C0-AMX-069

Canadian Bulletin of Fisheries and Aquatic Sciences 225

Management of Marine Fisheries in Canada

By

L. S. Parsons

Department of Fisheries and Oceans
Biological Sciences Directorate
200 Kent Street
Ottawa, Ontario, Canada K1A 0E6

NATIONAL RESEARCH COUNCIL OF CANADA
AND
DEPARTMENT OF FISHERIES AND OCEANS
Ottawa 1993

ISBN 0-660-15002-6
ISSN 0706-6503
NRCC No. 36175

Canadian Cataloguing In Publication Data

Parsons, L. S.
Management of marine fisheries in Canada

(Canadian Bulletin of Fisheries and Aquatic Sciences; no. 225)
Includes an abstract in French.
Includes bibliographical references.
ISBN 0-660-15002-6
DSS Cat No. Fs 94-225E

1. Fishery management -- Canada. 2. Fisheries -- Canada. 3. Fisheries policy -- Canada.
I. Canada. Dept. of Fisheries and Oceans. II. Title. III. Series.

SH328.P37 1993 333.95'6'0971 C93-099550-3

This publication is available from:

Subscription Office, Research Journals, National Research Council of Canada, Building M-55, Montreal Road, Ottawa, Ontario, Canada K1A 0R6

Enquiries: Tel.: 613-993-9084 FAX No.: 613-952-7656

Remittances should be made payable to the Receiver General for Canada, credit National Research Council of Canada.

Printed on recycled paper.

Correct citation for this publication:

Parsons, L. S. 1993. Management of marine fisheries in Canada. Can. Bull. Fish. Aquat. Sci. 225: 763 p.

CONTENTS

CHAPTER 13. RIDING THE ROLLER COASTER

CHAPTER 14. THE SOCIAL DIMENSION

CHAPTER 17. SCIENCE AND FISHERIES MANAGEMENT. I — THE DEVELOPMENT OF FISHERIES RESEARCH AND THE SCIENTIFIC ADVISORY PROCESS

CHAPTER 18. SCIENCE AND FISHERIES MANAGEMENT. II — FISHERIES SCIENCE: CHALLENGE AND OPPORTUNITY

CHAPTER 19. ENSURING COMPLIANCE: FISHERIES ENFORCEMENT

CHAPTER 20. SOME OBSERVATIONS ON MARINE FISHERIES MANAGEMENT IN OTHER COUNTRIES

CHAPTER 21. CONCLUSIONS

ABSTRACT

This volume describes and evaluates the impact of major changes in the management of Canada's marine fisheries in recent decades. In just 25 years, these fisheries have gone from underdevelopment to overcapacity. Regulatory interventions have mushroomed. These include Total Allowable Catches (TACs), allocation of access among fleet sectors, limited entry licensing and individual quotas. Major benefits flowed from the 1977 extension of fisheries jurisdiction to 200 miles. However, Canada's marine fisheries continue to be plagued by instability because of various problems and constraints. These include:

1. natural resource variability, often environmentally determined;
2. the common property problem and resultant overcapacity;
3. market fluctuations;
4. heavy dependence on the fisheries in isolated coastal communities;
5. recurrent conflict among competing users; and
6. conflicting objectives for fisheries management.

Various combinations of these factors have contributed to recurrent boom-and-bust patterns in many marine fisheries. Extended jurisdiction and various post-extension management innovations have not solved the problems of the fisheries sector. There is an urgent need to bring harvesting and processing capacity into balance with sustainable resource levels. Periodic fisheries crises and demands for government assistance can be expected to continue unless viable alternative economic opportunities can be developed in the coastal regions.

RÉSUMÉ

Ce document décrit et évalue les répercussions des principaux changements apportés à la gestion des pêches maritimes canadiennes durant les dernières décennies. En 25 ans seulement, ces pêches sont passées de la sous-utilisation à la surutilisation, et les organes de réglementation ont multiplié leurs interventions : introduction du total des prises admissibles (TPA), répartition des droits d'accès entre les secteurs de la flottille, limitation du nombre de permis et introduction de quotas individuels. En 1977, l'agrandissement de la zone de pêche jusqu'à 200 milles de côtes a procuré des avantages énormes. Toutefois, l'instabilité demeure une charactéristique dominante des pêches maritimes canadiennes en raison de différents facteurs, notamment :

1. l'instabilité des ressources halieutiques, souvent déterminée par des facteurs environnementaux;
2. la question de la propriété commune et de la surcapacité qui en a résulté;
3. les fluctuations du marché;
4. la forte dépendance des villages côtiers isolés vis-à-vis de la pêche;
5. les éternelles disputes entre concurrents;
6. la non-coïncidence des objectifs de gestion des pêches.

Les pêches maritimes ont connu une alternance constante d'expansions et de chutes rapides, en partie provoquée par différentes combinaisons de ces facteurs. L'agrandissement de la zone de pêche et l'introduction subséquente de nouvelles méthodes de gestion n'ont pas réglé les problèmes du secteur des pêches. On sent un besoin urgent d'adapter la capacité de capture et de transformation au niveau des ressources halieutiques disponibles. Il est probable que les crises périodiques qui frappent le secteur de la pêche et les demandes d'aide gouvernementale vont se poursuivre à moins que des initiatives de diversification économique viables ne soient mises sur pieds dans les régions côtières.

ACKNOWLEDGEMENTS

My thanks go, first of all, to Dr. Peter Meyboom, former Deputy Minister of Fisheries and Oceans, who provided me with the opportunity to undertake this study of Canadian marine fisheries management, and to his successor, Mr. Bruce Rawson, for granting me the time to complete it. I also owe a great debt of gratitude to Henry Lear and Ronnie Sanford, who assisted in all phases of this project from its conception to completion. Without their help, this book would not have been realized.

Numerous other people have contributed to this project. Glen Hodgins played a vital role in the early phases in accessing information from many sources within DFO, including searching for and retrieving files and records dormant and often forgotten. Staff of the DFO Library in Ottawa — Cuneas Boyle. Heather Cameron, Denis Lasalle, Jean Weerasinghe, Graeme Durkin and others — provided expert assistance in conducting bibliographic searches and responding to my never-ending requests for books, periodicals, etc. Leo Lanthier, Carol Logan, John Horsey, John Leclair of the Records Management group provided invaluable assistance in locating and providing files and records. I am also grateful to my colleague Dr. Bill Doubleday for his help in facilitating the logistic arrangements for this project.

Staff from numerous Branches of the various DFO Sectors in Headquarters and the Regions provided data and assistance of various kinds. It would be difficult to list every one who assisted.

I am indebted to Art Read of the Canadian Hydrographic Service who patiently drafted and redrafted all the illustrations.

I am also grateful to the individuals who commented on drafts of various portions of the manuscript. Greg Sheehy performed exceptional service in editing and helping to improve the manuscript. John Camp, David Cook. Gerry Neville, Jeannine St-Pierre and others in the Scientific Publications Unit were always helpful in guiding the book to completion. My wife Loretta reviewed several chapters and made valuable suggestions.

Colleen Meloche and Lisa Noall assisted in typing and proofreading the various drafts.

Finally, I would like to thank Dr. Arthur May, a former Deputy Minister of Fisheries and Oceans, who invited me to Ottawa in 1976 to participate in the preparations for extended fisheries jurisdiction and who encouraged me to move from fisheries science to fisheries management. This provided the base of knowledge and experience which stimulated me to write this book.

INTRODUCTION

Canada has abundant marine fish resources. Despite this, many of Canada's marine fisheries have been plagued by recurrent crises. These are rooted in the inherent natural variability of the fish resources and their common property nature. The problems are compounded by the vagaries of market fluctuations and a very large reliance on export markets.

The 1960's, 1970's, and 1980's were decades of major change in the management of Canada's marine fisheries. This book examines the nature of those changes and traces their evolution. In Chapter 1, I provide an overview of Canadian fisheries. The next three chapters examine: the jurisdictional basis for fisheries management in Canada, certain biological and economic aspects of fish and fishing which profoundly affect fishing and the management of fisheries, and the objectives of fisheries management. Changes in objectives over time and the failure to set clear objectives complicate the task of fisheries management. In Chapters 5 and 6, I describe common techniques of marine fisheries management and their application in Canada's marine fisheries. Chapters 7 through 9 examine the common property problem and Canada's attempts to address the "race for the fish" and the overcapacity dilemma.

Extension of fisheries jurisdiction to 200 miles in 1977 did not bring all of the fish stocks off Canada's coast under Canadian jurisdiction. Major transboundary stocks continue to be shared with Canada's fisheries neighbours and are fished beyond 200 miles. The transboundary migrations of these fish stocks complicate Canada's attempts to manage its domestic fisheries. Three chapters (10–12) are devoted to examining the evolution and impact of the 200-mile limit, Canada's international post-extension fisheries policy, and Canada's often stormy relations with its fisheries neighbours.

Debates have raged over the relative importance which should be given to social and economic factors in Canada's fisheries objectives, policy and management practices. In Chapter 13, I examine the roller coaster nature of the fisheries and the reasons for the boom and bust pattern which has characterized recent Canadian fisheries experience. In Chapter 14, I examine the social dimension of Canada's marine fisheries including incomes, employment and community dependence.

One of the central challenges of fisheries management has been the management of conflict between competing user groups. In Chapter 15, I examine the dynamics of group conflict and the elaborate fisheries consultative structures which were developed to reconcile conflicting interests. This chapter also describes institutional reforms currently being implemented to foster a closer partnership between industry and government in fisheries management.

One of the chief conflicts in fisheries is between competing users of fish habitat. The quality of the aquatic environment is crucial to the well-being of Canada's fisheries. In Chapter 16, I examine the importance of fish habitat, Canada's fish habitat policy, and habitat/environmental issues which affect the sustainability of fisheries.

The natural and social sciences have had a major influence on Canadian marine fisheries policy over the past several decades. Fisheries science is a crucial input to the management process. In Chapters 17 and 18, I trace the development of fisheries research and the scientific advisory process in Canada. I examine some of the major issues confronting fisheries science today and suggest priorities for future research including a greater emphasis on understanding and predicting the influence of the environment on the abundance of marine fish populations.

Over the past three decades a complex web of regulations has grown up to provide a framework for implementing fisheries policy. But fisheries regulations are only effective if complied with. In Chapter 19, I examine the factors which influence an individual's choice whether to comply with fisheries regulations and the relative effectiveness of alternative enforcement methods in securing compliance.

In the penultimate Chapter (20), I examine the marine fisheries management experience of several other developed countries. This chapter situates the Canadian experience in a world context and describes some lessons to be learned from others' experiences. Finally, in Chapter 21, I describe the characteristics of the fisheries system which have contributed to the recurrent boom-and-bust pattern and repeated demands for government financial assistance in many fisheries. I offer some conclusions about the effectiveness of Canadian marine fisheries management. This chapter analyses the attempts to achieve greater stability in the marine fisheries and suggests why these have been only partially successful. I conclude that Canada's marine fisheries will continue to be subject to periodic crises. Demands for government assistance will continue unless viable economic opportunities can be created in the coastal regions. I predict a continuation of the boom-and-bust pattern moderated somewhat by fisheries management intervention.

CHAPTER 1
CANADA AND THE FISHERIES

1.0 INTRODUCTION

Historians tell us that the fishery attracted the first Europeans to what is now Canada's Atlantic coast. It influenced patterns of settlement and was a vital part of commerce from the 16th century onwards. Fish and fishing were an integral part of the life style of Canada's native peoples for thousands of years prior to the arrival of the Europeans.

To today's city dweller, the fishery may appear to be a relic of the past and a drain on, rather than a contributor to, the national economy. However, to many Canadians in more than a thousand communities along the Atlantic, Pacific and Arctic coasts, the fishery provides employment and income in areas where alternatives are few.

2.0 THE HISTORICAL CONTEXT

Canada's fisheries began with the capture of fish for food and trade by Canada's aboriginal peoples. The commercial marine fisheries in Canada originated with John Cabot's discovery of Newfoundland in the summer of 1497. Still, it was not the English who began to exploit the new overseas fishery. Their interest at that time focused on Iceland. The French and the Portuguese established the fishery in the Northwest Atlantic during the early 1500's (Innes 1978).

It is beyond the scope of this volume to describe the evolution of Canada's commercial marine fisheries from 1500 to the present day. For a brief account of historical developments on the east and west coasts, see Gough (1988) and Gough (1993). For a more detailed account of the development of the Atlantic coast fishery over several centuries, see Harold Innis' treatise *The Cod Fisheries: The History of an International Economy* (Revised Edition 1978).

The early fishery on the Atlantic coast by European-based fishermen was for cod. Cod was king on the Atlantic for centuries and remains a key component of these fisheries in the late 20th century.

During the 17th and 18th centuries France and England vied for dominance of the fisheries of the New World, with England finally prevailing in the 19th century. Following the withdrawal of the French, settlers in Newfoundland and Nova Scotia fished in small open boats from shore.

The early 19th century also saw the emergence of new methods of fishing such as the purse seine and longlining. By the late 19th century steel vessels began to replace wooden ones. Canning technology fostered the development of the lobster fishery and the Bay of Fundy sardine fishery. The cod trap supplanted the hook-and-line method of fishing in Newfoundland. Dried salted cod production on the Atlantic peaked in the 1880's and then declined. This decline was the result of several factors: (1) the disappearance of wooden ships; (2) competition from other food products; and (3) the appearance of outlets for fresh fish. By the end of the 1800's the lobster fishery had grown in importance to rival the cod fishery.

On the Pacific coast a single species, the sockeye salmon, dominated the early history of the commercial fishery which commenced much later than on the Atlantic coast. Native populations had concentrated at locations on the rivers near the salmon spawning grounds. When the fur trade reached the Pacific coast in the early 1800's, these sites served as bases for the trade. The expansion of B.C.'s population following the gold rush in the 1850's provided a local market for fish. Commercial canning of salmon began on the Fraser River around 1870 and on the Skeena River shortly thereafter.

Completion of the transcontinental railway provided access to markets for salmon in eastern Canada and the United States. In 1887 cold storage plants for salmon were built on the Fraser River. This provided a basis for expansion of the salmon and the halibut fisheries. The halibut fishery in turn required bait which led to the development of a fishery for herring (Gough 1988).

Inland, the resources of the lakes and rivers were utilized by natives for thousands of years. As Europeans moved inland to pursue the fur trade and to colonize the interior, they too began to exploit the fish resources. Some areas were overfished early in the process of settlement.

Around the beginning of the twentieth century, a number of important changes occurred in the Canadian fisheries. The gasoline engine began to be used in the small-boat fisheries. Purse seining began in the British Columbia salmon and herring fisheries. In 1908 otter trawling was introduced to the Atlantic coast. Inshore fishermen opposed this latter development, fearing that an increase in the number of other trawlers would hurt their sector of the fishery. In response, the government severely restricted otter trawling, the intent being to maintain the largest possible labour force in the fishery. The trawler fleet was restricted to 3 or 4 vessels during the 1930's. This retarded the development of the Atlantic fishery, in particular the growth of a year-round industry based on fresh and fresh frozen fish.

The Second World War intensified the demand for fish products, keeping alive the salt fish industry. A filleting and freezing industry was established in Newfoundland. Government restrictions on the use of otter trawlers were relaxed.

The fishing fleets adopted new technology including radios, radar and sonar. Governments, both federal and provincial, placed increased emphasis on fisheries development, following the 1944 Report on the Atlantic Sea Fishery by Stewart Bates (Bates 1944). Modernization and expansion of the Atlantic fisheries was a major objective. More vessels and plants were built with government assistance. New fisheries were developed for species such as redfish, flounder, scallops, shrimps and crab.

The Fisheries Prices Support Board was established in 1947, to stabilize returns to fishermen in periods of market downturn. The Fishing Vessel Assistance Program was introduced in 1942. The objective was to help modernize the fleet. The Fishermen's Indemnity Plan was introduced in 1953 to provide fishermen with affordable commercial insurance. The Fisheries Improvement Loans Act was instituted in 1955 to assist fishermen in obtaining credit from private lending institutions. Unemployment insurance coverage was extended to self-employed fishermen in 1957 (Crowley et al. 1993).

With the development of refrigeration in transportation and storage facilities, the demand for frozen fish increased. The groundfish industry switched to frozen fish production. Several vertically integrated companies began to dominate the groundfish fishery and subsequently extended into the lobster, scallop and herring fisheries.

This expansionist trend came to an end in the late 1960's with the realization that the fishery resources were finite. Some stocks had been overfished domestically, e.g. B.C. herring. The build up of the foreign fishery on the Atlantic was reducing the abundance of important groundfish stocks. In just 25 years Canada's marine fisheries had gone from underdevelopment to overcapacity. From the mid-1960's onward the federal government began to wrestle with the twin problems of overfishing (conservation) and overcapacity (economic viability). The emphasis shifted from developmental assistance to regulatory control to address these problems. The late 1960's marked the transition to a new era of modern-day fisheries management. This involved new regulatory techniques, including limited entry licensing and catch quotas. Conflict intensified among various groups as each fought for a greater share of the fisheries pie. Government became increasingly involved in determining who could fish, where, when and how.

3.0 AN OVERVIEW OF TODAY'S CANADIAN FISHERIES

3.1 General

Canada has one of the world's largest fishing zones, the longest coastline, and an abundance of freshwater lakes and rivers. It is one of the world's top twenty fishing nations in terms of fish production. From 1974 to 1988, Canada ranked 16th in catch in 11 years out of 15. This placed it well behind the top four fishing nations, Japan, the USSR, China and the USA, but in a league with South Korea, Iceland, Mexico and Spain. Canada's share of the world catch was generally less than 2% through the 1970's and 1980's. Canada's reputation as a fishing nation derives from its ranking as the world's number one fish exporter in terms of value from 1978 to 1987 (FAO Statistics). Most of Canada's fish production is exported.

Canada has important fisheries on both the Atlantic and Pacific coasts and in the inland lakes. Overall, however, the fishery has accounted for only about 1% of the value-added in commodity producing industries in Canada since the 1970's. Fishermen and plant workers accounted for only 3% of Canadian employment in those same industries through the 1980's.

On a regional basis, the fishery makes a much more significant contribution to the value-added in commodity-producing industries. This is particularly true in the Maritime provinces and Newfoundland (Table 1-1a). In 1986, fisheries contributed 15.8 and 15.9% of the value-added in commodity-producing industries in Nova Scotia and Prince Edward Island and 20% in Newfoundland. Elsewhere the percentage contribution was 5% in New Brunswick and 3% in British Columbia. In Quebec, Ontario and on the Prairies the percentage was less than 1%.

In employment terms, the regional impact is even more striking in Newfoundland, Nova Scotia and Prince Edward Island (Table 1-1b). In 1985, 72.0% of the employment in goods-producing industries in Newfoundland was generated by the fisheries, compared with about 1% in Quebec and in the Prairie Provinces and only a fraction of 1% in Ontario.

These figures illustrate the large dependence on the fisheries as a source of employment and income in the Atlantic provinces, particularly in Newfoundland, P.E.I. and Nova Scotia. Indeed, Newfoundland is

TABLE 1.1(a). Contribution of fishing and fish processing to value added in commodity-producing industries, 1986 (Values in million dollars). Source: Statistics Canada. Catalog 15-203. 1990. Provincial Gross Domestic Product by Industry.

	Value Added				
Province	Fishing	Fish Processing	Sub-total	All Commodity Producing Industries	Fisheries as % of Total
Nova Scotia	292.3	246.6	538.9	3416.3	15.8
New Brunswick	69.3	82.3	151.6	3029.8	5.0
Prince Edward Island	41.4	23.4	64.8	406.5	15.9
Québec	68.0	40.8	108.8	36,011.6	0.3
Newfoundland	141.8	261.3	403.1	2014.0	20.0
Ontario	35.3	***	***	67,498.9	***
Prairie Provinces	19.3	***	***	33,844.9	***
British Columbia	264.5	165.6	430.1	15,557.7	2.8
Yukon and NWT	1.6	***	***	850.2	***
Canada Total	933.4	843.4	1776.8	162,629.7	1.1

*** = Confidential

TABLE 1.1(b). Employment in the fisheries compared with employment of all goods-producing industries, 1985 (Employment reported in thousands). Source: Statistics Canada. Catalog 35-250. 1990. Fish Products Industry.

	Fisheries				
Province	Fishermen	Fish Plant Workers	Sub-total	Goods Producing Industries	Fisheries as % of Total
Newfoundland	27	9	3	50	72.0
Nova Scotia	14	7	21	90	23.3
New Brunswick	7	4	1	68	16.2
Prince Edward Island	4	1	5	15	33.3
Québec	6	2	8	818	1.0
Ontario	2	—	2	1452	0.1
Prairie Provinces	6	—	6	612	1.0
British Columbia	18	4	22	320	6.9
Canada Total	84	27	111	3425	3.2

Note: The numbers of fish-plant workers obtained from Statistics Canada reflect only the numbers of plant workers in the larger fish plants with more than 200 workers.

noteworthy among the Canadian provinces in its chronic dependence upon the fisheries. At the community level fisheries activities are even more significant (Kirby 1982; Poetscke 1984; also see Chapter 14, Section 2.1. The Task Force on Atlantic Fisheries (Kirby 1982) concluded that more than one-quarter of the total population of the Maritime Provinces and Newfoundland lived in approximately 1300 small fishing communities. They estimated that at least 35% of the overall employment in these communities was provided by fishing and fish processing. In Newfoundland more than one half the population live in such communities.

Of the 600,000 people in the Atlantic who live in fishing communities, about 200,000 live in communities where fishing activity is the principal, if not the only, employer.

In British Columbia, the fisheries are relatively less important to the overall economy than in the Atlantic provinces.

In the Arctic, native peoples depend upon fish for food and income in isolated communities scattered across the northern part of Canada. Freshwater fisheries are important to native communities throughout Canada.

Canada's fisheries resources are both varied and abundant. These resources are generally classified into two major groupings — "finfish" and "shellfish". Finfish include the major marine species and most commonly utilized freshwater species. Shellfish include crustaceans such as lobsters and crabs, and molluscs such as clams, oysters and squid. Species are also commonly grouped according to their preferred habitat and life history patterns. Both finfish and shellfish, for example, are grouped into littoral, demersal (or groundfish) and pelagic species.

Littoral species are those which live inshore, within the 50–60 fathom contour adjacent to the coastline. Crustaceans and molluscs, with the exception of scallops, shrimp and squid, generally fall into this category.

Groundfish species live near the bottom. Pelagic species spend a large part of their life cycle in midwater or near the surface of the open ocean. Groundfish species include both common finfish such as cod, haddock, and flatfish as well as some crustaceans (e.g. crabs), and a mollusc (scallops). The major pelagic species are all finfish, e.g. herring, mackerel, tuna, except for squid. Redfish do not fit neatly into either of these categories but are considered to be groundfish.

Anadromous species constitute the other major category. These spend much of their adult life in the ocean but swim upstream into freshwater to spawn. The five species of Pacific salmon, which are anadromous, support extremely valuable fisheries in British Columbia.

Canadian commercial fisheries landings and landed value generally increased from the mid-1970's onward, with a decrease in the early 1980's. The rate of growth slowed in 1988 and landings and landed value dropped slightly in 1989. Total Canadian commercial landings increased from 969,000 tons in 1974 to 1.44 million tons in 1979. Landings decreased to 1.28 million tons in 1984. Thereafter landings increased to a peak of 1.65 million tons in 1988. The 1988 landings were 70% greater than the low point of 1974, and 10% above the 1984–1988 five-year average (Fig. 1-1). However, 1988 landings were only 14% higher than the 1960's high of 1.45 million tons in 1968, just prior to the decline in the Atlantic groundfish fishery in the early 1970's under the pressure of foreign overfishing (Table 1-2).

The nominal landed value of the Canadian commercial fisheries increased dramatically during the 1980's, more than doubling between 1980 and 1987. The major increase occurred during 1985–1988. The landed value from 1988–1990 was about $1.5–$1.6 billion (Table 1-2). The market value of production in 1988 was $3.2 billion, up from approximately $2.0 billion in 1984.

More than 80% of Canada's fisheries production is exported. The quantity of exports increased from 366,000 tons in 1976 to 613,000 tons in 1988, an increase of 67% (Fig. 1-2). The overall increase in exports from 1976 to 1988 was 67%.

The total value of exports in 1988 was $2.7 billion, approximately the same as in 1987. Export value increased from $600 million in 1976 to $1.1 billion in 1978, hovered around $1.6 billion from 1982 to 1984, and then almost doubled between 1984 and 1988 reflecting the market boom of the 1985–1988 period.

TABLE 1.2. Canadian commercial fish landings and landed values 1966–1990 (Quantity in metric tons and value in 000$).

	Atlantic		Pacific		Freshwater		Canada Total	
	Q	V	Q	V	Q	V	Q	V
1966	995,290	95,859	271,540	60,642	55,051	14,853	1,321,881	171,354
1967	1,040,667	98,059	159,812	48,959	48,434	11,831	1,248,913	158,849
1968	1,267,539	110,617	131,145	57,268	52,231	12,957	1,450,905	180,842
1969	1,207,549	114,828	88,418	47,381	54,547	15,660	1,350,534	177,869
1970	1,173,959	125,685	117,021	60,255	42,933	13,237	1,333,913	199,177
1971	1,094,744	128,803	113,368	58,588	41,511	13,132	1,249,623	200,523
1972	931,234	141,410	163,317	75,128	42,458	15,840	1,136,009	232,378
1973	888,478	167,553	183,827	130,409	45,529	19,095	1,117,834	317,057
1974	781,003	163,614	141,141	100,976	47,007	18,241	969,151	282,831
1975	805,345	184,524	132,916	79,681	42,479	20,944	980,810	285,149
1976	880,892	218,665	180,942	141,851	39,667	24,146	1,101,461	384,662
1977	1,003,074	282,536	204,310	166,250	47,289	31,091	1,254,713	479,877
1978	1,153,231	408,466	198,743	249,729	47,571	32,959	1,399,505	691,154
1979	1,237,702	499,558	155,216	327,224	49,152	46,868	1,442,130	873,650
1980	1,156,088	501,902	129,946	179,746	54,297	48,352	1,340,311	730,000
1981	1,194,557	550,177	183,117	232,976	49,956	57,125	1,427,650	849,278
1982	1,197,632	581,608	157,813	235,774	57,743	58,847	1,413,218	876,229
1983	1,108,439	617,394	191,543	203,011	48,818	48,464	1,348,800	868,869
1984	1,065,205	595,493	169,118	238,404	43,424	61,505	1,277,797	895,402
1985	1,187,937	683,044	213,844	371,946	48,078	58,361	1,449,849	1,113,351
1986	1,245,280	874,470	222,417	395,587	45,270	77,180	1,512,967	1,347,237
1987	1,265,430	1,116,868	251,336	442,296	50,800	88,500	1,567,566	1,647,664
1988	1,338,681	1,012,429	265,847	533,559	48,000	82,000	1,652,528	1,627,988
1989	1,271,713	959,775	283,395	453,664	51,199	82,690	1,606,307	1,496,129
1990	1,296,684	953,086	305,207	478,192	45,500	78,000	1,647,391	1,509,278

Source: DFO Statistics, Ottawa.

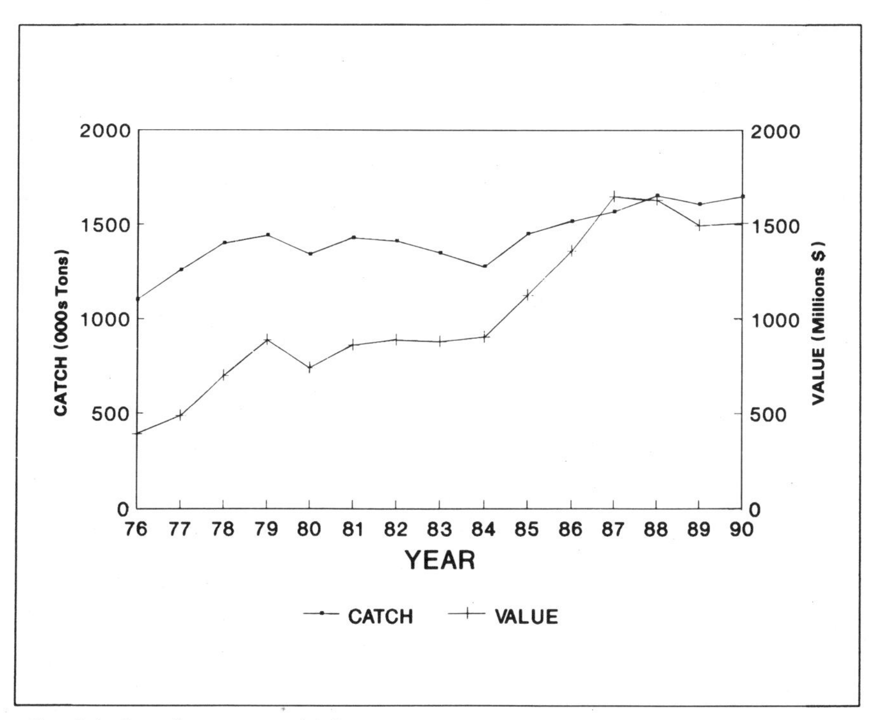

FIG. 1-1. Canadian commercial fish landings in quantity and value from 1976 to 1990.

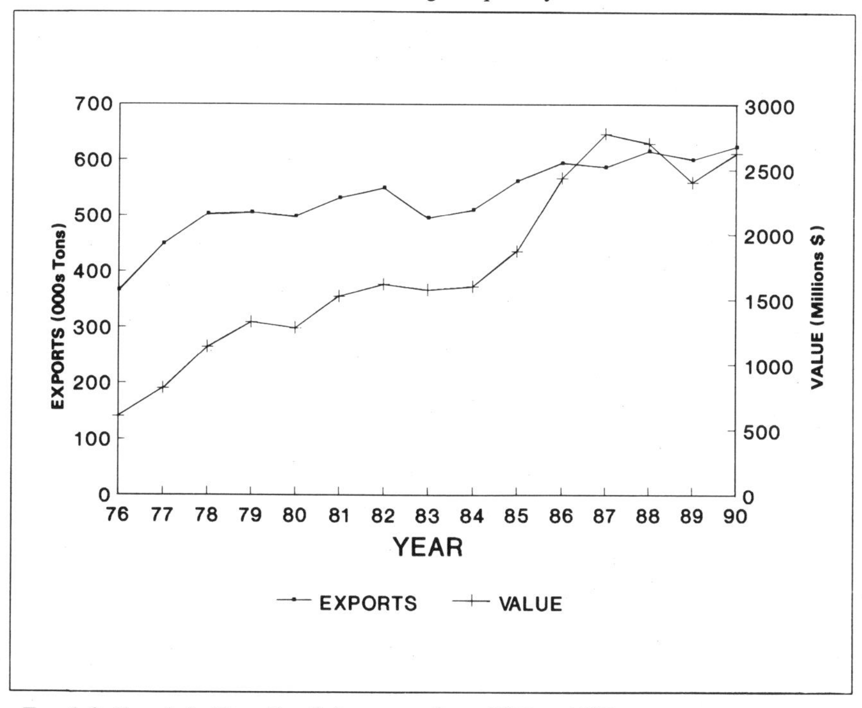

FIG. 1-2. Trends in Canadian fish exports from 1976 to 1990.

Despite attempts at market diversification through the 1970's and 1980's, the United States remained Canada's largest export market, accounting for 56% of total export volume and 52% of total export value in 1988 (Fig. 1-3). However, the percentage of Canada's fish exports going to the United States market declined from 65% of volume in 1976 to 59.5% in 1984 and 56% in 1988. By 1988 Japan was Canada's second largest

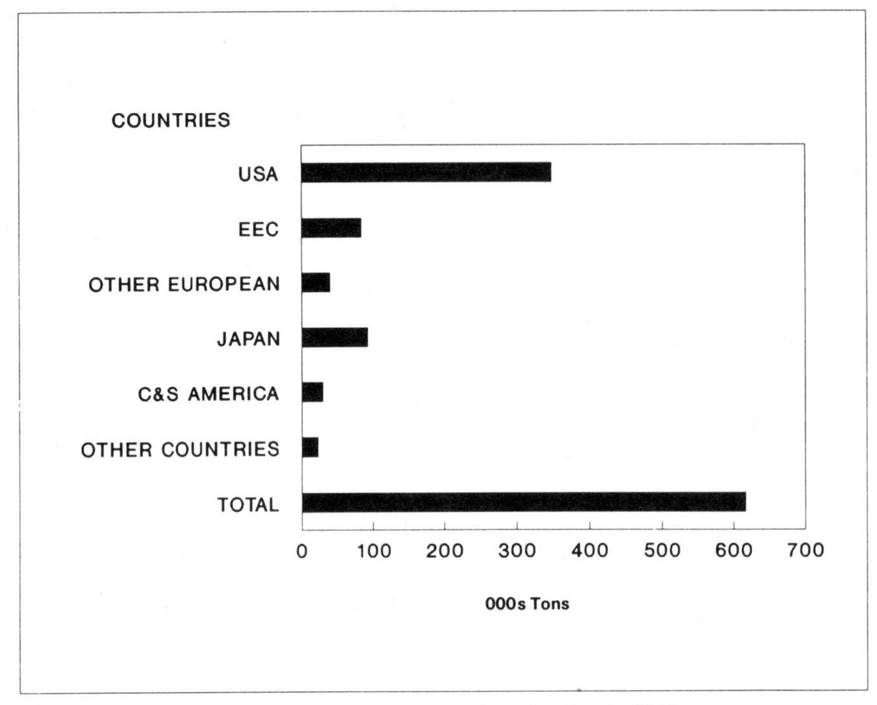

FIG. 1-3. Canadian fish exports by country of destination in 1988.

export market, absorbing 15% of exports by volume and 22% by value. This compared with only 5.5% of the volume in 1976 and 12.9% of the value.

Exports to the European Community in 1988 represented 14% of the Canadian volume and 17% of the value, unchanged from 1984.

Canada also imports significant quantities of fish and shellfish. In 1988, imports were 171,000 tons valued at $737 million. The United States is Canada's main supplier of fisheries products, accounting for 53% of the volume and 47% of the value of fish product imports. Fresh or frozen shellfish, particularly shrimp, and canned fish, particularly tuna, accounted for a very significant proportion of overall imports.

With exports and imports valued at $2.701 billion and $737 million, respectively, Canada's net balance in international fishery product trade was $1.964 billion in 1988.

In 1988, 95,000 fishermen were involved in Canada's commercial marine fisheries. This was roughly equivalent to the number of fishermen engaged in fishing in the late 1950's–early 1960's, following a dramatic decline to a low of 56,000 in 1974. In 1988, approximately 66,000 fishermen were engaged in the Atlantic commercial fisheries, 21,000 in the British Columbia commercial fisheries and about 8,000 in the freshwater fisheries.

According to Statistics Canada, the number of persons employed in fish processing fluctuated around 25,000 from 1982 to 1985. Because the numbers of fish plant workers reported by Statistics Canada include only the numbers of plant workers in the larger fish plants, i.e. those with more than 200 workers, these underestimate the actual number of fish plant workers in Canada. According to DFO statistics, the total number of fish plant workers in 1988 was 40,000. Approximately 33,000 were employed in the Atlantic fisheries, 6,000 in British Columbia and 1,000 in the freshwater fisheries.

Canada's commercial fisheries employ about 135,000 individuals in the harvesting and processing sectors combined. This number seems insignificant when viewed in terms of national employment statistics. However, regionally this employment is very important to about 1,500 coastal communities. In a great many of these, fishery-related employment determines whether or not a community survives or disappears.

The harvesting sector of Canada's marine fisheries is incredibly diverse and difficult to portray nationally. A multiplicity of types and sizes of fishing vessels and types of gear are used to harvest the wide diversity of species which support Canada's commercial marine fisheries. There is as much difference between the small-boat inshore cod fishery using cod traps in Newfoundland and the dragger fishery for scallops on Georges Bank as there is between the Gulf of St. Lawrence crab fishery by a midshore fleet using crab traps and the purse seine or troll fishery for Pacific salmon in British Columbia. The fishing vessels used range all the way from a small open boat powered by an outboard motor in the lobster fishery, which fishes on day trips, to the large powerful offshore otter trawler, which goes to sea for up to 2 weeks before landing its catch for processing. Modern factory freezer trawler technology is used for the northern shrimp off northeast Newfoundland-Labrador.

Geographically, there are three major Canadian fisheries — the Atlantic, the Pacific and freshwater. Over the past two decades the Atlantic fisheries have generally accounted for about 80% of the total commercial landings in Canada. The Pacific share of total landings has been around 15%.

In terms of landed value, the highly valuable Pacific salmon and herring fisheries shift the balance somewhat toward the Pacific. The Atlantic contribution to the total Canadian landed value has been around 60% during the past two decades. The Pacific share has been around one-third of the Canadian total. In 1988, the Atlantic accounted for 66% of the market value of fishery products, the Pacific 30% and the inland fisheries 5%.

Thus the Atlantic and Pacific fisheries dominate the Canadian commercial fisheries in landings, landed value and value of fishery products. Because of differences in values of particular species and products, the Pacific share of landed value and value of fishery products is higher than would be indicated by the relative share of landings.

3.2 The Atlantic Fisheries

The groundfish fishery remains dominant in landings and value (55.4% of volume and 37.1% of value). There are also very important shellfish fisheries, particularly lobsters, crab, scallops and shrimp (only 15% of volume in 1988, but 53% of value). Pelagic fisheries, principally for herring and capelin are significant in volume but less so in value (29.5 and 10%, respectively in 1988). Of the three major species groupings, groundfish constitute more than half the volume of landings but shellfish constitute more than half the landed value of the catch.

The relative contribution of various species to the Atlantic coast catch in volume and value in 1988 is shown in Fig. 1-4 and 1-5. Cod is dominant in volume (35.3%). Groundfish other than cod account for 20.1%. Herring ranks second overall in volume at 19.8%. Capelin is second among the pelagic species in volume at 6.8%. Scallops account for 5.8% of the overall volume and 38.6% of the shellfish fishery value. The principal scallop fishery is the offshore fishery on Georges Bank.

In terms of landed value, lobster and cod are the dominant species (26.3 and 23.8%, respectively). The next most valuable species are crabs and scallops (9.9 and 8.5%, respectively).

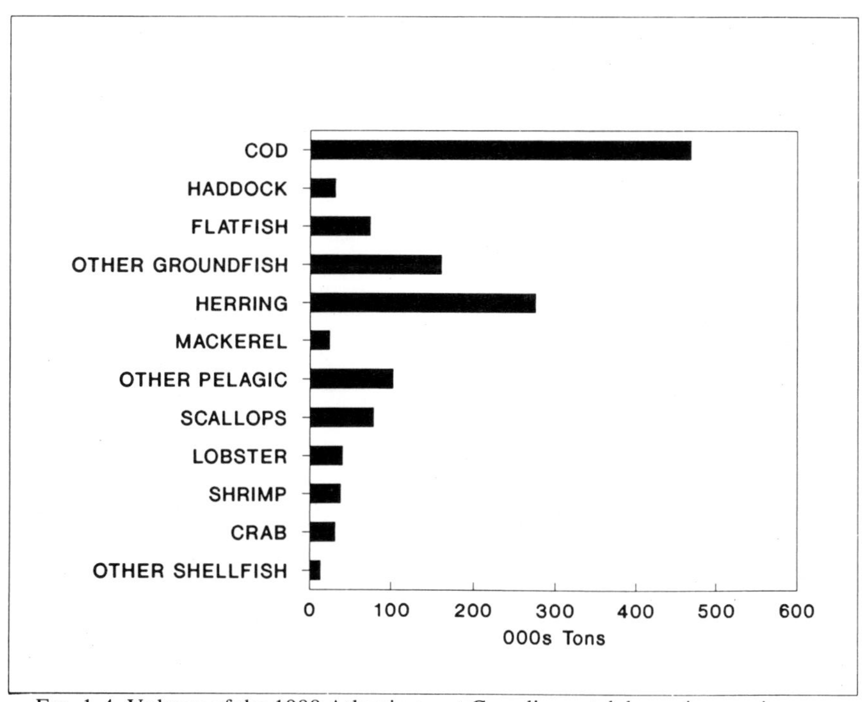

FIG. 1-4. Volume of the 1988 Atlantic coast Canadian catch by major species.

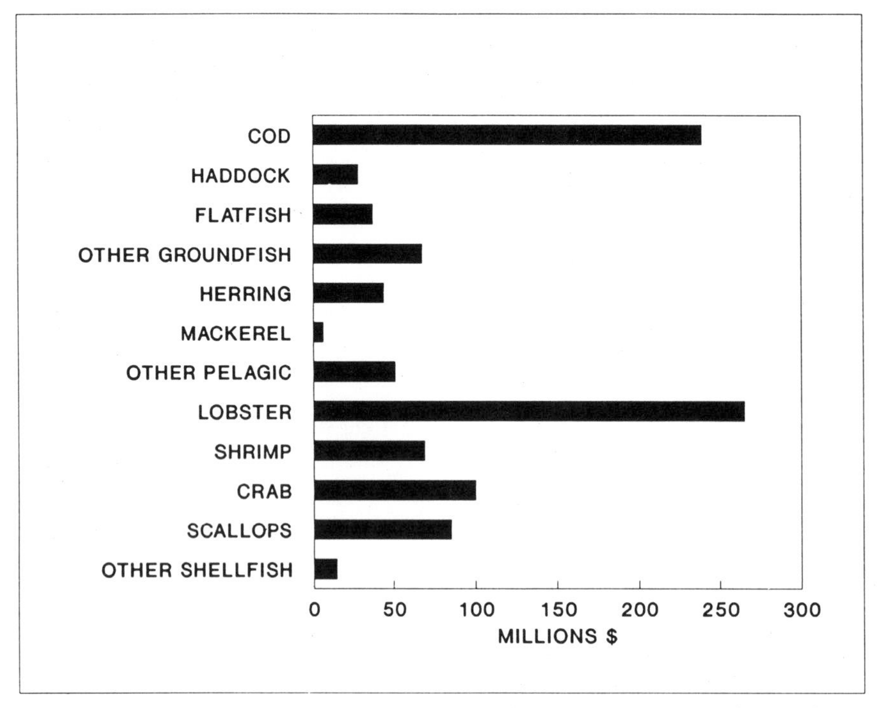

FIG. 1-5. Landed value of the 1988 Atlantic coast Canadian catch by major species.

This is a thumbnail sketch of the situation in 1988. There have been significant shifts in the relative contribution of various species to the total Atlantic catch and landed value over time. The relative contribution of groundfish has declined significantly since 1988 because of a decline in groundfish stocks, particularly cod.

Differences in the relative abundance of these species along the Atlantic coast contribute to significant differences in volume and landed value among the five Atlantic provinces. As indicated by fisheries' contribution to the value-added by commodities-producing industries, Newfoundland and Nova Scotia are the dominant Atlantic provinces in fisheries terms. Prince Edward Island ranked ahead of Nova Scotia in relative contribution of fisheries to the provincial economy.

Newfoundland and Nova Scotia together accounted for 78.8% of the Atlantic coast catch and 71.7% of the landed value of that catch in 1988 (Fig. 1-6). New Brunswick has consistently ranked third in volume and value but well behind Newfoundland and Nova Scotia. In 1988, New Brunswick accounted for 11.4% of the catch and 11.8% of the landed value. Quebec has consistently ranked fourth and Prince Edward Island fifth in recent years in catch and landed value. Quebec accounted for 6.6% of the catch and 9.9% of the landed value in 1988 and P.E.I. accounted for 3.2 and 6.6%, respectively.

Although Newfoundland and Nova Scotia are the most important of the five eastern fishing provinces, there are considerable differences in their fisheries. During the 1980's, Newfoundland led in volume, accounting for slightly more than 40% of the Atlantic total catch. Nova Scotia, however, has consistently had the largest landed value, accounting for more than 40%. Newfoundland has had the lowest ratio of landed value to volume of all five provinces.

The reason for these differences lies in the different mix of species harvested in each province. Newfoundland still largely depends on groundfish, and cod is king. Nova Scotia harvests a much greater variety of species and, in particular, is blessed with very valuable shellfish fisheries including lobsters and scallops. The price per pound for these species is much higher than that for groundfish and herring. Because lobsters are the most valuable species Atlantic-wide and Nova Scotia's share of the lobster catch in 1988 was 45.8% (53.1% of landed value), it is not surprising that Nova Scotia ranks first among the Atlantic provinces in fisheries value.

In contrast, Newfoundland accounted for 63.3% of the Atlantic cod catch and 57.0% of the landed value. Cod accounted for 53.5% of the total Newfoundland catch and 47.7% of the value of that catch. In Nova Scotia, cod accounted for 26.5% of the catch and only 18.8% of the landed value. Lobsters, on the other hand, accounted for 32.3% of the landed value of the Nova Scotia fishery.

In addition to the greater diversity of species and the greater preponderance of higher-valued species in the Nova Scotia fishery, there are other important differences between the fisheries of the eastern provinces.

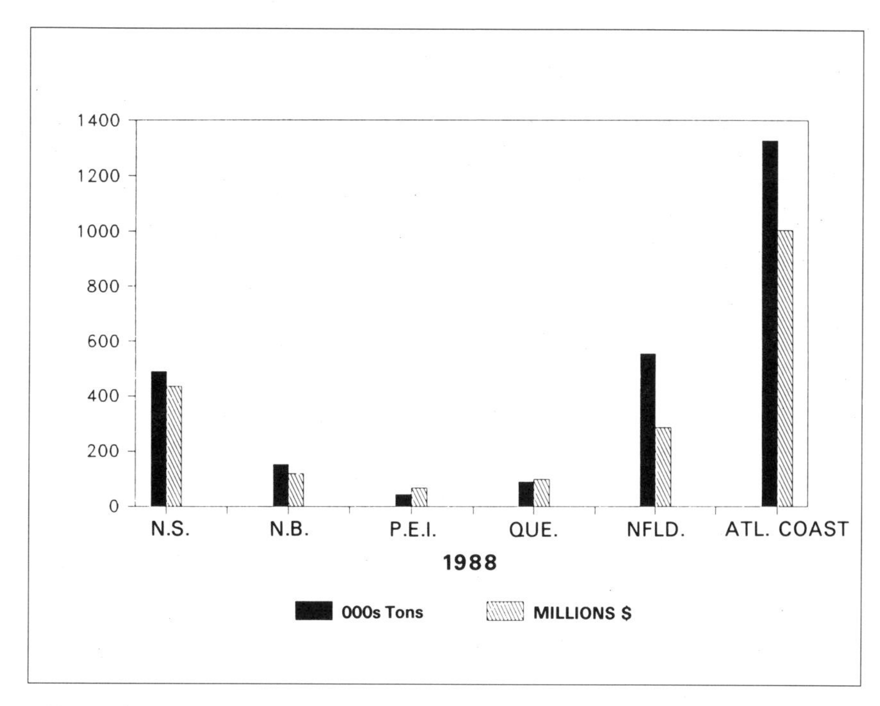

FIG. 1-6. Provincial distribution of catch and landed value in the Atlantic.

Much of the Newfoundland coast is ice-bound for a significant portion of the year. Hence, the inshore fisheries for groundfish experience much shorter seasons than those in Nova Scotia. More favourable weather conditions in southwest Nova Scotia and the Bay of Fundy make it more feasible to conduct a year-round fishery using relatively small fishing craft. One of the most important fisheries is a winter lobster fishery.

The fishing grounds adjacent to southwest Nova Scotia are also suitable for otter trawlers less than 65 feet long. Otter trawling by similar-sized vessels has not proved feasible close to the coasts of northeast Newfoundland and Labrador. Thus the groundfish fishery in Newfoundland has been more clearly divided into a seasonal inshore fishery by cod trap, gillnet and longline, and an offshore fishery by large otter trawlers based primarily in ice-free ports on the south coast of Newfoundland.

Nova Scotia has a more diversified groundfish fishery, involving a large midshore fleet (vessels 40 to 65 feet long) of powerful otter trawlers, or "draggers" as they are often called, as well as a fleet of offshore trawlers servicing major ports such as Lunenberg. It also has a specialized scallop dragger fleet and a large fleet of herring purse seiners which harvest the lion's share of the valuable Bay of Fundy herring stock.

Southern New Brunswick also shares in the valuable Georges Bank scallop fishery and the Bay of Fundy purse seine fishery. The fisheries of Northern New Brunswick are primarily carried out in the Gulf of St. Lawrence which is also ice-infested for much of the year. Vessels from that area fish valuable species such as crabs, lobsters and shrimp, as well as groundfish. Vessels from four and sometimes five provinces fish side by side on the same fishing grounds in the Gulf.

These interprovincial differences in the composition of the fisheries and dependence on them are reflected in differences in the numbers of fishermen and numbers and types of fishing vessels among provinces. The number of fishermen in the Atlantic provinces increased from a low of approximately 36,000 in 1974 to 66,000 in 1988. The number of fishermen had declined through the late 1960's-early 1970's as the fish stocks came under pressure from foreign overfishing. By 1979 the number had exceeded the previous high in the 1960's at 56,000. Through the early and mid 1980's the number of fishermen hovered around the 58,000 to 60,000 level. Historically, Newfoundland has had by far the largest number of fishermen, around 45% of the Atlantic total.

The relative landed value of the Newfoundland and Nova Scotia fisheries as a percentage of the Atlantic total (28.4% versus 43.3% in 1988) and the relative numbers of fishermen (45.0% versus 24.5%), demonstrate that there are many more fishermen in Newfoundland earning lower fishing incomes. This was borne out by the 1981 survey conducted by the Task Force on Atlantic Fisheries. The Task Force found that full-time fishermen earned the highest net incomes in southwest Nova

Scotia ($28,766) and the lowest in northeast Newfoundland-Labrador ($4,512) (Kirby 1982). These results have been confirmed in subsequent surveys by the federal Department of Fisheries and Oceans. Overall, fishing incomes are low for a majority of Atlantic fishermen. A significant proportion of the families of fishermen have incomes near or below the poverty line for rural Canada (see Chapter 14, Section 3.0). Higher incomes in areas such as Southwest Nova Scotia and southern New Brunswick are an exception to the general pattern.

The total number of Atlantic fishing vessels has generally remained stable at around 30,000 since 1980. However, the effective fishing power has increased enormously as larger, more efficient vessels have been built. There have been numerous attempts to constrain this growth in fishing capacity, but significant overcapacity exists today in many sectors of the Atlantic fishery (see Chapters 7, 8 and 9).

More than half (56.2%) of the total number of fishing vessels are based in Newfoundland. More than 90% of the Newfoundland fishing vessels are 35 feet or less in length. The effective fishing power of this fleet is constrained by the rather primitive fishing technology which can be employed by such vessels. Yet there are large numbers of fishermen who use such craft to fish. In some fisheries, e.g. lobsters, relatively small vessels can be used quite effectively to harvest the available catch.

The Atlantic processing sector is widely distributed along the coastline of the four Atlantic provinces and the Gulf of St. Lawrence shoreline of Quebec. In 1988, there were approximately 950 federally registered fish processing establishments.

These establishments vary in size, diversity of operation and type of proprietorship. Processing operations range from small one- or two-person curing or trucking-of-fresh-split-fish operations to large modern plants with several hundred employees. Processing activities are diverse, including fresh/frozen production, canning, salting, pickling and marinating. Proprietorship ranges from multinational public and private companies to independent fishermen, to a crown corporation.

While numerous processors contribute to the total output of the industry, three large vertically integrated firms — Fishery Products International, National Sea Products and Clearwater Seafoods — predominate. They own more than 80% of the offshore trawler fleet which caught about 38% of the total Atlantic groundfish catch in 1988. Also, through their own inshore plants and subsidiaries they account for an additional substantial proportion of the groundfish catch and that of other species. Overall, they account for more than 50% of production. Another significant participant in the processing sector is the Canadian Saltfish Corporation. This corporation is jointly supported by the federal and provincial governments. It has exclusive responsibility for marketing saltfish produced in Newfoundland-Labrador and along the north shore of Quebec.

The majority of processing establishments are located in Nova Scotia and Newfoundland (35 and 26%, respectively), with 20% in New Brunswick, 11.6% in Quebec, and 6.8% in Prince Edward Island. Despite the large number of processing establishments, concentration of production, particularly in the groundfish sector, is high.

Fish processing plants are often divided into two categories, year-round and seasonal. The year-round plants are owned primarily by the large vertically-integrated firms. The seasonal plants tend to be operated by small independent companies. There is, however, overlap between the two major groupings. Approximately 35% of the total number of plants operate year-round. A number of these large groundfish plants have either been closed or reduced to seasonal operations with the reduction in groundfish quotas in the early 1990s.

The operating period for seasonal plants varies depending on area, resource availability and product diversification. Plants providing a single product such as canned or frozen shellfish or cured herring are greatly dependent on resource availability and, as a consequence, often operate only for a few months. The ability of a plant to diversify into other products may extend its operating season to 6 or 8 months.

Fish processing capacity has increased considerably since 1977. Accurate measures of the extent of capacity growth are not available. However, the increase in the number of fish processing establishments provides a rough indicator. The total number of fish processing establishments increased from 519 in 1977 to 953 in 1988, an increase of 84% (Fig. 1-7). Studies prior to 1977 indicated that there was already overcapacity in the processing sector. The growth since 1977 has added to that overcapacity (see Chapter 13, Section 3).

There have been two growth spurts in the number of processing establishments since 1977 (Fig. 1-7). From 1977 to 1984 there was a 35% increase. Much of that growth occurred in Newfoundland where there was a 53% increase. Nova Scotia also experienced a significant increase (26%). During the Atlantic groundfish crisis of the early 1980's the number of processing establishments levelled off from 1981 to 1984. With the market-driven boom of 1984–88, there was a further significant increase of 31% between 1984 and 1988. This time the greatest increase occurred in Nova Scotia (39.3%) and the lowest percentage increase in Newfoundland (19.0%). New Brunswick, Quebec and Prince Edward Island all experienced an increase of 30–33%. Today there is significant excess processing capacity in most areas of the Atlantic fisheries (see Chapter 13).

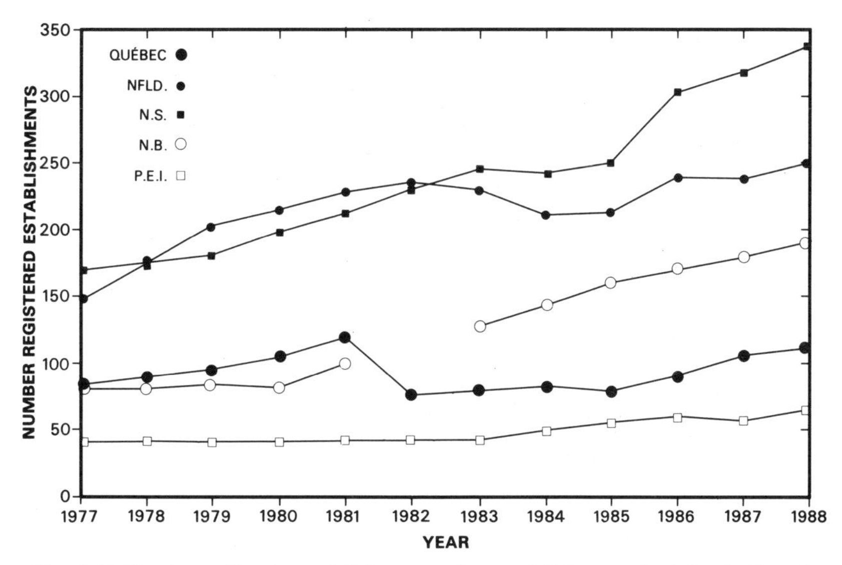

FIG. 1-7. Numbers of registered fish processing establishments in Atlantic Canada — 1977–88 (from data provided by Economics Planning and Analysis Division, Gulf Region, DFO).

The U.S. is Canada's primary export market, taking about 56% of Canadian exports in 1988. However, the U.S. is a much more important market for the Atlantic fisheries than for the Pacific. In 1988, 64% of Atlantic Canada's fish exports worth $1.175 billion went to the U.S. (Fig. 1-8). The EC was second with 16% valued at $275 million. Japan was third with 14% valued at $252 million. Other countries accounted for the remaining 7% valued at $137 million. The U.S. market was particularly important for Atlantic Canada's groundfish exports.

3.3 The Pacific Fisheries

Overall, the Pacific fisheries account for about 15% of Canada's commercial landings, about 30% of the landed value and about 30% of the value of fishery products. Pacific salmon (five species) accounted for 33% of the catch in volume in 1988 at 87,000 tons (Fig. 1-9). Herring accounted for 12% of the catch by weight and shellfish only 9%. Groundfish species as a group constituted 44.0% of the catch.

These species' relative contribution to the value differs significantly from their contribution to the catch. Salmon were king at 60% of landed value (Fig. 1-10). Herring was second at 15.6%, groundfish species third at 15.5%, and shellfish fourth at 8%.

Salmon have supported a commercial fishery off the coast of British Columbia for more than a century. Since 1966 the salmon share of the catch has fluctuated between 30% and 60%, reflecting fluctuations in the Pacific salmon catch. Landings in 1985 and 1986 were at record highs of 107,000 and 100,000 tons, respectively. Landings during the 1980's averaged 75,000 tons annually compared with 65,000 tons during the 1970's. Salmon generally account for 50 to 70% of the landed value of the catch. The landed value of the B.C. salmon fishery has increased substantially since the 1950's. Year-to-year earnings can be quite volatile, given particular combinations of high or low stock abundance and high or low prices. The industry has undergone periodic booms and busts, e.g. downturns in the late 1970's and a market downturn beginning in 1990.

The herring fishery went through a major transformation from the 1960's to the 1980's. Up until the federal government closed the fishery for conservation reasons in 1967, herring were primarily used for reduction to meal and oil. With the reopening of the fishery in the early 1970's, herring were utilized for roe for the Japanese market. The herring share of the catch increased from a low of 2–3% during 1968 to 1970 to 24% by 1972 and 48% in 1977 when a post-closure peak catch of 97,000 tons occurred. Subsequently, the frenzied roe fishery was brought under stricter controls. Herring constituted around 20% of the catch from 1980 to 1984 and then decreased to the 10–15% range. This was partly due to the above-average salmon catches during the second half of the 1980's. Herring constituted about 15–20% of the landed value through the 1980's.

The groundfish contribution to total landed value fluctuated in the 10–20% range through the 1970's and

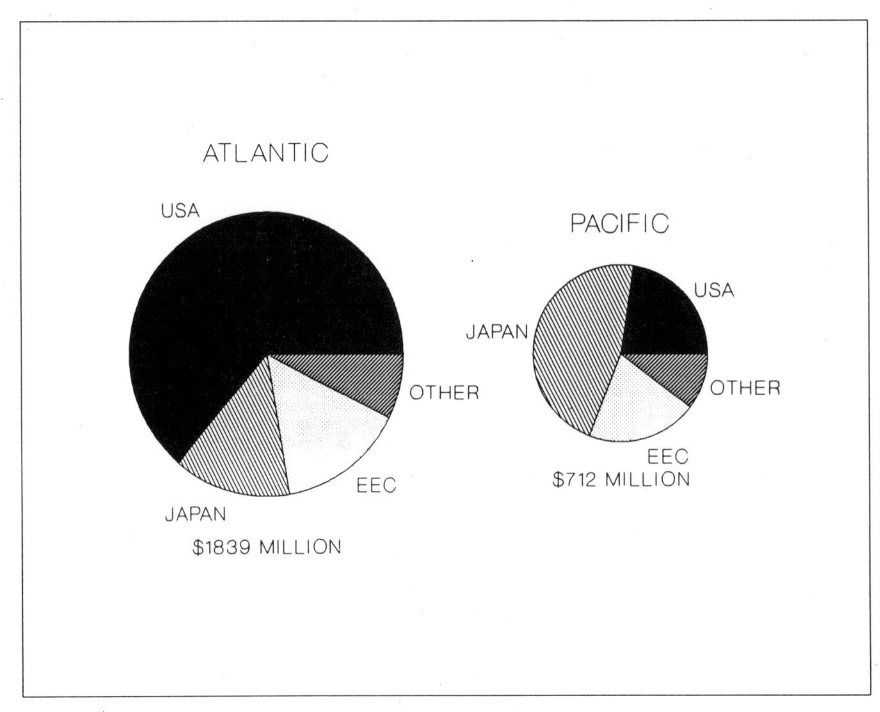

FIG. 1-8. Value of Canadian fish exports to various countries in 1988 in millions of dollars (from data provided by DFO Marketing Group, Ottawa).

1980's. Recently, groundfish have contributed a far greater percentage of the catch than of the landed value (44 and 15.6%, respectively in 1988). The species composition of the catch has changed significantly in recent years. The fishery can generally be divided into three components: the groundfish trawl fishery that harvests species such as rockfish, Pacific cod, hake, sole and a variety of other bottom fish, the Pacific halibut fishery and the sablefish (black cod) fishery. In relative order of economic importance the principal groundfish species exclusive of halibut are rockfish, Pacific hake, Pacific cod, sole and a variety of other species.

Historically, the groundfish fishery in British Columbia was directed almost exclusively at Pacific halibut, which for a long while was the third most significant fishery in British Columbia after salmon and herring. Pacific halibut accounted for as much as 24% of the total B.C. catch during the herring closure from 1968 to 1971. With extension of fisheries jurisdiction in 1977, Canadian halibut fishermen were prevented from fishing in U.S. waters after a 2-year phase-out period. From 1977 to 1988 halibut constituted only 3% or less of the total catch. The value of the halibut catch declined from 18% of the total landed value in 1970 to 5% by 1974 and fluctuated between 3 and 6% during the 1980's.

From 1986 to 1989 there were approximately 20,000 registered fishermen in British Columbia. This is the highest number since the 1950's and double the low point in 1972 when the estimated number of fishermen was 9,900. This total represents the number of personal fishing licences issued to commercial fishermen. It has been estimated that only 15,000 of these could be considered to be "active" fishermen (Fisheries Council of British Columbia 1989; Price Waterhouse 1990). About 10,000 individuals, two-thirds of the number of active fishermen, are involved in the salmon fisheries.

In 1988, there were 226 fish processing plants registered in British Columbia, most of which are seasonal. Thirteen were engaged in canning, 117 were cold storage plants and 96 were involved in packing and other types of processing. Ninety four of these plants were located on Vancouver Island, 113 on the Lower Mainland and 19 on the Central and North coasts. In value of finished goods, the plants on the Lower Mainland generated 70% of the total, those on Vancouver Island 10% and those on the Central and North Coast 20% (Price Waterhouse 1990). Processing activities are less dispersed in British Columbia than in the Atlantic provinces.

Monthly average employment in the fish processing sector was around 5,000 during 1987–89. Because of the seasonal nature of the industry, peak employment is considerably higher, around 8,000–11,000 during 1986–89. The direct person-years of employment in the fish processing sector were estimated at 5,600–5,700 in 1988 and 1989 (Price Waterhouse 1990).

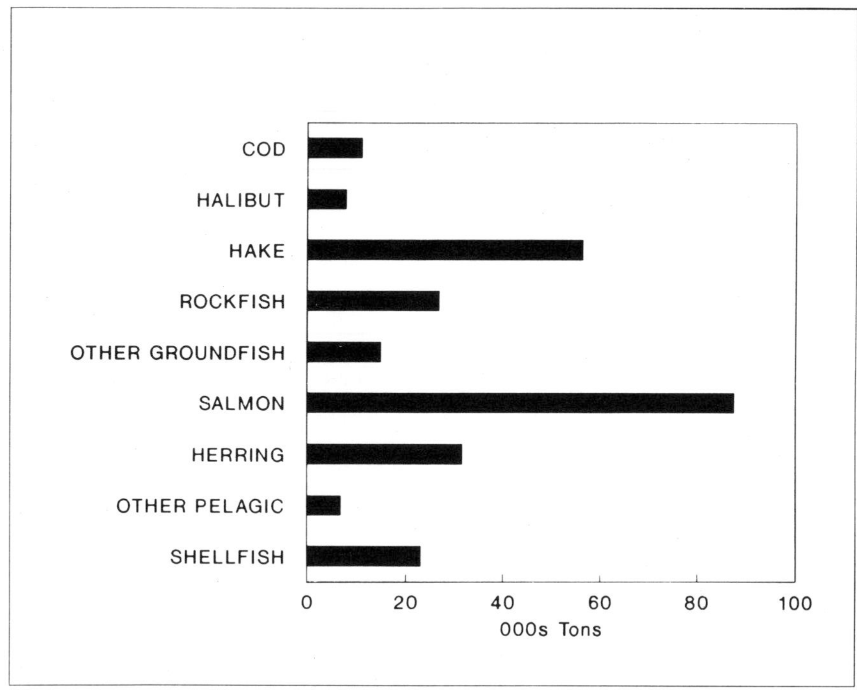

FIG. 1-9. Volume of the 1988 British Columbia catch by major species.

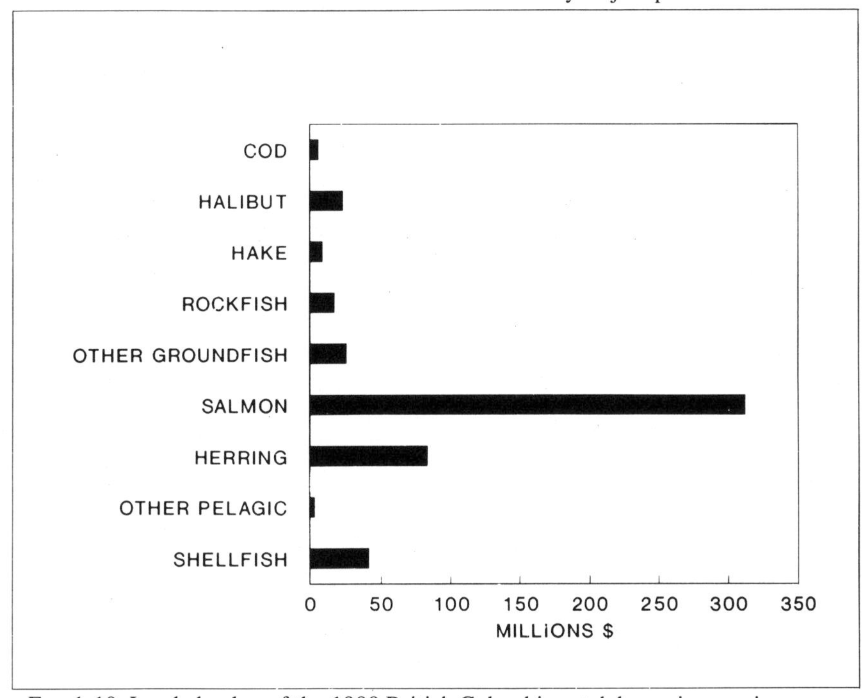

FIG. 1-10. Landed value of the 1988 British Columbia catch by major species.

Price Waterhouse (1990) concluded that at the height of the fishing season there are some 29,750 persons directly at work full and part-time in the fish harvesting, processing, sportfishing and aquaculture sectors in British Columbia. They estimated that the total fishing industry generates about 16,500 direct person-years of employment.

The British Columbia fishing industry is also heavily export-oriented although somewhat less dependent on exports than the Atlantic industry. About two thirds of

B.C. salmon products are exported, most of the herring roe and 75% of the groundfish production.

While both the Atlantic and B.C. fishing industries are export-oriented, there is a significant difference in the markets. Japan is the largest export market for B.C. fish products, accounting for approximately 46% of total exports in 1988 valued at $331 million. In addition to being virtually the only market for roe herring, Japan also receives about one quarter of B.C.'s salmon production. The U.S. was the second largest market in 1988 accounting for 23% of B.C. exports valued at $161 million. Salmon accounted for 51%, groundfish 21%, shellfish 18%, and halibut and herring 10% of B.C. exports to the U.S. In 1988, the EC ranked third with 21% of B.C. exports. The primary export to the EC is salmon.

3.4 The Inland Fisheries

Despite Canada's great abundance of freshwater lakes and rivers, its inland commercial fisheries produce only a very small percentage of commercial fish landings. Inland commercial landings in 1988 were 50,000 tons with a landed value of $91 million, accounting for 3% of total Canadian landings and 5.6% of landed value.

This, however, is not a true measure of the significance of the inland fisheries to Canadians. Recreational fisheries are a very important component of the total freshwater catch. Pearse (1988) estimated the total harvest of all Canadian freshwater fisheries at 174,000 tons per year. Of this, the commercial fisheries accounted for 28.7% at 50,000 tons per year, the recreational fisheries 66.1% at 115,000 tons and the native fisheries 5.2% or 9,000 tons.

The story of Canada's freshwater fisheries is generally one of decline with the advance of settlement by European immigrants in the 19th century. Numerous human activities destroyed or reduced fish habitat. These included lumbering, dams, agricultural clearance and run-off, railway and road building, mining and industrial and domestic pollution. Overfishing also became a problem as the human population increased.

Canada's commercial freshwater fisheries are dominated by the fisheries in the Prairie provinces, the Northwest Territories, and the Ontario fisheries including northern Ontario and those in the Great Lakes. Small commercial fisheries occur in freshwater in the Atlantic provinces, Québec and the Yukon. The primary species caught in these various fisheries include whitefish, walleye, northern pike, mullet, yellow perch, rainbow smelt, sauger, Pacific salmon, gaspereau, and the American eel (Pearse 1988).

For a description of Canada's freshwater fisheries, problems and suggested solutions, see Peter Pearse's 1988 *Rising to the Challenge: A New Policy for Canada's Freshwater Fisheries.*

3.5 The Arctic Fisheries

Although not substantial in quantity or landed value, fish and marine mammal resources contribute significantly to the livelihood and culture of Canada's native peoples in the Arctic. The Arctic coastal area is populated primarily by aboriginal people, particularly Inuit. About 40,000 people, 78% aboriginal, occupy 60 communities along the Arctic coast and in the MacKenzie Delta (Clarke 1993).

Arctic native peoples have harvested fish and marine mammals for thousands of years. Subsistence fisheries provide a major source of food and support the cultural life of these residents. As Clarke (1993) notes, the commercial and recreational fisheries and the commercial use of marine mammal products contribute one of the few sources of employment and cash to the economies of the area.

The species exploited are primarily anadromous fish and marine mammals. The anadromous fish migrate to estuaries, rivers and lakes and away from the Arctic seas in winter. Most marine mammals also migrate to areas where water and ice conditions are more favourable, e.g. southern Davis Strait and the Bering Sea.

The anadromous fishery harvests primarily Arctic charr. It also harvests other species such as whitefish, cisco and inconnu and Atlantic salmon. Although the exact amount of the subsistence harvest is not known, it probably constitutes the major portion of the total catch of anadromous fish (Clarke 1993). There is also a small commercial fishery for Arctic charr and some other anadromous species. This involves about 500 fishermen.

Several species of marine mammals occur in the Arctic — beluga, narwhal, bowhead, walrus, ringed seal, harp seal, and hooded, bearded and harbour seals in the eastern Arctic and Hudson Bay, and beluga, bowhead and the ringed and bearded seals in the western Arctic. About 6,000 natives participate in the subsistence harvest for marine mammals. The subsistence harvest of beluga, narwhal and walrus numbers in the hundreds. The harvest of seals has been in the low thousands since the collapse of the harp seal market following the European ban on the import of "whitecoat" pelts in the early 1980's.

Clarke (1993) concluded that the Arctic fisheries provided significant benefits to the local residents. He estimated the rough dollar value of benefits as $15 million for the replacement value of food from the subsistence harvest, $2 million as other consumer surplus benefits and $7 million as the value added to the Canadian economy. The most important contribution is the support that the Arctic fisheries provide to the aboriginal cultures, food and other materials for their use, and employment for 50 to 75% of the population.

3.6 The Recreational Fisheries

About 90% of Canada's recreational fisheries occur in fresh water. These fisheries are beyond the ambit of this book. There are, however, important marine recreational fisheries, e.g. for various species of Pacific salmon, bluefin tuna, and fishing in fresh water for anadromous species such as Pacific and Atlantic salmon. These anadromous fish are heavily exploited by commercial marine fisheries and also important to Canada's native peoples. Conflicts between commercial and recreational fishermen over use of this common resource have intensified in recent years.

Canada's coastal and inland waters can be divided into five areas from a recreational fisheries perspective. In the *Pacific* area, which includes the coastal and inland waters of British Columbia and the Yukon, three salmon species predominate in the marine recreational fishery — chinook, coho and to a lesser extent pink salmon. Chinook, coho and steelhead are also fished in inland waters during their spawning migrations. In the Yukon, the sports fishery catches Arctic grayling, trout and chinook.

In the *Northern* area, four salmonids — Arctic charr, Arctic grayling, lake trout and brook trout — are the major interest for anglers. The *Prairie* area is exclusively freshwater. In the *Southeastern* area of Ontario and Québec, freshwater species are also the primary attraction. Chinook, coho and pink salmon have been introduced into the Great Lakes system. Atlantic tomcod are fished through the ice in a Québec winter fishery on the St. Lawrence system.

In the *Atlantic,* Atlantic salmon is the primary ocean-going species of interest to anglers. They also fish, to a much lesser extent, for mackerel, pollock, cod and bluefin tuna.

Statistics on recreational fishing are difficult to obtain and those on the impact of recreational fishing are subject to widely differing interpretations. The Department of Fisheries and Oceans in cooperation with the provinces conducts a National Sportfishing Survey every 5 years. The 1985 Survey indicated that more than five and a half million Canadian and over 900 thousand non-resident anglers fished in marine and inland waters in 1985. They spent approximately $4.4 billion on recreational fishing (DFO 1987a). Figures on the marine recreational fisheries component of the national total are not available.

The 1985 Survey indicated that recreational fishing was on the increase. Approximately 500,000 more people fished recreationally in 1985 than in 1980, an increase of 10%. Anglers spent approximately $2.7 billion more on goods and services for fishing in 1985 than in 1980, an increase of 76% in constant dollars (DFO 1987a).

The majority of resident anglers live in Ontario (46% of the total angling population), Québec (20%) and British Columbia (12%). More than two thirds of tourist anglers fished in Ontario.

Ontario attracted almost half the total recreational fishing effort in 1985, measured by the number of days a person fishes or "angler days". Québec was second and British Columbia third.

The 1985 Survey indicated that anglers spent $4.4 billion in 1985 on goods and services directly for sport fishing. Of this, $2.5 billion was expended on vacation packages, accommodation, campsites, food, travel, boat rentals, fishing supplies and equipment, guide services and fishing licences. Another $1.9 billion was spent on major items purchased for sport fishing, such as special vehicles, camping equipment, boats, cottages and land.

These expenditures support about 40,000 jobs in the recreational fishing industry. This includes people employed with lodges, outfitters, manufacturers and retailers of sport fishing goods, boats and boating services, fishing guides and other related businesses.

The $4.4 billion figure estimated as the expenditure and investment in recreational fisheries in Canada in 1985 is not directly comparable to the $3.4 billion value of fishery products from the commercial fisheries in 1988. Attempts to compare these figures can be misleading.

One thing is clear. The recreational fisheries predominate in the inland waters of Canada, both in tonnage and value of the harvest. In marine waters, the commercial fisheries are dominant. This book focuses primarily on the marine commercial fisheries. Recreational fisheries for marine or anadromous species are dealt with but these form a small component of Canada's marine fisheries.

3.7 Aquaculture

In addition to the fisheries harvesting "wild" populations of fish, mammals and aquatic plants, Canada has a small but rapidly growing aquaculture industry. Aquaculture involves the husbandry or cultivation of aquatic animals or plants. It has been practised in some parts of the world for centuries.

Attempts to enhance production of wild species through fish culture in Canada date back to the 1850's. An extensive network of federal hatcheries was built up producing mostly salmonids. The 1890's saw attempts to enhance cod production through hatcheries in Newfoundland. Aiken (1993) concluded that a century of salmonid enhancement culture enabled Canada in the 1980's to rapidly establish a salmon sea cage culture industry. Federal hatcheries through their expertise and the production of smolts enabled the Canadian industry to establish itself in the world market before increasing global salmon production in the late 1980's began to soften the world market.

Canada's share of world aquaculture is small. Canada probably ranks about fiftieth in total aquaculture production (Aiken 1993). In 1988 total aquaculture production from all species was 18,000 tons. Over the past 15 years, the aquaculture industry in Canada has grown rapidly. In 1984 the annual value of the industry totalled approximately $7 million (DFO 1990a). By 1987 the total had grown to $62 million. By the end of 1988 the growth rate had accelerated again — by 76% — to $109 million. In 1987, aquaculture represented only 3% of the landed value of Canadian fisheries. During 1988, the figure doubled to 6%. Industry has forecast that the aquaculture share of the landed value of Canadian fisheries could rise to 25% by the year 2000. The recent rapid expansion has focused primarily on salmonids. During the 1980's the Canadian salmonid farming industry became the fourth largest in the world.

Canada's aquaculture industry has been described as "a mosaic comprised of small family operations, publicly owned and listed companies, and subsidiaries of large corporations involved in international seafood markets and foreign-owned enterprises" (DFO 1990a). The industry is diversified, with considerable variation between and within provinces in the species cultivated and the magnitude of the enterprise.

In the Atlantic provinces, there is a well established aquaculture industry based on Atlantic salmon, trout, mussels and oysters. In the 1980's the industry diversified into scallops. Nova Scotia is involved with all four of the primary species. New Brunswick is the leader in producing Atlantic salmon, taking advantage of favourable climatic conditions in the Bay of Fundy. Prince Edward Island leads in mussel production. Newfoundland has the smallest aquaculture sector of the four Atlantic provinces. Even there, however, there is growing interest in aquaculture, particularly of Atlantic salmon.

Ontario and Québec have long been major trout producers. Québec also produces mussels. These two provinces are now experimenting with other species. Ontario has been testing salmon and Arctic charr, and Québec has begun commercial production of salmon.

Aquaculture of trout, primarily rainbow trout, has been underway in the Prairie provinces since prairie "pot hole" culture began in Manitoba in 1968. Recently, there has been some commercial production of Arctic charr.

The aquaculture industry in British Columbia primarily involves salmon and oysters but also cultivates trout and marine plants. The most rapid aquaculture growth in Canada in the 1980's occurred in the British Columbia salmon farming. This industry mushroomed from four commercial farms in 1981 to 135 operating farm sites in 1989. In 1989 some rationalization occurred as a number of firms were placed in receivership. Lower salmon prices worldwide, partly as a result of overproduction by aquaculture in Norway, injected some realism into euphoric growth expectations.

B.C. farmed salmon production in 1989 was 12,400 tons, compared with 6590 tons in 1988. Estimated production in 1990 was 15,000 tons (DFO 1990a). These figures compared with a 1988 B.C. commercial salmon catch of approximately 85,000 tons. The value of B.C. salmon farm production rose dramatically from less than $1 million in 1980 to an estimated $72 million in 1989.

Factors which contributed to the expansion of Canadian aquaculture in the 1980's included:

- a growing worldwide demand for seafood, which is expected to continue;
- the need to meet this demand with production from new sources such as cultured fish;
- an increasing reliance on aquaculture products as a dependable year-round source of quality seafood; and
- widespread recognition of the high quality of Canadian cultured seafood products (DFO 1990a).

Other factors, however, may limit future growth of the Canadian aquaculture industry. Immediate concerns include the following:

- New sources of supply have lowered prices for some species;
- Processing, cultivation and distribution costs are reducing profits; and
- Competition within Canada and among countries is intensifying (DFO 1990a).

These are cost and market-related constraints. Another major constraint is the growing conflict between the emerging aquaculture industry and traditional fishermen, real estate developers and recreational interests. There has been growing opposition by other users of the coastal environment.

Despite these concerns, the Canadian aquaculture industry appears well positioned for future growth. In recognition of the potential of the Canadian aquaculture industry, DFO in 1990 released *Cultivating the Future: An Aquaculture Strategy for the 90s* (DFO 1990a). Underpinning this Strategy document was a recognition that the aquaculture industry was increasingly developing the capacity to address its own needs for supplies and services. The stated objective was to "encourage private-sector development by creating a climate for entrepreneurship and innovation."

One of the challenges of the coming decades will be to optimize benefits from both the wild fisheries and the cultured fisheries, while managing the conflict between them. This book focuses on the challenge of managing the wild fisheries. Aiken (1993) provides an overview of the aquaculture industry in Canada, its problems, challenges and opportunities. For further information consult Pritchard (1984), Wildsmith (1982), Pfeiffer and Jorjani (1986), Boghen (1989), Egan and Kennedy (1990), and the 1988 Report of the Standing Committee on Fisheries and Oceans entitled *Aquaculture in Canada* (Canada 1988a).

4.0 CONCLUSIONS

Canada's abundant fish resources and marine and freshwater habitats support fisheries in the many lakes and rivers scattered across the vast Canadian land mass, in coastal regions, and on the continental shelf within and outside Canada's 200-mile fisheries zone. These fisheries are diverse and dependent upon a renewable but variable resource base.

The two major marine fisheries occur along the Atlantic and Pacific coasts. Although the Arctic coastline is long, the Arctic supports primarily subsistence fisheries for Canada's native peoples. Inland the commercial fishery is small and overshadowed by the important recreational fisheries.

Canada's marine commercial fisheries on the Atlantic coast occur primarily in economically impoverished regions with few alternative employment opportunities. More than a thousand Atlantic coastal communities depend on the fisheries for employment and income. On the Pacific coast the degree of regional dependence on the fisheries is less than on the Atlantic. Nonetheless, the fisheries are an important component of the British Columbia economy.

Canada's commercial fisheries appear insignificant to the national economy. Regionally, however, they are the lifeblood of more than a thousand communities. Fishermen, plant workers and their families depend on the fisheries to put bread on the table. Some fishermen earn good incomes from fishing; others eke out a meagre existence. But fishing is much more than a source of income. It is a way of life.

From generation to generation.

CHAPTER 2

THE JURISDICTIONAL CONTEXT FOR FISHERIES MANAGEMENT IN CANADA

"By its text the Canadian Constitution appears to grant the federal government total control over the fisheries."

– David VanderZwaag, 1983

1.0 THE LEGISLATIVE BASE

1.1 General

The federal-provincial division of powers over fisheries in Canada is established in the *Constitution Act, 1867*. Unlike other natural resources, which were assigned under Sections 109 and 117 to the provinces, the exclusive legislative authority of the Parliament of Canada under section 91(12) of Part VI of the Act, entitled the "Powers of Parliament", included "sea coast and inland fisheries."

Section 92 of the *Constitution Act, 1867* set forth the exclusive powers of the provincial legislatures. The provincial powers were intended to be clear-cut, relating to matters of local concern. Sections 109 and 117 vested in the four original provinces, Nova Scotia, New Brunswick, Québec and Ontario the natural resources within their boundaries. Section 92(13) gave the provinces the authority to legislate with respect to property and civil rights. Furthermore, the provinces were given power over provincial public lands in s.92(5).

At first glance, it would seem that the *Constitution Act, 1867*, gives the federal government exclusive authority for the management of fisheries in Canada. Federal legislators and administrators took this view in the 30 years following Confederation. As an example, the first *Dominion Fisheries Act, 1868*, was wide-ranging in scope. That Act (s.2) authorized the federal Minister of Marine and Fisheries "where the exclusive right of fishing does not already exist by law, [to] issue or authorize to be issued, fishery leases or licenses for fisheries and fishing wheresoever situated or carried on..." The Act (s.19) also gave the Canadian cabinet complete power to make any regulations it saw fit "...for the better management and regulation of the sea-coast and inland fisheries...".

The Ministry of Marine and Fisheries used this authority to extend licensing and leasing of fishing rights to all parts of the new Dominion, including the British Columbia salmon fishery in 1882. Licensing and leasing was originally intended to play a central role in management for conservation purposes. By the late 1800's, however, licensing was being used mainly to identify fishermen for statistical purposes. Nonetheless, "for more than 30 years following Confederation, the federal government exercised unchallenged authority in fisheries matters" (Scott and Neher 1981).

Discontent with this broad view of federal powers grew throughout the early decades following Confederation. Fisheries was just one area of jurisdictional dispute between the federal and provincial governments (Thompson 1974).

1.2 Judicial Interpretation of the Division of Powers

During the past century the courts have rendered judgments on a number of fisheries disputes. Over time these judgements have clarified the authorities of the two levels of government pertaining to fisheries.

1.2.1 The Queen v. Robertson

The first case to go before the Supreme Court of Canada was *The Queen v. Robertson* (1882). This dispute involved the granting by the federal Minister of a 9-year salmon angling lease on a section of the Miramichi River in New Brunswick. The lease was challenged by persons claiming the exclusive right to fish that portion of the river. Mr. Robertson's lease had been ruled invalid by the courts of New Brunswick, and he sought compensation from the federal government. The Exchequer Court of Canada decided in Robertson's favour. The federal government appealed this decision to the Supreme Court of Canada.

The Supreme Court ruled that Mr. Robertson's lease was invalid because the federal government did not have the power to grant fishing privileges in a river which had belonged to a province or an individual. As mentioned earlier, under Sections 109 and 117 of the *Constitution Act, 1867*, the provinces retained ownership of provincial lands. Under section 92(5) they were given the power to make laws respecting "the management and sale of public lands belonging to the province and of the timber and wood thereon."

The court held that riparian owners (owners of the land bounding a stream) had an exclusive right to fish in nontidal streams. The court concluded that it was beyond the authority of the federal government to grant leases or licences to fish in nontidal portions of rivers.

> Chief Justice Ritchie stated:
>
> "to all general laws passed by the Dominion of Canada regulating 'sea-coast and inland fisheries', all must submit, but such laws must not conflict or compete with legislative power of the local legislatures over property and civil rights beyond what may be necessary for legislating generally and effectually for the regulation, protection and preservation of the fisheries in the interests of the public at large. Therefore, while the local legislatures have no right to pass any laws interfering with the regulation and protection of the fisheries, as they might have passed before Confederation, they, in my opinion, clearly have a right to pass any laws affecting the property in those fisheries, or the transfer of transmission of such property under the power conferred on them to deal with property and civil rights in the province, inasmuch as such laws need have no connection or interference with the right of the Dominion parliament to deal with the regulation and protection of the fisheries, a matter wholly separate and distinct from the property in the fisheries."

The court also determined that in New Brunswick there existed an exclusive right to fish in nontidal waters flowing through ungranted lands. This exclusive right to fish belonged to the Crown as trustee for the benefit of the people of the province.

Following this decision, Québec and New Brunswick began to lease salmon angling rights in the nontidal reaches of rivers within their jurisdictions. It appears that the federal government regarded this decision as pertaining only to the leasing of salmon angling privileges. Apart from this concession in Québec and New Brunswick, it continued to exercise complete control over fisheries.

1.2.2 The Ontario Fisheries Reference

Other provinces were becoming discontent with the federal government's control of fisheries. In the late 1800's, Ontario became a leading proponent of provincial rights. In 1885, Ontario adopted the *Ontario Fisheries Act*, an act to "regulate the fisheries" of the province. It appears that this was an attempt to capture revenues from the fisheries (Thompson 1974). In 1892, the Ontario government enacted legislation purporting to regulate the manner of fishing in the province as well as attempting to license and collect fees. This was a clear challenge to the federal government's attempt to exercise control over the inland fisheries. After being heard by the Supreme Court of Canada, this case went to the Judicial Committee of the Privy Council on appeal by both parties. The Privy Council decision in this case was a turning point in clarifying the powers of the federal and provincial governments with respect to fisheries.

The 1898 decision (the *Ontario Fisheries Reference*) distinguished between legislative jurisdiction and proprietary rights. The Judicial Committee determined that the existence of legislative power in a particular subject area does not imply any proprietary rights in that area and that all proprietary rights held by provinces at the time of Confederation remained intact. This judgment declared those sections of the *Fisheries Act* which empowered the Minister of the Department of Marine and Fisheries to grant licenses or leases conferring exclusive rights of fishing in nontidal waters to be beyond the powers of the Parliament of Canada:

> "...Insofar as s.4 of the Revised Statutes of Canada c.95 empowers the grant of fishery leases conferring an exclusive right to fish in property belonging not to the Dominion, but to the provinces, it was not within the jurisdiction of the Dominion Parliament to pass it."
>
> Their Lordships continued:
>
> "...Provisions prescribing the mode in which a private fishery is to be conveyed or otherwise disposed of, and the rights of succession in respect of it, would be properly treated as falling under the heading 'Property and Civil Rights' within s.92, and not as in the class 'Fisheries' within the meaning of s.91. So, too, the terms and conditions upon which the fisheries which are the property of the province may be granted, leased, or otherwise disposed

of, and the rights which consistently with any general regulations respecting fisheries enacted by the Dominion Parliament may be conferred therein, appear proper subjects for provincial legislation, either under class 5 of s.92, 'The Management and Sale of Public Lands' or under the class 'Property and Civil Rights'."

This decision confirmed that provinces may legislate on matters regarding property and civil rights in fisheries, such as transfers, rights of inheritance or conditions of leasing a *provincially-owned fishery*.

The Privy Council ruled that the provisions of Ontario's *Fisheries Act* which attempted to *regulate the manner of fishing* were invalid because these matters were clearly within the federal parliament's exclusive jurisdiction. Thus the exercise of provincial rights can be restricted by federal regulations governing the conservation and preservation of the fishery resource including such matters as type of fishing gear, limits on the amount of catch, closed seasons and the species and size of fish that may be caught. Federal jurisdiction on these matters encompasses all Canadian waters, both coastal and inland.

1.2.3 Federal-Provincial Negotiations 1898–1949

Although seemingly clear, the 1898 decision of the Privy Council led to confusion about the extent of provincial rights respecting fisheries. Federal-provincial conferences were convened and the federal government transferred to the provinces the administration of fisheries legislation for those fisheries which they considered to be subject to provincial law because of property rights. The federal government was not, however, prepared to recognize a provincial property right in the tidal or seacoast waters of the provinces. The initial administrative delegation to Ontario and Québec was confined to the nontidal fisheries.

Initial arrangements for the delegation of administrative control of the nontidal fisheries were concluded in 1899 with Ontario and Québec. Administrative control of all of the inland fisheries, both sport and commercial, was transferred to Ontario. The federal government also delegated the administration of fisheries in inland rivers and waters to Québec but it claimed and exercised jurisdiction over the fisheries of the Gulf of St. Lawrence below a line drawn from Cap Chat to Pointe des Monts.

The next fisheries jurisdictional confrontation involved British Columbia. At the turn of the century, the British Columbia government became interested in securing management control over the B.C. fisheries and, in particular, using licence fees as a source of revenue. British Columbia, in concert with the eastern maritime provinces, claimed property rights to the then-existing 3-mile limit. The issue was resolved temporarily in 1901. It was agreed that, pending a settlement of the question of jurisdiction over seacoast fisheries and the outcome of a federal provincial conference to be held in 1902, the federal government would retain control of the fisheries, essentially as before. It would, however, account for any portion of licence fees which might thereafter be agreed belonged to the province. Essentially, that agreement left British Columbia free to administer that portion of the inland commercial and sport (non salmon) fisheries outside of the so-called 'railway belt' which had been conveyed to the federal government by B.C.'s terms of union with Canada.

A major Federal-Provincial Conference on Fisheries in 1902 accomplished little to resolve the remaining jurisdictional questions. Perhaps the most significant outcome was the position adopted by the three Maritime Provinces. They agreed with Québec and British Columbia in arguing that the 1898 Privy Council decision gave the provinces property rights in the fisheries, not only in nontidal but in tidal waters to the 3-mile limit. However, they offered to lease any such right to the federal government for "adequate consideration". A subsequent Privy Council decision prevented this.

The agreement between British Columbia and the federal government foundered in 1907, when the province informed the federal government that it was not prepared to renew the agreement and would bring its *Fisheries Act* of 1901 into force. It also declared its intention to require licences from all fishermen. Section 8 of the *B.C. Fisheries Act* stated: "We may prohibit fishing in any waters of the province, excepting under authority a fishing lease, permit or licence, granted by authority of our Act." The B.C. government moved to enforce the Act, collecting licence fees both from fishermen and fish processors. This precipitated a reference to the Judicial Committee of the Privy Council in 1913-1914. The Privy Council (*Attorney General for British Columbia v. Attorney General for the Dominion of Canada — the B.C. Fisheries Reference*), concluded that the federal government had exclusive jurisdiction in tidal waters because a public right to take fish there existed dating back to the Magna Carta in 1215. The Council held that the provinces had no property rights and hence no jurisdiction over any aspect of fisheries in tidal waters. The Lord Chancellor, Viscount Haldane, explained the Council's decision as follows:

> "It was a right open equally to all the public, and therefore, when by section 91 seacoast and inland fisheries were placed under the exclusive legislative authority of the Dominion Parliament, there was in the case of the fishing in tidal waters nothing left within the domain of the Provincial Legislature."

The federal government took over the administration of the B.C. tidal fisheries and those in the 'railway belt' following the decision. Administration of oyster farming had been transferred by agreement in 1912 to the B.C. government and has remained with the province since then.

Like British Columbia, Québec had claimed jurisdiction in tidal waters to the 3-mile limit. Because of the 1914 decision, the federal government wanted to modify the provisions of the 1899 agreement whereby Québec had been administering certain fisheries in the Gulf of St. Lawrence. Québec, however, contended that the Magna Carta did not apply to that province and hence the 1914 decision did not apply to Québec fisheries. A further reference to the courts was necessary, with the Privy Council handing down its decision in November 1920. In the meantime Canada had transferred all of its inland fish hatcheries within Québec to the province. Thus the province had secured control of the inland fisheries and the hatchery system. (Similarly, federal fish hatcheries were transferred to the province of Ontario between 1912 and 1918).

In 1920, the Privy Council decided against Québec's position. While the Magna Carta did not apply in Québec, the rights of the public respecting fisheries had been established by statutes passed between 1763 and 1867. These statutes provided for a public right to fish in all navigable waters. The Justices concluded: "As the public right was not proprietary, the Dominion Parliament has in effect exclusive jurisdiction to deal with it."

The Council determined that the public right of fishing did not, however, extend to fixing 'fishing engines' to the soil. Therefore, in tidal waters, the bed of which may be owned by the province, and in which the fishing would be carried on by 'engines fixed to the soil' there was an element of conflict. (Fishing 'engines' in effect refers to fishing gear). In such situations there could be a form of concurrent federal and provincial jurisdiction over fisheries using fishing gear attached to the soil. (The *Quebec Fisheries Reference 1921*).

Superficially, the federal government had won more than in the earlier court decision because of the decision regarding "navigable waters," which would put some inland nontidal fisheries under federal jurisdiction. At that time virtually all fishing in Québec was carried out by fixed gear attached to the soil. Thus both levels of government claimed jurisdiction in tidal waters, placing fishermen in a difficult situation. For practical reasons, the administration of the fisheries in all tidal waters within the Province, with the exception of those around the Magdalen Islands, was delegated to Québec in 1922. The right to administer federal regulations governing the province's coastal fisheries was delegated to Québec by federal Order in Council (P.C.360 passed by the Governor in Council on February 13, 1922). Québec was delegated "under proper regulation by the Federal Government as to conditions under which such fishing may be carried on, the responsibility for the administration of all the coastal fisheries of the Province [as well as the river fisheries], with the exception of those about the Magdalen Islands." The intent was to avoid a double licensing system. The administration of fisheries surrounding the Magdalen Islands was not transferred to Québec at that time because of the different manner in which these fisheries were conducted. In 1943 the Orders in Council were amended to delegate to Québec administration of the fisheries of the Magdalen Islands (P.C.1890, on March 15, 1943). Thus, until the 1980's, Québec exercised management privileges over its coastal fisheries which were not available to other provinces.

Meanwhile, another form of fisheries confrontation had developed in British Columbia. In spite of the 1914 decision of the Privy Council, British Columbia continued to assert a presence in fisheries management. It introduced measures in 1910 to license canneries and placed a ban on new cannery construction in the northern areas of British Columbia. In 1911 the federal government introduced a requirement for federal cannery licences and a limitation on the number of licences in northern B.C. (Neher and Scott 1981). A section of the federal *Fisheries Act* of 1914 introduced licences and fees for fish canneries. The federal and provincial governments battled over this issue until a further decision by the Privy Council in 1929 declared that cannery licensing was outside federal jurisdiction (*Attorney General for Canada v. Attorney General for British Columbia — Fish Canneries Reference, 1930*). The federal government had contended that the licensing scheme was "necessarily incidental to the protection and conservation of fisheries", a view which was rejected by the Privy Council. The Council determined that section 91 (12) of the *BNA Act* did not give the federal government the power to regulate fish processing. In its judgement, the contentious sections of the 1914 *Fisheries Act* fell under the subject "Property and Civil Rights", a matter under provincial jurisdiction. Lord Tomlin said:

> "...Trade processes by which fish when caught are converted into a commodity suitable to be placed upon the market cannot upon any reasonable principle of construction be brought within the scope of the subject expressed by the words 'sea coast and inland fisheries'."

No part of section 91 gave the federal government the authority to licence fish canneries. The federal regulations encroached on the provincial right to legislate on the subject of property and civil rights.

From this decision it would appear that federal control over fisheries normally ends once the fish has been legally caught. Legally-caught fish become the property of the fisherman capturing them and are then subject to provincial law with respect to processing. However, fish processed in a province but destined for export are subject to the *Federal Fish Inspection Act.* A distinction must be drawn, however, between freshwater and tidal fisheries, specifically tidal fisheries located outside the boundaries of a province. As demonstrated by the court decisions discussed previously, provincial jurisdiction over property and civil rights is confined to the territorial boundaries of a province. For coastal provinces, this is normally the low-water mark. Thus fish caught in tidal waters remain under federal control until landed on provincial territory or taken out of Canadian fishing zones (B. Menoury, Ottawa, pers. comm.).

Around that time (1929–30), the federal government agreed that the natural resources of Manitoba, Saskatchewan, and Alberta and the "railway belt" in British Columbia should be administered by the provinces. In 1870, when the province of Manitoba was formed and in 1905, when the provinces of Saskatchewan and Alberta were formed, Canada had retained the right to administer the natural resources in these provinces. In 1929 and 1930, Canada entered into agreements with Manitoba, Saskatchewan and Alberta which contained the following provision for fisheries:

> "Except as herein otherwise provided, all rights of fishery shall.... belong to and be administered by the Province, and the Province shall have the right to dispose of all such rights of fishery by sale, licence or otherwise, subject to the exercise by the Parliament of Canada of its legislative jurisdiction over seacoast and inland fisheries."

These agreements were later confirmed by the *BNA Act, 1930.*

Also in 1930, an Act was passed which returned to B.C. all of its natural resources, the "Act Respecting the Transfer of the Railway and Peace River Block." This transfer included the administration of the inland fisheries within the 'railway belt'.

Newfoundland, as a separate colony of Britain, and then Dominion, had managed its own fisheries prior to Confederation with Canada in 1949. Alexander (1976, 1977) has described the management mechanisms in place prior to Confederation. Under the Terms of Union with Canada, the provisions of the BNA (later Constitution) Act were made applicable to Newfoundland. Thus the federal government obtained the exclusive power of legislating with respect to Newfoundland's seacoast and inland fisheries. There was, however, one restriction on the use of this legislative power. Term 22(2) stipulated that:

> "All Fisheries Laws.... shall continue in force in the Province of Newfoundland as if the Union had not been made, for a period of five years from the date of Union and thereafter until the Parliament of Canada otherwise provides, and shall continue to be administered by the Newfoundland Fisheries Board; and the costs involved in the maintenance of the Board and the administration of the Fisheries Laws shall be borne by the Government of Canada."

At the time of Union provincial authorities retained control of the inland sport fisheries. Canada assumed control of all other fisheries. Administration of the inland sport fisheries was transferred to the Federal Government in 1954.

1.2.4 Principles for Division of Responsibility

These developments in the interpretation of the *Constitution Act, 1867*, established the following principles for the division of powers regarding fisheries between the federal and provincial governments:

- The Parliament of Canada has exclusive legislative authority over both seacoast and inland fisheries, under section 91(12).
- In nontidal waters there is a right of property in fisheries. The provinces, pursuant to section 92(13), may enact legislation concerning such matters as conveyances, leases and succession to proprietary fisheries, subject to federal regulations on fishing seasons and manner of fishing and other conservation methods.
- In tidal waters there are no property rights in fisheries. Hence, provincial legislatures have no jurisdiction over fisheries in those waters except to the extent that provincial laws can infringe on provincial/federal jurisdiction via the constitutional doctrine of "necessarily incidental effect" (J. Meaney, DFO, Ottawa, pers. comm.).

2.0 THE POWER TO LEGISLATE FOR SOCIO-ECONOMIC AS WELL AS CONSERVATION OBJECTIVES

2.1 General

One important issue concerning the scope of federal legislative power for fisheries has only recently been clarified by the courts. It had long been held that the primary federal jurisdiction was restricted to the conservation and protection of fish stocks. Chief Justice Laskin expressed this view in 1976 in the case *Interprovincial Co-operatives v. the Queen*:

> "Federal power in relation to fisheries does not reach the protection of provincial or private property rights in fisheries through actions for damages or ancillary relief for injury to those rights. Rather; it is *concerned with the protection and preservation of fisheries as a public resource*, concerned to monitor or regulate undue or injurious exploitation, regardless of who the owner may be, and even in suppression of an owner's right of utilization" (Emphasis mine).

Thompson (1974) commented on the various court interpretations of section 91 and 92 of the BNA, 1867:

> "For a federal government scheme to be upheld as valid under section 91(12) of the *BNA Act 1867*, the seacoast and inland fishery, *it must not appear to relate to anything that is not strictly relating to the regulation, protection, and preservation of fish and fisheries* or to something that is not necessarily incidental to the regulation, protection, and preservation of fish and fisheries" (Emphasis mine).

While this view of the federal government's role may have held in the late 1800's and the early part of this century, by the middle of this century the federal government had become heavily involved with social and economic aspects of the marine fisheries. By the 1970's federal policies and regulations invoked specific social and economic objectives in allocating access to fishery resources (see Chapter 4). Nowhere was this more evident than in the 1976 Policy for Canada's Commercial Fisheries (DOE 1976a). The government stated:

> "The strategies adopted reflect a fundamental redirection in the government's policy for fishery management and development. Although commercial fishing has long been a highly regulated activity in Canada, the object of regulation has, with rare exception, been protection of the renewable resource. In other words, fishing has been regulated in the interest of the fish. In the future it is to be regulated in the interest of people who depend on the fishing industry. Implicit in the new orientation is more direct intervention by government in controlling the use of fishery resources, from the water to the table."

2.2 The Gulf Trollers Decision

The federal government's move to regulate the fisheries to meet social and economic objectives came under court challenge in the 1980's. Justice Collier of the Federal Court's Trial Division in 1984 issued an order quashing certain Public Notices issued by federal fishery officers in British Columbia varying "close times" for commercial salmon troll fishing in the Gulf of Georgia. (Gulf Trollers Association v. Minister of Fisheries, and Shinners 1984). Justice Collier endorsed the contention of the Gulf trollers that the powers of the Federal Minister and of Parliament over "seacoast and inland fisheries" were limited to conservation and protection of the resource. He concluded:

> "The respondent's decisions of April 16 were, to my mind, promoted by two disparate and pervading reasons: conservation, and socio-economic management allocations. The second purpose was, in my view, beyond permissible constitutional powers. The two considerations were inextricably mixed. In those circumstances the court cannot segregate. The decision must fall."

Both the fishing industry and government viewed Justice Collier's decision as a fundamental challenge to the powers of the federal government to regulate the fisheries for social or economic purposes. This precipitated a major debate. The federal government appealed the Collier decision to the Federal Court of Appeal. It also took action to amend the *Fisheries Act* to clarify the Minister's powers. On March 5, 1985, it introduced in the House of Commons Bill C-32, an Act to amend the *Fisheries Act*. This Bill included a new "Purposes" section for the *Fisheries Act*. Section 2.1 read:

"*Purposes*

2.1 The purposes of this Act are:

a) to provide for the conservation and protection of fish and waters frequented by fish;
b) to provide for the proper management, allocation and control of the sea-coast fisheries of Canada;
c) to ensure a continuing supply of fish and, subject to paragraph (a), taking into consideration the interests of user groups and on the basis of consultation, to maintain and develop the economic and social benefits from the use of fish to fishermen and others employed in the Canadian seacoast fishing industry, to others whose livelihood depends in whole or in part on seacoast fishing and to the people of Canada; and

d) to provide for the proper management and control of the inland fisheries of Canada and, subject to the constitutional jurisdiction of the provinces, for the allocation of those fisheries."

Thus by proposing a Purposes section in the *Fisheries Act* the federal government sought to clarify the scope of its legislative powers respecting fisheries to include allocation of fish, taking into account social and economic considerations. This proposal met stiff opposition from parliamentary critics and various industry groups. Testifying before the House of Commons Standing Committee on Fisheries and Forestry, Mr. Pierre Asselin, Director, Legal Services, Department of Fisheries and Oceans, explained the impact of Justice Collier's decision:

> "The decisions of Mr. Justice Collier, being decisions of a court of first instance, are not binding on other judges. They cast doubt on the Minister's authority to allocate, from a constitutional point of view. The best advice that my department could give the Minister in the circumstances is, since Mr. Justice Collier's decision is not a decision of a Court of Appeal and therefore not binding on other judges, then neither is the Minister bound by it and he can continue to allocate, in specific areas, until challenged. In the meantime.... we attempt to amend the *Fisheries Act* to clarify the Minister's authority to allocate, at least under the Act, so if a future challenge arises, the challenge will be squarely based on the constitutional issue and will prevent a court from ducking the constitutional issue by saying the Act is insufficient to give the Minister that authority" (Canada 1985a).

Fisheries and Oceans Minister John Fraser, in making the case for the amendment, stated:

> "Those who say, maybe no Minister should have power other than to make regulations directly and clearly linked only to conservation, fail to recognize the complexity of the situation.... If I administer this department looking only to the conservation of the stock, I will have a heck of a lot easier job, but I do not know that I am acting in the public interest. You can save the fish, but if you cannot manage the fishery and the people in it in such a way that they get a share, then what happens to the complex fishing society that is out there? You would literally end up in a situation where the mighty and powerful will get the first grab at whatever species there are and nobody else will get a look-in.... I do not know how else we can ask somebody to manage, under the rule of law, a complex situation and try to obtain two things: one, the conservation and enhancement of the stocks; and secondly, some kind of equitable sharing of the resource" (Canada 1985b).

There was prolonged debate in the Standing Committee, particularly on the question of amending the Act "to provide for regular formal consultation with traditional user groups in the commercial fisheries". The Bill that passed and became law stipulated that the changes to the *Fisheries Act* to clarify its purpose and scope would, unless replaced by further legislation, sunset (cease to be law) on January 1, 1987 (S.C. 1985, c.31).

Meanwhile the federal government pressed forward with its appeal of Justice Collier's decision to the Federal Court of Appeal. On November 3, 1986, the Federal Court of Appeal reversed the Collier decision thereby determining that the legislative power of Parliament with respect to seacoast fisheries is not confined to the conservation and protection of the resource (Re: Minister of Fisheries and Oceans and Gulf Trollers Association 1986). Justice Marceau stated: "The case.... raises a constitutional issue which must undoubtedly be dealt with." Referring to the court cases described previously, he observed:

> "The question to be answered is whether Parliament, in the exercise of its legislative competence under s.91(12) of the *Constitution Act, 1867*.... can establish close and open times for catching fish not only for the purpose of conservation but also for a purpose of a socioeconomic nature.... The words used by the judges in those cases to characterize and better describe the federal power were not intended to indicate authoritatively that Parliament's competence was confined to legislation necessary to conserve and protect the fishery to the exclusion of all other objectives. In fact, I never understood the distribution of legislative powers made by sections 91 and 92 between the central parliament and the provincial legislatures as having been devised with some regard to the purpose for which a power could be exercised. The distribution is made on the basis of classes of subjects..."
>
> Justice Marceau concluded:
>
> ".... It is the pursuit of allocative objectives in the management of the fisheries which is

objected to, but such allocation, even if considered independently of any idea of conservation, does not trench on any provincial power. The power conferred on Parliament in s.s.91(12) of the *Constitution Act, 1867*, is not qualified, in my understanding, by any inherent condition that it be used to pursue some specific objectives and not others. Parliament may manage the fishery on social, economic or other grounds, either in conjunction with steps taken to conserve, protect, harvest the resource or simply to carry out social, cultural or economic goals and policies. In fact, in my view, unless and until the party attacking legislation on division of power grounds identifies a possible trespass on a specific law-making power of the other level of government, the purpose for which a piece of legislation was passed is of no concern of the courts."

2.3 McKinnon v. Canada

In another recent case in 1987 (*McKinnon v. Canada*), a longline fisherman from Shelbourne County in Nova Scotia challenged the application of the federal government's Sector Management Policy. The challenge was made on several grounds, one being that it did not come within Parliament's authority to regulate the fisheries because that authority "is limited to laws which are directed at the preservation and protection of the fish stocks and laws necessarily incidental to that purpose." The Federal Court of Canada's Trial Division rejected this argument.

Justice Martin stated:

> ".... motivation, in this case, is irrelevant. Once it is determined that Parliament has the legislative authority to regulate any particular field it is not for the courts...to question Parliament's motivation or the wisdom of the legislation.... The federal authority to regulate the fisheries undoubtedly includes the right to determine the times during which fish may be caught and the means employed to catch them. It also includes, in my opinion, the right to determine the areas in which fishermen may or may not prosecute the fishery.
> "The fisheries.... is more than fish or the preservation and conservation of fish. It includes those who prosecute it, and the means, times and places of its prosecution. It is not only desirable but, in my view, essential that the federal authorities consider in their regulatory schemes or licensing systems for the fisheries, the fishermen and the social and economic impact on their livelihood of an orderly system for allocating the available fish stocks to the several groups, categories or classes of operators."

Thus, the courts in the Gulf Trollers and McKinnon cases confirmed the right of the federal government to regulate the tidal fisheries for social and economic as well as conservation objectives.

3.0 RECENT ATTEMPTS TO CHANGE THE CONSTITUTIONAL DIVISION OF POWERS RESPECTING FISHERIES

The late 1970's witnessed a resurgence of federal-provincial confrontation over the constitutional division of powers with respect to fisheries. Unlike the battles early in this century, this time attention focused, not on interpreting the constitutional powers as agreed in 1867, but rather upon changing the division of powers. Some provinces sought through constitutional amendment greater provincial control over the management of fisheries.

3.1 The 200-mile Limit and Demands for Expansion

Elsewhere we shall see that the 1970's were a decade of turbulence, particularly in the east coast fisheries. Foreign overfishing during the 1960's had depleted the principal resource, Atlantic groundfish. Existing international mechanisms had failed to come to grips with the resource decline in a timely and effective manner. Canada, through its participation in the Law of the Sea negotiations, had sought control over the fishery resources to the edge of its continental shelf. On January 1, 1977, Canada extended its fisheries jurisdiction to 200 nautical miles. This was done in the context of a developing consensus on a new Law of the Sea. Reacting to pressures to put in place a more effective management regime, Canada extended its fisheries jurisdiction to 200 nautical miles. It implemented stringent management controls on both foreign and domestic fishermen to rebuild fish stocks and provide a more stable resource base for the fishery. The coastal provinces and the fishing industry greeted the 200-mile fishing zone with considerable optimism.

Canada instituted a comprehensive management regime including the establishment of conservative catch quotas, allocation of quotas among competing fleet sectors and limited entry/licensing systems. At the same time, certain provincial governments and segments of the fishing industry called for expansion of the

Canadian fleet and processing capacity to reap the perceived benefits of the new zone. The governments of Newfoundland and Nova Scotia joined forces to advocate expansion of the fleet. Supported by Prince Edward Island, they developed a fleet development plan and presented it to the federal government late in 1977. Newfoundland saw the new zone as an opportunity to expand its fishery. The provincial government produced several major policy papers, setting forth its vision of how the benefits of the new zone might best be utilized (Newfoundland 1978a). One Newfoundland proposal was the "Primary Landing and Distribution Centre" concept. This envisaged joint ventures with foreign freezer trawlers which would land their product at one port facility (Harbour Grace) for distribution to plants dependent on the inshore fishery. The intent was to extend the operating season of those inshore plants. Meanwhile in Nova Scotia major processors were pressing the provincial government to support joint ventures and expansion of the offshore fleet.

The Newfoundland-Nova Scotia position paper called for fleet expansion on the grounds that the inshore fleet required major investment because "most vessels were of wooden construction and had to be replaced in order to maintain the primary economic base for the continued existence and future prospects of most small communities in the Atlantic provinces." It argued that the offshore fleet required major investment because a large proportion of the fleet would be obsolete or unserviceable within a decade. The estimated price tag for fleet replacement was $900 million (*Lunenberg Progress Enterprise*, January 4, 1978).

Federal Fisheries Minister Roméo LeBlanc met with the provincial ministers in Ottawa in October 1977. Federal scientists described the status of the fishery resource and the opportunity for stock rebuilding. They stated that, as the stocks rebuilt, catch rates for the existing fleet would improve. They also observed that the increase in the allowable catches of stocks traditionally harvested by Canada could be caught with the existing fleet. Thus, fleet expansion was not warranted.

In a November 1977 speech to the Yarmouth Board of Trade, Fisheries Minister Roméo LeBlanc cautioned that a $900 million fleet development scheme could not be justified (DOE 1977a). This Yarmouth speech was considered by many to be his reply to the provincial development proposal. But the provinces pressed for a formal federal reply.

3.2 Pressures for Greater Provincial Control Over Fisheries

At the February 1978 First Ministers' Conference, Newfoundland Premier Frank Moores linked fisheries development to the province's place in Confederation and made the point that fish was as important to Newfoundland as oil to Alberta and wheat to the Prairie provinces. While Nova Scotia Premier Regan indicated that he still accepted the legitimacy of federal jurisdiction over fisheries, Premier Moores launched the first salvo in the constitutional battle to follow:

> "It is generally accepted that ownership of resources is the tool provided by the Canadian constitution for the survival of the provinces as viable members of the Canadian confederation. In the case of fisheries, we as a Province have neither ownership nor control of a resource which is of vital importance, socially and economically.... We find that the present *de facto* control of the fishery by the central government is totally unacceptable" (CICS 1978a).

Newfoundland called for control of licensing policy to be delegated to the province for 5 years, with the province having the right to participate in setting and allocating quotas. Federal Minister LeBlanc responded forcefully:

> "Those who might urge a balkanizing of management into regional or provincial units should recognize that, while lines on maps may provide solutions where resources are geographically fixed, they do not readily contribute to solving the problems associated with managing migratory fisheries" (CICS 1978b).

Throughout 1978, federal/provincial acrimony over fisheries deepened as Minister LeBlanc continued to reject proposals for fleet expansion. By April 1978 Nova Scotia Fisheries Minister Dan Reid raised publicly the possibility of a new division of federal-provincial responsibilities. In July 1978 Newfoundland threatened to take the fisheries jurisdiction issue to court.

In early fall 1978 the Conservative Party came to power in Nova Scotia. The new Premier, John Buchanan, was soon criticizing the federal government's perceived inaction, calling Ottawa's follow-up on extended jurisdiction "pathetic". He called for greater federal-provincial cooperation (*Halifax Chronicle Herald*, October 27, 1978).

The battle was joined again at the October 1978 constitutional conference and the November First Ministers' Conference on the Economy. At the First Ministers' Conference, Nova Scotia criticized federal fisheries policy: "Only a dramatic reversal of present policy will prevent foreign fishing fleets and processors from continuing to reap the major rewards from our resources" (CICS 1978c).

Newfoundland Premier Moores pressed forward aggressively on the issue of constitutional reform. He contended that the division of powers was the most important issue in constitutional reform. Control over natural resources was a particular concern. Unlike the other provinces, Newfoundland did not have control of the natural resources most crucial to its future development — the fisheries and offshore petroleum. He stated:

> "My government is and should be held largely accountable for the condition of the fishery and we require the tools to execute these responsibilities.... We believe strongly that any changes in our constitution must include provision for a real and meaningful input by the Provinces into the management of our fishing resources and we are prepared to accept the costs proportionate to our involvement, as well as any revenues involved. Serious consideration should be given to attaining concurrent jurisdiction in this sector with provincial paramountcy" (CICS 1978d).

In the ensuing months Newfoundland continued to push the case for constitutional change. At the Federal-Provincial Continuing Committee of Ministers on the Constitution (CCMC) on November 23–25, 1978, Newfoundland proposed an amendment to section 91(12) of the *BNA Act* which would have given the federal government paramountcy in:

- — scientific and other forms of research;
- — setting Total Allowable Catches;
- — issuing quotas to foreign countries and licensing foreign vessels; and
- — all aspects of fish inspection.

Under Newfoundland's proposal, the provincial governments would have paramountcy in licensing domestic vessels and in establishing quotas within the Total Allowable Catches for provincial use and establishing the level of surplus for allocation by the federal government to foreign countries. It appears that the Newfoundland proposal was rejected at that time by the other nine provinces. (Pross and McCorquodale 1987). However, in October 1978, Alberta, in its paper on constitutional change titled *Harmony in Diversity: A New Federalism for Canada,* recommended that "seacoast and inland fisheries be a concurrent power in the Constitution, with provincial paramountcy".

At a CCMC meeting in December 1978, Newfoundland refined its proposal:

> The federal government would have paramountcy over:
> - — international negotiations;
> - — surveillance and international enforcement;
> - — basic and applied research;
> - — quality standards and inspection for export;
> - — licensing of foreign vessels (based on residual quotas).
>
> Provincial governments would have paramountcy over:
> - — determination of quotas after Total Allowable Catches had been established;
> - — division of those quotas;
> - — plans for harvesting;
> - — licensing of local boats and those of other provinces (Newfoundland 1978b).

Following these meetings, Newfoundland apparently concluded that Nova Scotia's support was the key to a successful strategy. Newfoundland pressed its case in a series of bilateral meetings. A tentative agreement was reached with Nova Scotia that the concept of concurrent jurisdiction should be embodied in the constitution (*St. John's Evening Telegram*, November 30, 1978).

At the February 1979 First Ministers' Conference Premier Buchanan put the argument for concurrent jurisdiction thus:

> "Unlike all other natural resources, the fishery is managed by federal authority. Since it is not possible, certainly not practical, to realistically pursue fisheries development separately from fisheries management, the province cannot develop its fishing industry. We are convinced.... that it is necessary at this time for Canada to accept concurrency in the fisheries as a constitutional fact of life" (CICS 1979a).

Although Newfoundland, Nova Scotia and Alberta were now calling for constitutional change with respect to fisheries, they did not make headway at that meeting. Prince Edward Island and New Brunswick felt that their claims to traditional grounds and quotas were threatened by the proposal. The federal government agreed to involve the provinces more directly in fisheries decision-making (through a regional council of fisheries ministers) but opposed further study of constitutional change. Although the Newfoundland-Nova Scotia proposal had some support from Ontario, British Columbia and Alberta, it was not supported by the federal government, New Brunswick and Prince Edward Island. In the meantime the constitutional treatment of inland fisheries had emerged as an issue in the CCMC. Minister LeBlanc had said that the federal government had no aversion to change. The problem was to avoid fragmented management, a threat to resource conservation. In the end, the Prime Minister concluded:

> "There is not a willingness on the part of the federal government to change the constitution yet in that regard. Some of you have been rather adamant on other subjects. This is one where I would like to show progress, but where I take the advice of the man who represents the fisheries and the fisherman just as much as some of you at this table" (CICS 1979b).

Shortly after the conclusion of the Conference, the Newfoundland-Nova Scotia alliance was sundered because of conflicting views over the allocation of the recovering northern cod stock. Federal Fisheries Minister LeBlanc met with Atlantic fisheries ministers on February 15, 1979, in Halifax to discuss the allocation of an additional 12,000 tons of northern cod. The Provinces were unable to agree on how it should be allocated. Newfoundland demanded that all of it go to Newfoundland companies.

When the meeting concluded there was apparently some expectation that Newfoundland and the Maritimes would receive 5,000 tons each with 2,000 tons going to Québec. However, the federal government ended up allocating 9,500 tons to Newfoundland, 2,500 tons to the Maritimes and 500 tons to Québec (*Halifax Chronicle Herald*, March 7, 1979).

In the weeks following the First Ministers' Conference various fishermen's organizations spoke out in support of continued federal jurisdiction over the fishery. The Executive Director of the Fisheries Council of Canada on February 26, 1979, stated:

> "The present constitution provides the best and most realistic division of powers between the federal and provincial governments and the Fisheries Council of Canada is strongly opposed to any change" (FCC 1979).

Jerry Nickerson, head of the Nickerson-National Sea Products conglomerate, and a leading critic of federal fisheries policy, stated that he was "quite unequivocal in his belief that the federal government should retain paramount authority in most areas of marine fishery jurisdiction" (*Halifax Chronicle Herald*, May 26, 1979).

3.3 The Northern Cod Fishery and The Constitution

By the summer of 1979 there were new players both at the federal and provincial level — a new federal Minister of Fisheries, James McGrath, and a new Premier of Newfoundland, Brian Peckford. The management and allocation of the northern cod resource in the post-extension era figured prominently in their relationship and ultimately impacted on the outcome of the September 1980 constitutional conference.

The federal government convened a Northern Cod Seminar in Corner Brook, Newfoundland in August 1979 to address the future management of this important resource. Newfoundland argued there that Northern Cod should be reserved for its own inshore and middle distance effort. Any offshore surplus should be allocated for landing in Newfoundland ports.

A November 1979 Discussion Paper on Bilateral Issues presented by Premier Peckford to Minister McGrath defined the major issue for bilateral discussion — the management and harvesting of the Northern Cod Stock. Premier Peckford contended "that the inshore fishery can and should take up to 85 percent of the Northern Cod, with any residual volume being taken by Newfoundland-based offshore effort to supply resource-short plants in this province" (Newfoundland 1979a).

He also reiterated Newfoundland's commitment to jurisdictional change respecting fisheries:

> "The ultimate fishery policy objective of the Province is constitutional change.... This objective will be pursued within the framework of the Continuing Committee on the Constitution."

But he then took action which may have thwarted the achievement of his constitutional objective. In December 1979, Premier Peckford told all "mainland" fish companies operating in Newfoundland that if they accepted (directly or indirectly) Newfoundland cod caught by freezer trawlers and landed on the mainland the Government would reserve the right to "consider them ineligible for any provincial assistance programs or to apply for any further processing licences" (Newfoundland 1979b).

This prompted the Nova Scotia government to consider the full implications of concurrent jurisdiction for Nova Scotia's fishery interests. It began to realize that the constitutional change advocated by Newfoundland would likely deny Nova Scotia fishermen access to the recovering northern cod resource off Newfoundland.

3.4 The September 1980 Constitutional Conference

Preparatory work for a major Constitutional Conference continued during the summer of 1980. The Continuing Committee of Ministers on the Constitution met in Vancouver in July. It concluded that "agreement or near agreement is possible regarding *inland fisheries, marine plants and sedentary species*" (CCMC 1980a). This option would have given the provinces exclusive jurisdiction over:

- — inland fisheries in nontidal waters;
- — aquatic plants in nontidal waters;
- — marine plants in tidal waters in or adjacent to a province;
- — sedentary species in tidal waters in or adjacent to a province; and
- — aquaculture within a province and in tidal water adjacent to a province.

Views differed with respect to diadromous species. Two alternatives were presented. Under one, the federal government would establish the total allowable catches for diadromous species in nontidal waters and their allocation between provinces. Under the other "the Parliament of Canada would have the power to exclusively make laws in relation to sea coast and inland fisheries for diadromous species" (CCMC 1980a). The federal government considered it necessary to retain control over diadromous species and in particular to provide for the protection of fish habitat and native peoples' fisheries as well as the prevention of fish diseases.

Regarding the *sea coast fisheries*, the Committee reported that "it has been possible to develop a 'best efforts' draft based on a revision of the Newfoundland proposal which provincial officials, two coastal provinces excepted, consider worthy of consideration by governments as a basis for consensus." It noted that "Federal representatives continue to oppose this approach maintaining it is unworkable." In the meantime they had developed a draft proposal for mandatory consultation (CCMC 1980b).

In the meantime the shift in the Nova Scotia position had become publicly apparent with the appointment of Edmund Morris as Fisheries Minister. On September 2, 1980, he announced that "Nova Scotia is not prepared to renounce two centuries of tradition by supporting concurrent jurisdiction of the fishery.... concurrent jurisdiction as they propose it is not a concept we are prepared to support.... In truth, concurrency means divided authority, not concurrency or togetherness" (*Halifax Chronicle Herald*, September 2, 1980). On September 4, Minister Morris attacked Newfoundland directly stating that Newfoundland's northern cod claim had been "a warning as to how they could conduct themselves if the provinces got together to decide allocation of various species. Newfoundland's demands, when they haven't had the authority to make demands, [have] been a great instructor as to what they would demand if they had the authority" (*Halifax Chronicle Herald*, September 4, 1980).

Thus, immediately prior to the September 8–12, 1980, Constitutional Conference, the federal government appeared willing to contemplate a transfer of inland fisheries to provincial jurisdiction, provided that native rights were protected and that the federal government would retain jurisdiction over fish diseases and control of pollution in international and interprovincial waters. The federal government was also prepared to contemplate the transfer of jurisdiction over aquaculture and immobile coastline species (such as oysters). For diadromous species (such as salmon) and for marine fisheries generally the federal government had rejected proposals for a change in jurisdiction. Instead, it had offered to make consultation mandatory under the Constitution.

All provinces were in favour of a transfer of jurisdiction with respect to inland fisheries although they disagreed about the federal caveats. Regarding aquaculture and immobile coastline species, the provinces seemed willing to accept a transfer of jurisdiction. On diadromous species, the provinces almost all advocated provincial jurisdiction. This would have had the most significant implications for British Columbia given the dominance of Pacific salmon in that province's fishery.

With the single exception of Nova Scotia, all provinces seemed to favour some form of jurisdictional change giving the provinces greater powers over marine fisheries. Québec and British Columbia favoured exclusive provincial jurisdiction; the rest favoured concurrent jurisdiction with more or less power left with the federal government. Prince Edward Island and New Brunswick apparently wanted a strong federal role in the context of concurrent jurisdiction, while Newfoundland favoured a greatly diminished federal role.

At the September 1980 Constitutional Conference, Premier Peckford pressed his case for change, stating: "The Government of Newfoundland insists that it is absolutely essential that there be a constitutional provision ensuring a role for the province in fisheries management. There is no credible argument which can be made to deny this right to coastal provinces.... The local aspects of fisheries are matters which properly reside in the provincial legislatures.... The Newfoundland approach does not seek to exclude the federal government in any area. The underlying principle is shared jurisdiction. Our proposal is an effective reflection of the basic principle that the government with the most immediate connection with an area of authority should have jurisdiction in that area." He noted that some progress had been made on the question of inland fisheries. With respect to sea coast fisheries, he asserted that: "In the discussions of our Ministers, all provinces except one agreed that there must, at least, be concurrent jurisdiction in this area" (Newfoundland 1980a).

The federal position was summarized by Prime Minister Trudeau as follows:

> "*On the seacoast and marine fisheries*, we have said no, we think it would be wrong to have concurrent jurisdictions, but we have offered to have much closer consultation — and have even offered to write a mandatory provision to that effect into the new Constitution.
>
> "*On inland fisheries*, we are ready to give this to the provinces because, generally, they are doing it now and doing it well. More important, most of these fisheries lie within a single province.... *On marine and aquatic plants*, we have said no to the provinces, because of our concern about protecting the plants which are so often a major part of the habitat of seacoast species.... *On sedentary species*, we are ready to give the provinces the jurisdiction they seek. *On the anadromous species*, we are not prepared to go along" (CICS 1980).

While Newfoundland had achieved a consensus of nine provinces on most issues and while the federal government had indicated flexibility on certain issues, in the end the matter was dropped from the agenda which led to the *Constitution Act, 1982*. The federal government was adamant in its belief that fragmented management of the marine fisheries would be a backward step for the conservation and management of fish stocks and for the fishing industry. It was supported by one major coastal province, Nova Scotia, and by virtually all fishermen's and processor's organizations which had spoken out on the issue.

3.5 Why the 1978–80 Provincial Attempt to Change the Division of Powers Regarding Fisheries Failed

It is interesting to speculate what might have happened had Premier Peckford not antagonized Nova Scotia by his strong claim to northern cod for Newfoundland. What if Nova Scotia had continued its support for concurrent jurisdiction and hence formed a consensus of ten provinces seeking change? Would this have resulted in some form of jurisdictional change respecting fisheries? The dispute over northern cod affected the dynamics of the constitutional discussions on fisheries, just as it was later to reach international prominence in the federal-provincial dispute over French fishing rights and the Canada-France boundary dispute.

Another factor too often discounted is the influence of personality in shaping policy. Fisheries Minister LeBlanc was a major force in resisting the provincial thrust for constitutional change in fisheries. Pross and McCorquodale (1987) commented:

> "Like most political scientists we are inclined to argue that whilst individuals often affect events and decisions, they seldom change the course of public policy. Yet Roméo LeBlanc, who held the position of Minister of Fisheries from 1974 to June 1979 and from February 1980 to September 1982 appeared to do that. He towered over the fisheries policy community influencing events far beyond his jurisdiction.... LeBlanc said he would give nothing away and he did not."

They go on to assert that the provincial attempt to change the division of powers respecting fisheries probably failed for four reasons:

> "First, it was too ambitious. Paramountcy implied competence and resource beyond what the public in general and the policy community in particular knew was, or could be, available to the provinces. Concurrency — in the form of institutionalized consultation and some sharing of functions — might have been politically feasible, but concurrency with paramountcy was a nonstarter. Second, the provinces were not able to demonstrate to one another or to the industry that they were capable of statesmanlike management of the resources. In this their downfall was accelerated by the third factor undermining the whole campaign, LeBlanc's skill in dividing the forces ranged against him.
>
> "Finally the talks fell victim to the forces which can be seen at play in the constitutional talks at large...This (general) insistence on consensus in the end simply resulted in intransigence and an 'all or nothing' attitude and prevented multilateral negotiations between the provinces to bridge the very real differences of interest and ideology which divided them. Once negotiations failed during the autumn

of 1981 and the federal government embarked on its option of unilateral action, the question of transfer of power to the provinces, except for the provisions of Part IV of the New Act, were lost in the struggles between patriation and the entrenchment of a Charter of Rights."

I would add these factors to this list:

1. The fisheries constituency (fishermen's organizations, processors, etc.) supported unanimously the position of federal paramountcy over the sea coast fisheries.
2. They did so because they agreed with the federal government that inherently fragmented management would be disastrous for the fishing industry. As LeBlanc said:
   "The other natural resources don't swim.... A fish caught in a five-way tug of war is going to be a funny-looking fish at the end" (DFO 1979a).
3. It is not only fish which move seasonally across political boundaries; the fishermen do so even more. The changes proposed by Québec and Newfoundland, in particular, would have required fishermen to be licenced by up to five different provinces in the Atlantic and by the federal government in international waters. The potential for conflict and confusion was enormous.

Meanwhile, Newfoundland continued its fight to change the division of powers respecting fisheries. The province reiterated its position on constitutional amendment in the 1980 document — Towards the Twenty First Century-Together (Newfoundland 1980b). This position was essentially the same one advanced in the discussions leading up to the 1980 Constitutional Conference.

3.6 Consolidation of Federal Powers Over Fisheries

In April 1983, Nova Scotia put forward a new position on Fisheries Jurisdiction (Nova Scotia 1983). This suggested that provincial jurisdiction or some form of delegated responsibility would be appropriate for:

- — inland fisheries, excluding diadromous species;
- — sedentary species (e.g. clams, oysters);
- — sea plants;
- — aquaculture; and
- — certain socioeconomic or contractual matters occurring within "these waters" (e.g. over-the-side sale) — meaning waters belonging to Nova Scotia.

The Nova Scotia position paper asserted that federal authority should be paramount in fisheries matters: (1) which affect more than one province, (2) which deal primarily with the protection and conservation of the common property fisheries resources "in the wild", or (3) which involve international agreements, interprovincial trade and international trade, navigation, shipping or harbour facilities. The position paper also explained Nova Scotia's support of the federal position on fisheries jurisdiction in 1980, namely, to prevent the balkanization of fisheries by restricting the mobility of the fleets. The document appeared to differ from Nova Scotia's 1980 position in one important respect. It differentiated between biological factors and socioeconomic factors and stated: "The increasingly dominant role of the federal government in this latter aspect of fisheries management can be seriously questioned." It suggested that these aspects are more properly the responsibility of the province. This position was put forward before the court decisions in the Gulf Trollers and McKinnon cases clarified the existing legislative powers of the Federal Parliament in this respect.

During the constitutional discussions of 1979 and 1980 Newfoundland had cited the 1922 delegation of the administration of coastal fisheries to Québec in advancing its case for a greater provincial role in the management of fisheries. This administrative delegation was terminated and the federal government resumed the administration of marine fisheries in Québec effective April 1, 1984. The Task Force on Atlantic Fisheries proposed the consolidation of "federal management of the fisheries in the Gulf of St. Lawrence by resumption of full federal responsibility for licensing and other aspects of marine fisheries management in Québec." The Task Force rationale was as follows:

> "The continued division of responsibility for administration of the harvesting sector in Québec creates (1) confusion among fishermen; (2) both duplication and gaps in essential activities and in obtaining essential information; and (3) dissatisfaction in other provincial governments as well as in the industry generally. The difficulty of obtaining the most basic information on the Québec fisheries itself leads to the conclusion that the status quo is not acceptable" (Kirby 1982).

The federal cabinet accepted this recommendation and proceeded to rescind the Order in Council delegating administration of the marine fisheries to Québec. A new full-fledged Québec Region of the Department of Fisheries and Oceans was established, with bolstered staff, along with a new research facility, the Maurice Lamontagne Institute at Mont Joli, Québec. Conflict soon erupted between the federal Minister of

Fisheries and Oceans, Pierre De Bané, and the Québec Minister responsible for fisheries, Jean Garon. When the dust settled, Québec was left with the administration of inland fisheries including anadromous species in non-tidal waters.

3.7 Meech Lake and Beyond

During the "Québec round" of constitutional discussions leading up to the Meech Lake Accord of 1987, Newfoundland again pressed for increased provincial jurisdiction over fisheries. By this time, however, the degree of support for Newfoundland's position was less than it had been in 1980. Nonetheless, the Meech Lake Constitutional Accord of 1987 included a provision (Section 50(11)) for constitutional conferences to be convened at least once each year. Section 50(2) provided that:

> "The conferences convened under subsection (1) shall have included on their agenda the following matters:
> (a) Senate reform;
> (b) roles and responsibilities in relation to fisheries; and
> (c) such other matters as are agreed upon."

To some degree, this was an important gain for Newfoundland. Certainly, Premier Peckford saw it as leaving the door open for future discussion of jurisdictional change.

Speaking on ratification of the Accord in the Nova Scotia legislature, Premier Buchanan stated that:

> "The Government of Canada has made it very clear, many times, that there will be no constitutional change as far as jurisdiction of fisheries is concerned, and over half the Premiers have already indicated that clearly, so...of course, there is no possible way of it ever happening" (Nova Scotia 1988).

Nova Scotia indicated that it would seek to entrench a provision for mandatory consultation in the constitution, when it came up for subsequent amendment.

In the discussions leading up to ratification of the Accord by New Brunswick, Premier McKenna indicated that federal jurisdiction in fisheries matters must be maintained. There was continued fishing industry opposition to any change in fisheries jurisdiction. In a submission to the Special Joint Committee on the 1987 Constitutional Accord in September 1987, the Maritime Fishermen's Union stated:

> "Our fishermen members are deeply concerned over the reference to fisheries in the Meech Lake Accord. They feel that anything that leads to increased provincial jurisdiction in the fishery constitutes a direct threat to their future livelihood.... We feel the real losers in the 'provincialization' of the fishery would be the inshore fishermen. Since the fish stocks themselves know no provincial boundaries, anything like transferring of licensing or quota management powers to the provinces would invite fish wars, increase effort on stocks, and eventually lead to their collapse" (MFU 1987).

Also testifying before the Joint Committee, Mr. Ron Bulmer, President of the Fisheries Council of Canada, stated that further discussions on fisheries roles and responsibilities would fuel conflicts between provinces and various fishing interests. He urged the committee to prevent the institutionalizing of annual debates over fisheries jurisdiction. The FCC also requested that there be no further devolution of current federal responsibilities in fisheries to the provinces (FCC 1987). Ultimately, the Meech Lake Constitutional Accord did not receive the necessary ratification by all the provincial legislatures within the 3-year time limit for ratification stipulated in the Accord. The Newfoundland legislature ratified the Accord but later rescinded its ratification when a new government took office. Manitoba failed to ratify the Accord.

The demise of the Accord plunged the country into a constitutional crisis, with the future of Canada and Québec's role in Canada as central issues. The question of "roles and responsibilities for fisheries" was soon lost in this intense debate.

Meanwhile, the new government of Newfoundland, headed by Premier Clyde Wells, altered somewhat Newfoundland's decade-long search for greater constitutional jurisdiction over fisheries. Instead, the Wells' government sought joint management of the fish resources adjacent to Newfoundland. Joint management was recommended by the 1990 Harris Panel on Northern Cod (Harris 1990) and the 1991 Report of the Maloney Commission of Enquiry into the Alleged Erosion of the Newfoundland Fishery by Non-Newfoundland Interests (Maloney 1990). Responding to the Maloney Report, Premier Wells stated:

> "The findings of the Maloney Commission strengthen our position that day-to-day management of fisheries in the waters around Newfoundland can only be effectively and efficiently achieved through a joint Canada-

Newfoundland Board that would serve to integrate our respective policies. However, I must emphasize that the Province is not seeking a change in legal jurisdiction, but is seeking shared management with the Federal Government" (Newfoundland 1991).

Intense constitutional discussions again consumed the time and attention of Canadians during 1991 and 1992. The federal government in constitutional proposals released in 1991 did not envisage any change in jurisdiction over fisheries. The Charlottetown Accord negotiated by the Prime Minister and the provincial Premiers in August 1992 contained no reference to fisheries jurisdiction. In any event, this Accord was rejected by Canadians in the national referendum of October 26, 1992. Constitutional issues were then put on the back-burner.

It is unclear whether or to what extent the question of fisheries jurisdiction might surface again in a future round of constitutional discussions. As late as the fall of 1992, the Newfoundland government was still pressing for joint management. Testifying before the House of Commons Standing Committee on Fisheries and Forestry on December 8, 1992, Newfoundland Fisheries Minister Walter Carter reiterated Newfoundland's proposal for a Joint Fisheries Management Board modelled after the Canada/Newfoundland Offshore Petroleum Board. The proposed board would have the power to set allocations, license fish harvesters, vessels and processing plants (Newfoundland 1992).

The federal government, however, was moving towards institutional change which would vest decision-making powers on licensing and allocation matters in new Atlantic and Pacific quasi-judicial boards, which would place more power in the hands of the fishing industry (see Chapters 8 and 15). Meanwhile the jurisdictional context for marine fisheries management in Canada remains as set forth in the Constitution Act, 1867, and as interpreted by the courts since then.

CHAPTER 3

SOME BIOLOGICAL AND ECONOMIC ASPECTS OF FISH AND FISHING

"The management of marine fisheries presents a complex mixture of biological, economic, social and political problems."

– John Gulland, 1974

1.0 INTRODUCTION

The complex interplay of biological, economic, social and political factors in the management of fisheries is well illustrated in the management of Canada's marine fisheries over the past century. The past three decades in particular have brought sweeping change, both in Canada and worldwide. The most dramatic change has been the establishment of 200-mile national fishing zones.

What is fisheries management? One definition is — everything done to maintain or improve fisheries resources and their utilization. In this sense, fisheries management ranges from managing the resource to undertaking activities to promote fisheries development, e.g. exploratory fishing, gear and fleet development, environmental maintenance, processing and marketing. Initial efforts at fisheries management focused on the conservation of wild fish stocks. In the modern era, however, there has been growing recognition that fisheries have to be viewed as a total system from the fish in the water to the fish on the table. This system includes the harvesting, processing and marketing sectors as well as the resource, and it combines economic, diverse social and political as well as biological factors.

Management of fisheries — like the management of government or business — consists of seizing opportunities and tackling problems. When we attempt to manage fisheries systematically, we try to replace chance with control, blind sailing with good navigation, uncertainty about the future with a reasonable degree of predictability.

John Gulland (1974) listed several main questions faced by fisheries managers:

"(a) How big is the resource and how many fish can be caught each year while maintaining the stock for the future?;

"(b) Given the potential catch, how should this be used for the greatest benefit of the country?; and

"(c) What action needs to be taken to achieve these objectives?"

Fisheries biologists have been preoccupied with answering the first question. They have generally done well despite some high profile exceptions. There is still considerable room for improvement: in forecasting stock fluctuations and yields, understanding multispecies interactions, and understanding the influence of the environment on fluctuations in the abundance and availability of fish populations. Fisheries economists have been concerned with the questions of "greatest benefit" and "best use". One of their chief preoccupations has been the "race for the fish" arising because of the *common property* nature of the resource.

In this chapter I will outline the biological basis for fisheries management, the theory of fishing and the economics of fishing. This will provide the background to examine the evolution of fisheries management in Canada over the past several decades.

2.0 SOME BIOLOGICAL CHARACTERISTICS OF FISH RELEVANT TO FISHERIES MANAGEMENT

There are two main branches of fisheries biology (Cushing 1968). One is the study of the *natural history* of the stocks, concerned with how the fish spawn, feed and grow. The other is the study of the dynamics of fish stocks — the rates at which fish grow, reproduce and die. This latter aspect includes estimating fish abundance and is commonly known as *population dynamics*.

A number of biological characteristics of fish influence their harvesting and the methods of fisheries management. These characteristics include:

2.1 A Renewable Resource

A fish population is self-renewing unless subject to excessive fishing pressure or environmental catastrophe. There are endangered species or populations, e.g. the lake sturgeon in Canada. Normally, however, commercial fish populations are resilient and can continue to yield substantial catches without fear of extinction. Nonetheless, there are instances where fish populations have collapsed (at least in commercial terms) due to fishing down the parent stock to a level at which it is unable to sustain a commercially viable fishery. Examples include the Grand Banks haddock and Georges Bank herring. The recent decline in northern cod abundance appears to be an example where adverse environmental factors contributed to a sudden, dramatic resource collapse.

2.2 Unit Stock

Fisheries scientists and managers talk of particular "fish stocks", e.g. the northern cod stock, the Gulf of St. Lawrence redfish stock. A *stock* is a group of fish that can be treated as a homogeneous and independent unit. A few members of the group may mingle with other groups, but most stay with their own group. A fish population is considered a separate stock when it can be fished without the fishing affecting other populations of the same species. Cushing (1968) described the ideal stock as one that "has a single spawning ground to which the adults return year after year. It is contained within one or more current systems used by the stock to maintain it in the same geographical area." A stock is thus "a large population of fishes distinct from its neighbours and differences between stocks should be detectable genetically. Stock unity is maintained by reproductive isolation from generation to generation." For a discussion of mechanisms determining the differentiation of marine fish populations, see Sinclair (1988a).

In the Canadian Atlantic, there are about a dozen different stocks of cod, each of which is treated separately for management purposes (Fig. 3-1). Atlantic mackerel occur in Canadian waters in summer and fall and migrate south to the Canada/U.S. transboundary zone or into U.S. waters to overwinter. These mackerel are treated as a single stock, even though distinct southern and northern spawning components have long been recognized. Greenland halibut, or turbot, which are distributed over an area from the southern Grand Bank to Greenland, are also considered to be a single stock.

The concept of a stock is perhaps most clearly defined for the various species of Pacific salmon. The ability of adult Pacific salmon to return to a distinct spawning stream has led to the genetic separation of stocks. Ricker (1972) estimated that there are at least 8000 stocks of salmon in the north Pacific. Harvesting plans for Pacific salmon must take this into account and at the same time deal with the complex problem of "interceptions" of various stocks in mixed fisheries.

2.3 Age

Although they are comparatively small creatures fish live for a surprisingly long time. In their unexploited state Atlantic cod live to about 20 years, Atlantic herring to about 15 years, American plaice to about 25 years, and redfish to as much as 50–60 years.

Salmon, capelin and silver hake are exceptions to the general rule that the marine species commercially exploited in Canada live for a relatively long time. Anadromous species (species which spawn in freshwater) have a complex life cycle. Atlantic salmon spend 2–3 years in freshwater before going to sea and return to spawn after 1 year at sea (at this stage they are known as "grilse") or 2–3 winters at sea (large salmon). Most return to spawn at 5 years of age. Among the five species of Pacific salmon, the smallest, the pink salmon, has an extremely short 2-year life cycle. Most chinook are 4–5 years old when they return to spawn. Sockeye spend from 1 to 3 years in a lake before migrating to sea. They return to spawn after between 3 and 6 years at sea, although generally the run-cycle in southern British Columbia is 4 years. Most Pacific salmon die after spawning. Capelin, for example, many of which spawn and die on coastal beaches, live for only 4 years or so. Also under fishing pressure the proportion of older fish in a fish stock is reduced. The age profile of a moderate or heavily exploited stock shows a greater proportion of younger age groups.

In temperate waters such as the Canadian Atlantic and Pacific, annual rings on otoliths (ear-bones) or scales can be readily distinguished, and the age of the fish thus determined. This is analogous to determining the age of a tree from the annual rings in its trunk. Cod, plaice and redfish, for example, are normally aged by counting the annual rings in the otoliths. Salmon, on the other hand, are aged by counting the rings in the scales. Figure 3-2 shows the pattern of annual rings in a cod otolith. Early European researchers (Hjort, Peterson) verified that these rings were in fact "annual" by tracing very dominant year-classes through the fishery over several years.

At any given time, a stock of fish is made up of several *year-classes* (fish born in a particular year). A knowledge of the age composition of a fish stock is essential to any study of population dynamics and to fisheries management. Such knowledge enables researchers to determine such factors as longevity, growth rates, and age at maturity.

2.4 Growth

Generally, fish grow quickly in juvenile life, more slowly after reaching sexual maturity and even more slowly in old age. (Fig. 3-3). Because they grow continuously and by a factor of several times even during the period of mature life, the *growth rate* has considerable implications for fishing and fisheries management. The age at which fish are harvested can affect significantly the potential yield.

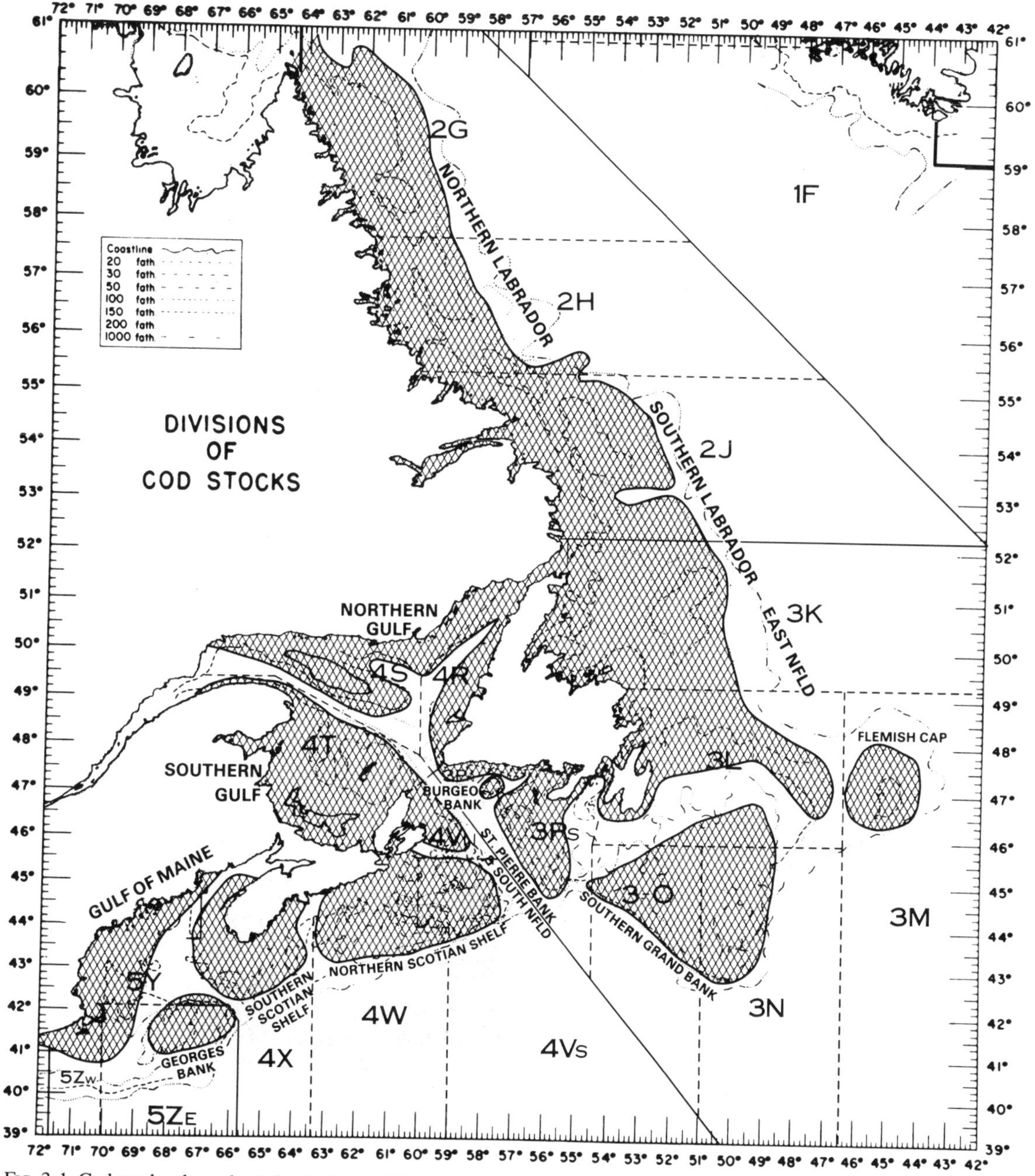

FIG. 3-1. Cod stocks along the Atlantic Coast of Eastern North America (adapted from Templeman 1962 and Pinhorn 1976).

Growth is at first slow then becomes faster and faster until an "inflection point" is reached. After this point growth slows more and more with time. There thus is an age at which the total weight of a year-class is at a maximum where increases due to growth of fish are greater than losses due to mortality (Fig. 3-4). Theoretically it would be best to exploit a fish stock by catching all members of each year-class at that age. This, however, is not feasible. Catches consist generally of a mixture of age groups.

Most growth estimates are based on the length of the fish. The relationship of length to weight remains fairly constant for a fish throughout its life. Thus length measurements generally provide a satisfactory basis for measurement of growth.

The growth rate of fish is affected by changes in various physical environmental factors, such as

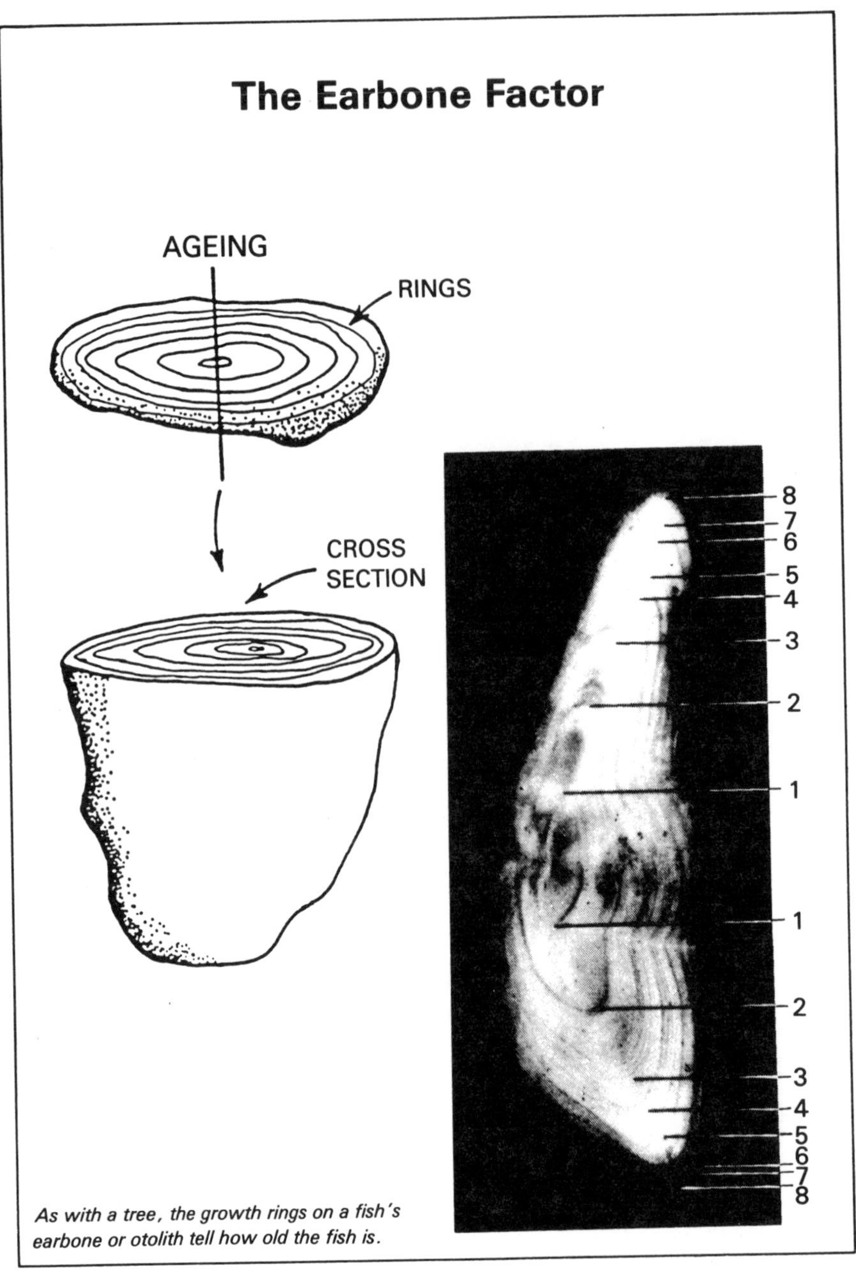

FIG. 3-2. Cross-section of ototlith of cod showing the pattern of annual rings.

temperature, and by such biotic factors as changes in availability of food and changes in stock density. North to south changes in growth rate for Canadian Atlantic cod are illustrated in Figure 3-5.

2.5 Fecundity

Marine fish produce large numbers of eggs, only a small portion of which survive to maturity. Cod, for example, are prolific. The number of eggs spawned depends upon the size of the female, larger fish producing more eggs. Female cod, 100 cm long, produce approximately 5,000,000 eggs. The size of eggs varies among species from about 1–2 mm diameter in cod, to 3–4 mm in salmon and halibut. The survival rate from eggs to adults for fish is quite small, in contrast with most vertebrates which produce relatively few eggs or young per female. Most of the mortality takes place

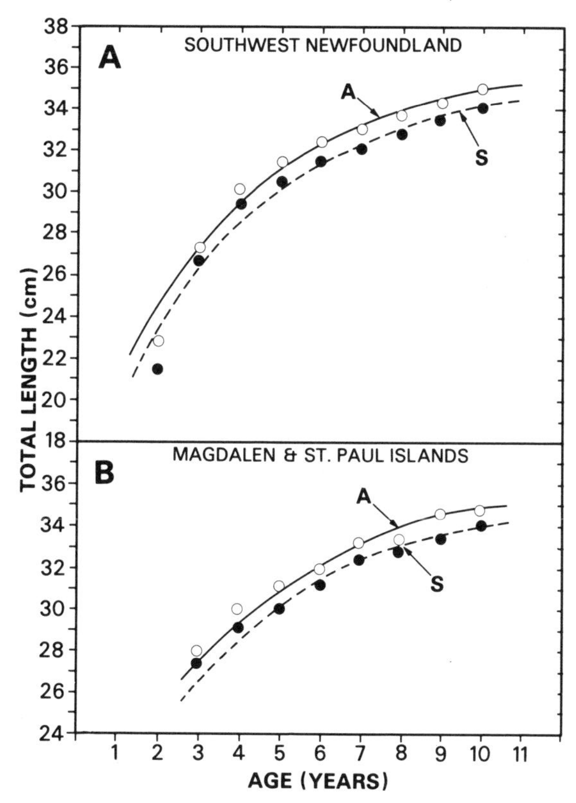

FIG. 3-3. Growth curve for spring spawning(S) and autumn spawning(A) herring (adapted from Parsons and Hodder 1975).

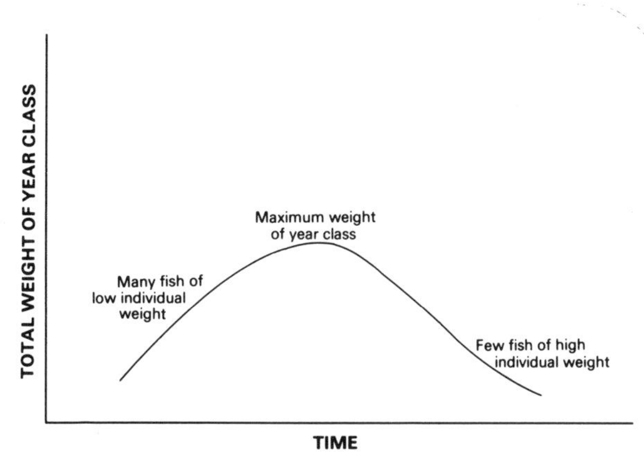

FIG. 3-4. Change in total weight of a year-class throughout its life (adapted from Armstrong 1978).

in the larval stage. In cod, the combined death rate of eggs, larvae, immatures and adults is, on average, 99.99% or higher.

Most marine species spawn in midwater. Herring lay their eggs on the bottom, some capelin stocks in the

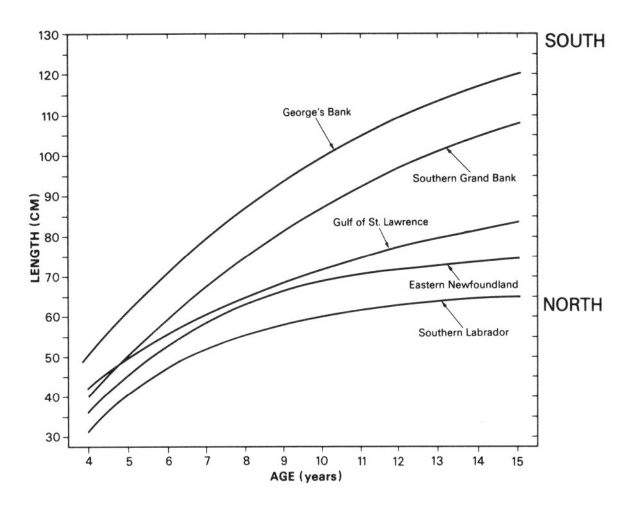

FIG. 3-5. North to south changes in growth rate for Atlantic Cod (adapted from Lear 1985).

intertidal zone on Newfoundland beaches and salmon on gravel in streams or lakes.

2.6 Natural Variability

There is significant year-to-year variability in the proportion of young which survive, leading to considerable fluctuations in year-class strength. In some instances, these fluctuations can be so pronounced that a large year-class or year-classes can sustain a fishery for several years. In Atlantic Canada there have been two well-known examples in recent decades where such year-classes have supported major expansion of particular fisheries. These were the southern Gulf of St. Lawrence-southwest Newfoundland herring and the Gulf of St. Lawrence redfish fisheries (Fig. 3-6). In the case of the southern Gulf-southwest Newfoundland herring, there are two spawning components, one which spawns in the spring and the other in the autumn. The 1958 year-class of autumn spawners and the 1959 year-class of spring spawners were exceptionally large. These large year-classes supported the development of a major purse seine fishery in the late 1960's-early 1970's, with catches peaking at 180,000 tons in 1970. Subsequent year-classes have been small in comparison, and the current fishery is only a fraction of that which occurred in the heyday.

Similarly, two large year-classes of Gulf redfish, those of 1956 and 1958, supported the recovery and the expansion of that fishery through the mid-1960's and into the early 1970's, with the catches peaking at 130,000 tons in 1973. These year-classes were then 17 and 15 years old respectively. This fishery has been supported by a series of infrequent large year-classes. Because redfish is a slow-growing, long-lived species,

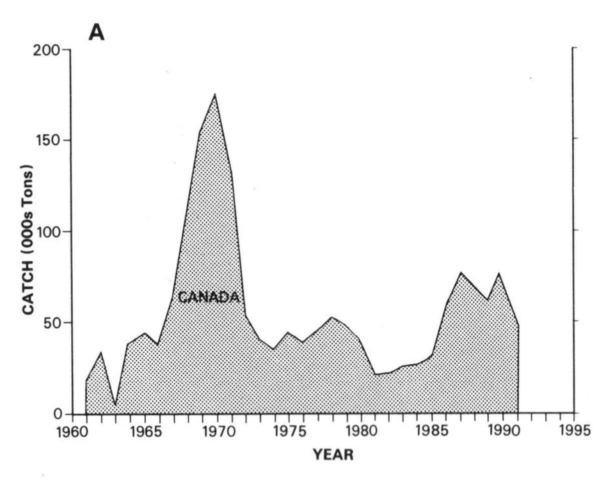

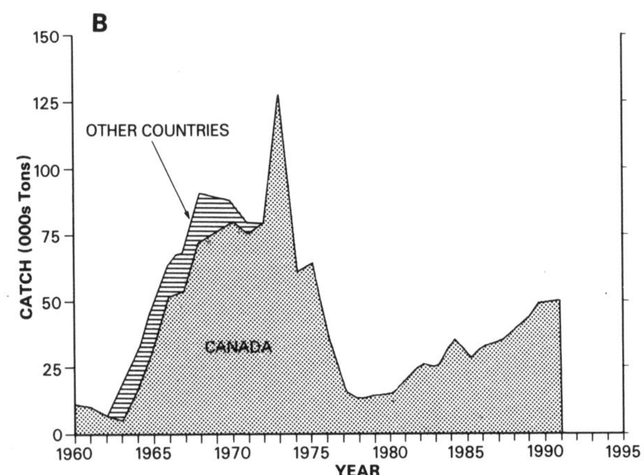

FIG. 3-6. The surge in catches of Southern Gulf herring (A) and Gulf redfish (B) when the large year-classes were being fished (adapted from Rivard et al. 1988).

the effects of such year-class fluctuations can be observed in the fishery for as much as 10 years (Fig. 3-6).

This phenomenon of infrequent large year-classes has been observed worldwide. Rothschild (1986) presented data on the ratio of maximum year-class strength to minimum year-class strength and the number of years between very large year-classes for various stocks. He observed that most year-classes are very low in abundance. These relatively low values are interspersed by the occurrence of very large year-classes for the species considered. He concluded that the average interval between very large year-classes is about 11 years.

This natural variability can confound any attempt to describe the response of a fish population to fishing (see Chapter 18). The search for a better understanding of the phenomenon is one of the major challenges facing fisheries scientists.

2.7 Migrations

Many commercial species of marine fish migrate great distances, hundreds, even thousands, of kilometres each year. Adults migrate between spawning and feeding grounds and back again in a regular cycle. Larvae drift from spawning grounds to settle in nursery areas. As they mature they migrate to the feeding grounds and join the adult stock (this is termed *recruitment*). Harden Jones (1968) depicted the migration pattern in the form of a triangle with spawning and nursery areas forming the base and the feeding area of the adult stock at the apex (Fig. 3-7). He hypothesized that life-history migrations, including egg and larval drift from spawning areas to juvenile nursery areas caused by surface-layer residual currents, were responsible for observed patterns of population structure. Cushing (1975) suggested that the larval drift from spawning ground to nursery ground occurs in the same geographical position from year to year and also is the period when the major natural control of population numbers takes place. He suggested that spawning grounds in temperate and higher latitudes tend to be fixed in position and in season and suggested that this may be an adaptation to variability in the influence of temperature on production. He advanced the "match/mismatch" hypothesis which suggested that the success or failure of a year-class may depend upon the match or mismatch of the timing of larval production to that of their food. Under this hypothesis ocean currents are a major force in determining year-class survival. Sinclair (1988a) took issue with the hypotheses of Harden Jones and Cushing and advanced a more complex hypothesis concerning the factors determining population regulation (see Chapter 18).

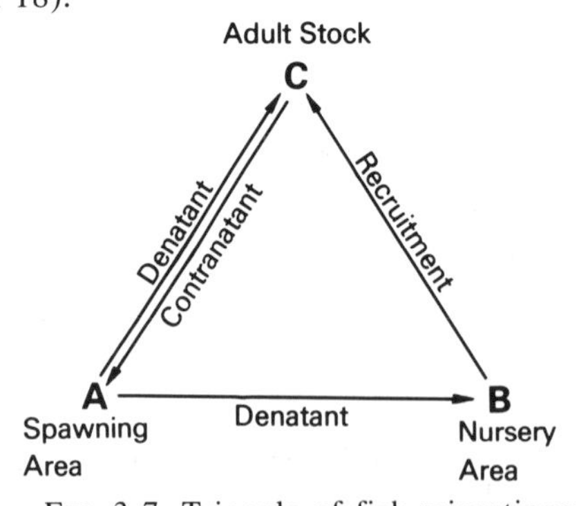

FIG. 3-7. Triangle of fish migrations (from Harden-Jones 1968). Denatant: downstream, passive drift; Contranatant: upstream, active migration to spawning area.

The migration patterns of various commercial fish species have important fisheries implications. Salmon migrate great distances, both in the Atlantic and Pacific oceans. Atlantic salmon produced in the rivers of New Brunswick and Quebec migrate along the south and east coasts of Newfoundland to feeding grounds at West Greenland, where they mingle with salmon from European rivers. In addition to the significant quantities caught in the West Greenland fishery, which intercepts salmon produced in North America and Europe, some salmon are caught in interception fisheries at Newfoundland during the return journey to their home streams. In addition, Canadian Atlantic salmon fisheries intercept a few fish of U.S. origin during their return migration from the West Greenland to northeastern U.S. rivers (Fig. 3-8).

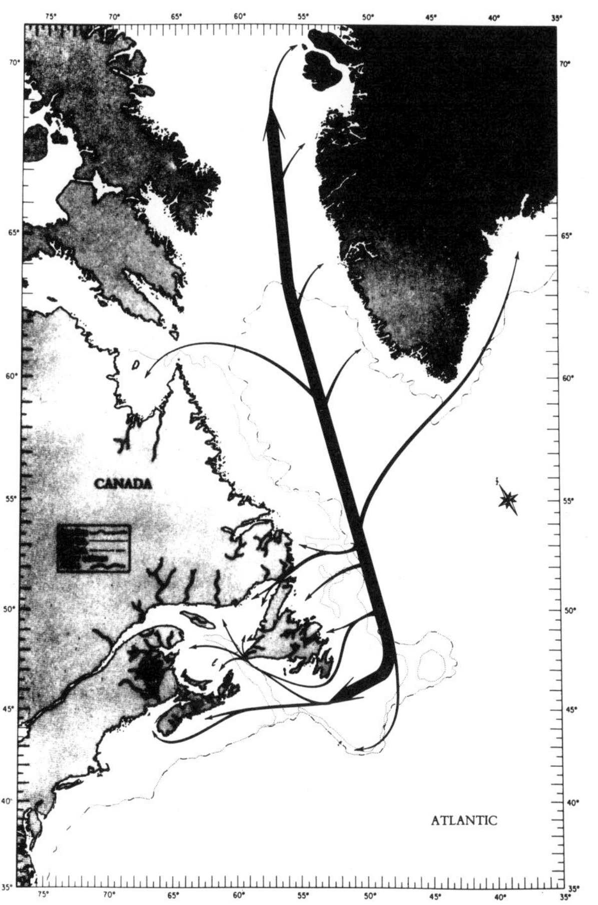

FIG. 3-8. Migrations of Atlantic Salmon (from Haig-Brown 1974).

The five species of Pacific salmon have complicated life histories. They migrate vast distances to the mid-Pacific to feed. North American stocks of all species are largely concentrated in the Gulf of Alaska, with their migration having a large counterclockwise pattern. Asian stocks also migrate counterclockwise. Stocks of North American and Asian origin intermix in the central North Pacific. While pink and coho salmon produced in British Columbia rivers appear to remain within the Gulf of Alaska, chinook, chum and sockeye migrate to approximately 165°–177° W with sockeye migrating as far as 178° E (Fig. 3-9 and 3-10).

Salmon of Canadian and U.S. origin were intercepted for decades in the Japanese high seas salmon fisheries. In addition, Americans intercept Canadian salmon off Alaska and Canadians intercept U.S. salmon in mixed fisheries along the British Columbia coast as these salmon return to the Pacific Northwest.

Some pelagic species also migrate long distances. Mackerel on the Atlantic coast are known to migrate from Long Island, New York to the northern tip of Newfoundland. Bluefin tuna, which occur in Canadian Atlantic waters during the summer and autumn, migrate there from spawning grounds in the Gulf of Mexico.

Groundfish species such as cod and flatfish migrate shorter, but nonetheless considerable, distances in the course of their annual spawning and feeding cycles. There are two cod stocks in the Gulf of St. Lawrence commonly designated as the southern Gulf and northern Gulf stocks. The southern Gulf cod stock occurs throughout the waters of the southern Gulf in summer but overwinters outside the Gulf off Cape Breton. The northern Gulf stock spends the summer and autumn in the eastern and northern Gulf along the west coast of Newfoundland and the north shore of Quebec but migrates south to overwinter along the western portion of the south coast of Newfoundland at the mouth of the Gulf (Fig. 3-1).

One of the best-known cod migrations is the inshore-offshore migration of the northern cod stock of eastern Newfoundland and Labrador. These cod winter on the offshore banks where they spawn in the spring (March–April). They migrate inshore in late June and July to feed on capelin in inshore waters. Not all of the stock comes inshore and the proportion which migrates inshore varies from year to year. This gives rise to considerable fluctuations in the availability of cod to inshore fishing gears, e.g. the cod trap, and in inshore catch. During the mid-1980's (1982–1986) inshore cod catches were relatively poor in relation to offshore catches and in comparison with the increasing catches during the late 1970's following declaration of the 200-mile limit. This led to considerable controversy about stock abundance and accusations that offshore fishing was "ruining" the inshore fishery. A Special Task Force was established by the Minister of Fisheries and Oceans in 1987 to investigate the reasons for this situation. The Task Force concluded that the cod stock had been increasing in abundance and that the decline was due to a combination of factors,

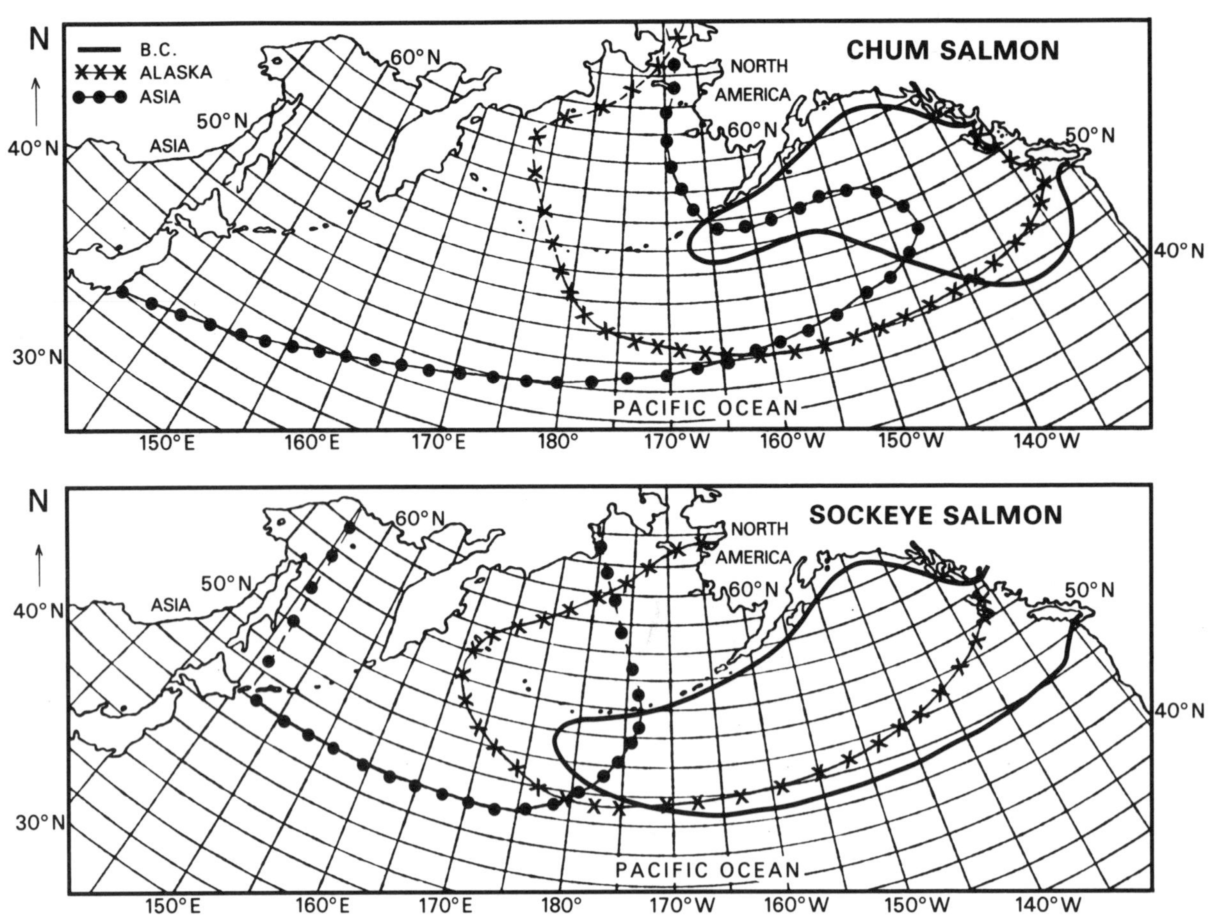

FIG. 3-9. Chum and sockeye salmon distributions in the North Pacific.

including colder-than-normal temperatures, availability of food capelin offshore and possibly less fishing effort inshore than exerted previously. Early in 1989 Canadian scientists revised downwards their estimates of the abundance of this stock. These revised estimates were confirmed by a Special Panel on Northern Cod early in 1990. During 1991 there was a sudden, drastic and unexpected decline in the spawning biomass of the northern cod stock, which ultimately lead to a 2-year moratorium on fishing for northern cod in 1992 (Chapter 18 and Lear and Parsons 1993).

Fish migrations can have a considerable impact on the availability of fish to different types of fishing gear and can pose problems for fisheries managers as stocks cross management boundaries (national or international) during their annual migrations.

3.0 BIOLOGICAL THEORY OF THE EFFECTS OF FISHING

Russell (1931) developed the simplest model describing the factors affecting stock size (Fig. 3-11). The *population biomass* (living weight) is increased by *growth* of individuals within the population and by *recruitment* (the addition of new young fish to the adult stock). *Fishing mortality* and *natural mortality*, i.e. all deaths not resulting directly from fishing, act to decrease biomass.

A stock which has never been fished is an *unexploited* or *virgin stock*. According to the *equilibrium theory of fishing*, in an unexploited fish stock the biomass is in equilibrium with its environment at any given time. The biomass is in equilibrium or balance because growth of individual fish in the stock and the additions through recruitment equal the weight of fish which die due to natural causes. This natural mortality may be due to predation, starvation, disease or old age. In this situation, the stock biomass remains unchanged and the net growth rate is zero. The stock biomass will remain at that level as long as the environmental factors which affect growth, recruitment and natural mortality do not change significantly. Should changes occur in any of these factors, then the stock biomass will, in theory, change to reach a new

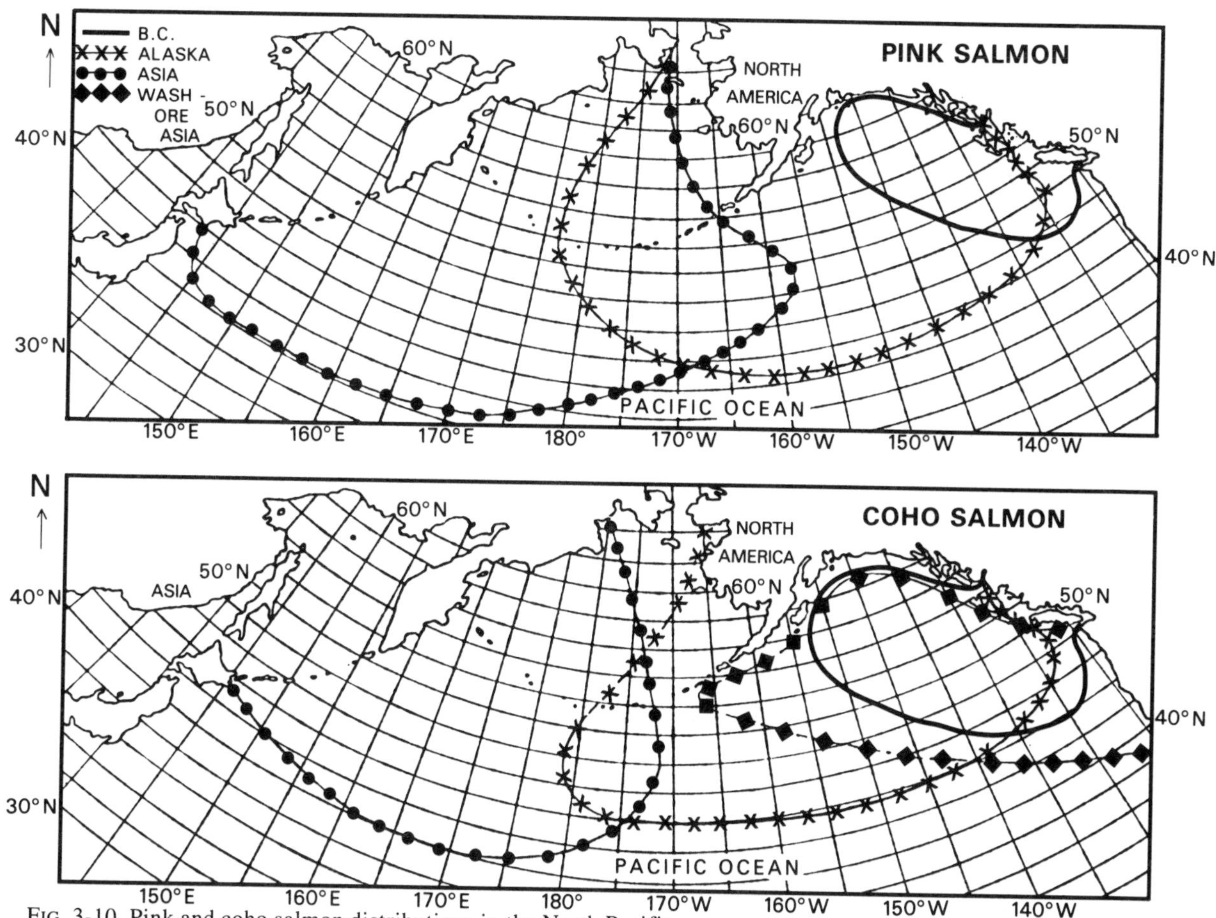

FIG. 3-10. Pink and coho salmon distributions in the North Pacific.

equilibrium level with the carrying capacity of the changed environment.

The rate of net growth is related to the size of the stock. It is zero both when the stock size is zero and when the stock is unexploited but in equilibrium with its environment. The net growth rate is highest at some intermediate stock size.

When fishing begins on a stock, the equilibrium is upset because the weight lost to the stock increases (due to fishing mortality) and the stock size begins to decrease. This results in an increase in the rate of net growth because there is more food for the fish remaining. When there is less competition for food, generally individual fish grow faster and fewer fish die from natural causes. In theory, the faster and more intensively a virgin stock is fished, the faster the remaining fish grow and reproduce. The net rate of growth of the stock increases even though the stock size is being reduced initially.

If the fishing pressure stabilized at this low level, the increased rate of net growth would balance the increased removal due to fishing and the stock would reach a new equilibrium at this level of stock size and fishing pressure. However, if the level of fishing continues to increase, the stock size will continue to decrease and the rate of net growth will continue to increase as the stock tries to replace the weight of removals. The catch will continue to increase until the stock size is reduced to some intermediate level at which the rate of net growth is at a maximum. This corresponds to a level of catch called the *Maximum Sustainable Yield (MSY)* (Fig. 3-12). Beyond this point the rate of net growth decreases, often rapidly, and the level of sustainable yield diminishes.

When fishermen begin fishing an unexploited stock, the total catch of an individual fishing vessel (tons of fish caught) will be high, and the catch rate (tons of fish caught per hour or day fished) will also be high. Since exploitation of the stock is light, the net growth rate is increasing. As the number of fishermen grows, the stock size decreases due to fishing mortality, and the net growth rate approaches its peak. This peak corresponds to the level of Maximum Sustainable Yield.

As increased fishing pressure brings the stock closer to the Maximum Sustainable Yield level, the total

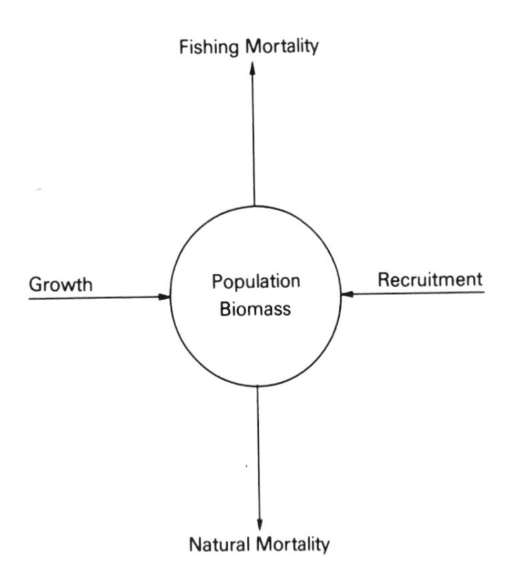

FIG. 3-11. Input-output model of exploitable population biomass (from *Marine Resource Economics* Vol. 1 p2 M.P. Sissenwine. Taylor and Francis, Inc. Washington D.C. Reproduced with permission. All rights reserved.)

catch increases only slowly, but the catch rate of an individual vessel will decrease significantly. At the level of MSY, the catch per unit of fishing effort, which was very high when fishing began on the lightly exploited stock, will be only half or less of that at the beginning. The inverse relationship between catch rate and increase in fishing effort is illustrated in Fig. 3-13. Catch rate (*catch per unit of effort* or CPUE) is often used as a measure of stock density.

This inverse relationship between stock density and fishing effort was first demonstrated by W.R. Thompson to the Pacific halibut fishery (Thompson and Bell 1934). The simple model of changes in stock size in relation to fishing pressure described above (the logistic curve) was applied by Hjort et al. (1933) to the analysis of whale populations and Graham (1935) to the stocks of plaice in the southern North Sea.

3.1 Surplus Production Models

This relationship between catch rate and fishing effort became the foundation for the development of the first of two main streams of fish population models — the surplus production model which is now commonly called the Schaefer model or method. Schaefer (1957) extended the first approaches by expressing the dependence of stock density on fishing effort as a parabola relating catch to stock where the MSY was obtained at half the virgin stock, as described above. Other curves are possible where the maximum occurs at some point

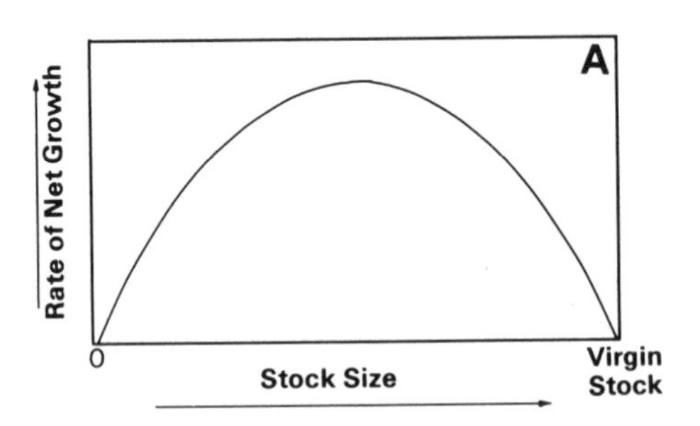

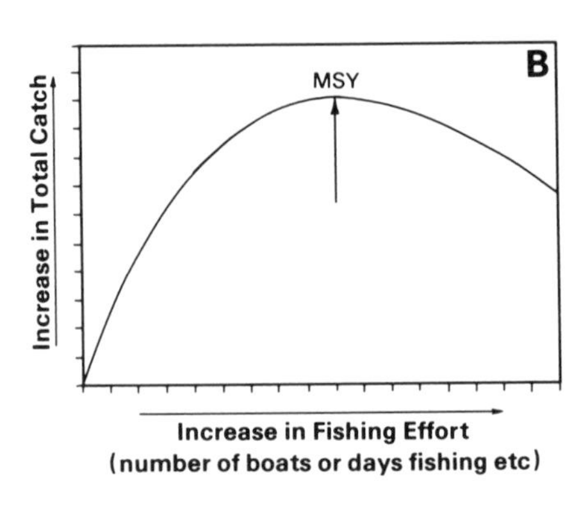

FIG. 3-12. A. Rate of net growth against stock size; B. Increase in total catch against increase in fishing effort.

other than half the virgin stock size. These general production models consider the fish stock as a whole and do not take account of the current or changing age composition of the stock. Over the years modifications have been made to the simplest form of the pro-

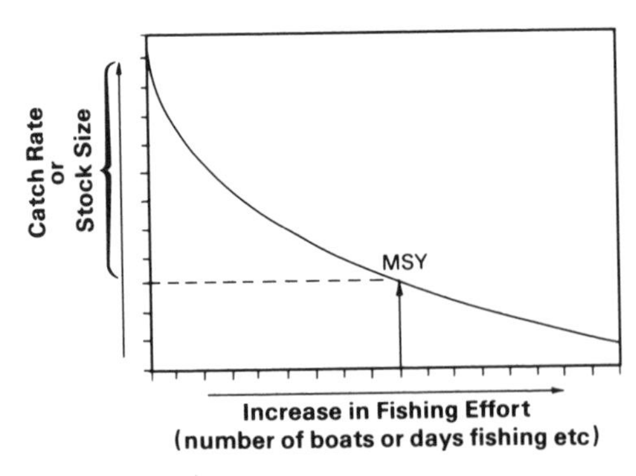

FIG. 3-13. Relationship between catch rate and increase in fishing effort.

duction models for ease of application or to achieve a closer approximation to reality.

The advantage of the general production model is its simplicity. The only data required are measures of the total catch and of total effort or catch per unit of effort. But this model has several disadvantages. The chief disadvantage is that it does not distinguish between the effects of recruitment, growth, natural mortality and fishing mortality. A second disadvantage is that the model ignores the *learning factor* in the development of a new fishery. Catch rates can be distorted as fishermen become familiar with new grounds. Catch rates can also be distorted by improvements in fishing technology. However, this model is particularly useful for species such as tuna which cannot readily be aged, or for species such as Atlantic redfish for which aging is difficult. The production model has also been used to estimate the MSY from a complex of species, such as the groundfish resource as a whole. Figure 3-14 illustrates its use to estimate the production of groundfish on the Scotian Shelf.

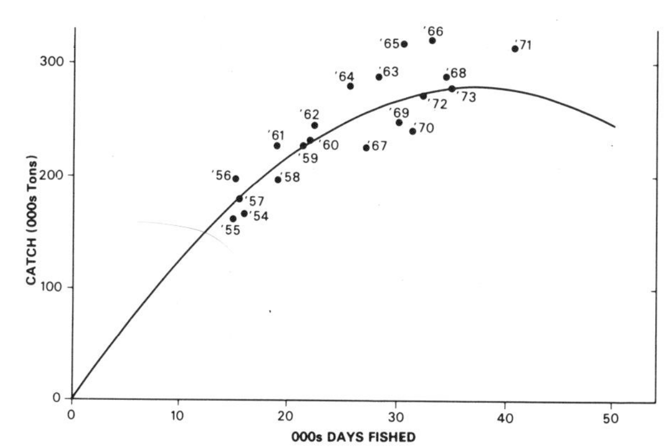

FIG. 3-14. An application of the Schaefer model to estimate groundfish production (excluding silver hake) on the Scotian Shelf (adapted from Halliday and Doubleday 1976).

3.2 Analytic Models

The second main stream of model development led to the analytic models. Analytic models recognize that: a fish stock includes a large number of individuals; the age structure of the stock varies over time; and recruitment, growth, natural mortality and fishing mortality all have important effects on stock dynamics.

Each stock is made up of year-classes. Analytic models examine what happens to a year-class from the time it recruits to the fishery until the last fish of that year-class has died. Five factors are taken into account (Gulland 1983a):

1. The numbers of young fish recruiting to the fishery each year;
2. The rate at which they are caught (which depends on the amount of fishing effort);
3. The range of sizes (ages) of fish that are caught (which depends on the selectivity of the fishing gear-e.g. mesh size);
4. The rate at which fish die from natural causes; and
5. The pattern of growth of the individual fish.

Russell (1931) first developed the analytic approach. He described the fish stock as changing from year to year because of increments of growth and recruitment and decrements of mortality (natural and fishing). Beverton and Holt (1957) were the first to apply the analytic model in their landmark monograph *On the Dynamics of Exploited Fish Populations.* They advanced the approach an order of magnitude by developing mathematical formulae which treated the stock as the integral of numbers and of weight from the age of recruitment to the age of extinction of a year-class. They devised methods of estimating total mortality and its components, natural and fishing mortality. They factored in the growth of fish by incorporating the von Bertalanffy growth equation into their model. The von Bertalanffy growth equation stipulates that growth in length or weight tends toward a maximum at infinite age (Fig. 3-3). Two constants are derived by fitting the curve to data, one being the maximum length or weight and the second being the rate at which the maximum is attained in age (von Bertalanffy 1938). Beverton and Holt (1957) incorporated the two mortality coefficients and the two growth constants to calculate the stock in numbers and in weight.

Beverton and Holt (1957) extended these models to account for the possibility that growth and natural mortality were density-dependent (dependent on stock size) during adult life. This in effect flowed from the model of balance described earlier. But their major contribution was in expressing the yield not as catch but as yield-per-recruit. The yield-per-recruit model describes what would happen in the real world if recruitment were constant. We have seen earlier that recruitment in fact varies considerably from year to year. Nonetheless, Beverton and Holt held the view that the yield-per-recruit models could be applied widely because the variability of recruitment was so high that no change in recruitment could be detected in relation to changing stock size. The yield-per-recruit model became widely used in the scientific analyses conducted for various international fisheries commissions. Even today much of the advice to fisheries managers is formulated in terms of yield-per-recruit.

Various mathematical methods have been developed since then to express the relationship among recruitment, growth, natural mortality and fishing mortality. They have also been used to provide means for estimating these parameters from fisheries data such as commercial catches, fishing effort and age structure of the population. One recent method which was developed simultaneously in the North Atlantic (Jones 1961; Gulland 1965) and in the Pacific (Murphy 1965) is called *virtual population* or *cohort analysis*. In this method, fishing mortality is estimated from the ratio of catch to stock. The sum of catches throughout the life span of a year-class is called the "virtual population". It represents a least estimate of stock. Ricker (1975) called this approach "sequential computation" of rate of fishing and stock size but the term proposed by Pope (1972) — cohort analysis — is now in common use. Basically, if the natural mortality rate is known or assumed, computations are made of the fishing mortality rate to which a year-class is subjected at successive ages, using its catch at each age. It is also necessary to know or assume a value for the fishing mortality F or total mortality Z for one age as a starting point. Generally, this value is assumed for the oldest age. The calculated values of fishing mortality converge toward the correct value as the analysis moves to progressively younger ages. From the ratio of catch to stock in the second-oldest age group a new estimate of fishing mortality is calculated. In this way the stock is estimated back through the life of the year-class and yearly estimates of fishing mortality are calculated.

Gulland (1983a) has described the underlying principles of the analytic models as follows:

> "These factors can be handled mathematically in various ways, but the basic principle is the same. The total life span can be divided into successive periods, during which the numbers caught will be proportional to the fishing effort (plus a factor depending on the selectivity of gear for the particular size of fish) and the numbers dying from other causes can be considered as being proportional to some natural mortality coefficient. The numbers at the beginning of the first period considered (i.e. starting with the age of recruitment) will be equal to the numbers of recruits. The numbers at the beginning of each subsequent period will be equal to the numbers in the previous period, less the numbers caught and the number dying from natural causes. In this way a brood of fish can be followed through its life in the fishery. The numbers caught can be found by simple addition. The weight caught is similarly calculated by multiplying the numbers caught in each period by the average weight of fish of that age, and adding up.
>
> "As described, the model enables one particular brood of fish to be followed through its life in the fishery. In the same way other broods can be followed, so that by looking at all the year-classes which are present in a fishery at a given time, the entire stock can be described in terms of the number of recruits in each of the previous years, and their history of growth, fishing and natural mortality. Most of the important characteristics of the stock, e.g. the number or weight of adult fish, can be computed by fairly simple arithmetic, provided the values of the parameters (growth, natural mortality and recruitment) are known."

Discussion of the technical derivation of these approaches and the variety of methods that can be used to estimate the critical parameters is beyond the scope of this book. The reader seeking further understanding of the methodology of population estimation is referred to the manuals by Ricker (1975), Gulland (1983b) and Hilborn and Walters (1992), where the various methods and their application are described in detail. An article by Pope (1982) describes the cohort analysis procedure.

3.3 Dependence of Recruitment on Parent Stock

We have seen that recruitment in marine fish stocks can be quite variable. Much has been written about the extent to which recruitment depends on the size of the parent stock as opposed to being determined environmentally. Graham (1935) believed that recruitment increased relative to stock size as a stock was reduced by fishing. He concluded that fishing would never reduce a fish stock enough to reduce recruitment either relatively or absolutely. Beverton and Holt (1957) felt that the variability of recruitment was so high that it could be ignored in terms of modelling fish stocks.

Experience in many fisheries has shown that recruitment to a stock at a low level can be as high as recruitment at a high stock level. However, experience has also shown that many of the catastrophic failures of fisheries have occurred as a result of recruitment failure. Cushing (1971) suggested that four great pelagic fisheries disappeared through *recruitment overfishing*: the Hokkaido herring, the Japanese sardine, the Norwegian herring and the Californian sardine. Clupeoid (herring-like) fishes are susceptible both to climatic changes and to excessive fishing. A large number of herring-like fishes have declined and/or collapsed under high fishing pressure. Gadoid-like fishes appear to be more resilient but are nonetheless also vulnerable to recruitment overfishing.

The Pelagic Fish Symposium held at Aberdeen, Scotland (Saville 1980) concluded that:

"In all cases considered during the symposium there is strong positive, or presumptive evidence that the stock collapse was, in the final analysis, due to reduced recruitment, generated by a decline in spawning stock. It would therefore seem evident that the first consideration in advising on stock management must be the maintenance of the spawning stock above the level at which there is a serious danger of development of a stock-recruitment relationship. Unfortunately this danger level can be specified with any degree of precision for relatively few stocks.... In the interim a very conservative policy should be adopted."

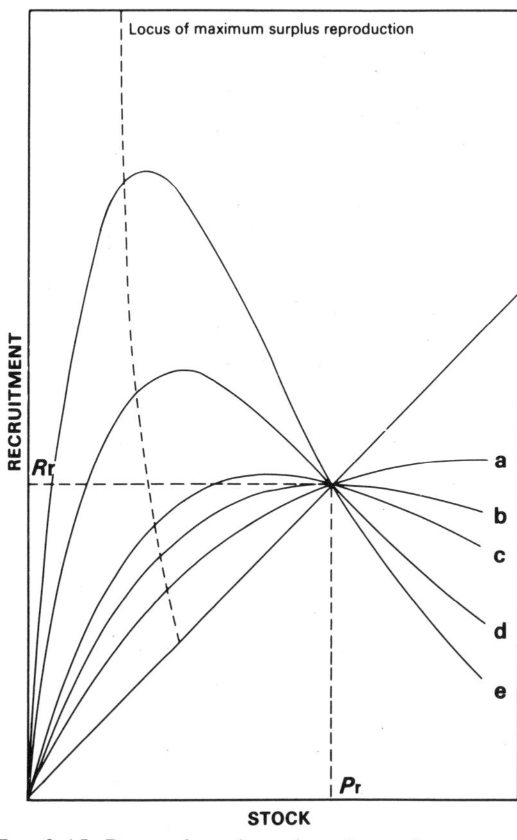

FIG. 3-15. Dome-shaped stock and recruitment curves from Ricker (1954). The point where the curves cut the diagonal is the replacement level of stocks and reproduction. (from *Marine Resource Economics* Vol. 1 p3 M.P. Sissenwine. Taylor and Francis, Inc. Washington D.C. Reproduced with permission. All rights reserved.)

Several models of the relationship between recruitment and parent stock have been proposed, most noteworthy being those by Ricker (1954) and Beverton and Holt (1957). These models assume that the mortality between egg and recruitment must include a density-dependent component. A large proportion of this mortality appears to be independent of stock density with much of the variability of recruitment determined by environmental factors. Ricker (1954) hypothesized that density-dependent mortality of young fish is generated by the aggregation of predators (or cannibals of older year-classes) on the stock of eggs or larvae. Under this assumption, the biological mechanism of cannibalism would lead to dome-shaped stock-recruitment curves (Fig. 3-15). Beverton and Holt (1957) hypothesized that density-dependent mortality results from density-dependent growth of young fish combined with size-dependent mortality. This assumption led to a different form of stock-recruitment curves (Fig. 3-16).

Cushing and Harris (1973) suggested that the nature of the stock-recruitment curve differs among various groups of fishes. They hypothesized that density dependence can be related to fecundity or the distance between larvae in the sea. Thus, density dependence would increase from herring to plaice to cod, with the curve of recruitment on parent stock bending from a near linear form to a flat-topped one and finally to a dome (Fig. 3-15). Cushing (1975) hypothesized that any stock and recruitment curve cuts the bisector at the point of stabilization, about which the virgin stock varies. In this hypothesis, a low recruitment reduces the stock below the point of stabilization, which generates a higher recruitment and so the stock is brought back to the

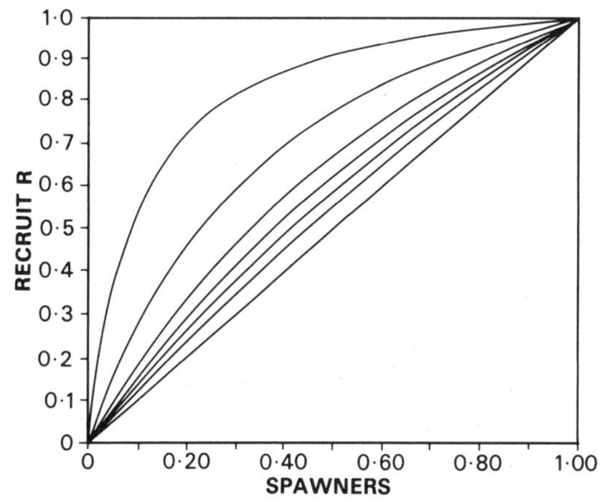

FIG. 3-16. Beverton and Holt stock and recruitment curves (adapted from Sissenwine 1984b). (from *Marine Resource Economics* Vol. 1 p4 M.P. Sissenwine. Taylor and Francis, Inc. Washington D.C. Reproduced with permission. All rights reserved.)

stabilization point; above that point in stock, the same process works in the opposite direction. Cushing further suggested that if the stock and recruitment curve for herring is nearly linear, there is little capacity for stabilization. Cod, with a markedly dome-shaped curve, would have a greater stabilizing capacity.

While a number of theoretical curves have been constructed, it is generally acknowledged that recruitment in fact tends to be highly variable. What are the implications of this for fisheries managers? Ricker's methodology has been applied to the Pacific salmon, a stock that spawns, recruits and dies in virtually a single age group. His equations have been used extensively and successfully in the study and management of Pacific salmon. Shepherd (1982) attempted to bridge the gap between the general production and analytic models for fisheries by developing a versatile functional form relating recruitment to spawning stock biomass. He contended that this form represents a conservative approach to safeguard against recruitment overfishing. He advocated that "suitable stock-recruitment relationships be explicitly incorporated into single-stock assessments of fisheries in a comprehensive way." This approach is now being incorporated by the International Council for the Sea's Advisory Committee on Fisheries Management (AFCM) in its assessments for stocks in the Northeast Atlantic.

Sissenwine (1984a) concluded that "spawning stock size alone is not an adequate basis for predicting recruitment." But is spawning stock size a relevant management concern? Obviously, some spawning stock is necessary or there will be no recruitment. Furthermore, there are numerous examples of recruitment failures e.g. European herring, Georges Bank herring, Flemish Cap cod, Georges Bank haddock.

Generalizations about stock-and-recruitment relationships are a perilous venture. This is illustrated by comparing the conclusions of Winters and Wheeler (1987), from a recent study of the recruitment dynamics of spring-spawning herring in the Northwest Atlantic, with the hypothesis of Cushing and Harris described earlier. Winters and Wheeler concluded that the recruitment patterns of the seven major spring-spawning herring populations in the Northwest Atlantic were determined largely by annual variations in overwintering temperatures and salinities associated with the Labrador Current. The patterns of recruitment success were similar among stocks. They found a high level of density dependence of recruitment. The level was higher than comparable estimates for many demersal species which Cushing and Harris hypothesized were more resilient to exploitation than herring. This was a surprising finding but not inconsistent with the high incidence of herring stock collapses due to overfishing. Burd (1985) noted that a prominent feature of herring is their ability to recover from extreme levels of overfishing. Winters and Wheeler suggest that this resilience is primarily due to the stock-recruit relationship of herring: a curve with a rapidly ascending left limb and a relatively low critical spawning stock size (Fig. 3-17). This characteristic accounts for the high frequency of herring stock collapses, as the sharply rising curve would tend to mask the imminence of recruitment failure, despite drastic reductions in adult stock size levels. They suggested that the most appropriate fishing strategy for herring is to maintain fishing mortality at a level of $F=0.2$ with the option of closure when parent stock size falls below the critical level (assumed to be approximately 20% of maximum observed egg production levels).

What does all this mean? Several conclusions seem reasonable:

1. Understanding the relationship between parent stock size and recruitment and the influence of environmental factors in determining recruitment variability remains one of the major challenges of fisheries science;
2. While the relationship between parent stock size and recruitment is unclear, a number of fish stocks around the world have collapsed due to recruitment overfishing; and
3. A prudent fish manager would set a conservative level of harvest to safeguard against recruitment overfishing, pending better understanding of the stock-recruitment relationships for particular stocks.

For a further discussion of the scientific challenge of understanding the mechanisms determining recruitment variability, see Chapter 18.

3.4 Yield-per-recruit

Because of the variability of recruitment and the difficulty of forecasting it, scientists commonly advise managers in terms of the *yield-per-recruit*, i.e. the average yield to be expected under any given pattern of fishing from an individual fish recruiting to the fishable population. If one has an estimate of the recruitment, this can be used to estimate the total yield. Assuming that recruitment will be at the average level of recent years (an often dangerous assumption), total yield and yield-per-recruit will change in a similar manner.

This advice is often couched in terms of the fishing mortality which will produce the maximum yield-per-recruit (F_{max}) or some variant of this. The nature of the yield-per-recruit curve varies among species (Fig. 3-18) and among stocks. In cases where the yield curve is flat topped (e.g. plaice) the value corresponding to F_{max} will occur at a ridiculously high level of little significance to management.

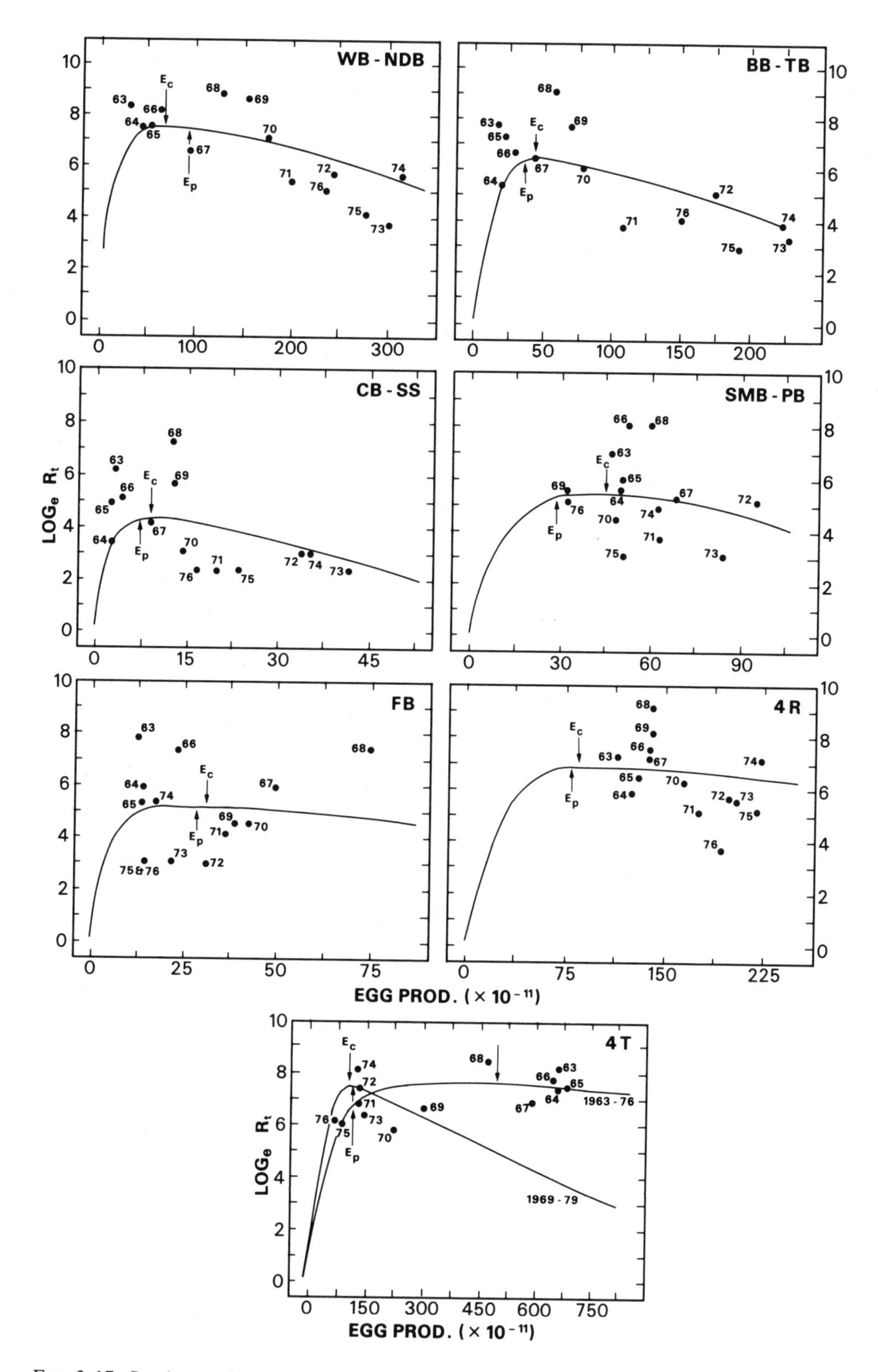

FIG. 3-17. Stock-recruitment curves for various stocks of spring-spawning herring in the Northwest Atlantic (from Winters and Wheeler 1987).

3.5 Species Interactions

Single species models have normally been used by fisheries scientists to describe the effects of fishing on fish stocks. The real world situation is considerably more complex.

Many fisheries harvest a number of species. The simplest case involves the incidental taking (*by-catch*) of other species in a fishery targeted primarily at one species. By-catch regulations are normally introduced to limit the take of those species caught as by-catch. In other instances, a fishery takes a great variety of species but is not directed at any one species. Calculating the sustainable yield(s) in such situations becomes more difficult.

The more challenging scientific problem relates to fisheries which are independent but which exploit biologically interdependent species. These could involve predator/prey species or species that compete for food. An example of the predator/prey situation is the cod and capelin fisheries of the Northwest Atlantic. Capelin, which are one of the major prey species of cod, were heavily exploited in the 1970's. The offshore capelin fishery on the Grand Banks was subsequently closed to fishing for 9 years because of the low level of the spawning stock.

Because of marked fluctuations in recruitment and the short life cycle of capelin, the Northwest Atlantic Fisheries Organization (NAFO) Scientific Council concluded in 1979 that capelin populations will fluctuate widely in biomass. They suggested a low exploitation rate of 10% to protect the spawning stock (NAFO 1979). In its 1982 report, the Scientific Council noted that capelin is an important food for predators, especially cod. The low exploitation of 10% would therefore provide a safety factor (NAFO 1982).

Despite concern that cod depend on capelin and that a reduction in capelin stocks would affect the inshore cod fishery, studies on the subject have been inconclusive (Akenhead et al. 1982; Lilly 1987a).

The opposite argument is made in the case of the Atlantic harp seal population, which has grown rapidly since exploitation was drastically reduced by the cessation of the large-vessel hunt. Fishermen argue that a growing harp seal population will reduce prey species such as cod. The Malouf Royal Commission on Seals and Sealing (Malouf 1986) concluded that:

> "The impacts of seals on fisheries are very real, but they are also difficult to express in reliable, quantitative terms. Fewest doubts concern the losses caused by removal of fish from, and damage to, fishing gear, which are

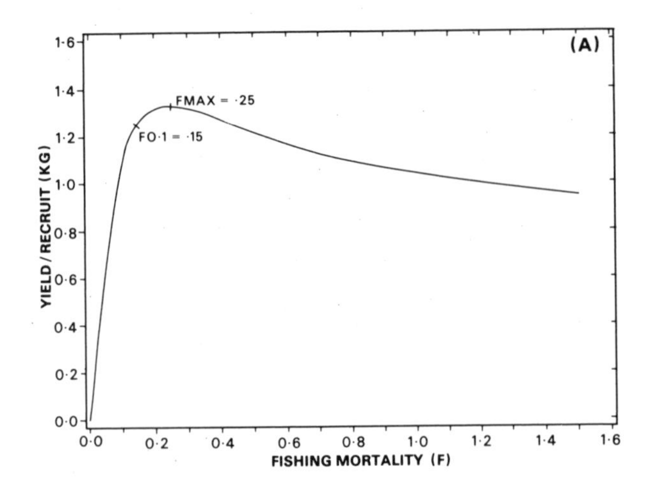

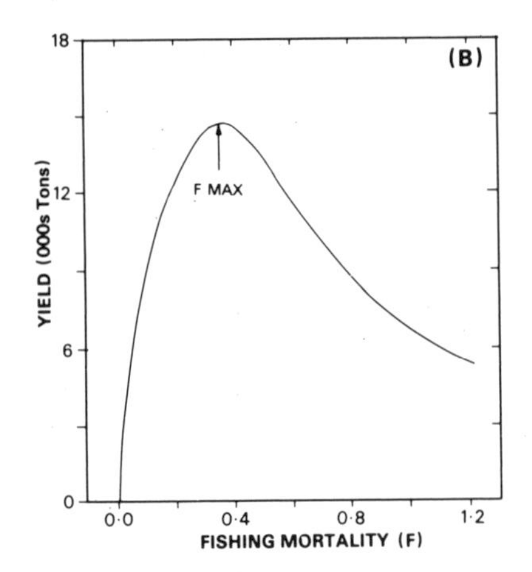

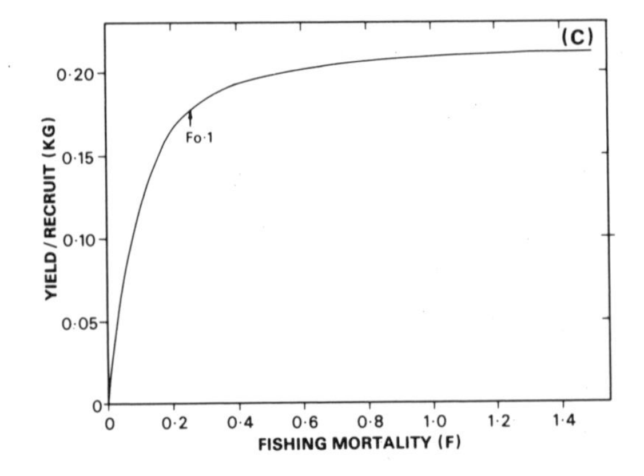

FIG. 3-18. Typical yield-per-recruit for cod(A), herring(B) and American plaice(C) in the Canadian Atlantic (adapted from Bishop et al. 1988; Winters and Wheeler 1987).

usually clearly visible. Losses from these sources are estimated to run to a few million dollars annually. Losses resulting from competition between seals and fishermen for commercial species of fish, and from the spread of parasites, are almost certainly much higher. Even the lower bounds of the ranges of losses, which are believed to be conservative, are substantial when compared with the total value of the fishing industry.

"Seals consume large quantities of commercial fish in Canadian waters and, consequently, cause a reduction in the catches of fishermen. On the Atlantic coast, roughly five million tonnes of a wide variety of fish and some crustaceans and molluscs are consumed, mainly by harp and hooded seals. Rather less than half of this amount is taken on or near commercial fishing grounds, off southern Labrador, Newfoundland, the Maritimes and Quebec. The consumption of commercial species is considerable. This must have some impact on catches, though the catch will not be reduced by exactly the amount of consumption of that species by seals. For heavily exploited species the reduction may be similar to or exceed the amount consumed.... On the Atlantic coast the value of this unavailable catch is undoubtedly very great. It is clearly significant in comparison with the total value of the current commercial catch."

May et al. (1979) discussed the way multispecies food webs respond to the harvesting of species at different levels in the food chain. They cited the commercial fisheries of the North Sea as an example of the effects of competition and prey-predator interactions among species. They suggested that the populations of large gadoids and other smaller fish increased with declines in the herring and mackerel stocks. This was because diminished stocks of herring and mackerel reduce predation on larval and juvenile stages of other fish species. They also modeled interactions among trophic levels in the Southern oceans (e.g. krill, cephalopods and sperm whales). These models simulated the effects of reductions in whales and possible reductions in krill through fishing upon other components of the ecosystem.

May et al. (1979) concluded that:

1. For populations at the top of the trophic ladder, Maximum Sustainable Yield (MSY) will remain a useful reference point for management. Such stocks should be kept at or above the greatest net annual increment;
2. For other populations ecosystem preservation requires that stocks be maintained at a level which will not reduce the population's productivity, or that of other populations dependent on it;
3. It is very important to keep in mind the different time scales for different population processes. The slowest time scales (i.e. that of the long-lived species at the top of the food chain) should be used in monitoring a harvesting regime; and
4. All estimates of population parameters may fluctuate in response to environmental variability. Harvesting levels should be set conservatively to guard against accidental overexploitation.

In the Canadian Atlantic, management of the southern Gulf of St. Lawrence cod stock was, for a while, based on scientific advice which took into account a multispecies model (Lett 1977, 1978). This model postulated relationships between mature cod biomass and production, and between cod biomass and mackerel biomass. Mackerel was seen as a primary determinant of cod recruitment. However, Doubleday and Beacham (1982) concluded that these interactions could be artifacts and that the model had little predictive capability.

Ambitious attempts have been made in recent years to model the ecosystem of the North Sea. Andersen and Ursin (1977) developed elaborate equations to describe the interactions among North Sea species under various harvesting assumptions. They suggested that sustainable yields could be increased by systematically depleting the stocks of the larger predatory fish (cod, haddock, saithe) and by fishing the younger age classes and smaller species that would then predominate. Such suggestions ignore the important economic interests of the existing fishing industry.

While some progress has occurred in understanding multispecies interactions, it is still not possible to use such models in setting harvest levels. Specific advice on this aspect of fisheries is unlikely in the near future.

Meanwhile fisheries managers wrestle daily with practical multispecies fisheries problems. The salmon fisheries along the west coast of North America provide a good example of the management challenge. As Healey (1982) pointed out, much of this complexity stems from the multispecies, multistock nature of these fisheries. The intermingling of species and stocks is such that there are virtually no single stock fisheries. Some fisheries involve mixtures of several species and hundreds of stocks. Healey (1982) concluded that the theory of mixed-stock fisheries (as it then existed) was of little value for managing the multistock salmon fisheries.

Considerable research is required to advance our understanding of multispecies interactions and their implications for fisheries management. For a further discussion of this challenge, see Chapter 18.

4.0 ECONOMICS OF FISHING

As a fishery intensifies, the catch rates of individual participants decrease even though the total catch will increase slowly as more fishing brings the stock closer to the level of MSY (Fig. 3-12 and 3-13). Under the Schaefer general production model, at the level of MSY individual catch rates will have decreased by 50% or more compared with the catch rates experienced by the first participant(s). This decline in catch rate has significant economic implications.

The work described earlier focused on the biological effects of fishing, laying the groundwork for regulations to conserve the resource. During the 1950's it became apparent that biological theory was insufficient because it did not take into account the economic effects of fishing. Burkenroad (1953) stated: "The management of fisheries is intended for the benefit of man, not fish, therefore, [the] effect of management upon fish stocks cannot be regarded as beneficial *per se*." H. Scott Gordon in his classic 1954 paper *The Economic Theory of a Common Property Resource: The Fishery* observed: "The great bulk of the research that has been done on the primary production phase of the fishing industry has so far been in the field of biology.... On the whole, biologists tend to treat the fisherman as an exogenous element in their analytical model." He then applied economic theory to the fishing industry demonstrating that the "overfishing problem" had its roots in the economic organization of the industry. He defined the optimum degree of utilization of a fishery as that which maximizes the *net economic yield*, the difference between the total cost, on the one hand, and total value of production, on the other. Scott (1979) credits Gordon and others with lifting fisheries economics out of its previous obscurity. The new paradigm considered the economic implications of fish as a *common property resource*. This concept has developed in the intervening years as a new and broader groundwork for regulating fisheries.

The problems of managing a *common property resource* are discussed in more depth in Chapters 7, 8 and 9. Here I will merely outline the economic aspects of the theory of fishing. This will build upon the general production biological approach described earlier. Again, we are discussing the relationship between the level of fishing effort and the amount of fish that can be harvested year after year without affecting the stock. These are at best only average relationships, which in reality are affected by many factors other than the level of fishing. The growth rate and equilibrium population size of even one fish stock is a complex phenomenon.

Two factors affect the cost of harvesting the resource: (1) costs of labour and capital, i.e. the cost of the fishing effort, and (2) the catch rate. When catch rates are high, less fishing effort is required to catch the same amount of fish. The revenue from fishing can be viewed as the volume of the catch multiplied by the price received for the fish. The biological curves relating the catch to the level of fishing can be used to explore the economic impacts of fishing. The curves can be converted by adopting the simplifying assumptions that the value of the catch is proportional to its volume and the costs of fishing proportional to the amount of

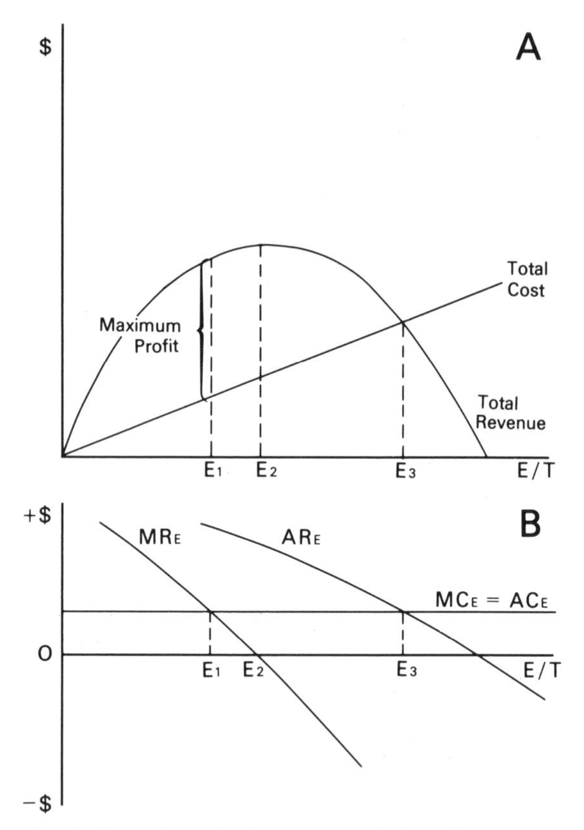

FIG. 3-19. A. The long-term relationship between total revenue and effort and total cost and effort.
B. The curves for marginal revenue (MR) average (AR) revenue and marginal costs (MC). (AC) is average (from Anderson, 1986).

E_1 is the level of fishing effort where the difference between the total revenue and total cost curves is at a maximum

E_2 is the level of fishing effort producing the Maximum Sustainable Yield

E_3 is the equilibrium level of fishing effort in an unregulated fishery, i.e. the open-access equilibrium fishing effort. This assumes that the price of fish and the cost of effort are constant. Figure 3-19a illustrates the long-term relationships between total revenue and effort and total cost and effort.

Because it has been assumed that the price of fish is constant, i.e. that the demand for fish products is perfectly elastic, the total revenue curve has the same shape as the curve of sustainable yield. The sustainable yield multiplied by the market price of fish (net of unit processing costs) produces the sustainable revenue from fishing. Also, assuming that the cost of fishing is proportional to the amount of fishing effort, the cost curve is a straight line.

At this point we need to consider other concepts — *marginal cost, marginal revenue and average revenue.* For this simplified model the curves for marginal revenue, average revenue and marginal costs are shown in Fig. 3-19b. Marginal revenue and the change in revenue resulting from a change in the level of effort slopes downwards because the marginal catch per unit of effort declines as fishing effort increases. Average revenue (revenue per unit at each level of effort) also slopes downward because of declining catch per unit of effort. The curve showing marginal cost (the change in cost due to a change in effort) is constant by assumption in this model and equal to average cost per unit of effort.

Without regulation the fishery will approach an *equilibrium* at the point E_3 where total revenue equals total costs (Fig. 3-19a). At this point the average revenue per unit of effort equals the average cost per unit of effort. To the left of this point, existing participants tend to expand their effort and new participants are attracted to the fishery because the average revenue per unit of effort is greater than the average cost per unit of effort. Beyond the point where total costs equal total revenues, the participants will experience losses. In an open-access fishery the effort will tend to increase below this point (E_3) and to decrease above it. Economists call this equilibrium level of effort in an open-access fishery the *open-access equilibrium yield.*

In this model the optimal level of effort will be E_1 (Fig. 3-19b) where the marginal cost equals the marginal revenue. Increasing the effort will decrease profits because costs will increase more than revenue. According to economic theory when the marginal cost exceeds the marginal revenue there is a loss to society. This occurs because the cost of harvesting the additional fish is greater than their value. If effort is reduced below E_1, the profit would decrease with revenues decreasing faster than costs. Economists have categorized E_1 as the optimal allocation of effort to the fishery since the marginal revenue just balances the marginal cost (Anderson 1986).

The difference between the total costs of fishing and the sustainable revenue is the "sustainable resource rent" (SRR) or "maximum economic yield" (MEY). Until the 1970's, fisheries economists believed that maximizing the sustainable resource rent or economic yield was the best policy. This may be a more conservative harvesting policy than the pursuit of MSY because the target biomass level is larger than that associated with the MSY. Standard theory of fishery economics postulates that fishing effort will expand until the sustainable resource rent is dissipated. At the open-access equilibrium, there is no further incentive for expansion because of the low catch per unit of effort and the high harvesting costs. In such a situation, the fishery would be considered to be overexploited both by biologists and economists.

This simplified model of fisheries economics is *static* — it does not take time into account. A fish population does not react immediately to a change in the level of fishing effort. Any adjustment may occur over several years. The rate that participants enter or leave a fishery is also more complex than suggested by the simple static model.

Economists have constructed *dynamic* models to describe the effects of fishing. These models take time into account explicitly. In effect, they apply the theory of capital and investment to fisheries matters. Dynamic analysis is concerned with the extent to which it is prudent to invest in a resource. Investing involves short-term sacrifices for long-term benefits while disinvesting has the opposite effect. Economists attempt to determine the economic yield from a marginal investment in the fishery resource and compare this with yields that society could expect from alternative investments.

Anderson (1986) pointed out that the simple static model of fisheries economics ignores potential year-to-year changes in stock size. He argued that sustained yield curves are not adequate because economic efficiency requires maximizing the *present value* of harvest, and harvest in one year affects future harvest because of its effect on stock size.

A very important factor in determining the actual location of the stationary optimum level of effort is the nature and size of the *discount rate.* Clark (1973) showed that with a zero discount rate, MEY occurs at a point such as E_1 where sustainable net revenues are maximized. In such a case, it is not rational to increase current net revenue at the expense of the future because a dollar of future revenue is equal in value to a dollar of present revenue. With an infinite discount rate, MEY is the same as the open-access equilibrium. With a positive discount rate, the stationary optimum level of effort occurs between the level that generates the highest sustainable revenues (MEY) and the open-access equilibrium level. The exact location depends upon the nature and magnitude of the discount rate. Anderson (1986) pointed out that optimum utilization, from an economic perspective, may not involve a stationary level of effort. The dynamic economic models suggest that in some cases "pulse" fishing or the complete destruction of a fish stock would be the economically optimal approach (Munro 1980).

In addition to the discount rate, the main economic factors influencing the level of MEY are the price of fish and the cost of catching them. In the static model of fisheries economics, both of these were assumed to be constant. In fact, both of these factors are subject to changes from year to year. Thus, the estimate of MEY will change in a static as well as a dynamic model.

Anderson (1986) summarized the complexities of this approach to fisheries economics:

> "The proper operation of a fishery requires that the sum of the present values of the yearly total returns be maximized. This means that the proper level of catch over time depends upon expected future prices and costs, intertemporal population growth rates, and the rate of interest. In cases where the price of fish is expected to fall drastically, this may call for very heavy fishing in the present even though future outputs will be adversely affected. When the fish stock is extremely slow-growing, or when there are economics of scale in the production of effort, very heavy fishing may be beneficial every five years or so, with little or no harvesting in between. Or if prices and costs are expected to remain relatively constant or to grow at constant rates, the proper output may be an equal amount each year, although with a positive discount rate it will not be the amount that achieves the highest net revenue in each period. In summary, maximum economic yield of a fishery can be more properly thought of as the optimal stream of yields over time. This stream can consist of a constant amount year after year, or yearly catch may vary widely. To make things more complex, this optimal stream can change over time with changing expectations as to cost, price, and rate of growth of the stock."

Munro (1982) asserted that conservation was the central preoccupation of biologists from the end of the Second World War until recent years. The central management criterion was "full utilization", usually expressed in terms of the MSY. He noted that economists from Gordon onwards objected to this approach to fisheries management because it focused solely on physical yields, ignoring economic consequences. From a traditional economic perspective, the goal of MSY seemed ill-conceived because an MSY policy leads to economic overfishing unless the cost of fishing effort is zero. Munro (1982), however, concluded, from dynamic economics analysis, that it is invalid to assert that an MSY policy will necessarily lead to economic overfishing. The conclusion from the static model that optimal management requires maximizing sustainable resource rent is economically valid only with a zero discount rate. If the discount rate is positive, then maximizing the sustainable resource rent would mean overinvestment in the resource from a societal standpoint. Munro (1982) suggested, from the perspective of dynamic fisheries economics theory, that an MSY policy could result in economic underexploitation of a resource.

Dynamic economic models have been criticized as so complex that they cannot be applied with our available data. Thus, their present utility in fisheries management is questionable. For further information on the subject, see Clark (1976), Munro (1982) and Anderson (1986).

5.0 THE CONCEPT OF THE MARGINAL YIELD AND $F_{0.1}$

The question of an appropriate fishing strategy has bedeviled fisheries biologists, economists and managers for decades. Gulland examined these issues from a slightly different perspective than the traditional fisheries biological and economic models. He developed the concept of the marginal yield by distinguishing between three types of fisheries — a very heavily fished stock, a very lightly fished stock and a moderately exploited stock (Gulland 1968). In a very heavily fished stock increased fishing will yield no significant increase in catch and will perhaps result in a decrease. Catches by additional vessels will be at the expense of those already fishing. In a lightly exploited stock, additional fishing would have a negligible effect on the stock and on the catches by existing vessels. For a moderately exploited stock, increased fishing would increase the total catch but the catch per unit of effort will decrease (Fig. 3-20).

The yield curve, as before, represents a general relationship between fishing effort and average long-term catch. An increase in fishing effort from E_1 to E_2 is represented by the line AB. The expected catch from this additional effort, as calculated from the present CPUE, would be BD. The actual increase in catch would, however, be BC. This increase in total catch, BC, is the *marginal total catch*. The ratio BC/BD is the *marginal*

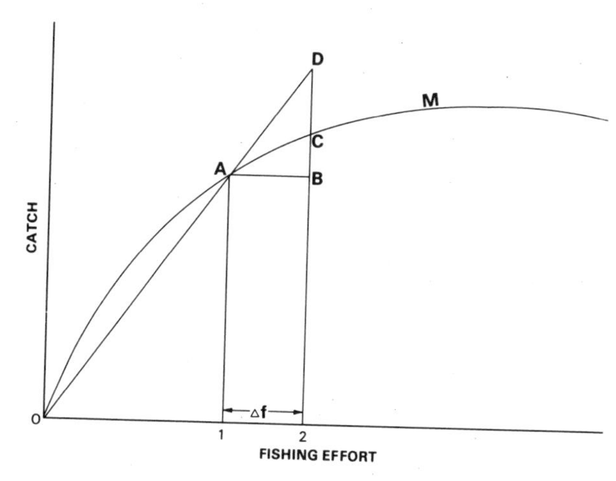

FIG. 3-20. A general relation between total catch and fishing effort where the marginal increase in yield, BC, is less than that expected from the catch-per-unit of effort (from Gulland 1968).

efficiency of the increase in effort. The marginal efficiency tends to unity near the origin, is zero at the point of maximum catch and negative for increases in effort beyond that which yields the maximum catch. For example, if the marginal efficiency is 40%, and at the present level of fishing a certain type of vessel catches on average 100 tons of fish a year, then the addition of one extra boat will only increase the total catch by 40 tons. If 100 tons is only a small fraction of the total catch, then the catch of the new boat will be only marginally less than 100 tons but the catch by all the other participants together would decrease by about 60 tons.

Where differences in year-classes are very large, it is likely that when a strong year-class enters the fishery the stock will increase whatever the magnitude of the catch (within limits). When a succession of strong year-classes is replaced by a run of poor ones, the stock may decrease even if fishing is cut back drastically. In such situations it is somewhat ridiculous to talk about a sustainable yield. However, when a stock is declining with strong year-classes being replaced by weak ones, managers are most likely to be seeking specific scientific advice.

In the simplest case, namely, that where recruitment is independent of parent stock abundance, management can only make the best of incoming recruitment by maintaining fishing at a level corresponding to the optimum position on the yield-per-recruit curve. In practice the yield-per-recruit curve may be quite flat or without a pronounced maximum (Fig. 3-18). In such situations determining the maximum is difficult. In any event, obtaining the theoretical maximum catch would require unwarranted increases in effort and costs. It makes little sense to use an extra 10–20% of fishing effort to obtain the last 1 or 2% of the possible maximum yield.

It is possible to calculate the fishing mortality which will maximize the net economic yield for a given set of circumstances, if the necessary biological and economic data are available. This can, however, be complicated by a variety of factors particularly when there are a number of countries, or different types of vessels from one country, participating in a fishery.

Gulland and Boerema (1973) proposed that a more objective method of calculating a desirable fishing mortality would be to consider the increase in total yield from adding one extra unit of effort. The marginal yield will equal the slope of the tangent to the curve of catch against fishing effort. It will always be less than the CPUE. The simple economic optimum occurs when the value of the marginal yield equals the marginal costs of a unit of effort. Gulland and Boerema (1973) pointed out that it would be undesirable to increase the amount of fishing beyond the level at which the value of the marginal yield is small compared with the costs of the extra unit of effort required to produce that yield. Considering what marginal yield might be regarded as small, they noted that an arbitrary figure used for advice on the management of Georges Bank Herring (ICNAF 1972a) was a marginal yield equal to one-tenth of the original catch per unit effort in the very lightly exploited fishery.

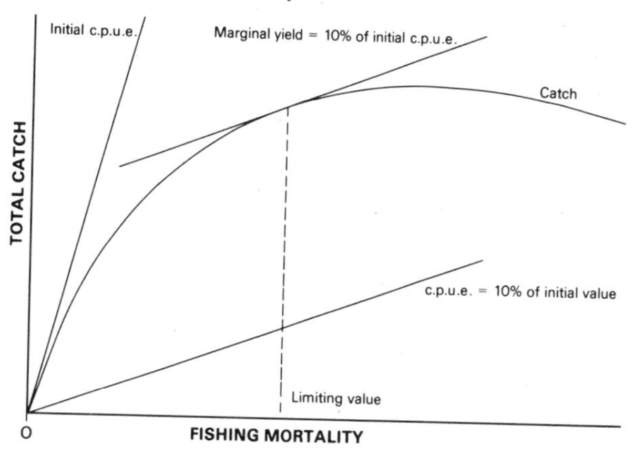

FIG. 3-21. An illustration of the situation where the marginal yield is 10% of the initial catch-per-unit of effort (from Gulland and Boerema 1973).

The calculation of this is illustrated in Fig. 3-21. There are two straight lines through the origin which show the catch per unit of effort in the lightly exploited fishery (the tangent to the catch curve at the origin) and a catch per unit of effort of 10% of this. The point where the tangent to the curve is parallel to this 10% line represents a limiting point beyond which any increase in fishing effort (mortality) would not be worthwhile.

This point has become widely known in the scientific and management literature as the $F_{0.1}$ *reference level* of fishing mortality. Although somewhat arbitrary, it was adopted by the International Commission for the Northwest Atlantic Fisheries (ICNAF) as the basis for setting the level of harvest for certain groundfish stocks in 1976. It subsequently became the standard reference point for scientific advice on management of Canada's Atlantic groundfish stocks.

6.0 CONCLUSIONS

In this chapter we have reviewed some of the basics of fish biology, and the biological and economic theories of fishing. We have seen that fisheries management is complicated by such biological factors as migration patterns, natural variability and multispecies interactions.

Traditional biological and economic models of fishing inadequately represent the "real world" of fisheries. They assume a stability in the fisheries system which is rarely observed in practice. Attempts to develop and apply more complex but more realistic models are confounded by a lack of reliable and comprehensive data on various facets of the fisheries system.

Despite these difficulties, fisheries managers must deal with an often unpredictable resource, a myriad of user groups, and conflicting objectives which are often vaguely articulated. In the next chapter we examine these objectives of fisheries management and their evolution over the past several decades.

CHAPTER 4

OBJECTIVES OF FISHERIES MANAGEMENT

"There is an infinite range of biological, social, economic and political objectives determined by local circumstances. Each of these objectives if followed could lead to a different pattern of harvest and exploitation."

FAO, 1979

1.0 INTRODUCTION

An ideal set of objectives for fisheries management is like the Holy Grail — desirable, much sought after, but elusive. Since the first significant attempts at fisheries management in the last century many conflicting objectives have been proposed. From the middle of this century to the mid-1970's, a debate raged between those who favoured a biologically based reference point for management (MSY) and those who advocated maximizing the net economic (MEY). From the mid-1970's onwards, both of these concepts have been replaced in Canada and the USA by the more ambiguous concept of Optimum Yield or Optimum Sustainable Yield. This is considered by its proponents to better accommodate the diverse, conflicting objectives involved in managing any fishery.

There are also different views about the scope and need for specificity of fisheries management objectives. Some observers, usually economists, contend that objectives must be very specific and measurable (e.g. Anderson 1983). Others (e.g. Gulland 1977; and Alverson and Paulik 1973) have taken the view that objectives can be general in nature.

In this chapter I examine the rise and fall of MSY as the basic objective for fisheries management in the world context, the attempts by fisheries economists to replace it with some version of MEY, and the recent shift to the more diffuse concept of Optimum Yield (OY). I then sketch the evolution of the objectives for fisheries management in Canada, with some observations about specific versus general objectives.

The view that fisheries management is concerned primarily with the biological conservation of fish resources has been generally replaced worldwide by a broader view of fisheries as an economic activity encompassing a broad spectrum of individuals, processes and interests. The fisheries system includes the resource base, the harvesting and processing sectors (primary and secondary industry) and the utilization of those resources in the marketplace.

A variety of models of the fishery system have been proposed. These range from the simple to the complex. The FAO ACMRR Working Party on the Scientific Basis of Determining Management Measures (FAO 1980) emphasized the complexities of the fisheries system but concluded that fisheries management structurally is the same as any other form of management.

Figure 4-1 depicts their concept of the management process. Defining objectives is the first step. Then data are collected and analyzed for managers. Managers use this information to examine options, and make and implement decisions. Results of those decisions are monitored and evaluated in the context of the original objective(s).

This often results in new objectives and management processes. Various models of fisheries systems which have been developed have one common feature — the need to define objectives as a first step in management. There is considerably less agreement on the nature and scope of objectives. The terms "goal" and "objective" have often been used interchangeably in the literature. Webster defines an "objective" as "something toward which effort is directed; an aim, goal, or end of action." Anderson (1983) defined management objectives as "the stated goal or goals to be achieved by fisheries policy." Halliday and Pinhorn (1985) noted that distinctions among definitions of strategies, objectives, tactics, issues, policies, regulating actions or mechanisms and management tools are "fertile ground for semantic arguments." They defined objectives as "the goals to be achieved by managing fisheries." I will use the term "objective" in this sense.

Various authors have emphasized the need for clear objectives for fisheries management (e.g. FAO 1979; Brewer and deLeon 1983). Anderson (1983) contended

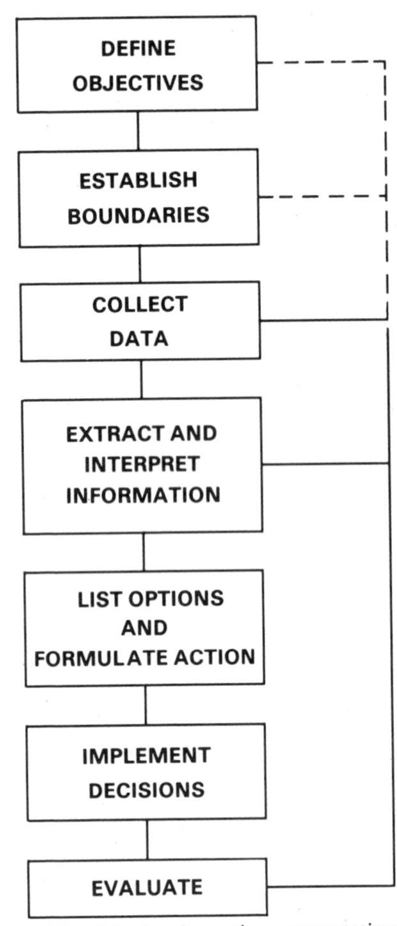

FIG. 4-1. A schematic representation of the components of a fishery management system.

that objectives must be "operational". He stipulated several important characteristics of operational objectives: "First, they must be stated such that criteria for success or failure can be quantified. The objective of improving fisheries may sound good, but it offers little help in comparing various policy alternatives. On the other hand, the goal of maximizing net revenues from the fishery is operational because earnings and costs can be measured and hence can provide an explicit means of comparing policy options. While some important aspects of fisheries, such as equity, industry structure, etc, are difficult to quantify, they can still be appropriate considerations for management objectives. However, unless the objective is stated such that there is a viable method of determining that it has been met, comparisons of regulatory techniques are not possible." Anderson also stressed the need to specify relative values of objectives to compare various regulatory options. A third important characteristic of operational objectives is that they include schedules within which they are to be achieved.

FAO (1984a) identified the inadequate formulation of objectives as one of the most general weaknesses of fisheries policy. They cited three reasons why objectives may not be adequate: (1) They may not have been sufficiently flexible; (2) Multiple objectives may conflict with one another; and (3) Objectives may be poorly defined, and hence subject to different interpretations.

Despite the benefits, the process of setting clear objectives is not easy. The diverse groups involved in any fishery will have a myriad of conflicting interests. The reconciliation of these conflicting interests and the articulation of a set of generally acceptable objectives pose a major challenge for fisheries managers. Making trade offs is a formidable but essential task.

2.0 EVOLUTION OF OBJECTIVES IN THE GLOBAL CONTEXT

2.1 Early History

Alverson and Paulik (1973) defined three groups of fisheries management objectives: (1) objectives concerned with maintaining the resource; (2) socioeconomic objectives; and (3) objectives reflecting national and international political interests. The concepts of fisheries conservation and management of fish stocks extend far back into human history. Egyptian, Greek and Roman literature shows evidence of laws to control fishing. Protection of spawning or mature fish was even linked to religion in some ancient societies.

Peter Larkin (1980) succinctly summarized man's earliest fisheries objective as the need to get something to eat for himself, his family or his clan. While there were early attempts to manage freshwater fisheries by manipulating lakes, ponds and streams, there is no mention in early writings of management of marine fisheries. In fact, as late as the last century, the stocks of marine fish were considered to be limitless. This was best reflected in Sir Thomas Huxley's famous statement in 1883 that the seas were inexhaustible. Ten years later, however, a British Select Committee recognized that depletion of marine fish stocks was indeed possible and recommended minimum size limits.

2.2 The Maximum Sustainable Yield Concept

The basic concepts of marine fisheries management evolved just before and immediately after World War II. These were summarized by Russell in a monograph on *The Overfishing Problem* (Russell 1942). The concept of Maximum Sustainable Yield had its origins in the simple biological models of stocks of single species of fish.

The apparently simple concept was attractive to international fishery managers. Throughout the 1940's and 1950's the theory and practice of MSY became widespread. It was embodied in the 1949 Convention setting up the International Commission for the Northwest Atlantic Fisheries (ICNAF). The aim of ICNAF was "to make possible the maintenance of a maximum sustained catch from the fisheries." MSY became the reference point for management in other international fisheries commissions during the 1950's and 1960's.

Larkin (1977) described the 10 years following World War II as the golden age for the concept of Maximum Sustainable Yield. Larkin summarized the MSY approach as follows: "...any species each year produces a harvestable surplus, and if you take that much, and no more, you can go on getting it forever and ever (Amen). You only need to have as much effort as is necessary to catch this magic amount, so to use more is wasteful of effort; to use less is wasteful of food."

During the late 1950's and 1960's the MSY concept came under assault. Many fisheries economists argued that the continuing pursuit of the maximum physical yield was not an appropriate objective for fisheries management. Initially, they pushed maximum economic yield (the static concept) as a more appropriate goal, modifying the concept by introducing dynamic models. But fisheries managers continued to support MSY until the mid 1970's. Economists persuaded fishery managers in some countries (e.g. Canada, Australia, New Zealand) to address the problems of open-access common property fisheries through such measures as *limited entry licensing*. However, very few managers accepted MEY as an appropriate objective. Ultimately, the shortcomings of MSY and some of the advantages of MEY became apparent. In the mid-1970's the biological and economic considerations implicit in these two approaches were subsumed in the more diffuse concept of optimum yield.

Given its almost universal acceptance in international fisheries management for some considerable time, why was MSY discarded in the 1970's? To answer that question, it is necessary to examine the advantages and disadvantages of the MSY concept. One of the primary reasons why MSY was the dominant objective for three decades was that fisheries managers and fishermen could grasp the basics of the concept. Also the idea of catching the maximum sustainable yield was readily acceptable to countries with divergent economic, social and political goals. For those countries concerned with the maximum production of food, MSY could ensure a continuing maximum supply of food from the sea. MSY also accounted for concerns about biological overexploitation, as these were then understood. At the same time MSY management minimized declines in fishing employment. MSY was thus an acceptable objective to most fishermen and governments.

The apparent simplicity of MSY is deceptive (Gulland 1968, 1977). It is too simple in at least four ways: (1) the great size of the system being considered; (2) the measures of input; (3) the measures of output; and (4) the trade-off between short-term and long-term interests. MSY does not account for species interactions, either in terms of the effects a fishery for one stock has upon other stocks or biological interactions among the species being fished. The latter could be either a predator-prey relationship or competition for food. Under the MSY approach, a particular stock is viewed in biological isolation, the only factor upsetting the equilibrium being the fishery for that stock. This of course is a gross simplification of what happens in the oceans. However inadequate our knowledge, these interactions should be taken into account.

The effect of natural variability in fish stock abundance also complicates the MSY concept. Under normal conditions the MSY approach may produce the desired results. Under adverse environmental conditions, stocks may collapse if the level of fishing is maintained at the theoretical MSY level. The theoretical relationship of sustainable yield to fishing effort does not hold in such circumstances. Environmental fluctuations can change the relationship so that the effort which produces MSY one year may have a considerably different impact in another year. This would result in overexploitation or underexploitation from a MSY perspective. For substocks of species, a general MSY can result in overexploitation and sometimes the extinction of less productive substocks. This has been demonstrated for Pacific salmon and herring and may well hold true for other species.

But the chief criticism of MSY was levelled by fisheries economists (Scott 1955; Gordon 1954; Christy and Scott 1965). MSY is a purely physical concept which considers only one output, the magnitude of the catch. It takes no account of the value of the catch or the cost of catching it. Economists and some biologists (e.g. Gulland 1968) have argued that, at the maximum, the curve of catch against fishing effort is flat. Hence, the marginal increase in catch by increasing the fishing effort by one quarter or one third is very small. It makes no sense to increase the effort by this amount to increase the catch by two to five per cent (see discussion of marginal yield in Chapter 3). Economists advocated the objective of obtaining the greatest net economic yield from a fishery. This would involve fishing at the point of the maximum difference between the value of the catch and the costs of catching it, hence the concept of Maximum Economic Yield (see previous Chapter).

2.3 The Maximum Economic Yield Concept

Gulland (1977) described MEY as a great advance on MSY because it recognized the importance of reducing inputs, for example, the costs of catching fish. Cost reduction is a management objective. A disadvantage of static MEY is that it is not uniquely definable (Gulland 1977). Few fisheries are valued solely in economic terms. Different countries view their fisheries from considerably different perspectives. Some are concerned with employment, others with food supply and still others are preoccupied with fisheries as a source of foreign exchange. Other factors, such as minimizing conflict between groups of fishermen or maintaining coastal communities, might be as important as minimizing the economic input for a particular country. The major drawback with static MEY is it is in fact static (Cunningham 1981). It fails to take time into account despite the fact that growth over time is extremely important with fish.

The concept of MEY shifted from the earlier static to a more dynamic concept in recent years. Static MEY aims to achieve the optimal allocation of resources to the fishery in any period by ensuring that price is equal to marginal cost. In contrast, the goal of dynamic MEY is to achieve the optimal resource allocation over time. Dynamic MEY seeks to maximize the discounted present value of future net returns.

Cunningham (1981) described the concept of dynamic MEY as theoretically sound but practically weak. Apart from the discount rate, the level of fishing effort corresponding to dynamic MEY depends on the price of fish and the costs of catching it. In theory information on price and cost would be required until infinity. In practice revenue and cost would be of little use beyond about 30 years. Thus information would be required for 30 years into the future. Cunningham (1981) concluded: "It thus appears probable that many, perhaps widely differing estimates, may be made of the future trend of revenue and costs, and it will be a serious problem to decide which estimate to believe." Crutchfield (1979) argued that dynamic MEY assumes "a level of information on a forecast basis which is not only unavailable at present, but almost certainly will be unavailable at any conceivable cost in the future." Despite the theoretical elegance of the approach, the dynamic MEY model is unlikely to be applied by fisheries managers in the near future.

2.4 Real World Problems

So far we have been discussing fisheries management objectives from biological and economic perspectives. From a biological perspective, MSY is simply not adequate. It provides a first index of production potential and is useful as a first cut at managing marine fisheries. But, it is not sufficient as a long-term management objective. Fisheries economics, on the other hand, has concentrated almost exclusively on allocative efficiency (Cunningham 1983). The distributional problems of management (the equity issues) are often assumed away or ignored.

In the real world, management decisions are often taken by politicians rather than biologists or economists. Politicians must consider the diverse and often conflicting interests involved in the fishery. As Larkin (1980) noted: "Fishermen have their lives to lead, and governments must consider national interests. The question is not so much how to harvest fish stocks biologically, or how to promote economic return, but rather how best to proceed in the face of current social circumstances." This leads to a series of questions and choices. "For example, if it is apparent that too few fish are being taken by too few fishermen to achieve either a biological or economic maximum return, should the fishery be curtailed or closed regardless of whether the fishermen will no longer have a living? Should the guiding principle be drawn from the literature of welfare economics? Should social welfare be maximized?... Should the attempt be made to maximize 'aesthetic values'?".

Fisheries management objectives must consider other important factors. These include income distribution and employment as well as conservation and economic efficiency. Managing a fishery by limiting open access, for example, increases income for those involved in the fishery by freezing or reducing the number of participants. Those excluded lose income as well as any non-monetary benefits from fishing. Rettig (1973) observed that, in general, society regards equity as avoidance of drastic changes in income distribution. Unless management schemes account for this, they are unlikely to be adopted.

Pontecorvo (1973) observed: "A bio-economic position dominates thinking about fisheries management simply because there is no body of social or political theory (relative to welfare maximization in economics or population dynamics in biology) to force a modification of either the biological or economic position. In these circumstances, which appear unlikely to change in the foreseeable future, political and social considerations can only be considered on an ad hoc basis...." Bishop et al. (1981) noted the need for a multi-objective approach: "The heavy emphasis on the efficiency goal in most of the economics literature has been challenged by both economists and other social scientists.... The concern has been that efficiency is not the only goal in fishery management and may be less important than other goals."

2.5 Emergence of the Optimum Yield Concept

By the mid 1970's, biologists, economists and fishery managers were all searching for an alternative to the MSY/MEY paradigms. From the ferment emerged the concept of Optimum Yield, also known as Optimum Sustainable Yield. The term OSY was first used in the 1958 Law of the Sea Convention where conservation was defined as "... The aggregate of the measures rendering possible the optimum sustainable yield from those resources so as to secure a maximum supply of food and other marine products...." In this context OSY was equated with MSY.

By the early to mid 1970's attention focused on a variety of factors of importance to fishermen and fisheries managers. A 1974 symposium sponsored by the American Fisheries Society reached a consensus that fisheries management required a new broader objective. The concept of Optimum Yield which emerged meant different things to different people (Roedel 1975). Wallace (1975) observed:

> "Such a concept requires taking into account all the factors mentioned — economic, sociological and biological — in determining the optimum level of harvest.... In general, the results will be a level of fishing that would normally be below MSY but would provide for the special needs of special groups and the specific requirements for conservation of the target species.... Economic, social, and biological values will serve as a basis for the statement of the objective. The optimal yield allows for these inputs, rather than being limited to maximizing net profits or maximizing sustainable yield."

Harville (1975) proposed that Optimum Yield incorporate a full array of interdisciplinary goals as a composite objective, including:

(a) objectives for maintaining resources, including sustainable yield, prevention of waste and protection and enhancement of environmental productivity;
(b) objectives for economic efficiency and maximizing the economic yield consistent with protection of the resource; and
(c) quality of life objectives for educational, recreational, aesthetic, and other social benefits.

Summarizing the Symposium, Roedel (1975) defined "optimum sustainable yield" as "a deliberate melding of biological, economic, social and political values designed to produce the maximum benefit to society from stocks that are sought for human use, taking into account the effect of harvesting on dependent or associated species."

Roedel suggested that these definitions:

(i) recognize non-extractive uses and values;
(ii) recognize the importance of quality to the sport fishing experience;
(iii) consider return on investment as a major criterion in setting harvest rates;
(iv) allow management based on traditional MSY if there is an overriding need for fisheries products; and
(v) temper all these factors with knowledge of the real world and what is acceptable to the body politic.

Larkin (1977), in *An Epitaph for the Concept of Maximum Sustained Yield*, criticized these definitions as "virtually meaningless" and "an eclectic mishmash that was all things to all people." He suggested that "the concept is potentially subject to abuse and would almost certainly be used primarily as a way of justifying a political course of action." He observed: "The chances that your optimum is my optimum are nearly zero.... natural systems are sufficiently diverse and complex that there is no single, simple recipe for harvesting that can be applied universally. When there is added in the complexity and variety of social, economic and political systems, the number of potential recipes is just too enormous to be easily summarized by simple dogma."

By the time Larkin's Epitaph for MSY was published, the USA and Canada had from differing perspectives already adopted some concept of optimum sustainable yield as the general objective for fisheries management.

The U.S. Fishery Conservation and Management Act of 1976, the so-called Magnusson Act, adopted the concept of optimum yield and defined it as the amount of fish:

"(a) which will provide the greatest overall benefit to the Nation, with particular reference to food production and recreational opportunities; and
"(b) which is prescribed as such on the basis of the maximum sustainable yield from such fishery, as modified by any economic, social, or ecological factor."

Spencer Apollonio (1983), the first Executive Director of the New England Fishery Management Council, described "optimum yield" as an elusive concept that allows social and economic considerations to be taken into account. He observed: "Experience has shown that it is difficult for councils to define their management objectives.... Casual examination of fisheries management plans shows, in many cases, very generally stated objectives.... Many of these objectives are so general that they amount, one suspects, to no clear objective.... The MFCMA, by setting undefined 'optimum yield' as the goal of fisheries management, in effect has given extraordinary latitude to the regional councils in setting policy. In practice, optimum yield is what the managers say it is."

For a further elaboration on the application of optimum yield under the U.S. Regional Council system, see Chapter 20.

In 1976, Canada fought in ICNAF for the abandonment of MSY and adopted a national optimum yield approach in the Policy for Canada's Commercial Fisheries (DOE 1976a). This document stated that the guiding principle in fisheries management no longer would be "maximization of the crop sustainable over time" but the "best use" of society's resources. "Best use" was defined as "the sum of net social benefits (personal income, occupational opportunity, consumer satisfaction and so on) derived from the fisheries and the industries linked to them."

In the next section I examine the factors which led to this shift in policy and the changes in Canadian fisheries management objectives since that time.

3.0 EVOLUTION OF OBJECTIVES IN THE CANADIAN CONTEXT

3.1 The Early Years: From 1857 to the Second World War

To a large extent, the evolution of fisheries management objectives in Canada to 1976 paralleled developments on the world scene.

The first comprehensive fisheries legislation in British North America was passed in 1857. That legislation incorporated implicitly two sets of objectives. The major principle for management of the river and estuary (primarily salmon) fisheries was conservation. The 1857 Act consolidated various local regulations to control fishing practices considered harmful to the resource. The principle of conservation was apparently first recognized in 1822 by the legislature of Upper Canada which legislated closed seasons to protect the breeding period of game fish. The 1857 Act created a body of fisheries overseers with enforcement powers. The violent opposition to those who sought after 1857 to enforce regulations indicated that earlier conservation measures had never been enforced.

The second major focus during this period was the fisheries of the Gulf of St. Lawrence. "The Gulf was a vast stretch of relatively unknown waters whose wealth in fish must have been regarded with the same optimism as that [with] which the lumberman scanned his seemingly inexhaustible forest wealth" (Hodgetts 1955). There was considerable international competition for those resources. Captain Pierre Fortin on the vessel *La Canadienne* policed the fisheries of the Gulf, (see Report of the Crown Lands Department, 1852), as "the magistrate in command of the expedition to protect the fisheries of the Gulf." Policy became "a matter of stimulating Canadian fishermen to take advantage of their natural riches before strangers better organized and equipped got too deeply entrenched" (Hodgetts 1955). In 1858 a system of fishing bounties was introduced to encourage the achievement of this objective.

The first *Dominion Fisheries Act, 1868,* gave the Governor in Council power to make "every such Regulation or Regulations as shall be found necessary or deemed expedient for the better management and regulation of the sea-coast and inland fisheries." Fisheries policy for the estuary and river fisheries continued to focus on conservation, using licensing, time, area and gear restrictions. Policy for the ocean fisheries continued to focus on development. The bounty system was suspended for a brief period following Confederation, but it was reinstituted in 1882. At that time it was funded by a cash settlement from a dispute over access by American vessels within the Canadian three-mile limit (Canada, 1873). Apart from the bounty system, the ocean fisheries policy was laissez-faire. For the ocean fisheries, "it appears highly desirable to abstain from interference as much as possible" (Canada, 1876).

The Commissioner of Fisheries for Canada in 1902 described the objectives of fisheries policy as follows:

> "Four main interests have been prominent in the forming of fishery regulations generally. These are: First, the interests of the fish. If there were no fish there would be no fisherman and no fishing industries. Hence, the preservation and fostering of the fish-supply in their native waters is imperative. Second, the interests of the fishermen as an industrial community. The body of fishermen have legitimate rights, which must be recognized by the state. The rights of labour cannot and ought not to be ignored; and the fishermen form an important part of the population in most countries.... Third, the interests of the state as a whole. The interests of the state, or as it is commonly expressed, the public interest, may not always coincide with the first or

second interest described above, indeed they may come into serious collision, and many authorities might be quoted to show that the public interest should be paramount and that all other interests should be regarded as of secondary importance.... Fourth, international interests, which may affect the comity of nationals and which have often reached a stage so crucial and perilous as to over-ride the interests of the fish, the fishermen and the nation, requiring these interests, indeed, to a large extent, to be sacrificed to avoid momentous and lasting evils, such as foreign unfriendliness or even war" (Prince 1902).

This was one of the first acknowledgements that fisheries policy must reconcile multiple conflicting objectives. The primary emphases for a number of decades to follow were conservation for some species and development of fisheries for other species or areas considered to be underexploited.

Needler (1979) noted: "The principal purpose of early fisheries regulations was to protect the fish stocks from overfishing and depletion. Attention was concentrated on those stocks where the danger appeared to be greatest — the fish and shellfish caught close to shore and the freshwater and anadromous species." He pointed out that some regulations were related more to marketing than to conservation, e.g. lobster closed seasons. For some species, e.g. lobster and scallops, conservation regulations included the establishment of fishing districts, fishing seasons and minimum size limits. These became increasingly detailed over the years.

The cod fishery was treated quite differently. There were no regulations at all in this or other groundfish fisheries until the 1940's. The exception was a limit on the number of trawlers in the 1920's and 1930's in response to pressure from hook-and-line fishermen. The MacLean Royal Commission of 1928 investigated the fisheries of the Maritime Provinces and the Magdalen Islands (MacLean 1928). It addressed nine specified issues and had a more general mandate "to enquire into the general condition of the fishing industry, how existing conditions of the fisheries and fishermen might be improved and how the industry might be further developed with expedition and efficiency." The Commissioners could not agree on the trawler issue. There were separate reports by four commissioners, on the one hand, and the Chairman, on the other, with the Chairman favouring the use of trawlers under certain strict controls. In response, the federal government limited the use of trawlers. It appears that no more than three operated during the 1930's.

The MacLean Commission reiterated the conservation objective but also advocated "more adequate returns to the fishermen and greater prosperity for the industry in general." To achieve this, "a larger amount of invested capital is needed, a larger expenditure by the Federal Government, temporarily at least, is urgently required, and more co-operation among fishermen and dealers is essential." The Commission concluded that the fisheries of the Maritime Provinces "are capable of great expansion."

Despite this note of promise, the Atlantic fisheries which had prospered during World War I declined through the Depression. Towards the end of World War II, the use of trawlers became more acceptable. The Quebec fishing industry changed from one based on salt codfish to one based on fresh and frozen fish. This transition was aided by government construction of cold storage facilities.

3.2 The Post-War Years

Stewart Bates, as part of the Nova Scotia Royal Commission on Provincial Development and Rehabilitation, produced in 1944 a major *Report on the Canadian Atlantic Sea Fishery* (Bates 1944). From a comprehensive examination of the fishing industry before and during the war, Bates called for structural changes to make the industry more competitive in world markets:

> "A fuller use of the fishery resources of these provinces is physically possible and necessary if the industry is not to return to its pre-war depressed state. But this revolution can be achieved only by new methods that will increase its efficiency and its productivity per man."

He found greatest potential for expansion in the frozen trade, particularly for groundfish. To achieve this, he recommended technological upgrading, new marketing arrangements and strengthened education, research and development. He concluded that, if expansion were to be rapid and integrated, financing would have to be primarily governmental. He also advocated closer cooperation within the industry to make it more competitive.

Following Bates' recommendations, the emphasis shifted in the post-war period from conservation of the nearshore, estuarine and river species to development and expansion of the Atlantic fishery, particularly groundfish. Interestingly, Bates stated: "No wise government would encourage (or permit) such expansion of private enterprise as would result in overfishing because that leads to reduction in income. The government aim would be expansion to the *optimum point,* but not beyond" (Emphasis mine). No definition of the *optimum point* was provided. By the late 1960's this note of optimism would be lost in the mists of history.

Escalation of foreign fishing.

The postwar years saw an expansion of fisheries and the use of new technologies worldwide. There were more powerful, long-range vessels with onboard processing and storage facilities, fish-finding instruments, new types of fishing gear and improved processing, storage and transportation of fish products. This led to expanded world consumption and a considerable increase in fishing effort. The world catch increased from about 20 million tons in the early postwar years to about 70 million tons in the early 1970's.

In postwar Canada, trawlers were acquired with government assistance. The Canadian government promoted industry expansion through a variety of subsidy and assistance programs to help fishermen modernize and upgrade their equipment (Crowley et al. 1993). These initiatives and the development of North American markets had killed the offshore schooner industry by the early 1960's and drastically curtailed the inshore saltfish industry.

New distant-water fleets began to appear in the Northwest Atlantic in the 1950's. Catches by these fleets increased rapidly in the 1960's. Total catches increased from about 2 million tons to a peak of 4.6 million tons in 1968 and remained at about this level through 1973. Groundfish catches peaked in 1965 at 2.8 million tons and declined steadily from 1968 to 1974.

At the beginning of the postwar period there appeared to be little need to regulate the fisheries of the Northwest Atlantic. Nonetheless, an International Convention for the Northwest Atlantic Fisheries was negotiated in 1949 and came into force in 1951. This initiative was prompted because of Canadian and U.S. concern about the fisheries off their Atlantic coasts. The International Commission for the Northwest Atlantic Fisheries (ICNAF) was established. By the early 1950's ICNAF was beginning to regulate the fisheries using mesh size controls. By the late 1960's, it began discussing more active intervention for the Northwest Atlantic as stocks declined with the huge escalation of distant-water fishing effort. The management objective embodied in the ICNAF Convention was the maximum sustained catch, i.e. MSY. But until the early 1970's there was no limit on the amount of fishing or catch. For more details, see Chapter 10.

Canada officially subscribed to MSY in the ICNAF context during the 1950's and 1960's. Domestically, however, it pursued modernization and fleet upgrading. This was intended to improve incomes of fishermen, to provide the groundfish processing industry with year-round fish supplies through expansion of the Canadian offshore fishing effort, and to compete with the foreign fleets. Small and medium-size firms consolidated into a few large, vertically integrated companies. The

Industrial Development Branch of the federal Department of Fisheries (and provincial counterparts) encouraged the expansion of Canada's offshore, midshore and nearshore fleets and the development of new fisheries (e.g. scallops, shrimps and crabs). This resulted during the 1960's in conflict between conservation-oriented scientists of the Fisheries Research Board of Canada and exploitation-oriented engineers and technologists in the Industrial Development Branch.

Barrett (1984) described the period from World War II to the early 1970's, when fishery policy promoted the industrialization and centralization of the industry, as "a 30 year honeymoon between the private and public sectors when large-scale development projects and expansion were seen to be synonymous with modernization and progress... the crisis in fish stocks from 1968 onward brought an end to this age of innocence."

The modernization thrust of the 1960's attempted to help the Canadian offshore industry compete with the distant-water fleets. This thrust did not meet its objective. The Canadian share of the groundfish catch in the Northwest Atlantic dropped from 34.5% in 1955 to 20.2% in 1965. While the Soviets increased their effort off the Canadian Atlantic coast more than fourfold between 1960 and 1965, the Canadian catch increased by only 6%.

The number of Canadian vessels (50 gross tons and over) on the Atlantic coast increased from 211 in 1959 to 558 in 1968. The average size of vessel also increased. However, the Canadian Atlantic groundfish catch grew by only 18% while the total groundfish catch by all countries increased 57%. Between 1965 and 1968 both the Canadian catch and the total catch levelled off. While this was occurring, provincial Departments of Fisheries, the Atlantic Development Board, and the Atlantic Provinces Economic Council all called for more federal assistance for offshore expansion. They believed that the foreign fleets would gradually withdraw because of Canada's proximity to the fishing grounds.

Hédard Robichaud, Minister of Fisheries between 1963 and 1968, fostered federal-provincial co-operation and industrial expansion through shared-cost programs. The federal government covered most of the cost. He encouraged the use of Pacific purse seiners to develop the east coast herring fishery, and new vessel and plant construction. Robichaud convened a special Federal-Provincial Conference on Fisheries Development in Ottawa in January, 1964. Following that conference, cooperative federal-provincial programs aimed at development of the Atlantic fisheries were expanded. While Atlantic fish stocks were beginning to suffer from excessive fishing pressure, Robichaud (1966) maintained that "our Atlantic Fisheries are expanding in a spectacular manner, but it is orderly expansion." Robichaud envisaged considerable growth in the Atlantic fisheries. This was to be achieved through diversification to other species, transformation of the Atlantic groundfish fishery from an inshore to an offshore fishery and investment of about $300 million over 10 years in modern fishing vessels.

Although preoccupied with development and expansion, Minister Robichaud recognized the importance of conservation: "This task demands of industry not only to expand production as much as economically desirable, but also to ensure, by observation of conservation practices, an adequate harvest for posterity" (Robichaud 1967).

3.3 An End to the Age of Innocence

Effects of excessive fishing pressure were evident in declining Atlantic groundfish catches. The total catch dropped from 2.8 million tons in 1965 to 1.6 million tons in 1974. The Canadian groundfish catch decreased from 620,000 tons in 1968 to 418,000 tons in 1974. By the mid-1960's, ICNAF scientists were warning that the amount of fishing must be restricted to maintain fish stocks at the MSY level (ICNAF 1964a; Templeman and Gulland 1965). Total Allowable Catches (TACs) or catch quotas for two haddock stocks for 1970 were the first steps in this direction. With an amendment to the ICNAF Convention which allowed for allocation of catch shares to member countries, the first national allocations were set for herring for 1972. The ICNAF Convention's objective was also changed to "optimum utilization", to accommodate economic and technical, as well as biological factors. By 1974 all of the major groundfish stocks of the ICNAF area were under a system of Total Allowable Catches and national allocations. These initial TACs were established at the MSY or F_{max} level (Chapter 3).

Neither Canada nor the USA believed that the single-species MSY approach was adequate (ICNAF 1974a, 1975a, 1976a). The USA achieved the adoption of a second-tier TAC and catch quota system for the southern part of the ICNAF area (Subareas 5 and 6). This measure was in effect from 1974–76 to control the overall level of exploitation (O'Boyle 1985). It represented the first significant departure from the single-species MSY approach. At Canadian insistence, ICNAF discussed fishing effort controls (ICNAF 1975a, 1976a). As early as May, 1972, Jack Davis, the Minister responsible for Fisheries, outlined in the House of Commons the Canadian approach in ICNAF to deal with overfishing. Two basic thrusts were involved: (1) conservation, and (2) a special preference for the coastal state. He observed: "The emphasis is on conservation. Most of the traditionally fished stocks in the Northwest Atlantic are either being over-fished or are under heavy fishing pressure. Result: a reduction in the supply of fish to our inshore fishermen. This is why Canada is urging a cut back in the offshore fishing effort by all nations and a special allocation, in Canada's favour, to improve the income of inshore fishermen in the Maritime Provinces, Newfoundland-Labrador and Quebec" (Canada 1972).

At the 1975 ICNAF meeting, Canada secured the adoption for 1976 of a regulation reducing fishing effort by non-coastal states by approximately 40% from that of 1972–73.

Canadian managers had come to regard the MSY/F_{max} approach as inadequate. The $F_{0.1}$ concept, described in Chapter 3, had been adopted for herring in the ICNAF area and Canadian managers began to press in 1975 for a strategy of fishing below F_{max} for the Atlantic groundfish stocks. They persuaded ICNAF to agree in 1976 on TACs set at the $F_{0.1}$ level for 1977. These were, in effect, transitional measures since coastal state jurisdiction had already been announced for 1977.

The stage had been set for this shift in management objectives by the ICNAF Standing Committee on Research and Statistics (STACRES) at the Seventh Special Commission in September, 1975 (ICNAF 1976b). The Committee pointed out the need to recognize limitations of the F_{max} reference point. These were greatest for fish stocks which: (1) had no clearly defined maximum for the relationship between catch per recruit and fishing mortality or, (2) if present, the maximum occurs at a relatively high value of fishing mortality rate. STACRES noted that in these situations setting TACs at the F_{max} level could severely reduce the stock size. This could also reduce the number of age-groups in the exploited stock, and cause large short-term changes in catch (and hence in the magnitude of the short-term changes which must be made in the TACs). Recruitment failures might also occur due to the generation of too low spawning stock sizes.

The Committee concluded: "In view of the possible large adverse consequences of setting the fishing mortality rate too high in cases where there is doubt about its adequacy, a more restrictive management system than that based on the F_{max} level of fishing mortality would be justified.... A lower fishing mortality rate than F_{max} can be set which would result in only a small loss in average catch but would achieve a substantially higher average stock biomass, greater stock stability due to the presence of a larger number of age-groups in the exploited phase, higher average catch per unit effort, and increased economic efficiency." $F_{0.1}$ was suggested as a possible reference level of fishing mortality to achieve these benefits (ICNAF 1976b).

At its 1976 Annual meeting, STACRES was more explicit. "Although a single management objective could not be recommended to cover all stocks because the objective may vary from stock to stock, TACs for 1977 should generally be recommended with the aim of achieving $F_{0.1}$ rather than F_{max}" (ICNAF 1976c).

3.4 The 1974–76 Groundfish Crisis and the Policy for Canada's Commercial Fisheries

While Canada was working within ICNAF to put means in place to rebuild the fishery, the Canadian Atlantic groundfish industry was plunged into crisis. The crisis was due to declining catch rates combined with a downturn in the U.S. market for groundfish markets (see Chapter 13). The severity of the crisis caused the federal government to institute a Groundfish Bridging Program involving $140 million in assistance over 3 years. Meanwhile the state of the industry was being examined by a Task Force headed by Mr. Fern Doucet, a Special Advisor to the Minister of Fisheries. The Doucet team, formed in December 1974, studied every aspect of the industry.

Speaking to the Fisheries Council in May 1975, Minister LeBlanc commented on the Task Force's findings: "Through ICNAF, we have provided at least some protection for the stocks...but the protection has not been enough to maintain the stocks at levels where they can make our fisheries pay.... Management practice has been concentrated too much on biological considerations and too little on tailoring our fleet to the capacity of the resources to sustain them." He called for broader fisheries management objectives (DOE 1975a).

A new set of objectives emerged from the fire of the 1974–76 crisis. These were articulated in the May 1976 *Policy for Canada's Commercial Fisheries* (DOE 1976a). This document took a broad view of the fisheries system. It indicated that the extension of fisheries jurisdiction would not by itself solve industry's problems: "Extended jurisdiction should be seen as being as much a challenge as an opportunity, and in facing up to that challenge we should always keep in mind that the fortunes of the fishing industry depend on more than fish. They depend on markets, on production costs, on the industry's built-in ability to compete, and on a myriad other factors."

One special feature of the fishery recognized in the 1976 Policy was its instability: "The fishing industry...suffers from unstable product prices and unstable material and service costs. The fisheries are also buffeted by forces of a special kind arising from the character of the resource base...year-to-year variability is much more difficult to deal with particularly when (as is commonly the case) the variations cannot be accurately predicted.... Resource-based fluctuations from time to time may be superimposed on the movement in costs and prices stemming from the other sources mentioned. The result can be a steep, and usually unforeseen, fall or rise in earnings and profits. Industry reaction to this form of uncertainty frequently has been to install sufficient catching and processing capacity to handle the peaks in supply, thereby inflating industrial overheads and reinforcing the inherent tendency toward over-expansion in the commercial fisheries."

The 1976 Policy identified the defects in the MSY approach to resource management and indicated that existing industrial and social development policy had skirted some basic problems. The Policy predicted that fundamental restructuring of the fishing industry was inevitable. It would occur in an orderly fashion, under government auspices, or through the operation of inexorable economic and social forces.

The document described two major shifts in policy:

1. The guiding principle in fishery management would no longer be maximum sustainable yield but the best use of society's resources. "Best use" was defined as the sum of net social benefits (personal income, occupational opportunity, consumer satisfaction and so on) derived from the fisheries and the industries linked to them.
2. Private enterprise would continue to predominate in the commercial fisheries. However, fundamental decisions about resource management and about industry and trade development would be reached jointly by industry and government.

The long-term viability of the industry was seen to depend on "getting rid of certain structural defects, notably catching and processing over-capacity, dispersal of processing facilities, and fragmentation of business organization." These changes should be gradual to minimize the disruptive impact of change: "Where adverse social side-effects, such as reduced employment opportunities can be kept within acceptable limits, restructuring should proceed. Where damage to the community would outweigh advantages in the short run the changes must be postponed."

Fewer people employed in harvesting was seen as essential. However, the document was quick to add that this would not mean drastic dislocation of the people dependent on the fishing industry. Rather, where it was feasible to expand, this should be accomplished without increasing fisheries employment.

Pacific salmon development required dealing with the threat from environmental degradation of their freshwater and estuarine habitat and artificially enhancing salmon numbers. For both Atlantic groundfish and Pacific salmon, necessary first steps included improving the international management regime and establishing systems to control access to the resource.

Copes (1980) described the 1976 Policy as "the first comprehensive statement of fisheries analysis and policy ever issued by a Canadian government. The document clearly confirmed the economic analysis that had emerged over the previous 15 years. It acknowledged the need to apply limited entry universally, to reduce significantly the excessive manpower of the inshore fishery, and to rationalize the dispersed and fragmented processing industry.... Conscious of the socioeconomic problems of the Atlantic region, this policy went well beyond the now familiar criterion of simple resource rent maximization. It seemed in tune with the more elaborate criterion of maximizing the combination of resource rent, producer's surplus and consumer's surplus, leaving room for the accommodation of any relevant social cost/benefit considerations."

Larkin (1977) was quick to characterize the 1976 approach as "typical, for it says (in only 302 words) that the goals are to maximize food production, preserve ecological balance, allocate access optimally, provide for economic viability and growth, optimize distribution and minimize instability in returns, ensure prior recognition of economic and social impact of technological change, minimize dependence on paternalistic industry and government, and protect national security and sovereignty — it being kept in mind that there is no priority implied in the order things are listed; that there are interactions in the objectives; and that trade-offs and compromise will be necessary." Larkin further observed: "These goals are striking in implying that there is no single optimum policy, for as we all know, one cannot optimize for two things at the same time, let alone a dozen. They are humorous because they so accurately reflect the real difficulties of managing human affairs."

While it lacked specificity, the 1976 Policy was the first comprehensive attempt to propose policy objectives for the entire fisheries system from the water to the table. It set the scene for ambitious management initiatives following Canada's extension of fisheries jurisdiction in January 1977. Canada took steps to reduce foreign fishing and to restrain expansion of the Canadian offshore fishing effort to provide for stock recovery. Considerable euphoria followed the 200-mile limit (see Chapter 2). There were grandiose provincial plans for fleet development and expansion and demands by Newfoundland for constitutional change for fisheries jurisdiction. Speaking for the processing sector, the Fisheries Council of Canada (FCC) attributed the 1974–76 crisis solely to resource overfishing. The Council believed the problem would be solved with extended jurisdiction and quota management. In a 1977 brief to a Cabinet Committee on Food Strategy, the FCC expressed fear that increased federal control over processing and marketing implied in the 1976 Policy would curtail free enterprise (*St. John's Evening Telegram*, December 7, 1977).

A major battle ensued between the processing sector, some provincial governments and the federal minister. Enjoying the benefits of a recent federal government bail-out, some companies felt ready to exploit the opportunities of the 200-mile limit. They wanted no truck-or-trade with LeBlanc's interventionist policies. The companies began a major expansion, particularly in the unregulated processing sector. This contributed largely to another crisis in 1981 and 1982 when the industry began to sink in a wave of recession savaging North America. Chapter 2 told part of the story of this period. The remainder will be addressed in subsequent chapters.

In the immediate post-extension era, fisheries policy under Minister LeBlanc became more regulatory. Catch quotas were widely instituted, with allocation of access through complex fishing plans and an elaborate licensing system (see Chapters 7 and 8). Catch rates and catches increased. Markets were buoyant, and the industry enjoyed several good years.

In 1981, DFO issued a more specific document titled *Policy for Canada's Atlantic Fisheries in the 1980's* (DFO 1981a). It stated:

> "$F_{0.1}$ will continue to be the principal reference point for scientific advice on the level of harvest. As a general rule, no TAC will be set above that reference level but in certain instances, TACs will continue to be set below that level in order to provide for stock rebuilding, larger average size of fish in the catch, improved catch rates and greater stability of catches."

The 1981 Policy restated a commitment to maintain and revitalize the inshore fishery. It favoured qualitative over quantitative development. The $F_{0.1}$ reference point established in the 1981 Policy continued to be the basis for setting the level of harvest for most Atlantic finfish stocks through the 1980's. But other aspects of the 1981 document were soon outdated as the industry plunged into another crisis, greater than that of 1974–76.

3.5 The 1981–83 Crisis and the Kirby Task Force

A severe recession gripped North America in 1981. In August of that year, industry leaders met with senior government officials to signal a crisis that would lead to fundamental industry restructuring. However, this was not necessarily the kind of restructuring that Minister LeBlanc had once intended.

By November 1981, two of the four major processing companies on the east coast were asking the federal government for financial assistance. The government decided to set up a task force. The Task Force on the Atlantic Fisheries was established in January, 1982 with Michael Kirby, Secretary to the Cabinet for Federal-Provincial Relations, as Chairman. The Task Force had two mandates. The first was to advise the government on long-term policies for a healthy fishery. The Task Force Report titled *Navigating Troubled Waters* addressed this issue (Kirby 1982). The second mandate was to advise the government on the requests for financial assistance from the Lake Group Ltd. of Newfoundland and H.B. Nickerson & Sons Ltd. of Nova Scotia. This resulted in an Atlantic Fisheries Restructuring Act. Several medium-size firms were consolidated into two larger ones, Fishery Products International (FPI) and National Sea Products (NSP). The federal government injected equity in both these firms and in the new Pêcheries Cartier, a smaller government controlled firm in Quebec. For details, see Chapter 13.

The Task Force made 57 recommendations including a new set of objectives for Atlantic fisheries management. Parsons (1993) provides details of these recommendations and the government's response.

The Task Force proposed three objectives for Atlantic fisheries policy in order of priority:

Objective 1: The Atlantic fishing industry should be economically viable on an ongoing basis, where to be viable implies an ability to survive downturns with only a normal business failure rate and without government assistance.

Objective 2: Employment in the Atlantic fishing industry should be maximized subject to the constraint that those employed receive a reasonable income as a result of fishery-related activities, including fishery-related income transfer payments.

Objective 3: Fish within the 200-mile Canadian zone should be harvested and processed by Canadians in firms owned by Canadians wherever this is consistent with objectives 1 and 3 and with Canada's international treaty obligations.

The Task Force gave the following rationale for these objectives:

> "It is essential to develop an Atlantic fishing industry that does not require regular or periodic government subsidies in order to survive.... Economic viability must therefore be the primary concern of government policy makers.... [This] does not mean that every

Conflicting technologies: cod trap (top), Atlantic trawler (bottom).

processing firm or harvesting enterprise now in the industry should become permanently economically viable. There will be casualties, there will be bankruptcies, and these should be allowed to occur. They are a normal part of the Canadian economic system."

The Task Force, however, believed that industry should not be allowed to develop solely based on economic efficiency. The social consequences of such an approach were unacceptable.

The second objective emphasized the need for the fishery to employ as many people as possible. This recognized that the Atlantic region was economically disadvantaged. The fishing industry was the only possible source of employment in much of that region. The Task Force had, however, rejected the view that maximizing employment should be the primary fisheries objective. As far as possible, the existing distribution of fishery-dependent population should be maintained. This was particularly true for Newfoundland. Consolidation of plants, for example, should be extremely exceptional rather than part of a "heartless economic rationalization plan."

The Task Force recognized that Objectives 1 and 2 would often conflict. Nevertheless, they concluded that "fisheries policy must reflect both economic and social realities." The "social dimension must take precedence when the issue at stake is increased profitability but not when the issue is the continued economic viability of a firm."

The Task Force observed that almost all the Atlantic fishing industry supported objective 3. It left the door open to favour objective 1 over objective 3 if new federal, provincial or private investment were inadequate.

The Minister of Fisheries and Oceans, Pierre De Bané, announced that the government accepted the objectives in the order of priority set out in the Task Force Report (DFO 1983a).

The objectives proposed in this Report represented a considerable shift from the earlier stated objectives for Canadian fisheries policy. The resource conservation objective was conspicuous by its absence. It had been dominant for most of the preceding century. When questioned about this, the Chairman stated that it was taken as "given" and need not be included in the list of objectives. An indirect reference to this in a section on the *Role of Ideology* stated: "Government has to play a major role in the industry at least as manager of the resource." Nonetheless, omitting conservation from the list of major objectives was significant. Also significant was the designation of economic viability as the primary objective. When Roméo LeBlanc was the Fisheries Minister, industry often saw the social objective — maximizing employment — as dominant.

Ranking the objectives was another distinctive feature, however one feels about the ranking presented. The Kirby Report did not provide operational objectives. There were, however, some general directions for administrators. These were more specific than earlier proposals for reconciling economic and social objectives. The major deficiency was the lack of reference to and ranking of the resource conservation objective. In reality, conservation remained a paramount objective of the Department of Fisheries and Oceans.

3.6 The Pearse Commission on the Pacific Fisheries

In January 1981 just prior to the Kirby examination of the Atlantic fishery, the government had established a Commission on Pacific Fisheries Policy. The Commission was to find ways to improve Canada's Pacific fisheries. Commissioner Peter Pearse issued his final report *Turning the Tide — A New Policy for Canada's Pacific Fisheries* in September 1982 (Pearse 1982). This wide-scope inquiry addressed fisheries resource management and conservation, industrial regulation, sportfishing policy, Indian rights, environmental protection, problems of the Yukon, intergovernmental arrangements, enforcement and research.

Chapter 1 of Pearse's report addressed policy objectives. He concluded that the problems facing the Pacific fisheries were "numerous, grave and very complicated." These included overfishing, conflicts among users, overexpansion of the fishing fleets, and the erosion of marine and freshwater habitat. He had been offered an astonishing variety of explanations for the problems afflicting the fisheries. These ranged from avaricious fishermen to abusers of the habitat, natural predators and incompetent managers. He concluded, however, that the "inquiry pointed inescapably to deficiencies of government policy." Among those deficiencies were "uncertain objectives." Existing policies were "neither coherent nor well suited to modern needs." Pearse acknowledged that the Department of Fisheries and Oceans had made improvements in regulating and managing the Pacific fisheries over the previous dozen years but concluded that "these innovations have taken place in a piecemeal fashion without a clearly articulated policy objective to guide them. The result has been unpredictable and inconsistent regulation."

Pearse proposed a framework of objectives:

1. *Resource Conservation*
   Fishery policy must first and foremost ensure that the resource is properly protected and, whenever advantageous, enhanced.

2. *Maximizing the benefits of resource use*
   This meant ensuring that the resources available for harvesting "make the highest possible contributions to the economic and social development of the people of Canada, especially of those resident on the Pacific coast of Canada, recognizing that this contribution may be realized in economic, recreational and other social forms." In Pearse's view, this meant that the resources should be allocated to those who could and would make the best use of them.
3. *Economic development and growth*
   Pearse's goal of promoting economic development and growth embodied at least two supplementary objectives:
   (i) to improve incomes in the fisheries. Returns to labour and capital were typically low and unstable; he envisaged much better potential returns from a rationalized industry.
   (ii) to develop economic opportunities for coastal communities and Indian people.
4. *Social and cultural development*
   Because fisheries policy bears heavily on certain groups, Pearse argued that it should be consistent with, if not promote, public objectives for those groups. "In addition to the special needs of coastal communities, I have taken into account the special economic problems of Indians and their unique dependence on fish for nutritional needs and cultural activities. I have also taken into consideration the need to preserve recreational opportunities and to protect the commercial fisherman's lifestyle."
5. *Returns to the public*
   Pearse saw the fisheries as a heavy burden on taxpayers. Relatively few commercial fishermen enjoyed earnings in excess of "reasonable" returns to their labour and capital. He envisaged that fleets better adjusted to the available resources could yield very substantial net gains. He combined recommendations for rationalizing the fishing fleets and improving their economic performance with recommendations to capture some of the gains for the public.
6. *Flexibility*
   The Pacific fisheries were remarkably susceptible to rapid change — in resource abundance, in markets, and in fishing technology and effort. Pearse argued that fisheries policy must accommodate continuing shifts in the external environment.

The Pearse Report contained very specific recommendations under the framework of objectives listed above. The objectives themselves were very general with no explicit ranking. The exception was resource conservation where the first obligation of the federal government "is to ensure that the resources are properly conserved, managed and developed". Furthermore, the objectives could hardly be described as operational.

Pearse's summary of proposed objectives for the Pacific fisheries was similar in many ways to those in the 1976 Policy for Canada's Commercial Fisheries. Pearse made specific proposals for economic rationalization of the commercial fishing fleet. However, his general statement of objectives did not adequately address the economic versus social conflict which Kirby addressed explicitly. He was more specific than Kirby in one respect: his commitment to resource conservation as the first obligation of the federal government. I discuss his specific proposals for restructuring the Pacific fishery in later chapters. See also Parsons (1993).

Unlike Kirby, Pearse did not have the advantage of having his recommendations considered by the federal Cabinet prior to the release of his report. Much of the generality of his report, including this section on objectives, was soon lost in the sea of controversy which swept over his very detailed proposal for a bidding/auction system for quota licenses.

3.7 The 1984 Agenda for Economic Renewal

In September, 1984, a new Conservative federal government took office. In his November 1984 statement, "An Agenda for Economic Renewal", Finance Minister Michael Wilson described the new government's vision for the fishery. This recognized that Canada has some of the world's richest marine and freshwater habitat containing some of the most productive fish stocks in the world. The challenge was to "transform this endowment into economic wealth." The Agenda emphasized economic viability as a dominant goal. But it recognized that many fisheries are "set in the context of regional economies in which alternative employment opportunities are scarce." Excessive fishing and processing capacity on both coasts needed to be addressed. However: "Government financial rescues are not the answer. If it is to be self-sustaining, the industry must attract new investment both domestically and from abroad, adopt new technology, enhance product quality and consistency, achieve a higher degree of product and market diversification, and improve its overall marketing performance." The document proposed that government facilitate these industry improvements but not hinder adjustments to market realities (Department of Finance 1984).

The Agenda for Economic Renewal recognized that planning for an improved Atlantic fishery must be part of an overall approach to economic development for Atlantic Canada. Any "consolidation of employment in the sector would have to be complemented by public and private initiatives to promote job opportunities in other sectors of the economy, as well as fair and generous adjustment programs for workers in the sector."

These themes were generally consistent with the Kirby Report's emphasis on economic viability. One striking difference was the emphasis in Objective 3 of the Kirby Report on Canadianization of the fishery. In keeping with the Conservative Government's general approach to foreign investment, the Agenda for Renewal emphasized attracting new investment "both domestically and from abroad."

3.8 Legislating Objectives

The new government introduced in the House of Commons Bill C-32 to amend the *Fisheries Act* (see Chapter 2). This Bill was a response to court challenges to the Minister's powers to allocate fish among user groups for reasons other than conservation. The legislation as passed introduced the following *Purposes* section to the Fisheries Act:

"2.1 The purposes of this Act are:

- (a) to provide for the conservation and protection of fish and waters frequented by fish;
- (b) to provide for the proper management, allocation and control of the sea-coast fisheries of Canada;
- (c) to ensure a continuing supply of fish and, subject to paragraph (a), taking into consideration the interests of user groups and on the basis of consultation, to maintain and develop the economic and social benefits from the use of fish to fishermen and others employed in the Canadian seacoast fishing industry, and others whose livelihood depends in whole or in part on seacoast fishing and to the people of Canada."

These changes to the Fisheries Act to clarify its purposes and scope sunset on January 1, 1987. This, in fact, occurred following judicial clarification of the Minister's allocative powers. Nonetheless, this amendment, however shortlived, represented an explicit statement of the government's objectives for fisheries management. The conservation and protection objective was clearly deemed of primary importance, applying both to fish and fish habitat. Also noteworthy was the objective "to maintain and develop the economic and social benefits from the use of fish" to three groups: (1) fishermen, (2) others employed in the fishing industry and (3) the people of Canada.

Most of the debate on this legislation centered on the Minister's power to allocate fish for other than conservation reasons. This section was too sketchy to be an operational statement of objectives. However, it was the first explicit recognition in legislation of economic and social objectives for fisheries management and three groups of clientele. The *Purposes* section did not provide a hierarchical ranking of objectives. Nor did it weight the economic and social goals. In this sense, it incorporated the general "best use" approach of the 1976 Policy for Canada's Commercial Fisheries. Missing was the emphasis on economic viability present both in the Kirby Report's statement of objectives and the 1984 Agenda for Economic Renewal.

3.9 The Task Force on Program Review

In September 1984, the Conservative Government created a Task Force on Program Review with the Honourable Eric Neilson, Deputy Prime Minister, as Chairman. There were two major objectives — better service to the public and improved management of government programs. Federal fisheries programs were examined by the Natural Resources Study Team, which published a report in September 1985 titled *Natural Resources Program: From Crisis to Opportunity.* (Canada 1985a).

The Natural Resources Study Team made numerous recommendations regarding fisheries programs. Discussion of the team's findings in this chapter will be limited to objectives for fisheries policies and programs. The Study Team concluded that only a few fishermen in Canada earn a good living. The large majority's insufficient income is supplemented by an elaborate system of social safety nets and assistance programs. The result was:

- (a) an industry which is inefficient and not economically viable in many areas of Canada;
- (b) overregulation of the viable part of the industry (to protect the less efficient segments) thereby seriously hampering its potential profitability and export competitiveness.

The Study Team observed:

"This untenable situation is most obvious to the Department of Fisheries and Oceans. This department wants to be a department of fisheries, not a department of social welfare or employment. Unfortunately, circumstances have driven the department into this realm of conflicting objectives. Its primary mandate of protecting the fisheries resource and maintaining a viable fishing industry is being dissipated by the numerous socioeconomic burdens placed on it."

A more appropriate direction for fisheries policy was suggested:

"The fishing industry of Canada can become profitable, not require extensive government assistance and be a net contributor to the national economy, but only if it is not burdened with the need to maximize jobs and provide an entry to the social welfare system. The problem of unemployment and low income in many parts of coastal Canada must be dealt with through other programs and departments."

The Minister's original responsibility under the Fisheries Act of protecting and enhancing fish stocks was emphasized. Over recent years this mandate had been extended to managing the fishery for social objectives (employment, income, regional development). The Minister's expanded mandate under Bill C-32, the authority for economic and social initiatives, was considered "most undesirable."

The Study Team recommended that the fisheries component of the DFO focus on:

(a) Conservation and enhancement of the natural fisheries resources in the wild;
(b) Basic inspection services to benefit fish consumers, leaving quality control to the private sector;
(c) Basic infrastructure such as wharves, breakwaters and dredging; and
(d) Marketing for export and trade.

A further recommendation: "The government must focus the task of DFO to maximize economic returns to Canada from the fish resource and allow the industry to operate in a business-like manner, including reducing the requirement to maintain jobs merely to permit entry into the social welfare system." This meant "focusing the task of DFO on its primary objective of managing and enhancing fish stocks and eliminating the conflicting and counter-productive objectives now being imposed on the department."

The Natural Resources Study thus came out squarely in favour of resource conservation and protection and economic efficiency as the appropriate objectives for fisheries management. Employment, income and community dependence should be left to natural forces of adjustment or be handled by other government departments. Following this report DFO began to reorient its policies and programs. The emphasis shifted from using Canada's fisheries as the primary means to address income and employment problems in coastal regions. Renewed emphasis was placed on the conservation objective. The Department's fisheries science and ocean science functions were integrated in one organization, focused more directly on fisheries management requirements (Chapter 17). Fisheries surveillance and enforcement were bolstered to deter foreign and domestic illegal fishing (Chapter 19). The Department also changed a number of its economic related programs. The Fishing Vessel Assistance Program (FVAP) was terminated. The Fisheries Improvement Loans Act (FILA) was not renewed. The Fishing Vessel Insurance Program was required to recover a portion of its costs, and the Department's Food Centre was transferred to the private sector (Crowley et al. 1993). Economic viability was also promoted through implementing and extending systems of individual quotas where feasible (Chapter 9).

3.10 Reconciling Objectives in Times of Crisis

During 1985 and 1986, the federal government vigorously pursued agreement with the provinces on means to address four specific "challenges" facing the fisheries sector: trade, income stability, resource considerations and fisheries development. At their Annual Conference in November 1985, First Ministers agreed that timely, cooperative efforts in these four areas were required of both orders of government and the private sector. At the November 1986 First Ministers Conference, Federal Fisheries Minister Siddon reported the conclusions of Fisheries Ministers: Conservation and effective management of Canada's aquatic resources for the benefit of the Canadian fishing industry should be reaffirmed as Canada's primary objectives in international fisheries relations.

Thus, from a variety of sources post-1984 it appears that the predominant fisheries objective was regarded to be resource conservation and protection. However, economic and social factors continued to be important considerations in fisheries policy. There was greater emphasis on promoting a commercially self-reliant industry. Fisheries policy continued to reflect concerns about community dependence and employment. However, there were no specific new policies or programs to these ends until the Atlantic Groundfish crisis of 1989–1993. Some programs seen to foster overcapacity in the harvesting sector, e.g. Fishing Vessel Assistance and Fisheries Improvement Loans, were discontinued. It appears that the three primary objectives were, in order of priority: (1) resource conservation and protection; (2) economic viability; (3) minimizing social disruptions.

The difficulty of reconciling these three objectives became evident in the 1989–93 downturn in the Atlantic groundfish industry. Scientific reassessment of certain stocks, e.g. Northern Cod, declines in Gulf of St. Lawrence and Scotian Shelf stocks and continued fleet and processing overcapacity combined to produce another Atlantic groundfish crisis (see Chapter 13). In addressing this crisis, the Government had to balance the need to adjust TACs for conservation with community, company and worker dependence upon a stable supply of fish.

Scientists advised that the $F_{0.1}$ TAC for northern cod was substantially less than previously estimated. The interim report of the Independent Review Panel on Northern Cod confirmed this. In response Fisheries Minister Tom Siddon stated:

> "Conservation and maximization of the fishery resource for the long-term benefit of Atlantic Canada's fishing industry is our ultimate goal. A management regime based on a fishing mortality level of $F_{0.1}$....remains Canada's fundamental conservation approach. However, if scientific advice recommends fishing levels well below those of previous years, there is a built-in management mechanism to move to $F_{0.1}$, over a longer period of time" (DFO 1989a).

Announcing the 1990 Atlantic Groundfish Management Plan, which reduced the Northern Cod TAC from 235,000 t to 197,000 t (significantly higher than the $F_{0.1}$ level), Minister Siddon noted:

> "Our actions are all aimed at managing the human and economic impact of declines in our fish stocks, and creating a fishery which is sustainable and viable in the long-term. The key element of the plan must be the conservation and sustainable development of the resource" (DFO 1990b).

In a more comprehensive response to the Atlantic Groundfish crisis, Federal Ministers Valcourt, Crosbie and MacKay announced on May 7, 1990, a 5-year $584-million Atlantic Fisheries Adjustment Program (AFAP). AFAP had three elements "aimed at ensuring a viable fishery in the long-term for Atlantic Canadians, while supporting individuals and communities in the fishery to adjust to the realities of declining fish stocks and plant closures."

The three elements were:
— Rebuilding the Fish Stocks ($150 million);
— Adjusting to Current Realities ($130 million); and
— Economic Diversification ($146 million).

These components supplemented a previously announced $130 million short-term response program and a $28 million commitment to augment aerial surveillance (DFO 1991m).

This Program recognized the need to rebuild the fish stocks while cushioning the impact of the adjustment required. Minister Valcourt commented:

> "Adjustment is already happening in the fishery. But we must manage that adjustment instead of having it forced upon us through a declining resource base, with all the human and community costs that this entails. We must take steps to link the size of the fishery to what can realistically be sustained by the abundance of the fish stocks."

Minister Crosbie stated:

> "We recognize that long-term diversification of the Atlantic economy is necessary to provide alternative employment opportunities to those who have traditionally relied upon the fishing industry for their jobs" (Canada 1990a).

Announcing a 3-year Atlantic Groundfish Management Plan on December 14, 1990, Minister Valcourt spoke of striking "a reasonable balance between conservation of the resource and stability for participants in the industry and for the economy of Atlantic Canada" (DFO 1990c).

The three primary objectives of resource conservation, economic viability and minimizing social disruptions were all intertwined in the Government's response to the 1989–1993 Atlantic Groundfish crisis. Conservation was an important objective. The conservation measures initially adopted were tempered by the need to maintain economic and social stability in the short-term. However, in response to subsequent scientific advice in 1992, Fisheries Minister John Crosbie first slashed the Canadian northern cod TAC to 120,000 tons and then in July 1992 imposed a 2-year moratorium on fishing for northern cod because of conservation concerns.

Responding to scientific advice in the fall of 1992, which indicated that several Atlantic cod stocks were at low levels, Minister Crosbie slashed TACs by as much as 70%, setting TACs for the major groundfish stocks in 1993 at the $F_{0.1}$ level. Announcing major adjustments to the multi-year Groundfish Management Plan on December 18, 1992, Minister Crosbie stated:

"[Rebuilding the groundfish stocks] will require measures that address the fundamental problems leading to their decline, to the extent those are within our control rather than determined by nature!

— Once and for all we will change the way we fish to conserve and protect small fish;
— We will establish a strong spawning stock base so that when environmental conditions change, the resource will grow in abundance and increase the sustainable harvest; and
— We will establish a better balance between the resource and the number of individuals and communities who are dependent on it." (DFO 1992i)

4.0 CONCLUSIONS

In Canada, as elsewhere, fisheries managers face multiple conflicting objectives which shift over time with changes in the philosophy of the players in power. Official objectives for fisheries management tend to be stated as a general framework, incorporating biological, economic and social dimensions, with the balance and relative weights attached to component subobjectives shifting from time to time.

Kirby (1982) attempted to provide an explicit, ranked set of objectives and in 1985 the Study Team on Natural Resources called for a return to paramountcy of the resource conservation objective. The political system, however, has been reluctant to embrace a clear, hierarchical set of objectives. Stated objectives tend to be vague. Because of the extreme diversity of the fisheries in Canada and radical shifts in recommendations which can occur within the space of only 2 years, it is not surprising that a national statement of objectives would incorporate a broad range of conflicting subobjectives. An example of this is the framework of objectives subsumed in the "best use" or "optimum yield" approach of the 1976 Policy for Canada's Commercial Fisheries. There is one exception to that fuzziness, the reemergence during the 1980's and early 1990's of resource conservation and protection as the paramount objective of fisheries policy in Canada. Following the 200-mile extension of fisheries jurisdiction, that objective was expressed on the Atlantic coast in the $F_{0.1}$ reference point for Atlantic finfish stocks. This is the closest Canada has come to an explicit replacement for MSY.

Attempts at a clear national statement of ranked fisheries management objectives, however desirable, are probably doomed to failure. The Task Force on Atlantic Fisheries came the closest to having such a set of objectives adopted by government. It is ironic that they omitted the one objective which has consistently garnered the greatest public and political support: that the federal government's primary fisheries obligation is to conserve and protect fish and their habitat. The choice between objectives is inevitably a political decision in a democratic society.

In this respect I agree with John Gulland, who observed:

> "A strict definition of the general aims of conservation and management, if it could be achieved, would clearly be of great value in setting detailed regulations.... The arguments suggest, however, that it is impossible to define a 'best' fishery policy because what is 'best' will vary from time to time. What is considered the 'best' policy at any one moment will be determined by the current preferences between different measures of input and output, the uses that can be made of different products from a given ecosystem, the weights given to long-term and short-term interests, as well as the knowledge of how the natural systems operate, and what will be the effect of alternative actions. These are bound to change.... Nevertheless, some general statement of purpose is highly desirable.... Insofar as a specific term or phrase is needed, it is likely that 'optimum utilization' or 'optimum sustainable yield' is at least as suitable as any other. Unlike some other terms, it implies some balance between different factors" (Gulland 1977).

Roedel (1975), summarizing the AFS Symposium on Optimum Sustainable Yield, recognized that OSY was a deliberately imprecise term but wisely observed: "It acknowledges the reality of political forces, no matter how much the purist may deplore their existence."

Should we abandon all efforts to develop clear, ranked and measurable objectives for fisheries management? No! But such efforts are more likely to succeed if they focus on specific fisheries. Sophisticated tools exist to assist managers in this process.

Healey (1984) attempted to demonstrate the application of decision analysis in an analytical model for optimum yield. He noted that neither the FCMA of the USA nor Canadian policy documents indicate how conflicting goals can be reconciled. He observed that some economists have attempted to equate OY with MEY. This was invalid because it substituted another single objective approach (maximizing return on capital and labour) for the old MSY approach. Healey pointed out that fisheries management had no acceptable analytic methodology for ranking, weighting, and combining the multiple objectives that should determine OY. He argued that *multiattribute utility theory* was an appropriate, well-developed methodology. He applied it to analyze the Gulf of Maine herring stock and the Skeena River salmon fishery.

I will not discuss Healey's analysis in detail but his conclusions are interesting. He identified four advantages of simple multiple attribute utility theory (MAUT) techniques for complex fishery management problems like OY:

1. The techniques are specifically designed to help with the practical solution of multiattribute decision problems;
2. The techniques mimic the natural decision making process so that they are intellectually appealing and understandable to the decision maker;
3. They provide a structural analytic framework for quantitatively evaluating alternative solutions; and
4. The techniques permit a broad range of types of information and levels of measurement to be brought together into a single objective decision rule.

Healey was correct in his assertion that determining OY requires value judgments. He observed that fisheries managers abhor the thought of value judgements, while in reality they make such judgments all the time. Healey argued: "The MAUT techniques not only provide a way to bring the value judgments out in the open, but also provide a way to absolve the manager of the criticism that he is imposing his values on the industry. Provided the manager is diligent about identifying the important stakeholder groups and capturing their attribute weights, he can reasonably argue that everyone's views were taken into account in determining OY."

An advantage of the optimum yield approach is the flexibility it provides to vary the weight given to biological, economic, and social factors according to the circumstances appropriate to a fishery. That flexibility has really not been tapped in any rigorous way to date.

Optimum yield must be defined in fishery-specific terms to achieve clear, weighted and measurable objectives for fisheries management. Practical applications of the techniques explored by Healey and other related techniques should be vigorously pursued. These techniques might advance the effort to formulate multiobjective decision frameworks for particular fisheries. These efforts are more likely to succeed than attempts to recast national fisheries management objectives in a more rigorous manner. Meanwhile fisheries managers must continue to grapple with multiple conflicting objectives as part of their daily business.

CHAPTER 5

TECHNIQUES OF RESOURCE MANAGEMENT

"The basic problem of fisheries management is to select the particular control measures which best achieve the objectives of management."

– L. Anderson, 1983

1.0 INTRODUCTION

Choices among regulatory measures in fisheries depend on the management objective. Just as there are multiple objectives for fisheries management, a wide variety of regulatory approaches have been applied in fisheries around the world. These have differed widely in effectiveness. Some techniques attempt to maintain productivity of the resource. Others attempt to resolve conflict among industry groups. Still others are intended to promote efficient use of the resource.

There are a variety of ways to classify these regulatory techniques. Crutchfield (1961) identified two categories: (1) regulations affecting fishing mortality, and (2) those affecting the size and age at which fish are taken. Templeman and Gulland (1965) distinguished between regulations which (1) affect the sizes of fish caught, or (2) affect the total fishing effort. Gulland and Robinson (1973) used a somewhat similar classification based on (1) control of the size caught, (2) control of the amount of fishing, and (3) controlling excess capacity. Their last category exemplified the drive for economic efficiency in the fisheries. Parsons (1980) distinguished two categories of regulation: (1) those controlling the catch composition, and (2) those controlling the total amount of fishing. Pearse (1979a) used a similar system: (1) regulating the composition of the catch, (2) regulating the size of the catch, i.e. the amount of fishing, and (3) promoting efficiency in the fishing process. His third category also recognized the increasing emphasis on economic efficiency.

Clark (1980) summarized these attempts to classify fisheries regulations:

> "Techniques of fishery management can be classified broadly into two types, according to whether they are mainly directed towards the control of the fish population so as to maintain a high level of productivity, or whether they also attempt to maintain economic efficiency of the fishing industry. The methods that have traditionally been used such as total catch quotas, closed seasons and areas, gear restrictions, and so on belong largely to the former category whereas methods such as licences, taxes or royalties and allocated quota systems are of the second type. In practice it may be appropriate to employ a combination of methods of both kinds."

Regulations of the first category are *open-access techniques* and those of the second category *limited-access techniques* (Anderson 1986). This chapter focuses on the *open-access techniques* aimed at control of the fish population to maintain a high level of resource productivity. *Limited-access techniques* are dealt with in Chapters 7, 8 and 9. The distinction between these categories is not clear cut. Any particular regulatory technique can have multiple effects and impact on more than one objective of fisheries management.

This chapter deals with regulations which control the catch composition and the amount of fishing, as well as some auxiliary methods which do not fit either category. These measures include:

- Mesh Size and other Gear Restrictions
- Size Limits
- Closed Seasons
- Closed Areas
- Catch Quotas
- Fishing Effort Controls.

All of these methods have been used extensively in the management of Canadian marine fisheries. The indirect or passive means of controlling fishing such as gear restrictions, size limits, closed seasons, and areas, have been used in managing the Atlantic lobster fishery for approximately a century. The more direct means of controlling fishing mortality such as catch quotas and fishing effort controls are of more recent vintage. In the Canadian context catch quotas were first applied in the Pacific halibut fishery by the International Pacific Halibut Commission in the 1930's and by Canadian management authorities in the Pacific herring fishery from the late 1930's onwards. Catch quotas, more commonly known as *Total Allowable Catches* (TACs), were first widely applied to the groundfish fisheries of the Canadian Atlantic in the early 1970's. TACs were used subsequently as a primary management tool for virtually all finfish species of the Atlantic coast. Fishing effort controls, through limitation of entry, were introduced in the Pacific salmon and Atlantic lobster fisheries in the late 1960's. They were extended subsequently to all major Canadian commercial marine fisheries in the 1970's. This chapter compares the passive and active methods of regulating fishing mortality and provides examples of their application in Canadian fisheries. The particular regulatory techniques applied in five Canadian fisheries — Pacific halibut, Pacific herring, Pacific salmon, Atlantic lobsters and Atlantic groundfish — are examined in greater detail in Chapter 6.

2.0 CONTROLLING THE COMPOSITION OF THE CATCH

2.1 Mesh Size Regulations

The usual intent of mesh size and minimum fish size regulations is to control the capture of small fish to avoid *growth overfishing*. Control of mesh size has been the most widely used technique to protect small fish, particularly in the international trawl fisheries of the North Atlantic. Other measures available include minimum fish size limits and closed areas or seasons. The latter two approaches may be used if the sizes of fish caught differ significantly between areas or seasons.

Mesh size regulations are intended to control size at first capture. The theoretical models assume knife-edge selection with all fish below a certain size escaping to grow to a larger size, while those above this size are liable to be caught at the full rate of exploitation. The length selection curves from experimental studies show a more gradual relationship between the percentage of fish retained at a particular length in relation to various mesh sizes (see Fig. 5-1 and 5-2).

Selection experiments to determine the effect of mesh size on the size of fish caught and released were begun in the Northeast Atlantic in the late 1800's. Extensive mesh selection experiments were conducted in the Canadian Atlantic in the 1950's and 1960's. These became the basis for the first 15 years of management of the groundfish fishery by the International Commission for the Northwest Atlantic (ICNAF). Mesh selection measures were intended to select a mean age (size) at first capture which maximized yield-per-recruit for a given species. The appropriate mesh size depends on the natural mortality and growth rate of that species, based on the yield-per-recruit theory of Beverton and Holt described in Chapter 3. During its early years ICNAF was preoccupied with reducing the wastage of small fish at sea. The first regulations restricting the size of mesh used in the codend of the otter trawl were introduced by ICNAF in 1953. In its first 15 years, ICNAF established some 20 regulations, all of which focused on minimum mesh sizes and trawl construction methods.

Mesh size regulations are based upon calculations of the 'mean selection length,' the size at which half the fish are retained and half are released (Fig. 5-3). This increases with mesh size and differs among different types of fishing gear, e.g. otter trawl and cod trap. In a mixed species fishery such as that for groundfish, each species tends to have a different mesh selection curve.

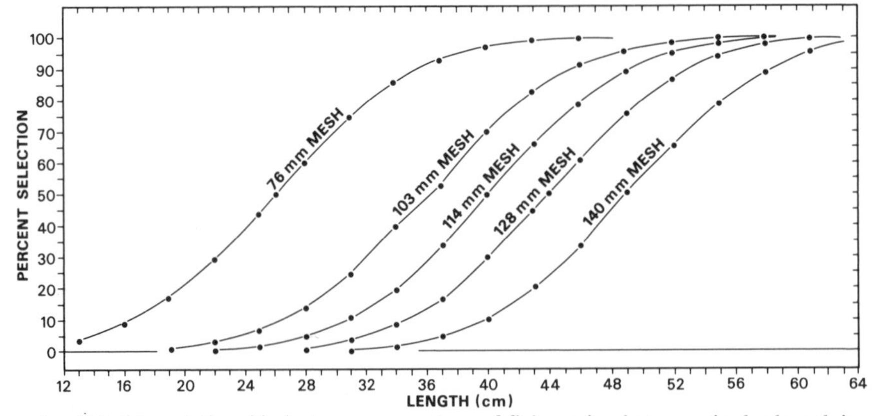

FIG. 5-1. The relationship between percentage of fish retained at a particular length in relation to various mesh sizes for cod in Subarea 3 of the Northwest Atlantic Fisheries Organization (NAFO) (from Hodder and May 1964).

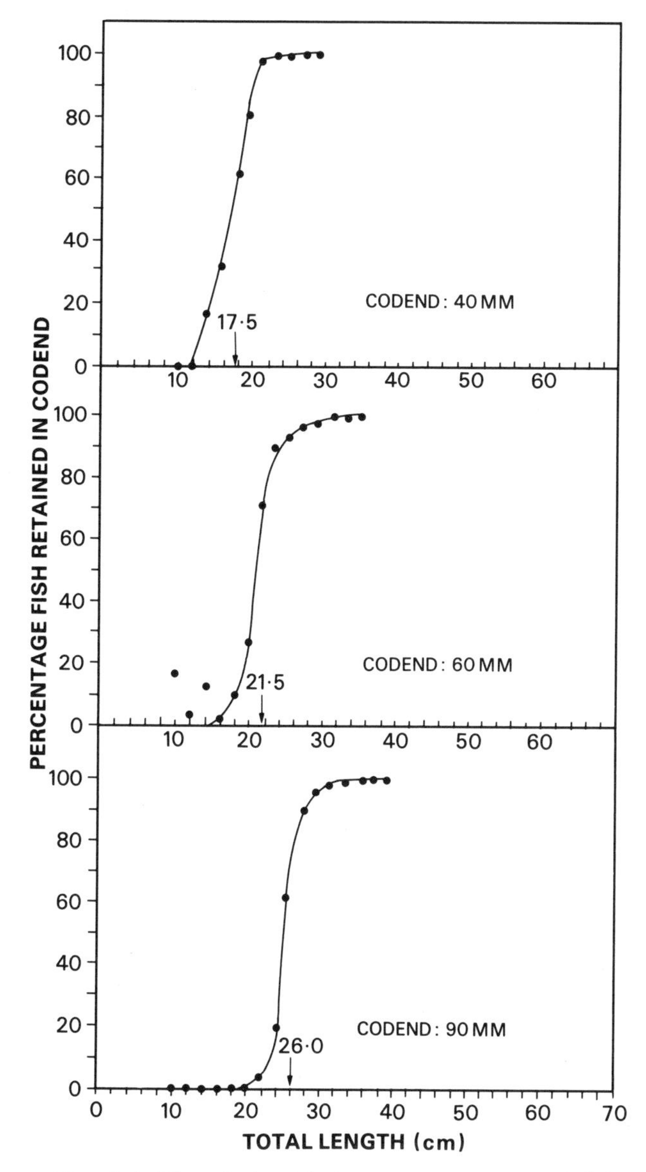

FIG. 5-2. The relationship between percentage of fish retained at a particular length in relation to various mesh sizes for silver hake on the Scotian Shelf (from Clay 1979).

Therefore, different mesh sizes were initially adopted for species such as cod, redfish, and American plaice. ICNAF introduced mesh size regulations when the level of fishing effort was moderate. As fishing effort increased in the 1950's and 1960's it became necessary to progressively increase the minimum mesh size. Following the introduction of the 200-mile limit, Canada implemented a uniform 130 mm mesh size for otter trawls on the Atlantic coast. This excluded smaller species such as redfish and silver hake. It was intended to help enforce the minimum mesh size. However, this resulted in an increase in the average mesh size actually used by approximately 10 %.

Any increase in mesh size has both short- and long-term effects. There are short-term reductions in catch because most of the smallest fish are released. In theory the fish that escape will grow larger and hence produce a greater yield when captured later. Other potential benefits include increases in stock size and future catch rates.

One problem with mesh size regulations is that catch rates will fall in the short-term. Hence, fishermen make short-term sacrifices for long-term gain. Another problem is that, although potential benefits can be calculated, it is difficult to demonstrate such benefits in practice. Changes in stock size or increased yields to fishermen may be masked by natural fluctuations in recruitment.

Mesh regulations have been widely used in Canadian fisheries and around the world. With the rapid growth of certain fisheries, changes in fishing techniques, and the development of "mixed" fisheries, particularly during the 1960's, minimum mesh size by itself was insufficient to regulate the fishery. Mesh regulation suffers from two major weaknesses when used as the primary management measure for otter trawl fisheries:

> First, most trawl fisheries take a mixture of species, with the optimum mesh size for different species often being very different.
>
> Secondly, while mesh regulation can in theory increase the total catch to some extent, the economic benefit will be short-lived unless the amount of fishing is controlled. Indeed, as soon as any benefit becomes apparent, there is incentive for additional effort to enter the fishery. Thus the benefits are dissipated.

Difficulties similar to those with mesh size controls in trawl fisheries occur in other fisheries. Although the size of fish caught in longline fisheries can be influenced by hook size, the relationship is not very exact. In purse seine fisheries for pelagic species it is virtually impossible to use minimum mesh size to regulate the size at first capture.

In addition to their use to achieve a desired type of selectivity, gear restrictions have been used to limit the use of more efficient fishing methods. A classic Canadian example is the virtual ban on groundfish trawlers on the Atlantic coast from 1928 until the Second World War. This was done because of the concerns of inshore fishermen that trawlers would destroy the fish stocks. Only three trawlers were in use from 1928 until the Second World War when this policy

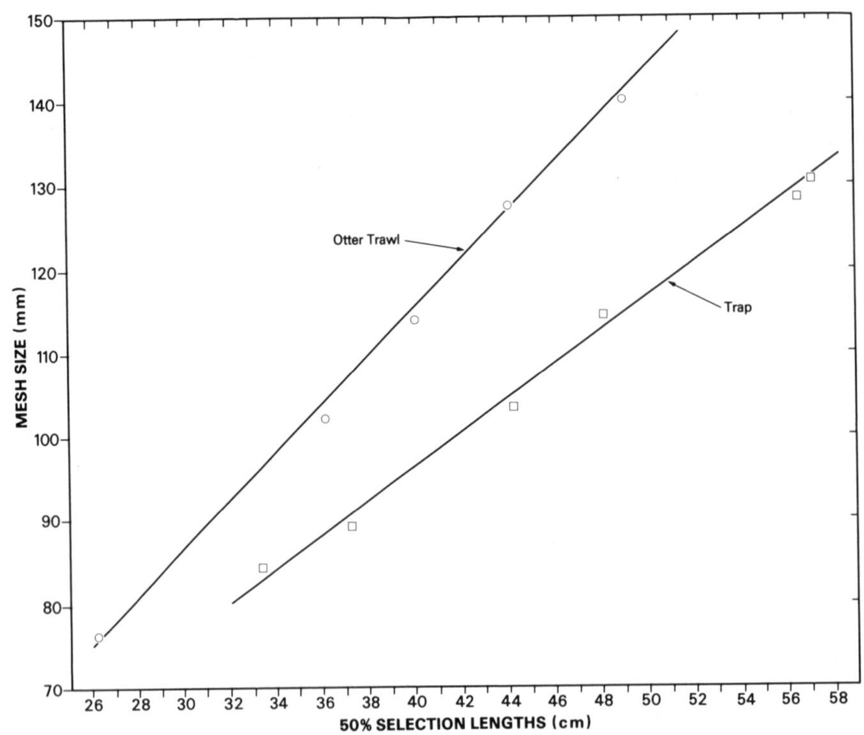

FIG. 5-3. Comparison of 50% selection lengths with mesh size for cod trap and otter trawl (from Bishop 1982).

was reversed. Other examples of the restriction of efficient gears are the prohibition of traps in the Pacific salmon fishery and the prohibition of trawling for Atlantic lobsters. The use of gear restrictions to limit new or more efficient technology leads to an upward regulatory spiral. Regulations may impose artificial inefficiencies on a fishery. However, fishermen will find ways to circumvent these constraints. Regulators then search for new ways to restrict this technology.

2.2 Minimum Size Limits

Another regulatory technique for controlling the capture of small fish is a minimum size limit. This can take two forms: (1) an absolute minimum size limit and (2) a regulation based on the average size of fish landed or on maximum count per unit of landed weight. The former has been applied for decades in the Atlantic lobster fishery. The latter has been applied in the Atlantic scallop fishery.

Garrod (1987) argued that minimum size limits in the Northeast Atlantic have usually been implemented to support minimum mesh size. They are not necessarily closely related to the age at maturity of the fish. The rationale for a minimum size limit in such a situation is that without it there would be continual pressure to reduce mesh size or to ignore the mesh regulation. A minimum size limit sometimes encourages fishermen to avoid areas where juveniles are known to congregate.

Minimum size limits have been used extensively for species such as lobster and scallops where they are particularly useful because specimens under the legal limit can be returned to the sea alive. Their use in the lobster fishery is discussed in the next chapter. They have been used less extensively in finfish fisheries, e.g. Pacific salmon, Atlantic herring, and more recently Atlantic groundfish. The 5-year Atlantic salmon management plan introduced in 1984 (Lear 1993) used a somewhat different approach. It prohibited capture of multi-sea-winter Atlantic salmon, as distinct from grilse, in the recreational fishery. A 63 cm *maximum* size limit was introduced as part of a complex package of management measures which affected all sectors of this fishery. There was considerable debate about the merits of such a measure. Many participants believed that salmon would die upon return to the water. However, studies have shown that salmon released as a result of the hook-and-release policy have a survival rate of 90% or more (CAFSAC 1987a).

Another variation on the minimum size approach is the average meat count regulation in the Atlantic scallop fishery. Caddy (1984) suggested that such regulations be applied under circumstances where:

1) Large numbers of small individuals are captured by unselective gear, e.g. dredges, making an absolute minimum size limit unenforceable, and where mechanical or manual sorting of the catch is not feasible; and
2) Although the gear is unselective, the species in question is either fairly well segregated by size in areas known to the fisherman, or survival of discarded small individuals at sea is known to be high, e.g. shellfish.

This variation on minimum size limits is also intended to increase the mean age of capture and protect juveniles from premature exploitation. The average meat count approach has been compared with the more traditional minimum weight (size) limit for the Georges Bank scallop stock. Mohn and Robert (1984) compared the effects of a minimum meat weight (MMW) of 11.3 g (equivalent to 40 scallops per pound) and an average meat weight or meat count (MC) of 30 scallops per pound. There was not a great deal of difference between 30 MC and 40 MMW. The 30 MC seemed to have an advantage in terms of conservation and stability of catch. They suggested that for each MC there would exist a MMW which would give approximately the same overall catch and stock conservation benefits. The MMW is an abrupt limit while the MC imposes a gradual limitation. In their view, blending tended to spread the effort over a number of ages resulting in greater catch stability. Naidu (1984) reached a different conclusion from a study of the meat count regulation for St. Pierre Bank scallops. He contended that the meat count regulation did not protect young scallops or delay the capture of those recently recruited.

The Canadian Atlantic Fisheries Scientific Advisory Committee (CAFSAC 1984a) considered the effectiveness of average meat count and minimum meat weight regulations in delaying exploitation of young, rapidly growing scallops. CAFSAC concluded:

> "Either approach to regulation could potentially achieve a reduction in exploitation rates of young scallops. However, in practice the average meat count regulation allows more potential for a significant departure from a gradual recruitment pattern.... The minimum meat weight at a particular count is potentially more effective in protecting young scallops than an average meat count corresponding to the same size. The implementation of a minimum meat weight regulation would be more consistent with the initial objectives of the meat count regulations."

Canada's Atlantic scallop fishery has been managed by an average meat count regulation since April 1973. In 1983 a meat count of 39/500 g was imposed; this was reduced to 33/500 g effective January 1, 1986. In the Atlantic scallop fishery this minimum size regulation was the primary regulatory tool for a considerable period.

Size limits have also been used in the Atlantic snow crab fishery. This fishery developed rapidly in the 1970's and early 1980's but began declining in the late 1980's. Size at first capture for snow crab was set with the objective of ensuring that no females were captured and males were only captured after being permitted to spawn at least once. This was to ensure full utilization of egg production potential. Halliday and Pinhorn (1985) concluded that:

> "Exploitation of the recruited population (males) at 50–60% (F0.1) appears to result in close to full fertilization and the overall strategy clearly avoids recruitment overfishing whereas it may or may not maximize recruitment, and hence usable fishery yields."

The subsequent decline of the Gulf Snow Crab stock raises doubts about this approach. Recently, the appropriateness of the minimum legal size of 95 mm CW for maximizing yield per recruit and for avoiding recruitment overfishing has been questioned (Jamieson and McKone 1988). There is evidence that most morphometrically mature males over 95 mm CW were heavily exploited as soon as they molted to legal size and therefore did not have the opportunity to mate before being caught (Conan et al. 1989). For a fuller discussion of management methods in this fishery, see Hare and Dunn (1993).

2.3 Effectiveness of Attempts to Control the Age at First Capture

Opinions vary on the utility of gear regulations and minimum size limits as management tools. Crutchfield as long ago as 1961 observed that "the history of gear regulation is a testimony to the viciousness of the infighting which develops when technology cuts across the interests of particular groups." He was more positive about the use of mesh size to regulate age at first capture: "The technique is limited in applicability to fisheries where gear selectivity is controllable and where the gear contacts a wide range of sizes in any fishing period. It is therefore most directly useful in achieving optimal average age of fish taken by trawling." Regarding size limits Crutchfield stated: "In most saltwater fisheries size limits do not, of themselves, afford much protection, since losses of undersized fish returned to the water are normally very heavy . There are, of course, some exceptions; and, in addition, size limits provide useful support to other, more effective, regulations designed to allow greater growth before capture."

Gulland and Robinson (1973) concluded:

> "Actions to control the sizes of fish caught generally have no unexpected economic implications. Measures such as control of the mesh size in trawls, or limits to the size of fish that can be landed or sold, allow fishermen to carry on their activities in virtually the same way as before regulations were introduced, except for modifications to their gear or changes in the proportions of their catch that they are permitted to land....Since such measures should make fishing more profitable, without additional management action the long-term effect is to attract further capacity into the fishery, and hence to reduce profitability back to the former level. Therefore, such measures, while economically attractive in the short run, give no assurance of long-term economic benefit."

An Expert Consultation on The Regulation of Fishing Effort convened by FAO in Rome in January 1983 concluded:

> "Size limits and mesh size regulations may provide a base level of protection to the stock in many fisheries, particularly if, without such control, the fish would be caught before they became mature and spawned. Minimum size limits, if they can be enforced, should at least prevent the stock from becoming excessively depleted even if other measures controlling fishing mortality are temporarily ineffective.... Gear restrictions and/or size limits can be important components of a management regime; they can be used with considerable flexibility and sophistication to achieve the economic and biological objectives sought. Under favourable conditions — especially in single-species fisheries where the restrictions are understood and accepted by the fishermen — they have been successful. Even in these circumstances they cannot resolve a number of problems, including the economic problem of overcapacity. In other conditions they may be of very limited value" (FAO 1984b).

In general, minimum mesh size and minimum size limit regulations are a useful adjunct to other elements in a comprehensive management plan. By themselves they are not likely to meet either biological or economic objectives for a particular fishery.

2.4 Closed Areas

Closed areas are spatial restrictions on fishing. These are usually intended to protect particular life history stages of a species. They work by protecting a stock from overexploitation at a point where it is particularly vulnerable to exploitation, decreasing by-catches of particular components of the resource or minimizing the likelihood of conflict between users of different types of gear. Area closures can involve long-term or seasonal closures. Closure of nursery grounds or the banning of fishing on spawning or prespawning concentrations are often advocated for conservation purposes when the real objective is to minimize gear conflict.

Canadian examples of area closures include:

1. The limitation of fishing for silver hake on the Scotian Shelf to the "silver hake box";
2. The seasonal closure of the Browns Bank haddock spawning area during the period March – May;
3. The prohibition of fishing for salmon along a portion of the southwest coast of Newfoundland;
4. The closure of certain nursery areas in the Pacific to halibut fishing;
5. The general prohibition on fishing for salmon in certain areas of the high seas on both the Pacific and Atlantic coasts;
6. The general prohibition of fishing by groundfish trawlers greater than 65 feet within 12 miles of the Atlantic coast; and
7. Numerous instances on both the Atlantic and Pacific coasts of prohibiting fishing by particular gears within certain specified areas, e.g. "boxes".

The "silver hake box" on the Scotian Shelf is an example of limiting fishing to a particular area to permit fishing by small-meshed gear for one species without impacting adversely on other species through the capture of juveniles of those other species. Prior to the 200-mile limit, there was a large-scale fishery by the East European fleets for silver hake on the Scotian Shelf using small-meshed gear. Evidence indicated that this fishery was capturing large quantities of juveniles of other species, e.g. cod and haddock, which was contributing to overexploitation of those species. Because of the by-catch problem ICNAF at its Annual Meeting in June 1976 set a 60 mm minimum mesh size and discussed the possibility of limiting the commercial hake fishery to midwater trawls. Additional studies carried out in 1976–77 identified three primary areas in which silver hake fishing with bottom trawl was concentrated. There were substantial by-catches of juvenile cod and haddock in these areas.

At the Ninth Special Meeting of ICNAF in December 1976, Canada announced that it would limit bottom trawling with small mesh gear to deeper water in the summer months when other species would not be concentrated in those areas and established a seasonal limitation on hake fishing (ICNAF 1977). Fishing silver hake with small-meshed gear would henceforth be limited to an area at the edge of the Scotian Shelf (Fig. 5-4). A silver hake season of April 15 to November 15 was also established. This regulatory proposal allowed fishing with small meshed gear (60 mm) south and east of a line which became known as the Small Mesh Gear Line (SMGL).

ICNAF adopted these measures in 1976. In 1977 Canada codified the ICNAF regulation in its Foreign Vessel Fishing Regulations. In addition, regulations were introduced to limit by-catches of haddock to less than 1% and that of other important commercial species to less than 10% of the total weight on board the vessel. Studies of the impact of the SMGL were conducted from 1977 to 1982. Waldron and Sinclair (1985) reported that geographic distributions of by-catch for cod, haddock, and pollock supported the location of the SMGL. They concluded that this method of managing the Scotian Shelf small-meshed fishery was adequate to minimize by-catch and yet permit access to the silver hake stock.

The closure of the haddock spawning areas on the Scotian Shelf in Division 4W to all trawling during the peak spawning season was discussed at the 1970 Annual Meeting of ICNAF. However, this was deferred largely because of objections that the closed area would interfere with fisheries for other species, particularly cod. At the same time, closed season-area regulations were introduced for Division 4X haddock. STACRES was requested to examine the impact on other fisheries. At its 1971 meeting STACRES advised that:

> "In general, such closures will create problems in the southern areas of ICNAF because of the many species. Closure of spawning areas would not be expected to result in any direct

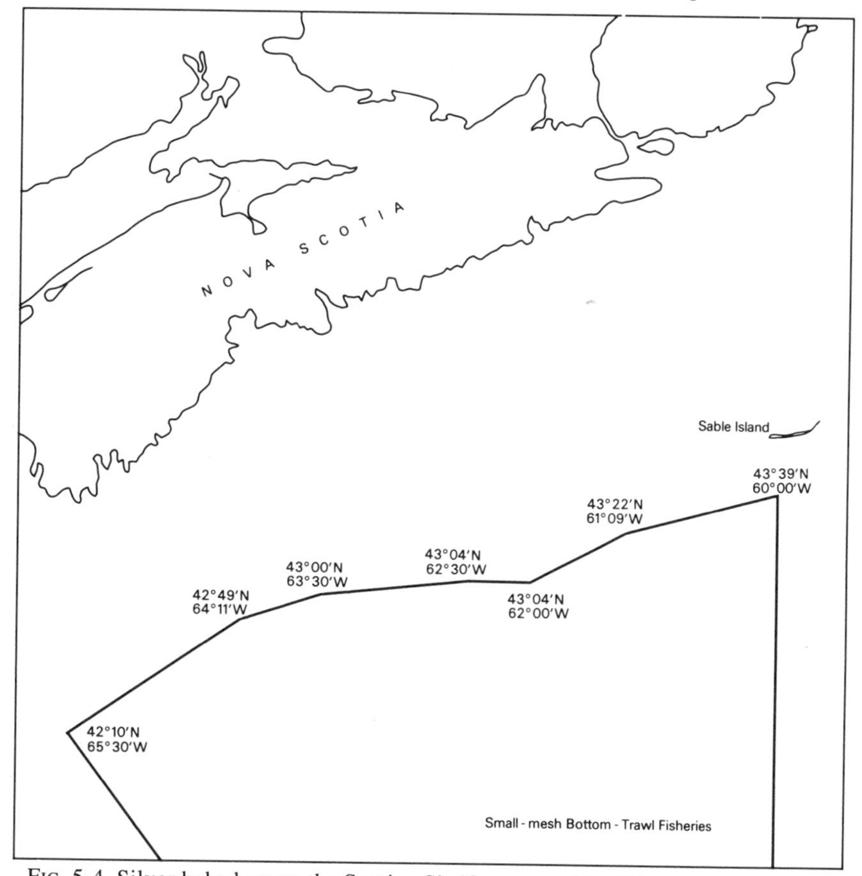

FIG. 5-4. Silver hake box on the Scotian Shelf, as agreed by ICNAF (1976).

significant biological benefits. It is utilized primarily to prevent catches from being concentrated in a short time span. The consequences of closures are thus primarily related to economic or administrative factors. They could be negative in the sense that the period of highest catch rate is closed. Also because fishermen are able to adjust their fishing strategy, such closures may not in fact achieve even the desired result.... The problem of spreading catches over the year is better solved by more direct methods such as seasonal quotas" (ICNAF 1971).

Nonetheless, the Division 4X closed area/season regulation was maintained. The closed area was reduced in size in 1972 and the time of closure extended to include May.

STACRES in 1975 examined the impact of the closed area and season regulations for haddock on the Scotian Shelf and concluded that the regulations had reduced fishing mortality during the months of closure. It observed that regulations in force for 1975 encompassed almost all of the area in which haddock concentrate and that this inevitably interfered with fisheries for argentine, silver hake, cod and pollock. STACRES observed: "minimization of haddock mortality is important to its management. Should present closed area/season regulations prove an unacceptable interference with other fisheries, the Commission should consider alternative methods of regulating haddock mortality" (ICNAF 1975b).

Despite this adverse impact on other fisheries, the closed area/season regulation was maintained. CAFSAC reviewed the spawning ground closure in 1981 and concluded:

> "The rationale behind establishing a closed area to fishing during the haddock spawning season (March–May) was that groundfishing may in some manner detrimentally affect recruitment to the stock through disturbance of the spawning act. Since investigation of this management scheme in 1970 no data have become available to suggest that a closed area has either a good or bad effect on spawning success and thus recruitment. Imposition of the closed area was coincident with application of quota management thus confounding biological interpretation" (CAFSAC 1981).

One of the most effective uses of a closed area regulation was the permanent closure of the Atlantic salmon fishery in southwestern Newfoundland in 1984. This was combined with a mandatory buy-back of fishing licences. These measures were part of a comprehensive package aimed at reversing the decline in spawning escapement. The commercial salmon fishery in this zone was closed because of high interception of mainland-origin salmon returning to mainland rivers to spawn (Lear 1993).

One highly visible application of the closed-area technique was the closure by the Pacific Halibut Commission of designated halibut nursery grounds. The Halibut Convention permitted the Commission to close all halibut fishing grounds populated by small, immature halibut. During the 1930's two nursery areas were closed, one off British Columbia and the other off southeastern Alaska. In 1967, the Commission designated a large section of the southeastern Bering Sea as a nursery area closed to all halibut fishing. These regulations were difficult to enforce (Bell 1981).

Abstention provisions or agreements preventing the fishing of salmon on the high seas are another type of closed area regulation. In the Convention establishing the International North Pacific Fisheries Commission, Japan agreed to abstain from fishing salmon off the coasts of Canada and the U.S. (exclusive of the Bering Sea and of the waters of the North Pacific west of a provisional line approximately along 175° W. Longitude). Both Canada and Japan also agreed to abstain from fishing salmon in waters of the Bering Sea east of a particular line. Subsequently, in the Commission meetings the interest of the United States and Canada lay in moving the Protocol line as far to the west as possible. This would mean that most salmon of North American origin would be east of the line. The Japanese fishing industry wanted to keep the line as far east as possible. Canada and the U.S. had prohibited net fishing by their nationals for salmon outside the "surf line." Following extension of fisheries jurisdiction, the "abstention" line was adjusted to provide further protection for salmon of North American origin (see Chapter 11).

On the east coast, the 1982 Convention for the Conservation of Salmon in the North Atlantic established the North Atlantic Salmon Conservation Organization (NASCO). Article 2.1 of the convention prohibits salmon fishing beyond the fisheries jurisdictions of the coastal States. Article 2.2 stipulates: "Within areas of fisheries jurisdiction of coastal States, fishing of salmon is prohibited beyond 12 nautical miles from the baselines from which the breadth of the territorial sea is measured, except in the following areas:

(a) in the west Greenland Commission area, up to 40 nautical miles from the baselines; and

(b) in the North-East Atlantic Commission area, within the area of fisheries jurisdiction of the Faroe Islands."

While allowing existing high seas fishing for Atlantic salmon to continue under regulations by the relevant Commission, this provision prohibited new fisheries for Atlantic salmon except close to the coasts of the States concerned. This effectively closed a large area of the North Atlantic to salmon fishing.

Other closed area regulations aim to minimize gear conflict. For example, the limitation on the area of operation of trawlers over a certain size is intended to prevent disruption of the fishing operations of inshore fishermen. There are numerous instances on both the Atlantic and Pacific coasts where specific fishing grounds or areas are closed to fishing by particular gear types.

Some closed area regulations reduce efficiency and increase costs for some fishing enterprises, with no biological advantage. These may close spawning grounds or close areas to specific gear types. In other instances, closing areas is clearly beneficial. Examples include the SMGL to minimize by-catches and prohibition of salmon fishing in certain areas to minimize interceptions.

2.5 Closed Seasons

The imposition of closed seasons is one of the most common techniques used to manage fisheries. There are a variety of reasons for closing fisheries for specific fixed periods during a year. The chief objective is to protect a stock from exploitation during part of its life cycle or during seasons of high vulnerability. There are two aspects to this:

1. To maintain exploitation rates at desirable levels if fishing effort is uncontrollable; and
2. To prevent fishing during particular seasons of the year.

Closed seasons of fixed duration to achieve the first purpose are unlikely to protect the stock or maintain a given stock size. Theoretically, such a measure should be enforceable. However in practice such measures may lead to trade in illegally caught fish. De Wolf (1974) cited instances of this in the Atlantic lobster fishery.

Closures of fixed duration during particular seasons may have purposes other than conservation. Quite often closed seasons are introduced for market reasons, e.g. when market prices are poor or when fish quality is poor. These considerations apply in the case of the Atlantic lobster fishery where the seasons were established in many instances to avoid competition with the period of high landings in Maine. The winter open season for the southwest Nova Scotia lobster fishery coincides with high prices and low product availability on the North American market. This is also a period when quality problems are at a minimum (see the next Chapter). Closed seasons are also used in other shellfish fisheries to minimize quality problems, for example, the capture of soft-shelled crabs.

Seasonal openings of variable duration are a primary management tool for Pacific salmon and Pacific herring. In both cases, fishing capacity is far larger than necessary to harvest any surplus to required spawning escapement. The roe herring fishery in B.C. is renowned for its short 15-minute open seasons. The B.C. salmon fishery is highly complex. Five species and a multitude of stocks intermingle on their return to the rivers to spawn. Thus the use of seasonal openings is fine-tuned to account for last minute estimates of stock abundance and permit sufficient fish to get up river to meet target spawning escapements. The closed/open season technique is an essential regulatory tool in the Pacific herring and Pacific salmon fisheries.

Restrictions on time and area fished have been criticized because both will increase the cost of fishing (they constrain the way vessels and gear can be used). See Anderson (1986). If area closures force fishermen to fish on less productive grounds, the cost curves for the individual fishermen and the fishery as a whole will also shift upward. Thus spatial and temporal restrictions on fishing often promote inefficiency because fishing is carried out at times and places where the catch per unit of effort is relatively low.

One of the touted advantages of closed areas and seasons is that they are relatively easy to enforce using patrol boats and airplanes. There are, of course, exceptions. Enforcement of closures, for example, as that during the haddock spawning season on the Scotian Shelf, is relatively expensive.

Closure of nursery grounds or the banning of fishing of spawning or prespawning concentrations can have serious implications for fishermen. The effects may be quite different for different groups of fishermen depending upon where, when and how they fish. Thus the political and social consequences of such prohibitions can be as important as the purely biological impacts (Gulland and Robinson 1973).

Area and season closures can be useful adjuncts to more direct methods of controlling fishing mortality. However, they are generally less useful as the primary methods of management. They are useful conservation measures, in cases such as the "silver hake box", to minimize by-catches of other species. They can also be used to optimize spawning escapement in species such as Pacific salmon and herring. DFO has in some instances closed certain spawning and nursery areas for groundfish as a conservation measure, e.g for haddock on the Scotian Shelf.

3.0 CONTROLLING THE AMOUNT OF FISHING

In recent years there has been an increasing emphasis on regulatory methods to control the amount of fishing. The primary objective of controlling the amount of fishing is to achieve some target level of fishing mortality. The choice of the target level will depend upon the management objective(s) for, and the nature of, a particular fishery. Examples of target fishing mortality cited in Chapter 3 were F_{max} and $F_{0.1}$.

Unfortunately, fishing mortality cannot be observed or controlled directly. Therefore, managers usually monitor and/or control two parameters directly related to fishing mortality: catch and fishing effort. These are related according to the following formulae:

$C = F\tilde{N}$

where $\tilde{N}$ is the average abundance of the population during the year, C is the catch and F represents fishing mortality, i.e. the rate at which fish are removed by fishing;

$F = qf$

where f is the fishing effort and q is the catchability coefficient representing the amount of fishing mortality induced upon the population by a unit of fishing effort.

The second equation indicates that the relationship between f and F is a straight line passing through the origin with slope q, a constant catchability coefficient. In practice, q is rarely constant. A particular amount of fishing effort will not always remove the same proportion of the stock.

Catch or nominal effort will be satisfactory measures of the fishing mortality if $\tilde{N}$ or q remain constant. Variations in $\tilde{N}$ or q mean that controlling the catch or effort at some particular value will not achieve the desired level of fishing mortality. If they fluctuate randomly with only a moderate degree of variation, this would not be serious. However, if there are significant trends in the variation of $\tilde{N}$ or q, the effects could be substantial. An obvious example would be an increase in fishing mortality, sometimes undetected. Thus catch limits generally have to be adjusted annually, particularly when there are significant year-to-year changes in population abundance. Regarding fishing effort, the fishing mortality exerted by a nominal unit of effort tends to increase with technical improvements in a fishery. This can be accommodated by adjusting the unit of effort to reflect increased fishing power provided the latter can be determined. However, it is difficult to standardize different vessel-gear combinations from a variety of countries or even within a single country. It is usually very difficult to determine the increases in efficiency of particular vessel-gear units.

Changes in abundance of the fish stock under study can often result in undetected changes in catchability (q). This has been observed for pelagic species such as herring. Reductions in stocks of these species may be accompanied by a shrinking area of distribution of the fish. Under such circumstances, monitoring of catch per unit effort (CPUE) and nominal effort may result in erroneous calculations of the fishing mortality being applied to the stock. Thus unadjusted fishing effort controls can have detrimental effects similar to those mentioned for unadjusted catch controls.

To account for these factors, the limit used to control fishing mortality, whether it be catch or fishing effort, must be regularly adjusted. Catch quotas are generally adjusted annually. Adjustments are based on the most up-to-date information on stock abundance. Limits on the amount of fishing effort are more difficult to modify because of difficulty of documenting changes in fleet efficiency. Catch quotas and effort controls have different advantages and disadvantages as regulatory measures.

3.1 Catch Quotas

In the 1970's and 1980's the setting of limits on total catch became the most common method of controlling the amount of fishing. These are now widely known as Total Allowable Catches (TACs). Catch quotas were used in the North Pacific halibut fishery as early as the 1930's. They were applied to Pacific herring shortly thereafter. Immediately after the Second World War they were introduced for Antarctic whales.

It became accepted in the 1960's that the amount of fishing in the North Atlantic should be controlled and that mesh size regulations were ineffective for this purpose. Catch quotas were chosen as the primary regulatory instrument because it was easier to implement national allocations under a system of catch quotas than under a system of effort limitations. Following their introduction by ICNAF in 1970, TACs were rapidly adopted as a management measure for finfish stocks in the North Atlantic. This was accompanied by national allocation of shares of the total quota.

Catch quotas can be readily understood by fishermen and the fishing industry. They can also be readily compared between countries and between fishermen. By comparison, effort quotas and allocations are complicated to devise and difficult to implement.

TACs as a regulatory method were reasonably quickly accepted in multinational fisheries and within particular countries. The rationale for establishing a particular level of TAC has not been as readily accepted. Internationally, this has been the case in the area beyond the Canadian 200-mile limit on the Atlantic coast. For complex political reasons, the European Community repeatedly objected in the late 1980's to TACs established there by the Northwest Atlantic Fisheries Organization (NAFO) (see Chapter 11). Within Canada debate has often raged about the scientific basis of TACs for certain stocks. Examples are the controversies surrounding the TACs for Gulf of St. Lawrence redfish, Gulf herring and northern cod during the 1970's and 1980's. There were ongoing mini-wars about the appropriate level of TAC. These battles were often couched in terms of challenges to scientific advice but the root cause was often conflict over how the resource should be shared among competing user groups.

Where the TAC system has been applied it has generally been accepted. It has worked reasonably well from national and international perspectives. Exceptions have taken three forms. Some countries or managers have satisfied competing user groups only by setting TACs larger than scientists recommended. A second problem has arisen from erroneous estimates of stock abundance. Sometimes this has resulted in dramatic adjustments in the level of TAC advised from one year to the next (see Chapter 18).

One major disadvantage is the need to ensure timeliness in the setting of catch quotas by minimizing the time lag between the collection and analysis of data and the quota period. In practice, this is often difficult to achieve. During the 1970's and 1980's catch quotas for most groundfish fisheries off the Canadian Atlantic coast were recommended 6 months in advance of the fishing season based on assessments performed 8 months earlier using data from the year previous. For example, the 1990 TACs would be recommended in May–June of 1989, based on 1988 data. Ideally, the assessment should be conducted and catch quotas established just prior to the next fishing season using data from the current year's fishery. Although this is possible for a few fisheries, it is the exception rather than the rule.

The effects of time-lags are least significant for long-lived fish. Failure to set the TAC "right" in one season can be compensated in the following season without significant loss in yield or overfishing. Also for long-lived species individual year-classes can be monitored as they recruit to the fishery. Thus potential yields can be predicted up to 2–3 three years in advance with some reasonable confidence.

Major problems can arise with short-lived species whose abundance fluctuates widely from year to year, e.g. capelin and silver hake. If a weak year-class enters the fishery, and the catch quota is not altered in time, a high fishing mortality could lead to overexploitation of the year-class. This could have major consequences for the stock. To avoid this, it is necessary to predict reliably the strength of recruiting year-classes and adjust the quota during the season. In the case of the Pacific salmon fishery, for example, adjustments are made weekly, even daily, by varying the duration of open/ closed seasons.

Errors in estimating stock abundance and identifying trends in fishing mortality can be substantial (see Chapters 18 and 20). One source of error arises from control by catch, rather than effort. To calculate TACs it is common to use a method called Virtual Population Analysis (VPA) (see Chapter 3). Results of the analysis depend critically upon the choice of the "terminal F": the fishing mortality assumed to occur in the last year for which data are available (Pope 1972). For example: In 1990 when the TAC for 1991 is being calculated on the basis of 1989 data, the value of F for 1989 could be under-estimated. This would mean that the stock in 1989 was less than was estimated. The catches during 1989 will in this case take a higher proportion than expected of this smaller stock. Hence, the extent of over-estimation of the stock at the beginning of 1991 and the TAC for 1991 would probably be considerably greater than the original degree of under-estimation of terminal F.

Gulland (1974) suggests that catch quotas are "comparatively easily controlled and enforced." In an earlier paper (Parsons 1980) I disagreed, pointing out that the number of days fishermen spend on the fishing ground is more readily controlled under a system of aerial patrols, supplemented by at-sea surveillance. This has been proven in managing fishing by foreign vessels in the Canadian zone. Experience over the past 15 years with the TAC system in the North Atlantic has shown that the most serious indirect effect of catch quotas is the falsification of data on the fishery. There have been numerous instances where the actual landings have been considerably larger than the reported figures. One glaring example was the underreporting of catches in the Bay of Fundy purse seine fishery for herring during the 1980's. Similar cases have been reported from the Northeast Atlantic where in several instances ICES scientists have advised that it was no longer possible to assess the state of the stocks (see Chapter 20).

Gulland (1984) observed that attempting to control the amount of fishing through catch limits is often a "second — or third best method." He analyzed management by catch quotas using four criteria — (1) maintaining the resource; (2) economic efficiency; (3) equity; and (4) transaction costs. In practice, knowledge of the resource is less than perfect. Thus the catch quota set will not be the correct value to achieve the desired fishing mortality. Its effectiveness relative to other types of control depends on the year-to-year variability of the stock and the degree to which this can be measured and predicted. Thus, catch quotas are highly suitable for whales. With stocks of long-lived fish, for which year-class variability can be monitored reasonably well, control by catch is suitable, "though probably no better than effort limits." With short-lived stocks, for which it is difficult to make accurate estimates of abundance in the following year, other types of control, e.g. on fishing effort, may be more effective. For pelagic schooling species, the catchability coefficient varies inversely with abundance. These are a special case: controls on nominal effort do not control the actual fishing mortality. Catch quotas may be used successfully if acoustic surveys can provide a direct estimate of the current abundance.

The impact of catch quotas on economic efficiency depends upon the extent to which the total catch is allocated among participants in the fishery. Global quotas which are not allocated to individual groups are clearly inefficient. They lead to a mad race to catch the quota. This can be overcome by various forms of allocation of the global quota. The relative equity of catch-versus-effort controls depends upon the manner in which these are applied. For a discussion of this, see Chapters 7, 8 and 9.

Acquiring the scientific knowledge base necessary to permit year-to-year adjustments in catch quotas is costly. Large amounts of data must be collected, costly research surveys conducted and scientists diverted from the broader questions of resource assessment and management. Costs of enforcement can also be high.

Catch quotas have been successfully applied in many fisheries to solve the conservation problem for single species. By themselves, they do nothing to address excessive fishing capacity. In many instances global catch quotas can exacerbate overcapacity unless other measures are introduced to address the problem.

3.2 Effort Controls

Superficially, regulating fishing mortality by control of effort appears to offer advantages over catch control. Target fishing mortality rates might be more directly achieved by controlling fishing effort than by controlling catch. In theory, once an appropriate level of fishing effort is determined, this level may be maintained without year-to-year adjustments dependent on annual stock assessments. Certain studies have suggested that during periods of low productivity fish stocks are less vulnerable to pronounced reductions in biomass when they are regulated by a constant fishing mortality (effort) than by a constant catch approach, e.g. Reeves (1974).

If the measures of effort were accurate, it would not be necessary to adjust quotas from year-to-year to account for year-class fluctuations, nor would it be necessary to monitor the strength of recruiting year-classes quite so closely. In practice a fishing effort regulation is intended to harvest a proportion of the stock. Determination of the level of fishing effort depends on the assumption that a given unit of fishing will harvest a constant proportion of the stock so that, once the proportion is determined, the number of units of effort is defined. This requires converting a target fishing mortality to a quantity of fishing by a standard unit and its reconversion to component units. Usually, this derivation depends at some point on comparing catches over time of different vessels. This system is particularly liable to error if the regulatory constraint changes fishing patterns, as is often the case. Studies in ICNAF from 1973 to 1976 demonstrated considerable changes in efficiency of the fleets of various countries over the previous 10–20 years. Actual fishing effort, and hence, fishing mortality was increasing even in cases where the apparent fishing effort remained the same.

Thus there are two aspects to selecting an appropriate level of fishing effort to achieve a target fishing mortality. The first is to detect and measure changes in efficiency. The second is to compare and calibrate effort by two or more fleets fishing the same stocks. Increases in efficiency may be obvious or subtle and can often only be derived by guesswork. Such guesswork is rarely an acceptable basis for adjusting effort quotas allocated to particular segments of a fishery. Another difficulty is that many of the adjustments made by scientists depend on hard-to-verify reports by fishermen. Also the number of hours fished per trip may change over time. This might not be accounted for when regulating the number of days on the fishing grounds, should that be the effort measure chosen. Regulating the number of fishing hours, a more accurate measure, would be impossible to enforce.

STACRES (ICNAF 1973a) examined the implications of regulating effort on the basis of hours fished, days fished, and days on ground. It concluded that "days fished" was the best measure of fishing mortality. Regulating "days fished" had two drawbacks however. One was the difficulty of monitoring in the international fishery the number of days being fished. The second problem is that fishermen from one country, or fleet sector, cannot always determine whether other vessels on the fishing grounds are in fact fishing. This might lead them to make erroneous conclusions about compliance with regulations. "Hours fished" suffered even more from both those drawbacks. "Days on ground" may lack the precision of days fished in relation to fishing mortality but it has a smaller margin of error than just regulating the number of vessels. "Days on ground" can be more easily monitored, are easily observable by fishermen, and thus could represent a more credible regulation than "days fished."

It is very difficult to regulate fishing mortality precisely by control of fishing effort. Effort control is most beneficial in situations where a precise cause and effect relationship is not needed. Direct control of the number of vessels and/or vessel days on ground could then be useful. The latter approach might be appropriate to halt declines in resource abundance pending the introduction of a more sophisticated system.

Effort regulation has been applied to marine commercial fisheries less extensively than catch regulation. In their review of possible conservation actions for the ICNAF area, Templeman and Gulland (1965) observed: "There must be some direct control of the amount of fishing. All methods of doing this raise difficulties, but that presenting the least difficulties is by means of catch quotas. There must be separate quotas for each stock of fish, e.g. for cod at West Greenland, and preferably allocated separately to each section of the industry."

Following the introduction of national catch quotas in ICNAF in 1970, the USA proposed effort regulation for the area off the northeastern USA. However, ICNAF rejected the proposal. A two-tier catch quota system was adopted instead. Canada secured a commitment from ICNAF in 1976 to a 40% reduction in fishing effort by distant-water states off the Canadian Atlantic Coast. Once Canada extended fisheries jurisdiction in 1977, it instituted a licensing system for foreign vessels. The number of vessels and vessel days on ground were regulated to provide effort control to supplement catch quotas.

Licensing has been widely used in domestic fisheries. Limited entry licensing is usually intended to allocate resources among user groups or to control capacity to increase economic efficiency. It is not generally sufficiently fine-tuned for directly regulating fishing mortality, the subject of the present chapter. The pros and cons of limited entry licensing are discussed in Chapter 7.

4.0 CONTROL OF MULTISPECIES FISHERIES

We have discussed various regulatory techniques for managing single species stocks. In practice trawl fisheries are seldom based on single species. Instead, they harvest a complex of resources of varying interest to different countries or fleet sectors within a country. In such mixed fisheries, a framework of individual single-stock catch quotas may be incompatible with TACs on species caught together. Thus it may not be possible to achieve the TACs of all simultaneously.

The degree of complexity of fisheries interactions ranges from the relatively simple, where fisheries on two different species are so distinct that they can be managed quite separately, to an extreme situation where the species are so evenly mixed on the fishing grounds and in the catches that no preference for one species or another can be detected. In the latter instance, the species complex should be managed as a unit. The complex of species can be so large that the only practical method is to treat the catch as if it came from a single species. This is true for some tropical areas. It is better to introduce simple management approaches early than sophisticated management too late.

For an intermediate degree of fisheries complexity, Garrod (1975a) suggested two alternatives: (1) to establish the TACs of individual stocks with minimum stock size constraints for particular stocks, or (2) to consider the resource complex in particular areas as a whole and adopt either catch or fishing effort regulations for the whole. Garrod (1975b) described the formulation of catch regulations based on designated stock size. This assumed that the first responsibility of management is to maintain a stock size that will ensure continuity of the resource. He concluded that this implied a variable catch and fishing mortality to maintain a constant stock (Garrod 1975a). This would necessitate defining a minimum spawning stock level.

Some stocks have been depleted to 10% of the unexploited level. Garrod suggested a 15% minimum spawning stock constraint. This proposal, however, was not widely supported. The most acceptable approach would be to adopt a spawning stock level which is known to be adequate. Given the uncertainties about stock and recruitment described in Chapters 3 and 18, this is unlikely to be easy for any stock. It will be impossible for some stocks.

Doubleday (1976) suggested that an equilibrium spawning biomass constraint at a biomass level about two-thirds of that of the virgin stock could provide an adequate biomass buffer to maintain stock stability in the face of large fluctuations in recruitment. In practice, with the exception of the salmonids, minimum or optimum spawning stock sizes have been established for very few species and stocks.

In the 1970's there were some attempts to use general production (Schaefer-type) analyses of catch in relation to fishing effort (see Chapter 3) for the groundfish fisheries of the Northwest Atlantic to estimate an overall Total Allowable Catch level or overall level of

By-catches in a mixed fishery.

fishing effort for managing these "assemblages" of species. Brown et al. (1976) estimated the projected maximum sustainable yield from Schaefer yield curves for the finfish resource off New England as 900,000 tons. The sum of the MSYs from individual assessment studies was 1,300,000 tons. This suggested that summing the MSYs from individual assessments might overestimate the total MSY because of species interactions and by-catch complications.

Pinhorn (1976) used the Schaefer general production model to estimate the groundfish MSY for Subareas 2 and 3 at 900,000 to 1,000,000 tons. He concluded that catch quota regulations for 1975 would not prevent continuing stock decline. A reduction in fishing effort of 30–40% might be necessary to reduce fishing effort to the MSY level. In a similar analysis, Halliday and Doubleday (1976) estimated the MSY of groundfish on the Scotian Shelf to be 250,000 tons (excluding silver hake). They concluded that the catch quota regulations for 1974 and 1975 could not halt stock decline. A reduction in fishing effort of 37% from the 1973 level was required to reduce effort to the MSY level.

Pope (1976) and Horwood (1976) considered the case of pronounced biological interactions where one species may increase to replace another which is heavily fished. They found that the overall MSY of the resource complex may only be achieved by a very specific mixture of fisheries aimed at the different species. Outside these limits the species composition of the complex, and catches, will depend heavily on the level of fishing mortality. However, the overall total catch may be fairly stable. If there are no biological interactions, attaining the MSY will depend on the species preference of the fisheries involved. If these vary with species abundance, the yield may be stable over a range of fishing. However, this may not be the theoretical MSY of the resource. The overall MSY will be associated with a particular species composition. Its attainment will depend on matching the fisheries exploiting the complex to the "optimal" species composition.

Thus general production models do not necessarily indicate the MSY of a complex. They may only indicate the MSY of fisheries with the species preference observed during development of the fishery. If the total level of exploitation were regulated by an overall catch or effort regulation, the "true" MSY is likely to be achieved only by a very mixed fishery. The level of catch may remain fairly stable over a wide range of levels though there is a progressive change in the species composition. Thus the resource composition may shift towards species which were initially of least value.

Hence, a single overall catch regulation might not achieve the MSY. It would not protect particular stocks but would allow a shift in the species composition. Equally a single overall effort regulation might not achieve the MSY but could tend to fix the species composition of the resource.

Of the various measures available, control of fishing effort is least suitable for modifying the balance between species. In practice fishermen will continue to seek the highly valued species. Closed areas and seasons could help differentiate between species. If one species is heavily exploited, it may be preferentially protected if there are areas or seasons where it is particularly abundant or easily caught. If the dynamics of the resource complex were known, an array of individual catch quotas might achieve the yield objectives. Since these dynamics are not fully understood, it is necessary to protect the individual species so far as possible. This could be augmented by an overall catch or effort control to provide a cushion against unknown interactions and preserve a satisfying resource configuration. An overall regulation cannot be regarded as a precise measure. However, it may be necessary because the exact sizes of all individual stocks are unknown.

Single species regulation is also imprecise. A catch quota will be most heavily influenced by the error in estimating numbers of young fish entering the fishery. Effort regulations will be influenced by year-to-year or within-year variations in the catchability coefficient of the units of effort. In both cases, it was suggested in the 1970's that the level of exploitation achieved by a regulation will have a coefficient of variation ± 20–30% of the intended level (Pope and Garrod 1975). Recent studies (see Chapter 18) suggest that the uncertainty can be greater than this. Multispecies fisheries introduce additional error because of undetected biological or fisheries interactions among the component resources. Where catch quotas are used, the overall TAC will, in general, be less than the aggregate of the individual species TACs because of the by-catch problem.

Where effort controls are used, the amount of effort directed to particular species could be expected to vary with short-term variations in abundance and species preference. This could change the catchability coefficient. As with errors in stock size, it is impossible to quantify the errors involved in projecting the present effect of fishing to their presumed effect in 2 years time. They may be very specific to each species complex and vary with time within it. With respect to such errors, neither method (TACs or effort control) is clearly superior.

Catch quotas present better opportunities for differential species management. One approach experimented with in the 1970's was a two-tier system consisting of catch quotas on individual species/stocks within an overall catch quota for all species/stocks. This approach was in effect during 1974–76 in Subareas 5 and 6 off the U.S. Atlantic coast. Another option for regulating multispecies fisheries is combined catch and effort regulation. The chief problem is the need for compatibility between the two regulatory approaches. Such a system was implemented off the Canadian Atlantic coast in the last years of ICNAF. These approaches to management of the Atlantic groundfish fisheries are described in Chapter 6.

5.0 CONCLUSIONS

In summary, a variety of regulatory methods have been devised for marine fisheries. The two primary methods involve controlling the age (size) at first capture, through minimum mesh size or minimum size limits, and controlling the amount of fishing, through catch quotas or fishing effort controls. Other auxiliary methods include closed areas and seasons. In practice, no one method is ideal to achieve resource management goals for any particular fishery. Generally, a combination of various regulatory methods are necessary depending upon the objectives for a fishery.

CHAPTER 6

RESOURCE MANAGEMENT OF CANADA'S MARINE FISHERIES

1.0 INTRODUCTION

The previous chapter outlined the various techniques available to regulate the harvest of marine fish stocks, to ensure continued productivity of the resource. All of these techniques have been applied in various Canadian fisheries over the past century. This chapter examines the application of these techniques using five selected fisheries — Pacific halibut, Pacific herring, Pacific salmon, Atlantic lobsters and Atlantic groundfish. These examples show how the appropriate mix of techniques varies from fishery to fishery.

2.0 PACIFIC HALIBUT

Pacific halibut (*Hippoglossus stenolepis*) occur on the continental shelf off western North America. There are major concentrations off Alaska and British Columbia north of Vancouver Island. Because of migrations, both Canadian and American fisheries exploit substocks which migrate across international boundaries (Cook and Copes 1987).

The commercial fishery for Pacific halibut commenced in 1888. By 1915 the total catch had reached 70 million pounds. It decreased to between 40 and 50 million pounds by 1920 (Fig. 6-1). This prompted fishermen to call for an investigation which resulted in the establishment of the International Fisheries Commission in 1923.

The International Fisheries Commission was established by a Convention between Canada and the United States for the preservation of the halibut fishery of the North Pacific Ocean and the Bering Sea. The Commission was renamed the International Pacific Halibut Commission in 1953. The Commission's authority was extended by modifications to the Convention in 1930 and 1937.

A 1922 Royal Commission in Canada emphasized the need to conserve the halibut resource, arguing that the stocks had declined significantly. The 1923 Halibut Convention provided for a 3-month winter closed season subject to modification after 3 years. The winter closed season was motivated mainly by economic rather than conservation considerations (Bell 1981). In

A large Pacific halibut about 2 1/2 metres long and weighing about 180 kilograms (Photo courtesy International Pacific Halibut Commission).

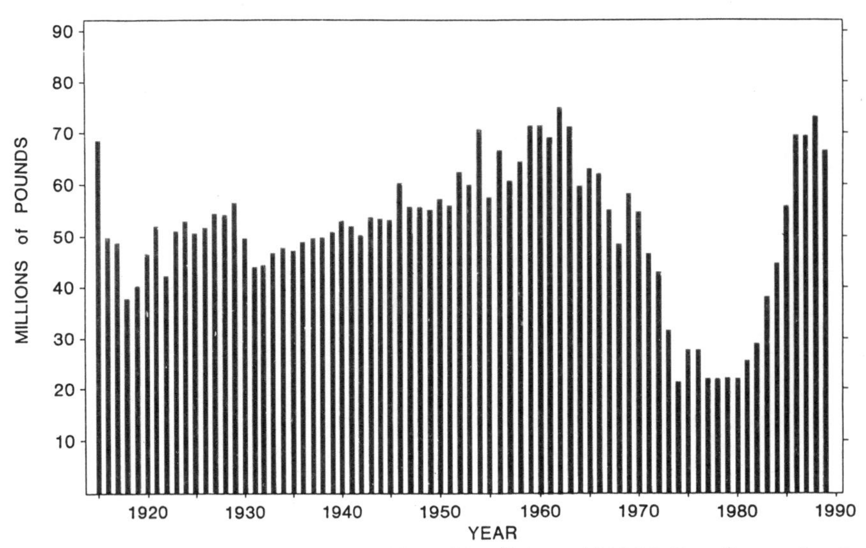

FIG. 6-1. Trends in catches in the Pacific halibut fishery: 1915-89 (from International Pacific Halibut Commission).

a 1928 scientific report to Canada and the USA, the International Fisheries Commission concluded that the winter closed season had not protected the resource. The stocks of halibut had continued to decline. The Commission requested authority to establish areas where the catch of halibut could be reduced by a predetermined percentage annually until there was stability of yield. It also recommended prohibitions on destructive gear and a two-week extension of the three-month season.

The provisions of the 1930 revised Halibut Convention were similar to the 1923 Convention. However, the 1930 Convention granted greater powers to the International Commission. The Commission could now divide the convention waters into areas and limit the catch of halibut taken from each area during the fishing season. It could also regulate the licensing and departure of vessels to collect statistics and fix the character and type of gear to be used. The Commission could also close areas populated by small immature halibut. Enforcement of the regulations was the responsibility of the participating governments.

Problems in enforcing catch quotas and controlling the incidental catch of halibut by vessels fishing for other species led to a revised Convention in 1937. This provided more effective control of the capture of halibut caught incidentally to fishing for other species in areas closed to halibut fishing. It also allowed prohibition of the departure of vessels to any area when it was considered that sufficient vessels had already departed to harvest the catch quota for that area. The provision for the control of halibut caught incidentally remained in effect until 1965. The departure date regulation was used only from 1937 to 1939.

A revised Halibut Convention signed in 1953 authorized the Commission to establish one or more open or closed seasons each year in any area. This enabled the Commission to establish supplementary seasons in areas where the short single season had resulted in underfishing. Areas where the halibut showed signs of overfishing during the single regular season could be closed and then opened at other times. The 1953 Convention also gave the Commission the power to establish size limits. It also specified that halibut stocks be developed and maintained to produce the maximum sustained yield.

Thus the regulatory techniques applied in the Pacific halibut fishery were extended by these various modifications to the original Convention. Bell (1981) described the events leading to these changes. More detail is provided in Crutchfield (1981).

The era of regulation established by the 1953 Convention continued with only minor modifications until 1978. The primary regulatory tool was catch quotas by area. These were supplemented by size limits, closures of nursery areas and prohibitions on the capture of halibut by "destructive gear", i.e. trawling.

Skud (1973) reviewed the management of the Pacific halibut fishery. He distinguished three periods in the fishery since the establishment of the Commission, illustrated by the trend in commercial catches (Fig. 6-1). During the initial period (1923-30), catches continued to decline, diesel-powered schooners were introduced, and gear and fishing methods were changing. Longlining from larger vessels replaced fishing from dories that delivered the catch to "mother-ships". During the second period (starting around 1930) the fleet converted to new methods which dominated the fishery for the next

20 years. Over 90% of the fleet fished with 13 ft. (4.0m) hook spacing and fishing efficiency improved with the use of electronic gear. The Commission took a more active regulatory role. During this period the catch and catch per unit of effort gradually increased.

The third period described by Skud was from 1950 to the mid 1970's. During this period there was a further gradual change in fishing techniques which continued to improve efficiency. The average annual catch increased consistently from 1935 to the early 1960's. After 1962 however, catches in each region declined until 1980. Stock abundance, estimated by catch per unit effort (CPUE), also showed a decline for each area from 1960 to 1980.

Catches which had been declining through the late 1920's reached a minimum in 1931. However, the abundance of the stocks increased after the Commission introduced measures to limit the catch.

Catches had declined from 70 million pounds in 1915 to about 45 million pounds by the early 1930's. From 1935 onward total catches, CPUE, and average size of fish caught increased consistently until catches finally peaked at 75 million pounds in 1962. A major groundfish fishery by foreign fleets began in the early 1960's in the Bering Sea and the Gulf of Alaska. These fleets caught substantial quantities of halibut incidentally while fishing for groundfish. The extent of this halibut bycatch was underestimated. Stocks declined to the lowest recorded levels by the mid-1970's (McCaughran 1989).

Commenting on the management initiatives of the International Pacific Halibut Commission (IPHC) until the early 1960's, Bevan (1965) stated: "The work of the Pacific Halibut Fisheries Commission provides an outstanding example of successful regulation by means of a quota. The commission's regulation has resulted in virtually full rebuilding of halibut stocks of the North Pacific." Crutchfield (1965) observed: "The halibut programme has been a conspicuous success — from a biological standpoint." Skud (1973) observed: "The maintenance of a viable fishery under intense exploitation for a 30-year period certainly speaks for the Commission's contribution. Many scientists have recognized IPHC's role as a classic example of fishery management based on scientific information."

By the mid 1960's commercial catches were declining. Estimates of stock abundance had been too high. This was partly due to certain assumptions about technological change which were subsequently shown to be false, e.g. the assumption that catch is independent of hook spacing. Re-examining existing data, Skud (1973) showed that the catch per hook increased as the spacing increased. Thus gear efficiency had been underestimated and the CPUE overestimated. This had resulted in overestimation of stock abundance.

In addition to errors in estimating abundance in earlier years, during the 1960's there were increases in efficiency of the longline gear and the percentage of the catch taken by small setline vessels and salmon trollers. There was also an increase in fishing effort by trawlers from Canada, the U.S., Japan and the USSR. This increased the incidental catch of halibut. Although the retention of halibut caught by Canadian and U.S. trawlers was prohibited, there was undoubtedly increased mortality.

With increased removals by the trawl fisheries, effort by the longline fleet was greater than should have been permitted. This occurred because the impact of the greater incidental catches was not immediately detected by CPUE. As the abundance decreased, the Commission reduced the catch limit, generally by 500 to 1,000 tons a year. But these reductions did not arrest the decline. In 1972 the limit was reduced by 6,000 tons after Skud's analyses showed that the stock had been overestimated during the previous decade.

Catches in Area 2 exceeded MSY in the mid-1950's and in Area 3 MSY was reached during the early 1960's. The incidental catch by trawlers accelerated the decline which resulted initially from fishing by the longline fleet. The mortality of halibut caught incidentally by trawlers and returned to the sea was estimated at 50% for Canadian and U.S. trawlers and 100% for foreign trawling operations (Hoag and French 1976).

The IPHC was established to combat biological overexploitation of the stocks. It was, to a large extent, successful from the 1930's to the early 1960's. The catches and CPUE of both Canadian and U.S. halibut fleets declined in the latter part of the 1960's and in the 1970's because of several factors. These included the increase in incidental catch, reduced recruitment of juvenile halibut probably related to changing environmental conditions, and apparent overfishing of the stocks by the longline fishery itself when the IPHC increased the catch quotas to test the estimated MSY. The problem of incidental catch by foreign trawlers was brought under control with the extension of fisheries jurisdiction by Canada and the U.S. in 1977. However, the problem of incidental catches by domestic trawlers remained. The 200-mile limit itself impacted significantly upon the operations of the IPHC and the halibut fleets. These impacts are described in Chapter 10.

In brief, major dislocation of the Canadian fleet occurred from Alaskan to B.C. waters. Canada placed the halibut fleet under limited entry. The USA placed strict bycatch controls on the foreign fleets. This reduced the foreign bycatch and produced more accurate estimates of bycatch.

Following reductions in the catch quotas in the mid-1970's, the stocks began to recover. The stricter controls on the foreign bycatch assisted in this rebuilding. The increase in the stocks continued through the 1980's. By 1986 the stocks in some areas equalled or surpassed the biomass level estimated to produce the MSY (McCaughran and Deriso 1988).

The total exploitable halibut stock biomass apparently peaked in 1986 and in the late 1980's declined by about 5–6% per year (IPHC 1989).

The catch quotas for Pacific halibut in the late 1980's were set to achieve a Constant Exploitation Yield (CEY). Based on an estimated optimal exploitation rate of 0.35, this yield is roughly a third of the exploitable biomass. The recommended allowable commercial catch for the directed setline fishery is determined after allowing for removals in the recreational fishery, bycatch and wastage.

The 1988 commercial catch of 74.6 million pounds was the second highest on record. The total catch in 1988 was probably the highest ever when bycatch, sport catch and waste are added to the commercial catch. In 1989 the Commission reduced the catch quota significantly to 64.65 million pounds. The actual exploitation rates for 1986 through 1988 were higher than the target CEY. According to Commission staff, factors contributing to these increased exploitation rates included decreases in recruitment, overestimates of stock biomass due to changes in the trend in abundance, and increasing bycatch mortality. The 1988 bycatch mortality was the highest observed since 1982 (IPHC 1989). In its 1989 Report the Commission noted that recruitment had dropped off dramatically in all areas. This lower recruitment, combined with exploitation above the recommended 0.35 level, indicated that the stock would continue its decline at a rate of about 5–15% per year over the next several years (IPHC 1990).

The management success of the Halibut Commission was limited because management for economic purposes was prohibited in the Convention(s). There was no provision to allocate catch quotas between the two countries or among fleets. Neither was there any provision for regulating the amount of fishing effort.

Crutchfield (1965) commented on the period 1933–1957:

> "The rebuilding of the halibut population led to a 300% increase in the number of boats participating in the fishery. But as the number of boats increased each boat faced higher and higher operating costs per unit of catch. Again, the end result was to increase costs until only minimum necessary returns could be earned.... The conservation authority is not inept, nor has it been unaware of these economic problems. Indeed, many of them were forecast years ago by members of the Halibut Commission staff. What is involved is not a failure of performance but a mistaken concept as to what is to be conserved and what is to be maximized."

These events have been repeated more recently. With the decline in catch and abundance that began in the early 1960's, landings decreased from 70 million pounds to a 1980 level of 20 million pounds. The number of units participating declined moderately until 1973 but increased again following a dramatic increase in real prices beginning in 1972. This occurred even though the expectations were that even deeper cuts in quotas would be required. Thus "entry into the fishery was increasing at a time when fishing mortality had reduced the halibut fishery to a near-crisis level" (Crutchfield 1981).

Following extension of jurisdiction by Canada and the United States, Canada attempted to restrict the size of its fleet through limited entry licensing. Individual vessel quotas were discussed in the early 1980's but not adopted until 1991. Although the U.S. took steps to curtail the foreign bycatch, an increase in the U.S. fleet occurred following the displacement of Canadian vessels from Alaskan waters. As the stocks began to improve, many new vessels were attracted to the fishery. Overall, there was a dramatic increase in the U.S. halibut fleet (McCaughran 1989).

The North Pacific Fishery Management Council discussed limited entry during the 1980's. The Council recommended a moratorium on halibut licences in 1983. However, this was rejected by the U.S. Office of Management and Budget. There were significant improvements in fishing technology during the 1980's, increasing the effective fishing effort. McCaughran (1989) pointed out that the fleet was far too large for the available catch. Fishing time had been reduced from many months to a few 24-hour openings, despite high stock levels.

A further problem was the rapid growth of the U.S. North Pacific groundfish fleet in the late 1980's (see Chapter 20). This fleet began to capture large quantities of halibut incidentally, increasing the bycatch mortality. Following the 1988 IPHC Annual Meeting, the Commission wrote to the Canadian and U.S. governments expressing concern about the problems created by the short, intense fishing seasons in Alaska that occurred because of uncontrolled fishing effort. The Commission stated that the long-term solution was for the North Pacific Fishery Management Council to reduce the number of participants in the fishery (IPHC 1989).

Following the 1989 Annual Meeting, a letter was sent to each government expressing concern about the increasing mortality of juvenile halibut in the U.S. groundfish fisheries in the Gulf of Alaska and Bering Sea (IPHC 1990). The total incidental catch in 1990 was estimated as 17.4 million pounds. Of that total, 15.8 million pounds were caught by U.S. fishermen and 1.6 million pounds by Canadian fishermen.

A 1979 protocol to the Halibut Convention altered the Commission's mandate to manage for optimum yield instead of maximum sustainable yield. This allowed social and economic factors to be considered. However, this did not give the Commission authority to regulate fishing effort directly.

Skud (1985) evaluated the use of closure regulations in the Pacific halibut fishery. He pointed out that, before effort became excessive, the seasonal and area closures served biological goals without hampering the economic efficiency. As effort increased, fishing time decreased well below 100 days in the 1950's. Biological goals were jeopardized and the economic efficiency of fishing operations were affected adversely. Conditions improved in the 1960's and early 1970's when fishing time increased on average to 150 days. With the increase of effort again in the late 1970's and during the 1980's, the problem became more acute than in the 1950's.

From the experience with seasonal and area closures in the halibut fishery, Skud concluded that regulatory measures which control effort are more critical than those concerned with gear and protection of juveniles. He observed:

> "In the halibut fishery, the catch or quota (and its attendant closures) has been the regulatory measure with the greatest impact. Without the quota, the present-day effort probably would have led to the inevitable depletion predicted by economist's models of uncontrolled fisheries."

The impact of the Commission's regulatory measures on the rebuilding of the halibut stocks has been disputed. The debate has focused on whether the fluctuations in abundance resulted from fishing/catch restrictions or changing environmental conditions.

The most famous controversy involved Thompson and Burkenroad. Burkenroad, in a series of publications from 1948–53, argued that the decline in the halibut stocks prior to 1930 could not have been the result of fishing alone. Furthermore, the subsequent recovery was so rapid and so major that it could not be accounted for by the Commission's management measures (Burkenroad 1948, 1953). Thompson maintained that the fishery was the dominant factor (Thompson 1950, 1952). The debate was inconclusive. It appeared that overfishing had been a major factor but that environmental changes unrelated to fishing may also have contributed to fluctuations in stock abundance. Skud (1975) reviewed this controversy. He concluded that observed abundance variations were not necessarily due to fishing levels alone.

Recent studies indicate that even after another 40 years of research it remains unclear whether environmental factors are the primary determinants of recruitment fluctuations. Parker (1989) studied the influence of oceanographic and meteorological processes on the recruitment of Pacific halibut in the Gulf of Alaska. He found that variations in certain oceanographic factors explained up to 66% of recruitment variability not accounted for by the effects of spawning stock size and density dependence.

Nevertheless, it appears that the regulatory measures of the IPHC, specifically catch quotas, have helped conserve the resource in the face of strong fishing pressure. Stauffer (1988) described Pacific halibut management as a "classic example of what can be achieved when biological assessments of the production from a fish stock are successful and well modelled, and the dynamics of the stock are relatively well behaved."

Despite the IPHC's relative success in achieving biological goals using catch quotas and auxiliary measures to control bycatch, a major problem of excess fishing capacity worsened during the 1980's. Stauffer (1988) summed up the situation:

> "The management process for halibut has successfully accomplished the goal of maintaining the productivity of the stock but has failed miserably in managing the fishery or the people harvesting the resource to achieve an efficient use of human and capital resources .... Until the institutions responsible for managing the halibut fishery are structured to manage on the basis of not only biological objectives but also social and economic objectives, the current shotgun nature of the fishery will persist".

3.0 PACIFIC HERRING

In British Columbia, each Pacific herring (*Clupea harengus pallasi*) population migrates in the fall from offshore feeding grounds to an inshore spawning area. Here they form dense pre-spawning aggregations, awaiting the ripening of their reproductive cells. Traditionally the fishery took place during the winter resting stage. From the early 1970's onwards, however, a major "roe" fishery has concentrated on capturing the herring when the eggs are at a very specific ripe stage.

The spawning season extends from mid-February to mid-April. Peak spawning occurs a little earlier in the southern part of B.C. than in the northern. The eggs are deposited mainly on vegetation in and immediately below the intertidal zone.

Most of these herring move offshore to feeding grounds on the continental shelf immediately after spawning. They reappear in inshore waters from one to three months before spawning in the following year. Some local stocks remain inshore year round. However, they generally mix to some extent with the migratory stocks at spawning time (Hourston 1978).

Herring were first caught commercially in B.C. in 1877. The fishery primarily involved purse seine gear until the 1970's. From a low of about 500 tons in 1900, the catch increased to about 30,000 tons in 1909. Between 1919 and 1927 the catch rose from 30,000 to 85,000 tons because of the expansion of the dry-salting industry. Between 1927 and 1935 the catch dropped abruptly to 30,000 tons. This was because of a major decline in the dry-salted market in the Orient (Fig. 6-2).

Between 1935 and 1938 the catch increased to a new peak of about 100,000 tons as a result of developments in the reduction industry. After the end of the Second World War catches increased sharply. Between 1953–54 and 1961–62 the catch averaged around 200,000 tons annually. The greatly increased catches resulted from more intensive exploitation, aided by the development of more efficient fishing methods. In the 1962–63 season the total catch reached a record high of 240,000 tons. Subsequently the catch declined to 181,000 tons in 1965–66 and to 133,000 tons in the 1966–67 season.

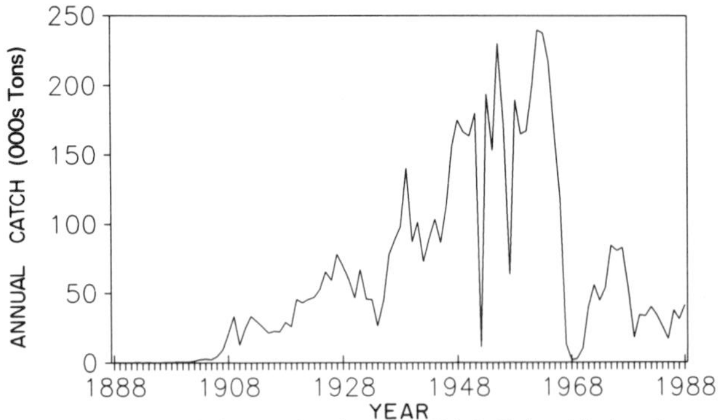

FIG. 6-2. Trends in catches in the British Columbia herring fishery: 1888-1988 (from Stocker 1993).

This decline in catch reflected a sharp decline in abundance which led to a closure of the reduction fishery for an indefinite period in 1968. Only traditional minor fisheries for local food and bait were allowed for the next 4 years.

Taylor (1964) described the early history of the fishery. For an account of the decline and recovery of the stocks in the mid 1960's and early 1970's, see Hourston (1978).

Closure of the major reduction fishery helped stocks recover. A roe fishery commenced in 1972 using seines and gill nets. This fishery caught an average of 48,000 tons over the next 4 years, peaking at 100,000 tons in 1976. During the 1970's and 1980's the roe herring fishery became a significant but unstable component of the B.C. industry (Stocker 1993). The landed value of the herring fishery averaged $60.5 million from 1981 to 1988. The 1988 value was $96.5 million with the roe

Pacific herring.

fisheries contributing 98.5% of the total. This occurred despite a decline in catch after 1976. From 1977 to 1989 the catch fluctuated between 14,500 and 74,000 tons, averaging 34,800 tons annually. MacGillivray (1986) described development of the roe herring fishery from the 1970's to the mid 1980's.

Pacific herring regulations have changed significantly over the past century. Initial regulations involved closures and mesh regulations to protect spawning and juvenile herring (Hourston 1980a). In addition to the closure of various areas of the coast to all types of fishing, certain small areas have been closed specifically to herring fishing. Closing these areas did little to limit herring catch. They generally served to protect certain local stocks used for food or bait, or they served to allay fears of local residents that extensive herring fishing might adversely affect local salmon fishing.

A closed season was in effect for many years to protect herring during spawning. Other measures included a 48-hour weekly closed period and an unofficial Christmas closure of 3–4 weeks.

The Pacific herring fishery was one of the first fisheries where catch quotas were instituted. They were introduced in 1936 to halt an apparent decline in herring abundance on the west coast of Vancouver Island and to prevent a similar decline on the lower east coast. Quotas were applied to the middle east coast in 1940 and to the upper east coast, and northern and central areas in 1941. Quotas on the west coast of Vancouver Island were removed in 1946–47. Prior to the 1968 closure of the overall fishery there had been no quotas for the Queen Charlotte Islands area.

Quotas were frequently adjusted during the fishing season. In many areas quotas were increased. Hence, they did not limit catch (Hourston 1980b). Limitations on catch on the west coast of Vancouver Island were initially an experiment to run for 5 years (1936–37 to 1940–41). By the end of the 5 years, there was still considerable uncertainty about the need for quotas. It was decided to retain the existing quotas in modified form.

Because of uncertainty about the effectiveness of catch quotas, the government decided in 1946 to conduct what became known as "the west coast experiment." The original proposal called for the removal of all restrictions (except for closed areas on and adjacent to spawning grounds) for 2 years, followed by complete closure in the third year. This was to apply for 12 years to the whole west coast of Vancouver Island. The scientists involved at that time argued that this approach offered the best opportunity to evaluate quotas in a short time.

Certain segments of the fishing industry objected strongly to the removal of restrictions on herring fishing, while others considered the lack of a fishery every third year a big economic disadvantage. Federal fisheries officials concluded that the proposal was not feasible. A compromise proposal emerged whereby quota restrictions would be removed in all years, but restrictions on the closing date of the season would be retained unchanged. Removal of the quota was expected to result in an increase in fishing effort. In 1946–47, the quota for the west coast of Vancouver Island was removed to compare the effects of unlimited fishing on this stock with those of fishing under quota on the lower east coast stock. No detectable difference was noted in the response of the two stocks. Taylor (1964) concluded that, even with large increases in fishing efficiency after 1946, the absence of quotas did not result in a sufficiently great increase in exploitation to affect recruitment. He concluded that at the level of exploitation occurring in 1960, catch quotas, at least on the west coast of Vancouver Island, were probably not necessary to ensure adequate reproduction. This, in effect, led managers to believe that the level of fishing was unlikely to reduce recruitment of herring stocks.

This reluctance to accept the possibility that herring could be overfished set the scene for the stock collapse of the late 1960's. This conclusion now seems incomprehensible in light of the experience of the intervening two decades with the collapse of pelagic stocks worldwide.

Hourston (1980b) attributed the decline of the B.C. herring stocks in the late 1960's to several factors. Eight of the nine major herring populations experienced sharp declines between 1963 and 1967. For each population, recruitment of the incoming year-class was weak in the year when the decline was initiated. At least two of the preceding and/or succeeding year-classes were also weak. Fishing efficiency had increased more than was realized at the time. Also, stock assessment procedures then in use could not detect the decline in spawning escapements until these were well advanced.

Catches remained high on the series of small year-classes recruited in the mid-1960's. Most of the mature adults and many age one and age two fish were removed by the fishery before the problem of depletion was recognized and the fishery closed.

Following the shock of the late 1960's, the management objective changed from maintaining the catch to preserving the resource, and from maximizing the tonnage caught to maximizing the benefits from the catch. Area-specific optimum target spawning escapements became the basis for management of the fishery with the commencement of the roe fishery in 1972. The runs to the individual fishing grounds were forecast based on average recruitment. Allocations to the food export, local food and bait, and spawn on kelp fisheries were deducted from the forecast run, as were spawning escapement requirements. Any surplus was made available for harvest (Hourston 1980a). Also, fishing capacity was limited through licence limitation and closely controlled openings and closures (with some seine fisheries as brief as 15 minutes). The fishery was opened and closed on the fishing grounds by the local manager. Under these circumstances, the control over the catch was largely limited to controlling the duration of the opening.

This "in season" management system prevailed through the 1970's. As Stocker (1993) has noted, this approach resulted in very large catches during years of strong recruitment, e.g. 1975–1977, and exacerbated biomass declines during periods of low recruitment.

By the early 1980's, the difficulty of determining the "optimum" spawning escapement for any herring population became evident. In 1982, the policy was changed. Effective in 1983, there would be a fixed exploitation rate harvest strategy of 20% of forecast run size. Scientists argued that a constant harvest rate strategy would...."provide less variable yields than fixed escapement policies, and permit informative variation in spawning stock" (Stocker et al. 1983).

This constant harvest rate strategy was combined with a system of area licensing introduced experimentally in 1981. This meant that fixed catch quotas were established annually for each licence area. Under the fixed quota constant harvest rate management system implemented in 1983, the size of herring stocks was estimated before the fishing season, and catch quotas set accordingly. The catch quotas were not subject to revision as the season progressed unless new evidence indicated that the stocks were significantly less abundant than anticipated.

This system combined elements of two regulatory techniques, catch quotas and short open seasons. In 1985 the approach was further modified by the introduction of the CUTOFF policy (Stocker 1993). The CUTOFF policy was intended as a safety measure. CUTOFF was defined as a level of biomass for each stock below which no catch is recommended. The intent was to minimize the risk of reaching very low biomass levels which could reduce reproductive potential. These CUTOFF biomass levels were utilized along the south coast of British Columbia in 1986 and in the Queen Charlotte Islands in 1988.

The management strategy in the late 1980's was based on a fixed harvest rate of 20% with predetermined CUTOFF levels for each stock. Hall et al. (1988) evaluated the alternative management strategies of constant escapement versus constant harvest for the roe herring fishery in the Strait of Georgia. They concluded that the constant escapement strategy provides highest average catches, but increases catch variance. The constant harvest rate strategy appeared to reduce the variance in catch with only a slight decrease in the average catch relative to the fixed escapement strategy. Hall et al. (1988) concluded that maintaining a minimum spawning biomass reserve combines the safety of the constant escapement strategy and the catch variance-reducing feature of the constant harvest rate strategy. Their analysis supported the use of the CUTOFF policy for British Columbia herring.

For more details on management practices in the British Columbia herring fishery, see Stocker (1993).

4.0 PACIFIC SALMON

There has been a large scale fishery for salmon in British Columbia since the late 1800's. The fishery has been based on the five species of Pacific salmon (sockeye [*Oncorhynchus nerka*], pink [*O. gorbuscha*], chum [*O. keta*], coho [*O. kisutch*] and chinook [*O. tshawytscha*]) which frequent British Columbia waters. There are considerable differences among species' life histories. In general, however, these species spawn in freshwater in the autumn and winter. The resulting juveniles migrate to the ocean where they put on most of their growth. The maturing adults normally follow well-defined migration routes back to their streams of origin ('homing'), where they spawn and then die. A "stock" of salmon generally consists of individuals from a particular spawning population, which is relatively distinct genetically (e.g. Ricker 1972). In practice a "stock" of salmon is usually the salmon of one species which inhabit a particular stream (Larkin 1988).

Management of Pacific salmon is complicated by the distinctive life cycles of the five species. Sockeye migrate to sea after spending one to 3 years in a lake. After 3–6 years at sea they return to their natal stream to spawn. Pink salmon, which are the smallest of the five species, live only 2 years. They migrate to sea as fry upon emerging from the gravel in the spring. They return to spawn as 2-year-olds. Coho generally spawn in coastal creeks close to the ocean. They have the simplest life history next to the pink salmon. After 1 year in the nursery stream, they spend 2 years in the ocean. They generally return to spawn in their third year. Chinook are the largest of the five Pacific salmon species and live the longest. They mature between 3 and 7 years of age. Most are 4–5 years old when they return to spawn. They are the least abundant of the five species in B.C. Most go to sea soon after hatching but some remain in freshwater for 1 or 2 years before migrating to the sea. Chum salmon, like the pinks, migrate to the sea almost as soon as they emerge from the gravel as fry. They return to spawn in their third or fourth year.

Pacific salmon migrating upstream to spawn.

The problems of salmon management are exacerbated by the existence of numerous stocks of each of these five species. Healey (1982) indicated that there are about 300 stocks of sockeye, 700 stocks of pink, 970 stocks of coho, 300 stocks of chinook and 880 stocks of chum in British Columbia. He estimated the relative abundance of the five species at that time as follows: sockeye, about 5–12 million, depending largely on the size of the Fraser River run; pink, about 10 million in odd-numbered years and 25 million in even-numbered years; coho, about 4 million; chinook, about 1–1.5 million; chum, ranging from 2–10 million.

Pink salmon, because of their unique life history, show extreme cyclic dominance. Cyclic dominance refers to a persistent temporal pattern of dominant, subdominant and off years. Some stocks of sockeye salmon, e.g. the Adams River stock, also exhibit cyclic dominance among brood lines. One of the most famous is the Adams River stock in the Fraser River which has a 4-year cycle (Fig. 6-3).

Pacific salmon from B.C. migrate great distances in the North Pacific. They generally migrate north during the spring and summer and south in the autumn and winter. Research on the ocean migrations has shown that the B.C. stocks do not migrate far west of 175° W latitude or south of 45° N longitude. Stocks intermix in the open ocean and during their migrations along the coast en route to the spawning grounds.

A mixture of gear types is used to fish these species at various times and places. Three basic gear types are used in the commercial salmon fishery: gill nets, purse seines and trolling gear (hook and line). There is also trolling along the coast by sports fishermen. Gill nets have traditionally been used mainly in or near the mouths of rivers or at the heads of inlets. But gill nets are also used effectively wherever salmon concentrate on their migrations towards the spawning grounds, e.g. the narrow entrances to the Strait of Georgia and in channels between islands. Purse seiners usually fish in straits or passages along migration routes where salmon are concentrated. Commercial vessels which use hook and line generally fish in open waters and sometimes range as far as 50 km offshore. The sports fishery is primarily a near-shore, sheltered water fishery with the Strait of Georgia being a primary centre of sport fishing effort.

In earlier years, fishing took place near the river mouths. With the advent of powered vessels, the fleets became more mobile. Hence the fisheries increasingly focused on mixed stocks in areas of concentration along migration routes.

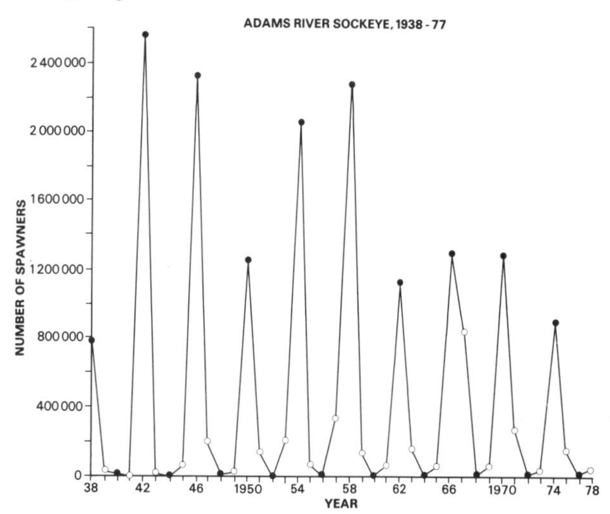

FIG. 6-3. Trends in the Adams River Sockeye stock (from Healey 1982).

In the past, pink, chum and sockeye were considered "net" species because gill-netters and purse seiners fished mainly for these species. The commercial troll fleet fished for coho and chinook, with the sport fishery also targeting coho and chinook. In recent years this pattern has broken down. Trollers also catch the so-called net species, seiners and gill-netters fish coho and chinook, and the sport fishery harvests greater numbers of coho and chinook.

Because of these complexities of salmon biology and fishing practices, regulations devised over the years for the B.C. salmon fisheries have also been complex. This reflects a compromise between biological objectives and social and economic goals. Management of Pacific salmon has focused on allowing some target or optimum number of salmon up river to spawn. The relationship between stock and recruitment is poorly defined (Larkin 1988; Healey 1993). However, it is generally assumed that recruitment is maximum at an intermediate stock size. Maximum sustained yield would occur with an optimum escapement that is commonly from one-third to one-half the equilibrium unfished population (Larkin 1988). Recruitment from any given stock size varies considerably. Thus the common strategy has been to harvest the surplus above the assumed optimum spawning escapement (Larkin 1988).

In the early history of the salmon fishery, managers usually allowed the fishery to continue until a certain quota was achieved. This quota was based on market rather than biological considerations. This resulted in the elimination of certain substocks of "early run" fish (Larkin 1988). Current Pacific salmon management is based largely on a strategy of regulating for an optimum spawning escapement.

Today managers attempt to regulate the harvest to remove the same proportion of fish over the whole time span of the migration of a stock through a fishery. The rate of harvest is estimated as a fishery begins each year and then adjusted as the run proceeds and more data become available. Weekly estimates of CPUE and escapement (by netting upstream) are used to attempt more precise management (Larkin 1988). Fisheries are restricted to only a day or two per week. Even a short fishery can exert a high rate of harvest.

As early as 1890, closed seasons, mesh size, other gear restrictions, vessel limitations and licence fees were introduced in the Fraser River salmon fishery. All these measures were intended to ensure an adequate spawning escapement. In the early 1900's the government granted exclusive rights to a limited number of canners, who were left to manage the fishery. When this did not work, the government intervened with conservation regulations. The Sanford-Evans Commission in the early 1900's recommended that the government stabilize the industry by preventing the excessive use of capital and labour. By the mid-1920's, however, the government had lowered licence fees and removed licence restrictions on canneries.

Following a disastrous year in 1927, attributed by some to additional fishing effort entering the fishery as a result of decontrol, the government imposed more restrictive area and season closures. Lyons (1969) reported that in 1927 the weekly restricted periods were extended. Salmon fishing areas were closed for at least seven consecutive days during the runs, and the fishing season closed earlier than usual. Regulations became increasingly restrictive in 1929, 1930 and subsequent years.

From the early 1920's onward area and season closures were used both to allow adequate spawning escapement and to allocate the catch among gear types. By the early 1950's, season and area closures were intended to constrain increasing fishing effort. In the Strait of Juan de Fuca the average days per week fished decreased from 5½ in 1951 to 3½ in 1959 (MacDonald 1981).

Increases in fishing pressure were making it difficult to achieve adequate escapement. Fishermen were also suffering financially. These considerations led to a study by Dr. Sol Sinclair in 1958–60 (Sinclair 1960). He recommended restrictive licensing. These measures were included in the Davis Plan of 1969 (see Chapter 7).

The primary mechanism to promote optimum spawning escapement for particular salmon stocks has continued to be closure of the fishery for a specific time period in a specific area. Fraser (1979) observed that over the years closed seasons, closed areas and restrictions on more efficient gear were all used to restrict the catch to a biologically sustainable maximum. Argue et al. (1983) detailed actions from 1950 to 1979 for the Strait of Georgia chinook and coho fishery. These included reductions in the daily sport bag limit in tidal waters, increases in the sport size limit, and shortening the open season for the commercial troll fishery. Regulations prohibited both sport and commercial trolling in certain areas. Allowable fishing time for the commercial net fishery was also significantly reduced.

Given the complexity of the B.C. salmon fishery, the detailed evolution of these various regulatory techniques is beyond the scope of the present chapter. Healey (1993) discusses the management of various components of the B.C. salmon fisheries. McDonald (1981) observed that regulatory actions over the years may have reduced fishing rates, at least in the short term, but this did not prevent the shift of effort and catch from terminal fishing areas to areas where more extensive mixing of stocks occurs. He stated: "The large number of stocks involved, together with intensive fishing on

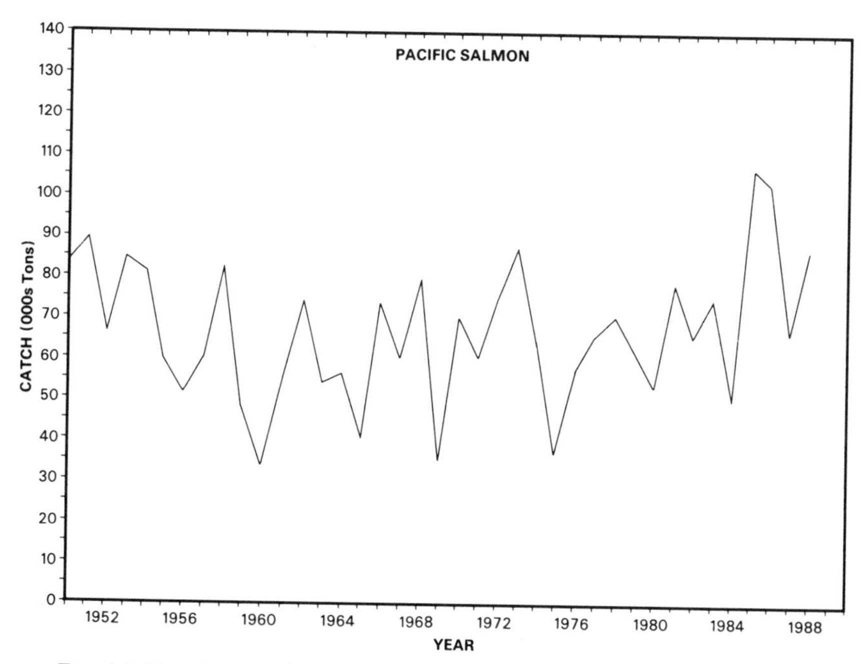

FIG. 6-4. Trends in catches in the British Columbia salmon fishery.

mixed stocks, has largely precluded the development of useful stock-recruitment relationships as a basis for forecasting abundance and for setting escapement goals. The catching power of the fleet has become so great and fishing time so short that the provision of hoped-for escapements is as much good luck as good regulation."

In a study for the Economic Council of Canada, MacDonald (1981) concluded that the relative constancy of salmon landings through the years despite intensive fishing pressure indicated that existing measures to conserve the resource had achieved some success.

Pearse (1982) reviewed stock trends for B.C. salmon and concluded: "First, in the aggregate our salmon stocks are well below their original levels of abundance. Second, while in the last two decades the overall decline has been arrested and for many stocks declines have been reversed (due mainly to improved fishery management), some stocks appear to be declining still. I am particularly apprehensive about the condition of many chinook and coho stocks. Third, the immediate cause of continuing declines and low levels of abundance is overfishing. And, finally, salmon stocks can undoubtedly be rebuilt substantially through better management, more careful regulation of catches and enhancement."

Pearse's observations were made in the context of low catches overall during the late 1970's and a major problem of fleet overcapacity. This generated proposals for fleet reduction to better match capacity to the available resource (see Chapter 8). By the mid-1980's overall catches of salmon in British Columbia reached record levels of more than 100,000 tons. The average catch through the 1980's was higher than normal (Fig. 6-4). Commercial landings in the 1930's averaged just under 20 million fish and in the 1940's and 1950's just over 20 million. They increased to an annual average of 23 million in the 1960's, 22 million in the 1970's and 26 million in the 1980's.

Some stocks of Pacific salmon have been eliminated. However, in general through the 1980's overall abundance was high. Salmon enhancement has contributed to this increase in productivity. But the relative contribution of enhancement efforts and natural recruitment fluctuation is unclear. Also, the overall higher catches have masked declining trends in the stocks of certain species, particularly chinook. Tougher conservation measures were introduced in the mid-1980's to arrest the decline of the major chinook stocks (see Chapter 7).

Overall, management of British Columbia salmon has maintained and, in some instances, enhanced the resource. However, the success of biological management has been uneven. The benefits have been dissipated by excessive fishing capacity.

Several authors have questioned current resource management practices. Larkin (1988) noted that errors in counting and recording escapements are large enough to preclude accurate estimates of production and trends in stock size. Walters and his colleagues have pressed the case for "adaptive management". This would involve deliberately manipulating catch and escapement to evaluate alternative harvest strategies (e.g. Walters 1986).

Historical harvest strategies have also been questioned in relation to cyclic dominance in sockeye salmon (Collie et al. 1990; Welch and Noakes 1990). These recent studies have examined harvest strategies and escapement for Fraser River sockeye. Because most Fraser River sockeye mature as 4 year olds, there are four relatively discrete populations in each stock. In the Adams River tributary of the Fraser, the dominant run is nearly 500 times larger than the off year runs. Because scientists have been unable to identify the cause of cyclic dominance, managers have tried to maintain this cyclical pattern (Welch and Noakes 1990).

There have been various suggestions, but no clear explanation, of the causes of this phenomenon. One suggestion is that negative interactions between different runs perpetuate it (Ricker 1950; Ward and Larkin 1964; Larkin 1971). Another is that higher harvest rates in years of weak returns helped create and maintain the cycle (Walters and Staley 1987). In the late 1980's a Fraser River Cyclic Dominance Working Group examined how to increase the total yield of Fraser sockeye by increasing escapement levels.

Welch and Noakes (1990) examined cyclic dominance and optimal escapement of Adams River sockeye. They concluded that, regardless of the degree of interaction between years, the best policy is always to equalize escapement. They suggested that an equal escapement policy could increase yields by at least 35% over that obtained by the current cyclic escapement pattern.

Collie et al. (1990) investigated harvest strategies for rebuilding the less abundant stocks of Fraser River sockeye. They suggested that the only definitive test of the mechanisms causing cyclic dominance would be an experiment to increase abundance of the small runs in cyclic populations. They concluded that stocks exploited together cannot be managed for stock-specific escapement goals. In the case of Fraser River sockeye, a fixed harvest rate rather than a fixed escapement policy appeared optimal. They concluded that 60–70% harvest rates, compared with the existing average of 80%, should greatly increase yield regardless of uncertainties about the appropriate population model. More experimentation appeared necessary to determine whether off-year populations would rebuild with reduced exploitation.

At the end of the 1980's fisheries managers began to experiment with variations on the traditional optimum spawning escapement policy. More general large-scale adaptive management experiments for Pacific salmon appear unlikely because of the potentially significant disruption of traditional harvest patterns.

5.0 ATLANTIC LOBSTERS

5.1 General

The lobster (*Homarus americanus*) fishery is the most important inshore fishery in Atlantic Canada. In 1988 Canadian landings totalled 41,000 tons with a value of $283 million. In landed value lobster is second only to salmon in Canada.

Canadian landings peaked at 47,600 tons in 1886 and reached a low of 12,200 tons in 1924. Secondary peaks in landings occurred at about 22,000 tons in 1924 and the 1950's. Landings rose substantially during the

Atlantic Lobster.

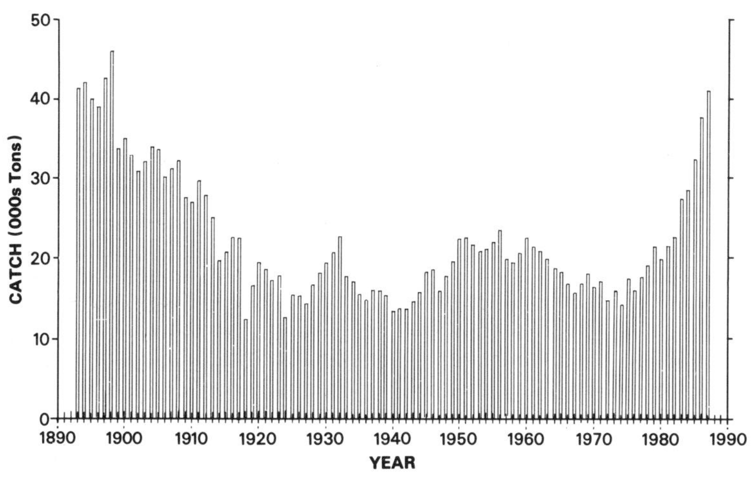

FIG. 6-5. Trends in Atlantic Lobster landings (from Rivard et al. 1988).

1980's from 20,000 tons in 1980 to 47,000 tons in 1990, the highest level of landings since the late 1800's (Fig. 6-5). While some of the upswing can be attributed to increased fishing effort, there has been a recent general increase in recruitment to the commercial stocks (Rivard et al. 1988).

Thus there has been an intensive lobster fishery in Atlantic Canada since about 1870. Landings increased rapidly in the early years of the fishery because the fishery was being prosecuted on previously unexploited stocks of large, old lobsters. In 1873, lobsters in New Brunswick and Nova Scotia averaged 1.1 to 1.4 kg. -two to four times as large as the average run today. The average size declined rapidly with expansion of the fishery. This decline gradually levelled off, and there has been no significant change in the size of lobsters caught in recent decades.

The lobster has two parts — the "body" which consists of the combined head and thorax covered by an unjointed shell, and an abdomen in six sections. The latter is commonly called the "tail." Lobsters grow by moulting, which involves the complete replacement of the shell. Lobsters moult as many as five or six times in their first season and three or four in their second. The time between moults increases as they grow. As lobsters approach commercial size, moulting occurs only once or twice a year. In some areas where the water is very cold commercial-sized lobsters may go two years between moults (Scarratt 1979). The size at maturity differs considerably among areas. Lobsters in the Gulf of St. Lawrence mature at about 250 grams compared with about 700 grams in the Bay of Fundy. This difference appears to be related to differences in summer water temperature.

These biological characteristics of Atlantic lobsters have influenced the regulatory regime. The lobster fishery has been one of the most closely regulated fisheries in Canada. Regulations were first imposed during the development of the fishery in the late 1800's. A wide array of regulatory techniques has been applied over the past century. Major regulatory measures include:

(1) Restriction of gear type;
(2) Minimum size limits;
(3) Closure of the season during moulting and mating periods;
(4) Limitation of the length of the season;
(5) Prohibition against taking egg-bearing female lobsters;
(6) Prescription of lath spacing in traps (to permit escape of small lobsters);
(7) Division of coastal areas into fishing districts;
(8) Allocation of fishing seasons by districts;
(9) Licensing of fishermen;
(10) Restriction of fishermen and their boats to one district in a season; and
(11) Limitation of the number of traps.

Limited entry licensing was introduced in the late 1960's and Licence Buy-Back programs were implemented to reduce fishing effort. In this chapter I will focus on the conservation-related regulations, leaving the discussion of the impact of the limited entry licensing to Chapter 8.

Lobster management is complicated by the division of the Atlantic coast into 25 lobster fishing districts. The specifics of the regulatory regime vary considerably among these districts.

Wilder (1965) described three phases in the fishery since its development. The first phase involved rapid expansion with the discovery of new areas, increases in the fleet and a steady increase in landings to about 47,600 tons in 1886. The second phase lasted until about 1918. It saw a steady decline in catch to about 12,000 tons with a significant decline in the average size of lobsters caught. From 1918 to the mid-1960's, Wilder considered the annual landings to be relatively constant, averaging 18,000 tons and ranging from 13,500 to 22,500 tons. He categorized this period as phase three, during which the fishery and stock remained essentially in balance. This phase lasted until the early 1970's (Fig. 6-5), following which there has been a steady upswing in landings to a high of 47,000 tons in 1990. This upsurge in the 1980's constitutes phase four in the history of the fishery.

The first regulations for the lobster fishery, established in 1873, prohibited capture of soft-shelled lobsters, egg-bearing females and lobsters under 1½ pounds (DeWolf 1974). Prohibiting capture of soft-shelled lobsters was a quality control rather than a conservation measure. This was replaced by a closed season shortly thereafter, and the weight limit was changed to a minimum length limit (9-inch total length). The "berried" lobsters regulation was retained. These measures — closed/open seasons, minimum size limit and prohibiting the retention of egg-bearing females — were categorized by Wilder (1958) as the three most important regulations to appear up to that time. They are major elements of the lobster fishery regulatory regime even today.

During the early period of the fishery the closure and size regulations were rarely enforced or obeyed (Scott and Tugwell 1981). The decline in lobster landings in the late 1800's and increases in efficiency around 1900 generated interest in more stringent measures to conserve the lobster stocks.

Scott and Tugwell (1981) described the evolution of the lobster regulatory regime:

> "Nine commissions were appointed in the fishery between 1887 and 1927. As a result of these commissions' investigations, regulations on the boundaries of fishing districts, lobster size, closed seasons and trap construction were variously modified, rescinded, reintroduced and further modified. More significantly, perhaps, lobster fishermen were required in 1918 for the first time to obtain licenses before they were permitted to fish, and in 1934 this restriction was extended by the introduction of a regulation prohibiting fishermen from fishing lobsters in more than one district in any one year. Size limits, fishing districts and closed seasons have been adjusted slightly in the past 40 years but the major changes in lobster fishing regulations have been in the area of factor input limitation, that is, gear, vessel and fisherman restriction. A major instance was that of 1945, when capital, vessel and gear mobility were drastically restricted by the introduction of a regulation which stated that no one could use in lobster fishing any boats, traps or other lobster fishing equipment that had been used during that year in lobster fishing operations in any other lobster district.
>
> "Later, enforcement difficulties caused that part of the regulation referring to traps and other equipment to be rescinded in 1959. Again in the 1960's, license and trap limitations were introduced with the aim of controlling fishing pressure and so increasing the net incomes of fishermen."

In the following sections I examine the impact and relative effectiveness of the main regulatory techniques.

5.2 Closed Seasons

The purpose for introducing closed seasons in the lobster fishery was not clearly stated. Wilder (1958, 1965) suggested that closed seasons were introduced to protect lobsters during the egg-laying, moulting and hatching periods. Later, seasons were adjusted to reduce fishing effort and so arrest a steady decline in landings during the early 1900's. Length and timing of the seasons were adjusted for marketing reasons and to permit fishermen to engage in other fisheries or other occupations. Scott and Tugwell (1981) argued that the closed seasons were introduced to prevent capture and canning of lobsters in poor condition and to prevent fishing in bad weather.

In any event, it appears that the fishery has simply adjusted to the shorter seasons by employing more men, boats and gear. Hence, the closed seasons have not reduced the overall rate of exploitation. A comparison of exploitation rates in long-season and short-season districts by Rutherford et al. (1957) indicated that short seasons had not limited catches. The rate of exploitation was often much higher in the areas with the shorter seasons.

There have been other secondary effects of the seasonal closures. Wilder (1958) contended that closed seasons had improved the quality of the lobsters. Less than 20% were caught from July to September when they are soft-shelled, difficult to hold and produce a lower meat yield. Rutherford et al. (1957) pointed out that existing seasonal closures could be justified for adjusting production to exploit demand and supply conditions in the U.S. market. The May–July peak in Canadian landings precedes the August–October peak in the Maine lobster catch. Winter production in southwest Nova Scotia occurs at a time when supplies are low and market prices high.

Enforcing lobster seasons has been costly and in some areas extremely difficult (Wilder 1965). Wilder (1965) noted that these seasons had been in effect for many years. Hence, any sudden major changes would have serious effects not only on the primary industry but on processing, storage, distribution and marketing. Earlier he had suggested that a feasible alternative would be a universal closed season from July to September. This would avoid the period of heavy U.S. production and would limit the sale of newly-moulted lobsters. Under such a system landings would be concentrated in the early fall. Scott and Tugwell (1981) supported Wilder's proposal of a universal open season between October and June.

5.3 Protection of Egg-Bearing Females

The retention and sale of egg-bearing females has been prohibited by regulation since 1873. Current regulations state that "no person shall at any time fish for, sell or have in possession any lobster with eggs attached." In the early years this regulation was largely ignored. Hatcheries were established in an attempt to ensure continued productivity of the lobster stocks. Following closure of the hatcheries in 1917, there was more emphasis on protecting egg-bearing females.

Scarratt (1964) showed that the relative abundance of first larvae in Northumberland Strait varies from year to year by about 4:1. There is great variability in the survival during the free-swimming stage, with the numbers reaching the fourth larval stage varying about 40:1. The relationship between the number of egg-bearing lobsters, the number of larvae surviving between the first and fourth stages, and the number of recruits to the fishery, is unclear. Wilder (1965) concluded that the abundance of egg-bearing females is probably not an important factor in limiting the commercial production which is determined largely by factors other than the abundance of egg-bearing females. That conclusion was based upon a comparison of the proportion of egg-bearing females in the southern Gulf and in southwest Nova Scotia in relation to the productivity of the fishery. At that time the relationship between offshore lobsters and larval drift to inshore grounds in southwest Nova Scotia was not understood. Pringle et al. (1983) concluded that most lobster stocks are recruitment overfished. Exploitation rates are between 70% and 90% and minimum legal carapace length is at or below the size at reproductive maturity. Some stocks have collapsed with catches at 5% of maximum level. Thus it seemed prudent to maintain the regulation protecting egg-bearing females. Most fishermen seem to support this regulation (DFO 1989b).

5.4 Trap Lath Spacing

The rationale for trap lath spacing regulations is similar to that for mesh size regulations. The intent is to release small (sublegal size) lobsters while the lobster traps are still on the fishing grounds. These should survive and grow to a larger size before capture, thus increasing the yield-per-recruit.

Experiments in the 1940's demonstrated that traps with 1⅜ inches (35mm) spaces between the two lowest side laths would permit the escape of up to 90% of lobsters below 7 inches total length without reducing the total catch of legal-sized lobsters (Wilder 1965). In areas with a 3⅛ inch carapace length minimum size limit, 1¾ inch spaces allowed 75% of the sublegal lobsters to escape without reducing the catch of legal-size lobsters.

Regulations requiring 1¼ inch and 1⅝ inch spaces came into effect in 1949 in the Maritime Provinces. Reaction to the regulations was mixed. Some fishermen were enthusiastic. Others were reluctant to alter their traps. Enforcement became extremely difficult because of lack of proof of ownership and alteration of traps at sea. These factors led to the rescinding of the trap lath spacing regulations in 1955 in the Maritime Provinces and Quebec. In Newfoundland regulations still require a 1 3/16 inch spacing.

Anthony and Caddy (1980) reported that studies in many U.S. states had shown that proper lath spacing would decrease handling damage to pre-recruits and increase the catch per unit effort of legal lobsters. The U.S. State-Federal Scientific Committee had recommended the use of such spacing in all traps.

More recently, escape mechanisms have been introduced for lobster traps in the southern Gulf of St. Lawrence. The escape mechanisms are plastic laths with circular or rectangular openings which allow undersized lobsters to escape, reducing the capture and handling of so-called "short" lobsters.

5.5 Minimum Size Limits

Minimum size limits for lobster have been in effect since the early years of the fishery. Early attempts to enforce such limits were unsuccessful and most of the regulations, except those in southern New Brunswick and Newfoundland, were abandoned. In 1932 a 9-inch total length limit was adopted in southern Nova Scotia. This was changed to a 3 3/16 inch carapace length limit in 1934 and a similar limit was applied in Newfoundland in 1939. In other lobster areas size limits were reintroduced in the early 1940's, with the carapace method of measurement gradually being adopted in all areas. The size limits range from 2½ inches to 3³⁄₁₆ inches (63.5 to 89 mm).

The minimum size limit regulation is intended to increase the size (age) at first capture and thus increase the yield per recruit. The "canner" industry in the southern Gulf depends on relatively small lobsters. There has been considerable debate over the years about the impact of size limits and in recent years the merits of increasing the existing minimum size limits. Wilder (1965) attempted to assess the impact of minimum size limits upon lobster landings and concluded:

> "(This) review has failed to provide consistent, convincing evidence that minimum size limits have improved landings significantly. It seems entirely possible that the greater stock associated with large, well-enforced size limits reduces growth and survival to the point where no improvement in yield results. There is, however, no indication that size limits actually reduce sustained yields. Where larger lobsters are worth appreciably more per pound, the greater value of the catch may more than balance the cost of enforcing appropriate size limits."

In an earlier paper, Wilder (1958) had concluded:

> "These observations provide strong support for minimum size limits.... Wherever size limits have been observed the commercial catch has improved. On the basis of the available evidence it is concluded that size limits are the most effective means of increasing the yield from the fishery."

Scarratt (1979) observed that:

> "Most areas would probably benefit by an increase in the minimum size limit so that all lobsters could breed at least once before they are caught."

In the 1970's biological research indicated that many local lobster stocks were growth overfished with exploitation rates between 70 and 90% (Anthony and Caddy 1980). A 1978 Canada-U.S. workshop on the Status of Assessment Science for N.W. Atlantic Lobster Stocks (Anthony and Caddy 1980) concluded that the legal minimum sizes for all stocks were at a level well below that producing maximum yield-per-recruit.

Pringle et al. (1983) observed that in some Canadian lobster districts the minimum legal size was below the smallest size at which females lay eggs. With the very high exploitation rates common in lobster fisheries recruitment could fail. They recommended increasing the minimum carapace lengths to 3 inches for warm-water stocks and 3.5 to 4 inches for cold-water stocks.

In the late 1980's-early 1990's a major dispute occurred between Canada and the United States resulting from U.S. efforts to prohibit the import of lobsters below the U.S. minimum size limit (see Chapter 13). The 1987 American Lobster Fishery Management Plan provided for increases in the U.S. minimum size limit to 3¼ inches in 1989, 3⁹⁄₃₂ inches in 1991 and 3⁵⁄₁₆ inches in 1992.

In November 1989, the U.S. Congress adopted an amendment to the Magnuson Fishery Conservation and Management Act. The amendment would prohibit from interstate commerce lobsters and lobster products smaller than the minimum possession limit in effect under the American Lobster Fishery Management Plan.

Canada initiated a Chapter 18 complaint in December 1989 under the Canada-U.S. Free Trade Agreement (FTA) against the U.S. restriction. A dispute settlement panel was established. The Panel submitted its report in May, 1990, ruling that the U.S. measure was not an unfair trade restriction. Following this ruling, members of the two countries' lobster industries attempted to reach a negotiated settlement. An industry-to-industry understanding was reached in July 1990, followed by a tentative agreement between officials of the two governments.

This agreement would have required Canada to increase its minimum lobster size in all areas of Atlantic Canada, except the Gulf of St. Lawrence, to the same level as that in U.S. federal jurisdiction. In return, the U.S. would have agreed to suspend for 3 years further scheduled increases in its minimum size requirement.

The fishing industry was split on the proposal. Because of strong opposition from Canadian lobster fishermen who would be most seriously affected by the size increase, the Canadian government decided that the agreement was not in Canada's interest.

The U.S. size limit increased to 3⁹⁄₃₂ on January 1, 1991, in accordance with the original plan. In mid-January the New England Fishery Management Council decided to ask the Secretary of Commerce to suspend the schedule of size increases beyond 3¼ inches, i.e. that in effect in 1990.

At the same time, the New England Council indicated its intention to amend the existing FMP to specify an "optimum target level of effort" for the fishery and management measures to achieve that target. This represented a move toward the more comprehensive Canadian management regime. Licensing and limitation of the number of traps per boat were introduced in the Maritimes lobster fishery in the late 1960's. These were accompanied by a Lobster Licence Buy-Back Program in the 1970's (see Chapter 8).

Scientific studies have argued for increasing the minimum size limit to conserve Atlantic lobsters (e.g. Campbell 1985, 1990; Fogarty and Idoine 1988). Direct reduction of fishing mortality is, however, favoured where possible.

The widespread increase in Atlantic lobster abundance during the 1980's is more likely due to favourable environmental factors than the management regime. A study of the Scotia-Fundy lobster fishery concluded that the increase in landings during the 1980's appeared "not to be directly attributable to management changes during the late 1970's and early 1980's, though they are no doubt important in sustaining both the recovery and higher landings." This same study concluded that "the minimum carapace size restrictions are the most important regulations for protecting lobster stocks and for maximizing the total yield in both weight and value from the fishery" (DFO 1989b).

The recent dispute with the U.S. makes it uncertain whether an increase in the minimum size would have long-term benefits. Opposition to the U.S. initiative was based on the belief that Canada should be free to choose management measures independent of the whims of another country. The recruitment pulse of the 1980's is not likely to be sustainable in the long term. The effectiveness of the current management regime will be tested again should stock abundance return to more normal levels.

6.0 ATLANTIC GROUNDFISH

The past two decades have seen sweeping changes in management of the Canadian Atlantic fisheries. Little has been documented about the major changes in the management of one of Canada's most important fisheries — the Atlantic groundfish fishery. This fishery accounts for 60% of catches and 45% of the landed value of fish from the Canadian Atlantic. Nationally, Atlantic groundfish accounted for 55% of catches and 37% of landed value in 1988.

Groundfish have been fished off the Atlantic coast of Canada since the time of John Cabot. Indeed, pursuit of the cod fishery was one of the primary factors leading to European settlement of the maritime regions of Atlantic Canada. Groundfish are demersal or bottom-dwelling fish including cod, haddock, redfish and flounders. They are fished by several types of fishing vessels and gears. Approximately 40–45% of the catch is by otter trawling involving large modern stern trawlers fishing offshore. The chief species, cod, is also taken in large quantities in cod traps in summer when it migrates inshore.

The groundfish fishery is really a complex of fisheries. It has waxed and waned over the years in response to natural fluctuations in abundance and availability (e.g. inshore cod migrations) and major external factors (e.g. market demand, world wars). The decades since the Second World War can be divided into two distinct periods. The first involved post-war expansion from about 1945 to 1970. This was fuelled by the Bates' Report of 1944 (Bates 1944). That report concluded that under-development in the fishing industry was the result of trawler restrictions in the 1930's and undercapitalization in both the primary and secondary sectors. During the 1950's and 1960's governments in Canada pursued aggressively modernization of the fishing industry. The second phase which commenced in the early 1970's involved major regulatory intervention to conserve fish stocks and limit fishing capacity.

During the 1960's there was a massive build-up of foreign fishing off the Atlantic coast by the USSR, Spain, Portugal and other countries. This led to over-

Atlantic cod, conerstone of the groundfish industry.

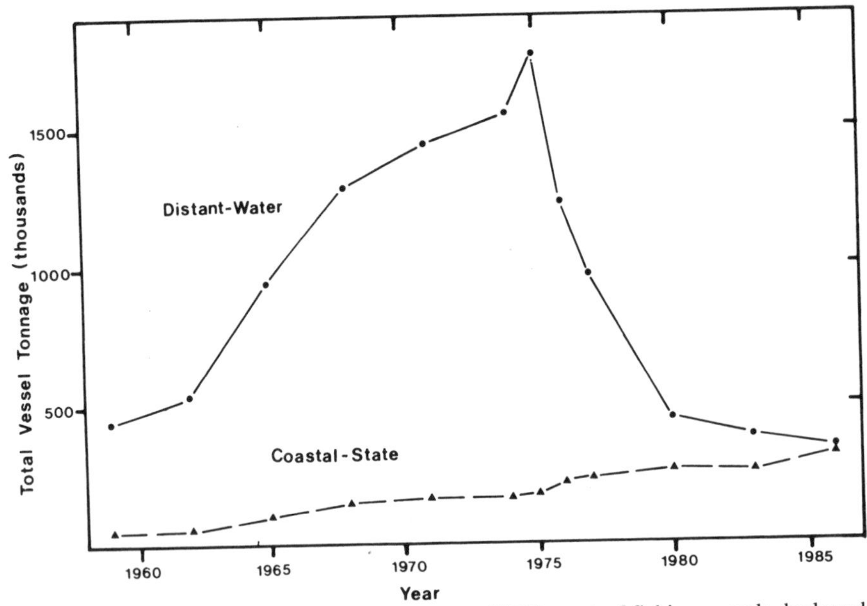

FIG. 6-6. Aggregate gross registered tonnage (GRT, tons) of fishing vessels deployed in the Northwest Atlantic by distant-water and coastal-state fleets (from Pinhorn and Halliday 1990).

fishing of the groundfish stocks and a decline in catch rates and economic returns from the fishery. This occurred when Canadian companies were building up their catching capacity, using the modern stern trawler. A temporary market downturn in 1968 was followed in 1973–74 by a more severe crisis in the groundfish fishery due to a combination of resource decline and market factors. This crisis increased domestic pressures to extend Canadian fisheries jurisdiction to encompass all fish stocks upon which Canadian fishermen depended.

Modern management of the Atlantic groundfish fishery began with the establishment of the International Commission for the Northwest Atlantic Fisheries (ICNAF) in 1949. Prior to that there had been restrictions within Canadian waters on types of fishing gear (e.g. the prohibition of trawlers) and local closed areas had been used to minimize gear conflicts among gear types, but there had been no real attempt to manage the groundfish resource as a whole. Indeed, given the distribution of that resource, it would not have been possible for Canada to play an active management role without international cooperation. At the time ICNAF was established, there was concern in Canada and the United States about the potential impact on North American fishermen of the increase in foreign fishing effort.

In the early days of ICNAF the level of fishing effort was still considered moderate. ICNAF concentrated initially on increasing the size (age) at first capture. In 1953 ICNAF introduced the first mesh size regulations. Some 20 regulations concerned with mesh size and trawl construction limitations were established during the first 15 years of ICNAF (Halliday and Pinhorn 1985). Beverton and Hodder (1962) assessed the impact of minimum mesh size regulations. They noted that the fishing effort in most areas was expected to increase and that the optimum mesh size increases with fishing effort.

Beverton and Hodder (1962) also considered the implications of a minimum size limit. For the main species of the ICNAF area they considered it unlikely that many discarded fish would survive. Thus a minimum size limit would only have conservation value if it served one of the following functions: (a) to discourage fishing on grounds which contain mainly small fish that the Commission wished to protect; or (b) to discourage fishermen from reducing their effective mesh size below that prescribed by regulation. They did not recommend minimum size limits, and ICNAF did not set general minimum size limits for groundfish. Beverton and Hodder (1962) described the fishing intensity on most of the stocks of cod and haddock, up to 1958 at least, as moderate and not as heavy as in some areas of the Northeast Atlantic.

Fishing effort and fishing mortality increased rapidly throughout the 1960's (Fig. 6-6). Thus, even where the minimum mesh sizes were being enforced, the mesh size was below the optimum for the level of fishing effort. Even though the minimum mesh sizes increased through the 1960's and early 1970's, the anticipated benefits of mesh size regulation were not achieved because of the greatly increased fishing effort.

The ICNAF Standing Committee on Research and Statistics (STACRES) realized by the mid-1960's that the amount of fishing would have to be controlled by means other than mesh size regulations. In 1964 STACRES advised the Commission that additional regulatory measures were required to arrest the dramatic expansion of fishing effort (ICNAF 1964b). Templeman and Gulland (1965) outlined the implications of alternative regulatory measures for the Commission. They warned: "For many if not most of the stocks of major importance the amount of fishing has now reached a level such that further increases in fishing will bring little or no increase in catch, and may even reduce the catch.... There must therefore be some direct control of the amount of fishing. All methods of doing that raise difficulties, but that presenting least difficulties is by means of catch quotas. There must be separate quotas for each stock of fish and preferably allocated separately to each section of the industry."

Templeman and Gulland (1965) also laid the groundwork for the shift in management objective to $F_{0.1}$ which occurred a decade later. They observed that it made no sense to try for the last possible pound of catch when 90% of the maximum yield could be taken with perhaps 50% of the maximum effort.

In 1967, the ICNAF Working Group on Joint Biological and Economic Assessment of Conservation Actions reported on assessments of selected cod and haddock stocks: at the high level of fishing intensity reached from 1962 to 1965, the fishing mortality rate was higher than required for the maximum sustainable catch per recruit (ICNAF 1968a). They suggested that the catch could be sustained over the long term, and perhaps even increased by 10%, with a 30 to 40% reduction in the fishing mortality rate. This catch would be accompanied by "a substantial increase in catch per unit fishing intensity, less year-by-year variability in total catches and an increase in the average size of fish in the catches."

The Working Group warned:

> "By 1965 fishing intensity on cod and haddock in the North Atlantic had reached the stage where the cost of harvesting the annual catch was substantially greater than necessary. Unless restrictions are imposed, this situation appears likely to persist and indeed to increase in severity.... The need for taking positive steps toward effort limitation in the North Atlantic demersal fisheries before the problems become still more complex is therefore urgent."

The Working Group compared fishing effort limitation and catch limitation. In the end they echoed the call by Templeman and Gulland (1965) for catch quotas allocated among countries. At the same time STACRES warned that the continuing growth in fishing effort increased the urgency for more effective control measures.

In 1968, STACRES reported that cod in Subarea 1 and haddock in Subarea 5 were "demonstrably overexploited." STACRES could not estimate the level of fishing mortality giving the maximum sustainable yield for most of the ICNAF stocks. However, it indicated the required scientific data were available to set catch quotas for Subarea 1 cod and Subarea 5 haddock. STACRES found no difference in effectiveness between limitation of effort and limitation of catch. However, it preferred limitation of catch to limitation of effort because catch was easier to determine than was effort (ICNAF 1968b). Thus, the Commission was nudged toward catch quotas.

In 1969, STACRES provided specific assessments for Subarea 1 cod and Subarea 5 haddock as a basis for catch quotas. It also warned that it was unlikely that the record catches of cod in the Convention Area in 1968 could be sustained. It drew attention to the escalating catches of herring. For Subarea 5 haddock, it considered closed seasons and/or closed area regulations as alternatives to catch limitation (ICNAF 1969a).

Meanwhile the 18th Annual Meeting of the Commission in 1968 had considered catch quotas. The USA proposed principles for establishing general catch quotas by species and by fishing areas and for allocating quotas among countries. It proposed to: (1) establish overall quotas only for species that were being demonstrably overfished; (2) adjust these quotas annually based on scientific forecasts; (3) allocate only 80% of the overall quota as country quotas, leaving the remainder available to member countries, countries with developing fisheries, or third parties; (4) allocate overall quotas among countries in a fishery in proportion to their average catch during a 10-year period; and (5) give coastal countries a special preference when allocating quotas (ICNAF 1968b).

These proposals were not discussed in detail at the 1968 meeting. They were referred to the Standing Committee on Regulatory Measures (STACREM). In April, 1968 the USA proposed an amendment to the ICNAF Convention to allow for national allocation of catch quotas.

At its 1969 Annual Meeting ICNAF agreed to Total Allowable Catches or TACs for haddock on the southern Scotian Shelf (Division 4X) and Georges Bank (Subarea 5) haddock. These were 18,000 tons and 12,000 tons respectively for 1970, 1971 and 1972 (ICNAF 1969b). These were global TACs in that countries were not allocated shares. Advisory bodies to ICNAF believed that a system of catch quotas would only work if accompanied by national allocations. The original ICNAF Convention permitted only overall catch limits. ICNAF, in 1969, adopted a proposal to amend the Convention to allow national allocations of catch quotas. The amendment also modified the MSY objective of the original Convention to allow for "appropriate proposals... designed to achieve the optimum utilization of the stocks..." This amendment came into effect in December, 1971.

At its 1971 meeting STACRES advised that:

1. The catch of harp seals was far above the sustainable yield;
2. Immediate action was needed to limit catches on all cod stocks in the ICNAF area;
3. Haddock stocks in Subareas 4 and 5 would probably continue to decline without more severe limitations; and
4. All herring stocks had decreased in abundance in the last year, and the Georges Bank stock was extremely low. Action should be taken to reduce catches and rebuild the stocks (ICNAF 1971).

The 1970 and 1971 Commission meetings focused on means for acceptable national allocations. There was no action on national quotas for the main commercial species. Immediately after the implementation of the Protocol amending the Convention, ICNAF held an extraordinary (Special) meeting in February 1972 about the herring fisheries. The Commission adopted global quotas and national allocations for herring in Subareas 4 and 5 (ICNAF 1972b). At the 1972 Annual Meeting ICNAF adopted catch quotas for most of the major groundfish stocks in the Canadian Atlantic effective in 1973 (ICNAF 1972c). At the 1973 Annual Meeting TACs were established for all groundfish stocks effective in 1974 (ICNAF 1973b).

Needler (1979), who was deeply involved as Head of the Canadian delegation in the negotiations leading to this breakthrough, described these events:

> "It was quickly agreed that if a total allowable catch is established for a fish stock, each country taking part in the fishery must have a separate national allocation, otherwise there would be a mad rush, with each country trying to maximize its share, leading to chaos and inefficiency in all stages of the fishery from catching to marketing. Delay resulted from the need to amend the Convention to give ICNAF a broad enough mandate to recommend national quotas but ICNAF moved quickly after receiving the authority in December 1971. Within 6 weeks TACs and national allocations were established for three major herring stocks, within 6 months a score of groundfish stocks were added, and a year later similar action was extended to most of the major stocks supporting international fisheries in the ICNAF area. This constituted a major ICNAF accomplishment and a first in international fisheries management. Thus was born the regime under which each year TACs and national allocations are set for the succeeding year."

The TACs established by ICNAF were largely based on STACRES' advice. The reference point for establishing these TACs was fishing at F_{max} or F_{msy}, depending whether the stock was assessed with a yield-per-recruit or a general production model. Individual countries were to enforce national allocations set by ICNAF. This enforcement was inadequate.

Catch rates, reflecting stock abundance, continued to decline dramatically from 1970 to 1975. They increased only slightly from 1975 to 1977 (Fig. 6-7). Thus it appeared the management system could not arrest the overall decline in stocks. A combination of factors may have contributed to this situation. Stock assessments were limited by insufficient data, the inaccuracy of catch reporting and time lags in obtaining data. Also, the use of MSY or F_{max} as the reference point for calculating TACs did not provide a sufficient safeguard against errors in assessment resulting from these factors.

Both Canada and the USA became convinced that single species catch quotas did not provide an adequate basis for managing the complex groundfish fisheries off the eastern seaboard of North America. At the January 1973 meeting of ICNAF the US proposed fishing effort allocations by country to put an overall limit on exploitation in the area. In June 1974, ICNAF adopted a two-tier quota system for Subareas 5 and 6 with a global catch quota which was generally 20% lower than the sum of the TACs for the individual stocks (ICNAF 1974a). The stated objectives of the two-tier quota system were:

(a) to compensate for bycatch mortality which was difficult to quantify and control by more direct means;
(b) to account for species interactions which were not satisfactorily taken into account in single species stock assessments; and
(c) to allow for recovery of the total biomass from the reduced level in the 1960's to a level giving the maximal or some optimal yield in a relatively short time frame.

O'Boyle (1985) has described the methodology of the two-tier system. Unfortunately the system was in effect for too short a time to determine its effectiveness. It was abandoned when the U.S. extended fisheries jurisdiction in 1977. By 1982 the U.S., under its new Regional

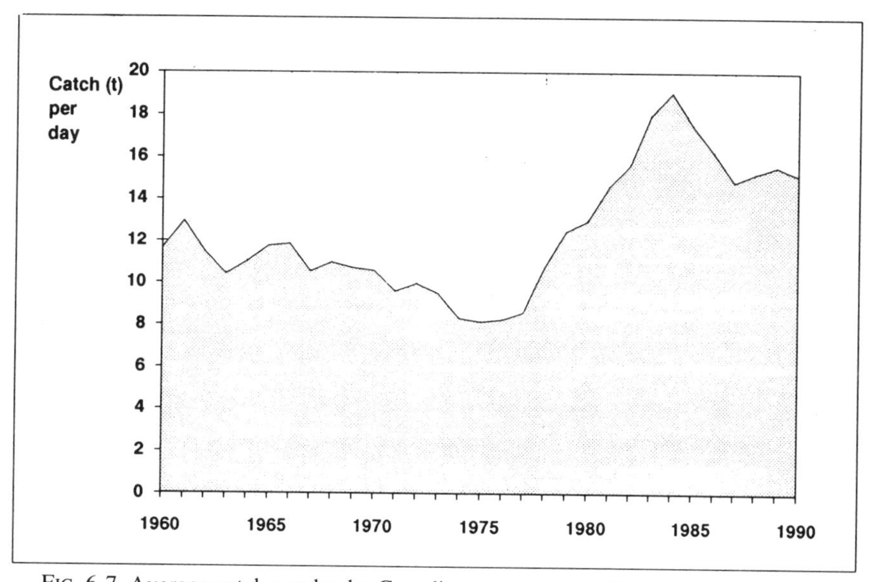

FIG. 6-7. Average catch per day by Canadian otter trawlers (from D. Rivard, Fisheries Research Branch, DFO, Ottawa).

Council Management System, had dropped even stock-specific TACs and reverted to the more indirect regulatory techniques (closed area/seasons, mesh size and minimum landed sizes) (see Chapter 20).

Canada adopted a different approach, seeking a combined catch and effort regulation. Canada first proposed a 40% reduction in fishing effort by noncoastal states in ICNAF in 1975 (ICNAF 1975c). ICNAF adopted 40% reduction in effort combined with stock-specific quotas in 1976 (ICNAF 1976a). The chief argument against this approach was that it could lead to substantial underutilization (e.g. Garrod 1975a).

If management action must target individual species, then the fishing activity must be directed. This can rarely be achieved through an effort regulation. Thus individual species and stocks must be protected by catch quotas. The individual species catch quota was not sufficient by itself in a mixed fishery. The real choice was whether an overall resource quota was set in terms of catch or effort. Garrod (1975a) argued: "The advantages of the overall catch limit lie in the negotiation of allocations and avoidance of the inequities that will be associated with fishing effort control due either to trends in fishing efficiency or incompatibility with species catch allocations; it would be extremely difficult to match the effort allocation to the catch allocation with any confidence or equity of both regulations for all countries....It would appear therefore that so far as the biological objectives of management are concerned, and, given adequate enforcement, neither overall catch nor effort regulation has an indisputable advantage, but the former has fewer disadvantages."

Garrod's reservations about a combined catch quota and effort regulation system may have been pertinent in the pre-200 mile limit era but held little relevance for coastal states following extension of fisheries jurisdiction. In 1976 for the area off the Canadian Atlantic coast, ICNAF adopted an overall effort limitation with a reduction of fishing effort by noncoastal states by 40% from that observed in 1972–73, superimposed on individual stock quotas (ICNAF 1976a). In 1977, Canada implemented within its 200-mile zone a system of licensing-effort control superimposed on individual stock quotas for foreign fishing. Based on the most recent catch rate information, analyses were made of the level of effort required to take national allocations of particular species quotas. Control was exercised through licensing of individual vessels and specifying the number of days permitted on ground. The chief reason for this approach as opposed to the two-tier catch quota system was the ease of enforcing such controls on number of vessels and days on the fishing grounds. Such an approach provided a safeguard against quota violations which were not present in the two-tier catch quota system.

Another major event in the last days of ICNAF was a shift in management approach away from the pursuit of Maximum Sustainable Yield to a more conservative approach to management, using $F_{0.1}$ as the reference point for setting TACs rather than F_{max} or F_{msy}. This shift occurred at a Special STACRES meeting in September 1975 and the STACRES Annual Meeting in 1976. At the 1975 meeting STACRES had identified the principal deficiencies of F_{max} as a basis for management (see Chapter 4) and advised TACs for some stocks at the

$F_{0.1}$ level (ICNAF 1976b). At the 1976 meeting, STACRES went further and proposed that the TACs for 1977 should be set to achieve $F_{0.1}$ rather than F_{max}. As pointed out previously, $F_{0.1}$ was an attractive alternative to F_{max} because it promised in the long-term 85-90% of the yield-per-recruit at F_{max} but required only about two-thirds of the fishing effort to catch it. The benefits envisaged from fishing at $F_{0.1}$ included:

1. Larger stock sizes and hence increased catch rates;
2. More age groups in the stock, reducing the probability of errors in TACs and the need to adjust TACs from year to year because of year-class fluctuations;
3. Larger average size of fish; and
4. Larger spawning biomass, increasing the probability of more frequent good year-classes (ICNAF 1976c).

Because of the perceived conservation and economic benefits, Canadian fisheries managers quickly accepted the strategy of fishing below F_{max}. They proposed in 1975 that ICNAF adopt such a strategy. At the 1976 ICNAF meeting Canada secured agreement on TACs at the $F_{0.1}$ level (ICNAF 1976a). This significantly reduced TACs for most stocks (e.g. the TAC for northern cod was reduced from 300,000 tons at F_{max} in 1976 to 160,000 tons at $F_{0.1}$ in 1977). These were transitional measures, however. ICNAF was aware that Canadian fisheries jurisdiction would be extended to 200 miles on January 1, 1977. There is little doubt that this influenced ICNAF's adoption of the $F_{01.}$ strategy. Thus the $F_{0.1}$ approach to management should be viewed as a Canadian coastal state initiative of the post-extension era.

The regime which had evolved in ICNAF during 1970-76 to a large extent set the scene for the Canadian system adopted in 1977. Pinhorn and Halliday (1990) summarized the situation:

> "The Canadian inheritance from ICNAF was a management system consisting of long-standing mesh size regulations, recently imposed TAC controls with national allocations, and 1-year-old fishing effort controls on non-coastal-nation fleets."

Following extension of jurisdiction Canada continued to use the TAC/allocation control approach, maintained mesh-size regulations, and implemented more rigorous fishing effort controls on foreign fleets fishing in Canadian waters. Canada also used $F_{0.1}$ as the scientific reference to manage finfish stocks on the Atlantic coast and maintained this approach through the 1980's.

For its domestic fisheries, Canada extended the national allocation approach of ICNAF in a complex allocation of access system for its domestic fleet. This was coupled with introduction of limited entry licensing for the Atlantic groundfish fishery during the late 1970's — first to the offshore groundfish trawler fleet and subsequently to the small-boat inshore fleet. The federal government did not wish to see the benefits of extended jurisdiction dissipated by expansion of fishing effort. As we saw in Chapter 2, the euphoria surrounding extension of fisheries jurisdiction generated considerable pressure from some provinces and segments of the fishing industry to expand the fleet. The federal government believed that the existing fleet would benefit from increasing stocks through increased catch rates. This would improve the economics of fishing. This aspect will be addressed more fully in the chapters that follow.

Halliday and Pinhorn (1985) pointed out that the actual reduction in fishing effort was less than the target of 40% below the 1972–73 level. It appears that a reduction by perhaps as much as one-third occurred between 1973 and 1976 in Subareas 2, 3 and 4. While the management approach adopted by Canada after extension of jurisdiction did not differ much from that developed in the last days of ICNAF, Halliday and Pinhorn (1985) concluded that measures taken were more effective. Catch rates reached a minimum in 1975 in the Canadian Atlantic area but returned by the early 1980's to levels prevalent in the 1960's. Groundfish abundance in all of Subareas 2, 3 and 4 decreased by more than half between 1967 and 1975. It remained stable between 1975 and 1976, then more than doubled between 1976 and 1984 (Pinhorn and Halliday 1990) (Fig. 6-8).

A central element of Canada's post-extension resource management strategy for the Atlantic groundfish fishery has been to maintain a moderate exploitation rate. Pinhorn and Halliday (1990) described post-extension trends in stock abundance and fishing mortality rates. Their analyses indicate that F_s for the Atlantic cod stocks in the 1960's were, in most cases, above 0.50. There was an increase in F in the early 1970's, with peak values for several stocks of about 1.0 during 1975–76. Yellowtail flounder on the Grand Bank had similar F_s before and after extension, significantly above $F_{0.1}$. It appeared that post-1977 F_s were probably less than $F_{0.1}$ for redfish and substantially below $F_{0.1}$ for Greenland halibut. Silver hake were fished close to $F_{0.1}$.

Pinhorn and Halliday (1990) concluded that Canada had not been able to reduce F to $F_{0.1}$ for the stocks most important to the Canadian fishery — Atlantic cod, haddock, pollock, and the major flatfish stocks. Except for haddock, post-1977 Fs were reduced to

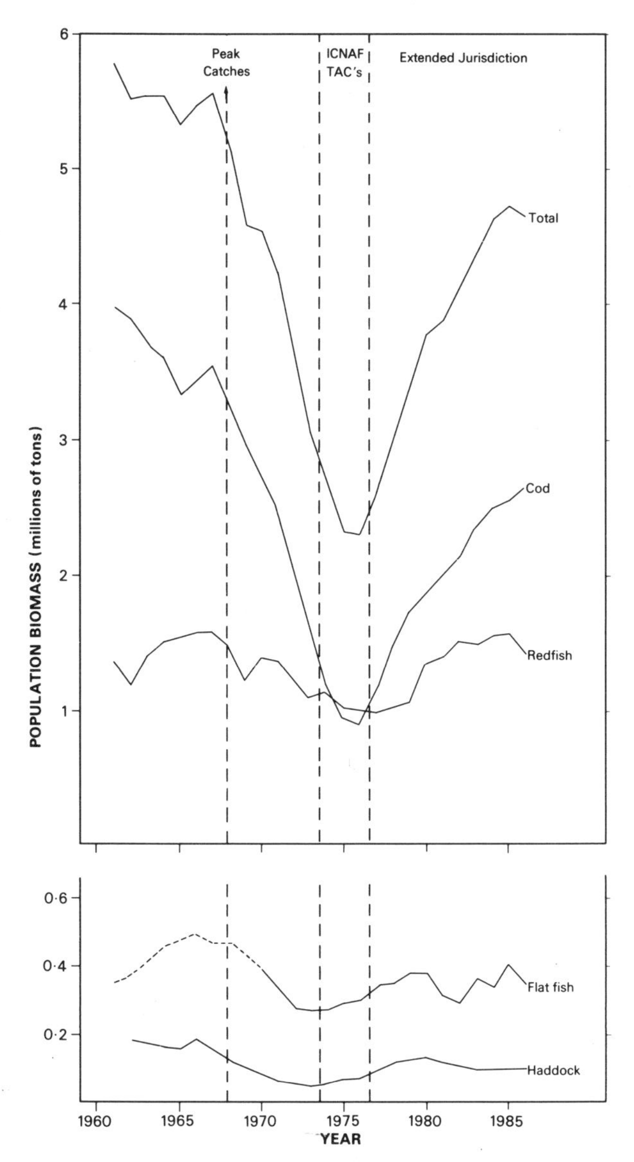

FIG. 6-8. Trends in biomass of Atlantic Groundfish: 1961–86 (from Pinhorn and Halliday 1990).

about F_{max}. The reasons for this are discussed in later chapters.

In the post-extension era catch projections from stock assessments tended to overestimate the predicted catch at $F_{0.1}$. This overestimation averaged about 30% in the early 1980's (Rivard and Foy 1987). The most widely publicized recent example was the dramatic reassessment in 1989 of the abundance of the northern cod stock (Lear and Parsons 1993). Despite this overestimation of stock abundance, it is clear that Atlantic groundfish off the Canadian coast increased significantly in abundance after 1976. Pinhorn and Halliday (1990) observed: "Canadian resource rehabilitation and conservation efforts appear to have met with a large degree of success overall."

However, in the late 1980s, the "resource bubble" burst. The TACs for the major northern cod stock were adjusted downwards, beginning in 1989. In 1991, there was a sudden, unexpected and dramatic decline in the abundance of northern cod. This led to a 2-year moratorium on fishing for this stock, beginning in July 1992. Meanwhile, cod stocks on the northern Scotian Shelf and in the Gulf of St. Lawrence experienced substantial declines in the late 1980s and early 1990s. Initially, government responded to these declines by adopting a 50% rule, i.e. adjusting the TAC by 50% of what would be required to achieve $F_{0.1}$. For some stocks TACs were set at twice $F_{0.1}$ as a proxy for F_{max}.

During 1992, however, it became apparent that the gradualist approach was not working. Therefore, for 1993, Fisheries Minister Crosbie announced the adoption of an immediate move to $F_{0.1}$ for the other major cod stocks in the Canadian zone. This necessitated draconian reductions in TACs, exacerbating a crisis in the Atlantic groundfish industry which had commenced in the late 1980s. As an example, TACs for southern Gulf cod and northern Scotian Shelf cod were reduced from 43,000 tons and 35,200 tons in 1992 to 13,000 tons and 11,000 tons respectively. These reductions, combined with earlier reductions for the haddock stocks, and the Grand Banks flatfish stocks managed by NAFO, plunged the industry into a deep crisis.

There appeared to be a combination of reasons for the resource decline in the late 1980s, early 1990s. Lower resource productivity was reflected in low average weights-at-age and low recruitment levels. This change in resource productivity stemmed, at least in part, from adverse environmental conditions. In addition, fishing mortality had remained high for many stocks. Anecdotal evidence indicated that there was widespread dumping and discarding of small fish and hence under-reporting of actual catches. Harvesting and processing capacity were generally acknowledged to be excessive.

A minimum fish size of 16 inches for cod, haddock and pollock had been introduced in 1988, primarily for market reasons. This was adjusted in 1990 to 17 inches on the Scotian Shelf, coupled with a proposed mesh size increase which was not implemented. By 1992 it became clear that this minimum fish size did not correspond to the effective mesh size in use. The 1993 Groundfish Plan, in addition to very low TACs, incorporated provisions for mandatory landing of all fish caught, i.e. the minimum fish size measure was abandoned. The various fleet sectors were required to produce acceptable harvesting plans demonstrating how they would avoid the capture and discarding of small fish.

It was hoped that these much tougher conservation measures, together with reductions in harvesting and processing capacity, would arrest the stock declines and rebuild the stocks over the next several years. The success of these measures would depend not just on the cooperation of the fishing industry but also on whether environmental factors facilitated an increase in resource productivity. For a further discussion of the cyclic boom-and-bust pattern in the Canadian Atlantic groundfish industry, see Chapter 13.

7.0 CONCLUSIONS

In this chapter I have examined the wide range of resource management techniques which have been applied to manage Canadian fish stocks. The five fisheries described briefly in this chapter illustrate the diversity of techniques used.

The great diversity of Canada's fisheries was illustrated in Chapter 1. Given this diversity and the different biological characteristics of the many species which support the major commercial fisheries, no single regulatory technique or set of techniques will suit all situations. On the east coast, catch quotas have become the dominant technique for resource management in the post-extension era. However, it is clear that catch quotas are not the preferred technique for all fisheries. Anadromous fisheries such as the multispecies, multistock Pacific salmon fisheries demand a different approach. In these fisheries the more traditional use of closed seasons and areas continue to be primary management tools. In invertebrate fisheries such as the valuable Atlantic lobster fishery, minimum size limits are an important regulatory tool.

Thus there is no uniquely appropriate method of regulating fisheries, whatever the management objective. Beddington and Rettig (1983) recognized this:

> "Successful programmes of fisheries management will involve a mix of regulatory and other devices. However, there is no point in seeking for a mixture which will provide some perfect solution for all time. Economic and social situations develop, the behaviour of the fish resources of the oceans are variable and to a considerable extent unpredictable, and hence management measures will have to alter with changing circumstances. It is therefore essential to recognize that management must be flexible, new measures to regulate effort may be needed and other ones abandoned. The management scheme that is designed to last forever will fail, perhaps as badly as the absence of management."

So far, this discussion of regulatory techniques has focused on methods to maintain or enhance productivity of fish stocks. We have seen from the examples in this chapter that attempts to conserve fish resources will fall short of the mark if they do not also address the economic problems which arise from the common property nature of fish resources. None of the measures discussed so far deal with the inexorable trend to overcapacity in an open access fishery. An effective regulatory framework must also address the common property issue.

CHAPTER 7

MANAGING THE COMMON PROPERTY I — ALLOCATION OF ACCESS

"When everyone competes for a share of a common, but limited, resource, the result is a zero-sum game; one man's gain is always another man's loss. All the conflicts over allocations — whether between provinces, between inshore and offshore fleets, or between individual fleets — are of this type. And the conflict will become more intense when there is no longer any new growth to allocate."

– M.J. Kirby, 1982

1.0 THE TRAGEDY OF THE COMMONS

Fishery resources are *common property*. Gordon (1954) first described clearly the problems this causes. He compared *open access* to the fishery to the use of common pasture in the medieval manorial economy:

> "There appears, then, to be some truth in the conservative dictum that everybody's property is nobody's property. Wealth that is free for all is valued by none because he who is foolhardy enough to wait for its proper time of use will only find that it has been taken by another."

Subsequently, the effects of open access to fish resources became known as the Tragedy of the Commons. Garrett Hardin (1968) developed the 'tragedy of the commons' idea to illustrate that unrestricted freedom to produce children would, in the long run, bring ruin to all. He dramatized this concept by a parable concerning unrestricted grazing rights in a hypothetical village commons, where pursuit of self-interest by the cattle owners led to collective tragedy.

Fish belong to no one in particular and everyone in general. Unlike other natural resource-based industries, few mechanisms are in place for participants to enjoy tenure over definable units of the fishery resource base (see Chapter 9). The common property characteristic of the fishery has been identified as one of the most important factors contributing to inefficiency and instability in the fishing industry (e.g. Kirby 1982; Pearse 1982).

Turning the Tide, the Report of the Pearse Commission on Pacific Fisheries Policy, discussed the implications for fisheries:

> "The perverse tendency for fishing fleets to overexpand is rooted in the way the commercial fisheries have traditionally been organized. Until very recently, fisheries throughout most of the world were open to unrestricted numbers of fishermen and fishing enterprises. Harvesting was, and still is, based on the 'rule of capture'.... In these circumstances, temporary profits will stimulate fishermen to expand their vessels' fishing capacity in order to increase their catch, and will attract new entrants into the fishery. So the fleet will expand even if it is already capable of taking the entire harvest.... An increase in the price of fish will set off a wave of investment in vessels and gear even when there are no more fish to catch. The result is the excess fishing capacity we observe in all of our major fisheries.
>
> "Several effects of this phenomenon should be noted. First, it threatens the stocks.... Second, the redundant capacity raises the capital, labour and operating costs involved in fishing, and so erodes the net returns the fishery could otherwise generate.
>
> "Third, such fisheries are unstable....
>
> "All of these effects — stock depletion, poor economic performance and instability —

> result from treating the resource (fish) as common property until they are caught, and are normal whenever resources are treated this way" (Pearse 1982).

Overexpanded fishing capacity is not the result of irrational behaviour on the part of fishermen. No individual fisherman can control the fishing effort of others. Thus he has a strong incentive to expand his fishing capability to obtain a larger share of the catch or to protect his share from competitors. The resulting excess capacity is irrational, but it is the result of rational responses of individual fishermen to economic incentives.

In addition to competition among fishermen, the common-property nature of the resource stimulates competition among processing companies for raw material and among provinces for employment opportunities. Processors are motivated in the same way as fishermen to obtain a larger share of the catch or to protect their share from competitors. This leads to increases in capacity and a preoccupation with supply of raw material for processing. Processors compete for fishermen in the same way as fishermen compete for fish. Provincial governments, in order to create jobs (and attract federal transfer payments), have over the years offered financial incentives to fishermen and processors. This has led to overcapacity both in the fleet and onshore.

Stillman (1975) pointed out that Hardin's parable involves three assumptions additional to the common-property approach:

1. The users must be selfish and they must be able to pursue private gain even against the best interests of the community;
2. The environment must be limited, and there must be a resource-use pattern in which the rate of exploitation exceeds the natural rate of replenishment of the resource; and
3. The resource must be collectively owned by society (common-property) and freely open to any user (open-access).

If these three conditions are fulfilled, Stillman postulated that the Tragedy of the Commons is inevitable. However, Berkes (1985) provided a contrasting analysis of Hardin's paradigm in the context of empirical case studies of fisheries worldwide. He rejected conventional wisdom, stating that the Commons paradigm is not *the* model of reality for *all* fisheries. He concluded that:

1. The fishery resources are almost never truly open-access; and
2. Individual interests are often subservient to collective interests of a community.

Nonetheless, the effects described by Gordon, Hardin and Pearse have been observed in most of the western world's major marine commercial fisheries. Within Canada over the past 25 years a complex web of measures has been developed to deal with the problem of the "race for the fish." These include *allocation of access* to the resource among user groups, *limited entry licensing* systems which have been extended to virtually all marine commercial fisheries in Canada, and, in the 1980's, the introduction of *individual quotas* or enterprise allocations. Individual quotas are a step in the evolution of property rights.

The previous two chapters dealt with various *open-access* regulatory techniques primarily for conserving the resource. This and the next two chapters deal with various limited-entry techniques used in Canada to address the common-property problem. Although allocation of access among fleet sectors has developed in parallel with limited entry licensing, I will first describe the evolution of the controlled access (allocation of access) regime in specific fisheries. This will illustrate how managers have attempted to limit the "race for the fish" by allocating portions of the available catch among fleets and gear types.

The following fisheries were chosen because they illustrate the diversity and complexity of the approaches which have evolved. They also exemplify the clash of conflicting interests in the fishery. These fisheries are the Atlantic groundfish fishery in general and certain components of that fishery in particular, and the Pacific salmon fishery, with particular emphasis on the chinook fishery.

2.0 ATLANTIC GROUNDFISH

2.1 Evolution of Atlantic Groundfish Fishing Plans

ICNAF established a Total Allowable Catch (TAC) regime for Atlantic groundfish in the early 1970's. Agreement on a system of global catch quotas became possible once it was agreed that these global quotas would be allocated among countries. The introduction of national allocations in ICNAF in 1971 was a break from the international common-property approach which had prevailed until then. Once Total Allowable Catches were established for particular species, with the total partitioned into national allocations, it was then up to each country to decide whether and how access would be controlled within its own allocation.

Distribution of the available resource among the participants in the Canadian groundfish fishery first became a pressing management issue in 1976 when Canada first imposed a TAC for redfish in the Gulf of St. Lawrence. This was a very low allowable catch compared with previous catches. It substantially reduced the quantity of traditional groundfish species available to the Canadian offshore fleet. This was coupled with substantial reductions in ICNAF-set TACs for stocks on the Grand Banks and the Scotian Shelf. The result was widespread concern that there was insufficient quota for the Canadian offshore fleet to operate year-round.

In June 1976, Fisheries Minister Roméo LeBlanc announced a series of interim measures to keep the east coast trawler fleet operating for the balance of 1976 (DOE 1976b). These measures included:

- a special financial allowance under the Temporary Assistance Program for groundfish catches made outside the Gulf of St. Lawrence by Gulf-based vessels;
- exploratory fishing projects for redfish in "distant waters" and for species which Canadian vessels had not traditionally fished, such as silver hake, mackerel, grenadier, squid and Scotian Shelf shrimp;
- assistance to defray increased costs for fishing for traditional species in new areas, for example, fishing for cod and turbot in areas off northeast Newfoundland and Labrador[1]; and
- a monthly limit of 500,000 pounds of redfish and/or flounders per vessel for the major portion of the east coast.

At that time DFO scientists attempted to project the rate of recovery of various stocks based upon a number of assumptions. The Gulf redfish stock was at a low level and would not rebuild until the early 1980's (DOE 1978a). The northern and southern Gulf cod stocks were expected to rebuild slowly. For those stocks, restrictive TACs had severely constrained fishing effort. Any increase in TACs as stocks rebuilt would be offset by an increase in catch rates and the normal activity of the existing fleet. Small boats would tend to fish cod longer each year as catch rates returned to past levels. Overall there appeared to be a need to divert offshore effort out of the Gulf.

Scientists anticipated increases in catches and catch rates for most traditional Scotian Shelf groundfish fisheries. These improvements would probably allow for an increase in effort levels beyond those expended in 1977, and absorb some effort displaced from the pollock or cod fisheries.

Special problems were foreseen for fleets and plants based at the Gulf entrance in Sydney Bight and on the southwest coast of Newfoundland (Burgeo and Gaultois), which at that time were supplied by relatively old side trawlers of limited range.

DFO scientists projected that the cod and flatfish stocks of the Grand Banks would increase over time under a $F_{0.1}$ management regime. This predated the problem of overfishing of the straddling stocks outside 200 miles in the post-extension era. In 1976–77 it was thought that the groundfish fisheries of the Grand Banks provided some hope for expansion of Canadian effort in the long-run.

Off northeast Newfoundland and Labrador, the dominant feature is the northern cod stock (Divisions 2J3KL) (Lear and Parsons 1993). The biomass of this stock was projected to increase under Canadian management with the TAC expected to increase significantly over the next several years, perhaps to 350,000 tons by the mid to late 1980's[2].

Thus prospects for the major groundfish fisheries of the Atlantic coast were described as:

1. Slow Growth: The Gulf, Scotian Shelf and Grand Banks;
2. Decline: Fisheries of the Gulf Entrance; and
3. Sustainable Growth: Cod stocks of northeastern Newfoundland.

In the short-term, this resource availability imbalance appeared to necessitate a dramatic shift in the fishing strategy of the Canadian offshore trawler fleet. To achieve this, federal managers developed a plan to apportion the available stocks among the various components of the offshore trawler fleet to keep trawler fleets and processing plants operating year round in 1977. The Atlantic Groundfish Management Plan for 1977 was the first attempt in Atlantic Canada to allocate or distribute stocks within an overall TAC.

In the Gulf the Plan provided a potential catch for the Canadian trawler fleet of 80,000 tons in 1977. From an actual 1975 catch of 125,000 tons, this was a decrease of 36%. On the Scotian Shelf TACs had been decreased substantially in the 4VsW cod stock (from 30,000 to 7,000 tons) and in the transboundary pollock stock (from 55,000 to 30,000 tons). This, combined with the decline in TACs for Gulf groundfish where Nova Scotia-based trawlers had in the early 1970's taken a high proportion of their catch, meant that the total groundfish available to this fleet in traditional areas would be about 40% less than their 1975 catches.

[1]This became the stimulus for the development of a Canadian offshore fishery for northern cod.

[2]This stock did increase substantially in abundance in the post-extension era but not to the levels expected and then subsequently declined. The reasons for this are discussed in Chapters 13 and 18.

For most of the groundfish species traditionally harvested by Canadians on the Grand Banks, St. Pierre Bank and in areas east and north of Newfoundland, the Canadian quota allocations had either stayed about the same or increased. The Canadian quotas off Newfoundland and Labrador in 1977 were 261,000 tons compared with 1976 quotas of 232,000 tons and 1975 catches of 148,000 tons for the same stocks. This major difference between catches and allocations was largely due to higher redfish allocations in areas not normally fished by the Canadian fleet and higher allocations of species not usually taken by this fleet in directed fisheries (e.g. turbot).

In summary, Canada's 1977 groundfish allocations under ICNAF totalled almost the same as in 1976. This was significantly less than in earlier years. Considerable quantities of potential catch had been lost off Nova Scotia and in the Gulf from stocks which the Canadian trawler fleet could have harvested. These losses had been made up by redfish so that the overall quantity of redfish available in 1977 was slightly higher than in 1976. But the harvesting of these quotas would necessitate an overall northward shift in fishing activity by the Canadian offshore trawler fleet.

Thus, as groundfish fishermen and fisheries managers on the Atlantic anticipated the 200-mile fisheries limit, they needed to determine how to distribute the available resource among competing groups. The total pie, the sum of the TACs for particular stocks, had to be divided among these fleet sectors. Otherwise, the offshore trawler fleet would harvest many of the available Canadian quotas in the Gulf and on the Scotian Shelf before inshore fishermen even got to sea in their small boats.

As a basis for allocating the available resource among fleet sectors and gear types, the first Groundfish Management Plan enunciated the following Management Principles (DOE 1976c):

"1. The overriding principle is that the fishery resources must be conserved and restored;
2. In principle the fishery resources are accessible to all Atlantic fishermen. Access to fishery resources will be allocated on the basis of a satisfactory trade-off between economic efficiency, the dependency of the fleets involved, and, in the short run, sharing the burden of shortages of resources amongst all those involved;
3. Coordinate the deployment of mobile fishing fleets, over the fishing grounds and the operating season;
4. Provide for the withdrawal of excessive catching capacity in congested fleet segments and in areas of low productivity, and for the best possible mix of fleet units; and
5. Utilize the fishery resources over the calendar year to the maximum degree possible."

The 1977 Groundfish Plan was a complex of quotas by vessel type and gear classes including allowances for inshore vessels, closed seasons, limits on numbers of trips per month and catch limits per trip. It represented a great number of compromises among the different sectors of the groundfish industry.

Management planning for the Gulf of St. Lawrence proved especially difficult. This was due partly to the severely depressed state of the Gulf redfish and southern Gulf cod stocks and partly to the varied interests demanding access to this fishery. French treaty rights were a further complication. The Gulf-based vessels were probably capable of catching most of the available stocks. Balanced against this was the argument that some Maritimes and Newfoundland-based vessels had traditionally caught up to 60% of their catches in the Gulf. They were very insistent that their right of access be protected. Quebec-based vessels were unwilling to fish outside the Gulf (DOE 1976c).

Attempting to balance these conflicting interests, the first Groundfish Plan provided incentives for offshore trawlers to fish northern cod in winter. Approximately $4.5 million was provided to encourage fishing in northern areas, fishing of redfish and turbot, and fishing of unutilized quotas in traditional areas. This incentive program was instrumental in developing an offshore Canadian fishery for northern cod. While the incentive program was being developed, one prominent fishing company executive remarked: "The feds are crazy if they think Canadian trawlers are ever going to be successful in fishing northern cod during the winter in the ice!" This was typical of the Canadian industry's resistance to change.

Overall, the first Groundfish Plan attempted to address the problem of resource shortage in the Gulf by pushing the more mobile Nova Scotia and Newfoundland-based trawler fleets out of the Gulf and encouraging the Gulf-based offshore trawler fleet to fish outside the Gulf. The Plan also attempted to accommodate the perceived lack of mobility of the side trawlers and lower-powered stern trawlers based in Cape Breton and on the southwest coast of Newfoundland. These vessels were given preferential access to local resources. The larger stern trawlers were given financial incentives to fish for northern cod off Labrador and other species elsewhere in the Canadian 200-mile zone.

Different gear types used in the Atlantic groundfish industry. (Longliner (top left), cod trap (top right), small dragger (bottom left), gillnetter (bottom right)).

Canadian trawler fishing in the ice for northern cod.

Displacing the larger trawlers from the Gulf and the northward shift in allocations caused other distributional conflicts. The Nova-Scotia-based trawlers had traditionally fished on Georges Bank, the Scotian Shelf and in the Gulf of St. Lawrence. The larger Newfoundland trawlers had become highly specialized in fishing flatfish on the Grand Banks. The Newfoundland interests became concerned that the Nova Scotia trawlers would start scooping up the flatfish quotas. There was also growing concern that the Nova Scotians would start fishing northern cod intensively, leading to a redistribution of "Newfoundland" fish to Nova Scotian plants. This presaged a major interprovincial conflict which played a key role in the outcome of the 1980-1981 constitutional discussions about fisheries (see Chapter 2).

These conflicts were reflected in the positions of the Newfoundland and Nova Scotia industry groups in discussions in the fall of 1976 about the 1977 groundfish plan. The Newfoundland groups advocated catch limits per trip, while the Nova Scotia groups strongly favoured closed seasons and limits to the number of trips per month. These pressures brought a mixture of controls, with catch limits per trip for yellowtail flounder on the Grand Banks and limits on numbers of trips and closed seasons for those stocks of primary importance to Nova Scotian vessels (e.g. non-Gulf redfish stocks).

In announcing the 1977 Groundfish Management Plan, Fisheries Minister LeBlanc observed that the Plan had three objectives: "to avoid conflicts between local and mobile fleets over scarce fishery resources, to ensure rebuilding of depleted fish stocks, and to stretch out available resources to keep the groundfish industry working all year long." LeBlanc noted that the intermediate and small boats, more than 10,000 of them, had only limited range. Hence the large trawler fleet had a "duty and opportunity of going farther afield" (DOE 1976d).

Principle 2 of the 1977 Management Plan talked of allocating access "on the basis of a satisfactory trade-off between economic efficiency, the dependency of the fleets involved, and, in the short run, sharing the burden of shortages of resources amongst all those involved." In fact, the achievement of an agreed "satisfactory trade-off" proved elusive.

Groundfish resource allocation issues engendered great public debate from 1976 to 1982 as the various interest groups jockeyed for the greatest share of the available resource. This was understandable since the resource allocation process involved distributing potential income from the fishery among various groups in society. Although the ultimate aim of allocating access to the resource was "best use", as set forth in the 1976 Policy for Canada's Commercial Fisheries, in practice a variety of factors had to be accommodated in allocating the resource among competing fleets. These included proximity to the resource, community dependence, and efficiency of fleet sectors. In the short term, there was a greater emphasis on such factors as proximity and dependence.

The situation in the late 1970's–early 1980's was characterized by conflict over how scarce resources would be shared among the competing fleet sectors and by squabbles over sharing the growth in catch resulting from stock rebuilding. The latter debate was exacerbated by the fact that the growth in fish stocks post-1977 was not uniform throughout the Canadian Atlantic. The greatest increase in volume following extension of fisheries jurisdiction was expected to occur in the major northern cod stock (Divisions 2J3KL) off eastern Newfoundland and Labrador. The bubble of anticipation burst in February 1989 with a scientific reassessment of the rate of increase in stock biomass in the 1980's. Until then, expectations ran high about the long-term potential catches from this stock. Even as late as 1986–87, scientists thought that the rebuilt stock could support catches at $F_{0.1}$ in the order of 300,000 tons per year. Not unexpectedly, the areas of minor growth (the Gulf) and anticipated major growth (northern cod) were the focus of much of the debate on allocation of access to the Atlantic groundfish stocks.

The allocation debate centered on: (1) the general inshore/offshore split; (2) access to the Gulf of St. Lawrence by large trawlers based outside the Gulf; and (3) interprovincial rivalry between Newfoundland and Nova Scotia about the shares of northern cod and where the northern cod catches were to be landed. In the late 1980's the "Gulf" provinces joined the debate over allocation of northern cod. Most of these conflicts focused on access to cod, specifically, Gulf cod and northern cod.

2.2 The Inshore/Offshore Split

The question of the inshore/offshore split was at the forefront of the debate. Offshore trawler companies frequently contended that there had been a considerable tilt in allocations to the inshore fleet following extension of fisheries jurisdiction. This argument was a reaction to policy pronouncements by Fisheries Minister LeBlanc in the late 1970's and to the displacement of the offshore fleet from fishing grounds in the Gulf. Minister LeBlanc in a number of speeches in 1977 and 1978 indicated that the inshore sector would be favoured over the offshore trawlers. Speaking to the Rotary Club of St. John's in May, 1977, he stated, with reference to northern cod (DOE 1977b):

> "Who gets first crack at these fish? Here I must say... that I have a clear bias for the inshore fisherman. Not because of some romantic regard, not because of his picture on the calendars, but because he cannot travel far after fish, because he depends on fishing for his income, because his community in turn depends on his fishery being protected."

Thus DFO gave priority to the inshore sector. Particularly in certain stocks, such as Gulf groundfish and northern cod, inshore fishermen were to benefit from stock rebuilding and regain ground lost because of the stock declines in the early 1970's. However, the priority given in allocations to the smaller and less mobile fleet sectors was attacked by opponents of LeBlanc's policies.

To examine the effects of this policy, it is necessary to differentiate among classes of vessels operating in the Atlantic groundfish fishery. The terms "inshore" and "offshore" are misleading. Many of the vessels classed as inshore, e.g. 65-foot draggers in southwest Nova Scotia, are quite capable of fishing most of the year and harvesting most of the species available on the Scotian Shelf. Operators of the large trawler fleet prefer to use a cut-off of 100 feet in length to distinguish between vessels which can fish anywhere along the Atlantic coast and those which must fish close to home port. Even this cut-off is not a completely valid distinction. Side trawlers longer than 100 feet could not, for example, fish the northern cod in winter. Early debates about the inshore/offshore split were distorted by differing definitions of "inshore" and "offshore" among the various interest groups. Gradually, however, the 100-foot cut-off was accepted as the basis for determining the inshore/offshore distribution of allocations.

There was very little change in the split of groundfish catches between the inshore and offshore over the decade 1971–80 (Fig. 7-1). Excluding redfish, there was little change in the percentage of groundfish catches taken by vessels greater than 100 feet. Redfish is an offshore species, the catches of which have been heavily influenced by natural large-scale changes in abundance in the Gulf of St. Lawrence. The percentage of the total catches taken by vessels over 100 feet hovered around 40% in the late 1970's, a slight decrease from around 44% in the early 70's.

A 1988 analysis by the Department of Fisheries and Oceans, for all groundfish species fished including redfish, showed slight year-to-year variations in the percentage shares of the catch by vessels greater and less than 100 feet. The percentage share of the catch taken by vessels longer than 100 feet averaged about 45% during the decade following extension of jurisdiction (Fig. 7-2). The offshore share decreased slightly in the early 1980's, but by mid-decade had returned to the late 1970's level (DFO 1988a).

For most groundfish stocks there was very little change in the percentage shares of allocations in the annual Atlantic Groundfish Management Plans for the inshore and offshore sectors over the decade 1977–86. There were, however, exceptions: northern cod, Grand Banks cod (Divisions 3NO), Gulf cod, Gulf redfish, Scotian Shelf haddock, Georges Bank haddock and cod and Scotian Shelf pollock.

From the mid-1970's onward a significant shift occurred in the species sought by the offshore fleet. After the collapse of the Grand Banks haddock stock in the 1960's, Canadian trawlers over 100 feet caught primarily redfish and flatfish. Following the decline in the Gulf redfish in the early 1970's, some of these vessels

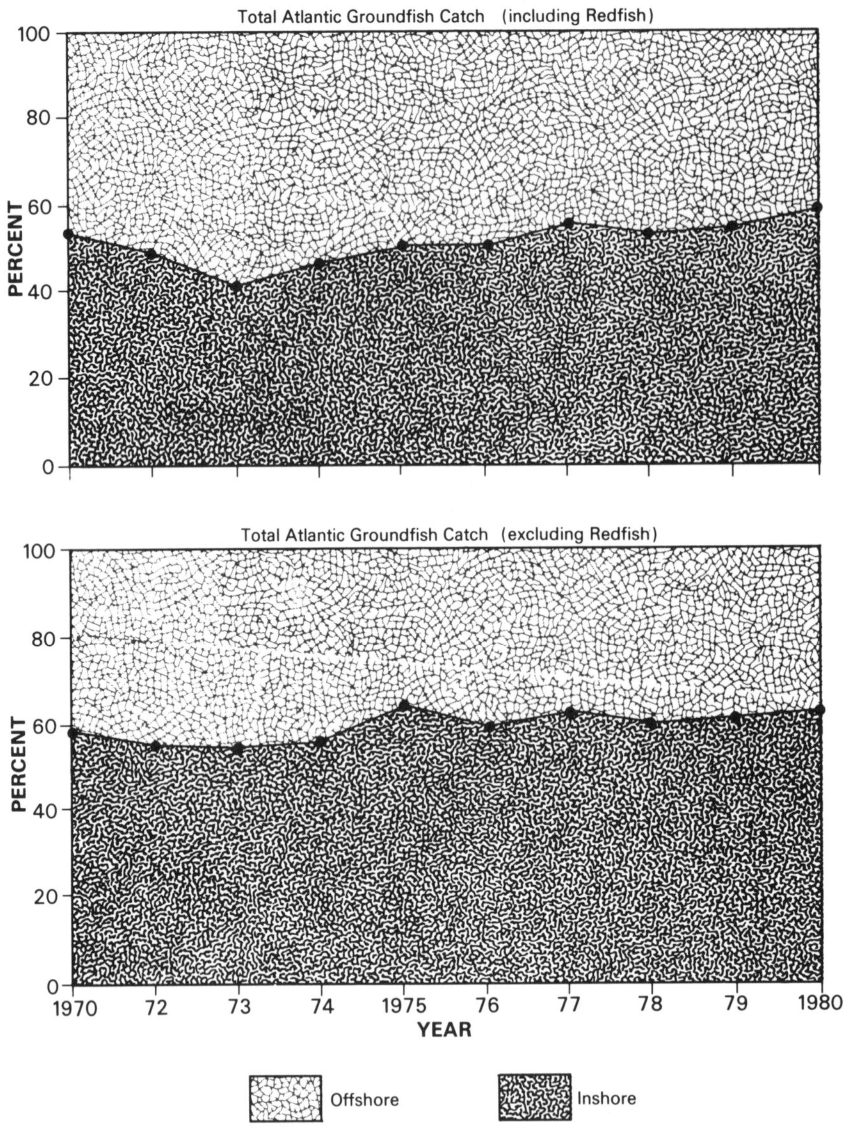

FIG. 7-1. Distribution of percentage shares of the Atlantic groundfish TACs between the inshore and offshore sectors: 1971–80.

turned for a couple of years to fishing redfish east and north of Newfoundland. They were assisted in this by government incentives under the Groundfish Vessel Dislocation Program. However, because of a downturn in the redfish markets, the trawler fleet caught only about 50% of their redfish quota in 1980 and 1981. This situation intensified their demand for a greater share of the TACs of cod, the species on which the inshore fishery had traditionally depended. The share of the total Canadian cod catch for the offshore fleet increased steadily from a low of 18% in 1975 to 42% in 1986. In absolute terms, the cod catch by this vessel class increased more than sevenfold from 27,000 tons in 1975 to 196,000 tons in 1986. From 1977 to 1986 the cod catch by the inshore fleet rose from 187,000 tons to 269,000 tons, an increase of 44%.

2.3 Access to Gulf Cod

A second major aspect of the groundfish allocation debate concerned access to Gulf of St. Lawrence cod stocks. The Gulf was considered to be an area of slow growth. Because of decreased TACs in the Gulf, access to Gulf cod by the non Gulf-based offshore fleet began to be constrained in the 1977 Groundfish Plan. The Plan set aside 18,500 tons of northern Gulf cod for

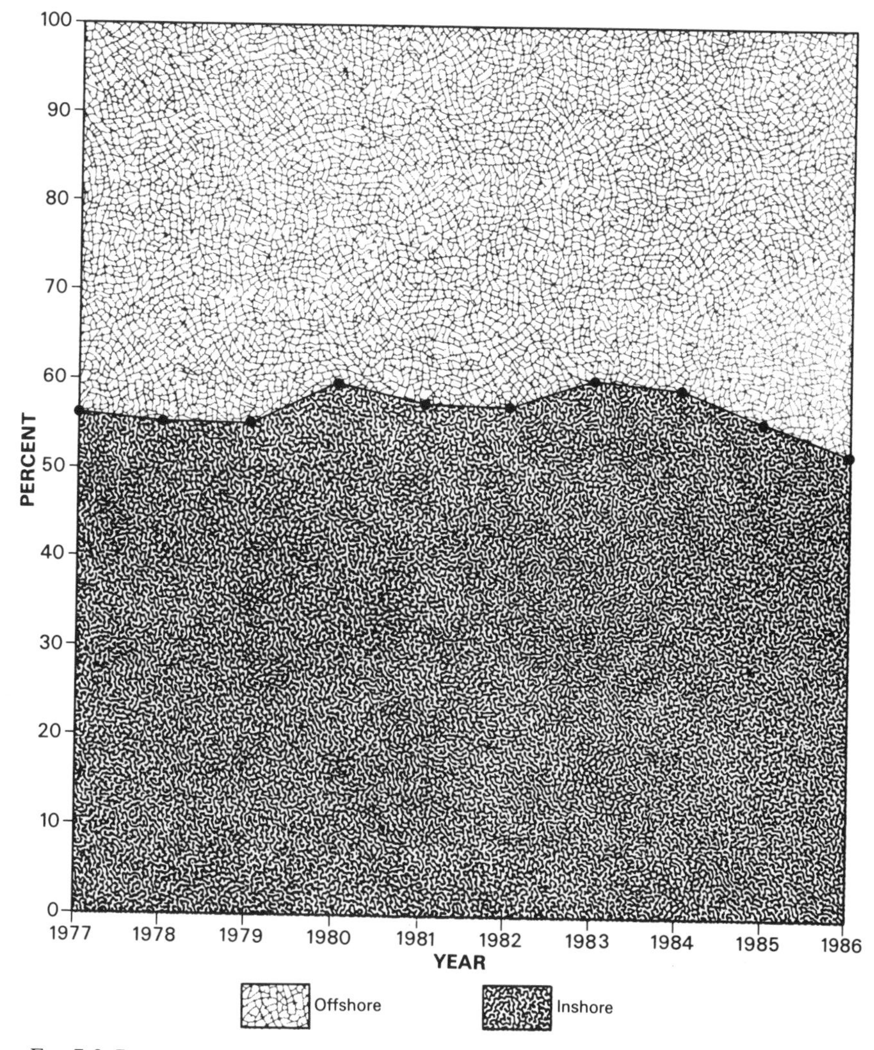

FIG. 7-2. Percentage of the total Atlantic groundfish landings, taken by the inshore and offshore sectors: 1977–1986 (from DFO 1988a).

all vessels over 100 feet and allocated only 1,000 tons of southern Gulf cod in Sydney Bight (4Vn) for the side trawlers based at plants adjacent to the Gulf. In the 1978 Groundfish Plan non-Gulf based vessels greater than 100 feet were restricted to 2,500 tons out of a TAC of 55,000 tons for northern Gulf cod and a portion of a Canadian quota of 3,200 tons in 4Vn out of a TAC of 27,000 tons for southern Gulf cod.

In the 1979 Groundfish Plan the TAC for northern Gulf cod was increased to 75,000 tons from 55,000 tons the previous year. The non-Gulf based offshore fleet was allocated 10,000 tons. With an increased TAC of 36,000 tons for southern Gulf cod, the non-Gulf based trawlers over 100 feet obtained access to quotas of 2,800 tons for stern trawlers and 4,000 tons for side trawlers. In announcing the 1979 Plan, Fisheries Minister LeBlanc alerted the non-Gulf vessels over 100 feet to "the distinct possibility that their cod allocation (10,000 m.t.) from this stock might be substantially reduced or eliminated in future years, in order to provide for adequate fishing opportunities for existing vessels in the Gulf" (DOE 1978b).

During 1979, a scientific reassessment of the southern Gulf cod stock resulted in a mid-season increase in the TAC from 36,000 tons to 46,000 tons. The new Fisheries Minister, James McGrath, decided to allocate an additional 6,000 tons to vessels both Gulf-based and non-Gulf based to fish in Division 4T, an area from which the non-Gulf based vessels had been excluded for some time. This ignited a firestorm of protests from Gulf-based interests. In the 1980 Plan, the TAC for southern Gulf cod was set at 54,000 tons, of which the non-Gulf based offshore fleet was allocated 9,100 tons, split equally between the side and stern trawlers. But there was no provision for them to fish in 4T. Specific reference was made to the difficulty of reconciling the interests of Gulf-based vessels with the requests for allocations for non-Gulf based vessels.

With the change of federal government in February 1980, Roméo LeBlanc returned as Fisheries Minister. During the election campaign he had made it clear that he supported the Gulf-based interests. He decided to bring the conflicting groups together in a September, 1980, Seminar on Gulf Groundfish held at Memramcook, New Brunswick. The central issue was access to the Gulf by fleets based outside the Gulf. Many different points of view were expressed in papers submitted by seminar participants (DFO 1980a).

Inshore fixed gear fishermen laid claim to at least the amount of fish they had caught in the 1940's and 1950's. This implied an increase in the inshore allocation to 50,000 tons. The midshore mobile fleet representatives argued that their fleet could not be viable unless their catch rates increased. This could only be accomplished by a reduction of effort. They called for prohibition of fishing in the Gulf by vessels over 100 feet. Representatives of the deep sea mobile fleet fishermen argued that, although the resource situation had improved, the high cost of travelling to distant fishing grounds in winter required continued access to the Gulf winter fishery to ensure fleet viability. They also contended that access to the Gulf was essential for some of the older side trawlers and smaller stern trawlers if they were to continue fishing.

Provincial representatives expressed widely divergent views. Newfoundland argued for concurrent jurisdiction through constitutional change (see Chapter 2). Regarding the Gulf, it argued that access to fish stocks must be accorded primarily to those regions which had historically fished them and which had become dependent on them. New Brunswick contended that non-Gulf based trawlers greater than 100 feet should be excluded from the Gulf. P.E.I. called for a more conservative approach to setting TACs, i.e. at a level lower than $F_{0.1}$, to improve the economics of fishing operations, increase the catch of inshore fishermen and improve product quality. Québec demanded that all increases in TACs be reserved for Gulf-based fishermen. Nova Scotia pointed out that it had a 250-mile coastline on the Gulf of St. Lawrence embracing 79 fishing communities supplied by inshore vessels. It argued that some of the Nova Scotia offshore fleet, because of age and limited catching capability, were very dependent upon stocks close to home. Nova Scotia demanded that the large offshore trawlers be permitted to continue fishing in Sydney Bight in the southern Gulf and in the northern Gulf. These diverse views are summarized in the Report of the Seminar (DFO 1980a).

Fisheries Minister Roméo LeBlanc had attended the Seminar and listened to most of the sessions. His conclusions were reflected in the 1981 Groundfish Plan. In announcing this Plan in December 1980, Minister LeBlanc noted that Newfoundland and Nova Scotia had been the major beneficiaries of the increase in groundfish landings in recent years. He pointed out that groundfish landings in Newfoundland had risen from 277,000 tons in 1970 to 374,000 tons in 1979, while landings in Nova Scotia had risen from 143,000 tons in 1970 to 231,000 tons in 1979. He stated: "Those provinces which are dependent — because of the nature of their groundfish fleets — almost exclusively on groundfish resources in the Gulf of St. Lawrence have not experienced a similar rate of gain in landings" (DFO 1980b).

In the Minister's view, the Gulf Groundfish Seminar had demonstrated that the Gulf resource was not sufficient to sustain the existing Gulf-based groundfish fleet of more than 10,000 vessels. Therefore, he announced that non-Gulf based stern trawlers longer than 100 feet, powered by engines greater than 1,050 brake horsepower (BHP) would not have access to Gulf cod stocks in 1981. Vessels between 1,050 BHP and 1,300 BHP were an exception. They would have access to 1,000 tons. Non-Gulf based vessels over 100 feet in length and less than 1,050 BHP, (i.e. side trawlers) were considered less mobile and not equipped to fish as far away as Labrador or the Flemish Cap. They were granted continued access to Gulf cod, thus assuring plants at the mouth of the Gulf a continued supply of groundfish. These trawlers were assigned an allocation of 10,000 tons in 1981.

Another significant change in the 1981 Groundfish Plan (in part a concession to offshore trawler owners) was that, for the first time, fixed gear groundfish fisheries such as traps, gillnets and longlines were placed under quota. These fisheries had previously been managed under an "allowance" system, with an allowance defined as a "provision for normal catch levels." In reality, an allowance was an estimate of the expected average catch. Thus, from the 1981 Plan onward all groundfish fleet sectors were now managed by allocated quotas. An exception was made for the Newfoundland small-boat fixed-gear fishery for northern cod which again reverted to an allowance system in 1982 because of the special problem created by the unpredictable fluctuations in inshore catches for this stock.

In February 1981, DFO released a discussion paper titled *Toward a Policy for the Management of Gulf Groundfish* (DFO 1981b). Allocation priorities were ranked as follows: "First priority of access must be given to the Gulf based fixed gear fleet. The second priority must go to the Gulf based mobile gear vessels under 100 feet, and if there is any surplus to their needs, the balance could be allocated to the deep-sea fleet." Non-Gulf based vessels over 100 feet but less than 1,050 BHP were, however, to continue to be allocated cod in 4Vn (Sydney Blight) and in the northern Gulf. The Discussion Paper also identified a problem of excess fishing capacity in the Gulf and concluded that capacity should be controlled through more stringent regulations on vessel replacement and on gear.

This policy set the stage for allocating Gulf groundfish over the next decade. The restriction on access to cod by non-Gulf based stern trawlers (vessels greater than 1,050 BHP) was maintained until the 1987 Groundfish Plan, when the "less than 1,050 BHP" designation was removed. This happened because of a general acceptance that, with enterprise allocations in effect, offshore companies should be free to choose which part of their fleet would harvest their enterprise allocation (Chapter 9).

Arguments have been advanced in various fora that Nova Scotia-based trawlers were disadvantaged by this policy. It is true that the offshore fleet based in Nova Scotia and Newfoundland had less access to Gulf cod than previously. However, the extent and impact of that reduction were often greatly exaggerated. In the 1970's, the big four trawler companies averaged about 18,000 tons of Gulf cod per year. In 1981, they were allocated 10,000 tons from the two Gulf stocks. They had gained far more elsewhere, both on the Scotian Shelf and off Newfoundland. The displacement from the Gulf did not affect seriously the overall operating pattern of the Nova Scotia fleet. Nova Scotia-based trawlers, for example, in 1980 took about 84% of their total catch within 250 miles (one day's steam) of home ports compared with 83% in 1978. During the 1980's, as the offshore allocation of northern cod increased, there was a greater shift northward.

The side trawlers and processing plants which had been established largely to exploit the major redfish stocks on the south coast of Newfoundland and in the Gulf were disadvantaged by the general shift in resource abundance. This was the reason for the special provision for vessels less than 1,050 BHP in the 1981 and subsequent Groundfish Plans.

In the first decade post-extension there had in fact been little overall change in the inshore/offshore balance. There had been some limited displacement from the Gulf but the effects had been more than offset by significant increases in the fish available to the offshore trawler fleet on the Scotian Shelf and off eastern Newfoundland-Labrador. The considerable improvement in the daily catch rates of the offshore fleet from 1977 onward improved the economics of trawler operations.

This situation changed with the resource downturn on the Scotian Shelf in the late 1980's and the drastic reductions in the Canadian northern cod TAC from 1989 to 1992. Further reductions for 1993 had a dramatic impact on the profitability of the offshore companies operating the offshore fleet (see Chapter 13).

2.4 The Struggle for Gulf Redfish

The non-Gulf based stern trawlers lost their access to Gulf redfish with the first TAC, at 30,000 tons, in 1976. The major fishery which had developed on the extremely abundant 1956 and 1958 year-classes peaked at 130,000 tons in 1973 but decreased rapidly thereafter. The federal government subsequently excluded non-Gulf based trawlers longer than 100 feet from the Gulf redfish fishery.

This launched the initial battle in the war over access to the Gulf stocks. Representatives of the trawlers based outside the Gulf strongly criticized establishment of a TAC of 30,000 tons. They claimed that the scientists' advice was at odds with the offshore fleet's good catch rates in January and February. The fishery had been supported by the 1956 and 1958 year-classes. However, immediate recruitment prospects were dismal, and between 1977 and 1980 the TACs had to be reduced further to a low of 16,000 tons. Approximately 6% of these TACs were allocated as incidental catches (i.e. by-catch) for non-Gulf based vessels over 100 feet fishing for northern Gulf cod. By 1981, catch rates indicated that the stock had stabilized and the TAC was increased to 20,000 tons. This included 1,500 tons for non-Gulf based vessels over 100 feet.

The 1982 TAC was set initially at 28,000 tons and subsequently increased to 31,000 tons. Of the increase, 1000 tons was allocated to the non-Gulf based vessels greater than 100 feet, bringing their share of the TAC to 3,500 tons. That represented 11.5% of the Gulf redfish TAC. For 1983 the TAC was held at 31,000 tons even though scientists indicated that it could be increased to 37,000 tons as a result of improved recruitment prospects.

In the 1983 assessment, CAFSAC confirmed that the 1970 and 1971 year-classes were abundant. It advised that the TAC for 1984 could be set as high as 75,000 tons fishing at the $F_{0.1}$ level (CAFSAC 1984b) but would decline to 40,000 tons over several years. This provoked a disagreement between Gulf-based and non-Gulf based interests on the level at which the TAC for 1984 should be established. The Gulf-based interests argued that the TAC should not be increased to ensure stability in catch over a number of years. They were also motivated by a strong desire to prevent the non-Gulf based stern trawlers from returning in force to the Gulf redfish fishery. Representatives of the non-Gulf based interests contended that the 1984 TAC should be established at 60,000 tons. They considered this a conservative level.

The Gulf redfish stock was known for extreme fluctuations in recruitment. Hence, the fishery would be dependent on the strong year-classes of 1970 and 1971 for several years. Federal officials therefore believed that a good case could be made for fishing below the $F_{0.1}$ level and spreading the yield from these year-classes over several years. This would provide some stability for participants in the fishery until the next strong year-class could be harvested. In addition to providing stability of catch, a TAC lower than the $F_{0.1}$ level would mean improved catch rates and hence improved economic viability. An explicit trade-off would have to be made between large catches and stable catches with higher vessel productivity. Scientists were asked to describe the implications of fishing for several years at constant catch levels of 40,000, 50,000 and 60,000 tons respectively.

Because of the controversial nature of any increase in the TAC and the possible return of non-Gulf based stern trawlers to this fishery, there were extensive consultations in various fora. These included the Atlantic Groundfish Advisory Committee; the Federal-Provincial Atlantic Fisheries Committee — Deputy Minister level — and the Atlantic Council of Fisheries Ministers — Ministerial level. Those discussions made it clear that the Gulf and non-Gulf based interests were diametrically opposed on whether to increase the TAC and how any increase should be shared. The Gulf-based interests argued against any significant increase in the TAC. This position was supported by fishermen and processor associations in the Gulf and the provinces of Quebec, New Brunswick and P.E.I. The non-Gulf based interests demanded that the TAC be increased to a minimum of 60,000 tons, with all of the increase to be allocated to them. This position was supported by associations of vessels and plant owners based outside the Gulf and the provincial governments of Nova Scotia and Newfoundland.

This was a classic case of conflicting interests, which the federal government had to reconcile.

The first question — the level of the TAC — was not so much how to measure the total benefits the resource could yield, but over what period the benefits should be realized. The alternatives were: (1) to maximize production in the short run; and (2) to spread the benefits over the longer run and promote a more stable fishery.

The second question — how to allocate the benefits between competing groups — was largely an equity issue: who should share in the benefits? This was particularly complex. There was a broad array of variables to consider. Many of these were not easily quantifiable. These variables included: access to other stocks and possible substitution of one species or stock for another, relative growth of different stocks, market demand for various species and price fluctuations, harvesting and processing costs.

The government appeared to have three major alternatives:

1. Minimal increase — a TAC no higher than 40,000 tons, the position advocated by the Gulf-based interests;
2. Increase the TAC to 60,000 tons — the level demanded by the non-Gulf based interests; and
3. Increase the TAC to somewhere between 40,000 and 60,000 tons. The scientific advice indicated a TAC of 50,000 tons could be sustained over the next several years thus providing stability of catch.

While officials sought a compromise, the interest groups lobbied federal Cabinet Ministers. This lobbying elevated the issue to the Cabinet level for decision.

On December 29, 1983, Fisheries Minister Pierre De Bané explained the government's decision:

> "It has been decided to establish the TAC at 50,000 tons for 1984 and 1985.... It is planned to increase this TAC level to 55,000 tons in 1986 and 1987 and to 60,000 tons in 1988. These provisional TACs would be subject to annual scientific review, with the proviso that the TAC in any year will not be established above the $F_{0.1}$ level. Any increase would only occur if confirmed by the latest scientific advice available at that time. The Gulf-based quota will remain at 40,000 tons for any level of TAC between 50,000 and 60,000 tons. For instance, should the TAC increase above 50,000 tons and then subsequently decrease, the Gulf-base share will be maintained at 40,000 tons. If the TAC should fall below 50,000 tons, the shares to all interests will be reviewed.
>
> "During the period 1984 to 1988, the Gulf quota of 40,000 tons will be set aside for Gulf based vessels, and also if necessary, chartered non-Gulf based vessels, all for landing at Gulf plants.
>
> "The remainder of the TAC will be reserved for non-Gulf based trawlers greater than 30.5 m (100 feet) in length. During September each year, a review of the Gulf redfish quota

will be undertaken and any anticipated uncaught amount from the Gulf quota will be made available to non-Gulf based vessels, on condition that an equitable amount, as determined by the Department of Fisheries and Oceans, be landed at Gulf-based plants" (DFO 1983b).

From 1984 to 1986, the actual TACs followed the De Bané Plan, 50,000 tons in 1984 and 1985, 55,000 tons in 1986. For 1987 the TAC was reduced to 50,000 tons because the scientific analyses indicated the TAC equivalent to $F_{0.1}$ would be 50,000 tons. Thus the reduction accorded with the proviso in the De Bané Plan. In 1988, the TAC was increased to 56,000 tons. The 6,000 ton increase was used to create a special program for Gulf-based groundfish plants short of resources for processing. For 1989 the TAC was increased again by 1,000 tons. This slight increase was allocated to small mobile gear vessels less than 65 feet based in the Gulf. This gear sector was also given first priority to contract for harvesting the 6,000 tons of Gulf redfish allocated to the Special Program for Gulf Plants.

Overall, the process of setting TACs and allocations for Gulf redfish from 1983 to 1989 illustrated the influence of conflict over allocations in determining the level of harvest. It also showed how the various fisheries groups sought to exert political pressure for an outcome favourable to their interests. "Best use" in this case was determined by a weighing of the political pros and cons of a given course of action.

2.5 Northern Cod

Another major focus of debate over groundfish resource allocation in the years from 1977 to 1992 was the northern cod stock off northeastern Newfoundland and Labrador. This stock is fished on the offshore banks during the winter as the fish congregate prior to spawning. It migrates inshore during the late spring and early summer to feed, where it is caught by fixed gear, particularly cod traps. Historically, the northern cod stock was the basis for settlement of the coast of eastern Newfoundland and Labrador. It continues to be the main support of hundreds of coastal communities there. These communities suffered great hardship because of the build-up of the foreign fleets and overfishing during the 1960's and early 1970's. The inshore cod catch declined from close to 200,000 tons in the 1950's to only 34,000 tons in 1974 (Fig. 7-3). Up to that time, there had been very limited fishing of this stock by the Canadian trawler fleet (a high of 20,000 tons in one year when the foreign fishery was at its peak). Following extension of fisheries jurisdiction the federal government took a conservative approach to managing this cod stock to rebuild the area's inshore fishery. Reported catches in the 1960's had peaked at approximately 800,000 tons in 1968. Scientific projections indicated that, if recruitment returned to the levels of the 1960's, the long term yield fishing at $F_{0.1}$ could be as high as 400,000 tons. This projection was subsequently reduced to 300,000 tons.

TACs for this stock were first introduced by ICNAF in 1973 at about 650,000 tons and continued at this level

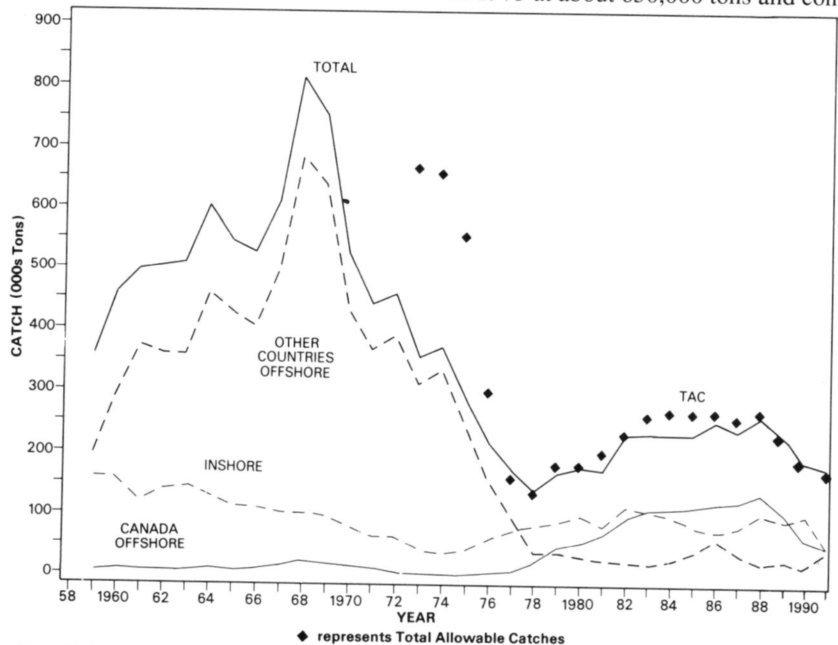

FIG. 7-3. Trend in catches of northern cod: 1959–91.

in 1974. The TAC was reduced to 550,000 tons in 1975. In retrospect, it was clear that these TACs were set too high. In 1976 the TAC was cut to 300,000 tons to reduce the fishing pressure on this stock (Pinhorn 1979). The TAC for 1977 was established at 160,000 tons, corresponding to the $F_{0.1}$ reference level. Following extension of fisheries jurisdiction the TAC for 1978 was set below $F_{0.1}$, 135,000 tons, to provide for more rapid rebuilding to restore the inshore fishery.

The stock began to increase in abundance after 1977. Catch rates in the winter offshore fishery increased very significantly. Canadian trawlers began to fish this stock offshore in 1976 and 1977 with the assistance of federal subsidies and grants under the Groundfish Vessel Dislocation Program and the Northern Fishery Incentive Program. Both Newfoundland and Maritimes-based vessels participated. Inshore catches increased from 34,000 tons in 1974 to about 90,000 tons in 1979. Catches by the Canadian offshore fleet increased from 1,000 tons in 1975 to 45,000 tons in 1980. Newfoundland-based offshore vessels caught 82% of the offshore total and Maritimes-based vessels caught 18%.

With these developments in the fishery and the anticipated increase in the stock over the next several years, this stock became very attractive to the Canadian offshore fleet. As representatives of the offshore fleet were battling to retain access to cod in the Gulf of St. Lawrence, they were also jockeying for a greater share of the anticipated TACs of northern cod.

Public debate about access to the rebuilding northern cod stock began to escalate with each interest group seeking the largest share of the increasing TAC. Given the great social and economic importance of this fishery, the federal Minister of Fisheries convened a major government/industry seminar on northern cod. The seminar was held in Corner Brook, Newfoundland in August 1979, with representatives of the provincial governments and the inshore and offshore fleets participating. The Rapporteur's report (DFO 1979a) presents conclusions of the Seminar:

> "(a) While there was no unanimity, there was strong agreement that the TAC on northern cod should be established at a conservative level, not only to enable the stock to rebuild rapidly but because there was uncertainty expressed by the scientists about their knowledge of the stock....;
>
> (b) There was general agreement that in granting access to the northern cod stock, the inshore fishery should be given priority and that there should be some measure of control on the size of the inshore fleet. The proportion of the stock to be allocated to that sector of the fishery should be about two-thirds of the total."

It was also agreed that a new middle distance fleet should replace part of the fixed gear fleet to lengthen the inshore fishing season. There were three other important points of agreement: (1) Neither factory nor freezer trawlers were required to catch the northern cod. (2) Some offshore northern cod caught in winter should be distributed to underutilized seasonal plants. (3) In the absence of a surplus, northern cod should not be used to obtain market access or tariff reductions. Although strong views were expressed for and against the use of northern cod to obtain conservation cooperation beyond the 200-mile zone, no consensus was achieved. Foreign access to non-surplus cod was an extremely controversial issue at that time (see Chapter 11).

The government of Newfoundland and Labrador advanced this position at the Seminar:

> "This stock must, to the extent possible, be harvested by inshore/middle distance effort based in communities along the east coast of the Island and along the coast of Labrador. We envisage this effort yielding approximately 300,000 of the 350,000 to 400,000 metric ton TAC projected for 1985.
>
> "If it can be demonstrated that a part of this resource cannot be harvested by inshore effort or, based on new knowledge that a part can be harvested offshore without adversely affecting the inshore fishery, that part must be reserved to offshore effort which is based in Newfoundland and which will land its catch in Newfoundland ports for distribution to processing plants which now operate on a seasonal basis. We would estimate that approximately 50,000 metric tons will be made available for this purpose.... Extension of the period during which raw material is available to the remaining plants must be the primary objective of any offshore effort that might be directed at northern cod" (Newfoundland 1979a).

One month after the Seminar, the federal government set forth its position in the discussion paper *Toward a Policy for the Utilization of Northern Cod*. The policy for northern cod management had these basic elements:

"(a) The first and over-riding priority in allocations is to the inshore fishery. The consensus from the seminar participants was that two-thirds of the TAC of northern cod should be set aside as an allowance for the inshore fishery;

(b) There must be some control on the number of units in the inshore fishery; and

(c) An amount not to exceed 10% of the TAC of northern cod or alternatively an amount of 20 to 30 thousand tons should be set aside for negotiating, through NAFO, a larger allocation to Canada of fish beyond the 200-mile zone. As the offshore fleet will be the principal beneficiary, this amount should come from the offshore allocation. The domestic offshore allocation would thus approximate 25% of the TAC" (DFO 1979b).

The province of Newfoundland was not happy with this policy. Premier Peckford subsequently stated that "the inshore fleet can, and should, take up to 85% of the Northern Cod, with any residual volume being taken by Newfoundland-based offshore effort to supply resource short plants in this province." He argued that only constitutional realignment of jurisdictional responsibilities could enable Newfoundland to ensure that the northern cod resource was managed for the full benefit of Newfoundlanders (Newfoundland 1979b).

In December 1979, Newfoundland attempted to force all offshore vessels fishing this stock to land their catch in Newfoundland for processing. Premier Peckford summoned the executives of National Sea Products and H.B. Nickerson to his office and told them that if they landed northern cod caught in "Newfoundland" waters at mainland plants, the Government would reserve the right to consider them ineligible for any provincial assistance programs or to apply for any further processing licenses. By adopting these tactics, Premier Peckford lost Nova Scotia's support for constitutional change respecting fisheries and hence undermined his constitutional initiative in September 1980 (see Chapter 2).

Nova Scotia firmly resisted the Newfoundland ultimatum to the Nova Scotia-based companies. On September 4, 1980 the new Nova Scotia Fisheries Minister Edmund Morris went on the attack. He said that Newfoundland's northern cod claim had been a warning about their behaviour if the provinces got together to allocate various species.

For 1980, the TAC for northern cod was 180,000 tons. The Canadian quota was 155,000 tons, of which 45,000 tons was allocated to the offshore fleet and 110,000 tons set aside as an "inshore allowance." For 1981, the TAC was increased to 200,000 tons, with a Canadian quota of 185,000 tons. Of the Canadian quota, 62,500 tons was allocated to the offshore fleet and the amount set aside for the inshore fishery was increased to 120,000 tons. The 1981 Plan stipulated that when 47,500 tons had been taken the directed offshore fishery would close. The remainder of the offshore quota would be used to support a flexible by-catch in the flatfish fisheries.

Both the inshore and offshore fisheries for northern cod are seasonal. The inshore fishery takes place in summer and early fall and the offshore fishery in winter and late fall. The offshore fishery primarily occurs from January to April. Catches usually peaked during the months of January and February, declining through March and April as the cod stocks began to disperse prior to their inshore migration. By May, catch rates were considered too low for trawlers to fish economically. They remained low on the offshore grounds throughout the summer and early fall. Catches did not begin to increase offshore until November, as the cod once again began to concentrate in areas 2J and 3K. The fishing "window", therefore, for offshore trawlers covered a maximum of approximately 6 months. Ice conditions and general environmental factors in many years also constrained the effective fishing period. In general, the effective fishing "window" is probably only 4–5 months. It was therefore not possible to conduct a year-round offshore fishery for northern cod.

The 1981 fishing season was unusual. The ice-cover was light and concentrated north of the prime fishing grounds. Wind and weather conditions were also more favourable than normal. These conditions plus the competitive forces of the open-access regime contributed to a tremendous race to catch the fish. By the end of January it was apparent that the offshore quota would be caught before the end of February.

Fisheries Minister LeBlanc discussed this matter with Atlantic Fisheries Ministers at a meeting in early February, 1981. The Fisheries Minister for Nova Scotia strongly opposed any change in the Groundfish Management Plan. Newfoundland Fisheries Minister Jim Morgan reiterated his government's position that 85% of the TAC should be set aside for the inshore/midshore fishery.

On February 9, 1981, Fisheries Minister LeBlanc announced that the offshore fishery for northern cod could be closed before the end of February — more than 60 offshore vessels were catching about 1,700 tons per day. The Minister noted that trawler owners the previous year had emphasized the importance of year-round operation for their vessels. "In spite of this, trawler owners have been directing excessive effort towards the offshore cod fishery. I think the result of this unrestrained activity is obvious," he stated. Mr. LeBlanc said he had discussed the issue with the provincial fisheries ministers from Québec and the Atlantic provinces. "Although not unanimous, it was the general agreement of my provincial colleagues that no action be taken to close the fishery before the quota was reached," he observed (DFO 1981c).

Much of the 45,000 tons of northern cod caught during the months of January and February 1981 was processed into cod blocks. The companies did not have sufficient capacity for fillet production during the seasonal peak offshore winter fishery. These effects of the 1981 "race for the fish" set the stage for the trial introduction of enterprise allocations for the 1982 fishery (see Chapter 9).

One of the major principles emerging from the Northern Cod Seminar and embodied in the 1979 federal Discussion Paper was that the inshore fishery had priority. This resulted in an allowance for the inshore fishery equal to two-thirds of the northern cod TAC. This priority was reflected in the 1980 and 1981 Groundfish Plans. For 1981, the inshore figure was identified as a "quota", implying that the inshore fishery would be closed when this catch level was reached. In 1982, an "allowance" was again set aside for the inshore fishery. It was decided to maintain this approach subsequently because the inshore fishery had fallen significantly short of its 120,000 ton quota in 1981. Thus it appeared preferable to manage it as an average over several years. The rationale was that this would allow the inshore fishery to take its full potential with existing effort levels in years of high cod abundance inshore. It would also ensure that the offshore fishery was not unduly constrained by overestimates of potential inshore catch. On a practical basis, this resulted in a 55:45 inshore/offshore split of the northern cod TAC by 1982. The Newfoundland government continued to argue its case for virtually exclusive access by Newfoundland-based vessels. However, it appeared to drop its original demand for a 85:15 percentage split in favour of the inshore fishery.

The second major element of policy that emerged from the Northern Cod Seminar was setting the TAC at a level below $F_{0.1}$. The level that was adopted in 1980 was F=0.17 (180,000 tons), compared with an $F_{0.1} = 0.20$. The TAC continued to be set at this level of fishing mortality until 1983 when, to conform to the scientific assessment, the TAC at F=0.17 would have had to be reduced. It was decided to move to the $F_{0.1}$ level for 1984. In 1986 and 1987 the TAC for this stock was again set below the calculated $F_{0.1}$ level. For 1987 the level was set 10,000 tons below the calculated $F_{0.1}$ level and for 1988 it was set 27,000 tons below the calculated $F_{0.1}$ value. For 1986–88 these lower levels in relation to the calculated $F_{0.}$1.values were established in response to concerns about declining inshore catches. (Subsequent scientific assessments established that the actual fishing mortalities during these years were more than double the target level).

The Task Force on Atlantic Fisheries in its 1982 Report devoted a chapter to the Northern Cod Stock (Kirby 1982). The Task Force assumed that the Canadian quota by 1987 would reach 380,000 tons. It proposed that the 1987 quota be allocated as follows (Initial 1982 allocations are shown for comparison):

	1982	1987
1. Inshore allowance	120,000 tons	145,000 tons
2. Existing trawler fleet — vessels over 100 feet	87,250 tons	145,000 tons
3. Resource-short plants	5,250 tons	50,000 tons
4. Other fixed and mobile gear	2,500 tons	40,000 tons
TOTAL	215,000 tons	380,000 tons

The major recommendation was that the allowance to existing inshore vessels and the allocation to existing trawler owners not be increased proportionately to the anticipated increase in the TAC. Instead, the Task Force proposed that a substantial proportion of the growth should be allocated to supply so-called Resource-Short Plants in the off-season and to support the development of a fleet of large "Scandinavian-type" longliners as an alternative to the small fixed gear vessels currently utilized in the inshore fishery. The intent of both the Resource-Short Plant Program and the fleet of longliners would be to reduce the seasonality of processing activities along the northeast coast of Newfoundland.

These recommendations were accepted by the federal government and shaped the allocations policy for northern cod for the next several years. The expected volumes did not materialize because the stock did not reach projected levels. Some allocations were established, however, for the Resource-Short Plant Program and the development of a Scandinavian longline fleet. The 1983 Groundfish Plan included an allocation of 10,000 tons of northern cod plus 17,000 tons of other stocks and species for the Resource-Short Plant Program and 3,000 tons for fixed gear vessels between 65 and 100 feet.

The federal government developed eligibility criteria and a list of eligible plants for the Resource-Short Plant Program. At the same time, 25% of the

RSPP was made available to plants based outside northeast Newfoundland and Labrador (Divisions 2J, 3KL). The northern cod allocation for the RSPP was maintained at 10,000 tons from 1983 to 1986, reduced to 9,300 tons in 1987 with the TAC reduction, and increased to 15,350 tons in 1988 when the TAC was restored to 266,000 tons.

Meanwhile from 1983 to 1987 there was growing concern that inshore catches which had reached 113,000 tons in 1982 had subsequently declined to a low of 72,000 tons in 1986. CAFSAC (1986a) concluded that the decline in inshore catches was the result of a combination of factors:

1. Abundance of cod inshore;
2. Catchability of cod; and
3. Fishing effort.

CAFSAC stated that the overall abundance of the stock and the extent of the spring migration to the inshore areas determines the abundance of cod inshore. The inshore migration was influenced by the prevailing water temperatures and by the relative supply of food, mainly capelin, in the inshore and offshore areas respectively. CAFSAC concluded that the decline in 1985 in inshore catches was likely due mainly to severe environmental conditions since the biomass of the stock had apparently not declined. In 1986, the low inshore catch was probably a result of an abundance of food capelin offshore and a diversion of effort to other species inshore (CAFSAC 1986a).

This explanation was not acceptable to inshore fishermen. They argued that the declining inshore catch was the result of overfishing of the cod stock offshore by Canadian offshore trawlers within the Canadian fishing zone and foreign fishing vessels outside the 200-mile zone on the Nose of the Grand Banks.

Fisheries and Oceans Minister Tom Siddon in July 1987 appointed a Task Force headed by Dr. Lee Alverson, a natural resources consultant, to investigate the reasons for the decline in the inshore cod catches. The Task Force was to determine whether the decline was due to more than natural causes and, if so, what management measures could be taken to control it. The Task Force (Alverson 1987) concluded that the decline in the inshore cod catch was due to a combination of several factors:

1. Changes in availability resulting from predator/prey and/or environmental relationships, coupled with a slower than anticipated rebuilding of the major stocks;
2. Uneven distribution of fishing on stocks or components of offshore stocks migrating to inshore fishing grounds;
3. Potential over-fishing in the southern areas by the inshore fishermen of separate inshore stocks of cod;
4. Redeployment of effort to other target species;
5. Possible effects of fishing on recruitment; and
6. A slower growth rate of individual cod.

The Alverson Task Force's explanation of the decline agreed, in large part, with that offered earlier by CAFSAC. The one new possibility introduced in the task group report was potential over-fishing of inshore stock components by inshore fishermen in southeastern Newfoundland.

Inshore fishermen were seeking lower TACs because of the poor inshore catches. At the same time, however, some Gulf-based interests, having successfully constrained the participation by Newfoundland and Nova Scotia-based trawlers in the Gulf fishery, turned their eyes enviously toward the stocks of the continental shelf, in particular, northern cod. In 1987, a consortium of Quebec and New Brunswick interests, La Société de Pêche Nova Nord (Nova Nord), submitted a request to the federal government for access to northern groundfish. The Consortium sought 34,500 tons per year initially, rising to 68,400 tons over 10 years. The Nova Nord consortium involved seven Quebec and five New Brunswick companies. Nova Nord argued that:

1. Its plants had not received an equitable share of the new resources resulting from extension of fisheries jurisdiction; and
2. Plant capacity utilization in the Gulf was low, and access to northern cod resources would alleviate the problem.

The irony of Gulf-based interests seeking increased access to northern cod was not lost on the Newfoundland government. It lobbied vigorously and successfully against the proposal during the development of the 1988 and 1989 Groundfish Plans. Late in 1988, the Newfoundland government issued an analysis titled *Northern Cod Under Attack: The Newfoundland and Labrador Perspective on the Allocation of Northern Cod* (Newfoundland 1988). It argued that Quebec and New Brunswick had experienced greater percentage increases in the landed value of fish catches than had Newfoundland since 1977 (Fig. 7-4).

The Newfoundland government claimed that its position on resource access had been misrepresented over the years: "As a province, we have never taken the position that other provinces should not enjoy access to fish resources in waters adjacent to Newfoundland and Labrador. First and foremost, we have always held that historical fishing practices and patterns should not be jeopardized. Furthermore, we have always maintained that other provinces could access a share of any given resource once the demonstrated needs of Newfoundland and Labrador have been met. Northern

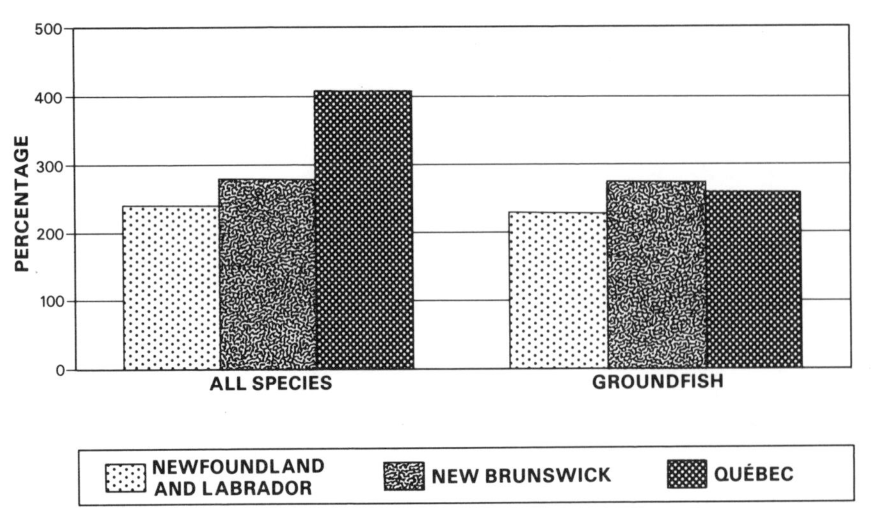

FIG. 7-4. Percentage changes in landed value of the Atlantic catch by province: 1977–1986 (from Newfoundland 1988).

cod, in particular, is not a species which is surplus to our Province's needs. Accordingly, the demands of the Nova Nord consortium must be dismissed outright."

The Alverson Task Group estimated a 5-fold increase in stock size from 1976–86 (an annual rate of increase of 13%). CAFSAC had estimated a 5.5-fold increase over the same period (an annual rate of increase of 15%).

In announcing the 1988 Groundfish Plan on December 30, 1987, Fisheries Minister Siddon stated:

> "The Task Group has confirmed the evidence developed by DFO scientists that the northern cod stock has increased substantially in abundance since 1976...The Task Group recommended a cautious approach to the setting of the 1988 TAC in order to accelerate the growth of the northern cod stock and provide some buffering against the combination of factors which have contributed to the decline in the inshore fishery. I have taken this advice into account in establishing the TAC for 1988 at 266,000t...." (DFO 1987b).

Regarding the allocation issue, Minister Siddon observed:

> "There are many new proposals for access to Northern cod based upon the expectation of a much higher TAC. However, the increase in the TAC of 10,000t will just allow me to reinstate the groups which suffered reductions in 1987.... Other proposals will have to be held in abeyance until the resource picture improves."

Inshore catches which had been decreasing from 1982 to 1986, increased slightly to 79,000 tons in 1987 and more significantly to 102,000 tons in 1988, the highest level since the previous peak of 113,000 tons in 1982. However, despite the increase in the inshore catches, there were some signs that the annual assessments had been optimistic. CAFSAC in January 1989 (CAFSAC 1989a) concluded that the stock had recovered since the very low period of the late 1970's but at a slower rate (10%) than thought previously. In particular, CAFSAC concluded that the indications of stock status provided over the years by the research vessel surveys, which were more pessimistic than the commercial fishery indicators, had more accurately reflected historical trends in the cod population. CAFSAC further observed that the almost constant increase in offshore commercial fishing success (catch per hour) between 1977 and 1983 was due to increases in efficiency as well as stock growth. The extent of stock recovery was less than the improvement in fishing success would indicate. In a dramatic conclusion, CAFSAC estimated that the fishing mortality in 1988 was probably about 0.44 and over the past several years had been between 0.4 and 0.5. This was far above the $F_{0.1}$ level of 0.2.

This analysis indicated that the biomass of cod aged 3 years and older increased from a low of about 450,000 tons in 1976 to about 1.2 million tons in 1984. It declined thereafter to about 1 million tons in 1988 (Fig. 7-5). The reduction in biomass after 1984 was attributed to fewer young fish entering the population in 1986 and 1987 than in the previous several years, rather than increased harvest of older fish. The immediate recruitment prospects were considered poor. CAFSAC concluded that fishing in 1989 at $F_{0.1}$ would generate a catch of 125,000 tons. This implied a reduction in fishing effort by more than 50%.

This dramatic change in the perception of stock size posed grave questions about the future management of this fishery. The management strategy since 1979,

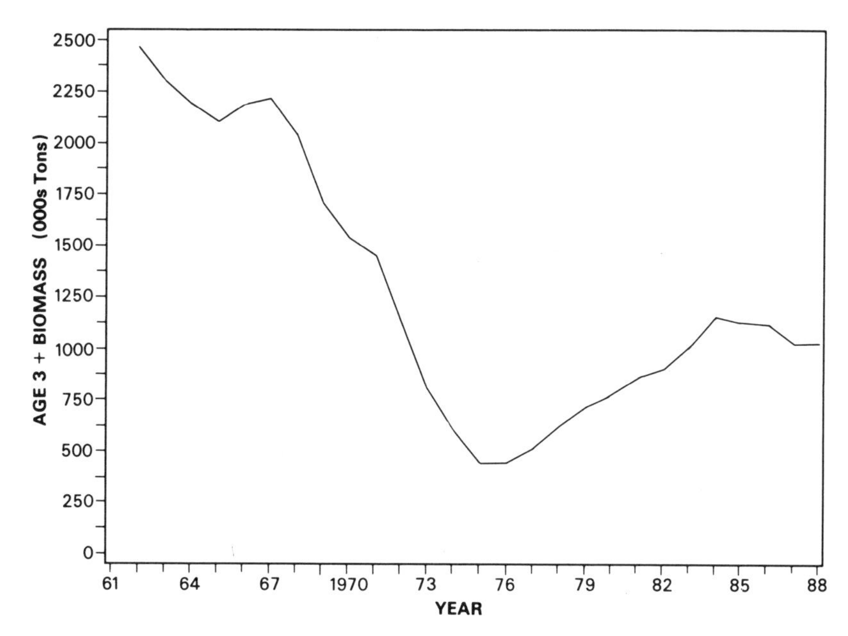

FIG. 7-5.Age 3+ Biomass for cod in divisions 2J3KL: 1961–88 (from CAFSAC 1989a).

confirmed by the Task Force on Atlantic Fisheries in 1982, had been based on the assumption that the TAC would have reached 300,000 tons or more by 1987–1988. If the 1989 CAFSAC assessment proved correct, then a radical reevaluation of the management of this stock was required, given that the inshore fishery in 1988 harvested 102,000 tons. The social and economic implications of a TAC of 125,000 tons appeared catastrophic.

Responding to the CAFSAC advice, Fisheries Minister Siddon on February 8, 1989, announced a downward revision of the 1989 TAC to 235,000 tons, a level which it was thought would stabilize the stock at its existing level (DFO 1989c).

On February 12, 1989, the Minister announced that the inshore sector would retain its allowance of 115,000 tons. The 31,000 ton TAC reduction would come from the other fleet sectors. This reduction was shared proportionately among the remaining user groups (DFO 1989d).

The Minister also announced the establishment of an independent review panel, headed by Dr. Leslie Harris, President of Memorial University. The Panel was to examine trends in stock abundance, the data used in assessing and forecasting catches, the mathematical methods used for stock assessment in Canada and other countries, and, the calculations leading to the 1989 advice.

The CAFSAC advice had shattered a decade of expectations. Thus verification of the assessment seemed prudent.

On May 15, 1989 the Independent Review Panel released an interim report (Harris 1989). The Panel confirmed the CAFSAC reassessment of the northern cod stock and concluded that F was between 0.35 to 0.55. The Panel recommended that the fishing mortality in 1990 be reduced to halfway between the existing level (0.45) and the $F_{0.1}$ level (0.20). This implied a TAC of 190,000 tons for 1990.

On January 2, 1990, Minister Siddon announced that the TAC was being reduced from 235,000 tons to 197,000 tons for 1990 (DFO 1990b).

On March 30, 1990 the final report of the Independent Review Panel was released (Harris 1990). It recommended an immediate reduction of F to 0.30 and as early as possible to 0.20.

The socioeconomic impacts of proposed reductions in the northern cod TAC were far reaching. The two major companies, Fishery Products International and National Sea Products, announced plant closures and vessel tie-ups. In response to the northern cod crisis and quota reductions elsewhere in the Atlantic, the government on May 7, 1990 announced a major Atlantic Fisheries Adjustment Program (AFAP) (Canada 1990a). This provided funding for expanded scientific research and enforcement, an industrial adjustment program for workers in communities facing plant closures, and a program for community economic diversification. For details, see Chapter 13.

The government chose not to reduce further the 1990 TAC because of the additional hardships it would produce. On December 14, 1990, Fisheries and Oceans Minister Valcourt announced a new multi-year Groundfish Plan (DFO 1990c). This included a 3-year program of provisional TACs for northern cod at 190,000 tons in 1991, 185,000 tons in 1992 and 180,000 tons in 1993. The inshore allowance was maintained at 115,000 tons for the 3-year period. The allocation of northern cod to the Resource-Short Plant Program was eliminated.

This period of relatively stable TACs was proposed in the expectation that the stock would rebuild. Fs at 0.45 were far above the revised $F_{0.1}$ value of 0.25 and approaching F_{max} (0.52). Thus any rebuilding would occur slowly. Hopes were pinned upon a forecast that the 1986 year-class was relatively better than average.

The 1979 allocations policy for the northern cod stock had been developed when expectations about the level of future TACs were high. When the inshore fishery failed to reach its allowance, the policy was revised to provide for further growth of the Canadian offshore fishery. New programs (the RSPP and a Scandinavian longliner fleet) recommended by the Task Force on Atlantic Fisheries had also been developed.

The reassessment of the status of northern cod in January 1989 had significant implications not only for the level of TAC but also for allocations policy. The offshore share of the TAC was reduced and the newer programs eliminated. The bitter federal-provincial and interprovincial debates over allocation of northern cod during the previous decade were overtaken by the changed perception of the amount of resource available. In retrospect, these shattered expectations illustrate the fallacy of relying on long-term resource projections when determining resource conservation and allocation policies.

Early in 1992 hopes for a quick rebuilding were dashed when CAFSAC advised that the biomass of cod ages 3 and older had declined to 800,000 tons, mostly because of the disappearance of older fish (CAFSAC 1992a). This indicated that under the 3-year plan fishing mortality would remain high. The spawning biomass for 1991 was estimated at 130,000 tons, amongst the lowest observed. CAFSAC expressed concern about the low estimate of the adult stock biomass. It advised that 1992 catches for the first half of the year should be restricted to the lowest level possible, to about 50% of catches during the first half of 1991.

In response, Fisheries Minister John Crosbie slashed the Canadian northern cod TAC to 120,000 tons. The offshore quota was reduced from approximately 72,000 tons in 1991 to 20,000 tons for the first 6 months of 1992. The implications of this reduction sent shock waves through the Atlantic fishing industry.

These shock waves intensified when new scientific advice was received in June and July 1992. Because of EC failure to curtail its fishery for cod on the Nose of the Grand Banks, Canada requested the NAFO Scientific Council to review the status of this stock. The Scientific Council at a Special Session, in June 1992, confirmed that the biomass of northern cod had declined sharply during 1991. It concluded that the stock was at a very low level. The age 3+ biomass at the beginning of 1991, between 520,000 and 640,000 tons, and the 7+ biomass (approximately the spawning stock biomass — SSB), between 72,000 and 110,000 tons, were at or approaching the lowest levels ever observed for this stock. It expressed concern about the low level of the SSB and recommended that fishing mortality should be reduced in 1992 from the level of recent years. While calculating a range of possible $F_{0.1}$ catches under various assumptions and using different mathematical models, the Council suggested it would be wise to consider the $F_{0.1}$ catch for 1992 to be at the lower value of 50,000 tons (NAFO 1992a).

Following receipt of the NAFO Scientific Council Report, Fisheries Minister Crosbie called upon all countries to observe the NAFO moratorium on northern cod outside the Canadian 200-mile zone. The European Community responded by announcing that its vessels would cease fishing for northern cod, at least until the fall of 1992.

At the beginning of July, 1992, CAFSAC released its final advice on the northern cod stock for 1992. CAFSAC concluded that the abundance of cod had declined drastically in the previous 12–18 months. It estimated the age 3+ biomass at the beginning of January 1992 to be between 527,000 and 696,000 tons and the age 7+ biomass to be between 48,000 and 108,000 tons, about 10% of the long-term average. All indications were that the stock had recently declined abruptly and was at or near its lowest observed abundance. CAFSAC advised the 1992 catches be kept to the lowest level possible (CAFSAC 1992b).

The exact cause of the sudden, drastic and unexpected decline during 1991 was not clear. Water temperatures off Labrador and northeastern Newfoundland were considerably below normal in 1991. The ice coverage was greater than normal and persisted for a record length of time, with ice present inshore into what would normally have been summer. Low ocean temperatures persisted throughout summer and early autumn and the cold intermediate layer of waters less than 0° C of the Labrador Current was at or near its greatest size (CAFSAC 1992b).

The entire stock was essentially composed of the 1986 and 1987 year-classes (5 and 6 year-olds in 1992). The following four year-classes appeared to be weak. Fisheries in the next few years would depend heavily on the 1986 and 1987 year-classes and would exploit them before they had made their full contribution to the spawning stock. CAFSAC recommended that all efforts be made "to enhance the possibilities of stock rebuilding."

The federal government reacted swiftly to the new scientific advice. On July 2, 1992, Fisheries Minister

John Crosbie imposed a 2-year moratorium on the northern cod fishery to continue until the spring of 1994. The rationale was that since 1990 a devastating decline in the northern cod stock, due primarily to ecological factors, had reduced the biomass by half and the spawning biomass by three-quarters. A 2-year moratorium appeared to offer the only chance for the spawning biomass to recover quickly to its long-term average, permitting resumption of an inshore fishery in the spring of 1994 (DFO 1992a).

The bitter allocations debate since 1977 was overshadowed by this sudden, drastic and unexpected decline in the stock. Both the offshore and inshore fisheries had been closed temporarily in the interests of stock rebuilding. This represented the most dramatic conservation action in Canadian fisheries history, with profound social and economic implications for Newfoundland and Labrador (see Chapter 13).

2.6 Sector Management

Let's return to the broader question of overall inshore/offshore allocations in the Atlantic groundfish fishery. With a few exceptions like northern cod, from 1977 to 1983 the inshore/offshore shares for each stock were determined annually on an ad hoc basis. The inshore Atlantic fisheries were initially managed under an allowance system. This changed in 1981 for fisheries other than northern cod. Sufficient catching capacity clearly existed in the inshore fleet to exceed the allowances in most years. The result was TAC overruns and further pressures to reduce the share of the TACs allocated to the offshore fleet. Once the inshore fisheries were placed under quotas, further measures were introduced to prevent the mobile-gear small vessels (less than 65 feet) from moving to other areas after taking their local quotas. The fleet of small draggers less than 65 feet in length based in southwest Nova Scotia had grown rapidly in number and catching capacity in the late 1970's. The quotas available to these vessels from local stocks were insufficient to keep that fleet operating year-round.

The threat of these vessels moving into the Gulf and rapidly catching the small vessel quotas there, led to the introduction of Sector Management on January 1, 1982. With Sector Management, vessels under 19.8 metres (65 feet) were to be regulated within three geographic sectors, "to bring more order and discipline into the harvesting operations of this fleet" (DFO 1981d, 1984a). The three sectors are southern and eastern Newfoundland and Labrador (NAFO Areas 0, 2, 3KLMNOPs), the Gulf of St. Lawrence (NAFO Areas 4RST and 3PN) and the Scotian Shelf-Georges Bank (NAFO Areas 4VWX and 5) (Fig. 7-6).

Sector Management recognized the differing needs within the inshore groundfish fisheries. A rationale for the policy was that Sector Management would enable a Regional Director General to make decisions respecting the small-vessel fleet specific to a sector. This could be done only if such decisions could not disrupt fishing patterns and practices in other areas. Licences for vessels less than 65 feet became nontransferable among Sectors. There were provisions for historic overlap so that vessels could maintain their traditional fishing patterns as much as possible. Sector management for the groundfish fishery was similar to the concept of management zones already in effect in the Atlantic salmon, herring and lobster fisheries.

Minister LeBlanc's announcement of the policy stated that the Sector Management approach was widely accepted by industry. However, there was vocal and vigorous opposition from some spokesmen for the southwest Nova Scotia small dragger fishermen, who felt most affected by the Policy. This resulted ultimately in a court challenge to the Minister's authority to allocate on social and economic grounds — the MacKinnon case referred to in Chapter 2. The Federal Court of Canada's Trial Division rejected the argument that Parliament's legislative powers were confined to the conservation and protection of fish stocks. Justice Martin concluded that: "The federal authority to regulate the fisheries includes the right to determine the areas in which fishermen may fish."

2.7 Southwest Nova Scotia Fleet Overcapacity and its Impact on Inshore/Offshore Sharing Arrangements

While this settled the question of the legality of the Sector Management approach, it did not restrain the appetite of the southwest Nova Scotia dragger fleet for more fish. During 1980 to 1983 there had been continuing battles over the sharing of the TACs. Both Gulf and southwest Nova Scotia inshore interests pushed for reduced shares of the offshore trawlers in those areas. They wanted to restrict the offshore trawlers to fishing on the Grand Banks, and for northern cod. This dispute led in the fall of 1983 to the introduction of Enterprise Allocations on a 5-year trial basis in the offshore Atlantic groundfish fishery (Chapter 9). For Enterprise Allocations to succeed, it was essential to establish a longer term sharing arrangement between the inshore and offshore components of the fleet. The EA Plan established by the federal government in December 1983 included a set of rules for sharing the Canadian quota between the inshore and offshore fleets (Parsons 1983). The 1984 EA Plan stipulated that:

1. The quotas for the inshore and offshore fleets be specified as proportions of the Canadian share of each groundfish stock. These ratios were for a 5-year period.

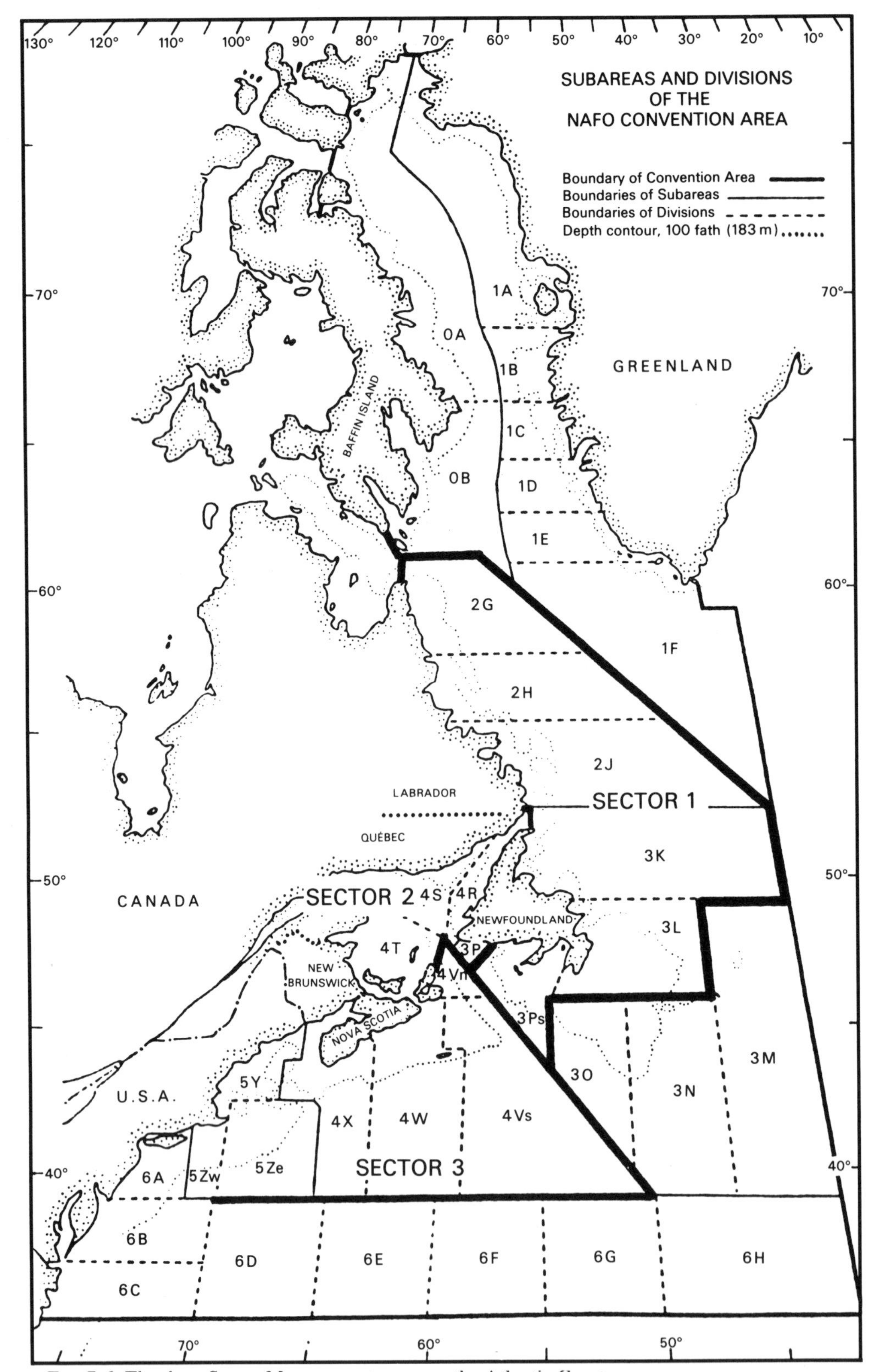

FIG. 7-6. The three Sector Management areas on the Atlantic Coast.

2. For 1984 and beyond, the percentages were based on percentages specified in the 1984 groundfish plan with the exception of the following stocks: Northern Cod, St. Pierre Bank Cod and Gulf Redfish.
3. For the Gulf redfish, the split was left to the Fisheries Minister to determine. Because of a low TAC for 3Ps cod, the offshore share was reduced to a by-catch allocation, but was to revert to the previous share if and when the TAC returned to the 1983 level. For Northern Cod, the share for the offshore fleet was to increase to meet by 1987 a target allocation consistent with that recommended by the Task Force on Atlantic fisheries.
4. Provision was made for adjustments in the percentage sharing arrangements by the federal minister when:
   (a) The scientifically-recommended TAC changed by 15% or more in any year, or cumulatively by more than 15% over a series of years during the 1984–88 period;
   (b) as a result of international events or negotiations, the Canadian share of the recommended TAC changed by 15% or more from that previously in force;

There was a provision for within-season quota transfers between the inshore/offshore sectors by mutual consent of the parties concerned. Such transfers were to be without prejudice to future years.

These provisions for a long-term sharing arrangement between the inshore and offshore fleets were a significant step forward in allocating access to the groundfish stocks. They provided a more sound basis for fisheries management planning. The actual percentage shares for particular stocks from 1977 to 1988 are given in Annex I of the report of the Government/ Industry workshop on Inshore/Offshore Sharing of the Atlantic Groundfish Resources, April, 1988 (DFO 1988a).

Subsequent to the 1984 arrangement, the imbalance between the available resource and fleet capacity created considerable pressure to change the inshore/offshore shares for Scotian Shelf stocks. This resulted in a number of temporary quota transfers from the offshore to the inshore for several stocks. For Georges Bank haddock, the inshore-offshore ratio was relatively stable from 1981 to 1984, averaging 27 and 73% of the Canadian quota respectively. Temporary transfers from the offshore to the inshore in 1985, 1986 and 1987 adjusted this arrangement with the inshore receiving on average, over the 3-year period, 57% of the Canadian quota. Similarly, for area 5Z cod in 1985, 1986 and 1987, the inshore share increased to about 74% of the Canadian quota. For Scotian Shelf-Georges Bank pollock, temporary transfers from the offshore to the inshore increased the inshore share from 40% in 1980 to 50% in 1987 and 1988.

In announcing the 1989 Groundfish Plan, Minister Siddon observed that there had been an overall decline in Scotian Shelf and Georges Bank TACs, of approximately 50% since 1984. Inshore and offshore sharing had been reviewed in preparation for longer-term renewal of Enterprise Allocations (see Chapter 9). The Minister announced that 12,000 tons of cod, haddock, pollock, redfish and flounder from the Scotian Shelf and Georges Bank would be transferred permanently from the offshore to the inshore under the 1989 Management Plan. This institutionalized the temporary transfers of the previous several years (DFO 1988b).

When the drastic reductions in cod TACs for 1993 were announced in December 1992, Fisheries Minister Crosbie reiterated the government's commitment to proportional reductions. The focus would be on rebuilding fish stocks, not on reallocating fish from one gear sector to another.

Over the years, the allocation principle in the Groundfish Plan had been restated:

> "Allocation of fishery resources will be on the basis of equity taking into account adjacency to the resource, the relative dependence of coastal communities and the various fleet sectors upon a given resource, and economic efficiency and fleet mobility" (DFO 1988c).

Behind this statement lay more than a decade of bitter conflict. It reflected a concerted effort to reconcile the widely differing interests in the Atlantic groundfish fishery.

3.0 PACIFIC SALMON

The major Pacific anadromous salmon fisheries are considerably different from the groundfish fisheries of the Atlantic coast. During the 1980's, however, they were buffeted by similar large scale conflicts over resource access.

The multispecies, multistock nature of the Pacific salmon fishery was described earlier. There are three major user groups in these diverse fisheries — commercial fishermen, recreational fishermen and native fishermen. The commercial category includes three major gear types — purse seine, gillnet and trollers — which also compete for the resource. Given the complexity of the Pacific salmon fisheries, this chapter cannot portray adequately all of the conflicting interests. Therefore, I will present the general context for allocating access. I will then examine in more depth one particular user group conflict — the battle over sharing declining chinook stocks. This battle was at the forefront of public debate

Pacific Salmon Fishery: Pacific Salmon Troller (top), Pacific Salmon Seiner (center) and Angler with Georgia Strait Chinook (bottom).

about Pacific salmon management from the mid-1980's onwards.

3.1 Pacific Salmon Allocations

Before 1985, there was no formal process to allocate the Pacific salmon catch. Although there was no explicit allocation framework, fishery managers emphasized maintaining 'traditional' fishing patterns. MacDonald (1981) observed that "a historically important management objective has been to maintain the share of each gear type in the total catch." This had been done by varying time and area closures among the three gear types to allocate a given share of the catch to each. The distribution of the commercial catch among gear types changed significantly from 1960 to 1977 (Table 7-1). Gillnet catches decreased from about 40% in the period 1951–65 to 23% in the 1977–80 period. The purse seiner share dropped slightly in the 1960's. By the late 1980's however, it had increased to about 58%. The share of the total catch taken by trollers increased consistently from 12% in 1951–55 to 26% in 1977–80 but then declined to 22% in the late 1980's.

There was considerable variability in the catch shares for particular species and stocks. For example, the Fraser River sockeye run had a variable catch history. The catch share for the troll fleet ranged anywhere from 19.7% of the run in 1978 to 39% of the run in 1982 when the troll fleet was used as a lever to bring the USA to the table in treaty negotiations. The gillnet share ranged anywhere from 14.8% in 1982 to 33.2% in 1974. Seiners caught between 46.2% and 62.5%.

Prior to the 1980's, fisheries managers conducted regular reviews of catches and the number, type and distribution of fishing vessels. Assuming that the efficiency of the various fleets remained constant, percentage shares among the different gear sectors were maintained or altered by varying fishing times.

The catch distribution among gear types in 1950 had changed significantly by 1980. This was caused by numerous changes in the relative efficiency of the different sectors and the relative changes in fleet composition. This reflected the following trends:

1. Larger and more efficient vessels, particularly seine and gillnet vessels, increased the mobility of the fleet;
2. Increased efficiency of seine vessels provided more opportunity to fish in bad weather and at night;
3. Troll vessels developed methods to target on non-traditional species, i.e. sockeye and pink salmon;
4. Larger troll vessels with onboard freezing capabilities were added to the fleet;

TABLE 7-1. Percentage composition of the B.C. salmon catch by species and gear type — 1960–1988.

	Sockeye	Coho	Pinks	Chums	Chinook	All Species
Gillnet						
1960–1963*	81.3	25.1	32.3	57.5	30.3	41.7
1964–1967*	75.8	25.0	29.9	53.8	25.5	39.7
1968–1971*	69.2	22.0	25.6	62.3	19.4	40.6
1972–1975*	64.0	18.3	19.6	51.1	15.4	38.6
1977–1980*	46.4	9.2	11.7	44.9	10.9	23.0
1981–1984*	38.6	5.2	10.7	40.1	9.6	19.5
1985–1988*	43.1	5.8	10.0	31.0	8.9	19.9
Seine						
1960–1963*	18.0	12.8	62.1	42.4	4.9	39.1
1964–1967*	21.9	11.4	59.3	45.7	5.6	33.1
1968–1971*	23.6	11.9	61.4	37.1	7.6	32.6
1977–1980*	46.1	14.5	67.4	52.8	13.0	49.5
1981–1984*	50.7	12.9	71.8	56.3	11.6	54.9
1985–1988*	43.1	11.8	74.4	62.4	12.8	58.3
Troll						
1960–1963*	0.7	62.1	5.6	0.3	64.8	19.2
1964–1967*	2.3	63.6	10.8	0.5	98.9	27.2
1968–1971*	7.2	66.0	12.9	0.6	73.0	26.8
1972–1975*	6.3	65.1	13.0	0.4	77.7	23.1
1977–1980*	7.5	76.3	20.9	2.3	76.1	27.5
1981–1984*	10.8	81.9	17.5	3.5	78.8	25.6
1985–1988*	13.8	82.4	15.6	6.6	78.3	21.8

* = four-year average.

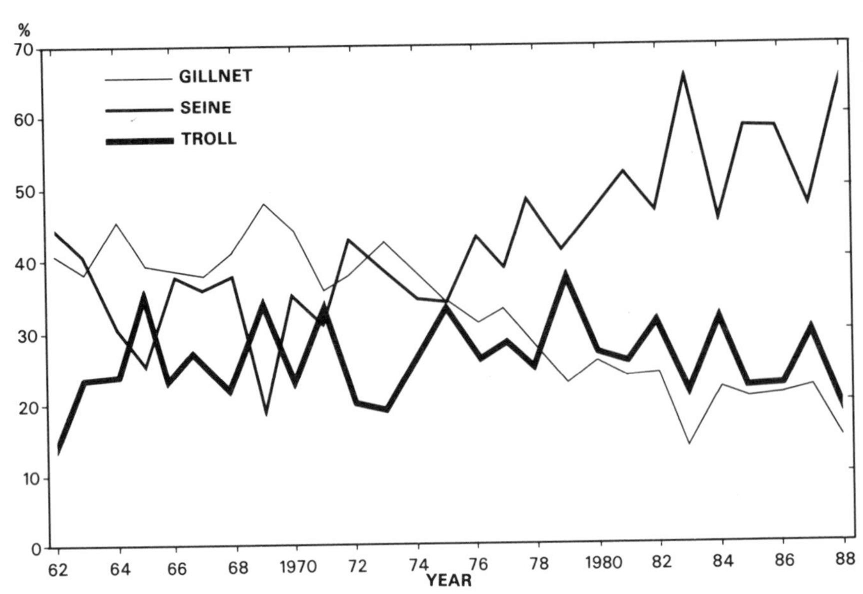

Fig. 7-7. Percentages of total salmon catch by gillnet, seine and troll gears: 1962–1988 (from data provided by Michelle James, Pacific Region, DFO).

5. Many gillnet vessels added trolling capability;
6. The number of seine vessels increased from 391 in 1968 to 592 in 1979; and
7. The number of gillnet vessels declined from 3760 in 1968 to 2441 in 1979.

Overall, seine catches had increased during the 1970's and 1980's while gillnet catches had declined substantially (Fig. 7-7). The seiners' share of the catch increased for all species. The most substantial gains were in the sockeye and chum fisheries. The troll fleet increased substantially its share of coho and chum.

MacDonald (1981) concluded that the use of area and time closures to ensure adequate escapement and to allocate the catch across gear types had induced fishermen to adjust to the reduced fishing allowable by increasing the fishing power of their vessels. Furthermore, in his view, the regulations were biased among gear types. Seiners were more restricted than gillnetters and gillnetters more than trollers. MacDonald conceded that the relative constancy of salmon landings through the years despite intense fishing pressure suggested that input controls had generally been effective in conserving the resource. He noted, however, that the success of biological management had been uneven:

> "Species primarily subject to the troll and recreational fisheries, which are not subject to extensive area and time closures, have experienced significant declines in stock levels. The chinook species, in particular, appears to be in considerable trouble."

In making this observation, he was reflecting public fears which had been front page news during the 1980 fishing season. By the late 1970's-early 1980's fisheries managers had recognized that the increased numbers and efficiency of the troll fleet had increased the proportion of salmon harvested offshore. By 1979 the Fishing Vessel Owners Association of B.C. had called upon the Department for controls on the troll fleet to counter its increasing efficiency in targeting pinks, sockeyes and chums, in addition to its proportionately high catch of commercially caught chinook and coho. In 1956 a "surfline" limit had been introduced to keep the "net" fleet closer to shore and protect the mixed stocks. Net fishermen argued that trollers operating outside the surfline had so improved their technology that they now threatened the mixed stocks as much as the net fishermen did when the "surfline" was first established.

An increased offshore harvest could upset the traditional balance of catch among gear types and increase the complexity of management. Action on this problem was delayed because the troll fleet was the only one which could effectively target U.S. stocks off the West Coast of Vancouver Island, an important factor in the Canada-USA interception negotiations (see Chapter 12).

Fishery managers recognized a need for chinook conservation measures. Such measures were needed across the board to protect the resource rather than simply redistribute the catch. In 1981, several measures were introduced to protect chinook. These primarily affected the troll fleet. While the net fleet had unexpectedly good catches of sockeye and pink salmon that year, trollers were not successful with the pink salmon run. But the gillnet fleet had by this time been forced out of most terminal fisheries with a considerable impact on their fishing operations. Beginning in 1981, DFO attempted to prevent further erosion of the gillnet share of the overall salmon catch. The 1984 fishing plan attempted to give the gillnet fleet more fishing time than seiners, for a "fairer" distribution of catch. Specific shorter seasons were proposed for the troll fleet.

Traditionally the troll season in the Gulf of Georgia had been from April 15 to September 30. The 1984 troll season for Georgia Strait was established as July 1 to August 30. This action prompted the Gulf Trollers Association to seek a court injunction preventing this change. The Gulf Trollers contended that the open season for chinook would be limited to a period when fishing chinook had been marginal for the trollers. Their primary target during July and August was coho. Appearing before Justice Collier of the Federal Court's Trial Division, the Trollers Association argued that new restrictions on chinook fishing were being imposed on the commercial fishermen with no corresponding restrictions on other groups in the fishing industry. They were particularly concerned about competition from sports fishermen. They further contended that the Minister's powers were limited to protecting and conserving the resource. They argued that he did not have the legislative authority to allocate fish on social or economic grounds. Justice Collier agreed with the trollers that allocation for socioeconomic purposes was "beyond permissable constitutional powers."

As detailed in Chapter 2, this decision was subsequently reversed by the Federal Court of Appeal. The appeal court determined that the legislative power of Parliament for seacoast and inland fisheries is not confined to conservation and protection.

The Gulf Trollers' court challenge to the Minister's allocative powers was a shock to the Department and the fishing industry. The industry came to recognize the need for a formal allocation process. The Minister's Advisory Council (MAC), representing various sectors of the B.C. fishing industry (see Chapter 15), recommended that the 1985 salmon fishing season be managed with traditional harvesting practices on the following basis:

That gillnet, seine and troll fisheries on sockeye, pink and chum be allocated to each sector on the basis of a percentage share of the total allowable catch using the 1981 catch distributions as a guideline, with the following provisos:

1. The allocations were to be a target for the 1985 Fishing Plan and for establishing openings;
2. The allocations should be organized in practical units so that catches could be shared by area, region and run;
3. When actual run sizes were different from the pre-season forecast, openings of the various sectors should be adjusted to achieve the targets for that allocation unit;
4. The fishing plan for the "outside troll" catch of Fraser pinks should be structured for a continuous troll fishery (Pacific and Freshwater Fisheries staff, DFO Ottawa, Personal Communication).

The areas referred to in the following discussion of Pacific salmon management are shown in Fig. 7-8.

The Pacific Salmon Treaty of 1985 established catch levels and ceilings for all salmon species to address Canada-USA salmon interception problems. It provided further impetus for a formal allocation process within Canada. As a result, in 1985 the Pacific salmon fishery was managed on the MAC-proposed allocation principles. The sockeye catch was divided 60:40 between the seiners and the gillnetters. The chum catch was divided 50:50 between the seiners and the gillnetters. Pinks, chinooks and cohos were to be managed in accordance with the Canada-USA Salmon Treaty. The troll catch of sockeye and chum salmon was not discussed.

In 1986, 1987 and 1988 these principles were further developed through consultations with the fishing industry, in MAC until 1987, and for 1988 based on advice from the newly established Commercial Fishing Industry Council (CFIC). These allocation patterns are shown in Table 7-2 (a-d).

To a large extent, these allocations were determined by agreement within MAC and subsequently the CFIC. Pacific salmon allocations for 1986 reflected historical harvest patterns with some minor exceptions. Allocations on the North Coast generally reflected the catch levels of the various gear types during the brood year 1982, with the exception of obligations resulting from the Canada-USA Pacific Salmon Treaty. On the South Coast, MAC recommended for Fraser-bound sockeye, allocations of 22.8% for trollers; 29.7% for gillnetters; and 47.5% for seiners. Trollers had sought 33% of the Fraser sockeye allocation. In announcing the 1986 Plan, Fisheries Minister Tom Siddon sympathized with the Gulf trollers: "I have decided to alter slightly the MAC recommendations for Fraser-bound sockeye to more closely recognize the reality and historical involvement of the Gulf trollers. Therefore, I have decided to allocate, for 1986, an additional 100,000 sockeye to Gulf trollers. With this increase..., the overall troll share of Fraser-bound sockeye will reach 24 percent" (DFO 1986a).

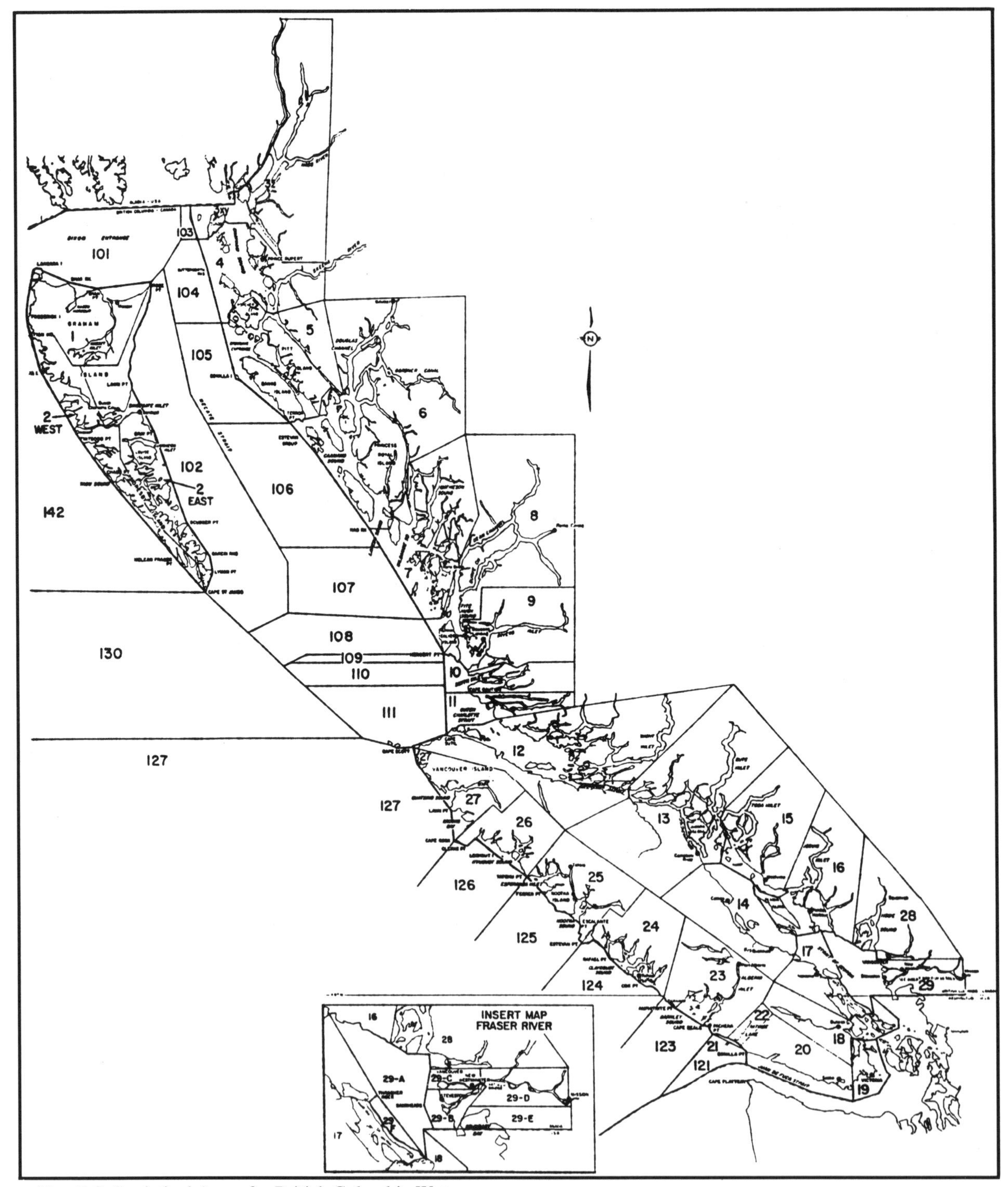

FIG. 7-8. Statistical Areas for British Columbia Waters.

For 1987, the Minister again followed recommendations of his Advisory Council. MAC was replaced in April 1987 by the Pacific Advisory Council (PARC), a policy advisory body (see Chapter 15). However, the Minister asked MAC for advice on 1987 salmon allocations.

The 1987 allocations generally reflected MAC's recommendations. However, MAC could not agree on

TABLE 7-2a. Sockeye salmon allocations for 1985 to 1990.

North Coast Troll

	1985	1986	1987	1988	1989	1990
Areas 1,3,4,5	Not to exceed 5%	Not to exceed 5%	Not to exceed 5%	Not to exceed 5%	Not to exceed 5%	Not to exceed 4%
Areas 6,7,8		Incidental catches not to escalate	Incidental catches not to escalate	Incidential catches not to escalate	Incidential catches not to escalate	Incidential catches not to escalate
Areas 9,10	The gillnet fishery and traditional troll fishery to coincide with net fishing time. The troll fishery should not escalate.					

North Coast Net

	1985	1986	1987	1988	1989	1990
Areas 1,3,4,5	76% GN 24% SN	76% GN 20% SN	75% GN 24% SN 1% TR	75% GN 25% SN	76% GN 24% SN	73% GN 27% SN
Areas 6,7,8	Incidental	Incidental	Incidental	Incidental	Incidental	Incidental
Areas 9,10	Traditional	Traditional GN, TR	Traditional GN, TR	Traditional GN, TR	Traditional GN, TR	Traditional GN, TR

South Coast

	1985	1986	1987	1988	1989	1990
Area 23	40% GN 60% SN	40% GN 60% SN	40% GN 60% SN	40% GN 60% SN	40% GN 60% SN	40% GN 60% SN

Fraser Bound

1985	1986	1987	1988	1989	1990
30.0% GN 68.0% SN 2.0% TR	29.36% GN 46.96% SN 20.26% TR-outside 3.42% TR-inside	34.5–35.0% GN 51.5–52.0% SN 10.5–11.0% TR-outside 1.5–3.5% TR-inside	38.4% GN 52.8% SN 4.8% TR-outside 4.0% TR-inside	33.0% GN 55.0% SN 10.0% TR-outside 2.0% TR-inside	28.0% GN 45.5% SN 22.7% TR-outside 3.8% TR-inside

Source: Pacific, Arctic and Inland Fisheries Operations.

TABLE 7-2b. Pink salmon allocations for 1985 to 1990.

North Coast Troll

	1985	1986	1987	1988	1989	1990
Area 1	Fishery was managed in accordance with Canada-U.S. Treaty levels until theArea 1 Treaty arrangement was negotiated in 1988. Canadian allocation in 1988 and 1989 was 1.7 million. In 1990 the maximum harvest was to be 1.95 million (Total of 5.125 million for 1990–1993)					
Areas 2–10	Maximum 5%	Maximum 4%	Maximum 5%	Maximum 4%	Maximum 4%	Maximum 4%

North Coast Net

	1985	1986	1987	1988	1989	1990
Areas 1,3,4,5	20% GN 76% SN	20% GN 76% SN	28% GN 72% SN	21% GN 79% SN	28% GN 72% SN	22% GN 78% SN
Areas 6–10	Maintain traditional fishing patterns with the primary objective toharvest available surpluses and ensure that escapement goals are met.					

South Coast

1985	1986	1987	1988	1989	1990
No Agreement for GN & SN 30% TR	OFF YEAR	9% GN 58% SN 29% TR-outside 4% TR-inside	9% GN 73% SN 9% TR-outside 9% TR-inside	9% GN 58% SN 29% TR-outside 4% TR-inside	OFF YEAR TRADITIONAL FISHING PATTERNS

Source: Pacific, Arctic and Inland Fisheries Operations.

TABLE 7-2c. Chum and coho salmon allocations for 1985 to 1990.

CHUM — North Coast Troll

	1985	1986	1987	1988	1989	1990
Areas 1–10	Traditional troll fishery should not escalate.					

CHUM — North Coast Net

	1985	1986	1987	1988	1989	1990
Areas 1–10	Traditional fishing patterns. Net fleet to maintain past catch shares.					

CHUM — South Coast

	1985	1986	1987	1988	1989	1990
Inside	N/A N/A	35% GN 65% SN	35% GN 65% SN	35% GN 65% SN	35% GN 65% SN TR, maximum 1%	35% GN 65% SN TR, maximum 1%
	50% GN	50% GN	50% GN	50% GN	50% GN	50% GN
	50% SN	50% SN	50% SN	50% SN	50% SN	50% SN
Outside	The troll fishery should not escalate.					

COHO — South Coast

	1985	1986	1987	1988	1989	1990
Outside	Not to exceed 1.75 million	Not to exceed 1.75 million	Not to exceed 1.8 million	Not to exceed 1.8 million	Not to exceed 1.8 million	Not to exceed 1.8 million

Source: Pacific, Arctic and Inland Fisheries Operations.

TABLE 7-2d. Chinook salmon allocations for 1985 to 1990.

CHINOOK — North & Central Coast

1985	1986	1987	1988	1989	1990
In accordance with Treaty levels, all gear catch not to exceed 263,000					Not to exceed 264,000
			Troll allocation 203,000	Troll allocation 203,000	Troll allocation 187,000

CHINOOK — South Coast

	1985	1986	1987	1988	1989	1990
Outside	Troll catch not to exceed 360,000					
Inside	50,000	50,000	50,000	31,000	31,000	31,000

Source: Pacific, Arctic and Inland Fisheries Operations.

the South Coast allocation of pink and sockeye. The Minister decided that the south coast pink catch would be allocated 4% for Gulf trollers, 29% for outside trollers, 9% for gillnetters and 58% for seiners.

For Fraser River-bound sockeye the Gulf trollers were allocated a minimum of 1.5% and a maximum of 3.5% of the total catch. The outside trollers were allocated 10.5 to 11.5%; gillnetters were allocated 35%; and seiners were allocated 52%. The result was a modest increase in access to pinks and sockeye for trollers without significantly affecting the net fleet. Allocations for the North Coast generally reflected the historic catch levels of troll, gillnet and seine gear types. A major factor in shaping the 1987 Plan had been a demand by the Gulf trollers for increased allocations of chinook, sockeye and pink salmon, plus an extended fishing season.

For 1988 the Minister sought advice from the newly established Commercial Fishing Industry Council. The CFIC was unable to agree on a comprehensive salmon allocation package. The major issues of contention were the allocation of pink salmon between the net fleet and the troll fleet in the North Coast area, the sharing of Fraser River sockeye and the Gulf troll chinook allocation.

Anticipating weak sockeye returns to the Fraser River system, the Minister of Fisheries and Oceans made these allocations: seiners 52.8%; gillnetters 38.4%; outside troll fleet 4.8%; and inside troll 4%. These allocations were to provide for a modest increase in the sockeye catch for Gulf (inside) trollers without adversely affecting the net fleets.

Returns of pink salmon to B.C.'s North and Central coasts were expected to be very good. Strong returns of pink and sockeye salmon were forecast for the Skeena River system. In keeping with the Pacific Salmon Treaty, the commercial troll fleet was entitled to a maximum of 1.7 million pinks in specified sub-areas of Area 1. The troll fleet was allocated up to 4% of the North Coast pink catch in areas 2 to 10, including any catch from Area 1 above the 1.7 million.

Over the previous several years allocation had been based on historic averaging combined with some adjustments to cut back on chinook and coho catches because of the Pacific Salmon Treaty. The Fishermen's Union (UFAWU) and the Fishing Vessel Owners Association (FVOA) continued to support traditional fishing patterns and historic catch as the basis for allocation. The Pacific Trollers Association (PTA) and the Northern Trollers Association (NTA) supported allocations based on the number of fishermen employed in each sector (DFO 1989e).

Essentially, the troll groups supported the concept of equal numbers of fish per fisherman. They ignored the issues of other labour employed in the industry and returns to capital (NTA 1989; PTA 1989). The Pacific Gillnetters Association (PGA) wanted their share to be based on an overall share gillnetters had at some time in the past.

Again in 1989 the CFIC was unable to agree on salmon allocations. Ultimately, the Minister had to arbitrate the allocation for Fraser River sockeye for 1989. Apart from that, status quo allocations based on brood cycle year allocations were maintained, with the exception of chinook which is discussed in the next section. Strong returns of Fraser River sockeye were anticipated. Fraser River sockeye were allocated as follows: seines, 55%; gillnets, 33%; outside Gulf of

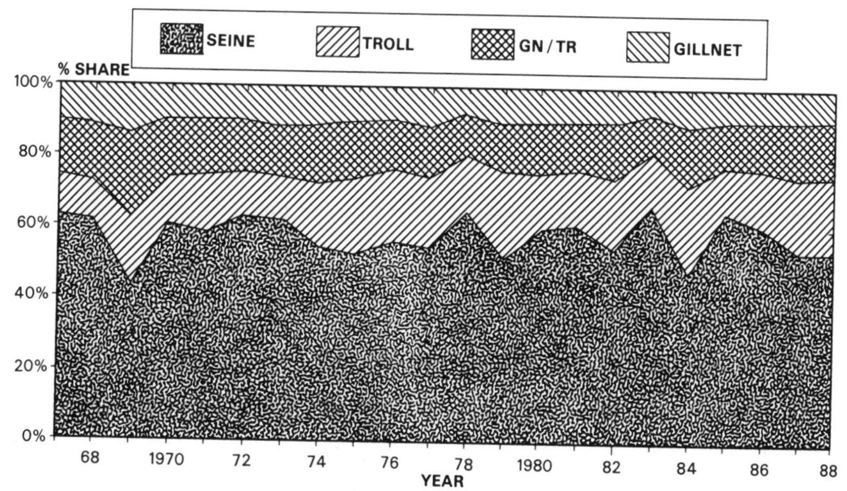

FIG. 7-9. Average gross salmon incomes per vessel by gear category for the Pacific Region. Average incomes are expressed in 1981 dollars (from data provided by Michelle James, Pacific Region, DFO).

Georgia trollers, 10%; and inside Gulf of Georgia trollers, 2%.

In comparison with 1988, the gillnet share was decreased and the share of the outside troll fleet doubled.

DFO argued that the share of landed value attributable to an average vessel by gear had not changed substantially over time (Fig. 7-9). The seine/non-seine split averaged 58/42% over the previous 22 years. On the basis of average gross earnings per vessel, seine vessels appeared not to have gained in relation to the other gear types. Greater changes in catch shares had occurred in the non-seine sector. Average gross salmon incomes of combination and gillnet vessels declined in comparison with those of single gear troll vessels (DFO 1989f).

The objectives for the 1989 allocations were:

1. Maintain industry stability by using cycle year historic allocations as the basis for allocation;
2. Adjust troll catches to maintain average troll earnings by allocating access to non-traditional troll species (i.e. sockeye);
3. Maintain a viable Gulf troll fishery without encouraging more trollers to enter that fishery;
4. Improve the relative earnings of gillnet vessels;
5. Maintain traditional fishing patterns; and
6. Ensure allocations are manageable (DFO 1989f).

In determining the 1990 allocations, DFO used the accommodations on North Coast sockeye and pink allocations that were considered but not recommended by a CFIC Subcommittee. These represented a change from 1986 allocations for sockeye and 1988 allocations for pinks. For North Coast sockeye, 1990 allocations were a return to more traditional patterns between the gillnet and seine fleets.

In announcing the 1990 salmon allocations, Fisheries Minister Valcourt stated that the Plan was based on the principle of fair allocation to each type of gear based on coastwide catches of all species (DFO 1990d). The anticipated commercial harvest of Fraser River sockeye was allocated as follows: seine, 45.5%, gillnet, 28%, and troll, 26.5 %.

This represented a significant increase for trollers from their 1989 share of 12%. The increase was "in recognition of a reduced share of the salmon resource for trollers in recent years, and in response to decreased access to chinook salmon in northern British Columbia waters" (DFO 1990d). This increased troll allocation was significant: a variation of only 1% in catch quotas meant approximately $1.2 million to fishermen.

Thus over the period 1988 to 1990 there was some shift in the pattern of Pacific salmon allocations toward a coastwide approach. Within the commercial sector a formal allocation process evolved during the late 1980's which resolved many allocation issues without ministerial intervention. From 1988 to 1990 the Minister was forced to intervene each year to decide Fraser River sockeye allocations. By 1990 discussion had begun in earnest on the development of a long term allocation plan. So far I have discussed the question of allocations among different gear types within the commercial sector. A major driving force in public debate about Pacific salmon allocations during this period was the question of allocating a declining chinook resource between commercial (specifically the troll fleet) and recreational fishing interests.

3.2 Chinook Quotas and Allocations

Chinook stocks in the coastal waters of British Columbia can be grouped into five major stock complexes on the basis of geographic location, life histories, ocean distributions and interception patterns in ocean fisheries. These stock complexes are:

1. North Coast (Areas 1–5);
2. Central Coast (Areas 6–10);
3. Strait of Georgia (Areas 11–19 and 28);
4. West Coast of Vancouver Island (Areas 20–27); and
5. Fraser River and Tributaries (Area 29) (see Fig. 7-8).

The Fraser River system has been further subdivided into four major sub-stocks (DFO 1987c).

By the late 1970's-early 1980's scientists warned that natural populations of chinook salmon along the west coast of North America had declined to a low level of abundance. Chinook were not only reduced in abundance but severely overexploited. British Columbia chinook escapements had decreased by over 40% since the early 1950's (Healey 1982), while catches by commercial and sport fisheries had increased considerably, especially in the 1970's. These trends were observed all along the west coast of North America (Fig. 7-10). Concern about this trend led to the establishment of an international program under the Pacific Salmon Treaty to rebuild chinook stocks to "optimum sustainable levels."

Riddell and Starr (1987) identified three causes for the decline of natural chinook populations. First, chinook were harvested primarily in large mixed-stock fisheries — commercial troll and recreational. There had been no active management of these fisheries because catch by stock was unknown. Second, the complex life history and migration of chinook meant that a year-class was harvested over several fisheries and years, resulting in a high cumulative exploitation rate. Third, the rapid increase in hatchery production in the United States since the early 1960's had masked the decline in production from natural populations.

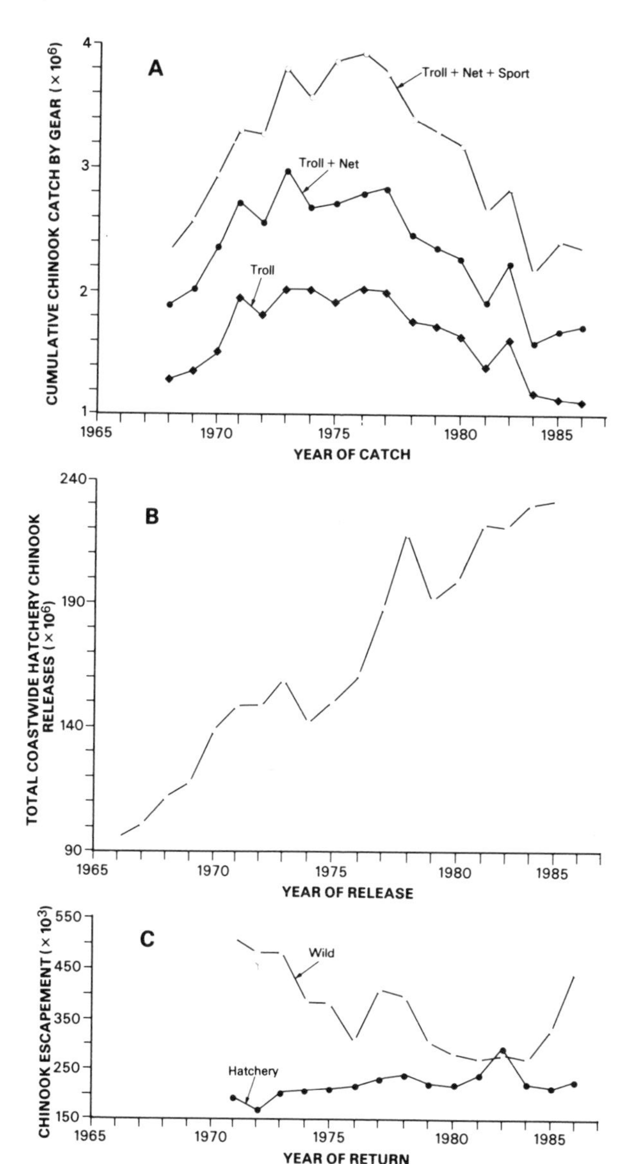

FIG. 7-10. Trends in Chinook catch escapement and hatchery releases from Oregon through Southeast Alaska (from 1986 Report of Chinook Technical Committee, Pacific Salmon Commission).

Apparently what occurred in the 1970's is that fishing effort increased in response to the apparent abundance of chinook, which was composed of more hatchery-produced fish and fewer natural chinook. This resulted in overexploitation of the natural populations which declined rapidly.

The major concern about chinook within British Columbia in the mid-1980's focused on the Georgia Strait stock(s). Georgia Strait is that body of water between mainland British Columbia and Vancouver Island. Because of its proximity to major urban centers, this area attracts large numbers of recreational fishermen, both resident and tourists. Argue et al. (1983) estimated that this sport fishery involved over 100,000 boats and 345,000 fishermen, with chinook salmon as the choice sport fish. Large numbers of chinook salmon were taken historically in the commercial troll fishery in Georgia Strait and a commercial net fishery at the mouth of the Fraser River. Small numbers of chinook salmon were intercepted in native food fisheries. The commercial fishery involved approximately 9,000 fishermen using about 4,400 troll, gillnet and seine vessels. The native Indian food fishery involved another 6,300 fishermen who fish primarily in lakes, rivers and streams draining into the Strait. The freshwater sport fishery involved another 50,000 fishermen. This combination of four fisheries, six basic gear types, and in excess of 400,000 participants in the harvesting sector alone constituted one of the most complex fisheries in North America.

So far in this chapter we have focused on the problems of allocating access to fish resources among different vessel sizes, gear types and among provinces in the commercial fisheries for various species. The fishery for Georgia Strait chinook also involved a substantial marine and freshwater recreational component.

Although the decline in Georgia Strait chinook stocks had become more pronounced in the 1980's, federal fisheries officials recognized there was a problem by the late 1970's. They took some action to arrest the decline of the natural stocks. In the late 1970's most terminal gillnet fisheries on chinook were eliminated. Gillnet fisheries in the Nass, Skeena and Fraser estuaries, together with scattered spring and early summer fisheries in many coastal fisheries, were eliminated. In terminal fisheries for other species, maximum mesh size restrictions were introduced to reduce incidental chinook catches.

Prior to 1964 the trolling season for chinooks was open almost year-round. In 1966 the season was shortened to 5½ months. Trollers were allowed to fish 7 days per week during the open period. Until 1981 all commercial trollers could operate in the Strait during open periods.

Historically, the marine sport fishery in B.C. had a 12-month season. Daily bag limits were first introduced in this sport fishery in 1950.

In October 1980, DFO announced a two-area troll licence system which would force trollers to choose each year between fishing in Georgia Strait or in outside waters. This measure was intended to decrease fishing pressure on the inside stocks through the removal of most large trollers from inside waters. Other measures announced then included: (1) the restriction of trollers coast-wide to six gurdies (mechanical reels) per vessel; (2) a requirement that trollers use barbless hooks; (3)

spot closures for both commercial and sport fishing. Spot closures involved closing off specific areas for certain periods to protect threatened stocks or heavy concentrations of juvenile salmon.

In February 1981, the Regional Director General of Fisheries and Oceans announced a series of further conservation measures for chinook salmon to reverse the trend of declining chinook stocks, especially those of the Fraser River system. Catches from these stocks had declined dramatically over the previous 10 years, from 230,000 chinook in 1970 to 110,000 in 1980.

The 1981 measures included lowering the maximum allowable gillnet mesh size in the Fraser River to reduce the incidental catch of chinook salmon when fishing for other species. The use of downriggers by sports fishermen was banned, to reduce the efficiency of the sport fishery. A Georgia Strait salmon sport fishing closure for 4 months was announced, to take effect on December 1981. Sports fishermen were to be licensed. The minimum size limit was to be increased and a one chinook per day catch limit imposed. In explaining the rationale for these measures, the Director General stated: "One of the guidelines in drawing up these measures was that any restrictions should be shared by those having an impact on the chinook stock — both commercial fishermen and the sports fishermen. The sports fishery in B.C. waters is now of such a magnitude that it cannot continue almost totally unregulated while restrictions are being imposed on those who depend on the fishery resource for their livelihood" (DFO 1981e).

These measures were controversial. In the winter of 1981 the Pacific Region Sport Fish Advisory Board suggested a seven-point conservation package:

(1) An 18 inch minimum size limit;
(2) The stringent and more frequent application of spot closures whenever required to protect juvenile and adult chinook salmon;
(3) Continuation of the Fraser River closure for sport fishing for adult chinook salmon;
(4) Commencing in 1981 a coastwide daily bag limit reduction to two chinook per day from four chinook per day during the four winter months (December–March);
(5) The implementation, in 1982, of a 30 chinook annual bag limit;
(6) A ban on the use of all 'meat' lines, such as a downrigger without a quick release mechanism, or other devices used to catch salmon in a non-sporting manner; and
(7) Where necessary, the extensive use of river mouth closures to protect adult chinook escapement.

Responding to the seven-point package in April 1981, Fisheries Minister Roméo LeBlanc indicated that he was willing to accept the Sport Fish Advisory Board recommendations "because they meet the primary objective of contributing 12,000 additional chinooks to the spawning grounds." He also announced that a limited directed commercial fishery would be allowed while sockeye were migrating through Area 29D on the Fraser River (DFO 1981f). The sports fishermen had obtained substantial changes to the package of regulations originally proposed for the 1981 season.

Total chinook catches by the three major gear types had increased from 322,000 in the early 1950's to 623,000 during the 1972–76 period. By then about one half of the total chinook catch in Georgia Strait was taken by sport gear (343,000 or 55% compared with 24% in 1962–66). The share taken by commercial trollers had increased from 28% during 1952–56 to

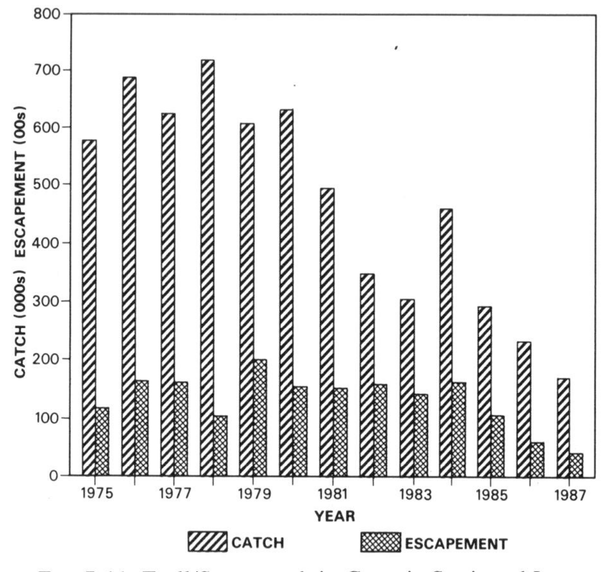

FIG. 7-11. Troll/Sport catch in Georgia Strait and Lower Georgia Strait chinook escapement. Source: DFO Pacific Region.

Fig. 7-12. Georgia Strait sport catch of age two chinooks. Source: DFO Pacific Region.

35% during 1957–66, then dropped to 29% by 1972–76. The catch by the other major gear type — Fraser gillnets — had decreased from 47% during 1952–56 to 16% during 1972–76. The marine sport catch had increased fourfold since the mid-1960's.

Meanwhile, the Strait of Georgia chinook stocks continued to decline. In the late 1970's-early 1980's there was a precipitous decline in the Georgia Strait chinook catch, from an average of 623,000 in the early 1970's to 291,000 by 1985 (Fig. 7-11). Concern over the state of the stocks increased when the sport catch of age-two chinook, a measure of upcoming recruitment, declined dramatically from 1983 to 1985 (Fig. 7-12). This apparent collapse in recruitment led to calls for more drastic conservation measures.

Alarmed by the continuing decline of chinook stocks, DFO on March 16, 1984, announced a significantly shortened troll season that implied catch reductions of about 20% for most fishermen. Since the U.S. was unwilling to agree to conservation measures recommended by Canadian and U.S. scientists, it had become necessary for Canada to look after its own stocks as best it could (DFO 1984b).

Chinook declines in the Strait of Georgia had necessitated a total early season closure. The troll season in the Strait of Georgia was reduced to July 1 to August 31 in place of the normal season of April 15 to September 30. The seasons for troll gear were also shortened in other areas.

These measures aimed to reduce the take of chinook salmon and increase escapements. In the Strait of Georgia, the closure amounted to the total elimination of the directed chinook fishery, since in July and August the trollers targeted on coho. During 1983 the Gulf trollers catch of chinook was approximately 125,000. Given the same availability of fish as 1983, it was anticipated that their incidental catch of chinook during the coho fishery would be approximately 25,000. These severe measures were taken with the intent of getting as many of the 100,000 chinook saved from the trollers' hooks as possible onto the spawning grounds.

At that time no further action was taken to limit the sport fishery. The reduced troll season and the lack of further action against the sport fishery led the Gulf trollers to seek a court injunction to set aside the changes in their fishing season, using the argument that the Minister did not have the power to allocate for socioeconomic purposes. This led to the Justice Collier decision (described in Chapter 2) which was reversed in the Federal Court of Appeal in November 1986. The trollers believed that the change to their season, essentially closing their directed fishery for chinook, could reallocate 100,000 chinook to the sport fishery. This would happen unless there were further restrictions on the sport fishery to allow the fish foregone by the trollers to get to the spawning grounds.

On April 19, 1984, Fisheries Minister Pierre De Bané responded to pressures from the trollers for additional measures to curtail the sport fishery: "This is not the time to decide on reallocations, what is at stake is the conservation of a resource that is crucial to all sectors." De Bané stated that, while the Seven-Point Program agreed upon in 1981 placed stringent restrictions on the sport fishery, it was clear that on a species basis there were areas where the sport sector accounted for a very large part of the catch and hence must carry a further burden (DFO 1984c). He indicated that further measures were required to reduce the sport catch of chinooks in the Gulf from about 200,000 fish in 1983 to 160,000 fish in 1984. However, no further action was taken involving the sport fishery in 1984. The sport fish catch in the Strait of Georgia by the end of November was more than 396,000, considerably exceeding the Minister's suggested target of 160,000.

On the same date, Minister De Bané called on all B.C. Native leaders to help ensure that conservation measures in the non-native sector would succeed in increasing spawning escapements. (For a discussion of Native food fishery issues, see Chapter 14).

In July 1984, the Gulf trollers sought to set aside the decision to have a 2-month troll season in the Gulf. Although Justice Collier agreed, he did not hand down his decision until August 31, the date on which the fishery was scheduled to close. Judge Collier, in his decision, did not recommend additional fishing time for the trollers, but merely commented that the department did not have the authority to curtail the troll fleet and thereby reallocate fish to the sport sector. He did not, however, give any instructions on what should be done to rectify the situation. Although the trollers won their interim action in the court, nothing changed.

Frustrated by their lack of success through the court, the trollers decided to mount a protest fishery in September. Their protest was motivated by the lack of remedial action by DFO to curtail the recreational fishery, "clearly in contempt of Judge Collier's decision." Michael Griswold, President of the Gulf Trollers Association, stated: "The total harvest of chinook in the Gulf of Georgia has now exceeded 400,000 fish, double the DFO pre-season safe harvestable limit. Further harvesting of the resource this year is intolerable to the commercial fisherman, the responsible sports fisherman and the people of Canada" (Gulf Trollers Association 1984).

The total 1984 Strait of Georgia chinook catch was approximately 500,000 fish in comparison with the earlier recommendation from the Canada-USA Chinook Technical Committee that the combined troll and sport catch be held to 225,000 chinook. This catch level was comparable to that taken in 1980 and 1981. It would appear that the delayed opening for trollers had improved the sport fishery, which in turn spurred an increased catch by attracting additional sports fishermen to the grounds.

By the end of 1984 the Strait of Georgia was beginning to resemble a sport fishing preserve. The Gulf trollers and the marine sports fishermen were locked in bitter battle over how the burden of chinook conservation should be shared. Meanwhile, the Canada-USA negotiations aimed at managing interceptions of salmon in mixed fisheries, after 20–30 years of on-again/off-again discussions, culminated in an historic accord. The Canada-USA Pacific Salmon Treaty, to take effect on March 18, 1985, was a comprehensive attempt to deal with the bilateral interception problems for all salmon species. The complex issues surrounding this treaty are dealt with in Chapter 12. Here I will touch only upon those aspects relating to chinook.

Provisions respecting chinook are contained in Annex IV, Chapter 3 of the treaty. Paragraph 1 provided for the establishment of a chinook salmon management program to meet these objectives:

(i) halt the decline in spawning escapements in depressed chinook stocks;
(ii) attain by 1998 escapement goals established in order to restore production of naturally spawning chinook stocks.

A Joint Chinook Technical Committee was established to evaluate annually the status of chinook stocks, evaluate management measures and recommend adjustments to management measures set forth in the Treaty.

Paragraph 1(d)(iii) stipulated that in 1985 and 1986, the total annual catch by the sport and troll fisheries in the Strait of Georgia should not exceed 275,000 chinook.

A catch ceiling of 275,000 chinook for the sport and troll fisheries in the Strait of Georgia for 1985 and 1986 was well below catches in preceding years. Hence, it was anticipated that adherence to the ceilings would reduce exploitation of Georgia Strait chinook enough for natural stocks to rebuild.

The immediate issue was how to allocate the 275,000 catch quota. Lobbying from the powerful sports interests was intense. They wanted to reduce the troll share to no more than an incidental catch of 25,000 chinook in the coho fishery. Gulf trollers pressed their case for a 45% share, based on historic catches.

In February 1985, Fisheries Minister John Fraser announced new measures for the sports fishery. The intent was to allow sport fishing throughout the year without exceeding the chinook quota. Daily bag limits of 2 chinook and a seasonal bag limit of 20 chinook were imposed. In addition, provision was made for widespread spot closures to protect chinooks at critical times and in critical area. This prompted U.B.C. Professor Carl Walters to warn that the fisheries minister had set such generous limits for sports fishermen that he would be forced to violate the treaty or close the fishery before the end of the summer. Walters said that it was highly likely that sports fishing catches would reach the annual treaty limit "well before the end of the summer, possibly as early as mid-July" (*The Vancouver Sun*, February 13, 1985).

Meanwhile, intense discussions continued with the two groups on how to allocate the catch. In the absence of an agreed allocation, the Gulf Trollers Association announced on March 12 that its members would start fishing for chinooks on April 15 with or without federal government approval — this was a sign of the growing battle with the sport fishing industry over access to the Strait of Georgia (*The Vancouver Sun*, March 12, 1985).

The Minister's Advisory Council was divided on the issue of chinook allocation in the Gulf. The commercial sector recommended the trollers be given 45%. But the sports fishing representatives had rejected the idea. In April the Pacific Trollers Association withdrew from MAC.

In June 1985, Minister Fraser announced that he would allocate 50,000 chinooks — 18% of the total catch — to commercial trollers and the rest to sport fishermen. Spot closures throughout Georgia Strait were announced June 6. Sports Fishery Advisory Board Chairman Danny Sewell responded: "I think the chinook conservation program is a very positive one. The closures are small and short enough to do the job. It's a win for salmon. It's a win for you and me. It's a win for the treaty and it's a win for future generations" (*The Vancouver Sun*, June 25, 1985).

1985 was an exceptional year for sockeye salmon returns. Although the original 1985 salmon plan made no provision for sockeye for the trollers, in August 1985 Fisheries Minister Fraser announced that the Gulf trollers would be allowed to fish Fraser River sockeye at times equivalent to those of Fraser gillnet vessels. The Minister indicated he had taken this action to allow Gulf trollers to share in the unexpected bonanza sockeye run (DFO 1985a).

The 1985 Strait of Georgia chinook catch was 291,000, down considerably from 480,000 in 1984. In January 1986, the Pacific Trollers Association, in a brief to the new fisheries minister, Tom Siddon, contended that the trollers had been disadvantaged by the Pacific Salmon Treaty. In their view, there was a definite linkage between the reduced West Coast of Vancouver Island troll coho catch quota of U.S.-bound fish and an increased Canadian share of sockeye bound for the Fraser River. The

Canadian troll fleet had been restricted in several instances with the benefits accruing to the net fleet and, in the case of Georgia Strait chinook, the recreational fishery. They requested greater allocations of the species traditionally caught by the gillnet and seine fleets. In their view the Salmon Enhancement Program had been used to enhance "net" species and chinook and coho beyond the reach of the majority of the troll fleet (PTA 1986).

In 1985, DFO had commented on the implications of the Chinook Annex of the Pacific Salmon Treaty as follows:

> "By signing this agreement, Canada is committed to a 15 year (3-cycle) rebuilding program for the most depressed naturally-spawning chinook stocks. In the next two years (1985 and 1986), all major coastal interception fisheries have been assigned catch ceilings. From 1987 onward, Canada is committed to continue the rebuilding process. During the rebuilding period, however, user groups will be guaranteed a minimum catch" (DFO 1985b).

In 1986 the Chinook Technical Committee reviewed catch and escapement data for the 1985 season and concluded that chinook stocks in the Strait of Georgia needed further protection if the 15-year rebuilding program were to be successful.

The committee concluded:

> "Without reductions in catch ceilings or an equivalent action, harvest rates on naturally spawning chinook populations would be expected to increase and prolong the rebuilding schedule" (Pacific Salmon Commission 1986).

For 1986 and 1987 Fisheries Minister Tom Siddon maintained the chinook quota and allocations in Georgia Strait at the 1985 level of 275,000, divided 225,000 for sport fishing and 50,000 for trollers. However, he provided some relief to the Gulf trollers by increasing the troll allocations of other species.Troller representations in 1986 and 1987 had some impact in securing increased access to other species in compensation for their small share of the Strait of Georgia chinook fishery.

The chinook catch fell to 226,000 in 1986 and declined further to 159,000 in 1987, the lowest level in the last two decades. By late 1987 scientists had concluded that harvest rates had not been reduced nor had Lower Georgia Strait stocks recovered. The assumptions on which the treaty catch ceiling was based had not materialized. Scientists had wrongly assumed that the survival of naturally spawned recruits from 1983–1985 brood years would improve to 1979–1982 levels and that survival of hatchery releases to recruitment would remain constant.

In December 1987, Fisheries Minister Siddon established a Georgia Strait Task Group of scientists and fishermen to study options to conserve and enhance chinook resources. DFO proposed to double base period (1977–82) escapements of Lower Strait of Georgia chinook by 1998 using a combination of management actions and enhancement. A variety of alternative management actions were discussed with the Task Group. The group believed that LGS stocks were severely depressed and could not meet rebuilding goals without more effective means of increasing escapements (Anon 1988a).

On March 7, 1988, Fisheries Minister Siddon announced a "tough but fair" conservation program for the declining LGS chinook stocks. The program consisted of a dual approach — conservation measures and enhancement. The initiatives included:

1. *Reduced Sportfishing Limits*
   The annual bag limit in the Strait of Georgia was set at eight chinook for 1988, down from the previous limit of 20. Outside the Strait, the limit remained at 30 chinook per year.
2. *Area and Time Closures*
   Spot closures for the sports fishery were to be continued.
3. *Commercial Fisheries Restrictions*
   The number of fishing days for Johnstone Strait net fisheries would be reduced in the following years. Troll fishing in some portions at the north end of Johnstone Strait would be closed in 1988 to reduce the harvest of lower Strait stocks. A limit of 35,000 chinook was set for trollers inside Georgia Strait, down from the 50,000 level in effect during 1985–87.
4. *Native Fisheries Restrictions*
   The Minister indicated that the native food fisheries would be adjusted commensurate with the reductions being applied to the sport and commercial sectors.
5. *Conservation Tagging*
   All fishermen — sport, commercial and native — would be required to affix a conservation tag to all chinook caught in B.C. waters. It would be an offence to be in possession of an untagged chinook salmon (DFO 1988d).

"Overall, we aim to reduce the chinook harvest rate in the Strait of Georgia by at least 20%, " Mr. Siddon stated, with this reduction being "applied equitably to all the user groups" (DFO 1988d).

The Gulf troll allocation was subsequently reduced to 31,000 for 1988, 1989 and 1990. By 1990 there were still no signs of rebuilding. Only time will tell whether the goals of this program will be achieved and whether the measures taken were sufficient to rebuild the Strait of Georgia chinook stocks.

Meanwhile, in 1989 and 1990, similar recreational/ commercial conflict arose over the allocation of North Coast chinook. From 1985, under the terms of the Pacific Salmon Treaty, the chinook ceiling for the North Coast of British Columbia was 263,000 fish. For several years the domestic allocations were set at: troll — 203,000; net — 40,000; and sport — 20,000.

The North Coast sport catch increased from 14,000 chinook in 1987 to 21,000 in 1988 and 34,000 in 1989. Early in 1990 DFO projected that the sport catch would rise to between 40,000 and 50,000 in 1990.

In January 1990, Minister Siddon informed the Northern Trollers Association of his intention to cap the recreational fishery. In June 1990, Minister Valcourt adjusted the allocations of North Coast chinook to recognize the growth in the recreational fishery that had occurred, but capping the expansion of this fishery. The new allocations for 1990 were: troll — 183,000; net — 35,000; and sport — 45,000 (up from 20,000) (DFO 1990d).

The commercial sector opposed this reallocation to the recreational sector, particularly the local Queen Charlotte Island trollers who heavily depended upon chinook. They believed the compensation in terms of increased access to sockeye salmon by trollers was inadequate. Meanwhile, the recreational sector opposed capping the growth of this fishery for 3 years. The Georgia Strait and North Coast chinook allocation battles illustrate the growing conflict between the commercial and recreational sectors in British Columbia. The conflict threatened to escalate further in the 1990's.

4.0 CONCLUSIONS

The two major fisheries — Atlantic groundfish and Pacific salmon — used as examples in this chapter represent the diversity of marine fisheries in Canada. They illustrate well the problems of allocating access to common property resources, and the process and approaches which have evolved in Canada to address these. Basically, resource allocation is concerned with dividing a limited "pie" among many conflicting interests. The pie is rarely large enough to satisfy those seeking shares. The common-property nature of the resource encourages overcapacity in the harvesting sector and quite often the processing sector as well. A lack of intervention at an early stage in the development of a fishery inevitably compounds the problem. Overcapacity generates enormous conflict as interest groups compete for their "fair share" of the resource. To most groups, "fair share" means meeting their needs at the expense of others.

Allocation of access to wild fish resources involves the distribution of wealth among those involved in the fishery. This raises difficult questions of equity. The evolution of the Atlantic Groundfish Plan is a case in point. Since their inception Plans have embodied as a principle that "allocation of fishery resources will be on the basis of equity, taking into account adjacency to the resource, the relative dependence of coastal communities and the various fleet sectors upon a given resource, and economic efficiency and fleet mobility." Similar phraseology occurs in the Basic Principles of the Atlantic Salmon Management Plan, the Northern Shrimp Management Plan and others.

The Northern Shrimp Plan talks about "fair access to, and equitable sharing of the Northern shrimp resource by all legitimate Canadian user groups, with particular emphasis on the needs of the people and communities most adjacent to the resource" (DFO 1990e). The Atlantic Salmon Management Plan emphasizes that "the social and cultural importance of fishing to native communities which have traditionally harvested the resource for their own consumption is recognized and will be given priority after conservation." It also states that "the limited fishery for Atlantic salmon will be managed so as to distribute the benefits most effectively among the largest number of Canadians" (DFO 1989g).

The saga of the Strait of Georgia chinook allocations was a story of confrontation between the respective needs and benefits of the commercial and recreational fisheries. The Atlantic Salmon Management Plan addressed this issue explicitly as follows: "In the maritime provinces, the importance of the recreational fishery will be given greater recognition based on the relatively larger potential benefits to be generated. However, there will be a continuing role for the commercial fishery. In Newfoundland and Labrador, it is recognized that there is much greater economic dependence upon the commercial fishery than upon the recreational fishery."[1] This was a more explicit recognition of the factors which shaped resource allocation decisions in the Strait of Georgia chinook battles on the Pacific coast.

Criteria for allocating access vary from fishery to fishery, but inevitably, priority is given to conservation. Native food fisheries generally follow conservation in priority. Determining how much is allocated to the commercial versus the recreational fisheries or among commercial groups is one of the thorniest aspects of fisheries management.

Regarding Pacific salmon, Healey (1982) observed:

[1]Despite this, in 1992 the Newfoundland commercial fishery for Atlantic salmon was closed for a minimum of five years.

"Allocation among gear types in the gauntlet fisheries is a significant problem for which there is no established policy guideline. Allocation among user groups is an equally controversial issue. Each...group stridently demands a greater share of the dwindling resource and all are politically active in pursuing their demands. In almost every instance the fishery manager is squeezed between the demands of both the gear types and user groups for more of a share in the catch and the need to let more salmon through the fishery to the spawning grounds."

Conservation, native food fisheries, adjacency to the resource and community dependence appear to be the dominant criteria in resource allocation decisions. Beyond these general considerations, the criteria for allocating access among various groups tend to be blurred. When the pie is shrinking, emphasis is placed on equitable sharing of the burden of conservation. This generally leads to sharing arrangements based upon some interpretation of historic participation, modified to take into account the criteria cited above. When the pie is increasing (eg. the fierce battles over northern cod allocations in the late 1970's-early 1980's) the allocation battles can be just as heated. Criteria for sharing are formulated through Seminars, consultative mechanisms or other fora and then modified over time with changing circumstances.

It is sometimes possible through extensive consultation to achieve consensus on resource allocation issues. But there are always problem stocks (e.g. when the allowable harvest is decreasing or increasing significantly) where agreement among the user groups is not possible. In the Canadian system the problem becomes one for the federal government. Fisheries managers are always involved. A large number of such issues, however, have historically ended up on the Minister's desk for decision. Unfortunately, there are no magic solutions for these problems.

Most resource allocation decisions appear to be based on subjective criteria rather than analytical frameworks. Economically efficient resource allocation has not been prominent in decision-making. Most decisions have been weighted by considerations of fairness and equitable sharing. This is not surprising given the complex nature of fisheries such as Atlantic groundfish and Pacific salmon and the role they play in regional economies. However, it distresses some observers of the fisheries scene. Rothschild (1983) observed:

"The pervasiveness of the allocation process — affecting both the magnitude of the net value of the fishery resource and the way in which this value is distributed among the producers, the processors and the consumers of fish — suggests that the quality of allocations, that is, the extent to which allocations produce desired societal benefits, is one of the most critical public-policy issues in fishery management."

He noted that there are good technical methods to develop optimal allocations, but these are underutilized. Rothschild (1983) acknowledged that the implementation of any allocation scheme depends upon its political acceptability.

Sprout and Kadowaki (1987), in an examination of the process and the problems in managing the Skeena River sockeye salmon, observed: "Consensus among interest groups on major allocative and other issues is rare, and no formal method has yet been devised to assign weights or values to varied or conflicting advice." They recognized that management necessarily involved "a good deal of subjective rather than analytical decision making", but suggested that there should be rigorous documentation of both analytical and subjective judgments.

Healey (1984), as mentioned in Chapter 4, explored the use of techniques of decision analysis to provide an analytical model for determining optimum yield. He recognized that, while fisheries managers feared value judgments, they made such judgments all the time. He argued that multiattribute utility techniques could be use to bring these values into the open and assist managers in dealing with the conflicting preferences of the various user groups.

Stroud et al. (1980) noted that the first approach to "best-use" allocations of fishery resources was taken by the ancestors of modern man on the basis of brute force. The strongest and most aggressive merely usurped the fishery resources to the extent both of their needs and their capabilities to repel competitors. Stroud et al. observed: "Thereafter, the method of allocation of fishery resources (has) changed only in form, not in principle. In recent times political strength has been substituted for physical strength. Although more subtle, the users having the greatest amount of political power determine the allocation."

Allocation of access is inextricably bound with concepts of equity and fairness. Hence, allocation decisions will likely continue to be based on subjective judgments on how the fish "pie" should be shared. There is merit, however, in Healey's suggestion of providing the decision makers with practical analytical tools. This sort of analysis should be extended to other species and areas.

In the end, conflict is the name of the resource allocation game. User groups are driven to seek greater shares of the pie because of the common property nature of the resource. The formal allocation-of-access frameworks which have been developed have been superimposed usually on fleet overcapacity situations. Thus, they have not solved the common-property problem. The resource is shared in discrete segments, but then there is competition within each sector for the greatest share of these segments just as there was competition in the open access situation for the greatest share of the global "pie". Thus, allocation of access among groups, by itself, does nothing to restrain the tendency to overinvest and build bigger and better boats to catch the fish.

Other approaches have been developed in parallel with allocation of access to control the level of participation in the fishery and to deal with the negative effects of the common property problem. The next two chapters examine the nature and impact of these initiatives.

CHAPTER 8

MANAGING THE COMMON PROPERTY II — LIMITED ENTRY LICENSING

"The economic health of the industry cannot be maximized until limitation of numbers of participating units is accepted as the major weapon against overfishing."

– J.A. Crutchfield, 1961

1.0 INTRODUCTION

Much has been written during the past 25 years about the problem of limiting fishing effort and the advantages and disadvantages of alternative forms of limiting entry to the fishery. Many fisheries economists have agreed that entry to particular fisheries must be restricted to avoid the problems of open access to a common-property resource. There has been less agreement, particularly in recent years, on the merits of specific means of restricting access.

Direct limitation of the number of licences to harvest has become the best known and most widely used method of limiting entry into a fishery. Some economists have advocated taxes or royalties, but this approach has rarely been implemented. During the 1980's individual quotas gained acceptance in the pursuit of economic efficiency in fisheries management. Various forms of individual quotas have been introduced in fisheries around the world, but this approach was still at an early stage of development at the beginning of the 1990's.

2.0 EVOLUTION OF THE LIMITED ENTRY CONCEPT

Licences were first introduced in Canada as a fisheries management tool in the post-Confederation years. According to Gough (1993), licences were first used "mainly to keep some sort of law and order in the fishery, by identifying participants and establishing that the fishery was a privilege under the Crown". Leases on salmon rivers served both conservation and orderly sharing purposes. However, court decisions in the 1880's and 1890's weakened the federal licensing power in inland waters. By World War I licences were used in many commercial fisheries. However, the number of licences was limited in only a few fisheries (Gough 1993).

Gough found little evidence of widespread application of limitation of entry on the Atlantic. He describes E.E. Prince's failed attempts in the early 1900's to apply limited entry in the lobster fishery. Prince argued for limited entry: "So long as the taking of lobsters on Canadian shores is a free fishery, so long will it be difficult to carry out the preservative measures that are desirable." Prince predicted that the lack of such measures would eventually exhaust the fishery (Canada 1913).

Not until the 1960's did limited entry become widely accepted as a fisheries regulatory instrument. *Limited entry* is any control of a fishery that curtails or restricts the addition of fishermen, fishing vessels or equipment. *Limited entry licensing systems* are the preferred tool for controlling entry into many countries' fisheries. Within Canada over the past 25 years such licensing systems have been extended to virtually all marine commercial fisheries. Although individual quotas were also introduced in selected fisheries during the 1980's (see Chapter 9), licensing remains the major tool for limiting entry employed in Canada. Papers describing other countries' experience with limited entry can be found in Rettig and Ginter (1978) and Pearse (1979). Cicin-Sain et al. (1978) made some worldwide comparisons of the experience with limited entry, based on the limited information available at that time.

Limited entry and allocation of access (the subject of Chapter 7) are closely linked. In Canada, these two approaches have evolved in parallel. Although I treat these issues separately, at times they become entwined to the extent that concern over group shares has coloured views on limited entry.

Crutchfield (1961) was an early advocate of restricting the number of units participating in a fishery. He contended that excessive use of labour and capital was inevitable unless a public agency intervened. At that time the arguments against restricted entry as the principal regulatory device centered on various problems of application. Crutchfield acknowledged that these were "by no means negligible... but a start must be made somewhere." A major problem was foreseen in eliminating an already excessive number of vessels and fishermen in many fisheries. Crutchfield (1961) proposed a gradual reduction of participants.

Initially he suggested that this be achieved by making licences non-transferable and issuing no new ones. Subsequently he concluded that this would involve serious inequities and proposed that licences should be transferable through auctions or government-funded buy-backs.

In Canada in the late 1950's to early 1960's fisheries experts recognized the need to restrict entry through licensing limitations. Although licensing had been employed in the west coast salmon fishery in the late 1800's, it had been used as an administrative rather than regulatory tool for improving the economics of the fishery. A form of limited licensing was introduced in the Atlantic lobster fishery in 1967. The first comprehensive program to manage a Canadian fishery by means of limited entry licensing was implemented in the British Columbia salmon fishery in 1968. In the decade thereafter, limited entry licensing was gradually introduced throughout Canada's marine commercial fisheries. The number and nature of these initiatives has been so diverse that a detailed examination would fill a book. I will therefore concentrate on the evolution and impact of this technique in selected fisheries.

3.0 THE PACIFIC FISHERIES

3.1 B.C. Salmon Fishery

3.1.1 General

Chapter 6 discussed the application of traditional open-access regulatory methods to the multispecies, multistock B.C. salmon fishery. From the 1880's to the 1950's increasingly complex arrays of gear, areas, season and other open-access control measures were used to attempt to conserve the resource, in the face of ever-escalating fishing capacity. This increase in fishing capacity threatened the survival of some salmon stocks. It also resulted in obvious economic waste through increasing the costs of fishing.

By the late 1950's this problem had become acute. In 1958, the federal government commissioned Dr. Sol Sinclair, Head of the Department of Agricultural Economics and Farm Management at the University of Manitoba, to study the problems of the salmon and halibut fisheries and recommend measures for sustaining these two fisheries. His study entitled *Licence Limitation — British Columbia: A Method of Economic Fisheries Management* was published in 1960 (Sinclair 1960).

Licence limitation in the British Columbia salmon fishery was not a new concept. Two earlier unsuccessful licensing schemes had been implemented, the first on the Fraser River between 1889 and 1892 and the second in the northern areas of the province between 1907 and 1917. The first attempt involved a specific limit of five hundred boats on the Fraser River. Three hundred and fifty of these licences were granted to cannery operators for company vessels and the balance issued to independent "outside" fishermen.

The major flaw in this initiative appeared to be the failure to recognize the resource rent or excess profit accruing to the licensees, which resulted in demands from other independent fishermen for the right to fish the river. Cannery capacity increased significantly. In 1892, the licensing system was terminated and by 1893 the number of vessels involved in the fishery had more than doubled (Fraser 1978).

The second attempt, in the northern area of B.C., involved a limitation in 1908 on the number of canneries. The cannery operators initially agreed on a voluntary system of boat allotments. The provincial government assumed responsibility for this in 1910. Again, the managers failed to foresee the creation of a resource rent, which benefited the cannery operators and resulted in a demand for more licences. Higher prices for canned salmon during World War I increased the pressure for new entry. New licences were issued, and in 1917 the federal government removed the restriction on cannery licences.

Sinclair (1960) drew on these earlier experiments as background. He suggested a gradual approach to the problem, with higher licence fees and a 5-year moratorium on licences. After the 5-year-period, the government would fix the number and type of licences needed for an economic harvest. These licences would be allotted by auction and would be transferable between fishermen.

Although fishermen's organizations had pushed the idea of a licence limitation program for years, the Sinclair Report received a frosty reception from the industry. Despite general agreement on the principle of limited licensing, views diverged on the form of the system and how to implement it. During the 1960's changes in the management system laid the groundwork for limited entry. However, politicians would not act in the absence of an industry consensus.

Pacific Salmon Seiners.

3.1.2 The Davis Plan

The political climate changed when a majority government was elected in 1968. Jack Davis, the new Fisheries Minister, called for a limited entry licensing plan, which was announced in September, 1968, as the Salmon Licence Control Program. The program took effect for the 1969 fishing season (DOF 1968a) and became known as the Davis Plan.

The Plan's objectives were:

1. To increase incomes of fishermen to the average regional wage;
2. To reduce the level of overcapacity by reducing the size of the fleet; and
3. To reduce the number of vessels to improve the management of the resource (Newton 1978).

Fraser (1978) described the primary objective as the reduction of overcapitalization and excess labour usage within the fishery:

> "This was intended to reduce the cost of production and create an economic surplus that would, first, raise the level of fishermen's remuneration and second, provide a certain return to government to compensate for the use of this public resource and the ever-increasing costs of resource management. Finally, it was to achieve these ends with minimal dislocation of capital and labour then employed."

The Davis Plan envisaged four distinct phases:

1. Freezing the fleet at a stable level;
2. Effecting a gradual reduction in fleet size;
3. Improving the standard of vessels in the fishery; and
4. Introducing economically optimal gear and area regulations.

Phase I

In the first phase the fleet was frozen by creating a specific salmon licence in place of the general fishing licence used previously. This licence applied to a vessel rather than an individual fisherman. Salmon vessel licences were to be issued under special circumstances including: 1. vessels with commercial salmon landings in 1967 or 1968; 2. a vessel under construction by September 6, 1968; or 3. a vessel replacing one with certain standards of production. Licences had to be renewed annually and could be maintained only if vessels fished at least every other year.

Vessels qualifying for salmon licences were placed in two categories, based on previous catch performance:

'A' category vessels were those with annual landings of salmon larger than 10,000 pounds of pink or chum salmon (or an equivalent amount of other salmon species);

'B' category vessels had some commercial landings of salmon but less than 10,000 pounds of pink or chum.

The cut-off between the categories was approximately $1,250 (1968 dollars). In Phase I, only 'A' category vessels could be retired and replaced by new vessels. 'B' category vessels could fish but could not be lengthened or improved and could not be replaced. Although Sinclair had proposed non-transferable licences for the first 5 years of any program, this was not done because it would hamper the sale of vessels as fishermen retired.

Various groups sought changes in the Salmon Control Plan. The United Fishermens Food and Allied Workers' Union (U.F.A.W.U.) opposed vessel-based licensing, claiming that it would increase the power of the big companies. In response, the federal government put a freeze on company ownership of class 'A' vessels in April, 1969. Commercial fishermen fishing for other species, and who had not fished salmon in the qualifying years of 1967 or 1968, contended that they traditionally fished for salmon during peak salmon production years. Initially, there was no provision stopping vessels licensed for salmon from fishing halibut, shellfish or groundfish. In response to criticism, the plan was changed to allow a vessel that participated in any fishery in 1967 and 1968 to obtain either an 'A' or 'B' category salmon vessel licence in 1969.

This modification seriously undermined the objectives of the licensing program. Although only a few vessels (160 at most) were added to the fleet, many of these were large groundfish trawlers and halibut longliners. This created a large pool of unused capacity which could fish salmon. Subsequently, it became common for the non-salmon vessel to be retired from the salmon fishery and replaced by a salmon vessel. The original vessel continued to fish non-salmon species.

Phase II

In January 1970, Minister Davis announced the elements of Phase II:

1. A substantial increase in licence fees;
2. The phase-out of category 'B' vessels from the fishery; and
3. A buy back program (DOF 1970).

'B' category vessel licences would be phased out over 10 years. The intent was to prevent an effective increase in fishing effort by these vessels.

The Buy-Back Program was a major feature of Phase II. Proceeds from the higher licence fees went to a special fund established in 1971 for purchase of 'A' category vessels to reduce the fleet. Any vessels purchased were offered for sale outside the fishery. Proceeds were channeled back to the fund. The Buy-Back program continued until 1973. By then 350 vessels had been retired, about 7% of the 'A' fleet. The salmon harvest was extremely good in 1973, escalating the value of class 'A' licences, and resulting in termination of the Buy-Back Program.

Another major provision of Phase II was a new ton-for-ton replacement rule in lieu of the boat-for-boat rule which had been in effect. In 1969 and 1970, 76 class 'A' vessels with a total capacity of 186 tons had been removed and replaced by vessels with a total capacity of 596 tons.

Effective June 26, 1970, the ton-for-ton replacement rule required that the replacement vessel be no larger in capacity than the original vessel. In August 1972, the replacement rules were again modified, limiting the length of the replacement vessel to that of the vessel removed.

Another problem addressed under Phase II was the hardship caused to native Indian fishermen by the increased licence requirements. Many Indian vessels were classified as category 'F' and hence subject to phase-out. To address this, in the 1971 season a special A-1 licence was created. Any native who owned an eligible vessel had two options:

1. Pay a licence fee of $10 per year and not be eligible for Buy-Back; or
2. Pay the regular licence fee and be eligible for all Buy-Back provisions.

The owner could sell his vessel to either native Indian or white fishermen, but, if sold to a white, the vessel would revert to class 'B' status unless all exempted licence fees were paid retroactively.

Later Phases of the Davis Plan

Two later phases of the Davis Plan had a marginal effect on salmon management. Phase III involved the introduction of higher vessel standards to improve the product and ensure the safety of participants. Standards imposed in 1973 applied to vessels for all species.

This had very little to do with the Salmon Licence Control Program and will not be considered further.

Phase IV, which was intended to bring gear and area restrictions in line with economic factors and conservation requirements, was never carried out. Thus further discussion of the Limited Entry Licensing Program in the B.C. salmon fishery will deal with initiatives under Phases I and II.

3.1.3 Impact of these Initiatives

The B.C. salmon limited entry program was one of the first comprehensive attempts to apply limited entry licensing for economic purposes. Thus it has been examined by a number of authors including Fraser (1978, 1979), Newton (1978) and Pearse and Wilen (1979).

Rather than re-invent the wheel, I will summarize those reviewers' findings. Pearse (1972) analyzed the program's priorities shortly after it was introduced. He speculated that it would hasten automation of the industry, that it would discriminate against companies with low opportunity costs, and it might be unable to control expansion of the fleet. Subsequent analyses indicated that these speculations were well-founded.

Fraser (1978, 1979) concluded that the program was largely successful in meeting its secondary objective, minimizing dislocation of the capital and labour already employed in 1969. Any vessel that depended at all on the fishery was granted a licence in the original scheme. During the second, "fleet-reduction", phase, class 'B' vessels were allowed a 10-year phase-out period. The Buy-Back program was voluntary.

Regarding the more basic objectives, Fraser (1978, 1979) observed that "on a superficial level it appears that the program is also a reasonable success." Several trends were observed 10 years after the program began:

1. The number of fishermen employed in the salmon fishery had declined (from about 9,600 in the pre-limitation period to about 8,600 in 1979);
2. The number of vessels active in the fishery also declined (from 6,639 in 1967 to 5,028 in 1975);
3. The composition of the fleet changed dramatically. The decline in fleet size occurred entirely in the gillnet and troll sectors of the fleet. Hence, the smaller rather than the larger vessels were eliminated;
4. The number of large vessels increased. Non-salmon vessels were converted to salmon vessels or replaced by new salmon seiners. The licences of a small seiner and any number of gillnetters and trollers could be combined or "pyramided" into larger seine vessels; and
5. There was also a marked increase in the number of combination gear vessels compared with single-gear vessels. As a result, the number of seiners and seine combination vessels increased from 369 in 1969 to 483 in 1975.

Fraser (1978, 1979) concluded that the labour input provisions of the program were effective. The average fishing vessel had, however, become larger and more capital intensive. Between 1968 and 1977, average engine horsepower increased by 47% for the gillnet fleet, 43% for the seine fleet and 36% for the troll fleet. Average vessel length for the three fleets increased by 6, 10, and 11%, respectively. Average net tonnage increased by 24, 11 and 17%. Thus Fraser concluded that the "average" vessel in the 1979 fleet was "larger, better equipped, and far more mobile than its counterpart in 1968."

Fraser estimated that the market value of the fleet had increased from $73 million in 1968 to a level of $273 million in 1977. Factoring out capitalized licence values, it appeared that the nominal capital employed in the industry had increased to $161 million by 1975. Allowing for a rapid inflationary spiral in boat-building and repair costs, Fraser (1979) estimated that the real capital employed had increased by $36 million, or 49%. Further increases have since occurred.

Over this period market conditions had changed significantly, with the value of landings increasing by about 35% (after adjusting for increases in the Consumer Price Index). Without licence limitation, this increase in landed value would have dramatically increased the fishing effort. The question was whether limited entry had constrained this trend. Fraser (1979) noted that since the 1973 season, salmon licences had been become quite valuable. An average licence sold for $17,000 in 1974. Although the licence values may have reflected unrealistic expectations among fishermen, high licence value was interpreted as a sign of success. Hence, Fraser (1979) concluded the program was "at least a qualified success with respect to its primary objectives."

Fraser (1979) noted a positive effect on the incomes of operators who were originally licensed in 1969 or purchased their way in during the first few years of the program. However, fishermen who entered the fishery after 1973 paid a high premium. New entrants had apparently gained little from the program. Fraser (1979) suggested that the original licence holders were the only clear beneficiaries. He examined means to make the licence program more effective, suggesting that the most realistic measure would be a change in the level and structure of the licence fees. He recommended a substantial increase in fees and a transition to a resource-value-based fee, such as a landings royalty. He concluded that limited entry alone would not protect the resource and create a more efficient industry.

Newton (1978) pointed out that a mistake was made at the start of the program by limiting entry into the salmon fishery only. This disrupted other fisheries. After 1973 halibut and groundfish vessels could not compete with the salmon fleet for crew, returns to capital, or other inputs. This was rectified ultimately by extending limited entry to the other fleets. Newton also noted that the anticipated unemployment due to limited entry never occurred. He observed that the processing companies' control over fishing licences was contained; they were limited to 12% ownership of the licensed fleet. Financing of the fleet shifted largely to the banks. Newton suggested that limited entry provided security of tenure, a limit on overcapacity, and improved financial prospects, all of which encouraged lending institutions to participate. This improved the fishermen's bargaining position with processors.

Newton (1978) used similar information as Fraser to conclude that limited entry programs could improve incomes of fishermen, facilitate resource management, and generate government revenues given flexibility in program administration. He stressed the need to evaluate limited entry programs over an entended period.

Pearse and Wilen (1979) also appraised the British Columbia fleet rationalization program. Their study questioned whether entry into the fishery had been restricted in terms of the value of labour and capital employed. They found that the labour engaged in the fishery decreased by about 16% between 1968 and 1975 and concluded that the real costs of labour had, at least, not risen. They also suggested that the program had been partially successful in checking the expansion of capital. The growth in redundant capital had not been halted however. The capacity of the fleet had continued to grow, even though it had far exceeded the fishery's needs when the program began. They speculated that the rate of capital accumulation was slower under the controls, even though the fleet's revenues continued to grow roughly as before. They thought that this might allow some resource rents to be realized.

Pearse and Wilen (1979) disagreed with Fraser's recommendation that there should be increased licence fees or a landings tax or royalty. They dismissed a levy on the catch as impractical. Instead, they suggested that the existing licences to engage in fishing be converted to licences to take specified quantities of fish. This would free fishermen to find the least-cost method of fishing.

Crutchfield (1979) suggested that the conclusions of Fraser (1979) and Pearse and Wilen (1979) were overly pessimistic. He noted the continued existence of very high prices for licences in the British Columbia salmon fishery. This meant that either there was a net saving in real factor inputs compared to that which would prevail in open access; or fishermen had continued to entertain, for 10 years or more, erroneous expectations. He found the latter possibility a "little hard to swallow." Rising salmon prices, in the absence of limited entry, would undoubtedly have induced further investment in new boats.

3.1.4 The Post-Davis Era — Mid 1970's to 1980

Sol Sinclair was commissioned by the Fisheries and Marine Service in the mid-1970's to develop a proposal for a revised Licensing and Fee System for the Coastal Fisheries of British Columbia. Sinclair concluded that the goals of the Davis Plan had been only partly achieved (Sinclair 1978).

After reviewing that experience and the three main methods of limited entry management tools, Sinclair proposed a comprehensive licensing and fee system for the entire British Columbia fishery. He recommended a royalty on the catch and capacity restrictions. This would consider length, hold capacity and horsepower. He also recommended the introduction of area licensing.

Sinclair remained skeptical of the merits of transferable licences between retiring and new entrants to the fishery. However, transfers had been permitted for the previous decade, and it was now difficult to cancel the practice. Instead, he suggested that further fleet reduction could be achieved by reestablishing the Buy-Back Program suspended in 1973.

Some elements of Sinclair's proposals were adopted over the next several years. For example, in March 1979, Fisheries Minister Roméo LeBlanc created a Fishing Vessel Licence Appeal Board for the B.C. industry. The Board was to review appeals in cases where licences had been refused through the department's normal licensing procedure and make recommendations to the Minister.

During 1977 to 1980 the Department made several changes to the vessel replacement rules established in Phase I and II of the Davis Plan. In July 1977, an attempt was made to control upgrading of the fleet by a regulation preventing transfer of a vessel licence from retiring gillnet or troll vessels to a new seine vessel. This meant that an old seine vessel had to be retired before a new one could be licensed. In June 1978, further action was taken to limit the size of newly licensed vessels. This regulation stipulated that if two or more vessels were replaced by a single vessel that vessel must be less than 50 feet long. In January 1979, the ton-for-ton replacement rule was revised. Henceforth, a replacement vessel was limited to a net tonnage less than or equal to the replaced vessel and could not be longer than the replaced vessel. In mid-1980, "pyramiding" of vessels was prohibited.

Meanwhile, other studies emphasized the problem of fleet overcapacity. In 1980, a weak salmon run coincided with a weak market price for salmon products. The adverse economic effects were compounded by a failure of the roe herring fishery, and a fishermen's strike. The combination of events created a financial crisis for the British Columbia fishing industry. Many vessel owners faced bankruptcy.

3.1.5 The Doucet-Pearse Report

In the autumn of 1980, Fisheries Minister LeBlanc requested Mr. Fern Doucet and Dr. Peter Pearse to examine the industry and recommend solutions for the problems. Their October 1980 report recommended a seven-point plan for immediate action:

1. A new pledge to salmon enhancement;
2. Stronger regulations and enforcement;
3. A renewal of buy-back operations;
4. Removal of "perverse" subsidies;
5. Licensing of sportsfishermen;
6. A landings charge for the commercial salmon fishery; and
7. Changes in vessel licence fees and restrictions (Doucet and Pearse 1980).

More specifically, they proposed a two-area system for the troll fishery. This would require troll vessels to fish either in the Gulf of Georgia or elsewhere. Thereafter licences would be amended to restrict trolling in the Gulf of Georgia to those who had chosen that area.

The Department of Fisheries and Oceans implemented the major recommendations of this study. In October 1980 additional resources were allocated to enforcement. A two-area troll licence system and spot closures of the troll fishery (where necessary) were implemented in 1981. For 1981 sports fishermen in the tidal waters of B.C. were licensed for the first time. All of the recommended steps regarding vessel licence fees and restrictions were taken. Commercial licence fees were doubled. The moratoria on vessel pyramiding and addition of new seine gear were kept in place (DFO 1980c).

One of the recommendations for immediate action was to reinstitute the salmon vessel Buy-Back Program. Pearse and Doucet considered that the Buy-Back approach directly addressed the overcapacity problem while giving fishermen a viable option to exit the industry.

In late October 1980 Minister LeBlanc met with representatives of the west coast fishing industry. Industry viewed renewal of the Buy-Back Program as a positive first step in solving its problems. On October 28, the Minister announced the renewal of Buy-Back operations "when funds could be made available" (DFO 1980c). On January 13, 1981, he announced the appointment of Dr. Peter Pearse as a Commissioner under the Inquiries Act to examine the industry and recommend long-term measures to control fishing (DFO 1981g). On February 12, 1981, the Minister announced a reinstitution of a modest Vessel Buy-Back Program with funding up to $3.5 million (DFO 1981h). This could have resulted in retirement of 40–50 fishing vessels. Because transactions had to be completed by the end of the fiscal year (March 31), only 42 offers were made. This resulted in 26 licensed salmon vessels being purchased for a total of $2.5 million.

In 1981, the Economic Council of Canada undertook a case study of the Pacific salmon fishery (MacDonald 1981). This was part of a broader examination of the public regulation of commercial fisheries in Canada. MacDonald reviewed the events in the fishery since the introduction of the Davis Plan and concluded:

> "The licence limitation programme and subsequent replacement restrictions did not tackle the fundamental problem in the fishery; they did not remove the *incentive* for fishermen to over-invest in harvesting facilities. Instead, they attempted to inhibit a fisherman's ability to over-invest by limiting his capital investment options. The increasing profit potential in the industry in the 1970's resulted in fishermen devising unanticipated methods to circumvent restrictions placed on them....."

After considering taxes (landings charges) and individual quotas, MacDonald (1981) concluded that "quotas appear to be the only viable alternative." I will examine the arguments for and against this in Chapter 9.

3.1.6 The Pearse Report

Meanwhile, Pearse in 1981 and 1982 examined the Pacific salmon fishery in-depth, with extensive public hearings and numerous background studies (see Parsons 1993). His 1982 report *Turning the Tide* concluded:

> "Our catches of salmon and roe-herring could be taken with fleets half their present size and at half the cost now expended in fishing.... The (Davis) plan has clearly failed in its main purpose, which was to control and reduce excessive fishing capacity."

Pearse proposed that limited-entry licensing systems be replaced by quota licences in "those fisheries where it is feasible to do so." In his view, this included all of the significant commercial fisheries other than the salmon and roe-herring fisheries. Chapter 9 describes his quota-licence proposals and discusses their fate.

Regarding the Pacific salmon and roe-herring fisheries, Pearse recognized that "at the present time, any system of individual catch quotas would.... be difficult for these fleets to adjust to and probably beyond the capability of the Department to administer. The stocks and available catch of these species.... are notorious for their wide and unpredictable year-to-year fluctuations, making it impossible to allocate individual quotas in advance with any degree of certainty." As an alternative, he proposed special measures to improve the limited-entry licensing system and reduce the fleets (Pearse 1982).

These included:

1. A program to reduce the salmon and roe-herring fleets to half their 1982 size over 10 years;
2. Allocating the catch among competing sectors of each fleet to ensure that all would share the benefits of fleet reduction;
3. Royalties on landings to capture some of the financial gains from fleet rationalization; and
4. New restrictions on vessel replacement which, coupled with levying royalties and eliminating subsidies, would dampen licensees' incentives to expand their fishing power.

Pearse proposed that all existing regular salmon licences and roe-herring licences be replaced in 1983 by new 10-year licences. The existing 'B' salmon licences should be renewed only until the year in which they were scheduled to expire. They would then be eliminated. Licences should be issued to persons or companies and designate the vessel to be used. Comprehensive gear licensing should supplement the existing licensing system. Pearse proposed also that, before the 1986 fishing season, each salmon licensee be required to select one of three zones (north, south and west) in which his licence would apply for the remainder of its term. This is known as *area licensing*.

Pearse proposed that the target fleet at the end of a 10-year transitional period begining December 1982 be 50% of the existing licensed capacity. He proposed that licences for this target fleet be allocated through a complex, phased, competitive bidding process. To reduce the fleet during the initial 10-year transition period, Pearse proposed a major voluntary Buy-Back program to be administered by a new Pacific Fisheries Licensing Board. He suggested that this be funded from four sources:

1. An initial grant from the federal government of $10 million;
2. A government payment each year equal to the royalties paid in that year on roe-herring plus one-half of the royalties paid on salmon. This amount should be doubled by means of a dollar-for-dollar matching grant from the federal treasury;
3. Payments from the federal government each year in amounts equal to the revenues from competitive bids for salmon and roe-herring licences; and
4. Borrowing by the Board against its anticipated revenues to a maximum of $100 million.

Regarding fleet replacement, Pearse recommended that no new vessels, except those already under construction at the time his report was released, should be eligible for any commercial fishing licence during the 10-year transitional period. Licensees would be permitted to replace their vessels with already-licensed vessels. This would be subject to the established foot-for-foot and ton-for-ton replacement limits.

Pearse's proposals clearly were comprehensive, bold and imaginative. But events during the decade following release of his report have shown that the proposals did not meet two of the three criteria enunciated by Sinclair (1978) as essential to the success of any restricted access program. Sinclair had suggested that any successful program would have to be:

1. Administratively feasible;
2. Politically acceptable; and
3. Publicly defensible.

It is debatable whether Pearse's proposals were administratively feasible. They engendered a storm of public debate over the next 5 years which made it clear that they were neither politically acceptable nor publicly defensible. In essence, the ambitious proposal for restructuring of the Pacific fishery sunk in a tidal wave of opposition and protest from the major interest groups in the fishing industry. See also Parsons 1993.

3.1.7 Events Post-Pearse

In April, 1982, following release of Pearse's Interim Report, Fisheries Minister LeBlanc established a Fleet Rationalization Committee chaired by Don Cruickshank. The mandate was to advise LeBlanc on (1) means to reduce fleet capacity, (2) the appropriate fleet size and mix after rationalization, and (3) means to strengthen vessel replacement rules. This Committee reviewed Pearse's final report before proposing that:

1. Fleet capacity be reduced through a voluntary Buy-Back program;
2. Further growth of the fleet be prevented by curtailing capitalization through a royalty on landings;
3. Fleet rationalization should achieve a 24% reduction in the roe-herring seine and gillnet fleets, a 24% reduction in the number of salmon seine vessels and a 12% reduction in the small boat salmon fleet; and

4. Vessel replacement rules be strengthened by restricting replacement vessel size to that of the existing vessel. This would involve a new measurement system based on the product of length, breadth and width (Anon 1982).

The Minister's Advisory Committee on the B.C. fisheries in 1983 essentially rejected Pearse's licensing proposals. These included the proposals that would have had all fishermen bid for their licence in an open auction and the establishment of a Crown Corporation to administer the licensing and appeal process.

Following extensive consultations with the industry, Minister Pierre De Bané and Senator Jack Austin in June 1984 announced a new policy for the Pacific fisheries (DFO 1984d). The government's proposals involved two steps:

1. Reducing the overcapacity of the fleet and strengthening it financially; and
2. Adopting a "more rational approach to harvesting."

The proposals aimed at reducing the fleet, by as much as 45%, through a Buy-Back of at least $100 million. This would be accompanied by a management system "that puts a final end to the 'race to the fish.'" Mr. De Bané announced the government's intention to implement area and gear licensing and individual fishing allocations. The proposals included a Buy-Back Corporation, with funding by government and financial institutions. Although a total amount of $100 million was mentioned, the details were vague "pending completion of negotiations."

Proposed initiatives included:

1. Individual fishing allocations would be introduced in selected areas in the troll and net fisheries in 1985 and extended to all salmon fisheries by the end of 1987;
2. Salmon vessel owners would be required to select one of four areas if using net gear, or one of three areas, if using troll gear (i.e. area licensing);
3. All salmon licence holders would be restricted to one gear type;
4. The salmon catch would be divided within the commercial fishery by gear type;
5. More fish would be caught in terminal fisheries;
6. Licence fees would be increased.

The government tabled in the House of Commons draft legislation to amend the Fisheries Act and create a new Pacific Fisheries Restructuring Act (Canada 1984). On July 9, 1984, the new Prime Minister John Turner dissolved Parliament and called an election. The Liberal Government was defeated and a majority Conservative Government swept to power in the early September election.

The previous government's proposals to restructure the Pacific fisheries were jettisoned. The new Government launched its own consultations with the industry about the problems of the Pacific fishery. A voluntary government-funded Buy-Back Program was considered but never materialized. The concept of individual quotas, which had been opposed by the major sectors of the industry, was dropped.

The B.C. salmon limited entry program, as modified over the years, neither reduced the overcapacity problem in the salmon fisheries nor eliminated the incentive to overinvest. However, it did provide some protection against a surge of new entrants. A DFO study of the British Columbia salmon fishery in 1987 indicated that, without the limited entry program, the 7,500 licensed vessels could have increased to 10,000 during the period of rapidly rising fish prices in the 1980's (Department of Fisheries and Oceans, Pacific Region staff, pers. comm.). Instead, the number was reduced to about 4,500 licensed vessels (Fig. 8-1). Despite this curb on expansion, in the late 1980's chronic overcapacity was still a major problem for Canada's Pacific commercial fisheries.

Taking inflation into account, between 1969 and 1988 the capital value of the seine fleet had increased by a factor of 3.6. The overcapacity problem was masked by higher-than-average salmon abundance and prices during the late 1980s. The industry during 1986 to 1990 has been described as "buoyant, prosperous and healthy". Increased capital investment, however, left the fleet more vulnerable to cyclical catch and declining prices. Vessel owners had invested heavily in their vessels, leaving them little cushion for decreases in gross income (DFO 1992j).

3.2 B.C. Roe Herring Fishery

Despite the shortcomings of the B.C. salmon limited entry program, this type of restricted licensing was subsequently adopted for the other major fisheries on the Pacific coast in response to similar problems of overcapacity. One fishery after another developed, overexpanded and was subsequently placed under restrictions on additional entrants through limited entry licensing. Federal officials introduced limited entry to the roe-herring fishery when it commenced in 1974. However, the licence was attached to the fisherman instead of the vessel. The groundfish trawl fishery was restricted in 1975 and the shrimp fishery in 1976, both by vessel licence programs. Licences in the abalone and roe-on-kelp herring fisheries were restricted in 1977. As of 1981, the only fisheries not subject to limited entry were

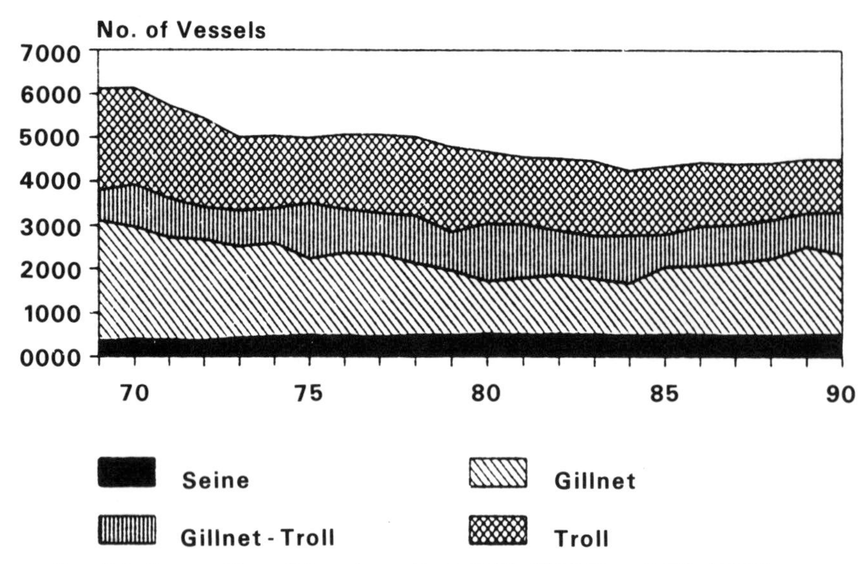

FIG. 8-1. Composition of the salmon fleet, 1969–1990 (from DFO, 1992j).

groundfish hook and line, crabs, smelts and sardines by gillnet or set net.

The remainder of this section will focus on certain features of limited entry in the roe herring fishery. This program differed in form and content from the salmon program. In 1974, when licencing was introduced, the fishery was a new one. Limited entry was introduced to prevent, rather than reduce, overcapacity and overfishing.

Following closure of the B.C. herring fishery for conservation reasons, the Department of Fisheries permitted an experimental roe herring fishery in 1971. This fishery grew in value very rapidly to $35 million wholesale in 1973 and $170 million in 1979. The commercial roe fishery takes place in a very short period during the 6 weeks between late February and early April, on and adjacent to the spawning grounds which are in and immediately below the intertidal zone.

When fishing recommenced, a relatively low catch-quota was established (see Chapter 6). By late 1973, it appeared that several thousand vessels would fish herring the following season. This led to a licensing program with three objectives:

1. To maintain fleet size below levels that might threaten herring stocks;
2. To provide economic returns at least sufficient to cover costs for fishermen; and
3. To provide a return from the resource to the Crown.

The situation at that time was quite different from that in the salmon fishery. Fleet capacity was not yet excessive, and the initial participants were doing quite well. The roe fishery was a bonanza. On the one hand, federal officials were concerned about a potential erosion of returns if capacity increased significantly. At the same time, the Minister of Fisheries appeared unwilling to deny access to any "bonafide" fishermen (Meyer 1976). Thus federal officials developed a licensing plan which attempted to reconcile conflicting objectives.

The 1974 licensing plan permitted initial entry by all applicants. Limitation was to be accomplished through setting fees considerably above previous levels. For example, fees were set at $2,000 per seiner and $200 per gillnetter compared with a fee of $400 for a salmon seiner in 1973.

To facilitate entry of natives into this fishery, DFO granted them free entry until January 15, 1977, while closing entry to all non-natives on January 15, 1974.

In contrast with the transferable vessel licence for salmon fishing, the herring licence was a non-transferable personal licence. There was also an owner-operator clause, which applied only to vessels first entering the fleet in 1974. However, the owner-operator restriction was circumvented through vessel leasing.

Fraser (1982) observed that the program could "not be considered an outstanding success" in controlling the number of vessels. The number of seine vessels in this fishery increased from 161 in 1973 to 252 in 1974. In 1974, 1,579 gillnet licences were issued, about six times more than the number of vessels participating in 1973. The number of *active* seine and gillnet vessels increased by about 40 and 400%, respectively.

Despite these increases, the program was not a complete failure. Fisheries officials anticipated that, without licence limitation, a fleet of 5,000 vessels would participate in the fishery. By 1976 the number of licensed seine and gillnet vessels had declined to 214 and 1,285 vessels, respectively (Fraser 1982). By 1976 the licensing program was considered "a qualified success" and the fleet "relatively manageable." Although average returns had declined, the fishery "still provided a very handsome return" (Fraser 1982).

The year 1976, however, marked a transition in the fishery. Shortfalls in production from China, the other major supplier to the Japanese market, left British Columbia producers with a virtual monopoly. This, with smaller catches of B.C. roe herring in 1978 and 1979, caused an upward price spiral. As a result, gold fever hit the fishing grounds. Capital costs of fishing increased substantially without an equivalent increase in output. Fishing pressure also increased, with the fleet rushing from opening to opening, threatening the stocks. The fleet operated at such a pace that even minor management blunders could be disastrous for the resource. Gillnet openings were down to one or two days in some areas, and fifteen-minute openings were common for the seine fleet. In 1978 and 1979 returns from the fishery were exceptionally high. But in 1980, the roe-market collapsed, depressing 1980 prices.

Pearse (1982) described the fishery at the end of the 1970's:

> "The roe herring fishery is extraordinarily hectic due to the unpredictable stocks and available catch, the massive and excessive fishing power, the need to limit the fishing time to the moment when the fish are about to spawn and the high values at stake. It is probably the most difficult of fisheries to manage; fishery officers, under extreme pressure and great uncertainty, have to try to restrict openings to a few minutes in many cases, during which fortunes have sometimes been made."

Pearse considered that the roe-herring licensing system had caused several serious problems: "The first and most obvious is that it failed to curtail the size of the fleet to the needed capacity. This was due partly to generous initial eligibility criteria." The attempts to make licences nontransferable had been subverted in practice. Pearse further observed:

> "Because the licenses apply to persons, who can change their designated vessels, there is little to restrain the growth in fishing power of the vessels used and hence the expansion of the fleet's fishing capacity. In this respect the licensing system has been even less effective than salmon licenses, being incapable of controlling the fleet size, which is its main purpose."

Pearse's proposals for licensing in the roe-herring fishery were similar to those for salmon.

Meanwhile, the Department of Fisheries and Oceans had made a major modification to the herring licensing system. In 1981 the Department introduced an experimental system of area licensing for the roe fishery. MacGillivray (1986) described the structure of area licensing and evaluated its impact.

Under the area licensing scheme, roe herring licences were restricted to one of three defined fishing areas. Each licence holder could choose one of those areas. The area choice applied for 1 year. Licences could not be transferred among vessels during the season except where a vessel was lost or destroyed. A ceiling was established on the number of vessels to be licensed for each area.

Subsequently, the program was modified to allow a vessel to fish more than one area. This was termed multiple licensing. In 1983, fixed catch quotas were introduced for each licence area. Under this system, catch quotas were established before the season and not to be revised during the season (see Chapter 6 and Stocker 1993). In 1985 the areas were redefined.

The major objective of area licensing was immediate improvement in the in-season manageability of the herring stocks. While the economic impact of area licensing was also considered important, area licensing by itself could not solve the fishery's economic problems (MacGillivray 1986).

MacGillivray (1986) reported that the majority of fishing industry groups were satisfied with area licensing. Support for area licensing increased from 53–67% in 1981 to 78–86% after the 1983 fishery.

MacGillivray (1986) concluded that area licensing had introduced some stability into the fishery. Fishermen and processors could plan more effectively for the season, and the frantic running around in anticipation of openings was greatly reduced.

The economic impact was also significant. Industry representatives identified major savings in operating costs and lower packing costs, as a result of restricted mobility (MacGillivray 1986).

Area licensing had two potential equity impacts: (1) on the distribution of catch among fishermen, and (2) potential unemployment resulting from multiple licensing. Area licensing had different effects on catch distribution for the gillnet and seine fleets. The risk of a seiner catching nothing was increased with area licensing. This created pressure for multiple licences. Area licensing resulted in a more equitable distribution of the gillnet catch per licence. According to the U.F.F.A.W.U. multiple licensing threatened jobs.

From a biological perspective, area licensing improved the management of the roe-herring fishery

(Stocker 1993). This occurred because of the reduction in the large concentrations of boats at major openings.

The limited entry licensing regime for the B.C. roe-herring fishery was imperfect. However, improvements in the 1980's addressed some of the concerns in Pearse's 1982 Report. There was a major impact on fleet size. By 1985 the herring fleet had been reduced by approximately 30%.

4.0 THE ATLANTIC FISHERIES

4.1 Atlantic Lobsters

Limited entry was introduced experimentally in the Atlantic lobster fishery in 1967. Subsequent developments in the limited entry program for this valuable fishery have not been documented as thoroughly as the British Columbia salmon program introduced about the same time. There has been no systematic assessment of the impact of the Atlantic lobster limited entry program.

4.1.1 The Fishery in the Early 1960's

The first serious examination of the economics of the Canadian Atlantic lobster fishery was undertaken in the mid-1960's. The results were published in Fisheries Research Board Bulletin 157 — *An Economic Appraisal of the Canadian Lobster Fishery* (Rutherford et al. 1967). There was a form of licensing prior to that time, but statistics on participation in the fishery were poor. Official statistics for the Maritime Provinces estimated the number of lobster fishermen at 16,000–17,000 in 1959 to 1963. This compared with 20,000–23,000 licences issued during this period. The count by fishery officers, the lower figures in the official statistics, was a more reliable indicator of the number of fishermen earning a living from lobsters (Rutherford et al. 1967).

Measures of fishing effort were inadequate. Although the number of traps on hand was a crude index of fishing effort, there appeared to be a steadily increasing trend in the number of lobster traps over the years. The number of traps increased from 1.25 million in 1918 to 1.6 million in 1925 and to 2.5 million in 1960. The number of traps declined slightly in Nova Scotia in the later years but increased in Quebec, New Brunswick and Prince Edward Island by about 25% during 1951–1960. The lobster catch fluctuated around an average of 19,500 tons a year during that period (Rutherford et al. 1967).

Although lobster fishing was a short-season operation, many operators apparently earned a large proportion of their incomes from the lobster fishery. Some individuals reported no other source of earnings. Survey results indicated that, to a large extent, movement of fishermen into and out of the lobster fishery was determined by the availability and attractiveness of alternative employment opportunities. Lobster fishermen were particularly opportunistic workers. A "hard core" of lobster fishermen continued to fish lobster through periods of prosperity and depression. One of the more interesting observations from the 1967 study was the considerable resentment among fishermen toward what they regarded as "unfair intrusion by newcomers into the lobster fishery." Rutherford et al. reported:

> "The fishermen are directly affected by additional boats and gear brought into operation on already over-crowded fishing grounds. They look with particular disfavour on the so-called 'moonlighters'. These are persons who are in occupations other than fishing but take time off from regular employment, or use their holidays or spare time after working hours, to fish for lobsters."

These observations foreshadowed a major issue in lobster licensing policy during the 1970's: eliminating moonlighters from the fishery.

Unlike the B.C. salmon and herring fisheries, the lobster fishery was prosecuted primarily by small boats from 22 to 48 feet in length. Most boats were open-decked. The average number of traps used during 1961 was 206 per enterprise with significant differences among areas. Gross income from lobster fishing averaged just $1,600 per enterprise for all areas. Net current income — the difference between receipts from the sale of lobsters and the current operating expenses — averaged $779.

Rutherford et al. (1967) considered fishing effort at that time to be "much in excess of what is required to take the maximum sustainable catch of lobsters." They suggested that reducing effort would increase the net economic yield of the fishery.

The 1967 study examined several alternatives for lobster fishery management to increase the net economic yield. Sinclair's (1960) proposal for licensing in the B.C. salmon fisheries was examined as an option for the lobster fishery. They rejected an arbitrary reduction in the number of fishermen but they favoured taxing fishermen by auctioning licences.

Rutherford et al. made no specific proposals to pursuing this "desired economic objective" of increasing the net economic yield from the fishery. Nonetheless, in 1966 and 1967 the federal government introduced a series of new management measures for the lobster fishery to improve the economic returns of commercial lobster fishermen.

4.1.2 Limited Entry in the Lobster Fishery

In March 1966, Fisheries Minister Robichaud announced limits on the numbers of lobster traps coupled with registration of lobster boats in Lobster Fishing District 8 in Northumberland Strait (DOF 1966). This was extended early in 1967 to other parts of the southern Gulf (DOF 1967a and b).

In June 1967, Robichaud announced similar licensing restrictions for the summer and fall lobster fisheries in District 8-Northumberland Strait. The Minister indicated that this was being done for economic reasons:

> "The legal lobster crop is harvested up to 85% each year, and we feel that fishermen can land the same number of lobsters with fewer traps and with less labour and capital costs. In the past several years the lobster fishing effort has intensified greatly" (DOF 1967c).

In February 1968, the Minister announced that all lobster fishing boats in the Maritimes and Newfoundland would have to be registered and that operators' lobster fishing licences in all Maritime districts would be limited in 1968 to those holding such licences in 1967 (DOF 1968b and 1968c). Thus, 1968 marked the first wide-scale application of licence limitation in the Maritimes lobster fishery.

In January 1969, new Fisheries Minister Jack Davis announced that the limit on the number of operators would be replaced by a limit on the number of boats that would be allowed to fish for lobster. The lobster licence would stay with the boat. If an owner sold his boat to another fishermen, he would withdraw from the fishery and the purchaser would be able to enter. Mr. Davis stated: "From now on a licensed lobster boat is the passport to the lobster fishery" (DOF 1969a). He was undoubtedly influenced by the approach adopted in the Pacific salmon limited entry program.

In February 1969, this program was supplemented by:

1. Minimum trap limits in all Maritime lobster districts; and
2. Termination of the provision allowing jointly owned and operated boats to fish additional traps (DOF 1969b).

The minimum trap limits were introduced "to distinguish between bonafide fishermen and others who catch lobsters on a casual, weekend, or part-time basis." No fisherman who fished fewer traps in 1968 than the lower limit for his district could add traps in the future. Thus two classes of licences were established. Class "B" licences were issued to all boats from which less than 100, 75 or 50 traps, depending on the district, were fished in 1968. Class "A" licences were issued to all other boats, i.e. boats or replacements from which more than the minimum number of traps were fished in 1968. Class "A" licences were transferable when a boat was sold. When a fisherman using a class "B" boat ceased fishing, the licence for that boat would not be renewed. Thus Class "A" licences became the "ticket of entry" for the lobster fishery.

Some lobster fishermen feared that the new licence limitation scheme might concentrate ownership of lobster boats (and, in effect, licences) in the hands of lobster buyers and processing companies. Minister Davis addressed this fear in April 1969, announcing restrictions on multiple licensing in the lobster fishery. No person or company could licence more lobster fishing vessels in 1969 than they had registered in 1968. New entrants into the lobster fishery could licence only a single vessel (DOF 1969c).

De Wolf (1974) reviewed the theoretical economic impact of alternative forms of regulating the lobster fishery. He observed:

> "Given our present inadequate knowledge of the behaviour of lobster stocks and the costs and returns of fishing enterprises, the most we can hope for is that we are moving in the right direction, i.e. towards the maximum net economic yield."

Net economic yield in the lobster fishery had been dissipated by excessive fishing effort. Trap limits and limited entry licences were an attempt to improve the economic conditions of lobster fishermen. De Wolf (1974) reported that licence limitation and trap limits did not immediately reduce the total number of traps fished. Indeed, the total number of traps fished apparently increased in some fishing districts.

The number of traps fished increased slightly (5.6%) between 1968 and 1972. There was a 8.7% reduction in the number of lobster fishing boats from 10,339 in 1968 to 9,441 in 1972. The number of class "B" vessels decreased from 1,106 in 1969 to 730 in 1972; this of course was an objective of the 1969 policy.

From an analysis of the number of traps fished in 1972 compared with the total number of traps that could be set if all class "A" boats fished the maximum possible for each district, De Wolf concluded that approximately 25% more traps could be fished under trap limit by class "A" boats than were fished in 1972 by all boats. This indicated that trap limits and licence limitation had not significantly reduced fishing effort.

De Wolf (1974) concluded:

> "Licence limitation and trap limits as introduced in the late 1960's will redistribute

incomes more equally, will not affect total fishing effort substantially, may lead to an increase in the total number of traps, may have an adverse effect on economic efficiency, and will increase the value of boats as the right to fish lobsters becomes capitalized."

4.1.3 Lobster Fishery Task Force

In March 1974, at a meeting of the House of Commons Standing Committee on Fisheries, Mr. Roméo LeBlanc, then MP for Westmorland-Kent, requested that a Task Force be established to investigate the lobster fishery. Such a Task Force was appointed in the spring of 1974, chaired by economist Gordon De Wolf.

Following extensive consultations with fishermen, lobster buyers and processors, and departmental managers, the Task Force made its recommendations in March 1975 (DOE 1975b). According to the Task Force, the lobster industry of Atlantic Canada was characterized by a large number of fishermen earning low incomes. There was low capital investment in fishing operations. This resulted in relatively easy entry even with licence limitation, excessive capacity in both the harvesting and processing sectors and little technological innovation.

The Task Force advocated maximum net economic yield as an ideal at which to aim. However, it concluded that there were many reasons why the lobster fishery could not be managed solely based on economic criteria. In many areas where there were few alternative occupations, a large displacement of fishermen could have serious social consequences and could substantially increase welfare payments. As a long term objective, the Task Force suggested that the level of fishing effort in the lobster fishery be decreased by reducing the number of fishermen and the amount of gear. It suggested a long-term integrated approach to licence all inshore fishermen to create a group of 'professional' fishermen. These would derive their livelihood from several fisheries and not be overwhelmingly dependent on one. Meanwhile, the Task Force foresaw that "the level of fishing effort in the lobster fishery will be greater than that dictated by economic criteria alone."

The Task Force supported trap limits as necessary measures to control fishing effort and thereby improve fishermen's incomes and economic efficiency. It concluded that the initial trap limits were much too high in relation to the number of traps actually fished. The Task Force recommended reducing the maximum trap limits then in force and extending trap limits to Newfoundland.

The Task Force noted that the number of class "A" lobster boats in the Maritimes had remained at about 8,800 since the introduction of limited entry for boats in 1969. Class "B" boats had declined from about 1,100 to 600. In Québec the number of licensed lobster fishermen increased from about 1,600 in 1969 to 1,725 in 1972 while the number of lobster boats increased only slightly, from 780 to 793. Apparently after 1972 there were substantially fewer fishermen. The number of lobster fishermen in Newfoundland decreased from 6,400 in 1969 to 4,800 in 1972 but increased significantly in 1973 and 1974.

The Task Force observed that the 1969 limited entry program for lobster boats had distorted the boat market. Licensed lobster boats were selling for abnormally high prices because the sale included payment for the privilege of fishing. Unlicensed boats, on the other hand, sold for abnormally low prices. It noted that the sale of lobster fishing privileges provided retiring fishermen with a high payment for their boat and gear, but the practice was not popular with fishermen. The sale of class "A" boats led to a situation where those best able to afford licences purchased them. The Task Force noted that some licences had been sold to individuals who engaged in the fishery more for recreation than for production. It concluded:

> "With governments attempting to increase incomes and employment in Atlantic Canada, it appears reasonable to use the valuable lobster resource solely to generate income and employment."

Nonetheless, the Task Force concluded from the limited available evidence that "the commercial fleet must be reduced by 25 to 50% if the fishery is to become economically viable." It also concluded that "the lobster fishery should be reserved for commercial fishermen in regions where such fishing will provide a reasonable standard of living to those engaged in it.... Persons fully employed in other occupations should be eliminated immediately."

In Quebec some reduction in the number of fishermen had already been achieved by eliminating those who did not meet specified criteria of eligibility. Most fishermen who attended the Task Force meetings had called for reductions in the number of participants. The most common suggestion was a government buy-back scheme. Although it saw the need to reduce the number of fishermen, the Task Force did not favour a buy-back. Any buy-back program should be voluntary. Neither boats nor gear should be purchased. There should be a uniform payment to be financed in part through higher licence fees. The Task Force's preferred approach was to reduce the number of lobster fishermen through attrition.

4.1.4 The Government's Response to the Lobster Task Force Report

The federal government responded quickly to the Report of the Lobster Fishery Task Force. On December 30, 1975, Roméo LeBlanc, Minister of State for Fisheries, announced new measures to exclude from the lobster fishery those who earned their living elsewhere to increase returns to 'legitimate' fishermen. The rationale for these new measures was:

> "Stricter licence controls against 'moonlighters' along with other measures will give a smaller fleet catching more lobsters per fishermen. The fishermen demanded strongly that we exclude 'moonlighters' from this fishery. This we intend to do. We have no intent to disturb the part-time fishermen with a real dependence on the lobster fishery or the man using his boat in more than one fishery. We want only to honour the demands of these legitimate fishermen by excluding a minority with no real stake in the lobster industry" (DOE, 1975c).

The package included these measures:

1. Those with full-time employment in year-round occupations such as school teachers, civil servants, professionals, or in occupations that coincided with the lobster season, would lose their privilege to fish lobster;
2. Local advisory committees would be created to help put these measures in place for each lobster district;
3. A temporary freeze was placed on the issuance of new licences and the transfer of licences for lobster fishermen and vessels pending announcement of further measures;
4. Fishermen in Newfoundland could fish no more traps than they fished in 1975, until surveys determined suitable trap limits for each district; and
5. Limited entry licensing would now apply in Newfoundland.

In September 1976, Minister LeBlanc announced three new categories of lobster licences for the Maritime provinces. In the intervening months since his December 1975 statement, fishermen in the Maritimes had been canvassed and 70% supported the policy of giving priority access to the lobster stocks to those persons most dependent on fishing. The new licence categories were:

Category "A" — full licence: This was reserved for those who depended on the lobster fishery, and who had no year-round employment nor any full-time seasonal job coinciding with the lobster season. With approval of the Department, the holder of an "A" licence could under specific conditions transfer it to another person.

Category "B" — limited licence: Although the holder of this licence had regular employment elsewhere, he had a claim to fish lobster because of his previous participation. Persons who had fished lobster from 1968 or earlier might qualify for this licence. A "B"-licence holder could fish 30% of the maximum number of traps allowed in his lobster district. Normally, a "B"-licence would remain restricted to the licensee and terminate when he left the fishery.

Category "C" — temporary licence: With full time employment elsewhere, with only a short history in this fishery, and with no real dependence on the lobster fishery, persons in this category were considered to have no claim to continue holding a licence. To allow transition time, for example, to sell off boats and traps, those licence holders were allowed limited fishing privileges for up to two years. No transfer of "C"-licences would be allowed (DOE 1976e).

Problems arising from this policy resulted in some modifications over the next couple of years. Initially, it was intended that Category "A" licences would not be issued to persons employed full-time outside the lobster fishery or part-timers working concurrently with the lobster fishing season. However, an exception to this rule was made in cases where the individual was otherwise qualified and did not earn more than $10,000 outside the fishing industry. This was later changed to minimum wage plus 25%. Provisions were also made to upgrade a Category "B" licence to Category "A" if the licensee's employment status changed to meet the regulatory requirements for Category "A". With such an upgrade (to a Category "A-1") the fishermen could fish 100% of the trap limit. A "Category A-1" licence differed from a Category "A" licence in that it was not transferable for a period of two complete lobster seasons after which time it became a normal Category "A" licence.

4.1.5 The Lobster Vessel Certificate Retirement Program

The Lobster Fishery Task Force had proposed reducing participation in the lobster fishery by 25 to 50%.

During the consultations fishermen had suggested this be accomplished by a government-funded voluntary buy-back program. In 1976 the Minister of Fisheries for Prince Edward Island, with the support of P.E.I. lobster fishermen, called for a project on P.E.I. to buy back lobster licences. The result was the P.E.I. Lobster Vessel Certificate Program established in January 1977. It was jointly sponsored by the federal government and the P.E.I. Department of Fisheries, with funding on a 90:10 basis. It was to continue until March 31, 1980. The objective was to establish "a higher and more stable income for lobster fishermen through a voluntary retirement of vessel certificates." The program was open to any resident of P.E.I. who had a lobster fishing vessel certificate for a vessel operated by a lobster fisherman with a Category "A" licence for the years 1977, 1978 or 1979. Compensation was determined by multiplying the applicant's average annual documented landings by the average landed value per pound on P.E.I. during the previous year, plus an additional 20% to compensate for unrecorded sales. The compensation was to be a minimum of $2,000 but not more than $6,000.

The P.E.I. pilot project had by its midpoint led to the voluntary withdrawal of over 50% of its target objective of 400 Category "A" lobster licences.

Spurred by the initial success of the P.E.I. pilot project, in July 1978, Fisheries Minister LeBlanc announced a 3-year program "to establish a higher and more stable income for lobster fishermen in New Brunswick and Nova Scotia" (DOE 1978c). The Lobster Vessel Certificate Retirement Program, more commonly known as the Lobster Licence Buy-Back Program, was open to any resident of New Brunswick and Nova Scotia holding a Category "A" Lobster Vessel Operators Licence. Anyone who surrendered his Vessel Certificate would be ineligible to fish lobster within 3 years of his withdrawal. Thereafter, he could return only by replacing another Category "A" licence holder.

The compensation provisions of the program were identical to those of the P.E.I. pilot project. In total, $5,159,000 was expended to remove 1,569 of a total of 6,941 Category "A" lobster vessel certificates from the fishery. The vast majority of participants were Nova Scotia residents (86 percent). The average payment per fishermen was $3,284. The response to the program was such that it was extended to December 31, 1981 (DFO 1981i)

The majority of those who opted to participate in the program were older fishermen (47% were over age 54). A DFO evaluation concluded that the program benefited mainly fishermen who wished to retire. Since 63% of the prices paid were at the minimum end of the range, it appeared that many sellers were inactive or so-called "back-pocket" licence holders (DFO 1983c).

The timing of the program was opportune. Catches had declined just prior to the program. Hence fishermen were inclined to take the opportunity to get out of the fishery. This increased the level of participation in the program.

Subsequently the stocks improved and catches doubled during the 1980's. The lobster fishery entered a boom period. Licences which had a low value during the poor years became more highly prized. The resurgence in lobster stocks was not a result of any effort reduction resulting from the buy-back program but rather a general Atlantic-wide lobster recruitment pulse (Chapter 6).

Although there is no clear evidence that the Buy-Back Program increased the average income, the program removed potential fishing effort. About 22% of the 1978 total of 6,941 Category "A" lobster licences were removed. Even if a large portion of these were relatively inactive fishermen, these licences could have been sold to potentially more active fishermen. Without the buy-back program the resurgence in the lobster stocks and catch rates in the 1980's would undoubtedly have attracted more effort into the fishery. This would have dissipated the benefits of the improved catches. Thus I conclude that the Lobster Buy-Back Program had a positive impact on the earnings of fishermen who remained in the lobster fishery.

This conclusion is supported by a 1989 analysis of the Scotia-Fundy lobster fishery (DFO 1989b). This study revealed that net incomes of the typical lobster vessel (combined shares of skipper-owners and crew) increased in Lobster Fishing Area 32 from $1,200 in 1978 to $8,100 in 1987 (a sevenfold increase). In Lobster Fishing Area 34 there was a two and one-half fold increase. The results for other areas fell between these extremes. Adjusting the 1978 figures by a factor of 1.75 to account for inflation, skippers in seven districts made significant gains (Fig. 8-2). Skipper incomes remained fairly stable in real terms in western Nova Scotia and southern New Brunswick. Crew shares had increased substantially in all areas. They almost doubled in several areas and tripled in others. Returns to vessels (profits) also increased in all areas.

This study (DFO 1989b) indicated that the impact of limited entry had been dramatic. The number of fishermen would have been much higher without it, costs would have been higher and incomes would have been lower. The study estimated the effect that the buy-back program had on the fishery:

> "If the 1,300 withdrawn licences were active in 1987, as most would have been in this high landing period, they would have produced the same $139 million gross value but the total net income would have reduced to

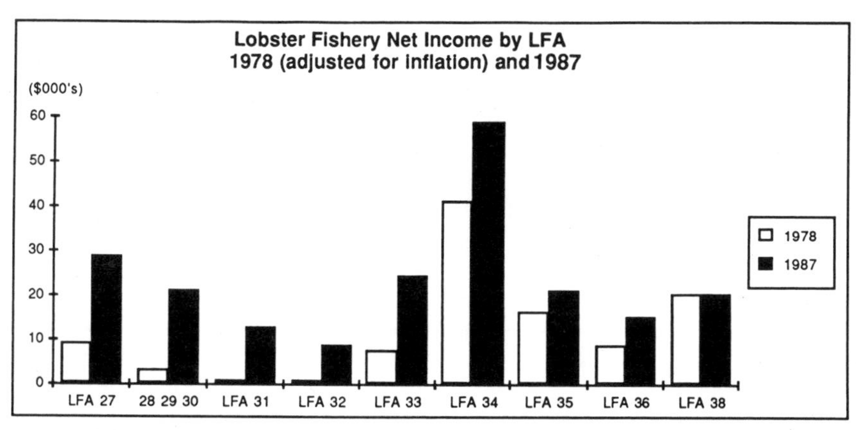

FIG. 8-2. Lobster fishery net income by Lobster Fishing Area in 1978 and 1987 (from DFO 1989b).

> $89 million due to higher costs. The number of fishermen would have been greater, at least 8,300 in total. This would have reduced the average [gross] return for skippers and crew to a maximum of $10,500" compared with the average $14,400 gross return actually experienced."

This supports my general conclusion that limited entry has had a beneficial impact in the Atlantic lobster fishery.

4.2. Atlantic Groundfish

4.2.1 Developments Prior to Limited Entry

Entry restriction by licensing was extended in 1971 to the offshore lobster and scallop fisheries and to herring purse seiners. In each instance, entry restrictions were to constrain catching capacity in fisheries where the resource base was declining.

In the Atlantic groundfish fishery, limited entry licensing was not introduced until 1974 and even then only incrementally. Different sectors of this fishery were gradually brought under entry restrictions until 1980. The Atlantic groundfish fishery was not restricted to waters under Canadian control. The policy thrust of the 1950's and 1960's had been to develop the groundfish fleet, particularly the offshore trawler fleet. This was done to compete with the build-up in the foreign fishery for groundfish off the Canadian coast.

During this period Canada was competing in the open-access international 'race for the fish'. In the post-Second World War era Canada had embarked on a program of modernization and diversification of the fleet. Most efforts were devoted to enlarging the near-shore, mid-shore and offshore sectors, and encouraging these fleets to utilize new species. Aggressive federal-provincial industrial development programs sought new fishing opportunities. These programs experimented with new types of vessels, gear and navigational and fish finding technology. They supplemented federal and provincial vessel subsidy programs and provincial vessel loan programs. In addition, there were considerable expenditures on port infrastructure, wharves, community stages, and fresh water supply systems for fish processors.

These measures markedly changed the size and composition of the Canadian fleet. Declining catch rates in the inshore cod fisheries from the late 1950's onward led to significant declines in vessels under 10 tons and in total numbers of fishermen. Inshore fishermen left the industry or adapted to larger vessels. The number of fishermen declined from 49,000 in 1965 to 39,000 in 1973. At the same time, the near-shore, midshore and offshore fleets increased. From 1964 to 1973, vessels in the 35–50 foot length category increased from 1,644 to 2,496; vessels in the 50–75 foot length category increased from 496 to 702; vessels in the 75–100 foot length category increased from 101 in 1964 to 144 in 1970 but declined to 102 by 1974. Vessels greater than 100 feet in length increased from 126 in 1964 to 235 in 1970. There had been a significant increase in capacity in these vessel classes prior to 1964. For example, the number of trawlers greater than 100 feet in length increased from 26 in 1947 to 156 in 1965.

The total number of vessels declined slightly during the 1970–74 period with most of the decline occurring in the small boats category. The number of vessels longer than 75 feet also declined during this period because of the decline in the Atlantic herring fishery.

Between 1974 and 1979 the number of vessels increased again to the 1964–65 level. The increase was due to the growth in the numbers of nearshore and middle-distance vessels-vessels between 35 and 100 feet in length. The major change occurred between

1976 and 1979; in 1976 there were 3,304 such vessels but by 1979 the number had increased to 5,872 vessels.

The increase in the number of vessels between 35 and 100 feet was fuelled by high expectations of benefits from the 200-mile limit. The expansion was encouraged by vessel construction subsidies from provincial and federal governments. In many instances, fishermen replaced existing vessels with larger, more productive and expensive vessels expecting that much more fish would be available. While replacement rules at that time restricted increases in length, they did not preclude major increases in vessel tonnage and horsepower.

4.2.2 Evolution of Limited Entry Controls

Prior to 1973 offshore trawlers, draggers and Danish seiners were required to obtain fishing licences. However, there was no restriction on entry to these fisheries. By 1973 fishermen and resource managers were becoming increasingly concerned about the rapid expansion in the number and catching capacity of the Atlantic fleet. There appeared to be prospects for an expanded Canadian share of the North Atlantic catch in the future. However, a serious short-term imbalance between catching capacity and available resources appeared imminent. The federal government established a Licensing Policy Review Committee to review existing licensing arrangements and develop new policies.

In October 1973, a Government Industry Policy Development Seminar on licensing was convened in Halifax. Participants discussed proposals for a new licensing policy for the Atlantic fisheries. In opening remarks to the seminar participants, Mr. Ken Lucas, Senior Assistant Deputy Minister of the Fisheries and Marine Service, observed:

> "At this moment.... there is no government policy on licensing for the Atlantic fisheries.... There (has been) a real upsurge in interest in the Atlantic coast fisheries.... Those people who are sitting in the fishing economy began to realize that they were sitting on a resource which had terrific potential for growth and for making money.... There were a lot of people talking about expanding their investments, expanding their fishing fleets, and so forth" (DOE 1973a).

He identified two reasons for this:

1. The market price for fish was increasing at a rate faster than the cost of living because of an imbalance between demand and supply; and
2. With the Law of the Sea discussions and the expectation that coastal states would be extending their jurisdictions seaward, there were good prospects of increasing Canada's supply of fish.

Minister Jack Davis had announced that anyone who decided to enter the fishery after August 13, 1973 did so at his own risk. There could be no assurance that a licence would be issued for the vessel which the person might purchase or build. Subsidy payments for vessel construction were frozen until new licensing guideline could be formulated (DOE 1973b).

In introducing new licensing policy proposals, Lorne Grant, Chairman of the Atlantic Licensing Policy Review, noted that the fisheries for the various Atlantic species could be grouped into three categories:

> "There are certain fisheries that will stand expansion and greater efforts, there are other fisheries where the situation is stabilized, and there is a third area where we have too many people in too many boats chasing too few fish....
> "The catch rates per ton of vessel on the Atlantic coast have declined during the past decade. This is particularly true of herring, groundfish, scallops and lobster. There are only limited exceptions to this generalization. We have to work harder, invest more capital and invest more labour to catch the same number of fish that we were catching a decade ago" (DOE 1973c).

On November 14, 1973, Environment Minister Jack Davis announced what was termed "a new fishing fleet development policy for Canada's Atlantic Coast." The stated aim of the policy was "to match fleet size to fish stocks by instituting a more selective subsidy program for vessel construction and by establishing a new licence control program" (DOE 1973d). The policy consisted of the following elements:

1. All commercial fishing vessels on the Atlantic coast would be registered by the Fisheries and Marine Service;
2. All operators of fishing vessels would be licensed;
3. Vessel operators must be Canadian citizens or landed immigrants to be eligible for licences;
4. Entry permits would be required for the following fully exploited fisheries: lobsters, scallops, salmon, herring and snow crab (most of these fisheries were already under limited entry control);
5. In fisheries which had not been previously controlled, all skippers and operators of vessels engaged in fishing during the previous 5 years

would be granted registration and entry permits for those vessels and fisheries;

6. The freeze on subsidies was lifted to encourage vessel modernization throughout the fishing industry. Subsidies would in future be provided for new vessels designed to fish unexploited stocks, for conversion of existing vessels to increase productivity and for replacement of existing vessels;
7. Subsidies would not be provided for construction of vessels to be used in those fisheries where excessive fishing capacity exists (i.e. the limited entry fisheries listed in [4]); and
8. New vessel registrations in the groundfish fishery would be encouraged for underexploited species and areas. Replacement and conversion of existing vessels would also be encouraged. In the traditional groundfish fishery, the moratorium on new vessel registrations would continue for an additional 9 months.

Subsequently, the number of offshore trawlers was frozen and entry into the offshore fishery restricted. Licences were introduced for fixed gear (e.g. gillnets, longlines) fisheries in the Maritime Provinces in 1974 but entry controls were not rigorously applied. Over the next 7 years, similar measures were adopted for the other groundfish fleet components as follows:

1. June, 1976 — Licences for otter trawling by vessels under 65 feet were limited;
2. November, 1978 — A moratorium was placed on entry to groundfish fishing by vessels under 65 feet in all of Subarea 4 and Division 3P (the east and north coasts of Newfoundland and Labrador were not included because of the anticipated growth in the northern cod stock);
3. June, 1979 — The moratorium was relaxed to allow entry by vessels using baited gear only and to remove restrictions on personal fishing licences on inshore vessel registrations for residents of Labrador;
4. March, 1980 — A complete freeze was placed on entry of inshore groundfish vessels anywhere on the Atlantic coast;
5. May, 1980 — A limit was placed on the issuing of otter trawl licences to vessels under 65 feet in eastern and northeast Newfoundland and Labrador (Divisions 2J-3KL); and
6. June, 1980 — A complete freeze was placed on personal commercial fishing licences.

The 1980 measures were taken to arrest the tremendous influx of additional people and vessels to segments of the inshore fisheries where entry was not limited. The lack of entry restrictions for this sector of the groundfish fishery had allowed significant growth in the number of vessels less than 65 feet. This contrasted sharply with the offshore sector where replacements were fairly tightly controlled following 1973.

As of 1973–74 no additional unrestricted groundfish trawling licences for vessels over 65 feet were made available. Only licences already issued were valid. There was a "reserve" bank of licences which had been issued for vessels which had been subsequently retired from the fishery but not replaced. That "reserve" was frozen for a number of years and eventually eliminated. Active vessels could be replaced but each replacement required Ministerial approval. Large otter trawl vessels could be replaced on a one-for-one basis with the new vessel not exceeding 125% of the length of the replaced vessel. A single vessel could also replace two or more vessels provided the length of the single vessel did not exceed 80% of the combined length of the replaced vessels. These replacement guidelines were intended to limit increases in fishing capacity. Very few replacements occurred. Effectively, the harvesting capacity of the offshore fleet was constrained to the level existing in 1974. For example, between 1973 and 1980 only 14 otter trawl vessels over 100 feet were acquired. All were replacements on a one-for-one basis.

Because entry restrictions for groundfish vessels less than 100 feet were introduced later than for the offshore fleet, the number of vessels in this size category increased significantly. Vessel capacity also increased. In July 1976, the following vessel replacement rules were adopted for vessels less than 65 feet:

1. Vessels 45 feet and less — open access to all gear types;
2. Vessels between 45 feet and 65 feet were placed under limited entry, with licences issued only to a vessel which had been licensed in the previous two calendar years (1974 and 1975); and
3. A replacement vessel in the 45–65 foot class could not exceed 125% of the length of the vessel being replaced up to a maximum of 64 feet 11 inches.

The July 1976 policy led, over time, to a large overcapacity problem as a result of the development of a new fleet of "jumbo" 45-foot and 65-foot vessels in the Maritimes, particularly in southwest Nova Scotia. Fishermen circumvented the length ceilings by building wider vessels.

In November 1978, a 6-month freeze on new entrants into the inshore groundfish fishery was announced (DOE 1978d). This became known in bureaucratic circles as the "warm freeze", because senior Maritime Region officials interpreted this as a freeze on only otter trawlers over 45 feet. As a result several hundred new otter trawl licences were issued in 1979 to individuals who already had at least one other fishing licence. This led to a substantial expansion of the small dragger fleet. This, combined with existing replacement rules for vessels less than 65 feet in length, was a recipe for disaster. The enormous overcapacity problem in the southwest Nova Scotia small vessel fleet of the late 1980's can be traced in part to those policies and the manner in which they were implemented.

4.2.3 The Levelton Licensing Study

By 1978 there were a number of problems with the application of Atlantic licensing policy. Management approaches varied greatly among Regions and among fisheries. Because of its piecemeal development, the licensing system was administratively cumbersome and frequently misunderstood by the public and fishermen. There was little opportunity for public input to licensing decisions.

In 1978, Fisheries Minister Roméo LeBlanc commissioned a special study of these problems. The study, headed by former federal Fisheries executive Cliff Levelton, placed particular emphasis on the groundfish fishery. Fishermen, processors and provincial government representatives were consulted extensively over several months.

The Levelton Report, titled: *Toward an Atlantic Coast Commercial Fisheries Licensing System*, released in June 1979, described the existing Atlantic licensing system as "complicated and confusing and not easily understood by those in the fishing industry or even by those administering the system" (Levelton 1979). Levelton recognized a considerable divergence of views within the fishing industry about appropriate objectives for the licensing system. He proposed the following overall objectives (in order of priority):

1. To promote economic viability of fishing operations;
2. To promote just and equitable distribution of access to the fisheries resource;
3. To assist in directing fleet development in line with changing conditions in the fishery;
4. To aid in fisheries management and in conservation of the fisheries resource; and
5. To facilitate administration, information gathering and enforcement.

The Report noted that the views of the Atlantic coast industry on limitation of entry ranged from satisfaction with existing programs to complete dissatisfaction with limitation of entry *per se*. Even the supporters of Atlantic coast limited entry programs criticized the distribution of licences among local areas or groups of fishermen. This problem was usually caused by the way in which limited-entry programs were put in place. Some, for example, were based on past performance in some period prior to limitation of entry. This had inevitably resulted in unequal distribution of licences among communities, fishermen, areas or provinces. Another contributing factor was the fact that limited-entry programs were very seldom, if ever, instituted in developing fisheries, but were usually implemented when a fishery already had excessive numbers of people and vessels involved.

The Levelton Report reiterated the advantages of limited entry and made the following recommendations:

1. That limitation of entry be maintained in those Atlantic coast fisheries where it was already in place and that expansion in those fisheries take place only as the overall situation permitted;
2. That in all Atlantic coast fisheries licensing take the form of a licence issued to the eligible owner (individual or corporation), or lessee in the case of a leased vessel, to use a specified registered fishing unit for a specified type or types of fishing activity;
3. That there be a simple registration of all commercial fishing units, renewable annually;
4. That categorized personal fishing licences be issued, providing for differentiation of regular fishermen, apprentices/learners and casual fishermen;
5. That free or open transfers of licences not be permitted but that limited transfers, to allow continuation of existing enterprises by other members of those enterprises, should be permitted through the licensing authority. Otherwise, licence holders wishing to leave the fishery should be required to relinquish their licences to the licensing authority;
6. That criteria for vessel replacement in limited entry fisheries be defined and that such criteria focus on catching capability and economic returns in each fishery;
7. That participation requirements be defined for all limited entry fisheries;
8. That when vessels under 100 gross tons owned by fish buying corporations were sold or replaced, the first right of refusal for the licence should go to independent fishermen, with a fixed time period for the exercise of this option;
9. That when any vessel is repossessed all relevant fishing privileges held by the owner should revert to the licensing authority and that independent fishermen be given first right of refusal within a specified time period to take up these privileges;

10. That acquisition of freezer trawlers be permitted through replacement or conversion of existing licensed units and that requests for licences for freezer or factory freezer trawlers to fish solely for non-or under-utilized species be favourably considered provided the vessels were owned and crewed by Canadians;
11. That the licence fee structure be revised such that licence fees would cover administrative costs;
12. That an improved administrative framework be implemented, including more direct involvement by fishermen, a clearer and more uniform interpretation of policy, and properly applied program evaluation procedures.

The Levelton Report took a middle-of-the-road approach on most licensing issues. Most of the recommendations regarding the licensing system were accepted and implemented by DFO. By 1982 all Atlantic fisheries were placed under limited entry. A fisherman licence categorization system was introduced in 1981. Local licensing committees were established as a forum for consultation on licensing policies.

Licence fees were increased significantly in 1981 and 1982. This apparently served to deter a number of marginal participants. For example, in Newfoundland the number of personal fishing licences issued decreased from 35,000 in 1980 to 28,500 in 1981 (This may also have been due to a decline in the squid fishery, E. B. Dunne, St. John's, pers. comm.). The ratio of full-time to part-time licences in the various regions (13:15 in Newfoundland, 11:8 in the Maritimes) indicated that the increase in apparent numbers in the inshore fishery in the late 1970's was biased by the number of licences issued to part-time fishermen. The number of full-time fishermen in Newfoundland in the early 1980's was not greatly different from the number which existed in the 1960's and remained much the same at the beginning of the 1990's.

4.2.4 Categorization of Atlantic Fishermen

The major legacy of the Levelton Report was a revamped licensing system which involved categorized personal fishing licences differentiating between full-time and part-time fishermen. This was a refinement of the universal licensing of commercial fishermen introduced in 1973. Since that time fishermen's groups had pressed for licensing which distinguished between full-time or professional fishermen and those with other primary occupations. Classification or categorization of fishermen as proposed by the Levelton Report was not a new concept. The approach was used in the original licensing of the commercial Atlantic salmon fisheries in 1971 and the lobster fishery in 1968.

The program was described as follows:

— a method to determine who was primarily dependent on the fishery for a livelihood;
— a step in the development of professional status for fishermen;
— a method of regulating the entry of newcomers;
— a means to assist in determining (i) who should receive limited fishery licences (new or transfers); and (ii) who should receive subsidies and other financial assistance (DFO 1980d).

Fisheries Minister LeBlanc, in a speech to the Newfoundland Fishermen, Food and Allied Workers' Union in St. John's in November 1980, elaborated on his view of the new licensing system:

> "This system is not intended to take anything away from anyone, but is necessary to provide... protection for fishermen... and to develop policies for the differing needs of each group.
> "A fisherman will be classified full-time if he fishes consistently during the fishing season in his area and has little other income except on a limited basis from such things as farming or logging.
> "Fishermen will be given a greater voice in licensing decisions through the establishment of local and regional licensing appeal committees. These will be your committees, with a majority membership of fishermen. They will be empowered to hear appeals of fishermen dissatisfied with the licence assigned to them" (DFO 1980e).

Although Minister LeBlanc had stated that the "system was not intended to take anything away from anyone," in the same speech he announced that in 1981 only full-time fishermen with a previous history in the herring fishery could fish herring with fixed gear in Newfoundland. He also announced that only full-time fishermen with a previous history in the cod trap fishery could draw for cod traps in the local cod fishery committees.

The categorization program was strongly supported by the fishermen's union in Newfoundland. However, it was resisted in some quarters, particularly by individuals who expected to fall in the part-time category. The Newfoundland provincial government expressed reservations about the program (*St. John's Evening Telegram*, November 12, 1980).

In a speech to the Maritime Fishermen's Union in Halifax on February 26, 1981, Minister LeBlanc backpedalled somewhat on the purpose of the program:

> "We have no hidden surprises up our sleeve. No fisherman, be he full-time or part-time, is going to wake up one morning to learn we have taken away his fishing privilege....
> "The classification of fishermen is simply a means for us to determine the extent to which fishermen are involved in the fishery. It is not designed to be a tool to eliminate fishermen" (DFO 1981j).

The Task Force on Atlantic Fisheries (Kirby 1982) recommended continuing and improving the categorization program to tailor policies and programs for each group.

The full-time/part-time categorization was subsequently used by DFO to reduce effort in the Atlantic salmon fishery. A major element of the 1985 Atlantic Salmon Management Plan was a mandatory licence buy-back program for part-time commercial salmon fishermen in Newfoundland and Labrador. Cancelling the part-time salmon licences was aimed at "reducing the fishing effort for conservation purposes," specifically to reduce the interception of Maritimes-origin salmon, and "at the same time allow the remaining full-time Newfoundland fishermen who depend on the resource for their livelihood greater opportunity to be economically viable" (DFO 1985c).

The privileges of full-time and part-time fishermen were clarified in the 1989 *Commercial Fisheries Licensing Policy for Eastern Canada* (DFO 1989h). Section 10(2) of that policy stated:

> "A full-time fisherman will not lose any fishing licences on being downgraded to part-time with the exception of the current restriction in the salmon fishery in the province of Newfoundland and the herring fixed gear fishery in the Newfoundland Region."

Section 10(3) stated:

> "When a specific species is being over-fished, one or more of the following restrictions may be imposed on part-time fishermen:
> (a) restriction on reissuance of licences to another person;
> (b) restriction on obtaining additional licenses;
> (c) stricter vessel replacement restrictions; and
> (d) a licence moratorium."

4.2.5 The Factory Freezer Trawler Controversy

There was considerable pressure from certain provinces and some companies immediately after the 200-mile limit for expansion of the offshore groundfish harvesting capacity. However, federal Minister of Fisheries, Roméo LeBlanc, resisted this pressure (see Chapter 2). By 1979 pressure was being exerted again by one of the major companies, National Sea Products, for permission to acquire factory freezer trawler technology. The Nova Scotia Fish Packers Association in a brief to the 1979 Northern Cod Seminar argued that the choice of appropriate technology should be left to the private sector. In their view, government's role was to ensure that the total licensed "fishing power" was compatible with resource availability. They contended that licences for Canadian factory freezer trawler vessels should be available immediately (Nova Scotia Fish Packers Association 1979).

The Report of the Northern Cod Seminar (DFO 1979b) recorded the differing views of the participants but provided no conclusions. In September 1979, the Department convened another seminar on the development of a policy for squid and other underutilized species. Again, Nova Scotia companies argued that freezing-at-sea technology was necessary if Canada was to exploit these nontraditional species.

On November 30, 1979, Fisheries Minister James McGrath announced a policy for the licensing of freezer and/or factory freezer trawlers (DFO 1979c). The main elements were:

1. Canadians could acquire vessels capable of freezing at sea for harvesting non-traditional species.
2. For those who wished to fish traditional groundfish species with a freezer trawler, licences would be granted to replace vessels currently licensed and active in the groundfish fishery. The replacement vessels would also be licensed for non-traditional species. (This was a continuation of existing policy).
3. For new applicants wishing to fish squid, silver hake, grenadier, capelin, argentine and offshore mackerel only, a maximum of four additional licences would be issued for 1980. Applications involving consortia of fishermen and/or processors would be considered. Charters would be permitted for 1980 only. Such vessels, whether Canadian owned or chartered, could be no longer than 200 feet.

A new 200-foot maximum length rule was added to the existing trawler replacement criteria. Vessels would be permitted to carry the filleting equipment necessary for processing non-traditional species but would not be permitted to fillet any traditional groundfish species. The latter restriction was intended to protect existing on-shore processing employment.

Of the four licences granted, three licensees chartered foreign vessels for the permitted one year but did not acquire permanent vessels. Hence, the licences subsequently were revoked. One licensee, Mersey Seafoods, did attempt fishing nontraditional species with a new vessel. The President of this firm estimated that in his first year he caught 50 tons of silver hake at a harvesting cost of $35.00 per lb. (William Murphy, Nova Scotia, pers. comm.). However, through combining licences from other retired vessels, the company secured a groundfish licence for the vessel and subsequently did very well freezing northern cod at sea. The major proponent of factory freezer trawlers, National Sea Products, did not acquire one of the four non-traditional species licences in 1980.

In 1980, DFO forcasted Atlantic groundfish trawler supply requirements for 1981 to 1985 (Carpentier et al. 1980). At that time the fleet of groundfish trawlers 100 feet or more in length consisted of 5 vessels based in Quebec, 58 vessels based in the Maritime Provinces and 81 vessels based in Newfoundland, for a total of 144. There were 8 more vessels which had been approved for replacement as groundfish vessels. However, these were not then engaged in the groundfish fishery.

DFO anticipated that a substantial number (47) of the active vessels would become obsolete and need to be replaced during the 1981–85 period. The replacements would likely be substantially larger, more productive and much more expensive to acquire and operate. Four possible replacement scenarios were analyzed. These analyses indicated that projected fish abundance in 1985 would permit only a modest replacement schedule: 3–5 vessels per year under the existing 125% replacement length rule.

An examination of alternate vessel replacement criteria (foot-for-foot replacement) indicated that a 'foot-for-foot' replacement guideline would reduce the amount of fish the fleet required (Carpentier et al. 1980). It would also improve fleet viability. DFO introduced a "foot-for-foot" replacement guideline in June 1981.

However, National Sea Products did not give up the fight to acquire a factory freezer trawler. By 1985, the company was under new management. With a new federal fisheries management regime involving enterprise allocations for offshore groundfish, it was pressing its case with renewed vigour. In June 1985 National Sea Products applied for a licence to operate a factory freezer trawler. NSP argued that it required factory freezer trawler technology "to meet the growing demand for a superior quality product from an aggressively competitive marketplace" (NSP 1985).

DFO in August 1985 released a Discussion Paper on Factory Freezer Trawlers (DFO 1985d). The DFO Discussion Paper considered the implications of FFT technology for the offshore Atlantic groundfish fishery, including the key economic and social aspects. The paper suggested several options for government:

(a) continue the existing ban on processing at sea;
(b) permit factory trawlers with restrictions on the number, on the areas of operation, and/or quotas and/or species; and
(c) permit factory trawlers without restrictions.

These options were reviewed for effects on: (i) resource base, (ii) economic viability, (iii) employment, (iv) marketing, (v) anticipated provincial positions, and (vi) international implications. The tentative conclusions were:

1. There was an adequate resource base within existing enterprise allocations to accommodate domestic freezer trawlers.
2. Factory trawlers could operate more profitably than wetfish trawlers or freezer trawlers and their associated onshore processing plants.
3. At that time a few used West German factory freezer trawlers were available at low prices. These could be a good addition to any firm's fleet.
4. Replacement of wetfish trawlers with factory freezer trawlers would normally result in a net loss of shore-based employment. The net loss would be less if a FFT harvested unutilized enterprise allocations.
5. A change in policy would likely result in applications for more FFTs. Plant closures might occur depending on the number of FFTs licensed and their fishing plans. Such closures could remove the economic base of certain communities.
6. Many U.S. buyers believed that freezing at sea provided "consistently high quality products." This resulted in a growing preference for these products.
7. There were market opportunities in Japan, Europe and other countries for headed and gutted frozen at sea products (turbot, redfish). These could be exploited by freezer trawlers.
8. With stringent handling and processing controls, premium quality groundfish could be produced from any harvesting technology.

In consultations on the FFT issue, fishermen's groups, particularly the NFFAWU, generally opposed introducting FFT technology to the Canadian fishery. "What National Sea Products wants to do is take the inshore fish plants and move them out to sea", said one union member. Richard Cashin, NFFAWU President, stated: "This is not a case of Newfoundland versus Nova Scotia. This affects the inshore and deep sea fishery in all of the Atlantic provinces. They are considering buying technology which is outdated in Europe. If FFTs are allowed in, where are the quotas for our on-shore plants going to come from when Fishery Products are telling us they don't have enough in their quotas to keep plants in Grand Bank, Burin and St. Lawrence operating at full capacity?" (*St.John's Evening Telegram*, November 7, 1985).

The major opposition came from the province of Newfoundland. In September 1985 Fisheries Minister Rideout released a Discussion Paper entitled *Appropriate Offshore Fish Harvesting Technology: An Assessment of the Detrimental Effects of the Use of Factory Freezer Trawlers* (Newfoundland 1985a). Newfoundland advanced seven arguments against changing the policy respecting FFTS:

1. The introduction of FFTs would violate the agreement of September 1983 between Canada and Newfoundland concerning the Restructuring of the Newfoundland fishery. Clause number twelve stated: "Factory trawlers will not be permitted to harvest northern cod" (see Chapter 13).
2. It was mistaken and dangerous to assert that FFT technology was necessary to supply a top quality product to export markets.
3. FFTs would lead to significant net loss of onshore processing jobs and a precarious future for many Atlantic communities.
4. The unnecessary utilization of distant water technology such as FFTs would negate Canada's geographic advantage of proximity to its marine resources.
5. Issuing an FFT licence would probably prejudice Canada's position in an arbitration to interpret key language in the 1972 Agreement between Canada and France (see Chapter 12).
6. FFTs would imperil the market-driven strategy behind the Enterprise Allocations system.
7. Any policy supporting FFTs would be economically inefficient.

The Newfoundland government strongly emphasized Clause 12 of the Restructuring Agreement. It contended that by including this clause the Federal Government bound itself to the existing policy, prohibiting the filleting of traditional groundfish at sea, as a condition for both parties to accept the Restructuring Agreement.

In November, 1985, National Sea Products' application to operate one FFT in the Atlantic groundfish fishery was approved for a 5-year introductory trial period. Acting Fisheries and Oceans Minister Erik Nielsen announced that a maximum of three FFT licences would be granted to selected east coast fishing companies. The licences would be for a 5-year introductory period. One licence each had been reserved for National Sea Products and Fishery Products International while the third licence was reserved for a company or consortium from the remaining offshore groundfish companies. Mr. Nielsen stated:

> "The decision to grant the licences for the introductory period does not signify a dramatic shift in the philosophy of the Government of Canada toward the traditional Atlantic coast fishery. The inshore small boat fishery and the wetfish trawler fishery will continue to be the backbone of the Atlantic groundfish industry. This decision merely allows companies to investigate further the role of FFTs in their operations and will permit the government to clarify the important socio-economic aspects associated with the introduction of factory trawlers" (DFO 1985e).

The companies receiving FFT licences would have to meet certain conditions:

1. At least 50% of a factory trawler's catch had to be made up of previously under-utilized enterprise allocations.
2. No more than 6,000 tons of a company's northern cod allocation could be harvested by a factory trawler in any one year.
3. Factory trawlers could not operate in the Gulf of St. Lawrence or the Bay of Fundy.
4. To operate a factory trawler, a company must retire or convert equivalent length and capacity from its present fleet.
5. A company had to outline an acceptable plan to minimize the socio-economic/community impact.
6. FFTs must be registered immediately as Canadian vessels and crewed fully by Canadians within a 2-year period.
7. FFTs could not harvest species such as capelin where an economically efficient fishery had already been developed.

Addressing the specific argument raised by the Province of Newfoundland about Clause 12 of the Restructuring Agreement, Mr. Nielsen said:

> "The creation of FPI and its economic rejuvenation was the overriding intent of the agreement. The federal government remains committed to the intent of this agreement and the quick approval of the $76.7 million equity injection, as requested by the

FPI Business Plan, is evidence of this commitment."

Reacting to the federal government's decision, Premier Peckford stated in the House of Assembly:

> "The violation of Clause 12 is a serious breach of trust. It casts into doubt the security of all agreements thereby creating a serious precedent with implications for all Provinces in Canada.
> "This is indeed a sad day for the people of this Province. We have again seen a vivid demonstration of the lack of understanding by the Federal Government of the fundamental importance of this particular cod resource to improving our fishing industry and the way of life it supports. It is a clear justification for our long-stated policy that this Province must have a larger jurisdictional role in the management of our most important resource" (Newfoundland 1985b).

As events turned out, only National Sea Products exercised the option to acquire a Factory Freezer Trawler. FPI did not do so, and none of the other eligible companies could develop a proposal which met the conditions. The NSP factory freezer trawler Cape North commenced fishing in 1986. To acquire the Cape North licence NSP had to retire equivalent catching capacity (five vessels) from its fleet.

The FFT decision was a compromise between the strongly held views of the Newfoundland government, on the one hand, and the clearly articulated position of National Sea Products supported by the province of Nova Scotia, on the other. The door was opened to introduce this technology in the Canadian offshore fishery. The results of the 5-year trial indicated that the Cape North had become an important profit center for NSP and allowed them to develop new markets for frozen-at-sea products. The socioeconomic impacts were found to be minimal and limited to a number of communities around Lunenberg (D. Tobin, DFO, Ottawa, pers. comm.).

On June 10, 1991, Fisheries Minister John Crosbie announced that the FFT program initiated in 1985 as an experiment had been made permanent. National Sea Product's licence would continue subject to a review of the vessel's operating conditions every 5 years. The other two licences announced in 1985 continued to be available, one set aside for Fishery Products International, and the other for a consortium of independent operations (DFO 1991a).

The factory freezer trawler Cape North (Photo courtesy of National Sea Products).

It is doubtful that the use of factory freezer technology will increase in the 1990's. Undoubtedly, policy decisions respecting the use of such technology will continue to be constrained by the pressure to maintain onshore processing employment in coastal communities with few alternative employment opportunities.

4.2.6 Addressing Overcapacity in the Inshore/Nearshore Fleet

The Department of Fisheries and Oceans took steps in 1980 to address overcapacity in the inshore-nearshore fleet sector by cancelling back-pocket licences for otter trawlers less than 65 feet in length. Because of the "warm freeze" the number of otter trawler licences for vessels less than 45 feet long had more than doubled in the Maritimes between 1977 and 1979. Faced with this situation, DFO decided to reissue 1979 licences only on the basis of proven landings in 1979 or evidence of purchase of the appropriate gear for otter trawling. 814 of these inactive licences were cancelled by the end of 1980 (Pierre Comeau, DFO, Ottawa, pers. comm.). This action helped prevent an already serious overcapacity problem from becoming even worse.

Recognizing the growing overcapacity in the small vessel fleet, DFO revised its groundfish vessel replacement policy in June, 1981 (DFO 1981k). The revised policy restricted the replacement of all vessels longer than 35 feet to a foot-for-foot, hold-for-hold basis. Following consultations with industry, this replacement policy was modified in August 1981 such that vessels between 35 feet and 64 feet 11 inches could be replaced within five-foot intervals: i.e. 35'–39'11", 40'–44'11", 45'–49'11", up to 60'–64'11". This created limits at five-foot intervals which could not be exceeded by replacement vessels. It was estimated that, over time, this could result in an increase in fleet capacity of between 8 and 14%.

Vessels between 65 feet and 100 feet long could be replaced only on a foot-for-foot basis with a 5% tolerance factor. The hold capacity of any replacement vessel 35 feet and over could not exceed that of the vessel it was replacing by more than 10%. For vessels less than 35 feet in length, replacements could not exceed 34 feet 11 inches.

A 1987 DFO analysis indicated the following trends in effectiveness in curbing the growth in fishing capacity:

Gulf Region — Average overall length of vessels in the under 65 foot class had grown by 10%, tonnage by a negligible amount, while brake horsepower had decreased. The number of vessels declined by about 30%, mainly within the under-35 foot class.

Newfoundland Region — The number of registered vessels had declined with all of the decrease occurring in vessels operated by "part-timers." The number of vessels operated by full-time fishermen had increased by about 4%, almost entirely in the under 35-foot class. A 9% increase occurred in the gross tonnage of registered "full-time" vessels. There had been a substantial increase in the average brake horsepower of almost all vessel categories.

Scotia-Fundy Region — There was a decrease in the number of vessels licensed to fish groundfish between 1982 and 1986. There were, however, 165 banked licences which could be activated. The average length and net tonnage for each vessel length group had remained relatively stable but the average gross tonnage increased by 2–8% and the average brake horsepower by about 6% over that period. A relatively small number of groundfish vessels over 25.5 GRT, representing about 12% of the fleet, accounted for most of the Scotia-Fundy catches by vessels under 100 feet. Of these 323 vessels, only 15 were constructed between 1983 and 1986; 207 had been built between 1978 and 1982 (DFO 1987d).

In 1981 Sector Management had been introduced for this class of vessels to better balance the fish quotas and fishing effort within each geographic sector. The 1987 Vessel Replacement Policy study suggested a three-sector approach in place of an overall Atlantic-wide policy of curbing capacity;

- *Reducing Capacity* on the Scotian Shelf and the south coast of Newfoundland (Division 3P) where excessive capacity was threatening economic viability for the entire groundfish fishery;
- *Diversifying Capacity* in the Gulf to allow adaptive vessel design to meet the changing biological and economic conditions of the diverse Gulf fisheries;
- *Expanding Capacity* in northeastern Newfoundland where the inshore-nearshore fleets had been unable to harvest the available quotas. (This suggestion was made in expectation that the northern cod stock would continue to increase in abundance).

It appeared that since the first steps were taken in 1973 to constrain fishing capacity and overcapitalization the following cycle was occurring: A problem was recognized, control measures were developed, then these were subsequently relaxed to allow for "flexibility", efficiency and diversification. Relaxation would usually be followed by a series of licence freezes, licence cancellations and new replacement criteria.

The 1987 DFO study indicated that the 1981 groundfish vessel replacement guidelines had permitted a "modest" increase in the size and capacity of fishing vessels within the designated 5-foot intervals and the 10% carrying capacity restrictions. The guidelines had also permitted increases in effective capacity resulting from improvements in vessels, equipment and fishing gear. The Study concluded that a volume-based replacement system was potentially more efficient and equitable than existing guidelines. The suggested approach involved a cubic measure derived from a combination of the length, breadth and depth of a vessel (DFO 1987d).

The growth in overcapacity in some segments of the inshore/nearshore fleet, particularly in southwest Nova Scotia, was apparently the result of action taken too late combined with administrative blunders which subverted the intent of the policy (e.g. the so-called "warm freeze" in the Maritimes). The replacement rules introduced in 1981 appear to have put a "cap", with a certain margin of slippage, on the growth of the fleet. But, as in many other limited entry fisheries worldwide, it was a case of closing the barn door after the horse had escaped.

The extent of the overcapacity problem in southwest Nova Scotia became apparent by the mid-1980's. Quotas were caught and fisheries closed down early in the season. This brought pressure on the federal government to increase TACs and reallocate fish from the offshore to the inshore fleet. As a result, there were a number of temporary quota transfers in 1985, 1986 and 1987 and a permanent transfer of 12,000 tons of groundfish from the offshore to the inshore under the 1989 Groundfish Management Plan. The 1987 Groundfish Plan included a whole array of measures to slow down the fishery, improve quality and increase returns to the industry. These measures included seasonal quotas and trip limits and steps to enhance quality, like bleeding and gutting at sea. Because of overcapacity on the Scotian Shelf, Minister Siddon declared a moratorium for 1987 on the issuance of inactive groundfish licences, with a 1-year freeze on combining these licenses. (DFO 1986b). On March 16, 1987, the Minister asked the Chairman of the Atlantic Regional Council (ARC), Mr. Gilles Theriault, to meet industry representatives to discuss alternatives to the moratorium (DFO 1987e).

ARC recommended that a small committee representing all major sectors of the industry address the issue over the next 6–9 months and develop recommendations for the 1988 season. ARC also favoured lifting the moratorium if it was proven that this would not result in increased effort. On April 16, 1987, Minister Siddon lifted the moratorium while fishermen in the region worked on the overcapacity problem (DFO 1987f).

In November, 1988, the Scotia-Fundy Groundfish Capacity Advisory Committee reported:

> "Total effort is currently higher than the resource can sustain. The impact of too much effort has been to maintain a fishing mortality, (F), level which is not only far in excess of the target which is the underlying basis of the management plan, but is, in fact, above F_{max}. Fishing effort is double the target level. This has happened despite intense efforts to control effort by means of seasonal quotas, trip limits on catch, and so on. Reductions in effort by these measures appear to have been offset by increases in numbers of trips and in the size of the active fleet. It is estimated that inactive capacity in the groundfish fleet is as large as the active capacity. In other words, the fleet has the potential to exert four times the effort required at $F_{0.1}$" (Anon 1988b).

To reduce harvesting capacity, the Committee recommended a package of measures:

1. Officially splitting the fleet into "generalists" and "specialists." The specialist dragger boats made up only about 20% of the fleet but harvested about 80% of the catch.
2. Reducing the specialist fleet by 25–40% to bring its harvesting capacity more in line with the resource supply. This was to be accomplished through an industry funded buy-back of licences.
3. Containing but not reducing the capacity of the generalist fleet.
4. Increasing the minimum trawl mesh size from 130mm to 152mm and increasing the legal minimum landed size for cod, haddock and pollack toward 20 inches.
5. Using trip limits, fleet tie-ups, seasonal and area closures, and combinations of limits on catch per trip and number of trips to control fishing effort.

In July 1989, Minister Siddon established a Task Force to examine the Scotia-Fundy Groundfish Fishery, headed by Jean Haché, Director General, Scotia-Fundy Region of DFO. This was prompted by the closure of most of the groundfish fishery to the inshore mobile gear (dragger) fleet, resulting in plant closures, layoffs and general hardships. The Task Force in its December 1989 Report (DFO 1989i) tended to confirm the conclusion of previous studies: the excessive catching capacity of the inshore fleet, particularly the mobile gear sector, was the major underlying problem.

The Task Force identified two major trends — declining stocks and increasing fishing capacity. It concluded that harvesting capacity and the accompanying overinvestment must be reduced as quickly as possible.

To accomplish this, it recommended that fishermen in Scotia-Fundy be given a choice of participating in one of three fleet combinations. *Group A* was to include small boats operating on day trips, consisting only of fixed gear fishermen. These boats would carry on a traditional, seasonal fishery of less than 6 months per year. It would not be subject to premature closure. The major management control would be a trip limit in the range of 1,500 kg.

All inactive licences were to be placed in Group A. None of these currently inactive licences would be transferable. This group was estimated at 900 active and 1,000 inactive licences.

Group B was to consist primarily of fixed gear fishermen often operating on multi-day trips. This group would operate under a fleet quota with the fishery subject to closure when their quota was caught. About 400 licence holders would likely opt for this category.

Group C would contain the highly capitalized, high capacity mobile gear fleet. Individuals in Group C, (approximately 400 licence holders) would be allowed to choose a fleet management system subject to DFO approval. These choices would include a strictly enforced group quota, individual quotas, or individual transferable quotas. The group could also collectively opt for some type of self-funded licence retirement scheme.

In announcing the 1990 Groundfish Plan, Minister Siddon stated that boat quotas would be introduced for the mobile gear sector in southwest Nova Scotia to deal with the overcapacity problem (see Chapter 9).

4.3 Licensing Recommendations of the Task Force on Atlantic Fisheries

The Report of the Task Force on Atlantic Fisheries critized existing licensing and replacement rules as "artificial and arbitrary and....often not successful" (Kirby 1982).

The Task Force recommended the following approach to licensing:

1. Continuation and improvement of the process of categorization of fishermen;
2. Adoption of the following licensing principles: (i) the licence would pertain to the individual as a quasi-property right (the licence would be on the man, not the boat); (ii) the licence would specify either a limitation on the catch (an 'enterprise allocation' or a 'quota licence') or on the catching capacity of the fishermen's vessel and gear (an 'effort-related licence' as now exists in, for example, the lobster fishery); (iii) the licence would be divisible and transferable (that is, it could be sold or traded) subject to certain conditions, with the transfer process being supervised by a quasi-judicial board; and
3. Establishment of a quasi-judicial Atlantic Fisheries Licence Review Board to review and hear appeals for the existing licensing system, as well as for the proposed system of enterprise allocations and quota licences.

The government's response to the Task Force Report (DFO 1983a) indicated that the recommendation to implement a system of quota or effort-related licences would be phased in over the next several years, beginning with the offshore fleet (see Chapter 9).

The government rejected the recommendation for a quasi-judicial Review and Appeal Board.

4.4 The 1989 Commercial Fisheries Licensing Policy for Eastern Canada

Changes had been made in the Atlantic licensing system following the Levelton Report in 1979. By 1985 there was a clear need to clarify and standardize licensing policies for the various Atlantic fisheries.

An internal consolidation resulted in a December 1985 Discussion Paper on Commercial Fisheries Licensing Policy for Eastern Canada. This served as the basis for consultations with all segments of the Atlantic fishing industry over the next couple of years.

DFO issued a Proposed Licensing Policy document in May 1988. This was the subject of further consultations. Finally, in January 1989, Fisheries and Oceans Minister Tom Siddon issued the *Commercial Fisheries Licensing Policy for Eastern Canada* (DFO 1989h). This document consolidated all federal licensing policies for the commercial fisheries in Atlantic Canada. The Minister emphasized that the majority of the policies contained in this document were already in effect. Atlantic Fisheries Ministers endorsed this policy statement at their meeting on December 15, 1988.

The definition of a "licence" differed significantly from the transferable licence approaches proposed by the Economic Council of Canada, the Pearse Commission and the Task Force on Atlantic Fisheries. Section 4(1) of the Policy stated:

> "A 'licence' grants permission to do something which, without such permission, would be prohibited. As such, a licence confers neither property nor any proprietary or contracted rights which can be legally sold, bartered or bequeathed. Essentially, it is a privilege to do something, subject to the terms and conditions of the licence."

Section 4(2) defined a "fishing licence" as:

> "An instrument by which the Minister of Fisheries and Oceans, pursuant to his authority under the Fisheries Act, grants permission to a person (an individual or a company) to harvest a certain species of fish, subject to the conditions attached to the licence. This is in no sense a permanent permission; what the licensee essentially acquires is a limited fishing privilege rather than any kind of absolute or permanent 'right'.
> "A fishing licence grants private access to, and use of a common property resource-such a resource cannot be alienated to 'private ownership' (private property) without extinguishing its 'common property' nature."

By so defining a fishing licence, the government clearly rejected the views of those who had advised over the previous decade that the resource should be privatized.

The 1989 document also set forth the following objectives for the licensing system (Section 5):

- — to aid conservation of the fisheries resource;
- — to promote the stability and economic viability of fishing operations;
- — to promote equitable distribution of access to the fisheries resource;
- — to promote orderly fleet development by controlling the number, size and types of new vessels; and
- — to facilitate data collection for administration, enforcement and planning purposes.

The evident intent was to continue to use licensing to pursue a broad range of conservation, economic and social objectives.

Section 11(8) stated that a licence holder is not required to own the vessel he registers and operates under licence. This was a change in policy. Fishermen would no longer have to purchase or own the vessel from which they fish.

The vessel separation policy promoted by former Minister Roméo LeBlanc, whereby licences for vessels under 65 feet would be issued to individuals, not companies, continued in force.

The Department's policy on foreign ownership and involvement with licensed fishing companies was more clearly defined. The 1989 Policy (Section 15) provided that:

1. If a foreign interest acquires over 49% of the common (voting) shares of a Canadian owned company which has fishing licences, the licences cannot be retained by that firm.
2. If such a firm having a subsidiary in Canada which has fishing licences is taken over by another foreign-owned firm from the same country, the licences may be retained as part of the continuing Canadian operation (i.e. no net increase in foreign ownership).
3. If a foreign interest purchases a minority ownership interest in a Canadian firm or establishes a jointly owned subsidiary with a Canadian firm in which the Canadian firm owns more than half the voting shares, licences will not have to be surrendered.

The 1989 policy changed the treatment of licences in the case of repossession by a government lending agency. The Provinces of Nova Scotia and New Brunswick had argued for years for a change in the previous policy which gave the licence holder up to 2 years to place the licence on another vessel. The 1989 Policy stipulated that the licence would be held by the Department for 60 days. If the loan had not been settled within 60 days, the licence privilege could be cancelled or reissued to an eligible recipient not selected by the original licence holder. The owner of a vessel repossessed by a provincial loan agency could thus no longer place the licence on another vessel or benefit through the transfer of the licence.

There were significant changes to vessel replacement rules. Holders of groundfish or shrimp licences using vessels between 65 and 100 feet cannot replace such a vessel with a vessel whose Length Over All (LOA) is greater than 105% the LOA of the original vessel, or with a hold capacity that exceeds 110% of the hold capacity of the original vessel.

Under the offshore enterprise allocations regime a vessel may not be replaced by a vessel with a greater catching capacity than the capacity of the original vessel. In this instance, catching capacity was to be determined by the amount of fish that can be carried in a vessel's fish hold and on the average catch by that vessel during its best 3 years of fishing.

The most significant change involved the replacement rules for vessels between 35 and 65 feet long. Replacement rules based on vessel length and hold capacity were changed to an overall measure of vessel capacity, a cubic number derived from length, width and depth. This new approach, a result of extensive technical work and parallel initiatives on vessel replacement rules in other countries, was intended to provide fishermen with more flexibility in choosing a replacement vessel while also providing the government with a more effective mechanism to control fleet capacity.

These changes, which took effect April 1, 1989, were another step toward limiting the growth in fleet capacity.

5.0 THE PACIFIC COAST COMMERCIAL LICENSING POLICY

In September 1990, DFO issued a Discussion Paper on its commercial fishing licensing policy for the Pacific Coast (DFO 1990f). This summarized existing licensing policies. Like the Atlantic Licensing Policy, it defined a fishing licence as a privilege rather than a right. The basic objectives of commercial fishing licensing in the Pacific were stated as follows:

1. To facilitate responsible management and conservation of fisheries resources;
2. To promote the stability and economic viability of fishing operations;
3. To facilitate cost recovery for the management and enhancement of the resource;
4. To facilitate necessary data collection for regular administration, enforcement, and planning purposes.

Objectives 1, 2 and 4 were similar to objectives stated in the Atlantic Policy. References to "equitable distribution of access" and "orderly fleet development" were missing from the Pacific licensing policy.

6.0 LICENCE REVIEW AND APPEAL PROCESSES — NATIONAL

Throughout the 1970's and 1980's, fishermen were concerned about their right to appeal the Department of Fisheries and Oceans' decisions on licensing matters. Prior to 1980, fishermen had to appeal decisions to Departmental committees. If they were not satisfied by the rulings of those committees, they could appeal directly to the Minister. In rendering his decisions the Minister sought advice from bureaucrats, often including those who had turned down the earlier appeal. This was neither a fair nor just procedure.

Fisheries Minister LeBlanc, when announcing the program for categorization of fishermen in 1981, called for independent review committees, consisting of fishermen and others familiar with the fishing industry. This process worked reasonably well. But the final route of appeal to the Minister was still not independent of the bureaucracy.

In 1982 the Task Force on Atlantic Fisheries criticized this process. It noted industry's great concern about the lack of an independent referee:

> "Fishermen believe that licensing decisions are influenced by politics, presumably because the Minister is seen to be open to pressure from the many groups lobbying for changes in policy."

The Task Force proposed the establishment of a quasi-judicial Atlantic Fisheries Licence Review Board. The major function of the proposed Board would be to act as arbiter and referee, within policy guidelines established by the Department.

The Task Force emphasized that the specific structure of such a Board was not as important as ensuring fairness in the review and appeal process:

> "It is essential that licensing decisions be subject to review and that the whole licensing process be fair and be seen to be fair. To this end, decisions regarding granting or withholding licences should be subject to open public review, with all interested parties free to make representations. Review board findings should also be public, with the reasons for decisions documented and published.... There must be a registry of licenses, including any terms and conditions that may be attached to them, maintained and open for public scrutiny."

The Economic Council of Canada had earlier recommended that licensing should be administered by a body separate from the agency responsible for managing the resources because "fishery officials should be as insulated as possible from decisions about who is to participate, so as to depersonalize and depolitize the choice of gear and fisherman" (Scott and Neher 1981).

Peter Pearse in *Turning the Tide* proposed that a Pacific Fisheries Licensing Board be created as a Crown Corporation:

> "The board should be given responsibility for administering commercial fishing licences within the general policy set out in the Fisheries Act and regulations, and should be responsible to the Minister for ensuring that licensing policy is applied uniformly and consistently. It should be responsible for conducting competitions for new licenses, carrying out the licence retirement program, maintaining a public record of licence holdings, deciding licence appeals, and advising the Department on needed changes in licensing policy" (Pearse 1982).

Pearse reviewed the existing licensing appeal process as it pertained to the Pacific fisheries. He concluded that the existing Pacific Region Licence Appeal Board was "largely independent of the Department." Despite this, he observed that many fishermen distrusted this elaborate appeal system. This distrust was apparently caused by the unnecessary secrecy of the appeal process. The board never disclosed the grounds on which appeals were made or its rationale for ruling on them.

Pearse's proposal for a crown corporation to administer Pacific licensing, including the review and appeal process, was more radical than Kirby's proposal for a quasi-judicial Licence Review Board. Pearse's proposal was rejected by the Pacific fishing industry and by the government.

Responding to the recommendations of the Task Force on Atlantic Fisheries, Minister De Bané announced that the government had rejected the recommendation for a quasi-judicial body. Instead, the government would establish a consultative advisory committee on licence decisions and appeals. It would also establish a procedure for Members of Parliament to be involved in licence decisions (DFO 1983a).

An Atlantic Fisheries Licence Review Board was established in June, 1984 "to provide fishermen with an independent body as a last level of appeal when they are dissatisfied with federal licensing decisions" (DFO 1984e). Minister De Bané said that the new Review Board would be "completely independent of the government process and therefore in a position to judge impartially." However, the Board never met. The federal election of 1984 intervened, and following the election the new government put the Board on hold while it considered its options.

In July 1986, Fisheries and Oceans Minister Tom Siddon appointed a "new Atlantic Fisheries Licence Appeal Board" (DFO 1986c). The Board would hear appeals from fishermen who were dissatisfied with departmental licensing decisions. The Board would be a last level of appeal for fishermen. Its decisions would be recommendations to the Minister. In presenting their cases to the Board, fishermen would have the right to be accompanied by a representative. In addition to hearing licensing appeals, the Board would consider general problems with licensing policy and make recommendations to the Minister and the Atlantic Regional Council. The Board would also advise the Minister on selection processes and eligibility criteria for new licences where an existing fishery may be expanded or established or a new one established. The Board would not consider requests for new licences in limited entry fisheries where no new licences are being issued. Its jurisdiction did not include the offshore fleet.

With the rejection of Pearse's recommendations the existing west coast Licence Appeal Board process set up in 1979 by Minister LeBlanc was maintained. Pearse had acknowledged that "the Appeal Board is largely independent of the Department." The Board's structure and functions changed little after 1984. It was the only avenue to appeal a licensing decision by DFO Pacific Region staff.

Thus by 1990 the final appeal process was essentially similar for both the Atlantic and Pacific fisheries. The Minister made the final decisions, but he received independent advice for both coasts.

7.0 THE MINISTER'S DISCRETIONARY LICENSING POWERS

For decades Section 7 of the Fisheries Act set forth the legislative authority for licensing as follows:

> "7. (1) Subject to subsection (2), the Minister may, *in his absolute discretion*, wherever the exclusive right of fishing does not already exist by law, issue or authorize to be issued, leases and licences for fisheries or fishing, wherever situated or carried on.
>
> (2) Except as otherwise provided in this Act, leases or licences for any term exceeding 9 years shall be issued only under the authority of the Governor General in Council" (R.S.C., 1985, C.F.-14 as amended) (Emphasis mine).

Licensing policies and procedures are defined in regulations made pursuant to Section 43 of the Act which provides that:

> "The Governor General in Council may make regulations for carrying out the purposes and provisions of this Act and in particular, but without restricting the generality of the foregoing, may make regulations:
>
> "(f) respecting the issue, suspension and cancellation of licences and leases; and
>
> "(g) respecting the terms and conditions under which a lease or licence may be issued."

Three major external reviews in the early 1980's, (the Economic Council, the Pearse Commission, and the Kirby Task Force) suggested that licensing be done by an independent body. Nonetheless, until 1991 successive governments from different political parties preserved the discretionary powers of the Minister. Indeed, most ministers jealously guarded this prerogative. At times, however, this discretionary power was a political liability rather than an asset.

In October 1991, Fisheries Minister John Crosbie stated that the existing system requiring the Minister to make all the decisions was "simply archaic" and "too political" (DFO 1991b). He announced the government would establish independent Boards for the Atlantic and Pacific fisheries. These Boards would allocate quotas and issue licences. The intent was to have the new Boards established by 1994. This would significantly alter the process for decision-making on licensing and allocation issues (see Chapter 15).

8.0 CONCLUSIONS

Virtually all commercial marine fisheries in Canada were under limited entry licensing by the late 1970's. Limited entry has been used for various objectives. This chapter has shown that limited entry licensing has had mixed success in curbing overcapacity and overinvestment in Canadian fisheries.

The B.C. salmon licensing program has been studied extensively. Most analysts have questioned the utility of limited entry licensing to curb overcapacity and overcapitalization. Experience in the inshore-nearshore groundfish fishery on the Atlantic coast, particularly the overcapacity in southwest Nova Scotia, supports that view.

On the other hand, the experience in the Atlantic lobster fishery has been more positive. In this fishery, limited entry licensing has constrained additional entry when lobster abundance increased substantially. Limited entry licensing and stringent vessel replacement guidelines also constrained the growth of capacity in the Atlantic offshore groundfish fleet since the mid-1970's. Fish stocks recovered following extension of jurisdiction. Limited entry licensing, combined with a conservative Total Allowable Catch regime, improved catch rates and reduced costs of fishing. Until certain groundfish stocks declined at the end of the 1980's, there was little evidence of overinvestment in offshore harvesting capacity.

Limited entry licensing has usually been introduced in Canadian fisheries where harvesting capacity has already become excessive. Restrictive action is often not taken sufficiently early in the development of a fishery. Ad hoc implementation over time contributes to the growth of overcapacity because of the incentive to beat the controls before they become tougher. In a few instances, such as the Atlantic Northern Shrimp fishery, limited entry was imposed in a developing fishery. This example suggests that limited entry licensing applied at an appropriate juncture in the development of a fishery can generate positive benefits. However, it must include tight controls on vessel replacement.

Several authors have argued that limited entry creates an elite group of licence holders, hurting rural communities and other prospective fishermen. Warriner and Guppy (1984) addressed the 1968 Davis Plan's influence on regional disparities. They concluded that, contrary to expectations, the Pacific fleet had not urbanized. In the salmon fishery, the proportion of the fleet based in small coastal centres had grown, with corresponding losses to urban centres. Over time some of the historical disparities in earnings between urban and rural fishermen had been reduced.

Sinclair (1983, 1987) suggested that limited-entry schemes had contributed to the partial achievement of the economic objectives of licensing in certain restricted fisheries. But this had a high social cost. In his view, state policy had created a local elite. This led to bitter resentment among those excluded from lucrative fisheries. He cited the shrimp and cod fisheries of northwest Newfoundland where the incomes of inshore dragger skippers and their crews were up to ten times those of typical inshore fishermen.

Davis and Thiessen (1988) argued that limited entry licensing created scarcity: there are too few licences issued to meet the demand. This would inflate the economic value of licences and increase the difficulty of licence acquisition for new entrants. They contended that limited entry licensing brought inequality of opportunity in the fisheries.

If indeed the effects described by Sinclair and Davis and Theissen are occurring, they are important matters of public policy. The problem should be addressed more systematically. In advocating limited entry to deal with overcapitalization, economists ignored the social impact. References to such problems were generally dismissed by arguing that other occupations have limited entry or closed shops. To become a doctor, plumber, or lawyer one has to pay one way or another for entry.

Crutchfield (1979) put this line of argument cogently:

> "I cannot generate much concern over the allegation that limited entry licensing schemes discriminate against potential new entrants. Their only 'loss' would appear to be the opportunity to share the capital gains accruing to those grandfathered into the system. Thereafter, any current or potential operator of a fishing unit faces precisely the same cost and revenue functions, and new entry is possible on the same basis as it would be in most other natural resource industries: purchase or lease of a right to harvest. It is surprising to find the fisheries singled out for special castigation because of the burdens imposed on potential new entrants. Why not equal concern over the plight of the young man who wishes to enter retailing, only to find that the cost of purchasing or leasing good retail sites has risen astronomically in recent

years? A quick glance at a semilog plot of farmland prices yields the same concern with respect to potential new farmers."

Licences have certainly acquired a capital value. In many cases this can be quite substantial (e.g. Wilen 1988). During the consultations on the 1989 Atlantic Coast Licensing Policy, there was general recognition that selling and trading of licences was occurring, even though officially such activity was not permitted. While some groups expressed concern, the federal and provincial governments maintained the status quo: Officially the licence is a privilege which confers no property right. However, governments acknowledged that the licence holder can in fact determine the recipient of a transferred licence when he retires from the fishery.

In many fisheries there was substantial overcapacity before limited entry licensing was introduced. One of the fundamental challenges is what is to do in such overcapacity situations. Limited licence buy-back schemes have been tried in certain fisheries (e.g. B.C. salmon, Atlantic lobster, Atlantic salmon), with varying degrees of success. The B.C. program was too short to reduce fishing effort significantly. It was terminated in 1973 because changing circumstances in the fishery elevated licence prices to a level where the program was no longer feasible. A limited reactivation of the program in 1981 had little effect on the problem. Proposals for a large-scale buyback program, proposed by Pearse in 1982 and announced by Fisheries Minister De Bané in 1984, were never implemented. An Atlantic Salmon Licence buy-back program in 1984, 1985 and 1986 reduced the number of active participants in the Atlantic salmon fishery in a resource crisis. This did not, however, succeed in restoring Atlantic salmon stocks. In 1992, a 5-year moratorium on commercial salmon fishing on the island of Newfoundland was introduced, coupled with a further voluntary buy-back of Atlantic salmon licenses (Lear 1993).

Copes and Cook (1982) described three major problems that have emerged in buy-back programs:

1. The expectations trap;
2. Capital stuffing; and
3. Transitional gains trap.

The "expectations trap" arises because a buyback program generates expectations of improved incomes. Thus it raises the price owners demand for their licence or licensed vessel, either from another potential buyer or from a government buy-back authority. This can kill a buy-back program as it becomes too expensive. The B.C. salmon fishery provides an example. "Capital stuffing" occurs when the increased capacity of individual vessels adds more capacity in total than is removed through the buy-back. The "transitional gains trap" occurs when limited entry raises expected fishing incomes and vessel operators and operators are allowed to sell their licences. The anticipated stream of additional earnings becomes capitalized in the value of a licence. Under such circumstances only the first generation of licence holders really benefits from limited entry.

Copes and Cook (1982) suggested that capacity might be more effectively reduced if licences were not transferable, but terminated when a fisherman retired. This would provide for automatic attrition of the fleet. It is, however, politically difficult to change from a transferable to a non-transferable system.

The authors also suggested that a politically tough-minded approach might make a regular buy-back program workable. This could involve very high levies on the fleet for a longer period of time to finance a buy-back program. The duration and size of the levies would have to be substantial to reduce expectations of profit and make it easier to buy out licences. Despite these shortcomings, I remain convinced that properly designed buy-back programs can reduce gross overcapacity.

Another major problem is to limit capacity growth in a fishery where there is a perceived incentive to invest in bigger and better boats. The 1981 "foot-for-foot", "hold-for-hold" restrictions in the Atlantic groundfish fishery had a significant impact in slowing but not halting capacity growth. The overcapacity problem which exists now generally preceded the imposition of those vessel replacement rules. The latest innovation of restricting replacement on the basis of a cubic measure offers real promise of curbing further growth in the fleet.

Wilen (1988) pointed out that the focus of economists' concerns (overcapitalization or excess financial capital) tends to differ from that of fisheries managers (excess physical fishing capacity). Although the fishing process allows more substitution of inputs than was anticipated when limited entry programs were introduced, he concluded that substitution possibilities on the vessel are linked to and limited by gear characteristics. If regulations constrain net depth, length, and other gear characteristics, there are often few competitive options to pursue.

Wilen (1988) also noted that where entry has not been limited the results have been detrimental: strong competition between fishermen and "racing" to catch fish, redundant and inefficient investment, and increasingly severe biological controls (e.g. the Pacific halibut fishery). Conversely, where limited entry programs were introduced, they have resulted in rents, sometimes substantial, emerging in licence values. Meanwhile, fishermen continue to try to outflank their competitors. This is offset by fisheries managers' attempts to prevent increases in fishing capacity.

In my view, the failure of existing input restrictions to constrain capacity has been a matter of too little action too late and administrative/political missteps in phasing in restrictions. The recent volumetric/cubic measure approach is a significant advance.

Despite its shortcomings, limited entry licensing has proven useful in managing Canadian fisheries. The very high price that limited entry licences now bring to those leaving the fishery is strong evidence that limited entry licensing has improved fishermen's incomes in many fisheries.

Most countries concerned with rationalizing their fisheries have chosen limited entry licensing as a primary tool for fisheries management. Crutchfield (1981) suggested several reasons for this:

> "It is the easiest to introduce since it must, of necessity, be used in parallel with other methods of regulation and involves the least disturbance to existing ways of organizing and operating a fishing venture. In addition, its tightness can be varied from a very modest moratorium approach to one in which significant reduction of unnecessary inputs is undertaken as a matter of policy. Whatever the route chosen, a limited entry programme can be phased in a manner which minimizes the amount of compulsion that must be exerted to trim the level of fishing effort to desired levels."

Regarding the efficacy of limited entry, Wilen (1988) concluded:

> "On the one hand, it is clear that fisheries have been successfully controlled where managers have been willing to employ the instruments of their disposal with enough vigour to combat increasing effort. On the other hand, these traditional means of regulating fisheries have done nothing to alter the incentives to compete for shares of the fixed resource."

A major challenge lies in reducing fishing capacity where significant overcapacity exists. Apart from buyback programs under limited entry licensing, two other approaches have often been suggested as more effective rationalization schemes. These are output controls-taxation and a form of quasi-property rights (individual catch quotas). These are examined in Chapter 9. Wilen (1988) suggested that some forms of area licensing could also reduce the incentive to overinvest.

Despite shortcomings, limited entry licensing continues to be a valuable method to pursue economic rationalization and associated fisheries management objectives. Most limited entry programs are relatively recent. Thus there is a paucity of data on which to judge their relative success. Improving our understanding of the effects of limited entry programs and establishing criteria for evaluating the impact of such programs should be a priority.

CHAPTER 9

MANAGING THE COMMON PROPERTY III — INDIVIDUAL QUOTAS

"An overall quota of fish must be the principal weapon to control a fishery, but with it there must be individual quotas for boats and plants as well, depending on circumstances. What you are then doing, in effect, is to give particular people exclusive licence to take part of the fish resource, in the same way that an oil company is allocated an offshore block or a mining company is given an ore concession. This idea applied to fisheries makes people throw up their hands in horror. But I am sure that in 20 years this will be the accepted method of running fisheries."

– G.H. Elliott, 1973

1.0 INTRODUCTION

When G.H. Elliott made this prediction at the 1973 Technical Conference on Fishery Management and Development, sponsored by the U.N. Food and Agriculture Organization (FAO), economic theorists were still debating the merits of licensing versus taxation as regulatory instruments for controlling investment and capacity in world fisheries. Less than a decade later many economists were touting quantitative rights or individual quotas as preferred regulatory instruments, e.g. Pearse (1979a), Scott (1979). Within just a few years of Elliott's prediction, individual quotas were introduced experimentally in a few fisheries (e.g. Bay of Fundy herring fishery in the mid-1970's). A decade after the 1973 FAO Conference, enterprise allocations (a form of individual quotas) were introduced in Canada's offshore Atlantic groundfish fishery (Parsons 1983). Within 15 years individual quotas were being tested in numerous fisheries around the world.

The impetus to seek alternative methods for economic rationalization of fisheries arose from the perceived deficiencies of limited entry licensing. Economists and managers alike began to realize that the initial experiences with limited entry licensing were not all positive. Limited entry had not removed the incentive to build bigger and better boats to compete for a larger share of the available resource. For a while several economists favoured taxation over limited entry licensing (Anderson 1977; Scott 1962; Strand and Norton 1980). For a discussion of the pros and cons of taxation as a fisheries regulatory instrument, see Clark (1980), Stokes (1979), and Cunningham et al. (1985).

Taxes or royalties have not been used to regulate any Canadian marine fishery. Indeed, the approach has rarely been used anywhere for this purpose (Clark 1988).

Most economists have argued, however, that royalties or landings charges are a legitimate means for governments to capture part of the resource rent in prosperous fisheries. Within Canada, landings charges have long been discussed (but never implemented) as a device to recover costs of salmonid enhancement in the B.C. salmon fishery.

2.0 EVOLUTION OF THE INDIVIDUAL QUOTAS CONCEPT

The federal Department of Fisheries and Oceans first considered individual quotas to regulate fisheries because the various forms of input controls then in use (such as limited entry licensing and vessel replacement rules) dealt with the effects of the common property problem rather than the cause. During the 1960's, several economists had discussed the concept of introducing property rights to deal with the cause of the problem, but the use of output controls in the form of individual quotas was first seriously considered in the 1970's. In a 1971 background paper for a Canadian Fisheries and Marine Service seminar on fisheries policy, Peter Pearse made no mention of individual quotas as a possible regulatory instrument (Pearse 1971). Later, however, Pearse became an advocate of individual quotas. The first serious proposal for individual quotas for fishermen was raised in a paper by Christy (1973). Christy identified considerable difficulties with limited entry licensing and a system of taxes or user fees that would discourage entry to the fishery. He suggested a third approach — fisherman quotas. In the early 1970's ICNAF in the Northwest Atlantic had established national quotas to divide up international Total Allowable Catches and to end competition among countries for the available catch. Just as national quotas were intended to eliminate the incentive to overcapacity among countries, so a system of individual quotas at the national level might end the "race for the fish" among individual vessels or operators within a country. Christy saw fisherman quotas as a logical extension of the national quotas approach in the international fishery. He suggested that the greatest advantage of a system of fisherman quotas would lie in its flexibility and the freedom for fishermen "to innovate as they wish and avoid the onerous and cumbersome constraints of detailed regulations on size of vessel and kind of gear." In addition, "it would appear to permit significant reductions in the costs of management and actually provide for some coverage of these costs by the users of the resource."

In 1972 individual quotas were introduced in the Lake Winnipeg fishery. In 1975 a fisherman's quota system was adopted by Michigan for its Lake Superior fisheries. In 1976 individual quotas were introduced in the Bay of Fundy herring fishery.

From 1979 to 1982, a number of economists seized on individual quotas, quasi-property rights, or quantitative rights as the means for economic rationalization of fisheries. Scott (1979) and Moloney and Pearse (1979) came to see output controls, specifically quantitative rights, as ideal instruments for regulating commercial fisheries. By 1981 DFO was discussing enterprise quotas with companies in the Canadian Atlantic offshore groundfish industry. A large scale experiment with enterprise allocations (company quotas) was introduced in that fishery for 1982. Following the restructuring of the offshore groundfish fishery in 1983, a system of enterprise allocations was adopted as a 5-year experiment in this fishery beginning in 1984. This was subsequently adopted for ongoing management of the offshore groundfish fishery.

Meanwhile, Scott and Neher (1981) had made an impassioned plea for widespread adoption of individual and exclusive rights of access. They termed these "usufructuary rights." Their envisaged rights system embodied these elements: marketable, divisible quotas specific for species, time, location and gear. Their plea was followed by Pearse's (1982) proposal for 10-year quota licences and an auction system for the Pacific fisheries. The Task Force on Atlantic Fisheries (Kirby 1982) also recommended a system of quasi-property rights for the Atlantic fisheries, in the form of individual licences specifying either a limitation on the catch or on the catching capacity of the fisherman's vessel and gear. These quota licences would also be divisible and transferable, i.e. marketable.

Different terms have been used for variants of this approach: quota licences, enterprise allocations, individual transferable quotas (ITQs), and quantitative, stinted or usufructuary rights. For this chapter "individual quotas" will be the generic term for this approach to the problems arising from the common property nature of the fishery. By 1983 the economic community had clearly embraced individual quotas as the means to deal with open access fisheries. Governments, both in Canada and elsewhere, were experimenting with the approach in selected fisheries.

The rationale for an individual quota system is that it would eliminate the incentive to overinvest and end the competitive "race for the fish". Under an individual quota system, the available catch (e.g. Total Allowable Catch for a particular stock) is divided among individual fishermen, fishing units or fishing enterprises before the fishing season. Each individual, unit or enterprise is assigned a fixed share of the TAC, either as a specific quantity or as a percentage of the TAC. This is done for one year or for a longer period. Percentage shares of the TAC in a given season translate into specific quantities once the TAC for that season is determined.

In advocating property rights to the catch, Scott (1979) suggested that the quota system would resemble some existing limited entry licence systems, except that a quota would permit the holder to land a certain quantity of fish. Quotas would be transferable, marketable and of fixed duration. Among the advantages he foresaw were:

1. A fisherman would know how much fish he could catch for a given period and hence could choose the pace at which he caught his quota. This would probably spread the catch more evenly over the season, to suit the market.
2. Individual quotas could be located and dated to discourage excessive fishing under high-cost, high-concentration conditions.
3. Individual quotas could promote flexibility in the fishery.
4. An individual quota system would have lower transaction and administration costs than other methods.

Moloney and Pearse (1979) also advocated regulating commercial fisheries by providing fishing enterprises with transferable rights to harvest specific quantities of fish. They contended such a regime would maximize resource rents and permit the gains to be distributed flexibly. This would be superior to most other proposals for rationalizing fishing. They saw several disadvantages with Christy's earlier proposal for "fisherman quotas". One was that quotas would be defined as a percentage of the total catch. Moloney and Pearse argued that this would restrict flexibility in their allocation and increase the difficulty of reducing the catch. In fact, this feature of Christy's proposal has been incorporated in most of the experiments with individual quotas in Canada's marine fisheries. Their chief criticism of Christy's proposal was that his quotas were not transferable. Again, many individual quota experiments involve nontransferable quotas.

The system of quantitative rights proposed by Moloney and Pearse (1979) incorporated these elements:

1. Rights of the holders to catch specific quantities of fish;
2. These rights would be freely transferable and divisible;
3. The rights should be allocated by an open auction; and
4. The regulatory authority could enter the market for rights, buying or selling at the market price in response to changing market requirements.

This system, they believed, would generate economic incentives to minimize costs or maximize rents. All rents would be capitalized in the value of the quantitative rights. Furthermore, there would be no incentive to adopt the kind of wasteful technologies associated with competition among fishermen for larger catch shares.

Stokes (1979) suggested that an individual quota system: (1) transfers competition for shares from the fishing grounds to the market, and (2) encourages voluntary reductions in fishing effort. He recognized that enforcing an individual quota system could be difficult. Stokes (1979) warned that the gains from underreporting the catch and upgrading (taking only the best quality fish and discarding the rest) could be substantial. He emphasized the need for voluntary compliance, reinforced by peer pressure. Such a system could only succeed if supported by fishermen.

The proposals of Scott and Neher (1981), Pearse (1982), and Kirby (1982) were similar in advocating quasi-property rights in the form of quota licences. All envisaged that such quota licences would be divisible and transferable (marketable). While strongly advocating quasi-property rights, Scott and Neher (1981) recognized that acceptance by fishermen was central to the success of such a system. They urged caution in introducing changes. They suggested that the government select two fisheries on the east coast and one on the west coast as 'model fisheries' to test such a system. Candidate fisheries should: (1) be relatively new fisheries with modest investments to overcome the artificial impediments already imposed by control schemes currently in place; (2) employ adaptable gear so that any redundant capital could be absorbed elsewhere; and (3) employ men and boats with well-established records of participation. The last would ensure that historical rights could be established with a minimum of fuss.

Pearse (1982) argued forcefully in favour of transferability:

> "While the objections to transferability are weak, its benefits are substantial. Transferability permits flexible reallocation of fishing privileges to enable the industry to adjust and to provide an avenue for new participants. It will be particularly valuable in promoting fleet rationalization where licenses provide catch quotas because it will enable licensees to adjust their rights to the most economical amount for their fishing units. The more flexibly the quotas can be divided and combined, the more they will encourage this kind of rationalization."

The Task Force on Atlantic Fisheries (Kirby 1982), while adopting Pearse's terminology and rationale for quota licences, were less categoric in espousing the approach. They foresaw:

> "A system by which fishermen would be allocated individual annual catch quotas that, *within certain limits*, they could catch by whatever means they wished. The initial allocation of these quotas might be based on recent historical performance. The quotas would normally be expressed as a percentage of the TAC for a given stock. They could be transferred or sold among fishermen *subject to restrictions the government might wish to impose....* Once a fisherman had a guaranteed quota, he could then plan to catch it at least cost." (Emphasis mine)

However, the Task Force saw some practical limitations for such a quota licence approach. A quota licence concept might be impractical for vessels under 35 feet, but possible in some cases for vessels between 35 and 65 feet. For the smaller boats the Task Force suggested licensing for a certain amount of fishing capacity or catching potential. Overall, they recognized the need for a great deal of consultation, analysis and experimentation with pilot projects.

By the end of the 1980's, individual quotas had been widely introduced in Canada's Atlantic fisheries. Some systems had become permanent. Others were still experimental. On the Pacific coast, the concept had been tried only in small-scale fisheries. The reasons for this difference will be elaborated upon later. In the following sections I examine the nature and relative success of these experiments with individual quotas.

3.0 THE ATLANTIC FISHERIES

3.1 Bay of Fundy Purse Seine Herring Fishery

Before economic theorists fixed during 1979–1982 on the concept of individual quotas as a preferable regulatory instrument, a system of individual vessel quotas had already been introduced in the Bay of Fundy purse seine herring fishery, commencing in 1976.

The Department of Fisheries and Oceans introduced individual vessel quotas in 1976 as part of a complex restructuring of the Bay of Fundy herring fishery. This was precipitated by a severe crisis in the industry. Major changes had occurred in the Atlantic herring fishery in the 1960's. A large number of reduction plants were constructed to process herring to fish meal and oil. These plants required large volumes of raw material. This necessitated the use of purse seiners capable of catching large quantities of herring. The fleet expanded rapidly through conversion of existing vessels, new construction, and the diversion of several large herring seiners from British Columbia in 1966 and 1967.

Expectations were high that the fishery could sustain greatly increased catches. In 1966, the Industrial Development Branch of the federal Department of Fisheries projected catches of one million tons annually (Anon 1967). This was never achieved.

The purse seine catch from the Bay of Fundy increased from 12,000 tons in 1963 to 200,000 tons in 1968 but by 1971 had decreased to 67,000 tons. The fishery was placed under a catch quota (TAC) in 1972 under the auspices of ICNAF. This increased competition leading to increased catching capability. By 1975 the entire quota could be caught in a few weeks. That year the allowable catch was 100,000 tons, with 57 seiners fishing. With the drop in price from $60.00 to $30.00 per ton, the average gross income for a seiner was only $50,000. The financial prospects for 1976 for vessel owners and crew looked bleak.

Mitchell and Lennon (1975) concluded that the problems resulted from resource scarcity, overcapacity in

Atlantic Herring Fishery.

both the primary and secondary sectors of the industry, and competition for herring between the food processing and reduction industries.

In November 1975, the Minister asked his Special Adviser, Mr. Fern Doucet, to meet with the herring fishermen to develop a long-term solution to the industry's problems. These discussions made it apparent that the income of the fishermen could only be increased by increasing the value per ton of herring. This was because the volume of herring available to the fleet was essentially fixed at about 60,000 tons. The use of herring for reduction had to be cut back in favour of utilizing it for human consumption (F. Doucet, Ottawa, pers. comm.).

By this time the fishermen also recognized that solving their problems depended on their ability to organize effectively as a bargaining group. The Bay of Fundy processors did not compete with each other for catches. This resulted in artificially low prices to fishermen.

The fishing season had become very short since 1972. Once the need for expanded utilization of herring for food was acknowledged, it also became clear that the season should be lengthened. To achieve this, it was necessary for the purse seiners to function as a fleet. According to Doucet, this reasoning led to vessel quotas. Through negotiation, the fishermen's committee agreed on a minimum vessel quota of 1,500 tons and a maximum of 2,500 tons for 1976.

Following Doucet's negotiations, Fisheries Minister LeBlanc approved a series of measures to address the problem:

1. A $750,000 program of deficiency payments for the purse seiners;
2. A ban on the use of whole herring for fish meal (This was already prohibited by Section 29 of the Fisheries Act unless special Ministerial permission was granted. 1976 marked the first time in the history of the Bay of Fundy purse seine fishery that such permission was not granted);
3. Processing companies were compensated for losses on capital investment in reduction plants;
4. A purse seiners' cooperative, the "Club" or Atlantic Herring Fishermen's Marketing Co-operative (AHFMC), was formed, to control harvesting and marketing. This included the authority to negotiate prices with processors and to facilitate sharing of catches. The cooperative would also control the days on which purse seiners could fish and the plants to which they delivered their catch;
5. Introduction of voluntary vessel quotas including daily and weekly catch limits to match the catch with processing capabilities;
6. Permission for fishermen to sell a portion of the available catch in unprocessed form over-the-side to Polish factory trawlers, at prices considerably higher than those paid by Canadian processors; and
7. Government assistance to:
   (i) Aid fishermen in acquiring ownership of the purse seine fleet;
   (ii) Foster vessel conversion to improve fish quality; and
   (iii) Encourage processing firms to increase capacity for human food production (DOE 1976f, 1976g).

These initiatives represented a major transformation of the Bay of Fundy purse seine fishery from reduction to processing herring for food. This process was facilitated by the decline of Northeast Atlantic herring stocks, which opened up the European market to Canadian herring food products.

The "over-the-side" sale to Poland served two purposes. It provided an alternative market to fishermen while the processors were converting their facilities to food production. It also served to stimulate more competitive prices from the domestic processors.

The AHFMC in its first year of operation had as members 27 purse seine captains out of a total active fleet of 53 vessels. Through control of the landing pattern of its members, the AHFMC strongly influenced the outcome of the 1976 fishery. It consciously chose to keep the official catch to a level of 60,000 tons. Landed value increased from $2.8 million in 1975 to $4.0 million in 1976 with a decline in landings from 84,000 to 60,000 tons. By 1978 the value of the purse seine fishery alone was $11.8 million. The total landed value increased from $5.2 million in 1975 to $20.0 million in 1978. Kearney (1984) estimated that real incomes of Bay of Fundy purse seine fishermen on the smaller vessels doubled between 1975 and 1978.

The success in 1976 induced all purse seiners to join the Co-op the following year. Doucet has given the individual vessel allocation scheme much of the credit for the successful restructuring of the industry. Unfortunately, the spirit of harmony did not prevail. In 1979, 14 purse seiners left the Co-op to form the Southwest Seiners Association.

The initial 3 years of the Bay of Fundy project were undoubtedly successful. The alternative market offered by over-the-side sales was the glue that made the Co-op system work. In 1979 processors convinced the government that they could now handle all of the available herring for food production at prices comparable to those offered by the foreign buyers. Thus permission for an over-the-side sale was denied. Nor was over-the-side sales permitted in 1980. By this time the Co-operative marketing approach was coming apart at the seams.

Kearney (1984) observed that the AHFMC could not compensate for the lack of this alternative market. By 1980 the system had broken down with members reestablishing individual marketing arrangements with the processors.

The period 1980–83 has been described by the Regional Director General for DFO at the time as follows:

> "The collapse of the fisherman managed system and rejection of its apparent advantages after only two years induced imposition of more direct control by government. That control included continuation of the vessel-quota system. By 1980 both volumes of herring and prices for catches had declined substantially below the 1978–79 peak when prices exceeded $250 per ton. Prices had fallen to less than $100 and many vessels began forming ties to the processing sector in order to secure a market outlet. Mutual cooperation among skippers had collapsed and Departmental attempts to enforce vessel quotas proved to be largely unsuccessful.... The loss of negotiating power on the part of purse seine fishermen, coupled with problems in controlling individual vessel quotas, created a nightmare condition in the fishery" (Crouter 1984).

In 1981 the government again intervened, reintroducing over-the-side sales to bring the seiners back together under one organization. This was unsuccessful because by now processors had re-exerted their traditional dominance over the purse seine fleet. Fish quality deteriorated with a high proportion of small fish in the catch. This reduced the volume of herring that could be marketed and the price paid for it.

The Task Force on Atlantic Fisheries focused on the problems in the groundfish industry. However, it also examined the plight of the herring purse seine fleet. The Task Force concluded that the herring resource on the Atlantic coast could no longer support the purse seine fleet at its existing size (63 vessels, plus an additional three licences from recently sunk vessels). TACs in the Gulf and western Newfoundland had fallen to very low levels, and there was no prospect of a seine fishery there in the foreseeable future. The Task Force recommended that the Atlantic purse seine fleet be reduced by at least 20 to 25 vessels for the remainder to be viable (Kirby 1982).

The Task Force considered government-funded Buy-Back to reduce the fleet. It concluded that it could not, in the existing economic climate (the recession of the early 1980's), justify a publicly funded licence or quota Buy-Back program when businesses in other sectors were going bankrupt without compensation. It recommended a rationalization program with these elements:

1. Immediate implementation of transferable vessel quotas for seiners, with the initial allocation distributed based on relative catches in the previous 3 years;
2. Establishment of a 5-year Buy-Back program for boat quotas funded by industry levies on domestic purchases and over-the-side sales. These quotas would be sold back to the remaining operators; and
3. Establishment of stringent measures to prevent misreporting of landings, for example, by requiring landings to be made only in the presence of a fisheries officer and by suspending or cancelling licences for misreporting.

The Task Force suggested that such a system did not require totally unrestricted trading in enterprise or boat quotas. For public policy reasons it might be desirable to restrict holdings of quotas so that no person, including processors, could have an interest in more than a fixed percentage of the TAC. This might be necessary to ensure adequate port market competition and reasonable prices to fishermen.

Of the 65 vessels involved in 1982, 49 were from the Scotia-Fundy Region and 16 were from the Gulf. These now belonged to three associations, the AHFMC, the Southwest Seiners and the Gulf Seiners Association. By this time these organizations had little if any control over the fishing activities of their members. Following extensive consultations with the industry on the Task Force recommendations, the government in August 1983 announced a restructuring of the Atlantic herring purse seine fleet (DFO 1983d).

The program was intended to permit voluntary reduction of the fleet by combining individual vessel quotas over the next 10 years. The major elements of the program were:

1. Separation of the fleet: Vessels based in Scotia-Fundy were given exclusive access to the Bay of Fundy and Chedabucto Bay (Divisions 4WX) areas while purse seiners based in the Gulf of St. Lawrence would have exclusive access to the Gulf and a portion of Sydney Bight (Divisions 4RST and Subdivision 4Vn); (Fig. 9-1).
2. A single fleet quota was developed for Divisions 4W and 4X;

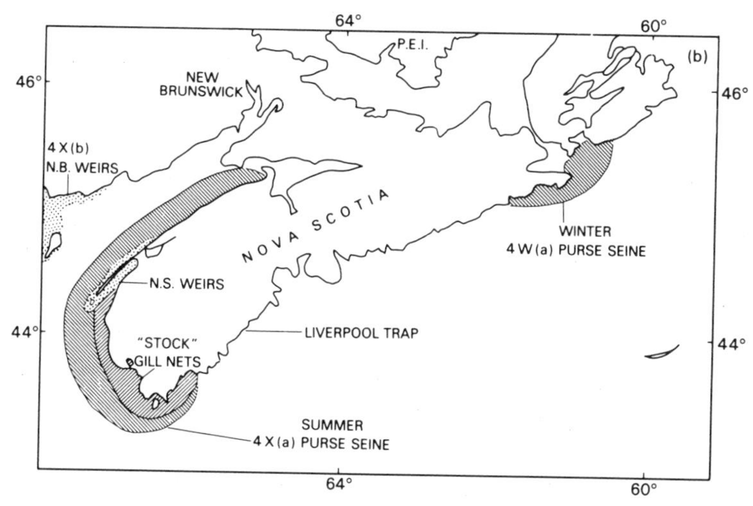

FIG. 9-1. Location of the components of the herring fishery in the 4WX management unit (from Sinclair et al. 1985). The different shadings indicate the various herring fisheries in 4WX.

3. Provisions for fleet reduction: Licence holders were granted the right to transfer vessel quotas through sale and purchase as a mechanism to remove vessels from the fleet, with a constraint that no single vessel could acquire more than 4% of the purse seiner allocation;
4. The following guarantees were given by the Minister of Fisheries and Oceans:
   — the program would last at least 10 years;
   — individual vessel quotas would remain in effect for the duration of the plan;
   — the Scotia-Fundy purse seine fleet was guaranteed at least 80% of the 4WX TAC; and
   — no new licences would be issued for at least 10 years.
5. The 49 vessel Scotia-Fundy fleet was categorized into 27 Class A licences (each entitled to 1.6% of the TAC), 15 Class B licences (each entitled to 2.7% of the TAC) and 7 Class C licences. The basis for the classification was:
   Class A: Fishermen-owned vessels with no history of participation in Division 4W during the period 1980–83 (essentially the non-mobile Bay of Fundy fleet of smaller vessels);
   Class B: Fishermen-owned vessels which had participated in Division 4W during the period 1980–83 (the larger, more mobile fleet); and
   Class C: Vessels owned by herring processors, both mobile and non-mobile. In most cases, Class A vessels were less than 65 feet and Class B and C vessels over 65 feet in length; and
6. Certain restrictions were placed on the purchase of quotas:
   — Class A could purchase Class A;
   — Class B could purchase Class B;
   — Class C could purchase Class C;
   — Class A and B could purchase Class C providing certain criteria were met; and
   — the seller of a quota was required to divest himself of his entire quota and exit the fishery permanently, surrendering his purse seine licence to DFO.

Initially, this system appeared to function reasonably well. By December 1983 five Scotia-Fundy licence holders and one Gulf licence holder had sold their quotas. This reduced the size of the Scotia-Fundy fleet to 44 and the Gulf fleet to 15 (DFO 1983e).

Within a year or so problems became apparent. The quota assigned to each vessel was, in most cases, far less than the vessel's catching capability. Underreporting again became widespread. In April 1985, recognizing that the actual catch was far larger than the reported catch, Fisheries and Oceans Minister John Fraser accepted the recommendation of a joint industry/government consultative committee for a 1985 allocation of 100,000 tons for the purse seine fleet, with the inshore fishery being managed on an allowance (DFO 1985f). In theory, the tripartite package deal among DFO, the purse seiners and the processing industry would provide greater access by DFO officers to catch information at the offices of the purse seiner associations. The industry agreed to respect the management plan and provide accurate catch information.

In March 1986, Fisheries and Oceans Minister Tom Siddon indicated that, as a result of the 1985 agreement, misreporting had been reduced from an estimated 77% in 1984 to between 10 and 15% in 1985 (DFO 1986d).

Regarding the early period of individual vessel quotas in the Bay of Fundy purse seine fishery, Crouter (1984) observed that individual quotas had the following effects:

> "1. They made the impact of fleet rationalization less severe but retarded the development of an efficient fleet so that we are left with a fleet segment consisting of small, inefficient vessels and another segment of over-capitalized fishing craft.
>
> "2. They favoured the survival of fishermen's associations and provided an opportunity for greater involvement by fishermen in day-to-day management but they could not preserve the initial unity among the fishermen.
>
> "3. They promoted individual ownership of vessels among fishermen but they did not retard ownership and control by herring processors."

A thorough evaluation will not be possible until the end of the 10-year period. However some preliminary conclusions about the 1983 plan are warranted. The plan was structured quite differently from earlier ones, and it appears to have had a modest impact on fleet rationalization in its initial 5 years.

The number of purse seine licences in the three categories (Classes A, B and C) was reduced between 1983 and 1988 from 49 to 40 (Table 9-1). This represented an 18.4% reduction in the licences over 5 years. Seventeen licence holders had increased the size of their quotas and 11 had reached the 4% maximum permitted an individual licence holder. Approximately 20% of the TAC was transferred among licence holders through these transfers. The fishing power of the remaining vessels had increased (MacFarlane 1989).

It is likely that continued misreporting of catches limited the degree of rationalization. By cheating, a quota owner could acquire additional supply without having to purchase additional quota. Thus there was clearly a reduced incentive to purchase and combine quotas. With the breakdown in the initial voluntary vessel quota system in 1979 and 1980, Doucet (Ottawa, pers. comm.) estimated that total landings were underreported by about 40%.

Crouter (1984) assessed the pre-1983 experience with individual vessel quotas:

> "Our experience with vessel quotas has been that each and every fisherman will attempt to 'cheat' on his quota and processors will promote that attempt through collusion in falsifying records."

Kearney (1984) asserted that the Bay of Fundy purse seine fishery had a history of underreporting the annual catch by about 30%. The most comprehensive analysis of misreporting was undertaken by Mace (1985). She estimated that underreporting was relatively stable at about 25% in the pre-vessel quota period (1973–75). It declined to about 15% during the Co-operative approach to vessel quotas (1976–79), and then increased steadily from about 30% in 1980 to about 44% in 1984.

The largely unsuccessful attempts to deter overfishing and limit individual catches to the quota allotted had taxed to the limits the enforcement manpower of DFO. Enforcing the vessel quota system had been time-consuming and costly. MacFarlane (1989) reported that the regulations governing catch reporting in place to the end of 1988 were ineffective and "frustrated the efforts of enforcement officers to obtain accurate records of herring removals and sales."

MacFarlane (1989) assessed the economic performance of Scotia-Fundy herring purse seiners from 1983 to 1987. He found that the profitability of these vessels in 1986 and 1987 had improved significantly compared with the situation in 1983. This occurred, not because of fewer vessels, but because of an increasing biomass, an increase in the TAC, a new market for roe-herring, and a continuing, if declining, supplement from over-the-side sales.

TABLE 9-1. Licence status of the Scotia-Fundy Herring Fleet, 1983 and 1988. For definitions of Classes A, B and C, see text.

	1988		1983	
	Number of Licences	Percentage of Purse Seine Quota	Number of Licences	Percentage of Purse Seine Quota
Class A	24	43.2	27	43.2
Class B	12	43.2	15	40.5
Class C	4	13.9	7	16.3
Total (A+B+C)	40	100.3	49	100.0

The earnings of seiner skippers and crewmen in 1986 and 1987 were as much as three times the 1983 earnings. By 1988 six new replacement vessels had been added to the fleet. There were significant expenditures on upgrading the technical capability of various seiners between 1983 and 1987. MacFarlane (1989) suggested that these investment decisions were not consistent with the cost minimization expected to occur under an individual quota regime. These expenditures increased the capacity of the individual enterprise overall.

Despite these problems, MacFarlane (1989) concluded that by 1988 the fleet of purse seiners was smaller, more efficient and more profitable. He anticipated that the tightening of regulations for catch reporting and compliance with quotas would "do much to reverse the abuse of earlier years."

3.2 The Offshore Atlantic Groundfish Fishery

3.2.1 Introduction of the EA Program

The first major trial of individual quotas in the Atlantic fisheries was the introduction of enterprise allocations in the offshore groundfish fishery during the period from 1982 to 1984. An enterprise allocation system was applied experimentally in 1982, suspended in 1983 because of the uncertainty during restructuring of the offshore groundfish industry, and reintroduced for a 5-year trial period beginning in 1984.

The "race for the fish" which then prevailed in the offshore fishery provided the impetus for the introduction of enterprise allocations to this fishery. The immediate trigger was the competitive fishery for northern cod: in January – February 1981, the offshore quota of northern cod was caught in 6–7 weeks (this would normally take several months). Associated with this were glut conditions at the processing plants, a large proportion of poor quality fish going into low-grade products, and consequent adverse market impacts. This "race for the fish" continued throughout 1981. Offshore quotas were caught rapidly in the northern Gulf, Sydney Bight and St. Pierre Bank. There were fleet tie-ups in August with plant closures, and employees were temporarily out of work.

The January rush for northern cod led Fisheries Minister Roméo LeBlanc to predict layoffs "because of the actions of trawler owners who have failed to bring a sense of orderly harvesting into this fishery" (DFO 1981). The trawler owners were unable to agree on a method of slowing down the fishery. This particular experience illustrated the effects of offshore companies' efforts to maximize their share of the available catch.

The offshore sector was a major component of the Atlantic groundfish fishery. It accounted for approximately 50% of total Atlantic groundfish allocations and catches. In 1981 there were about 140 trawlers over 100 feet in length active in this fishery. Eighty-one vessels were based in Newfoundland and 54 in Nova Scotia. The majority of these were operated by four major companies — H.B. Nickerson, National Sea Products, Fishery Products and the Lake Group. There were 14 vessels based in Nova Scotia operated by smaller companies (e.g. Atmar Marine, Mersey Seafoods, Connors Brothers). There were a few vessels (5) based in the Gulf of St. Lawrence.

In 1981, 82 of these vessels, or 61% of the fleet, had been built between 1958 and 1967. Assuming a useful life of 20 years, it was possible that all these vessels would have to be replaced in the next 5 years. The potential impact on the catching capacity of the fleet was enormous.

Any significant increase in capacity could only worsen an already difficult situation. During 1981, DFO undertook several studies in cooperation with the offshore companies. These included constructing a simulated computer model of the fleet. This helped convince company representatives of the need for a more rational approach to harvest the offshore groundfish share of the TACs.

DFO officials had already begun thinking about the possible application of individual or company quotas. Senior DFO officials had agreed in 1980 that individual quotas be tried in selected Atlantic fisheries, e.g. northern shrimp and offshore scallops. Because of the race for the fish in the 1981 offshore groundfish fishery, DFO developed a proposal for the introduction of company quotas in this fishery.

Some authors have suggested that the move to enterprise allocations in 1982 was primarily the product of initiatives taken by the four large offshore companies (e.g. Gardner 1988). On the contrary, some company officials were reluctant to embrace the concept of individual company allocations.

Despite reluctance on the part of the senior management of some companies to test new approaches, DFO officials continued to encourage the offshore companies to agree on a system of company quotas on a trial basis for the 1982 fishery. Failing this, the companies faced much more stringent controls on effort aimed at spreading out the fishery on a year-round basis. On September 22, 1981, DFO requested the four major offshore companies to develop a set of principles to introduce enterprise allocations in 1982. The companies were also asked to determine among themselves the division of the offshore quota into enterprise allocations.

Company representatives met on several occasions during the fall of 1981. The four major companies by now agreed that an implemented trial of enterprise allocations in 1982 would be useful and on the general principles to govern the trial. However, they could not agree on the stock by stock allocations. Nevertheless, three of the companies, The Lake Group Ltd., National Sea Products Ltd. and H.B. Nickerson and Sons Ltd., did agree on a proposal for enterprise allocations in 1982. This was submitted as a Working Paper to DFO in November, 1981 (Moores et al. 1981).

The central theme of the three-company proposal was that "the 1982 allocations should reflect and preserve, to the greatest practical extent, the recent shares achieved by each enterprise of the offshore catch of each species." Tentative enterprise allocations were calculated for every stock based on catch performance, largely over the 1977–1980 period.

The four companies disagreed on how to divide the offshore quota of northern cod. Fishery Products Ltd. argued that the 1982 quotas should be determined by some fixed formula applied stock by stock to the 1971–1980. This would have given Fishery Products a larger share of northern cod than had been the case since 1979. (Fishery Products Ltd. share had been 82.2% in 1976 but by 1980 had declined to 50.0%). The other three companies argued that 1982 enterprise allocations "must take account of the forced eastward shift of the Nova Scotia and Gulf entrance trawler fleets, and, in particular, the expulsion from the Gulf and Sydney Bight in 1981 of vessels over 1,050 horsepower."

Thus the major obstacle to implementing enterprise allocations on a trial basis in 1982 was the disagreement on how the northern cod quota should be shared. Given the anticipated growth in the northern cod stock, this was a crucial decision. DFO agreed to arbitrate on this issue. Following presentations by company representatives, it decided to allocate the northern cod offshore quota as follows:

TOTAL	FP	LG	HBN	NSP
Tonnage: 83,750	39,500	11,000	7,500	25,750
Percentage:	47.16	13.13	8.96	30.75

With this decision the agreement to put in place enterprise allocations on a trial basis for 1982 fell into place.

The principles and procedures for the 1982 experiment were described in a document titled *Rules for the Administration of Enterprise Allocations for the Atlantic Offshore Groundfisheries during the 1982 Trial Period* (DFO 1981m). The enterprise allocations trial applied to the four large offshore companies. Separate allocations were set aside for trawlers over 100 feet owned and/or operated by other than these companies ("others" later to be known as the "Independent Offshore Group", the IOG). Fishing for these allocations by vessels in the "others" group was on a competitive basis as before.

Reaction to the 1982 trial was positive. Both the government and the industry participants realized the program was a significant improvement over competitive fishing. Industry felt that the program should be extended into 1983. However, such a system required a stable policy and financial environment so that each new modification could be tested and its implications understood. Such stability was missing as the 1983 season approached. The four companies involved in the trial enterprise allocation program were in severe financial difficulty. The Task Force on Atlantic Fisheries was examining options for restructuring the industry to make it more viable. Because of the uncertainties surrounding the impending restructuring, the new Minister of Fisheries and Oceans, Pierre De Bané, decided to postpone formalizing enterprise allocations until decisions were made on the reorganization of the industry.

In December 1982, DFO officials met with company representatives and requested the submission of "orderly fishing plans". The companies agreed that, in lieu of enterprise allocations, they would follow such fishing plans, which would be submitted to the Department quarterly for "fine-tuning" and monitoring. In effect, the trial enterprise allocations program was continued into 1983 unofficially. (Telex, Minister De Bané to the offshore companies, December 30, 1982).

By the fall of 1983 the form of restructuring was becoming clearer (see Chapter 13). Negotiations commenced with industry representatives, including the Independent Offshore Group, to put in place a 5-year trial Enterprise Allocations Scheme beginning in 1984. DFO developed, through a series of consultations with industry representatives, a framework for Enterprise Allocations. This policy (Parsons 1983) included several elements: *first*, a statement describing how the Total Allowable Catch (TAC) would be set; *second*, a process to establish the Canadian portion of the TAC; *third*, a clear definition for deriving offshore quotas as a percentage of the total Canadian portion of groundfish TACs; *fourth*, principles for assigning enterprise allocations within the offshore quota; and, *fifth*, rules for administering enterprise allocations (involving vessel replacement, sale or transfer of quota among enterprises and licence fee structures). The first 5 years, 1984–88 inclusive, would be a transitional period during which the traditional approaches would be replaced with new guidelines based on experience with the new system.

The 1984–88 Enterprise Allocations trial was built upon the foundation laid in 1982. However, there were several new elements. Key among these was the establishment for the first time of inshore/offshore shares on a percentage basis (see Chapter 7). Establishing these percentage shares was a prerequisite for a 5-year EA trial for the offshore companies. A second major step was the decision to establish enterprise allocations as percentages of the offshore quotas, rather than as absolute volumes.

Laubstein (1990) defined an enterprise allocation in the Atlantic offshore groundfish fishery as "a specified amount of a particular species and stock of groundfish which an individual fishing enterprise may harvest during a given period of time — the right to harvest a specific quantity of a given groundfish stock." This is correct in so far as it goes but ignores the important point that the establishment of percentage allocations for particular species and stocks was a key element of the 1984–88 trial. From the TAC, the percentage allocated to the offshore, and the percentage allocated to an individual enterprise, the allocation for a given enterprise for a particular year and stock was determined.

Another major element of the 1984–88 trial program was the application of enterprise allocations within the IOG for the first time in 1984. These enterprises agreed by mutual consent to establish their allocations in volume terms for 1984 only, expecting that EAs within this group would be formalized in percentage terms for subsequent years. The introduction of EAs in the IOG was a significant step forward. To achieve this, an increased allocation from the offshore quota of certain stocks was necessary to facilitate the transition. This involved protracted negotiations with the Offshore Vessel Owners Working Group. Agreement was secured on a modest increase in the IOG northern cod allocation from 8,500 tons in 1983 to 12,500 tons in 1984. Twelve percent of the Canadian offshore quota for northern cod was also earmarked for the IOG for each of the next 5 years. These measures made possible industry-wide acceptance of the Trial Program for Enterprise Allocations for 1984–88.

Restructuring of the major companies, and the addition of the IOG as participants, resulted in allocations for nine major species and 34 stocks for 18 companies with a total of 200 EAs. With changes in the total TAC for groundfish and changes in various species and stocks, the overall share of the various companies fluctuates over time. The volume and percentage of enterprise allocations of the two major companies and the IOG as a group changed from 1982 to 1990 (Table 9-2). Together, Fishery Products International and National Sea Products account for about 80% of the offshore groundfish allocations with 13 smaller companies sharing the balance. FPI's overall percentage of the total enterprise allocations increased from 40.4% in 1984 to 43.7% in 1987 while NSP's percentage decreased from 43.7% to 38.3%. With decreases in the TACs for the flatfish stocks on the Grand Banks and northern cod, FPI's overall share decreased to 37.5% in 1990 while NSP's share increased to 39.5%. Actual volumes decreased substantially for both companies from 1986 to 1990 and declined even further with the closure of the northern cod fishery during 1992.

The 1984–88 EA program did not confer substantive property rights upon the offshore companies. Percentage allocations were established for 5 years but the actual enterprise allocations were determined annually. Although the 1983 policy framework (Parsons 1983) envisaged the sale or permanent transfer of EAs and included draft provisions for such sales or transfers, implementation of these was deferred pending further consultation with the industry and the provinces. The provinces were generally opposed to such provisions and no sale of allocations was permitted during the 1984–88 period. However, EAs were divisible and temporarily transferrable during the fishing season. Such transfers amounted to about 200 individual transactions involving a total of about 25,000 tons of groundfish each year. This was approximately 7% of total annual groundfish landings by the offshore fleet. In effect, an EA constituted a harvesting right for any particular enterprise, i.e. the right to harvest a specified share of the common property resource, not a property right.

TABLE 9-2. Allocations of groundfish under the Enterprise Allocation Program, by company, 1982–1990 (000s of tons).

Year	NSP MT	%	FPI MT	%	Other Companies MT	%	Total EA's MT
1982	195.3	45.3	178.8	41.4	57.5	13.3	431.6
1983	203.8	43.8	194.3	41.7	67.7	14.5	465.8
1984	208.6	43.7	193.0	40.4	75.9	15.9	477.5
1985	189.8	40.3	198.5	42.1	82.3	17.5	470.6
1986	182.4	38.7	205.2	43.6	83.5	17.7	471.1
1987	172.3	38.3	196.9	43.7	81.3	18.0	450.5
1988	170.6	39.6	183.3	42.5	77.1	17.9	431.0
1989	152.3	39.1	156.3	40.1	80.8	20.8	389.4
1990	123.2	39.5	117.0	37.5	71.5	22.9	311.7

The 1984–88 EA trial program allowed some freedom in the choice of technology. The foot-for-foot replacement rules were dropped in favour of limits applying only to the catching capacity of the offshore vessels. The catching capacity of replacement vessels could not be greater than the capacity of the vessels replaced, interpreted as the average catch of that vessel for the previous 3 years. Two or more older vessels could be replaced by one larger one provided that the catching capacity of the replacement unit did not exceed the catching capacity of the replaced vessels.

Several changes were made to the program in 1986. Companies within the IOG were assigned EAs as percentage shares. A formula for sharing stock declines among companies was established, which provided for declines to be shared in proportion to companies' percentage shares by stock. The major change in 1986 was the introduction of access fees to replace the traditional individual vessel licence fee system. The existing vessel licence fee system was not related to the value of enterprise allocations a company was entitled to fish, nor to any rent the resource might generate. The vessel licence fees ranged from $400 for a side trawler to $2,500 for a larger stern trawler. This system of fees was intended to recover the administrative costs of licensing.

The new system of access fees introduced in 1986 was related to the relative values of the species and the magnitude of the enterprise allocations of a particular species assigned to each company. Species were assigned an index value relative to that assigned to cod.

A predetermined amount to be collected was established and then the access fee by species for each company was determined according to the following formula:

$$\frac{|\{ EA_s \times I_s \}|}{|\{ EA_t \times I_s \}|} \times V$$

where

$\{EA_s$ is the sum of the allocations (tons) of species S;
I_s is the index number for species S;
$\{EA_t$ is the total allocation (tons) of species S for all companies; and
V is the predetermined amount to be collected.

For 1986, the amount to be collected was set at $289,300, equivalent to the amount paid as licence fees in 1985. In 1988 this was increased to $2.1 million as part of the government's cost recovery program. Subsequently, the amount collected declined to $1.4 million in 1991 and 1992, reflecting the decline in offshore groundfish allocations as Total Allowable Catches for certain stocks were reduced.

3.2.2 Impact of the Offshore Groundfish EA Program

Laubstein (1990) described the principal advantages of fishing under the EA system:

1. It eliminated the "race for the fish" which often leads to market glutting, poor product quality, overcapitalization and other negative effects of competition for the limited common property resource;
2. It provided for greater operational flexibility and coordination, and thus for a more efficient, cost-effective and integrated overall process of harvesting, processing and marketing;
3. It promoted greater market responsiveness by the fishing industry in terms of better product quality, more effective coordination of fish supplies with market demand and product development in general;
4. It provided for more effective long-term planning by the fishing enterprises in terms of capital expenditure, and market development and acquiring equity; and
5. It reduced the need for government regulation of the fishery since a full-fledged EA scheme is largely self-regulating.

It is difficult to quantify the impact of the EA program during the 1984–88 period because its effects are compounded by other factors which influenced the industry. During this period there were significant increases in groundfish prices and decreases in interest rates and fuel prices. There were also favourable exchange rates, changes in the geographic availability of resources, the major restructuring and refinancing of the offshore companies. And there were major changes in management structure and senior personnel. All of these factors contributed to the resurgence of the offshore groundfish fishery.

Nonetheless, the EA trial program for the offshore fishery improved company practices. The major motivation for introducing the EA program was to end the competitive "race for the fish". The individual companies no longer had any incentive to maximize their share of the offshore quota. With EAs, companies gained flexibility as to when, how and if to harvest their allocations during a given year. This probably contributed to a spreading out of offshore landings throughout the year, resulting in a more consistent supply of groundfish to the market and a better match of harvesting patterns with processing capability. Gardner (1988) argued that other constraints prevented a more marked change in fishing patterns.

The elimination of the incentive to maximize shares is illustrated by the reduction in the number of offshore fishery closures during the trial EA program. In 1981, prior to the first trial of EAs, there were 80 closures of the offshore fishery (where the quota was caught before the season ended and the fishery closed by DFO). By 1987, there were virtually no closures in the offshore fishery. This contrasted with more than 200 fishery closures in 1987 in the inshore fishery which was still largely subject to competitive fishing pressures.

The EA program contributed to fleet rationalization through improved trawler utilization, vessel modernization and capacity reduction. Between 1982 and 1987 the number of vessels in the offshore groundfish fleet decreased from 140 to 129 and the overall fishing effort in terms of total annual fishing days decreased by close to 10% (Laubstein 1990). The fleet reduction of the "big two", FPI and NSP, was 14 and 16%, respectively. This apparent reduction in harvesting capacity and effort occurred when buoyant market conditions would have exacerbated overcapitalization under normal competitive fishing pressure. However, these results have to be interpreted cautiously. With replacement of the less efficient side trawlers by stern trawlers, a drop in the number of sea days does not necessarily indicate a real decrease in effective fishing effort.

Gardner (1988) concluded that the number of active vessels of these two companies had dropped from 123 in 1982 to 99 in 1987. Overall harvesting capacity (using carrying capacity as a proxy) declined from 28,940 to 25,660 tons, or by 11%.

The number of companies within the Independent Offshore Group decreased from 18 in 1982 to 11 in 1988. However, the number of vessels remained at 16 and overall capacity of the IOG fleet increased.

Another significant benefit of EAs was a more uniform utilization of fish processing capacity. The companies involved in the EA program apparently became market-driven rather than volume-driven, as was the case in the pre-EA era. They placed increased emphasis on fresh fish sales and the production of premium products. Harvesting operations were integrated with the companies' processing and market operations.

Gardner (1989) used National Sea Products for a case study of the impact of EAs. He concluded that, "Virtually all aspects of NSP, from harvesting through processing to marketing, have been affected by the EA system." The most profound change appeared to be one of orientation, with market demand rather than pressure from competing fleets forming the basis for fleet operations. NSP had apparently shifted its objectives from share maximization to quality and efficiency. NSP had a four-part strategy to achieve these objectives: 1. improving the mix of vessels by retiring redundant capacity (mainly the less efficient side trawlers) and adding the factory freezer trawler; 2. quality-based investment in the remaining fleet; 3. modifying fishing practices; and 4. introducing a quality-based pricing system (Gardner 1989).

EAs in the offshore groundfish fishery had definite positive effects during the 5-year trial introduced in 1984. Although EAs constituted constrained harvesting rights rather than property rights as envisaged by economic theorists, their benefits were clear. They ended the race to maximize shares of the offshore quota, with its negative effects in terms of fish quality, gluts and production of low-grade market products. There was some fleet rationalization both in terms of a more efficient fleet configuration and some reduction in harvesting capacity. The shift from a volume-driven to a market-driven fishery and the integration of harvesting, processing and marketing operations helped foster a more efficient fishery with minimal job loss. Perhaps the most significant advantage was some stability in all facets of the offshore fishery.

This system could be introduced in an orderly fashion because of extensive consultation among federal and provincial governments and industry. Industry's recognition that a new approach was required and their willingness to commit to a trial scheme of enterprise allocations were major factors in its success. The government proceeded cautiously, making modifications as required. There was no massive overcapacity problem in the offshore sector at the time the program was introduced. While there was potential for a significant increase in harvesting capacity as vessels were replaced, this was averted by taking preventive action. The relatively small number of participants (enterprises) also facilitated implementation of the EA system.

Industry generally complied with the program's requirements. Although misreporting was less extensive than in the Bay of Fundy purse seine fishery, it was nonetheless significant (Fraser and Jones 1989). In the early years of the trial, there were reports of high-grading through excessive discarding of small fish at sea. CAFSAC (1986a) estimated that discards in the offshore fishery for northern cod increased from 7.2% by number in 1981 to 24.4% in 1986. Government intervened by placing observers on 50% of the offshore trawlers. However, there were continuing reports of discarding.

In 1987 a wide-ranging review of the EA system was undertaken. This resulted in widespread agreement by both the federal and provincial governments and the industry participants that the program should be continued beyond 1988. On December 30, 1988, Fisheries Minister Tom Siddon announced that the program of enterprise allocations in the offshore groundfish fishery would become permanent in 1989 (DFO 1988e).

Perhaps the most significant aspect of the Minister's statement was his clarification of what enterprise allocations would mean in the future:

> "Groundfish remains a common property resource. What the offshore companies gain under the EA program is a harvesting privilege for a specific share of the resource subject to scientific advice and Ministerial decision."

3.3 Western Newfoundland Inshore Otter Trawl Fleet

The first experiment with enterprise allocations in the Atlantic inshore groundfish fishery occurred in the otter trawl fleet based at ports along the west coast of Newfoundland. There are two major cod stocks in the Gulf of St. Lawrence, the northern Gulf cod which frequents Divisions 4RS and Subdivision 3Pn and the southern Gulf cod, which frequents Divisions 4T and Subdivision 4Vn in the winter. The inshore otter trawl fleet of western Newfoundland consisted of about 100 vessels ranging in length from 40 to 65 feet in the 1980's. It fished primarily the northern Gulf cod (approximately 30,000 tons) and shrimp (6,000 tons) in areas adjacent to their home port. From 1984 to 1989 the cod fishery by this fleet involved a system of enterprise allocations. Quotas elsewhere in the Gulf which this fleet was entitled to fish were fished competitively with vessels from other provinces.

Prior to the introduction of enterprise allocations, vessels of this fleet competed in the classic race for the fish, commencing fishing along the southwest coast in February and moving northward along the west coast in late spring and summer. The highest catch rates were experienced in winter to early spring. This, coupled with higher-than-average prices at that time of year, triggered a mad scramble to maximize individual shares of the available quota. This resulted in seasonal gluts, poor quality and a short harvesting and processing season. Vessels tended to operate only 8–10 weeks a year.

The Department of Fisheries and Oceans issued an additional 39 inshore otter trawl groundfish licences in 1981–82. This was to placate certain fixed gear fishermen. However, it reduced the catches and incomes of existing otter trawler operators. Competition stimulated acquisition of larger, more mobile vessels capable of extended fishing trips. By 1983 it became apparent that the economic viability of the existing fleet could not be maintained unless the resource was more equitably distributed. The fleet was divided into four sectors and separate quotas established for the winter and summer fisheries. This set the stage for consideration of individual vessel quotas.

The vessel owners were members of one organization, the Newfoundland Fishermen, Food and Allied Workers Union (NFFAWU). This made it easier to suggest enterprise allocations. The Union was initially concerned that crew members might be adversely affected if vessel owners could freely transfer individual allocations. However, the parties did agree on a rudimentary system of enterprise allocations.

A 3-year pilot project was introduced (for 1984, 1985 and 1986). The program was voluntary and had no provision for transferability in the initial phase. The program's objectives were:

1. To improve the quality of fish landed by the fleet;
2. To extend the harvesting season, thus maximizing employment potential in the processing sector;
3. To reduce the competitiveness that existed in the fishery in 1983 and reduce the risk of small vessels competing with larger vessels;
4. To assist vessels within the fleet to become economically viable over the long term; and
5. To reduce or eliminate fish processing gluts (DFO 1983f).

The four fleet sectors established in 1983 were maintained. These sectors were:

- E-2 : owners/operators of vessels holding both shrimp and groundfish licences and residents of Division 4R
- D-2 : owners/operators of vessels holding only groundfish licences and resident in Division 4R
- D-3 : owners/operators of vessels which acquired mobile groundfish licences in 1981-82
- D-4 : owners/operators of vessels holding groundfish licences and resident in Subdivision 3Pn

Different levels of enterprise allocation were negotiated for each category. Because of the late entry of some vessels into the fishery and the poor performance by some over the preceding 2–3 years, individual fishermen/vessels were given allocations based not on individual historical performance but on the average landings in their length class by fleet sector.

The 29,500 ton quota for otter trawlers less than 65 feet was subdivided by fleet sector:

	Quota	No. of Licensed Vessels	No. of Active Vessels
D-2	5,713 tons	19	17
D-3	8,930 tons	38	35
D-4	2,712 tons	8	7
E-2	12,165 tons	42	40
Total	29,500 tons	107	99

There were different allocations for various length intervals within each sector in accordance with the wishes of fishermen within that sector.

A large majority of fishermen favoured equal quotas for vessels of similar size. However, owners of highliner vessels opposed this approach. Hence, it was agreed that any surplus fish (uncaught quotas) would be reallocated at various times during the year. Superimposed on this complex array of EAs was a set of seasonal restrictions, which ensured that 60% of the quota allocated to vessels in the northern areas could not be harvested until after May 1.

Fishermen were generally wary of the concept of transferability. Therefore, the buying, selling or trading of EAs was not permitted during the trial period. Fishermen reacted favourably to the first year. As a result, DFO eliminated the former fleet quota distinction. In 1985 allocations were standardized by vessel size.

In 1986 the Department intended, in consultation with fishermen, to develop a permanent program. However, reductions in the Canadian quota for 4RS-3Pn cod from 86,500 tons in 1985 to 78,600 tons in 1986 and an accompanying decrease in the portion set aside for the EA project from 29,500 to 26,500 tons had a significant impact on the next few years of the EA program. With further TAC reductions projected, (the Canadian quota dropped to 73,900 tons by 1989 even with the phase out of the Metropolitan France fleet from the Gulf), fishermen were reluctant to make a long-term commitment to a permanent EA program. Because the results of the 3-year pilot project were encouraging, fishermen agreed to continue the pilot program for 1987 and 1988. The approach was essentially that adopted in the pilot project.

There was one significant change in 1987. New provisions allowed fishermen who had lost their vessels due to natural causes, or whose vessels were not operational for reasons beyond their control, to trade 100 tons of their individual allocations to other fishermen. This represented a significant portion of individual vessels' allocations. Seasonal restrictions were maintained.

Weekly trip limits of 100,000 lb were also maintained. Uncaught quotas could still be reallocated within the fishing season, with owners/operators relinquishing any fish remaining in their allocations by August 15th.

This Western Newfoundland experiment with EAs for the inshore otter trawler fleet combined some of the features of a theoretical individual-quota regime with more traditional management measures. The latter included seasonal restrictions, trip limits, and limits on transferability, and choice of technology. Despite the obvious limitations, fishermen liked the program. The major advantage lay in reducing the conflicts which had prevailed in the competitive fishery. Quality was improved, the fishing and processing seasons lengthened and gluts reduced (Fraser and Jones 1989).

The fishing season lengthened from an average of 9 weeks in 1983 to 20 weeks in 1986, despite a quota reduction of 3,000 tons between 1985 and 1986. The average operating costs of the fleet declined from $27,309 to $21,035 between 1983 and 1985, with average total expenses (including gear and maintenance) declining from $51,924 to $42,740 (Fraser and Jones 1989).

There was no provision for transferability during this program except for limited trades in exceptional circumstances in the latter years of the experiment. Fishermen remained reluctant to accept the concept of transferability.

This program evolved slowly over time in accordance with the willingness of the participants to accept change. The program was managed by a committee with representatives of the vessel owners, processing sector and both levels of government. Thus the fishermen themselves to a large extent shaped the program.

The program's structure prevented significant fleet rationalization. Some steps were taken to tighten licensing restrictions to encourage fleet reduction. During the pilot project seven vessels were withdrawn from the fishery. A buoyant market, with high fish prices, made it attractive to remain in the fishery.

High-grading was one major problem which was not publicly acknowledged at the time. This involved the discarding of small cod and consequent underreporting of the catch. Although there has been no quantification of the extent of this problem, there have been widespread reports of significant underreporting. One senior Newfoundland government official termed the project "an unmitigated disaster" because of the perceived level of discarding and underreporting. Although no data are available on the extent of underreporting or misreporting, anecdotal evidence indicates that the underreporting problem was substantial (J. Jones, DFO, pers. comm.). Fishermen also reported that up to 100% of the fleet was using small mesh gear or legal size gear with liners and discarding large quantities of small fish at sea.

By the fall of 1988 fishermen in western Newfoundland and southern Labrador began to discuss with DFO major changes to the program. They were prepared to contemplate a more permanent program with some form of transferability. By this time discussions were underway regarding the establishment of enterprise allocations throughout the Gulf of St. Lawrence small groundfish dragger fleet. Revision of the western Newfoundland program was subsumed in a broader Gulf-wide initiative (see next section).

3.4 Other Atlantic Fisheries

The perceived benefits of enterprise allocations in the offshore groundfish fishery and in the inshore otter trawler fishery of western Newfoundland fostered the acceptance of enterprise allocations by participants in certain other Atlantic fisheries. Enterprise allocations were introduced in the offshore lobster fishery in 1985, in the offshore scallop fishery in 1986, in a developing offshore clam fishery in 1987 and in the northern shrimp fishery in 1987. The introduction of EAs in the northern shrimp fishery was facilitated by a boat quota system in effect in this fishery since the early 1980's. An EA program has also applied since 1987 to offshore bluefin tuna and swordfish taken as by-catch while fishing for unregulated tuna species.

These programs have not been in place long enough to determine their effectiveness. One common feature is the relatively small number of enterprises involved in each of these fisheries. The experimental offshore clam fishery involves only four enterprises. The offshore scallop fishery which is a longstanding fishery of major significance in Nova Scotia was placed on EAs in 1986 for a 3-year trial period. At the start of the program 10 enterprises were involved. This was reduced to seven. The northern shrimp fishery originally involved eleven licensees when limited entry was introduced into this developing fishery in 1978. A twelfth licence was later issued in 1980 to Northern Quebec Inuit. In 1987 four new licences were issued, for a total of 16 licences. At the same time a 2-year experimental EA program was introduced.

The offshore scallop EA program was modelled on the offshore groundfish EA program. Enterprises were assigned percentage shares of the TAC. An individual enterprise's share was weighted 50% by the relative share of the total number of licences held by the enterprise and 50% by the relative catch history of the various enterprises. In the northern shrimp EA program each of the 16 licences was assigned a 1,000 ton share of the TAC. The offshore scallop and northern shrimp programs were similar to the offshore groundfish trial EA program in prohibiting permanent transfers but permitting intraseason transfers between enterprises. Existing vessel replacement provisions were maintained. One interesting feature of the offshore scallop program was a specific list of recommended penalties for violations of scallop fishing regulations, agreed to by the participants.

The EA program in the offshore scallop fishery may have increased the economic benefits from the resource (DFO 1991c). EAs helped reduce fishing costs and increase fishing returns. Fishing effort declined while total scallop landings increased. Between 1983 and 1987 the number of active fishing vessels dropped by more than 16% and the catch per sea day almost doubled. This increased gross and net revenue per vessel and average crew earnings.

Individual vessel quotas have been applied in the inshore snow crab fishery of Cape Breton since 1979 and in the inshore snow crab fishery along the north shore of the Gulf of St. Lawrence (Quebec and Newfoundland) since 1986. In April 1990 a pilot individual vessel quota program was introduced for the midshore crab fishery in the Gulf of St. Lawrence. This program responded to stock problems and declining snow crab prices. The principal objectives were more orderly harvesting and resource conservation.

In the Atlantic groundfish fishery, enterprise allocations were extended on a trial basis to the midshore fleet (vessels between 65 and 100 feet long) in 1988, for certain stocks, and in 1989 to mobile groundfish vessels less than 65 feet throughout the Gulf of St. Lawrence. The process for establishing EAs in the middle distance groundfish fishery closely resembled the approach used in the offshore groundfish fishery. EAs for 1988 were established as percentages of the middle distance gear sector quotas in key stocks. The share of the 65–100 foot fleet quota available to each enterprise in the fixed gear portion of the fleet was determined on the basis of past fishing performance (catch history) and the number of licences held by each enterprise. EAs for the mobile gear sector were based on a formula using the average catch histories of each vessel during the period 1982 to 1986. Vessels (licensees) possessing both a groundfish and shrimp licence or a groundfish and crab licence were not assigned enterprise allocations. This experimental approach to EAs applied only to certain key stocks. Several fleet sector quotas for other than designated EA stocks remained open to competitive fishing by all vessels in the particular fleet. Provisions for transfers were similar to those in effect in the offshore groundfish fishery. Enterprises participating in both the offshore groundfish EA program and the middle distance groundfish EA program were permitted to make temporary transfers between the enterprise's offshore groundfish allocations to its middle distance allocations but not vice versa.

The introduction in 1989 of a trial program of enterprise allocations for mobile groundfish vessels less than 65 feet in the Gulf of St. Lawrence in effect extended the western Newfoundland program to the rest of the Gulf. The new program involved 192 vessels from western Newfoundland, Quebec, New Brunswick, Nova Scotia and Prince Edward Island. Virtually all participants supported the program. This program emphasized the need for adequate monitoring and control. Fishermen and processors in the Gulf agreed to a system which included a special agreement for monitoring, reporting, penalties, revalidation of licences and an audit program. For 1989 DFO paid the direct costs of monitoring and control. Each enterprise agreed to sell its catch only to those fish buyers who had confirmed in writing to DFO that they would accept independent audits of their operations. (The agreement also provided for sanctions including reductions of EAs in subsequent years). These provisions exemplify the concern about possible misreporting of catches and the increased enforcement costs required to monitor a complex EA program. An audit process was developed to detect possible collusion between fishermen and buyers. This was to be tested first in the winter fishery in southern Newfoundland. DFO also increased its observer coverage on that fleet to better monitor the discarding at sea.

The reaction to this program in 1989 was favourable. In January 1990 a ten-year EA program with provision for permanent transfers and combining of licences within certain constraints was implemented. A dockside weighmaster program was established to monitor catch reporting. A registered company operating at arms length from local buyers, fishermen and processors was established to hire, train and deploy DFO certified weigh-masters in an attempt to ensure accurate reporting of landings. For the 1990 fishery funding was provided by the Job Development Program of CEIC. Thus was born the first Individual Transferable Quota (ITQ) system for groundfish in Canada (DFO 1990g).

In 1991 a mandatory dockside monitoring program was extended to all fleets of groundfish trawlers, midshore crabbers, shrimp vessels and large herring seiners in the Gulf. The costs of this program were to be recoverable from fishermen, buyers and processors. For the 1991 fishery the estimated cost of the program was $410,000 for processors and $660,000 for fishermen (B. Rashotte, DFO, Ottawa, pers. comm.). Because of declines in the northern Gulf Cod quota, fishermen and buyers in western Newfoundland strongly opposed the cost recovery aspect of the dockside monitoring program.

Under the 1991 Groundfish Plan individual quotas were introduced for the majority of groundfish draggers under 65 feet in the Scotia-Fundy Region (DFO 1990h). Included were about 450 of the 2800 vessels licensed for groundfish. Main features of the new system were:

- Individual allocations were based on historical catch: an average of the best two of the 4-year period 1986–1989.
- Operators with poorly documented or no catch in the base period could opt for minimum allocations of 7 tons for vessels under 45 feet and 50 tons for vessels between 45 and 65 feet long.
- The program applied only to cod, haddock and pollock.
- Draggers based in southwest Nova Scotia and southwest New Brunswick were obliged to take part.
- Generalists (small "day" draggers) could opt out of the program.

Mindful of the incentives to misreport, DFO introduced closer monitoring and stricter penalties to support this individual quota system. The new Commercial Catch Monitoring System was similar to that already implemented in the Gulf of St. Lawrence fishery for groundfish and snow crab. Some of the main features included:

- use of a new combined log book at sea and purchase slips;
- construction and staffing of four operations Centres to receive hails and process catch and landings data;
- a requirement for fishermen to "hail" their catch to the operations centres from sea or before any groundfish is offloaded;
- port monitors to provide 100% coverage for groundfish landings including checking the accuracy of scales, visually inspecting the catch and watching the offloading;
- use of independent audit teams to examine the books and records of fishermen, companies and plants.

This monitoring system applied to all groundfish vessels using mobile gear and to fixed gear vessels greater than 45 feet.

Thus in the late 1980's the concept of enterprise allocations was extended to a variety of fisheries in Atlantic Canada with a "go-slow" approach. These trials should clearly demonstrate the advantages and disadvantages of EAs for intermediate sized vessels. These programs have been in operation for only a short time. Thus it is premature to comment on their effectiveness in achieving the benefits theoretically possible under such a management scheme. However, by 1992 there was widespread concern about pervasive dumping and discarding of small fish in many of the groundfish fisheries.

4.0 THE PACIFIC FISHERIES

Unlike the Atlantic fisheries where innovative approaches to enterprise allocations have been introduced in several major fisheries since the early 1980's, individual quotas on the Pacific coast have been tried only in five small-scale fisheries — abalone, herring spawn-on-kelp, herring food and bait, geoduck and sablefish — and one major fishery — halibut. The geoduck, sablefish and Pacific halibut programs are very recent (1989 for geoduck, 1990 for sablefish and 1991 for halibut).

4.1 Abalone

Abalone are snails with a low spiral shell. The large muscular foot on which the abalone crawls is the part eaten. The main stocks are subtidal, and since they live on rocky shores and reefs, they can be effectively harvested only by diving. The fishery for abalone is small-scale; 42 tons were harvested in 1985 worth slightly less than half a million dollars. Catches and landed values declined after 1978 when 433 tons worth $1.9 million were harvested. The catch in the late 1980's was only one tenth of that in 1978. During the same period the landed price increased more than three-fold from $3.15 per kilogram in 1976 to $10.52 in 1985 and then doubled to $22.42 in 1988. This was the result of the rapid depletion of accumulated, slow-growing stocks (Bates, MS 1985).

Prior to 1976 the small abalone fishery was largely unregulated. The average annual catch between 1965 and 1971 was less than 8 tons. Landings increased dramatically to 273 tons in 1976, a fivefold increase over 1975. This sparked the introduction of limited entry licensing in 1977. Individuals were permitted only one licence, which was nontransferable. The effort and catch continued to increase in 1977. Some areas were closed and reductions in the length of the fishing season implemented. A further increase in the catch in 1978 led to a decision to implement a Total Allowable Catch of 227 tons in 1979. The TAC was subdivided into a portion to be fished competitively (113 tons) and a portion allocated equally among the 26 licence holders. In 1980 it became necessary to reduce the TAC. The entire TAC was allocated to individual quotas, with each licence allotted a nontransferable quota of 4.5 tons.

This individual quota program was continued in subsequent years, with TACs and individual quotas being progressively reduced. To ensure compliance with the individual quota regime, each licence holder was required to check in with the appropriate District office of DFO 24 hours prior to off-loading the catch. A fishery officer inspected the landings and checked the weight before the next fishing trip could begin.

Precipitous declines in the resource and catch during the 1980's render suspect any attempt to assess the benefits of the individual quota system in this fishery. It has been suggested that recruitment may be fluctuating independently of stock size. There are insufficient data to carry out stock assessments (Breen 1986; Sloan and Breen 1988).

An internal DFO study (Jacobson, DFO, Pacific Region, pers. comm.) estimated that illegal fishing effort on this resource was equal to the existing legal fishing effort. Of this, it was estimated that 60% of the illegal effort was due to poaching by non-licence holders and 40% to over-harvesting of individual quotas. Individual quotas in this fishery proved no more successful at halting or reducing illegal harvesting of abalone than the previous management regime.

Unfortunately, this experiment with individual quotas was aborted in November 1990 when DFO announced closure of the abalone fisheries at the end of the 1990 season. The Director General of the Pacific Region of DFO, Pat Chamut, stated that "serious depletion" of abalone stocks required closure of the commercial fishery and a ban on harvesting by recreational and native fishermen (DFO 1990i). The abalone quota had remained at 47 tons from 1985 to 1990.

Scientific studies indicated low levels of recruitment. Stock levels appeared to be only 20% of the levels observed during the 1970's and an even smaller fraction of pre-fishery abundance.

Chamut indicated that the closure was likely to last for 5 years or "until abalone stocks are rehabitated" (DFO 1990j).

4.2 Herring Spawn-on-Kelp

Pacific herring spawn in the intertidal zone. The fishery which has evolved since the recovery of stocks in the early 1970's currently has three major components. These are a food herring fishery (mid-November to mid-December), a roe fishery (March to mid-April) and herring spawn-on-kelp fishery (mid-March until the end of May).

Development of the commercial spawn-on-kelp fishery has been closely regulated since its inception in the early 1970's. Prior to that, native Indians along the British Columbia coast traditionally harvested herring spawn on marine plants as a source of food. The rate of development of the commercial fishery was tightly controlled to maintain high product prices and stabilize market share.

Limited entry licensing was introduced to this fishery in 1975. From 1975 to 1977 spawn-on-kelp licences were awarded through a rating system that favoured applicants who had previous experience in catching and holding live herring, residents of remote coastal communities, and Native Indians. In 1978, the number of licences was increased to 28 with the issuance of a few new licences to Indian bands. Thereafter the number

Herring Spawn-on-Kelp.

of licences was frozen at 28 until 1991. These licences are nontransferable privileges extended to individuals or bands. Each licence entitles the holder to produce a fixed amount of product measured in cubic metres. A holder of a spawn-on-kelp licence cannot hold a roe herring licence. Spawn-on-kelp must be harvested in enclosures.

The annual amount that could be produced by each licence holder varied between 1975 and 1987, from 5.44 to 9.07 tons. A quota of 9.07 tons applied in all but 4 years. The spawn-on-kelp licence stipulates that the amount of the individual quota can be adjusted from year to year.

Following a reduction in 1986 to 5.44 tons, by 1987 it was possible to increase the individual quota to 7.26 tons because herring abundance had returned to the average level for the decade. Market conditions for spawn-on-kelp improved from 1985 onwards. Prices paid per pound for spawn-on-kelp increased from $19.25 in 1986 to $24.35 in 1987, for example. By 1988 the spawn-on-kelp fishery had a landed value of $12.7 million. The fishery had become quite profitable for the existing individual quota licence holders, who grossed an average annual revenue above $450,000. With fees for spawn-on-kelp licences of $10 for native Indians and $2000 for non-natives, the fishery had become quite lucrative.

The potential for expansion of this segment of the B.C. herring fishery had been recognized for several years. Pearse (1982), recommended that the spawn-on-kelp fishery be expanded. This was supported by the Fleet Rationalization Committee. Subsequently, a market analysis in late 1984 suggested that B.C. production should be expanded to take advantage of a growing market with prices increasing for high quality spawn-on-kelp, and to maintain Canada's competitive position relative to production by Alaskan fishermen.

In the late 1980's there was considerable interest in expanding this fishery. By 1988 the Department had received about 260 requests from individuals for licences and 30 requests from Native Bands. Existing licence holders became concerned that their individual quotas would be reduced or that an increase in total supply would result in lower prices and diminished profitability for their operations. Given these factors, DFO considered in 1988 and 1989 maintaining the individual-quota level at 7.26 tons and issuing an additional 10 licences by increasing the overall B.C. spawn-on-kelp quota. Market analyses indicated that this would not reduce returns to existing licensees.

The question of how to allocate any additional licences was controversial. Suggestions included issuing new licences by lottery, auctioning new licences with revenue going to the Crown, and issuing new licences to native bands only. In 1991 the number of licences was increased by 10, with all of the increase going to native bands. Each was entitled to harvest 7.26 tons of herring roe. To achieve this amount of roe, about 90 tons of herring are harvested by each licence holder. The total 1990 allocation of 2360 tons represented about 6% of the total herring fishery.

Overall, the individual quota system in the spawn-on-kelp fishery has been relatively successful. Landed values are consistently high. The product is of consistently high quality. The harvesters appear to be cost efficient with increased net earnings over time. Also a high level of native participation has been achieved (DFO 1991c).

One negative note is the existence of an illegal spawn-on-kelp fishery worth up to $3 million, or roughly equivalent to an additional 8 licences.

4.3 Herring Food and Bait

Individual quotas were introduced in this fishery in 1985. In 1990 the TAC was 1125 tons, approximately 3% of the overall TAC for the B.C. herring fishery. The season is relatively short, about 6 weeks during November and December. The fishery serves markets for human food and frozen commercial fishery bait.

Two forms of individual quotas are used. In the northern sector, processing companies are given enterprise allocations. These totalled 225 tons in 1990. In the southern sector there was a lottery among vessel owners/licence holders for 20 units of 45 tons each, totalling 900 tons in 1990. The licences and quotas are non-transferable.

When this program was established in 1985, the objectives were to promote more orderly and market sensitive harvesting and to assure equality of opportunity to participate. A recent study (DFO 1991c) concluded that:

— fleet capacity was under control;
— conservation concerns had been effectively addressed;
— operators were "fishing to market";
— the use of the lottery for the southern sector had assured equality of opportunity to prospective participants.

4.4 The Geoduck Fishery

Geoducks are a species of large clams noted for their extremely large protuberant "foot" up to 1 m long, with which they burrow deeply into the substrate. Like abalone, geoducks are harvested by divers operating from vessels; the catch is delivered fresh (unprocessed) to shore facilities. Divers direct a jet of water from a surface pump through a nozzle into the substrate to loosen mud or sand around the geoduck.

The British Columbia geoduck fishery began in 1976 with landings of 44 tons. Landings reached 5,370 tons by 1985 and levelled off at around 5,000 – 5,300 tons in 1986 and 1987 (Fig. 9-2). The increase in landings between 1976 and 1980 appears to have been the result of increases in fishing effort. Higher landings from 1984 onwards resulted from the discovery of new fishing grounds and more intensive exploitation of geoduck grounds along the North Coast of British Columbia. Geoduck prices (average price per kilogram) were $0.85 during 1984–86, increased to $1.08 in 1987 and doubled to $2.09 in 1988. Landed value increased from $4–$5 million per year during 1985–1987 to more than $12 million in 1989. Average vessel income was estimated at $163,000 in 1988.

Between 1977 and 1982 vessels harvesting geoduck were issued permits by DFO. The number of permits issued peaked at 110 in 1979. Limited entry licensing

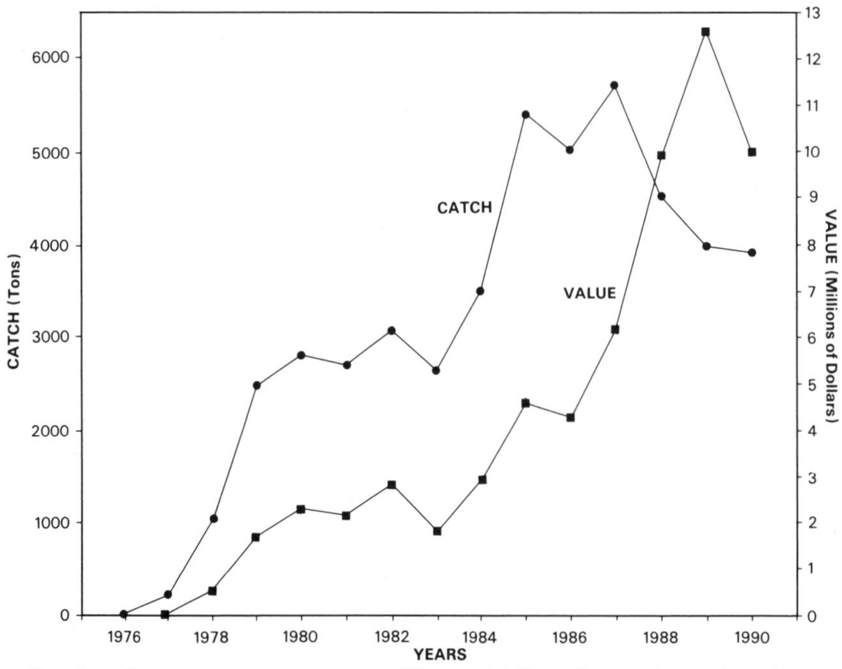

FIG. 9-2. Catches (tons) and values (millions of dollars) for geoduck: 1976–90.

Pacific Geoducks.

was implemented in 1983, with an initial 53 licences. This was subsequently increased to 55 licences as a result of successful licence appeals. Geoduck licence holders are also entitled to harvest horse clams.

Until 1989 the chief tools used for management of the geoduck fishery were limited area licensing, area quotas and control of seasonal fishing time. Under this management system fishermen had an incentive to catch as large a share of the allowable quota as quickly as possible. This increased harvesting costs and reduced profitability. Quality was also adversely affected.

Almost all geoduck licence holders are represented by the Underwater Harvesters Association (UHA). In April, 1988, the UHA asked DFO to introduce individual quotas in the geoduck fishery. The main elements of the proposal were:

- A coastwide TAC of at least 10,000,000 pounds (4536 tons) compared with the 1987 TAC of 4082 tons;
- This coastwide quota to be divided equally among all geoduck licence holders; and
- The geoduck licence would become a personal licence and the individual quotas would be transferable (Underwater Harvesters Association 1988).

DFO undertook an evaluation of the proposed system (Turris 1988). The assessment was positive, with one major reservation. Because geoduck stocks are sedentary and geoduck beds scattered along the coast, it was considered necessary to set appropriate harvest levels for different areas through area catch quotas. Some DFO officials favoured the introduction of an area licensing system similar to that applied in the roe herring fishery. The restriction on mobility would reduce harvesting costs, facilitate enforcement and minimize the incentive to discard. However, area licensing by itself would retain many of the economic disincentives of the existing management system.

The Department indicated to the UHA that it preferred a management plan combining Enterprise Allocations and Area Licensing. The UHA responded by informing DFO that 44 geoduck vessel owners (80%) would support this modified proposal.

Another DFO concern was the anticipated increased cost of monitoring and enforcing an EA/Area Licensing scheme. These increased costs were estimated as $230,000 in the first year. The UHA, determined to see this new approach adopted, responded by agreeing that industry would pay the additional management and enforcement costs.

The Department, in consultation with the UHA, developed a modified plan for Enterprise Allocations and Area Licensing in the geoduck fishery. The major elements of the plan were:

1. Each licence holder would receive an enterprise allocation equal to 1/55 of the coastwide quota;
2. The coast would be divided into three areas and licence holders must select annually the area from which they would fish their quota;
3. The number of licence holders permitted to fish in an area would be such that the sum of the licence holders' quotas for that area would not exceed the area quota;
4. Designated ports and a validation procedure would be implemented; and
5. Industry would provide the necessary funding to pay for contract observers.

This plan was adopted for a 2-year trial in 1989 and

1990 and extended for 1991, 1992 and 1993. For 1989, each licensee was assigned a quota (EA) of 160,000 pounds of geoduck clams (round, whole weight). Some flexibility in replacing current vessels with larger ones was provided. This was subject to the proviso that, should the EA/AL system be dropped after the trial period, fishermen must fish from the same vessel they fished from in 1988 or from a vessel of equivalent length to the vessel they fished from in 1988. Licence holders were not permitted to split their quotas and transfer the split quota to other licence holders. However, one vessel could fish multiple quotas.

A significant feature of the plan was the agreement by the UHA to contract port inspectors to collect biological and fisheries data on the geoduck fishery. These port inspectors record information on catch composition and weight as well as information on fishing equipment and methods used. The requirement for the industry to pay the cost of port inspectors was a step toward cost recovery in the fishing industry.

4.5 Pearse's Proposals for Quota Licences for the Other B.C. Fisheries

Pearse (1982) proposed the adoption of quota licence systems for every major and minor fishery in British Columbia except for the salmon and roe-herring fisheries. In *Turning the Tide* he mentioned 13 fisheries specifically. Of these, four (abalone, herring spawn-on-kelp, herring food and bait and geoduck) were brought under quota licences during the 1980's; two of these (abalone and herring spawn-on-kelp) were already being managed by quota licence systems at the time of Pearse's report.

For fisheries other than salmon and roe herring, Pearse proposed:

1. Where quota licensing replaced an established limited-entry system, all owners of licensed fishing vessels eligible to fish for the relevant species, reporting landings in 1980 and 1981, would be issued initial 10-year quota licences.
2. The quantity of fish authorized under each licence would be determined from the licensee's reported landings in 1980 and 1981 and the total allowable catch.

Quota licences would be issued for specific zones, typically the north, south and west zones. Initial quota licence holders would be required to select the zone or zones where their licences would apply.

These procedures would govern initial allocations of quota licences in each fishery. In the long term, Pearse envisaged issuing new 10-year quota licences through competitive bidding. Approximately one-tenth of the total allowable catch for a species would be allocated each year. The decade following the introduction of each quota licensing system would involve transition to a full competitive bidding regime. These quota licences would be issued to persons or companies and be freely transferable. Competitive bidding for new licences and renewal of existing licences were essential features of Pearse's proposal.

Apart from the quota attached to a licence, similar procedures were proposed for the issuance and renewal of licences in the salmon and roe-herring fisheries. In the case of the salmon and roe-herring fisheries, a substantial licence Buy-Back program was proposed (see Chapter 8). Pearse believed limited-entry licensing to be inherently weak. He also believed quota licences to be superior as a means of allocating fishing privileges and promoting orderly fleet development. However, he had recommended that "limited entry licences be retained for the time being for the salmon and roe-herring fisheries because the special characteristics of these two major fisheries makes individual catch quotas impractical." Nonetheless, he suggested that, if the salmon and roe-herring fleets could be reduced to a more manageable level over a 10-year period, individual quotas would be more feasible, especially for herring.

Initially, it appeared that Pearse's proposal for quota licences would be tried on a pilot basis for halibut, one of the major Pacific fisheries. The declaration of the 200-mile fisheries zone in 1977 led to the exclusion of U.S. halibut fishermen from Canadian waters in 1979 and Canadian fishermen from Alaskan waters in 1980. The impact on Canadian fishermen was substantial because their access to halibut stocks was more limited than in the past. Limited entry was introduced in the halibut fishery in 1979. Due to a liberal appeal system, the halibut fleet quadrupled to over 400 vessels. This increased the problem of overcapacity in the halibut fleet.

Following Pearse's interim report in 1981, halibut fishermen appeared receptive to his proposal for individual quotas based on historical catch performance. Fisheries Minister LeBlanc responded favourably to Pearse's recommendation. He would institute quota fisheries for halibut and food and bait herring as soon as possible. To deal with "technical problems", he established a Halibut Vessel Quota Implementation Committee. This Committee consisted of five experienced halibut fishermen chaired by a DFO representative (DFO 1982a).

Reporting in April 1982, the committee made recommendations about the size of individual quotas, transferability, fishing gear and seasonal fishing schedules. It also recommended an appeal system and precautions to ensure effective enforcement (Gibson 1982). The Committee recommended implementation of an individual quota system in time for the 1983 fishery.

In August 1982, Minister LeBlanc accepted the recommendations of the Halibut Vessel Quota Committee, authorizing a program for the 1983 fishery. Within a month opposition had begun to develop to the proposed individual quota system. In a letter to LeBlanc, Mr. Don Cruickshank, the Chairman of the Fleet Rationalization Committee established by the Minister in April 1982, expressed the Committee's "great concern regarding the proposed implementation of the quota system for the British Columbia halibut fishery." He identified six concerns:

1. The spiralling effect on licence values;
2. Difficulty of enforcing restrictions on the concentration of quotas;
3. Questions about whether the quota should be placed on the vessel or on the fisherman and the implications for transferability and security for vessel financing;
4. Possible shifts in the marketing pattern of halibut from frozen to fresh by the lengthening of the catching season through quotas;
5. An increase in direct sales by fishermen to the public would result in the harvest of more halibut (than) permitted by quota holders;

In September 1982, Peter Pearse released his final report. It incorporated the quota licence proposals discussed above. A negative reaction quickly developed in the industry to Pearse's quota licensing proposals. On November 26, 1982, the Prince Rupert Fishermen's Co-operative Association telexed the new Fisheries Minister, Pierre De Bané, urging him "not to institute the individual quotas for halibut as proposed in both of Dr. Pearse's reports" (Telex, Ken Harding to Minister De Bané dated November 26, 1982).

At the same time the Fleet Rationalization Committee submitted its report to the Minister. They opposed "the application of personal quotas on any fishermen." Halibut fishermen who had failed to fish halibut in 1980 and 1981, and would receive little or no quota, began to oppose the proposal.

The Minister's Advisory Council (MAC) on January 21, 1983, passed a resolution requesting that halibut quotas be delayed until MAC could develop alternatives for those fisheries for which Pearse had recommended quotas. DFO postponed a decision on implementation until consultation on the Pearse report was completed. Halibut fishermen were informed that a final determination on individual quotas would be made following review of the Pearse recommendations. Until then the halibut fishery would be regulated as in the past.

Thus a pilot project which might have led to a wider experimentation with individual quotas in the British Columbia fisheries in the 1980's was killed. The quotas were calculated, the regulations passed, but the storm of controversy generated by Pearse's final report made it difficult, if not impossible, to proceed. The quota licence proposals in his final report were sufficiently radical that they prevented action on the more modest system which was taken to the brink of implementation. Almost a decade passed before the idea could be resurrected and implemented.

Debate raged within the industry over the next couple of years on the fate of Pearse's ideas for restructuring the Pacific fishery, particularly his quota licence proposals. On February 18, 1983, MAC advised the Minister to reject Pearse's recommendations for auctioning licences. They also suggested that the proposal for a Crown Corporation to administer licensing and appeals be rejected. At a Vancouver press conference following the MAC meeting, Minister De Bané rejected the idea that established fishermen would have to compete in an auction for the right to remain in the industry. He also indicated that the proposal for a Crown Corporation to administer the licensing of fishermen would not be implemented. "I am not prepared to delegate to an outside body the decision-making authority entrusted to me by Parliament," De Bané said (*The Vancouver Sun*, February 19, 1983).

Meanwhile the problem of overcapitalization in the major Pacific fisheries remained. MAC was given until November 30 to develop proposals for fleet reduction, licensing and allocation of fishing rights among seiners, gillnetters and trollers. On September 28, 1983, MAC presented a general discussion paper titled *A First Step Towards Rationalizing the West Coast Fisheries*. That paper stated:

> "After careful and exhaustive deliberations, the Minister's Advisory Council has determined that a Buy-Back program is not only viable, but it is the *only* option that will provide some stability in the industry while allowing us to deal with the issues of licensing and allocation, which are part of the long term solutions necessary for a stable and profitable industry to develop on the west coast...
>
> "The first and foremost objective of a Buy-Back program is to relieve the pressure on an overfished resource and increase escapements to the spawning grounds by buying out excess fishing capacity. The second objective is to

increase the gross income of fishermen, so they can meet expenses, service their debts and obtain a reasonable net income" (MAC 1983).

As mentioned earlier, in June 1984, Minister De Bané announced a new Policy for the Pacific Fisheries (DFO 1984d). The De Bané Plan proposed a reduction of the fleet by as much as 45%. This would be achieved with a Buy-Back program of at least $100 million (see Chapter 8).

The De Bané Plan also proposed that individual fishing allocations be introduced starting in 1985. This would be done in selected demonstration areas in the Pacific salmon troll and net fisheries. A certain percentage of the total allowable catch would be assigned to individual fishermen. De Bané predicted that a full system of individual fishing allocations would apply in all salmon fisheries by 1987 (DFO 1984f). These allocations within each area and for each gear type would be transferable. Legislation tabled in the House of Commons to implement these proposals died when a federal election was called for September 1984.

The British Columbia fishing industry reacted angrily to the restructuring proposals. Jack Nichol, President of the United Fishermen and Allied Workers' Union, said the scheme was "throwing fishermen an anchor to sink them instead of a life raft to keep them afloat in hard times" (*The Globe and Mail*, June 13, 1984). Al Meadows, President of the Pacific Trollers Association said that the legislation had received near-universal condemnation: "Its compulsive nature and dictatorial intent we find particularly galling, especially in light of the considerable consultation that took place between the Minister and his advisory council" (*The Vancouver Sun*, June 27, 1984). Throughout the process, Meadows added, Minister De Bané and his staff were aware that the industry would never accept regulatory measures such as area licensing, individual quotas and single-gear licensing. David Elliott of the Pacific Trollers Association said that Minister De Bané was told repeatedly by his industry committee that quotas and area licensing were repugnant to fishermen (*The Vancouver Sun*, June 19, 1984). A British Columbia Fishermen's Coalition demanded the retention of common property rights for all fishermen.

With the election of a new Government in September 1984, the proposal to introduce individual quotas in the Pacific salmon fishery was dropped. The resurgence of Pacific salmon stocks and a surge in prices during the mid-1980's dissipated the pressure to reduce overcapacity in the salmon fleet. Meanwhile in other fisheries, e.g. sablefish, halibut, the effects of overcapacity became more evident. This led in 1990 and 1991 to individual vessel quotas in these fisheries.

4.6 The Pacific Halibut Fishery

After the exclusion of Canadian halibut fishermen from U.S. waters, DFO controlled the amount of halibut harvested off British Columbia by limiting the number of fishing licences (435 in the late 1980's) and limiting the number of fishing days. Following the aborted attempt to introduce individual quotas in the early 1980's, the Canadian annual halibut catch more than doubled from about 2,500 tons during 1980–1983 to more than 5,400 tons in 1987 and 1988. The catch then dropped to 3,600 tons in 1990.

Meanwhile the number of allowable fishing days decreased from 65 in 1980 to 6 days in 1990 (Fig. 9-3). In 1989 the fleet in just 11 days harvested twice the catch

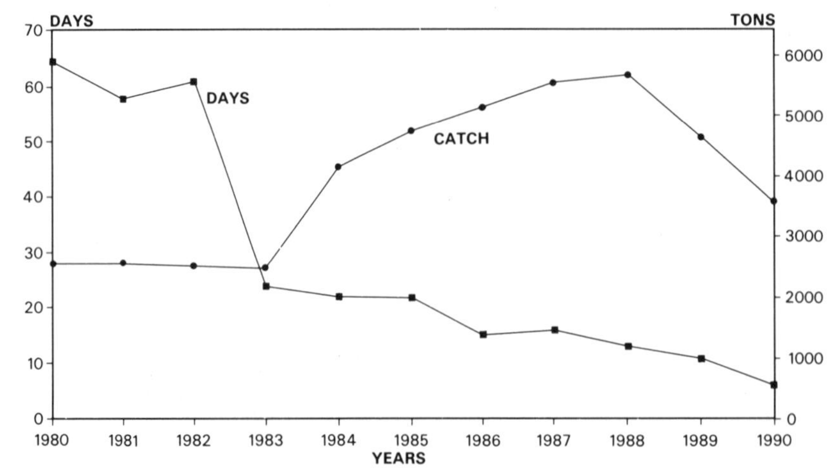

FIG. 9-3. British Columbia halibut fishery catch (tons) and length of fishing season (days): 1980–90 (from data provided by David Reid, Pacific Region, DFO).

taken in 65 days in 1980. This meant that the halibut fishery at the end of the 1980's was a short, highly intense fishery. Supply gluts and poor product quality resulted in lower landed prices. Fishing costs were continually increasing as fishermen tried to maintain their share of the catch through gear and vessel improvements. Fishermen were under pressure to compromise safety by fishing in bad weather and overloading their vessels. One of the greatest problems was that a slight miscalculation of fishery openings by fisheries managers could result in a catch that greatly exceeded the TAC, with detrimental implications for resource abundance.

In the spring of 1989 a small group of halibut licence holders from the Pacific Trollers Association (PTA) and the Pacific Coast Fishing Vessel Owners Guild (PCFVOG) asked DFO to consider using individual vessel quotas in managing the halibut fishery. This was prompted by the problems arising from the short intense fishery.

DFO began extensive consultation with all those involved with the halibut fishery. This included the release in September 1989 of a Discussion Paper on Individual Vessel Quotas (DFO 1989j). In November 1989 halibut licence holders were surveyed about their support for the concept of Individual Vessel Quotas (IVQs). Seventy-seven percent liked the concept.

Halibut licence holders then selected 18 representatives to a Halibut IVQ Advisory Committee (HAC). DFO also invited five non-licence holder representatives (1 UFAWU, 2 Processors, 1 Native and 1 PRFCG). Their mandate was to recommend a structure and monitoring system for a halibut IVQ program. A proposal from this Committee for halibut IVQs was mailed to halibut licence holders for them to vote on. In August 1990, 294 halibut licence holders responded, with 70% voting yes and 30% voting no.

The proposal supported by the majority of the interested parties incorporated these elements:

- IVQs were to be based 70% on the historical performance of the licence holders, and 30% on the overall vessel length for the specific licence;
- The IVQ program was to be implemented for a 2-year trial period commencing in 1991;
- The fishing season was to last 9 months, from March 1 to November 30;
- During the 2-year trial the IVQs would be non-transferable;
- An appeal process would be established to deal with those fishermen contesting their IVQ;
- A comprehensive monitoring and enforcement program would be implemented;
- The halibut license holders would pay all costs necessary to manage and enforce the program.

In November 1990 Fisheries Minister Bernard Valcourt announced a 2-year trial program of individual vessel quotas in the Pacific halibut fishery, incorporating the provisions discussed above (DFO 1990j). Mr. Valcourt emphasized the need for proper monitoring and enforcement of IVQs, with all costs to be borne by halibut licence holders. These costs were estimated to be $750,000.

The Minister also announced a Halibut Advisory Board to advise and assist DFO in managing the fishery and an Appeal Board to hear appeals from halibut licence holders who disagreed with their IVQ allocation.

Although many groups supported IVQs for the halibut fishery, the UFAWU was a notable exception. The UFAWU believed that IQVs:

- would result in the privatization of a common property resource;
- would be concentrated into the hands of processors and big business;
- would result in speculation and raise the price of entry into the fishery;
- would reduce employment;
- were unenforceable and would result in increased illegal fishing.

Apart from the concerns raised by the UFAWU, the generally acknowledged potential problems included high grading (culling of lower valued species), incentives for illegal fishing, dislocation of labour and increased fishing pressure on non-IVQ fisheries. Nonetheless, the introduction of IVQs in the halibut fishery was the first experiment with individual quotas in a major fishery on the Pacific coast. The results will have significant implications for the future application of individual quotas in Canada's Pacific fisheries.

4.7 Sablefish

The introduction of individual quotas in the 1990 sablefish (blackcod) fishery was partly a spin-off of discussions launched with halibut fishermen in 1989 about IQs in that fishery.

Limited entry licensing had been in effect in the sablefish fishery since 1981, with the number limited to 48 licences. Despite limited entry, fleet efficiency increased dramatically during the 1980's. In 1989 the sablefish fleet caught in 14 days (4719t) more than it caught in 245 days in 1981 (2628t) (Fig. 9-4). The landed value increased from $5.5 million in 1981 to $19.0 million in 1989. Fishing capacity had increased more than thirtyfold since 1981 (D. Reid, DFO, Vancouver, pers. comm.).

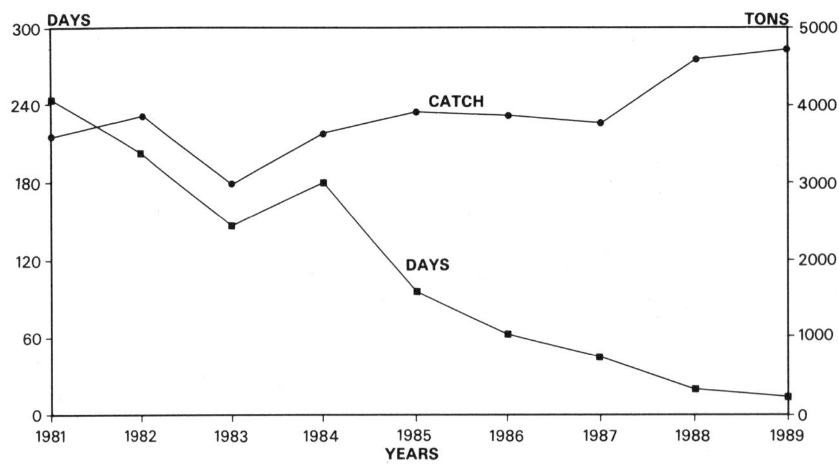

FIG. 9-4. Sablefish fishery catch (tons) and length of fishing season (days): 1981–89 (from data provided by David Reid, Pacific Region, DFO).

The results of this excessive capacity and progressively shorter seasons were an unsafe race to the fish, landings gluts, poor fish quality, and increasingly unmanageable fisheries. The fleet was greatly overcapitalized.

Late in 1989 it appeared that the 1990 sablefish fishery would last no more than 8 days. In October 1989 the Pacific Coast Blackcod Fishermen's Association proposed to DFO that IVQs be used in the 1990 sablefish fishery. More than 90% of sablefish licence holders agreed to a proposal by DFO (DFO 1989k) for individual vessel quotas, stricter enforcement requirements, and cost recovery.

The 1990 Pacific Sablefish Management Plan incorporated a 2-year trial of individual vessel quotas. Quotas were allocated among the licence holders with 70% of the quota based on historical catch (i.e. the best annual catch of 1988 or 1989) and 30% on the licensed vessel's overall length. These quotas could be transferred for one season only but reverted to the original licence holder at the end of the season. Quotas were non-divisible; they could be transferred only as a block.

A new monitoring system was instituted to guard against cheating. Landings were monitored by mandatory logbooks and sales slips. A small number of ports were designated as landing sites. Quota monitoring was contracted out to a private firm, with the costs ($250,000 in 1990) paid for by licence holders. Observers were stationed at the designated landing sites to weigh and validate all landings.

Licence holders reacted favourably during the first year of the program. The 1991 Sablefish Plan tightened up the monitoring, recording and reporting requirements for licence holders and land-based observers at the designated landing sites. The number of designated landing sites was increased from 5 to 7. Biological sampling of the developing year-round fishery was increased.

In 1990, 18 licences were combined to give an effective fleet size of 30 vessels. This was a promising beginning to an experiment introduced at the sablefish licence holders' own request. The trial was extended into 1992 and 1993.

5.0 CONCLUSIONS

During the 1980's the concept of individual quotas as a management tool for Canada's marine fisheries was widely debated and tested in several major fisheries on the Atlantic coast and, to a lesser extent, on the Pacific coast. A number of additional experiments were launched at the beginning of the 1990's. Experimentation is still underway but there is sufficient experience and information available to draw some tentative conclusions.

Initially, EAs or IQs were not the result of discussion and planning but rather of a crisis in particular fisheries or fleet sectors. In the early stages the desire was as much to solve a problem as to introduce long-term changes in support of desired objectives.

The success of enterprise allocations in the Atlantic offshore groundfish fishery indicates that a system of individual quotas can have considerable benefits. Chief among these has been the dampening of the incentive to race for the fish to maximize an enterprise's share of the total allowable catch. Experience in several fisheries has confirmed that individual quotas provide flexibility as to when, how and if an enterprise will harvest its allocation during a given year. Both the Atlantic offshore groundfish fishery and, to a lesser extent, the Bay of Fundy herring fishery have provided some evidence that individual quotas foster fleet rationalization. Another major benefit has been the transformation from a volume-driven to a market-oriented fishery.

However, there are some disadvantages to individual quotas. Chief among these are the problems of high-grading (through discarding at sea), misreporting and underreporting of catches and the consequent difficulty of ensuring compliance with individual quota management regimes. There has been massive underreporting in the Bay of Fundy herring fishery, discarding of small fish in many sectors of the Atlantic groundfish fishery, including widespread underreporting in the western Newfoundland small otter trawler fishery and underreporting in the British Columbia abalone fishery. All these cases attest that this problem is far greater than was envisaged by the economic theorists who originally promoted quantitative rights as the ideal regulatory instrument. Virtually all of the early proponents of individual quotas ignored this problem or downplayed its potential impact on the effectiveness of an individual quota system. The Canadian experience suggests that overcoming this problem is one of the major challenges in designing individual quota systems. Mechanisms to ensure compliance must be designed into IQ systems, if they are to have any hope of success. Now that the magnitude of this problem has been recognized, managers are trying to minimize the incentive to cheat. Steps here include placing observers on offshore groundfish trawlers, designating ports of landing and industry-funded port inspectors in the recently-introduced individual vessel quota systems in British Columbia. In the Gulf of St. Lawrence and southwest Nova Scotia, DFO has included a complex system of designated landing points, observers, plant audits and penalties in the recently-introduced small trawler experiments .

This problem is serious enough to wreck any attempts to rationalize fisheries through the use of individual quotas. There is still, however, insufficient recognition that the incentive to cheat or underreport in an individual quota system is probably as large and as problematic for successful fisheries management as the incentive to maximize one's individual share of the total catch in an open access or limited entry fishery. Because of this incentive to cheat, individual quota management schemes are inappropriate for situations where landings occur at a large number of geographically dispersed ports. This means, for example, that individual quota systems would be ineffective for the small-boat (vessels less than 35 feet) fisheries on the Atlantic coast where thousands of vessels use numerous, scattered, and, in many cases, relatively isolated landing points.

If an individual quota system is to succeed, fishermen must be willing participants in the design of such a scheme. This is vividly demonstrated by the Atlantic offshore groundfish trawler experiment where the offshore companies helped design the original system and subsequent modifications to it. The success of this effort over the past decade has resulted in large measure from ongoing dialogue among the companies, the provinces and federal government officials. The design and evolution of the enterprise allocations system for this fishery has been a cooperative endeavour, which culminated in its permanent adoption in 1989. Similarly, it was possible to introduce enterprise allocations in the otter trawl fishery in western Newfoundland because fishermen recognized the need for change and participated on an ongoing basis in the design and modification of the EA system. This was also true of the initial experience in the Bay of Fundy purse seine fishery and the subsequent modifications in 1983. Most recently, it has been possible to break the logjam on the Pacific coast and introduce a trial enterprise allocation system in several fisheries (e.g. geoduck, sablefish and halibut) because the participants themselves requested that such systems be put in place.

Attempts to impose individual quota systems against the wishes of the majority of the participants in a fishery have failed miserably. The most dramatic instance of this was the rejection of Pearse's proposals for quota licences in fisheries other than salmon and roe herring and the rejection of his proposals for 10-year term licences with competitive bidding in an open auction for the Pacific salmon and roe-herring fisheries.

Another feature that has fostered the success of some of the Atlantic trials of EA systems has been government's willingness to take a cautious approach, learning from initial experience and modifying schemes as they evolve, i.e. *adaptive management*. Another facet of this has been the recognition that one theoretical scheme cannot be force-fitted to all fisheries and circumstances. Instead, on the Atlantic coast, government has been willing to design EA systems to meet the needs and circumstances of particular fisheries. A major shortcoming of Pearse's proposals was that he attempted to design an ideal, uniform scheme for most fisheries, which incorporated a major feature — 10-year licences open to competitive bidding — which was unacceptable to the participants in those fisheries. The Task Force on Atlantic Fisheries advocated a general principle of a move to quota licences. However, it left the details to be worked out by government officials and the industry. This difference largely explains why we had numerous trials of individual quota systems in various Atlantic fisheries during the 1980's and early 1990's. On the Pacific coast, on the other hand, real experimentation with individual quotas did not take place until the 1989–1991 period.

There is one essential difference between most of the individual quota systems now in place and the model advocated in the early 1980's by Scott and Neher (1981), Pearse (1982) and Kirby (1982). These three major reports all advocated transferable, divisible and marketable quota licences. With a few exceptions, the individual quota systems adopted for Canada's marine fisheries over the past decade have not incorporated provisions for transferability and marketability. A major exception was the modified plan adopted in 1983 for the Bay of Fundy purse seine fishery. In the Bay of Fundy example, limits were placed on how much of the Total Allowable Catch could be held by one licensee (4 percent) and transferability was restricted within certain categories of vessels. More recently limited transferability has been adopted in the otter trawl fishery less than 65 feet in the Gulf of St. Lawrence. Permanent transfers are allowed in only two of the 10 fisheries under individual-quota regimes on the Atlantic coast.

Frequently, transferability is resisted by fishermen who fear the concentration of the privilege to fish in the hands of companies and large scale operations. The result is that individual quotas in Canada's marine fisheries represent constrained harvesting rights rather than quasi-property rights. Minister Siddon described the offshore groundfish EAs as "a harvesting privilege for a specified share of the resource subject to scientific advice and Ministerial decision." Although a DFO Working Group in 1990 recommended that the concept of transferability should be established during the development of all IQ regimes (P. Sutherland, DFO, Winnipeg, pers. comm.), progress towards free transferability and marketability in individual quota systems for Canada's major marine fisheries is likely to be slow. Although ITQs were supposed to be an alternative to the existing management system, in most instances they have been imposed on top of the existing system. Few of the rules or regulations in place prior to the introduction of ITQs have been relaxed.

Individual quota systems are clearly not suitable for all fisheries. Since stability is one of the major benefits offered by individual quotas, it follows that individual quotas are inappropriate for species/stocks/fisheries where the available catch fluctuates widely from year to year because of fluctuations in resource abundance or availability. This means that individual quotas may not be an appropriate management tool for fixed gear fisheries such as the inshore cod trap fishery in Newfoundland which is subject to substantial year-to-year variation in resource availability. Similarly, individual quotas are probably inappropriate in the case of the Pacific salmon fishery where there are widespread year-to-year fluctuations in abundance and availability of the resource to particular types of gear and at different points along the complex migration routes of the five Pacific salmon species.

Not all economists have been sanguine about the much-touted merits of individual quotas as the ideal regulatory instrument to eliminate the common-property "race for the fish". Copes (1986) recognized that: "When it comes to promoting individual quota management, its proponents often fail to apply the sharp insights gained in exposing the deficiencies of limited entry licensing. There is no reason to assume that fishermen, where confronted with the rules of individual quota management, will lose either their ingenuity at circumvention or their incentive to promote individual interest at the expense of collective interest." In an insightful analysis, he explored a variety of problems that should be anticipated with the introduction of individual quota management. He identified problems in the following areas: (1) Quota Busting, (2) Data Fouling, (3) Residual Catch Management, (4) Unstable Stocks, (5) Short-Lived Species, (6) Flash Fisheries, (7) Real Time Management, (8) High-Grading, (9) Multi-Species Fisheries, (10) Seasonal Variations, (11) Spatial Distribution of Effort, (12) TAC Setting, (13) Transitional Gains Trap, and (14) Industry Acceptance.

Most of the potential problems he raised have been touched on already in discussing Canadian marine experience with individual quotas. There are a couple of additional points worth noting. Copes observed that individual quota management is inherently unsuited to fisheries where the catch is residual to a managed escapement target. This provides a cogent argument why individual quotas are inappropriate in the management of the Pacific salmon fisheries. He also observed that individual quotas make little sense in the case of "flash fisheries" such as the British Columbia roe herring fishery where a "race for the fish" is necessitated by the nature of the fishery itself.

Unlike most of his colleagues, Copes also pointed out certain problems associated with making individual quotas transferable. The benefits of transferability go to the initial group of licence holders. They are able to "capitalize the stream of future benefits and extract them from those succeeding them, who must purchase these rights at full value." As a result succeeding generations enjoy no net benefits, and will likely fall to the lower levels of net income that government action was intended to overcome. Copes observed: "If it is also important to bring about a long-run improvement in income levels for succeeding generations in a particular fishery, the ITQ approach is likely to prove unsuitable." This is a problem but not just with ITQs. It was probably more of a problem with the initial introduction of limited entry. ITQs simply represent a refinement in the equalization of licences. Nonetheless, given the chronic problems of unemployment and low incomes in the Atlantic provinces, it is not surprising that Fisheries Ministers from the Atlantic provinces have opposed the concept of transferability of individual quotas.

Based on the Canadian experience over the past decade, it seems clear that individual quotas are an effective management tool for the management of some, but not all, fisheries. Some of the envisaged benefits, particularly the reduction of the incentive to maximize the individual share of the catch, are achievable without free transferability or marketability of individual quotas. The reduction of gluts, the landing of better quality fish, the reorientation from a volume-driven to a market-focused fishery are all tangible benefits of the individual quota approach as it has been applied in a number of Canadian fisheries.

Some benefits in terms of fleet rationalization are also being realized, although not to the extent anticipated, e.g. the offshore groundfish fishery and the Bay of Fundy purse seine fishery. These are most likely to occur in situations where enterprises own more than one fishing vessel and hence can match their capacity to that required to harvest their enterprise allocation e.g. offshore groundfish, offshore scallops. This will be more difficult to achieve in situations where all participants in a fishery own single vessels. Even there, it is possible to replace vessels with smaller ones to minimize individual harvesting costs. Under the latter circumstances, however, the economically optimum fleet mix is unlikely to emerge.

In a recent review of rights-based fisheries management in Canada, Crowley and Palsson (1992) concluded that there had been efficiency gains from granting "rights" and ensuring resource rent capture. Second, enforcement of rights is costly and demanding. Pressures to cheat on quotas can only be avoided by costly enforcement. Third, quotas have been traded under EAs despite legal discouragement. They further concluded that the absence of divisibility and ready transferability had possibly slowed but not prevented the consolidation and rationalization of rights-based program fisheries. They observed:

> "The overall lesson from the Canadian experience is that a system of property rights can yield substantial benefits, even if the rights are attenuated. It is not a disaster if the original design includes extraneous objectives such as employment maximization or regional segregation of quotas."

As with limited entry licensing, one of the chief constraints on individual quotas is the difficulty of introducing them in a fishery where there is already grossly excessive capacity. One reason why it was possible to introduce them relatively smoothly in the Atlantic offshore groundfish fishery was that the mismatch between harvesting capacity and available offshore quota was relatively slight at that time. The mismatch was relatively slight because 10 years before EAs, government froze entry into the fisheries and limited replacement. Then in 1980 it cancelled all banked (inactive) licences. Thus it had achieved some reduction in capacity prior to the introduction of EAs in 1982. In the Bay of Fundy purse seine fishery, on the other hand, the considerable overcapacity in the harvesting sector led to major problems of quota busting and data fouling.

Individual quota management seems best suited to fisheries where:

1. The resource is relatively stable;
2. The number of enterprises is small, perhaps tens or a hundred participants rather than thousands;
3. The number of landing points is relatively small and easily accessible to enforcement personnel, and there are clearly defined marketing channels;
4. Enterprises are formally organized in effective associations that can speak and negotiate for the members;
5. Participants recognize the negative effects of the "race for the fish" under open access conditions and are willing to experiment with innovative approaches which counter traditional notions of survival of the fittest; and
6. There is a voluntary commitment to comply with an individual quota regime and to assist in its enforcement.

Individual quota systems have to be carefully tailored to the different characteristics of particular fisheries. In appropriate fisheries and with appropriately designed compliance mechanisms, individual quotas constitute a useful addition to the array of fisheries management tools which have been traditionally employed to manage fisheries. Individual quotas are not, however, a panacea for the myriad of problems arising from the common property nature of fish resources.

CHAPTER 10

THE INTERNATIONAL DIMENSION
I — THE EVOLUTION AND IMPACT OF EXTENDED FISHERIES JURISDICTION

"I wish to announce today the government's decision to extend the fisheries jurisdiction of Canada out to 200 miles from the coast. The state of our fishery resource and the situation of our fishermen, of our fishing industry, and of our coastal communities, make this action imperative. There will be no fishery resource left to protect if action is not taken now — because the fish stocks will be so depleted as to disappear as a resource of commercial significance. Not only the fish but our Canadian fishermen too are an 'endangered species."

– Allan J. MacEachen, House of Commons, June 4, 1976.

1.0 INTRODUCTION

Canada's decision to extend its fisheries jurisdiction to 200 miles from the coast, effective January 1, 1977, on the Atlantic and Pacific coasts and March 1, 1977, in the Arctic, was part of a worldwide movement to enclose a substantial portion of the global oceans commons. Declaration of the 200-mile Canadian fisheries zone is the most significant event in Canadian fisheries history. Indeed, the global move to a 200-mile limit for fisheries jurisdiction during the 1970's is perhaps the most significant event in the history of world fisheries. It marked the transition from a regime where states could manage fisheries within only a small band of 3 to 12 miles from their coasts, to a situation where coastal states are responsible for managing substantial portions of the world's marine fish resources.

For several centuries it was generally accepted that the high seas were free to all. The doctrine of freedom of the high seas has come to be associated with the publication in 1609 of *Mare Liberum*, a work by the Dutch jurist Hugo Grotius. Grotius, in his work, developed the argument for a principle of freedom of the high seas, maintaining that some things, such as food, must be "exhausted" to be utilized. Thus ownership was necessary to use such things. Other things, such as air and running water, were not exhausted by use. These should remain free for all to use. The sea fell into the second category. Grotius argued no person or nation could appropriate any part of the high seas for exclusive use.

This brought the Dutch into conflict with the British, who, while acknowledging a freedom of navigation, reserved the nearshore fisheries area for their own use. John Selden, in a publication entitled *Mare Clausum: of the Dominion or ownership of the Sea* (1635) contended that the resources of the sea were just as exhaustible as land resources. He argued for the right to restrict foreign fishing off British shores. This led to the concept of a territorial sea and the principle of reserving areas close to the coast for limited or general purposes. Another Dutch jurist, Cornelius Van Bynkershoek (1702), subsequently took the view that the dominion of the coastal state over the sea extended to the point at which the coastal state's power effectively ended. This became interpreted as the distance within 3 nautical miles (Hollick 1981).

Thus evolved the concept of a territorial sea and the high seas. The doctrine of freedom of the high seas was reflected in the Treaty of Paris of 1856.

2.0 EVOLUTION OF EXTENDED FISHERIES JURISDICTION

2.1 The 1930 Hague Conference

The conflict between the concepts of coastal state rights over some portion of the seas and the freedom of the high seas continued into and intensified in the 20th century. The Hague Conference for the Codification of International Law in 1930 considered the matter of the territorial sea and contiguous zone. At that time the U.S. and Britain claimed a 3-mile territorial sea, the Scandinavian countries claimed 4 miles, and Russia under the Czars had laid claim to a 12-mile territorial sea. The maritime countries represented at the Conference agreed that the territorial sea formed part of the territory of coastal states and that the high seas were free to all. The Conference, however, failed to agree on the width of the territorial sea and on the nature and breadth of any contiguous zone. A 3-mile limit for the territorial sea was supported by Britain and the Commonwealth countries, including Canada. Others argued for limited jurisdiction beyond 3 miles. Britain, with the support of the Commonwealth countries, blocked any attempt to recognize any form of jurisdiction beyond the territorial sea. Nonetheless, this debate planted the seeds for the later concept of a separate fishing zone (Hollick 1981).

Following the Second World War Grotius' approach began to be questioned. His assumptions proved invalid. During the next two decades it became apparent that the resources of the oceans were far from inexhaustible. Advances in technology made it possible to deplete fish stocks through overfishing. Mineral and petroleum resources were discovered on the continental shelf and the deep seabed and techniques for their extraction developed. Pollution by passing merchant vessels became a significant concern.

2.2 The Truman Proclamations — 1945

These developments led to unilateral action by several countries from the mid 1940's to the late 50's. To the later chagrin of the United States, these actions had their origin in the Truman Proclamations of 1945. There were two proclamations, both issued on September 28, 1945 (Truman 1945a and b). One, the continental shelf proclamation, stated:

> "The Government of the United States regards the natural resources of the subsoil and seabed of the continental shelf beneath the high seas but contiguous to the coasts of the United States as appertaining to the United States, subject to its jurisdiction and control" (Truman 1945a).

The second proclamation, concerning fisheries, stated:

> "The Government of the United States regards it as proper to establish conservation zones in those areas of the high seas contiguous to the coasts of the United States wherein fishing activities have been or in the future may be developed and maintained on a substantial scale" (Truman 1945b).

The evolution of these policies in the last days of the Roosevelt administration is described in Watt (1979) and Hollick (1981).

2.3 Unilateral Actions by Latin American States

The Truman Proclamations were clearly expansionist in nature. They provoked a series of unilateral claims by a number of Latin American states. Mexico was the first to take action. In October 1945, the President of Mexico claimed as national territory the continental shelf adjacent to Mexico to a depth of 200 m, and its resources. In December 1945, an amendment to the Mexican Constitution laid claim to the waters covering the continental shelf as the property of the nation. This, however, was never promulgated by the Executive (Mexico 1945). In 1946 Panama and Argentina made somewhat similar proclamations (Panama 1946; Argentina 1946).

Chile, on June 23, 1947, was the first to assert jurisdiction over a 200-mile zone. Hollick (1977) described the origins of this claim. In August, 1947, Peru also proclaimed a 200-mile zone to protect the abundant fisheries off its coasts (Peru 1947).

All of these claims cited the Truman Proclamations as precedent. By 1950 U.S. officials had realized that the Truman Proclamations had led to a substantial attack on the long standing principles of the freedom of the high seas and the 3-mile territorial sea (Watt 1979).

The U.S. temporarily stemmed the flood of new claims by strong protests to Argentina, Chile and Peru in mid-1948. But in the early 1950's further claims were made. Six Latin American countries issued new or modified claims in 1950.

Opposed by the U.S., Chile, Ecuador and Peru (the CEP countries) took action to coordinate their claims. This led to the Santiago Declaration of August, 1952, in which the CEP countries asserted sole sovereignty and jurisdiction over a zone including the sea floor to a distance of 200 miles from the coasts or from islands (Anon 1952).

2.4 Canada's Position — 1945–1958

The waters of the Northwest Atlantic had been fished by Europeans for centuries. Following the Second World War and the Bates' Report, Canada focused on developing its offshore fishery. Meanwhile, foreign fishing off the Atlantic and Pacific coasts began to increase in the post-war period.

Britain had made several attempts to address overfishing in the Northeast Atlantic. These were intended to cover the entire North Atlantic. A Convention on the Regulation of Meshes of Fishing Nets and Size Limits of Fish was adopted in London in 1937. It was signed by 10 west European countries but never came into effect. Britain tried again in 1943 and 1946. At the 1943 meeting there were differences over the size of the proposed convention area. While Canada supported a North Atlantic convention, the U.S. wanted separate treatment of the fisheries of the Western and Eastern Atlantic, with a division at approximately 40° west longitude (Hollick 1981).

Britain backed off. Between 1943 and 1946 it persuaded Canada to consider separate conventions for the two areas. In April 1946, another Convention for the Regulation of the Meshes of Fishing Nets and Size Limits was adopted, applying to the Northeast Atlantic and Arctic (Cushing 1972).

Concerned about the possible shift of fishing effort to waters off their coasts, Canada and Newfoundland considered proclamations similar to the Truman Proclamations. Britain discouraged this. Meanwhile, concerned about the developments in Latin America, the U.S. sought to establish a multilateral conservation regime for the Northwest Atlantic. The U.S. invited Canada, Denmark, France, Great Britain, Iceland, Newfoundland, Norway, Portugal and Spain to a January, 1949 meeting in Washington to discuss a draft convention.

Canada expressed reservations about the scope of the proposed commission, preferring to limit its powers to investigation and recommendations. Also the U.S. draft had adopted Newfoundland's narrow territorial waters. This led to fears in Canada that it would be difficult to extend these limits outward later (Hollick 1981).

U.S. officials persuaded Canada to participate with the understanding that the commission's powers would be limited to investigation. The Conference agreed on a Northwest Atlantic Fisheries Convention, which entered into force on July 3, 1950. That Convention provided for an International Commission for the Northwest Atlantic Fisheries (ICNAF), which met for the first time in 1951. ICNAF had 10 members by 1953. Other countries joined over the next two decades.

The Northwest Atlantic Fisheries Convention recognized the traditional distinction between the territorial sea and the high seas. ICNAF's authority extended only to the high seas beyond the jurisdiction of coastal states. Canada entered a reservation to the Convention:

> "Any claims Canada may have in regards to the limits of territorial waters and to jurisdiction over fisheries particularly as a result of the entry of Newfoundland into Confederation will not be prejudiced" (UN 1953).

One of the conditions attached to the entry of Newfoundland into Confederation was that the Canadian government apply the headland-to-headland rule for the measurement of the territorial waters along the coasts of the new province. Prime Minister Louis St. Laurent stated in the House of Commons on February 8, 1949:

> "We intend to contend and hope to be able to get acquiescence in the contention that the waters west of Newfoundland constituting the Gulf of St. Lawrence shall become an inland sea," i.e. part of the territorial waters of Canada (Canada 1949).

No action was taken on this matter until the 1960's. Meanwhile domestic pressure was growing for Canada to abandon its traditional adherence to the 3-mile territorial sea. This was exacerbated by the fact that Canadian trawlers on the east coast had been prohibited since 1911 from fishing within 12 miles of certain portions of the coast (except Newfoundland) while foreign trawlers could fish in the zone between 3 and 12 miles from the coast.

On the Pacific coast, bilateral management of the Pacific halibut fishery took place through the International Pacific Halibut Commission, first established in 1923 (see Chapter 6). Canada's international interests were largely concerned with reducing the interception of Canadian salmon by Japan and the United States.

Fraser River sockeye salmon were also jointly managed through the International Pacific Salmon Commission. This commission was established by the Canada-U.S. Sockeye Salmon Fisheries Convention, ratified in 1937. Multilateral management of Pacific salmon was undertaken through the International North Pacific Fisheries Commission (INPFC), involving the U.S., Japan and Canada. The Truman Proclamation of 1945 respecting fisheries had, to a large extent, been motivated by the threat posed by Japan to the Alaskan salmon fisheries in the pre-war years (Hollick 1981). The U.S., Japan and Canada agreed in Tokyo after World War II to the International Convention for the High Seas Fisheries of the North Pacific Ocean. That convention introduced the principle of *abstention*. Under this principle, any of the contracting parties might be requested to abstain from participating in a fully utilized fishery, if such fishery had been subjected to an extensive conservation program by one or both of the other parties. This applied to Japan, which agreed to abstain from fishing certain species in certain areas. Canada and the U.S. agreed to implement necessary conservation measures for halibut, herring and salmon stocks in specified parts of the convention area off their coasts. During renegotiation of the Convention in 1951, Japan agreed to refrain from fishing Pacific salmon east of 175° W longitude, the *abstention line*. This provided substantial protection for Canadian salmon stocks from Japanese fishing (Bell 1981).

Despite these measures, in 1952 the United Fishermen and Allied Workers' Union proposed in a submission to the Minister of Fisheries that Canada adopt a 9-mile territorial sea to provide fisheries protection (UFAWU 1952). (Subsequently this union advocated a 12-mile territorial sea). The Minister of Fisheries, James Sinclair, reacted by appointing a committee, headed by the Dean of the Faculty of Law of the University of British Columbia, to review the situation and make recommendations.

On July 30, 1956, the Prime Minister stated in the House of Commons that "we think the twelve-mile limit should be recognized", with due recognition of historic fishing rights. On August 13, 1956, Minister Sinclair indicated that Canada favoured a 12-mile territorial zone (Canada 1956a and b).

The First United Nations Conference on the Law of the Sea, UNCLOS I, was held in 1958. This was followed by a Second Conference, UNCLOS II, in 1960. More than a decade later the Third UN Conference on the Law of the Sea, UNCLOS III, which extended over a decade from 1973 to 1982, produced a new Law of the Sea.

2.5 UNCLOS I

At UNCLOS I in March 1958, Canada proposed a 3-mile territorial sea and a 9-mile contiguous zone. This came with the proviso that, in the contiguous zone, a coastal state would have "the same rights in respect of fishing and the exploitation of the living resources of the sea...as it has in its territorial sea" (Canada 1958a). The Chairman of the Canadian delegation explained that, while Canada was sympathetic to the Latin American push for wider fishery jurisdiction, it was unlikely that all countries would agree on more than a 12-mile contiguous zone. The proposed fishing zone was to reserve "a reasonable coastal belt for the use of fishermen of the coastal states... many of (whose) communities may largely depend for their livelihood on the preservation of the fishing stock in the nearby seas" (Canada 1958a).

The Canadian proposal reflected the desire of coastal states to secure jurisdiction over the fisheries resources adjacent to their coasts. Nonetheless, it was tied to the concept of a narrow 3-mile territorial sea. Thus it attempted to reconcile the views of those who wished greater control of fisheries adjacent to their coasts and those who opposed extension of the territorial sea because of interference with the concept of the freedom of the high seas. However, Britain and the United States changed their position, proposing a 6-mile territorial sea and a 6-mile fishing zone beyond the territorial sea. The hook was that traditional fishing rights would be recognized in the outer 6 miles. This prompted Canada to shift ground. A Canadian proposal for a 6-mile territorial sea plus a 6-mile fishing zone was voted on, along with three other major proposals, in Plenary. There was insufficient support for passage of either of the proposals (Gotlieb 1964).

The Canadian delegation had also worked for adoption of the principle of abstention. This proposal failed to obtain the necessary majority. This was not, however, a major setback for Canada because of the protection afforded to salmon by the abstention line in the Pacific.

UNCLOS I did produce concrete results in the form of four Conventions:

1. The Territorial Sea and Contiguous Zone (which did not specify an outer limit);
2. The High Seas;
3. The Continental Shelf; and
4. Fishing and Conservation of the Living Resources of the High Seas.

The first two Conventions merely confirmed the existing law of the sea with some minor modifications. The Continental Shelf Convention confirmed the thrust toward coastal state rights over exploration and exploitation of the natural seabed and resources of the continental shelf. It contained a provision giving coastal states jurisdiction over the sedentary living resources of the seabed to the edge of the continental shelf. It defined the limits of the continental shelf as a depth of 200 m or a greater depth where exploitation is possible. The High Seas Fisheries Convention was never ratified by most nations actively involved in high seas fishing.

Despite the failure of UNCLOS I to define the territorial sea and contiguous zone, Canadian politicians and officials described the outcome of the Conference in positive terms (Canada 1958b; Ozere 1973).

2.6 UNCLOS II

In 1960, another Law of the Sea Conference was convened to resolve the dispute about the width of the territorial sea and the nature of a contiguous fishing zone. Canada again proposed a 6-mile territorial sea and a 6-mile contiguous fishing zone. The United States modified its earlier proposal. The new proposal incorporated the previous provision that states which had fished regularly in the contiguous 6-mile zone during the preceding 5 years could continue to do so in future. As modified, fishing rights would be limited to the species, quantities and areas fished during the preceding 5-year period. Canada and the U.S. then co-sponsored a compromise proposal which provided for a 6-mile territorial sea, a 6-mile fishing zone outside this and a 10-year phasing-out period for those countries which had fished in the outer 6 miles during the preceding 5 years. While there were other proposals, e.g. for a 12-mile territorial sea, the Canada-U.S. joint proposal garnered the most favour. It was not adopted, however, because it fell one vote short of the necessary two-thirds majority (Gotlieb 1964).

2.7 Canadian Actions — 1963–1964

With the failure of these multilateral efforts, Canada embraced a cautious unilateral approach during the 1960's. There was a rapid development in state practice from 1958 to 1964. A number of additional countries claimed a territorial sea of 12 miles. Another group of states established fishing limits beyond the limits of their territorial seas. In another application of the concepts discussed at UNCLOS II, Britain agreed with Iceland, Norway and Denmark during 1959-61 on various forms of phase-out agreements (Gotlieb 1964).

In a January 1963 submission to the Government of Canada, the Fisheries Council of Canada (FCC), urged the government to declare certain bodies of water as Canadian national waters, and adopt the straight baseline principle, from which the breadth of territorial seas and the exclusive fishing zone would be measured (FCC 1963).

On June 4, 1963, Canada announced its intention to take unilateral action. The Prime Minister stated:

> "The Canadian government has decided to establish a 12-mile exclusive fisheries zone along the whole of Canada's coastline as of mid-May 1964, and to implement the straight baseline system at the same time as the basis from which Canada's territorial sea and exclusive fisheries zone shall be measured" (Canada 1963).

The Canadian Parliament passed a new *Territorial Sea and Fishing Zone Act* in 1964. This act created a 9-mile fishing zone beyond the 3-mile territorial sea and provided enabling legislation for the closure of other areas to fishing through the use of straight baselines (Canada 1964). The use of headland-to-headland baselines provided the opportunity to significantly extend Canada's internal waters. However, in terms of the breadth of the territorial sea, the 1964 Act was a step backward from the "six plus six" formula which Canada had pursued at the 1960 LOS Conference.

The potential impact of the 1964 Act was negated by the passage of an Order in Council which allowed many foreign vessels to continue fishing in Canadian contiguous zones pending negotiation of agreements with foreign governments. U.S. fishing vessels could continue to fish in the contiguous zones on both the Atlantic and Pacific coasts. Vessels from Britain, Denmark, France, Italy, Norway, Portugal and Spain could continue fishing on the east coast, pending the negotiation of agreements with these countries. It was recognized that vessels of France and the U.S. would be allowed to continue fishing in the areas, subject to appropriate conservation regulations. However, the Canadian Government intended that fishing by the other countries would be phased out once specific arrangements had been negotiated.

Gotlieb (1964) described passage of the 1964 Act as "a decision of historic importance in the evolution of Canadian policy with respect to its adjacent waters." He was optimistic about a speedy implementation. He was mistaken. Negotiations with the various countries did not proceed smoothly, and agreements were not reached until after further unilateral action by Canada in 1970. Not until October 1967 was the first set of geographical co-ordinates of points to establish straight baselines issued by the Governor in Council. Those points established baselines only along the coast of Labrador and the eastern and southern coasts of Newfoundland. In 1969, a second list of geographical co-ordinates was published establishing baselines along the eastern and southern coasts of Nova Scotia and the western coasts of Vancouver Island and the Queen Charlotte Islands (Legault 1974).

Despite these delays Legault (1974) described the 1964 Act as a "turning point in the evolution of Canada's maritime policy and maritime claims." Until the 1958 and 1960 Conferences on the Law of the Sea, Canada had followed "a path of negotiation, arbitration and bilateral and multilateral agreement." While Canada did not abandon that path in the 1960's and 1970's, the 1964 Act added unilateral action as one of the tools of Canadian maritime policy.

Meanwhile fisheries developments during the 1950's and 1960's stimulated Canada's emergence as a coastal state in the 1970's.

2.8 Fisheries Developments in the Northwest Atlantic — 1945–1970

When ICNAF first met in 1951, there was no pressing conservation concern off the Canadian Atlantic coast. West European countries had fished there for centuries. Efforts were underway in Canada to develop an offshore Atlantic groundfish fishery by Canadian trawlers. The offshore trawler fleet grew to about 160 vessels, owned by a few major companies. Federal and provincial governments collaborated to promote post-war fisheries development, through exploratory fishing, vessel-building subsidies and grants for the construction of new processing plants.

Prior to 1947 the Northwest Atlantic fisheries were relatively stable. France, Spain and Portugal had been fishing in the area for centuries. Newfoundland, Canada and the USA began to develop offshore fisheries. The European fisheries were concentrated on the Grand Banks.

Cod was the primary species fished both by Europeans and Canadians. In Newfoundland, which was not yet part of Canada, "fish" meant "cod." Canadian and U.S. fishermen were, however, beginning to land other groundfish species such as haddock and pollock, and pelagic species such as herring and mackerel. Canadians began to fish redfish in the early 1950's.

Total catches off the Canadian Atlantic coast during this period have been estimated at 500,000–600,000 tons (Regier and McCracken 1975).

From 1947 to 1957 there was an orderly expansion of the fisheries in the Northwest Atlantic. There was no major change in the countries fishing off the Canadian Atlantic coast, except for new minor fisheries by Britain and Norway. European fishermen fished more in the Gulf of St. Lawrence and off Nova Scotia. European vessels continued to concentrate on cod and usually discarded other species. The Spanish, however, fished haddock during the mid-1950's when this species was abundant on the Grand Bank and St. Pierre Bank (Regier and McCracken 1975).

There were significant changes in the species sought by the Canadian fishery. Offshore trawlers fished for redfish and various flatfish species that they had not fished previously. Haddock became an important species in the Canadian fishery. The Canadian offshore fishery shifted from a hook and line dory fishery to offshore vessels using otter trawls. The Europeans increased their use of otter trawls and pair trawls (Regier and McCracken 1975).

Landings from the area off the Canadian Atlantic coast increased from 700,000 tons to 1,200,000 tons, with cod still comprising about 70% of the total.

The period 1958-1968 was one of dramatic, uncontrolled expansion in the Northwest Atlantic fisheries. The most significant event was the arrival of the USSR fleet in the Northwest Atlantic and rapid expansion of its activities. The USSR was followed by other Eastern European countries (Poland, Romania, German Democratic Republic and Bulgaria). Other West European countries (the Federal Republic of Germany, Britain and Norway) also joined in the rapidly escalating fishery.

The arrival of the USSR fleet added a new dimension to the fishery. The USSR sought any species or stock abundant enough to feed the appetite of its massive fleet. It engaged in what became known as "pulse fishing": a large amount of fishing effort was directed at a particular species in a given area until it was reduced to a low level of abundance. Then the fleet moved on to another species or another area (Regier and McCracken 1975).

The USSR's catch in the Convention Area increased from 17,000 tons in 1956 to 370,000 tons in 1962 and then more than doubled to 853,000 tons in 1965. It increased to a peak of 1,357,000 tons in 1973. This high level of catches was sustained by a remarkable shift in the species composition of the catch over time. The main species sought shifted over time from redfish, to cod, to silver hake, and herring and then in the 1970's to capelin and mackerel (Fig. 10-1).

During this period the fishery diversified. A winter fishery for northern cod developed off the east coast of Newfoundland and Labrador. A fishery for herring developed offshore. Flatfish fisheries expanded on the Grand Banks. In addition to traditional groundfish species, the USSR developed fisheries for such species as argentine, grenadiers and silver hake. Canadian fishermen also diversified, with more intensive fishing of herring, redfish, flatfish and pollock.

This era involved rapid changes and technological improvements in the methods of fishing. The major developments were the introduction of stern trawlers/factory trawlers for processing fish at sea, purse seining and midwater trawling for pelagic species, and more efficient fish-finding techniques (Regier and McCracken 1975).

The introduction of stern trawler factory trawler technology to the Northwest Atlantic during the 1960's changed dramatically the scale and nature of the area's offshore fishery. By 1970 there were about 900 freezer trawlers and factory trawlers over 1,000 tons in the world's fishing fleets. Of these about 400 belonged to the USSR, 125 to Japan, 75 to Spain, 50 to West Germany, 40 to France and 40 to Britain (Hjul 1972).

New technology and greatly increased fishing effort brought rapid growth in catches until 1968. Total Northwest Atlantic nominal catches increased from about 1,800,000 tons in 1954 to a peak of 4,600,000 tons in 1968. Much of that increase occurred from 1960 to 1968 (Fig. 10-2). Catches in ICNAF (NAFO) Subareas 2-4, which approximate catches off the Canadian coast, increased from 1,400,000 tons in 1962 to a peak of 2,700,000 tons in 1968. Canadian catches peaked at about 1,300,000 tons in 1968.

During this period expansion was uncontrolled. The tonnage of vessels fishing in the Northwest Atlantic (excluding those under 50 tons) increased dramatically from about 400,000 tons in 1959 to a peak of around 1,500,000 tons in 1974 (Fig. 10-3). Despite a continued increase in fishing effort from 1968 to 1974,

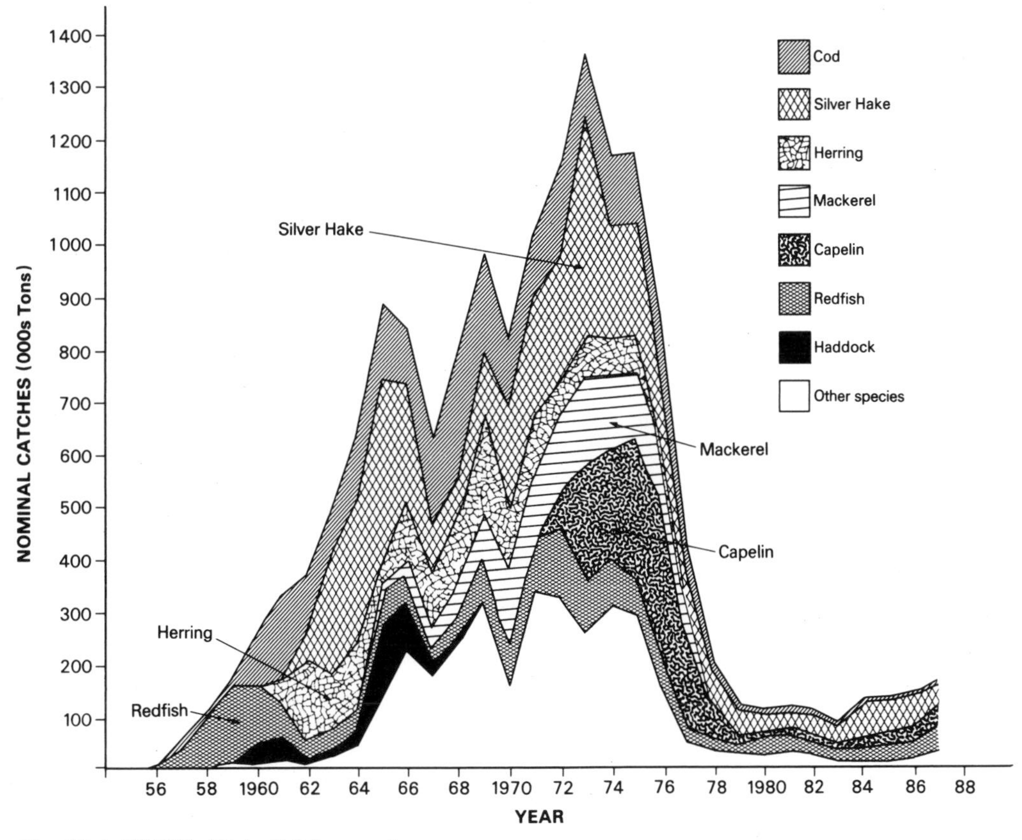

Fig. 10-1. USSR's "Pulse" fishery patterns.

total Northwest Atlantic catches declined to about 4,200,000 tons in the early 1970's. In the area off the Canadian coast (Subareas 2-4) catches declined steadily from the 1968 peak of 2,700,000 tons to less than 1,500,000 tons by 1977. Canadian catches from the Northwest Atlantic, which had also peaked in 1968, declined from 1,300,000 tons to less than 850,000 tons in 1974 and 1975.

Off the Canadian Atlantic coast groundfish had been the major component of the catch, particularly in the offshore fishery. Groundfish catches off the Canadian Atlantic coast peaked at 2,000,000 tons in 1968, then declined to less than 1,000,000 tons by the mid-1970's. The Canadian groundfish catch peaked at 620,000 tons in 1968 and then declined to a low of 418,000 tons in 1973.

The declines in catch did not reflect fully the extent of decline in the stocks because of the continued increase in fishing effort until 1975. Canadian offshore trawlers had been experiencing declines in their catch per day fished from a peak of around 12–13 tons per day during the early 1960's to about 10 tons per day in 1972 (Fig. 10-4). By 1974–75 the catch per day fished had declined to about 8 tons. This reflected a precipitous decline in the abundance of the major groundfish species, particularly cod.

Inshore fishermen were also affected by increased fishing pressure offshore. The most dramatic illustration of this was the decline in the inshore cod fishery along the northeast coast of Newfoundland and Labrador. Catches by inshore fishermen from this stock declined from 159,000 tons in 1959 to 97,000 tons by 1969 and then plunged to a low of 35,000 tons in 1974 (Fig. 10-5). This had a severe impact on the social and economic fabric of the hundreds of coastal communities whose fishermen depended on northern cod for their livelihood (Lear and Parsons 1993).

The decline in catches and catch rates led to pressure on the Canadian government to bring foreign overfishing under control. The chief avenue available to Canada in the late 1960's was multilateral action through ICNAF. The evolution of the ICNAF regulatory regime has been described in Chapters 5 and 6. ICNAF initially emphasized study of fish stocks in the ICNAF area. The first regulatory efforts aimed at minimizing wastage of small fish by imposing minimum mesh

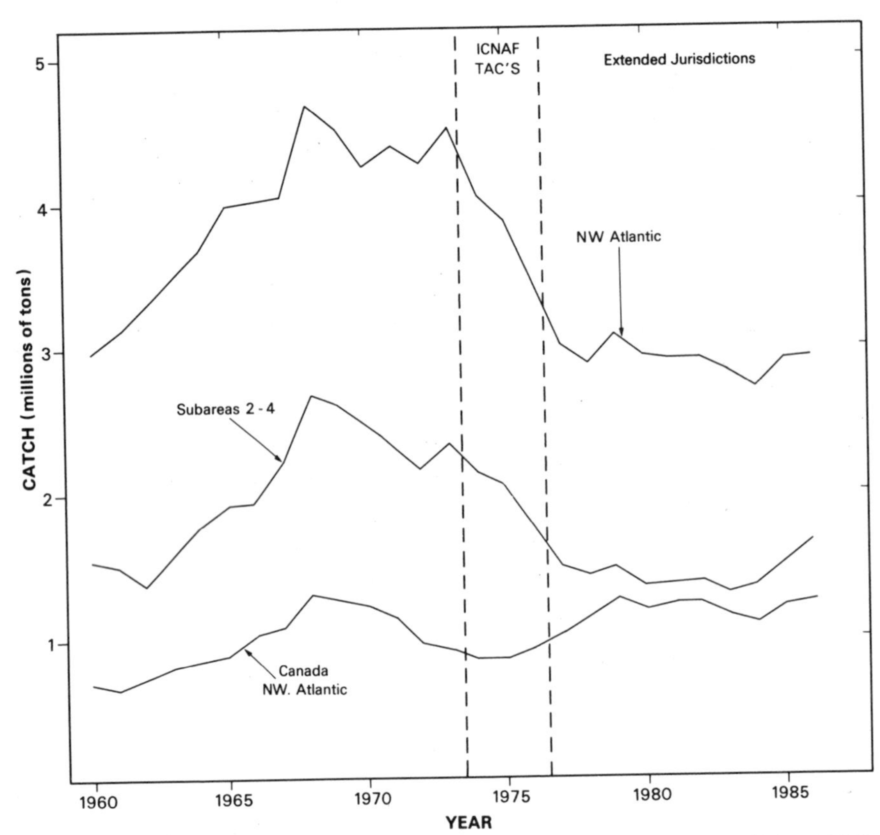

FIG. 10-2. Northwest Atlantic nominal catches (from Pinhorn and Halliday 1990). Subareas 2-4 are ICNAF/NAFO statistical areas of the Canadian Atlantic Coast.

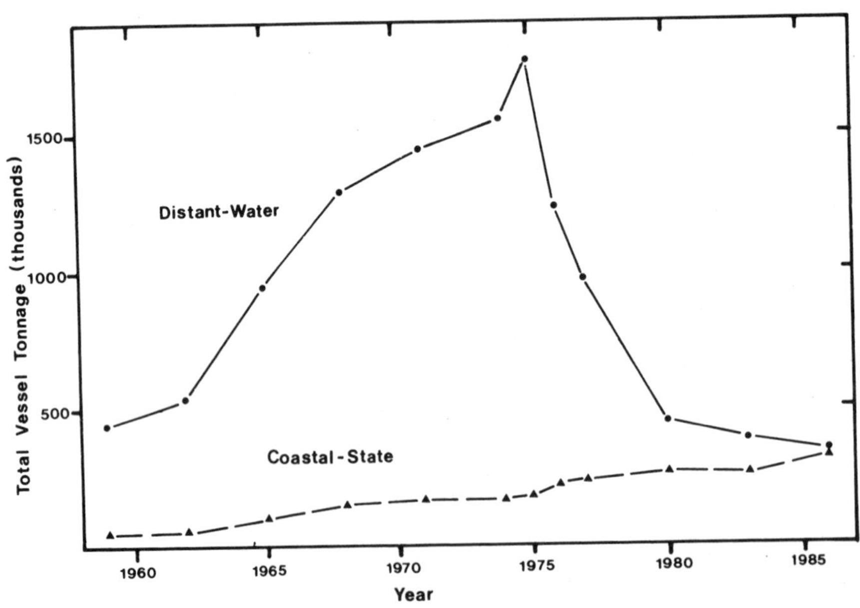

FIG. 10-3. Aggregate gross registered tonnage (GRT, tons) of fishing vessels deployed in the Northwest Atlantic by distant-water and coastal-state fleets (from Pinhorn and Halliday 1990).

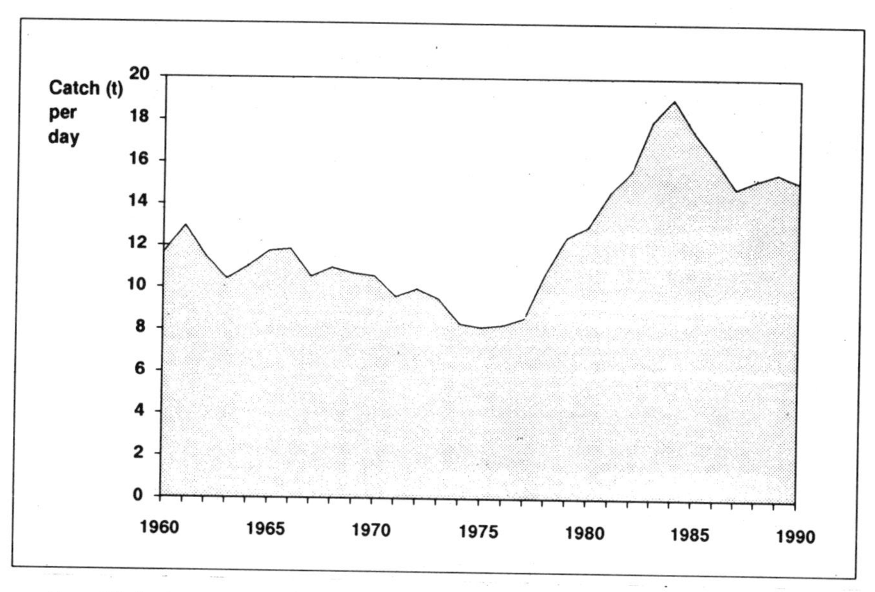

FIG. 10-4. Average catch per day by Canadian otter trawlers (from D. Rivard, Fisheries Research Branch, DFO, Ottawa).

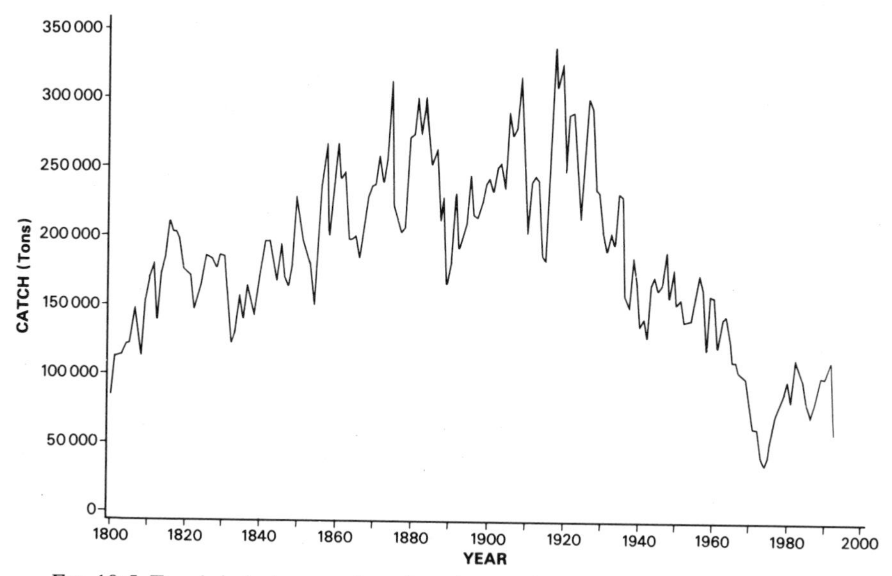

FIG. 10-5. Trends in inshore catches of northern cod: 1800–1991. In July 1992 a 2-year moratorium was introducted.

sizes in the codend of otter trawlers. Some 20 such regulations were adopted during the first 15 years of ICNAF.

In the mid-1960's ICNAF's scientific body, STACRES, advised that mesh size regulations were inadequate to control the amount of fishing. Additional regulatory measures were required to arrest the dramatic escalation of fishing effort. Templeman and Gulland (1965) called for direct control of the amount of fishing and advocated catch quotas as the most effective means of achieving this. In 1967, the ICNAF Working Group on Joint Biological and Economic Assessment of Conservation Actions called for "positive steps toward effort limitation." It too recommended catch quotas allocated among countries (ICNAF 1967). By 1968 Canada and the U.S. were collaborating to persuade ICNAF to adopt a catch quota system.

Initially, ICNAF was hampered by a provision in the

original convention stipulating that all Panel member states which would be affected by a proposal adopted by the Commission must agree to the proposal before it could apply to any member (there were five Panels each dealing with Subareas of the Convention Area). This procedure required active unanimity (Finkle 1974). This obstacle was removed in theory in June 1964 when the Commission adopted a Protocol to the Convention which modified the original provision so that proposals became effective for all contracting governments 6 months after the date of notification from the depository government. Any contracting government affected by a proposal could present an objection within the 6-month period. A proposal then became effective for all contracting governments, except governments which had presented an objection. Ratification of this Protocol by all members did not occur until 1969.

ICNAF also introduced an enforcement regime allowing mutual inspection of catches and gear at sea. By a Protocol of June 7, 1963 which did not become operative until July 1, 1971, the Commission adopted a Scheme of Joint International Enforcement (ICNAF 1974b). Under this Scheme each contracting government could appoint inspectors with authority to board a foreign vessel fishing in the convention area. The inspector would report any findings to the authorities of the flag state of the vessel concerned and to ICNAF. The flag state was left to investigate and impose penalties. Thus the Joint Enforcement Scheme had limited effectiveness. However, during the mid-1970's Canada used it to detect violations and pressure offending parties to take action. ICNAF agreed at its 19th Annual Meeting in 1969 to establish Total Allowable Catches (TACs) for haddock on the southern Scotian Shelf and Georges Bank. These were global TACs (ICNAF 1969b). It was clear, however, that for catch quotas to be widely embraced national allocations were needed. In 1969, ICNAF adopted a proposal to amend the Convention to permit national allocation of catch quotas. The amendment also modified the MSY objective to allow for proposals to achieve the optimum utilization of the stocks." This amendment took effect in December 1971 (ICNAF 1974b).

Immediately following this, a Special Meeting of ICNAF was convened in February 1972. This resulted in global TACs and national allocations for herring in Subareas 4 and 5. At the 1973 Annual Meeting in Copenhagen, TACs were established (for 1974) for 56 finfish stocks in the Convention Area (ICNAF 1973b).

By then Canada, faced with the growing conservation threat off its Atlantic coast, had taken bolder unilateral action building upon the 1964 Territorial Sea and Fishing Zones Act. It was also seeking increased coastal state jurisdiction within UNCLOS III.

2.9 Fisheries Developments on the Pacific Coast

Early bilateral and multilateral initiatives dealing with Pacific salmon have already been mentioned. These include the International Pacific Salmon Commission and the abstention line to reduce Japanese interception of salmon of North American origin. The latter measure meant fewer salmon from streams of Canadian origin were being caught on the high seas. This was distinct from interception in U.S. fisheries to the north and south of British Columbia.

The groundfish fishery off the B.C. coast was being prosecuted close to the coast in the Strait of Georgia, Queen Charlotte Sound-Hecate Strait and off the west coast of Vancouver Island (Fig. 10-6). Japanese and USSR vessels began taking groundfish off B.C. in 1966, mainly rockfishes, with an initial catch of about 45,000 tons. The Japanese then turned to blackcod using set line, and the USSR to Pacific hake using midwater trawl. Canadian and foreign fishermen fished different species. The scale of the foreign fishery was far smaller than off the Atlantic coast.

As Canada approached the end of the 1960's, the dominant Pacific fisheries international concerns were protection of Canadian salmon on the high seas and the development of mutually satisfactory salmon interception arrangements with the Americans. One of Canada's major international fisheries objectives had become the elimination of all high seas fishing by other countries for anadromous species originating in Canada. But the main pressure for extending Canada's fisheries jurisdiction seawards was being generated by the overexploitation of the groundfish resources on the Atlantic coast.

2.10 Unilateral Proclamation of Fisheries Closing Lines — 1970

In an October, 1968, speech to the American Commercial Fish Exposition in Boston, the Minister of Fisheries and Forestry, Jack Davis, spoke of Canadian concerns about the management of species beyond the existing limits of coastal state jurisdiction. He argued that, while the exclusive fishing zone had been pushed out to the 12-mile limit, the whole of the continental shelf should be subject to better management and conservation practices. He expressed concern about the survival of species like cod and haddock. The assault by the "new, large and expanding fishing fleets from Russia, Poland and East Germany" was of great concern. He spoke of leading an international crusade to make sure that "we are not going to destroy the supply of some of these higher species for all time to come" (DOF 1968d).

By 1970 some 57 states claimed a territorial sea of 12

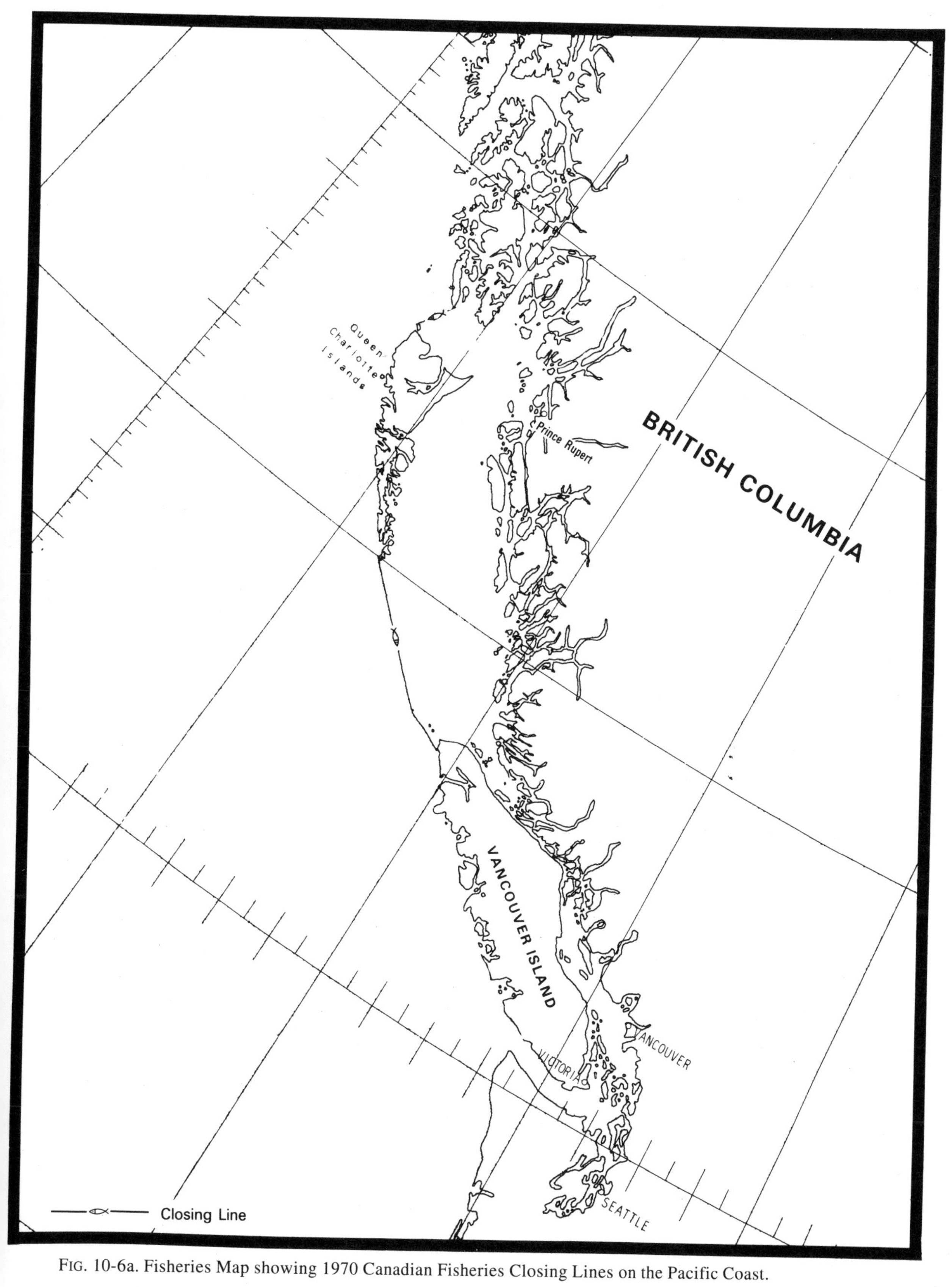

FIG. 10-6a. Fisheries Map showing 1970 Canadian Fisheries Closing Lines on the Pacific Coast.

FIG. 10-6b. Fisheries Map showing 1970 Canadian Fisheries Closing Lines on the Atlantic Coast.

miles or more (Johnson 1977). In April 1970, the Canadian government brought before Parliament two significant pieces of legislation — the *Arctic Waters Pollution Prevention Act* and an amendment to the *Territorial Sea and Fishing Zones Act*. The latter amendment established a 12-mile territorial sea (effectively eliminating the 9-mile contiguous zone of 1964). It also authorized the establishment of new fishing zones. On December 18, 1970, the Minister of Fisheries and Forestry announced that "fisheries-closing-lines were being drawn across the entrance to the Gulf of St. Lawrence, the Bay of Fundy, Queen Charlotte Sound and Dixon Entrance-Hecate Strait" (Fig. 10-6a and 6b) (External Affairs 1970).

Minister Davis informed the House that the Government would conclude negotiations to phase-out fishing by countries which had traditionally fished in the areas concerned, namely, Britain, Denmark, France, Italy, Norway, Portugal and Spain (External Affairs 1970). Canada had also signed an agreement on reciprocal fishing privileges with the U.S. The activities of United States fishermen in the areas concerned would not be affected by the promulgation of the fisheries closing lines (Canada-USA 1970). The fisheries closing lines approach pertained to fisheries jurisdiction rather than the complete sovereignty which states exercised in their territorial seas.

Following promulgation of the fisheries closing lines on the east and west coasts, phase-out negotiations were intensified and agreements reached by 1972 with all of the countries concerned. These agreements provided for the phase-out of Danish, Portuguese and Spanish fishing in the Gulf of St. Lawrence by the end of 1976 and U.K. fishing in the Gulf by the end of 1973. Fishing in the outer 9 miles of the territorial sea would cease by the end of 1978. The Norwegian agreement established a different arrangement for that country's sealing operations. This allowed Norwegian sealing in the Canadian territorial sea on the east coast on an occasional and strictly regulated basis. These agreements essentially terminated fishing within the Gulf or the territorial sea in 1978. No agreement was negotiated with Italy because it had discontinued fishing off Canada's Atlantic coast (Legault 1974).

The phase-out for France was different. It recognized that country's historic fishing rights. These rights were supported by French ownership of the islands of St. Pierre and Miquelon, immediately adjacent to the Canadian coast. The 1972 agreement with France was to become a thorn in Canada's side in the late 1980's. It provided for:

1. The phase-out of fishing by Metropolitan France trawlers in the Gulf by May 1986;
2. Continued fishing in perpetuity by a maximum of 10 trawlers registered in St. Pierre and Miquelon, with a maximum length of 50 m. These could fish along the Newfoundland and Nova Scotia coasts, except the Bay of Fundy, and in the Gulf on an equal footing with Canadian trawlers (reciprocal rights were provided for Canadian trawlers along the coasts of St. Pierre and Miquelon);
3. An arbitration mechanism to settle disputes concerning the implementation of this agreement; and
4. Establishment of the territorial sea dividing line between Newfoundland and St. Pierre and Miquelon.

For a fuller treatment of this agreement and its implications for Canada, see Chapter 12.

2.11 Coastal State Preference Within ICNAF

In addition to these unilateral and bilateral actions, Canada was now asserting itself within ICNAF. It sought preferential rights for the coastal state and a more effective conservation regime.

The ICNAF Convention had been amended to allow for national quotas. Canada and the U.S. pressed for more effective management action. ICNAF responded by adopting between 1972 and 1974 Total Allowable Catches for all commercially important stocks within the Convention Area. Canada advocated special status for coastal states in a May 1972 statement on Northwest Atlantic Fisheries Policy. In this statement to the House of Commons, the Fisheries Minister outlined Canada's approach at the upcoming ICNAF meeting in Washington. He identified two basic thrusts: conservation and a special preference for the coastal state. Regarding conservation, he indicated that Canada would be urging a cut back in the offshore fishing effort by all nations. It would also call for "a special allocation in Canada's favour, to improve the income of inshore fishermen in the Maritime provinces, Newfoundland-Labrador and Québec." The Minister described a 40-40-10-10 sharing formula: "40% to be allocated to each nation in proportion to its latest 3-year catch statistics; 40% reflecting its catches in the preceding 10-year interval; 10% additional to the coastal state; and 10% to look after contingencies" (Canada 1972).

At its 1972 Annual Meeting ICNAF adopted this formula to set national allocations for 1973. Mocklinghoff (1973) noted that, when a large reduction of total catch became necessary, the shares of coastal states were reduced to a lesser degree than the shares of other states. This represented *de facto* acceptance of coastal state priority. Although the Canadian argument that it be permitted to harvest as much as it could catch was not fully accepted by ICNAF until 1975, the principle of coastal state priority had been accepted in a tangible form in 1972.

While Canada was working within ICNAF to assert coastal state priority, it was also active at the United Nations. In concert with other coastal states, Canada was pressing the U.N. to ensure that coastal states would be assigned increased jurisdiction and management over the living resources adjacent to their coasts in a new Law of the Sea.

2.12 UNCLOS III

UNCLOS I and II had failed to reach agreement on the breadth of the territorial sea and the nature and scope of a contiguous fisheries zone. By the end of the 1960's, so many states had claimed a territorial sea of 12 miles that this was now regarded as customary international law. The explosive growth of distant water fisheries during the 1960's had reemphasized the need for coastal states to have greater control over the living resources off their coasts. There was also growing interest in the mineral resources of the deep sea.

The process leading to the Third Law of the Sea Conference was launched by Ambassador Arvid Pardo of Malta in an eloquent speech to the First Committee of the UN General Assembly. He spoke of the common heritage of mankind, calling for "an effective international regime over the seabed and the ocean floor beyond a clearly defined national jurisdiction" (UN 1967). In December 1970, the UN General Assembly adopted a Declaration of Principles Governing the Sea-Bed and Ocean Floor Beyond the Limits of National Jurisdiction. This declaration stated that the seabed, its subsoil and its resource in the area beyond national limits were: "the common heritage of mankind.... The exploitation of its resources shall be carried out for the benefit of mankind as a whole, irrespective of the geographical location of states, whether landlocked or coastal, and taking into particular consideration the interests and needs of the developing countries" (UN 1970a). The declaration called for an international regime to be established to govern all activities involving resources, including the orderly and safe development and rational management of the area, and the equitable sharing of benefits.

Another resolution called for a new conference on the Law of the Sea to be held in 1973. The range of issues would include the international regime for the seabed beyond national jurisdiction plus "the regime of the high seas, the continental shelf, the territorial sea (including the question of its breadth and the question of international straits) and contiguous zones, fishing and conservation of the living resources of the high seas (including the question of preferential rights of coastal states), the preservation of the marine environment (including inter alia the prevention of pollution), and scientific research" (UN 1970b). It was agreed that the Seabed Committee would meet during 1971 and 1972 to draft a treaty encompassing all these issues.

The Seabed Committee, expanded to 86 members, worked assiduously during 1971 through 1973 in an attempt to produce draft treaty articles. Canada's interests in the work of the Seabed Committee and subsequently in UNCLOS III were broad and ranged far beyond fisheries. I deal here only with the fisheries-related aspects of the Canadian position. An overview of the broad array of issues and Canada's evolving position concerning these can be found in Johnson and Zacher (1977), Buzan (1982), Hage (1984) and DeMestral and Legault (1979-80).

Canada took a leading role in the Seabed Committee and in UNCLOS III in seeking a new Law of the Sea to reflect the needs of coastal states, a group which Canada came to lead for a while at UNCLOS III.

Domestic concern about the overexploitation of fish resources, about potential pollution of the marine environment and related matters helped to shape the Canadian position. In January 1971, the Fisheries Council of Canada submitted a Brief to the Standing Committee on External Affairs and National Defence (FCC 1971).

The FCC advocated that the coastal state should have jurisdiction over the living resource of the Continental Shelf. Anadromous species were the one significant exception to this approach. "For these", the FCC stated, "we must have universal agreement that, a) anadromous species belong to the nation in whose territory they spend their fresh water phase, and b) no one will fish for anadromous species on the high seas." The FCC Brief presaged the key fisheries elements of the position Canada was to adopt at UNCLOS III.

Domestic pressure continued to mount in 1971, culminating in the formation of an east coast lobby group, the Save Our Fisheries Association (SOFA). Headed by Gus Etchegerry of Fishery Products Ltd., the Association was made up of some 14 groups representing the fishermen's union, processors, municipalities, development associations and other organizations affected by the fishing industry. This Association, formed in Newfoundland, sought and received support from industry representatives in the Maritime provinces. It proposed that Canada make its position absolutely clear at the 1973 Law of the Sea Conference that as a coastal state Canada would not accept "anything less than full management control over the main resources on the continental shelf by the end of 1973" (*The Evening Telegram*, St. John's. October 12, 1971).

Also in the fall of 1971 the Nova Scotia Federation of Fishermen called on the federal government "to assert control over fisheries to the edge of the continental shelf to protect the livelihood of Maritime fishermen" (*The Evening Telegram*, September 27, 1971). These domestic pressures influenced both the Canadian position in ICNAF, where Canada staked out a claim to preferential rights, and Canada's emerging Law of the Sea position.

Early discussions of the Preparatory Committee at the United Nations reflected widely divergent views. These ranged from the claims of many Latin Americans, to a territorial sea of up to 200 miles, to the views of the distant water states, who wished to minimize any extension of coastal state jurisdiction or powers. Canada initially favoured a substantial increase in coastal state powers over fisheries adjacent to its coast, and leaned toward preferential rather than exclusive rights for the coastal state. It described its approach as *functional*. This meant the granting of authority to coastal states to carry out limited and specialized functions as opposed to sovereignty. Canada's position was advanced in terms of "custodianship" and the "delegation of powers" to the coastal state (Legault 1974).

Canada's Law of the Sea position was influenced by the fact that it has a large continental shelf, with an area amounting to almost 40% of its land mass. This was the second-largest continental shelf in the world, second only to that of the former USSR. This continental shelf is narrow on the west coast but extends for several hundred miles off the Atlantic coast, up to 300–400 miles offshore in the vicinity of the Grand Banks. These factors influenced Canada's position. It did not see a 200-mile zone, along the lines proposed by the Latin Americans, as sufficient to protect its interests on the east coast because important fish stocks would straddle the outer 200-mile boundary. The driving force for the west coast fisheries was the protection of salmon on the high seas.

For these reasons Canada adopted initially a *functional* approach which differentiated between groups of fish species based on their ecology, distribution and migratory behaviour. The four categories were sedentary, coastal, anadromous and wide-ranging species. This approach was described in a statement to the Preparatory Committee by Dr. Needler, the Deputy Representative of Canada, on August 6, 1971:

> "With regard to fish stocks in coastal areas, the Canadian delegation [proposes] a resource management system under which the coastal state would assume the responsibility and be delegated the required powers, for their conservation and management — as custodian for the international community under internationally agreed principles.... Most of the stocks which live out their lives in relatively shallow areas adjacent to the coast would come within the ambit for such a delegation of responsibilities and powers to the coastal states.... The special interests of the coastal state would require provision to be made for preferential rights for the coastal states in the harvest of those species of particular socio-economic importance to the coastal population.
>
> "A sound management system for salmon would require the delegation to the coastal states of the sole right to harvest the salmon bred in their own rivers.... Canada [favours] the prohibition of the fishing of salmon on the high seas.
>
> "The management of [wide-ranging ocean species, particularly large pelagic fish and marine mammals] is best approached through international commissions" (Canada 1971).

Under this approach, the coastal state would continue to enjoy exclusive sovereign rights to the sedentary species of the continental shelf.

Canada had difficulty in selling the species approach for fisheries and preferential rights over living resources to the edge of the continental shelf in the Preparatory Committee. There were a number of reasons why the initial Canadian position did not find favour with many other countries. Under a continental shelf/species approach, a substantial number of African, Caribbean and Asian states with narrow continental shelves would gain little. The African coastal states wanted a 200-mile limit and fixed boundaries, following a commitment made to their landlocked states in 1972. Another problem for developing countries was that highly migratory species, e.g. tunas, were often the major resource off their coasts. On the other hand, the linkage of fisheries jurisdiction with the continental shelf appeared similar in certain respects to that of the Latin American territorialist countries. This led to opposition from the Maritime states. By 1972 it was clear that coastal states, both developed and developing, were in favour of a much wider jurisdiction over fisheries in a defined zone, most likely 200 miles.

In November 1973, External Affairs distributed a Paper titled *Third United Nations Conference on the Law of the Sea,* as background for consultations on the Canadian position prior to the commencement of the Conference. This paper maintained the species approach and the claim to preferential rights over living resources to the edge of the continental shelf (External Affairs 1973).

UNCLOS III stretched over nearly a decade, from 1973 to 1982. Major steps forward on the fisheries issues occurred at the first substantive session in Caracas in the summer of 1974. The major substantive issue addressed at Caracas concerned the limits of national jurisdiction. Canada and most other coastal states came to Caracas supporting expanded coastal state jurisdiction out to 200 miles. While there were some Latin American states which were sympathetic to broad-margin states like Canada, the African states were insistent that the boundary should be fixed at 200 miles for all states. The crucial turning point at Caracas was the shift in position of three major Maritime powers, the USSR, Britain and the United States. These three states, along with Japan, had been among the leading opponents of extended fisheries jurisdiction. They had been under considerable external and internal pressure to accept the growing trend to a 200-mile economic zone. In 1974, all three announced that they would support 200-mile economic zones and national jurisdiction over the resources of the continental shelf where it extends beyond 200 miles.

Sanger (1987) suggested that the matter was settled before Caracas. He saw the action by the maritime states as part of a bargain: agreement to a 12-mile territorial sea and a 200-mile economic zone in return for freedom of navigation through straits. He argued that the USSR and the United States competed in 1973 to be the first to offer the bargaining terms. Hollick (1981), on the other hand, contended that each country specified a number of conditions that must be met for their support of a 200-mile zone. Because they had not acted in concert, no reciprocal concessions were won from the coastal states.

Even though important differences of opinion still persisted, more than 100 states spoke out at Caracas in favour of a 200-mile exclusive economic zone. The surge towards the 200-mile zone was now irresistible. While Canada was pleased with the emerging acceptance of expanded coastal state jurisdiction over the living resources adjacent to its coasts, it continued to lobby for coastal-state control over stocks between 200 miles and the edge of the continental shelf, and for a special provision for anadromous species.

Canada mounted a massive lobby at Caracas to educate other non-salmon states about the unique nature of salmon and the need for special treatment for anadromous species in the new Law of the Sea. Each delegation was presented with a beautifully bound copy of *The Salmon*, a book especially written for this purpose by the well-known Canadian author and sports fishing enthusiast, Roderick Haig-Brown. Sanger (1987) described how Haig-Brown and Leonard Legault of the Canadian LOS delegation collaborated on this book. It eloquently portrayed the plight of the Pacific and Atlantic salmon, their high seas migrations and the unique contribution of the state of origin to the survival and development of salmon stocks. The book concluded by seeking recognition of the rights of origin on the following terms:

1. That anadromous stocks should be fished only by coastal states and only in areas under their jurisdiction, subject, however, to any appropriate arrangements between neighbouring states of origin where there was intermingling of their respective stocks;
2. That the conservation of anadromous stocks required comprehensive management throughout their migratory range, and that the state of origin had a special interest in such management; and
3. That a coastal state which, in its own area of jurisdiction, fished for anadromous stocks originating in another state should take into account these conservation and management requirements and consult with the state of origin in this regard (Haig-Brown 1974).

This publicity campaign, which included a visit by fisheries representatives from developing countries to British Columbia in 1974 to witness the salmon run, generated sympathy and support for the Canadian position.

At the Geneva session in 1975 detailed fisheries provisions for the zonal concept were negotiated. Following the tabling of a proposal by a group led by Kenya, the coastal state group really coalesced. Canada, while maintaining its case for preferential rights beyond 200 miles, participated actively in drafting the fisheries provisions for a 200-mile zone, thus abandoning its earlier species functional approach (A. De Mestral, McGill University, Personal Communication). At Geneva committee chairmen were asked to prepare a set of draft articles based on their best assessment of a possible compromise position. Those three sets were tabled during the last plenary meeting as the Informal Single Negotiating Text (ISNT) as a tool to further the negotiating process at the next session.

In terms of the fisheries issues, the maritime states had by now accepted the concept of an *exclusive* economic zone, rather than one involving preferential rights. The general direction of the ISNT was satisfactory to Canada, with the exception of the question of preferential rights for stocks beyond 200 miles. Canada secured agreement on a draft article on anadromous species. While it did not ban high-seas salmon fishing, it said the state of origin had ownership rights, and it provided a basis to limit high-seas salmon fishing. Johnson (1977) suggested that the salmon article outweighed the failure to get a preferential rights article for fish stocks bordering the zone. This was because salmon stocks were both economically and politically more significant than the groundfish fishery beyond the 200-mile limit on the east coast. This was a short-sighted view because overfishing beyond 200 miles became an issue of major significance to Canada during the 1980's (see Chapter 11).

Meanwhile Canada made further attempts to secure adequate provisions for conservation of stocks which overlap the 200-mile limit and the area beyond 200 miles. An article in the 1975 negotiating text provided that interested states would seek to agree on the measures to ensure the conservation of such stocks. In the end only Canada and Argentina were interested in strengthening this article. Sanger (1987) observed that the campaign by Canada and Argentina started too late, in 1980, after the margineers had won the battle for a broad definition of the continental shelf. This campaign on straddling stocks focused on an addition to Article 63, which treated the question of co-operation between two or more coastal states in conserving stocks straddling their zones. Some 16 countries co-sponsored an amendment in 1981 but there was significant opposition from Japan, the USSR, Spain, the U.S. and the EEC. During the final negotiating session, in April 1982, the straddling stocks amendment was shunted aside because it became linked procedurally with the USSR position on a compromise on the question of innocent passage by warships through the 12 mile territorial sea. Instead of being decided on its own merits, it was dropped to break a deadlock on another issue (Sanger 1987).

By this time events had in many respects overtaken the Law of the Sea Convention. As Hollick (1981) observed, the UNCLOS III negotiations fell broadly into two phases-the period up to 1976 and that from 1977 on. Up to and including 1976 the negotiations concerned the whole range of issues. After 1976, the problem of seabed mining became the only significant unresolved issue.

Buzan (1982) described Canada's policy, in terms of the fisheries provisions of the Law of the Sea Convention, as highly successful. DeMestral and Legault (1979-80) attributed Canada's success at the Law of the Sea Conference to a combination of factors. These included:

1. An early identification of genuine, tangible interests in a broad range of issues, involving thorough discussion with interested parties within Canada;
2. The delegation spoke for a country united on the issues under negotiation;
3. This support was manifested by the attendance of Cabinet Ministers at the negotiations;
4. The formation of a capable delegation which was a closely coordinated multidisciplinary team: in many instances, members of the delegation had direct operational responsibility for the issues under negotiation;
5. Strategic decisions on the forging of alliances with other states with compatible interests. These included new negotiating relationships transcending Canada's traditional western-bloc alliances;
6. Canada's ability to seek compromises among divergent interests, despite its own clearly identified interests;
7. Clear skills of advocacy and negotiation on the part of Canada's representatives. This was well illustrated by the success on the anadromous species provision, despite the fact that only fifteen states had a genuine economic interest in salmon, with major conflicting interests at stake; and
8. A recognition that "the overriding objective on any negotiation should be to reach agreement and to promote that international community interest which lies in the accommodation of conflicting national interests."

2.13 Events Leading to Canada's Unilateral Extension of Fisheries Jurisdiction to 200 Miles — 1974–1976

By 1975 essential agreement had been achieved at the LOS Conference on the concept of the 200-mile exclusive economic zone. It was clear, however, that it would be several years before all issues could be resolved and a new Law of the Sea Convention adopted. Canada, therefore, came under increasing pressure to act unilaterally to declare a 200-mile zone.

This pressure was heightened by the failure of apparently successful initiatives within ICNAF to produce immediately tangible results. By 1974, a global Total Allowable Catch system had been adopted. However, the TACs were too high to arrest the stock decline underway since the late 1960's. Dr. Arthur May, who was later to become Canada's Deputy Minister of Fisheries and Oceans, criticized the ICNAF quota system at the May 1973 meeting of the Fisheries Council of Canada in Charlottetown:

> "The present ICNAF quota scheme... must only be regarded as a short term solution for Canadian east coast problems. It does assure us of at least a status quo, a maintenance of our recent catches. But it does the same for other nations. It tends to preserve a distant fishing presence that otherwise might be more difficult to maintain, and the greatest benefits tend to go to the nations with the highest recent catches, the ones which caused the problem in the first place by virtue of the rapid build up in their fleets" (DOE 1973e).

The FCC's January 1974 Brief to the Standing Committee on External Affairs and National Defence, while supporting the basic Canadian position at the UNCLOS III, had further proposed that:

> "If the Law of the Sea Conference fails to establish a fisheries Convention which Canada can accept, Canada should unilaterally extend her exclusive fishing zones and management authority" (FCC 1974).

In June 1974, conflict developed on the east coast because foreign fleets (the USSR and Norway) were fishing capelin in a small area very close to the Newfoundland coast. On June 4, Minister Davis called for an immediate end to "large-scale Soviet and Norwegian fisheries which threaten capelin runs to inshore areas along the Newfoundland coast" (Anon 1974a).

The USSR assured Canada on June 12 that it would comply with ICNAF quotas and make accurate catch reports to ICNAF. On July 19, Canadian enforcement officials informed reporters that the results of surveillance of the USSR capelin fleet by fisheries officials from the Canadian destroyer *Algonquin* indicated that the USSR fleet had exceeded its quota of 85,000 tons of capelin (*Globe and Mail*, July 20, 1974). By July 23, it appeared that the Soviet fleet had stopped fishing for capelin (*Globe and Mail*, July 24, 1974). This incident was the prelude to a major confrontation with the USSR in 1975 (Anon 1974b).

The Minister of Regional Economic Expansion, Don Jamieson, told reporters in late June, 1974, that Canada would act unilaterally if the LOS Conference did not accept Canada's claim to a zone of 200 miles or to the limit of the continental shelf, whichever was greater. Mr. Jamieson said that "if the conference comes apart at the seams we can't let things go on the way they are now" (*Globe and Mail*, June 22, 1974).

However, in a December 1974 speech to the Rotary Club in St. John's, Mr. Jamieson, the senior Atlantic Cabinet Minister, excluded unilateral action because "we would find it extremely difficult and enormously costly to enforce a unilateral declaration against countries that might... challenge Canada's action." In the closing part of his speech he indicated that unilateral action was an option which "cannot be rejected forever" (DREE 1974).

By May 1975 the Fisheries Council was calling for unilateral action. On May 9, the Secretary of State said in the House that "unilateral action is one of the policy options open to the government, and it certainly is a lively option at present" (Canada 1975a). However, on May 12, Mr. MacEachen indicated that the upcoming ICNAF meeting would deal with a Canadian proposal calling for a considerable reduction in fishing effort: "It would seem premature to consider any action of any kind until the results of that meeting are clear" (Canada 1975b).

Canada pressed at the June 1975 ICNAF meeting in Edinburgh for a 40% reduction in foreign fishing effort off the Canadian Atlantic coast. ICNAF did not approve this request, claiming that further study was required.

ICNAF's reluctance to accept Canada's proposal resulted in further calls in the House for unilateral extension. The Newfoundland and Nova Scotia legislatures in late June, 1975, passed unanimous resolutions supporting unilateral action. Responding to questions in the House on July 22, Mr. MacEachen stated:

> "The Law of the Sea Conference was not a failure... the single negotiating text which was produced at the conference met almost all of the significant objectives which Canada had in the field of fisheries.... In the meantime, however, we as a government are considering as one of our options for future action the question of unilateral action.... It is not clear...that a simple declaration of unilateral action will, in fact, deal with the problem...I would want to be certain that taking that course would bring substantial benefit within a reasonable period to the east coast fishery" (Canada 1975c).

The following day, July 23, Minister LeBlanc announced that Canada's Atlantic ports would be closed to Soviet fishing vessels effective July 28. Mr. LeBlanc indicated that this action was necessary because the Soviet fishing fleet was consistently overfishing ICNAF quotas. Repeated attempts by Canada to bring these practices to a halt had met with no satisfactory response from Soviet authorities (DOE 1975d).

In announcing the closure of Atlantic ports to the USSR fleet, Minister LeBlanc indicated that the Spanish and Portuguese fleets had also been involved in 'violations' using nets with undersized mesh. They had also discarded large tonnages of species they did not want without keeping records as required by ICNAF. He indicated that Canada was making a direct approach to the Spanish and Portuguese governments and that "if the performance of their fleets does not improve immediately, our ports will be closed to them as well."

Spanish and Canadian officials met on August 6 and 7. In a Joint Communiqué on August 8, "both sides recognized that it was imperative to ensure strict fulfilment of obligations assumed under [ICNAF], particularly in light of serious declines in the stocks." The Spanish officials indicated that they had recently met with representatives of the Spanish fishing fleet to improve compliance with ICNAF regulations. It was agreed that under the ICNAF Scheme of Joint International Enforcement steps would be taken to enable Spanish fisheries inspectors to work with Canadian inspectors in securing compliance with ICNAF regulations (External Affairs 1975a).

Provincial Premiers, meeting in St. John's, passed a resolution on August 22, 1975, calling on the federal government to proceed with a unilateral declaration to give Canada effective control over all commercial fishing within 200 miles off Canada's Atlantic coast (*Globe and Mail*, August 23, 1975).

USSR and Canadian officials met in Ottawa from August 25 to 27 to address the USSR overfishing/ Canadian port closure issue. In a Joint Communiqué issued on August 28:

> "Both sides recognized that it was imperative to ensure strict adherence to and implementation of measures agreed within [ICNAF], particularly in light of the urgent need to maintain and restore the stocks" (External Affairs 1975b).

The officials agreed to recommend that their governments establish a "joint Fisheries Consultative Commission" to review problems, exchange information, cooperate on enforcement and help prevent damage to fishing gear. The two sides further agreed "to ensure the prompt discontinuance of a fishery, when the national quota allocation for the stock in question has been taken."

Prior to the September 1975 Special ICNAF meeting Canadians held further meetings with officials of Portugal, Norway, Poland and a high-level meeting with the USSR . LeBlanc announced the reopening of Canada's Atlantic ports to the Soviet fishing fleet on September 29, 1975.

The Soviet Union agreed to enter into a bilateral agreement with Canada in the future "covering fishing in an extended 200-mile Canadian fishing zone." The Minister for External Affairs described this as "the most important single development in the fisheries field that we have been working on yet" (External Affairs 1975c).

Through these bilateral meetings, Canada laid the groundwork for acceptance of its conservation and regulatory proposals at the Montreal Special ICNAF meeting.

At the September ICNAF meeting Canadian proposals for a 40% reduction in foreign fishing effort for groundfish species, for more stringent catch limits on six fish stocks in critical condition, and for Canada to be allocated higher percentage shares of the overall catches, were accepted (DOE 1975e).

In addition to the fishing effort regulation, another major result was the adoption of the $F_{0.1}$ approach for setting TACs for certain critical stocks — the first sign of the abandonment of the MSY approach of the past (see Chapter 4). This ICNAF meeting in September 1975 was a major turning point. Canada subsequently directed its efforts towards unilateral extension of fisheries jurisdiction in the context of the consensus which had emerged at the 1975 Geneva session of LOS.

In the summer of 1975 Canada headed down the path of unilateral extension but with acceptance by the major fishing countries negotiated in advance. This was intended to minimize the potential problems of compliance with an unexpected unilateral extension. The confrontation with the Soviet Union in July-August 1975 and the closure of Canadian Atlantic ports to the Soviet fleet, with accompanying threats of similar action to Spain and Portugal, were pivotal events. These set the scene to negotiate bilateral agreements which secured acceptance in advance of a Canadian 200-mile limit.

Following the ICNAF meeting, Canadian negotiators intensified their efforts to put in place the framework of bilateral agreements that would smooth the transition to a Canadian 200-mile limit. Norway was the ideal partner with which to start this series of agreements. Norway and Canada shared a common approach to coastal state jurisdiction, and Norway had only a small fishery in the Canadian zone. Canada had none in the Northeast Atlantic.

In December 1975, the Secretary of State for External Affairs announced an agreement between Canada and Norway on fisheries (External Affairs 1975d). The agreement would govern continued fishing by Norwegian vessels in areas "to be brought under Canadian jurisdiction beyond the present limits of the Canadian territorial sea and fishing zones off the Atlantic coast." The Agreement permitted Norwegian vessels to fish in the area concerned, under Canadian authority and control, for resources "surplus to Canadian requirements." The key provision was Article II, which served as the prototype for later agreements with Poland, Spain, Portugal and the USSR. This Article provided for continued fishing by Norwegian vessels in an extended Canadian fishing zone "for allotments, as appropriate, of parts of total allowable catches surplus to Canadian harvesting capacity." It stipulated that Canada would set TACs, determine the allocation of surpluses, and licence Norwegian vessels to fish in the Canadian zone.

Similar agreements were reached with Spain on February 20, 1976, Portugal on March 12, Poland on March 25, and the USSR on May 19.

The conclusion of the agreement with the USSR set the stage for the announcement of a decision which had been taken in principle by Cabinet in February 1976. The Secretary of State for External Affairs, Allan MacEachen, announced in the House of Commons on June 4, 1976, the Government's decision to extend Canadian fisheries jurisdiction out to 200 miles from the coast, effective January 1, 1977. He noted that the U.S. and Mexico were taking similar action effective in 1977 (Canada 1976).

Mr. MacEachen noted that the agreements with Norway, Poland, the USSR, Spain and Portugal, in addition to the existing 1972 agreement with France, covered the major foreign fisheries off Canada's Pacific coast and "more than 88% of the foreign catch in that part of the ICNAF Convention area to be incorporated within Canada's 200 mile fisheries zone."

On June 8, 1976, in a speech at the opening of the 1976 Annual Meeting of ICNAF in Montreal, Fisheries Minister LeBlanc informed ICNAF members that "Canada is committed to allowing others to fish for stocks which may be surplus to Canadian capacity. For many stocks there can only be a surplus if the stocks are rebuilt. It is the process of rebuilding that the Government of Canada is preparing to devote itself now" (DOE 1976h).

He stated that for 1977 Canada would determine within its 200-mile zone the conservation measures to be applied, the vessels which would be allowed to fish, and the allocations they would be allowed to take. As an interim measure, for 1977 only, Canada would give effect to regulations adopted within ICNAF with Canada's concurrence, by adopting and enforcing such regulations under Canadian law. He indicated that Canada might also adopt additional regulatory measures for 1977. However, these too would take into account decisions within ICNAF and would be consistent with agreements reached at the ICNAF meeting with Canadian concurrence.

Minister LeBlanc also signalled Canada's intention to seek changes in existing ICNAF arrangements to ensure continued multilateral cooperation in managing stocks beyond 200 miles. Canada served notice of its intention to withdraw from ICNAF on December 31, 1976, to preserve its options, but indicated that it would not necessarily proceed with withdrawal.

At the 1976 ICNAF meeting Canada secured the further reductions in TACs that it considered necessary and the adoption of the $F_{0.1}$ reference level as the guide post for management. For many stocks inside the Canadian 200-mile zone, including most of those off Nova Scotia, only Canadian fishermen could fish in 1977. For groundfish traditionally fished by Canadian fishermen, the TAC for all nations combined was decreased from 956,600 tons in 1976 to 668,500 tons in 1977, a reduction of 30%. Foreign fleets would absorb nearly all the quota reductions on these stocks. Their total share would decrease by 47%, with reductions as high as 68% for some countries (ICNAF 1976a).

For these same groundfish stocks, the Canadian quota increased very slightly to 339,600 tons in 1977 from 336,000 tons in 1976. However the Canadian percentage share of the TACs increased from 35% in 1976 to 51% in 1977.

ICNAF agreed to meet again in Tenerife, Spain, in December, 1976, to consider TACs and national allocations for seven stocks deferred pending further scientific information. It also established a working group whose members, independent of their governments, would suggest possible future arrangements for international fisheries cooperation in the Northwest Atlantic.

The December Special Meeting of ICNAF adopted, by a large majority, amendments to the ICNAF Convention proposed by Canada. These amendments recognized Canada's right to manage fisheries within its 200-mile zone effective January 1, 1977. The Commission would no longer have any management functions within this zone. It would continue managing fisheries beyond the 200-mile zone. The meeting also agreed that action be taken as soon as possible in 1977 to develop new multilateral arrangements for managing stocks beyond 200 miles (ICNAF 1977). ICNAF was thus the first international commission to adapt itself to extended fisheries jurisdiction.

3.0 THE NEW REGIME OF EXTENDED FISHERIES JURISDICTION

3.1 The Legal Context

Legal implementation of the Canadian 200-mile fisheries zone was a simple step. The *Territorial Sea and Fishing Zones Act* of 1964 provided the necessary enabling legislation. The 200-mile zone was given legal effect through an Order in Council, dated November 2, 1976. Two new Canadian fishing zones were proclaimed, Zone 4 on the Atlantic and Zone 5 on the Pacific coast effective January 1, 1977 (Canada 1977a) and Zone 6 in the Arctic on March 1, 1977 (Canada 1977b) (See Fig. 10-7). The Order in Council contained the geographic coordinates defining the zones in which Canada would be exercising fisheries jurisdiction. The Canadian claim extended up to a 12-mile territorial sea around St. Pierre and Miquelon and overlapped the U.S. claim published in the Federal Register on November 4, 1976. It raised obvious boundary delimitation issues. The Canadian Order-in-Council referred to boundary delimitation talks with the U.S., France and Denmark. It stated that the limits of the Canadian fishing zones as defined in the order were "without prejudice to any negotiations respecting the limits of maritime jurisdiction in such areas." Chapter

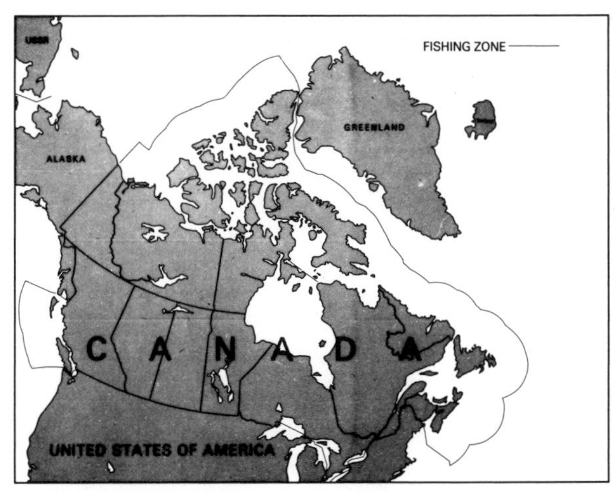

FIG. 10-7. Map showing the Canadian 200-mile fishing zone.

12 describes the evolution of Canadian fisheries relations with Denmark, France and the U.S. in the post-extension era.

Following the worldwide adoption of the 200-mile zone concept in 1976 and 1977, efforts continued to secure a comprehensive Law of the Sea Convention. The fisheries provisions had been largely settled by 1976. Negotiations continued for the remainder of the decade on two main issues: the regime for exploitation of the mineral resources of the deep seabed, and formation of an International Seabed Authority for the area beyond national jurisdictions over the continental shelf. The election of President Reagan in 1980 precipitated a crisis, as the U.S. government undertook a wide-ranging review of its Law of the Sea policy. This resulted in a tough new U.S. stance at the Law of the Sea negotiations. As a result, the U.S. voted against the Convention, even though substantial concessions were made by other parties in the last weeks of UNCLOS III. Despite U.S. opposition, the Convention was signed by 119 states and organizations in December 1982. The Convention was open for signature for 2 years. During that period the number of signatories increased to 159. These did not include the United States, Britain and West Germany. However, the European Economic Community signed the Convention on behalf of its members with respect to fisheries and marine pollution. The Convention will enter into force when 60 countries have ratified it. As of December 1992, 52 countries had ratified the Convention.

Thus, at the beginning of the 1990's, the Law of the Sea Convention and the 200-mile exclusive economic zone concept were still not codified in international law. However, the 200-mile zone was widely adopted in the late 1970's. Thus there is little doubt that the 200-mile zone is now customary international law. Scovazzi (1985) argued:

> "Owing to widespread acceptance, within a few years the 200-mile fishery zone acquired the status of a customary rule of international law and many coastal states, great maritime powers included, completed their shift towards extended marine jurisdiction by proclaiming exclusive economic zones."

Evensen (1985), who played a major role in negotiating the Law of the Sea Convention, stated:

> "The concept of the 200-mile economic zone has gained such worldwide acceptance in the practice of states, jurisprudence and international law literature that it in all probability today must be considered as forming part of the established principles of international law basically by reason of its general acceptance by the international community. It is one of the most conspicuous examples of how the technological revolution created a legal and political vacuum which was filled by this international law concept in a surprisingly brief period."

Some states have proclaimed 200-mile exclusive economic zones. Others have proclaimed 200-mile fisheries zones. Canada fell in the second category. It utilized the existing *Territorial Sea and Fishing Zones Act* to proclaim three new fishing zones. Thus Canada's 200-mile fishing zone falls short of the Law of the Sea Convention's provisions respecting an Exclusive Economic Zone. According to the 1982 LOS Convention, the coastal state has sovereign rights in an Exclusive Economic Zone to explore and exploit, conserve and manage the living and non-living resources of the waters above the seabed and the seabed and its subsoil. In the Exclusive Economic Zone, the coastal state also has jurisdiction to establish and use artificial islands, installations and structures, to conduct marine research and protect the marine environment. While Canada has exercised many of the provisions pertaining to an Exclusive Economic Zone, it has not legally proclaimed such a zone.

Early in UNCLOS III there was considerable debate about whether the new 200-mile zone would be regarded as territorial sea or high seas. In the end, it was agreed that the 200-mile Exclusive Economic Zone would be *sui generis,* i.e. unique. It would combine elements of both the territorial sea and the high seas. The coastal state exercises certain sovereign rights in the Exclusive Economic Zone (or Fishing Zone). However, it also has obligations regarding the rights of other states (e.g. freedom of navigation and overflight). Thus, the Canadian 200-mile zone is "neither fish nor fowl" but a different form of animal combining features of both.

The new Canadian 200-mile zone encompassed 503,000 square miles on the Atlantic coast and 129,000 square miles on the Pacific coast under Canadian fisheries jurisdiction. Because the continental shelf off the Pacific coast is very narrow, fishing occurred in only about 40% of this area. In January 1977 Canada's immediate challenges were to control foreign fisheries in this new zone and continue efforts to rebuild stocks which had been overfished under the previous multilateral fisheries arrangements.

A Task Group on Extended Fisheries Jurisdiction had been established within the Fisheries and Marine Service of the Department of Fisheries and the Environment in April 1976. It was to develop plans for the management of both foreign and domestic fisheries in the new zone. The group developed a regulatory regime to control foreign fishing. This was implemented through new sets of Coastal Fisheries Protection Regulations and Foreign Offshore Fishing Regulations. It developed plans for surveillance and enforcement of the 200-mile limit to ensure compliance with the new Canadian regulatory regime, and for a significantly increased research effort, particularly in the area of resource assessment.

Canada had extended its jurisdiction in the context of the developing consensus at UNCLOS III. The final text of the Law of the Sea Convention, signed in 1982, sets forth in Section V the provisions respecting the exclusive economic zone. Articles 61 and 62 define coastal state rights and obligations under this regime. These do not differ greatly from the Revised Single Negotiating Text under discussion at the time Canada extended its fisheries jurisdiction.

Article 61 dealt with conservation of living resources. Article 62 addressed utilization of these resources. These articles established the framework for setting total allowable catches, the factors to be taken into account, the coastal state's right to determine its harvesting capacity and provisions for determining the allocation of surpluses. Article 62 also established the obligation of other States to comply with the conservation measures and other laws and regulations of the coastal State. The scope of such regulations was suggested.

3.2 Regulation of Foreign Fishing

These two Articles (see Appendix I) established the context in which Canada would operate as a coastal State. Canadian rules governing foreign fishing were established in the Coastal Fisheries Protection Regulations under the *Coastal Fisheries Protection Act*, (R.S.C. 1985, c.c.33) and the Foreign Fishing Vessel Regulations, under the *Fisheries Act* (R.S.C. 1985, C.F-14). The Coastal Fisheries Protection Regulations covered the basic control aspects of foreign fishing, including licensing, reporting procedures and vessel identification requirements (C.R.C. 1978, C.413 (as amended)). The Foreign Fishing Vessel Regulations contained operational restrictions on foreign fishing including quotas, gear restrictions and closed areas (C.R.C. 1978, C.815 (as amended)).

Canada treated 1977 as a transitional year. Accordingly, it adopted the TACs and national allocations agreed to in ICNAF during 1976. Each foreign fishing vessel and each service vessel required a licence to operate in the Canadian 200-mile zone, although there were certain exemptions for French and American vessels. Canada had passed transitional regulations exempting French vessels from the licensing provisions of the Coastal Fisheries Protection Regulations in ICNAF Division 3Ps, in the region of St. Pierre and Miquelon. The licensing exemption applied to U.S. vessels in the area south of 63°N and seaward of 3 miles.

These foreign fishing licences specified what kind of fish could be caught and where, the quantity that could be caught and a variety of operating conditions. Each country was required to submit a fishing plan in the early autumn for their proposed fishery in Canadian waters during the following year. These plans had to conform to the national allocations assigned by Canada. The plan had to include the numbers of fishing and support vessels. It also had to identify type of vessel, fishing capacity, area to be fished, species to be fished, time of year when fishing was planned, and amount of fishing effort to be applied, (number of days on ground, number of days fishing).

These plans were then reviewed by Canadian technical experts to ensure that the planned amount of fishing effort was reasonable in light of anticipated catch rates and the available allocation. This meant that Canada was applying both a catch quota and a fishing effort regulatory system to the foreign fleets. This made it easier to enforce the regulatory system through monitoring of days on ground and days fishing. These fishing plans were reviewed in bilateral meetings with each nation. These discussions often focused on the determination of the appropriate catch rate for a species, type of vessel and area. Following approval of the fishing plans, each country was required to submit licence applications with a fishing plan for each vessel.

To facilitate surveillance and enforcement, vessels were required to report in by radio 24 hours before entering and 72 hours before leaving the Canadian 200-mile zone. Vessels had to report location, catches, discards and other operational information by radio at least once a week. Vessels also required certain types of markings for identification.

During the transition year 1977 no licence fees were charged. The fee system introduced in 1978 combined access fees for all fishing vessels and service vessels, which amounted to $1.00 per gross vessel ton, with a fishing fee of 8 cents per gross ton per day for each day on the fishing grounds. (These fees were increased in later years).

In the first year of extended jurisdiction 551 foreign fishing vessels were licensed on the Atlantic coast and 46 on the Pacific, with about 160 support vessels.

3.3 Surveillance and Enforcement

Canada faced the challenge of demonstrating that it could police effectively the 200-mile zone. It had been preparing for these responsibilities for some time. In March 1976, the federal Cabinet had approved a 5-year plan for fisheries surveillance and enforcement involving the coordinated deployment of vessels of the Fisheries and Marine Service, aircraft and ships of DND and vessels of the Department of Transport. An additional $4 million per year was diverted to offshore surveillance and enforcement for a total of $12 million per year commencing in fiscal 1976–77.

The additional funds were intended to enable:

1. An increase in Tracker aircraft patrols from 2000 to 4000 hours per year;
2. A doubling of offshore ship time to 1,650 sea days on the Atlantic and 495 days on the Pacific;
3. Air surveillance once weekly of sensitive fishing areas (Flying hours would total 3,750 per year on the Atlantic coast and 480 hours on the Pacific); and
4. The boarding of vessels offshore — with foreign vessels to be boarded at least four times per year, and Canadian vessels at least twice per year. The objective was to board and inspect one third of the foreign fleet and one-sixth of the Canadian fleet each month.

At that time the Department of Fisheries and the Environment had eight vessels deployed on the Atlantic coast and three on the Pacific coast. However, not all of these were capable of true offshore patrols. Sixteen Tracker aircraft plus three squadrons — six aircraft each — of long range Argus were used on the Atlantic coast and three Trackers plus another squadron of six Arguses on the Pacific.

Canada was able to maintain a credible presence in enforcing the 200-mile limit in the early years of extended jurisdiction. During the first year of extended jurisdiction 900 foreign and 170 Canadian vessels were inspected. Fourteen convictions were obtained against foreign captains for violating Canadian regulations. The major challenges to Canadian authority came later, in the 1980's. These involved incursions of foreign vessels, particularly Spanish, into the Canadian zone on the Grand Banks. This resulted in beefed-up surveillance and new enforcement approaches. There were also problems on the east coast with incursions of American vessels into the Canadian zone following the boundary delimitation on Georges Bank by the International Court of Justice in 1984. Chapter 19 discusses the evolution of Canada's approach to surveillance, enforcement and compliance of both the foreign and domestic fleets in the years following extension in more detail.

3.4 Fisheries Research Post-Extension

Another major component of the Canadian response to the 200-mile limit was to strengthen the Canadian capability for fisheries resource assessment research. Resource assessment was now the responsibility of Canada as the coastal state. To rebuild the resource, a more sophisticated Canadian capability was required to estimate, and predict trends in, fish stock abundance. The failure of many international commissions to manage the fisheries effectively was partly due to inadequate efforts by member countries to assess adequately the resource on a continuing basis. The level of sampling of the Atlantic offshore commercial fish catch in the mid-1970's was only 40% of that required to meet the minimum level of sampling identified by ICNAF's scientific committees (DOE 1976i). Similarly, resource survey activity had been far less than the minimum required. The data presented by certain countries were highly suspect, and appeared to have been manipulated by those countries to suit a particular end.

The Task Group on Extended Fisheries Jurisdiction concluded that Canada as the coastal state had to develop an independent capability to assess the fish resources off its coasts. This required significant increases in scientific activities aimed at assessing the resource potential, monitoring resource fluctuations and predicting trends in fish abundance.

In 1976, considerable scientific effort was directed toward participating in the preparations for an orderly transition to the 200-mile fishing zone in January 1977. This included the analysis of the research that would be required to manage the new zone effectively, the acquisition of the necessary additional funding and putting in place appropriate structures for implementation.

Proposals were made to double commercial catch sampling capability. It was also estimated that 1,500 ship days were required for various kinds of resource survey activity on the Atlantic coast. The Department at that time operated only two research trawlers capable of

Tracker aircraft engaged in fisheries surveillance (Courtesy Department of National Defence (IH 74-196).

Fisheries patrol vessel, Cape Freels.

offshore fishing operations, neither of which was ice-strengthened. Overall, it was calculated that the equivalent of an incremental five vessel years were required: (1) to conduct the required resource inventory surveys; (2) to conduct surveys with specialized gear to determine the abundance of precommercial-age fish for fish species which recruit to the fishery at an early age but the young of which are not taken in regular inventory surveys; (3) to conduct egg and larval surveys to provide a basis for describing the early life history of several major fish species and enable the construction of population models combining early life history stages with adult production and insights on stock recruitment relationships; and (4) to conduct specialized surveys devoted to evaluating and improving existing survey techniques and research methodology (DOE 1976i).

By 1979–80 the government had allocated an additional 216 person-years and $23.0 million to conduct the additional research considered necessary (W.G. Doubleday, Ottawa, Personal Communication).

This more than doubled Canadian resource assessment activities in the offshore area. A large proportion of this increase was used to expand Canadian direct-survey capability at sea. This included the long term charter of a 263-foot ice-strengthened offshore trawler, the *Gadus Atlantica*, capable of conducting research off Newfoundland-Labrador during the winter, and another large stern trawler, the *Lady Hammon*, capable of bottom and midwater trawling.

Another major element of the extended jurisdiction management regime involved observers on foreign vessels licensed to fish in the Canadian zone (see Chapter 19). These personnel had a dual scientific and enforcement role. Foreigners were levied a charge of $125 per day to pay for observers. Observers studied fishing methods, gear, catches and other aspects of the fishing operation. They also collected biological samples and obtained detailed information on by-catches and discards.

To assist in managing the foreign fleet, Canada developed a comprehensive computerized data management system called FLASH (Foreign Licensing and Surveillance Hierarchical Information System). This system processed information from foreign fishing licences, data acquired through surveillance and inspection activities and weekly catch and position reports submitted by foreign vessels. Computer terminals in St. John's, Halifax and Vancouver were linked to a central computer in Ottawa. The information in the central computer was updated daily in the Regions. Regional offices could determine from the system the status of any vessel, the current fishing activity, or the level of catch within any quota. Surveillance air crews and inspectors going to sea on the patrol vessels were briefed using reports produced from FLASH.

With increased surveillance, enforcement, and research, Canada appeared well positioned to manage the foreign fishing activities in the new Canadian zone.

4.0 IMPACT OF THE CANADIAN 200-MILE FISHERIES ZONE

4.1 General

Canada's extension of fisheries jurisdiction to 200 miles appeared to set the scene for an expanded and more properous Canadian fishery. Expectations were high in many sectors of the fishing industry that the 200-mile zone signalled the beginning of a bonanza. Several

Fisheries research vessel, Gadus Atlantica.

of the coastal provinces, e.g. Newfoundland and Nova Scotia, and many in the fishing industry, advocated a rapid expansion of Canada's harvesting and procesing capability to take advantage of the opportunities the new zone would make available. The federal government was almost alone in advising a cautious approach to the development of an expanded Canadian fishery in the new zone.

Canada did benefit considerably by its extension of fisheries jurisdication. The benefits came through increased management authority over a vast area with major fish resources, through the gradual displacement of foreign fisheries from the Canadian zone and through the development of Canadian fisheries in areas and for species not previously utilized by Canada. The most important benefits came from the rebuilding of fish stocks which had been overfished in the pre-extension era.

4.2 Increased Management Authority Over a Vast Area and Major Fish Resources

Canada was a major beneficiary of the Law of the Sea negotiations. With the second largest continental shelf in the world and one of the largest fishing zones after extended jurisdiction in 1977, Canada saw great promise in implementation of the LOS Convention. Johnston (1985) argued that Canada "might be regarded as the country which had the most to gain, in relative, if not absolute resource terms, from the new law of the sea." Canada's new 200-mile fishing zone was one of the largest in the world. Sanger (1987) ranked Canada seventh in terms of area of the new 200-mile zones. Alexander and Hodgson (1975) ranked Canada fifth behind the United States (1st), Australia (2nd), Indonesia (3rd) and New Zealand (4th). Johnston (1985) calculated that, "if one adds in those Arctic Ocean areas which, technically considered, might be regarded as falling under the UNCLOS III regime of internal waters, Canada is probably to be ranked third or fourth among the world's largest gainers of surface area, and second or third, with the depth dimension of ocean space." By any measure Canada has an extremely large 200-mile fishing zone.

The introduction of this zone had a dramatic impact. The immediate effect was to bring the major fish resources within 200 miles of the Canadian coast under more effective control. Although a few stocks on the Nose and Tail of the Grand Banks straddle the boundary and could be fished by foreigners outside the Canadian 200-mile limit, more than 90% of the stocks of commercial significance off the Canadian Atlantic coast occurred within the Canadian zone.

The major impact in terms of bringing fish resources under Canadian management was on the Atlantic coast. The Atlantic continental shelf extends beyond 200 miles on the Nose and Tail of the Grand Bank whereas on the west coast the continental shelf is very narrow — only 35–40 miles in breadth. The 200-mile fisheries zone brought the important Atlantic groundfish fishery under Canadian management, with the exception of the few stocks straddling the boundary on the Grand Bank and three relatively unimportant stocks on the Flemish Cap beyond 200 miles (see next chapter). Prior to extension the Atlantic groundfish fishery was managed by ICNAF, with varying portions of various stocks occurring within the Canadian 12-mile territorial sea.

The impact on the Atlantic herring fishery was much less significant since most stocks, with the exception of that on Georges Bank, spent most of their life cycle, and were fished, within Canadian internal waters. Similarly, most of the major shellfish resources, e.g. lobsters, crabs, occurred within Canadian internal waters prior to the 200-mile fisheries zone. The major exceptions were the Georges Bank scallops and the shrimp stocks off Labrador and northeast Newfoundland.

Another resource affected significantly by extended jurisdiction was the major capelin stocks off northeast Newfoundland-Labrador and on the Grand Bank.

On the Pacific coast, the resource implications of the 200-mile zone were much less significant. The major Pacific salmon fishery occurred within Canadian or U.S. internal waters prior to extension. Interceptions of Canadian-origin salmon on the high seas were not affected directly by the establishment of the 200-mile fisheries zones. Interception of Canadian-origin salmon by U.S. fishermen and U.S.-origin salmon by Canadian fishermen continued to be a problem until agreement on a Pacific Salmon Treaty was achieved in 1985. This Treaty did not solve all the interceptions problems but it provided the framework for the negotiation of long term and annual sharing arrangements.

B.C. herring were fished exclusively by Canadian fishermen in Canadian waters prior to extension. Their management was not affected significantly by establishment of the 200-mile zone. Similarly, the shellfish fisheries in British Columbia were prosecuted close to the coast.

The major B.C. groundfish fishery for halibut was affected substantially by extended jurisdiction. Although the International Pacific Halibut Commission continued to manage the resource, Canadian fishermen were excluded from fishing off Alaska after an initial phase-out period. Similarly, U.S. fishermen were excluded from fishing groundfish off British Columbia.

The foreign fisheries for B.C. groundfish, although small in comparison with those on the Atlantic coast, had occurred adjacent to but outside Canada's 12-mile territorial sea. Thus the 200-mile fisheries zone brought these resources entirely under Canadian management for the first time.

Overall, the most significant impact of the 200-mile fisheries zone occurred on the Atlantic coast, where most of the major groundfish resource was brought under Canadian management. Problems occurred with the management of those stocks which straddled and occurred beyond the 200-mile fisheries limit. These are discussed in the next chapter.

4.3 Impact on Foreign Fisheries in the Canadian Zone

4.3.1 General

Canada had gained control over considerable fish resources previously subject to international high-seas exploitation. With Canada's rights to manage and exploit these resources came certain obligations. Canada had extended its jurisdiction in the context of the developing consensus on the Law of the Sea and had undertaken to manage the fisheries of the extended zone in accordance with the general principles being developed at UNCLOS III. Article 62 of the LOS Consolidated Negotiating Text provided that the coastal state would set the Total Allowable Catch and determine its capacity to harvest the living resources of the exclusive economic zone. Where the coastal state did not have the capacity to harvest the entire allowable catch, it shall.... "give other states access to the surplus of the allowable catch."

4.3.2 Atlantic Coast

Over time, foreign fishing in the Canadian zone diminished substantially compared with the level of effort and catches in the late 1960's and early 1970's. Part of this reduction was achieved multilaterally in the year or two prior to extended jurisdiction. However, the major impact became evident after extension (see Fig. 10-8).

After extension there appeared to be surpluses to Canadian harvesting capacity in many stocks brought under Canadian jurisdiction. Foreign fleets were given access to these surpluses under specific terms and conditions of benefit to Canada (see Chapter 11). Over the following decade the situation changed dramatically. As the Canadian fleet demonstrated it could take most of the Total Allowable Catch for most species, the surpluses diminished. Canadian harvesting capacity replaced foreign fleets in many important stocks.

This occurred as early as 1977 and 1978 on the Atlantic coast for a number of fisheries from which the foreign fleet had formerly harvested a substantial portion of the TAC. Examples were cod, haddock, pollock and herring on the Scotian Shelf, American plaice and yellowtail flounder on the Grand Banks, and flounders generally on the Scotian Shelf. Early in the new regime the foreign allocations of cod, redfish and flatfish off Newfoundland and Labrador were also substantially reduced and the Canadian shares increased.

As Canadians began to redirect fishing effort to species and areas where they had previously fished little, the surpluses available for allocation to foreign

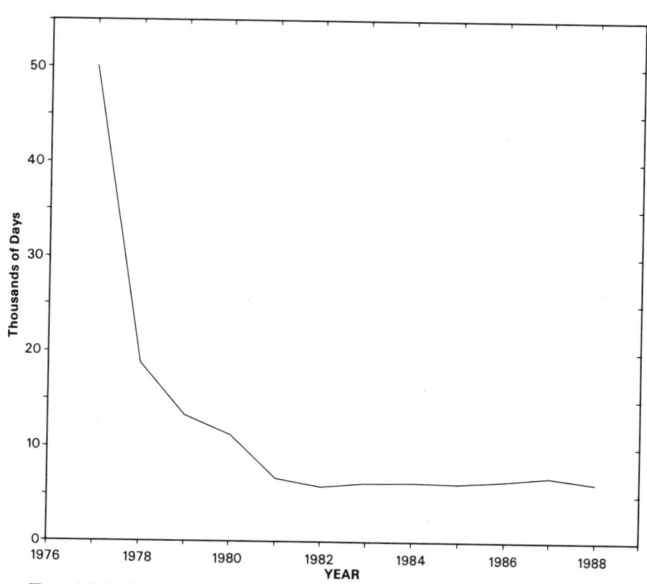

FIG. 10-8. Trend in number of days fished by foreign fishing vessels in the Canadian zone: 1977–1988.

countries in the Canadian zone diminished. One striking example of this involved the northern cod stock. By 1981 there was no longer any surplus. The question of allocating non-surplus cod to foreigners became an important policy issue. During the period from 1981 to 1983, 25,000 tons of "non-surplus" northern cod were reserved for foreign allocation.

A key dimension of the foreign allocations debate revolved around the criteria for allocation of the available surpluses (and temporarily specified quantities of non-surplus fish) among countries wishing to fish in the Canadian zone. Although Canadian allocations policy evolved over the first decade of extended jurisdiction, one consistent element was the search for cooperation in conservation of the stocks which straddle the Canadian 200-mile limit. Canada had been unsuccessful in the last days of UNCLOS III in securing recognition of a coastal State right to manage stocks straddling the 200-mile limit. This failure bedeviled Canadian fisheries management in the 1980's. Foreign overfishing of stocks beyond the 200-mile limit became a matter of increasing concern to Canada in the late 1980's. The attempts to deal with this problem through the establishment of a new multilateral fisheries organization, the Northwest Atlantic Fisheries Organization (NAFO), and various bilateral negotiations, are described in Chapter 11.

A primary objective of Canadian policy since 1977 has been to have Canadian fishermen harvest whatever fish in the Canadian zone can be harvested profitably by Canadians. At the same time, growth in harvesting capacity was constrained to foster improved profitability of existing operators. The objective of having Canadian fishermen harvest the lion's share of fish available within the Canadian zone was largely achieved in the 1980's.

Canada's share of Atlantic coast finfish allocations increased from 18.3% in 1974 to 44.9% in 1978. By 1984 Canada had secured 77% of the total allocations within and outside 200 miles. The remaining 23% consisted primarily of stocks entirely or partially outside Canadian jurisdiction and stocks within the Canadian 200-mile zone which were of little interest to Canadian fishermen. The latter included silver hake, roundnose grenadier and argentines (DFO 1985g).

4.3.3 Pacific Coast

The impact of the 200-mile limit on foreign fisheries off the Canadian Pacific coast was considerably different. Prior to extension there had been relatively small-scale foreign fisheries (with the exception of U.S. fishermen fishing for salmon and halibut) in the Canadian zone compared with the Atlantic.

The major impact of extended jurisdiction on the B.C. salmon fishery was upon the patterns of interception of U.S. origin-salmon by Canadian fishermen and Canadian-origin salmon by U.S. fishermen. Historically, the quantity of Canadian salmon taken by U.S. fishermen exceeded considerably the quantity of U.S. salmon taken by Canadian fishermen. In the years immediately prior to extended fisheries jurisdiction that difference was reduced because of the capture by Canadian fishermen of U.S. hatchery fish produced in Oregon and Washington. Copes (1981) observed that "the advent of the 200-mile limit probably has had an adverse impact on Canada's position in the interception contest." U.S. fishermen could intercept considerable quantities of Canadian-origin fish off Alaska, Canadians could continue to intercept U.S. origin salmon migrating to Washington and Oregon, but after 1980 Canadian fishermen were barred from fishing salmon off Alaska and Washington. The question of relative balance of interceptions had been argued for decades. Years of negotiation finally resulted in the Canada-USA Pacific Salmon Treaty in 1985 and the establishment of the Pacific Salmon Commission. The implications are examined in Chapter 11.

The two next most valuable B.C. fisheries are herring and halibut. Canada's access to and management of herring was virtually unaffected by the 200-mile limit. The Canadian halibut fishery, on the other hand, was adversely affected. The halibut fishery had been in decline in the pre-extension period. The 1977 Canadian Pacific halibut catch was about 25–30% of the catch from 1956 to 1965.

In the decades prior to 1977 Canadian fishermen had taken the bulk of their catch in waters off the U.S. coast. For example, from 1956 to 1965 Canada caught an annual average of 14,110 tons of halibut. Of this amount, 5,870 tons (41.6%) was taken in what became the Canadian zone in 1977 and 8,240 tons (58.4%) was taken off the U.S. coast. During the same period U.S. fishermen caught less than 10% of their catch off the Canadian coast. In the decade just before extension the portion caught by Canadians off the U.S. coast dropped to about 50%.

The 200-mile zone resulted in reduced access to each other's zones. The allowable Canadian halibut catch in the U.S. zone was reduced significantly in the late 1970's. Initial attempts to negotiate a mutually satisfactory reciprocal fishing arrangement were unsuccessful. In June 1978, Canadian fishermen were prevented access to the U.S. zone and vice versa. In February 1979, a phase-out arrangement was negotiated for the Pacific coast whereby Canadian fishermen were permitted to harvest two million pounds of halibut off Alaska in 1979 and one million pounds in 1980. U.S.

fishermen were allowed to harvest 14.3 million pounds of groundfish off British Columbia over the same 2-year period.

As a result, the Canadian fleet had to reduce its operations because it now had access to a more limited portion of the halibut resource. DFO implemented a "halibut relocation plan" in an attempt to reduce the size of the fleet. There were three aspects to this plan:

1. Vessels with several licences were encouraged to fish species other than halibut;
2. Vessels were offered a grant for gear conversion to fish black cod, an underutilized species; and
3. A minimum landing requirement (3,000 pounds of halibut in 1977 or 1978 on gear other than troll) was imposed for the issuance of licences to continue fishing for halibut. This resulted in the denial of halibut licences to about 400 part-time halibut fishermen.

These measures reduced the fishing capacity of the fleet by about 20% compared with a 58% reduction in the catch available to the Canadian fleet (Copes and Cook 1982). Flexibility in the appeal process, however, led to approval of additional licences with 422 vessels licensed by 1981. Prior to extended jurisdiction and limited entry, less than 100 vessels had relied primarily on the halibut fishery in Canadian waters. Dislocation from U.S. waters resulted in considerable excess capacity in the post-extension era. The Canadian catch of Pacific halibut decreased from 5,200–5,300 tons in 1977–78 to 3,100–3,200 tons during 1981–83. It subsequently increased to around 6,000 tons in 1987 and then declined again to 3,500 tons in 1990 (Fig. 10-9).

The one component of the British Columbia fishery that clearly benefited from the 200-mile limit was the groundfish fishery (excluding halibut). This fishery harvests a variety of species, including Pacific cod, black cod, pollock, flatfishes (other than halibut), rockfish, hake and dogfish. U.S. fishermen fished groundfish off British Columbia for decades, with the U.S. fishery accounting for 47% of the catch in 1964. The foreign fishery (other than that by the U.S.) commenced when the USSR started fishing groundfish off British Columbia in 1965 (Table 10-1). The USSR was joined by Japan in 1966. These were the only two foreign countries fishing groundfish until 1975 when Poland, the Republic of Korea and the German Democratic Republic started fishing these stocks. The GDR fished for only 1 year, but Poland and the Republic of Korea continued fishing there in 1976.

The catch by U.S. fishermen was roughly as large as that by Canadian fishermen during 1964–1970 but averaged less than half the Canadian catch during 1972–1976. The Canadian catch more than doubled from 11,600 tons in 1970 to 26,000 tons in 1976. The total groundfish catch increased from 24,600 tons in 1964 to 109,000 tons in 1966, peaked at 124,900 tons in 1969 and thereafter fluctuated between 60,000–80,000 tons. The distant water fleets accounted for 73.8% of the catch in 1969. Their share declined to about 50% during the 1974–1976 period.

In 1977, Canadian scientists advised that some of the groundfish stocks, e.g. ocean perch, had been overfished. TACs were set at relatively low levels to rebuild the

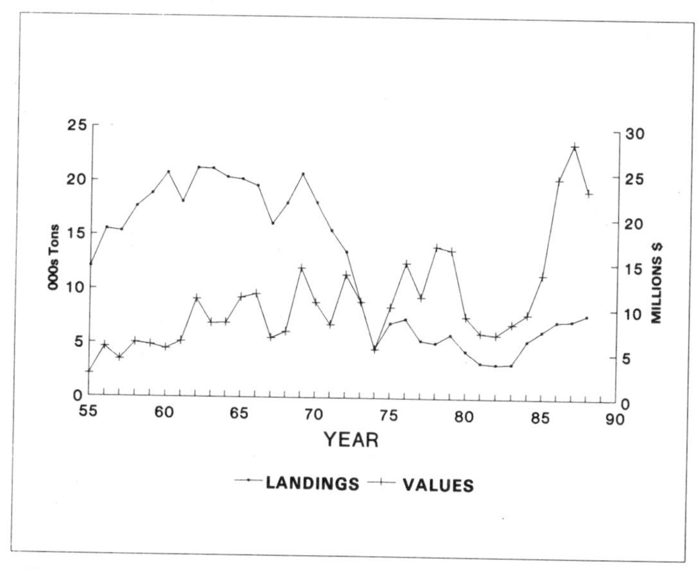

FIG. 10-9. Canadian landings (000's tons) and landed value (millions of dollars) of Pacific halibut: 1955–1988.

TABLE 10-1. Groundfish catches off British Columbia.

Year	Canada	USA	USSR	Japan	Poland	Korea	GDR	TOTAL
1964	13,156	11,451	—	—	—	—	—	24,607
1965	18,458	16,385	14,000*	—	—	—	—	48,843
1966	23,594	19,252	60,000	6,386	—	—	—	109,232
1967	15,814	16,293	21,373*	16,270	—	—	—	69,750
1968	17,781	12,543	54,135	19,988	—	—	—	104,447
1969	16,778	16,009	72,519	19,606	—	—	—	124,912
1970	11,626	13,705	25,577	14,497	—	—	—	65,405
1971	14,476	11,753	6,008	8,678	—	—	—	40,915
1972	20,980	10,975	8,328	17,842	—	—	—	58,125
1973	22,163	12,450	16,207	15,676	—	—	—	66,946
1974	19,854	10,854	2,651	28,351	—	—	—	61,710
1975	23,952	8,853	3,814	16,327	26,273	1,301	2,000	82,520
1976	26,040	8,877	4,264	11,891	6,739	2,358	—	60,169
1977	27,635	10,635	560	8,134	3,327	216	—	50,507
1978	33,597	NA	700	3,364	589	—	—	38,250
1979	42,718	NA**	0	3,637	4,263	—	—	50,618
1980	51,354	NA**	78	817	4,456	—	—	56,705
1981	54,385		227	187	3,189	—	—	57,988
1982	55,490		0	2,237	10,357	—	—	68,084
1983	64,193		0	0	13,177	—	—	77,370
1984	67,415		0	0	13,203	—	—	80,618
1985	58,397		0	0	10,533	—	—	68,930
1986	82,018		8,138	0	15,604	—	—	105,760
1987	118,196		11,737	0	9,716	—	—	139,649
1988	117,223		1,330	0	26,386	—	—	144,939
1989	129,681		13,322	0	18,254	—	—	161,257
1990	139,685		0	0	3,439	—	—	143,124

* = Rough estimate.

** = Under a bilateral agreement U.S. fishermen were allowed to catch 14.3 million pounds of groundfish off British Columbia over the 2-year period 1979 and 1980.

Source: 1964–1977 — (Copes 1981).

1978–1990 — Department of Fisheries and Oceans, Biological Sciences Branch, Pacific Region.

stocks to permit a more viable Canadian fishery. However, since the Canadian fleet was unable to harvest the TACs of all species, Japan, Poland and the USSR were allocated various quantities of hake, sablefish, dogfish and rockfish in the first year of extended jurisdiction. Foreign allocations totalled 31,000 tons, consisting of 20,000 tons of hake, 3,000 tons of black cod (sablefish), 3,000 tons of rockfish and 5,000 tons of dogfish. Half the total consisted of hake allocations (7,500 tons each) to Poland and the USSR. Total foreign allocations decreased to 20,200 tons in 1978 and 14,750 tons in 1979.

Foreign quotas for rockfish and dogfish were eliminated in 1978, leaving Pacific hake as the only significant foreign groundfish fishery in the Canadian zone on the Pacific coast. Through the 1970's and 1980's joint ventures/direct sales, whereby Canadian fishing vessels would catch hake for delivery at sea to foreign factory trawlers, were used to develop a Canadian fishery for this species. Foreign allocations were used as an inducement to attract foreign partners for these ventures. Foreign allocations of Pacific hake fluctuated during the 1980's ranging from a low of 10,000 tons in 1981 to highs of 37,000 tons in 1987 and 41,000 tons in 1988.

Foreign allocations dropped to 4,000 tons in 1990 and zero in 1992.

4.4 Rebuilding of Stocks Overfished in the Pre-extension Era

For the Canadian domestic fisheries, the period immediately following extended jurisdiction brought short-term pain with the expectation of long-term gain. Canada established low Total Allowable Catches for groundfish, for example, to begin rebuilding the resource. The amount of Atlantic groundfish available to the Canadian fleet in 1977 was insufficient to keep the fleet and onshore processing plants operating year-round. To deal with this problem, Canada introduced the first Annual Groundfish Plan and a complex system of allocating access among fleet sectors. This was intended to spread the burden of conservation among all participants in the industry. Chapter 7 describes the Fishing Plan process and the evolution of the domestic allocation of access regime.

Early in the first year of extended jurisdiction, Fisheries Minister LeBlanc cautioned against expecting too much too soon from the 200-mile zone:

> "The limit means nothing unless we use it right. Reaping its full advantages will take time and care.
> "The new zone offers only potential, we cannot take its promise for granted" (DOE 1977b).

LeBlanc spoke of plans to use the resource better, to help "those who suffered most from the years of overfishing: the small man, the inshore man, with his own boat." He also spoke of pursuing prosperity by increasing not just the catch but the return per pound of the catch. This would necessitate upgrading fish quality.

The federal government talked about rebuilding the resource and the inshore fishery and increasing the value of the catch. However, it was under considerable pressure from the Atlantic provinces to expand the fleet. LeBlanc said:

> "I have been urged to expand the fishing fleet, for example, with huge new freezer trawlers. Perhaps this would help the shipbuilders but I see no need yet of a huge new catching capacity...when our existing fleet is only beginning to get out of trouble" (DOE 1977b).

A few months later Minister LeBlanc again addressed the question of fleet expansion. He pointed out the recovery of the groundfish stocks would occur at different rates in different areas:

> "Only over time will the lowered fishing effort increase the amount of fish and the catch rates by our fishermen... [Our] plans can hardly include any vast fleet expansion. In almost no area of the Atlantic is our groundfish fleet getting anywhere near its catch rates of five years ago" (DOE 1977a).

By this time Newfoundland and Nova Scotia were at work on their fleet development plan which called for an expenditure of $900 million on new fishing vessels (expansion and replacement).

LeBlanc's rhetorical response was:

> "Do we want to double a fleet that is getting half loads? I hardly think so" (DOE 1977a).

There was considerable antagonism between the federal government and certain provincial governments on this issue (see Chapter 2 and Chapters 7 and 8). On the one hand, the provincial governments, and some large processors, saw a need to build more and bigger vessels to harvest the resources of the extended zone. On the other, the federal government believed, based on available scientific advice, that the increasing groundfish resource in future years would result in improved catch rates, which should first be utilized to restore the profitability of the existing fleet. This controversy led to the first attempt by federal scientists to project anticipated catches over the next several years. Given the uncertainties involved, this was an exercise fraught with considerable potential for error. The scientific projections indicated that a significant increase in TACs could be expected. Increasing stock abundance would result in increased catch rates for the existing fleet which would be capable of harvesting most of the increase in TACs of traditional finfish (DOE 1978a).

Dr. Arthur May as Acting Assistant Deputy Minister, Fisheries Management, of the Department of Fisheries and the Environment, gave a speech to the Canadian Labour Congress, on January 13, 1977. He described the Canadian approach to management of the new zone thus:

> "*The main thrust of the new 200-mile management regime is to rebuild the resource so as to provide increased opportunities for Canadian fishermen....* In order to rebuild these resources, we are applying stringent conservation measures to ensure that we do not replace foreign overfishing with Canadian overfishing. The 200-mile limit will not bring

about an overnight miracle, but the long-range future is bright.... It will take between 5 and 10 years of strict management to bring about restoration of the resource. The severe conservation measures presently being implemented will begin that process" (DOE 1977c) (Emphasis mine).

By and large the federal government succeeded in containing the capacity of the Canadian offshore fleet (see Chapter 8), although technological improvements undoubtedly let to some increase in fishing efficiency. There was some increase in the inshore sector to permit a rebuilding of the inshore fishery. As forecast, the Canadian fleet harvested significantly increased quantities of fish as the stocks improved. Canadian allocations of species under TAC management on the Atlantic coast increased from 409,400 tons in 1977 to 1,091,000 tons in 1984 (DFO 1985g). Canadian catches of traditional groundfish species rose from 467,000 tons in 1977 to a peak of 744,000 tons in 1982, an increase of 59%. Canadian Atlantic groundfish catches fluctuated between 700,000 and 750,000 tons until 1989, when they declined to 656,000 tons. Preliminary catch statistics for 1990 indicated a further decline to 603,000 tons. With the drop in the northern cod catch, there was a decline to less than 600,000 tons in 1991. Overall, however, Canadian catches of traditional groundfish were 50% higher in the 1980's than they were in 1977. The Canadian total Atlantic groundfish catch of 756,000 tons in 1987 was 80% higher than the low point of 1974.

Perhaps a more valuable measure of the impact was the increase in the catch rates of the large Canadian otter trawlers (Fig. 10-4). As mentioned earlier, Canadian offshore trawlers increased their catch rates to more than 15 to 19 tons of fish per day by 1982, compared with 8 tons per day in 1975. This reduced harvesting costs, at a time when other costs of fishing (e.g. fuel) were increasing. It appears likely that some of this increase in catch rate was due to technological innovations and changes in area and season fished (e.g. a learning curve as Canadian trawlers gained experience in fishing northern cod in the winter offshore).

Doubleday et al. (1989) examined trends in stock abundance up to 1983. Pinhorn and Halliday (1990) conducted a comprehensive study of changes in stock abundance on the Canadian Atlantic coast over the past three decades up to 1986. They derived biomass indices for the various species from a variety of sources. Their analyses indicated that total cod biomass declined to approximately one quarter of its early 1960's levels by 1975–76 and about tripled thereafter (Fig. 10-10). Cod was the major determinant of trends in groundfish catches and apparent trends in groundfish biomass. The abundance of groundfish off the Canadian Atlantic coast (Subareas 2–4) decreased by more than 50% between 1967 and 1975, remained stable between 1975 and 1976, then more than doubled between 1976 and 1984 (Fig. 10-10). Patterns of decline and recovery for stocks inside the Canadian zone and straddling the boundary on the Grand Banks were broadly similar. On the other hand, stocks completely outside 200 miles on the Flemish Cap had different trends, with neither cod nor redfish showing recent abundance increases (Fig. 10-11).

Pinhorn and Halliday concluded that fishing mortalities on cod stocks in the 1960's were, in most cases, about F=0.50, above F_{max} for most stocks. F increased in the early 1970's to about F=1.0 for several stocks in 1975–76. After 1977, Fs for cod were, on average, lower in comparison to the early 1970's but well above the target fishing mortality of $F_{0.1}$. Most cod stocks from 1977 to the late 1980's were fished at or near F_{max}.

Their analyses suggested that the overall biomass of commercial pelagic species halved between 1970–75 and 1978–80 from about 8 to 4 million tons. It may have increased back to the early 1970's level by 1985. This is largely a reflection of trends in capelin biomass. Capelin accounted for more than half of the total commercial pelagic biomass throughout much of the period under study. Mackerel declined in abundance throughout the 1970's by about half, then recovered substantially in the 1980's. Herring declined from about 2.5 million tons in 1970 to about 700,000 tons in 1978–82. It then increased to more than 1.5 million tons after 1984.

The exploitation rate for capelin appears to have been less than the target rate. The major fluctuations in population apparently resulted from fluctuations in recruitment caused by environmental factors (Leggett et al. 1984). On the other hand, exploitation rates played an important part in the dynamics of herring. Fs on the southern Gulf of St. Lawrence and Scotian Shelf stocks were generally above $F_{0.1}$, although Fs on the southern Gulf stock were around $F_{0.1}$ in the mid-1980's.

Pinhorn and Halliday (1990) concluded that "Canadian resource rehabilitation and conservation efforts appear to have met with a large degree of success overall." They observed that fishing mortality likely increased substantially after 1963, peaked sometime in the mid-1970's and became substantially lower after extension of jurisdiction. This is sufficient to explain the reversal of groundfish abundance trends.

One disturbing conclusion from their analysis was that Canada had not succeeded in reducing F to $F_{0.1}$ for the stocks most important to the Canadian fishery. Except for haddock, post-1977 Fs have been reduced to about F_{max}. Pinhorn and Halliday did not comment on the appropriateness of $F_{0.1}$ as the management objective. However, it is worth noting that one of the arguments advanced by STACRES for a target fishing mortality of

$F_{0.1}$ was that it would provide a margin of safety against errors in estimating stock abundance. Had Canada been using a target fishing mortality of F_{max} in the post-extension era, it is quite likely that Fs would have been well above F_{max}. The increases in stock abundance and catch rates witnessed in the post-1977 period would probably not have occurred. This supports the cautious approach of striving to fish at $F_{0.1}$.

Pinhorn and Halliday (1990) concluded that a dramatic reversal of "exploitable biomass" trends occurred on the Atlantic coast coincidentally with Canadian extension of fisheries jurisdiction in 1977. They observed:

"These results are consistent with the Canadian strategy of maintaining a moderate exploitation rate and with the Canadian

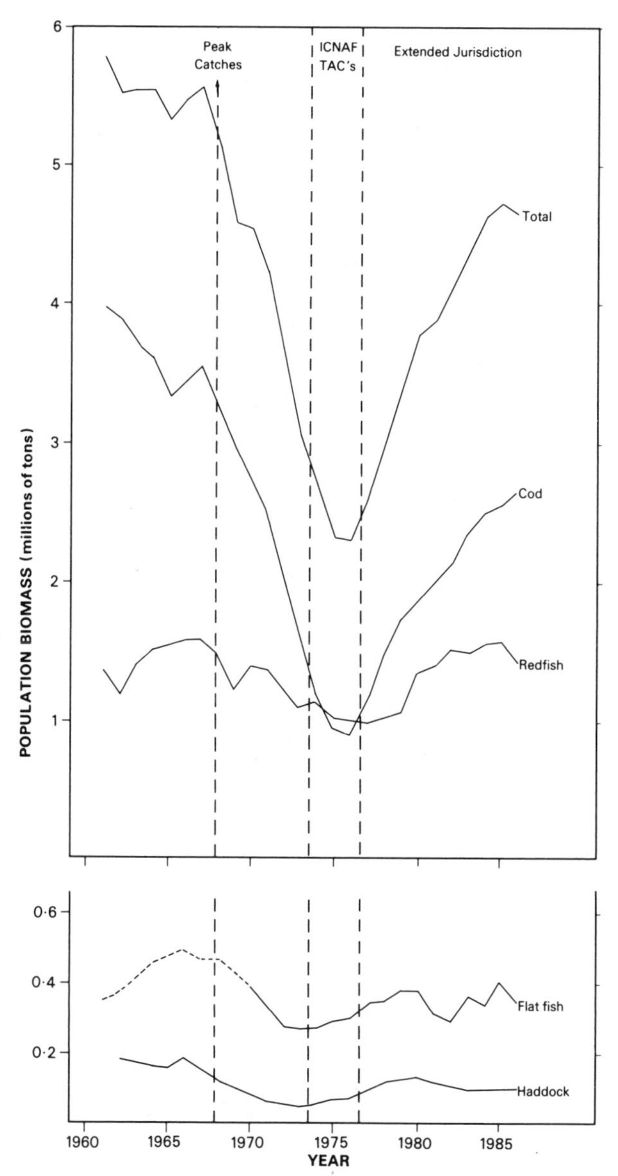

FIG. 10-10. Biomass of exploited age groups of fully exploited groundfish species in subareas 2-4 in 1961–1986, in total and for important species (from Pinhorn and Halliday 1990).

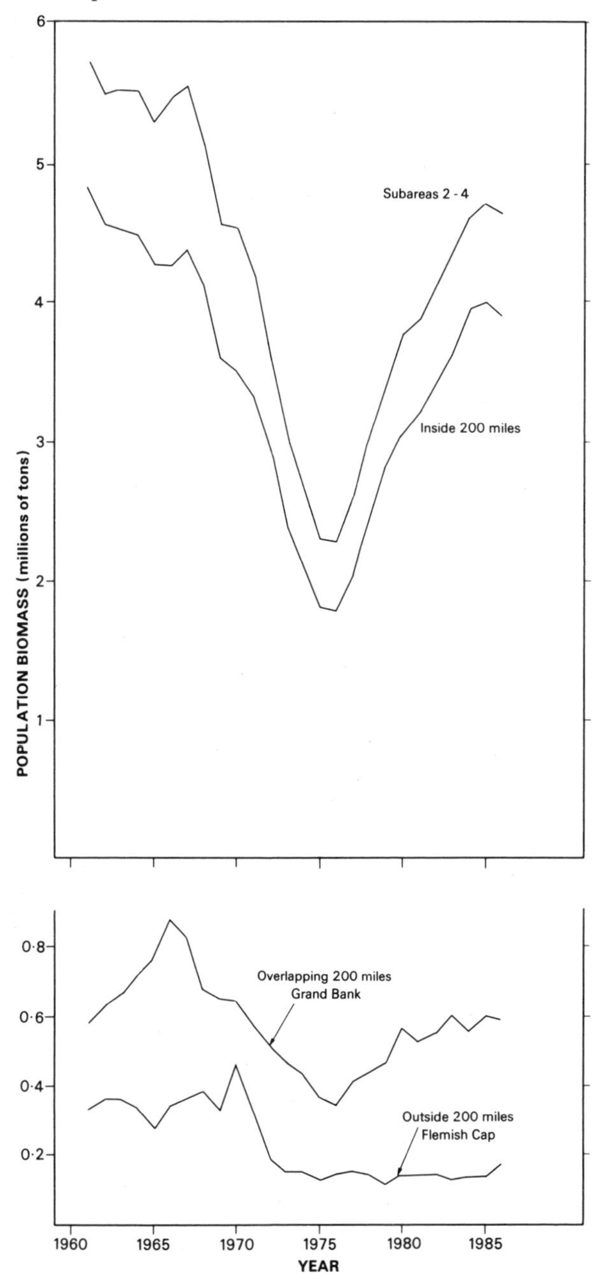

FIG. 10-11. Biomass of exploited age groups of fully exploited groundfish species in subareas 2-4 in 1961–1986, in total and for different jurisdictional regimes (from Pinhorn and Halliday 1990).

> ability to control fishing within its zone, through surveillance and observer activities, much more effectively than was apparently achievable under the international control schemes in effect prior to 1977."

Nonetheless, they noted that these trends were also consistent with an alternative hypothesis. Koslow et al. (1987) ascribed these trends to a periodicity in resource productivity, associated with large scale environmental events and mediated through recruitment. This would imply that there was an underlying cycle in the production of groundfish stocks and that the international fishery of the late 1960's and early 1970's simply amplified the declining trend in resource abundance. Similarly, under this hypothesis, the Canadian management regime amplified the increase in biomass which resulted from an increase in resource productivity coincident with extension of jurisdiction. For a more detailed discussion of the impact of environmental influences on fish productivity, see Chapter 18.

In any event, Canadian groundfish fishermen clearly benefited from improved catch rates following the 200-mile limit. Increases in catch rates and catches, combined with the anticipation of future increases, triggered a wave of euphoria in certain sectors of the Atlantic industry. There were the provincial proposals for fleet expansion, which were resisted by the federal government. Some "leakage" occurred but this was moderate compared with what might have occurred had the federal government also embraced the Klondike attitude of many in the industry and provincial governments.

There were significant changes in the areas, seasons and species fished by Canadian offshore trawlers. Fortunately there was no major dislocation on the east coast because of the boundary claims of Canada's three neighbours. Canadian vessels had done very little fishing on the Greenland side of Davis Strait (Subarea O). Fishing in the area claimed by France continued into the 1980's with minimal disruption. Following the International Court of Justice decision in 1984, Canadians were excluded from the U.S. portion of Georges Bank. Most Canadian fishing on Georges Bank had been conducted on that part which became Canadian waters as a result of the ICJ decision (Halliday et al. 1986).

One major consequence of extended jurisdiction on the Atlantic coast was a significant build-up in onshore processing capacity, which was under the control of provincial governments. Groundfish freezing capacity increased 150% from 1974 to 1980 (Kirby 1982). Private entrepreneurs, with assistance from provincial governments and the federal Department of Regional Economic Expansion (DREE), put in place more processing capacity than was warranted by the current or anticipated resource supply. Much of this excess processing capacity was debt-financed. DFO and DREE signed an agreement in 1981 whereby DREE would limit any future assistance to measures aimed at quality and productivity improvements. But this came too late to avert the build-up of processing overcapacity. With the surge in groundfish market demand in the mid-1980's, there was a further expansion of the already excessive onshore processing capacity.

The debt load associated with excessive plant capacity combined with the general economic downturn in the recession of the early 1980's to bring the large vertically integrated companies heavily reliant on groundfish to the brink of financial collapse. This led to the Task Force on Atlantic Fisheries in 1982 and the financial restructuring of the four major processors in 1983 (see Chapter 13 for a discussion of these events).

Foreigners (with the exception of France) had already been phased out of the Gulf of St. Lawrence. Thus Newfoundland and Nova Scotia were the major beneficiaries of the 200-mile zone, because of the displacement of foreign fishing effort. It was off their coasts that the greatest stock recovery was anticipated. This led in the post-extension era to interprovincial bickering. Pressure from the Gulf provinces of Quebec, Prince Edward Island and Nova Scotia led to displacement of large trawlers based in Newfoundland and Nova Scotia from the Gulf. By the late 1980's the Gulf provinces, through the Nova Nord consortium, were seeking increased access to fish in the 200-mile zone outside the Gulf.

As a result of the 200-mile zone, new fisheries were developed in areas and for species not previously fished by Canadians. An example of the former was the development of the offshore trawler fishery for northern cod (see Chapter 7). An example of the latter was the development of a Canadian fishery for Pacific hake. This involved over-the-side sales to foreign factory trawlers. For a discussion of the use of foreign vessels in an attempt to develop Canadian fisheries for new species, see Chapter 11.

More recently, Rivard and Maguire (1993) analysed the evolution of the $F_{0.1}$ strategy for the management of Canada's Atlantic groundfish stocks. In addition to documenting trends in catches, which correspond to those mentioned earlier in this chapter, they examined the trends in eight groundfish stocks for which analytical assessments were available. The catch from these stocks accounted for about 60% of the groundfish landings from traditional fishing grounds in 1990.

The total biomass of the eight stocks studied increased from approximately 1 million tons in 1975 to a peak of 2.4 million tons in 1984, decreased to 1.8 million tons in 1987 and increased slightly thereafter (Fig. 10-12). In

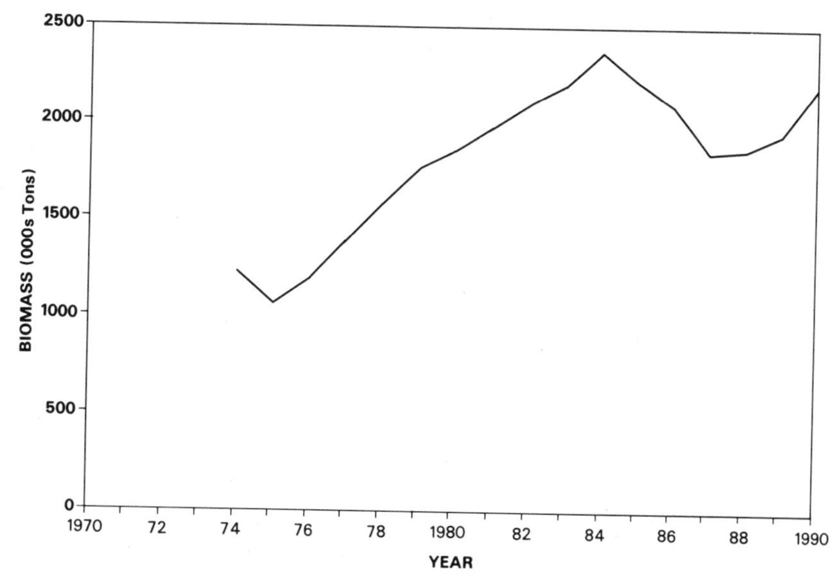

FIG. 10-12. Trend in stock abundance for eight groundfish stocks with a detailed analytical assessment for 1991. The biomass estimates given here correspond to the total for the eight stocks (from Rivard and Maquire 1993).

1990, the total biomass of these stocks was estimated at 2.2 million tons. Rivard and Maguire's estimates indicate that the total biomass had doubled since the extension of fisheries jurisdiction. Of the eight groundfish stocks studied, all showed a general recovery after extension of fisheries jurisdiction. They noted, however, that the northern Gulf cod stock had experienced a fast increase in the late 1970's — early 1980's, followed by a rapid decline from the mid-1980's onward. While the decline had stabilized by 1990, its level remained below that of the mid-1970's.

Overall, Rivard and Maguire concluded that both the total catches and the Canadian catches had improved in relation to the mid-1970's. The increase was particularly strong for cod, despite the decline noted in the late 1980's. All indicators suggested a significant improvement in relation to the mid-1970's despite the decline in many stocks in the late 1980's. Rivard and Maguire noted, however, that fishing mortalities had not been reduced to the $F_{0.1}$ level for any significant amount of time on any of the stocks reviewed. However, the high levels of fishing mortality observed in the late 1960's and early 1970's seem to have been avoided and, on average, the level of fishing mortality was somewhat lower than that exerted prior to 1977. This conclusion was similar to that reached by Pinhorn and Halliday (1990), based on data up to 1986.

It should be noted that the analysis by Rivard and Maguire did not take into account the sudden, drastic and unexpected decline in northern cod abundance during 1991, resulting in a two-year moratorium commencing in July 1992 (see Chapters 7 and 13). Nor did it take into account further dramatic downturns in certain other cod stocks in the early 1990's. Despite these events, however, the general conclusions reached by Pinhorn and Halliday (1990) and Rivard and Maquire (In Press) still appear valid. It appears likely, however, that the downturn beginning in the late 1980's resulted from productivity changes linked to environmental changes as well as excessive fishing mortality.

4.5 Costs and Benefits of Extended Jurisdiction

Doubleday et al. (1989) attempted a rough quantification of the direct costs of management and a qualitative indication of some of the benefits to Canada of extended jurisdiction on the Atlantic coast. They estimated an approximate total annual expenditure directly related to managing the Atlantic zone of extended fisheries jurisdiction of about $65 million in 1984. After allowance for overheads and excluding unrelated programs, the annual cost of managing the zone of extended fisheries jurisdiction on the Atlantic coast was estimated to be about $75 million Canadian in 1984, representing an increase of about $45 million 1984 dollars over pre-extension program costs. (The pre-extension base in 1978 dollars was $20 million for surveillance and enforcement and research annually in the area subject to extended fisheries jurisdiction, corresponding to a 1984 total of about $30 million).

Quantification of the net benefits of the fisheries management program in the 200-mile zone is, of course, exceedingly difficult. Doubleday et al. (1989) suggested that, to a significant degree, the increase in catch rates and groundfish catches within the Canadian 200-mile zone could be attributed to the program of management. They noted increases to 1984 of 300,000 tons in Canadian Atlantic catches with a landed value of $100 million. Employment in harvesting had increased from 41,000 to 53,000 and in processing from 15,000 to 22,000. They also noted the beneficial impact of the increase in catch per unit effort of Canadian trawlers.

Overall, Doubleday et al. (1989) concluded that extended jurisdiction had made it possible to stabilize and increase employment and to make some contribution towards reducing the costs of harvesting. It had not overcome structural problems in the fishing industry and the Atlantic coast regional economy generally. This left the fishing industry very vulnerable to the twin forces of inflation and high interest rates in the early 1980's. Despite this, the authors concluded that the benefits of extended fisheries jurisdiction in Atlantic Canada had been "real and substantial and justify the expenditures needed to achieve them."

Since 1984 there have been some changes in the expenditures on various aspects of the fisheries management program in Canada's 200-mile fisheries zone. Although general expenditure reductions have been implemented in some programs of the Department of Fisheries and Oceans, significant increases in program resources were allocated to offshore surveillance and enforcement and some areas of research (e.g. northern cod) as a result of the Atlantic Fisheries Adjustment Program in May 1990 (see Chapter 13). In the early 1990's there was a substantial drop in the catches and landed value of the Atlantic groundfish fishery, compared with the mid-1980's, due to a downturn in resource abundance. If capacity is brought more in line with the available resource in the 1990's, this is likely to result in a significant reduction in employment in harvesting and processing.

I have made no attempt to quantify the net benefits based on this later data because of the same difficulties faced by Doubleday et al. (1989).

5.0 CONCLUSIONS

The 200-mile fisheries zone brought the bulk of the fisheries resources off Canada's coasts under Canadian fisheries management in 1977. This provided Canada with the authority to manage these resources in accordance with Canadian fisheries objectives. Canada gained considerable benefits by taking measures to rebuild depleted fish stocks and utilize the fisheries resources of the new zone.

In the decade following extension, Canada increased significantly its share of the allowable harvest off both the Atlantic and Pacific coasts. The presence of foreign vessels in the Canadian zone diminished dramatically in the 1980's as the quantity of fish allocated to other countries was reduced substantially.

The 200-mile fishing zone meant that, on both coasts, Canada now had sovereign rights to manage the resource. Instead of being an international common property, the resource became national common property. Canada moved on a variety of fronts to deal with the common property problem, as described in Chapters 8 and 9. The 200-mile zone increased resource availability to the Canadian fleet on the east coast through displacement of foreign fishing effort. It made it possible for Canada to put in place a conservative resource management regime aimed at stock rebuilding. It fostered the introduction of innovative approaches such as enterprise allocations. It also shifted the responsibility for success or failure in management squarely to the shoulders of Canada as the coastal state.

The 200-mile zone made possible a more effective resource management regime than existed formerly under the international commission approach. Thus Canada gained greater security of supply. But it did not solve all of the problems of the fishing industry. Expectations were high at the time of extended jurisdiction; the future appeared full of promise. In 15 years some of these expectations were dashed, others shattered. Following the crises of the early 1980's on both coasts, the realization dawned that the 200-mile limit was not a cure-all for Canada's fisheries ills. But an upswing in industry fortunes in the mid and late 1980's again weakened the industry resolve to work with government to tackle fundamental problems. When another crisis hit the Atlantic groundfish industry in the early 1990's these fundamental problems remained unresolved. Some of these are addressed in the following chapters.

In summary, the 200-mile limit yielded considerable benefits for Canada but it also brought major challenges. Tackling these challenges remains a task for the 1990's and into the 21st century.

CHAPTER 11

THE INTERNATIONAL DIMENSION

II — CANADIAN INTERNATIONAL FISHERIES POLICY FOLLOWING EXTENSION OF JURISDICTION

1.0 INTRODUCTION

In the previous chapter, I described the impact of the 200-mile limit on Canada's domestic fisheries. This chapter focuses on the evolution of Canada's international fisheries policy in the post-extension era. As the era of the 200-mile limit dawned, the future appeared bright for Canada's fisheries. Canada had gained control over considerable fish resources previously subject to multi-country high-seas exploitation under the pre-1977 rules of international fisheries management. With Canada's rights to manage and exploit these resources came certain obligations. Leonard Legault handled Canada's fisheries negotiations during the transition to extended jurisdiction. He described the basic premises of the new regime as follows:

> "One of the most basic elements of this consensus (UNCLOS) is the principle of optimum utilization, which means simply that coastal states will not play dog in the manger — will not allow fish stocks under their jurisdiction to die of old age and go to waste — but rather will allow other countries access to such portions of the stocks as may be surplus to the coastal state's harvesting capacity, as determined by the coastal state and subject to terms and conditions established by the coastal state.... The principle of optimum utilization in no way represents a limitation on our sovereign rights. On the contrary. We will determine each year the TAC's for the various species of our 200-mile zone; we will determine what is the Canadian harvesting capacity in respect of these species; we will determine what may be surplus to our harvesting capacity; and we will determine who will have access to any such surplus and under what terms and conditions" (Legault 1977)

Canada had extended its jurisdiction in the context of the developing consensus on the Law of the Sea. It intended to manage the fisheries of the extended zone following the general principles being developed at UNCLOS III. Article 61 of the LOS Consolidated Negotiating Text provided that the coastal state would set the Total Allowable Catch "to maintain or restore populations of harvested species at levels which can produce the maximum sustained yield, as qualified by relevant environmental and economic factors, including the economic needs of coastal fishing communities." Article 62 provided that "...the coastal state shall determine its capacity to harvest the living resources of the exclusive economic zone." Where the coastal state does not have the capacity to harvest the entire allowable catch, it shall... "give other states access to the surplus of the allowable catch."

In the post-extension era important policy issues arose concerning the definition and use of "surpluses". Immediately following extension, Canadian fishermen faced a shortage of traditional groundfish species in traditional fishing areas and began to fish farther afield. Despite this, there were species and stocks where the Total Allowable Catch exceeded the harvesting capacity of the Canadian fleet. As Canadians began to redirect fishing effort to species and areas where they had previously fished little, the surpluses available for allocation to foreign countries in the Canadian zone diminished.

One striking example of this involved the northern cod stock. Foreign countries had harvested 310,000 to 395,000 tons from this stock between 1970 and 1974. Following extension, the amount available for foreign allocation decreased from 93,140 tons in 1977 to 20,700 tons in 1980. By 1981 there was no longer any surplus. The question of allocating non-surplus cod to foreigners became an important policy issue. Between 1981 and 1983, 25,000 tons of non-surplus northern cod were reserved for foreign allocation.

Key dimensions of the foreign allocations debate were the criteria for allocating surpluses (and temporarily specified quantities of non-surplus fish) among countries wishing to fish in the Canadian zone. Although Canadian allocations policy evolved over the first decade of extended jurisdiction, one consistent element was the search for cooperation by other countries to conserve the stocks which straddle the Canadian 200-mile limit.

In the previous chapter we saw that Canada had two key interests which were not fully satisfied by the Law of the Sea Convention. The first involved protection of stocks which straddled the 200-mile limit on the east coast. Canada had not succeeded in securing recognition in the LOS Convention of a coastal State right to manage straddling stocks. Foreign overfishing of stocks beyond the 200-mile limit became a matter of increasing concern to Canada in the late 1980's.

The second key interest was protecting anadromous stocks of Canadian origin beyond 200 miles. On the east coast, salmon stocks of North American (primarily Canadian) and European origin were exploited by several European countries at West Greenland. On the west coast, there was continuing concern about the potential for increased exploitation of Canadian-origin salmon on the high seas. There were also difficult and prolonged negotiations to establish a framework for managing Canadian and U.S. interceptions of Pacific salmon originating in each other's waters (see Chapter 12). The Atlantic salmon problem was addressed through a new North Atlantic Salmon Conservation Organization (NASCO) and the Pacific high-seas-interception problem through renegotiation of the International North Pacific Fisheries Commission (INPFC). More recently there have been attempts to negotiate a broader agreement involving countries who fish salmon in the North Pacific or catch salmon incidentally in drift nets used to fish other species.

A primary objective of Canadian policy since 1977 has been to have Canadian fishermen harvest whatever fish in the Canadian zone they can take profitably. At the same time, growth in harvesting capacity was constrained to improve profitability of existing operators. The objective of having Canadian fishermen harvest the lion's share of fish available within the Canadian zone was largely achieved in the 1980's. Canada's share of allocations of finfish (groundfish, pelagic) in the Northwest Atlantic (NAFO Subareas 2-4) increased from 10% in 1974, to 32% in 1977, to 77% in 1984 and to 84% in 1990.

The following sections address: (1) the evolution of Canada's foreign allocations policy; (2) the attempts to protect the straddling stocks on the Atlantic coast through bilateral fisheries relations and NAFO; and (3) the attempts to protect Canadian-origin salmon on the Atlantic coast through NASCO and on the Pacific coast through INPFC and new controls on drift netting.

2.0 FOREIGN ALLOCATIONS POLICY

2.1 Phase I — Allocations For Access

Foreign allocations policy evolved through three phases between 1977 and 1990. Phase I, which lasted from 1976 to 1982, can be characterized as "stability and allocations for access." During this period significant surpluses were available for allocation. These surpluses went to countries which had fished traditionally off the Canadian coast. In light of the anticipated decrease in the surplus, Canada considered it inappropriate to make allocations to "new flags", countries that had not previously fished in the area. Usually, allocations were given only to countries which had entered into a bilateral fisheries treaty with Canada.

In Phase I, Canada emphasized continued good relations with countries which had fished traditionally off the Canadian coast. Expectations were high that considerably increased quantities of fish would be available to Canada in the mid-1980's, as the stocks rebuilt. Some sectors of the industry were concerned that Canada might not be able to market its increased catch. This brought pressure to diversify markets, leading to the "allocations for access" policy. Canada sought commitments from the countries receiving allocations that they would initiate or increase imports of processed Canadian fish products, or improve access to their markets for such products. This approach in the late 1970's was termed the "commensurate benefits" policy.

Fisheries Minister Roméo LeBlanc was a strong proponent of trading fish for new or improved market access. In a May, 1978, speech to the Fisheries Council of Canada, he described the rationale for Canada's foreign allocations policy:

> "In return for access, we ask fishing fees.... We ask protection for high-seas salmon.... We have asked that countries fishing in the Northwest Atlantic incorporate our special interest in fisheries outside the zone into a new, binding treaty.... Finally, in return for access we ask market advantages for our fishery products. Markets for the increasing harvests of traditional species such as cod. Markets for increased harvests of underutilized species: silver hake, grenadier, capelin and the like.
>
> "In future negotiations over foreign quotas in our zone, those who give us the most will get the most" (DOE 1978e).

Canadian negotiators outlined this *commensurate benefits* approach to ICNAF member countries assembled in Bonn, Germany, in 1978. In a meeting convened by Canada in Bonn, but separate from the ICNAF meeting, the Canadian spokesman, Dr. A.W. May, informed ICNAF members that allocations in the Canadian zone would no longer be based strictly on historical participation. Parts of the surpluses would continue to be allocated in accordance with traditional fishing patterns. However, portions of the TACs for particular stocks would be reserved for separate allocation to countries with bilateral agreements with Canada. Allocations would be based on cooperation on fisheries matters, including the purchase of Canadian fisheries products.

Following the 1978 ICNAF meeting, Canada attempted to secure firm commitments from various countries to purchase specific quantities of Canadian fish products in 1979. The total reserves announced in Bonn amounted to 109,950 tons (of which 72,500 tons was capelin). Several countries made offers of purchases against the reserves. Specific commensurate benefit allocations for 1979 were made to the following countries: Romania, Poland, Portugal, Cuba, and the USSR.

A surplus of 25,000 tons of northern cod was also declared in the fall of 1978. Of this amount, a total of 7,500 tons was allocated to Canada's bilateral partners based on their traditional presence, but not their traditional level of fishing, for this stock. The remaining 17,500 tons was allocated in return for commensurate benefits. A significant portion of this was allocated to Portugal (8,000 tons) because of its traditional large role in the cod fishery and its expressed willingness to expand market opportunities for salt cod from Canada.

The market inroads secured by this approach were insignificant, with the exception of Portugal. At the time of the 1979 joint ICNAF/NAFO meetings, the Canadian commensurate benefits policy was maintained but the emphasis shifted to the principle of "a satisfactory trading relationship." Canada expressed its intention to continue to distribute a number of surpluses based on countries' catches prior to 1979. These would go to states which had bilateral agreements with Canada. Licences to fish in the Canadian zone for these allocations depended on five conditions:

1. Support for Canadian proposals for ICNAF/NAFO TACs, and Canadian allocation requests for straddling stocks and stocks beyond 200 miles;
2. Adherence to ICNAF/NAFO regulations in the area beyond 200 miles;
3. Willingness to accept Canadian scientific observers in the area beyond 200 miles;
4. Non-use of flags of convenience to avoid ICNAF/NAFO obligations; and
5. Maintenance of a satisfactory trading relationship with respect to the purchase of Canadian fish products.

Towards the end of Phase I, attention focused on the utility of "non-surplus" allocations in return for market preferences. In 1980 Canada offered to enter into long-term agreements with the European Community (EC), Spain and Portugal in which specific allocations in the Canadian zone would be made in return for specific market commitments for Canadian fish products. This resulted, after protracted negotiations and major controversy within Canada, in a Long-Term Agreement (LTA) with the EC. Negotiations with Spain were fruitless. This failure was linked to overfishing of the straddling stocks. It shaped the late 1980's confrontation between Canada and the EC over management of stocks on the Nose and Tail of the Grand Banks. Portugal chose not to consider a long-term agreement, but opted instead for year-to-year agreements on allocations and commercial cooperation. Ironically, it maintained "a satisfactory trading relationship" with Canada up to its accession to the EC in 1986.

By 1981 there was no longer a surplus of northern cod, the species of primary interest to the western European countries. Following the Northern Cod Seminar in 1979, the federal government allocated 25,000 tons of northern cod for foreign countries. In 1981 the Canadian offshore sector harvested its quota in just 2 months, and it clamoured for more fish. The industry argued that the LTA with the EC was failing to produce expected benefits and that it was counterproductive to negotiate with Spain. It generally opposed allocating "non-surplus" fish for any purpose.

2.2 Phase II — Reward for Past Performance

In response to industry concerns, in August 1982 DFO convened a government-industry seminar at Oak Island, Nova Scotia, to review international fisheries relations policy. This Seminar provided input to the Task Force on Atlantic Fisheries. The Task Force recommended a new approach to foreign allocations which was accepted by the federal Cabinet. The Task Force recommended that the government:

1. Allocate non-surplus resources to foreigners as part of agreements for reciprocal fishing rights by fishing vessels across international boundaries (e.g. with Greenland in the Davis Strait);
2. Allocate resources that are currently surplus to Canadian harvesting capacity (e.g. squid) and a

> fixed amount of 'non-surplus' resources (e.g. cod) preferentially to those countries that maintain a satisfactory fisheries relationship with Canada (including fisheries trade and conservation). Allocations of non-surplus resources should be made after the fact — that is, in a subsequent year as a reward for satisfactory behaviour in the previous year, rather than as an incentive. In particular, the government should not negotiate access by foreign vessels to non-surplus resources in return for access to markets (Kirby 1982).

The Task Force viewed the allocation of non-surplus fish as "expensive as well as unwise from a marketing point of view." Because market access was inherently more uncertain and more subject to manipulation than catching an allocation of fish in the Canadian zone, allocations for access was "a one-sided bargain." Therefore, the Task Force suggested that non-surplus allocations be strictly limited in total, be made unilaterally based on good performance and be offered to selected nations "after the fact" rather than in return for a promise of access to markets. The Task Force left open the possibility of seeking benefits, including greater access to markets, in return for allocation of surplus resources.

The Reward approach of Phase II was reasonably successful with Portugal (which was allocated surplus and non-surplus fish) until it joined the EC in 1986. Some progress was also made with the Eastern Bloc countries. Conservation cooperation was maintained and there were some gains in market access.

There were two major shortcomings of the Phase II approach. Continued allocations of non-surplus northern cod to Portugal meant that there was less fish available to the Canadian fleet than it could harvest. This policy also continued the link between market development and fish allocations in the Canadian zone. This posed a threat to trade expansion should allocations have to be reduced. Also during the 1982–86 period the problem of overfishing outside 200 miles by countries that did not have access to Canadian waters began to escalate. The Reward for Past Performance policy could not arrest that trend because surpluses were decreasing, even for the countries that were cooperating with Canada.

2.3 Phase III — Cooperation on Conservation

In August 1985, DFO convened another government — industry seminar on international fisheries policy in Oak Island, Nova Scotia (Oak Island II). Seminar participants recognized that there was a growing conservation problem outside 200 miles. They concluded that the allocation of non-surplus cod to the EC under the Long-Term Agreement and to Portugal annually had not benefitted Canadian suppliers. Since 1981 Spain had been fishing freely outside 200 miles. It had blocked imports of Canadian fish products to its market in an attempt to gain access to Canadian waters. The EC, with the accession of Spain and Portugal to the Community in 1986, would probably use access to its market to seek preferential allocations in the Canadian zone. Once Canada began allocating non-surplus fish in 1981, there was increasing pressure to increase non-surplus allocations.

These developments led the Canadian industry and government to conclude that the time had come to stop non-surplus allocations and to delink allocations and access to markets. Seminar participants recommended that in future Canada allocate only surplus resources in accordance with the Law of the Sea Convention. Instead of rewarding "good behaviour" outside 200 miles, the emphasis should be placed on punishing "bad behaviour" by cutting off access to surplus allocations inside the zone.

The Oak Island II recommendations were a significant departure from Oak Island I and the Task Force on Atlantic Fisheries. The attempt to improve market access in free-market countries through the use of fish allocations was abandoned. The federal government accepted these recommendations as the basis for a modified international fisheries relations policy, which emphasized Cooperation on Conservation. This represented a new phase — Phase III — in foreign allocations policy, which was in effect from 1986 to the early 1990's.

> The essential elements of the new policy were:
>
> 1. Apart from existing treaty commitments (e.g. LTA and the Canada-France 1972 Agreement), Canada would no longer grant non-surplus allocations;
> 2. Surplus allocations would be granted only to countries which cooperated in conservation efforts both inside and outside the 200-mile zone; and
> 3. Commercial cooperation would no longer be a factor except to the extent it could be secured in agreements with East Bloc countries (DFO 1986e).

The essential underpinning of this approach was to keep trade matters separate from fisheries allocations.

Throughout the post-extension era Canada maintained good fisheries relations with Japan, the East Bloc and the Scandinavian countries. However, relations with Spain, Portugal and the EC had become increasingly strained from 1981 to 1985. With the accession of Spain and Portugal to the EC in 1986, the problem of foreign overfishing outside 200 miles intensified.

Cooperation on conservation was threatened by new policies adopted by the enlarged EC. Canada attempted to protect its special interest in the straddling stocks but found it increasingly difficult to do so. To understand how the promise of 1977 had culminated in the international fisheries confrontation of 1985–1992, it is necessary to examine Canada's attempt to protect its interest in the straddling stocks following extension through the formation of NAFO. It is also necessary to trace the evolution of its fisheries relations with the EC, Spain and Portugal from 1977 to 1992.

3.0 NORTHWEST ATLANTIC FISHERIES ORGANIZATION (NAFO)

Although the bulk of the marine fish resources off Canada's coasts fell within Canada's 200-mile fisheries zone, seven stocks on the Grand Bank straddled the 200-mile limit and could be fished by foreign fleets outside 200 miles. Another three stocks on Flemish Cap lay entirely beyond 200 miles.

These represented 10 of the 56 groundfish, pelagics and squid stocks under quota management off the Canadian Atlantic coast (CAFSAC 1986b). In terms of the overall Canadian Atlantic fishery, their importance was significantly less than appeared from just the number of stocks. Excluding squid, a variable resource fished only occasionally outside 200 miles, in 1978 Canada's catch from the remaining nine stocks was 78,962 tons, representing 9% of the total Canadian Atlantic catch of groundfish and pelagic species (ICNAF 1985).

Nonetheless, Canada felt overfishing of these stocks could pose a significant problem unless adequate mechanisms were in place to ensure that management of those stocks was consistent with measures taken by the coastal state. It therefore moved quickly to secure agreement on a new multilateral forum for management of stocks beyond 200 miles off its Atlantic coast.

At its December, 1976, Special meeting, ICNAF adopted an amendment to its Convention stipulating that the Commission would function only in waters outside national fisheries jurisdiction. This was an interim measure pending new institutional arrangements to accommodate the new jurisdictional realities (ICNAF 1977).

Canada convened Preparatory Conferences in Ottawa in March and June 1977, which considered a convention drafted by Canada. This draft envisaged an organization with management powers in the area beyond 200 miles and a scientific consultative role for the convention area as whole, including the 200-mile zones of coastal states.

Canada's major objective in these negotiations was to secure recognition of its special interests in the area outside but immediately adjacent to the Canadian 200-mile zone on the Atlantic coast. Canada sought to embody in the new convention provisions that conservation regulations for the Grand Banks and Flemish Cap stocks would have to be consistent with Canadian regulations within 200 miles, and that Canada as the coastal state would be entitled to preferential shares of the TACs for these stocks. At the Preparatory meetings few countries expressed support for this position.

In several bilateral agreements signed post-extension (in 1977 and early 1978) with Bulgaria, Cuba, the GDR and Romania, Canada was able to incorporate recognition of "the special interest of Canada, including the needs of Canadian coastal communities" in the resources beyond and immediately adjacent to the Canadian 200-mile zone.

Compromise on a new multilateral convention was achieved in October, 1977, when Canada agreed to the phrase: "The Commission shall seek to ensure consistency" between its conservation measures for straddling stocks and those taken by the adjacent coastal state. Further compromise occurred at a Group of Experts meeting in May, 1978, on the wording presently found in Article XI.4 of the NAFO Convention: "Commission members shall give special consideration" to the adjacent coastal state.

NAFO began its work in 1979, in conjunction with the last meeting of ICNAF. The new organization consisted of three bodies: a General Council, a Scientific Council and a Fisheries Commission. The General Council administers the internal affairs of the organization including the relations among its constituent bodies. The Scientific Council is a forum for consultation and cooperation for "study, appraisal and exchange of scientific information and views relating to the Fisheries of the Convention Area and to encourage research". Its primary function is to provide scientific advice to the Fisheries Commission and to coastal States where requested to do so. Canada seeks advice from the NAFO Scientific Council about the status of certain stocks within the Canadian zone. The Fisheries Commission is responsible for the management and conservation of the fishery resources of the Regulatory Area.

The NAFO Convention Area covers the Northwest Atlantic (Fig. 11-1), including waters under national fisheries jurisdiction. The Regulatory Area is that part of the Convention Area which lies beyond coastal States' fisheries jurisdiction. In practical terms, this means the Nose and Tail of the Grand Banks and Flemish Cap. The Convention applies to all fisheries resources except for: salmon, tunas and marlins, cetacean stocks and sedentary species of the Continental Shelf.

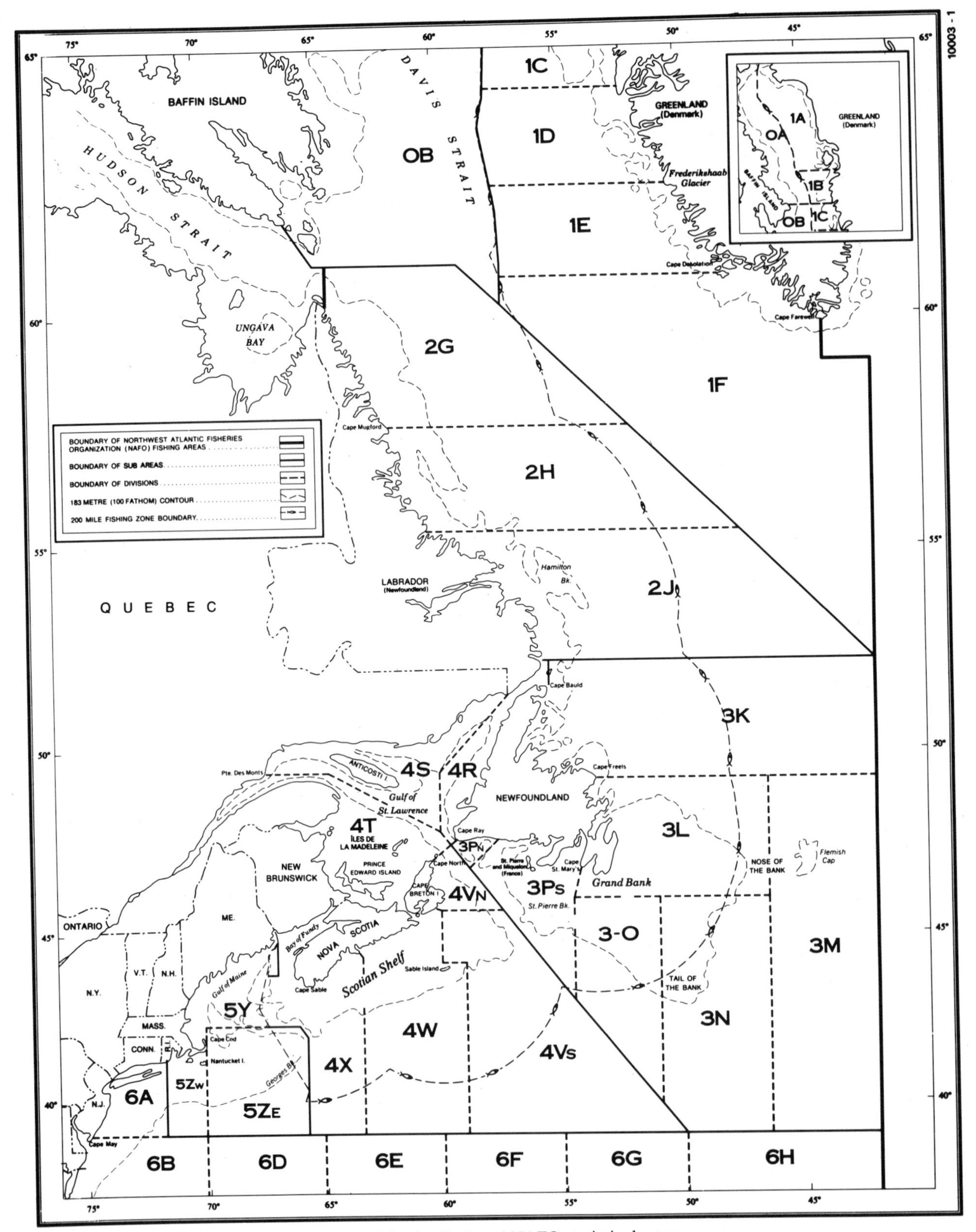

FIG. 11-1. NAFO area, Canadian 200-mile fisheries limit and NAFO statistical areas.

Henceforth, when referring to NAFO I will mean the Fisheries Commission, unless otherwise specified.

Nine countries signed the new *Convention on Future Multilateral Cooperation in the Northwest Atlantic Fisheries* on October 24, 1978, thus establishing NAFO. NAFO presently has 14 members comprising the major participants in the Northwest Atlantic fishery. In the 1980's other countries including Mexico, Panama, Venezuela, the USA and Korea have fished in the Regulatory Area but have not become NAFO members.

NAFO meets annually in September. The chief purpose is to establish management measures, particularly TACs and national allocations, for stocks in the Regulatory Area for the following year. These measures are adopted by a majority vote of the members of the Fisheries Commission, taking into account the advice from the Scientific Council, which meets annually in June. Within 60 days, contracting parties can lodge written objections to any or all of the measures adopted by the Commission. If a member objects to any particular measure, it is not legally bound by that measure. For example, if it objects to the TAC and national allocations for a particular stock, it can ignore the TAC and national allocations for that stock established by NAFO. If one member lodges an objection, there is a further period during which other members can object to the measure in question. The measure becomes binding only for those members who have not objected. This objection procedure, typical of international fisheries conventions, has become the Achilles' heel of NAFO. From 1985 onward the EC used the objection procedure repeatedly for a large number of stocks. This imperilled the effectiveness of NAFO.

Following the formation of NAFO, Canada pursued three objectives:

1. Ensuring that management measures adopted for the Regulatory Area were consistent with Canada's approach for adjacent stocks within its zone;
2. Ensuring that TACs were based on conservation requirements; and
3. Maintaining for Canada preferential allocations of the stocks beyond 200 miles, particularly the straddling stocks, thus implementing the "special consideration" provision of the Convention.

With a few exceptions, Canada persuaded NAFO to follow the conservative management regime it adopted for stocks within its zone. Except for Flemish Cap stocks which occur entirely beyond the Canadian zone, NAFO consistently established TACs at the $F_{0.1}$ level or its equivalent. Thus, Canada and the Fisheries Commission agreed on TACs for the seven straddling stocks. Similarly, Canada's identification of its needs for shares of the straddling stocks were largely accommodated by other NAFO members.

This approach characterized NAFO's deliberations from 1979 until 1985. The main difficulty from 1977 to 1983 was that Spain, a major participant in the fishery in the Regulatory Area, was not a member of NAFO. It was fishing freely, ignoring the ICNAF/NAFO quotas. This, plus increased activity by nonmembers, undermined NAFO's conservation efforts. Spain joined NAFO in 1983 and began to challenge its management approach and quota-sharing arrangements. When NAFO continued the conservative management approach and sharing arrangements which had been in effect since 1977, Spain invoked the objection procedure and continued to fish freely in the Regulatory Area. Thus, Spain's membership in NAFO from 1983 to 1985 had no effect on its fishing activities.

Up to and including the Sixth Annual Meeting, NAFO had generally followed the lead of Canada as the coastal state in setting TACs at the $F_{0.1}$ level or its equivalent. Canada's proposals were adopted for the seven straddling stocks. Spain challenged the Canadian management approach following its accession to NAFO in 1983. However, it had no support from other members.

Canada had removed the capelin fishery in Division 3L from NAFO in 1982. The decision which stocks would be considered as overlapping was made at the 1978 ICNAF meeting in Bonn. It was maintained, with the 3L capelin modification, into 1984. The most significant removal from the list of stocks considered by ICNAF at its 1977 meeting was northern cod (Divisions 2J, 3KL). From 1978 to 1984 Canada's right to manage this stock without reference to the NAFO Fisheries Commission was unchallenged.

NAFO's dynamics changed dramatically at the Seventh Annual Meeting in Havana in September, 1985. With the imminent accession of Spain and Portugal to the European Community, the EC changed its NAFO position radically. This threw NAFO into a crisis. To understand the confrontation between the other members of NAFO and the EC at the 1985 meeting and subsequent fisheries relations between Canada and the EC, it is necessary to examine Canada's bilateral fisheries relations with Spain, Portugal and the EC from extension of fisheries jurisdiction to the point where Spain and Portugal became members of the EC.

4.0 FISHERIES RELATIONS WITH SPAIN

Spanish fishermen had fished cod on the Grand Banks off Newfoundland for nearly 400 years. Annual Spanish catches off the Canadian Atlantic coast exceeded 200,000 tons from 1963 to 1972. This was prior to the imposition of catch quotas by ICNAF in 1973 (Fig. 11-2). By 1976, the year prior to extension, reported Spanish catches had declined to 47,000 tons. Spanish allocations of all species combined in 1976 was 82,000 tons. Following extension, Spain received combined allocations from ICNAF and Canada of 29,430 tons in 1977, 19,600 tons in 1978 and 16,360 tons in 1979.

Fisheries relations between Canada and Spain from 1976 to 1986 were governed, at least in theory, by the bilateral Fisheries Agreement of 1976. Under this agreement, Canada was committed to provide Spain with "allotments, as appropriate, of parts of total allowable catches surplus to Canadian harvesting capacity" (Article III). Canada and Spain undertook to "cooperate directly or through appropriate international organization to ensure proper management and conservation of the living resources of the high seas beyond the limits of national fisheries jurisdiction, including areas of the high seas beyond and immediately adjacent to the areas under their respective fisheries jurisdiction" (Article IV). Article VI stipulated that the two countries "shall promote future bilateral cooperation on... expansion of markets for fish and fish products originating in Canada" (Canada-Spain 1976).

Bilateral fisheries relations from 1977 onward were tense. They deteriorated rapidly in 1981 in light of Canada's rapidly declining surplus of cod. Cod was of primary interest to Spanish fishermen.

After the 1977 ICNAF meeting, Spain objected to quotas beyond 200 miles and threatened to withdraw from ICNAF. Following bilateral discussions in Madrid in December, 1977, Spain withdrew its objection. Discussions were launched on enhanced Canadian access to the Spanish market. In January, 1978, Spain agreed to promote increased purchases of dry salted cod from Canada. In return, Spain was allocated additional quantities of cod in the Canadian zone. Spain's total cod allocation in the Northwest Atlantic that year was 19,600 tons.

By May–June of 1978 Canada was alarmed by evidence of Spanish overfishing beyond 200 miles. Inspections under the ICNAF Scheme of Joint Enforcement indicated that Spain had exceeded its Flemish Cap cod quota (2090 tons) by 50% as early as May. Canada requested Spain to discontinue fishing. Canada had, in effect, already paid with increased cod allocations in the Canadian zone for the withdrawal of the Spanish objection to the ICNAF quotas. Thus

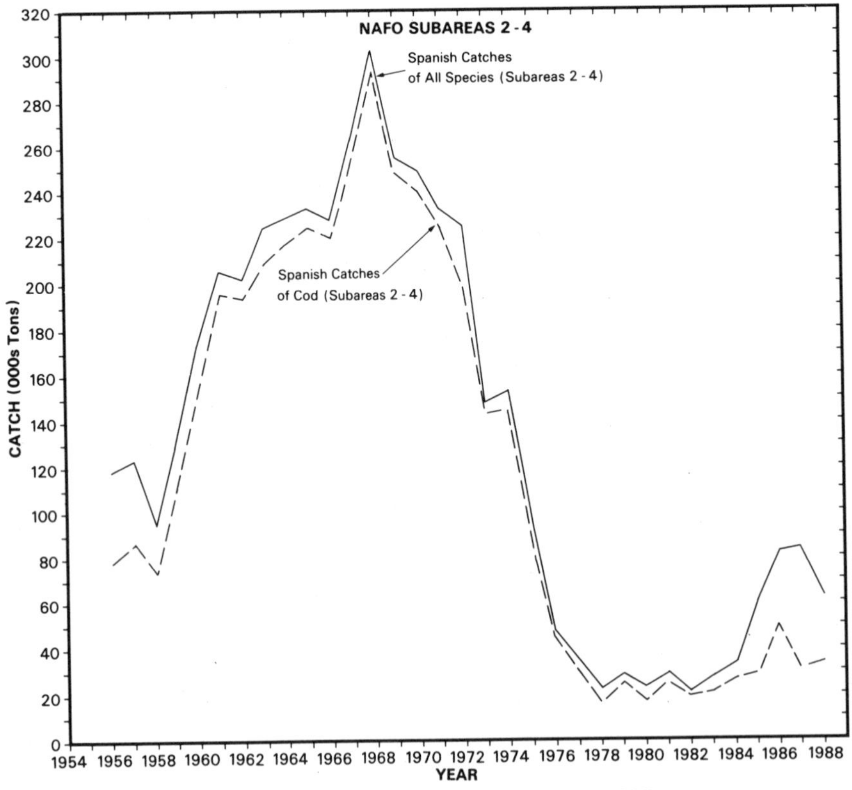

FIG. 11-2. Spanish catches in subareas 2–4 from 1956 to 1988.

Canadians were perturbed by Spain's failure to abide with ICNAF conservation measures. Spanish officials indicated that they would have great difficulty enforcing accepted quotas. They claimed that fisheries was a very "underdisciplined" sector of the Spanish economy.

By the fall of 1978 Spain did not appear to be meeting its commercial commitments for salt cod. Nonetheless, Canadian industry continued to advocate using "government clout on commensurate benefits to maintain an open door to Canadian products" (R. Bulmer, November 30, 1978).

In January, 1979, Canada and Spain agreed on the allocations (including 5,000 tons of northern cod) and commercial commitments for 1979. By May of 1979 Canada was again becoming concerned about Spanish overfishing outside 200 miles. Canadian inspections under the Scheme of Joint Enforcement indicated significant underreporting by Spanish vessels.

Matters came to a head in late May when Spanish and Canadian officials met to discuss the overfishing problem. This resulted in a June, 1979, Memorandum of Understanding. Spain agreed that since Spanish vessels had already "reached their level of catches authorized under the 1979 ICNAF allocation of quotas," it would issue an immediate order to all vessels to leave the ICNAF/NAFO area. It would also take "appropriate legal action" against all Spanish vessels which might have incurred infractions in 1979 (Canada-Spain 1979).

At the 1979 ICNAF meeting, Spain was the only country to vote against the Canadian proposal for 3NO cod for 1980 (ICNAF 1979). In consultations with industry shortly thereafter, Canadian officials recognized that control of Spanish fishing effort outside 200 miles, Spanish discards of flatfish, and the Spanish use of flags of convenience would remain problematic.

Spanish fleet owners clearly believed that, without substantially increased allocations within the Canadian zone, they would be "better off" in an unregulated fishery beyond 200 miles on the Grand Banks and Flemish Cap.

In November, 1979, Spain lodged an objection to the 1980 ICNAF TACs and allocations of cod in 3M and 3NO. In January, 1980, Spain notified NAFO of its rationale for not joining and indicated that it would be unilaterally establishing its own quotas for the NAFO area. By the end of February negotiations had resumed (DFO 1980f).

Canada and Spain concluded a Memorandum of Understanding on March 7, 1980, regarding their fisheries relations for 1980. The principal elements of this MOU were:

1. Spain would consider joining NAFO;
2. Spain and Canada agreed to the placing of scientific observers on each other's vessels in the area beyond 200 miles;
3. Spain agreed to attempt to curtail the activities of Spanish joint ventures in NAFO waters, as well as non-NAFO vessels (flags of convenience) operating in NAFO waters;
4. Spain agreed to issue import permits for specified quantities of Canadian cod and to give priority to squid imports from Canada; and
5. Spain would receive the following: 10,450 tons of cod allocations from Canada and 9,560 tons from NAFO and a NAFO squid allocation of 2,250 tons (Canada-Spain 1980).

By July, 1980, there was again evidence that Spanish vessels were falsifying their log books. By the fall of 1980 Spain was expressing interest in a new long-term fisheries agreement with Canada. It hoped this would provide for an allocation in the order of 40,000 tons of cod per year up to 1986. At that time Spain's existing allocations in Canadian and NAFO waters were 20,000 tons. Spain's request amounted to 10% of the TAC of cod in Canadian waters. Support in the Canadian industry for a long-term agreement had diminished dramatically in the absence of a surplus in the northern cod stock. Canada had given preference to Spain in 1979 and 1980 in an attempt to reach agreement. No other country received as high a cod allocation in Canadian waters, and Spain was second only to Japan in terms of squid allocations.

Responding to the views of the Canadian fishing industry, Minister LeBlanc offered Spain and Portugal a 2-year phase-out agreement from northern cod. Each country would get 5,000 tons in 1981 and 3,000 tons in 1982. Conditions for such an agreement would include:

1. A commitment to abide by NAFO regulations;
2. Commercial guarantees; and
3. Recognition that after 1982 there would be no northern cod allocation to these countries.

The 2-year phase-out approach was chosen because it would allow only limited access to non-surplus cod while providing time for each country to adjust its fishery.

Spain flatly rejected the "phase-out" proposal, threatening trade retaliation. By this time Spain's policy was to use access to the Spanish market to protect its fleet. The Spanish government criticized the TACs set by Canada as "excessively low". Phase-out of northern cod allocations would be met with phase-out of access to the Spanish market. By March of 1981 import licences required for access to the Spanish market were being withheld.

Fisheries relations between Canada and Spain were very strained throughout 1981. Under the 1976 Agreement, Canada was not obligated to allocate northern cod to Spain without a surplus. The "phase-out" offer was intended to ease the Spanish fleet out of that fishery. While in Canada this was considered generous, the Spanish were incensed that their fleet would not be considered for northern cod allocations after 1982. They disagreed with Canada's basis for establishing the cod TACs. Canada's conservative approach to stock recovery was seen as a cynical and self-serving use of science to exclude foreigners from Canada's cod fishery. Spain claimed that the "phase-out" offer was a breach of the 1976 Agreement. They retaliated with an unrestricted fishery outside the 200-mile zone and blocked the import of Canadian cod and squid to their market.

Since Spain had not yet joined NAFO, it was legally entitled to unrestricted fishing on the high seas. However, its failure to cooperate in conservation and its interference with fisheries trade directly contravened the 1976 Agreement. The trade actions also contravened the 1954 Trade Agreement between Canada and Spain, the two countries' principal bilateral trade agreement.

Canada protested these contraventions of treaty obligations. It rejected the Spanish contention that Canada breached its obligations with its conservative ($F_{0.1}$) approach to cod TACs. This situation provoked public debate in both countries. In Canada, the Spanish fishing fleet was seen as rapacious and undisciplined. In Spain, Canada's fishing policies were seen as self-serving measures to destroy the Spanish cod fleet.

At its June, 1981, Special Meeting, NAFO asked Spain to cease fishing for cod in Divisions 3M and 3NO immediately and to abide by NAFO conservation measures (NAFO 1981a). Spain refused to leave the NAFO area following the resolution adopted by NAFO in June. At its September meeting, NAFO removed the special reservation of quotas for Spain (NAFO 1981b).

Early in 1982, the Spanish government began to show some interest in ending the confrontation with Canada. At February, 1982, negotiations, Canada offered Spain for 1982 allocations of surplus cod in 2GH, squid in Subareas 3 and 4 and silver hake in Divisions 4VWX, a total of 11,250 tons of surplus allocations. Since there was no surplus of northern cod, Canada did not offer allocations from this stock.

Spain rejected this offer outright on the grounds that no agreement was possible without a northern cod allocation. The Canadian negotiator invited his Spanish counterpart to outline an agreement which Spain would accept. The result was a "draft ad referendum memorandum of understanding" (DARMOU) which outlined the minimum acceptable to Spain.

The principal elements of the DARMOU were:

1. Canada would offer 5,000 tons of non-surplus northern cod;
2. Canada would offer 11,250 tons of surplus Canadian stocks;
3. Spain would join NAFO by 1983 subject to Congressional approval and thereafter allow Canadian observers on Spanish vessels outside the 200 mile zone;
4. Spain would provide Canada with a cod fishing plan covering its proposed activities outside the 200 mile zone during the interim period prior to NAFO accession. It would invite Canadian surveillance and enforcement on the high seas to ensure the Spanish fleet's compliance with that plan;
5. Spain would restore a satisfactory trading relationship (Canada-Spain 1982a).

The Spanish fleet representatives indicated that the DARMOU was a "do or die" proposition. Should there be no agreement with Canada, they would use every political means to oppose Spanish accession to NAFO. They would never again cooperate with Canada, and they would use all possible leverage to block Canadian imports. While Spain would prefer to establish a long-term agreement with Canada prior to entry into the EC, it would "not forget" once in the EC if Canada remained intractable.

Federal fisheries officials consulted the Canadian industry about the results of the February negotiations. The industry consensus was that Canada should not make allocations to Spain until Spain would follow NAFO regulations and abide by NAFO quotas outside 200 miles. Canada should remain open to negotiate non-surplus allocations with Spain. As prerequisites, Spain must join NAFO and adhere to its NAFO quota and regulations.

Canada informed Spain that it would accept the DARMOU except for the conservation provisions which were considered insufficient. Canada urged Spain to apply for and abide by NAFO cod quotas in 3M and 3NO, instead of persisting with its unilateral fishing plan.

Spain expressed willingness to discuss this. However, it sought compensation in the Canadian zone for any fish it would otherwise have caught in the NAFO Regulatory Area. Canada could not satisfy such a request. However, at negotiations in May, the DARMOU of February was modified to provide for application by Spain to NAFO for cod quotas in 3NO and 3M. Based on these quotas, Spain would fish in those areas, "in full compliance with NAFO regulations, including the acceptance of observers and inspections on all Spanish vessels operating beyond 200-miles." The head of the Canadian delegation agreed to recommend an additional allocation to Spain of 1,000 tons of cod in area 2GH (Canada–Spain 1982b).

However, by this time the overwhelming mood among the Atlantic provinces and fishing industry was against giving in to the perceived "blackmail" by Spain.

At this point in the year 5,000 tons of northern cod for Spain could only be found by increasing the TAC. Since there were already Canadian vessels tied up for lack of quotas, this would be a dramatic exception to established principles. Such an agreement might have been possible early in 1982. However, stocks were fully allocated and Spain had again flaunted the NAFO regulatory regime. Thus by midsummer the window of opportunity had closed.

During 1983 and 1984 there were repeated attempts to seek some accommodation with Spain. These were fruitless.

In August, 1983, Spain joined NAFO. At the September, 1983, NAFO meeting, Spain sought large allocations of virtually all species. These requests were not acceptable to other NAFO members. The Spanish fleet continued to fish in NAFO waters after joining NAFO in 1983 even though Spain had not been allocated any NAFO quotas for 1983 (NAFO 1983). Following the 1983 NAFO meeting, Spain lodged objections to the 1984 NAFO quotas for 3M cod, 3M redfish and 3LN redfish.

In June, 1984, Canadian industry officials contended that the government had been too forthcoming in its dealings with Spain. They cited Spain's disregard for conservation in the NAFO area and the discriminatory closure of the Spanish market to Canadian fish products. The industry advised against further concessions. In the end there was no agreement between Canada and Spain for 1984. Spain's refusal to cease fishing for cod in 3L had by now emerged as one of the greatest impediments to an agreement. Canada was becoming increasingly concerned about the possibility of increased foreign fishing for northern cod in 3L. NAFO had up to this time accepted that this stock was managed by Canada.

In making offers of non-surplus cod to Spain in 1983 and 1984, Canada had made an exception for 2 consecutive years to the policy of Reward for Past Performance. By late 1984, the Canadian industry and the provinces were unanimous in recommending that no non-surplus resources be offered to Spain in 1985. The Spanish had not "performed" in conservation or market access in 1984. The federal government took this advice. To comply with the terms of the 1976 Agreement, Canada would make available certain surplus allocations (5,000 tons of silver hake and 4,000 tons of squid) provided Spain complied with the conservation and market access provisions of the Treaty. Spain was unlikely to meet these conditions since it had already lodged objections to the 1985 NAFO quotas for 3M cod and redfish and 3NO cod and 3LN redfish.

The Spanish market had essentially been closed to Canadian fish products since 1981. Spain had fished above the level of quotas reserved for Spain by NAFO before it joined NAFO in 1983. It had objected to the major quotas and continued to fish freely after it joined NAFO in 1983. By these actions Spain had tried to pressure Canada to offer larger quantities of cod.

Spain and Portugal were expected to become members of the EC on January 1, 1986. The 1976 Canada-Spain agreement would be automatically renewed unless terminated on June 10, 1986. Canada had obtained negligible benefits from the treaty from 1981 onwards. Thus, on June 28, 1985, the government announced that the Canada/Spain Fisheries Agreement would lapse (DFO 1985h).

Thus ended the era of attempts to reconcile Canadian and Spanish interests following Canada's extension of fisheries jurisdiction. With the accession of Spain and Portugal to the EC in 1986, the residue of the Canada-Spain confrontation dramatically altered Canada-EC fisheries relations.

5.0 FISHERIES RELATIONS WITH PORTUGAL

Canada's post-extension fisheries relations with Portugal were markedly different from those with Spain. There were problems but these did not overshadow the generally positive nature of the fisheries relationship.

Like Spain, Portugal had conducted a major fishery for cod off the Canadian Atlantic coast for several centuries. In the years prior to the 200-mile limit, Portugal's cod catch had declined from 171,000 tons in 1971 to 32,000 tons in 1977. From 1977 to 1982, Canada's fisheries relations with Portugal were marked by attempts to secure a larger share of the Portuguese market for cod while securing cooperation in conservation.

Efforts to obtain commercial benefits from Portugal were largely successful. Under the commensurate benefits approach, specific purchases in return for allocations of cod and other species led to a rapid increase in Canadian fish exports to Portugal to a level of $18 million in 1980 (DFO 1980g).

In 1980, Canada approached Portugal about a long-term agreement but the Portuguese showed little interest. Like Spain, Portugal was initially offered in 1981 a 2-year "phase-out" from non-surplus cod allocations. Canada offered Portugal 5,000 tons of northern cod for 1981 and 3,000 tons for 1982. In return Portugal was to:

1. Acknowledge that the 1981 northern cod allocation would be outside the framework of the 1976 bilateral agreement (i.e. non-surplus) and that, after 1982, where there was no surplus, there would be no allocations to Portugal;
2. Continue its membership in NAFO, abide by NAFO rules and not object to any NAFO regulation involving fishing in Divisions 3LNO;
3. Participate in the NAFO scientific observer scheme; and
4. Address the Portuguese aspects of the "flags of convenience" problem.

Canada for its part would wherever possible favour Portugal in any foreign arrangement programs, such as over-the-side sales or the resource-short plant program.

In addition to the non-surplus cod offer of 5,000 tons, Portugal was offered 11,720 tons of surplus fish of other species in the Canadian zone. In return Portugal was requested to purchase from Canada a minimum of 50% of its dry salted cod imports and a minimum of 40% of its wet salt cod imports.

Initially, Portugal appeared to accept the "phase-out" approach. However, it soon reversed its policy, insisting that there be no mention in the 1981 agreement of a phase-out of non-surplus allocations. While negotiations continued, the Portuguese private sector began buying large amounts of Canadian cod, particularly wet salt cod. This was very important for the Canadian industry in an otherwise bleak year. Consequently, Canada offered Portugal 5,000 tons of non-surplus northern cod plus the surplus allocations. Canadian officials agreed that any reference to "phase-out" be dropped from the agreement (Canada-Portugal 1981).

This marked the beginning of a special fisheries relationship with Portugal which continued to be allocated non-surplus cod on a year-to-year basis. Portugal fulfilled Canadian expectations by increasing its purchases from an average of less than $2 million per year in the late 1970's to a high of $62 million in 1982.

For 1982 Canada initially allocated Portugal 5,000 tons of non-surplus northern cod and 9,320 tons of surplus allocations of various species.

Subsequently, Portugal was allocated for 1982 an additional 700 tons of northern cod and 700 tons of non-surplus Scotian Shelf cod. Canadian officials decided to allocate a small quantity of Scotian Shelf cod to Portugal because they believed that, if non-surplus allocations were to continue, they must come from more than one stock. The additional allocations to Portugal were Canada's goodwill gesture recognizing the large Portuguese purchases in 1982.

As 1983 approached, Portugal stood to benefit from Canada's new policy on Reward for Past Performance. During bilateral discussions in February, 1983, Portugal assured Canada that it would be the major supplier of wet salt cod. Canada offered Portugal a 25% increase in its non-surplus cod allocations for a total of 6,500 tons of northern cod and 1,500 tons of Scotian Shelf cod. Atlantic Canadians were particularly sensitive to such an increase given the difficulties then confronting the Atlantic fishery. Later in the year Portugal was allocated 1,000 tons of 2GH cod for a total Canadian cod allocation of 9,000 tons (Canada-Portugal 1983).

In 1983, Canadian fisheries officers found evidence of Portuguese overfishing of their 3M redfish quota. Portugal's market and conservation performance in 1983 did not meet Canadian expectations. Purchases dropped to about half the 1982 level. However, with 1983 sales of about $40 million, Portugal remained a significant market for Canadian salt cod.

To deal with the problem of overfishing its 3M redfish quota, Portugal agreed at the 1983 NAFO meeting to trade its 850 ton 3LN redfish quota for 1,300 tons of 3M redfish from the Canadian allocation (NAFO 1983).

Despite Portugal's relatively poorer conservation and commercial performance in 1983, Canada continued allocations to Portugal in 1984 at the 1983 levels. These allocations were the maximum Canada would provide. These would be maintained over time if Portuguese performance was also consistent.

Fisheries Minister De Bané visited Portugal in January, 1984. Mr. De Bané instructed the Canadian Saltfish Corporation to give preference to Portugal when securing foreign vessels to supply the Resource-Short Plant Program (RSSP). This was in addition to 3,000 tons already negotiated. If foreign vessels were to participate in the RSPP program in future years, Portugal would have the right of first refusal. In return, Portugal indicated that Canada would be the major supplier of cod to the Portuguese market. It would supply 55% of Portugal's annual import of salt cod (Canada-Portugal 1984).

Despite these positive signals, the Portuguese continued in 1984 to overfish Flemish Cap redfish. Because of the trade of 1,300 tons of the Canadian 3M redfish allocation for 850 tons of the Portuguese allocation of 3LN redfish, Portuguese vessels had no quotas for the latter transboundary redfish stock and hence did not fish redfish in that area. At the 1984 NAFO meeting, Portugal worked hard to reverse this trade and regain access to the 3LN redfish stock. However, other NAFO members did not support this effort (NAFO 1984a). Subsequently, Portugal lodged an objection to its 1985 NAFO redfish quotas for 3LN and 3M.

Canadian sales of salt cod to Portugal in 1984 were about 17,000 tons of wet salt cod equivalent. This comprised about 50% of the imports of wet salt cod equivalent in 1984 compared with the Portuguese offer of 55%. These sales were about the same level as in 1983 but fell short of the purchase commitment made by Portugal in January.

Salt cod production was down in both Newfoundland and Nova Scotia in 1984. This contributed to the failure to achieve the level of the purchase commitment. Accordingly, Canadian industry recommended that the Portuguese not be penalized for their failure to fulfil the 1984 purchase commitments.

The 1985 agreement between Canada and Portugal was a roll-over of the 1984 agreement, with similar allocations and commercial cooperation provisions.

In 1985, Portugal overfished the NAFO "others" quota for 3LNO American plaice, a stock of considerable importance to Canada, and 3M cod, American plaice and redfish. It also overfished its Canadian northern cod allocation with a total catch inside and outside the Canadian zone above 9,000 tons. Thus in 1985 the conservation threat from Spanish activities on the straddling stocks in 3L and 3NO was heightened by the escalation of overfishing by Portugal. Late in 1985 Canadian officials informed Portugal that continued overfishing would harm the countries' relations.

Portugal was to join the EC in January, 1986. Nonetheless, a separate Canada-Portugal agreement was possible for 1986 provided it was concluded before January 1, 1986. Negotiations took place in December, 1985. Canada was prepared to maintain the 1985 level of allocations to Portugal if it committed its vessels to adhere to the NAFO conservation regulations.

Portugal was, however, caught in the vice of its impending accession to the EC. The EC had already instructed Portuguese authorities to object to the 1986 moratorium on cod fishing in 3L adopted by NAFO in September 1985 (see Section 7). Canada's one condition for granting allocations was that Portugal comply with the NAFO regulations. It became apparent at the December, 1985, discussions that Portugal, because of its 1986 membership in the EC, must follow the EC's lead on fisheries matters. Hence, it could not comply with Canada's condition for allocations. Canadian officials informed Portugal that, should it object to the NAFO regulations for 1986, Portuguese vessels could not fish in Canadian waters. They would also be denied port privileges and prohibited from participating in foreign arrangements.

Thus a relatively satisfactory fisheries relationship with Portugal ended with its accession to the EC.

6.0 FISHERIES RELATIONS WITH THE EC

Canada's fisheries relations with the EC can be divided into two distinct periods:

1. From 1977 to 1984 relations were characterized by attempts to seek mutually beneficial fisheries arrangements; and
2. From 1985 onwards the EC confronted Canada's fisheries policies and attempted to sabotage NAFO's effectiveness as a multilateral management forum for stocks beyond 200 miles.

Immediately prior to and in the period following extension of Canadian fisheries jurisdiction, Canada negotiated a number of long-term fisheries agreements. These included all the countries fishing for resources surplus to Canadian requirements within the Canadian zone, except for the EC (and, because of maritime boundary disputes, the USA). The pre-extension agreements benefitted Canada by providing advance recognition of its 200-mile limit. The post-extension agreements were to provide other benefits. These benefits included protection for Canadian salmon, recognition of Canada's special interest in stocks beyond and immediately adjacent to the Canadian 200-mile zone, and commercial cooperation.

In initial post-extension discussions, the EC had difficulty with all three objectives. Consequently, in June, 1978, an interim fisheries agreement was negotiated for 1978 and 1979. This agreement was provisional because of delays in ratification by the EC. The interim agreement benefitted Canada by keeping the door open to a long-term agreement incorporating commercial benefits. It also provided interim protection for Atlantic salmon of Canadian origin fished off West Greenland.

In December, 1979, Canadian and EC officials agreed to recommend that the 1978–79 interim agreement be "rolled over" for 1980, with minor amendments. This was done in April, 1980, and negotiations commenced towards a long-term agreement (Canada-EEC 1980a). Canadian officials wanted to eliminate EC tariffs favouring Norway, Iceland, Portugal and the Faroes. Participants in the European Free Trade Agreement (EFTA), as well as the Faroes, had secured preferential tariffs from the EC on a variety of products to compensate for the loss of duty-free markets in the U.K. and Denmark when these countries joined the EC in 1972.

The wide-spread introduction of 200-mile zones resulted in the EFTA preferences assuming a value far beyond that anticipated when those bilateral agreements were signed. The rapidly increasing importance of the EC member states as import markets for fish and fish products enhanced the export opportunities of the EFTA countries and accentuated the injurious effects of the tariff preferences on third-country fish exporters such as Canada. As an example, Canada had had a negligible share of the EC market for cod products. Between 90 and

95% of EC import requirements of cod blocks and fillets were met by those countries enjoying tariff preferences.

Canada offered the EC specific allocations of cod and squid each year for 5 years. In return, it requested equal treatment with Iceland and Norway, its major competitors in the EC market.

By mid-1980 the Canada-EC negotiations on a Long-Term Agreement (LTA) had become intertwined with internal EC negotiations on a Common Fisheries Policy (CFP). The United Kingdom, where most of the Canadian exports would likely go, would receive only a small share of the allocations in the Canadian zone. It pushed hard for a favourable deal on the CFP and tried to extract maximum benefits in return for concessions to West Germany (FRG) on the proposed agreement with Canada. Meanwhile, the Canadian industry was becoming more ambivalent about the proposed agreement. While the industry preferred reduced tariff barriers, it began to question the use of long term guaranteed allocations of Canadian fish as leverage for tariff improvements. Also it did not favour the Tariff Rate Quota (TRQ) approach proposed by the EC.

While initially in favour of an LTA, the industry by the fall of 1980 opposed allocating non-surplus fish in return for improved market access. There were two reasons for this. First, industry needed all the fish it could catch. Second, it was prepared to take its chances expanding its exports to the EC at existing tariff rates.

Government officials, on the other hand, thought that there would be market difficulties for cod and other products in the 1980's. Minister LeBlanc considered the allocation of a certain amount of non-surplus fish as a reasonable cost given the benefits sought. Thus, he authorized the negotiations to proceed.

On November 29, 1980, negotiators agreed to recommend to the two countries a long term agreement between Canada and the EC, coupled with an exchange of letters on tariff reductions and fish allocations. In summary, the EC agreed to establish tariff rate quotas (TRQs) as follows:

1. Frozen round cod and redfish at 3.7% with the TRQ starting at 5,000 tons for 1981–82 and rising to 6,000 tons in 1983 through 1986 for an average of 5,666 tons per year;
2. Frozen cod in blocks and fillets at 4–6% for an overall quota of 16,000 tons in 1981 rising to 24,000 tons in 1986 for an average of 20,000 tons;
3. Whole salted cod with no tariff with no quantitative limit for 1981 through 1983, with a limit of 4,000, 5,000, and 6,000 tons in 1984, 1985 and 1986, respectively;
4. Salted cod fillets with no tariff without quantitative limit for 1981 through 1983, with a limit of 2,500, 3,500 and 4,000 tons, respectively, for 1984, 1985 and 1986; and
5. Herring flaps in vinegar at 10% with a TRQ of 3,000 tons in 1981 rising to 7,000 tons in 1986 for an average of 5,166 tons.

In return, Canada would extend the following allocations to the EC:

1. 8,000 tons of northern cod in 1981 and 9,500 tons from 1982 through 1986;
2. 6,500 tons of 2GH cod from 1981 through 1987; and
3. 7,000 tons of squid from 1981 through 1986.

The draft agreement also provided that the salmon quota of 1,190 tons would be maintained at West Greenland in 1981 and would continue until the end of 1983 or until an international Atlantic Salmon Convention came into force, whichever came first (Canada-EEC 1980b).

At a meeting of the Atlantic Groundfish Advisory Committee in December, 1980, the Canadian processing industry and the Government of Newfoundland expressed strong opposition to the Agreement. Inshore fishermen in the Maritimes, who stood to loose nil in the way of allocations, supported the Agreement in the hope that improved market access would benefit their sales to processors. The province of Nova Scotia initially supported the Agreement on the grounds that it involved surplus fish. (By January 1981 it was opposing the Agreement). In their view, the TAC for northern cod could be raised above 200,000 tons in 1981 to make the Agreement possible. The Newfoundland Fishermen, Food and Allied Workers' Union opposed the Agreement. Overall, the weight of opinion was against the Agreement on the grounds that Canada should not attempt to buy market access with non-surplus fish (AGAC 1980).

Within the European Community discussions regarding a Common Fisheries Policy dragged on through 1981. The EC's decision on ratification of the LTA was linked to those discussions. The Community finally approved the LTA on September 29.

On December 30, 1981, the Canada/EEC Long-Term Fisheries Agreement (LTA) was signed in Brussels (Canada-EEC 1981a and b). In announcing the Agreement, Ministers MacGuigan and LeBlanc indicated that Canada's benefits from the Agreement would involve substantial reductions of the tariff rates levied by the EC on cod, herring and redfish products of special interest to the Canadian fishing industry. This was expected to significantly improve the Canadian industry's competitive position in the European market. The fishing rights and tariff rate quota benefits under the Agreement would extend from 1982 to 1987 (Canada 1981a).

The Canadian government came to regret this decision as events quickly proved that industry fears were justified. Immediately following the signature of the LTA, the European Commission introduced a tariff rate quota system that reduced the benefits Canada was to receive under the agreement. Under this regulation, the UK, the major market in the EC for frozen cod products, was allocated a very small proportion of the total initial tariff quota. It was exempted from the EC's normal practice of drawing on a "reserve" once its initial quota was depleted. As a result, almost 50% of Canadian exports of frozen cod fillets and blocks during the first 8 months of 1982 did not enjoy the benefits of the lower tariff rates provided in the LTA. This percentage continued to increase as UK authorities were no longer allowing imports of frozen cod fillets and blocks at the reduced LTA tariff rates. Sales which might have materialized were also lost because of the inability of UK importers to obtain the lower LTA tariff rates. The EC was failing to honour its full obligations under the agreement.

Canada's initial response was to honour its own obligations while warning the EC that it would review performance under the LTA later in the year. Canada would react appropriately if it continued to be denied benefits under the Agreement.

At the beginning of 1983, Canada was still dissatisfied with the EC's administration of the LTA tariff rate quota system and sought compensation for benefits lost in 1982. On January 28, 1983, Fisheries Minister De Bané announced an interim allocation of 2,000 tons of northern cod to the EC pending further review of the degree of EC compliance with the LTA in 1982. The Minister described the 2,000 ton allocation to West German vessels as a "goodwill gesture" by Canada. He indicated that, following completion of the review, additional permits would be issued for smaller allocations than the EC expected. This was because the Community had not lived up to its obligations under the LTA in 1982 and had already indicated that it would not fulfil its obligations in 1983 (DFO 1983g).

In February, 1983, the EC agreed to compensate Canada for impairment of its benefits in 1982. Discussions commenced on the level and form of such compensation. Meanwhile, it appeared that the maximum benefits which Canadian exporters could expect to receive in 1983 would be approximately 70% of the benefits due under the LTA. Accordingly, Canada decided to grant the EC only part of the fishing allocations to which it normally would have been entitled in 1983.

On December 13, 1983, Canadian and EC officials settled on the future implementation of the LTA. The main features of the new arrangements, which came into force on January 1, 1984, were:

1. A further reduction of 2% (from 6 to 4%) in duties for frozen cod fillets, covering potential sales of 51,000 tons of cod between 1984 and 1987;
2. A guaranteed level (53%) of access to the LTA tariff benefits in the United Kingdom, the principal EC market for cod;
3. A more liberal interpretation of the "further processing" requirement in the agreement, allowing direct sales to restaurants and catering operations to benefit from the reduced tariffs provided in the LTA; and
4. Restoration of the full Canadian fishing allocations specified in the LTA for EC vessels (DFO 1983h).

At consultations on December 9, 1983, most of the Canadian industry supported this package deal. The Provinces interested, particularly Newfoundland, remained firm opponents of the LTA on the grounds that improved access to markets should not be paid for by fishing allocations in the Canadian zone.

In the winter of 1985 a new element entered into play to disrupt Canada-EC fisheries relations. After fishing their northern cod quota in the Canadian zone, FRG vessels began fishing northern cod in 3L on the Nose of the Grand Banks outside the Canadian 200-mile limit, an area of increasing conflict between Canada and Spain. Canada requested the EC to stop the FRG fishery immediately because the 9,500 EC LTA cod quota for 2J3KL may already have been exceeded.

The Community failed to act on Canada's request and the FRG continued fishing in 3L outside 200 miles. By mid-April it appeared that FRG vessels had caught 19,000 tons outside plus 7,400 tons inside the Canadian zone for a total of at least 26,000 tons. The FRG fishery continued until a visit to Europe by Fisheries Minister John Fraser in May. Minister Fraser pressed the Canadian case with EC authorities in Brussels, and with the FRG in Bonn. Immediately following these discussions, the German Deep Sea Fishery Association instructed vessel captains still fishing 3L to leave the area (DFO 1985i).

This new development presented Canada with a significant resource management problem. The FRG vessels would likely be back on the Nose of the Grand Bank in the autumn of 1985. In 1985, Portuguese vessels had, for the first time since extension of jurisdiction, also fished on the Nose. Spanish vessels had been fishing there for several years after the 1981 impasse. The cod being fished outside 200 miles in 3L was part of the northern cod (2J3KL) stock complex. NAFO had since

1978 tacitly accepted Canadian management of this stock. The TAC for the northern cod stock was established each year by Canada, with advice from the NAFO Scientific Council, as had been the case for several years.

The estimated catch by the EC in the first half of 1985 was at least 25,000 tons compared with a LTA allocation of 9,500 tons. The lack of Community cooperation in responding to the problem posed by the FRG fishery had reinforced the negative perceptions of the LTA already widely held by the Canadian fishing industry and the provinces.

Canada attempted to resolve this matter bilaterally with the EC. The EC argued that Community vessels while on the Nose of the Grand Bank had been fishing in international waters and were not prohibited by the LTA from doing so. The LTA applied only to the Canadian zone. If the EC were to restrict its fishery beyond Canada's zone, then it must be compensated. The EC rejected Canada's right to manage the entire cod stock in NAFO Divisions 2J3KL.

EC negotiators by this time were being heavily influenced by the need to find fish for Spain and Portugal once they became members of the Community in 1986. Suspecting that Canada did not want the management of the northern cod stock brought before NAFO in September, the EC chose to raise the matter there to press Canada for significant concessions.

7.0 NAFO, THE EC AND CANADA — 1985–1992

The 1985 NAFO meeting was a stormy one. The EC challenged the Canadian position that 2J3KL cod should continue to be managed as one stock by Canada. It tried unsuccessfully to get NAFO to agree to set a TAC for 3L cod for 1986. Canada proposed that the Scientific Council be asked for advice on management in 1986 of various stocks. This would include certain questions to the Council regarding cod in Divisions 2J3KL. Because of the lack of information about cod in Division 3L outside the Canadian zone, Canada proposed a temporary restraint on fishing for 3L cod outside 200 miles. It would set a TAC of 266,000 tons for 2J3KL cod in the Canadian zone in accordance with the scientific advice for this stock.

The EC described the Canadian proposal as "creeping jurisdiction". The EC proposed there be no limit on fishing cod in Division 3L pending scientific advice. Canada proposed a one-year moratorium for 1986. The Canadian proposal was adopted (NAFO 1985a).

In the course of the heated discussions on this subject, the EC reversed its previous practice of supporting the scientific advice. Instead, it attacked the principles which NAFO used to establish TACs. When the Scientific Council report was presented, the EC delegate queried why the Council continued to recommend management at the $F_{0.1}$ reference level rather than providing a broad range of management options.

Supported by Spain and Portugal, the EC voted against proposed TACs for various stocks. The EC delegate stated that the EC would formally object to all NAFO decisions including the 3L moratorium (NAFO 1985a).

Thus, the 1985 meeting marked a major turning point for NAFO. The EC had hoped to force either Canada or the international community to give them more fish to accommodate the FRG, Spain and Portugal. For the previous 6 years, the EC had followed normal NAFO practices. Indeed, in the case of 3M cod, it had been a voice of conscience in favour of following scientific advice. At the Havana meeting, the EC reversed itself and attempted to destabilize NAFO. Through lengthy debate on the scientific advice and the $F_{0.1}$ management approach, it attempted to prevent NAFO from establishing conservation measures for 1986.

Because of the EC's insistence on a range of management options to consider, Canada agreed to formulate a broader request to the Scientific Council for advice in 1986.

Further Canadian efforts to reach a bilateral accommodation with the EC in 1985 failed. However, owners of the West German distant water fleet assured Canadian officials that they would comply with all NAFO regulations for 1986. This would include the moratorium on fishing cod in 3L outside 200 miles, provided they would be licensed to fish their share of the EC's LTA quotas in Canadian waters in 1986.

Early in 1986, the EC began refusing boardings by scientific observers under the NAFO Scientific Observer Scheme. Portuguese and Spanish vessels returned to fish on the Nose early in 1986, disregarding the NAFO moratorium and exceeding the EC's 9,500 ton quota under the LTA for 1986. While technically legal, the Spanish and Portuguese fishing for cod in 3L was inconsistent with the EC's conservation obligations under the LTA.

On June 13, 1986, Fisheries Minister Tom Siddon announced that, apart from obligations under existing treaties, Canada would make no new non-surplus allocations to foreign fleets. Using allocations to seek improved market access had proved ineffective. This would no longer be done except in the case of East Bloc countries where market forces did not operate in the conventional fashion. In the case of free market countries, "we will link allocations to one condition alone: good conservation behaviour." The Minister stated: "We are sending a very clear message to foreign fishing nations: in return for a licence to fish in our waters, and permission to use our ports, you must cooperate with our conservation efforts and observe the fishing limits that we and NAFO have established" (DFO 1986e).

In late June 1986, the EC Council of Fisheries Ministers decided to give notice of termination of the bilateral arrangement with Canada on scientific observers in the NAFO Regulatory Area. It would also withdraw from the NAFO Joint Enforcement Scheme effective one year after the date of notification to NAFO. The withdrawal from the NAFO Joint Enforcement Scheme was potentially the most damaging. This would reduce control and knowledge about EC catches in the NAFO Regulatory Area. The EC had also objected to the NAFO quotas for 1986 and established its own levels (see Table 11-1).

Thus, as the 1986 NAFO meeting approached, Canada and the EC appeared headed for a further confrontation. The EC would try to make the 1986 meeting a turning point. Its objectives included:

1. Terminating NAFO's long term management approach based on $F_{0.1}$;
2. Establishing NAFO management of the 2J3KL cod stock, or that portion of it occurring in 3L, and establishing NAFO quotas to be fished outside the 200 mile limit; and
3. Legitimatizing the large catches of Spain, Portugal and the FRG.

As part of its efforts to destabilize NAFO, the EC, in discussions on 3M cod, made a proposal: rather than set a TAC each Party should agree not to increase its current fishing effort. The total fishing effort would remain constant. (The estimated 1985 catch was 27,000 tons, far above the TAC of 12,965 tons). The EC proposal was rejected and a Danish proposal to maintain the TAC adopted.

On other stocks the EC noted the lack of management alternatives in the Scientific Council Report. It refused to support any TAC at the $F_{0.1}$ or equivalent level (NAFO 1986a).

In response to the Fisheries Commission request for information on cod in 3L, the Scientific Council had stated that:

> "The maximum proportion of the entire Divisions 2J3KL cod stock estimated to occur in the Regulatory Area is less than 10 percent in winter and less than 5 percent on average throughout the year" (NAFO 1986b).

Based on this information, Canada proposed that directed fisheries for this species in the area not be permitted in 1987. Canadian officials argued that the 2J3KL stock was a single stock. Less than 5% of it would normally be found outside the Canadian zone. Furthermore, the TAC was fully subscribed within the Canadian zone. The EC responded that it could not accept the principle of a stock occurring in the international waters of the Regulatory Area being "fully subscribed in the Canadian zone." The Canadian proposal was adopted with six votes in favour, three against and one abstention.

A major debate at the 1986 meeting concerned the EC's intent to withdraw from the Joint Enforcement Scheme and its request for a working group to develop a modified scheme. The Commission decided to set up such a Working Group to report to the next annual meeting. This decision was made on the understanding that "all NAFO Contracting Parties will fully ensure that their fleets will fully comply, at least until 15 November 1987, with a control regime corresponding to the provisions of the Scheme of Joint International Enforcement" (NAFO 1986a).

The EC objected to the measures adopted by NAFO for 1987 (Table 11-1). Despite the EC's actions in fishing on the Nose of the Grand Bank in 1985 and 1986, Canada decided to honour its obligations and provide the EC with its allocations under the LTA in the Canadian zone in 1987. The EC provided nothing in return. In April, 1987, Canadian officials notified the EC that the letters on allocations and tariff reductions which accompanied the LTA would not be extended beyond December 31, 1987, and that there would be no non-surplus allocations after that date (DFO 1987g).

The LTA had proven of little value to Canada in improving market access. The tariff rate quota concessions to Canada under the LTA were available to all GATT members. The manner in which the EC administered tariff quotas undermined any advantage to Canada. Overall, Canadian fish exports to the EC declined from 98,000 tons to 61,000 tons between 1977 and 1985 (Table 11-2). Canada was greatly disadvantaged by preferential treatment of Icelandic and Norwegian products in the EC. This treatment insulated Iceland and Norway from competition. There was no reason for Canada to renew the allocations for access arrangements.

At the 1987 NAFO meeting, Canada and the EC again locked horns. In a surprise development, Denmark, supported by Canada and Norway, proposed that the Commission follow the scientific advice for 3M cod and set the TAC at zero. This proposal was adopted by a slim majority.

A Canadian proposal to continue the moratorium on fishing for cod in 3L in the Regulatory Area was again adopted, with five votes for, one against and three abstentions.

The Commission considered the Report of the Working Group on Joint International Control amid

TABLE 11-1. Comparison of EC unilateral quotas, EC NAFO quotas, EC catches reported to NAFO for NAFO Groundfish stocks and 2J3KL cod for 1986–1991 and EC catches estimated by Canada for 1990–1991 (Quantities in tons).

STOCK	1986			1987			1988			1989			1990				1991			
	EC UNILAT QUOTA	EC NAFO QUOTA	EC CATCH REPORTED TO NAFO	EC UNILAT QUOTA	EC NAFO QUOTA	EC CATCH REPORTED TO NAFO	EC UNILAT QUOTA	EC NAFO QUOTA	EC CATCH REPORTED TO NAFO	EC UNILAT QUOTA	EC NAFO QUOTA	EC CATCH REPORTED TO NAFO	EC UNILAT QUOTA	EC NAFO QUOTA	EC CATCH REPORTED TO NAFO	EC CATCH CANADIAN ESTIMATE	EC UNILAT QUOTA	EC NAFO QUOTA	EC CATCH REPORTED TO NAFO	EC CATCH CANADIAN ESTIMATE
3M COD	7,500	6,465	11,079	7,500	6,465	6,441	0	0	562	0	0	547	0	0	637	22,100	*6,465	*6,465	4,280	7,600
3NO COD	26,400	14,750	30,470	26,400	12,345	21,885	26,400	14,750	19,816	26,400	9,220	20,937	7,000	6,860	6,823	13,300	*5,016	*5,016	6,509	9,500
3M REDFISH		3,100	11,571		3,100	22,648	12,000	3,100	7,247	12,000	3,100	13,062	12,000	7,750	13,680	21,500	*7,750	*7,750	10,111	8,800
3LN REDFISH		0	23,388		0	28,186	20,000	0	12,699	20,000	0	6,346	6,000	0	7,311	5,600	6,000	0	10,945	7,850
3M A. PLAICE		350	2,789		350	5,106	3,000	350	2,549	3,000	350	3,405	500	350	459	800	*350	*350	1,603	450
3LNO A. PLAICE		700	21,161		610	17,014	9,000	510	9,828	6,820	385	11,595	500	317	662	12,000	*328	*328	972	
3LNO YELLOW-TAIL		300	5,952		300	1,213	5,000	300	3,205	1,670	100	1,278	200	100	119	150	*140	*140	246	**16,500
3NO WITCH		0	3,788		0	2,957	4,000	0	2,888	4,000	0	1,990	1,200	0	1,155	500	1,000	0	1,097	
TOTALS	33,900	25,665	110,198	33,900	23,170	105,450	79,400	19,010	58,794	73,890	13,155	57,882	27,400	15,377	30,846	75,950	27,049	20,049	35,763	50,700
2J3KL COD	68,560	0	61,985[1]	76,400	0	35,392[1]	84,000	0	26,559	58,400[3]	0	35,594	32,000	0	23,758	21,800	27,000	0	22,835	41,900
GRAND TOTALS	102,460[2]	25,665	172,183	110,300[2]	23,170	140,842	163,400	19,010	85,353	132,290	13,155	93,476	59,400	15,377	54,604	97,750	54,049	20,049	58,598	92,600

1. In 1986 and 1987, Canada allocated 9,500t of 2J3KL cod to the EC under the Canada–EC Fisheries Agreement.

2. EC did not set unilateral quotas for all stocks.

3. Amended by EC in July 1989. Set initially at 84,000t.

* EC accepted NAFO decisions on these stocks.

**Includes 3LNO Plaice and Yellowtail and 3NO Witch.

Source: International Fisheries Directorate, Department of Fisheries and Oceans, Ottawa.

controversy over the EC's actions in withdrawing from the existing Scheme. EC vessels were now refusing boardings by Canadian NAFO inspectors unless they were accompanied by a flag state inspector. However, the EC would not provide details of the "equivalent" scheme it had put in place. The delegates could not agree on this issue. The EC invited the Fisheries Commission to meet in Brussels early in 1988 to discuss this matter further (NAFO 1987).

The Fisheries Commission met in Brussels in February 1988. It agreed on a revised scheme of control to be titled the Scheme of Joint International Inspection. This took effect on June 10, 1988 (NAFO 1988a).

On the surface some progress was being made toward resolving differences within the NAFO framework. But on the problem of overfishing beyond 200 miles, major obstacles remained.

Meanwhile, the problem of foreign overfishing had been examined by a special Canadian federal-provincial Task Force. The Task Force recommended that Canada try to strengthen NAFO to meet conservation objectives. Thus, the Canadian government attempted to persuade NAFO members of the wisdom of conservation and long term management and of the negative impacts of overfishing.

Canada made a major effort at the 1988 NAFO meeting to focus other NAFO members' attention on the long term implications of overfishing in the Regulatory Area and the need for action by NAFO. The Scientific Council advised for 1989 significantly lower TACs for 3NO cod, 3LNO American plaice, and yellowtail flounder (NAFO 1988b).

The Scientific Council Report expressed particular concern about overrun of TACs and uncontrolled fisheries. Regarding American plaice and yellowtail flounder, the Scientific Council indicated that there were nursery areas for juvenile plaice and yellowtail on the Tail of the Bank outside 200 miles. Uncontrolled fishing there, both by members and non-members of NAFO, threatened these stocks, which are of considerable importance to the Newfoundland-based trawler fleet.

Despite these scientific warnings, the EC continued to push NAFO to establish TACs at reference levels above $F_{0.1}$. For 3NO cod, for example, the EC proposed a TAC of 40,000 tons, in contrast with the Canadian proposal for a TAC at the $F_{0.1}$ level of 25,000 tons. The Canadian proposal was adopted, as were Canada's proposals for other straddling stocks (NAFO 1988c).

A Canadian proposal to continue the moratorium on fishing cod in 3L outside 200 miles was again adopted (NAFO 1988c).

In the General Council the Canadian delegate, Dr. Peter Meyboom, launched a major offensive on overfishing outside 200 miles. He pointed out that, 10 years before, the Northwest Atlantic fishery was a focus of hope for the future. Canada and other countries had faced a challenge to conserve and manage transboundary and other Northwest Atlantic stocks. He noted that Canada had largely succeeded in rebuilding stocks within its zone through strict regulations and quota controls. For the area beyond 200 miles, Dr. Meyboom observed:

> "Sadly, the recovery in the Canadian fishing zone has not been matched in the NAFO Regulatory Area. While initial efforts produced positive results in some stocks, the catch rates and biomass for others have remained low or have even declined further from their 1977 levels.
>
> "In the last three years, conditions in the Regulatory Area have taken a strong turn for the worst — to the extent that the progress that had been achieved for some stocks since 1978 is now in jeopardy. In short, NAFO is headed toward a resource crisis. One cause of that crisis is unregulated fishing leading to overruns of quotas and TACs and blatant disregard for other management measures" (DFO 1988f).

He pointed out the excessive catches stock by stock (Table 11-3). The message with respect to the EC's flagrant abuse of the NAFO objection procedure was clear.

The evidence was overwhelming that the EC's extensive use of the objection procedure was eroding the authority and effectiveness of NAFO. The EC's actions had placed the future of NAFO in jeopardy. That had been the EC's objective since 1985. The question was — would the other members of NAFO let this occur?

In response, the General Council adopted a resolution calling on "all Contracting Parties not to abuse the objection procedure against the regulatory measures adopted by the Fisheries Commission." All countries except the EC voted in favour (NAFO 1988d). Efforts continued through the fall to get the EC to adhere to the NAFO measures for 1989 but to no avail. In December, 1988, Minister Siddon denounced the EC's decision to establish unilateral fish quotas more than 12 times higher than its quotas from NAFO (DFO 1988g).

In 1989, Canada stepped up its efforts to eliminate foreign overfishing outside 200 miles. An intensive campaign involving Canadian Ministers, parliamentarians, and the fishing industry was launched to convince the EC and its member states to stop overfishing in the Northwest Atlantic. This involved high level political

TABLE 11-2. Canadian Exports of Fishery Products to the European Community — 1977 to 1988 (Quantities in Tons)

COUNTRY	1977	1978	1979	1980	1981	1982	1983	1984	1985	1986	1987	1988
United Kingdom	15,719	22,532	23,181	33,983	39,866	26,436	32,647	26,469	22,792	24,000	20,020	17,355
Belgium-Luxembourg	5,368	5,200	5,490	5,577	5,140	4,095	3,783	2,915	1,967	3,100	3,634	2,441
France	13,790	17,033	16,016	15,424	15,042	12,768	12,531	14,136	10,871	12,900	12,383	10,618
Fed. Rep. of Germany	49,885	40,841	34,567	35,346	32,176	19,080	13,231	9,957	11,181	7,500	8,976	17,278
Italy	1,497	2,600	4,835	4,087	4,394	2,575	3,650	2,927	3,787	3,700	3,809	3,814
Netherlands	8,064	10,807	5,908	5,068	3,107	3,379	2,790	2,173	2,941	4,500	4,466	3,591
Denmark	2,858	4,625	9,434	5,790	3,328	6,621	4,459	5,658	6,171	6,100	8,148	10,241
Ireland	265	706	518	580	159	980	1,017	463	216	500	410	405
Greece	N/A	N/A	N/A	1,608	1,367	667	692	1,024	900	900	1,049	1,017
TOTALS	97,446	104,344	99,949	107,463	104,579	77,001	74,800	65,722	60,826	63,200	62,895	66,760
Portugal	1,220	155	2,724	11,200	19,330	28,597	15,634	11,000	11,800	9,700	12,652	14,750
Spain	34	887	3,785	9,188	5,097	778	315	80	70	600	1,615	1,962
GRAND TOTAL	98,700	105,386	106,458	127,851	129,006	106,376	90,749	76,802	72,696	73,500	77,162	83,472

TABLE 11-3. NAFO TAC overruns 1985–1991.

SPECIES	1985			1986			1987			1988			1989			1990			1991		
	TAC	EXCEEDED BY	%	TAC	EXCEEDED BY	%	TAC	EXCEEDED BY	%	TAC	EXCEEDED BY	%	TAC	EXCEEDED BY	%	TAC	EXCEEDED BY	%	TAC	EXCEEDED BY	%
3M COD	12,965	1,000	8.0	12,965	2,000	15.0	12,965	Not Exceeded	N/A	0	2,000		0	40,000		0	32,000[1]		13,000	Not Exceeded	N/A
3NO COD	33,000	4,000	12.0	33,000	18,000	55.0	33,000	6,000	18.0	40,000	3,000	7.5	25,000	8,000	32.0	18,600	10,400[2]	56.0	13,600	15,400[3]	113.0
3M REDFISH	20,000	184	1.0	20,000	9,000	45.0	20,000	16,000	80.0	20,000	3,000	15.0	20,000	38,000[3]	190.0	50,000	33,000[3]	66.0	50,000	5,000[3]	10.0
3LN REDFISH	25,000	Not Exceeded	N/A	25,000	17,000	68.0	25,000	19,000	76.0	25,000	18,000[4]	72.0	25,000	9,000	36.0	25,000	4,000[4]	16.0	14,000	11,000[3]	78.5
3M A. PLAICE	2,000	Not Exceeded	N/A	2,000	1,800	90.0	2,000	3,600	180.0	2,000	800	40.0	2,000	1,500	75.0	2,000	Not Exceeded	N/A	2,000	Not Exceeded	N/A
3LNO A. PLAICE	49,000	5,000	10.0	55,000	6,000	11.0	48,000	5,000	10.0	40,000	1,000[5,6]	2.5	30,300	13,700[5,6]	45.0	24,900	7,100[5,6,7]	28.5	25,800	13,200[3]	51.0
3NO WITCH	5,000	4,000	80.0	5,000	4,000	80.0	5,000	3,000	60.0	5,000	1,000	20.0	5,000	Not Exceeded	N/A	5,000	Not Exceeded	N/A	5,000	Not Exceeded	N/A
3LNO YELLOW-TAIL	15,000	14,000	93.0	15,000	16,000	107.0	15,000	1,000	7.0	15,000	1,000[5]	6.6	5,000	5,000[5,6]	100.0	5,000	9,000[5,6,8]	180.0	7,000	8,100[3]	115.5

NAFO SCS DOC. 91/19

1. Includes estimates for non-members and Contracting Parties.
2. Includes estimates for non-members and Contracting Parties (10,600t).
3. Includes estimates of unreported catch.
4. Includes estimated catch for non-members who do not report to NAFO.
5. Includes a percentage of the "Flounder not specified" catch reported to NAFO by South Korea.
6. Includes estimates of catch based on surveillance reports.
7. Includes estimates for non-members and Contracting Parties (8,100t).
8. Includes estimates for non-members and Contracting Parties (5,100t).

N/A = Not applicable

Source: International Fisheries Directorate, Department of Fisheries and Oceans, Ottawa.

Distribution of foreign fishing effort beyond Canadian 200-mile fisheries zone in 1992.

discussions with EC leaders. A public information campaign was begun in Europe to show European decision makers the environmental damage being caused by EC overfishing. Canada also worked for international recognition of the special rights of coastal states to manage fish stocks that straddle 200-mile limits. The government appointed Mr. Alan Beesley Ambassador for Marine Conservation. He was to coordinate Canada's efforts to curtail foreign overfishing. (This role was later assumed by Mr. Randoph Gherson).

At the 1989 NAFO Annual Meeting, Canada tabled a paper which documented how the EC's use of the objection procedure had resulted in unilateral changes in the relative shares of NAFO members in the annual catches of the NAFO-managed stocks. The General Council adopted a Resolution calling "for compliance with the NAFO management framework in place since 1979, and compliance with NAFO decisions in order to provide for conservation and maintain the traditional spirit of cooperation and mutual understanding in the organization" (NAFO 1989a).

The EC adopted a less confrontational approach at the 1989 meeting. It abstained on most NAFO management proposals rather than voting against them. The TACs for 1990 for the 10 stocks managed by NAFO were based on advice from the NAFO Scientific Council. They were consistent with Canada's approach inside the 200-mile zone. The moratorium on fishing for cod in 3L in the Regulatory Area was continued for 1990,with the EC again opposing the proposal (NAFO 1989b).

In December, 1989, the EC's Council of Ministers established unilateral quotas of 27,400 tons for 1990 compared with EC NAFO quotas totalling 15,377 tons. This compared with EC unilateral quotas for the same stocks totalling 73,890 tons in 1989. While the discrepancy between the EC's unilateral quotas and the EC's NAFO quotas was narrowed, much of this was a "paper" reduction since the EC's reported catch had declined from 110,000 tons in 1986 to 58,000 tons in 1989. On the key issue under dispute, the fishery for northern cod in 3L (not included in the figures above), the EC unilateral quota of 32,000 tons was roughly equivalent to the average of the reported catches from 1987 to 1989 (Table 11-1).

For three stocks there was apparent progress on paper — 3NO cod, 3LNO American plaice, 3LNO yellowtail flounder. For these the 1990 unilateral quotas amounted to 7,700 tons compared with NAFO assigned quotas of 7,277 tons and recent catches by the EC fleet from these stocks of 20,000–30,000 tons. NAFO had reduced significantly the TACs for these stocks for 1990 because their abundance had been reduced by overfishing.

In 1990, Canada renewed its public relations efforts to persuade the EC member states to abide by NAFO decisions. It also worked to improve NAFO's enforcement mechanisms and to strengthen actions by NAFO members against non-members fishing in the Regulatory Area.

In May, 1990, EC Vice President and Fisheries Commissioner Marin visited Ottawa and met with senior Canadian Ministers concerned with the overfishing issue. The Marin visit set the scene for a more cooperative approach to the 1990 NAFO meeting. The EC voted with Canada and the majority in setting TACs and allocations for seven of the 10 NAFO stocks. However, it abstained on three stocks (3M cod, 3LN redfish and 3NO witch flounder) (NAFO 1990a).

NAFO again agreed to continue the moratorium on fishing cod in 3L, with the EC voting against.

At the 1990 meeting, Canadian officials distinguished between the Grand Banks straddling stocks and the Flemish Cap stocks. For the Grand Banks stocks, the Canadians continued to press for NAFO decisions consistent with the management measures adopted by Canada as the coastal state. Canada accepted less conservationist-oriented measures for the Flemish Cap stocks (NAFO 1990a).

NAFO established a Working Group to develop short-term measures to improve surveillance and enforcement in the Regulatory Area. The General Council adopted unanimously a resolution against fishing in the Regulatory Area by countries which are not members of NAFO. Fishing by these countries had increased in the late 1980's (Table 11-4). The resolution called on NAFO and its members to take steps to eliminate such fisheries. A Standing Committee on Fishing Activities of Non-Contracting Parties was created to:

— compile all available information on such activities;
— compile information on landings and trans shipments of fish caught in the Regulatory Area by non-Contracting Parties;
— recommend solutions to the problem (NAFO 1990b).

In December, 1990, the EC adopted quotas for the NAFO Regulatory Area for 1991. It accepted the NAFO quotas for the EC for seven out of 10 stocks. It reduced its unilateral quota for cod in 3L slightly to 27,000 tons. It established unilateral quotas for two straddling stocks — 3LN redfish and 3NO witch — even though NAFO had not set aside any share for the EC from these stocks.

While the EC had moved some distance towards adoption of NAFO quotas, its unilateral quotas, for 3L cod in particular, remained a major stumbling stock to ending the Canada-EC-NAFO confrontation.

In February, 1991, Canada released estimates of 1990 catches by EC vessels in the Northwest Atlantic outside the 200-mile limit. EC vessels were estimated to have caught 76,000 tons of NAFO-managed groundfish stocks (not including northern cod). This was five times greater than the quotas, totalling 15,377 tons, voted by NAFO members for the EC. These catches were also far above the unilateral quotas totalling 27,400 which the EC had set for 1990 (DFO 1991d).

Canada proposed an emergency decision by NAFO to bring a new "hail" system into effect. In an April, 1991, emergency vote NAFO agreed to implement the new sys-

TABLE 11-4. Unregulated fishery by non-members of NAFO.

Country	Number of Vessels								
	1984	1985	1986	1987	1988	1989	1990	1991	1992
Cayman Islands	0	1	1	1	1	1	1	0	0
Korea	1	1	1	1	3	5	6*	3	2
Mauritania	0	0	1	0	1	1	0	0	0
Malta	0	0	0	0	0	1	1	0	0
Panama (Korean-crewed)	0	4	3	4	5	5	2	2	1
(European-crewed)	4	4	5	8	15	19	22	27	25
St. Vincents	0	0	0	0	1	1	1	1	0
USA	0	14	15	9	11	14	9	0	0
Mexico/Chile	6	6	4	6	4	0	0	0	0
Venezuela	0	0	0	0	0	0	2	2	2
Honduras	0	0	0	0	0	0	0	2	2
Sierra Leone	0	0	0	0	0	0	0	1	1
Morocco	0	0	0	0	0	0	0	1	1
Vanuatu	0	0	0	0	0	0	0	0	1
Total	11	30	30	29	41	47	44	39	35

* May include a squid fishing vessel registered in Taiwan.
Source: International Fisheries Directorate, Department of Fisheries and Oceans, Ottawa.

tem. Under the "hail" system, fishing vessels were required to provide notification by radio whenever they entered or left the NAFO area, or moved from one NAFO sector to another outside Canada's 200-mile limit. This system was intended to make it more difficult for any vessel of a NAFO member to falsify its catches. Canadian Fisheries Minister Crosbie described the new system as "a step in the right direction" (DFO 1991e).

Canada continued to press for compliance by all countries with the NAFO moratorium on fishing for cod in 3L and for an end to the foreign overfishing of the flatfish stocks beyond 200 miles. In 1991, due to unusual environmental conditions, a greater proportion of northern cod than usual became available in the Regulatory Area outside 200 miles. According to Canadian surveillance estimates, the total foreign catch of cod in 3L was 49,000 tons, about 42,000 tons of which was taken by EC vessels.

At the 1991 NAFO meeting the EC again accepted the NAFO decisions for six stocks. It again established a unilateral quota for cod in 3L at 26,300 tons, in continued defiance of the NAFO moratorium.

Meanwhile, Canada was attempting to secure international agreement on the need for changing some aspects of the international fisheries management regime for straddling stocks and stocks beyond 200 miles, as part of the negotiations leading up to the June 1992 United Nations Conference on Environment and Development. At a May, 1992, International Conference on Responsible Fishing, held in Cancun, Mexico, Fisheries Minister John Crosbie called for recognition of the special interest and responsibility of coastal states in the conservation and management of straddling stocks on the high seas. He noted that unrestrained fishing of a straddling stock in an area beyond 200 miles can render useless any measures taken within 200 miles to manage that stock. Mr. Crosbie pointed out that "inadequate surveillance and control is the Achilles' heel of international fisheries management, rendering all else useless" (DFO 1992b).

To give practical effect to the principle of sustainable development in high sea fisheries, 40 countries, including Canada, sponsored a proposal on high sea fisheries at UNCED in June 1992. These countries were successful in securing agreement that an intergovernmental conference under United Nations auspices be convened as soon as possible to promote effective implementation of the Law of the Sea Convention on straddling fish stocks and highly migratory fish stocks. The conference, to be held in 1993, was to identify and assess existing problems related to the conservation and management of such fish stocks and consider means of improving cooperation on fisheries among states, and formulate recommendations. This Conference offered some prospect of developing new approaches for the management of straddling stocks.

At UNCED, the EC, at the last moment, withdrew its objection to the proposed conference. The report from the June, 1992, Special Meeting of the NAFO Scientific Council on northern cod, confirming the very low level of this stock (Chapter 7), was instrumental in changing the EC's position.

Shortly thereafter, the EC, which had been experiencing poor catch rates, announced that its vessels would cease fishing for cod in 3L, at least until the fall of 1992. The estimated catch by its vessels to the end of June was 10,700 tons. The 2-year moratorium introduced by Canada in July, 1992, placed additional pressure on the EC and other countries to comply with the NAFO moratorium in 3L. It also placed additional pressure on Canada to secure such compliance.

In June, 1992, following a visit by Minister Crosbie, Panama agreed to apply sanctions to vessels registered in Panama that fished contrary to NAFO conservation decisions, and to allow registration in future only to vessels that agree to comply with NAFO decisions.

Meanwhile, further declines were observed in key straddling stocks. The NAFO Scientific Council, at its June, 1992, meeting, reported that the American plaice stock in Divisions 3LNO was at a level far below historic levels — a decline of about 50% from 1984 to 1991. The spawning stock biomass had declined from about 180,000 tons in the early 1980's to approximately 50,000 tons in 1991. The Scientific Council recommended that fishing mortality be reduced substantially in 1993 and suggested a $F_{0.1}$ catch of 10,5000 tons (compared with a TAC of 25,800 tons for 1992) (NAFO 1992b). The continuing decline in this stock, which had been vitally important to plants on the south coast of Newfoundland, reinforced the need for a new regime for the management of fisheries beyond 200 miles.

Following the June 1992 Report of the NAFO Scientific Council, the EC over the next couple of months closed its fisheries outside 200 miles as its quotas for 1992 were reached. A major breakthrough occurred at the September, 1992, NAFO meeting when the EC, for the first time since 1984, agreed to abide by all NAFO conservation decisions for 1993. At this meeting NAFO unanimously adopted a ban on fishing for northern cod outside 200 miles in 1993 and agreed on improvements to the surveillance and control system to take effect in 1993, including a pilot project for a NAFO observer program. Commenting on the EC's action, Fisheries Minister Crosbie stated:

> "This year, the EC seems to have embraced conservation as the guide to managing its fishery in international waters off Canada's Atlantic coast. That does not mean that the EC agreed

to all measures Canada proposed at the recent NAFO meeting....the EC did not. That does not mean that there are no further important steps that the EC should take...there are. That does not mean there will be no more differences of view with the EC on high seas fisheries management...there will be. What it does mean is a critical shift in our fisheries relations with the EC, away from confrontation toward co-operation. This is important progress." (*The Labradorian*, September 28, 1992).

Discussions continued between Canadian and EC officials during the autumn of 1992. On December 21, 1992, Fisheries Minister Crosbie and the Secretary of State for External Affairs, Barbara MacDougall, announced that Canada had reached agreement to end the serious and long-standing fisheries problems with the EC. Highlights of the Canada-EC Fisheries Agreement included:

— Agreement to comply with NAFO conservation and management decisions, including quotas;
— Agreement to support management and conservation measures in NAFO, in conformity with Article XI of the NAFO Convention, which provides that NAFO decisions for straddling stocks outside 200 miles should be consistent with management decisions taken by the coastal State for the same stocks inside 200 miles;
— The EC will ensure that catches by its fleets do not exceed quotas, through co-ordinated inspection activities with Canada and the introduction of a licensing scheme to match fishing effort with quotas;
— Co-operation between Canada and the EC to end fishing by non-NAFO fleets;
— Co-operation on the development of joint proposals to reform NAFO by adding a dispute settlement mechanism to avoid abuse of the objection procedure;
— Renewed access by EC vessels to Canadian ports;
— Access to surpluses in Canadian waters by the EC on a comparable basis to those provided to other countries.

With respect to northern cod, the agreement provided that Canada will establish the TAC for this stock. Canada and the EC would support adoption by NAFO of a procedure under which five percent of the northern cod TAC will be distributed to NAFO members after the moratorium on this stock ends. NAFO would allocate this five percent share to Contracting Parties in accordance with a distribution key to be established at the 1993 NAFO meeting. The EC and Canada would propose that the EC receive two-thirds of the five percent to be allocated by NAFO. Canada would largely forego any claim to a portion of this five percent, but would have for itself all of 95 percent of the TAC. A key provision was agreement by the EC that NAFO management decisions outside 200 miles for this stock should be consistent with Canadian management decisions inside 200 miles (Canada 1992).

Commenting on this agreement, Minister Crosbie stated:

"This agreement has the potential to go a long way toward ending the calamity of foreign overfishing outside 200 miles. It has the potential to make [NAFO] once more, as it was in the years following extension of fisheries jurisdiction in 1977, an effective means to conserve and manage fisheries for straddling stocks in international waters. Central to this is that the EC, as it did until 1986, now accepts and will abide by NAFO quotas and other conservation and management decisions."

The concerted effort launched by Canada in 1989 to combat foreign overfishing has resulted in a major change in the Canada-EC fisheries relationship. The decline in resource abundance from the late 1980's to 1992 had emphasized the need for a cooperative approach to conserve the stocks. The problem of foreign overfishing outside 200 miles was not ended by this agreement. There was still a significant fishery by non-NAFO members. But clearly the agreement with the EC represented a major step forward.

Minister Crosbie reiterated the Canadian Government's intentions to pursue efforts to bring fishing by non-NAFO members under control, and to pursue the legal initiative, under the auspices of the United Nations, to secure a more satisfactory legal regime for the management of straddling stocks. Whether the 1993–94 UN Conference on High Seas Fishing would provide the necessary impetus for such changes remains to be seen.

8.0 FOREIGN ARRANGEMENTS

8.1 General

One of the major objectives of post-extension Canadian fisheries policy was Canadianization of the fishery in the Canadian zone. However, there were species and stocks which the Canadian industry was not yet equipped to utilize given existing harvesting and processing technology and markets. To assist in the development of fisheries for such species and stocks, bridging mechanisms were necessary. These mechanisms could have two functions: 1. permitting the Canadian industry to test the technical and economic feasibility of new fisheries without premature commitments to substantial permanent investment; or 2. providing temporary harvesting or processing capacity from foreign sources.

These bridging mechanisms have been termed "foreign arrangements" or "cooperative arrangements". They included partnerships between Canadian fishing interests, be they government, company, union, fishermen's organization or private individual, and a foreign party (government or state trading organization, company or individual). Arrangements utilized since 1977 fall into three general categories:

1. Developmental Charters;
2. Over-the-Side Sales (Direct Sales) and Over-the-Wharf Sales; and
3. The Resource-Short Plant Program (RSPP).

The latter two programs have been in existence since extension in one form or another. They have involved the use of foreign fishing vessels either to purchase and process fish harvested by Canadian vessels or to harvest fish for delivery to Canadian processing plants.

8.2 Developmental Charters

Under the Developmental Charters program, a Canadian company could charter a foreign vessel to harvest an allocation of non-traditional species (particularly squid and silver hake). This was initiated in 1977 and continued in 1978 and 1979. The program recognized that new technology was required to harvest these non-traditional species. Questions about methods of catching, type of gear to be used, handling of fish on board and freezing requirements needed to be answered. On-shore processing techniques needed to be developed.

Canadian industry saw a major opportunity to develop a domestic offshore squid fishery. In the late 1970's, squid, which exhibits cyclical fluctuations in abundance, was abundant off the Canadian east coast. Canadians at that time had little experience in the harvesting, processing or marketing of squid, with the exception of a sporadic inshore fishery in Newfoundland. Canadian squid landings rose dramatically from 3,000 tons in 1975 to 38,000 tons in 1977. The bulk of 1977 landings (30,000 tons) was caught in the Newfoundland inshore fishery. About 8,000 tons were caught off Nova Scotia by foreign vessels for landing to and (theoretically) processing by Maritimes plants. In 1978 and 1979, allocations were made to various companies in Newfoundland and the Maritimes and permission given to enter into "charter" arrangements with foreign companies. In 1979, Canadian squid landings rose to 160,000 tons, effectively flooding the market. This resulted in a build-up of inventories which lasted into 1980.

A limited version of the charter program continued in 1980. No charters of this nature were permitted from 1981 onward. The chief reason for discontinuing the program in 1980 was that it had resulted in windfall profits for certain Canadian companies which had made no serious effort to gear up for handling this species. In any event, the offshore squid fishery off the Canadian Atlantic coast has been virtually non-existent since 1981 because of low resource abundance.

There have been other instances where genuine foreign charter arrangements were used as a transitional device to develop a Canadian fishery. One of the best examples was the development of the fishery for northern shrimp off Labrador. This fishery grew from zero in the mid-1970's to a lucrative multi-million dollar fishery by the early 1980's. During the first year or two all Canadian licensees were permitted to charter foreign vessels to fish under Canadian licence. In the case of the three licences granted to Labrador based interests, this practice of "royalty charters" was permitted to continue until the present time. Other new shrimp licensees have also been permitted to "charter" foreign vessels for an initial transition period. In recent years similar charters of foreign vessels have been permitted for other underutilized species, e.g. turbot.

8.3 Over-the-Side (Direct) Sales

Foreign vessels were first used to assist in the development of the Canadian fishery in 1976. This began with the first program of direct or over-the-side (OTS) sales in the Bay of Fundy. That program was part of the effort to convert the purse seine herring fishery from a meal fishery to a food fishery. As part of a broad array of initiatives, fishermen were permitted to sell 12,000 tons of unprocessed herring directly to Polish processing vessels at a price higher than Canadian buyers would pay.

In 1978, the direct sales concept was broadened on the Atlantic coast to species such as mackerel and squid. Since that time over-the-side sales have been a continuing, and for a while extremely controversial, feature of the Atlantic fisheries scene. In the late 1970's over-the-side sales were also initiated on the west coast for Pacific hake.

Except for Atlantic herring and Pacific hake, the quantities involved have been generally small. Nevertheless, the subject of over-the-side sales became extremely controversial during 1979–81 on the Atlantic coast. Processors contended that over-the-side sales were undermining the processing industry. In the late 1980's, an over-the-side program was worked out involving both fishermen and processors in an attempt to develop an expanded Canadian fishery for Atlantic mackerel, an underutilized species in Canadian waters.

One of the most successful and least controversial OTS sales programs in Canada has been that involving Pacific hake. Serious efforts to develop a Canadian Pacific hake fishery began in 1979 using over-the-side sales as a developmental mechanism. Experimental amounts of Pacific hake were brought to shore plants during a 1978 Canada-Poland cooperative arrangement. The Fisheries Association of B.C. agreed that shore processing of offshore hake stocks was not yet feasible. This was due primarily to low hake prices, high domestic labour costs and the difficulty of transporting hake to processing plants without critical loss of quality.

Organizations representing most of the groundfish processing industry in British Columbia formed a Pacific Hake Consortium to negotiate and coordinate over-the-side sales with foreign interests, commencing in 1979. From a 35,000 ton TAC, 22,000 tons were set aside for this purpose. Each year thereafter the Hake Consortium negotiated over-the-side sales arrangements on behalf of B.C. fishermen. The USSR and Poland have been the dominant participants in the Pacific hake fishery. Japan became involved in 1988.

Landings increased from 5,191 tons in 1977 to 40,800 tons in 1983, the first year landings exceeded 40,000 tons. By 1988 landings reached 90,600 tons and the TAC for 1989 was established at 98,000 tons (Table 11-5). The amount allocated for foreign arrangements increased from about 30,000 tons in 1983 to about 53,000 tons in 1988.

Over-the-side sales played a key role in the development of a Canadian fishery for Pacific hake. This was clearly a successful use of foreign capacity to assist in Canadian fisheries development. Success was due to a congruence of interest among Canadian processors, fishermen and foreign buyers. All parties, Canadians and foreigners alike, derived economic benefit from a non-traditional, highly perishable species that would have otherwise remained unharvested.

8.4 Resource-short Plant Program (RSPP)

This program developed initially as a means of utilizing some northern cod for landing during the fall and winter at Newfoundland plants which would have otherwise operated only seasonally. In 1979, deliveries were made to plants in both Nova Scotia and Newfoundland. From 1980 to 1982, only Newfoundland plants were involved in the program. Subsequently, the concept was broadened to include other species not being fully utilized by Canadian vessels.

The use of foreign vessels to catch and land the fish at Canadian plants was an integral component of the program from its inception until the late 1980's. The Task Force on Atlantic Fisheries identified the seasonal peaking of the inshore catch as a major problem requiring substantial processing capacity which would be seriously underutilized in the off season (Kirby 1982). The Task Force recommended that specific allocations from the growing northern cod stock be set aside for supplying resource short plants, particularly on the northeast coast of Newfoundland. The Task Force set out some guidelines. It suggested that 10,000 tons of northern cod in 1983, increasing to 50,000 tons by 1987, as well as quantities of certain other species, be delivered to the plants by a self-financing fishing company or consortium.

Following extensive consultation during 1983, Fisheries Minister Pierre De Bané announced in December 1983, that the RSPP would be implemented early in 1984. A total of 11,000 tons of cod, 7,900 tons of redfish, and 8,000 tons of turbot was set aside for 1984. Some 60 processing plants were eligible to participate in the program. The Minister stated that Canadian offshore fishing vessels owners would be given first opportunity to catch the entire allocation. If processing plants were unable to secure the services of a sufficient number of Canadian fishing vessels, foreign-owned vessels would be permitted to participate in the RSPP program to a maximum of 50% of the total RSPP allocation in 1984 (DFO 1983i).

In January, 1984, Minister De Bané made an arrangement with Portugal whereby the Portuguese would have first right of refusal on the foreign component of the program. The arrangement for 1984 involved 5,000 tons of northern cod being landed by Canadian vessels and 5,000 tons by Portuguese vessels. The entire foreign share was caught but some of the Canadian vessels were unable to deliver on their commitment. The Minister allowed the use of foreign vessels to harvest the remaining 2,600 tons of cod and 4,000 tons of turbot, on the basis that all of the landings would be 100% processed in Canadian plants.

The use of foreign vessels was intended to be phased down. The 1984 experience indicated that Canadian vessels were not available during the winter months to land 50% of the 10,000 tons of RSPP cod. It appeared likely that the problem of non-availability of Canadian vessels would become more severe if the northern cod RSPP allocations were to increase.

By 1987, two industry consortia of eligible resource-short plants — Newfound Resources Ltd. for the Newfoundland-based plants and Marque Resources

TABLE 11-5. Total landings of Pacific hake by the foreign and domestic fisheries off British Columbia during 1977–1992 in metric tons.

	Country							
Year	Canada	USSR	Japan	Poland	Korea	Greece	China	Total
1977 National	—	2708	1931	552	—	—	—	5191
Joint Venture	—	—	—	—	—	—	—	
1978 National	—	700	3364	589	—	—	—	6467
Joint Venture	—	—	—	1814	—	—	—	
1979 National	302	—	3637	4263	—	—	—	12435
Joint Venture	—	1131	—	3102	—	—	—	
1980 National	96	78	817	4456	—	—	—	17662
Joint Venture	—	4300	—	4560	—	3355	—	
1981 National	3283	227	189	3258	—	—	—	24116
Joint Venture	—	7232	—	4998	—	4929	—	
1982 National	2	—	2234	10246	—	—	—	32152
Joint Venture	—	9391	—	10279	—	—	—	
1983 National	—	—	—	13177	—	—	—	40834
Joint Venture	—	14192	—	13465	—	—	—	
1984 National	—	—	—	13202	—	—	—	42108
Joint Venture	—	19692	—	9214	—	—	—	
1985 National	1092	—	—	10534	—	—	—	24932
Joint Venture	—	—	—	13306	—	—	—	
1986 National	1999	8137	—	15605	—	—	—	55877
Joint Venture	—	16642	—	13494	—	—	—	
1987 National	5106	11738	—	9717	—	—	—	74638
Joint Venture	—	18866	—	26348	2863	—	—	
1988 National	779	13330	—	26385	—	—	—	90676
Joint Venture	—	22818	1050	26314	—	—	—	
1989 National	2448	13907	—	19658	—	—	—	102364
Joint Venture	—	27068	12980	26303	—	—	—	
1990 National	3806	—	—	3976	—	—	—	77075
Joint Venture	—	19918	16848	32527	—	—	—	
1991 National	15500	—	—	6043	—	—	—	97797
Joint Venture	—	—	8143	61483	—	—	6628	
1992 National	19308	—	—	—	—	—	—	88965
Joint Venture	—	—	23723	39843	—	—	6091	

Source: B. Foulds, Fisheries Operations Directorate, Ottawa. Landings do not include Strait of Georgia catches. 1991 and 1992 data are preliminary.

Ltd. for plants in the Maritimes and Quebec — were in operation. The original criteria for 1987 called for harvesting of the RSPP quotas by Canadian vessels only. However, in September 1987, the federal government announced that as "an interim measure aimed at ensuring the full utilization of the Resource - Short Plant Program during the remainder of the 1987 season", the two consortia were authorized to engage foreign - flag vessels to harvest the remaining 1987 RSPP quotas (DFO 1987h).

The 3-year agreements with Newfound Resources and Marque Resources expired at the end of 1988. On December 30, 1988, Fisheries Minister Siddon announced that the agreement with Newfound Resources would be renewed for a 3-year period because this consortium was (finally) Canadianizing its harvesting operation. The Newfoundland-based consortium was acquiring its own vessel. Minister Siddon stated the agreement with Marque Resources would be renewed for one year and only on the condition that the consortium submit a harvesting plan using Canadian vessels (DFO 1988b).

Thus more than 10 years after the first ad hoc program was authorized for the use of foreign vessels to catch and land northern cod at Newfoundland seasonal plants, it appeared that a Canadian capacity would be put in place to supply offshore fish, particularly northern cod, to these plants on a continuing basis. The Resource-Short Plant Program had evolved considerably over that time. The landings under this program helped sustain these plants. They were particularly helpful during years of a poor inshore fishery during part of the 1980's. While Canadianization had been an objective from the beginning, it was not easily achieved.

Unfortunately, the otherwise bright prospects for the Newfoundland consortium of Resource-Short Plants were considerably dampened by the revised scientific assessment for northern cod for 1989. This program had been developed based on expectations that up to 50,000 tons of northern cod would be available for Resource-Short Plants by 1987. This was not to be the case.

With the reductions in the northern cod TAC in 1989, through 1991, the northern cod allocation to the RSPP was first reduced and then in 1991 eliminated. While the allocations of other species continued, the loss of the northern cod allocation removed an essential underpinning of the program.

8.5 Impact of the Foreign Arrangements Programs

Foreign arrangements, while often controversial, played a much smaller role in the development of the Canadian fishery post-extension than had been originally envisaged. The two major programs–direct sales and the RSPP — had some beneficial impacts but these have been largely unquantified. Clearly, direct sales were instrumental in maintaining the viability of the Bay of Fundy herring fishery in the face of market fluctuations. Direct sales also assisted the development of a Canadian fishery for Pacific hake.

The RSPP had the potential to extend the operating period of formerly seasonal plants in northeast Newfoundland. It could have supported the processing capacity necessary to handle the seasonal peak in inshore catches. By the late 1980's this program seemed to be achieving the original program objectives, with a Canadianized harvesting capacity, in Newfoundland. The loss of the northern cod allocation in 1991 threatened the continued viability of this program.

Overall, foreign arrangements were only a minor component of Canada's efforts to Canadianize the fisheries in its 200-mile zone.

9.0 PROTECTION OF CANADIAN-ORIGIN ANADROMOUS STOCKS BEYOND 200 MILES

A second major Canadian objective at UNCLOS III was the protection of anadromous stocks of Canadian origin beyond 200 miles.

Article 66 of the LOS Convention (see Appendix I) recognized the primary interest and responsibility of States of origin for anadromous stocks. However, paragraphs 4 and 5 made it plain that further international agreements (either bilateral or multilateral) were essential to give effect to this provision. Following declaration of its 200-mile zone, Canada began negotiations to protect its salmon stocks beyond the 200-mile fisheries zones on both coasts.

9.1 Atlantic Salmon

Salmon are highly migratory species, travelling thousands of miles and in some instances through the national waters of several countries as well as the high seas. In the case of the Atlantic salmon, investigations in the late 1950's revealed large concentrations of salmon off the West Greenland coast and off the Faroe Islands. Substantial fisheries developed on these concentrations. Scientific research established that the salmon concentrations off West Greenland originated both in North American and European rivers, with about half originating on each side of the Atlantic. The North American component consisted mostly of salmon from Canadian waters with a very small portion originating in U.S. rivers. Small quantities of U.S. origin salmon were also taken in the Canadian commercial fishery, particularly off Newfoundland and Labrador. Despite the small numbers involved, the U.S., particularly recreational salmon groups, maintained a keen interest in the exploitation of Atlantic salmon by fisheries outside U.S. waters.

The fishery off West Greenland increased significantly in the 1960's and 1970's. This prompted efforts by Canada, the U.K. and the United States to reduce the catches in this fishery. ICNAF in 1969 adopted a resolution which recommended a complete ban on fishing for salmon throughout the ICNAF Convention Area (ICNAF 1969b). This would have stopped high seas fishing off West Greenland had not West Germany (on a matter of principle) and Denmark and Norway, the countries most involved in the fishery, registered formal objections to the proposed ban. This meant that the ICNAF resolution had no effect. The 1970 ICNAF meeting made some progress when the participants in the fishery agreed to restrict fishing each year to the 4 months from August to November. During this period, the fishery could not exceed the 1969 level of catch or number of vessels (ICNAF 1970).

Efforts continued over the next 2 years to obtain a ban on high seas fishing. In 1972 the U.S. and Denmark agreed (U.S.-Danish Atlantic Salmon Agreement of 1972) to the phase out of high seas fishing (beyond 12 miles) by vessels from Europe by 1976. They also agreed on a quota of 1,000 tons for the nearshore fishery by native Greenlanders. This occurred after the U.S. Congress passed a Bill empowering the government to prohibit import to the U.S. of fish products from any country "fishing against the dictates of conservation."

In 1972, ICNAF agreed that all fishing for salmon off West Greenland outside 12 miles would cease by January 1, 1976. The catch by non-Greenlanders off West Greenland would be gradually reduced during the 1972–75 period (ICNAF 1972c). For future years the catch by native Greenlanders would be limited to approximately 1,100 tons. In 1975, ICNAF agreed that from 1976 onward only native Greenlanders could fish salmon there, with their annual catch limited to 1,190 tons (ICNAF 1975a). This was an increase from the 1,000 tons contained in the U.S.-Danish Agreement of 1972 and the 1,100 tons agreed at the 1972 ICNAF meeting.

By 1978 Canada, the U.S. and certain European countries began to focus on how to ensure continued controls on the West Greenland fishery once ICNAF terminated in 1979. The Convention establishing NAFO specifically excluded salmon from its mandate.

The second International Atlantic Salmon Symposium, held in Edinburgh in 1978, called for a new international convention dealing specifically with Atlantic salmon (Went 1980).

Canada negotiated an extension of the West Greenland quota of 1,190 tons in its bilateral agreements with the EC in the early 1980's. The interim arrangements for 1980 and 1981 and the Long-Term Agreement signed in December, 1981, covered the gap between the termination of ICNAF and the coming into force of a new Salmon Convention in 1984. Canada alone paid the price for maintaining the West Greenland quota during those years.

Several countries, among them Canada, the U.S., Iceland and the EC, helped draft a new Atlantic Salmon Convention during the early 1980's (NASCO 1988a).

The Convention applies to "the salmon stocks which migrate beyond areas of fisheries jurisdiction of coastal States of the Atlantic Ocean north of 36°N latitude throughout their migratory range" (Article 1.1). The primary purpose of the Convention was to control interceptions of salmon when they are in waters other than the state of origin. The Convention prohibits fishing of salmon beyond the areas of fisheries jurisdiction of coastal States. Within the areas of fisheries jurisdiction of coastal States, salmon fishing is prohibited beyond 12 nautical miles except in the following areas:

1. Off West Greenland up to 40 nautical miles from the baselines of the territorial sea; and
2. In the Northeast Atlantic, within the area of jurisdiction of the Faroe Islands.

The Convention created a North Atlantic Salmon Conservation Organization (NASCO). The organization consists of:

1. A Council, and
2. Three regional Commissions:
   - a North American Commission
   - a West Greenland Commission
   - a North-East Atlantic Commission

The Council of NASCO, like the General Council of NAFO, is essentially a coordinating and administrative body. The main business of NASCO is conducted in the three Regional Commissions. Their membership is:

North American Commission (NAC)
— Canada and the U.S.
West Greenland Commission (WGC)
— Canada, Denmark (on behalf of Greenland), the EC, and the U.S.
North-East Atlantic Commission (NEAC)
— Denmark (on behalf of the Faroe Islands), the EC, Finland, Iceland, Norway, Sweden and the USSR.

The commissions provide a forum for consultation and cooperation among the members in conservation, restoration, enhancement and rational management of salmon stocks subject to the Convention. They can propose regulatory measures for fishing in the area of fisheries jurisdiction of a member country of salmon originating in the rivers of other parties. The functions of the North American Commission are more specific. Article 7 stipulates that it provides a forum for consultation and cooperation "on matters relating to *minimizing catches* in the area of fisheries jurisdiction of one member of salmon originating in the rivers of another Party." Also it is authorized "to propose regulatory measures for salmon fisheries under the jurisdiction of a member which harvest amounts of salmon *significant to another Party in whose rivers that salmon originates*." (Emphasis mine). All three Commissions make recommendations to the Council concerning scientific research.

One key feature of NASCO which has had a profound effect on its deliberations is the requirement (Article 11) that decisions of a Commission be based on unanimous votes of those present and casting an affirmative or negative vote. This means that any member can veto any particular decision of a Commission of which it is a member. This requirement for unanimity makes it easy for members to block measures which they oppose.

NASCO held its first Annual meeting in May, 1984. This occurred while Canada was taking radical conservation measures for its own Atlantic salmon fisheries to arrest a decline in salmon abundance. Canada's domestic 5-year plan was painful and far-reaching (Lear 1993). Canadian officials, therefore, approached the first West Greenland Commission (WGC) meeting seeking a substantial reduction in the West Greenland quota of 1,190 tons, in effect since 1976.

ICES reported to the West Greenland Commission that the low catch of 310 tons at West Greenland in 1983 had three possible causes:

"(a) low sea temperatures which may have affected the catch rates and/or availability of salmon;
"(b) reduced stock abundance in Canada, and reduced abundance of the spring-run component in Scotland; and
"(c) possible reduced fishing effort" (ICES 1984).

In light of the ICES advice and the extremely low catch in 1983, Canada proposed limiting the 1984 West Greenland fishery to 310 tons, the size of the 1983 catch. Canadian officials described the difficult resource management decisions which the Canadian government had just made. Canadian scientists had estimated that these regulatory measures would reduce Canadian commercial catches in 1984 by 60–80% in the Maritime Provinces and by as much as 60–70% in certain key fishing areas of Newfoundland.

The Canadian proposal was put to a vote, with Canada and the U.S. voting in favour and the EC voting against.

At a later session of the West Greenland Commission in July, the U.S. put forward a compromise proposal agreed between the Canadian and EC delegates. Thus the Commission adopted a TAC of 870 tons for the West Greenland fishery for the 1984 season. A major factor in achieving this reduction was the package of stringent regulatory measures Canada had adopted restricting its commercial and recreational salmon fisheries. Canada indicated that a similar conservation program would also be in effect in 1985 (NASCO 1984a).

At the second annual meeting of the WGC in June 1985, Denmark represented Greenland which had withdrawn from the EC. The EC's position changed from that of 1984 because it now represented only states of origin. The prognosis for North American origin salmon continued to be poor. Thus Canada argued for a further quota reduction at West Greenland commensurate with Canada's conservation measures for 1985. The EC delegate also argued for a lower TAC for 1985.

The Danish delegate responded that there must be a guarantee that when stocks recovered there would also be an increase in the West Greenland TAC. Canada proposed a TAC of 600 tons, based on an estimation of the impact of Canadian measures implemented in 1985. The EC supported this proposal, Denmark opposed it and the U.S. abstained. Denmark opposed the proposal because the EC had not taken conservation measures in the home rivers of its member states. There were further proposals but no agreement (NASCO 1985a).

No agreement was reached on quotas or other management measures in any of the three commissions of NASCO in 1985. Subsequent to the meeting, Greenland unilaterally established a quota of 852.3 tons, with an August 10 opening date, equivalent to the 870 ton quota agreed the previous year, with an August 15 opening.

At the third annual meeting of the WGC in 1986, various proposals for a lower and higher TAC than the previous year were put forward and rejected. Finally the EC proposed a TAC of 850 tons for the 1986 and 1987 seasons, with an opening date of August 1. If the opening date were changed, the catch limit would have to be adjusted accordingly. The proposal was adopted with Denmark and the EC voting for the proposal and Canada and the U.S. abstaining. Basically, this measure carried forward the TAC level adopted at the first meeting of the WGC in 1984 (NASCO 1986a).

The 1986 meeting also saw considerable debate over the relative rights of the States of origin and "host" states. However, there was no agreement about Greenland's right to share in the harvest of the North Atlantic salmon resource.

For 1986 the actual West Greenland TAC was adjusted to 909 tons based on an opening date of August 15. Since a 2-year TAC had been adopted at the 1986 meeting, there was little substantive discussion at the fourth annual meeting of the WGC in 1987. The opening date implemented for the 1987 fishery was August 25 with an adjusted quota of 935 tons.

At the 1988 meeting, the Canadian representative expressed concern about the increasing proportion of fish of North American-origin in the West Greenland fishery in 1986 and 1987. He also noted that the numbers caught at West Greenland were about the same as before the 5-year Plan.

Canada proposed a TAC of 740 tons, with an opening date of August 1 for both 1988 and 1989, with the harvest of North American salmon not to exceed 292,000 fish for the 2 years combined. This proposal was supported by the EC and the U.S. but Denmark vetoed it. Denmark then proposed a TAC of 1,000 tons for each of 1988 and 1989. This was opposed by Canada, the EC and the U.S.

The Danish Chairman of the Commission then put forward a draft emergency regulatory measure. He proposed that the total catch of salmon at West Greenland for the years 1988, 1989 and 1990 not exceed 2,520 tons. In any year the catch should not exceed the average annual catch (840 tons) by more than 10%. This was linked to an opening date of August 1. This proposal was adopted (NASCO 1988b).

Canada had not achieved the level of reductions it desired. However, there were still significant gains from the situation prior to the establishment of NASCO. As it turned out, the TACs adopted did not limit the catch, which declined from 893 tons in 1988 to 337 tons in 1989 and 227 tons in 1990.

NASCO was unable to reach agreement on a TAC for West Greenland for 1991 and 1992. The catch was 438 tons in 1991 and less than 200 tons in 1992.

Meanwhile Canada, in the North American Commission of NASCO, had taken significant steps to reduce Canadian interceptions of U.S.-origin fish. The U.S. had sought reductions in interceptions of U.S.-origin salmon by Canada since the first NAC meeting in May 1984.

At the first NAC meeting the U.S. proposed that Canada establish an overall quota of 1,706 tons. This would include a subquota of 938 tons for the Newfoundland and Labrador fishery. The U.S. also proposed that season off Labrador and eastern Newfoundland be delayed until the end of July. These proposals were unacceptable to Canada given the small numbers of U.S. fish being harvested and the size and importance of the Canadian fishery (NASCO 1984b).

Canada had severely restricted fishing for Atlantic salmon for 1984 as part of a 5-year rebuilding plan. These measures significantly reduced total Canadian catches of MSW salmon. At the 1984 NAC meeting the Canadian representative said that these measures should also reduce Canadian interceptions of U.S.-origin salmon. Further measures should await scientific advice. The two countries agreed to consult ICES (NASCO 1984b).

At the February, 1985, meeting of the NAC, the U.S. withdrew its quota proposal but again proposed a restricted season. Regarding the U.S. proposal to close the Newfoundland fishery in June and July, Canada noted the economic dependence on that particular fishery by fishermen of the area. Most of their income came in this period (NASCO 1985b).

At the June, 1985, meeting of NAC, the Canadian representative noted that the 1985 Atlantic Salmon Management Plan contained measures not in effect in 1984 that would further reduce the interception of U.S.-origin salmon. These additional measures included closing the commercial salmon fishery in the Maritimes and removing 683 part-timers from the fishery in Newfoundland and Labrador. Canadian scientists estimated that the 1984 and additional 1985 management measures could reduce the interception of U.S. salmon by 26% (NASCO 1985b).

Canada rejected a U.S. proposal for further seasonal restrictions. The Canadian representative stressed the agreement at the June, 1984, meeting of the WGC that the burdens and benefits of salmon conservation should be fairly shared. Further measures were unacceptable because of the failure of the West Greenland Commission to agree on appropriate conservation measures (NASCO 1985b).

In announcing the 1985 Atlantic salmon management plan on April 25, 1985, Fisheries Minister Fraser had linked the requirement for a reduction in the West Greenland fishery to Canadian consideration of the U.S. request:

> "We will examine the U.S. request carefully in light of that scientific advice once it is received taking into account the extent to which interceptions of Canadian-origin salmon in the West Greenland fishery can be further minimized this year" (DFO 1985j).

In February, 1986, ICES reported to the North American Commission that the harvest of U.S.-origin salmon in Canada averaged about 1,400 per year, with about one-quarter of those taken in the Newfoundland and Labrador commercial salmon fishery between September 1 and December 31. To put these interceptions in perspective, the catch of U.S. salmon in the Canadian Atlantic salmon fishery amounted to 5 tons on average compared with a Canadian average catch of 2,000 tons, thus representing 0.3% of the total Canadian catch. ICES also said that the measures taken in the 1984 and 1985 Management Plans reduced interceptions of U.S.-origin salmon in 1985 by 11–40% (ICES 1986).

At this meeting of NAC, Canada indicated that the Newfoundland salmon fishery would be closed from October 15 onward. The Canadian representative stated that this would reduce the interception of U.S.-origin salmon by as much as 58%. The U.S. acknowledged that this action combined with the measures already taken would reduce interceptions of U.S. salmon by more than 30%. The Commission adopted the Canadian proposal as a regulation (NASCO 1986b).

Canada's domestic salmon management plan established quotas for the Newfoundland and Labrador salmon fishery for the first time in 1990. It set a global TAC of 667 tons, subdivided by area, plus an allowance of 80 tons for northern Labrador. This compared with a 1989 catch of 937 tons and as much as 1,524 tons in 1987. Quotas decreased commercial catches in almost all Newfoundland salmon fishing areas. The total catch (586 tons) was 40% lower than the average for the previous 5 years. The lower catches were not due entirely to these management measures because catch quotas were not reached in five of the 11 quota areas (CAFSAC 1991).

On the international front, discussions occurred within NASCO during the late 1980's on the possible buy-out of salmon quotas at the Faroes and West Greenland. Norway, Iceland and the U.S. pushed for a compensation fund under NASCO for the annual buy-back of some of the quotas allocated by NASCO to the Faroes and Greenland. In June, 1990, NASCO established a working group to consider such a fund.

In April, 1991, an historic agreement was signed between the Faroese Fishing Vessel Association and a multi-national committee. The agreement provided for the Faroese commercial salmon quota to be relinquished for the years 1991–1993 through a buyout. The Committee for the Purchase of Open Seas Salmon Quotas paid $582,400 U.S. for the 1991 Faroese quota of 550 tons (Anon 1991). This was reportedly funded 55% by the Norwegian government, 6% by Iceland, 17% by Ireland, 22% by England together with several institutions and volunteer groups.

NASCO provided a useful forum for international cooperation in managing Atlantic salmon during the 1980's. The reduced quotas for the West Greenland fishery were still too high for Canadians. However, they had been a step in the right direction.

In March, 1992, Fisheries Minister John Crosbie announced that the commercial salmon fishery on the island of Newfoundland would be closed for a minimum of 5 years. This action was taken after the fourth consecutive year of declining commercial catches in Newfoundland and Labrador to just 433 tons in 1991. This was 39% less than the lowest catch during the 1980's, in 1984. This fishing moratorium would not apply in Labrador with its high dependence on salmon and limited prospects for fisheries diversification. Coincident with the moratorium, the federal and provincial governments announced a jointly-funded offer of cash payments to commercial salmon fishermen who would voluntarily retire their salmon licences. Total cost of the retirement program was expected to be about $40 million. The governments indicated that recent restrictions on the commercial sector, such as reduced seasons and quotas, had not rebuilt the salmon resource and that fishermen would have to face further restrictions in the future (DFO 1992c). The Faroese agreement in April, 1991, and the new initiatives by Canada in 1992 affecting the Newfoundland commercial salmon fishery, signalled a new era of Atlantic salmon management in the 1990's.

9.2 Pacific Salmon

The five species of Pacific salmon migrate long distances, throughout the North Pacific ocean. There are two dimensions to the problem of interceptory fisheries on the Pacific Coast. Canadian and U.S.-origin salmon intermingle in the waters along the Pacific coasts of both countries. Salmon of North American and Asian origin intermingle on the high seas of the North Pacific.

Canada-U.S. interception negotiations culminated in the Canada-U.S. Pacific Salmon Treaty in 1985 (see Chapter 12). Interceptions of North American-origin salmon on the high seas had traditionally been addressed through the International North Pacific Fisheries Commission (INPFC) and the abstention line, east of which the Japanese would not fish salmon.

The International Convention for the High Seas Fisheries of the North Pacific Ocean came into force in June 1953, following ratification by the U.S., Canada and Japan. The key management principle in the Convention involved the concept of *abstention*, set forth in Articles III, IV and V. This stipulated that one or more of the Contracting Parties would abstain from fishing "any stock of fish which the Commission determines reasonably satisfies all the following conditions:

> "(i) Evidence based upon scientific research indicates that more intensive exploitation of the stock will not provide a substantial increase in yield which can be sustained year after year;
>
> "(ii) The exploitation of the stock is limited or otherwise regulated through legal measures by each Party which is substantially engaged in its exploitation, for the purpose of maintaining or increasing its maximum sustained productivity, such limitations and regulations being in accordance with conservation programs based upon scientific research; and
>
> "(iii) The stock is the subject of extensive scientific study designed to discover

whether the stock is being fully utilized and the conditions necessary for maintaining its maximum sustained productivity" (Article IV 1.6).

In the Annex to the 1953 Convention, Japan agreed to abstain from fishing, and Canada and the U.S. agreed to carry out necessary conservation measures, for halibut and herring in the area off Canada and the U.S. Japan also agreed to abstain from fishing salmon off the coasts of Canada and the U.S. exclusive of the Bering Sea and the waters of the North Pacific Ocean west of 175° West Longitude and a line passing through the western extremity of Atka Island.

In a Protocol to the Convention, the U.S., Canada and Japan agreed these abstention lines would be considered provisional. They would be subject to confirmation or readjustment depending on the results of a scientific research program to be conducted under the auspices of the INPFC.

The designation of 175° W longitude as the eastern boundary of Japanese salmon fishing operations was arbitrary (Jackson and Royce 1986). The Protocol to the Convention stimulated a major research program to determine the migration patterns and areas and degree of intermingling of salmon from the three countries. Research results demonstrated intermingling of some species and stocks occurred west of the abstention line of 175° W Longitude. The U.S. advocated moving the line far enough west that the number of salmon of North American-origin west of the line would be negligible. Japan argued that a more equitable position was one in which the number of Asian salmon on the east side of the line was roughly equal to the number of North American salmon on the west side of the line. Discussions on this continued for years, with no agreement by the three countries until the various extensions of fisheries jurisdiction in 1977.

All three countries extended their fisheries jurisdiction 200 miles from their coasts in 1977. This had significant implications for the existing North Pacific Fisheries Convention. The location of certain Alaskan islands meant that the U.S. fisheries zone extended considerably to the west of the Protocol line of 175° W Longitude. North and south of the Aleutian Islands, the U.S. zone encompassed a considerable portion of the area in which Japanese motherships had traditionally fished salmon. The stocks of halibut, herring and salmon subject to the Abstention provisions of the INPFC now lay almost entirely within the U.S. and Canadian 200-mile zones. The Protocol Line lay within the U.S. 200-mile zone except for the southward extension of the line in an area where there had been little salmon fishing.

Salmon of North American origin were still accessible to foreign fisheries in the central Bering Sea, where Alaskan salmon and Canadian salmon from the Yukon River migrated beyond 200 miles. They were also vulnerable to foreign fishing in the central Gulf of Alaska.

Canada and the U.S. both signed fisheries agreements with Japan in 1978. Meanwhile the U.S. had given notice that it would terminate the North Pacific Fisheries Convention one year from February 1977, pending the results of efforts to renegotiate the Convention. Negotiations were initiated in 1977 and culminated in a new Protocol to the 1953 agreement which entered into force on February 15, 1979.

The 1953 Convention essentially protected salmon and herring of Canadian origin on the high seas. INPFC reports indicated that a few thousand sockeye salmon and a similar number of chinook salmon of Canadian origin were taken annually in the Japanese salmon fishery westward of the Abstention Line. United States salmon, especially those originating in Bristol Bay, Alaska, had not received the same degree of protection and were caught in large numbers annually by the Japanese (INPFC, various years).

One major Canadian concern in renegotiating the INPFC was to protect Canadian salmon of Yukon River-origin in the central Bering Sea. Canada also valued the scientific work carried out by member states through the existing convention. Until the IPNFC was established, little was known about the fish stocks of the North Pacific ocean. Early in the negotiations, Canada proposed that the Commission:

"a) provide for scientific studies and for coordinating the collection, exchange and analysis of scientific data regarding anadromous species; and

"b) pending the establishment of a wider-based scientific body dealing with these matters, provide a forum for cooperation and consultation among the Contracting Parties with respect to the study, appraisal and exchange of scientific information and views relating to the fisheries of the Convention Area, including environmental and ecological factors affecting these fisheries, the promotion of scientific research designed to fill gaps in knowledge pertaining to these matters, and the compilation and dissemination of statistics and records."

It became clear early in the negotiations that the principle of abstention could no longer serve as the basis for a new agreement. The U.S. objectives were twofold:

1. To greatly reduce interceptions of North American origin salmon by Japanese landbased and mothership salmon fisheries; and
2. To provide for research on marine mammals, particularly the Dall porpoise, to determine the effect of the Japanese fishery on marine mammal populations, and to reduce or eliminate the incidental catch of marine mammals (USA 1978).

Japan sought to protect its salmon interests without jeopardizing its access to the large groundfish resource now in U.S. waters. Early in the negotiations the U.S. proposed that the Japanese mothership fishery not fish east of 170° E Longitude and the landbased fishery not east of 165° E Longitude. Japan rejected these measures which would have intensified Japanese fishing pressure on salmon of USSR origin.

The 1979 Convention took the form of a Protocol amending the 1953 Convention but the text was completely revised and the functions of the Commission modified. The essence of the 1979 amendment was contained in the Annex, which moved the "reference" line from 175° W Longitude to 175° E Longitude. The U.S. used access to stocks such as Alaska pollock to persuade the Japanese to move their salmon fisheries substantially to the west. Except as otherwise specified in the Annex, Japan agreed to cease salmon fishing east of 175° E Longitude, unless temporary fishery operations in this area were approved by all Contracting Parties.

These arrangements offered almost complete protection to salmon of B.C. origin and provided additional protection to Canadian chinook and coho originating in the Yukon River.

The Commission's focus narrowed almost exclusively to research. The scope of the Commission's research was expanded to include nonanadromous species as well as anadromous species pending the establishment of a broader organization for multilateral scientific research in the North Pacific. (A North Pacific Marine Science organization, nicknamed PICES, was later negotiated in 1991). A separate 1979 Memorandum of Understanding between the U.S. and Japan established a 2-year research program on marine mammals caught incidentally in the Japanese salmon fisheries. This Program was subsequently renewed. This MOU arose from the U.S. Marine Mammal Protection Act of 1972.

Readers interested in further details on the establishment and evolution of this Convention and the INPFC should consult Jackson and Royce (1986).

Following the reductions in fishing area and effort negotiated under the 1979 Amendment to the North Pacific Fisheries Convention, approximately 170 Japanese salmon driftnet vessels entered a developing fishery for squid. This component of the Japanese squid fishery was restricted to the area in international waters, while the squid jigging fleet was restricted to the area closer to Japan.

In the early 1980's, soon after Japan established its high seas driftnet fishery, the Republic of Korea and Taiwan also began driftnetting for squid. The Korean and Taiwanese squid fishery expanded in the late 1980's, covering a wider area than the Japanese fleet.

In 1986, amendments to the Annex of the 1979 INPFC Convention further restricted the Japanese driftnet fishery for salmon. This phased out fishing by the mothership fleet in certain areas and moved the eastern boundary of the landbased fishery to 174° E. Longitude. These amendments also provided for intensified scientific studies in the landbased fishery area south of 46° North Latitude. This research was to determine continent of origin of the salmonid stocks in this area. Increased enforcement patrols by the Japanese were also required. Observers were placed on fishing and patrol vessels. Canada participated in this observer program. The Japanese directed salmon fisheries did not then harvest significant numbers of Canadian-origin salmon, except for small numbers of B.C.-origin steelhead, Yukon-origin chinook and possibly Yukon-origin chums.

In the late 1980's, Japanese fishing effort directed to salmon was considerably less than in previous decades. The number of high seas vessels in the landbased fleet declined from a high of 374 in 1972–74 to 157 in 1988. The Japanese mothership fishery was restricted to one mothership in 1988 (with 43 catcher boats) compared with the 1950's peak of 16 motherships and 460 catcher boats.

By the second half of the 1980's there was growing concern both in the U.S. and Canada about the potential incidental catch of salmon, marine mammals and birds in the driftnet fisheries for squid by Japan, Korea and Taiwan. Canada's concerns were threefold:

1. The incidental or deliberate interception of Canadian-origin salmonids;
2. The incidental killing of marine mammals and birds;
3. The impact of lost or discarded driftnets which continue to fish indiscriminately, killing marine life.

During the 1980's, Canada for a while licensed a small-scale experimental high-seas-type squid driftnet fishery using Canadian and Japanese commercial vessels. This fishing was conducted both inside and outside Canada's 200-mile fisheries zone but much farther east than the squid fisheries by Asian fishermen. In 7 weeks of fishing in 1987 the experimental vessels caught 639 salmonids, 44 marine mammals and 112 sea birds. In November, 1987, concern about the detrimental impact of this type of fishing led the Minister of Fisheries and Oceans to suspend the Canadian experiment. He placed a moratorium on such driftnets inside the Canadian zone (DFO 1987i).

That year Canada also raised this matter in the INPFC, seeking a commitment to international cooperation to address the uncontrolled high seas driftnet fishing. Minister Siddon called on all driftnetting nations to accept INPFC observers on their vessels to gather data on the incidental capture of salmonids, marine mammals and other marine life in the driftnet fisheries for squid.

In 1986, Canada began research on the potential impact of the squid driftnet fisheries. On a 1988 cruise the Canadian research vessel, the *W.E. Ricker*, identified a potential for salmon interceptions in the high seas squid driftnet fisheries. A follow-up survey, involving a chartered vessel, was carried out in 1989.

Late in 1987, Canada, the USA and Japan strengthened existing regulations to prohibit the importation of salmonids taken in the INPFC area other than by INPFC members. In December, 1987, the U.S. passed the *Driftnet Impact, Monitoring and Assessment Act*. This Act required that by June 29, 1989, nations driftnetting in the North Pacific implement agreements for bycatch monitoring satisfactory to the United States. By the summer of 1989 Japan had fulfilled these requirements.

In a 1988 meeting, INPFC initiated an eight-point action plan to deal with marketing illegally-caught salmon. This was a follow-up to action in 1984. That year the U.S., Canada and Japan restricted salmon imports from Taiwan following reports of a significant harvest of salmon by Taiwan in 1983. Surveillance in the late 1980's indicated that the Taiwanese driftnet fishery was the greatest potential threat in terms of salmon interceptions. Some Taiwanese vessels which were ostensibly driftnetting for squid conducted a targeted salmon fishery north of Taiwan's regulated squid driftnet area.

Even though Taiwan in 1985 prohibited the retention of salmonids, salmon caught by Taiwanese vessels was continuing to appear on world markets. U.S. authorities seized approximately 1.6 million pounds of Taiwanese-caught salmon in 1986 and 1987. Canada and the U.S. participated in 1989 in a "sting" operation involving salmon originating in Taiwan.

The magnitude of the fisheries and their distance from land made it difficult to obtain accurate information on the catch of nontargeted species in driftnet fisheries for species such as squid in the North Pacific. Hence, Canada and the U.S. developed observer programs in the late 1980's to gather more accurate information on the incidental catch of other species particularly salmonids in the driftnet fisheries for squid. A pilot scientific observer program began in 1989 with observers from Canada, the U.S. and Japan on 32 Japanese squid driftnet vessels. While the observer program was in effect, these vessels caught about 6,000 tons of squid. Bycatches included pomfret, albacore tuna, blueshark, dolphins, Dalls porpoise, northern fur seal and more than 9,000 seabirds. Only 79 salmonids of all species were recorded as bycatch. The observer program was expanded in 1990 to cover 74 Japanese squid driftnet vessels.

The General Assembly of the United Nations in 1989 and 1990 called for a moratorium on all driftnet fishing by June 30, 1992. This was to continue until effective conservation measures were taken to prevent detrimental impacts from such fishing practices (UNGA Resolutions 44/225 and 45/197). Canada co-sponsored these resolutions.

Canada hosted an international meeting in Sidney, British Columbia, in June, 1991, to review scientific data on the impact of high seas driftnet fisheries in the North Pacific. Reflecting the widespread public and intergovernmental interest in the driftnet issue, INPFC held a Special Symposium on Driftnet Impacts in November, 1991, in Tokyo.

In December 1991, the United Nations adopted a resolution (UNGA Resolution 46/645) banning drift-net fishing worldwide by the end of 1992. The UN Resolution had no provisions for enforcement but the three countries that opposed such a ban in the past — Japan, South Korea and Taiwan — agreed to abide by it. The resolution was adopted by consensus.

In 1990, discussions were initiated on creating a replacement for INPFC to incorporate the USSR, the U.S., Canada, Japan, and possibly other countries. These culminated in the signature in Moscow in February, 1992, by Russia, the U.S., Canada and Japan of a new Convention for the Conservation of Anadromous Stocks in the North Pacific. This Convention would apply to the waters of the North Pacific north of 33° North Latitude beyond the 200-mile zones of the signatories. The Convention provides for:

1) the prohibition of directed fishing for anadromous fish in the Convention Area;
2) minimizing the incidental take of anadromous fish in fisheries for other species (NPASC 1992).

The Parties agreed to take "appropriate measures, individually and collectively... to prevent trafficking in anadromous fish taken in violation of the prohibitions provided for in this Convention, and to penalize persons involved in such trafficking."

The Parties also agreed to cooperate in taking action "for the prevention by any State or entity not party to this Convention of any directed fishing for and the minimization by such State or entity of any incidental taking of anadromous fish by nationals, residents or vessels of such State or entity in the Convention Area."

The Convention provided for the establishment of a new North Pacific Anadromous Fish Commission to replace INPFC. The objective of this Commission is to promote the conservation of anadromous stocks in the Convention Area. The Commission could also "consider matters related to the conservation of ecologically related species in the Convention Area."

Thus, at the beginning of the 1990's new mechanisms were put in place to address the problem of the interception of Pacific salmon on the high seas of the North Pacific.

10.0 CONCLUSIONS

The 200-mile zone brought the bulk of the fisheries resources off Canada's coasts under Canadian fisheries management in 1977. This provided the authority to manage these resources in accordance with Canadian fisheries objectives. Canada benefitted considerably by taking measures to rebuild depleted fish stocks and utilize the fisheries resources of the new zone.

In the decade following extension, Canada increased significantly its share of the allowable harvest off both the Atlantic and Pacific coasts. The presence of foreign vessels in the Canadian zone diminished dramatically in the 1980's as the quantity of fish allocated to other countries was reduced substantially.

Canada's foreign allocations policy evolved through three phases — Stability and Access to Markets, Reward for Past Performance, and Cooperation on Conservation. This mirrored the changing international scene. From 1977 to 1982, Canadian catches of valuable groundfish species increased substantially. This left only nontraditional species to allocate to foreign countries. With the exception of Portugal, Canada's attempts to improve market access in return for allocations in the Canadian zone proved unsuccessful. The most explicit attempt to link allocations to market access was the Long Term Agreement with the European Community. This quickly proved a failure. The manner in which the Community implemented the LTA frustrated the achievement of Canada's market objectives.

The Reward for Past Performance approach adopted in 1982–83 was a little more successful. A distinction should be made between "market access", involving binding rules on access, and the simpler concept of developing sales into new markets. Real "toeholds" were achieved in the Polish and USSR markets (B. Applebaum, DFO, Ottawa, pers. comm.). By 1986 Canada moved to delink fish allocations from attempts to improve market access. By then Canada's attempts to rebuild fish resources within the Canadian zone on the Atlantic coast were partially threatened by overfishing of the stocks straddling the 200-mile limit on the Nose and Tail of the Grand Banks.

Management of these straddling stocks emerged during the 1980's as the major international fisheries problem off Canada's Atlantic coast. Through the establishment of NAFO, Canada had pushed for management of the straddling stocks beyond 200 miles consistent with Canada's approach within its 200-mile zone. Until 1985, NAFO generally accepted Canadian management proposals for these stocks. A notable exception was the noncompliance by Spain with quotas set aside for it by ICNAF/NAFO. Spain fished freely outside 200 miles from the late 1970's until it joined the European Community in 1985 and continued to do so thereafter as part of the EC. A major turning point in international fisheries relations occurred at the 1985 meeting of NAFO. The European Community, in an attempt to secure fish for Spain and Portugal outside the Northeast Atlantic, abandoned its previous conservationist approach within NAFO. It embarked upon a path of confrontation, challenging NAFO and Canadian approaches to management. It voted against many NAFO decisions on TACs and national allocations and repeatedly used the objection procedure so as not to be bound by these NAFO decisions. Instead, it established its own unilateral quotas. Until 1990, these were considerably larger than NAFO allocations to the EC.

In particular, the Community ignored the NAFO-agreed moratorium on fishing for cod in Division 3L outside 200 miles. This became the major point of contention between Canada, NAFO and the EC as confrontation escalated. In 1989 and 1990 Canada launched a major public relations and diplomatic initiative to persuade European leaders to abide by NAFO decisions. From 1989 to 1992 the EC gradually moderated its position, in response to Canadian initiatives. Finally, in 1992, it agreed to a moratorium in 1993 on fishing for cod on the Nose of the Grand Banks (Division 3L) and also agreed to abide by all NAFO conservation and management decisions. This set the scene for a Canada-EC agreement in December 1992, which provided for Canada to establish the TAC for northern cod. Canada and the EC would propose to NAFO a procedure under which 5% of the TAC would be distributed to NAFO members after the moratorium on this stock ends. Canada would harvest 95% of the TAC. This agreement did not solve the problem of fishing outside 200 miles by nonmembers of NAFO. Canada renewed its efforts to seek a new legal regime for fishing on the high seas, consistent with management initiatives by coastal States within 200 miles.

Applebaum (1990) examined the straddling stocks problem in the context of existing international law. He concluded that "sub-laws" deduced from current international law oblige states fishing on the high seas to constrain their fisheries for straddling stocks to accommodate the relevant coastal state's special interest in these stocks. Applebaum suggested that it is time to reexamine the international legal principles involved. He recognized, however, that obtaining international recognition of these "sub-laws" would be a problem.

The June, 1992, United Nations Conference on Environment and Development agreed to convene as soon as possible, under UN auspices, an intergovernmental conference to promote effective implementation of the Law of the Sea Convention on straddling stocks and highly migratory fish stocks. It remains to be seen whether this conference will facilitate the development of new approaches for the management of straddling stocks.

Regarding the other major Canadian objective at UNCLOS III, protection for Canadian salmon on the high seas, a new international organization, NASCO, has since 1984 provided some protection for salmon of Canadian-origin harvested at West Greenland. Given the unanimity provisions of the Convention establishing NASCO, the degree of agreement during the 1980's was remarkable. The agreement in 1991 for a buy-out of the Faroese salmon quotas for 1991 to 1993 may be followed by a similar arrangement for the West Greenland fishery. At the beginning of the 1990's, Canada's Atlantic salmon stocks continued to experience poor returns in many river systems. This was despite the drastic Canadian conservation measures from 1984 onwards and the reduction in the West Greenland quota.

On the Pacific coast, the problem of interception of Canadian-origin salmon in the high seas directed fishery for salmon by Japan, was largely eliminated during the 1980's as the Japanese fishery was moved farther westward and curtailed. The development of a major high seas driftnet fishery for squid, by Japan and various South Pacific nations, posed a new but unquantified threat through the incidental capture of salmonids. Two international meetings in 1991 on the impact of these driftnet fisheries helped to quantify the interceptions of salmon in fisheries targeted at other species. Implementation of the 1991 UN Resolution calling for a ban on drift-net fishing worldwide starting in 1993 should represent a significant step forward to eliminate the potential threat to salmon, mammals and sea birds. The promise by Japan, South Korea and Taiwan to abide by this resolution offers hope for the future.

Overall, in the post-extension era problems of international fisheries management continued to be of critical importance to the effective management of Canada's marine fisheries. So far we have been discussing multilateral international fisheries issues. Many of the most critical issues concern Canada's bilateral fisheries relations with its neighbours. These are addressed in the next chapter.

CHAPTER 12

THE INTERNATIONAL DIMENSION
III — CANADA'S FISHERIES RELATIONS WITH ITS NEIGHBOURS

"Before I built a wall I'd ask to know
What I was walling in or walling out,
And to whom I would like to give offense
Something there is that doesn't love a wall ...."

Robert Frost, Mending a Wall

1.0 INTRODUCTION

When Canada extended fisheries jurisdiction, it laid claim to management rights over a vast new territory. However, its 200-mile zone did not fully encompass stocks occurring in the Canadian zone. In addition to stocks straddling the boundary between the Canadian zone and the high seas, a number of stocks also straddled the boundaries or claimed boundaries between Canada and its neighbours. As part of the worldwide trend, Canada's neighbouring coastal states, the USA, France (St. Pierre and Miquelon), and Denmark (Greenland), also extended fisheries jurisdiction in 1977. In a number of areas there were major overlaps between the 200-mile zone claimed by Canada and those claimed by its neighbours, particularly the USA and France.

One of the boundaries, between Canada and Denmark in Davis Strait, had in effect been settled in 1973 with the delimitation of the continental shelf in that region. The major difficulty here concerned the management of certain transboundary or shared stocks (Fig. 12-1).

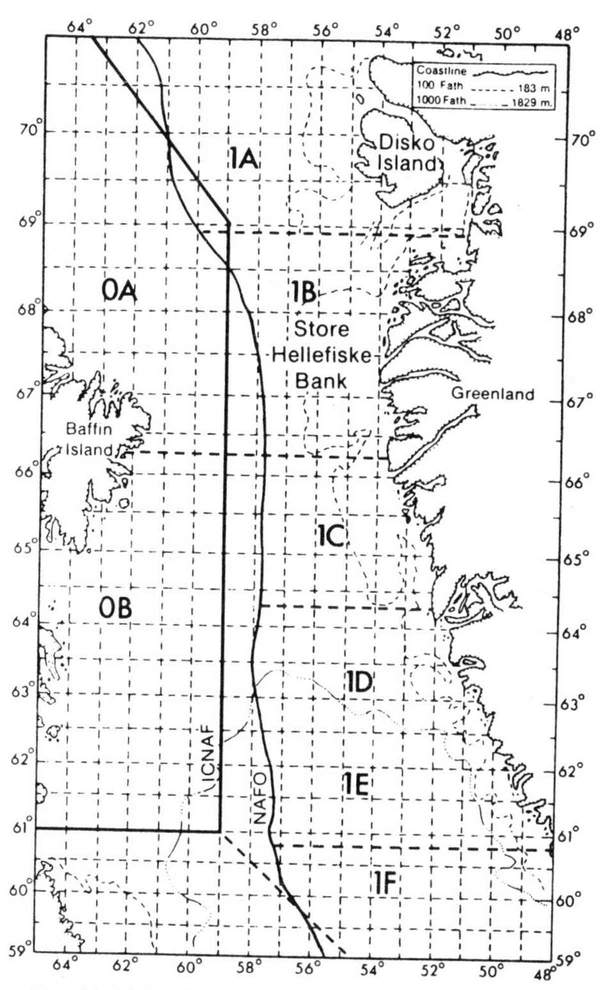

FIG. 12-1. The NAFO Subarea 0/1 boundary line in Davis Strait. The NAFO boundary line corresponds to the Canada-Greenland boundary.

On behalf of St. Pierre and Miquelon, France laid claim to a significant part of the waters claimed by Canada to the south of Newfoundland. In this region the claims overlapped in an area where there are significant fish resources exploited by the Canadian industry (Fig. 12-2). Relations were relatively amicable until the year before the scheduled May 1986 phase-out of Metropolitan France fishing vessels from the Gulf of St. Lawrence under the 1972 Canada-France Fisheries Agreement. In 1985, the boundary dispute and the management of shared stocks became a major bilateral dispute between Canada and France.

Relations with the USA following extension of fisheries jurisdiction were a web of conflicting interests.

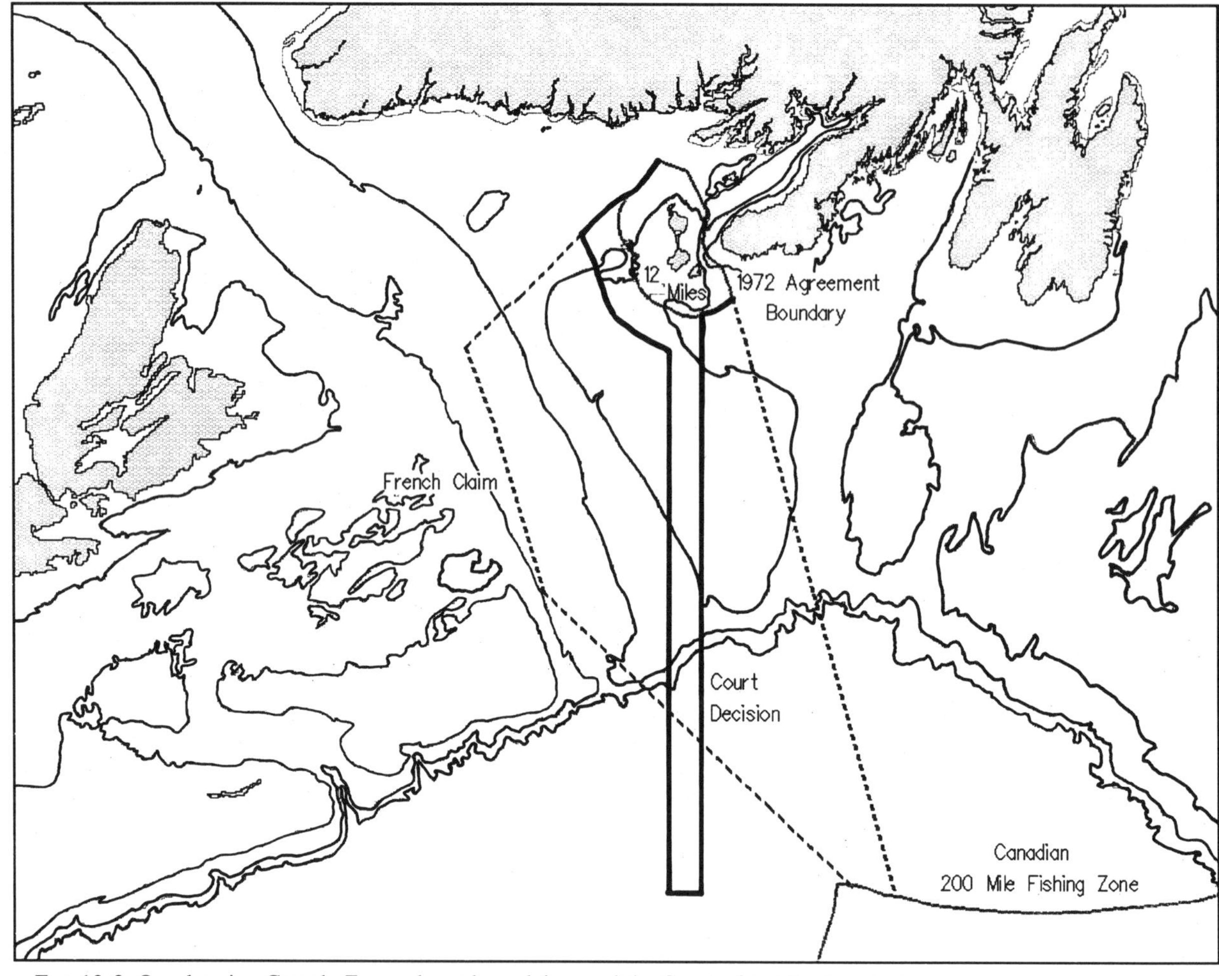

FIG. 12-2. Overlapping Canada-France boundary claims and the Court of Arbitration decision.

There were four conflicting boundary claims: Georges Bank-Gulf of Maine on the east coast, in the Beaufort Sea, and Juan de Fuca Strait and the A-B line in Dixon Strait on the west coast (Fig. 12-3A, B, C). Of these overlapping claims, the Georges Bank-Gulf of Maine dispute was the most significant in fisheries terms. The conflicts in this area were resolved by the International Court of Justice at The Hague in October 1984. The problem of managing shared stocks on the east coast has not been satisfactorily resolved.

On the west coast and in the Beaufort Sea the boundary disputes remain unresolved. Conclusion of the Canada-USA Pacific Salmon Treaty in 1985 and the establishment of the Pacific Salmon Fisheries Commission in 1986 brought significant progress on one of the major Canada-USA fisheries management problems. But problems remain. During the late 1980's major disputes arose over fisheries trade issues, particularly the export of unprocessed salmon caught in British Columbia waters and the implementation by the USA of a minimum size limit for lobsters imported from Atlantic Canada.

2.0 THE LAW OF THE SEA — BOUNDARIES AND SHARED STOCKS

Article 74 of the 1982 Law of the Sea Convention provides that:

> "The delimitation of the exclusive economic zone between States with opposite or adjacent coasts shall be effected by agreement on the basis of international law.... in order to achieve an equitable solution."

Article 63 of Part V of the LOS Convention deals with shared stocks occurring within the exclusive economic zones of two or more coastal States. Paragraph 63 (1) states:

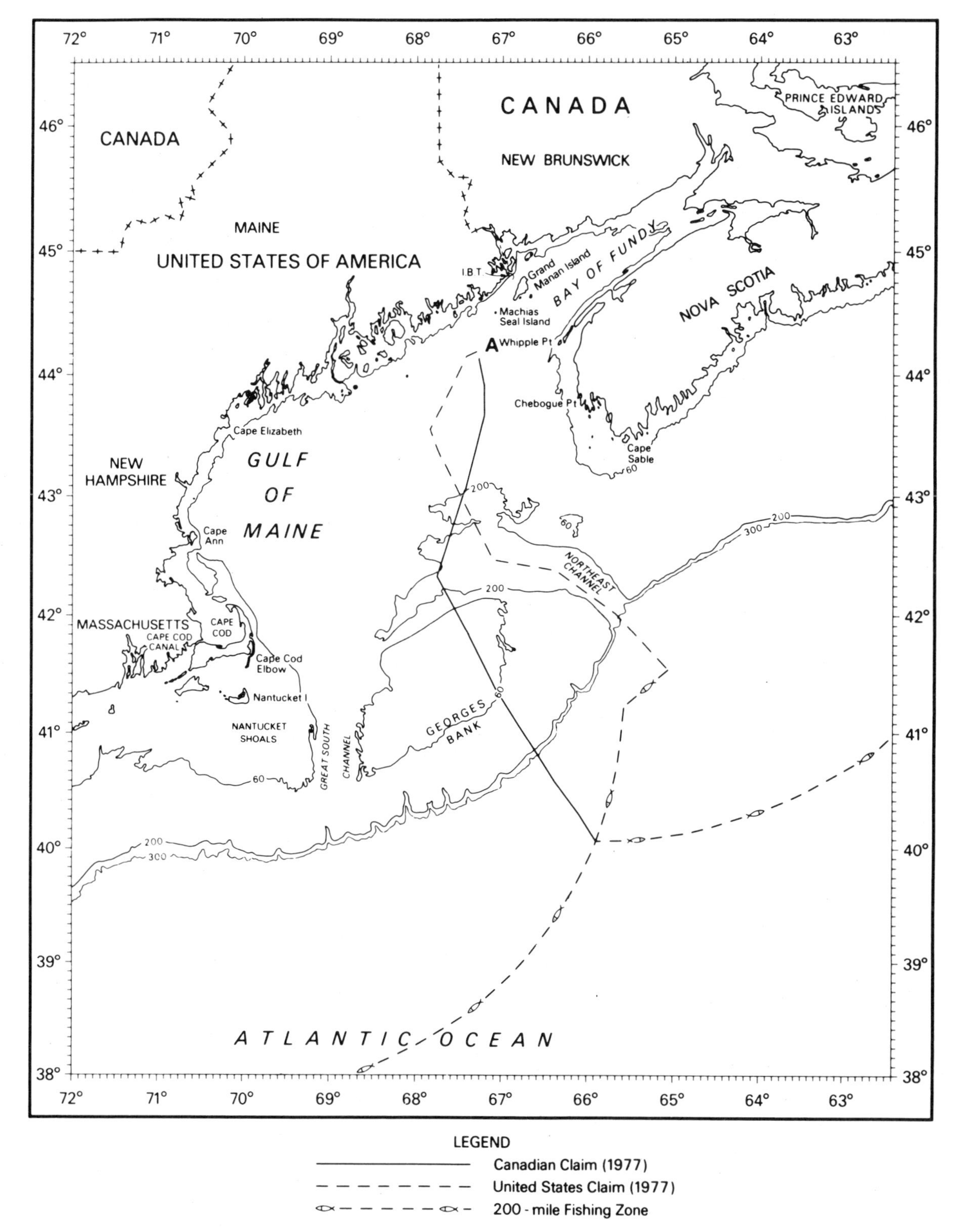

FIG. 12-3A. Canadian and United States claims of 1977 in the Gulf of Maine (from *Ocean Development and International Law* Vol. 16 p63 J. Cooper. Taylor and Francis, Inc. Washington D.C. Reproduced with permission. All rights reserved.)

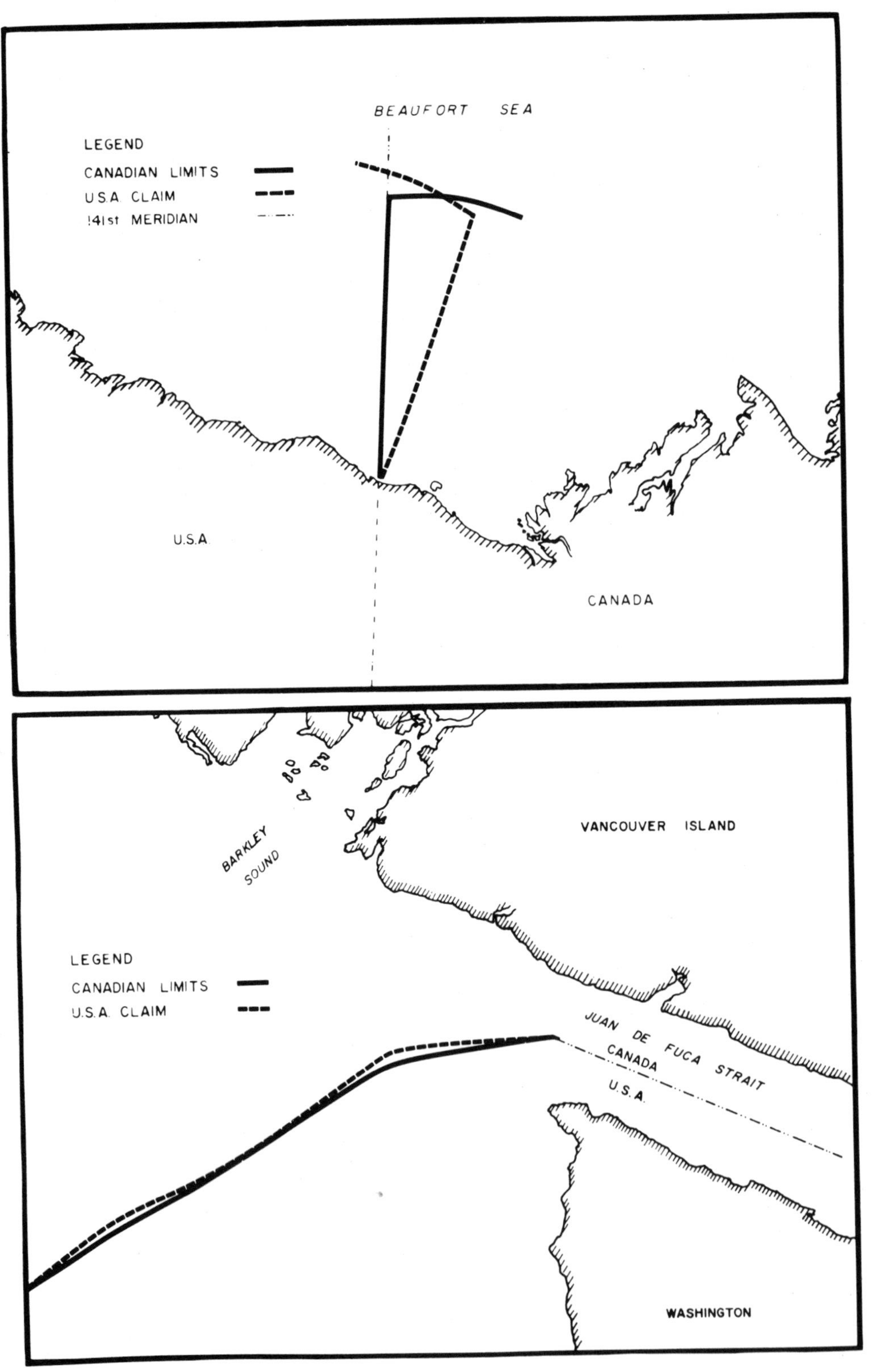

FIG. 12-3B. Canadian and United States claims in the Beaufort Sea (above) and off Juan de Fuca Strait (below) (from Wang, 1981).

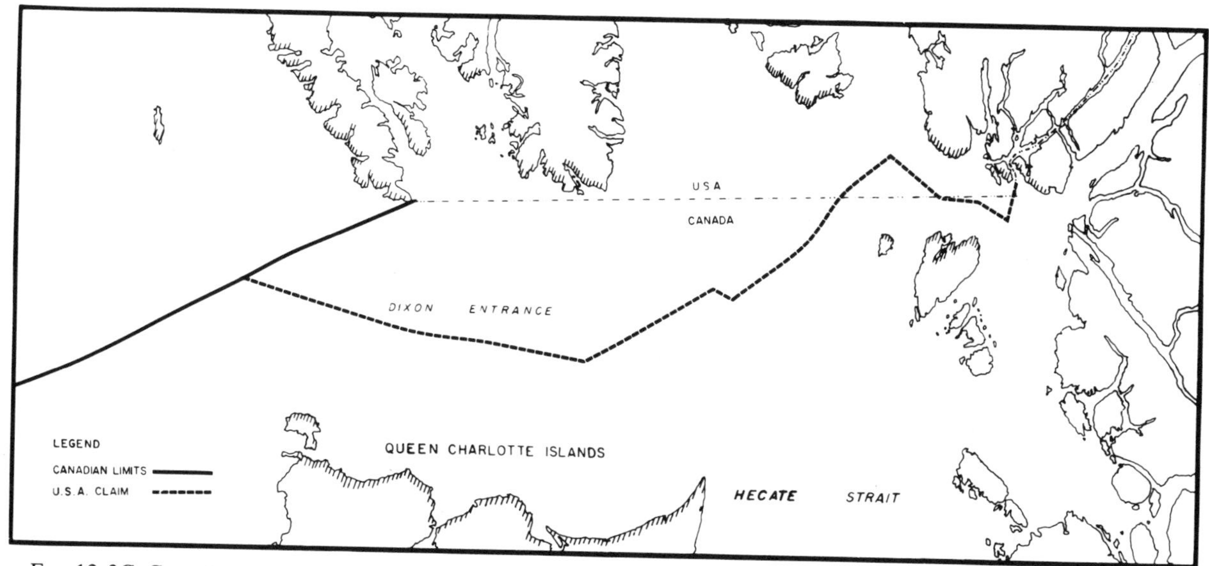

FIG. 12-3C. Canadian and United States boundary claims, inside and off Dixon Entrance (from Wang, 1981).

"Where the same stock or stocks of associated species occur within the exclusive economic zones of two or more coastal States, these States shall seek, either directly or through appropriate subregional or regional organizations, to agree upon the measures necessary to coordinate and ensure the conservation and development of such stocks."

Although this provision directs adjacent coastal states to "seek to agree" upon such measures for transboundary stocks, it does not compel them to reach agreement or to submit to the dispute settlement provisions of Part V if they fail to do so. In the absence of an agreement, the States concerned are free to manage the resources in their respective zones independently of each other. In this case fisheries management is governed by Articles 61 and 62.

Canada and its neighbours have sought to agree on the boundaries of their exclusive economic or fishery zones through negotiation. Where this has failed they have in certain instances referred boundary delimitation to the International Court of Justice or to a mutually agreed independent tribunal constituted for the purpose. Canada and the USA did the former in the case of the Georges Bank — Gulf of Maine. Canada and France did the latter for St. Pierre and Miquelon. Three of the disputed boundaries between Canada and the USA are still unresolved.

There have been a number of attempts at joint management of transboundary stocks. These have not generally been effective.

3.0 DENMARK (GREENLAND)

3.1 The Situation Pre-extension

Davis Strait is a relatively shallow body of water between Canada and Greenland linking the Labrador Sea in the south with Baffin Bay in the north. On December 17, 1973, Canada and Denmark signed an *Agreement Relating to the Delimitation of the Continental Shelf Between Greenland and Canada*. This agreement, ratified on March 13, 1974, was negotiated in the context of the 1958 Convention on the Continental Shelf and delimited the continental shelf in Davis Strait between the two countries for "the purpose of exploration and exploitation of the natural resources of the continental shelf" (Canada-Denmark 1973).

Davis Strait contains a mixture of water of Arctic and subarctic origin. In general, Arctic marine waters are not rich in fish. Dunbar (1951, 1970) has suggested that this is due to the extreme vertical stability of Arctic waters which limits the movement of nutrients from the bottom to the surface layers, hence inhibiting primary production. Water masses of Arctic-origin dominate the western part of Davis Strait along Baffin Island. Along the West Greenland coast, warmer subarctic waters are dominant. NAFO Subarea 1 consists principally of subarctic water, as does Division OB of Subarea 0 on the Canadian side. Division OA of Subarea 0 consists principally of Arctic-origin water (Fig. 12-1).

Cod is abundant enough off West Greenland to support a significant fishery but not so on the Canadian side of Davis Strait. Off West Greenland cod fluctuates between periods of high and low abundance. This is determined largely by changes in environmental conditions. The West Greenland cod stock is not transboundary. Management of Atlantic salmon, which is transboundary, was addressed in Chapter 11. The remainder of this section focuses on three other species which are transboundary in Davis Strait — roundnose grenadier, shrimp, and Greenland halibut.

The roundnose grenadier, a long-lived, slow-growing and late-maturing fish, occurs on both sides of the North Atlantic at depths from 350 to 2500 m. The greatest concentrations in the Northwest Atlantic occur from 50°N off northeast Newfoundland to 66°N in Davis Strait. There appear to be two distinct stocks in the Northwest Atlantic and the Northeast Atlantic (Grigor'en 1972; Savvatimsky 1969; Parsons 1976). The Northwest Atlantic stock is transboundary in Davis Strait. A fishery for this species, primarily by the USSR in Subareas 0 and 1, was first reported in 1968, with a catch of 6,000 tons. Catches averaged 6,500 per year from 1968 to 1977. Prior to 1977, this species was managed by ICNAF. ICNAF established a precautionary TAC of 10,000 tons in 1975, which was increased to 14,000 tons for 1976. Subsequently, the TAC was reduced to 8,000 tons for 1977. Following extension of fisheries jurisdiction, it became apparent that the ICNAF breakdown of catches by Subarea did not correspond to the new 200-mile zones. The dividing line between Subareas 0 and 1 was adjusted in 1979 to correspond to the boundary line separating the two fishing zones in Davis Strait (Fig. 12-1), with Subarea 1 being the zone off Greenland and Subarea 0 the zone off Canada.

In the early years the greater proportion of the catches were reported from Subarea 0. But from 1972 onward, based on the new boundary, the trend was toward increased catches in Subarea 1. Atkinson et al. (1982) postulated the probable location of the concentrated fishery for roundnose grenadier in an area bounded by 62°30' — 64°15'N and 58°00'W — 59°00'W (Fig. 12-4).

A second deepwater species of commercial significance in Davis Strait is the Greenland halibut. It occurs at depths to 1,500 m along the west coast of Greenland, in the deepwater fjords and off Baffin Island. Catches increased to 14,000 tons in 1972, peaked at 25,000 tons in 1975 and subsequently declined to 13,000 tons in 1977. Catches averaged 11,000 tons per year from 1969 to 1977. During this period the USSR accounted for 80% of the catch in Subarea 0 and 48% of the catch in Subarea 1 with Danish vessels taking about 20% of the catch in Subarea 0 and 40% in Subarea 1. There was considerable year-to-year variation in the proportion of the catch taken in each Subarea but overall about 25% was taken in Subarea 0 and 75% in Subarea 1, based on the old boundaries. With the 1979 adjustment, Subarea 0 became somewhat larger and Subarea 1 somewhat smaller than under the ICNAF boundaries.

ICNAF established a precautionary TAC of 20,000 tons for Greenland halibut in Subareas 0 and 1 in 1976 and 1977. The TAC was increased to 25,000 tons in 1978, based on a USSR research vessel survey which suggested that this stock was relatively large. Atkinson et al. (1982) concluded that commercial concentrations occurred in the area of the boundary south of 69°N.

The third species of commercial interest, the northern shrimp, *Pandalus borealis*, has been the most important to both Greenland and Canada in the post-extension era. Shrimp occur in both zones and in the boundary area. Based on information available in the mid-1970's, two stock complexes were recognized: 1. exploited offshore and inshore stocks off West Greenland and straddling between Subareas 1 and 0; and

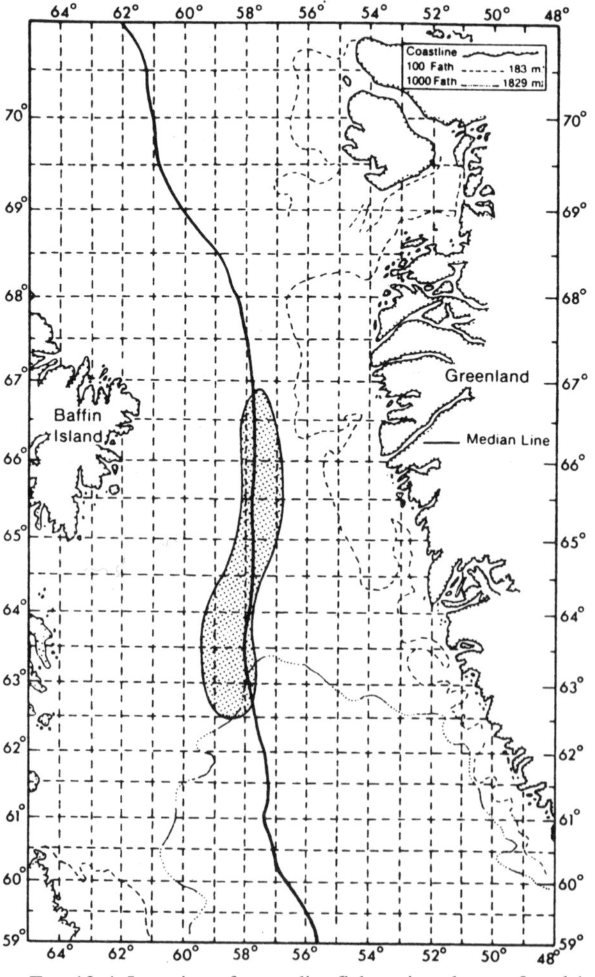

FIG. 12-4. Location of grenadier fishery in subareas 0 and 1 (from Atkinson et al. 1982).

2. unexploited stocks off Baffin Island. The offshore stock overlapped both subareas with the greater portion available in Subarea 1 (Fig. 12-5).

The shrimp catch in the offshore portions of Subareas 0 and 1 increased from less than 1,000 tons before 1972 to 43,000 tons in 1976 and then decreased to 34,000 tons in 1977. The inshore fishery off West Greenland was relatively stable from 1972 to 1980, averaging 7,000 – 8,000 tons. Canadian interest in the offshore shrimp fishery in Davis Strait began in 1979 following the development of a domestic shrimp fishery in the Canadian zone off Labrador.

Thus, prior to extension of jurisdiction in 1977 there was a well-developed fishery for several species by vessels based in West Greenland. There was also an offshore fishery for roundnose grenadier and Greenland halibut by the USSR. Vessels from several countries participated in the developing offshore shrimp fishery. Canadian vessels had virtually no interest in the offshore stocks of Greenland halibut and roundnose grenadier. Canadian interest in the offshore shrimp fishery materialized only in the late 1970's, following extended jurisdiction.

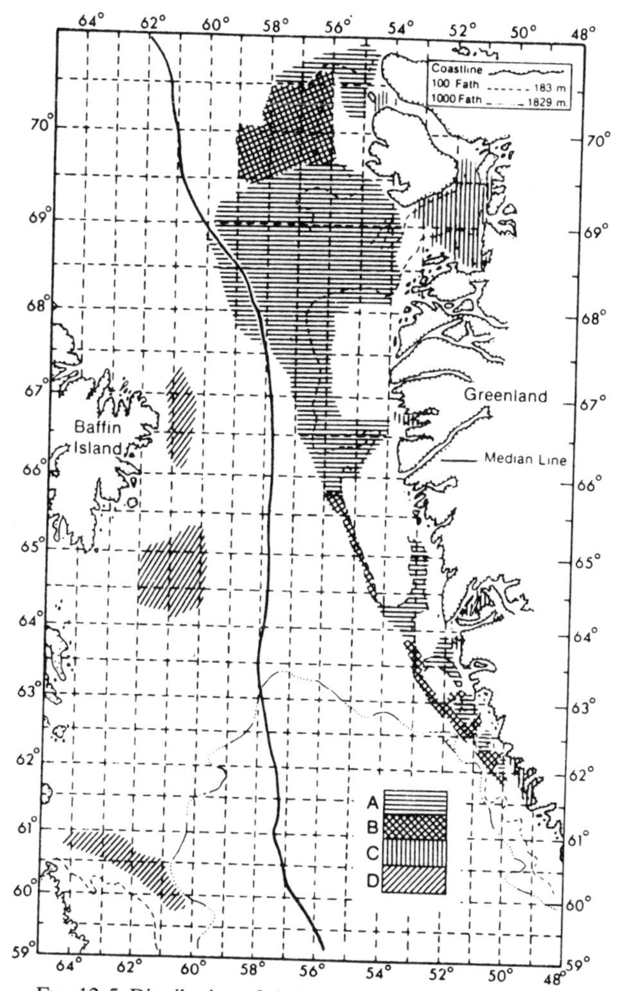

FIG. 12-5. Distribution of shrimp grounds in Subareas 0 and 1 (A = offshore grounds; B = potential grounds and nursery areas; C = major inshore grounds; D = areas where shrimp have been found off Baffin Island) (from Atkinson et al. 1982).

3.2 Post-Extension Arrangements between Canada and the European Community

With the advent of extended jurisdiction, the ICNAF regime in Davis Strait was replaced by arrangements between Canada and the EC, on behalf of Denmark (Greenland). Although Denmark had declared a 200-mile fishing zone in 1977 around Greenland, this did not give Greenland the right to manage the fishery in the extended zone. Since Denmark was a member of the EC, the fishing zone off Greenland was also a fishing zone of the EC.

Denmark joined the EC in 1973, but with a majority (over 70%) of Greenlanders voting against accession. In theory, all member states of the EC had the right to fish in Greenland waters. Because a common fisheries policy was not yet in place, the Act of Accession permitted Denmark to restrict fishing access to a 12-mile zone around Greenland to vessels from Greenland and those which had traditionally fished the area and operated from Greenland ports. Thus access to the inshore fisheries of Greenland (12 miles) was restricted to Greenlanders and Danes until December 31, 1977. From then until December 31, 1982, it was restricted to Greenlanders only.

In 1978, Greenland gained Home Rule status. The *Greenland Home Rule Act* permitted the delegation to the Home Rule authority of jurisdiction over "fishing in the territory." However, this merely transferred from Copenhagen to Greenland the jurisdiction over fishing in the 12-mile-zone which was already restricted to Greenlandic inshore vessels. The EC exercised control over the offshore fishery in Subarea 1 beyond 12 miles. Thus Canada had to "seek to agree" with the EC, not with Denmark or Greenland directly, on the joint management and sharing of the transboundary resources in Davis Strait.

Initially, Canada and the EC agreed to jointly establish TACs for the transboundary stocks based on scientific advice from the ICNAF Standing Committee on Research and Statistics until 1978. Thereafter they sought advice from the NAFO Scientific Council. The TACs for 1977 were global for Subareas O and 1, with no breakdown between the Canadian and Greenland fishing zones. In 1977, Canada allocated 11,200 tons of Greenland halibut in Subarea 0 to the GDR and the

USSR. Canada also allocated 4,900 tons of roundnose grenadier in Subarea 0 to the GDR and the USSR. For shrimp, Canada made no allocation as very little fishing activity had been reported in Subarea 0.

For 1978, annual bilateral meetings were initiated to establish TACs for each of the three species for Subareas 0and 1 and to divide these TACs between the two parties. Canada and the EC then suballocated these national shares of the TACs either to domestic or foreign vessels.

The 1978 arrangements were:

	TAC	EC	CANADA
Greenland Halibut	25,000 t	24,500 t	500 t
Roundnose Grenadier	8,000 t	7,500 t	500 t
Shrimp	40,000 t	39,000 t	1,000 t

The EC allocations included third party allocations.

Canadian interest in fishing shrimp in Davis Strait increased significantly during 1978. Following exploratory fishing and scientific study, the Minister of Fisheries issued 11 shrimp licenses in June 1978 for a new shrimp fishery off Labrador (DOE 1978f). The initial TACs for the Labrador shrimp stocks totalled 7,100 tons. By 1981, this had increased to 11,650 tons. In addition, in 1978 1,000 tons of the Subarea 0 and 1 shrimp TAC was available to these Canadian licensees, 650 tons of which could be harvested in the EC zone.

Over the next 3 years the acquisition of Canadian vessels by some of the licensees increased pressure on Canadian negotiators to secure access to larger quantities of shrimp in the EC zone. Subarea 0 and the area off Labrador were generally ice-covered in the winter and spring. But it was possible to fish shrimp year-round in Subarea 1. The major offshore shrimp stock in Davis Strait straddled the boundary between Subareas 0 and 1. The fishery on the EC side often extended into the Canadian zone. This indicated a potential for a reciprocal fishing arrangement. Satisfying Canada's needs, however, required increasing its share of the overall TAC and allowing Canadian vessels to fish this share in Subarea 1. An economically viable year-round Canadian fishery appeared to require access to a significant quantity of shrimp in Subarea 1.

Canada succeeded in increasing its share of the shrimp TAC from 1,000 tons in 1978 (of a TAC of 40,000 tons in 1978) to 2,000 tons in 1979 (of a TAC of 29,500 tons). 1,750 tons of this could be fished in Subarea 1. To secure this increase Canada agreed to relatively small shares of the Greenland halibut and roundnose quotas. These two species were of marginal interest to Canadians.

The 1979 arrangements were:

	TAC	EC	CANADA
Greenland Halibut	20,000 t	18,000 t	2,000 t
Roundnose Grenadier	8,000 t	7,500 t	500 t
Shrimp	29,500 t	27,500 t	2,000 t

The agreement for 1979 also provided that all allocations to third parties would be determined by agreement between the two parties. The EC ignored this provision and made allocations to third parties in the EC zone. Canada, on the other hand, sought EC consent before allocating the 1,500 tons of Greenland halibut and 300 tons of roundnose grenadier to be fished entirely within Canadian waters. Canada harvested its entire shrimp quota in 0+1 in 1979.

Canada approached the negotiation of management arrangements for 1980 seeking an increased share of the shrimp TAC and improved access to the EC zone. Negotiations for 1980 involved two phases: a November 1979 session in Ottawa and a January 1980 session in Brussels. Canada proposed percentage splits of the TAC based on estimates of the proportion of the stocks available in each zone. The EC emphasized the lack of an agreed scientific basis for such a sharing arrangement at that time.

Canada was willing to maintain its existing (1979) quotas for Greenland halibut and roundnose grenadier but sought a 3,500 ton Canadian quota for shrimp. Canada had another condition: as the ability of Canadian vessels to fish shrimp in this area increased, Canadian quotas should correspondingly increase to equal the proportion of the shrimp stock found on the Canadian side of the boundary line. The EC rejected this proposal. It argued that it required its full quota of 27,500 tons of shrimp for their own use and for allocation to third parties (The Faroes and Norway).

When negotiations resumed in January, the EC responded that it could agree only on an allocation of 2,500 tons of shrimp to Canada (up from 2,000 tons in 1979) with access to the EC zone. Otherwise, there could be no joint management with each party free to fish in its own zone but without access to the adjoining zone.

Canadian industry representatives indicated a need to fish on the EC side of the boundary line while the Canadian side was covered with ice (approximately the first 5 months of the year). Thus Canada accepted the EC proposal for shrimp. As a trade-off the Canadian allocations of Greenland halibut and roundnose grenadier were increased to 3,500 tons and 800 tons respectively. Allocations of these species could be made by either party to third parties without the consent of the other coastal state. The agreement also provided that Canadian and Canadian-chartered vessels would have access to the EC zone to catch a maximum of 2,000 tons of the 2,500 ton Canadian shrimp quota. Similarly, EC vessels could catch a portion of their total allocation of 27,000 tons of shrimp on the Canadian side of the boundary line.

During 1980 seven Canadian or Canadian-chartered vessels fished the Subareas 0 and 1 shrimp stock. The Canadian quota was reached in July. A Canadian patrol vessel was sent to Davis Strait to patrol the line and inspect vessels fishing on the Canadian side.

Canada and the EC had available at the November 1980 negotiations the Report of the Canada-EC Scientific Working Group on Joint Stocks in Subareas 0 and 1 (Anon 1980). The Group reported that:

1. It was unable to determine the relative distribution of roundnose grenadier biomass between the two zones;
2. The evidence for Greenland halibut, while inconclusive, suggested that as much as 40% of the stock occurred on the Canadian side of the boundary line;
3. The shrimp biomass appeared to be distributed 17% in Subarea 0 and 83% in Subarea 1.

Canada argued that the allocation to each Party should reflect the abundance of shrimp in each zone, i.e. 17% to Canada and 83% to the EC. This would have given Canada an allocation of 5,000 tons. The EC countered that, while stock distribution was an important factor, catch distribution should be the key factor in determining allocations.

The delegations agreed on the following proposed management agreement for 1981:

1.

	TAC	EC	CANADA
Greenland Halibut	25,000 t	20,000 t	5,000 t
Roundnose Grenadier	8,000 t	6,400 t	1,600 t
Shrimp	29,500 t	27,000 t	2,500 t

2. Provisions for reciprocal access were the same as in 1980;
3. Canada agreed to allocate to the EC 750 tons of redfish and/or Greenland halibut in the Canadian zone off Newfoundland;
4. Each Party would authorize fishing by vessels of the other Party from January 1, 1981 based on the Agreed Record (Canada-EEC 1980c).

While the delegations agreed at the conclusion of the meetings, the EC refused to sign the Agreed Record. EC officials explained that the negotiator had no authority to sign the 0 plus 1 agreement without an agreement on French allocations in the area off Newfoundland and St. Pierre and Miquelon. In late December the Community informed Canada that there would be no access for shrimp in Community waters in Subarea 1 without both sides approving the Canada-EEC long-term agreement (LTA).

In early January 1981, events took another twist when the EC negotiator informed his Canadian counterpart that the EC required additional shrimp allocations for Norway, Metropolitan Danish vessels and French vessels. Given the requirements of the Greenlanders, this would not be possible unless the TAC were increased. The EC accordingly suggested that the TAC be increased from 29,000 tons to 33,000 tons. It proposed amending the unsigned November agreement accordingly. The Canadian allocation would increase from 2,500 to 2,800 tons and the EC allocation increase from 27,000 to 30,200 tons.

Canadian authorities met with Canadian industry representatives on January 13 to inform them that the EC was linking the fate of the November Agreement on the Davis Strait stocks for 1981 with the fate of the Canada-EC LTA. They were also informed of the EC's request to modify the November agreement to increase the TAC and allocations for shrimp (Minutes, Northern Shrimp Advisory Committee, January 13, 1981).

Northern shrimp fishing interests believed the EC was "negotiating in bad faith." They opposed increasing the TAC beyond the level recommended by the NAFO Scientific Council because this would compromise the principles under which Canada normally established its TACs.

The licensees strongly advocated a hardline position by Canada on this issue, in view of the shifting EC position. Otherwise, they saw little hope in future negotiations with the EC about the Davis Strait stocks.

Accordingly, Canada informed the EC in mid-January that there was no scientific basis to increase the TAC of 29,500 tons for shrimp agreed in November. Canada argued that the EC's refusal to ratify the November agreement had already seriously prejudiced Canada's benefits under that agreement (access to shrimp in Subarea 1 as of January 1, 1981).

Despite this, Canada indicated that it would implement that agreement, provided licences were issued by January 31, 1981 to Canadian fishermen to fish for shrimp in Subarea 1.

Canadian vessels were standing by to commence fishing in early January. The decision to permit implementation as late as January 31 was a major concession for Canadian interests.

In early February, Canada informed the EC that, in the absence of a joint management agreement, Canada was establishing the following 1981 TACs for Subarea 0:

1. Canadian TAC of 5,000 tons of shrimp;
2. TACs of 12,500 tons for Greenland halibut and 4,000 tons for roundnose grenadier, at 50% of the levels of the previously agreed TACs for the entire area.

The EC proceeded to establish a shrimp TAC of 30,000 tons and TACs of 20,000 tons for Greenland halibut and 6,400 tons for roundnose grenadier in Subarea 1. The net result was a combined TAC of 35,000 tons for shrimp. The combined TACs for Greenland halibut and roundnose grenadier were also higher than those previously agreed on the basis of scientific advice but these were less likely to be harvested.

Canada protested the Community's actions but to no avail. In 1981, 12 licence holders operated in the Canadian northern shrimp fishery. All 12 operated vessels in Davis Strait. They harvested 4,500 tons from mid-June to mid-December. This proved that Canada could harvest its claimed share of the overall TAC for this stock.

As the 1982 fishing season approached, an agreement between Canada and the EC became less likely, because of the scheduled February 1982 referendum on Greenland's withdrawal from the EC. It was widely anticipated that Greenlanders would vote to leave the EC. For 1982 the EC established a shrimp TAC for Subarea 1 of 29,800 tons, of which 27,850 tons was reserved exclusively for Greenland fishermen. Combined with the Subarea 0 TAC of 5,000 tons established by Canada, this resulted in an overall 1982 TAC of 34,800 tons (see Table 12-1).

3.3 Greenland Withdraws from the European Community

On February 23, 1982, Greenlanders voted to withdraw from the EC. This required amending the European Community Treaties to exempt Greenland. Greenland would then have overseas country or territory (OCT) status. A principal reason for Greenland's request to withdraw from the EC was to gain greater control over its fisheries. Greenlanders resented the presence of other member states' vessels and the authority of the EC to set quotas for Greenland waters. The EC contributed to this resentment by occasionally reducing Greenland quotas to provide allocations for other member states (Harhoff 1983).

Greenland's request led to protracted negotiations which often centered on the fishing regime that would prevail after withdrawal and future financial assistance. The terms of Greenland's withdrawal from the EC were settled in February 1984. Greenland withdrew from the EC on January 1, 1985.

Hence, from 1982 to 1984, it was virtually impossible for Canada and the EC to establish joint management arrangements for Davis Strait. From the Canadian perspective this would have required a reduction in the EC and Greenland quotas for shrimp. This was not possible in the context of the negotiations underway for Greenland's withdrawal from the Community. Thus, Canada and the EC continued to establish unilateral quotas for their respective zones. In the case of shrimp, the overall TAC continued to exceed the scientific advice, which remained at 29,500 tons through 1984. The overall offshore shrimp catch averaged about 37,000 tons during the 1981–84 period with consistent overruns by the EC of its TAC for Subarea 1 (Table 12-1).

Canada's catch of shrimp in Subarea 0 fluctuated over this period (2,700 tons in 1983, 4,000 tons in 1983, 2,300 tons in 1984).

Greenland assumed management responsibilities for Subarea 1 in 1985. The first management-related discussions since 1980–81 took place when Canadian officials visited Nuuk in November 1986. Canada's primary interest continued to be access to Subarea 1 for shrimp in the winter months and ensuring that the overall TAC followed scientific advice. In these exploratory discussions, both sides agreed that more focused research was needed in Subareas 0 and 1. This research should assess the shrimp biomass, its geographic distribution as well as the appropriate level of removals. Both agreed to ask the NAFO Scientific Council for advice.

Despite these discussions, separate management regimes have applied to the two sides of the Davis Strait boundary since 1981. Despite fears that this could hurt the stocks, this did not occur. The overall TACs for shrimp were higher than the NAFO Scientific Council advised from 1981 to 1986 (Table 12-1). The Council had consistently advised a TAC of 29,500 tons for the offshore stock from 1979 to 1984. However, at its January 1985 meeting, the Council changed its advice. The stock appeared stable. Higher-than-advised yields had been realized during this period of stability. Thus the Scientific Council advised that the overall TAC for the offshore grounds in Subarea 1 and the adjacent parts of Subarea 0 in 1985 not exceed 36,000 tons, the level of average catch during 1979–84. The Council emphasized that this advice was not based on any evidence of an increase in stock size since 1979 but rather on a reevaluation of a longer data series (NAFO 1985b).

This advice would have permitted the overall TAC of 1981 to 1984 of around 35,000 tons to be maintained.

TABLE 12-1. Shrimp TAC's and Catches in Subareas 0 + 1 (Quantities in tons).

	1978	1979	1980	1981	1982	1983	1984	1985	1986	1987[a]	1988[a]	1989[a]	1990[a]
Catch Subarea 0	122	1,129	874	5,284	1,812	5,413	2,142	2,640	2,995	6,140	6,087	7,235	6,177
Subarea 1	34,347	33,458	43,278	39,516	42,515	41,354	41,241	51,396	60,134	57,641	54,455	60,849	63,431
Offshore (S of 71°N)	26,747	25,958	35,778	32,016	35,015	33,854	33,741	39,547	41,589	40,020	37,562	43,899	46,260
SA 0 + 1 Offshore Catch (S of 71°N)	26,869	27,087	36,652	37,300	36,827	39,267	35,883	42,187	44,584	46,160	43,649	51,134	52,437
SA 0 + 1 Advised Offshore TAC	40,000	29,500	29,500	29,500	29,500	29,500	29,500	36,000	36,000	36,000	36,000	44,000	50,000
SA 0 + 1 Effective Offshore TAC	40,000	29,500	29,500	35,000[b]	34,800[b]	34,625[b]	34,925[b]	42,120[c]	42,120[c]	40,120[c]	40,120[c]	40,120[d]	44,975[d]

[a] Provisional data.
[b] Includes TAC of 5,000 t in Subarea 0.
[c] Includes TAC of 6,120 t in Subarea 0.
[d] Includes TAC of 7,520 t in Division OA.

TABLE 12-2. France and Canada cod quotas and catches 1977–1987 in Division 3Ps from 1977 to 1987.

			FRANCE			CANADA		
Year	TAC[1]	15.6% of TAC	EC/France Quota	% of[2] TAC	Catch	Quota	% of[2] TAC	Catch
1977	32,000	4,992	5,000	15.6	2,986	26,900	84.1	29,259
1978	25,000	3,900	3,900	15.6	4,834	21,100	84.4	22,342
1979	25,000	3,900	3,900	15.6	6,349	21,100	84.4	26,657
1980	28,000	4,368	4,400	15.7	4,617	23,600	84.3	32,951
1981	30,000	4,680	4,700	15.7	7,807	25,300	84.3	31,085
1982	33,000	5,148	5,170	15.7	6,964	27,830	84.3	26,938
1983	33,000	5,148	5,170	15.7	9,856	27,830	84.3	28,595
1984	33,000	5,148	7,940	24.1	11,221	27,830	84.3	25,729
1985	41,000	6,396	10,000	24.4	18,508	34,600	84.4	32,859
1986	41,000	6,396	—	—	25,981	34,600	84.4	31,738
1987	41,000	6,396	26,000	63.4	24,502	34,600	84.4	33,049

Source: French and Canadian catches are from NAFO Statistical Bulletins for 1977–85 and NAFO SCS. Doc. 87/20 for 1986 and NAFO SCS. Doc. 88/18 for 1987.
[1]TAC as established by Canada.
[2]% is of the TAC as established by Canada.

Instead, it resulted in an increase in the overall TAC to 42,120 tons in 1985. Canada increased the TAC for Subarea 0 to 6,120 tons corresponding to 17% of the newly-advised TAC of 36,000 tons and Greenland established a new TAC at the 36,000 ton level. This TAC was maintained for 1986.

For 1987 the Scientific Council again advised a TAC of 36,000 tons. The TACs for 1987 and 1988 were reduced slightly to 40,120 with Greenland reducing its TAC by 2,000 tons.

At its June 1988 meeting, the Scientific Council concluded that the recent increasing trend in catch rates did not indicate changes in stock abundance. Despite increasing catches in the last 3 years, catch rates had not declined and data on size distribution showed no changes which could be related to fishing pressure. In this light the Council presented two options. Both envisaged some increase in TAC. The first option was to limit catches for a number of years to the levels observed in 1985 and 1986 (about 44,000 tons). The second involved increasing the TAC to 50,000 tons, the average catch for the 1985–87 period. This level would be maintained for several years, to determine whether the recent high catches could be sustained (NAFO 1988e).

Since Greenland had greatly exceeded its 1987 TAC of 34,000 tons, even the highest of the options meant a significant reduction in its catch. The actual TACs established for 1989 were 37,775 tons for Subarea 1 and 7,520 tons for Subarea 0 for an overall TAC of 45,295 tons.

The Canadian share of the overall TAC for 1988 was 15% and for 1989 16.6% . For the first time since 1981 the global TAC approximated the minimum (44,000 tons) advised by the NAFO Scientific Council. For 1990, the Scientific Council advised that catches should not exceed 50,000 tons.

Superficially, the absence of an agreed management regime has not adversely affected this transboundary stock. However, there is sufficient uncertainty in the scientific advice to warrant caution. In 1987, for the first time Greenland did not claim the entire TAC for Subarea 1. Canada consistently favoured establishing a TAC for Subarea 0 at 17% of the scientifically advised TAC. The Greenland share of the TAC decreased to 83.4% in 1989, compared with the period up to 1984 when it claimed the entire TAC for its zone. This appeared to offer some promise for future cooperation on management.

Canadian vessels were denied access to a winter fishery in Subarea 1 from 1981 onward. However, the Canadian fleet managed to compensate for this within the Canadian zone. About half of the 16 licence holders in 1990 also had groundfish licences and fished groundfish for part of the year. Those with access only to shrimp had diversified their area and season of fishing within the Canadian zone. Fishing in Subarea 0 was confined to the summer and fall but the fleet developed a significant shrimp fishery year-round within the Canadian zone. This was possible because the fleet has found economically fishable concentrations of shrimp in the winter to the south off northeast Newfoundland and southern Labrador. This reduced the impetus to seek reciprocal access to the Greenland zone. The Canadian catch in Subarea 0 from 1981 to 86 averaged 3,300 annually, but except for two years was less than 3,000 tons. In 1987 and 1988, with 16 licences in operation, Canada took all of the 17% of the TAC that it claimed.

Regarding the other two transboundary stocks, the absence of a joint management regime since 1981 has had no adverse impact on the stocks or the fisheries of the two coastal states. Roundnose grenadier is of little commercial interest to the fleets of either Party. There has been no directed fishery for roundnose grenadier in Subareas 0 and 1 since 1978. There were incidental catches of 3,000, 6,000, 7,000 and 2,000 tons in 1978, 1979, 1980 and 1981 respectively. Since then the catch has been less than a thousand tons annually.

The TAC for the third transboundary stock in Davis Strait, Greenland halibut, was maintained at 25,000 tons from 1982 onward. The catch did not exceed 10,000 tons annually between 1980 and 1990. For a while Canadians had no interest in fishing this stock in Davis Strait. The TAC for the large stock of Greenland halibut off the Canadian coast to the south in Subarea 2 and Divisions 3K and 3L increased from 55,000 tons during 1981–84 to 75,000 tons in 1985 and to 100,000 tons from 1986 onward. The average total catch from 1982 to 1987 was 23,500 tons, with the catch not exceeding 28,000 tons in any individual year. The Canadian catch averaged 14,700 tons during this period.

Then in 1990 a foreign fishery for this stock developed outside 200 miles in the Sackville Spur. At the same time Canadian - chartered vessels began to harvest more Greenland halibut in Subarea 2 and Subarea 0. It is likely that there will be greater Canadian interest in fishing this species in Subarea 0 in future. This is less likely for roundnose grenadier.

The saga of joint management and its demise in Davis Strait revolved around the offshore shrimp stock. The absence of a joint management regime since 1981 is regrettable. However, this had little impact on either the stocks or the fisheries in this area. By the early 1990's interest in a bilateral agreement with Greenland had revived.

4.0 FRANCE (INCLUDING ST. PIERRE AND MIQUELON)

4.1 General

Canada's fisheries relations with France in the post-extension era centered on two major issues:

1. The application and interpretation of the 1972 "Agreement on Mutual Fisheries Relations" between Canada and France; and
2. Overlapping maritime boundary and fishing zones claims to the south of Newfoundland following the proclamation of 200-mile zones by Canada and France in 1977.

The islands of St. Pierre and Miquelon off the south coast of Newfoundland are French territory. Thus, in the context of this chapter, France is Canada's neighbor because Canada has distinguished between its relations with St. Pierre and Miquelon, which is described as a neighbour in Article 4(1) of the 1972 Agreement, and Metropolitan France, which is not. St. Pierre and Miquelon are a "department" of France, and the 1972 Agreement concerned both the Metropolitan French fleet and the fleet based in St. Pierre and Miquelon. Thus this section deals with both sets of relations.

The islands of St. Pierre and Miquelon are located less than 25 km southwest of the Burin Peninsula in Newfoundland (Fig. 12-2). In addition to the two main islands of St. Pierre and Miquelon, there are some 10 other islets and rocks. These islands were claimed by Jacques Cartier for France in 1536. Normans, Bretons and Basques used the islands as a base for fishing cod from the 1600's onwards. Following wars between England and France, St. Pierre and Miquelon were ceded to Britain in 1713. France regained the islands in the Treaty of Paris in 1763. Britain invaded the islands several times in the remainder of that century but finally recognized France's sovereignty over the islands in the second Treaty of Paris in 1815.

Like that of the neighbouring island of Newfoundland the economy of St. Pierre and Miquelon is closely linked to the sea. Some fishing was carried on by inhabitants of the islands using schooners in the 1800's and early 1900's and the Metropolitan France fleet fishing in the Northwest Atlantic used the islands as a base.

4.2 The 1972 Agreement

As part of its decision to declare the Gulf of St. Lawrence Canadian waters (see Chapter 10), Canada in 1972 negotiated an Agreement with France on "their mutual fishing relations" (Canada-France 1972). In 1971, France was one of the most active countries fishing in Canada's territorial sea and the Gulf of St. Lawrence. France claimed the right to fish in these areas based on treaty rights, in particular the 1904 Convention between the United Kingdom and France Respecting Newfoundland and West and Central Africa (France-Great Britain 1904).

This Convention entitled France to fish "in the territorial waters on that portion of the coast of Newfoundland comprised between Cape St. John and Cape Ray, passing by the north" on "a footing of equality with British subjects." The right to fish along the "treaty shore" included port privileges to obtain supplies and shelter on the same conditions as the inhabitants of Newfoundland and the right to fish at the mouths of rivers. Another clause in the Convention stipulated that the exercise of these fishing rights was subject to the same regulations that might be applied in the future to British subjects concerning "the establishment of a closed time in regard to any particular kind of fish, or for the improvement of the fisheries."

France also claimed a general treaty right to fish in the Gulf of St. Lawrence outside the "treaty shore" based on the Treaty of Paris of 1763, the Treaty of Versailles of 1783 and two Treaties of 1814 and 1815 ending the Napoleonic Wars. Canada rejected the latter claim, maintaining that it had been superseded by the 1904 Convention.

Canada began negotiations with the French in May 1971 with the following objectives:

1. To secure French acquiescence to the extension of Canada's territorial sea to 12 miles and the designation of the Gulf of St. Lawrence and the Bay of Fundy as exclusive Canadian fishing zones;
2. To secure agreement on the termination of France's treaty rights under the 1904 Convention;
3. To secure agreement on the boundary in the area between Newfoundland and St. Pierre and Miquelon; and
4. To secure agreement on the phasing-out of French traditional fisheries.

The Canadian delegation was almost entirely successful in achieving its objectives. France accepted Canada's extension of its territorial sea to 12 miles. It accepted the designation of the Gulf of St. Lawrence and the Bay of Fundy as Canadian fishing zones. This was subject to phase-out provisions from the Gulf for the Metropolitan French fleet and a right for a limited number and size of vessels from St. Pierre and Miquelon to continue to fish along the coasts of Newfoundland and Nova Scotia and in the Gulf.

In Article 1 of the 1972 Agreement, France renounced "the privileges established to its advantage" by the 1904 Convention. It was agreed that "the present agreement supersedes all previous treaty provisions relating to fishing by French nationals off the Atlantic coast of Canada."

The Agreement also defined in the area between Newfoundland and the islands of St. Pierre and Miquelon the limit of the territorial waters of Canada and of "the zones submitted to the fishery jurisdiction of France" (Article 8 and Annex to the 1972 Agreement).

A brief resumé of these negotiations is given in the Memorial submitted by Canada on February 22, 1986, to the Arbitral Tribunal on the Dispute Concerning Filleting within the Gulf of St. Lawrence by trawlers based in St. Pierre and Miquelon, hereinafter referred to as the La Bretagne Arbitral Tribunal. Canada's objective was the phase-out of French vessels from the Gulf. While France accepted that Metropolitan vessels would be phased-out of the Gulf, it sought provision for future fishing by Metropolitan France vessels elsewhere in Canadian waters. Furthermore, it sought a special provision permitting trawlers from St. Pierre and Miquelon to continue fishing in the Gulf after the phase-out of fishing by Metropolitan France vessels. France argued that only on the basis of a concession permitting St. Pierre and Miquelon to continue fishing indefinitely could France renounce the rights they were to have enjoyed in perpetuity along the "treaty shore" of Newfoundland pursuant to the 1904 Convention. They argued that this was necessary for the economic survival of the islands (Canada 1985b).

The 1972 Agreement contained several provisions. Article 3 provided that: "Fishing vessels registered in Metropolitan France may continue to fish from January 15 to May 15 each year, up to May 15, 1986, on an equal footing with Canadian vessels, in the Canadian fishing zone within the Gulf of St. Lawrence, east of the meridian of longitude 61 degrees 30 mins west."

Article 4 provided that: "In view of the special situation of Saint-Pierre and Miquelon and as an arrangement between neighbours:

"a) French coastal fishing boats registered in Saint-Pierre and Miquelon may continue to fish in the areas where they have traditionally fished along the coasts of Newfoundland, and Newfoundland coastal fishing boats shall enjoy the same right along the coasts of Saint-Pierre and Miquelon; and

"b) A maximum of ten French trawlers registered in Saint-Pierre and Miquelon, of a maximum length of 50 metres, may continue to fish along the coasts of Newfoundland and Nova Scotia (with the exception of the Bay of Fundy), and in the Canadian fishing zone within the Gulf of St. Lawrence, on an equal footing with Canadian trawlers; Canadian trawlers registered in the ports on the Atlantic coast of Canada may continue to fish along the coasts of Saint-Pierre and Miquelon on an equal footing with French trawlers" (Canada-France 1972).

Article 6 provided that Canadian fishery regulations "shall be applied without discrimination in fact or in law to the French fishing vessels covered by Articles 3 and 4."

The 1972 Agreement was also significant in that it anticipated a possible further extension of fisheries jurisdiction by Canada. In return for the renunciation by France of its privileges under the 1904 Convention, Article 2 stipulated that: "The Canadian Government undertakes in the event of a modification to the juridical regime relating to the waters situated beyond the limits of the territorial sea and fishing zones of Canada on the Atlantic coast, to recognize the right of French nationals to fish in these waters subject to possible measures for the conservation of resources, including the establishment of quotas. The French Government undertakes for its part to grant reciprocity to Canadian nationals off the coast of Saint-Pierre and Miquelon" (Canada-France 1972).

4.3 The Overlapping Claims

At the time of the 1972 Agreement Canadian negotiators foresaw major benefits from the Agreement, particularly the phase-out of Metropolitan French vessels from the Gulf. However, other provisions of the Agreement came back to haunt Canada in the 1980's. The Canadian 200-mile zone which took effect on January 1, 1977, encompassed the waters around St. Pierre and Miquelon except for a territorial sea of 12 miles. Canada contended that, because of the special geographic circumstances of the islands of St. Pierre and Miquelon, they were not entitled to a 200-mile economic zone. Canada argued that they were entitled only to a 12 mile territorial sea. France, on the other hand, on February 25, 1977, proclaimed a 200-mile economic zone around St. Pierre and Miquelon based on the principle of equidistance (France 1977) (Fig. 12-2). France maintained that St. Pierre and Miquelon was entitled in principle to a full 200-mile exclusive economic zone. However, the actual French claim consisted of a smaller area, based on equidistance from the nearest coasts of Newfoundland, Nova Scotia and St. Pierre and Miquelon. The zone claimed by France was 13,480 square nautical miles in area. It extended 180 nautical miles south-southeast of the islands. This resulted in a considerable area of overlap between the two claims.

In November 1976, Canada and France agreed on interim measures for 1977 to avoid conflicts pending delimitation of their respective areas of jurisdiction around St. Pierre and Miquelon (DOE 1976j). The arrangements applied within ICNAF Subdivision 3Ps, a large triangle covering most of the area off the south coast of Newfoundland. It was agreed that the fishermen of each country could continue fishing without licences in this area for their traditional fisheries and for their ICNAF quotas. Enforcement was based on the ICNAF principle of flag-state enforcement. The fishery officers of each country could board vessels of the other country but would not conduct arrests or prosecutions. If a violation of a regulation were detected, a report would be sent to the authorities of the other country. Fishing by third countries was limited to a single USSR allocation agreed to in ICNAF in 1976 for 1977. From 1978 until 1981 these arrangements were renewed with the additional proviso that no third party fishing would be allowed in 3Ps.

Negotiations to delimit the continental shelf in the context of the 1958 Law of the Sea Convention on the Continental Shelf began in 1967 and continued intermittently thereafter. Attempts to negotiate the boundary intensified following the 1977 establishment of a Canadian 200-mile fishing zone and the French claim to a 200-mile EEZ around St. Pierre and Miquelon. In the negotiations which commenced in 1978, Canada contended that St. Pierre and Miquelon was only entitled to a 12-mile territorial sea. France sought a full EEZ to be determined by the equidistance principle. Various modifications to these positions were considered during the 1978–81 period but basically the two parties remained far apart. By late 1981 the boundary negotiations reached an impasse. Given the divergent positions, the French Foreign Minister proposed in November 1981 that negotiations be suspended for one year to allow time for reflection on both sides. He also proposed that the interim fisheries arrangements be continued for this period to avoid conflict. The alternative was to place the matter before some form of arbitration tribunal. Canada agreed to suspend negotiations until May 1982. Canada did not, however, agree to extend the interim fisheries arrangements, which were allowed to lapse. This was done to induce the French to modify their 200-mile claim.

From 1977 to the early 1980's, Canada and France agreed that the major resource which occurred partly in the disputed area, the 3Ps cod stock, would be shared with 15.6% going to France and 84.4% to Canada. This formula originated in ICNAF in 1976 as part of the transitional arrangements leading to coastal state management.

In 1977, Canada and France used this proportional split to divide the 1978 3Ps cod TAC 84.1% to Canada, 15.6% to France and 0.3% to others. France, either on its own or for a period through the EC, confirmed the 15.6% approach from 1977 to 1982. While initially there were annual discussions, Canada refused any form of joint management. Instead, it unilaterally established TACs and Canadian and French quotas until 1982 (Atlantic Groundfish Management Plans, various years). The EC, acting on France's behalf, established a quota for France in 1982 which was 15.6% of the TAC established by Canada, thus confirming France's acceptance of the formula which had been in effect from 1977.

4.4 Conflict Over Application of the 1972 Agreement

Meanwhile, during the late 1970's conflict arose between Canada and France about France's entitlement in the Gulf of St. Lawrence under the 1972 Agreement. The traditional French fishery in the Gulf involved two major cod stocks. These were the northern Gulf cod in Divisions 4RS and 3Pn and the southern Gulf cod in Divisions 4T and 4Vn (January–April)(see Fig. 12-6). During 1972 and 1973 there was no catch regulation for these stocks. For 1974 ICNAF established TACs and national quotas for 4T and 4Vn (January – April cod) but not for 4RS and 3Pn cod. It was not until 1977, following the emergency conservation measures for Gulf redfish in 1976 and the development of the first Annual Groundfish Plan, that Canada began to regulate the French Gulf fishery pursuant to the 1972 Agreement. As a result of discussions with France in 1976, Canada agreed to allocate France 30% of the 4Rs3Pn cod TAC, a figure which was based on the ratio of catches by France from the total stock in the 5-year period from 1967 to 1971 (Canada-France 1977). Article 5 of the 1972 Agreement required that Metropolitan French fishing vessels not substantially increase their fishing effort or direct their effort to species other than those they had traditionally exploited in the 5 years immediately preceding the 1972 Agreement.

The 5-year reference period was applied globally to the French fleet, even though Article 5 had no direct application to the SPM fleet. Canada left France to divide its quota between Metropolitan French and SPM vessels.

In subsequent annual discussions, France argued that "equal footing" implied guaranteed shares with France benefitting from a proportionate share of the phase-out of other countries from the Gulf. Canada maintained that "equal footing" dealt only with regulatory conditions and did not apply to the sharing of allocations. In 1978, Canada abandoned its approach of the previous two years, whereby it gave the French 30% of the 4Rs3Pn cod TAC. Canada argued that a fixed percentage entitlement was inconsistent with the ceiling on fishing effort mentioned in Article 5 (Canada-France 1978).

In 1980, Canada proposed that the French entitlement be restricted to a portion of the large-vessel catch rather than the TAC. The French opposed this approach.

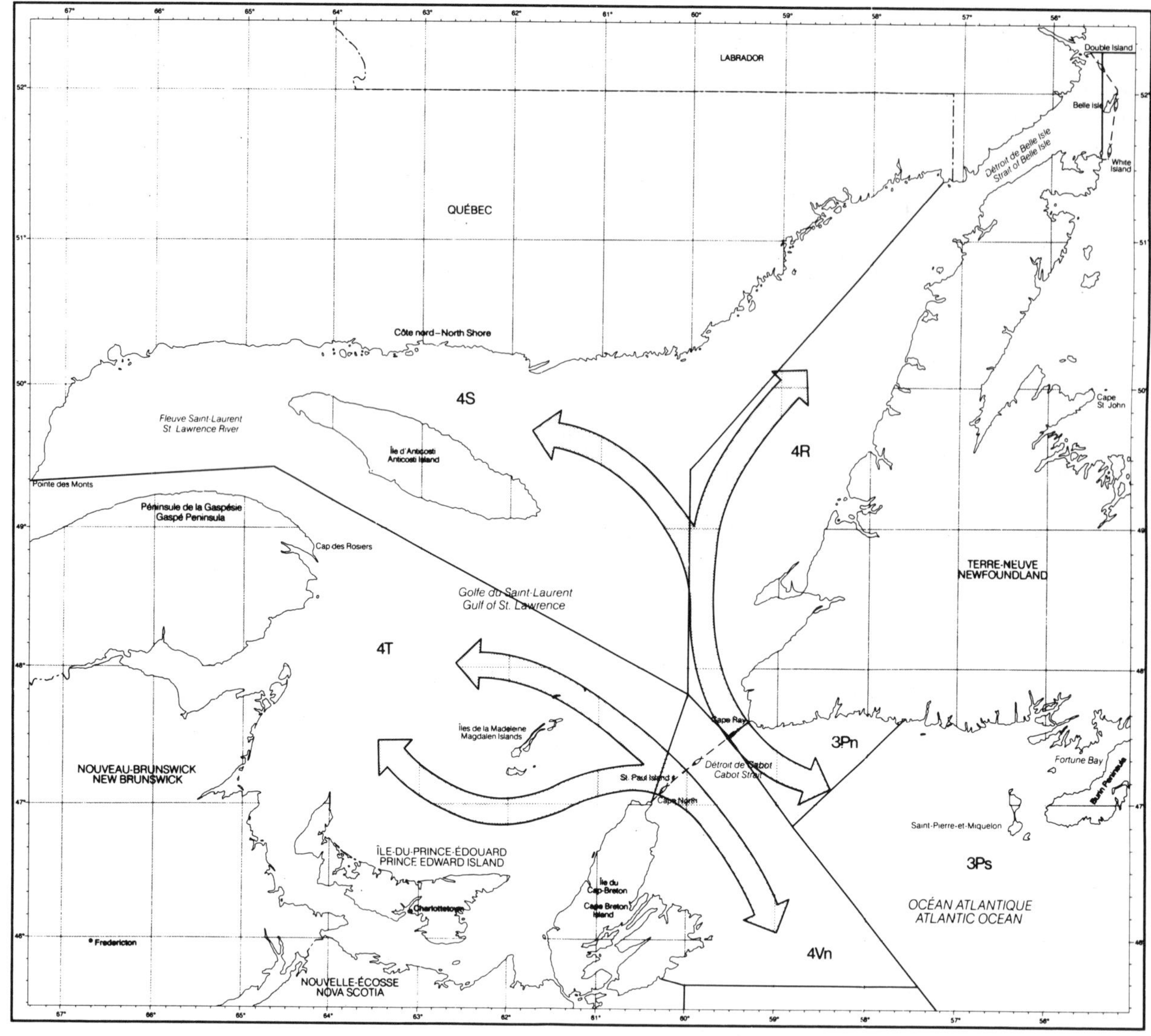

FIG. 12-6. Distribution of northern and southern Gulf cod stocks (from Canada 1985d).

The French portion of the Gulf cod TACs had been reduced from 27% in 1977 to 21% in 1978 and 19% for 1979. This occurred because the increase in the French quotas was minimized while the TACs for the Gulf cod stocks were increasing. The French expressed strong dissatisfaction with Canada's application of the 1972 Agreement. In 1979, they suggested invoking the dispute settlement procedures of the Agreement (Article 10).

This led to discussions in 1980 to review differences of interpretation of the treaty. In October 1980, these differences were resolved by an agreement (a Procès-Verbal) which stated that:

1. Pursuant to Articles 3 and 4 of the 1972 Agreement
   a. French vessels would be granted, from 1981 to 1986 inclusive, annual allocations of 19,960 tons of 4RS3Pn cod and 540 tons of 4T4Vn cod;
   b. These allocations would not be reduced except in proportion to any reduction in the 4RS3Pn TAC below 75,000 tons and the 4T4Vn TAC below 54,000 tons;
   c. Each year Canada could determine the portion of the French allocation of 4RS3Pn cod to be taken in 4TVn up to a maximum of 8,000 tons;

d. If the 4TVn allocation exceeded 540 tons in any year, the entire allocation could be fished outside the Gulf in 4Vn.

2. The parties agreed that Article 2 of the 1972 Agreement assured French vessels of a right to fish in the Atlantic coast zone of extended Canadian fisheries jurisdiction and Canadian vessels of a right to fish in the waters under French jurisdiction off St. Pierre and Miquelon. These rights would be exercised on the basis of quotas.
3. The parties agreed that, "In view of the geographic location of St. Pierre and Miquelon, particular provisions inspired by principles of good neighbourliness, establish certain reciprocal rights. These rights are set out in Article 4 of the Agreement and remain effective beyond 1986" (Canada-France 1980).

This agreement provided France with stable cod allocations in the Gulf up to 1986 at 20,500 tons. This appeared to settle matters in the Gulf pending the phase-out of the Metropolitan French fleet on May 15, 1986.

4.5 Resumption of Boundary Negotiations

Meanwhile, little progress was made on the boundary and fisheries arrangements issues for the area off St. Pierre and Miquelon. Following the decision in the fall of 1981 to suspend negotiations for one year, the interim fisheries (enforcement) arrangements were allowed to lapse. Over the next year enforcement incidents raised the profile of the issue on both sides. In November 1981, officers from a French patrol vessel boarded two Canadian fishing vessels 100 miles off the coast of St. Pierre and Miquelon. This boarding was widely perceived as an attempt to force Canada to make concessions in the boundary dispute. A further incident occurred in March, 1982, when Canadian fishery officers were inspecting two French vessels fishing the French Gulf cod quota in Division 4Vn, clearly in Canadian waters outside the disputed zone. The officers ordered the vessels into a Canadian port for further inspection of suspected violations of fisheries regulations. When the captains refused, the officers attempted to arrest them. The vessel captains resisted arrest and proceeded to St. Pierre with the Canadian officers on board. Following representations to the French authorities, the vessels proceeded to Halifax and the captains were charged with underreporting their catch.

These incidents sparked lively public debate both in Canada and on St. Pierre and Miquelon. Officials on St. Pierre requested the French Government to dispatch a gunboat to the area to protect French fishing fleets in the disputed zone.

During 1983 and 1984 relations deteriorated. In August 1983, France sent a seismic vessel to carry out research in the disputed area. A French naval vessel was also dispatched to patrol these waters for a brief period. During the autumn of 1983, Canadian officers carried out a number of inspections of French vessels. The vessels' captains permitted these boardings but French authorities protested each incident. Canadian authorities instructed Canadian captains to refuse boardings unless threatened with force. In January 1984, Canada and France agreed to forego inspections in the disputed area pending further negotiations in May 1984.

The May 1984 discussions made no progress toward resolving the dispute. There were several further boarding incidents in 1984. In January 1985, matters heated up considerably when three Nova Scotia scallop vessels were boarded by French fishery officers while they were tied up in port at St. Pierre. Canadian fishermen were again instructed to resist boardings outside the 12-mile territorial sea around SPM, unless threatened with force.

4.6 The La Bretagne Arbitration

In January 1985, matters took a new twist when the French trawler *La Bretagne*, registered in St. Pierre and Miquelon, was denied permission to operate as a factory freezer trawler in the Gulf. On January 4, 1985 the vessel had been authorized to fish outside the Gulf. By an amendment to its fishing licence on January 24, the *La Bretagne* was authorized to fish in the Gulf subject to the following conditions noted on the licence:

> "In accordance with the current Canadian prohibition against the filleting of traditional groundfish species at sea by Canadian vessels, the La Bretagne is permitted to process groundfish species in the Gulf of St. Lawrence to the headed and gutted form only."

On January 29, France protested the attachment of this condition to *La Bretagne's* licence, on the grounds that it violated the terms of the 1972 Agreement. Canada denied this. In February 1985, it was agreed to refer the dispute to arbitration in accordance with Article 10 of the 1972 Agreement.

In its Memorial (legal argument) submitted to the Arbitral Tribunal, Canada argued that it had the exclusive right under international law to regulate the fishery in the Gulf of St. Lawrence. This right was qualified only to the extent provided for in the 1972 Agreement. Canada contended that the primary purpose of the 1972 Agreement was to phase French vessels out of the Gulf fishery, as had been done with vessels of other countries fishing there. Canada further contended that, under the 1972 Agreement, it granted France an exceptional right to maintain a limited SPM trawler fishery in the Gulf in recognition that SPM was a "neighbour". The essence of the Canadian case was that Article 4(6) provided for the continuation of fishing in the Gulf by SPM trawlers on an "equal footing" with Canadian trawlers.

Canada further asserted that Article 4(6) limited the nature, number and length of the SPM trawlers that could "continue to fish" in the Gulf:

> "As Canadian vessels are prohibited from filleting at sea in the Gulf, Saint-Pierre-et-Miquelon registered vessels must abide by the same restrictions. Any other interpretation would defeat the express terms of Article 4(6) by putting Saint-Pierre-et-Miquelon registered vessels in a preferred position in the prosecution of the Gulf fishery and not on an "equal footing" with Canadian vessels" (Canada 1985b).

Finally, Canada argued that the development of an SPM factory freezer trawler fleet in the Gulf could transform the SPM fishery into something that Canada and France had not intended when they concluded the 1972 Agreement. Canadians feared that "in effect, France would bring back into the Gulf under the cloak of Article 4(6) the very fishery that is to be excluded under Article 3." If permitted to operate factory freezer trawlers from SPM in the Gulf, France could replace the original three wetfish trawlers with 10 FFTs and use this fleet as the basis for a claim to retain the 20,560 ton cod quota it had been allocated during 1980–86.

The arbitral tribunal decided by two votes to one that the 1972 Agreement did not entitle Canada to forbid the French trawlers registered in Saint-Pierre and Miquelon to fillet their catch in the Gulf of St. Lawrence.

France had argued that the term "fishing regulations" did not include regulations respecting processing at sea. The Tribunal concluded that "the authorities of one of the Parties may not use fishing regulations to subject the other Party's vessels to regulations unconnected with the purpose of fishing." The Tribunal further observed:

> "It is thus clear... that when the 1972 Agreement was concluded there was no regulation of fish processing in Canadian law.... One cannot therefore presume that, as Canada insists, the French Government could know in 1972 that fish processing on board foreign ships was covered by Canadian fishing regulations. Therefore, when the Parties used the term "Canadian fisheries regulations" they could hardly have had in mind at that time regulations covering fish processing on board fishing vessels."

The Tribunal also observed that:

> "When in the exercise of the full sovereignty to which it is entitled over its own nationals, Canada forbids them to engage in any given activity this does not entitle it to apply a similar provision in its fishing zone to French nationals enjoying fishing rights there that are recognized by treaty, for it cannot invoke its full sovereignty in a zone where the jurisdiction it exercises is only functional" (La Bretagne Arbitration 1986).

In a dissenting opinion, the Canadian nominee to the Tribunal, Donat Pharand, disagreed with the majority on the scope of "Canadian fishery regulations" envisaged by Article 6, the "equal footing" clause in Article 4(6) and the "arrangement between neighbours" which "characterizes the whole of Article 4." Pharand pointed out that in the Fisheries Case of 1910, a five member Tribunal concluded that processing was already subject to the regulatory authority of the coastal State in the same way as the liberty to fish itself.

Pharand also argued that the Tribunal's decision was contrary to the "equal footing clause" and could result in a preferential treatment of St. Pierre and Miquelon vessels in certain circumstances.

In an incisive critique of the Tribunal Award, American lawyer William T. Burke (1988) observed:

> "The decision by the Tribunal in the *La Bretagne* case has little substance that makes it worthy of consideration or adoption.... The opinion and underlying rationale are flawed, deliver general pronouncements which raise serious questions, and reach conclusions unsupported by international law.... Whatever the reasons for their decision, if the narrow and rigid perceptions of fishery management and law in this opinion were typical of decision-makers generally, there is cause for apprehension about the success of instituting modern fishery conservation and management measures around the globe when foreign

> fishing activities are affected. If these perceptions are more widely adopted, important interests of coastal states could be seriously harmed or destroyed entirely. These perceptions may lead to confusion or misunderstanding about the economic social and political goals and methods of fishery management and about a coastal state's authority to adopt appropriate regulations."

The "flawed and inexplicable" *La Bretagne* decision came at a particularly awkward time for Canada. It came in the midst of renewed attention to the boundary negotiations and fish quota negotiations relating to the phase-out of the Metropolitan France fishing fleet from the Gulf as of May 15, 1986. One consequence was that Canada became extremely wary of any suggestion by France to invoke again the arbitration process under Article 10 of the 1972 Agreement.

4.7 France Increases Fishing Effort in 3Ps and Seeks Transfer of Gulf Quotas Outside the Gulf

In addition to the original five (and recently six) trawlers registered in SPM entitled to fish in the Gulf, France had been normally sending seven or eight factory trawlers from Metropolitan France. These were to fish 17,000 tons of cod in the Gulf (the French share in 1980 minus an amount for SPM trawlers). These French vessels also fished cod in 3Ps as well as part of the allocations Canada had been providing to the EC off Newfoundland under the LTA. The French share under the LTA was 1,545 tons of 2J3KL cod, 200 tons of 2GH cod, and 2,400 tons of squid in Subareas 3 and 4. French allocations in the Canadian zone also included small allocations of transboundary stocks managed by NAFO but caught inside the Canadian 200-mile zone.

For several years (up to 1982) France had adhered to the 15.6% of the 3Ps cod TAC it was assigned by Canada. In the early 1980's, however, it departed from this practice. In 1983, the EC again set the French quota in 3Ps at 5,170 tons or 15.6% of the TAC established by Canada. However, France from 1982 onward began increasing its share of the overall cod catch in 3Ps. Presumably the aim was to increase pressure on Canada for a favourable boundary and fish quota deal. The French cod catch in 3Ps increased almost fourfold from 1982 to 1986 (Table 12-2). France greatly exceeded its own quotas from 1983 to 1986. This escalation by France was clearly contrary to the spirit of restraint called for by international law in the case of boundary disputes.

Canadian industry and negotiators had always expected that, as the 1986 phase-out from the Gulf approached, France would seek other opportunities for the Metropolitan French fleet. Hence the French would be more inclined to settle for a minimal zone around SPM. However, they did not behave as anticipated.

A series of negotiations in 1985 showed little chance for a negotiated settlement of the boundary issue. The Canadian attempt to use the May, 1986, deadline for the Metropolitan French fleet in the Gulf as leverage for a minimal zone around SPM had not shown much promise. The French sought a transfer of the Gulf quotas for Metropolitan French vessels (about 18,000 tons) outside the Gulf after 1986, using Article 2 of the 1972 Agreement as the rationale. The French also wanted the right for up to 10 SPM trawlers to fish up to capacity in the Gulf in perpetuity. In addition, they claimed an increased quota in 3Ps and were in fact setting the scene for this by increasing their catches.

Negotiations on the boundary continued at sessions in May and October 1985. But there was little or no progress towards a boundary settlement. In February 1986, Prime Ministers Mulroney and Fabius said that if no agreement was achieved at one last meeting, the dispute would be referred to international judicial settlement. At this meeting it became clear that there was no possibility of a negotiated boundary settlement.

Thus, from a Canadian perspective, the issue was one of negotiating the terms for referral of the boundary dispute to an international tribunal and for interim fisheries arrangements, pending a boundary decision.

The French were by now making unreasonable fish quota demands, from a Canadian perspective. They had increased their cod catch in 3Ps to 26,000 tons. They were seeking for the SPM fleet in the Gulf 12,000 tons (1,200 tons per vessel for each of 10 trawlers). For the Metropolitan French fleet they sought an equivalent amount outside the Gulf to what they had been catching in the Gulf, approximately 18,000 tons. This meant that overall the French wanted at least 50,000 tons of fish, almost exclusively cod. This compared with the approximately 30,000 tons they had been taking in the Canadian zone and off SPM prior to 1983. They were accepting no reduction of the latter figure despite the termination of their Metropolitan Gulf fishery. Furthermore, the French wanted a substantial quantity of 2J3KL cod as part of the "transfer" of Metropolitan French vessels from the Gulf.

Canada would not consider new quotas for France in the context of the 1972 Agreement pending settlement of the boundary around St. Pierre and Miquelon. It considered that the fish that France was harvesting in the disputed zone satisfied any Canadian obligation under Article 2 of the Agreement.

In November, Fisheries Minister Tom Siddon noted that Canada had offered the French "substantial allocations" of fish in Canadian waters on an interim basis, pending a boundary delimitation. "The allocations we have offered do not, however, include fish to replace the metropolitan catch in the Gulf of St. Lawrence, which is now terminated" (DFO 1986f).

4.8 The January 1987 Agreement and its Aftermath

On January 23 and 24, 1987, Canadian and French officials met in Paris. Those discussions resulted in the signature on January 24 of an Agreed Record which included the following elements:

1. Recognition that the two countries had not yet reached agreement regarding the application of the 1972 treaty in 1987. Canada had unilaterally established fishing quotas for France in December 1986 and January 1987. France had protested such action in January 1987.
2. Agreement that the representatives of the two Parties would meet before March 15, 1987, to begin negotiations on:
   a) A Special Agreement (Compromis) to submit to compulsory third party settlement the boundary dispute;
   b) A Procès-Verbal establishing the annual fishing quotas for French vessels in Canadian waters for 1988 through 1991. *These quotas would include cod quotas in NAFO Divisions 2J3KL* (Emphasis mine).
3. The negotiations were to be pursued to a successful conclusion as quickly as possible. The Procès-Verbal was, if possible, to be finalized before September 30, 1987 and the Special Agreement (Compromis) before December 31, 1987. The two agreements were to come into force on the same date. Each would be contingent on the other.
4. Canadian and French scientists would meet shortly to prepare a report on the state of the cod stock and the annual total allowable catch limits in 3Ps in conformity with specific terms of reference.
5. The two parties would maintain (pending a boundary decision) the arrangement in the form of diplomatic notes from June, 1984, whereby each Party agreed to abstain from regulating vessels flying the flag of the other Party in the disputed zone (Canada-France 1987a).

This agreement was reached between officials in Paris, with the blessing of both Governments, but in the absence of Canadian provincial government and industry representatives. Once the Agreement became public in Canada, it touched off a storm of protest. There was intense parliamentary and national media attention to the Canada-France dispute and the so-called "secret deal" whereby Canada promised France quotas of northern cod after 1987.

The Government of Newfoundland adamantly opposed providing even one ton of northern cod to the French. The Atlantic fishing industry was outraged by the decision to exclude advisers from the meeting between the two countries which led to this agreement.

Premier Brian Peckford of Newfoundland termed the January 1987 agreement a "secret sell-out of northern cod". He stated:

> "Every Newfoundlander and Labradorian and every fair minded Canadian must view the recent Canada/France Fisheries Agreement to be a secret betrayal of our legacy and a secret betrayal of our very right to have a meaningful voice in decisions that affect our very livelihood" (Newfoundland 1987).

In an attempt to "patch-up the quarrel", Deputy Prime Minister Mazankowski apologized to Peckford for the breakdown in communications (House of Commons Debates, January 30, 1987).

However, Premier Peckford demanded that Ottawa scrap the agreement with France and risk a diplomatic rupture to enforce its boundary claims off Newfoundland (*The Montreal Gazette*, February 4, 1987). Peckford persuaded eight of his fellow Premiers to meet in Toronto to discuss the impact of this agreement upon federal-provincial relations. The Premiers called upon the Federal Government to "review" the Interim Agreement with France which they considered to be inconsistent with the basic principles of federal-provincial relations. The Premiers said that Provinces must have confidence that "they will be consulted and involved in decisions which affect their vital interests" (*The Evening Telegram*, February 10, 1987).

At a meeting of the Canada-France Fisheries Advisory Committee, Canadian negotiators emphasized that no specific amount of northern cod had been offered to the French. The real negotiations were yet to take place.

In its response to the Newfoundland government's attack following the January agreement, the Canadian government in a brochure entitled "Canada-France, Our Fish, Our Boundaries" gave this explanation of events:

> "During the week of January 13–16, in Ottawa, Canada and France held negotiations, with Newfoundland, Nova Scotia and the fishing industry representatives partici-

pating. The provincial and industry advisers supported offers to France which included cod in the northern zone 2GH. This cod is a surplus resource" (Canada 1987a).

On January 16, negotiations broke down because France was not satisfied with these proposals.

"As a result of renewed contact and new instructions, Canadian officials travelled to Paris and signed an interim agreement on January 24. This was done without the participation of the Government of Newfoundland and Labrador and without the fishing industry and unions. This was inexcusable because the Government of Newfoundland and Labrador had been promised participation in these negotiations. The Government of Canada has apologized to the Government of Newfoundland and Labrador and will ensure that the Province, fishing industry and unions are fully involved to the end of these negotiations" (Canada 1987a).

The Prime Minister sent a letter to the Honourable John Crosbie, Newfoundland's Minister in the Federal Cabinet, which said that:

1. Mr. Crosbie would take a lead role with the Secretary of State for External Affairs and the Minister of Fisheries in all future aspects of the negotiations;
2. The outcome of the process must be in the best interests of the people of Newfoundland and Labrador and Atlantic Canada or there would be no deal;
3. The participation of the Government of Newfoundland and Labrador, the fishing industry and union representatives was to be continuous;
4. Any further agreement to allocate cod to France in Canadian waters would depend on:
   a) France accepting binding arbitration of the boundary dispute; and
   b) Agreement on a reduction of French fisheries in the disputed zone in 3Ps during arbitration (*The Evening Telegram*, St. John's, Newfoundland, February 12, 1987).

On February 12, 1987, Ministers Siddon and Crosbie stated that Canadian fisheries surveillance authorities were monitoring catches by French vessels in 3Ps. Preliminary information indicated that France was fishing above its cod quota of 6,400 tons. The Ministers warned that, if France overfished its quota, port privileges for the French fleet would be withdrawn (DFO 1987j).

Fisheries Minister Siddon met with the Atlantic Council of Fisheries Ministers in Moncton, on February 16, to discuss the Canada-France negotiations. At that meeting the federal government and the five eastern provinces agreed that talks with France should continue. These talks should find a way to send the boundary dispute to third party arbitration and to set interim fishing quotas between 1988 and 1991. Newfoundland agreed to continue to participate in all meetings at the Ministerial and official level concerning Canada/France maritime boundary and fisheries issues. Premier Peckford remained adamant, however, that there must be no allocation of northern cod to France.

In remarks following the meeting of Atlantic Fisheries Ministers, Fisheries Minister Siddon refused to back down on the northern cod issue, saying that northern cod had not been taken off the negotiating table (*The Chronicle Herald*, February 17, 1987).

On March 17, 1987, Ministers Siddon and Crosbie announced "new measures to limit French overfishing off the south coast of Newfoundland." The French fisheries in 3Ps had been declared closed and Canadian ports declared closed to French fishing vessels (Canada 1987b).

4.9 Negotiations on Boundary Arbitration and Interim Quota Arrangements

Two sets of negotiations were undertaken in 1987. One aimed at having a boundary reference treaty or Compromis concluded before December 31, 1987. The other sought a Temporary Fisheries Agreement for 1988–91 by September 30, 1987.

France would not finalize the Compromis unless it could obtain satisfactory interim fisheries arrangements. During the spring of 1987, France realized that Canada would not accept its interpretation of the 1972 Agreement as the basis for a quota package. France was also dissatisfied with closure of Canadian ports and Canada's enforcement activities in 3Ps. France cancelled scheduled June 24–25 fisheries negotiations and demanded arbitration of quotas. Canada said no to arbitration and asked France to return to the negotiating table.

Just prior to this, Canadian and French authorities received the report of the bilateral meeting of scientists on the state of the 3Ps cod stock. This stock had been reduced from a biomass of ages 3 and older of about 230,000 tons in 1959 to about 70,000 tons in 1975. The assessment indicated that it had recovered since 1982 because of two periods of strong year-classes (1972–74 and 1980–84). The overall TAC (combined TACs of Canada and France) for 1987 was 60,000 tons. The scientific projections indicated that fishing at $F_{0.1}$ in 1988 and 1989 would produce catches of 37,000 and 45,000 tons respectively. The catches at F_{max} in 1988 and 1989 would be 62,000 and 69,000 tons respectively (Canada-France 1987b).

For Canadians, the escalation in French catches had pushed the total catch to unacceptably high levels. This had been cushioned by an unusual increase in spawning and survival during the years before the French "overfishing" began. France, on the other hand, contended that the stock was healthy and could sustain the existing level of catch.

Meanwhile, Canada told France that its pre-conditions to resume the fisheries negotiations would not be met. Canada re-affirmed its commitment to the process set forth in the January 24 agreement. In a meeting between Prime Minister Mulroney and French Premier Chirac in late August, France agreed to resume negotiations. Before negotiations resumed in September, Canada appointed a new Chief Negotiator, Mr. Yves Fortier, in an attempt to break the stalemate in the negotiations.

Federal-provincial relations were again strained in September when Newfoundland government representatives walked out of the negotiations between Canada and France. This happened when the federal government placed an offer of non-surplus northern cod on the negotiating table, as promised in the controversial January 1987 agreement. Premier Peckford argued vigorously against offering any northern cod to the French. Unless the Canadian government changed its mind, Newfoundland representatives would not participate in the discussion. Meanwhile, representatives of the Newfoundland processors and the NFFAWU continued to participate. (Newfoundland government representatives later returned to the Canada-France Advisory Committee but stayed away from negotiating sessions where a quota of northern cod was being discussed).

On October 9, France again broke off negotiations and formally requested arbitration of quotas under Article 10 of the 1972 Agreement. Canada again refused to proceed to arbitration and informed France that, because of the level of French overfishing in 3Ps, France was not entitled to quotas in Canadian waters in 1988.

On November 6, negotiations resumed on the compromis but not on fish quotas. These continued on November 30. France, however, formally linked agreement on a compromis to resolution of the quota dispute. Canada offered no quotas to France for 1988. France formally accused Canada of breaching the 1972 Agreement. Canada responded, advancing legal arguments why it was not in breach and also refusing to go to arbitration on quotas. Canada also indicated that it might suspend the 1972 Agreement. France announced a cod quota of 26,000 tons for French vessels in the disputed zone.

Informal discussions continued between the Canadian and French negotiators through the winter and spring of 1988. In March 1988, the two Parties began informal negotiations on mediation of the dispute. The lack of quotas in Canadian waters outside the disputed zone increased the pressure on France to resolve the dispute. Catches in the disputed zone were poor in the early months of 1988. Thus France had lost the advantage it had held in 1987 of being able to add its catch in the disputed zone to French quotas in Canadian waters.

The informal discussions between negotiators resulted on April 30, 1988, in an agreement providing for mediation to assist direct talks between the two countries. The mediator was to be appointed jointly by May 7, 1988. Otherwise, the Secretary General of the UN would be asked to appoint a mutually acceptable mediator.

The deadline for the naming of a mediator passed without a mediator being named because France on May 5, 1988, arrested a Canadian vessel, the *Maritimer*, for fishing without a licence in French territorial waters. This occurred while Canadian negotiator Yves Fortier was en route to Paris. Inshore vessels (less than 65 feet long) from Newfoundland and SPM had traditionally fished in the waters of the other country without licences. In March 1987, France had adopted new fisheries regulations which required that all domestic and foreign vessels "trawling, dragging or setting nets" be licensed. Prior to arrest, however, France had given no notice it would apply these regulations to inshore vessels fishing in the territorial waters of SPM. Canadian regulations required all foreign (including French) vessels to carry Canadian licences. However, these had never been enforced for the smaller vessels from SPM.

In May, France acknowledged that the affected Canadian fishermen had not known about the decision to apply licensing regulations to inshore vessels using certain types of nets in the 12-mile French territorial sea. France, therefore, suspended the application of its licensing requirements for 3 months (Canada 1988b).

The election of a new French government on May 6, 1988, renewed Canada's hopes for resumption of negotiations. France's new Prime Minister Michel Rocard blamed the breakdown in negotiations on the "chauvinistic national policy" of his predecessor. For the first time it appeared that France wanted an overall settlement of the dispute. As a goodwill gesture, the French prosecutor in the *Maritimer* case requested that the charges be dropped. However, the judge declined and convicted the Captain of illegally fishing in French waters, but he did not impose a fine.

Negotiations were resumed in June. Through a series of negotiation sessions on the Compromis and interim fisheries arrangements, the two sides moved closer to resolving the dispute.

Matters were complicated at the end of June by the release of the NAFO Scientific Council Report on 3Ps cod. Unlike the earlier "optimistic" bilateral assessment of 1987, this report indicated that the level of fishing mortality since 1984 had been above 0.5. This was considerably above newly calculated values for $F_{0.1}$ of 0.15 and F_{max} of 0.27. This study indicated that fishing must be reduced to rebuild the 3Ps cod stock. The Council advised that the TAC for 1989 at $F_{0.1}$ would be significantly lower than the combined TACs for 1988 (20,500 tons compared with 60,000 tons) (NAFO 1988e). Commenting on this report on July 7, Ministers Siddon and Crosbie stated that reducing French catches of cod in 3Ps continued to be a fundamental Canadian objective (Canada 1988c).

Over the summer, attention focused on three major quota issues:

1. The appropriate TAC and shares for Canada and France of 3Ps cod;
2. The entitlement for SPM trawlers in the Gulf of St. Lawrence; and
3. The package of quotas for Metropolitan French trawlers in Canadian waters outside the Gulf and, more specifically, the amount of northern cod to be allocated to France.

Informal discussions were suspended because France challenged the NAFO Scientific Council report. Progress had been made on all the quota issues except for 3Ps cod. Given the latest scientific advice, Canada and France were unlikely to agree on an "appropriate" roll-back of the French fishery in this area. Thus the issue of mediation surfaced again. France was eager to pursue this because it now wanted the matter resolved in time to receive quotas in the Canadian zone for 1989. The Canadian strategy of containing the French fleet to the disputed zone in 1988 had worked because French catches there had dropped off (to 12,000 tons compared with 26,000 tons in 1987). The total French catch for 1988 would be considerably less than what France had fished in 1987.

4.10 Mediation and Agreement

On September 21, 1988, Ministers Clark, Crosbie and Siddon announced that Canada would proceed with non-binding mediation to resolve the fish quotas dispute with France. The mediation process would help the two countries reach an acceptable solution on interim fish quotas for French vessels off Canada's Atlantic coast, pending international adjudication of the boundary dispute (Canada 1988d).

Enrique Iglesias, President of the Inter-American Development Bank and a former Foreign Minister of Uruguay, was named as mediator. He was to try to facilitate agreement.

The mediator's role was to attempt to bridge the gap between the conflicting positions of the two Parties. To this end he made proposals in an attempt to bring the two parties to common ground. The mediator was given additional time, with his mandate ending on March 23, 1989. His efforts were clearly helpful in facilitating a settlement of the dispute.

On March 30, 1989, Canada and France signed the two agreements which had been envisaged in the January 1987 interim agreement. One agreement established a court of arbitration to delimit the maritime boundary between France and Canada. The court of arbitration had five members, one appointed by Canada, one appointed by France and three appointed by both countries. Mr. Jimenez de Arechaga of Uruguay chaired the Tribunal (Canada-France 1989a).

The second document was a Procès-Verbal containing the agreement relating to fisheries for the years 1989-91. This agreement established fish quotas to be allocated annually to French fishing vessels (Table 12-3). Excluding 3Ps, France was granted 26,450 tons of fish in the Canadian zone. Of this, 11,450 tons was cod and 15,000 other species for which quantities surplus to Canadian requirements had been identified. Since the 2GH cod was still considered to be a surplus stock, the bottom line involved 6,950 tons of non-surplus cod (4,000 tons of Gulf cod for trawlers based in St. Pierre and Miquelon and 2,950 tons from the controversial northern cod stock).

Regarding the northern cod quota, the 2,950 tons for 1989 corresponded to a TAC of 235,000 tons. The Agreement provided for adjusting the figure in proportion to changes in the offshore allocation part of the Canadian TAC for 1990 and 1991, with a proviso that an "upward adjustment shall not raise the annual quota beyond 3,800 tons" (Canada-France 1989b).

To protect the legal position of the two parties, the cod quotas in 3Ps were not covered in the formal agreement (Procès-Verbal) but rather were the subject of an exchange of notes verbales of the same date. Early in 1989, France had unilaterally stated its intention to fish 26,000 tons of cod in the disputed zone in 1989. In accordance with the mediator's suggestions, France reduced its unilateral quota by 10,400 tons to 15,600 tons for 1989. The French agreed to reduce further their unilateral quota to 15,000 tons in 1990 and 14,000 tons in 1991. While a significant reduction from the quotas France had claimed in 1986–88, this amount was still far above what Canada considered to be France's customary share.

Canada agreed to permit French vessels to fish up to 15.6% of the Canadian established TAC throughout

TABLE 12-3. Canada-France Settlement on Quotas (except 3Ps) (March, 1989).

Species	Area	Quota
Cod	4RS, 3Pn	
	4TVn (Jan.–Apr.)	4,000t
	2GH	4,500t
	2J3KL (1989)	2,950t
	2J3KL (1990–1991)	2,950t[1]
Redfish	2 + 3K	2,000t
	4VWX	1,500t
Greenland halibut	2GH	2,000t
	2J3KL	3,000t
Silver hake	4VWX	4,000t
Witch flounder	2J3KL	500t
Squid	3 + 4	2,000t

Source: Canada-France (1989b).

[1]This figure shall be adjusted upward or downward in proportion to changes in the offshore allocation which is part of the Canadian TAC for 1990 and 1991. Upward adjustment shall not raise the annual quota beyond 3,800t.

subdivision 3Ps. Once that quantity was caught, French vessels would be confined to the disputed waters in 3Ps.

By the exchange of notes, the two Parties maintained their respective legal positions while providing a *modus vivendi* to refer the boundary dispute to the special arbitration tribunal.

Minister Crosbie commented on the agreement:

> "We have achieved our major objectives. The settlement provides for a significant rollback of French catches in the disputed zone off the south coast of Newfoundland and for international adjudication of the boundary dispute. After intense negotiations we managed to limit, to the extent possible, the quotas granted to France for the 3 years while the boundary is being decided. Resolving the boundary is an essential step to protecting the fish stocks which are vital for the fishermen of the south coast of Newfoundland" (Canada 1989a).

The Procès-Verbal and exchange of notes also provided additional conservation measures. Canada and France would provide each other regularly with catch information, weekly for cod and monthly for other species. They also agreed not to change fundamentally the intensity, nature or method of fishing for other key species in the disputed zone. Subject to the right of pursuit, they agreed not to enforce regulations for vessels of the other country in the disputed zone. They also agreed to joint inspections.

4.11 The Canada-France Boundary Decision

On June 10, 1992, the court of arbitration for the delimitation of maritime areas between Canada and France rendered its decision. The line drawn by the court is compared with the parties' claims in Fig. 12-2. The court gave France a zone extending 24 miles southwest of the islands and a narrow corridor approximately 10 miles wide extending directly south to a point 200 miles from the islands. The Court relied on geographical factors to delimit the boundary (Canada-France 1992).

The Court decision gave St. Pierre and Miquelon a zone of 3,607 square nautical miles (12,372 km^2). This represented 20.4% of the French claim, 22.4% of St. Pierre Bank and 10.4% of area 3Ps. In 1985, Canada had offered France a zone of 2,484 square nautical miles (8,519 km^2).

The Court concluded that the proposed demarcation would not have "a radical impact on the existing pattern of fishing in the area." A preliminary fisheries impact analysis undertaken by Canada indicated that only 8.8% of 3Ps cod, 2.9% of 3Ps redfish, 9.9% of 3Ps American plaice, 11.3% of 3Ps witch and 8.4% of 3Ps haddock was located within the new St. Pierre and Miguelon zone. Very little of the Canadian groundfish catch had been taken historically within this zone (D. Rivard, Fisheries Research Branch, DFO, Ottawa, pers. comm.).

Only one resource of interest to Canada was affected significantly by the boundary decision. That was Iceland scallops, approximately 49% of which fell within the St. Pierre and Miquelon zone. This, however, represented only a very small portion of the resources of interest to Canada in the area south of Newfoundland.

A significant portion of the area awarded to France was in the deep waters of the Laurentian Channel, which contains virtually no fish resources.

The Canadian government and industry reacted positively to the court decision. Fisheries Minister John Crosbie described the boundary decision as being "in Canada's favour." "All of the key fishing areas for cod and other groundfish, which are critical to the livelihood of fishermen and plant workers off the south coast of Newfoundland, are located within the Canadian zone" Mr. Crosbie stated (DFO 1992d). Also most of the French 3Ps catches in recent years had been taken in waters that were now under exclusive Canadian jurisdiction.

With the boundary decision handed down, Canada and France still faced the question of the future interpretation of the 1972 Agreement. The ambiguous language of the 1972 Agreement regarding the right of French nationals to fish in Canadian waters, which had haunted Canada during the 1980's, now had to be addressed. The package of interim quotas agreed to on March 30, 1989, was to terminate in September, 1992.

Shortly after the boundary decision, negotiations commenced in July 1992 on future arrangements in the transboundary area and the future application of the 1972 Agreement. Canada's position in those negotiations had been strengthened considerably by the relatively small zone awarded to France.

Negotiations during the summer of 1992 failed to produce an agreement. On October 9, 1992, Minister Crosbie announced that Canada would be establishing unilateral quotas for France in the Canadian zone for 1993, to fulfill its obligations under the 1972 Agreement. An impasse resulted when France refused access to its zone by Canadian scallop fishermen. In January, 1993, fishermen and plant workers from St. Pierre and Miquelon conducted a protest fishery in the Canadian zone by two trawlers based in St. Pierre and Miquelon. These were arrested and brought to St. John's. This incident reflected the dissatisfaction of St. Pierre and Miquelon inhabitants with the Arbitration decision and the failure of Canada and France to agree on "satisfactory" quotas for SPM vessels in the Canadian zone (DFO, 1992k).

5.0 THE UNITED STATES

Canada and the USA had cooperated well on the east coast in the period leading up to the 200-mile limit, attempting in ICNAF to reduce the level of third party fishing in the Northwest Atlantic. These endeavours augured well for Canada-USA fisheries cooperation following extension of jurisdiction in 1977. However, by the time Canada extended its fisheries jurisdiction, it had bilateral agreements with all of the major countries fishing off the Canadian coast, except for the USA.

5.1 Overlapping Boundary Claims

The USA, under the Fishery Conservation and Management Act (FCMA) of 1976, extended its jurisdiction to 200 miles on March 1, 1977. These virtually simultaneous extensions of jurisdiction by Canada and the USA resulted in overlapping boundary claims in four areas. These were the Gulf of Maine on the east coast, off Juan de Fuca Strait and inside and off Dixon Entrance on the west coast, and in the Beaufort Sea (Fig. 12-3a, b and c).

The largest and the most significant of the overlapping claims involved the Gulf of Maine. Here the overlap was some 8,648 square nautical miles (29,663 square km), with Canada claiming about 35% of Georges Bank (Fig. 12-3a). The Canadian claim was based upon the principle of equidistance incorporated in the 1958 Convention on the Continental Shelf. The USA, on the other hand, claimed all of Georges Bank, with the original USA boundary claim running along the Northeast (Fundian) Channel. The U.S. claim was based on the "special circumstances" qualifier in the 1958 Convention. For further details on this dispute and its resolution, see section 5.8.

On the west coast, there were two overlapping boundary claims. To the south off the Strait of Juan de Fuca, Canada and the United States both proclaimed boundary lines based on equidistance. The area of overlap was relatively small (Fig. 12-3b). The equidistance line cut through a valuable fishing ground, the Prairie bank, which had been fished by Canadians. The Province of British Columbia argued that the Juan de Fuca Canyon was a natural division between the continental shelves of Canada and the USA and that the area to the north of the Canyon was the natural extension of the shelf off Vancouver Island. This argument was somewhat analogous to the U.S. claim to all of Georges Bank.

The boundary dispute in the Dixon Entrance area can be traced to the 1903 Alaska Boundary Tribunal which interpreted the rights of Canada and the USA under a 1825 treaty between Great Britain and Russia. The Tribunal established the so-called 'A–B line', running from Cape Muzon, Alaska (point A) almost due east to point B at the mouth of the Portland Canal. Canada since that time has regarded the A–B line as the international maritime boundary, with all the waters of Dixon Entrance and Hecate Strait being Canadian waters. This interpretation means that at two points the USA would be deprived of a territorial sea of 3 miles. The United States has maintained that the A–B line divides only the land territories of the two countries and not their maritime waters. In 1977, the USA published a boundary claim based on equidistance. Outside the Dixon Entrance both Canada and the USA published lines based on the equidistance argument. The area of overlap in the Dixon Entrance area was about 700 square nautical miles (2,500 square km) (Fig. 12-3c).

In the Beaufort Sea Canada's boundary claim again originated in an 1825 treaty between Great Britain and Russia. The argument was similar to that for Dixon Entrance, with Canada projecting the land boundary, the 141st meridian of west longitude, as the maritime boundary line. The U.S. claim was based on the equidistance argument. Canada used the 141st meridian west as the boundary of Canadian waters for issuing oil exploration permits and used it as the boundary in the 1970 Arctic Waters Pollution Prevention legislation (Fig. 12-3b).

Overall, Canada maintained that the maritime boundaries should be based on equidistance except where there was an applicable treaty, e.g. Dixon Entrance and the Beaufort Sea (Wang 1981). The USA based its claims on equidistance in all cases except the Gulf of Maine. There it cited "special circumstances."

5.2 Interim Fisheries Agreements

Previous extensions of fisheries jurisdiction from 3 to 12 miles had had little effect on the activities of Canadian and U.S. fishermen off each other's coasts. However, each of the new 200-mile zones encompassed areas in which fishermen from the other country had fished for centuries. An interim reciprocal fisheries agreement for 1977 was submitted to Congress by President Jimmy Carter one day before the U.S. 200 mile zone took effect on March 1. This agreement replaced a 1973 reciprocal fishing agreement which had granted fishing privileges to the nationals of both countries "in certain areas off the coasts of the United States and Canada."

In the 1977 Agreement the United States and Canada agreed to permit fishermen from the other country to continue fishing in the newly proclaimed 200-mile zones. The Agreement provided that "Fishing by nationals and vessels of each party in the zone of the other shall continue in accordance with existing patterns, with no expansion of effort nor initiation of new fisheries" (Canada-USA 1977). On the east coast, the two countries agreed to apply the 1977 ICNAF allocations, even though the USA was no longer a member of ICNAF. Fishing for herring "by nations and vessels of one party in the zone of the other" was restricted to the area beyond 12 nautical miles from the coast. On the west coast, catch quotas for U.S. and Canadian fishermen were set for ocean perch, black cod and shrimp. Halibut fishing was to continue in accordance with the recommendations and regulations of the International Pacific Halibut Commission. There were also a number of temporary measures respecting fishing for Pacific salmon. These were to allow more time to negotiate a comprehensive Pacific salmon treaty. The 1977 Agreement waived foreign licensing requirements. Also, to avoid prejudice to the conflicting boundary claims, it was agreed that 'flag state' enforcement would apply in the boundary areas. That is, each country was responsible for enforcing regulations against its own vessels.

5.3 Attempts to Negotiate a Comprehensive Agreement

Wang (1981) described the 1977 Agreement as "a stop-gap measure, valid until the end of 1977, designed to maintain the status quo." Both Parties recognized the need for a more comprehensive agreement. Accordingly, in August 1977, President Carter and Prime Minister Trudeau appointed special negotiators to negotiate a comprehensive agreement. This agreement was to deal with:

"1. Maritime boundaries delimitation,
"2. Complementary fishery and hydrocarbon resource arrangements as appropriate, and
"3. Such other related matters as the two governments may decide" (Canada 1977c).

The Canadian special negotiator was Mr. Marcel Cadieux, Canadian Ambassador to the United States and former Under-Secretary of State for External Affairs. The U.S. negotiator, Mr. Lloyd Cutler, was a well-known Washington lawyer, who subsequently became the White House counsel. In the negotiations that ensued, each special negotiator was assisted by government officials and industry advisers. Following a series of initial discussions, on October 15, 1977, the special negotiators submitted a Joint Report to the two governments. The report concluded Phase I of the negotiations. It recommended principles to resolve the fisheries and hydrocarbon issues, to facilitate progress on boundary delimitation during Phase II (External Affairs 1977).

The two sides had agreed on a text of "Proposed Principles for a Joint Fisheries Commission to be Established by Convention between Canada and the United States." This included provision for three categories of stocks to be managed differently:

- Transboundary stocks to be subject to joint management by a Fisheries Commission;
- Stocks considered appropriate for jointly agreed management as a unit, with management based on proposals submitted by the country with the primary interest;
- Stocks occurring clearly off the coasts of only one country, which would be managed by the coastal state with consultations within the Fisheries Commission.

Similarly, principles were agreed for sharing hydrocarbon resources in the boundary areas.

These basic principles for a comprehensive agreement were approved by the two governments and the Special Negotiators were instructed to continue their negotiations and develop the detailed terms of a settlement. The approach was to design resource access and allocation arrangements which would be considered as fair no matter where the boundary was. The conflicting boundary claims, however, gave rise to different perceptions of what was fair and reasonable in fisheries entitlements and shared access zones for hydrocarbon resources.

In late November 1977, the two negotiators requested an extension of their mandate to the end of January 1978. There would have to be further interim arrangements to allow reciprocal fishing pending completion of the negotiations. On March 28, 1978, in another joint report, the Special Negotiators indicated that some difficult problems remained to be resolved. The governments postponed formal negotiation on a long term agreement until the summer. Meanwhile, attention focused on the need for a renewal of interim arrangements for reciprocal fishing in 1978.

Problems had arisen in implementing the 1977 Agreement. This interim arrangement had been an executive agreement and hence had not been submitted to the U.S. Senate for approval. Under the U.S. Fishery Conservation and Management Act, the U.S. Regional Councils could establish regulations inconsistent with the Canada-USA interim agreement.

5.4 The 1978 Interim Agreement and its Suspension

On April 11, 1978 Ambassador Cadieux and Mr. Cutler signed an Exchange of Notes. These Notes provided an interim reciprocal fisheries agreement for 1978 (Canada-USA 1978). This agreement entered into force on a provisional basis pending legislative approval by the U.S. Congress. The basic elements were the same as for 1977: maintenance of existing fishing patterns with no initiation of new fisheries and no expansion of fishing effort. The 1978 agreement also provided for prior notification and consultation on proposed regulatory measures of one country which might affect the fishing privileges of the other.

Canada also insisted on a provision whereby, if a dispute were not resolved through consultation, the party whose interests were negatively affected could "take reciprocal action with regard to the activities of the fishing vessels of the other Party to an extent sufficient to re-establish the balance of fisheries interests between the two Parties."

The 1978 interim agreement included new provisions respecting fishing for Pacific salmon. Because of differences in the minimum size limit for chinook salmon (26 inches in Canada compared with 28 inches in the U.S.), the U.S. agreed that Canadian salmon troll vessels off the coast of Washington State, north of 47 degrees 55 minutes North Latitude, could have chinook salmon between 26 and 28 inches on board, provided these salmon had been caught in the Canadian zone.

The Agreement provided a further measure respecting Pacific salmon. Canada would consult with the United States about the conservation need to close Swiftsure Bank to all salmon fishing from April 15, 1978 through June 14, 1978 (Canada-USA 1978).

This Agreement also enlarged the fishing area in U.S. waters in which Canadian salmon troll vessels could fish. Immediately upon signature of the Agreement, the U.S. requested Canada to close the Swiftsure Bank to all salmon fishing after April 14. Canadian fisheries personnel closely monitored Canadian fishing activities through April to ascertain whether a large number of immature salmon were being caught. Canada concluded that there was no conservation problem except in one quarter of Swiftsure Bank which was closed. The U.S. repeated its call for a closure of all of Swiftsure Bank. Canada asked for clarification of the scientific basis for the U.S. position. The answers provided did not satisfy Canadian managers.

Tensions on the two sides of the boundary were rising. The U.S. in late April revoked the permission for Canadian salmon fishermen to fish in the enlarged area provided for in the 1978 Agreement. The Special Negotiators met on May 11 and 12 in an attempt to resolve the matter. Canada agreed to close Swiftsure Bank until further notice. On May 21, Mr. Cutler informed his counterpart that the U.S. would restore access subject to agreement on coordinated salmon troll regulations. Canada indicated it would consider this for 1979, or as part of a long-term salmon interception agreement.

Meanwhile, all was not quiet on the eastern front. The U.S. fishery for scallop and pollock in the Georges Bank-Gulf of Maine area was essentially unrestricted. Canada considered U.S. haddock catches to be excessive. If the U.S. continued to fish at these levels, stocks would decline, hurting Canadian fishermen. At a May 26 meeting the U.S. indicated that, because this fishery was now managed by the New England Regional Council, it was unable to restrict the U.S. scallop and pollock fisheries during the 1978 fishing season (Wang 1981).

These differences resulted in the suspension of the 1978 Interim Agreement. On June 1, Canada notified the United States that it would discontinue implementation of the 1978 agreement as of noon on June 4 (External Affairs 1978).

The United States retaliated. Fishing by each side in the undisputed fisheries zone of the other ended in June, except for halibut fishing which was allowed to continue. Ironically, the U.S. Senate approved the 1978 Fisheries Agreement a few weeks after it became defunct.

5.5 Another Attempt at a Comprehensive Agreement

Negotiations aimed at a comprehensive agreement resumed in June, with a sense of urgency surrounding the fisheries issues. The negotiations were not starting from scratch, as certain principles had been agreed in the

October 1977 Joint Report. Ambassador Cadieux outlined the tasks confronting the negotiators:

1. To agree on percentage allocations for a number of important fish stocks on the east coast, and on the west coast to develop the terms and conditions relating to the conduct of reciprocal fishing;
2. To close the gap between the maritime boundary positions in the Gulf of Maine/Georges Bank area, off Juan de Fuca, in the Dixon Entrance area and in the Beaufort Sea; and
3. To find the best means of promoting cooperation in Canada-USA fisheries relations until a permanent agreement was reached (External Affairs 1978).

Separate Atlantic and Pacific working groups were established to review specific stock allocation issues. A drafting group was also established. A series of meetings were held during the summer and fall. On September 15, 1978, Canada, taking into account a decision by the International Court of Justice involving a maritime boundary between Britain and France, extended its claim roughly 25 miles farther west. The U.S. protested Canada's extension of its boundary claim.

During the fall, the negotiating task was reduced. First, the negotiators abandoned efforts to work out shared-access zones for hydrocarbon resources. Ambassador Cadieux and Mr. Cutler were subsequently instructed to give priority to east coast fisheries and boundary issues. Wang (1981) indicated that among the factors which led the two governments to give priority to the east coast fisheries and boundary problems was the large size of the disputed area and its greater potential for confrontation. Another factor was the attractive prospect of offshore hydrocarbon resources in the Gulf of Maine.

Furthermore, the major issue on the west coast, the problem of salmon interceptions, was different in nature. Separate but parallel negotiations had been underway for years. These were to be intensified. Agreement on this issue eluded negotiators until 1985.

5.6 The March 1979 Agreements

Early in 1979, agreement was achieved on some fronts. Four agreements were signed in Washington on March 29, 1979, two dealing with the Atlantic coast and two with the Pacific (External Affairs 1979).

The two agreements for the Pacific coast provided for a 2-year phase-out of reciprocal rights for halibut and groundfish. This was to give fishermen time to adjust to the fence being erected between the Canadian and the U.S. zones.

In the Halibut Protocol on March 29, 1979, Canada and the United States agreed to maintain the International Pacific Halibut Commission. An Amended Annex to the Convention set stock quotas for April 1, 1979 to March 31, 1981. Specifically, Canadian vessels were permitted to continue to fish halibut off Alaska outside 3 miles during this period. There was a total catch ceiling of 3 million pounds of halibut, consisting of two million pounds from April 1, 1979 to March 31, 1980, and one million pounds from April 1, 1980 to March 31, 1981.

Pending determination of the maritime boundaries between Canada and the United States on the west coast, it was also agreed that:

1. enforcement would be carried out by the flag state;
2. neither Party would authorize fishing for halibut by vessels of third parties; and
3. either Party could enforce the Convention for halibut fishing by vessels of third parties.

The second agreement pertaining to the west coast stipulated that U.S. fishermen could not fish for groundfish in the Canadian 200-mile zone, except to catch 6,500 tons of groundfish from April 1, 1979 to March 31, 1981. This quota was subdivided into two annual allotments of 3,250 tons each.

Thus, by an Exchange of Notes between U.S. Secretary of State Cyrus Vance and Canadian Ambassador Peter Towe, on March 29, 1979, Canada and the United States entered into a 2-year arrangement whereby Canadian fishing for halibut in U.S. waters and U.S. fishing for groundfish in Canadian waters would be phased out. Although this agreement hurt the Canadian halibut fleet, it was implemented smoothly.

The impact of the dislocation on the Canadian halibut fishery and the subsequent development of the Canadian fishery for Pacific groundfish were described in Chapters 10 and 11.

On March 29, 1979, two agreements were also signed respecting the east coast. One of these, an Agreement between Canada and the United States on East Coast Fishery Resources, was probably the most comprehensive and detailed fisheries agreement ever negotiated by the Government of Canada. The second agreement was a Treaty between the United States and Canada to submit to binding dispute settlement the delimitation of the maritime boundary in the Gulf of Maine. The provisions of this treaty and its outcome are discussed in section 5.8.

5.7 The East Coast Fishery Agreement

There were five major elements to the East Coast Fishery Agreement (Canada-USA 1979a).

5.7.1 Management Regime

The Agreement proposed a management regime for some 28 fish stocks or groups of stocks of mutual interest to Canadian and American fishermen. These stocks were grouped in three management categories, based on their geographical distribution and the relative interests of the two countries.

Annex A included those stocks which would be jointly managed, with each side having an equal voice, because the stocks were transboundary, migrating extensively throughout the waters of both countries. Examples were mackerel, pollock and cusk.

Annex B included those fish stocks where one country was considered to have the primary interest, generally based on past fishing performance. The country of primary interest would propose management measures. These would be subject to certain criteria and a review process which involved the other country. Examples included scallops, cod and herring. The other country could object only if the proposed measures were "clearly inconsistent" with the Governing Management Principles set forth in the Agreement.

Annex C included those stocks found primarily in the undisputed waters of one country to which the other country had been given an entitlement by the treaty. Examples included redfish, loligo squid, and haddock. These would be managed by the country in whose zone they occurred. Management measures for these stocks would be subject to consultation but the "other country" could not invoke formal dispute settlement procedures if it disagreed with these measures.

5.7.2 The Fisheries Commission

The Agreement proposed a bilateral Fisheries Commission to make recommendations to the two governments for Annex A and Annex B stocks. In addition, the Commission was to be a forum for consultation between the two sides for Annex C stocks. Each country was to have one vote, with decisions by the Commission requiring the agreement of both sides.

5.7.3 Shares and Access

Annexes A, B and C of the Agreement established the negotiated shares the fishermen of each country could catch from each stock. The Annexes also specified the terms of access each country's fishermen would have to each stock. In many cases, the fishermen of both countries would not have access to a stock throughout its entire range. The management category, percentage shares and access provisions for each stock are summarized in Table 12-4. As examples, Canadian fishermen would have been entitled to 73.35% of the scallop catch. U.S. fishermen, on the other hand, could have 98.4% of Division 5Y cod, 99% of Subarea 5 redfish and 35% of Scotian Shelf redfish.

These entitlements were the product of hard bargaining. They involved trade-offs among areas, species and groups of fishermen. They were not frozen but could be adjusted every 10 years. Also, there was a provision to adjust the entitlement to take into account the results of the impending maritime boundary delimitation. But the "winner" in the boundary delimitation would not take all. Entitlements could be modified only within certain maximum limits.

5.7.4 Dispute Settlement Procedures

Since Commission decisions would require the affirmative vote of both sides, there was clearly a potential for disagreement. The Agreement established a two-level dispute settlement procedure. The Co-Chairmen would attempt to resolve any disputes between the two sides. If the co-chairmen resolved a dispute referred to them, their decisions would be binding. If they could not resolve a dispute, it would be referred to a third party arbitrator jointly appointed by the two countries. The decisions of the arbitrator were to be binding.

5.7.5 Review Procedures

Although the Agreement was to be permanent, it included a provision for review and adjustment of the shares of the stocks at ten-year intervals (except for the stocks of loligo squid, redfish in Subareas 3 and 4 and lobster). Either side could request such a review. If agreement could not be reached on adjustments to entitlements, the question would be referred to an Arbitrator.

If a country had more than a 50% share of the stock, the Arbitrator could adjust the original negotiated entitlements, up to a maximum of 10% of the TAC. If a country had less than 50% , the maximum adjustment would be 5% of the TAC. The Arbitrator was to take into account the delimitation of the maritime boundary and "the social and economic impact of any proposed changes on coastal communities."

The dispute settlement procedure was to be expeditious. If the Co-Chairmen could not resolve a dispute within 15 days, the dispute would automatically go to the Arbitrator who had to decide "urgent" matters within 15 days.

The East Coast Fishery Agreement was linked to the boundary adjudication agreement also signed on March 29, 1979. Neither could come into force without the other. Ambassador Cadieux explained:

TABLE 12-4. Management category, percentage shares and access provisions for each stock covered in the 1979 East Coast Fisheries Agreement.

SUMMARY OF FISHERIES AGREEMENT

SPECIES	AREA	MANAGEMENT CATEGORY	PERCENTAGE SHARES CANADA/U.S.	ACCESS PROVISION
COD	5Z	Primary (USA)	17.0% / 83.0%	Throughout area
	5Y	Exclusive (USA)	1.6% / 98.4%	" "
	4X (offshore)	Exclusive (Canada)	92.5% / 7.5%	" "
	4VW	Exclusive (Canada)	98.6% / 1.4%	" "
Haddock	5	Primary (USA)	21.0% / 79.0%	" "
	4X	Exclusive (Canada)	90.0% / 10.0%	" "
	4VW	Exclusive (Canada)	90.0% / 10.0%	" "
Redfish	5	Exclusive (USA)	1.0% / 99.0%	" "
	4VWX[a]	Exclusive (Canada)	65.0% / 35.0%	" "
	4RST[a]	Esclusive (Canada)	90.0% / 10.0% of quota allocated to Canadian non-Gulf-based vessels	" "
	3-0[a]	Exclusive (Canada)	600 metric ton U.S. quota	" "
Pollock	4VWX + 5	Joint Management	74.4% / 25.6%	Reciprocal fishing in 4X and 5Z only
Silver hake	5Ze	Primary (USA)	10.0% / 90.0%	Throughout area
Red hake	5Ze	Primary (USA)	10.0% / 90.0%	" "
Argentine	4VWX + 5Ze	Primary (Canada)	75.0% / 25.0%	" "
Cusk	5Ze	Joint Management	66.0% / 34.0%	" "
White Hake	4VWX	Primary (Canada)	94.0% / 6.0%	U.S. access limited to 4
	5	Primary (USA)	6.0% / 94.0%	Canadian access limited to 5Ze
Other groundfish	3 + 4	Exclusive (Canada)	99.0% / 1.0%	To cover bycatches in areas of specific entitlement
	5	Exclusive (USA)	1.0% / 99.0%	To cover bycatches in areas of specific entitlement
Scallops	5Ze	Exclusive (Canada) in area East of 68° 30'W longitude	73.35% / 26.65% of the full area	Throughout area
Lobster	5Ze	Joint management — disputed area pending determination of boundary sett. Primary — each country's respective side of	During Joint Management, no increase in either country's fishery. After boundary sett. each country establishes own level of harvest in its waters.	Limited to dispute area until determined and to respective side of boundary after boundary sett. unless otherwise agreed. Any access to the other country's water would be established on a reciprocal basis.
Squid (*Illex*)	3 + 4	Primary (Canada)	100.0% / 0.0%	No fishing in disputed area by either country pending boundary sett., except by mutual agreement. After determination of boundary each country's fishery limited to its own waters.
	5 + 6	Primary (USA)	0.0% / 100.0%	No fishing in disputed area by either country pending boundary sett., except by mutual agreement. After determination of boundary each country's fishery limited to its own waters.
Squid (*Loligo*)	5Z + 6[a]	Exclusive (USA)	9.0% / 91.0%	Not yet determined
Herring	5Z + 6[b]	Primary (USA)	Canada to receive 2000 MT quota for first 3 years of agreement. During next 3 years 2000 MT, if Total Allowable Catch (TAC) less than 21,000 MT. Between 21,000 and 45,000 MT Canada to receive 50% of increase in TAC, until it receives 33.3% of TAC. After 6 years, Canada to receive 33.3% of TAC regardless of TAC level.	Reciprocal access between 68° 30°W longitude and 65° 00°W longitude.
	5Y[b]	Primary (USA)	0.0% / 100.0%	No Canadian access except that portion of 5Y in the area of Grand Manan Banks.
	4WX[b]	Primary (Canada)	100.0% / 0.0%	No U.S. access
Mackerel	3, 4, 5, + 6	Joint Management for setting of TAC. Primary management for regulation of domestic fisheries two sides will consult in advance of establishment of regulations and allocations to third parties	40.0% / 60.0%	Limited to each country's own waters.

[a] The arrangement for Canadian access to loligo squid and U.S. access to redfish off Nova Scotia was only for a ten-year duration, at which time it was to be renegotiated.

[b] The management categorization of three herring stocks was to be reviewed at the end of 3 years and could be altered if the two sides agree adequate data were available to support a change. In any case, at the end of six years the management categories would be reviewed and, if necessary, their determination submitted to dispute settlement.

Source: Canada. Department of External Affairs. 1979.

> "They stand or fall together. The rationale for this arrangement was that the two governments might be more disposed to agree to a boundary settlement if there was a shock absorber, an insurance policy, to reduce the import of an adverse court or arbitral decision" (Cadieux 1980).

The value of the insurance policy was related to its duration. U.S. fishermen had pushed during the negotiations for a short-term agreement, perhaps 10 years. Under this scenario, once the boundary was determined, there would be an initial period of adjustment following which the chips would fall where they would. Canadian negotiators, supported by industry advisers, sought stability and certainty of access to certain stocks no matter where the boundary might be. The risks of an adverse boundary decision did not seem evenly balanced. If the U.S. boundary claim prevailed, Canada would lose access to all of Georges Bank. If the Canadian boundary claim prevailed, the United States would lose access to only 35% of the Bank (Wang 1981).

While reaction to the proposed Agreement was reasonably positive in Canada, in the U.S. it became increasingly negative over the next two years.

The proposed treaties required U.S. Senate ratification. Supporters of the treaty in the U.S. were primarily the owners of the larger vessels in Maine, Boston and Gloucester. They believed that the U.S. concessions were a fair swap to regain access to groundfish in Canadian waters. But their support was soon overwhelmed by a tide of protest. By the middle of May a coalition of the opposing groups had been formed, called the American Fisheries Defense Committee. It sought to prevent Senate ratification of the proposed East Coast Fishery Agreement. The committee was composed primarily of scallop fishermen, lobster fishermen, mid-Atlantic coast fishermen and related fish processors. The Committee did not have the support of Maine and other groundfish fishermen.

The AFDC argued that the best approach would be to settle the seabed boundary first. Then, based on an established boundary, a management regime for transboundary stocks could be negotiated (AFDC, May 15, 1979).

The AFDC's argument against ratification centered on a perceived imbalance of entitlements and "an impractical management bureaucracy" that would result from the proposed Fisheries Commission. Perhaps the chief complaint of the AFDC was against the scallop shares. The AFDC contended that the proposed treaty would halt the recent revival of the American scallop effort on Georges Bank, just when increased investments promised to help the United States regain its traditional predominance. The effect would be to idle major new investments in scallop vessels. It argued that the East Coast Fisheries Agreement with Canada would reduce the U.S. scallop industry's share of the total Georges Bank catch from about 31% in 1978 to 26.65% for at least 10 years (AFDC 1979).

In rebuttal, the U.S. State Department defended the Agreement, arguing that during the 6 years (1971–76) preceding establishment of the 200-mile fisheries zones, U.S. fishermen had not harvested more than about 15% of the major scallop stock in the Georges Bank fishery. Their 13-year average on the Bank (1964–76) was only 22.7%, and the annual U.S. percentage steadily declined during that period. The 26.65% thus represented a substantial increase over the 1971–76 average which Canada originally insisted was the appropriate U.S. share of the fishery (USA 1979).

The AFDC's "impractical management bureaucracy" argument claimed that the Fisheries Treaty would undermine the authority of the new U.S. Regional Council system. The U.S. State Department pointed out that the position of the U.S. members of the Fisheries Commission would be based upon fishery management plans developed by the Regional Councils, with full public participation. The Regional Councils would control the U.S. vote.

While U.S. officials and Mr. Lloyd Cutler testified before Congress and argued for treaty ratification, the members of the AFDC were spreading their message of doom and gloom.

These lobbying efforts had their first major impact in the June, 1979, hearings of a subcommittee of the House of Representatives Merchant Marine and Fisheries Committee. Following testimony by various groups interested in the treaty, this Subcommittee concluded that settling the fisheries question before the boundary dispute was contrary to U.S. interests. The Agreement was complex and unworkable, and should either "be renegotiated or be abandoned in favour of more suitable, and perhaps less formal arrangements" (Rhee 1980).

The lobbying delayed consideration in the Senate for more than a year. The Senate Foreign Relations Committee held a preliminary hearing on April 15–17, 1980, at which several New England Senators raised concerns about the Fisheries Treaty. In Canada, the House of Commons passed a unanimous resolution on April 23, 1980, urging the U.S. Senate to take early action to ratify the agreements.

For a while Senator Kennedy of Massachusetts was, behind the scenes, looking for possible amendments to the Agreement so that he could support its ratification. The U.S. National Fisheries Institute testified in favour of the treaty. Their support was contingent upon one amendment. The NFI said that it was "not appropriate to have a permanent arrangement which does not provide for alterations in economic, social and legal factors." Accordingly, the treaty should be amended to provide for renegotiation beginning January 1, 1984, or when the boundary adjudication was resolved, whichever was earlier, without any percentage limit on changes in either nation's entitlements (National Fisheries Institute 1979).

For Canadians such an amendment would have removed the treaty's utility as an insurance policy. Canada repeatedly told the United States it would not consider amendments to the Agreement as negotiated.

While the U.S. administration continued to support early ratification of the Fisheries Treaty without change, it was not seriously lobbying in support of the Agreements. Following election of a new President in November 1980, the prospects for Senate ratification diminished considerably.

The Senate never voted on the treaty. By the spring of 1981 it was clear that the treaty could not be ratified without substantial amendments. There were suggestions that the fisheries agreement be withdrawn and that both countries proceed with the boundary treaty. This was contrary to those provisions of the agreements of March 1979 which expressly linked the two treaties. Canada initially resisted delinkage. In a March 6, 1981 letter to the Chairman of the Senate Foreign Relations Committee, President Reagan said that the U.S. government would use discretion in enforcing U.S. laws "against Canadian fishing vessels in all maritime areas now claimed by Canada." In the same letter, President Reagan withdrew the Fisheries Agreement from consideration by the Senate. He stated that "it would be best to uncouple the two treaties and proceed with ratification of the Boundary Settlement Treaty" (USA 1981). The Senate adopted a resolution to this effect, and June 3, 1981, President Reagan signed the Boundary Settlement Treaty. On November 20, 1981, Canada and the U.S. exchanged instruments of ratification for the technically altered boundary settlement treaty. This referred the Gulf of Main dispute to the International Court of Justice at the Hague.

VanderZwaag (1983) listed 10 factors which helped kill the 1979 East Coast Fisheries Agreement:

1. Natural Delays,
2. The U.S. Presidential Election,
3. Strong New England Fishing Industry Lobby,
4. Fear of Subverting the New England Council,
5. Fear of a New Layer of Regulations,
6. Belief that the Scallop Allotment was inequitable to the United States,
7. Dislike for the Redfish — Loligo Squid Trade-off,
8. Willingness by U.S. Fishermen to Gamble,
9. Belief in the U.S. Legal Position, and
10. Belief in the Indestructibility of the Scallop Resource.

All of these obstructed ratification of the treaty by the U.S. Senate. However, I believe the primary factors were the perception that the U.S. scallop entitlement was too low and that the treaty arrangements would hamper the growth of the scallop fleet. These were coupled with a belief that the U.S. would gain all of Georges Bank in the boundary adjudication. While a vocal group of U.S. fishermen was willing to gamble on the outcome, they assumed the risk was small. Another major factor: during the two years of delay, the U.S. management system off New England had moved in a direction which made joint management increasingly untenable. This was because of the considerable differences between Canada's management approach and that of the New England Regional Council.

In a 1986 Speech to a Conference on East Coast Fisheries Law and Policy, Ed Wolfe, Deputy Assistant Secretary of State for Oceans and Fisheries Affairs, alluded to this when he remarked:

> "Whatever may be said of the 1979 agreement, it is clear that it was ambitious. The agreement made a frontal assault on the management of a vast range of stocks.... In retrospect, the 1979 agreement may have been too ambitious too soon after 200 miles. The Regional Councils were yet in their infancy, and both countries were jealous of their newly won management prerogatives in their respective zones.... These different management regimes do not readily lend themselves to coordinated approaches" (Wolfe 1987).

5.8 The Gulf of Maine Boundary Adjudication

5.8.1 The Context

On March 29, 1979, Canada and the United States had agreed to refer the Gulf of Maine boundary dispute to "a Chamber of the International Court of Justice, composed of five persons." This use of the Chamber procedure was the first in the history of the ICJ.

In the Special Agreement the Chamber was requested to decide "the course of the single maritime boundary that divides the continental shelf and fisheries zones of Canada and the United States" from a point (Point A) nearly 40 nautical miles southwest of the international boundary terminus in Grand Manan Channel to a point within a triad encompassing the terminal points of the boundary lines claimed by the two parties and the intersection of the 200 mile arcs used to determine the outer limit of their continental shelf and fisheries jurisdiction (Fig. 12-7). By the Special Agreement, Canada and the United States had agreed that the single maritime boundary would serve for all jurisdictional purposes (Canada-USA 1979b). This was the first time that an international tribunal had been asked to determine a single maritime boundary for both the water column and the continental shelf.

The formal proceedings commenced in January 1982 when the Chamber was constituted. The Chamber rendered its Judgement on October 12, 1984.

The Gulf of Maine dispute originated in May, 1964, when Canada began issuing oil and gas exploration permits for the eastern portion of Georges Bank. These permits were based on Article 6 of the 1958 Convention on the Continental Shelf. They were granted up to a hypothetical equidistance line between U.S. and Canadian baselines.

In November 1969, the United States informed Canada that it would not recognize Canadian permits for any part of Georges Bank. However, there was no specific U.S. claim at that time. Negotiations to delimit the boundary began in July 1970. Canada argued that the continental shelf boundary should be determined by an equidistance line, according to the Continental Shelf Convention. When Canada ratified that Convention, it had added a reservation that "the presence of an accidental feature such as a depression or channel on a submerged area should not be regarded as constituting an interruption in the natural prolongation." The U.S. objected to this. During negotiations on the Gulf of Maine boundary it invoked the "special circumstances" provision of Article 6 and proposed a boundary line along the Northeast or Fundian Channel.

With the declaration of 200-mile fishing zones in 1977, this dispute broadened to include the water column as well as the continental shelf. The Order in Council proclaiming the Canadian 200 mile zone gave coordinates of a lateral equidistance line for the Gulf of Maine. The U.S. Notice setting out the limits of its continental shelf and fisheries jurisdiction set forth a boundary which began at the international boundary terminus and followed the line of deepest water through the Gulf of Maine, exiting through the Northeast Channel (Fig. 12-3a).

During the 1977–79 negotiations, Canada informed the U.S. on November 3, 1977, that , based on the decision in the Anglo-French Continental Shelf Case of June 30, 1977, it intended to modify its claim to account for the distorting effect of Cape Cod, Nantucket Island and Martha's Vineyard on the equidistance line. A revised Canadian claim was formally promulgated by Order in Council on January 25, 1979 (Canada 1979). This was characterized as the "equitable equidistance line."

The U.S. maintained its 1977 claim up to the submission of its Memorial on September 27, 1982. In its Memorial the U.S. radically altered its claim, abandoning the Northeast Channel line and replacing it with a line drawn perpendicular to the general direction of the U.S. coast from point A into the triangle. This line was adjusted to avoid dividing Browns Bank and German Bank on the Scotian Shelf (USA 1982). The final claims before the Court were the adjusted perpendicular line of the United States and the adjusted equidistance line of Canada (Fig. 12-7).

5.8.2 The Canadian and U.S. Arguments

Neither side spared effort or expense in submitting its case to the ICJ Chamber. This documentation has proved invaluable to several scholars who commented subsequently on the arguments and the Court's Judgement. McRae (1983) provided a summary of the written proceedings. The more significant commentaries include those of Legault and McRae (1984), McHugh (1985), McDorman et al. (1985), Legault (1985), Schneider (1985), Pharand (1984), Clain (1985), Terres (1985), Legault and Hankey (1985), Cooper (1986), DeVorsey (1987) and Johnston (1988). The observations which follow are drawn from these sources and the Court documentation.

Both parties maintained that maritime boundaries are to be determined "in accordance with equitable principles." However, they diverged greatly in the specific arguments they advanced for their boundary claims.

The Canadian Case
Canada argued based on five propositions:

1. Equitable principles must be identified and applied based on the applicable law;
2. The boundary should respect the basis of coastal state title;
3. The boundary should respect the basic purposes of the rights and jurisdiction in issue;
4. The boundary should take account of legally relevant circumstances; and
5. The result of the application of equitable principles must itself be equitable in light of all the relevant circumstances (Canada 1982a).

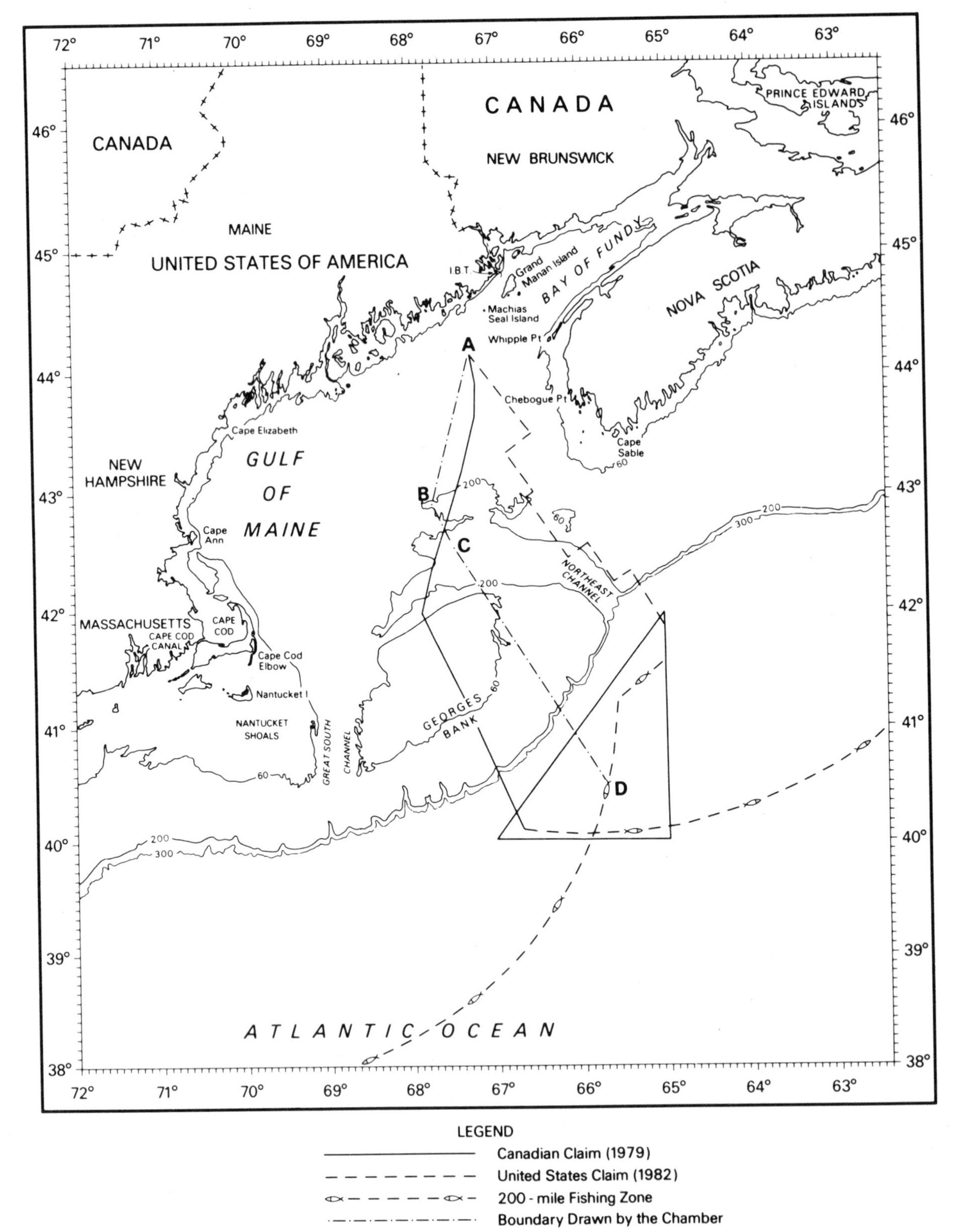

FIG. 12-7. The claims of 1979 (Canadian) and 1982 (U.S.) and the boundary drawn by the Chamber in the Gulf of Maine (from *Ocean Development and International Law* Vol. 16 p77 J. Cooper. Taylor and Francis, Inc. Washington D.C. Reproduced with permission. All rights reserved.)

Canada identified two main sources of applicable law. First, it asserted that Article 6 of the 1958 Convention on the Continental Shelf, and its application in the Anglo-French Award, was directly relevant to the delimitation of the continental shelf in this case because this Convention was binding between the parties. Second, Canada contended that the unity of the law respecting delimitation of the shelf and of the exclusive economic zone was reflected in the parallel wording of Articles 74 and 83 of the 1982 Law of the Sea Convention.

Canada's second main argument revolved around the view that the boundary must respect the legal basis of coastal state title. It argued that the older principle of natural prolongation had been replaced by a new emphasis on distance and proximity as the legal basis of title. This in turn should lead to a new emphasis on the equidistance principle.

Canada's third major argument was that the basic purposes of the rights and jurisdiction at issue should have a substantial bearing on the determination of a single maritime boundary. Canada argued that the central purpose was an economic one, reflected in the concept of an exclusive economic zone. Coastal states had a special dependence upon the resources off their coasts. Canada contended that the economic dependence of a coastal state on the sea area adjacent to it should have special weight. Canada stated that this "human dimension" was crucial because the economy of southwest Nova Scotia depended on its established fishery on Georges Bank. Canada contrasted the relative economic dependence of the parties upon the Georges Bank fish resources and argued that the placement of the single maritime boundary should maintain established fishing patterns.

Finally, Canada pointed to the conduct of the parties since Canada first issued oil and gas exploration permits for the eastern portion of Georges Bank. The United States had by its inaction effectively recognized Canada's proposed equidistance line. Technically, Canada claimed that the U.S. had acquiesced in Canada's issuance of exploration permits up to an equidistance line. The U.S. could therefore, not deny the applicability of equidistance in delimiting the boundary. Furthermore, the conduct of the parties regarding fisheries, particularly the negotiation and signature of the 1979 East Coast Fishery Agreement, confirmed Canada's established interest in Georges Bank and its status as a coastal state in the area.

The United States Case

The United States identified four "equitable principles" which it considered should be the basis for the Court's decision:

1. The boundary should respect "the relationship between the coasts of the parties and the maritime areas in front of those coasts";
2. The boundary should facilitate conservation and management;
3. The boundary should minimize the potential for international disputes; and
4. The boundary should take account of the relevant circumstances of the area (USA 1982).

With respect to the first principle, the U.S. advanced three subsidiary principles: non-encroachment, proportionality, and natural prolongation. Encroachment would occur if the method of delimitation left to one State an area that is off, or in front of, the coast of another. The proportionality test as envisaged by the U.S. required "that a delimitation take account of the relationship between the extent of the maritime area appertaining to the States concerned and the lengths of their respective coastlines."

The U.S. argued in favour of geographical natural prolongation. It took a macro view of the North American coastline, identified a single general direction of the coast and concluded that the coasts of Canada and the United States which accorded with the general direction are "primary" coasts. The U.S. contended that the coast of Nova Scotia facing onto the Gulf of Maine was a "secondary" coast. The U.S. described the Northeast Channel as "a significant break in the surface of the continental shelf." The U.S. also advanced an ecological argument that the Gulf of Maine is divided into three main ecological regions, two of which, Georges Bank and the Scotian Shelf, are divided by the Northeast Channel. The U.S. also argued historical rights in the area. It claimed to have mapped and surveyed the area, provided electronic and other aids to navigation, and conducted research, provided search and rescue services and defence. The U.S., asserted that these activities demonstrated its predominant interest in the Gulf of Maine area.

5.8.3 The Chamber's Decision

The Chamber drew a single maritime boundary in the Gulf of Maine area in its judgement of October 12, 1984, by a majority vote of four to one with Judge Gros dissenting. Judge Schwebel, while voting with the majority, preferred a somewhat different line (International Court of Justice 1984).

One of the first issues the Chamber had to address was whether it could draw a single boundary line for the continental shelf and the zones of fisheries jurisdiction. The Chamber was clearly uncomfortable with the concept of a single maritime boundary. On balance, however, it concluded that "there is certainly no rule of international law to the contrary, and in the present case, there is no material impossibility in drawing a line of this kind." For a further discussion of the implications of the decision to draw a single boundary line, see Legault and Hankey (1985).

The Chamber rejected virtually all of the arguments of the two parties in its boundary decision. It proceeded to draw its own line based on its judgement of the applicable law.

The Court reviewed the Continental Shelf and the Law of the Sea Conventions and previous international arbitration decisions. It interpreted international law for maritime boundary delimitation as follows:

> "Delimitation is to be effected by the application of equitable criteria and by the use of practical methods capable of ensuring, with regard to the geographic configuration of the area and other relevant circumstances, an equitable result."

The nuances of other maritime boundary delimitation cases are described by Pharand (1984) and Johnston (1988).

In its restatement of the fundamental norm of international law on this matter, the Chamber introduced a significant modification by singling out "the geographic configuration of the area" and referring only generally to "other relevant circumstances." In effect, the Chamber set aside previous notions of equitable principles, special circumstances and equidistance and replaced them with the concepts of "equitable criteria" and "practical methods." The Chamber concluded that the so-called equidistance principle was not a rule of law but rather merely a practical method which could be used in delimitation. Similarly, the Chamber rejected the U.S. contention that preference should be given to "primary" as distinct from "secondary" coasts as a rule of law.

Other arguments advanced by the parties respecting the maintenance of existing fishing patterns and fostering single management of living resources as rules of customary international law were also rejected. The Chamber conceded that, under certain circumstances, these could be equitable criteria.

The Chamber rejected criteria based on geomorphology or considerations relating to fish stocks as inappropriate. The former were relevant only to the continental shelf and the latter only to the water column. It concluded that "neutral" criteria were required. These must inevitably derive from geography, particularly the geography of coasts within the delimitation area:

> "One should aim at an equal division of areas where the maritime projections of the coasts of the States between which delimitation is to be effected converge and overlap."

The Chamber had to choose practical methods to implement the equal division approach. The line was drawn as three distinct segments (see Fig. 12-7). The first portion, in the inner Gulf of Maine, the Chamber saw as an area where the coastlines of the two states were adjacent to each other. In the second part, the outer Gulf of Maine, the Chamber considered the coastlines to be opposite. The third portion was seaward of the Gulf of Maine itself. For a discussion of how the lines were drawn, see Cooper (1986).

The Chamber then proceeded to test whether the resulting line was equitable. The various political and economic considerations, including fishing activities, hydrocarbon exploration, and research, which had been advanced by the parties, were not deemed equitable criteria. In the end, the Chamber asked itself whether the proposed line would "unexpectedly be revealed as radically inequitable, that is to say, as likely to entail catastrophic repercussions for the livelihood and economic well being of the population of the countries concerned."

The Chamber concluded that:

> "There is no reason to fear that any such danger will arise in the present case on account of the Chamber's choice of delimitation line or, more especially, the course of its third and final segment."

5.8.4 Legal Reaction to the Judgement

The Gulf of Maine boundary decision has proven fertile ground for legal experts, and has stimulated a large number of papers. Many of these criticize the Court's reasoning. Some commentators have criticized the Chamber's primary reliance on geography and its dismissal of resource allocation and management factors. McDorman et al. (1985) had serious doubts about the objectivity of the geographic approach the chamber used.

Clain (1985) described the Gulf of Maine decision as "a disappointing first in the delimitation of a single maritime boundary." He suggested: "The decision is destined to become nothing more than a footnote in the history of maritime delimitation."

Other commentators have come to a radically different conclusion and see the Chamber's decision as a significant step in the evolution of maritime boundary delimitation. In a 1985 article, Legault, who was Canada's Agent in the Gulf of Maine Case, observed:

> "Perhaps the three most important contributions of the Gulf of Maine case are that it clears away a lot of underbrush to let us see both the forest and the trees; it brings us back to the cardinal role of geography in maritime delimitation, and it sheds new light on the idea of proportion and disproportion in the relations between two coasts and the sea areas they attract within their jurisdiction" (Legault 1985).

In a spirited defence of the Chamber's decision, Collins and Rogoff (1986) disagreed with the various criticisms of the decision. They contended that the judgement was not a "splitting the difference" decision but rather was grounded in law and equity. They also contended that the Chamber's stress on geographical factors was correct, whether viewed from the perspective of theory, precedent or policy.

5.9 Post-1984 Fisheries Relations in the Gulf of Maine

Whatever the legal merits of the Chamber's decision, the line it drew in the Gulf of Maine was binding upon Canada and the United States by prior agreement. On October 26, 1984, the new boundary line went into effect. Canadian fishermen were prevented from fishing in U.S. waters and vice versa. Fishermen and managers of the two countries now had to assess the implications of the location of the new boundary and consider future management arrangements since the line divided many key stocks on Georges Bank.

Initial reaction to the decision was mixed. The official government position was that the decision was fair. An American official described the decision as a compromise: "We wanted 100% of Georges Bank, they wanted half of it, one would say that the court went for a split-the-difference sort of decision." The Canadian government expressed satisfaction with the result, saying that it "confirmed Canadian jurisdiction over Georges Bank" (*The New York Times*, October 13, 1984).

U.S. industry reaction was less favourable. On October 15, the *Boston Globe* reported that the Georges Bank ruling troubled fishermen. It quoted a Gloucester captain: "It's going to be the end for some of us. The Canadians will have our fish, the bankers will have our boats." Another fisherman stated: "It's going to hurt the draggermen, the lobstermen and the scallopers." Another observed: "The only hope is the U.S. government can negotiate some kind of access to the lost grounds for us. But that's going to take some time."

Peter Doeringer, a Boston University economics professor predicted that New Bedford scallop fishermen would be the big losers: "Basically, Canada picked up the richest prize in the disputed area. They got roughly half of the Georges Bank scallop beds."

Lucy Sloan, Director of the U.S. National Federation of Fishermen, indicated that the reaction of her members to the news ranged from outrage to suggestion of federal trade retaliation against Canadian fish imports.

Overall, it appeared that Canadian fishermen were reasonably satisfied with the outcome and U.S. fishermen were angry and disappointed. This was confirmed when some weeks later the U.S. industry's North Atlantic Fisheries Task Force asked the Department of State to request that Canada agree to a one-year moratorium on enforcement of fisheries law in the former disputed area. On November 28, 1984, 14 U.S. Senators and Congressmen wrote to U.S. Secretary of State George Schultz proposing an "interim agreement" with these elements:

1. Restoration of joint fishing in the area formerly in dispute exactly as was permitted just prior to the World Court decision;
2. One year in duration; and
3. Agreed upon without concessions or conditions.

In late November 1984, Senator George Mitchell of Maine visited Ottawa and proposed the one-year moratorium to Prime Minister Mulroney and External Affairs Minister Clark. Canadian fishermen were generally opposed to the proposal. On the advice of Canadian industry through the newly formed Gulf of Maine Advisory Committee (GOMAC), a government/industry committee formed to advise on transboundary issues, Canada rejected the U.S. proposal for a one year moratorium. Ironically, the U.S. fishermen who had scuttled the 1979 East Coast Fisheries Agreement were now attempting to ignore the Gulf of Maine boundary decision. These U.S. fishermen had gambled on winning all of Georges Bank and lost. Having made their bed, they would now have to lie in it.

The reaction of Canadian fishermen was partly influenced by a perception that Canada had benefited reasonably well from the boundary decision. This was confirmed by DFO scientists' analyses for GOMAC (GOMAC, 1984). Initial analyses indicated that the following proportions of key fishery resources lay on the Canadian side of Georges Bank: Scallops 54%, lobsters 21%, cod 24%, haddock 38%, pollock 27%, argentine 48%, and redfish 12%. The scallop figure was lower than the proportion of the total catch of Georges Bank scallops Canadian fishermen would have been entitled to under the proposed 1979 East Coast Fishery Agreement (73.3%) and a considerably lower proportion than they were catching in the late 1960's and early 1970's (80%). However, it gave Canadian fishermen the potential to catch more in absolute terms and a greater proportion of the total catch than they had harvested over the previous few years (5,900 t average during 1979–83 with a landed value of $65 million). This analysis was later updated by Halliday et al. (1986).

For groundfish, Canada had gained compared to its entitlement to under the 1979 Agreement. Canadian fishermen were excluded from the area they used to fish west of the line. However, this was more than offset by the removal of U.S. effort east of the line.

With these considerations, the industry advised that Canada not rush into new arrangements with the U.S. for the Gulf of Maine area. In analyzing options for future cooperation, the Gulf of Maine stocks were grouped into three categories:

1. Stocks for which independent management was possible (e.g. lobster, scallops);
2. Stocks for which some form of transboundary cooperation with the U.S. was desirable but not urgent (e.g. cod, pollock); and
3. Transboundary stocks which required immediate bilateral action (e.g. haddock and herring).

In the spring of 1985, the Canadian industry advised against discussions on reciprocal fishing with the U.S. There was no significant Canadian interest in fishing on the U.S. side of the line. It also recommended discussions aimed at a conservation agreement for haddock and herring. Before Canada could act on this advice, the U.S. industry filed a countervail petition with the U.S. International Trade Commission. This alleged unfair competition from Canadian fresh groundfish imports to the U.S. Notwithstanding this protectionist action, Canada wanted discussions.

On November 14, 1985, Canada submitted to Washington a formal proposal for consultations on the management of Gulf of Maine herring and Georges Bank haddock. The New England Fisheries Management Council responded unfavourably. On February 7, 1986, the U.S. counter-proposed that consultations at the technical level be initiated through existing channels of scientific and regional meetings. The State Department suggested that scientific consultation on haddock and herring be on the agenda for a bilateral scientific meeting in early April 1985. At this meeting scientists shared information on the status of these stocks but did not produce a joint assessment. Subsequently, in March 1987, there was a consensus of concern for the haddock stock. However, U.S. scientists indicated that they could not formally participate in a joint assessment. Nonetheless, the separate assessments by scientists from both countries urged managers to act to restore the declining stock.

Meanwhile, it was evident that the U.S. fisheries in the Gulf of Maine area were experiencing serious difficulties. In an article in the *Boston Globe Magazine* on October 18, 1987, journalist Jerry Ackerman described the U.S. fisheries as being in deep trouble. He stated that 10 years after New England fishermen thought their troubles were over with enactment of the 200-mile limit to keep foreigners out, times were worse than ever. He quoted various fishermen as stating the problem was no fish due to "overfishing". "By whom? The Russians? No, the Russians are gone.... Now, these men say the problem seems to be the New England fishermen themselves." He quotes a young fisherman as saying: "It seems like when we go out and we get near the Canadian border out there, right on the other side is where the fish are. None on our side."

This observation was borne out by the fishing practices of many U.S. fishermen who fished as close as possible to the boundary line and in many instances fished illegally on the Canadian side of the line. Following the boundary decision, 25 U.S. fishing vessels were charged with fishing illegally in the Canadian zone in 1985 and 1986. In 1986, the Canadian government increased fines and armed patrol vessels and fisheries enforcement officers. This appeared to reduce the number of incidents of illegal fishing with charges dropping to 16 in 1987 and 3 in 1988.

Despite this, New England fishermen resisted repeated proposals for tough conservation measures to speed recovery of these fish stocks.

Ackerman quoted Vaughn Anthony of the National Marine Fisheries Service's Northeast Fisheries Center at Wood's Hole as saying the status of haddock and cod was especially ominous. He cited haddock landings in New England falling almost 75% in just 5 years, from a 1982 peak of 20,000 tons to 5,390 tons. Cod was down from 58,000 tons in 1982 to 30,000 in 1986. Anthony (1990) later documented the decline in the New England stocks post-extension (see Chapter 20).

The New England groundfish fisheries are undoubtedly in a poor state, with catches at or near historic lows for 15 of 17 stocks. Catches per unit of effort had declined significantly since extension. A Report to the New England Fishery Management Council's Demersal Finfish Committee by its Technical Monitoring Group in June 1988 concluded that the most of the management measures were not working:

> "The stocks that have been primary targets of the fishery are all at record low levels of abundance, following general increases after

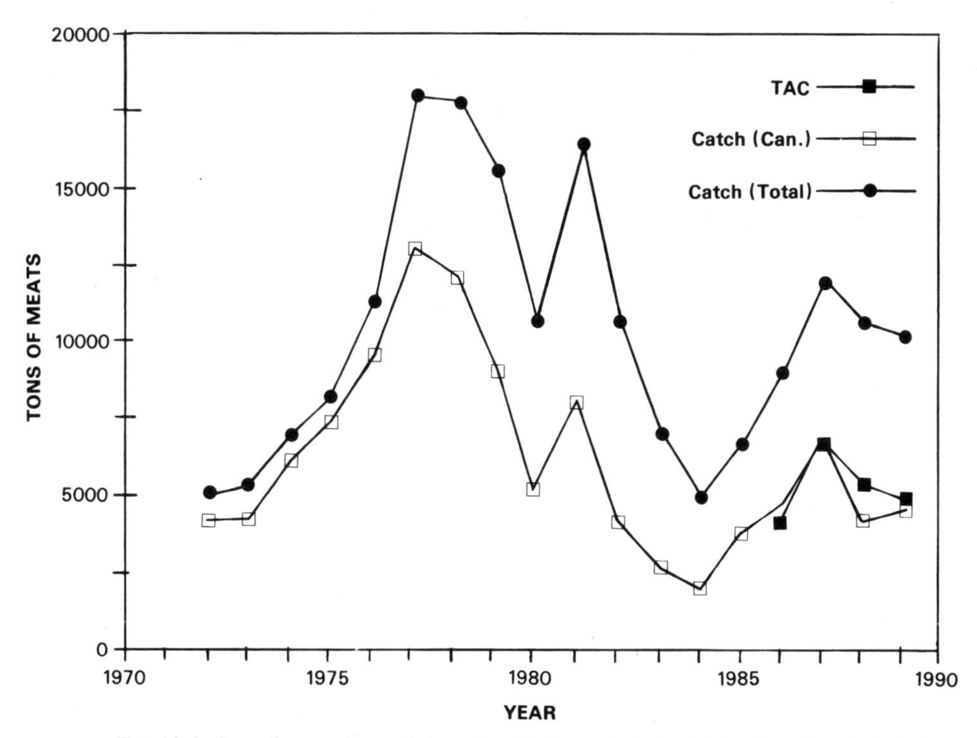

FIG. 12-8. Canadian catches and Canadian TACs, and total catch by Canada and United States for Georges Bank scallops.

> the implementation of the FCMA. The levels of fishing mortality for many stocks are similarly at record high levels" (USA 1988).

Regarding impact on the fishery, the Report noted that the total number of days fished by the larger vessels was at record high levels. The levels of fishing mortality were at record high levels for most stocks where this had been estimated. The total catch per unit effort had declined substantially for all vessel sizes. The Technical Monitoring Group recommended that "the transboundary nature of the stocks needs to be more fully considered within the management system."

The Gulf of Maine decision gave to Canada approximately one-sixth of Georges Bank, but the line cut through the fishing grounds for many commercial fish stocks. The chief resource of interest to Canada has been scallops, with the major scallop fishery taking place on Georges Bank. Scallop catches reflect trends in resource abundance which are heavily influenced by the irregular occurrence of strong year-classes. During the last 30 years four strong year-classes, those of 1957, 1972, 1977 and 1982 produced major peaks in landings (CAFSAC 1989b). Total catch and the Canadian catch increased from 1984 to 1987 due to the presence of the 1982 year-class but declined from 1984 to 1989 (Fig. 12-8). Canada's share of the total catch increased from 38.8% in 1984 to 56.4% in 1985 and to 58.3% in 1987. Canada's share decreased to 41.5% in 1988 and 45.1% in 1989. While it is difficult to generalize from this short period, it would appear that the boundary line did provide Canada with access to about 50% of the scallop resource. However, the combined Canada and U.S. fishing mortality is higher than F_{max} and substantially higher than $F_{0.1}$ and CAFSAC has advised that the fishing mortality be reduced to avoid growth overfishing.

For cod on the northeast part of Georges Bank, CAFSAC (1989c) concluded that catches since 1978 had resulted in fishing mortalities 2 or 3 times $F_{0.1}$. This means that fishing effort would have to be reduced by 1/2 or 1/3 to approach $F_{0.1}$. CAFSAC noted, however, that "reducing effort and catches by Canada to a level consistent with an $F_{0.1}$ management strategy may not result in substantial long-term increases in yield to the Canadian fishery because the benefits of reduced fishing effort could be negated by increased effort by the U.S.A in response to increased catch rates. The 5Zj and 5Zm management unit includes catches by both Canada and the U.S.A and it will be necessary to develop consistent management." CAFSAC advised that the 1989 Canadian quota of about 8,000 tons for Div. 5Z be maintained in 1990 for cod in unit areas 5Zj and 5Zm.

CAFSAC (1989c) could not provide an exact assessment for Georges Bank haddock. However, it concluded that "stock abundance and biomass are among the lowest since 1963 and recruitment in recent years has been highly variable and low. The 1983, 1985 and 1987 year-classes are the strongest since the 1978 year-class but less than a third its size.... The current fishing mortality rate and this associated catch of about 6,000 t are about 2 to 3 times greater than a harvest coincident with $F_{0.1} = 0.25$ management." CAFSAC advised that the Canadian catch of 5Zj and 5Zm haddock be reduced to less than 3,000 t in 1990 to approach $F_{0.1}$. In the case of both cod and haddock, CAFSAC suggested that "Canada and the U.S.A should pursue equivalent measures to reduce fishing effort."

On a more positive note, since 1984 there has been evidence of a recovery of the Georges Bank herring stock. However, CAFSAC (1989d) did not believe that this recovery was sufficient yet to support a fishery and advised that there be no directed fishery for herring in Division 5Z in 1990. The Georges Bank herring stock once supported a substantial fishery. If it could be rebuilt to former levels, this would provide significant opportunities for Canadian herring fishermen.

Overall, these Canadian scientific analyses indicated that the absence of consistent management measures for the transboundary stocks since 1977 had reduced the cod and haddock stocks on Georges Bank. Given the sedentary nature of the scallop resource, Canada has managed to hold its own since 1984 despite an intensive fishery on the U.S. side of the line. The sedentary nature of scallops has made it possible to improve the economics of the fishery through TACs and enterprise allocations commencing in 1986. The apparent recovery of Georges Bank herring in the absence of a fishery is encouraging. But this promise could be dashed by a sudden increase in fishing for this species on Georges Bank.

More than 15 years after extended jurisdiction and more than eight years after the Gulf of Maine boundary delimitation, it is time for Canada and the United States to rethink their management of the transboundary stocks on Georges Bank. There is a need for a consistent approach to certain key stocks. The major stumbling block continues to be the difference in the approach to management on the two sides of the Gulf of Maine boundary. On the surface, it would appear that U.S. fishermen are hurting the most as a result of the lack of consistent management. But Canadian fishermen too are suffering from the decline in certain transboundary stocks.

Reciprocal access still holds no appeal for Canadian fishermen. The activities of U.S. fishermen on the Tail of the Grand Banks subsequent to the Gulf of Maine decision, with catches around 5,000–6,000 tons in 1985 and 1986 and 3,000 tons in 1987 and 1988, are an attempt to seek out new opportunities. For Canada, they represent part of the conservation threat posed by fishing by non-NAFO members beyond 200 miles. U.S. fishermen have attempted to improve their market share by urging the curtailment of Canadian imports in the name of conservation. This is not conducive to the cooperation the situation demands.

Among the available options, the status quo of unilateral management by each country will not lead to stock rebuilding. Joint management of the transboundary stocks along the lines proposed in the 1979 Agreement appears to be only a remote possibility. Reciprocal fishing arrangements do not appear necessary. There is, however, clearly a need, as pointed out by recent scientific reports from within both countries, for cooperation or consistent management of selected transboundary stocks.

Whether this will be achievable in the short to medium-term depends upon the outcome of discussions within the New England Fishery Management Council on potential fishing effort controls. Without fundamental changes in the New England attitude to fishing effort control, progress on consistent or cooperative management for the Gulf of Maine stocks may be a long time coming. There appears to be a new understanding within the U.S. that the New England fisheries management system is in trouble, that there is a need to address the underlying problems, and that the U.S. is suffering from the lack of consistent management for transboundary stocks (see Chapter 20). Whether this will result in new initiatives in the 1990's remains to be seen.

5.10 The Canada-U.S. Pacific Salmon Treaty

5.10.1 Biological Context

Salmon from Canada and United States intermingle extensively during their migrations along the coasts of both countries and with stocks of Asian origin on the high seas. The attempts to curtail interceptions of North American salmon in the North Pacific high seas fisheries were described in Chapter 11. Here I deal with the negotiations aimed at minimizing the interception of Canadian and U.S. origin fish by nationals of the other party.

The different species of salmon exhibit widely different migratory patterns. Sockeye and chum salmon migrate great distances within the Gulf of Alaska, but pink salmon stay closer to their rivers of origin. Coho is the least migratory species. Many coho remain close to their home streams throughout their lives. Chinooks, on the other hand, stay close to the shore but migrate long distances in a north-south direction.

Pacific salmon begin their migrations toward their home streams in the late spring and summer of their last sea year. Since the feeding grounds lie chiefly to the north and west of their home rivers, the return migrations to home rivers are generally southeastward or eastward. The returning salmon generally come close to the coast some distance to the north and west of their home rivers. For example, sockeye and pink headed for the Fraser River arrive in inshore waters off Vancouver Island, moving southward to return to their home streams via Juan de Fuca Strait. Others arrive at the coast farther north and pass through Johnstone Strait at the northern end of Vancouver Island. Sockeye, pink and chum headed for the major rivers of northern British Columbia arrive first in Alaskan coastal waters, then migrate along the Alaskan shoreline and enter Dixon Entrance en route to their home rivers. Salmon originating in rivers of Washington and Oregon states move along the British Columbia coast during their return migration.

These migration patterns allow U.S. fishermen to intercept Canadian salmon off Alaska and Canadian fishermen to intercept U.S.-origin salmon from Washington and Oregon off British Columbia. Interceptions also occur at the northern and southern boundaries of British Columbia as Canadian salmon return home through Dixon Entrance and Juan de Fuca Strait.

5.10.2 The Interception Negotiations

The Canada-U.S. Pacific salmon interception problem is a century old. The earliest disputes in the late 1800's concerned U.S. catches of Canadian-origin sockeye returning to the Fraser River to spawn. From this beginning the scope and magnitude of interceptions broadened considerably to include interceptions in all of the areas mentioned above.

For most of the present century U.S. fishermen were the major interceptors. The problem started on the Fraser River where U.S. fishermen caught salmon in the approaches to the Fraser at the northern entrance to Puget Sound. The Canadian fishery was centered in the estuary and the river itself. Negotiations to address the interceptions problem for Fraser River sockeye were protracted and often frustrating. An International Fisheries Commission formed in 1908 by Canada (represented by Great Britain) and the United States developed regulations which were not enforced. This Commission ceased to function.

Following the famous Hell's Gate slide on the Fraser in 1913 and a decline in the sockeye salmon runs, negotiations were resumed. An agreement signed in 1929 came into force in 1937. Finally the International Pacific Salmon Fisheries Commission (IPSFC) began to regulate the Fraser River fishery in 1946. The scope of this Agreement, the Fraser River Sockeye Treaty, was broadened in 1957 by means of a Protocol which included pink salmon on the southern approaches to the Fraser under a management regime similar to that for sockeye.

Also in 1957, Canada and the United States informally agreed to voluntarily establish administration lines (the 'surf lines'), with a prohibition on fishing by nets seaward of these lines. These 'surf lines' did not restrict the developing Canadian troll fishery which took place off the coasts of Alaska, Washington and Oregon as well as British Columbia. The Canadian troll fishery intercepted coho and chinook returning to U.S. rivers. At the same time Canadian fishermen were not pleased with the 50% share allocated to the U.S. under the Fraser River Sockeye Treaty. It was costing Canada a considerable amount to preserve salmon habitat in the Fraser, with U.S. fishermen reaping 50% of the benefit.

Following the extension of fisheries jurisdiction by Canada and the USA to 12 miles in 1964 and 1970, the two countries negotiated a reciprocal fishing agreement in 1970. This was re-negotiated in 1973. Under these agreements, access by troll fishermen of both countries was curtailed within the 3–12 mile zone of the other country. Trolling by U.S. fishermen was eliminated in the 3–12 mile zone north of the middle of Vancouver Island. Trolling by Canadian fishermen was eliminated off the coasts of Alaska, Oregon and California. Because Canadian fishermen trolled more off the U.S. coast than vice versa, the agreements had a more significant impact on the Canadian than on the U.S. fishery. By 1970 both countries had become concerned about stock declines and were planning major salmon enhancement initiatives. The 'piecemeal' approach to salmon interceptions no longer was satisfactory. A Reciprocal Fishing Agreement in 1970 provided that, within one year, the two sides would begin negotiations "regarding all matters of mutual concern related to the fisheries for Pacific salmon."

In June 1971, the two countries agreed upon some general principles to guide future negotiations:

1. Each country should reap the benefits of its efforts to maintain or increase the stocks of salmon;
2. Each country should fish the salmon bound for its own rivers and should avoid intercepting salmon bound for their rivers of origin in the other country;
3. To the extent that interceptions continued, there should be an "equitable balance". This meant that the total value of salmon bound for Canadian rivers intercepted by the United States should be as nearly as possible equal to the total value of salmon bound for U.S. rivers intercepted by Canada;

4. This "equitable balance" should be achieved where possible by reducing rather than increasing interceptions;
5. Each country should try to adjust the techniques and economics of its fisheries to reduce interceptions; and
6. These adjustments should take into account the overriding requirement of conservation (Canada-USA 1971).

Negotiations up to 1976 failed to produce a detailed agreement based on the principles. In many respects, these principles were inconsistent. The negotiations got stuck on certain key issues including:

1. Whether and to what extent the Alaskan salmon fisheries should be limited;
2. Return of the Fraser River to Canadian management;
3. Achieving an "equitable balance" in the value of interceptions, given that the measurement of the "value" of interceptions was exceedingly difficult;
4. Conflicting views over whether U.S. fisheries on the stocks of transboundary rivers, specifically the Stikine and the Taku, should count as interceptions in the same manner as interceptions of salmon returning to the coastal rivers of British Columbia.

Following extension of fisheries jurisdiction in 1977, a new impetus was given to the negotiations by the overall decision to suspend reciprocal fishing privileges in 1978. Canada attempted to break the negotiating impasse in 1977 when it proposed that the negotiations emphasize cooperation in development rather than continuing to concentrate on the control of interceptions. However, negotiations again stalemated until 1980. By then fisheries managers on both sides were afraid to take necessary conservation action for fear of prejudicing the salmon-interception negotiations. The United States also faced considerable salmon management problems arising from judicial decisions allocating 50% of the salmon catch to Treaty Indian Tribes in the Northwest States (for details, see Yanagida 1987). This complicated the operations of the IPSFC which was being asked to regulate to meet internal U.S. allocation objectives.

Canada was catching an increasing share of the sockeye and pink salmon returning to the Fraser River. This occurred because a greater percentage of the Fraser River-salmon returned to the Fraser inside Vancouver Island through non-Convention waters. Canadian fishermen also initiated commercial fisheries on the transboundary Stikine and Taku Rivers, which had previously been fished almost exclusively by Alaskans.

Progress was finally made at a negotiating session in Lynnwood, Washington, in October 1980. The negotiators proposed the development of "a general framework Convention which would include a series of binding principles and a series of specific provisions related to an initial salmon interception limitation scheme, management of stocks bound for transboundary and Fraser Rivers and technical resolution procedures." The negotiators envisaged that a "Commission (with appropriate subsidiary Panels) would be formed immediately upon ratification to implement the Convention." Once the Commission was established, "it would be necessary for the Parties to negotiate further detailed implementation provisions regarding specific fisheries and approaches to management, development, research and monitoring" (Canada-USA 1980).

It was proposed that interim arrangements be developed to control the intercepting fisheries in the 1981 season. On June 19, 1981, interim fishing arrangements were agreed for certain key fisheries for the 1981 and 1982 seasons. These were implemented in 1981 in a spirit of mutual goodwill.

In mid-1982 the negotiators submitted a draft "Framework Agreement". This included proposed principles for cooperation on stocks subject to interception. It also proposed that the two countries negotiate annexes to the Treaty, to specify management measures for the intercepting fisheries during 1983 and 1984 (DFO 1982b).

Interim fishing arrangements were again implemented in 1982. Negotiations on the comprehensive agreement continued. A complete agreement was initialled on December 22, 1982 and submitted to the two governments for adoption (DFO 1982c).

Over the winter, however, major opposition to the treaty developed in Alaska. The new Governor requested a special Advisory Committee to examine the draft treaty. On February 21, 1983, Canadian Fisheries Minister De Bané noted that there was opposition to the treaty in both countries (DFO 1983j).

The Minister expressed concern that important matters addressed in the treaty would simply go unattended without agreement between the countries, particularly the rebuilding of natural chinook stocks of northern and central British Columbia.

Minister De Bané also stated that the historical 50:50 sharing arrangements for Fraser River sockeye and pink salmon were no longer appropriate. The process of reapportioning the catch of sockeye and pink salmon from the Fraser must continue whether the new treaty was adopted or not.

In late February 1983, the Governor of Alaska indicated that he would not endorse the draft treaty, mentioning seven problem areas and urging further negotiations. Consequently, the draft Treaty never reached the Senate. The Alaskan action put a considerable strain upon Canada-U.S. Pacific salmon fisheries relations throughout 1983. No interim fishing arrangements were put in place. On June 27, 1983, Minister De Bané stated that a further attempt to rescue the draft treaty had failed. He pointed out that a number of Alaskan concerns had been met but two remained. These were the chinook harvest levels for both sides in 1983 and the sharing arrangements for the salmon stocks spawning in the transboundary rivers, the Stikine and Taku. Canada had offered to reduce substantially its chinook catch from the level specified in the treaty. On the second issue,the Alaskans were demanding that Canada refrain from developing commercial fisheries for Canadian fish in the Canadian sections of the Stikine and Taku Rivers. Minister De Bané stated that this was unacceptable:

> "It would mean that the United States rather than Canada would be given the primary interest in and responsibility for salmon spawned in Canadian rivers" (DFO 1983k).

In 1983, the U.S. percentage of the Fraser River sockeye run dropped to a modern record low of 12%.

Negotiations resumed in November 1983, with new negotiators on both sides. The U.S. pressed the Alaskan demands and negotiations broke down again in January 1984. However, other forces were being brought to bear within the United States. The Pacific Northwest Treaty Tribes were seeking through the courts to include the Alaskan fisheries in the judicially decreed 50:50 sharing arrangements between Treaty Indians in the Northwestern states and other U.S. fishermen. In the Northwest states, a Pacific Salmon Treaty Coalition was formed and lobbied for resumption of the negotiations (Yanagida 1987).

Negotiations resumed again in early December 1984. The United States appointed former Congressman Edward Derwinski to provide political oversight of the negotiations. Canada appointed former Cabinet Minister Mitchell Sharp to perform a similar function. On December 15, 1984, agreement was reached. The Treaty itself was virtually identical to the draft Treaty of December 22, 1982, which had been earlier rejected. However, the Annexes to the Treaty were considerably different. The U.S. Senate approved the Treaty and it entered into force on March 18, when Prime Minister Mulroney and President Reagan exchanged instruments of ratification at the "Shamrock Summit" in Quebec City. For a more comprehensive historical and legal overview of the Pacific Salmon Treaty, see Jensen (1986).

5.10.3 The Pacific Salmon Treaty

The Pacific Salmon Treaty consists of the Treaty itself, a series of Annexes and a Memorandum of Understanding (Canada-USA 1985). The Treaty sets forth the objectives, principles, and procedures and provides for the establishment of a Pacific Salmon Commission as an ongoing mechanism for cooperation between the two Parties. The Annexes contain short-term commitments by the two Parties regarding the regulation of particular fisheries and the collection, exchange and analysis of information on the salmon stocks and the fisheries. The Memorandum of Understanding records the Parties' interpretations of certain terms of the Treaty. Later in 1985, Canada and the United States exchanged Diplomatic Notes covering the phase-out of the International Pacific Salmon Fisheries Commission and the division of work among the two countries and the Fraser Panel of the Pacific Salmon Commission.

The Preamble to the Treaty recognized "the interests of both Parties in the conservation and rational management of Pacific salmon stocks and in the promotion of optimum production of such stocks." The second paragraph of the Preamble reflects the provisions of Article 66 of the 1982 Law of the Sea Convention concerning anadromous stocks, recognizing "that States in whose waters salmon stocks originate have the primary interest in and responsibility for such stocks." This provision is given concrete expression in the body of the Treaty which stipulates that Canada and the United States have the right (subject to consultation within the Commission) to set the management objectives for the harvesting of such stocks wherever they migrate.

Article 1 provides a list of definitions of terms used in the Treaty. Of these, the definition of "overfishing" is interesting. It is defined in purely biological terms as "fishing patterns which result in escapements significantly less than those required to produce maximum sustainable yields." It is noteworthy, given the move to the optimum yield concept in both countries in the 1970's (see Chapter 4), to see the concept of MSY enshrined in a 1985 Canada-USA treaty.

Stocks subject to the Treaty are defined as Pacific salmon stocks which originate in the waters of one Party and

a) are subject to interception by the other Party;
b) affect the management of stocks of the other Party; or
c) affect biologically the stocks of the other Party.

This broad definition means that the Commission's attention could be attracted not only by interception but also by other non-exploitation activities, e.g. habitat degradation.

Conspicuous by its absence is a definition of the preambular term "optimum production".

Article II details the structure of the Pacific Salmon Commission and its Panels. The PSC has two national sections, a Canadian Section and a United States Section. It has 16 members, with four Commissioners and four alternates each from Canada and the United States. Unlike the earlier IPSFC, each Section has only one vote in the Commission. Thus the Commission can only take a decision or make a recommendation if both countries agree.

Article II also provides for three Panels as set out in Annex I. These are:

1. A Southern Panel for salmon originating in rivers south of Cape Caution, except for Fraser River sockeye and pinks;
2. A Northern Panel for salmon originating in rivers between Cape Caution and Cape Suckling in Alaska;
3. A Fraser River Panel for Fraser River-origin sockeye and pink salmon fisheries in southern British Columbia and northern Puget Sound.

The Panels are to "provide information and make recommendations to the Commission." For cases where fisheries intercept stocks for which more than one Panel is responsible, the Panels meet jointly.

The Pacific Salmon Commission does not regulate the salmon fisheries but provides regulatory advice and recommendations to the two countries. The basic role of the Commission is:

1. To conserve the Pacific salmon to achieve "optimal production"; and
2. To divide the harvests so that each Party reaps the benefits of its investment in salmon management and development.

In addition to the three Panels, the Commission is served by a Committee on Research and Statistics and a Committee on Finance and Administration. While the Commission may eliminate or establish committees as appropriate, it can only recommend to the Parties the elimination or establishment of Panels. In practice, the Commission has established a multiplicity of technical groups (e.g. Data Sharing, Yukon River, Chinook, Coho, Southern Chum, Northern Boundary, Transboundary). Each of these reports to the Commission or to one of the Panels or both but not, as might be expected, to the Committee on Research and Statistics.

Article III of the Treaty establishes Principles to guide the Parties and the Commission. These are:

"1. With respect to stocks subject to this Treaty, each Party shall conduct its fisheries and its salmon enhancement programs so as to:
   a) prevent overfishing and provide for optimum production; and
   b) provide for each Party to receive benefits equivalent to the production of salmon originating in its waters.

"2. In fulfilling their obligations pursuant to paragraph 1, the Parties shall cooperate in management, research and enhancement.

"3. In fulfilling their obligations pursuant to paragraph 1, the Parties shall take into account:
   a) the desirability in most cases of reducing interceptions;
   b) the desirability in most cases of avoiding undue disruption of existing fisheries; and
   c) annual variations in abundance of the stocks."

The two approaches suggested in paragraph 3, subparagraphs (a) and (b), are likely to usually conflict. Thus, the principles are equivocal in that they reflect the dichotomy between Canada's search for an "equitable balance" and the U.S.'s desire to protect traditional fisheries.

A Memorandum of Understanding, which forms an integral part of the Treaty, provided further insight into how the "equity" principle is to be applied. Imprecision in the measurement of interceptions and differences in the evaluation of benefits would necessitate a gradual implementation of programs to achieve a balance in interceptions. However, efforts were to be made in the initial short-term arrangements to achieve some balance.

Canada and the United States have differed about the means to achieve that balance. Canada has emphasized equity in terms of fish, while the United States has suggested compensation through cash contributions to enhancement programs. Secretary of State George Schultz, in a letter to the U.S. Congress, indicated that the "so-called equity principle is intended to provide for each Party to receive compensation benefits of unspecified form or quantity of fish."

Article IV stipulates procedures for the Commission and the Parties to ensure that their fisheries are conducted in accord with the Treaty.

The process in practice operates as follows:

Step 1: Each country provides technical information to the Commission on the conduct of its fisheries, pre-season expectations and enhancement activities, which is:

Step 2: Analyzed by bilateral technical committees, which then report to:

Step 3: Panels, which use these reports to develop their fishery recommendations. From here the various area plans are:

Step 4: Sent to the Commission for consideration. At this stage, the Commissioners meet to review and conclude negotiations on the plans, which are then:

Step 5: Transmitted to the Governments of Canada and the United States for final approval and regulatory implementation (Pacific Salmon Commission 1988).

According to Article IV, the State of origin develops the management requirements for stocks originating in its rivers. A technical dispute settlement process (Article XII) can be invoked. However, the management proposals by the State of origin are not subject to challenge by the other Party or the Commission unless they contravene the Treaty.

Another key part of the Treaty is Annex IV which contains agreed interception targets and certain management measures for stocks subject to the Treaty for the first few years after its entry into force. Annex IV contains 6 chapters dealing with specific stocks and a general obligation in Chapter 7 which stipulates that: "with respect to intercepting fisheries not dealt with elsewhere in this Annex, unless otherwise agreed, neither Party shall initiate new intercepting fisheries, nor conduct nor redirect fisheries in a manner that intentionally increases interceptions."

The specific chapters deal with interception and information collection and exchange arrangements for the Transboundary Rivers, Northern British Columbia and Southeastern Alaska, Chinook salmon, Fraser River Sockeye and Pink Salmon, Coho Salmon and Southern British Columbia-Washington Chum Fisheries. These contain provisions too detailed to summarize here. The Chinook Salmon Chapter provides an example of the provisions. It recognized that escapements of many naturally spawning chinook stocks originating from the Columbia River northward to southeastern Alaska had declined in recent years and were substantially below goals set to achieve maximum sustainable yields. The Parties agreed to:

"(a) instruct their respective management agencies to establish a chinook salmon management program designed to meet the following objectives:
(i) halt the decline in spawning escapements in depressed salmon stocks;
(ii) attain by 1998 escapement goals established in order to restore production of naturally spawning chinook stocks, as represented by indicator stocks indicated by the Parties, based on a rebuilding program begun in 1984."

Pending the recommendation of a Joint Chinook Technical Committee, the two Parties agreed to specific catch limits for chinook for 1985 and 1986 for various areas. These included 526,000 chinook in northern and central British Columbia and southeast Alaska divided equally between the Parties, a troll catch of no more than 360,000 chinook off the west coast of Vancouver Island, and a total annual catch by the sport and troll fisheries in the Strait of Georgia of 275,000 chinook. As described in Chapter 7, the Strait of Georgia chinook provision led to a controversial domestic resource allocation debate within Canada.

One of the most contentious issues in the negotiations was the treatment of fisheries on stocks originating in the Canadian sections of the transboundary rivers, i.e. rivers that drained to the sea through Alaska. These included the Yukon, Stikine, Taku and Alsek. After long and bitter negotiations, the two Parties agreed on the provisions contained in Article VII and Chapter 1 of Annex IV. Article VII provided that "whenever salmon originate in the Canadian portion of a transboundary river, the appropriate Panel shall provide its views to the Commission on the spawning escapement to be provided for all the salmon stocks of the river if either section of the Panel so requests." Then, on the basis of the views provided by the Panel, the Commission was to recommend spawning escapements to the Parties. Article VII also provided that "enhancement projects on the transboundary rivers shall be undertaken cooperatively."

Chapter 1 of Annex IV respecting the transboundary stocks stipulated specific arrangements for the Stikine and Taku Rivers for 1985 and 1986.

Basically, the provisions for the Stikine and Taku allowed for the maintenance of a symbolic Canadian commercial fishery on the stocks originating in the Canadian portions of these rivers. The Parties also undertook to rebuild chinook escapements to specific target levels by 1995.

The Yukon, which is a transboundary river, supports a substantial U.S. commercial fishery, mainly by local native people, in the lower reaches of the river. Subsistence fisheries by natives are important on both sides of the border. There is also a small Canadian commercial fishery. The U.S. opposed the inclusion of the Yukon fisheries in the Pacific Salmon Treaty because the U.S. fishery was conducted mostly by natives who had little interest in the overall Canada-U.S. negotiations. Article VIII of the Treaty proposed separate negotiations commencing in 1985 to develop an organizational structure to deal with Yukon River issues.

5.10.4 Impact of the Treaty

Initial reaction within Canada to the Salmon Treaty was mixed. The UFAWU condemned the Treaty and called on the government to scrap it. Tourist operators dependent on the sports fishery for chinook claimed it would destroy their industry. The *Vancouver Province* on February 3, 1985, described the Salmon Treaty as "A Boom for B.C." B.C. Editor Malcom Turbull wrote: "Short-term pain for long-term gain. That in a clamshell is the simple assessment of new Canada-U.S. West Coast salmon treaty that climaxed 15 years of sporadic negotiations."

In an article in the *Times-Colonist* on January 30, 1985, journalist Alec Merriman stated that the government "gave the whole shaft to more than 300,000 British Columbia sport salmon anglers and their $120 million-a-year sport-fishing industry.... Fisheries Minister Fraser has sold [sports fishermen] down the river by signing a treaty that all but wipes out the sport salmon fishery and the huge industry it supports."

Sports criticism of the treaty altered when Minister Fraser announced provisions for the sport fishery which permitted catch limits for Georgia Strait chinook of two per day and an annual limit of 20.

But other voices called for reason. The *Nanaimo (B.C.) Times* on February 2, 1985, stated:

> "Those whipping themselves into a frenzy over the proposed Canada-U.S. fishing treaty forget what the alternative is.... Without a treaty fishermen on both sides of the border are indirectly encouraged to catch as many salmon as they can. After all, don't the fish that get away end up in the holds of a competitor? Why would either government undertake a serious enhancement effort while so much of the benefit goes to the other country's economy?"

The impact of the Pacific Salmon Treaty is not yet clear. In the first 4 years of Treaty implementation, some progress was made. In terms of the transboundary rivers, in 1987 the 1985–86 treaty entitlements were scrapped. Canada then managed the rivers to increase the Canadian harvest of sockeye. In 1988, this Chapter of the Annex was renewed for 5 years with an improved system of sharing for Stikine sockeye, an increase in the share from the Taku River, and a commitment to cooperative sockeye enhancement of Stikine and Taku stocks. Thus, the Treaty arrangements for these rivers had been improved from a Canadian perspective.

In the northern boundary area, the percentage of the B.C./Alaska catch of Canadian sockeye stocks taken by Alaska did not increase but the absolute catch increased due to higher abundance of Canadian stocks.

Declining chinook stocks were of considerable concern to both Parties in the early 1980's. As of 1988, preliminary results from the Chinook Technical Committee suggested that the rebuilding program was being maintained although some stocks (e.g. those with fall migration timing) had not responded as well as others. Some, notably Lower Georgia Strait and West Vancouver Island chinook, continued to decline through 1987. The chinook interceptions balance now appeared in Canada's favour compared with pre-Treaty years, with each side now intercepting about 30% of the other's Total Allowable Catch.

Regarding Fraser River sockeye and pink salmon, Canada gained the authority to set escapement targets to accelerate rebuilding of these stocks. Canada controlled management targets. However the mandate provided in the Treaty for the Commission to consider harvests of Fraser-origin salmon no matter where they are caught appears to have reduced Canada's discretion in regulating its own fisheries.

Sockeye salmon abundance increased substantially during the latter half of the 1980's. The Pacific Salmon Treaty incorporated an 8 year harvest sharing arrangement for Fraser sockeye and pink salmon that was intended to:

1. Reduce the U.S. harvest share of the TAC from about 35% to about 27% for sockeye and 33% for pink salmon;
2. Cap the total U.S. catch in the 1989–92 period at 7 million sockeye (less than 15% of the TAC) and 7.2 million pink salmon.

While the Pacific Salmon Treaty provided benefits, problems emerged during the implementation of the Treaty. Although Canada was able to harvest more sockeye per year than it would have without the Treaty, the U.S. exceeded its 4 year ceiling of 7 million sockeye. In 1992 the U.S. deliberately exceeded its 1992 entitlement of 360,000 Fraser River sockeye by 337,000. This occurred at a time when several chapters of the Treaty were about to expire at the end of 1992 and were subject to renegotiation. These included the chapters dealing with the U.S. fisheries for Fraser River sokeye and pinks, chinook ceilings in northern and southern B.C. and southern Alaska, coho and chum ceilings in Washington State and southern B.C. and the northern transboundary rivers (the Stikine and Taku).

Canada was dissatisfied with the U.S. overfishing of its Fraser River entitlements. Many Canadians felt that further controls on the Alaskan fisheries were required. In 1992, there were record interceptions of Skeena and Nass River sockeye in the Alaskan fisheries. Overall, there was a continuing imbalance in interceptions on a coastwide basis with the U.S. reaping the benefit.

At the commencement of the renegotiations, Fisheries Minister Crosbie drew attention to the "large imbalance of interceptions on a coastwide basis that the U.S. has enjoyed at Canada's expense" and indicated that Canada was determined to move the Treaty toward its original intent of a balance of interceptions (DFO 19921). To this end, he appointed Yves Fortier, former Canadian Ambassador to the United Nations, to handle the negotiations for Canada.

Despite these difficulties, the Pacific Salmon Treaty was one of the brighter aspects of Canada-U.S. fisheries relations in the post-extension era. The full impact of the Treaty will only be apparent over the next two decades or so. This will depend, to a large extent, on whether the renegotiation of the various chapters which expire at the end of 1992 is concluded to the satisfaction of both Parties.

5.11 Other Canada-U.S. Fisheries Relations Issues

During the 1980's there were a number of disturbing developments relating to the trade in fish products between Canada and the United States. Repeated countervail investigations on the East Coast under Section 332 of the 1930 Tariff Act placed a strain on the bilateral fisheries relationship. In the late 1980's, a ruling under the General Agreement on Trade and Tariffs (GATT) rejecting the Canadian requirement that Pacific salmon and herring be landed in Canada for processing touched off a major bilateral fisheries dispute. For a discussion of these events, and their impact on the Canadian fishery, see Chapter 13.

6.0 CONCLUSIONS

The history of Canada's fisheries relationships with its neighbours in the post-extension era is a chequered one. Only three of Canada's six maritime boundaries with its neighbours had been resolved 16 years after extension. One of these, that with Denmark (Greenland), was delimited by mutual agreement prior to the 200 mile zone. The second, the Gulf of Maine boundary, was achieved through third party adjudication (Even there, the boundary in the inner Gulf of Maine, Bay of Fundy area around Machias Seal Island remains unresolved). A third, that with France (St. Pierre and Miquelon), was recently decided by third party arbitration. The remaining three boundaries with the U.S.A — Juan de Fuca Strait, Dixon Entrance and the Beaufort Strait — are less significant in fisheries terms and immediate resolution is less critical.

In general, the attempts to delimit the maritime boundaries through bilateral negotiation suggest that third party adjudication of one form or another is the preferable, if not the only, route to pursue. In the case of the Gulf of Maine and the area to the south of Newfoundland, there was clearly too much at stake for either party to back away from its claim. Because of the sovereignty linkage, neither country could be seen to yield. The decision in the Gulf of Maine case raised some interesting questions about the criteria for a multipurpose single maritime boundary.

Regarding the management of transboundary resources, again the record is chequered. The initial promise of reciprocal fisheries arrangements with the EC, on behalf of Denmark/Greenland, in Davis Strait, gave way quickly in the face of greed and differences of approach to the management of transboundary resources. The EC's refusal to abide by its agreement set the stage for a breakdown of the reciprocal fisheries arrangements. Because of the limited nature of the resources involved here (shrimp being the only one of mutual interest), the separate management regimes established on the two sides of the boundary do not appear to have hurt either the living resources or the fisheries. Recent developments in the northern shrimp fishery and the management approaches of the two parties suggest that some convergence of management approaches may be possible in the next decade, with a return to cooperative management.

Attempts to establish transboundary resource management arrangements with the United States have been less than satisfactory. The 1979 East Coast Fishery Agreement would have established a comprehensive management framework for the major transboundary resources. The efforts by various groups in the U.S. east coast industry to derail the Agreement have come back to haunt them through a boundary decision assigning a significant portion of Georges Bank to Canada. Developments since then suggest that both sides may have been better off had the 1979 Agreement been ratified.

The resource declines of the past few years underline the folly of attempting to manage transboundary resources in a manner which ignores their transboundary nature. The fishermen of both countries are suffering as a consequence, but U.S. fishermen more so. The different management approaches of Canada and the New England

Regional Council are a considerable obstacle to cooperative management. Unless the New England fisheries management regime evolves to limit fishing effort, there is little prospect of improving the current divided approach. There have been signs that some re-thinking is occurring within the New England industry. One can only hope that this will lead to some form of cooperative management for selected stocks.

Except for the impact on Canadian halibut fishermen, the termination of reciprocal fishing arrangements on the Pacific coast does not appear to have hurt non-salmon fishermen. The halibut fishermen suffered by their displacement from traditional fishing grounds off Alaska. However, maintenance of the International Pacific Halibut Commission has fostered continued cooperation in resource management.

For Pacific salmon, the 1985 Treaty offered the promise of a new era of cooperative management between Canada and the United States. Initial signs were encouraging that the regime under this Treaty would foster stock rebuilding and promote a more equitable balance in salmon interceptions. However, problems emerged in the early 1990's which threatened the successful renegotiation of various Treaty chapters which expired at the end of 1992. The future impact of the Treaty hinges on the outcome of those negotiations.

Canada's fisheries relations with France (both Metropolitan France and St. Pierre and Miquelon) in the post-extension era were complicated by the existence of the 1972 Agreement. This was signed before the global wave of 200-mile zones and indeed in advance of the widespread adoption of a management system relying on catch quotas. Relations with France were reasonably amicable in the initial years following the proclamation of 200-mile zones by Canada and France. The considerable overlap in the claimed areas of fisheries jurisdiction did not pose any immediate problems because of a gentlemen's agreement on catch quotas, national shares and flag state enforcement in the disputed zone. Relations deteriorated in 1984 and 1985 when France began to ignore previous understandings and increased significantly its cod catches in Subdivision 3Ps. This was triggered, to a large extent, by the imminent phase out (May 15, 1986) of Metropolitan France trawlers from the Gulf of St. Lawrence. The lack of a defined boundary set the scene for confrontation. Again it was not possible to arrive at a negotiated boundary delimitation. Referral of the boundary dispute to third party arbitration and agreement on an interim package of fisheries quotas calmed the waters temporarily. The boundary decision of June 10, 1992, was clearly in Canada's favour. The 1972 Agreement plus the boundary decision set the scene for negotiations on post-boundary arrangements.

These experiences, taken together, have demonstrated that the goal of cooperative management of transboundary resources is more elusive than the drafters of the 1982 Law of the Sea Convention envisaged. Where the fisheries management systems of two countries differ radically, this task is even more difficult. This is the case with Canada and the United States on the East Coast. But the era of 200-mile fishing zones and EEZs is still in its infancy. The lessons learned from the initial experiences should help to forge new approaches to cooperative management in the decades ahead.

CHAPTER 13

RIDING THE ROLLER COASTER

"The living provided by the fishery was continually buffeted by the vagaries of nature and world markets. People whose existence depended upon the fishery took as a matter of course the roller coaster nature of the industry as boom and bust alternated every few years.... There have been good years to be sure, but they have been part of a cyclical boom and bust pattern that has not captured the potential of the resource with any semblance of stability."

– M. Kirby, 1982

1.0 INTRODUCTION

Many observers of Canada's marine fisheries have noted the characteristic boom and bust patterns which have plagued the fisheries for decades, indeed centuries. This comment by the Task Force on Atlantic Fisheries was directed towards the Atlantic groundfish fishery. Similar phenomena have been documented for other species and other geographic areas. There is no dominant pattern which describes these up-and-down surges for all species and areas. Given the diverse nature of the Canadian fisheries and the often-unpredictable fluctuations in the resource base, the periods of prosperity and downturn differ from fishery to fishery. There is an underlying instability which continues to plague Canada's marine fisheries, despite the best efforts of governments, Royal Commissions, Task Forces, and other groups to identify causes and suggest solutions.

This boom and bust phenomenon has perhaps been most evident in the Atlantic groundfish fishery. Most of this chapter will examine this fishery and governments' attempts to control the roller coaster. Before discussing Atlantic groundfish, however, I will examine briefly the boom and bust phenomenon in certain other Canadian fisheries.

2.0 EXAMPLES OF FLUCTUATIONS IN VARIOUS FISHERIES

Various fisheries have experienced relatively rapid surges of prosperity followed by equally rapid downturns. These surges are due to a number of factors, either acting in isolation or in combination. One of these is the often volatile nature of the resource base. Despite decades of scientific investigation, it is still not possible to predict with certainty long-term changes in fish abundance or availability. Such changes can have a dramatic effect upon the year-to-year fortunes of fisheries for particular species.

Perhaps one of the most dramatic examples of this over the past several centuries has been the repeated "failures" in the inshore cod fishery along northeast Newfoundland and Labrador. Long before the advent of offshore trawling for northern cod, the inshore fishery was plagued by substantial variability in catches (Fig. 13-1). This variability was sometimes enhanced by changes in fishing effort, for example, during the two world wars. However, there has been an underlying variability in abundance and the availability of this cod stock to inshore fishing gear. This has persisted to the present day. This fishery, which until the 1950's, was based on salting of fish onshore, continues to alternate between years of feast and years of famine. Superimposed on the variability in resource availability were market swings. The impact of this variability upon inshore fishermen has been dampened in recent years by income support programs such as Unemployment Insurance for Fishermen (see Chapter 14). Nonetheless, the boom and bust phenomenon is still very much evident. This is well exemplified by the sudden, drastic and unexpected decline in the abundance of northern cod in 1991, resulting in the imposition of a 2-year moratorium on fishing for northern cod commencing in July 1992.

Dramatic year-to-year fluctuations in resource abundance have shaped the development and decline of various fisheries. On the east coast, two prominent examples were the development in the late 1960's and early 1970's of a large-scale fishery for redfish in the Gulf of St. Lawrence and the purse seine fishery for southern Gulf herring. Since its inception in the 1950's, the Gulf redfish fishery has depended on very few,

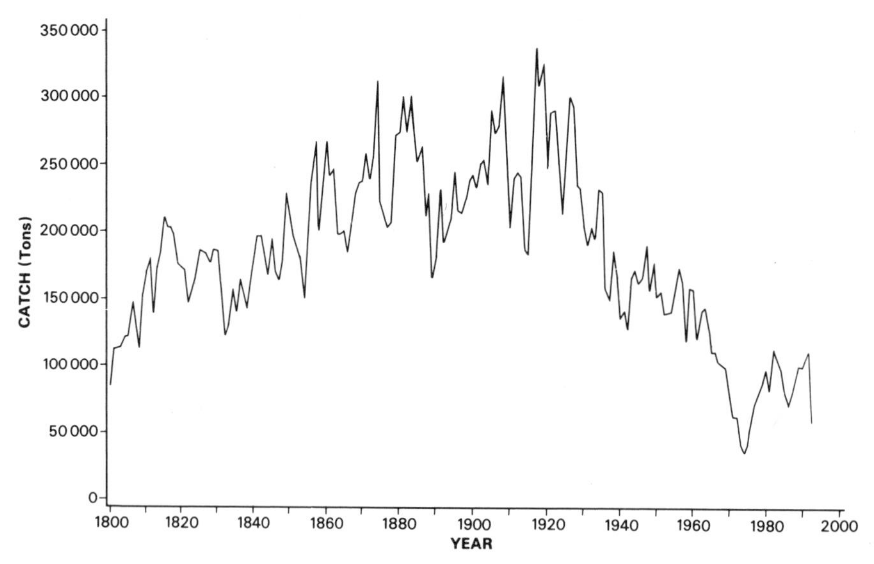

FIG. 13-1. Trends in inshore catches of northern cod: 1800–1991. In July, 1992, a 2-year moratorium was introduced.

infrequent large year-classes. The largest of these were the year-classes of 1956 and 1958 which sustained a dramatic expansion of the Gulf redfish fishery from less than 10,000 tons in the early 1950's to 130,000 tons in 1973. This was followed by a sharp downturn to less than 20,000 tons in the late 1970's (Fig. 13-2). Subsequently, the fishery recovered in the 1980's to the 30,000–40,000 ton range based upon moderately successful year-classes of the early 1970's. For about 10 years, the Gulf redfish fishery provided the main resource supply for certain plants on the south coast of Newfoundland and in Nova Scotia. Following the near-collapse of this fishery in the mid-1970's, the Canadian offshore trawlers which had fished this stock began to pursue new species and areas. During this period Canadian trawlers began to fish northern cod in the winter off northeast Newfoundland and Labrador, with government financial assistance under the Groundfish Vessel Dislocation Program.

Similarly, a purse seine fishery for southern Gulf herring developed rapidly in the late 1960's, based upon the large 1958 year-class of autumn spawners and the 1959 year-class of spring spawners. The federal and provincial governments encouraged development of this fishery following the closure of the B.C. herring fishery in 1967. Some purse seiners from British Columbia moved to the Atlantic coast to fish herring in the southern Gulf, along southwest Newfoundland and in the Bay of Fundy. The boom in the southern Gulf-southwest Newfoundland fishery lasted 5 to 6 years. This was followed by a collapse in the early 1970's after the fishing down of the two large year-classes. Since that time this fishery has fluctuated on a smaller scale (Fig. 13-3).

The squid fishery in the late 1970's was another dramatic Atlantic coast example of a short-term boom. Traditionally, there had been a small scale squid fishery by Newfoundland inshore fishermen, primarily for use as bait. Squid occurred irregularly in Newfoundland coastal waters. During the 1950's, landings ranged from 800 to 8,200 tons but were generally in the 3,000 to 7,000 ton range. During the 1960's, landings ranged from 24 tons to 10,800 tons. During the early 1970's, landings ranged from 26 tons in 1972 to 3,300 tons in 1975 but thereafter increased rapidly to a peak of 113,000 tons in 1979. This was 10 times the previous high prior to 1975. From 1977 to 1980 the squid fishery

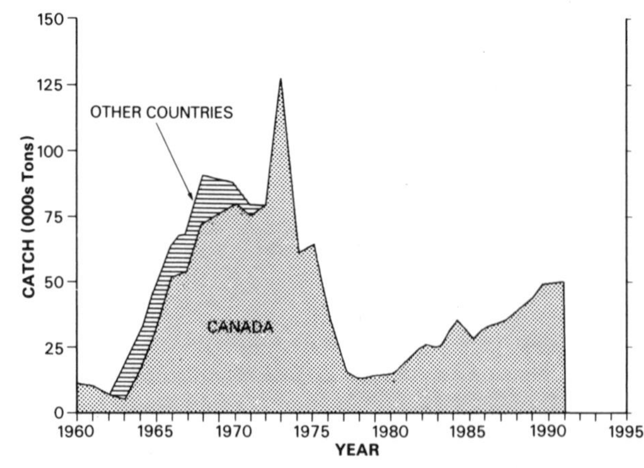

FIG. 13-2. Redfish catches in the Gulf of St. Lawrence (4RST): 1960–1991.

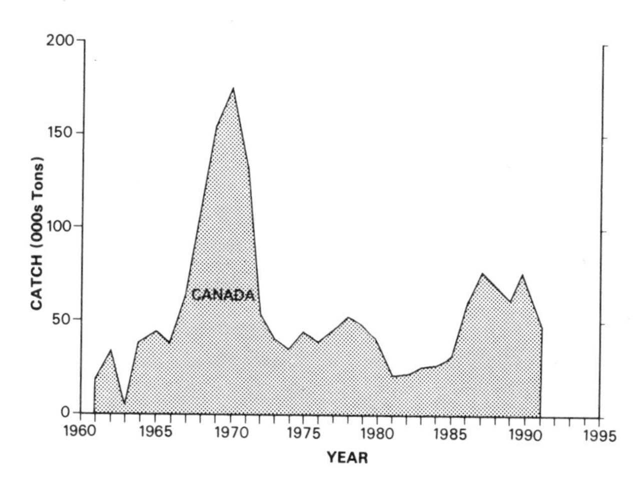

FIG. 13-3. Herring catches in the Southern Gulf of St. Lawrence (4T): 1960–1991.

attracted a great deal of attention as the traditional Newfoundland inshore fishery increased and a substantial offshore fishery developed on the Scotian Shelf. The chief market was Japan. The Atlantic industry developed "squid fever" as everyone saw the potential for big bucks from a resource previously considered insignificant.

There was considerable interest in developing an offshore Canadian squid fishery. So-called "developmental charters" were permitted, ostensibly for the transfer of technology to Canadian companies. In reality, squid allocations were sold to foreign companies, primarily Japanese, often with not much more than a phone call. The Canadian intermediary pocketed a tidy profit. In 1980, the government decided to put an end to these windfall-profit situations by discontinuing the developmental charters program. Shortly thereafter, squid abundance declined dramatically and the fishery collapsed. By 1983, the catch was only 13 tons. Since that time catches have remained at a very low level. The landed value for this fishery, which reached $1 million for the first time in 1976, increased rapidly to $30 million in 1979. By 1982, the landed value had plummeted to $2 million and since then it has remained at its former level, below $1 million. The rapid development and demise of this fishery was based on an unusual short-term surge in the abundance of squid in Canadian Atlantic waters. It is a classic case of a short-term boom with no lasting economic impact.

On the west coast, the valuable Pacific salmon has shown considerable variability in landings with pronounced peaks and troughs. Beamish and Bouillon (1993) have examined the historical Canadian catch of pink, sockeye and chum salmon for the period from 1925 to the late 1980's. They found that total catches were high from 1925 to the early 1940's, were slightly lower up to the early 1950's, increased briefly to the mid-1950's, and remained relatively stable until the early 1980's. There were two periods of high catches in the late 1930's and in the 1980's. The pattern of pink salmon catches was quite variable but trended upward in the late 1970's. Average chum salmon catches were relatively stable from 1928 to the mid-1950's when the average catch declined. There was some recovery in the 1980's. Average sockeye salmon catches did not change until a small increase in the mid-1950's; average catches were lower from the late 1950's until the mid-1960's. Average catches increased from the mid-1960's until the late 1970's and again beginning in the early 1980's. Beamish and Bouillon (1993) suggested that climate and the marine environment may play an important role in determining salmon production (see Chapter 18).

Given the complexity of this fishery for five species and the species' differing cycles of abundance, the downturn in the fishery for some species is sometimes masked by the upswing in the abundance of others. Landings have varied from 90,000 tons in 1951 down to 34,000 tons by 1960, up to 74,000 tons in 1962, back to 41,000 tons in 1965 and so forth. Landings were 50,000 tons in 1984 and 100,000 tons in 1985. Similarly, despite a generally increasing trend in salmon prices, landed values of Pacific salmon also fluctuated substantially from 1972 to 1988. Landed value doubled from $47 million in 1975 to $92 million in 1976, increased to $160 million in 1979, then dropped to $117 million in 1980. It increased again to $165 million in 1972 and dropped sharply to $111 million in 1983. From $111 million in 1983, landed value increased dramatically to $247 million in 1985. Clearly the prices and volumes of landings have fluctuated widely, contributing to an ongoing instability in the Pacific salmon fishery (Fig. 13-4). This underlying instability has been masked, to some extent, by the general upsurge in resource abundance of various salmon species throughout the North Pacific in the latter half of the 1980s.

Landings of Pacific herring also have pronounced peaks and troughs. During the days of the purse seine fishery for reduction, landings dropped from 180,000 tons in 1950 to 86,000 tons in 1952, then increased to 223,000 tons in 1956. Following a series of ups and downs, landings increased to a peak of 259,000 tons in 1963, then dropped to 53,000 tons in 1967. DOF then closed the fishery because of concern about the state of the stocks. Following the re-opening of the fishery for food purposes in the early 1970's, landings again fluctuated, increasing to 97,000 tons in 1977 but dropping sharply to 25,000 tons by 1980. During the 1980's, landings fluctuated in the 25,000 to 40,000 ton range (Fig. 13-5). The 1977 peak of 97,000 tons (since 1970) was associated with the so-called Klondike days of the Pacific roe-herring fishery.

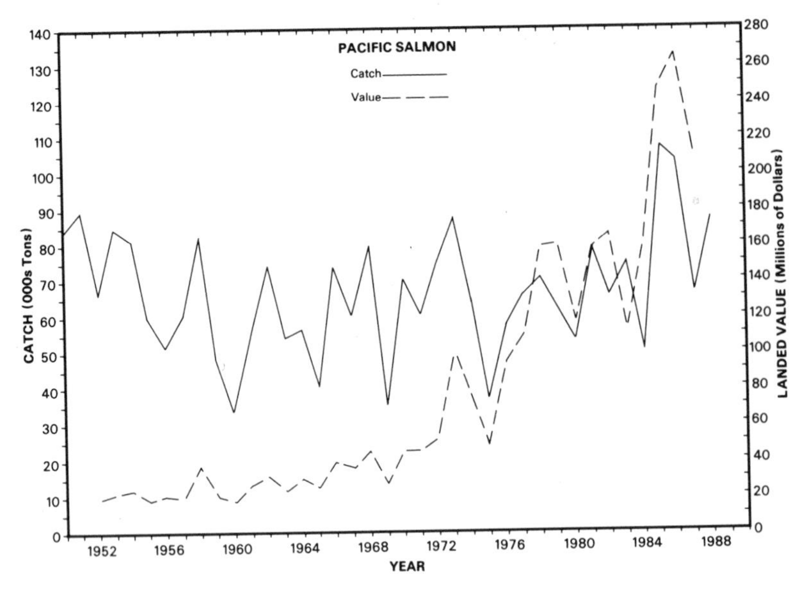

FIG. 13-4. Landings and landed value of Pacific Salmon in British Columbia: 1950–1988.

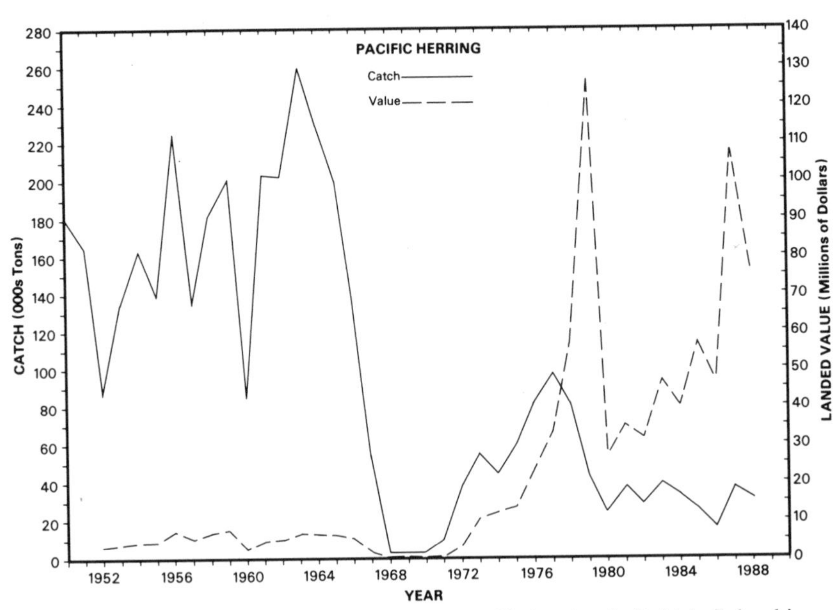

FIG. 13-5. Landings and landed value of Pacific herring in British Columbia: 1950–1988.

Resource variability is only one of the factors contributing to instability in the Canadian fishery. Another major factor has been dramatic swings in the market demand and prices for certain products. The wild swings in the market for roe herring in the mid-1970's are a good example. With the revival of the herring fishery in the 1970's, a major new fishery product, herring roe, developed for export to Japan. British Columbia became Japan's principal supplier with the collapse of Chinese herring roe production. Traditionally, salmon had accounted for more than 80% of the market value of the B.C. fishery. By 1979, with the development of the roe herring fishery, salmon contributed just over 50% of the value and herring roe almost 35%. Herring landings had peaked in 1977 but landed value peaked sharply in 1979. This occurred because of the dramatic increase in prices received for roe herring in 1977, 1978 and 1979. In 1978, Japanese companies paid as much as $3,000 a ton, cash on the fishing grounds, for roe herring. In 1979, price resistance developed in Japan. This led to a drop in the prices paid to B.C. fishermen for roe herring. Faced with dramatically reduced prices, in 1980 union fishermen refused to fish for herring. Consequently, landings dropped to 25,000 tons and the market value of herring dropped to less than 20% of its previous high. 1980 was also a poor year for salmon, with landings dropping to 54,000 tons and landed value to $117 million. As a result, the overall market value of the B.C. fishery declined by almost 40%. Until 1985, it remained less than two-thirds of the previous high.

These circumstances in the B.C. herring and salmon fisheries, combined with an overinvestment in the fleet which occurred in the late 1970's, led to a crisis for fishermen. There were excessive inventories of Pacific canned salmon. Following the bonanza in 1979 in herring, with gillnet vessels averaging gross earnings of $50,000 and seine vessels, on average, about $258,000, the 1980 roe herring prices collapsed from $1.34 per pound to 52 cents per pound. As a result, fishermen faced a severe economic squeeze by 1980. While market prices and earnings were decreasing, costs of production increased and interest rates soared. This meant that fishermen could not repay their investment in vessels, gear and equipment. From 1980 to 1984, fishermen were confronted by loan defaults, boat arrests, repossession and forced sales. The Pearse Commission recommended reducing fleet overcapacity. The government's response to this situation, and the recommendations of the Pearse Commission, is described in Chapter 8.

Thus, boom conditions of the late 1970's were followed by a bust in the early 1980's. Fishermen who survived experienced an upswing again from 1985 onward. These ups and downs on the west coast were somewhat similar to upheavals in the Atlantic groundfish fishery over the past four decades.

3.0 THE ATLANTIC GROUNDFISH FISHERY

3.1 The Decline of the Saltfish Industry

The fishery for Atlantic groundfish is the dominant fishery on the Canadian Atlantic coast. It is export-oriented and in recent decades has been particularly dependent on the U.S. market. For centuries the primary production was saltfish. Ryan (1986) traced the origins and early history of the saltfish industry in Newfoundland. Alexander (1977) described the decay of the Newfoundland saltfish trade from the mid-1930's to the mid-1960's.

Ryan (1986) documented fluctuations in Newfoundland's saltfish exports during the century from 1814 to 1914. He described the history of

Saltfish drying on flakes.

Newfoundland's saltfish industry during this period as "fluctuations around a declining trend", on a per capita basis. Periodic catch failures, inferior curing and developments in market countries and among competitors combined to hinder the production and sale of saltfish from Newfoundland.

Nova Scotia's saltfish industry developed in the late eighteenth century. The total production and export of saltfish from Nova Scotia increased significantly during the second half of the nineteenth century but declined considerably after the 1880's. There was a further more gradual decline in the early twentieth century (Ryan 1986).

By the early 1880's the Newfoundland saltfish industry was also in trouble and faltered further in subsequent decades. Ryan (1986) concluded that the problems of Newfoundland's saltfish industry resulted from the nature of the fishery, with its relatively short season, and from the nature of the economy. At the same time, there were external (market) factors which influenced the relative success of the industry.

The saltfish industry in both Newfoundland and Nova Scotia experienced considerable problems during the years between the two World Wars. The situation was exacerbated in Newfoundland by the virtual collapse of the economy and the establishment of a Commission of Government. The problem in Nova Scotia was mitigated because Nova Scotia fishermen and merchants were a much smaller portion of the labour force and contributed less to the provincial economy. Also, a fresh-frozen industry was beginning to develop.

Stewart Bates, in his 1944 Report on the Canadian Atlantic fishing industry, concluded that the industry would have to be re-oriented to fresh and frozen markets in central Canada and the United States. He suggested that the saltfish industry should be integrated with fresh and frozen operations to provide an outlet for poor-quality fish caught by trawlers and a "safety net" when fish gluts occurred (Bates 1944). Bates recommended modernization and the use of new technology.

In Newfoundland, the Fisheries Post-War Planning Committee also concluded that the saltfish trade to European markets would decline and be replaced by a fresh/frozen trade to the U.S. market (Newfoundland 1946). The Commissioner for Natural Resources, P.D.H. Dunn, stated that the new industry would have to be based on the production of frozen fish. This would require a large capital investment in new catching and processing methods. Dunn indicated that many existing communities would have to be closed down and their populations relocated to a few major ports (Alexander 1977).

3.2 Groundfish Trends — 1945 to 1965

In 1944, the federal government enacted the Fisheries Price Support Act (S.C. 1944-45, c.42). That Act provided for a Board with up to five members including a chairman and vice-chairman. Section 9 of the Act provided the following instructions for the Board:

> "In prescribing prices...., the Board shall endeavour to ensure adequate and stable returns for fisheries by promoting orderly adjustments from war to peace conditions and shall endeavour to secure a fair relationship between the returns from fisheries and those from other occupations.

The Act provided a working fund of $25 million (removed in the early 1980's) and established methods to support the prices of fisheries products. These methods were:

(a) the purchase and sale of any fisheries products; and
(b) the making of deficiency payments to producers of fisheries products equal to the amount by which the average prices of such products fell below prescribed prices during a specified period.

The Annual Reports of the Fisheries Prices Support Board (Various Years) generally contain a short section on economic conditions in the Canadian fisheries for the year under review. These reports outline trends in market conditions and prices during the 1950's and 1960's. The policy emphasis in the post-war years turned to the modernization of fleet and processing plants, and the development of an offshore trawler fishery to supply fresh and frozen fish primarily to the U.S. market. This thrust flowed from the Bates Report of 1944. Government gradually withdrew support for the traditional saltfish industry in favour of building up the capability to supply the frozen groundfish market in the U.S.

Alexander (1977) traced the decline in the saltfish trade during the period from 1945 to 1965. He described the policy of the federal and provincial governments during this period as one of neglect for the saltfish industry.

The immediate postwar years involved adjustment to peacetime conditions. The Fisheries Prices Support Board considered the years 1948 to 1954 to be generally "good" years for the industry. By the end of 1951, however, there were signs of a downturn.

In 1952, catches and landed value dipped. This was largely due to poorer results in British Columbia. Generally, 1952 was a good year for the fishery in the Maritimes, Québec and Newfoundland. The FPSB noted that, since 1947, annual North American consumption of groundfish fillets had grown from 0.6 pounds to 1.2 pounds per capita. This doubling of consumption had permitted an increase in Canadian production from 21,000 tons to 47,000 tons. The build-up had been based on greater availability of haddock and increased fishing for redfish and flatfish, rather than an increase in cod production.

The 1952 heavy production of fresh and frozen fillets resulted in large inventories of frozen products towards the end of the year. In 1953, the Atlantic catch was lower than previously. In Newfoundland, effort and production in the saltfish industry diminished in contrast with increased activity in the fresh and frozen segment of the industry. This was reflected by a decrease of over 7000 in the number of inshore cod fishermen, as well as a reduction in the number engaged in the Labrador fishery. There was expansion of effort in the fresh and frozen segment of the industry throughout the Atlantic.

Prices to fishermen for the major groundfish species were "at rather depressed levels" from late 1952 until the late summer of 1953, largely because of lower prices for fresh and frozen fish on the American market. The U.S. fresh and frozen market improved during the later months of the year and returns to fishermen improved (FPSB 1953–54)

This dip in the Atlantic frozen groundfish industry from late 1952 to mid-1953 marked the first downturn since the Second World War. The Fisheries Price Support Board did not consider the situation serious enough to intervene.

1954 proved to be an excellent year, on both coasts. On the Atlantic, all provinces except Québec had increased landings. Overall, there was an increase over 1953 landings of approximately 68,000 tons. This was due to substantially higher landings of cod and haddock.

In 1955, there was a substantial decline in landings of cod in Newfoundland, largely the result of a disappointing trap fishery on the northeast coast. Nonetheless, the output of Atlantic groundfish in general in chilled and frozen forms increased.

By 1956, the FPSB observed that "a review of the statistics of 1956 production and income from coast to coast.... indicate[s] a widespread measure of stability and growth."

From the FPSB Reports, there would appear to be little evidence of any serious crisis in the frozen segment of the Atlantic groundfish industry during the early to the mid-1950's. There was a slight downturn in 1952–53, but this was relatively modest and insufficient to warrant intervention by the Fisheries Prices Support Board. There were problems during the 1950's but these related generally to a general decline in the saltfish industry and "poor years" in the Newfoundland inshore cod trap fishery. The Labrador floater fishery came to an end in 1954. Market returns were not covering production costs and federal Fisheries Minister James Sinclair told the Newfoundland Fisheries Board that "specific areas of production or specific products will not be denied consideration of price support but when consideration is given it will be based on price changes and not on the basic low income of the area in question" (Minutes of the Newfoundland Fisheries Board, July 5, 1954, quoted in Alexander 1977).

The Newfoundland saltfish industry was declining during these years. Fishermen in 1953 appealed for assistance to the Fisheries Prices Support Board. The Board bought $950,000 worth of saltfish for relief shipments to Greece and Korea. At the end of the year it also provided a deficiency payment of $1.12 per quintal.[1] Saltfish markets were again poor in 1954, but a request for assistance to the FPSB was refused. Parliament initiated a Salt Assistance program, administered by the FPSB, which provided a 50% rebate on salt. This program continued until March 1969, as a small, permanent subsidy to fishermen salting their catch. The FPSB continued to report inadequate prices and incomes of fishermen in the saltfish industry. In 1958, the Board recommended a community-stage building program in lieu of price assistance. Apart from this, the Board did not intervene for the saltfish industry until 1968 when markets were depressed by a series of currency devaluations. Alexander (1977) observed:

> "The saltfish trade was never regarded as a jewel in the crown of the Federal Department of Fisheries. The inclination was to regard it as a primitive industry producing an obsolescent product."

Jones (DFO, Moncton, pers. comm.) has pointed out that the 1958 recommendation of the FPSB to provide assistance to build community stages had important implications later for the fish processing sector. This, coupled with the advent of comprehensive regional development programs in the late 1960's, resulted in the construction of numerous community stages in Newfoundland. Most of these community stages later became, with the help of regional development programs, fish plants, as the saltfish trade declined and the industry shifted to a fresh frozen trade.

Regarding the fresh and frozen groundfish industry, the FPSB reports indicate that the years 1957 to 1965 were generally favourable.

For 1959, the FPSB observed:

> "Later in the year a serious weakness occurred in the United States market for groundfish fillets and blocks as increased supplies, including fish from European sources, tended to oversupply the market. Storage stocks both in Canada and the United States reached excessive levels toward the end of the calendar year although improved demand early in 1960 resulted in some liquidation of stocks and a somewhat stronger tone to the market."

[1]Quintal — 112 pounds of split, salted, dried cod.

For 1960, the FPSB reported:

> "Major export markets for fishery products showed considerable strength throughout 1960 although as the year progressed, the renewed strength of the fresh and frozen market was offset by weaknesses in some important markets for salted and cured fish. This was a reversal of the experience of 1959."

In its report for 1961, the Board observed:

> "During 1961 the North American market for groundfish products-fillets, sticks and portions continued its recovery from the difficult position that had been encountered in the winter of 1959–60."

Thus, again the so-called crisis of 1960–61 was in reality a minor market downturn in the winter of 1959–60. There is little evidence to support the suggestion that the slight market downturns in 1952–53 and 1959–60 represented significant groundfish crises comparable to those which occurred in the late 1960's, 1974–75, 1981–83 and 1989–93.

On the other hand, it is possible that these market downturns may have been just as severe but, because of structural changes in the industry, the effects of the later downturns were much more severe.

The years from 1961 to 1965 were good years for the Atlantic groundfish industry, even though the seeds for later crises were being sown through the build-up of a massive foreign fishery off the Canadian Atlantic coast. Resulting declines in stock abundance and catch rates were not felt until the late 1960's and early 1970's. Between 1960 and 1963 the value of exports of fishery products increased by 25%. Part of this was due to depreciation of the Canadian dollar, but a significant part was due to higher unit prices in the U.S. and other markets.

1964 brought a new record high in volume and value. The value of fishery exports jumped 17% over the previous year. Again, this resulted primarily from higher unit prices in the market rather than an increase in the volume of the catch.

Market demand had been growing slowly but steadily, especially in North America. The combination of a relatively stable supply and rising demand was increasing prices. In 5 years the export value had increased by approximately 50%.

3.3 The 1967–69 Groundfish Crisis

In 1966, new records were achieved both in quantity and value of landings. In Newfoundland, landings surpassed all post-Confederation records. The relatively poor cod trap fishery was offset by higher landings from the growing trawler fleet. Landings of most other groundfish species were substantially higher than in 1965. However, there was a significant price decline in the U.S. market for some products in the fall of 1966, particularly Atlantic groundfish. This resulted from larger than usual supplies of groundfish fillets and blocks available in the U.S., combined with an apparent slackening in the rate of growth of consumption of fish sticks and portions. High inventories depressed prices. The price of frozen blocks and consumer packs of groundfish fillets declined sharply. By April 1967 prices had dropped 25% from the mid-1966 peak.

In 1967, the problem which had emerged late in 1966 worsened. Nationally, the upward trend of the previous several years in landings and value of the fishing industry was reversed. Despite the increased investment in trawlers, Atlantic groundfish landings declined 7% from a combination of poor market prices and scarcity of some species. Landed value also declined by 8%.

The production of frozen groundfish fillets and blocks decreased by 11% from 1966 to 1967. This resulted primarily from weak market conditions in the United States where most of the Canadian production was sold. The over-supply situation which began to develop in the U.S. market during the summer of 1966 contributed to weak market conditions during the remainder of 1966 and throughout 1967. All types of groundfish fillets and blocks were affected, but cod blocks were affected most. The wholesale price of cod blocks at Boston which was firm at around 29 cents (U.S.) in the first part of 1966 began to decline in July and reached a low of 21 cents in March 1967 (Fig. 13-6). The price began to recover in May and by the end of 1967 it was around 26 cents (U.S.) (Fig. 13-6). U.S. imports of groundfish fillets decreased by 7% from 1966 to 1967.

In 1968, the Canadian supply situation changed with fishermen on the Atlantic coast landing a record catch of 1.134 million tons, up 214,500 tons from 1967. The landed value was also up but only by $12.1 million. Most of the quantity increase was due to higher catches in the expanding herring purse seine fishery. Groundfish landings also increased somewhat, but the Fisheries Prices Support Board noted that the catch per unit of effort of the trawler fleet "continued to decline."

The market again weakened in the U.S., particularly for cod and redfish. The price of cod blocks decreased from 26 cents (U.S.) per pound in January 1968 to 24 cents in May and 21 cents by July, similar to the low of March 1967 (Fig. 13-6). Thus the recovery of late 1967 was short-lived.

These market trends in 1966–68 constituted the first significant crisis in the frozen groundfish segment of the Atlantic fishing industry. In May 1968, the federal government commenced deficiency payments to fish-

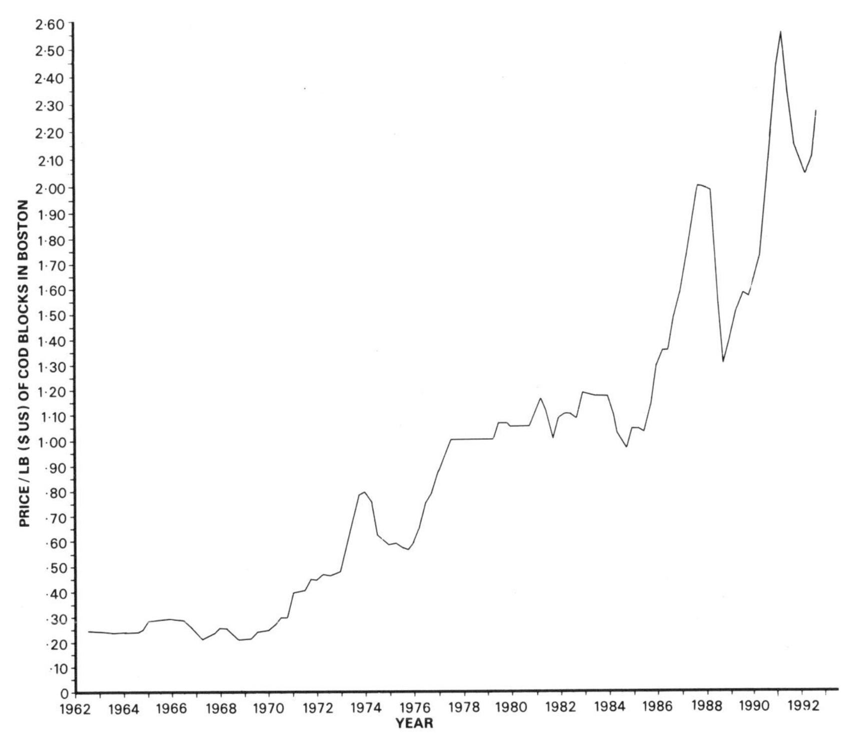

FIG. 13-6. Trends in cod block prices in the U.S. market: 1962–1992.

ermen through the producers of selected frozen groundfish products in the Atlantic. These payments, totalling $4.2 million, were intended to maintain prices to fishermen at or above the previous year's levels. Support was restricted to three species — one cent per pound on cod and one half cent per pound on flounder and redfish.

In 1969, Atlantic groundfish landings declined slightly. The production of groundfish fillets and blocks was about 27% lower in 1969 than in 1968. Prices continued to be weak during the first 9 months of 1969. In May 1969, the FPSB launched a frozen groundfish stabilization program to forestall distress selling in normal markets and to raise the market price of selected groundfish products to cover basic costs. In late spring the price for cod blocks was around 21 cents (U.S.) per pound. The FPSB purchased 7,847 tons of cod blocks (approximately 26% of total production) and 109 tons of fillets at prices ranging from 23 1/2 to 26 cents per pound. The entire inventory was returned, at cost, to the industry before the end of the fiscal year. By the end of 1969 the market price for cod blocks had increased to 26–27 cents per pound (Fig. 13-6). The initial cost of this program was $4.2 million, which was recovered from the industry when the inventory was returned to them.

The crisis in the saltfish industry was more critical. This period was the beginning of the end of this segment of the industry (Alexander 1977; Ryan 1986). The 1967–69 groundfish crisis led to the establishment of the Canadian Saltfish Marketing Corporation in 1970. However, by this time Newfoundland's saltfish production was down to 10,000 tons. The Saltfish Corporation failed to revive the saltfish industry (see Crowley et al. 1993).

3.4 A Brief Recovery — 1970 to 1973

By 1970 the Atlantic groundfish industry appeared to have recovered from the market downturn which began in 1966. Groundfish landings were declining due to foreign overfishing on the groundfish stocks (see Chapter 10). The production of total fillets and blocks decreased from 114,760 tons in 1969 to 107,049 tons in 1970. There were no marketing difficulties as demand exceeded supply, resulting in higher prices. The consumption of groundfish in the U.S. increased from 261,272 tons in 1969 to 282,590 tons in 1970. During the previous 10 years, the U.S. groundfish market had grown at the rate of about 6.5% per year. There was increasing competition for the U.S. cod market. By 1970, Canada supplied about 30% of U.S. imports of cod fillets compared with over 90% of the U.S. imports of flounder and ocean perch fillets. Iceland and Norway had increased significantly their share of U.S. imports of cod blocks and fillets over several years.

Prices of almost all groundfish products increased gradually in the early months of 1970. From September to December of 1970, prices of cod blocks jumped from 30 cents (U.S.) per pound to 40 cents (Fig. 13-6).

During 1971 and 1972 prices of most groundfish products again experienced significant increases. Cod block prices increased from 40 cents (U.S.) per pound in January, 1971, to 48 cents per pound by the end of 1972. In 1973, prices for groundfish products increased substantially. Cod blocks rose to 80 cents per pound, an increase of 31 cents per pound during the year. This was a dramatic increase, when one considers that the price at the beginning of 1970 was around 27 cents per pound (Fig. 13-6). Market demand was growing at a time when supply was decreasing. Landings of cod, the most important Atlantic groundfish species, decreased 19% from 1972 to 1973. Because of the greatly increased market prices, the landed value increased by 13% in spite of the decline in landings. Overall, the market value of all Canadian fishery products increased to $700 million in 1973 compared with $500 million in 1972, an increase of 40%.

3.5 The 1974–76 Groundfish Crisis

The Atlantic groundfish industry had gone from bust in the late 1960's (1966 to 1969), needing government assistance to survive a weak market, to a boom in 1973. But the boom was short-lived. Catch rates of Canadian otter trawlers had been declining since the mid-1960's (Fig. 10-4). The decline was precipitous between 1972 and 1975, when the catch per day fished dropped by 20–25% over 2 years. The declining catch rates increased significantly the cost of catching a pound of fish. Furthermore, in 1973 a major oil crisis occurred when the Organization of Petroleum Exporting Countries abruptly raised the world price of oil. This increased significantly the fuel costs of Canadian Atlantic groundfish fishermen, particularly in the offshore sector (Fig. 13-7). In 1974, the cost of catching fish with large offshore groundfish trawlers had increased to 10–12 cents per pound compared with 5–6 cents in the late 1960's.

Faced with rapidly increasing costs, the industry was ill-prepared for the market downturn of 1974. As

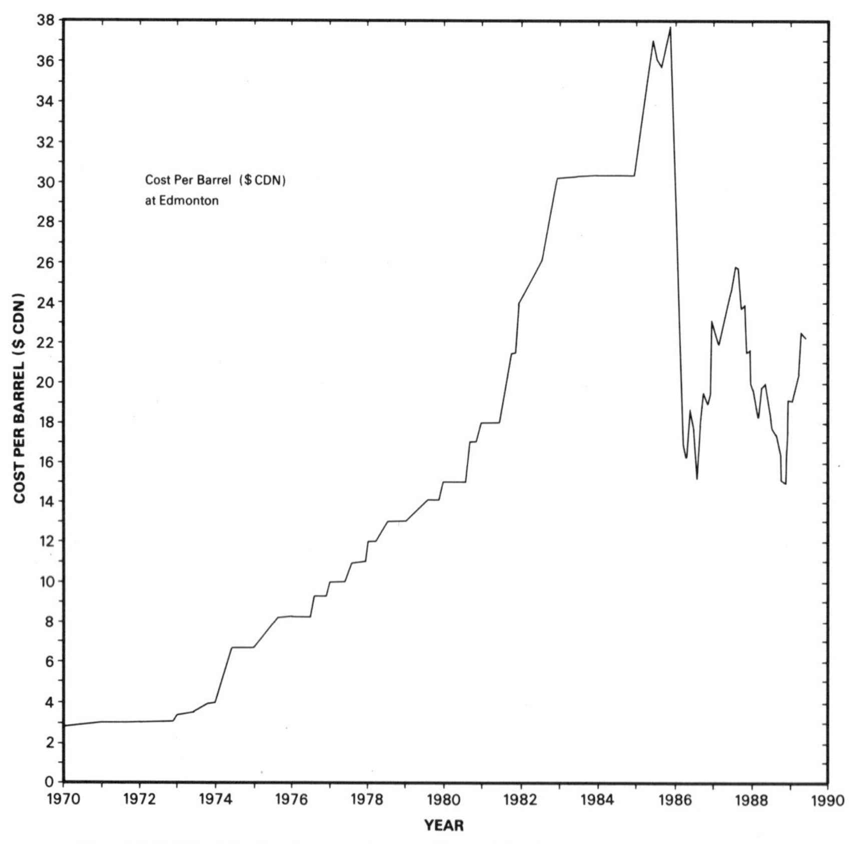

FIG. 13-7. World oil price trends as reflected in the cost per barrel at Edmonton during 1970–1989 (data provided by Energy Mines and Resources Canada).

one indication, cod block prices which had been 80 cents (U.S.) per pound at the end of 1973 decreased to 60 cents by July 1974. Groundfish export prices on the average declined by about 20%. In 1974, total U.S. groundfish consumption decreased 20% from the 1973 record consumption. A significant build-up of inventories occurred in the U.S. and exporting countries.

By mid-1974 the boom conditions of 1973 were replaced by a major crisis in the Atlantic groundfish industry. The industry faced a major cost/price squeeze. Industry leaders appealed to the government for assistance to ensure their survival. The government responded in July 1974, authorizing working capital loans and assistance for inventory financing and product promotion at an estimated cost of $10 million. This program covered all frozen groundfish products as well as canned and frozen lobster and crab meat. This program was later extended to the end of March, 1975, at an estimated additional cost of $4.5 million (DOE 1974a). At the same time, Fisheries Minister LeBlanc established a Task Force led by Mr. Fern Doucet, Chairman of the Freshwater Fish Marketing Corporation. The task force would examine the causes of the groundfish crisis and recommend long-term measures to rehabilitate the industry.

On December 20, 1974, Minister LeBlanc announced three more measures of assistance "as a first step towards the rehabilitation of the Canadian groundfish industry." These were to keep the groundfish fleets operating during the winter while longer term solutions were sought. The $20 million package included:

1. $14 million for short-term deficiency payments (conditional cash grants) on frozen groundfish production for the period January 1 – April 30, 1975;
2. $3 million to salvage frozen groundfish inventories by canning and making this canned product available as Canadian food aid products; and
3. $3 million for working capital loans to proprietors of plants affected in early 1974 by ice conditions in northeastern Newfoundland, Labrador and the lower North Shore of Québec (DOE 1974b).

In April 1975, the Doucet Task Force presented its recommendations. These called for new initiatives in resource management, harvesting, processing and marketing. In announcing a new $51 million "bridging" program of assistance on April 23, 1975, Fisheries Minister LeBlanc commented on the Task Force's conclusions:

> "Some of these recommendations call for better management of stocks and a reduction of effort in overcrowded fisheries, both national and international; fleet rationalization and development; entry control; an increase in efficiency in both primary and secondary industries; more orderly marketing; and a stabilization plan that would replace the bridging program as soon as possible. A recommendation that there should be no increase in the number of vessels fishing groundfish species and stocks, which are at present fully exploited, is to be implemented immediately" (DOE 1975f).

The new "bridging" program, which took effect on May 1, 1975, included for the first time direct assistance to fishermen. The program had several components:

1. $27 million for deficiency payments directly to groundfish fishermen of 2.5 cents per pound for first quality fish as landed;
2. $14 million for conditional grants to processors of first-quality frozen groundfish fillets and fillet blocks (eight cents per pound of finished product) and first quality fresh groundfish fillets within Canada, provided processors maintained the basic price to fishermen paid on July 1, 1974;
3. $600,000 for deficiency payments to crab fishermen ($300,000) and to crab processors ($300,000);
4. A collection of other small-scale deficiency payments to freshwater fishermen; and purchase of lobster inventory, mackerel, herring, gaspereau and groundfish for food-aid and development programs.

In a speech at the Annual Meeting of the Fisheries Council of Canada in Halifax on May 5, 1975, Minister LeBlanc observed:

> "When I visited this city last fall, some of you told me that, in effect, you might fail to survive the winter. The groundfish industry was sick. Within six weeks of that meeting, the Federal Government applied the miracle drug: money. We announced a $20 million program to keep the industry — companies and fishermen — going through the winter. We'd given the fisheries a $15 million injection only three months before that. And two weeks ago, I announced another $50 million program. It has been enough to make one wonder: Are we dealing here with a hypochondriac, or an invalid, or an addict.... Crisis has followed crisis, partly because of the patient's own bad habits...Let me admit that the fishing industry's bad habits have been only one cause of the trouble it is now in" (DOE 1975g).

LeBlanc went on to identify foreign overfishing as one

of the major contributing factors. He saw a need for increased coastal state authority for management and conservation of the stocks off Canada's coasts. He also identified destructive competition in the marketplace as an underlying problem.

He spoke of the need for more offshore control, for better resource management and improved quality of fish at point of landing and leaving the plant. He called also for the licensing of export groups, starting with groundfish and crab. He said that the government was considering a stabilization program whereby fishermen and processors would contribute in good times and receive benefits in bad times.

In October 1975, Minister LeBlanc announced stricter quality requirements would be applied under the Bridging Program. To continue receiving the program's benefits, fishermen had to improve their fish handling practices. Quality requirements were also applied as a condition of grants to processing plants (DOE 1975h).

In the latter half of 1975, market conditions improved somewhat. The high inventories held in Canada at the beginning of 1975 were reduced to more normal levels by year-end. While the demand and prices for cod blocks remained sluggish, demand and prices increased for most groundfish products during the last 6 months of 1975. The market for flounder and redfish strengthened. The U.S. consumption of groundfish fillets and cod blocks increased by 17% over 1974.

The Bridging Program was reviewed quarterly with changes in eligible species and rate levels. In January 1976, the rate to processors for certain products was reduced. Effective February 16, 1976, redfish, pollock, haddock and Greenland turbot were removed from the list of species eligible for the deficiency payments to fishermen and the conditional grants to processors. The 1975 landings of these species had exceeded those of 1974. The Bridging Program included a provision that groundfish species would become ineligible for the program when their landings reached the level of the previous year.

Despite these improvements, Fisheries Minister LeBlanc established a $44 million temporary assistance program (TAP) effective April 1, 1976. This successor to the Bridging Program provided deficiency payments to fishermen of 2 cents per pound for first quality groundfish landed, gutted or bled, and 1½ cents per pound for first quality groundfish landed, not gutted or bled. This component of TAP was designed to support fishing enterprises facing fewer fish and rising costs (DOE 1976k).

TAP also provided conditional grants to groundfish plants on fillet products produced from first quality groundfish. Processors were eligible for a conditional grant of 6 cents per pound for frozen fillets produced from bled and gutted groundfish and 3 cents per pound for fillets produced from groundfish which had not been gutted or bled. Assistance to processors was conditional upon their maintaining July 1974 prices to fishermen as a minimum.

In May 1976, the federal government released its new Policy for Canada's Commercial Fisheries (DOE 1976a). It recognized that overfishing, rising costs and a market downturn had combined to produce the 1974–75 Atlantic groundfish crisis. However, the document also blamed certain structural defects in the industry. The perceived defects included too much capital and labour for the available fish, and inefficient distribution of industrial capacity. Industry's overcapacity had resulted from competition among fleet owners and fish buyers for a dwindling supply of groundfish, abetted by government loans and subsidies for vessel construction.

There was also overcapacity in the processing sector resulting from competition for supply and buoyant market conditions, again abetted by public assistance programs. This competition for supply in the face of a scarce resource led to deterioration in fish quality.

The 1976 Policy postulated that "fundamental restructuring of the industry is inevitable. It will come about either in an orderly fashion under government auspices or through the operation of inexorable economic and social forces." The document proposed two major policy shifts:

"1. The guiding principle in fishery management no longer would be maximization of the crop sustainable over time but the best use of society's resources. 'Best use' is defined by the sum of net social benefits derived from the fisheries and the industries linked to them.

"2. While private enterprise, individual, cooperative and corporate, would continue to predominate in the commercial fisheries, fundamental decisions about resource management and about industry and trade development would be reached jointly by industry and government."

The strategies to accomplish the new policy objective included:

1. Obtaining national control of fishery resources throughout a zone extending at least 200 nautical miles from the Canadian coasts;
2. Applying systems of entry control in all commercial fisheries;
3. Facilitating price differentiation according to quality of fish landed;
4. Promoting the consolidation of the export-marketing of fishery products;
5. Developing programs to mitigate the effect of

the instability inherent in the commercial fisheries on the net revenue of fishing enterprises; and

6. Providing, through the adaptation of existing programs and/or the design of alternative programs, for the relief of chronically income-deficient fishermen.

Market conditions improved considerably for most Canadian fishery products during 1976. The overall market value increased by 42% compared with that of 1975. There was a strong upward trend in prices for a wide range of groundfish products. Demand was buoyant throughout the year for fishery products generally and for groundfish fillets and fillet blocks in particular. Cod block prices in the Boston market increased from about 60 cents (U.S.) per pound at the end of 1975 to the 87–88 cents range by the end of 1976 (Fig. 13-6).

3.6 The 200-mile Limit and Apparent Prosperity — 1977–79

Despite these improvements in the market place, levelling off of costs and a halt to the resource decline, by mid-1976 parts of the industry continued to operate at a loss. Operating losses had, however, been reduced by government financial assistance and the changing resource, cost and market conditions. Canada's extension of its fisheries jurisdiction to 200 miles brought the bulk of the offshore groundfish resource under its control. This provided the opportunity to rebuild the resource and promised a brighter future. However, the government concluded there was a short-term need to continue the subsidy program. $41 million was provided for 1977–78 under the same conditions as applied in 1976–77. Except for some problem areas (side trawlers and difficulties associated with specific plants), the industry was returning to profitability.

Canadian Atlantic groundfish catches reached 515,000 tons in 1977, up 10% from 1976 and 23% from the low of 418,000 tons in 1974. There were substantial increases in catches of cod, haddock and turbot. Redfish catches, on the other hand, were down because of the substantial decline in the Gulf of St. Lawrence fishery. Overall, the increase in gross returns to groundfish fishermen was close to 30% above 1976 levels (FPSB 1977–78).

The emphasis in fisheries program planning began to shift from interim support to long term management and development. When the Bridging Program was extended for 1976–77 as the Temporary Assistance Program, unused funds were to be applied under the 1976 Fisheries Policy to specific programs to rehabilitate the fisheries. In 1976–77, this cost $4.6 million. Part of the money went to the Atlantic Groundfish Vessel Dislocation Program, to encourage vessels to fish new species and/or areas. In 1977–78, a total of $14.3 million was allocated for Fisheries Rehabilitation measures. These included the conversion of Maritimes groundfish side trawlers to scallop dragging, incentives for fishing in northern areas, the charter of a freezer trawler, and fish chilling assistance.

The assistance program of 1975–76 and 1976–77 was a rescue operation and has been credited with helping the core of the Atlantic groundfish fishery survive until the resource supply and market conditions improved. Overall, during this period more than $170 million in special federal aid had been authorized just to keep the industry alive. That aid amounted to about $8,500 for every fishing enterprise or $2,900 for every fisherman across Canada. The bulk of the money, however, went to the Atlantic fisheries.

In November 1977, Fisheries Minister LeBlanc observed that the industry was beginning to emerge from the crisis which commenced in 1974. He noted that the Atlantic fisheries were not yet a stable industry:

> "The reasons are many: open access, the scattered and insular nature of fishing communities, the cyclical nature of the harvests, the dependence on one main market, the American, the unpredictable competition from foreign suppliers with strong and consolidated marketing agencies, all this helped build a perpetual insecurity.
>
> "These weaknesses — instability and low income — are historic; they far preceded the last two decades, when foreign fishing began to damage our traditional fish stocks and brought on our extension of jurisdiction. The weakness of the Atlantic fishery industry is not a phenomenon of 200 miles, it is a phenomenon of 200 years.
>
> "Along with the transfusion of money to keep major parts of the fishing industry, especially the groundfish industry, alive,...came a change of government attitude. Before, the basic goal of fisheries management was to protect the fish; we regulated on behalf of the resource. The new approach [means] that we will regulate on behalf of people" (DOE 1977a).

LeBlanc mentioned three fundamental changes. The first involved the limitation of entry into the fisheries because open access had "probably been the single biggest cause of the industry's chronic troubles" (see Chapter 8). The second involved the abandonment of the Maximum Sustainable Yield (MSY) approach because "at times it will be better to catch fewer fish than we could, because less fishing pressure makes fish more abundant in the water, easier to catch, more stable in their yield, and more profitable" (see Chapter 4). The third fundamental change was the 200-mile limit.

Following extension of jurisdiction, the federal government limited entry in virtually all fisheries, adopted the $F_{0.1}$ approach to set the harvest level for Atlantic groundfish stocks, and promoted orderly restructuring of the industry. On the marketing side, the government encouraged Canadian exporters of groundfish to form exporting groups to cope with fragmentation. During 1977–78, the Fisheries Council of Canada developed a proposal for a voluntary organization to bring together many of the companies which exported fish. This crystallized in the Canadian Association of Fish Exporters (CAFE) in 1978. Currently, CAFE's activities fall into four areas: market intelligence, promotions, government liaison, and sales coordination.

The need for restraint in harvesting and processing capacity and orderly marketing was soon forgotten as the stocks and markets continued to recover. Fishermen, processors and some provinces were soon swept up by the euphoria of the 200-mile limit and the perceived benefits of displacing foreigners in many fisheries. The federal government, however, cautioned that it would take 5 to 10 years for the groundfish stocks to rebuild and the rate of growth would differ among areas.

Federal scientists suggested that the TACs for the groundfish species traditionally fished by Canada would increase by 40% between 1977 and 1985 (DOE 1978a). However, this potential varied considerably among areas. The projections for future growth fell into three categories:

1. Areas of slow growth — The Gulf, Scotian Shelf and Grand Banks.
2. Areas of decline — Fisheries of the Gulf Entrance.
3. Area of rapid and sustainable growth — Cod stocks to the east and north of Newfoundland.

Scientists also estimated that the existing fleet could catch considerably more than it did in the mid-1970's. Indeed, it required increased catch rates and increased volume to be viable. Resource projections indicated that:

> "Much of the increase in catch of traditional groundfish can be harvested by the existing fleet, which will experience improved catch rates. Overall, the catch rate index for cod was projected to increase by between 40 and 70% between 1977 and 1985, depending upon the stock" (DOE 1978a).

The scientists noted that "predictions of stock status in the 1980's are to a large extent best guesses." Nevertheless, Canadian Atlantic groundfish catches did increase substantially in the post-extension era, peaking in 1982 and 1986 (Fig. 13-8).

In addition to harvesting overcapacity, there was considerable processing plant overcapacity or underutilization. A study of processing capacity in Newfoundland in the mid-1960's indicated that the utilization rate for freezing plants was only 37% based

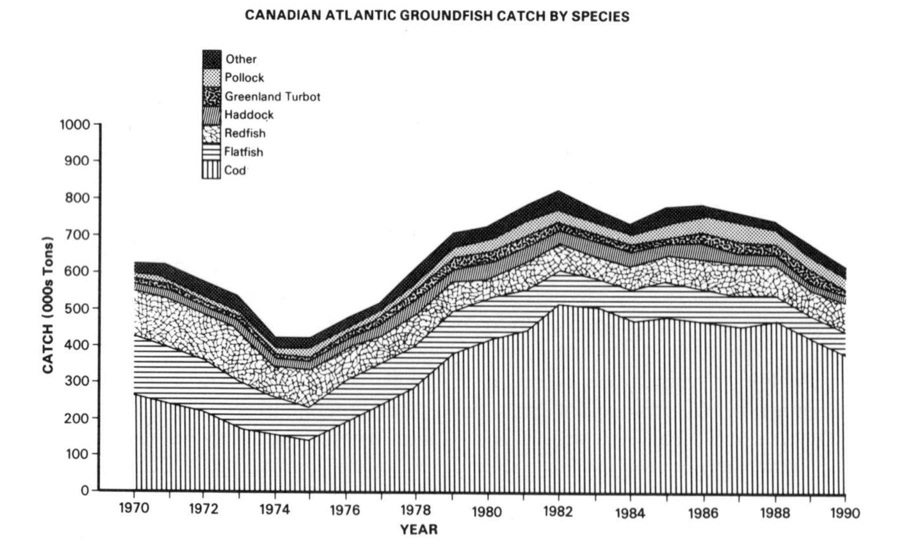

FIG. 13-8. Trends in Canadian Atlantic groundfish catches, 1970–1990.

on a standard year of 250 eight-hour days (Mensinkai 1969). Blackwood (1974), in a study presented at a conference sponsored by the Federal-Provincial Atlantic Fisheries Committee, on the Utilization of Atlantic Marine Resources, indicated that capacity utilization was only 27% for the Atlantic provinces as a whole.

These analyses suggested that the first priority was to better utilize existing harvesting and processing capacity. Despite this, some segments of the fishing industry and the provinces of Nova Scotia and Newfoundland began to develop ambitious plans for fleet and onshore plant expansion. The $900 million fleet development proposal of Newfoundland and Nova Scotia, referred to in Chapters 2 and 6, generated considerable pressure on the federal government to permit expansion of the offshore fleet. Two of the five major vertically integrated companies involved in the Atlantic groundfish fishery launched a major propaganda offensive criticizing the cautious approach of the federal government. They called for major changes in policy to allow fleet expansion. In a booklet titled "Where Now? A Discussion of Canada's Fishery Opportunity and the Considerations Involved in Realizing It", H.B. Nickerson & Sons Limited and National Sea Products Limited declared:

> "Canada cannot reap the potential benefits of the 200-mile management zone if its fleet does not have the capacity to harvest the fish stocks available during the next decade. If it is to take the initiative in fleet replacement and development, the industry requires comprehensive policies that will allow the industry to proceed with confidence in:
>
> - The replacement of the inshore fleet;
> - The replacement of the offshore fleet;
> - The expansion of the inshore fleet and the offshore fleet as fish stocks permit;
> - The development of a new generation fleet of Canadian trawlers, with freezer capability and possibly some on-board processing" (Nickerson and NSP 1978).

At the 1978 meeting of the Fisheries Council, Minister LeBlanc confronted the fleet expansion issue head-on:

> "The present groundfish fleet of larger vessels has the capacity to take half again its present catch, and provide better incomes — if we increase the fish in the water and the catch rates. If we do it the other way around — increase the fleet first — we are like a man with an exhausted woodlot, who instead of planting more trees to get more growth, spends all his money on more chain saws to cut the shrubs. Massive fleet expansion at the moment would be a Titanic undertaking — and I use the word advisedly" (DOE 1978e).

LeBlanc called on the Fisheries Council to provide a clear and responsible voice on fleet development:

> "I would like to see you join me in resisting suggestions that fleets should be vastly expanded, that plants be vastly enlarged — in other words, to resist the temptation of exaggerated expectations. I see no faster road to disaster than forgetting the very simple lesson that the biology cannot keep up with the technology — that the wealth of the oceans cannot yet match the greed of man."

Unfortunately, Minister LeBlanc's prophetic words fell on deaf ears. While he was able to prevent expansion of the offshore trawler fleet, he moved too late to prevent the development of overcapacity in the inshore fleet (see Chapter 8). Meanwhile, there was a significant expansion of onshore processing capacity, particularly in Newfoundland. Several large companies, particularly H.B. Nickerson & Sons Limited, took on heavy debt for this expansion. The number of federally registered fish plants in the Atlantic provinces increased from 519 in 1977 to 700 in 1981. In Newfoundland the number of plants increased from 147 in 1977 to 225 in 1981 (DFO/DREE 1982). Much of the excess capacity was concentrated in a few relatively large plants.

While the industry led the push for plant expansion, assisted by the banks and provincial governments, the federal government was not blameless. A 1982 DFO/DREE Report on Processing Capacity observed:

> "In examining the reasons for such an increase in fish processing capacity, two factors emerge: federal and provincial government incentives to invest in the fish processing industry and the optimism within the industry that was generated by provincial governments following the extension of jurisdiction in 1977. The federal government, through DREE, has assisted 260 enterprises in the establishment of new capacity or the expansion of existing capacity since the inception of its Regional Development Incentives Act (RDIA) in 1969. The total assistance granted since 1969 is in excess of $46 million on a total eligible capital investment of $157 million. In addition to RDIA, other federal programs such as Canada Works have contributed to the increase in fish processing capacity. In

addition to federal programs, most provincial governments have assistance programs for the establishment of fish processing plants. The overall effect of these government support programs or subsidies has been to encourage investment in excess of the normal level" (DFO/DREE 1982).

This expansion occurred while the resource was recovering. The Canadian Atlantic groundfish catch increased by approximately 100,000 tons in 1978, a 19% increase over 1977. It continued to increase to 1982 when it peaked at 820,000 tons, almost double the level of 1974. This increase was primarily driven by cod catches which rose from 146,000 tons in 1975 to 517,000 tons in 1982, a 350% increase in just 6 years (see Fig. 13-8). More importantly in terms of fleet profitability, the catch per day fished of the Canadian offshore trawler fleet increased sharply from approximately 8.5 tons per day in 1977 to almost 14 tons per day in 1981 and 1982, the highest catch rates since the 1950's.

Market conditions for Atlantic groundfish products were also favourable through the late 1970's. As an example, U.S. cod block prices increased from 60 cents (U.S.) per pound at the end of 1975 to $1.00 at the end of 1977. They reached $1.05 at the end of 1979 and stabilized at the $1.05–$1.15 range during 1980 and 1981 (Fig. 13-6).

3.7 The 1981–83 Crisis

Despite these favourable factors, the Atlantic groundfish industry again plummeted into crisis by 1981. This occurred during the severe economic downturn and Great Recession of the early 1980's. The seeds of this crisis had been present for some time. But, since there was no ongoing monitoring of the economic pulse of the fishing industry, it came as a shock to government. Detailed financial monitoring had begun during the Bridging and TAP programs. However, this was abandoned because industry stopped cooperating once the post-extension recovery was in full swing.

By late 1981 it became apparent that two of the five vertically integrated groundfish companies faced a severe financial crisis. These were the Lake Group (LG) in Newfoundland and H.B. Nickerson & Sons Ltd. (HBN) with plants in both Nova Scotia and Newfoundland. Although the crisis was emerging as early as 1980, there was no public evidence of this. The only publicly-traded company, National Sea Products (NSP), had reported profits of almost $20 million in 1978 and 1979.

There were difficulties in the market resulting from the following factors:

1. The U.S. dollar was appreciating against most other currencies and the Canadian dollar followed close behind. Thus Norwegian and Icelandic producers had a competitive advantage over Canadians in the U.S. market. At the same time, U.S. buyers, facing good supplies and exchange rate premiums, could dictate lower prices. These factors created a price situation where profit margins were zero or negative for Canadian suppliers but were still attractive to Scandinavian suppliers.
2. High interest rates in both Canada and the U.S. had reduced the ability of wholesalers and distributors to hold inventory and had shifted this burden to the processors.
3. The relative prices of substitute protein sources such as beef and poultry had fallen, weakening demand for fish products. The block market was flat.
4. With the flat market, and with higher energy, interest, raw material and labour costs, the processors experienced another cost/price squeeze. This and the shortage of cash flow resulted in some distress selling, further depressing the market.

These factors had been developing for months. They resulted in plant closures, vessel tie-ups and requests for assistance beginning in mid-August 1981. Approximately 4,500 plant workers and 1,000 fishermen had been affected by the closure of year-round plants. Also, approximately 3,600 workers had lost income from early closure of seasonal plants.

Despite higher volumes and values of landings in 1981 compared with 1980, the groundfish industry was beset by high production costs and sustained severe losses for the second consecutive year. Following a request by the large processors, the Fisheries Prices Support Board in December, 1981, provided $15 million of temporary assistance to sustain the frozen groundfish processing industry and stabilize fishermen's incomes. This program applied to export sales of frozen fillets and fillet blocks of cod, haddock, redfish and flatfish made between September 1, 1981 and December 15, 1981. It also applied to unsold frozen inventories of Canadian processors from 1981 Canadian production (up to 3,200 tons of redfish and 4,600 tons of flatfish product) (FPSB 1981–82).

Representatives of the Lake Group, H.B. Nickerson & Sons and the Bank of Nova Scotia (BNS), the major industry creditor, informed the federal government that this was only a drop in the bucket: major segments of the industry would be bankrupt without massive government financial assistance.

Fisheries Minister LeBlanc viewed the new crisis as the result of inherent structural and behavioral characteristics of the industry. These magnified the economic distress created by market difficulties. These factors included:

1. The common property nature of the resource and the way access to it is regulated (see Chapter 6).
2. The processing sector's preoccupation with volume and through-put. This led to excessive investment in processing capacity, competition between companies at the port market level regardless of quality considerations, and inconsistent quality in processed product.
3. Fragmentation, lack of marketing innovation and undisciplined competition in export markets contributed to price vulnerability in the face of adverse market conditions or general economic downturns.
4. A history of poor decision-making and unwise investment patterns by the industry.

Recognizing that the banks would force restructuring if the government did not act, Minister LeBlanc advocated a solution involving:

1. Separation of the offshore fleet from the integrated companies and consolidation of its operations;
2. Establishment of a marketing mechanism for offshore-caught fish, through which inshore-caught fish could also be sold, and through which seasonal plant owners could have access to fish to extend their operations;
3. Restructuring of processing operations to improve productivity, while at the same time assuring that the negative social impacts of restructuring were minimized; and
4. Establishment of a consolidated marketing company to channel most, if not all, Atlantic fish products to export markets.

LeBlanc argued that the major companies were cash short and had a crippling debt burden. A substantial infusion of equity would be required over the coming months. In particular, Minister LeBlanc favoured government taking a majority position in the Lake Group.

There appeared to be widespread agreement that the Atlantic fishery had severe structural problems. In addition, there was some support for the proposed quality improvement initiatives, reduced harvesting and processing capacity, and company quotas. DFO had already set such quotas experimentally for 1982. However, many bureaucrats in the central agencies and some of LeBlanc's colleagues rejected his suggestions that the government intervene directly in the industry by taking equity in certain companies. An ad hoc Committee of Ministers on the Atlantic Fisheries was established in conjunction with the decision to approve $15 million of assistance through the Fisheries Price Support Board. This Committee was a response to rapid deterioration of the offshore groundfish industry. Financial collapse of the Lake Group and H.B. Nickerson & Sons Ltd. seemed imminent. The Lake Group was all but bankrupt and had closed most of its processing plants. The Bank of Nova Scotia had in effect taken over the company and wanted to close permanently the plants in Grand Bank and Gaultois.

Against this background, on January 8, 1982, the Prime Minister announced the formation of the Task Force on Atlantic Fisheries under Michael Kirby. It was to "inquire into and report upon the current conditions and future direction of the Atlantic Coast fisheries" and to "report to the ad hoc Committee of Ministers on how to achieve and maintain a viable Atlantic fishing industry, with due consideration for the overall economic and social development of the Atlantic provinces" (Canada 1982b).

The federal government recognized that the Lake Group was in great jeopardy. The Task Force was instructed to negotiate with the shareholders, the provincial government and the Bank of Nova Scotia and recommend the nature and extent of federal government assistance to keep that firm in business for one year. After that the government expected to determine its role in the long-term development of the industry. Following the Task Force's negotiations, the federal government agreed to provide up to $13 million to the Lake Group, as a loan guarantee, as part of a one-year stabilization plan for the firm. This assistance was conditional upon the resumption of normal fishing and processing by the Lake Group, including the Gaultois and Grand Bank operations.

The Task Force intended to examine the structural problems of the industry during 1982. This would involve extensive consultations with all key players in the industry. The Task Force would then make policy recommendations to the government by the end of the year. Should the government decide to assist the industry the Task Force would negotiate with the affected parties.

The Task Force met with H.B. Nickerson & Sons Ltd. and the Bank of Nova Scotia to address a Nickerson request for assistance. In February 1982, the Task Force met with a coalition of processors and fishermen representatives from the Atlantic provinces. They proposed an assistance program similar to that of 1974–77 but were unable to estimate the level of assistance required.

The Task Force agreed there was a major problem. Unfortunately, there was little evidence of the extent of the problem. Studies were being undertaken to provide this necessary information.

In addition to the loan guarantee to keep the Lake Group operating in 1982, the Task Force negotiated the continued operation of the Fishery Products Ltd. plant at St. Anthony, Newfoundland. In January 1982, Fishery Products had announced the permanent closing of this plant. It had not requested either federal or provincial assistance to maintain the operation. The plant was in a relatively isolated part of northeastern Newfoundland, and the Task Force considered it important to the area's economy. There were few other employment opportunities for the plant workers and no significant alternate outlets for local fishermen's catches. Accordingly, the Task Force arranged the incorporation of a new company, St. Anthony Fisheries Ltd. (SAFL) to operate the plant at St. Anthony from June 1, 1982 to December 31, 1982. The Canadian Saltfish Corporation would manage the operation. The federal government gave SAFL a $2 million recoverable grant and $8 million of loan insurance to operate the plant. The operation was successful and earned a small profit.

H.B. Nickerson and Sons Ltd. (HBN) submitted four requests for assistance to the federal government between November 1982 and April 1983. All sought guarantees through separate proposals for each of the Atlantic provinces. The HBN Nova Scotia proposal called for the purchase by National Sea Products Ltd. (NSP) of all of HBN's fleet and some of its plants in Nova Scotia. NSP would pay approximately $50 million and assume a debt of $17 million. The Task Force rejected the Nova Scotia proposals because they would get the Bank of Nova Scotia out of the HBN financial situation without suffering any losses. HBN was now insolvent. If it went bankrupt, the Bank of Nova Scotia would probably lose $30 to $40 million.

In October 1982, the Task Force suspended discussions with the Bank of Nova Scotia about HBN and the Lake Group pending further analysis of financial information. The Bank's proposal by this time centered on the federal government's injecting a large amount of equity into a restructured company. Meanwhile, on November 16, 1982, a third major company, Fishery Products Ltd. (FPL), said that it too was insolvent and needed help. Discussions involving FP, the Canada Development Corporation, which owned 40% of FP, and the Bank of Nova Scotia had failed to produce a satisfactory refinancing plan. Following preliminary discussion with the Task Force, CDC and the bank reached an agreement allowing Fishery Products to continue operating until year-end. By this time, Pêcheurs Unis du Québec was also in severe financial difficulty even though it had not yet sought financial assistance.

Meanwhile, the Task Force had completed its general analysis of the Atlantic fishery and drafted a report containing major policy recommendations. The report was released in February 1983, following Cabinet consideration of its recommendations.

In its report the Task Force analyzed the financial condition of the processing sector based on a study by Woods Gordon of some 100 processing enterprises. The Woods Gordon study revealed that in 1981 the enterprises surveyed incurred a consolidated loss of $57 million, following a loss of $22 million in 1980. The operating income of National Sea Products, one of the strongest performers in the industry, peaked at $21.0 million in 1978 on sales of $217 million. By 1980 NSP lost $4.5 million on sales of $274 million. In 1981, its loss increased to $7.9 million.

While the financial downturn was widespread, it was most severe in the frozen groundfish and herring sectors. Simplistically stated, the crisis was the result of costs rising more rapidly than revenues. Unlike 1974, fish abundance was increasing and hence resource supply was not a factor in this crisis. At that time, 1978 was the most profitable year since extension of jurisdiction and 1981 the least. Between 1978 and 1981 net income of the sample companies dropped by 10.6 percentage points of sales. Pre-tax income declined by 14.2 percentage points of sales. The principal contributing factors were a decline of 7.1 points in gross margin and an increase of 5.4 points in interest cost. This meant that margins in 1981 were too low and interest costs too high.

The ratio of fixed assets to long-term debt decreased between 1978 and 1981. This reflected an increase in the industry's already excessive reliance on debt rather than earnings or new investment to finance growth. Short-term bank financing tripled between 1978 and 1981. These loans were to meet cash obligations in the face of operating losses. Short-term interest expenses in 1981 amounted to 4.8% of sales compared with only 0.8% in 1977. The ratio of equity to assets declined steadily from 26.9% in 1977 to 3.7% in 1981. The east coast industry had a grossly inadequate equity base.

The Task Force examined fluctuations in market prices over the previous 20 years. It concluded that the poor performance of the fishing industry could not be blamed on weak prices over the previous decade. They found that the cyclical variations in fish prices were not markedly different from those of food products generally, even though the phasing and frequency of peaks and troughs might not coincide.

The Task Force summed up the cause of the 1980–81 downturn thus:

> "The extremely strong price performance of fish products between 1969 and 1978, interrupted only in 1974–75, compounded by the declaration of the 200-mile limit and the rapid increase in Atlantic coast landings, created extraordinary optimism in the industry, in governments and in the financial commu-

nity. The optimism together with the inevitable competition among stake holders to be the first to take advantage of the 200-mile limit, accounts for the surge in investment and employment in the fishery between 1977 and 1980. The new claimants on the fishing dollar – lenders, fishermen, plant workers and suppliers — represented costs that soon grew to offset extraordinary increases in market prices, leaving the industry floating once more in a sea of red ink" (Kirby 1982).

Put more simply, the problem was unbridled greed which led to debt-financed overexpansion and plunged the industry into crisis when interest rates soared in 1980 and 1981. Tables 13-1 and 13-2 reveal the impact of soaring interest rates and debt on the 1980–81 crisis.

Total interest expenses for plants in the Woods Gordon sample increased by $57.9 million between 1978 and 1981. Sixty-eight percent of this was due to short-term loans and 32% to interest on long-term debt. The Task Force concluded that 44% of the interest expense between 1978 and 1981 was due to a higher than normal increase in the bank loan balance, and 46% was due to higher rates. Each of these contributed about $17.5 million in higher interest costs.

The areas of greatest financial distress were northeast Newfoundland, the south coast of Newfoundland and eastern Nova Scotia. In each of these areas a greater than average decline in gross margin was combined with a steep rise in interest expenses, most of it on short-term loans.

Interestingly, the Task Force found a marked difference between the situation of plants belonging to the Big Five vertically integrated companies and those of the independent firms. In 1978, the Big Five performed as well or slightly better than the independents. By 1981 they had slipped dramatically because of a larger decline in gross margin and a larger increase in interest expense. The Task Force attributed the difference in interest expense to the higher investment undertaken by the Big Five after 1978 and to the financing with short-term loans of this group in 1980 and 1981.

The Task Force offered the following overall observations on the economic condition of the processing sector:

"A grossly inadequate equity base has plagued the fishing industry for many years. The financing of assets has depended far too much on debt. Financial risk in the industry is excessive. At a short-term interest rate of 15 percent, and assuming a target return on equity of 19 percent, the industry would require an average gross margin of around 19 percent.... Lower interest rates (and reduced inflation) would lower the return required on equity and reduce the gross margin targets. Significantly higher returns would be required in good years to offset inevitable downturns....

"To have met the targets in 1981, the plants in the Woods Gordon survey collectively would have had to reduce the average short-term bank loan balance by $140 million and attract $250 million in equity capital. The principal financial challenge to the processing industry is to bring shareholders' equity to adequate levels. This will require sustained profitability" (Kirby 1982).

TABLE 13-1. Average rates of interest (percentages), 1979–1981.

	1978	1979	1980	1981
Bank Prime*	9.7	12.9	14.3	19.3
Long Term**	9.3	10.2	12.5	15.2
Sample average (short term)	8.6	11.2	11.0	19.8
Sample average (long term)	8.8	9.3	9.4	11.3

* Bank of Canada Review
** Government of Canada bonds over 10 years. These rates were merely indicative of trends and were not a source of capital for the industry.
Source: Kirby (1982).

TABLE 13-2. Components of interest expense growth, 1978–1981.

	Growth of Business	Excess Loans	Higher Rates	Joint Effect	Total increase
Short-term ($ millions) (% increase)	$ 4.2 11 %	$10.9 28 %	$11.7 30 %	$12.8 32 %	$39.6 100 %
Long-term increase ($ millions) (% increase)	$ 9.7 53 %	$ 2.7 15 %	$ 5.1 28 %	$ 0.8 4 %	$18.3 100 %

Source: Kirby (1982).

To a large extent, the Task Force had merely quantified what was already known before its formation. The question was: what was the government going to do, if anything, about the desperate financial problems of the Atlantic groundfish processing industry? The Task Force recognized that, without financial assistance, a number of major companies would go bankrupt. It also acknowledged that fishermen had financial problems. In many cases their incomes had not been sufficient to repay their vessel loans, to acquire new gear, or to maintain adequate working capital.

The Task Force considered the need for special new programs to provide financial assistance to fishermen and/or processing companies. It recommended against a new general program of financial assistance for either fishermen or processors. The most pressing problem was the near collapse of several major processors with extensive operations in Newfoundland. The Task Force observed that, if these major processors were forced out of business, the effect on fishermen, plant workers, communities and the Atlantic economy would be devastating. The Task Force concluded that a large injection of new equity capital was required. Additional debt would not solve the problem. It noted that attracting equity capital from the Canadian private sector at that time would be difficult in light of the fishery's recent performance. For companies to become viable, some reduction of over-capacity was considered inevitable. The Task Force stated:

> "A number of plants appear unlikely to be profitable under any foreseeable circumstances and they will have to be closed."

Having rejecting a program of general assistance, the Task Force was, however, negotiating to restructure those offshore processing companies which were then virtually insolvent.

The Task Force more or less dismissed the financial problems of fishermen, noting that "they are not in a significantly worse position than comparable groups like farmers." This was at odds with the Task Forces's own conclusions about fishermen's incomes.

3.8 Restructuring the Atlantic Groundfish Fishery — 1983–84

When the Task Force Report was released in February 1983, its Chairman, Michael Kirby, was already engaged, with a small core group, in negotiations to restructure the Big Five vertically integrated groundfish companies.

With the completion of its report, the Task Force as such was disbanded. Negotiations to restructure the insolvent companies were handled by a Federal Negotiating Team. The members were Michael Kirby, Halifax lawyer David Mann, Price Waterhouse Senior Partner Jack Hart and a former Vice-President of H.B. Nickerson & Sons, Peter Nicholson. Kirby (1984) has defined "restructuring" as "the orderly reorganization and refinancing of a group of insolvent or nearly insolvent companies with the objective of creating new enterprises which have a good chance of long-term viability."

Initially, the restructuring process focused on four insolvent companies — The Lake Group (LG), John Penny & Sons (JP), H.B. Nickerson & Sons (HBN) and Fishery Products (FP). These four companies employed approximately 20,000 plant workers and fishermen in some 50 communities in Nova Scotia and Newfoundland. They processed approximately 25% of the fish produced in the Atlantic provinces. When restructuring negotiations began in earnest at the beginning of 1983, these companies had suffered a combined net loss of $44 million in 1981 and $48 million in 1982. Their negative net worth at that time was estimated at $97 million. This situation would continue to deteriorate as the negotiations proceeded. By the end of 1983, these four companies plus National Sea Products (NSP), which also ran aground in 1983, recorded losses of more than $150 million.

At the end of 1982, the major creditor of the first four companies, the Bank of Nova Scotia (BNS), had loans outstanding totalling $222 million. The provinces of Newfoundland and Nova Scotia had outstanding loans of $20 million and $19 million, respectively. The Federal Negotiating Team retained Price Waterhouse of Toronto to evaluate corporate data and develop detailed financial models of the existing companies and of these companies restructured in several different configurations.

3.8.1 Restructuring Options

In considering restructuring options, the negotiating team worked from new federal objectives for the Atlantic fishery. The three objectives, ranked in order of priority, had been recommended by the Task Force and adopted by Cabinet (see Chapter 4). The first objective was for the fishery to become economically viable on an on-going basis. The second was maximization of employment consistent with the primary objective of economic viability and at a reasonable level of income for those engaged in fishing and fish processing. The third objective was that the fishing industry be Canadian owned.

While these objectives were generally endorsed, various players in the restructuring scenario differed significantly on the relative importance of the first two. The debate crystallized on whether certain fish plants in Newfoundland should remain open.

Initially, four basic options to reorganize the existing companies were considered:

Option 1:	Two companies	— (a) HBN/NSP and (b) LG/JP/FP
Option 2:	Three companies	— (a) HBN/LG/JP, (b) NSP and (c) a modified FP, minus certain of its plants.
Option 3:	Two companies	— (a) HBN/LG/JP/FP and (b) NSP.
Option 4:	One company,	— a single entity formed by combining the two in option 3.

In his 1984 discussion of the restructuring, Kirby downplayed the significance of the various reorganization alternatives. In fact, there was a major battle within the federal government during the winter of 1983 over which option would best meet federal policy objectives. The Federal Negotiating Team preferred option 1 and other players, e.g. DFO, preferred some version of option 3. The real question was whether to create two strong interprovincial companies or one strong interprovincial company. The latter option would combine NSP with the best of HBN's assets, leaving a weaker sister in Newfoundland.

The various parties involved attempted to persuade federal ministers of the merits of particular options. The management of the existing companies made a submission in which they argued that option 1 was the most viable and argued against option 3 and option 4. This presentation was orchestrated by the Federal Negotiating Team through the Bank of Nova Scotia. The bank and the government of Nova Scotia also supported option 1. The government of Newfoundland expressed some concern about option 1 and appeared interested in another combination, FP/HBN and NSP/LG/JP. The NFFAWU (Richard Cashin) spoke in favour of separation of the trawler fleet from processor ownership and some extension of the role of the Canadian Saltfish Corporation. He opposed options 1 and 4 and expressed a lesser preference for option 2, followed by option 3.

The supporters of option 1 argued that there was a "naturalness" to the proposed company combinations (NSP/HBN and FP/LG/JP) building on existing strengths. They also contended that the available financial projections suggested that option 1 was best overall from a business point of view with good long-term prospects for both companies.

Those who opposed option 1 argued that it would create a "weak sister" company (FP/LG/JP) in Newfoundland that could not compete with a strong NSP/HBN. One argument was that the species mix of FP/LG/JP would be too heavily weighted to groundfish. This could lead to earnings instability given the periodic fluctuations in groundfish price and supply.

3.8.2 The Selected Option

The federal Cabinet ultimately chose a modification of these options involving formation of two interprovincial companies — NSP/HBN less the Riverport, Nova Scotia scallop fleet and certain Newfoundland facilities; and FP/LG/JP plus part of the Riverport scallop fleet. This reorganization option essentially involved the combination of the Lake Group, John Penny & Sons, and Fishery Products with Nickerson's Newfoundland facilities and the Nickerson scallop fleet on the one hand, and NSP combined with Nickerson's groundfish fleet and the Canso plant on the other hand. The addition of the scallop fleet to the Newfoundland-based company was intended to provide it with a broader species mix and to broaden the base of the company beyond one province.

The federal government considered five ways of refinancing the troubled trawler companies. Kirby (1984) summarized these:

1. Provide loans and guarantees to existing owners;
2. Provide grants with few strings attached;
3. Find a private sector investor to put up the required funds for one or more companies as a whole;
4. Stand aside and allow events to take their normal course, almost certainly via bank receivership and disposal of assets; and
5. Make an equity investment in the companies, implying some degree of government voting participation.

The government endorsed the reasoning in the Task Force Report. The Task Force had said that an injection of equity was required and that, without Canadian private sector interest, governments would probably have to provide this. At that time there appeared to be no private investors willing to risk the necessary capital.

Thus, in March 1983, the federal government faced essentially two options:

1. Allow BNS to put the companies into bankruptcy and let events take their own course; or
2. Restructure the trawler companies with federal government investment in return for shares.

According to Kirby (1984), the federal government rejected the bankruptcy option for the following reasons:

"First, it would be profoundly disruptive in the marketplace. The liquidation of inventories would have ruined the U.S. market for dozens of independent processors as well as what ever remained of the large companies. It would have taken years to get this market back for Canadian fish products.

"Second, hundreds of small unsecured creditors would have been wiped out if the offshore trawler companies had gone into bankruptcy.

"Third, there would have been a loss of continuity of employment for over 40,000 Atlantic Canadians. In human terms, this element of the cost of bankruptcy would have been too high.

"Fourth, there seemed little likelihood that the offshore fishery would be reconstituted out of bankruptcy and remain in Canadian hands without very substantial government assistance."

For these reasons, the federal government decided to invest in a restructured offshore industry, if lenders (particularly BNS and the two provincial governments) would write off an amount roughly equivalent to the loss they would have incurred if bankruptcy occurred. This decision triggered a rocky series of negotiations.

In Newfoundland, plant closures quickly became the key issue. On March 28, 1983, while the federal Cabinet was still debating the reorganization options and the nature of federal investment in restructuring, a People's Conference was held in St. John's, Newfoundland. This was attended by Ministers of the Newfoundland government and representatives of community and labour groups. The conference adopted a resolution which stressed the need for the federal and provincial governments to ensure that all fish plants dependent on the deep sea fishery be reopened and kept open. It also called for public ownership of FP, LG, JP and HBN. The resolution was transmitted to the Honourable Donald Johnston, Chairman of the federal Committee of Ministers.

The FNT, based on its financial analyses, thought initially that five plants should be closed, namely, Fermeuse, Gaultois, Grand Bank, Burin and St. Lawrence. Kirby (1984) described the fundamental conflict with Newfoundland:

"In Newfoundland, the fundamental issue was the relative priority to be placed on the federal government's two fishery policy objectives of economic viability and employment maximization. The first — economic viability — is abstract, longer-term and appears to reward a small capitalist elite. The second — employment maximization — when expressed as a struggle to "save" Burin, Grand Bank and St. Lawrence is concrete, immediate and of benefit to hundreds of local citizens. The fact that ultimately these jobs and thousands of others will be threatened if the company failed to be profitable was never accepted because it was next year's problem, not today's."

In an interview in *Atlantic Business* magazine (April 1983), Dr. A.W. May, Deputy Minister of Fisheries and Oceans, suggested assistance would be in the form of an equity investment in the companies by government. He stated: "Loan guarantees would be of no help to the industry, because the companies would just be getting further in debt. Cash grants are out of the question because it would be unfair to companies that have managed to stay afloat in the recession." Dr. May indicated that, regardless of the type of assistance that would be provided, there would be layoffs and plant closures. "Twenty percent of the processing capacity in Eastern Canada could shut down tomorrow and nobody would miss it. I'm not saying that's going to happen. Obviously, in cases where communities would disappear if plants closed, the government will keep them open. But there will have to be some belt-tightening," he said (Woodworth 1983).

The options initially analyzed by Price Waterhouse assumed that at least five existing plants would close or merge. These plants were: Grand Bank, Burin, Gaultois, Fermeuse, and St. Lawrence. Price Waterhouse suggested that certain other plants, Ramea, Harbour Breton and St. Anthony, would continue operating primarily for social reasons regardless of their economic viability. Newfoundland questioned the rationale for such distinctions. It argued that the long term viability of the industry would only be secured by:

1. Increased throughput from greater resource utilization and expanded markets;
2. Strengthened management and improved productivity; and
3. Quality enhancement for fish products to attract top dollar in the world markets.

To get over the short term difficulties, the Province proposed that the federal government join the Newfoundland government "in approaching the Newfoundland Fishermen, Food and Allied Workers and other interested parties with a view towards developing a Social Compact with a minimum of 2 years duration. The Compact would be designed to preserve employment and the viability of fisheries operations through Government assistance and through participation by the workforce in the form of cancelled and reduced employee benefits along with a program to improve and reward production" (Newfoundland 1983a).

The general attitude of the NFFAWU had been summed up by Secretary-Treasurer Earle McCurdy when he said:

> "The processors built too many plants, some of them in the wrong place, and now they want us to pay for their mistakes" (Woodworth 1983)

Around this time the relative roles of Mr. Kirby and Fisheries and Oceans Minister De Bané were unclear. In early May 1983, Mr. De Bané met in Newfoundland with representatives of the communities that might be affected by plant closures and left the impression that the government accepted some financial responsibility for their future. Premier Peckford suggested that Ministers Morgan and De Bané meet prior to his meeting with the Chairman of the federal Committee of Ministers.

3.8.3 The Negotiations with Newfoundland

On May 17, 1983, Ministers Morgan and De Bané, supported by officials including Mr. Kirby on the federal side, met in Toronto. The outcome was a Memorandum of Understanding on Fisheries Restructuring signed by the two Ministers. The Memorandum incorporated a proposal, subject to ratification by the two governments, that:

1. A new company be formed, principally from a combination of FP, LG, JP;
2. The fleet, the marketing arm, and certain subgroups of plants would be operated as divisions of this new company;
3. The feasibility of separate Newfoundland offshore fleet, processing and marketing companies would be studied subsequent to the formation of the new company;
4. The governments would agree on disposition of existing shareholder's interests and jointly negotiate a re-capitalization plan;
5. The governments would jointly negotiate a "Social Compact" with the NFFAWU, recognizing the need for fishermen and workers to contribute to the revitalization of the companies and which could include representation of the union on the Board of Directors of the new company;
6. A Northern Fisheries Development Corporation (NFDC), including the plant at St. Anthony and plants on the Labrador coast, would be established.

Attached was a listing of plants to be included in the new company, including the Riverport scallop fleet.

A key clause in this Memorandum (Paragraph 6) dealt with plant rationalization. It stated:

"(a) the plants at Ramea, Harbour Breton and Gaultois will be reopened in consideration of the extreme dependence of these communities on the fishery;

"(b) the plant at Fermeuse will operate as an inshore plant and will receive fish under the Resource Short Plant Program;

"(c) the operations of Burin will be merged with those at Marystown and operations at Grand Bank will be merged with Fortune and St. Lawrence will remain closed;

"(d) the future of the plants at Hermitage and Belleoram will be assessed in light of their impact on inshore fish supply to Harbour Breton and Gaultois" (Canada-Newfoundland 1983a).

The memorandum further provided for:

1. The conversion to equity in the new company by Newfoundland of monies owed it in return for which the Province would receive shares in proportion to its contribution;
2. The conversion to equity by the federal government of its outstanding loan guarantees in respect of the Lake Group;
3. The new equity required to create an economically viable company would be provided by the federal government;
4. The two governments to establish a multi-million dollar Burin Peninsula Development Fund to diversify the economic base and provide new employment opportunities for the people in the area; and
5. The establishment of a Resource Utilization Task Force along the lines proposed by Newfoundland.

The surprising element of this agreement was Clause 6, where Minister Morgan agreed to the merger of Grand Bank and Fortune, on the one hand, and Marystown and Burin, on the other, and the continued closure of St. Lawrence. Premier Peckford overruled Morgan and returned to the negotiating table on May 18 seeking an amendment to Paragraph 6. Subparagraph 6(c) was replaced with:

> "The management of the new company will decide upon the future of the plants at Grand Bank and Burin. The two governments agree that, irrespective of their shareholding in the company, or their number of Directors on the Board, they will support without reservation the decision of management regarding Grand Bank and Burin. Pending formation of the new company and the decision of new management, the plant at Grand Bank will continue to operate and Burin will remain closed" (Canada-Newfoundland 1983b).

The federal Cabinet ratified the amended Memorandum as the basis for an agreement with Newfoundland. But Premier Peckford sought further amendments. In particular, he wanted a provision that the Burin plant be reopened based on a Social Compact with the NFFAWU and the results of the work of the Resource Utilization Task Force. After this Task Force submitted its report, there would be a review of the prospects for viability of the Grand Bank and Burin plants.

The federal government stood by the position agreed to on May 17, and as amended and then approved by Cabinet on May 18. Premier Peckford met with federal Ministers in Ottawa on May 24. At a meeting between Mr. Kirby and Mr. Morgan on the weekend of May 28–29, Kirby informed Morgan that the federal government would not reconsider the decision that Burin remain closed pending formation of the new company. On May 31, Premier Peckford informed Mr. Johnston that the Province had reached its bottom line, namely, both plants must remain open for at least 3 years.

On June 30, Premier Peckford confirmed that the future of the plants at Grand Bank and Burin was the only issue preventing a restructuring agreement. The Premier indicated that Ottawa was saying they had to close and the province was fighting the move:

> "No plant should be arbitrarily closed without a reasonable chance to succeed and achieve profitability" (Newfoundland 1983b).

Ignoring the De Bané-Morgan agreement of May 18, the Premier indicated that the federal government was proposing that the future of Grand Bank and Burin should be decided upon by the management of the new company to be established after the restructuring process was completed. The Premier stated: "If we were to accept this position, we are confident that the death knell will have been sounded for these industries in Grand Bank and Burin."

The Premier thus placed the federal government on the defensive, portraying it as hell-bent to destroy the communities of Grand Bank and Burin. The federal government's response was to announce on July 4 that it would unilaterally restructure the industry. At a Press Conference in St. John's, Minister De Bané stressed that the Premier had broken off negotiations after he rejected two agreements that Mr. Morgan had signed and after "I had made nearly a dozen further amendments to those agreements in order to try to reach a compromise agreeable to the Premier."

Minister De Bané indicated that long before it met with Newfoundland the federal government had decided that "the plants at Ramea, Harbour Breton and Gaultois and Harbour Breton must reopen and that the cost this would impose on the companies would be fully offset as part of the federal government's contribution to restructuring."

He concluded:

> "It seems crystal clear to us that further negotiations with the premier would be fruitless. No matter how many concessions we make in search of an agreement, they will never be enough unless we capitulate completely. And this we will not do" (DFO 1983).

He announced that the federal government would proceed to restructure the industry and pay the full cost. With the assistance of the Bank of Nova Scotia, a new company would be formed around a merger of FP, LG, and JP. This company would be funded with more than $75 million in cash from the federal government, a substantial conversion of debt to equity by the BNS and contributions of various kinds by others who then had equity in the companies being merged. The federal government would be the major shareholder. The new company's management would decide the future of the plants at Grand Bank, Burin and St. Lawrence. If management decided to merge Burin with Marystown or Grand Bank with Fortune or not to re-open St. Lawrence, then a labour adjustment package would be provided to employees who could not find re-employment. Until management decided the fate of these plants, the status quo would prevail.

To support its position, the federal government released copies of the May 17 and 18 Memoranda of Understanding and subsequent correspondence between the two governments. A propaganda war ensued. The federal government presented its case in a brochure titled *Restructuring*. There was major opposition in Newfoundland to the federal move, centered on the future of the three threatened fish plants. By mid-July, Newfoundland was seeking to reopen the negotiations on restructuring. Premier Peckford, however, said that the province would resume negotiations only on condition that the plants at Burin, Grand Bank and St. Lawrence be opened before the negotiations resumed (*St. John's Daily News*, July 19, 1983).

On August 9, Premier Peckford telexed the Prime Minister seeking his personal intervention in the restructuring issue. On August 18, the Premier released the text of the May presentation to the federal government, entitled "Restructuring the Fishery." He reiterated: "When you take into consideration each and every component of our submission to the federal government, it becomes abundantly clear that there exists no acceptable rationale for permanently closing down even one fish plant in this province" (Newfoundland 1983c).

The Premier called on the public of the Province, especially those directly affected by the threatened plant closures, to pressure the federal government to return to the negotiating table.

3.8.4 The Umbrella Agreement with the Bank of Nova Scotia

Meanwhile, other forces were at work. In mid-June, Michael Kirby had negotiated an umbrella agreement with the Bank of Nova Scotia on the restructuring of the east coast fishery.

The agreement dealt with FP, LG, JP, HBN and NSP. It was agreed that $355 million was required to refinance the companies by conversion of existing debt to equity and by infusion of new cash. This would be financed as follows:

— BNS (roll-over to equity)	$144 million
— Federal government (cash equity)	$110 million
— Newfoundland (roll-over to equity)	$30 million (tentative)
— Nova Scotia (roll-over to equity)	$38 million (tentative)
— CDC (roll-over to equity)	$34 million (tentative)

Two holding companies were to be formed. One, referred to provisionally as Hold Co X, would have as its only asset about 80% of the common shares of the new NSP (NSP plus some assets of HBN). The other, Hold Co Y, would have as its only asset 100% of the common shares in the new Newfoundland-based company.

Of the projected federal cash injection of $110 million, $35 million would go to the restructured NSP and $75 million to the new Newfoundland company. Of the $277 million owed by the companies to BNS at December 31, 1982, the bank would, as a result of restructuring:

(a) provide operating loans of $107 million to the new companies;

(b) receive $26 million in cash from the refinancing; and

(c) convert $144 million of debt (including $25 million of NSP shares) into equity in the restructured companies (Canada 1983a).

This agreement with BNS made possible the announcement in July of unilateral federal action in Newfoundland even though this was not compatible with the original assumption that the Newfoundland government would roll-over debt to equity. As the FNT moved to implement the July 4 announcement, it ran into difficulties with shareholders of the existing companies, FP and LG. In particular, it was not possible to reach agreement with the Canadian Development Corporation (CDC), the major shareholder in Fishery Products. When an impasse was reached in mid-August, the BNS moved to protect its position by calling for payment of its loans on August 25 (Lake Group) and August 26 (Fishery Products).

3.8.5 Reaching an Agreement with Newfoundland

On September 2, Fisheries Minister De Bané restated the Government of Canada's commitment to assist in restructuring the Newfoundland deep-sea fishing industry along the lines announced on July 4. He said that the receivers for the Lake Group and Fishery Products had informed the government that they intended to operate the companies with as little disruption as possible of normal operations.

Meanwhile, the government had no legal authority to make a direct equity investment and hold shares in the proposed fishing companies. Cabinet, therefore, authorized the necessary legislation to proceed as soon as possible.

While agreement had been reached with the Lake family for the purchase of all of the issued and outstanding shares of LG, JP and Caribou Fisheries Ltd. (the associated U.S. marketing arm), Fishery Products Ltd had taken steps to frustrate any attempt by the bank to place it in receivership. It sought the protection of the U.S. courts for itself and Fishery Products Inc, its American affiliate, under Chapter 11 of U.S. bankruptcy laws. FP then claimed that as the matter was before the U.S. court, BNS could not place it in receivership. FP also filed an action against the federal government, BNS, Price Waterhouse, the receivers (Clarkson Company Ltd) and Michael Kirby alleging that they had conspired to expropriate the company's property without fair compensation.

Against this background, the federal team, led by Dr. Kirby, met with Newfoundland negotiators on September 15–17. This resulted in a September 26, 1983, agreement between Minister De Bané and Premier

Peckford. The agreement provided that a new company be formed from the assets of FP, LG, JP and North Atlantic Fisheries Ltd. (owned by the province and HBN). The company would be financed by a cash contribution of $75.3 million from the federal government to purchase equity in the company, the conversion of $31.5 million of debt to equity by Newfoundland and the conversion of $44.1 million debt to equity by the Bank of Nova Scotia. The company would be owned 60% by Canada, 25% by Newfoundland, 12% by BNS and 3% by employees (tentative). The Board of Directors was to consist of five appointed by the federal government, three appointed by the provincial government, one by BNS and one (tentative) by the company's employees. The Chairman and Chief Executive Officer was to be jointly appointed by both governments. Both governments expressed their desire to return the business to the private sector as soon as possible (Canada-Newfoundland 1983c).

On the contentious issue of the Burin Peninsula plants, it was agreed that:

1. The plant at Burin would become an upgraded and expanded secondary processing or cooked-fish plant and a trawler refit center;
2. The plant at Grand Bank would remain open as a primary processing plant for at least 18 months to give the management of the company time to assess its economic viability;
3. The plant at St. Lawrence would become an inshore feeder plant and handle overflow offshore landings if they became available;
4. The plant at Fermeuse would become an inshore plant and would be eligible for the Resource Short Plant Program;
5. The plants at Harbour Breton, Gaultois, Ramea and St. Anthony would remain open for the foreseeable future; and
6. A multi-million dollar Burin Peninsula Development Fund would be established.

The two governments would jointly negotiate a "social compact" with the NFFAWU, to allow employees to purchase shares in the new firm. In return, they would have a representative on the Board of Directors.

Regarding the future of the industry, the Agreement provided that:

> "Plant closures, mergers, mechanization or trawler transfers resulting in a significant permanent change in employment in excess of 100 people, or one-half the workforce, as the case may be, associated with any single plant location would be subject to the approval of both Governments.
>
> "In the event that one of the Governments opposes the action contemplated, then the Government opposing that action shall assume the additional costs associated with the continuation of the existing level of operations" (Canada-Newfoundland 1983c).

Both sides had yielded somewhat to reach an agreement. The Newfoundland government had achieved its objective of keeping all plants open for the time being. Much remained to be done to put the restructured company into operation. Additional unforeseen funding would be required to make this company economically viable. But the foundation had clearly been laid for a new company, later named Fishery Products International, which became one of the world's largest fishing companies.

3.8.6 Negotiations with Nova Scotia

Ironically, the agreement to merge the Newfoundland companies had been expected to take much longer to achieve than the anticipated smooth integration of HBN assets in Nova Scotia with National Sea Products. Negotiations with Nova Scotia and NSP had been underway for more than a year. But little real progress had been made until the summer of 1983. The FNT had hoped to reach agreement with Nova Scotia, then inform the Board of NSP and HBN, then obtain the shares of HBN. Following that, it intended to obtain the agreement of NSP shareholders to amalgamate NSP and HBN. Finally, it would form a merged operating company which would receive equity from the NS Hold Co. (as per the June 24 letter of understanding with BNS) in return for common shares.

But, as in Newfoundland, matters would not unfold neatly in accordance with this scenario. As early as June 1983, when Kirby and his associates outlined the proposed refinancing scheme to provincial officials, it became evident that Nova Scotia would not roll over debt to equity. The province took this position because:

1. Their security was more than adequate (on several vessels and the Canso plant);
2. They anticipated that other independent fish companies in the province would resent the creation of a super-company with government equity; and
3. They were concerned about the precedent for other troubled businesses in the province.

Nova Scotia was requested to convert existing debt to equity in the form of preferred shares. The federal government would provide new cash (about $35 million) and would purchase a further $70 million of senior preferred shares from BNS over 5 years. As part of this proposal, the BNS and the federal government agreed to sell to the private sector within 5 years the shares in NSP/HBN held by the Nova Scotia Hold Co. The Province was informed that, if this plan could not be agreed, the likely alternative would be to fold the HBN Nova Scotia assets into the Newfoundland company, with no federal involvement in the refinancing of NSP.

In early August, three directors of NSP met with federal government officials to inform them of the company's financial condition. NSP had announced a second quarter loss of $1.9 million compared with a profit of $1.1 million in the second quarter of 1982. NSP losses were mounting; it had lost $2.6 million during June and forecast a July loss of $4.0 million. The company had an inventory of $130 million, about $30 million greater than normal. Management was now forecasting a pre-tax loss for the year of about $17 million. This compared with a previously budgeted pre-tax profit in 1983 of $17 million. The Board disagreed with management's estimate of the loss and forecast a loss as high as $27 million. This would dissipate more than half of NSP's shareholders equity by year-end. The company faced a critical cash shortage. The bank line of credit, which stood at $10 million, was being drawn down at a rate of $4 million per week.

On August 4, management presented the Board with a plan to reduce inventory levels through a cutback in production. This would still leave a pre-tax loss of $9–$10 million, despite closures and cutbacks.

The Board decided that these measures were insufficient in light of its more pessimistic estimate of likely losses. It decided to create a special subcommittee, consisting of three senior executives from outside the fishing industry, to recommend remedial measures, including the reduction of both the numbers and the wages of employees and the possibility of permanent plant closures. However, the NSP Board indicated that it would still consider restructuring proposals by the federal government and BNS.

The NSP Board also made the Province of Nova Scotia aware of NSP's deteriorating financial situation. Despite this, Nova Scotia still opposed in mid-September converting its $50.6 million debt in NSP and HBN to equity. It offered instead to give the restructured company a 5-year holiday on interest and capital payments. The result of this, under the proposed refinancing scheme, would have been to raise the federal percentage of the common shares of the Nova Scotia company above 50%. The federal government would then have controlled both the new companies, rather than just the one in Newfoundland. NSP's deteriorating financial situation called for additional investment in the $25–$50 million range.

3.8.7 The Port Hawkesbury Agreement

Federal-provincial negotiations resumed in late September. On September 30, 1983, in Port Hawkesbury, Premier Buchanan and Minister De Bané announced that the federal and Nova Scotia governments had agreed to cooperate in helping to restructure NSP and certain assets of HBN "with the aim of creating a commercially viable, privately owned offshore fishery" (Canada-Nova Scotia 1983).

The terms of the agreement included:

1. The Canso plant, the Wedgeport plant, HBN's 14 trawlers and two HBN scallop vessels would be amalgamated into NSP.
2. The restructured NSP would be refinanced by the conversion of some debt to equity by both the Bank of Nova Scotia and the Province of Nova Scotia, and the injection of new cash for equity by the federal government.
3. Both governments committed to divest their equity to the private sector as soon as the long run commercial viability of NSP was assured.
4. Existing shareholders and financial institutions were not to be bailed out.
5. The governments pledged that the company would be managed on a strictly commercial basis.
6. Plant closures, sales, mergers and trawler transfers would be subject to the approval of both governments.

The two governments stated that "an equity interest in NSP in return for new federal government cash, and conversion of some bank and government debt to equity, is the only alternative for solving" the deteriorating financial condition of NSP.

Anticipating negative reaction from the many independent fish processors in Nova Scotia, the Premier and Mr. De Bané stressed that "it would be completely wrong to characterize restructuring as nationalization of the offshore industry. From the outset, there will be private shareholders of the restructured company and...it is expected that its shares will continue to be traded on the stock market."

With the Port Hawkesbury Agreement, the stage appeared set to establish two companies which would be among the largest fish processing operations in the world. The Nova Scotia company would have sales of roughly $450 million (1983) and assets of $335 million. It would operate 24 processing plants with 11,000 full and part-time employees, supplied by 50 trawlers and 9,000 fishermen. The Newfoundland company would have sales of $395 million (1983) and assets of $220 million. It would employ 12,000 people, full and part-time, in 33 plants, processing fish supplied by 66 trawlers and 10,000 fishermen. Together, the two restructured companies would provide some or all of the income of more than 40,000 plant workers and fishermen. In 1983, the combined sales of the component companies totalled a little more than half the value of all fish processed on the east coast. Between them the two restructured companies marketed about 80% of the groundfish and significant, though lesser, shares of other major species.

Independent fish processors in Nova Scotia reacted negatively to the announced restructuring of NSP. On October 1, the *Halifax Chronicle Herald* quoted several industry spokesmen saying that the restructuring of Nova Scotia's largest fishing companies would transform the fishery into a social welfare system and stifle entrepreneurial spirit. "The destruction of the industry has been charted today," said one spokesman. Ernest Cadegan of Comeau Seafoods Ltd. stated: "Those of us who have been left out of this sweetheart deal felt very much like sacrificial lambs."

Allan Billard, Executive Director of the Eastern Fishermen's Federation, said that the federal and provincial governments should have allowed Nova Scotia's two biggest fish companies to sink into receivership if necessary and be bought by the pieces by small companies (*Moncton Times Transcript*, October 3, 1983).

On October 18, the federal government put its restructuring proposal to the Board of Directors of NSP. Mr. De Bané stated publicly that the federal government would hold a controlling interest in NSP. Under the $90 million restructuring proposal, a consortium of the federal and provincial governments and the Bank of Nova Scotia would have a 75% interest in the restructured firm. Mr. De Bané said it would be great if the private sector or the provincial government could take a greater piece of the new company. Minority shareholders (Nickerson's had held 51% of the shares) would have the option to retain their percentage of the restructured firm through additional stock acquisition. "If they want to put up more money, they can keep their position constant. If they don't, of course they will be diluted," Mr. De Bané said (*The Halifax Chronicle Herald*, October 18, 1983).

On October 19, the *Toronto Globe and Mail* carried the headline "Fish firms fear merger means nationalization." It reported that independent fish processors in Nova Scotia feared the proposed consortium takeover of NSP and its majority shareholder HBN might lead to federal nationalization of a large proportion of the East Coast industry. This view, expressed by Basil Blades of Sable Fish Packers Ltd., was shared by the owners of other small and medium-sized companies as well as some officials of the Nova Scotia Government. The *Globe and Mail* quoted an unnamed Nova Scotia official as suggesting that the proposed takeovers would ultimately allow Minister De Bané "to achieve an earlier expressed desire to see some form of nationalization of the industry."

By late October, the *Chronicle Herald* was reporting that the Nova Scotia fishing industry regarded the Newfoundland restructuring agreement "with fear and loathing — loathing the agreement and fearing the Nova Scotia industry will be stuck with a similar agreement" (October 27, 1983). The Nova Scotia industry was concerned about the apparently low priority given to economic viability, about government control of the company, and about the requirement to keep money-losing operations open to provide employment. The general reaction in Nova Scotia was quite different from that in Newfoundland. Whereas the Newfoundland government, communities and labour groups had demanded federal involvement to restructure the industry, in Nova Scotia there was a deep-rooted suspicion of the federal government. There was some willingness to accept federal cash but only if it came with no strings attached.

Responding to this situation, Premier Buchanan moved to distance himself from the Port Hawkesbury agreement. His tactic was to accuse the federal government of failing to deliver on implied allocation commitments in the September 30 agreement. At first, there was wrangling with federal officials over the nature of enterprise allocations, with Nova Scotia attempting to force its view upon the other provinces and players in the industry. However, Nova Scotia's initial objections to the federal proposal for 5-year enterprise allocations were allayed when federal officials struck a delicate compromise package with all the industry participants involved. The next issue was access to Gulf redfish (see Chapter 7). The TAC and allocation of Gulf redfish was an extremely controversial issue which required resolution at the Cabinet level.

Under the Port Hawkesbury agreement, Nova Scotia was to convert about $22.5 million of debt to equity and reschedule the balance of its present loans to NSP and HBN (another $22.5 million) with 5% interest and waiver of principal for 5 years. But the Province balked at formalizing this agreement without:

1. Explicit assurance of 'equitable' allocations of Gulf fish stocks and northern cod; and
2. An assurance that neither government would acquire a controlling position in the restructured Nova Scotia based company.

The minority shareholders of NSP were considering the consortium offer but were concerned about dilution of their existing position and the degree of government control of the company. Meanwhile, the financial condition of NSP continued to deteriorate rapidly and critically. The projection of year-end loss had been raised to $22 million.

3.8.8 The Atlantic Fisheries Restructuring Act

To acquire the authority to invest in the restructured companies, Minister De Bané introduced a proposed Atlantic Fisheries Restructuring Act in Parliament on October 31, 1983. He sought rapid passage of the legislation so that the two restructured companies could be in place for the 1984 fishing season.

Speaking on second reading of Bill-C170, De Bané reiterated:

> "This is *not* nationalization, this is *not* the beginning of nationalization. We are *not* creating crown corporations. All of the governments involved see their participation as temporary. We are convinced that the interests of the industry and the communities which depend on them will be best served by returning the Atlantic fisheries to the private sector as soon as possible. This is our stated and explicit aim. Indeed, we would be ready to do so immediately — except that, at this low point in the fortunes of the industry, no private investors are willing to take the risk alone" (Canada 1983b).

The Standing Committee on Fisheries held extensive hearings on this proposed legislation. There were several interesting developments during the Committee hearings:

1. The June, 1983, letter of agreement with the Bank of Nova Scotia was provided to the Committee;
2. The Minister of Fisheries and Oceans assured the Committee that he would not hold the shares in the two new companies;
3. The Committee proposed amending the bill by adding a clause on divestiture.

The Atlantic Fisheries Restructuring Act became law on November 30, 1983 (1980–81–82–83 c.172 s.9). The *Purpose* of the Act was:

> "to facilitate the development of viable Atlantic Fisheries that are competitive and privately-owned through the restructuring of fishery enterprises."

To carry out this purpose, the Act empowered the Minister of Fisheries and Oceans to acquire shares or interest in any fishery enterprise.

Subparagraph 4(2) required the Minister to dispose of any interest in a fishery enterprise "as soon as practicable after the enterprise becomes, in the opinion of the Minister, economically viable on a continuing basis."

3.8.9 "The Private Sector Solution" in Nova Scotia

Meanwhile, the federal government and BNS had on October 31, on the strength of the Port Hawkesbury agreement, acquired all of the shares of HBN. When the deal with Nova Scotia fell apart during the fall of 1983, the federal government and BNS were left with HBN and 56% of the common shares of NSP.

As late as December 8, Premier Buchanan was publicly expressing hope that the restructuring agreement for Nova Scotia would be completed. The Premier said that agreement would depend on whether provincial demands were met: "We must have agreement that the federal government will not get control of the [restructured National Sea] company. We must have assurances it will be returned to the private sector within 5 years of the agreement" (*The Chronicle Herald*, December 9, 1983). However, behind the scenes, the dynamics of the situation were changing. Mr. David Hennigar made a proposal to the Restructuring Committee of NSP on November 30. Hennigar, Director of Scotia Investments Ltd., represented shareholders with 12.5% of NSP shares. Hennigar had been holding informal discussions with Bill Morrow, Jerry Nickerson and representatives of the Nova Scotia government throughout the fall. Hennigar's proposal to the NSP Restructuring Committee included the conversion of NSP debt into "financial difficulty" preferred shares and $10 million of equity to be raised by the issue of new shares to existing shareholders. Conditions included retaining the existing senior management of NSP and reducing the voting position of the government and bank to minority status.

On December 19, Hennigar put his formal proposal before the NSP Restructuring Committee. It had these elements:

1. The Province would defer principal payments on existing loans of $33 million to NSP for 5 years amortized from that date over the next 15 years;
2. NSP would offer to purchase HBN assets for $18 million — current debt to the Province;
3. The Province would convert its HBN debt into NSP 25 year preferred shares and would provide up to $5 million for the renovation and upgrading of the HBN assets;
4. The Consortium would agree to sell at least half of its current share position to existing minority shareholders via a rights offering at $7 per share; the Consortium shares were to be resold to the market in 5 years; and

5. The Banks would convert $75 million of existing loans to "financial difficulty term preferred shares" with interest at 1/2 prime + 1%, these shares to be repayable in 1987, 1988 and 1989.

The Board of NSP agreed to decide at a special Board meeting in January, 1984, which of the two proposals — the Hennigar proposal or the Consortium proposal — it would recommend to shareholders.

The Bank of Nova Scotia and the Royal Bank of Canada rejected the Hennigar proposal. The banks' rejection of the Hennigar proposal was made public in early January. Peter Nicholson, a member of the FNT, emphasized that the minority shareholders were putting no new money into the company (*The Globe and Mail*, January 4, 1984).

On January 6, 1984, just prior to the special NSP Board meeting to consider the Consortium and Hennigar proposals, Premier Buchanan and Ministers MacEachen and Regan met with the BNS, the Royal Bank and the Hennigar group. As a result of this meeting:

1. Nova Scotia reserved its position on the Consortium proposal while the federal government stated it would not proceed without provincial participation;
2. The BNS made clear its rejection of the December Hennigar proposal;
3. The federal government reserved its position on certain aspects of the Hennigar proposal; and
4. The Royal Bank agreed to support the Hennigar proposal provided government guaranteed $75 million of term preferred shares of NSP, and BNS agreed to relinquish control of NSP.

Because of this impasse, the NSP Board, at its meeting on January 9, could recommend neither of the restructuring proposals because each contained elements vetoed by other parties. On January 12, the BNS called for payment of its loans to NSP (approximately $75 million). The Royal Bank had until January 26 to tell the BNS whether it would participate in the call of NSP loans or pay off BNS and assume its position as creditor of NSP. If the RBC participated in the loan call, this would have pushed NSP into receivership.

By this time the issue of restructuring NSP was generating intense political debate in Nova Scotia. The Province and the media characterized the issue as a struggle to save private enterprise. The federal government was left with little choice but to seek an improved "private sector solution." On February 6, Minister De Bané announced that the Directors of National Sea Products had approved a comprehensive plan for restructuring and financing the company. The Hennigar group had increased its direct equity contribution to NSP from zero to $20 million. The Toronto-Dominion Bank purchased $75 million of financial difficulty term-preferred shares without any federal government guarantee. The federal government contributed $10 million in equity to purchase preferred shares of National Sea. The HBN assets in Nova Scotia (less the 12 scallop vessels and associated facilities at Riverport which went to the Newfoundland company) were passed to NSP. The Government of Nova Scotia converted $18 million of HBN debt to equity in NSP and added $7 million for a total $25 million of NSP preferred shares. In total, NSP received an equity injection of $55 million. The private sector retained control of the company with private shareholders holding 66% of NSP's common shares, the federal government 20% and the BNS 14% (DFO 1984g).

The June 24 agreement with the BNS was reworked with the federal government agreeing to purchase, at year 10 and at the Bank's option, one-half of the $29 million of the Bank's Class B preferred shares in the Newfoundland company. Downstream, this meant an additional cost to the federal government of $15 million. Additionally, the federal government had made a significant contribution to restructuring NSP by ensuring that the HBN assets being acquired by NSP would carry no debt owing to the BNS. This re-arrangement of the federal-BNS agreement allowed the revised private sector proposal to proceed. The federal government was still committed to pay the BNS $70 million in equal instalments over the next 5 years plus interest of 5%. In return, the BNS released HBN from almost $100 million of loan obligations. The federal government, in return, was to receive the first $70 million of proceeds from the eventual sale of the approximately 2.6 million common shares of NSP pledged to the BNS as security for its loans to HBN. This agreement with NSP in February 1984 set the stage for the emergence of two large fishing companies from the ashes of the previous Big Five.

3.8.10 Federal Intervention in Québec

The federal government also intervened in Québec to purchase the assets of the insolvent Québec cooperative Pêcheurs Unis. This was done without the cooperation of and, indeed, in the face of concerted opposition from, the Québec government. This was part of a larger confrontation between federal Fisheries Minister De Bané and the Québec Minister of Agriculture and Fisheries, Jean Garon. In response to one of the recommendations of the Task Force on Atlantic Fisheries, the federal government rescinded the Canada-Québec 1922 agreement and resumed the responsibility for administration of the marine fisheries in Québec (see Chapter 2). The assets of the former Pêcheurs Unis were acquired by a new company, Pêcheries Cartier, which was federally dominated but also involved the main financial institutions, the Caisses Desjardins and the Banque Nationale du Canada. The federal government injected $23.6 million in equity and the two financial institutions agreed to convert $2.0 and $1.0 million, respectively, of their debt into equity in the new corporation (DFO 1984h).

3.8.11 A Hefty Bill for the Federal Government

Overall, the federal government through its investment in the new Fishery Products International and National Sea Products had intervened in a major way to prevent the collapse of the Atlantic offshore groundfish fishery and to weather the latest crisis in the industry. The hope was that these restructured companies would be economically viable and strong enough to withstand any future downturns.

The extent of federal investment turned out to be much higher than originally envisaged. Some of it was upfront equity injection; some involved downstream purchase of BNS shares. Furthermore, the federal government injected additional capital into FPI in 1985 following the development of a 5-year business plan by the new management under the leadership of Chief Executive Officer Victor Young, formerly head of Newfoundland Hydro. The Annual Reports to Parliament on the *Atlantic Fisheries Restructuring Act* (DFO, Various Years) reveal the following expenditures overall:

1983–84:	$ 38.5	million
1984–85:	$114.275	million
1985–86:	$ 93.595	million

Parliament appropriated a total of $246.37 million from November 30, 1983 to March 31, 1986, to cover the federal investment in the three restructured companies. No funds were appropriated for this purpose in 1987–88 and 1988–89. Additional expenditures will be incurred in future years for the purchase of BNS shares in FPI and NSP.

According to the Annual Reports, the expenditures to the end of fiscal year 1985–86 were:

Pêcheries Cartier	$31.5	million
Fishery Products International	$167.575	million
National Sea Products	$ 44.3	million

The federal government invested $167.6 in FPI from 1984 to 1987. When FPI was privatized in 1987, the federal government received $104.4 million. This left a net expenditure on the FPI restructuring of $63.2 million. Despite the large expenditure of public funds to rescue the offshore Atlantic groundfish fishery in the early 1980's, the federal government received little credit for this initiative.

3.9 The Glory Years — 1985–1988

Federal intervention put the industry on the road to recovery. The groundfish resource had recovered significantly since 1977 (Fig. 13-8). When restructuring was completed in 1984, groundfish markets were soft. Cod blocks, for example, had been around $1.17 U.S. per pound through 1983. These dipped to about $1.00 in the latter half of 1984 and first half of 1985 (Fig. 13-6). However, in the second half of 1985 the market began an upswing which continued until mid 1988. The price of cod blocks, for example, doubled during this period. The per capita consumption of fish in the U.S. increased significantly in the mid-1980's, peaking at 15.4 pounds in 1987. In the latter half of 1988, high inventories, slow seafood sales, and lower prices for alternative sources of protein combined to decrease seafood prices.

The upswing in the U.S. market in the mid-1980's fuelled a recovery in the Atlantic groundfish industry. But the restructured companies, which constituted the major component of the industry, did not show a profit again until 1985 in the case of NSP and 1986 for FPI. David Hennigar, architect of the "private sector" restructuring of NSP, became Chairman of the company. In June 1984, he was told by Gordon Cummings of Woods Gordon Inc. that he might as well have let the federal government nationalize the company. Cummings had reviewed NSP's books and projected a loss of $15 million in 1984, compared with a management projection of $5 million. Hennigar hired Cummings as NSP's executive vice president and chief operating officer. NSP sold six of its small plants and made other changes on the road to recovery.

The market upswing beginning in 1985 helped the recovery. In 1985, NSP reported a net operating income of $7.2 million. This increased to $21.8 million in 1986 and $27.6 million in 1987. NSP then launched a program of diversification and acquisition abroad. In 1987, the company acquired 64.6% of Pacific Aqua Foods Ltd., an aquaculture company in B.C. Early in 1988, the company agreed to purchase Bretagne Export S.A., a major seafood trading company, and La Surgelation Lorientaise S.A., a seafood processing facility, both located in France. During 1987 and 1988 the company modernized vessels through containerization, completed a replacement fresh fish processing plant in Louisbourg and made additions at Canso. NSP also acquired Canada's first Atlantic coast factory freezer trawler in January 1986 (see Annual Reports, National Sea Products).

FPI, on the other hand, took slightly longer to get on its feet. It reported a net operating loss of $35.0 million in 1984 and $20 million in 1985. In April 1987, the federal and provincial governments sold their shares in FPI to private investors. Following privatization, the company was owned 88.5% by private shareholders, 8% by the Bank of Nova Scotia, and 3.5% by the FPI employees. FPI reported a net operating income of $22.3 million in 1986 and $31.0 million in 1987. From 1985 to 1987, the company invested almost $90.0 million in rebuilding its fleet, modernizing its plants and upgrading its port facilities (see Annual Reports, Fishery Products International).

The years 1985 to mid-1988 brought a boom for the Atlantic groundfish industry. This period was probably the most prosperous in history for the Atlantic fishery. Prices and earnings reached record highs and catches were maintained at high levels. Unfortunately, it is not possible to quantify the financial returns in the industry because the Woods Gordon surveys, which had been initiated by the Task Force on Atlantic Fisheries, were only maintained for a year or two. Despite the Task Force recommendation that these surveys be repeated regularly, they were discontinued. Apparently this was because the processing sector refused to cooperate once the good times returned. This was similar to the experience of 1975–1977 when detailed financial information was made available for the duration of the government assistance programs but dried up when the assistance programs ended.

3.10 Crisis Again? 1989–1993

3.10.1 Another Crisis in the Making

While prices and earnings were high from 1986 to 1988, resource problems were beginning to emerge. The catch rate of Canadian trawlers peaked in 1984 (Fig. 10-4) at 19 tons per day fished and declined thereafter to 16 tons per day fished in 1986. The increase in groundfish catches from a low in 1974 to peaks in 1982 and 1986 had been supported by a large increase in cod catches from 146,000 tons in 1975 to the 500,000 ton range from 1982 to 1988 (Fig. 13-8). Cod catches as a percentage of the groundfish catch increased from 35% in 1975 to 66% in 1983 and remained at more than 60% from 1982 to 1988. The biomass of cod off the Canadian Atlantic coast increased three-fold from 1975 to 1984 but then declined slightly to 2.6 times the 1975 level in 1988 (Fig. 13-9).

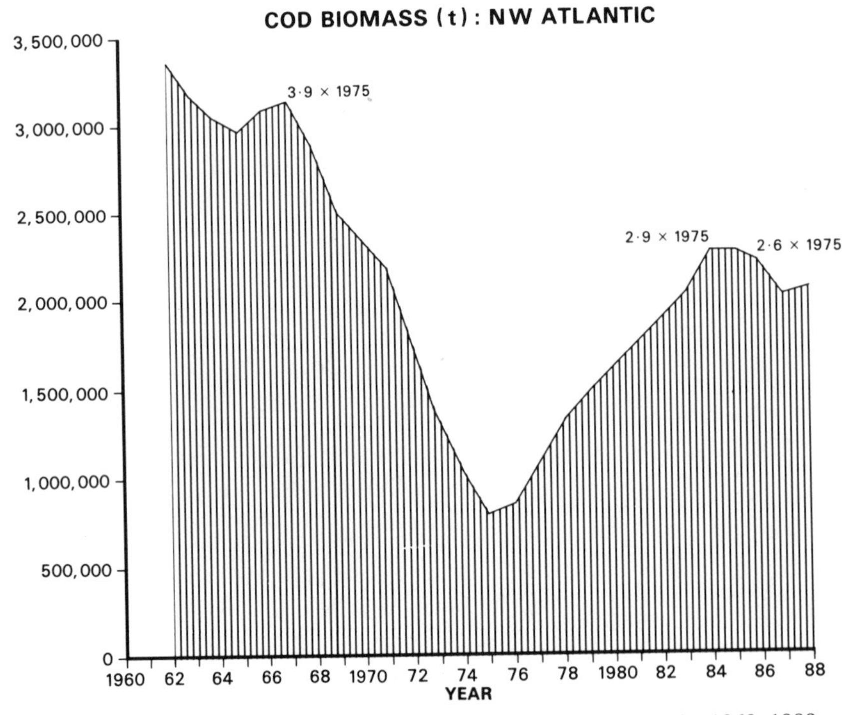

FIG. 13-9. Trends in cod biomass (tons) in the Northwest Atlantic: 1962–1988 (from Fisheries Research Branch, DFO, Ottawa).

More recent assessments indicated that fishing pressure remained high for haddock stocks and certain cod stocks in the 1980's and that the management target of fishing at $F_{0.1}$ was not attained for many stocks (Pinhorn and Halliday 1990). Overall, the resource picture which had been bright in the early 1980's appeared less so by 1989. Scientists revised their estimates of the size of the northern cod stock and advised a substantially lower TAC for 1989 (see Chapter 7). The TACs for other cod stocks had declined significantly from the mid-1980's to 1989. Haddock TACs had been sharply reduced from 1986 onward (Figs. 13-10 and 13-11).

Thus, by 1989, it appeared that the industry could expect lower, not higher, catches for the next several years. The reasons for this levelling off in some stocks and decline in others were not clear. However, it

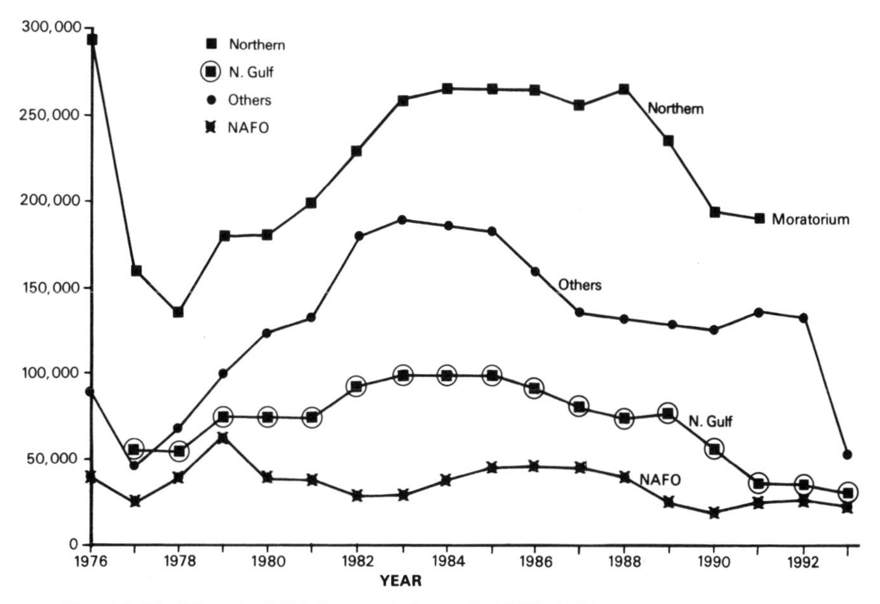

FIG. 13-10. Historical TACs (tons) for cod: 1976–1993.

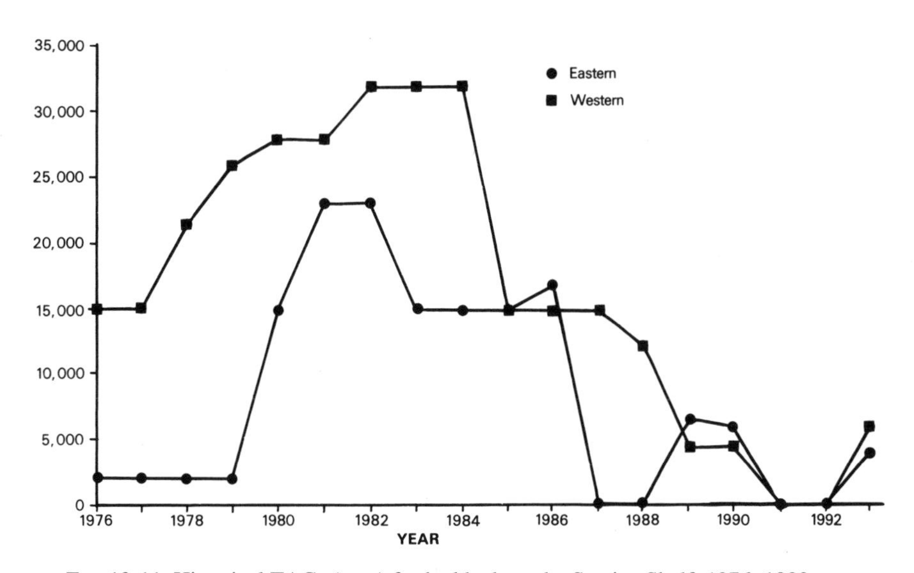

FIG. 13-11. Historical TACs (tons) for haddock on the Scotian Shelf: 1976–1993.

appeared that the stocks had been less productive in the 1980's. This was manifested in fewer young fish on average and slower growth rates, i.e. a decline in average weight at age (Fig. 13-12).

For a while, rising prices masked stock declines, e.g. Scotia-Fundy groundfish (Fig. 13-13). Also, the industry and governments had not learned from the 1981–83 crisis. The Task Force on Atlantic Fisheries had recommended restraining growth and reducing capacity in the industry. Nevertheless, significant growth in capacity continued through the 1980's. The number of registered fishermen Atlantic-wide increased only marginally between 1983 and 1988 (from 57,000 to 66,000). While the number of registered fishing vessels in Atlantic Canada remained about the same, technological change brought significant growth in catching capacity. The vessels, in many instances, became larger and more efficient. Perhaps the most dramatic example of this occurred in Scotia-Fundy.

The 1989 Report of the Scotia-Fundy Groundfish Task Force (DFO 1989i) noted that the fishing power of vessels in the Scotia-Fundy groundfish fishery had increased dramatically since 1980. The newer vessels, especially the wide, deep "jumbo" draggers, had larger hold capacity, bigger engines, more sophisticated fish-finding equipment and more efficient gear. During the 1980's the smaller, less powerful vessels were usually replaced with larger ones (Fig. 13-14). The average annual catch of the newer "jumbo" vessel was 200 tons compared with 23 tons for the conventional dragger.

Two surges of boat-building occurred in the post-extension era – from 1977 to 1982 and during the late 1980's. The high rate of vessel replacement in the immediate post-extension era (1977–82) was spurred by rising catches. Much of the 60–65' dragger fleet was built during this period. As catches levelled off in the mid-1980's, few new boats were built. The market surge from 1985 to mid-1988 led to a new round of vessel replacement. According to the Haché Task Force, vessel replacement rules and other factors made the fibreglass "jumbo 44'11" vessel the new replacement choice. When groundfish prices decreased in 1988, and quotas were reduced further, fishermen's earnings declined. The new vessel replacement guidelines issued in the spring of 1989 "drastically reduced future potential for fishermen to upgrade lobster boats into powerful groundfish vessels" (DFO 1989i). Background studies done for the Task Force indicated that fishing mortality for the primary species of interest – cod, haddock, and pollock — was more than double $F_{0.1}$ during the 1980's. Fishing capacity was probably four times the level required to harvest the available resource economically. As a result, haddock, pollock and cod stocks on the Scotian Shelf all declined during the late 1980's. The haddock decline commenced in the early 1980's.

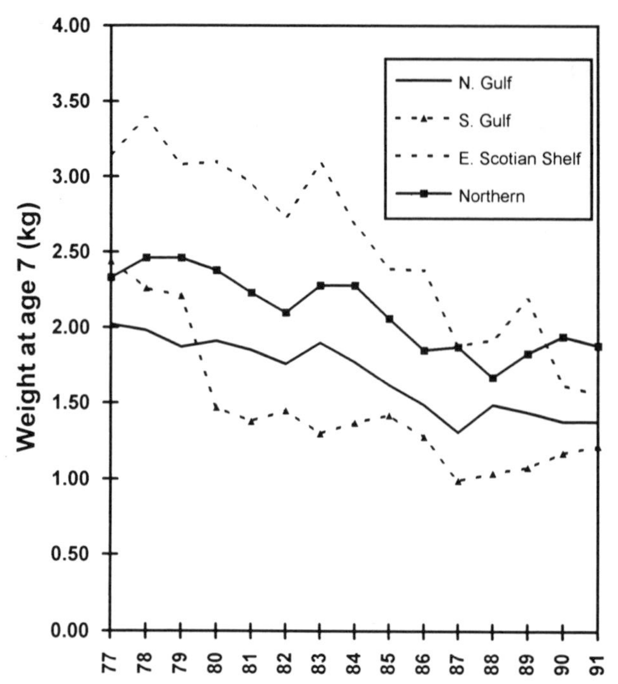

FIG. 13-12. Weight of an age 7 cod.

Despite the warnings about existing overcapacity in the processing sector, the number of federally registered fish plants in Atlantic Canada, which had increased from 519 in 1977 to 700 in 1981, increased further to 953 in 1988. This amounted to a 35% increase between 1977 and 1981 and a 36% increase between 1981 and 1988. Most of the growth in the latter period occurred between 1984 and 1988, when the number of fish processing establishments increased from 727 to 953. The greatest rate of growth in the post-1981 period occurred in Nova Scotia and New Brunswick (Fig. 13-15).

This increased capacity in the harvesting and processing sector increased the competition for raw material. Given this expansion of capacity in conjunction with the resource declines of the late 1980's, the boom could not last. The market downturn in mid-1988 combined with decreased TACs for certain key stocks (for example, northern cod in 1989) set the scene for another downturn in the groundfish industry, approximately 5–6 years after the trough of the last crisis.

The two major companies dependent on groundfish, NSP and FPI, which had been so recently restructured, again faced stormy seas. NSP went from a net income of $24.8 million in 1987 to a loss of $5.8 million in 1988. FPI's net income went from $58.0 million in 1987 to $16.8 million in 1988. For 1989 both companies reported losses — NSP of $32.4 million and FPI of $22.2 million.

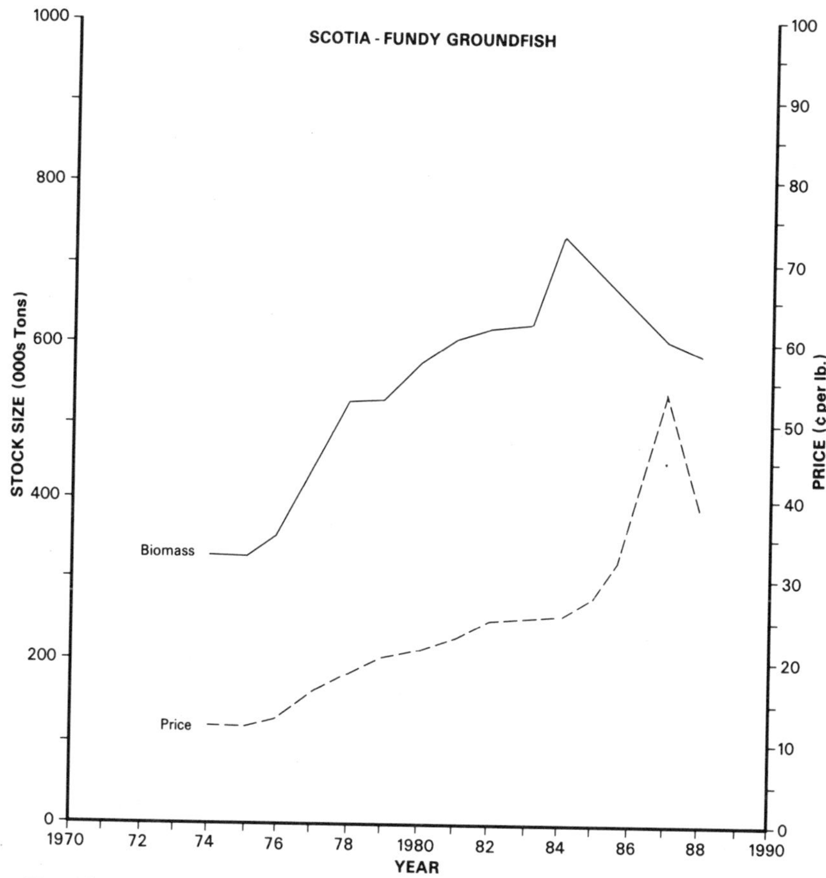

FIG. 13-13. Rising prices mask stock decline for Scotia-Fundy groundfishery (from Market Analysis Group, DFO, Ottawa).

The decline in revenues was attributed to cuts in quotas and a downturn in fish prices. NSP's and FPI's revenues were cut further by a strong Canadian dollar, which reduced the profit margin on exports to the U.S.

Both companies tried to lower costs in 1988 and 1989 by reducing excess capacity through rotating layoffs or permanent closures of some fish plants. One Toronto analyst observed that NSP was worse off than FPI because of NSP's relatively heavy debt load. With virtually no debt, FPI did not face heavy servicing charges. Much of NSP's debt was taken on in the effort to diversify beyond Atlantic Canada. While financial analysts considered this a wise long-term move, it left little room to manoeuvre in an emerging crisis. "National Sea Products is in serious financial straits," one analyst said. "They need a serious injection of equity" (Financial Post, November 6, 1989).

Both NSP and FPI began talking of retrenchment as soon as it became apparent that they were facing quota reductions in northern cod. FPI was also affected by the decline in flatfish TACs on the Grand Banks due to for-

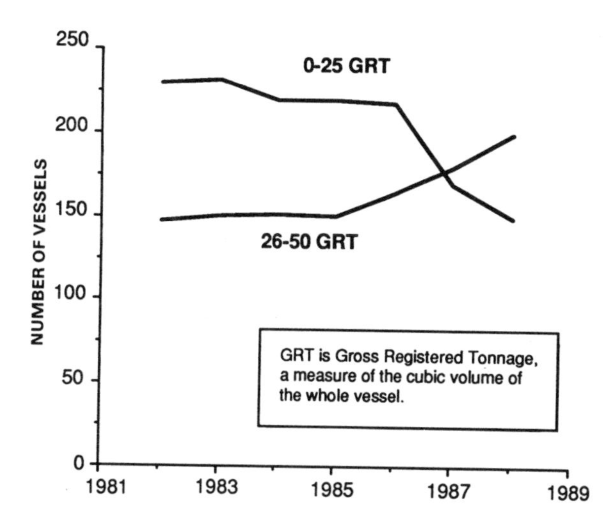

FIG. 13-14. Trends in the number of low capacity (<25 gross tonnes) and high capacity (26–50 gross tonnes) vessels in the Scotia–Fundy Region's Mobile Gear Fleet under 45 feet in length (from DFO 1989i).

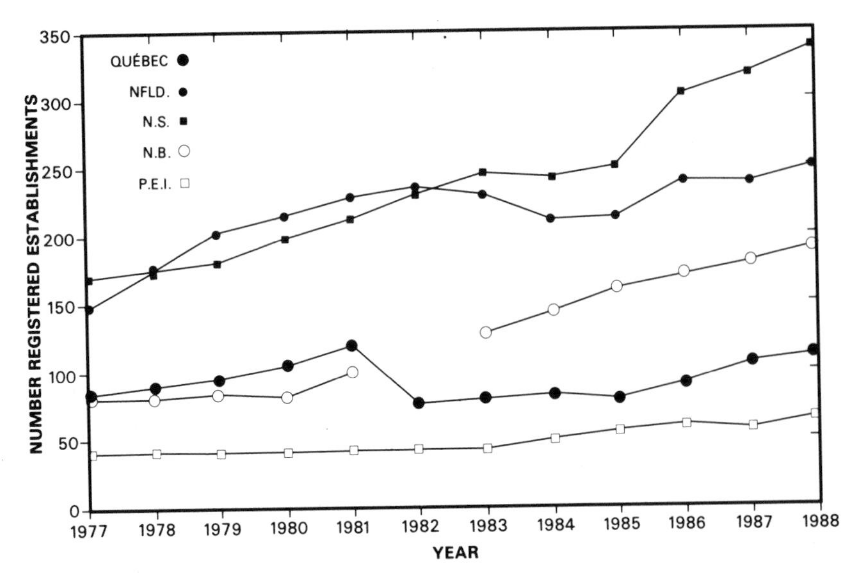

FIG. 13-15. Numbers of registered fish processing establishments in Atlantic Canada — 1977–1988 (from data provided by Economics Planning and Analysis Division, Gulf Region, DFO).

eign fishing. NSP was hurt by severe quota reductions in haddock and other species on the Scotian Shelf and by continued pressure from the overcapitalized small dragger fleet to reduce its Scotian Shelf quotas even further.

Following the 1984 restructuring, another major company had emerged in the east coast fishery in addition to NSP and FPI. The Clearwater group of seafood companies, only a small player as late as 1984, grew enormously, with sales increasing from $80 million to $700 million in just a few years. Clearwater's President John Risley built a multi-million dollar empire from the back of a rented truck in just 12 years. Clearwater started to blossom when it began exporting live lobsters to Europe. In the mid-1980's it diversified and bought a number of small and medium-sized fish enterprises including groundfish operations in Nova Scotia and even extended its operations into British Columbia. Early in 1989, Clearwater became the focus of controversy when it announced the closure of a fish plant in Port Mouton that it acquired as part of the acquisition of C.W. McLeod Ltd.

The company was blasted by politicians, plant workers and the media for closing the plant. In response, President John Risley predicted that at least eight more fish plants, employing hundreds of Nova Scotian workers, would be shut down by their owners by the end of 1989 (*The Chronicle Herald*, March 29, 1989). Clearwater came under attack because 57% of the non-voting preferred shares were held by Hillsdown Holdings of the United Kingdom. Ownership in terms of voting shares rested with Canadian John Risley and his brother-in-law Colin McDonald. Hillsdown PLC had the majority of equity and non-voting shares.

Labouring under the burden of a reported $300 million debt, in mid-August 1989 Clearwater Fine Foods Inc. was restructured. Under this arrangement Clearwater purchased all existing Clearwater shares held by Hillsdown. In return, Hillsdown purchased Risley's and McDonald's shares in Clearwater, USA, including the subsidiaries of the U.S. company. This followed the earlier purchase by Hillsdown of all of the shares in Clearwater U.K.

Mr. Risley stated: "the balance sheet we're left with is one which allows us the comfort and capacity to weather future storms if they're out there" (*The Toronto Globe and Mail*, August 22, 1989).

The newly restructured company expected annual sales of $140 to $150 million, compared with $500 million before. In view of the emerging downturn, Clearwater's actions appeared a prudent response to the realities of the fishery.

By early 1989 it was clear that the industry was facing another groundfish crisis. The boom in the mid-1980's was being followed by another downturn. The federal government appointed a Task Force on Northern Cod, headed by former Associate Deputy Minister of Fisheries, Ken Stein, reporting to a special Cabinet Committee, as well as the Harris Panel on Northern Cod. Fisheries Minister Siddon had also appointed a Scotia-Fundy Groundfish Task Force headed by Regional

Director-General Jean Haché. These groups attempted to determine the extent and potential impact of the resource decline and overcapacity problems.

3.10.2 Trends in the Market for Groundfish

Looming in the background was a market downturn which began early in 1988 but lasted only until about 1990 (Fig. 13-6). Overall, Canada was still heavily dependent on the U.S. market. By the end of the 1980's almost 80% of the Canadian fish production was exported, almost identical to that at the beginning of the decade. For all species combined, Canada's export markets in 1988 were, in order of importance, the United States (56%), Japan (15%), the European Community (14%), and others (15%). By 1990 the U.S. accounted for over 58% of Canadian fish exports.

The share of Canadian groundfish production consumed within Canada increased significantly from 1980 to 1989 (Fig. 13-16). Over 80% of groundfish exports went to the U.S. market. The per capita consumption of fish in the U.S. which had risen to a peak in 1987 levelled off at the end of the 1980's. Canadian groundfish faced competition from new species such as orange roughy and Alaska pollock. Between 1984 and 1988 U.S. imports of groundfish blocks increased from 139,000 tons to 183,000 tons. The volume imported from Canada increased from 45,000 to 59,000 tons.

During this period there was an increasing demand for non-traditional species. Alaskan pollock was available in plentiful, year-round supply from U.S. waters. Consumers accepted the species as a substitute for traditional groundfish. From 1982 to 1987 Canada's overall share of the U.S. groundfish import market declined.

3.10.3 Report of the Harris Panel on Northern Cod

The Canadian Atlantic Fisheries Scientific Advisory Committee's January 1989 report on northern cod, (CAFSAC 1989a) was a bombshell which shattered

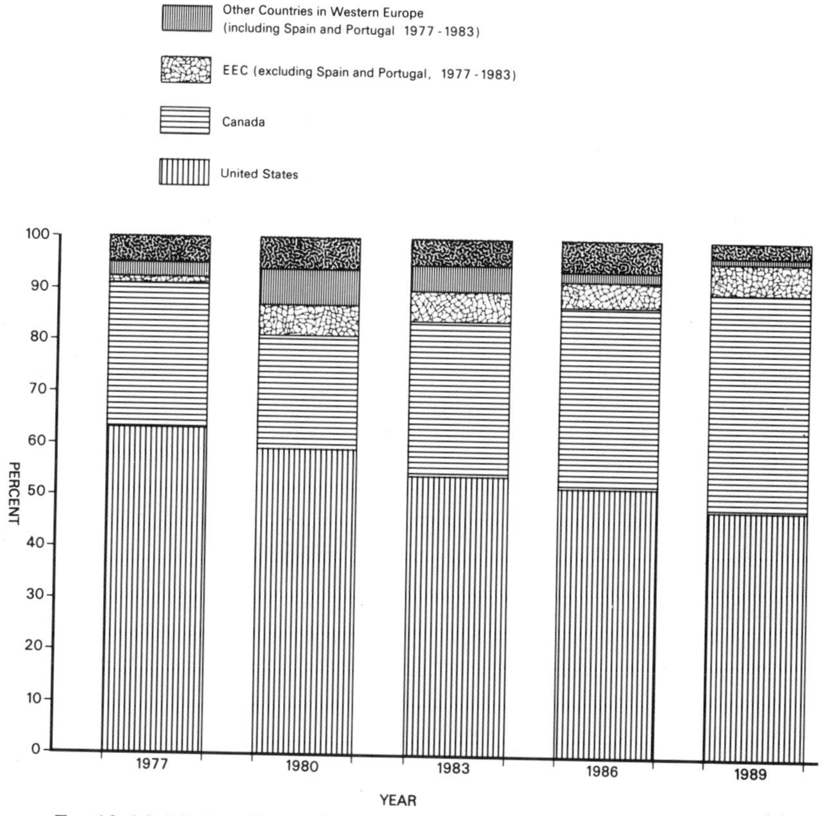

FIG 13-16. Market Shares for Canadian Groundfish (from data provided by Market Analysis Group, DFO).

expectations that had buoyed up the fishing industry in Newfoundland and to a great degree in Atlantic Canada since 1977. CAFSAC calculated that the catch corresponding to $F_{0.1}$ in 1989 would be 125,000 tons, compared with an earlier advised $F_{0.1}$ catch of 293,000 tons for 1988. CAFSAC concluded that previous assessments had underestimated fishing mortalities and that mortalities in recent years had been about double the $F_{0.1}$ level.

On February 8, 1989, Fisheries Minister Siddon announced a revised provisional TAC for northern cod of 235,000 tons, a reduction of 31,000 tons from the 266,000 ton TAC set in December, 1988. Because of the dramatic change in the scientific assessment of this stock, the Minister established an independent northern cod review panel chaired by Dr. Leslie Harris, President of Memorial University.

The Panel submitted an interim report on May 15, 1989 (Harris 1989). The Panel concluded that the most recent CAFSAC analysis was in the right ballpark. It recommended that the fishing mortality in 1990 be reduced to a value halfway between the existing level (0.45) and the $F_{0.1}$ level (0.20). This implied a 1990 TAC of 190,000 tons. On January 2, 1990, Minister Siddon announced that the Canadian northern cod quota would be reduced from 235,000 tons in 1989 to 197,000 tons in 1990.

On March 30, 1990, Bernard Valcourt, the new Minister of Fisheries and Oceans, released the final report of the Independent Review Panel on Northern Cod. The Panel made a total of 29 recommendations under several broad headings (Harris 1990). The Minister accepted the basic principles of the report as well as its major recommendations aimed at rebuilding fish stocks through intensified scientific research, new management measures and improved communications with fishermen and the fishing industry. The Panel recommended an immediate reduction of fishing mortality to 0.30 and as early as possible to 0.20. For details of the other Panel recommendations, see Lear and Parsons (1993).

3.10.4 The Scotia Fundy Groundfish Task Force

In July 1989, in response to quota reductions on the Scotian Shelf, which were exacerbating the effects of an already severe overcapacity problem, Minister Siddon appointed a Scotia-Fundy Groundfish Task Force chaired by DFO Regional General Jean Haché. This action was triggered by inshore plant shutdowns and layoffs resulting from closure of most of the groundfish fishery to the inshore dragger fleet in June 1989. The offshore industry was also reeling from repeated decreases in annual groundfish quotas on the Scotian Shelf. Two plants had closed permanently. National Sea's large plants closed on 2-month rotations to spread out the effects of the shortage of fish.

The Haché Task Force reported in December 1989 (DFO 1989i). It expressed concern about the three major groundfish species — cod, haddock and pollock — but especially haddock (Fig. 13-17). The haddock stock had been exploited at two to four times target levels since 1984. Scientists warned that, if this trend continued, the stock might collapse. The Task Force reconfirmed previous studies which had concluded that, over the short to medium term, the excessive catching capacity of the inshore fleet, particularly the mobile gear sector, was the major obstacle to a turnaround in the groundfish fishery.

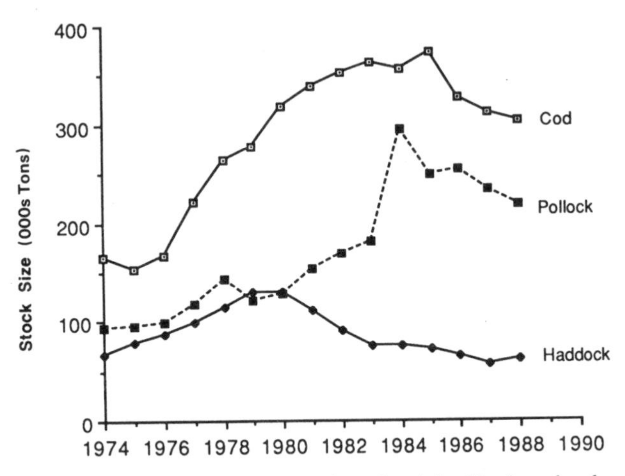

FIG. 13-17. Trends in stock size of cod, haddock and pollock on the Scotian Shelf (from DFO 1989i).

The Task Force concluded that harvesting capacity and overinvestment must be reduced as quickly as possible. It made 30 recommendations on a broad range of subjects — conservation, catch monitoring, enforcement, science, international, communications, fleet management. Most of these recommendations are addressed in other chapters. For a discussion of the recommendations to deal with the dragger overcapacity problem, see Chapters 8 and 9.

The Report emphasized that tough choices were required to address the imbalance between the available resource and harvesting and processing capacity.

3.10.5 Industry Response to the 1989–1992 Crisis

The two major vertically integrated companies involved in the offshore Atlantic groundfish fishery,

NSP and FPI, both experienced major losses in 1989. The offshore and inshore landings processed by NSP had declined each year from 1985 to 1989 with a cumulative drop of 31% (43,100 tons). This reduced catch resulted in significant plant overcapacity with the reduction in catch representing the annual processing capabilities of two large plants. During 1988, NSP experimented with rotating plant shutdowns which, with a strike early in 1989, resulted in 28 months of plant shutdowns (the equivalent of 2.3 of NSP's nine plants being closed year-round). (See Annual Reports for 1988 and 1989, National Sea Products).

NSP determined that these temporary closures did not significantly reduce its operating costs. It decided to permanently close its Lockeport plant in Nova Scotia on October 1, 1989. Also, in response to the quota reductions, NSP announced on December 11, 1989, the indefinite closure of two more plants, Canso, Nova Scotia, and St. John's, Newfoundland, and the reduction of wetfish operations at its North Sydney plant. These closures involved the loss of employment for 1500 people.

Subsequently, NSP reached an agreement with the provincial and federal governments in 1990 to operate Canso at reduced volumes and to examine the feasibility of converting the St. John's plant to shrimp processing. NSP also decommissioned four of its offshore trawlers in 1989 and another four trawlers in 1990.

FPI also experienced a major loss in 1989. This resulted from a combination of a soft market, the exchange rate, and a drop in catches. Its groundfish landings decreased by 15% between 1988 and 1989. At the end of 1989, it estimated that its 1990 quotas would be 26% less than 1988 levels. The capacity utilization rate in FPI's main processing plants dropped from 65% in 1988 to 55% in 1989 and was projected to reach 45% in 1990.

Early in 1990, FPI addressed this overcapacity problem by rationalizing its operations. On January 5, 1990, it responded to the 1990 Groundfish Management Plan with plans to reduce both harvesting and processing capacity. The proposed measures included:

1. The permanent retirement of 13 trawlers;
2. The processing plants at Gaultois, Grand Bank, and Trepassey would be permanently closed but not until the end of their operating period in 1991. (The plants would continue to operate in 1991 because the Newfoundland government had agreed to finance an extended notice period).
3. Staff cuts throughout the company (FPI 1990a).

In response, International Trade Minister Crosbie announced that the federal government would spend up to $130 million over the next 4 years to assist the individuals and communities affected by the plant closures and layoffs announced by FPI and NSP. This amount was intended "to respond to the immediate needs of the people and communities affected" (Canada 1990b). The immediate focus was to utilize the federal government's Industrial Adjustment Program. This was to be supplemented by additional support for older plant workers through separate measures that would apply to those 50 years of age and older. The short-term measures also included the creation of Community Diversification Funds.

Both NSP and FPI rebounded in 1990, partly as a result of consolidating and streamlining their operations. NSP sold its plants in Canso and Burgeo to a new company named Sea Freez on November 2, 1990. This deal was facilitated by the federal government's allocating 126,000 tons of underutilized species to Sea Freez and providing a loan guarantee of 85% of $15 million. Sea Freez traded its allocations of silver hake and capelin to the USSR for more valuable groundfish such as Barent's Sea cod and shrimp harvested by Soviet vessels outside the 200-mile limit. In any event, this reduced NSP's processing capacity. On sales of $608 million in 1990, NSP reported a loss of $2.2 million. It attributed the improvement to a number of factors, including the restructuring of operations in Canada and the U.S., a reduction in processing capacity and strong market prices (NSP 1990).

FPI rebounded from a 1989 loss of $22.2 million on sales of $349.8 million to a net income of $11.8 million on sales of $535 million in 1990. These results were attributed to cost reductions, higher sales and new products from the acquisition of Clouston Foods Canada Ltd. This was an international seafood trading and marketing firm with subsidiaries in Massachusetts and Washington (FPI 1990b).

1991 brought further changes to FPI's operations. In addition to closing several sales offices, FPI closed its plant in Gaultois on May 24, 1991 and its Grand Bank plant on May 31, 1991. The Trepassey plant was closed at the end of August or September. These closings reduced the number of FPI employees by approximately 1300. That left the company with approximately 6,400 employees and 15 plants operating at about 65% of their combined capacity in Newfoundland, Nova Scotia and Massachusetts. In mid-1991 efforts were underway to have other companies operate these closed plants.

In 1991, FPI essentially broke even partly due to the acquisition of the profitable Clouston Foods Canada, Ltd. NSP, on the other hand, had an operating loss of $7 million in 1991. During the first 9 months of 1992, FPI recorded a net operating loss of $3.5 million compared with an operating loss of $1.5 million during the same period of 1991. NSP had a net operating loss of $6.1 million for the first 9 months of 1992.

The latest crisis, which commenced in late 1988-early 1989, was not yet over. The Canadian offshore fleet experienced lower catch rates in fishing for northern cod in 1991 and had great difficulty in locating fishable concentrations of commercial size fish. At the beginning of 1992 the Canadian offshore fishery for northern cod was severely curtailed as part of the 120,000 tons conservation ceiling established in February, following CAFSAC advice that the stock had declined abruptly in 1991.

3.10.6 The Federal Government's Response to the Crisis in 1989 and 1990

In response to the social and economic dislocation caused by the quota reductions for northern cod and other species, and overcapacity problems, the federal government announced an Atlantic Fisheries Adjustment Program in May, 1990. AFAP had three elements "aimed at ensuring a viable fishery in the long-term for Atlantic Canadians, while supporting individuals and communities in the fishery to adjust to the realities of declining fish stocks and plant closures" (Canada 1990a).

These elements were:

— Rebuilding the Fish Stocks
— Adjusting to Current Realities
— Economic Diversification.

This expenditure of $426 million was additional to the previously-announced $130-million in the short-term response program and $28-million for aerial surveillance. Thus, the total federal contribution would be $584 million over the next 5 years.

Under the Rebuilding the Fish Stocks element about $150 million was targeted for expanded research on northern cod, other cod stocks off southwest Newfoundland and in the Gulf of St. Lawrence, Scotia-Fundy groundfish and Gulf snow crab. There was to be a greater focus on the ecology of cod in relation to its ocean environment and the impact of trawling on spawning grounds and nursery areas. Greater involvement of fishermen in DFO scientific research was emphasized. Research on harp and grey seals was to be increased. Surveillance and enforcement activities were to be expanded, and penalties increased through amendments to the Fisheries Act.

An Industrial Adjustment Service, (IAS), funded at $130 million, was to benefit laid-off workers (or workers identified for lay-off) in seven communities facing plant closures resulting from the northern cod reductions. The communities were: St. John's, Gaultois, Grand Bank, Trepassey, Canso, North Sydney and Lockeport. In addition, the IAS was to be used to help four other groups adjust to quota reductions: FPI trawlermen, inshore fishermen, the Scandinavian longliner fleet and the Burin refit centre.

The Economic Diversification Element, funded at $146 million, was to promote community economic diversification focusing on fishery dependent communities affected by quota reduction and stock depletion. The aim was to: 1. provide alternative employment opportunities; 2. encourage the exploration of underutilized species and stocks by assisting with marketing and fisheries development; and 3. assist aquaculture development.

Commenting on the program, Minister Valcourt focused on the need for adjustment and diversification:

> "Adjustment is already happening in the fishery. But we must manage that adjustment instead of having it forced upon us through a declining resource base, with all the human and community costs that this entails. We must take steps to link the size of the fishery to what can realistically be sustained by the abundance of the fish stocks.
>
> "To deal with the adjustment that is necessary principally in the Northern Cod fishery but also in other key Atlantic fisheries, jobs need to be created in other sectors of the economy....promoting diversification within the fishery will provide a broader base for a viable fishery over the long-term."

Minister Crosbie also observed:

> "We recognize that the long-term diversification of the Atlantic economy is necessary to provide alternative employment opportunities to those who have traditionally relied upon the fishing industry for their jobs" (Canada 1990a).

In June 1990, Minister Valcourt requested Mr. Eric Dunne, Director General of DFO's Newfoundland Region, to carry out consultations and develop recommendations on how best to implement the recommendations of the Harris Review Panel. The Report of the Dunne Implementation Task Force on Northern Cod was released in October 1990 (DFO 1990k). Regarding the level of TAC, the Task Force concluded that a decision was required on the short-term objective: whether (1) to continue the status quo and keep the biomass constant or (2) to start rebuilding by reducing the TACs, or (3) to rely solely on future recruitment to improve the stock status.

Various options and their likely consequences were examined, ranging from a drastic (50%) immediate cut in the TAC — which would allow a rapid increase in stock size but virtually eliminate the offshore harvesting/processing sector — to maintaining a constant catch of 200,000t. The latter would minimize further disruption in the industry but would probably allow only minimal stock growth. The Task Force concluded that "stock rebuilding must be started with an immediate, even if modest, reduction in catch....Even a small reduction in catch will improve stock status faster than any of the other available management measures."

The Task Force recommended that multi-year TACs be used to reach $F_{0.1}$. Multi-year plans should be continued thereafter to maintain stable catches and reduce pressure for major increases in the TAC.

On December 14, 1990, Minister Valcourt announced a new Multi-year Groundfish Plan (DFO 1990c). The TAC for northern cod was set at 190,000 tons in 1991, 185,000 tons in 1992 and 180,000 tons in 1993. This period of relative stability in TACs was proposed in the expectation that the stock would rebuild. This rebuilding would, however, occur very slowly. Hopes for the stock rebuilding were based upon an apparently stronger-than-average 1986 year-class.

This Multi-year Groundfish Plan set 3-year TACs for most stocks, generally at the same level as in 1990. Exceptions included northern cod, northern Gulf of St. Lawrence cod and southern Gulf of St. Lawrence cod. To address the state of the Gulf of St. Lawrence cod stocks, new measures were introduced including lower TACs, stricter observer coverage and dockside monitoring, and experiments with increased mesh size to reduce the harvest of juvenile fish.

To address the overcapacity problem in Scotia-Fundy, individual quotas for cod, haddock and pollock were allocated to the majority of groundfish draggers under 65 feet in length (see Chapter 9). This was accompanied by a new Commercial Catch Monitoring system to curb misreporting and under-reporting of catches.

Canada's Atlantic groundfish quotas were reduced by 16% between 1988 and 1990. However, DFO estimated that by the end of 1990, total groundfish exports would only have decreased by 8% (DFO 1990c). Canadian fishermen were using more of their total groundfish quota, thus reducing the impact of quota reductions on supplies in the market place.

Demonstrating the volatility of markets as well as the resource base, prices for most groundfish products rose dramatically during 1990. For example, the price of cod blocks in the U.S. market increased 50% during 1990. Canada's position in the U.S. market appeared secure vis-a-vis major competitors such as Iceland and Norway. However, there was significant new competition from a variety of nontraditional groundfish, species such as Alaska pollock and cod, farm-raised catfish, South American hake and from new fish exporters.

The Atlantic groundfish industry began to emerge from the 1989–1990 downturn. This downturn was largely a result of adjustments in resource supply, combined with exchange rate pressures. Again the federal government had stepped in with major financial assistance. Initiatives undertaken during the previous crises of the mid-1970's and the early 1980's had not resolved the underlying dilemma of fisheries management. It was not yet clear whether initiatives under AFAP would have a lasting impact through capacity reduction or provision of alternative employment opportunities through economic diversification.

3.10.7 The Deepening Crisis of 1992 and 1993

Just as it appeared that the Atlantic groundfish industry had survived its third major downturn in less than two decades, the crisis deepened again early in 1992. Fisheries Minister John Crosbie, in response to preliminary CAFSAC advice for 1992 on northern cod (CAFSAC 1992a), slashed the Canadian northern cod TAC to 120,000 tons. This resulted in widespread temporary closures of fish plants in Atlantic Canada while the industry assessed the impact of the latest quota reductions.

In June and July, 1992, the NAFO Scientific Council and CAFSAC concluded that there had been a sudden, drastic and unexpected decline in the abundance of northern cod during 1991. The total biomass had been reduced by half and the spawning stock biomass by three-quarters. The spawning stock biomass at the beginning of 1992, estimated to be in the range of 48,000 to 108,000 tons, was near or at least the lowest level ever observed and probably only 10% of the average spawning biomass (540,000 tons) from 1962 to 1990 (Fig. 13-18). While no single factor was identified as the cause of the sudden drastic decline, the primary factors suggested were ecological. Extreme environmental conditions in 1991 (the worst ice coverage and the coldest ocean temperatures in 30–40 years) were thought to be the primary factors contributing to the abrupt decline.

Because the stock now consisted essentially of the 1986 and 1987 year-classes (5 and 6 year-old fish in 1992), with the prospects that the four succeeding year-classes were weak, CAFSAC advised that catches be kept to the lowest possible level and all efforts be

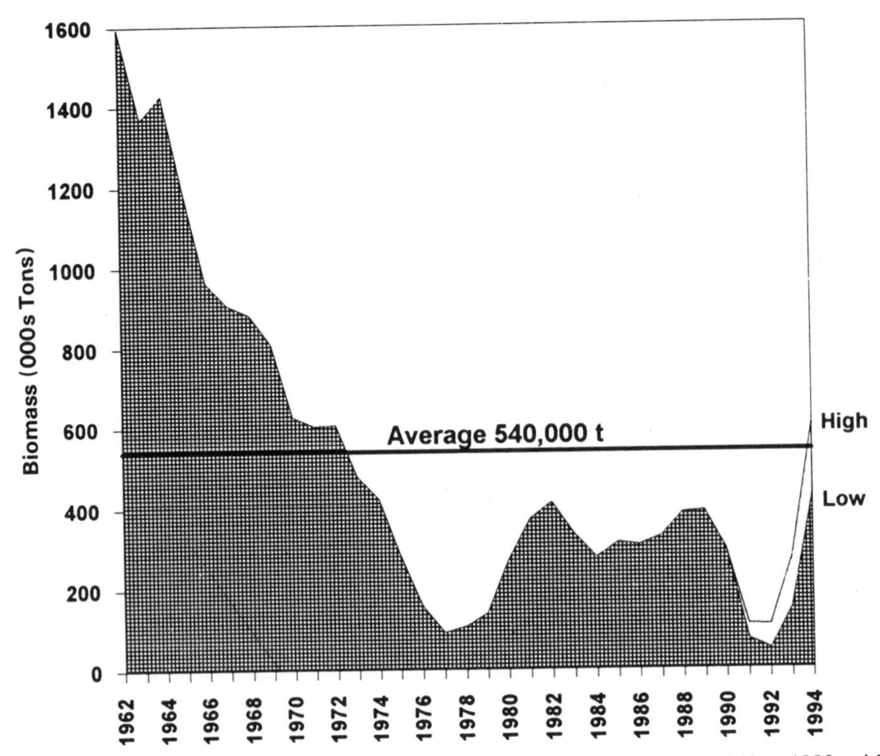

FIG. 13-18. Trends in the spawning stock biomass of northern cod from 1962 to 1992, with projections of the impact of the two-year moratorium on spawning stock biomass in 1993 and 1994 (DFO, 1992a).

made to enhance stock rebuilding (CAFSAC 1992b). The federal government reacted swiftly to this advice. On July 2, 1992, Fisheries Minister Crosbie announced a 2-year moratorium lasting until the spring of 1994 (DFO 1992a). This included the dramatic and unprecedented closure for 2 years of the Newfoundland and Labrador inshore fishery for northern cod, affecting 19,000 people directly dependent on this fishery. Although such an action would have been unthinkable previously, the moratorium was generally considered necessary by industry participants, including many inshore fishermen.

Inshore fishermen, however, condemned the federal government's initial emergency assistance payments of $225 a week to fishermen and plant workers affected by the moratorium. On July 17, 1992, Minister Crosbie announced an increase in compensation and a comprehensive adjustment program for northern cod fishermen and plant workers. New income replacement benefits, to a maximum of $406 per week with a minimum of $225 per week, took effect on August 1, 1992. The income replacement payments would be based in most cases on individuals' average unemployment insurance benefits over the previous 3 years (DFO 1992e).

Between August 1 and December 31, 1992, fishermen and plant workers would be asked to choose among a variety of alternative skills development programs. These included training for work outside the fishery as well as professionalization and certification within the fishery. Those who opted for these programs would continue to receive the full income replacement payments up to the maximum of $406 per week throughout the northern cod moratorium to the spring of 1994. Eligible individuals who chose not to enter one of these programs would revert to basic payments of $225 a week for the remainder of the moratorium. Older fishermen could choose early retirement through extension of an existing program for plant workers and trawlermen between the ages of 50 and 65.

The program also included a plan to retire licences from fishermen choosing to leave the fishery. This option would be available to the 4,000 active and 2,000 inactive groundfish licence holders in the northern cod fishery. Minister Crosbie indicated that, when the moratorium ended, priority of access to the resource would be given to full time, professional fishermen.

With respect to the processing sector, Minister Crosbie stated that it was up to the Newfoundland government, which has jurisdiction over processing plants, to decide the level of processing capacity and the distribution of fish plants. However, he indicated that the federal government "believes that capacity should be matched to the limits of the resource and be distributed along the coast in a way that provides reasonable access by fishermen for the sale of their catches."

The federal government recognized the need for the processing sector to maintain a core of capacity and employment. To this end, the federal government indicated that it would assist in identifying other sources of fish to be processed at northern cod plants during the moratorium. These sources could include Greenland halibut harvested in Davis Strait and fish purchased on the international market for processing in core plants. Details were to await the results of a Peat Marwick study of the processing industry.

The stated goals of the Northern Cod Recovery and Adjustment Program (NCARP) were as follows:

1. Protection and rebuilding of the northern cod stock;
2. A sustainable fishery for that resource once the moratorium ended;
3. A professional work force of fishermen and plant workers deriving adequate incomes from harvesting and processing northern cod;
4. Processing and harvesting capacity matched to the limits of the resource; and
5. Processing plants distributed along the coast to provide reasonable access to fishermen to sell what they harvest.

In response to questions from the media, Minister Crosbie indicated that he viewed the northern cod crisis as "an opportunity as well as a tragedy." He also affirmed his belief in the importance of the inshore fishery " which has the priority in the fishery scheme of things." He also stated that the historic share for the inshore sector must be maintained.

According to Minister Crosbie, the cost of the income replacement and adjustment program would be well in excess of $500 million. Various sources estimated the likely cost at $750–800 million. Once again the federal government had to intervene with costly measures to deal with a crisis in the Atlantic fishery. This time it appeared that at least one sector of the Atlantic groundfish industry might emerge from this crisis with a better match between the available resource and harvesting and processing capacity. The social and economic implications for the Newfoundland economy were profound.

Meanwhile, other parts of the Atlantic groundfish industry were hit hard in the autumn of 1993 when CAFSAC advised drastic reductions in the TACs, for certain other cod stocks. These included:

1. a reduction in the TAC for southern Gulf cod from 43,000 tons in 1992 to 13,000 tons in 1993;
2. a recommendation that catches in the 4Vn (May–November) stock be limited to the lowest possible;
3. a reduction in the TAC for 4VsW cod from 35,200 tons to 11,000 tons; and
4. significant reductions for Georges Bank cod and haddock.

Taken together with the northern cod moratorium, prospects were grim for the Atlantic groundfish industry in 1993 and immediately thereafter. The Canadian groundfish catch, which had been declining from the mid-1980's, would decline further. Canadian cod catches had declined substantially from 510,000 tons in 1983 to 307,000 tons in 1991. Catches of cod and haddock in general and flatfish catches on the Grand Banks appeared likely to decline further before the stocks began to recover.

During 1992 it had become apparent that the gradualist approach to TAC reductions, e.g. 50% rule, was not working. Therefore, for 1993, Fisheries Minister Crosbie announced the adoption of an immediate move to $F_{0.1}$ for the other major cod stocks in the Canadian zone. This resulted in draconian reductions in TACs, exacerbating the crisis in the Atlantic groundfish industry which had commenced in the late 1980's. These reductions plunged the industry into an even deeper crisis in 1993.

These very low TACs were accompanied by much tougher conservation measures aimed at reducing the capture and discarding of small fish. It was hoped that the tougher conservation measures, together with reductions in harvesting and processing capacity over the next several years, would arrest the stock declines and rebuild the stocks over the next several years. To a large extent, this would depend on whether environmental factors facilitated an increase in resource productivity.

The 1989–1993 crisis represented another chapter in the continuing saga of the Atlantic groundfish industry, riding the roller coaster from boom to bust again and again.

4.0 OTHER MARKET AND TRADE FACTORS AFFECTING THE CANADIAN FISHERIES

4.1 Countervail

Looking more broadly beyond groundfish, there were a number of other market and trade factors which had the potential to destabilize other important sectors of the Canadian fisheries. On the east coast, U.S. fishermen have periodically complained about the perceived subsidization of the Canadian fishing industry. They have threatened countervail actions under Section 332 of the Tariff Act. In 1978, following complaints from U.S. fishermen, the U.S. Tariff Commission held countervail hearings involving over $200 million in Canadian fish exports. This was the amount that U.S. interests claimed benefitted from subsidies such as the Fishing Vessel Assistance Program (FVAP), the Temporary Assistance Program (TAP) and various DREE grants. U.S. fishermen maintained that Canadian subsidies were putting them at a disadvantage. They petitioned the Secretary of the Treasury to impose countervail duties on Canadian seafood exports.

Canadian negotiators impressed upon the U.S. Administration the detrimental effects such duties would have on Canadian industry located in areas already suffering from high employment. Under the U.S. Trade Act of 1974, the Secretary could waive countervailing duties on certain products. In return, the exporting country must reduce, or take steps to reduce, its subsidy programs. U.S. authorities linked the Canadian request for a waiver to the suspension of the interim fishing agreement. In a major concession, Canada agreed to end the Temporary Assistance Program on October 1, 1978. In return, the Secretary of the Treasury granted the waiver.

Again in November 1983, the U.S. Trade Representative, responding to petitions from the New England fishing industry, asked the U.S. International Trade Commission to undertake an investigation under Section 332. In a December 1984 report, the Trade Commission confirmed that the Canadian industry had much greater government assistance than the U.S. industry. The Commission said Canada's share of the fresh groundfish market had been increasing, and the price at dock-side of Canadian fish was lower because of the concentrated buying power of the larger government supported companies. However, the Report also noted that the build-up in the New England industry under open access conditions resulted in rising costs, declining worker productivity and lesser resource availability in the New England fishery.

The North Atlantic Fisheries Task Force decided to file a countervailing duty case with the Secretary of Commerce and the International Trade Commission. Meanwhile, the Task Force suggested that a trade-off could be made in return for renewed access to the Canadian side of Georges Bank. Canada rejected this. Following extensive representation by U.S. and Canadian interest groups, the U.S. International Trade Commission ruled on the latest countervailing duty petition. In April 1984, it imposed a 5.82% duty on imports of certain species of fresh whole groundfish from Canada.

Action had also been taken against Canadian saltfish. In 1984, the International Trade Commission determined that Canadian salt cod firms were selling their products in the U.S. market at less than fair market value and that these pricing policies retarded the development of a U.S. salt cod industry. Anti-dumping duties were imposed against a number of Canadian salt cod exporters. A final U.S. ruling in September, 1987, resulted in duties on U.S. imports of dried saltfish from Canada being reduced to zero for the Canadian Saltfish Corporation and other appellants.

For a detailed critique of the use of American trade law in the Atlantic groundfish case of 1985–1986, see Rugman and Anderson (1987). The authors' careful

Fishery officer measuring lobster carapace length.

review of the U.S. countervail process concluded: "The ITC Commissioners are prepared to vote material injury based on the superficial data and inconclusive evidence produced by the ITC." They cited the dissenting report of Chairperson Stern who observed: "The Canadian product has been alleged to suppress the returns to U.S. fishermen. No conclusive evidence has been presented to support, or, conversely, to disprove this theory. Rather, an equally possible result of imports is that they have kept the average U.S. price *affordable* at the retail level for most consumers, thus protecting the market share won by fish in recent years. This result is not injurious, but rather desirable."

These actions did not have an immediate impact on the Atlantic groundfish industry. However, they indicated the potential negative impact of U.S. countervail law on the marketing of Canadian groundfish in the U.S. Canadians have generally perceived the American approach to these issues as dominated by political considerations rather than the merits of any particular case.

4.2 Non-tariff Barriers

The United States has taken other non-tariff actions, on the grounds of conservation, with a potential negative impact on the marketing of Canadian products in the U.S. During the early 1980's, the U.S. introduced its own meat-count regulations for scallops and threatened to apply these to scallops imported from Canada. Canada circumvented this by introducing an inspection program to certify that Canadian-caught scallops exported to the U.S. had been caught in accordance with Canadian conservation regulations.

Similarly, in October 1987, at the request of the New England Regional Council, the U.S. Secretary of Commerce prohibited landings of round cod and haddock under 19 inches. This also applied to imported fish. Canada threatened to complain to the GATT unless this action were reversed. In the end, however, Canada introduced its own minimum size limit for groundfish.

These were but two examples of the U.S. using minimum fish size requirements as non-tariff barriers to trade. More recently, a major dispute resulted from U.S. efforts to prohibit the import of lobsters below the U.S. minimum size limit.

Lobster is the second most important species for Canadian sea food exports to the U.S. It is second only to cod and constitutes about 15% of Canadian fish exports to the U.S. The 1987 American Lobster Fishery Management Plan proposed increases in the U.S. minimum size limit to 3¼ inches in 1989, 3 9/32 inches in 1991 and 3 5/16 inches in 1992. A proposal to prohibit the import of lobsters smaller that this was considered but dropped following representations from Canada.

In 1988, DFO, as a result of consultations with fishermen, decided that the minimum size limit should be increased along the Atlantic coast to mirror U.S. increases. The department initiated regulations for a size increase. In 1989, the various Atlantic lobster advisory committees reversed their earlier recommendations for a size increase because of industry concern over possible negative impacts on catches and markets. DFO accepted this advice and withdrew the proposed regulations.

In September 1989, Senator George Mitchell introduced a Bill in the U.S. Senate amending the Magnuson Fishery Conservation and Management Act to prohibit from interstate commerce lobsters and lobster products smaller than the minimum possession limit in effect under the American Lobster Fishery Management Plan. Congress passed this bill which became law on December 12, 1989.

Canada initiated a Chapter 18 complaint in December 1989 under the Canada-USA Free Trade Agreement (FTA) against the U.S. restriction. A dispute settlement panel was established. In May, 1990, the Panel ruled that the U.S. measure was not an unfair trade restriction. Following this ruling, members of the two countries' lobster industries attempted to negotiate a settlement. An industry-to-industry understanding was reached in July 1990, followed by a tentative agreement between officials of the two governments.

This agreement would have required Canada to increase its minimum lobster size in all areas of Atlantic Canada, except the Gulf of St. Lawrence, to the same level as that in U.S. federal jurisdiction. In return, the U.S. would not increase its minimum size requirement for 3 years.

Canadian lobster fishermen who would be most seriously affected by the size increase were strongly opposed. They persuaded the Canadian government to reject the proposed agreement.

The U.S. size limit increased to 3 9/32 on January 1, 1991, following the original plan. In mid-January the New England Fishery Management Council asked the Secretary of Commerce to take emergency action to suspend size increases beyond 3¼ inches, i.e. that in effect in 1990.

At the same time, the Council indicated it would develop a more comprehensive amendment to the existing FMP to specify an "optimum target level of effort" for the fishery and management measures to achieve that target. This was a move toward the more comprehensive Canadian management regime (see Chapter 6).

Canada has suggested that the U.S. use of technical barriers to trade such as the possession limit requirements regarding size limits for scallop, groundfish and lobster imports are inconsistent with the FTA (DFO 1988h). Fisheries Minister Tom Siddon, in remarks to the Senate Standing Committee on Fisheries, suggested that "these kinds of standards will be eliminated except where they are clearly in harmony between the two countries and are needed from a quality, health or management perspective" (Canada 1988e).

Under the FTA the continued use of technical standards that restrict trade are subject to the binational dispute settlement process.

4.3 The West Coast Salmon and Herring Dispute 1987–1990

Such technical barriers to trade continue to threaten the export of Canadian fish products to the U.S. market. On the west coast, however, recent developments in the context of GATT and the FTA respecting export restrictions on fish threatened to disrupt the supply of fish to the Canadian processing industry. From 1986 to 1988 Canada and the U.S. were locked in dispute over Canadian regulations prohibiting the export of unprocessed salmon and herring. Sections 6 and 7 of the *Pacific Commercial Salmon Fishery Regulations* required that all sockeye and pink salmon be processed prior to export. They prescribed processing and packaging requirements for salmon and salmon eggs destined for export. Section 24 of the *Pacific Herring Fishery Regulations* required that all herring be processed prior to export. It established processing and packaging requirements for herring, herring roe, and herring spawn on kelp destined for export.

Certain U.S. processors in Alaska complained that unprocessed salmon and herring were being shipped from Alaska to British Columbia but that Canadian regulations prohibited similar trade going the other way. During the fall of 1986, the United States and Canada held discussions which failed to resolve the matter. The U.S. believed that these Canadian regulations impaired U.S. rights under GATT. Therefore, it complained to the Council of GATT that this "processing in Canada" requirement violated GATT. In February 1987, the U.S. requested a Panel to examine the matter.

The main thrust of the Canadian argument was that the restrictions on export of unprocessed herring and salmon were a multi-purpose measure. Without them, separate regulatory measures would be needed to deal with landing, inspection, weighing, sorting, grading and collecting data. Canada argued that one purpose of the export restrictions was to maintain quality standards. These regulations were also part of the conservation framework because they assisted in the process of statistical information gathering which was necessary for the determination of conservation measures.

On November 20, 1987, the GATT Panel concluded that the export prohibitions on certain unprocessed salmon and unprocessed herring were contrary to GATT rules (GATT 1987). The Panel found that the Canadian prohibitions "could not be deemed to be primarily aimed at the conservation of salmon and herring stocks and at rendering effective the restrictions on the harvesting of these fish." The Panel recommended that Canada bring its measures into conformity with GATT.

Canada had argued that its measures restricting processing were justified in the context of international fisheries treaties and the 1982 LOS Convention. It contended that "the concept of conservation had evolved to include socioeconomic as well as biological dimensions." The Panel, however, considered that its mandate was limited to the examination of Canada's measures in the light of the relevant provisions of the General Agreement.

The reaction to the GATT Panel decision was negative from both B.C. fish processors and the union representing plant workers and fishermen. They suggested removing the prohibition on exports of unprocessed salmon and herring could cost up to 5,000 jobs (The *Toronto Globe and Mail*, December 19, 1987). Industry lobbied the government to oppose the Panel's decision when it came before the GATT Council.

Industry also pressured the federal government to address the issue through the Free Trade Agreement (FTA) then under negotiation. Canada tried to include in the FTA a provision allowing the Canadian West Coast herring and salmon export restrictions to continue. Given the GATT Panel decision, the U.S. would not agree to this (The *Toronto Globe and Mail*, February 10, 1988). In Article 1203(c) of the FTA Canada got an exception for "controls by Canada on the export of unprocessed fish" originating in the five eastern provinces.

This exemption provided no relief on the west coast issue. On March 21, 1988, Minister for International Trade, Pat Carney, informed the House of Commons:

> "It is the Government's intention to allow adoption of the GATT Panel Report and to dismantle the GATT – inconsistent export restrictions by January 1, 1989. But, at the same time, we intend to enact new regulations which will require that salmon and herring caught off the Pacific Coast be landed in Canada in order to ensure accurate catch reporting, inspecting, grading and quality control. While Americans will have access to unprocessed fish landed onshore at designated landing stations along the coast, they will not be allowed to buy fish directly from Canadian fishermen 'over the side' at sea. Overall, the landing requirement will improve management of the fisheries while preserving

the livelihood of coastal communities" (Canada 1988f).

On March 22, 1988, Canada informed the GATT Council that it considered that the Panel finding went too far. However, Canada would remove the measures which the Panel found inconsistent with GATT.

On April 25, 1989 Canada amended the *Pacific Commercial Fishery Regulations* and the *Pacific Herring Fishery Regulations* as follows:

1. It revoked Sections 6 and 7 of the *Pacific Commercial Salmon Fishery Regulations* which required that all salmon be processed prior to export;
2. It revoked Section 24 of the *Pacific Herring Fishery Regulation* which required that all herring be processed prior to export;
3. It introduced a requirement that all Pacific salmon and herring taken under a commercial fishing licence be landed in B.C. at a landing station. A landing station was defined as a building in British Columbia or a barge permanently affixed to the shore that was licensed as a fish processing plant or a fish buying station under the Fisheries Act of British Columbia;
4. It amended Section 14 of the Fish Inspection Regulations to permit the washing, icing and boxing of finfish, lobster and crab intended for export at locations other than an establishment registered under the *Fish Inspection Act*;
5. It required operators of landing stations to complete a record of each off-loading of salmon and herring and make that record available to the Department; and
6. It required operators of landing stations to make their salmon and herring available for biological sampling.

The Regulatory Impact Analysis Statement accompanying those regulations gave the following rationale for the new regulations:

> "Changes to the export restrictions without a requirement for landing would inevitably result in less accurate catch data and a limited biological sampling program, thereby seriously impairing Canada's ability to properly conserve and manage Pacific salmon and herring stocks. Accordingly, a landing requirement is imposed to ensure that the Department can maintain the ability to obtain accurate catch data and have access to salmon and herring stocks for the purposes of biological sampling."

Canadian officials consulted the U.S. government before introducing the regulations. The U.S. believed the new regulations had the same intent as the export regulations which had been ruled inconsistent with GATT.

In late May 1989, Canada and the United States agreed to establish a binational dispute settlement Panel under Chapter 18 of the FTA. The Panel, chaired by a Canadian (Donald M. McRae), submitted its final report in October, 1989. The issue was whether the Canadian landing requirement was incompatible with Article 407 of the FTA and, if so, whether the requirement was subject to an exemption under Article 1201.

The U.S. argued that the new Canadian landing requirement was an export restriction contrary to GATT. It contended that such a measure was a restriction because it imposed additional burdens on U.S. buyers. Buyers would need extra time for transporting the fish. They would have extra costs in landing and unloading, possible dockers fees, and product deterioration resulting from off-loading and reloading.

Canada argued that the landing requirement was not a restriction on the "exportation" or "sale for export" of herring and salmon to the U.S. within the meaning of GATT. It contended that U.S. buyers were free to purchase unprocessed herring and salmon under the same terms and conditions as Canadian buyers. Canada's chief argument was that the landing requirement was "primarily aimed" at conservation. Canada argued that a landing requirement provided the best information for conservation purposes because it was inherently more accurate than "hails" of the catch, it allowed for consistent verification and enforcement measures, and it provided access to 100% of the catch for biological sampling.

The U.S. argued that the landing requirement served no useful conservation objective. Hence it was not "primarily aimed" at the conservation of herring and salmon stocks. It considered the requirement a disguised restriction on international trade.

The Panel concluded that, "as presently constituted, the Canadian landing requirement is a restriction on 'sale for export' within the meaning of GATT Article XI(1)." The Panel also concluded: "Because it is applicable to 100% of the salmon and herring catch, the present Canadian landing requirement cannot be said to be "primarily aimed at" conservation and thus cannot be considered a measure "relating to the conservation of an exhaustible natural resource" within the meaning of GATT Article XX(g) and hence not a measure subject to an exception applicable under Article 1201 of the Free Trade Agreement" (Anon 1989).

However, the Panel "was also of the view that Canada could bring its landing requirement within Article XX(g) by structuring it" differently. In the Panel's view, one way that a landing requirement could be considered "primarily aimed at" conservation would be "if provision were made to exempt from landing that proportion of the catch whose exportation without landing would not impede the data collection process. Although any such proportion would have to be determined on the basis of the actual data and management needs of each fishery, or group of related fisheries, the Panel was of the view that the 10–20% range could provide appropriate guidance."

Given the wording of the Panel's decision, it is perhaps not surprising that both parties claimed victory. Superficially, the Panel upheld the U.S. claim. However, it did not rule that a landing requirement *per se* contravened GATT and the FTA but only a landing requirement applying to 100% of the catch. The Panel clearly stated that Canada could comply with Article XX by permitting up to 20% of the catch to be exported directly from the grounds.

The majority of the B.C. industry were disappointed with the decision. The Fisheries Council of British Columbia and the United Fishermen and Allied Workers' Union advised the Government to ignore the report. The British Columbia Minister of International Business publicly asked the federal government to disregard the report. Not to do so would undermine Canada's conservation regime.

In February 1990, the Canada-U.S. Trade Commission resolved the dispute. Canada would essentially retain its landing requirement for salmon and herring. However, 20% of the total allowable catch would be exempted from the requirement that salmon and herring be landed in British Columbia during the period March 1, 1990, to February 28, 1991. Twenty-five percent would be exempted in the 3-year period from March 1, 1991, to February 28, 1994. Canada would continue to control the export of roe herring to all destinations (Canada 1990c).

Thus, this dispute was finally resolved. However, market impediments and trade disputes such as this, which dominated the bilateral fisheries relationship on the west coast during the late 1980's, can have a destabilizing effect on the Canadian fishing industry.

5.0 CONCLUSIONS

The Canadian fishing industry in general, and the Atlantic groundfish industry in particular, have experienced alternating booms and busts every few years. Unpredictable fluctuations in fish abundance, excessive dependence on the U.S. market in the case of groundfish, and a general vulnerability to economic recessions, have all combined to produce periodic crises in the industry. In the case of Atlantic groundfish, there have been alternating periods of boom and bust with major crises occurring in 1967–69, 1974–76, 1981–83 and another commencing in late 1988-early 1989. While these crises have been spaced about 6–7 years apart, the factors leading to the downturns have varied from one crisis to the next. The 1967–69 downturn resulted from a downturn in the U.S. groundfish market. The major crisis of 1974–76 resulted from a combination of low resource abundance, rising costs and a market downturn. Together these factors threatened the survival of the industry. The 1981–83 groundfish crisis was the result of overexpansion in the processing sector financed by excessive debt in the euphoria following the 200-mile limit. When high interest rates of 20% plus struck during the Great Recession of the early 1980's, the offshore groundfish sector became insolvent. In 1989, downward adjustments in the TACs for many stocks, continued overcapacity, exchange rate pressures, and a temporary softening in the U.S. market combined to produce the third major Atlantic groundfish crisis in less than two decades.

Following each crisis, the industry has experienced boom conditions for a period of 3–4 years during which the problems of the past are rapidly forgotten. Everyone rushes to capitalize on the opportunities of the moment, forgetting that there is a tomorrow. Major government financial assistance in 1974–76 through the Bridging and Temporary Assistance Programs, and in 1983–84 through the financial restructuring of the large offshore companies, failed to buffer the industry against the shocks of subsequent downturns and failed to provide stability. Hence, the industry was again plunged into crisis in 1989 when TACs had to be adjusted downwards because of resource fluctuations and new perspectives on the status of certain stocks. This crisis deepened with the sudden, drastic and unexpected decline in abundance of the northern cod stock in 1991 and the introduction of a 2-year moratorium on fishing for northern cod commencing in July, 1992. Again, the federal government stepped in with major financial assistance, in the order of $800 million, to avert a social and economic catastrophe in Newfoundland and Labrador. Other sectors of the Atlantic groundfish industry were hit hard by drastic reductions in the TACs for certain other cod stocks in the Groundfish Plan for 1993.

Overcapacity in the harvesting and processing sectors resulting in excessive costs, and an excessive dependence on singular markets, render the industry vulnerable to resource downturns, market changes and any general economic downturns. While stability has been a goal of fisheries management since the 1970's, it will remain an elusive dream unless the problems of overcapacity can be addressed effectively and costs reduced. Even then, natural resource fluctuations will continue to produce unpredictable upswings and downturns at different times and in different fisheries. Existing management strategies aimed at catching a constant fraction of a variable number make it inevitable that catches will fluctuate as stock abundance varies. Even with relatively low exploitation rates, stock abundance will vary as a result of large-scale, environmentally-induced changes in year-class survival from time to time.

Bringing harvesting and processing capacity more in line with the productive capacity of a variable requires reconciling conflicting objectives. In 1983, the federal government adopted for the Atlantic fishery the three objectives proposed by the Task Force on the Atlantic Fisheries — economic viability, maximization of employment and Canadianization of the fishery. To promote economic viability, it is necessary to reduce capacity in the harvesting and processing sectors. Capacity reduction would lead to a reduction in employment in coastal communities dependent on the fisheries. The major Atlantic fishery crises of 1974–76 and 1981–83, led to calls for rationalization of capacity. This did not happen except for some subsequent rationalization of offshore groundfish harvesting capacity made possible by the introduction of enterprise allocations. It is not yet clear whether the response to the 1989–1993 crisis will significantly reduce capacity. The determination to implement a fundamental restructuring of the Atlantic groundfish industry seemed clearer in 1993 than at any time in the past.

A number of factors have hampered economic rationalization. These include the dependence of more than a thousand isolated coastal communities on the fishery, low fishing incomes in such communities, and the lack of alternative employment for fishery workers (see Chapter 14). Meanwhile, the Canadian fisheries seem likely to continue to ride the roller coaster, with alternating periods of boom and bust, exhibiting different patterns in different fisheries.

CHAPTER 14
THE SOCIAL DIMENSION

"When we think of optimum biological management and optimum economic yield, we must consider also optimum social yield. That is, how can we best satisfy and serve the most people?... When fish are counted, it's people that count."

– Roméo LeBlanc, October 22, 1974

1.0 INTRODUCTION

For decades battles have been fought between governments, between fishermen and processors, and within the academic community over the ill-defined concept of the 'social' versus the 'economic' fishery. Economists (e.g. Copes 1972, 1983) have waxed eloquent about the ills of the social fishery and the need for 'rationalization', i.e. reducing the numbers of fishermen in the inshore fishery, particularly in Newfoundland. Proponents of this approach have advocated, for example, relocating fishermen from numerous, isolated fishing communities along the Newfoundland coast to growth centres. Provincial and federal-provincial resettlement programs in Newfoundland during the 1950's and 1960's attempted to accomplish this but ultimately failed because of the lack of alternative employment opportunities. Copes (1972) generated a storm of controversy when he proposed:

> "As the Newfoundland economy has shown few signs of developing [new employment opportunities outside the fishery] on an adequate scale, it seems almost certain that greater reliance will have to be placed on outmigration from the province to solve both the general problem of unemployment in the province and the problem of underemployment in the Newfoundland outport."

Copes' proposal to depopulate large portions of Newfoundland made him a favourite target for those who wanted to maintain the rural fabric of Atlantic Canada. Other social scientists, e.g Wadel (1969), criticized these initiatives on the grounds that, instead of improving fishermen's employment opportunities and incomes, they destroyed livelihoods.

Over recent decades there has been a widespread perception that the inshore fishery is simply a "social" fishery while the offshore fishery is an "economic" one. This belief has prevailed despite the fact that over the last 20 years governments have intervened in a major way at least twice to rescue the offshore companies. While these companies have earned profits for only five or six of these years, the rate of return on investment of many inshore fleets (e.g. lobster, shrimp, crab and inshore draggers) has been generally positive over the same time period (Jim Jones, DFO, Moncton, pers. comm.). Jones (pers. comm.) has argued that the average cost of harvesting fish offshore has, except for 1986 and 1987, been well above that in the inshore fishery.

The 1986 Royal Commission on Employment and Unemployment in Newfoundland observed that "an invidious distinction has been made between the so-called 'economic' (read profitable) and so-called 'social' (read state-dependent) fishery" (Newfoundland 1986). It criticized a major assumption of Canadian fisheries policy, that there are too many fishermen chasing too few fish, and the further inferred assumption that "only through state intervention to restrict the number of entrants can the fishery be saved from a state of perpetual poverty."

The Royal Commission noted, however, that fisheries policies had not been consistently guided by such assumptions. Rather, federal and provincial governments, have had to "temper the logic of the industrialization/common property approach to fisheries management in order to protect Newfoundland outport families. Resettlement was curbed; unemployment insurance, make-work programmes, and various forms of subsidization to fishermen were introduced." The Commission disagreed with the distinction between so-called "economic" and "social" fisheries and argued for a new initiative to develop an economically viable inshore fishery.

The social/economic issue has not been confined to the question of the appropriate strategies for the inshore fishery. During the 1981–84 debate about restructuring the offshore Atlantic groundfish fishery, the federal and provincial governments argued over whether certain offshore fish plants should be kept operating. Plant workers blocked the sailing of offshore trawlers to emphasize the need to preserve employment opportunities within particular communities, e.g. Grand Bank and Burin.

The debate has not been confined to Newfoundland. Parts of Québec and northeastern New Brunswick and PEI have demonstrated the same emotional attachment to rural communities and to the fishing and processing opportunities on which they depend. Scattered throughout the Canadian Arctic are other communities whose existence is tied to local resources.

2.0 COMMUNITY DEPENDENCE

Tens of thousands of Canadian fishermen in more than a thousand, often isolated, rural communities depend upon the fishery as their sole means of gainful employment. They earn from the fishery chronically low incomes, buttressed by a special program of unemployment insurance for fishermen. For many, the total of their income is below the poverty line even for rural areas. Yet they cling to the fishery for a variety of reasons. Prominent among these is the lack of alternative employment opportunities locally or even within their province. But other factors, such as work satisfaction and attachment to the rural lifestyle, play a part.

These coastal communities were established because of the fisheries. They are part of the nation's social and cultural heritage. It is understandable that their inhabitants and governments are concerned about the welfare and preservation of these communities.

2.1 Atlantic

There are significant differences in the economic role of fisheries within the various regions of Canada. It would be difficult to overstate the socioeconomic significance of the fishing industry to Atlantic Canada. The industry plays a much smaller role in British Columbia's economy. Nationally, the commercial fishery contributes less than 1% to the Gross Domestic Product (GDP), but in the Atlantic it makes a major contribution to a struggling regional economy. In Newfoundland, the fishing industry's contribution to the Gross Provincial Product is the highest in Canada, around 15%. Next highest are the fishing industries in P.E.I. and Nova Scotia which contribute about 13% and 11% to their respective GPPs. In British Columbia the fishery contributes less than 1% to the GPP.

In 1988, the Atlantic fishing industry provided jobs for approximately 65,000 registered fishermen and 40,000 plant workers. The dependence on jobs in the fishing industry ranges from a high of 25% on the south coast of Newfoundland to around 2% in large cities such as Halifax. The fishing industry provides over 10% of all jobs in Atlantic Canada.

At the community level, fisheries activities are even more significant. Based on the 1971 Census, McCracken and MacDonald (1976) identified 702 communities as dependent to some degree on fishing or fish processing. This was out of a total of 964 communities along the Atlantic coast in Newfoundland, PEI, Nova Scotia, New Brunswick and the North Shore and Gaspe areas of Québec. The communities which had some dependence on fishing were distributed as follows:

Newfoundland	273
Prince Edward Island	94
Nova Scotia	87
New Brunswick	136
Québec	112
	702

There were often several landing points and settlements to each community as defined by the Census. In the case of Newfoundland, for example, there were 745 landing points and 615 settlements reported, and approximately 1,400 other settlements in the other Atlantic coast provinces for a total of about 2,000.

The Task Force on Atlantic Fisheries (Kirby 1982) concluded that more than one-quarter of the total population of 2.1 million in the four Atlantic provinces (Newfoundland, Nova Scotia, PEI and New Brunswick) lived in 1,339 small fishing communities. They identified small fishing communities as:

1. Any community with a population of fewer than 2,500 people and having at least 5 fishermen using it as their home port or usual port of landing; and
2. Any community with a population between 2,500 and 10,000 people, if the total of fishermen plus plant jobs exceeded 1% of the population. (A total of 28 towns met their second criterion).

The provincial distribution and population ranges of these small fishing communities are shown in Table 14-1. While the majority of these communities had populations of less than 500 people, many were larger. Newfoundland had the largest percentage of its provincial population living in small communities, while New Brunswick had the smallest (Fig. 14-1). More than half of these small communities had essentially single sector economies with fishing and fish processing employing 30% or more of the labour force. A single-sector community, as defined by the Department of Regional Economic Expansion, had at least 30% of its labour force employed in a single industry.

Fisher family putting away the catch.

Mending nets on an offshore trawler.

TABLE 14-1. Size and location of small fishing communities.

Population Range	Nfld.	N.S.	N.B.	P.E.I.	Qué.	TOTAL
5000–9999	3	3	0	1	1	8
2500–4999	13	3	4	0	4	24
1000–2499	43	20	19	6	16	104
500–999	98	34	23	3	24	182
200–499	193	102	29	17	37	378
100–199	123	97	28	18	11	277
>100	74	71	22	17	9	193
Unknown*	81	34	31	3	24	173
TOTALS	628	364	156	65	126	1339

*Population figures were not available for the communities labelled "unknown", probably because they were too small to be included in the census as discrete communities.

Source: Kirby (1982).

Some 42,000 of the 46,800 jobs identified in fish processing plants in 1980 were in small fishing communities. These jobs comprised 22.2% of the labour force in these communities. This was considered an underestimate of total plant employment because it omitted a large number of very small operations in the small communities.

Of the 23,000 full-time licence holders who fished in 1981, the TFAF estimated that at least 18,000 lived in the small fishing communities. It concluded that the fishing industry was the source of at least 62,250 direct jobs in these small fishing communities in Newfoundland and the Maritimes. This meant that at least 35.3% of the overall employment in these communities was provided by fishing and fish processing.

Poetschke (1984) attempted a more rigorous analysis of community dependence on fishing in the Atlantic provinces. His study focused on small coastal fishing communities, not all fishing communities. His definition of small coastal fishing communities was essentially the same as that of the TFAF.

Poetschke observed that 90% of the small Atlantic fishing communities had a population of less than 1,000 and 75% had fewer than 500 inhabitants. Only 137 had a population of over 1,000.

One-third of the 600,000 people living in Atlantic fishing communities (including Québec) lived in single-sector fishing settlements. At least 40% of all Atlantic fishing communities were single-sector. Almost 80% of the known single-sector communities were in Newfoundland (55%) and Nova Scotia (24%). One-quarter of the Atlantic fishing communities were extremely dependent on fishing with about 60% of the labour force working directly in fishing and fish processing.

About one-third of the communities were plant communities. These were relatively large with an average population of 822, more than twice the average population of 369 in non-plant communities. One-half of those who lived in plant communities lived in single-sector towns. On the other hand, over 80% of the people living in non-plant communities lived in towns that were not single-sector. The degree of community dependence on the fishery increased with plant size (Table 14-2).

TABLE 14-2. Job range of plants by community degree of dependence (calculations in percent).

Degree of Dependence	Number of Jobs				
	<25	26-50	51-150	>150	Total
45+	13.0	20.0	41.9	52.4	29.0
30–44	16.4	20.0	17.4	15.9	17.2
15–29	34.9	33.3	24.4	19.5	29.0
0–14	35.6	26.7	16.3	12.2	24.7
Total	100.0	100.0	100.0	100.0	100.0

Source: Poetsche (1984).

Only 60% of the smallest plants were in single-sector communities compared with 90% of the largest plants. However, the degree of dependence varied inversely with the population of the plant community. About 85% of plant communities with a population under 500 were highly dependent on fishing compared with about 50% of communities with a population between 1,000 and 5,000. The plant communities most dependent on fishing were those with the largest plants and populations under 1,000.

Dependence also decreased with increasing population among non-plant communities. Only 2% of non-plant communities with a population between 1,000 and 5,000 were single-sector compared with about 40% of communities with populations less than 500.

Overall, three-quarters of the very small fishing communities (populations less than 100) were single-sector compared with about 30% of those with populations between 1000 and 5000.

These data from the TFAF and Poetschke's analyses illustrate the overall importance of fishing to the Atlantic region. Of the 600,000 people who lived in fishing communities, about 200,000 people lived in communities where fishing activity was the principal, if not the only, employer. When jobs created by the multiplier effect of the fishing sector are added to direct fishing employment, the dependence on the Atlantic fishery is considerably greater than is indicated by direct employment figures alone (Poetschke 1984).

Many plant communities have developed a more complex infrastructure than non-plant settlements and have, in many instances, become regional service and employment centres. Poetschke hypothesized that people in plant communities are harder hit by economic shocks than those in non-plant settlements because greater dependence on fishing means greater vulnerability under difficult economic conditions. Similarly, the larger the plant, the greater the dependence on the fishery. This explains the intense debate generated by possible closure of large plants during periodic crises in the Atlantic fisheries.

Poetschke suggested it would be good long-term fisheries policy to encourage smaller plants because communities with smaller plants are less dependent and more flexible, though just as efficient as those with bigger plants. This interesting suggestion runs counter to the industrialization/centralization initiatives in Newfoundland of the 1950's and 1960's.

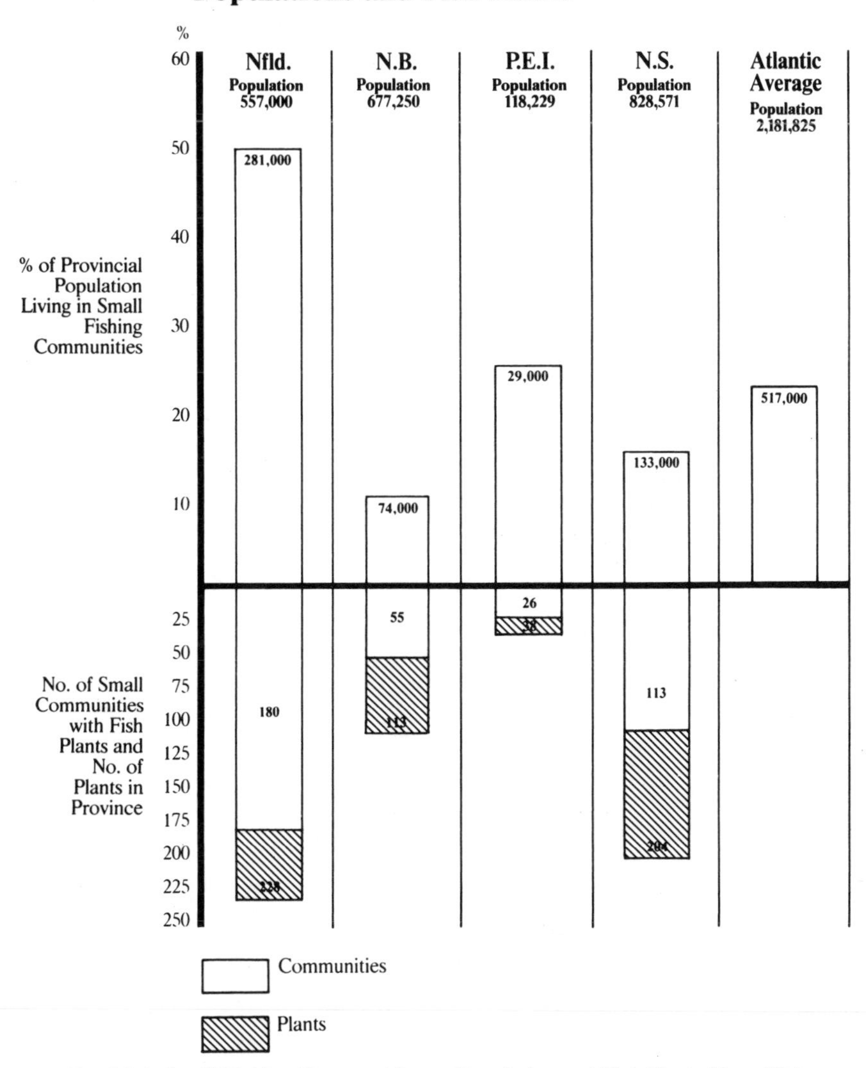

FIG. 14-1. Small Fishing Communities — Population and Fish Plants (from Kirby 1982).

2.2 Pacific

British Columbia's situation is quite different from that in the Atlantic. In British Columbia, the fisheries have considerably less importance in the overall economy. There are no thorough recent studies of relative community dependence on the fishery in British Columbia. Sinclair (1971) assessed the economic and social significance of the west coast commercial fishing industry to selected isolated coastal communities of British Columbia. Marchak et al. (1987) studied fishing and fish-processing industries in British Columbia and described the relative importance of fishing within British Columbia. The comments that follow are drawn largely from these two sources.

The major B.C. resource-based industries are mining and forestry. Even agriculture ranks ahead of fisheries. In 1988, for example, the landed value of the fishery was $533 million compared with a Gross Domestic Product of $1.9 billion for mineral production and $1.7 billion for logging and forestry.

These figures, however, do not account for the recreational fishery and its contribution to tourism. Expenditures attributable to sportsfishing in British Columbia in 1985 were $144 million, the third largest in the country behind Ontario and Québec (DFO 1987a).

Marchak et al. (1987) put the commercial fishery in perspective:

> "The commercial fishery is important for other reasons, however. It employs labour in coastal communities which have few or no alternative employers and provides a commercial livelihood for vessel owners and crew. It has some industrial linkages of benefit to B.C. shipbuilders and marine equipment suppliers, especially in the harvesting sector. While it has never been the major industry in B.C., it did have a more prominent position in the early history of the province and a paramount position in the precapitalist history of the coastal region. The economics and cultures of the coastal native peoples were built on the salmon runs, and one could not possibly describe the west coast of Canada without reference to fish and fishing. Yet in relative terms, the industry today has low priority on the provincial agenda."

At work in salmon cannery.

Sinclair (1971) pointed out that, in the interior of B.C., mining and agriculture were predominant with logging playing a comparatively minor role. In the coastal region, logging and fishing were of primary importance with mining playing the minor role. He also observed that fishing in some areas was the primary economic activity and that its position in the economy of the coastal region was more significant than in the provincial economy as a whole. In earlier times the salmon fishery helped to establish B.C.'s coastal communities. Canneries, the major processing operation, were highly labour intensive. This encouraged workers and their families to live near the processing plant.

In more recent times, however, the role of the salmon fishery in developing and maintaining remote coastal communities has diminished. This has resulted from two factors:

1. The relative decline in the importance of fishing to B.C.'s expanding economy; and
2. The impact of technological advances in fishing combined with the fact that successive generations tended to settle in large urban centres.

The acquisition of faster and better-equipped fishing vessels permitted fishermen to transport salmon over long distances. Fishermen were thus able to move from small coastal communities, close to main fishing locations, to the southern urban centres considerably removed from prime fishing areas. Changes in the processing sector, however, had a more dramatic influence on the coastal communities of British Columbia. Prior to improvements in the transport of raw salmon, the major constraint in the salmon processing industry was the need to obtain, locally, large quantities of good quality salmon for processing. With the advent of improved refrigeration techniques and faster boats, canneries were better off near populated areas, where labour and market outlets were more readily available during the short fishing season. Thus salmon canneries concentrated in two main coastal areas, the Lower Mainland and the North Coast.

In addition to geographical concentration of canneries, there has been since the late 1930's a dramatic decline in the number of salmon canneries on the coast (outside the Fraser District) and since 1960 a gradual decline in numbers for British Columbia as a whole (Fig. 14-2). The number of canneries declined from a widely dispersed coastal system of 94 canneries in 1917 to the point where there were only canneries in two main locations. This, of course, hurt communities which relied on the fishing industry.

The coast of British Columbia is now relatively unpopulated. Many of the communities dependent on fishing are gone. Outside the southern Lower Mainland area, the Coastal population is concentrated in the northern transportation centre of Prince Rupert or a few lumber or mineral-dependent towns: Nanaimo, Courtenay, Campbell River, Kitimat, Port Alberni, and Powell River. Only a small proportion of the coastal residents rely on fishing; an even smaller number rely on fish processing. In the mid-1980's there were only five canneries outside the Fraser (Lower Mainland) district where once there were more than sixty. Three of

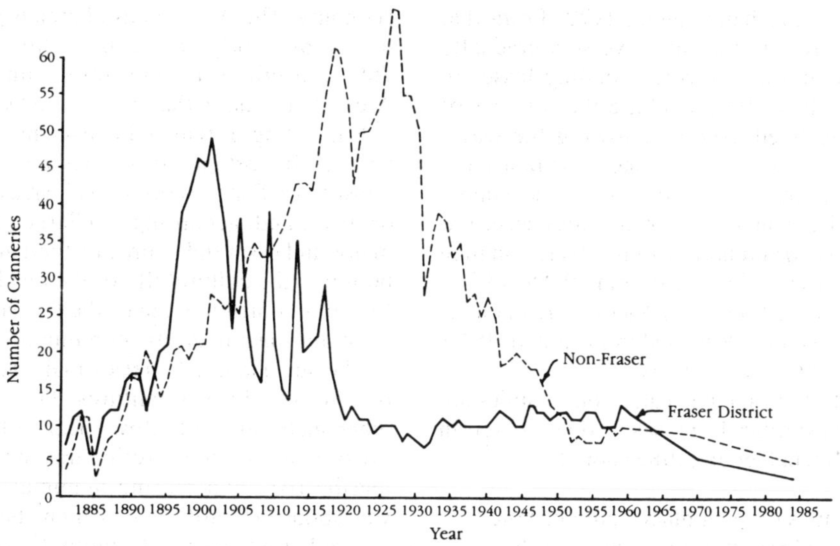

FIG. 14-2. Coastal Canneries in British Columbia from 1881–1984 by Region (from Marchak et al. 1987).

these were in Prince Rupert with only two located along the 4,500 miles of coast between the Fraser and Skeena rivers. Whereas 86% of all fish processing workers were located outside the Fraser district in the mid 1920's, by the mid-1950's the percentage had declined to 42% because of plant closures. Today, most of the remaining settlements along the coastline are native villages.

For a discussion of the causes of the decline of these coastal cannery communities, see Marchak et al. (1987). Outside Vancouver and Prince Rupert, the fish processing sector is now almost non-existent. A significant proportion of fishermen, however, still reside in rural communities.

Marchak et al. (1987) compared the relative prosperity of fishermen living in rural areas with those in urban centres. They classified communities as:

1. Remote — small, isolated communities often largely dependent on fishing and without significant service sectors or industrial differentiation;
2. Towns — more populous hinterland communities generally dependent upon large-scale mineral or log processing, with more services and acting as hinterland distribution centres; and
3. Coastal urban centres — Prince Rupert and the Vancouver and Victoria areas.

Generally, fishermen from the rural areas received less income from fishing than their urban counterparts but received greater returns on invested capital.

During the 15-year period from 1967 to 1981, the most significant change in the coastal fishing economy was an apparent reversal of a long-standing trend to increasing urbanization of the fleet. This trend, which began about 1950, appeared to have halted about 1972. From that year onward, the overall number of vessels gradually increased in rural areas, with corresponding losses in Vancouver and Prince Rupert. While the number of salmon vessels declined overall following the Davis Plan, the decline occurred at a slower rate in the rural areas than in the urban centres. From 1979 to 1981, the number of salmon vessels based in remote communities decreased by about 11.6%, but the proportion of the overall salmon fleet located in such areas increased from 10.9 to 13.3%. In the Prince Rupert and Vancouver-Victoria areas, on the other hand, the proportions decreased from 12.2 to 10.3% and from 59.0 to 52.6% respectively.

Losses to the fleets based in isolated communities are more serious than similar decreases in fleets based in urban centres. Marchak et al. (1987) noted:

> "The economics of such places are near the threshold level for survival and, unlike urban economies, less able to withstand further losses to economic inputs in any form. Hence, decreases in local fleets will have disproportionate consequences for such communities."

Another factor threatening these coastal communities is the differential impact of allocation policies upon the various fleet sectors. Rural-based fishermen in the salmon fishery, for instance, are more active in the troll than in the net fishery. Seine vessels tend to be based in urban areas, while rural fishermen engage more in combination gillnet and trolling. Recent restrictions on the troll fishery (see Chapter 7) are likely to impact more upon fishermen based in rural communities because they dominate the troll fishery.

Pinkerton (1987) concluded:

> "In sum, fishing dependent regions tend to be exceptionally vulnerable. Villagers often have few job opportunities and marketable skills apart from fishing. Important features of their social organization are often conditioned by fishing. They cannot always protect or maintain local stocks, which may be overfished by the larger and more mobile members of the British columbia fleet. Local fishers tend to lose fishing licences. Processing plants and packing and collecting services, without which it is difficult for small-scale fishers to operate, have often been removed."

2.3 Arctic

The Arctic is quite different from the Atlantic and Pacific coasts in the role the fishery plays in the regional economy. The Arctic coastal area is still occupied primarily by aboriginal peoples. About 40,000 people, 78% aboriginal, occupy 60 communities along the Arctic coast and in the MacKenzie Delta (Clarke 1993).

Harvesting of fish and marine mammals has always been an important subsistence activity in the Arctic. Subsistence fisheries occur in all areas where people live or travel and, according to Clarke (1993), "provide a major and essential source of food and a major contribution to the cultural life of the residents." The Arctic fisheries contribute one of the few sources of employment and cash in Arctic communities.

The subsistence fisheries make an essential contribution to traditional cultures. They provide food and other materials and allow local self-sufficiency. The gross value of the Arctic subsistence anadromous, marine fish, and marine mammal fisheries in 1987 was about $15 million, of which about $6 million was derived from the anadromous fishery and $8.5 million from the marine mammal fishery (Clarke 1993).

Arctic natives hauling in a Narwhal.

Overall, the most significant aspect of the fisheries for Arctic communities lies in their cultural and social benefits. Clarke concluded:

> "Although the Arctic fisheries are small compared to the Atlantic and Pacific fisheries, they play an integral and major role in the lives of northern Canadians. Harvesting fish and marine mammals is part of the traditional culture of the native peoples and provides a considerable portion of their food. The commercial and recreational fisheries provide one of the few sources of cash and employment in northern communities. The fisheries provide employment or occupation for 50–75% of the population, have an estimated value of about $15 million as the replacement value of food from the subsistence harvest, $1 million as other consumer surplus benefits and $6.3 million as the value added to the Canadian economy."

2.4 General

Overall, the extent of regional and community dependence on the fishery differs considerably among the Atlantic, Pacific and Arctic coasts of Canada. On the Pacific coast, fishing is a relatively minor component of the economy. The number of coastal communities dependent on fishing has declined dramatically over the past half-century. Today there remain some rural communities and Indian villages where fishing is an important contribution to the way of life and as a means of livelihood for many individuals. In general, however, the fishery is based in large towns and metropolitan areas, particularly for the processing sector but also, to a lesser extent, for the harvesting sector.

On the Atlantic coast, the situation is quite different. The fishery was the basis for the establishment of hundreds of small coastal communities and remains the chief, and, in many cases the only, source of livelihood for the majority of such communities. The degree of dependence of these communities upon the fishery cannot easily be overstated. On the contrary, too often planners have glibly envisaged an industrialized fishery based in large growth centres. The botched attempts at resettlement in Newfoundland are but one testimony to the failure of such policies to recognize the harsh realities of the rugged Atlantic coastline. The people who live in these communities are attached to fishing as a way of life and depend upon it for a living. They have few alternative employment opportunities. The bright lights of distant cities have enticed some with dreams of prosperity. But these dreams have often been shattered.

These coastal communities have played and will continue to play a vital role in the Atlantic economy and social structure. But the nature of life in these settlements, where people eke a meagre existence from the tempestuous sea, is often harsh. Residents who earn their living from the sea in many instances live on the edge of poverty, with chronically low incomes, and little opportunity to improve their economic situation.

3.0 FISHERMEN'S INCOMES

3.1 Atlantic

Large numbers of Atlantic fishermen live near or below the poverty line for rural Canada. This is not a new phenomenon. Bates (1944) observed that for fishermen in 1939, "the net income remaining for living was inadequate, no matter how it is estimated."

Unpublished data from the 1970's indicate average net fishing incomes of Atlantic coast fishermen in 1973 were approximately $2,550. Inshore fishermen averaged $2,050 and offshore fishermen $4,460. Comparable estimates for 1975 were $2,900, $2,270 and $4,970, respectively (T. Peart, unpublished data). The 1973 data showed the lowest average incomes were in Newfoundland and Prince Edward Island. The highest average net income was earned by Nova Scotia fishermen. When one considers that the Newfoundland figure was an average for inshore and offshore fishermen, it is clear that the lowest average net income was that of Newfoundland inshore fishermen.

In 1973, income from fishing averaged only 50% of total income. The other half consisted of:

U.I. Benefits	19%
Non-fishing earned income	31%

Total average cash income from all sources in 1973 was $5,112 for those Revenue Canada classified as fishermen.

Overall, 60% of Atlantic fishermen had total incomes less than $5,000. The poorest fishermen were those in P.E.I. and Newfoundland.

The Task Force on Atlantic Fisheries in 1982 undertook a detailed survey of the 1981 incomes of Atlantic-coast fishermen, using the DFO categorization of full-time and part-time fishermen. Of the 48,434 individuals issued licences in the four Atlantic provinces in 1981, approximately 23,400 earned their living as full-time fishermen. In addition, about one-quarter of part-time fishermen had fishing revenue comparable to full-timers. From this, the Task Force estimated that a total of nearly 28,000 active fishermen derived their primary source of earnings from fishing.

The TFAF noted wide variations in the seasonal patterns of fishing along the east coast, due to such factors as local weather and ice conditions, seasonal availability of fish species, catch quotas and types of fishing gear. They estimated that full-time fishermen devoted 29.4 weeks to fishing work during the year compared with 14.5 weeks for part-timers.

There was considerable regional variation in the weeks worked. Including preparation time, the average duration of fishing activity was 41.7 weeks in southern New Brunswick and 39.4 weeks in western Nova Scotia compared with only 23.5 weeks in northeast Newfoundland and Labrador where seasonal ice cover and rough weather limit the length of the inshore fishing season.

The average net fishing incomes of full-timers were more than four times those of part-timers (Table 14-3). The median income for full-timers was $6,500 and for part-timers $840. Average incomes were distorted by the fact that the top 10% of full-timers earned $23,350 or more and the top 10% of part-timers $6,000 or more.

TABLE 14-3. Revenue, costs, average and net fishing incomes for full-time and part-time fishermen.

Revenues, Costs and Average Fishing Incomes	Full-time	Part-time
Gross Revenues	$22,452	$3,703
Total Costs	10,545	920
Net Income (before tax)	$11,907	$2,783

Net Fishing Income for Full-time and Part-time Fishermen	Full-time	Part-time
25% (bottom quarter)	$ 2,731	$ 000
50% (median)	$ 6,500	$ 840
75% (third quarter)	$14,680	$3,100
90% (top decile)	$23,350	$6,000

Source: Kirby (1982).

Considerable variability in average and median incomes among areas was observed (Fig. 14-3). Full-time fishermen earned the highest net fishing incomes in western Nova Scotia ($28,766) and the lowest in northeast Newfoundland and Labrador ($4,512). Part-time fishermen earned the highest net incomes in southern Newfoundland ($8,107), working on the offshore trawlers, and in western Nova Scotia ($5,334).

Both types of fishermen benefitted significantly from income supplements such as unemployment

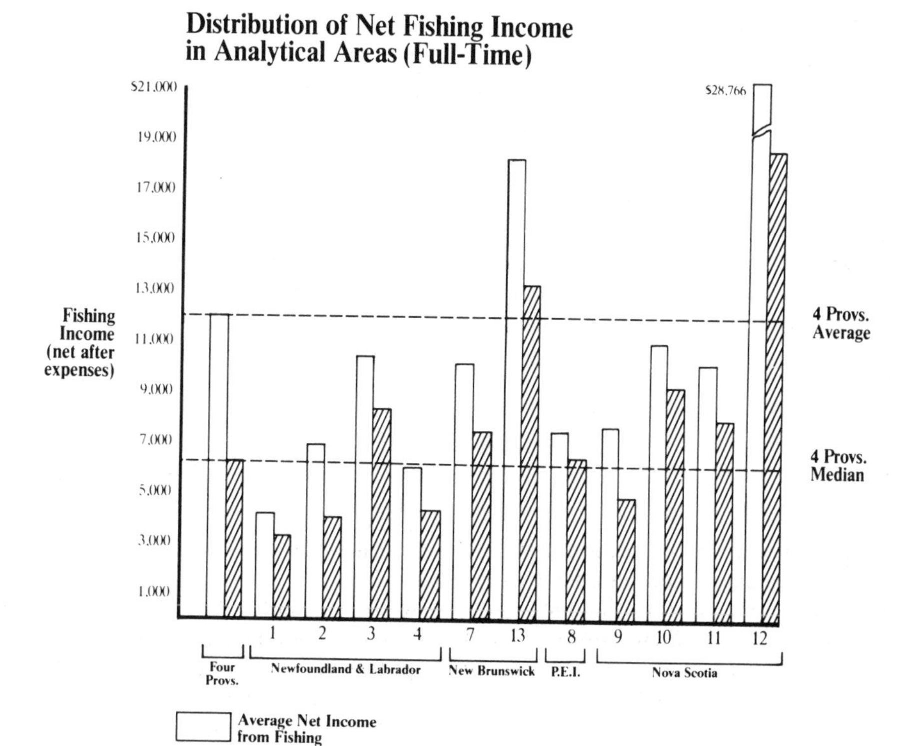

FIG. 14-3. Distribution of Net Fishing Income in various areas of the Atlantic coast for full-time fishermen (from Kirby 1982).

insurance, family allowance and pension benefits. These totalled an average of $2,910 for full-time fishermen and $2,206 for part-time fishermen. Unemployment insurance contributed the bulk of this ($2,466 for full-timers and $1,483 for part-timers) (See Unemployment Insurance, section 4.0).

Overall, full-timers received an average of $3,837 from non-fishing sources and part-timers $8,409. Among full-timers, only the top 10% received more than $6,300 from non-fishing sources. On the other hand, three quarters of part-timers did not rely on fishing for most of their income. The top 10% received more than $19,176 from non-fishing sources.

Previous observers had contended that fishermen supplemented their cash incomes extensively by hunting game, raising vegetables and small-scale farming. This was known as 'income in kind'. Faris (1972), for example, argued that this was an important contribution to the household economy for fishermen in Cat Harbour (Lumsden) on the northeast coast of Newfoundland. The Task Force was unable to assign a cash value to 'income in kind'. However, it concluded that the amount of these income supplements had been greatly exaggerated. Their survey indicated that fewer than a third of fishermen's households had vegetable gardens and fewer than one-quarter reported hunting large game.

Recently, Doug House, Chairman of the Newfoundland Royal Commission on Employment and Unemployment, criticized the TFAF underestimation of the value of household production. House et al. (1986) observed:

> "It completely ignores the most important components of household production in Newfoundland outports. First and foremost among these is housing. People build their own homes, and in concert with their relatives and their friends, they install their own plumbing, wiring, insulation and other services. Secondly, and of increased importance since the rise in world oil prices, people collect their own firewood to heat their homes. Thirdly, people do their own repairs and maintenance.... Fishing, hunting, gathering and growing food for the household's own consumption, the *only* item considered by

Kirby, constitutes only a small part (around 10 percent) of household production."

House et al. (1986) estimated that household production contributes at least a quarter of a typical fisherman's economic output.

All sources agree that the total net incomes of fishermen are low. The TFAF figures for the average net income from all sources in 1981 were $15,791 for full-timers and $11,182 for part-timers. Again there was considerable variability among areas. From these observations, the TFAF concluded:

> "A sizeable majority of full-time fishermen on Canada's east coast have total incomes below the recognized poverty line for rural residents."

To address the poverty question appropriately, it is necessary to take into account household incomes. Average total household incomes for full-time fishermen were lowest in northeast Newfoundland and Labrador ($14,319) and highest in southwest Nova Scotia ($35,882). Among part-timers, average total household incomes were lowest in Prince Edward Island ($12,002) and highest in southern New Brunswick ($25,952).

Table 14-4 shows the accepted poverty line figures from 1980 to 1988 for rural households of various sizes. The official rural poverty line for 1981 for a 4 person household was $12,035. Given that the average size of an Atlantic coast fisherman's household was 4.1 people, almost one-third of the households of full-time fishermen had total incomes below the poverty line. Among part-time fishermen, 40% of households had incomes below the poverty line. More significantly, total household incomes for a high proportion of full-time and part-time fishermen in northeast Newfoundland and Labrador, eastern Newfoundland, Prince Edward Island, the Gulf side of Nova Scotia and central Nova Scotia were well below the poverty line for rural Canada. Without the supplementation of incomes from non-fishing sources (particularly Unemployment Insurance and earnings by other household members), the majority of east coast fishermen's households would be at or below the poverty line.

Fish processing plants are an important source of non-fishing employment for members of fishermen's households. In the Atlantic as a whole, processing plants created as many jobs and almost as much income as fishing. In 1980, nearly 48,000 jobs depended on fish plants. Given the high turnover rate, many more than 48,000 individuals benefitted from plant employment during the year. Many fish plant jobs are seasonal. Thus, the total number of person-years of employment was estimated at 31,000 in 1980. About 20% of fishermen's households had one or more members working in fish plants in 1981. The links between fishermen's households and plant employment were strongest in Newfoundland and in northeast New Brunswick. In 1981, the average income earned by a fish plant worker who was also a member of a fisherman's household was $4,520. Clearly, fish plant employment by wives or other residents of fishermen's households was an important factor in determining the proportion of households whose total income fell below the poverty line.

TABLE 14-4. Statistics Canada Revised Low Income Cut-offs, 1980–1988, for family sizes of 3, 4 and 5 persons residing in rural areas.

	Family Size		
Year	3-Person	4-Person	5-Person
1980	$ 9,256	$10,699	$12,441
1981	10,412	12,035	13,995
1982	11,537	13,336	15,507
1983	12,203	14,106	16,403
1984	12,734	14,720	17,117
1985	13,244	15,310	17,803
1986	13,785	15,936	18,531
1987	14,389	16,634	19,343
1988	14,979	17,316	20,136

Source: National Council of Welfare, April, 1989.

More recent snapshots, from the 1984 and 1988 DFO surveys of Atlantic fishermen, confirm the findings of the TFAF (Table 14-5). In 1984, total net incomes ranged from $9,935 in Newfoundland to $18,028 in Nova Scotia. The distribution among provinces was similar to that observed for 1981, with total incomes being lowest in Newfoundland and highest in Nova Scotia. Other employment income was generally between $1000 and $2000, except for Nova Scotia where it was slightly higher. Overall, the net total income (before taxes) was up only slightly from 1981. This is not surprising since the early 1980's were recession years and the "restructuring" of the Atlantic fishery was not completed until 1984–85.

The figures for 1988 show net total fishing income was up by 50% or more from 1984. This reflects the market boom in 1985 to mid-1988. In 1988, other employment income was in the $2000–$3000 range, also reflecting an approximate 50% increase. Even in 1988, however, the vast majority of active fishermen (62.7%) had a net fishing income of less than $10,000 and a total net income from all sources of less than $20,000 (Table 14-6; Fig. 14-4).

TABLE 14-5. Income summary, all active fishermen, by province, 1984 and 1988.

	Nova Scotia	New Brunswick	P.E.I.	Québec	Newfoundland
			1984		
Gross Fishing Income	$21,373	$16,749	$15,768	$11,298	$ 8,703
Fishing Costs	(8,438)	(7,113)	(7,375)	(5,813)	(3,380)
Other Employment Income	2,397	1,760	1,014	1,645	1,214
Regular UI	555	1,134	1,294	1,254	713
Fisherman's UI	2,141	2,751	3,262	2,237	2,685
TOTAL INCOME	$18,028	$15,281	$13,963	$10,621	$ 9,935
			1988		
Gross Fishing Income	$28,406	$24,221	$22,302	$24,981	$10,979
Fishing Costs	(9,823)	(11,126)	(8,418)	(13,299)	(3,385)
Other Employment Income	3,223	2,501	1,488	1,974	2,294
Regular UI	876	3,493	2,036	957	1,133
Fisherman's UI	3,671	3,979	4,500	5,384	4,492
TOTAL INCOME	$26,353	$23,068	$21,098	$19,997	$15,513

Source: DFO Statistics
1984 and 1988 Survey of Atlantic Fishermen by the Department of Fisheries and Oceans.

TABLE 14-6. 1988 Income Distribution Among Active Fishermen in Atlantic Canada (DFO, 1992m).

		Average $ per year				
Net Fishing Income Range	Number of Fishermen	Gross Fishing Revenue	Net Fishing Income	Fishermen's UI	Regular UI	Total Net Income, All Sources
Under $10,000	30,860	7,298	3,794	3,805	1,903	13,053
$10,000–20,000	8,773	21,544	14,803	5,139	698	22,299
$20,000–30,000	4,526	34,035	25,173	4,183	707	31,043
$30,000–40,000	2,394	49,224	34,784	3,731	101	40,495
$40,000–50,000	1,206	68,692	45,503	4,211	565	51,424
$50,000 plus	1,440	143,367	83,559	4,180	333	88,898
Total/Average	49,198	$19,825	$12,588	$4,095	$1,412	$20,853

The 1984 and 1988 surveys did not continue the collection of information on household incomes. Looking just at the average net total incomes, in 1981 the figure for full-time fishermen ($15,791) was equivalent to the rural poverty line for a family of six. The average for part-time fishermen was equivalent to the rural poverty line for a family of three. In 1984, the average net total income (full-time and part-time fishermen combined) was equivalent to the rural poverty line for a family of two in Newfoundland, a family of three in Prince Edward Island, and a family of four-five in Nova Scotia. In 1988, the average net total income was equivalent to the rural poverty line for a family of three in Newfoundland, a family of five in Prince Edward Island, and higher than that for a family of seven or more in Nova Scotia.

Although the comparisons are inexact, it appears that fishermen's incomes improved in relation to rural poverty line figures in the mid-1980's. This reflects the general economic upturn in the Atlantic fishery (see Chapter 13) related to considerably increased prices for fish products in the U.S. market. The 1986–1988 period was the most prosperous in history for the Atlantic fishery. Fishermen's incomes improved somewhat as a

Net Fishing Income Distribution, Active Fishermen

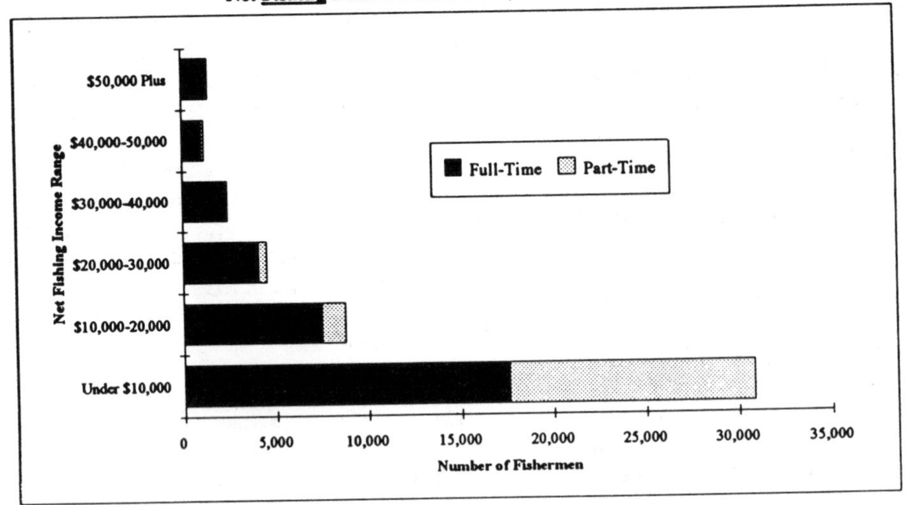

Total Net Income (All Sources), Active Fishermen

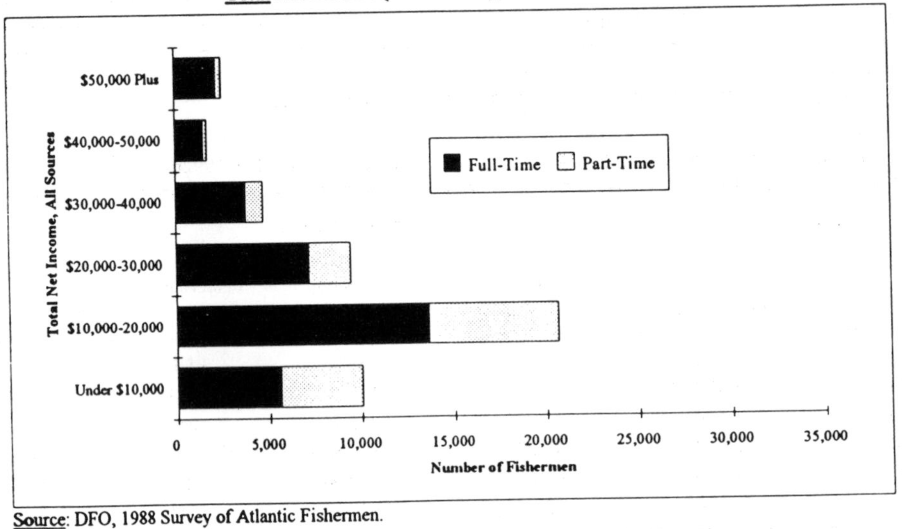

Source: DFO, 1988 Survey of Atlantic Fishermen.

FIG. 14-4. Net Fishing Income and Total Net Income Distribution of Active Fishermen in Atlantic Canada (DFO, 1992m).

result of the general prosperity of the fishery.

Recent results confirm the TFAF findings of considerable variability among areas in fishermen's incomes related to species fished, length of the fishing season and gear type. Fishermen who hold restricted licences for lucrative species such as lobster and licences for purse seine or small mobile gear tended to outperform the average substantially. Nova Scotia fishermen continued to earn the highest incomes on the Atlantic coast. Similarly, Newfoundland inshore fishermen with short seasons and limited species availability continued to earn the lowest incomes. In 1984, the average total income in Nova Scotia was double that in Newfoundland. In 1988, the average total income in Nova Scotia was 70% greater than that of Newfoundland fishermen. The average gross fishing income in Nova Scotia was almost triple that in Newfoundland but that was offset somewhat by lower fishing costs in Newfoundland and higher average transfer payments (Economics Branch, DFO, Ottawa, pers. comm.).

The average annual income figures mask the fact that some fleet segments provide high incomes to captains and crews. There were considerable differences in the average net incomes (before taxes) for various components of the groundfish fleet in the Scotia-Fundy Region (Table 14-7). Incomes for captains of longliners and

TABLE 14-7. Scotia-Fundy Groundfish: Annual incomes (before tax) from fishing in 1988.

	Captains	Crew
Inshore Generalist	$15–30,000	N/A
Inshore Specialist		
Longliners & Gillnetters:		
— 35–44 ft.	$26,000	$14,500
— 35–64 ft.	$24,500	$16,000
Draggers:		
— 35–44 ft. (NE)	$31,000	$15,000
— 35–44 ft. (SW)	$53,000	$29,000
— 45–64 ft. (SW)	$74,000	$28,000
Offshore		
Trawlers (>100 ft.)	$70–90,000	$24–50,000*

*Trawler crews range from deckhands to ship's officers.

Source: DFO Survey, Unpublished data, 1989.

gillnetters were in the $25,000 range and for captains of draggers between 45 and 64 feet in the $74,000 range. Crew members on the small draggers earned between $15,000–$30,000 and on the offshore trawlers between $24,000–$50,000. These incomes compare favourably to incomes earned by skilled and semi-skilled workers generally in Canada. However, these Nova Scotia fishermen's incomes tend to be the highest in Atlantic Canada. They are considerably higher than those earned in the inshore fishery in Newfoundland where incomes are chronically low.

A 1986 study undertaken for a First Ministers Conference concluded that fishermen generally have lower incomes than individuals employed in other economic sectors. Newfoundland fishermen were worse off than workers in other sectors in that province and Nova Scotia fishermen better off than workers in other sectors in that province. The ratio of the average fishermen's income to the provincial average income is plotted in Fig. 14-5. In Newfoundland, the average fishermen's income was only 75% of the overall provincial average in 1985. In that year the average fishermen's income was 30% higher than the overall provincial average in Nova Scotia and 17% higher in New Brunswick (Grady and MacLean 1989).

3.2 Pacific

As we have seen earlier, the fishing industry in British Columbia is highly dependent on salmon and herring. These species exhibit considerable variability in landings and landed value. In the late 1970's, fishermen's incomes in British Columbia were the highest in Canada, and over 50% higher than the overall average income for the province. In 1980, a year of crisis, their incomes fell to less than 80% of the provincial average. Their incomes remained below average until 1985 when they increased to 5% above the provincial average.

Marchak et al. (1987) presented data from 1981 to illustrate the wide disparity in the earnings of different groups within the British Columbia fishery (Table 14-8). These figures were divided by gear type into three categories: (1) highliners — fishermen whose earnings fall in the top 25% of their gear type; (2) average earners — the middle 50% of earners in each gear type; and (3) lowliners — those fishermen earning in the bottom quarter of their gear type. The extremes

TABLE 14-8. Fishing incomes and earnings of various sections in the British Columbia fishery in 1981.

		Average Fishing Incomes and Earnings			
Gear-Type		Top 25%	Middle 50%	Bottom 25%	Overall
Troll	Gross $	$ 56,847	$ 28,629	$13,536	$ 31,910
	Net $	20,008	6,973	3,116	9,267
	Sample Size	31	62	31	124
Gillnet	Gross $	$ 41,453	$ 20,330	$ 7,459	$ 22,437
	Net $	17,473	7,519	858	8,360
	Sample Size	12	23	12	47
Combinations	Gross $	$ 84,780	$ 46,613	$20,973	$ 49,860
	Net $	23,410	18,355	1,819	15,377
	Sample Size	35	65	35	135
Seine	Gross $*	$218,433	$110,522	$52,197	$121,965
	Net $	91,977	35,106	−4,341	39,126
	Sample Size	15	35	15	65

*Boat and net share only, crew shares are excluded. Only seiners fishing both herring and salmon are included.

Source: Marchak et al. (1987).

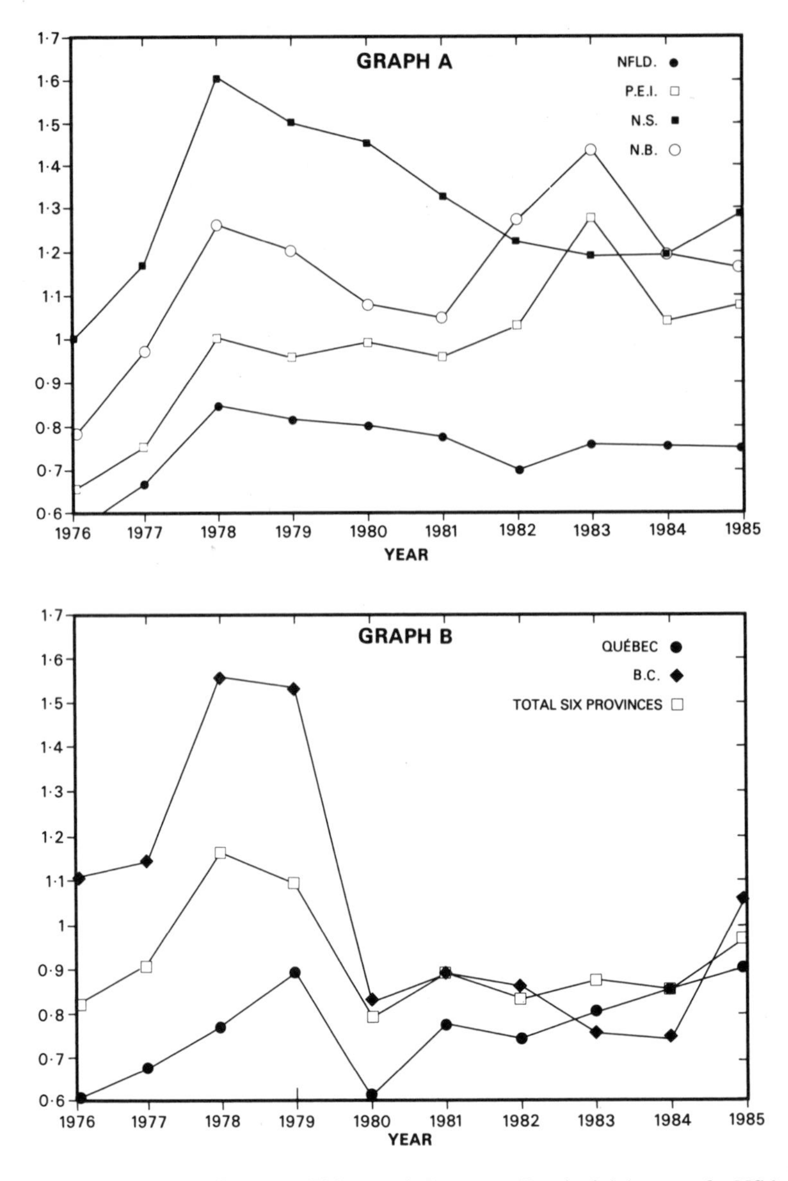

FIG. 14-5. A. Ratio of Average Fisherman's Income to Provincial Average for Nfld, P.E.I., N.S. and N.B.; B. Ratio of Average Fisherman's Income to Provincial Average for Quebec, B.C. and average of six provinces (from Grady and Maclean 1989).

in earnings are illustrated by comparing the ratio of earnings for highliners relative to lowliners. For trollers, the ratio was 6.42:1, with highliners receiving 6 times as much as lowliners. For fishermen with combination gears, the difference was more than twelvefold. For gillnetters it was twentyfold.

Viewing this another way, for trollers the top 25% of earners received 54% of the net fishing income. The bottom 25% of earners received only 8.4% of the total troll-fleet earnings. The pattern of inequality was similar for the other gear categories. Marchak et al. (1987) observed that the earnings disparities within gear sectors in the B.C. fishery exceeded the general levels of income inequality found in Canadian society as a whole.

Similarly, there were large disparities between fishermen in different gear categories. For highliners, the average earnings of seine boat owners exceeded by a factor of 5.3 the average earnings of their highliner counterparts among gillnetters.

3.3 Some Other Considerations

Barrett and Davis (1984) contended that the incomes analysis by the Task Force on Atlantic Fisheries was incomplete at best, since it failed to examine income distribution based on organizational factors such as sector participation (inshore/offshore), enterprise ownership (owners/non-owners) and crew position (captain/crewman). Davis and Thiessen (1986) and Thiessen and Davis (1986) also believed that income distribution among Atlantic Canadian fishermen could be adequately understood only after accounting for these organizational dimensions. They reanalysed the socio-economic data collected by the TFAF and concluded that the east coast fisheries could not be described as a single entity (hardly a startling conclusion). Nor was it useful to separate them by area or by full-time/part-time involvement. (Given the huge differences among areas, their suggestion that it was not useful to examine incomes by areas is highly questionable).

Their reanalysis was based on an inshore/offshore split determined by fleet and fishing effort characteristics. All fishing employing boats 42 feet or longer and using any combination of groundfish gillnet, longline, otter-trawl, purse-seine, Danish-Scottish seine, scallop-dredge, crab, or shrimp-fishing gear was classified as offshore. In addition, all boats at least 45 feet in length and using either lobster gear or herring/mackerel gill-nets were classified as offshore. All others were designated inshore. Their definition differs from the inshore/offshore classification used by DFO in the groundfish fishery. The latter is based on size of vessel, with those over 100 feet long being designated offshore. Given the mobility of draggers in the 45–65 foot and 65–100 foot range, the DFO designation is now questionable. Davis and Thiessen were attempting to differentiate between fishing effort in day or shore fisheries from enterprises which are geographically mobile and targetting continental shelf marine resources.

Offshore fishermen reported, on average, their enterprises grossed $55,541 in 1981 while inshore fishermen reported an average gross income of $8,861. On average, inshore enterprises realized a net income of $4,866 while offshore enterprises earned $25,194 after costs were deducted. Annual costs absorbed approximately 45% of inshore gross incomes and 55% of offshore gross incomes. From this, Davis and Thiessen concluded that offshore fishing was less efficient operationally than inshore fishing. This conclusion was based on the observation that inshore fishing generally cost substantially less and retained proportionately more income for participants than offshore fishing per unit of fishing effort.

Their reanalysis also indicated that ownership and crew position had a profound impact on income distribution in both the inshore and offshore sectors. In the inshore sector, owning the enterprise provided fishermen with almost 95% more gross income than that of non-owners ($7,746 compared with $3,984). Ownership had an even greater impact on the distribution of gross income in the offshore sector, with owners receiving over three times the gross incomes of non-owners ($71,202 compared with $17,210). Both non-owners and owners in the offshore fishery received average gross incomes far larger than those reported by inshore owners.

Crew position also strongly influenced gross income distribution. In the inshore sector, owner-captains reported an average gross income of $7,744 while non-owning captains received an average of $5,329. Non-owning crewmen, on the other hand, reported an average gross income of $3,298, only 42.6% of the owner-captains' gross income and 61.9% of the average gross income of non-owner captains. In the offshore sector, captain-owners received an average gross income of $77,756. Non-owning crewmen received an average of $13,983, 17.9% of that of owner-captains. Non-owning offshore captains received average gross incomes far greater than those of owning and non-owning crewmen, ($57,635 compared with $22,122 and $13,983 respectively).

These results suggested that fishing sector, the distribution of vessels and equipment ownership, and the position held in the fishing crew had a significant impact on fishing income.

From these analyses, Davis and Thiessen agreed that the plight of the majority of Atlantic fishermen was indeed grim. However, while agreeing with the TFAF contention that many in the industry had poverty-line incomes, they suggested that this was largely a problem of inshore fishermen. Despite the differences between the inshore and offshore sectors as they defined them, the level of earnings in both sectors suggested "a difference in order rather than a difference in kind." Davis and Thiessen concluded:

> "Most men participating in the Atlantic fisheries earn low to extremely low incomes. In fact, these data suggest that few have the capacity to generate sufficient earnings solely from fishing to fulfill annual income needs, let alone satisfy the cost of purchasing and maintaining a vessel and equipment. Given this, it is little wonder that federal and provincial government transfer payments, and subsidization is central for the ability for most fishermen to participate in the industry."

There are other income disparities in the fishing industry. Guppy (1987a), in a study of the British Columbia fish processing sector, revealed sexual and ethnic divisions in job allocations. Women were generally assigned jobs that were routine and monotonous, stationary and seasonal. Historically, the temporary, seasonal shoreworker jobs were undertaken by indigenous Indian and immigrant Asian women. In contrast, monthly rated men (those with high seniority and relatively good job security) tended to be non-Indian Canadians or northern Europeans.

These differences translate into differences in earnings. Women in 1981 received $9,700 in average annual income from fish processing work compared with an average of $17,400 for men. The bulk of this discrepancy occurred because women worked fewer months per year than men. Guppy explained this phenomenon as the result of creating "women's work", a limited range of jobs for which there is a large pool of applicants. Connelly and MacDonald (1983) observed, for the east coast fish processing industry: "Women have always constituted an available cheap labour reserve and...they have consistently responded to any and every wage-labour opportunity open to them."

Guppy (1987a) presented data on the average earnings in B.C. fish plants by gender and ethnic background. Incomes differed by gender but also by ethnic group, with Anglo-Canadians and Europeans receiving higher incomes. Anglo-Canadian and European men earned the top annual salaries largely because they were employed in the fish plants throughout the year.

Lamson (1986) reported the results of a study of women and fish plant jobs in Atlantic Canada. This study highlighted the importance of womens wages to the household economies of Atlantic coastal communities and the workforce reliability of female plant workers. Despite degrees of job dissatisfaction, the majority of female plant workers interviewed had remained in the same job for three or more years because they needed the money above all else. Many of these workers travelled by car 20 minutes or more to their place of work and quite a few travelled 45 to 50 minutes each way to work. This information contradicted the myth that fish plant employees work simply because convenient wage opportunities are available locally. Instead, Lamson's study confirmed the findings of the TFAF that the work of other members of fishermen's households is an essential component of household income. It was the household, rather than the individual, that was perceived as the central economic unit. The importance of women's work was indicated by their wages' estimated contribution to the total household income (i.e. 10–60%). The role of female plant workers is vital to the economic survival of households in those coastal communities where fishermen's incomes are chronically low.

4.0 UNEMPLOYMENT INSURANCE

4.1 General

The survival of many coastal fishing communities and fishermen's households has been dependent upon the availability of Unemployment Insurance Benefits over the past three decades. Unemployment Insurance for Fishermen has become an integral part of the economics of coastal fishing communities.

Unemployment Insurance for Self-Employed Fishermen was introduced in 1956, with fishermen first becoming eligible in April 1957. The Forget Commission of Inquiry on Unemployment Insurance described the origin of UI for Self-Employed Fishermen:

> "This was essentially a political decision, motivated by social rather than economic considerations. The purpose was to render support to incomes of self-employed fishermen in the inshore fishing industry, particularly on the Atlantic, and to the many coastal communities there that depend upon the fishery for their survival" (Forget 1986).

Currently there are three UI programs which cover fishermen:

1. Fishermen who are employees of a fishing enterprise are covered by the regular UI program;
2. Those who are self-employed are covered by two special fishermen's programs:
   a) one for individuals involved in seasonal fishing; and
   b) one for year-round fishermen.

In general, self-employed workers are not covered by UI because they do not have "a contract of service". Fishermen are an exception to this exclusion (Forget 1986).

4.1.1 Regular Unemployment Insurance

Regular UI benefits are available to fishermen who are employees of fishing enterprises. As of 1988, to qualify they had to have worked in insured employment for 10–14 weeks out of the previous 52. The number of weeks varied according to the regional unemployment rate. In 1988, the required number of weeks was 10 in Newfoundland, Prince Edward Island, coastal Québec, New Brunswick, British Columbia and parts of Nova Scotia. New entrants to the labour force required at least 20 weeks of insurable employment to claim benefits.

In the initial phase, individuals were entitled to 1 week of benefit for every week of insurable employment in their qualifying period, up to a maximum of 25 weeks. Those individuals who had more than 26 weeks of insurable employment could claim 1 week of benefit for every 2 weeks of employment in excess of 25, up to a maximum of 13 weeks. Individuals were also eligible for additional benefits based on the unemployment rate in their region. For every 0.5% increment in the regional unemployment rate above 4%, an additional 2 weeks of benefits could be claimed, to a maximum of 32 weeks. In 1988, individuals could claim the full 32 weeks in Newfoundland, Prince Edward Island, coastal Québec and parts of New Brunswick, Nova Scotia and British Columbia. This meant that in many areas fishermen could work for 10 weeks, collect regular UI for 10 weeks and regionally extended benefits for 32 weeks.

4.1.2 Year-round Fishermen's Unemployment Insurance

To qualify for year-round fishermen's UI, applicants had to meet all the requirements for regular UI. They also needed at least 20 weeks insurable employment as fishermen, with at least 6 insurable weeks of fishing in the last three consecutive quarters. Their last job had to be on a designated year-round fishing vessel. For a week to count as an insured week, the fishermen must have fished for at least 15 hours and have earned at least 20% of maximum insurable earnings. In the late 1980's, 10% of fishermen fell into this category. Under regular fishermen's UI, the employer's portion of the premium was paid for by the fish processors.

4.1.3 Seasonal Fishermen's Unemployment Insurance

Up until 1983, the benefit rate was based on the average of the last 10 weeks. Until 1983, benefits could only be claimed in the winter months (November 1 to May 15). To qualify for seasonal UI, fishermen must have the usual 10–14 weeks of insurable employment, including at least 6 insurable weeks of fishing. As under the regular UI, the benefit rate is 60% of insured earnings. For every 6 weeks of insurable employment, individuals can receive 5 weeks of benefit payments. They can also receive regionally extended benefits.

4.2 Cost of UI

In 1987, payments under Fishermen's UI totalled $223 million or 2% of payments made under regular UI. Federal government contributions to Fishermen's UI totalled $201 million in 1986/87 or 7% of its contributions to the regular UI program. The government paid a substantially higher proportion of Fishermen's UI than it did of the regular program.

The payments under Fishermen's UI differ significantly among regions. In 1988, 37.7% of the payments ($80 million) went to Newfoundland, 20.7% ($44 million) to Nova Scotia and 17.8% ($38 million) to British Columbia. New Brunswick, Québec and Prince Edward Island received 8.6%, 8.4% and 6.8% respectively of the overall expenditures under the Fishermen's UI program. Details on payments for the 1972–1988 period are given in Table 14-9.

4.3 Impact of Unemployment Insurance on the Incomes of Fishermen and their Households

The Task Force on Atlantic Fisheries data showed that benefits received through the Unemployment Insurance program were a significant source of cash income for fishermen in all geographical areas. Contrary to popular perception, not all fishermen received UI benefits. In 1981, for example, 63% of full-time fishermen on the Atlantic coast and 52% of part-timers received UI payments.

On average, full-time fishermen received $2,466 in UI benefits in 1981, while part-time fishermen averaged $1,483. UI payments averaged 16% of total net income (before taxes) for full-time fishermen and 13% for part-time fishermen. There was significant regional variation in the contribution of UI payments to net income. In northeast Newfoundland and Labrador, where earned incomes were lowest, UI payments made up 32% of average total net incomes for full-timers. This contrasted with western Nova Scotia where UI payments made up only 6% of average total net incomes for full-timers.

Nationally, in 1983, there were 35,000 claimants for Fishermen's UI, 7,500 claimants for Regular UI and 17,000 fishermen who made no UI claims for a total of 60,000. Of those who claimed fishermen's UI benefits, 79% of their earnings were derived from fishing, 3% from other sources of earnings and 18% from UI payments. Of those claiming regular UI, 38% of their earnings were derived from fishing, 37% from other sources of earnings and 25% from UI. When those who did not make UI claims were taken into account, fishermen derived 72% of their earnings from fishing, 12% from other sources of earnings and 16% from Unemployment Insurance (DFO, unpublished data).

The percentage of income derived from fishing varied from lows of 63% in Newfoundland and 65% in British Columbia to a high of 80% in Nova Scotia. The percentage of income derived from UI payments varied from lows of 9% in the inland provinces and 11% in Nova Scotia to highs of 20% and 23% in Prince Edward Island and Newfoundland respectively (Table 14-10).

When UI is viewed solely in the context of net fishing income, its importance as an income supplement looms

TABLE 14-9. Fishermen's Unemployment Insurance by province and national totals 1972–1988.

	Newfoundland		P.E.I.		Nova Scotia		New Brunswick		Québec		British Columbia		TOTAL
Year	Total Payments $'000	% of Total	Total Payments $'000	% of Total	Total Payments $'000	% of Total	Total Payments $'000	% of Total	Total Payments $'000	% of Total	Total Payments $'000	% of Total	Total Payments $'000
1972	4471	23.1	1592	8.2	4811	24.9	2079	10.8	1087	5.6	5280	27.3	19320
1973	4638	24.3	1312	6.9	4696	24.6	1966	10.3	1078	5.6	5424	28.4	19114
1974	5026	23.2	1634	7.6	5684	26.3	2127	9.8	1411	6.5	5741	26.6	21623
1975	4423	19.7	2066	9.2	6375	28.4	2130	9.5	1502	6.7	5978	26.6	22474
1976	6465	23.5	1923	7.0	7333	26.7	2646	9.6	2036	7.4	7101	25.8	27504
1977	13207	28.3	3432	7.4	9991	21.4	4938	10.6	3574	7.7	11466	24.6	46608
1978	18948	30.9	5035	8.2	12411	20.2	6732	11.0	4814	7.9	13372	21.8	61312
1979	24608	35.5	5900	8.5	13413	19.3	7342	10.6	6008	8.7	12137	17.5	69408
1980	33039	41.0	7244	9.0	16251	20.1	8585	10.6	7076	8.8	8484	10.5	80679
1981	36980	41.0	7439	8.2	18321	20.3	8652	9.6	7917	8.8	10885	12.1	90194
1982	42033	38.6	8686	8.0	23734	21.8	10295	9.5	10058	9.2	14118	13.0	108924
1983	52536	38.2	10450	7.6	29479	21.4	12315	8.9	12030	8.7	20791	15.1	137601
1984	57313	36.1	12063	7.6	36226	22.8	14762	9.3	13306	8.4	25137	15.8	158807
1985	63414	36.3	13063	7.5	38442	22.0	16016	9.2	14422	8.2	29553	16.9	174910
1986	71954	35.2	15928	7.8	43698	21.4	18746	9.2	16553	8.1	37736	18.4	204615
1987	80082	36.6	15978	7.3	45156	20.6	20282	9.3	17637	8.1	39891	18.2	219026
1988	80184	37.7	14479	6.8	44043	20.7	18242	8.6	17762	8.4	37818	17.8	212528
TOTAL	599321	35.8	128224	7.7	360064	21.5	157855	9.4	138271	8.3	290912	17.4	1647647

Source: Grady and MacLean (1989).

TABLE 14-10. Distribution of income from fishing versus that from UI by province in 1983.

	Number of Fishermen	% Income from Fishing	% Income from UI
Newfoundland	19,000	63%	23%
Prince Edward Island	3,500	72%	20%
Nova Scotia	13,000	80%	11%
New Brunswick	5,000	76%	15%
Québec	3,500	77%	18%
Inland	4,000	69%	9%
British Columbia	12,000	65%	14%
Canada	60,000	72%	16%

Source: Unpublished data, Economics Branch, DFO.

larger. From the 1984 survey of Atlantic fishermen's incomes, fishermen's UI as a percentage of net fishing income was 29%, on average. 1988 figures confirmed that UI had continued to be an important income supplement, particularly in Newfoundland (Table 14-5).

When figures are broken down intraprovincially, the degree of dependence on UI payments for income supplementation in particular areas is even more evident. Even within provinces there were significant differences. In northeast Newfoundland and Labrador, around 42% of income was derived from UI payments. On the south coast of Newfoundland (the trawler ports), on the other hand, income from fishing accounted for 77% of overall income and UI payments accounted for 17%, close to the national average for 1983. Within New Brunswick, there was also a considerable disparity in the contribution of UI payments to overall income — 24% on the Gulf shore compared with only 9% on the Fundy shore. The latter area has a year-round fishery and access to lucrative species such as scallops.

Guppy (1987b) observed an increasing dependence by fishermen in British Columbia on unemployment insurance. As a group, fishermen relied far more than other workers on UI payments. Nearly one-half of fishermen reported receiving some UI income, while one in four primary-sector workers and only about one in 10 farmers reported UI income. In addition, the average annual amount of such benefits was higher for fishermen ($2,800) than for other groups ($1,500). The number of fishermen in British Columbia receiving UI benefits rose from 3,176 in 1980 to 6,140 in 1984 — an increase of 93% in just 4 years.

Various studies over the years have identified shortcomings in the Fishermen's UI program and called for changes.

The Task Force on Atlantic Fisheries concluded that funds transferred to fishermen through the unemployment insurance program were vital to the Task Force's second objective for fisheries policy — that those employed in the fishery receive "a reasonable income as a result of fishery-related activities." The Task Force rejected any notion that the amount of this transfer should be reduced or eliminated. However, it questioned whether a better program could be designed to stabilize and supplement incomes. It made proposals for the development and testing of pilot programs (discussed in the next section).

Meanwhile, as a "transition measure", it recommended that the Canada Employment and Immigration Commission amend the regulations for Fishermen's UI to provide benefits based on the "best 10 weeks fished" for fishermen who fish at least 15 weeks. The Task Force also recommended greater flexibility in defining the fishing season to allow those who fish exclusively during the winter to qualify for benefits. It proposed restricting entry to the UI program, so that persons who fished less than 6 weeks would not qualify for Fishermen's UI. It also proposed revising the UI rules to permit boat-building during the benefit period for personal commercial use. More significantly, the Task Force proposed that the government adopt a 'sunset' provision in the UI regulations for self-employed fishermen. This meant that the entire program would no longer be in force after April 1, 1988, "provided that [a] production bonus system...and [an] income stabilization scheme can be implemented fully as replacements, with general approval from participants in the industry."

The federal government accepted the proposals to amend the Fishermen's UI regulations but rejected the call for a 'sunset' provision.

The 1986 Forget Commission concluded that the fishermen's UI program is "manifestly inconsistent with the principles of social insurance in that contributions from participants constitute only a small percentage of the total outlay and therefore substantial funds are needed from general revenue." It noted that in recent years over 90% of the benefits for fishermen had been paid by the government out of the Consolidated Revenue Fund.

More seriously, the Commission questioned the adequacy of the program:

"The program has proven to be ineffective and inadequate in terms of its objective of providing support to the incomes of needy fishermen. It does not address the tremendous diversity within the industry and the often inescapable fluctuations in the level of income of fishermen."

The Commission proposed that income supplementation be a major element in a reformed and improved income security system for the entire country. It suggested that the Fishermen's UI program be modified without waiting for a comprehensive nation-wide program. Specifically, it proposed that the money currently devoted to regionally extended benefits and to fishing benefits provide the basis for negotiating a program of income supplementation with each of the interested provinces. The resulting income supplement would be available to all workers in need.

A couple of other studies have addressed the impact of UI on the fishery. Ferris and Plourde (1982) modelled the effect of seasonal unemployment insurance on the size and efficiency of the Newfoundland inshore fishery. They concluded that the program generated a form of UI dependency, as seasonal UI benefits come to form an integral rather than supplementary part of inshore incomes. Furthermore, they suggested that UI could not permanently raise the relative levels of inshore incomes. Without entry controls, the program would provide at best temporary gains. Over the longer run, it would only expand the problem.

Gaskill et al. (1987) suggested that Newfoundland communities work share in order to maximize UI payments at the community level.

The Newfoundland Royal Commission concluded:

"The UI system encourages short-term make-work projects rather than long-term economic development, undermines work initiatives, discriminates against self employment and discourages the formation of sound work habits and attitudes. In its place we need a new income security system which provides every household with a minimum level of support, supplements low incomes in a way that rewards people for working, and provides income maintenance between jobs through a modified programme of unemployment insurance" (Newfoundland 1986).

When it was introduced, Fishermen's UI was supposed to be a temporary program, pending the development of a more satisfactory means of income support. No satisfactory alternative has yet been devised and it survived intact changes in 1989 to the Unemployment Insurance Program.

On April 11, 1989, the Honourable Barbara MacDougall, Minister of Employment and Immigration, announced a new Labour Force Development Strategy. Ms. MacDougall described this initiative as "a shift away from passive income support towards programs and services that will help Canadians get back to work as quickly as possible" (EIC 1989a) This Strategy included some changes to the UI program. One change meant that in areas where there are more job opportunities, claimants would have to work longer to qualify for benefits and the duration of their benefits would be shorter. The Minister announced the reallocation of $1.3 billion primarily to training initiatives and to make UI fairer (EIC 1989b).

Significantly, the Fishermen's Special UI program was left intact. Fishermen would still require the same number of weeks of insurable earnings to qualify for unemployment insurance and could draw benefits for the same number of weeks as before. Fish plant workers, however, who fall under the Regular UI program, were affected by the changes in the eligibility criteria for that program. Although fish plant workers were affected, the impact was diminished by the fact that the UI eligibility criteria continued to be linked to regional unemployment rates. In Newfoundland, for example, fish plant workers continued to qualify for UI with 10 weeks of insurable earnings.

Meanwhile, the Unemployment Insurance Program and Fishermen's UI, in particular, continue to supplement inadequate incomes in most of Canada's coastal fishing communities.

5.0 INCOME STABILIZATION

In recent years there has been considerable discussion about the variability as well as the inadequacy of fishermen's incomes and the perceived need for income stabilization measures. The Task Force on Atlantic Fisheries noted: "There is a considerable element of instability in pursuing a living from fishing. Fishing is a hunting occupation, and the 'beast' that is chased may outwit its pursuers or may not appear at all. Difficult weather and ice conditions or unpredictable species behaviour will play havoc with the revenues of even the most skilful fishermen" (Kirby 1982).

Although it did not examine in detail the issue of income stability as opposed to income supplementation, the Task Force considered several options for income stabilization. These included:

1. Catch Insurance;
2. Expansion of the mandate of the Fisheries Price Support Board, with improved funding; and

3. A Gross Revenue Stabilization Fund.

In terms of catch insurance, the TFAF foresaw a system to guard against severe declines in annual catches, similar to crop insurance in the agricultural sector. They envisaged 'risk rating' based on individual enterprises or groups of enterprises, in specific geographical areas. Baseline data on catches by gear types and species in clearly defined coastal areas would be measured over 3 to 5 years for all registered enterprises. This would establish a base against which to measure deviations in catch. While the Task Force believed that such a catch insurance scheme could be directly related to risks in species availability, weather effects, or other natural phenomena, it recognized that it was extremely difficult to develop a practical model. Feasibility studies in the past on this type of program had predicted high administrative costs, requiring continuing government subsidy. The Task Force also recognized that it would be difficult, perhaps impossible, to determine whether reasonable efforts were made by a fishing enterprise that reported a failure in catch.

As part of the response to the 1974–76 Atlantic fishing industry crisis and the restructuring of the Bay of Fundy herring fishery, the federal government funded a pilot catch insurance project involving herring weir fishermen in the Bay of Fundy. This project was abandoned as unworkable after a couple of years of trial.

The option of an expanded Fisheries Price Support Board would have had the Board intervene in a more active and anticipatory manner. Ad hoc deficiency payments during price declines would be payable to fishermen, processors or both on the Board's advice, with the Board having a revolving fund (as it did prior to 1981). The perceived advantages of this proposal were protection of fishermen's and processor's incomes from unusually severe price declines and because, of its ad hoc nature, avoidance of industry dependence on a regular price subsidy. The perceived disadvantages were that it could be subject to countervail action by the United States and was likely to result in some transfer of subsidy to processors when given to fishermen and vice versa.

A Gross Revenue Stabilization Fund, as envisaged by the TFAF, would guard against severe declines in gross enterprise revenues. In order to smooth out the peaks and troughs of gross revenue, a compulsory program would be designed to provide payouts when average revenues declined by a fixed percentage. Benefit levels would depend on actual revenues of individual enterprises, or groups of enterprises, compared with moving 5-year averages of gross revenues.

The Task Force foresaw a number of advantages to such a system:

1. Greater sensitivity to economic realities than a catch insurance scheme, because it could reflect major changes in either catch or market price;
2. Ease of administration compared with a catch insurance scheme, because it would be based on individual revenues reported through the existing taxation system;
3. Being largely self-financing through levies on fishermen's incomes in good years;
4. Avoidance of problems associated with net revenue stabilization schemes, which include an element of subsidy on capital cost and promote over-investment; and
5. Sufficient flexibility that pay-outs could be set at levels low enough to avoid creating disincentives to fish, and pay-outs would not be included in average gross revenue calculations for future years.

The chief perceived disadvantages were that baseline statistics on individual incomes, species and regional variations would require 3–5 years to establish, and verification of actual fishing effort would be needed to adjudicate claims.

The Task Force recommended that the government develop a gross income stabilization plan to smooth the high and low points in individual gross incomes over a rolling 5-year period. This plan would require participation by all fishermen, with funding from their contributions and from the federal government. It called for a detailed analysis and a possible pilot project in 1985. The government accepted in principle this recommendation, as well as a recommendation for a production bonus system to supplement fishermen's incomes by rewarding desirable fishing practices. Nothing much was done, however, to implement these two recommendations during the 1983–85 period.

The issue of income stabilization programs for the fishing sector was raised at the First Ministers' Conference in Halifax in 1985. First Ministers agreed that:

> "Alternatives to the current regime, such as income stabilization or catch insurance programs, must be explored to address income fluctuations in the harvesting sector of the industry" (DFO 1986g).

In response to this request, DFO analysed the potential for fisheries stabilization programs, based on agricultural income stabilization programs, in 1986. Agricultural stabilization programs provide deficiency payments to farmers following hardship years. The programs apply commodity-wide, generally on a national basis, and all eligible farmers receive the same per-unit rate of support. Payments are determined by comparing a particular year to the previous 5-year average. In most cases, deficiency payouts are made when earnings during a marketing year fall below 90% of the previous 5-year average. The difference is paid to farmers based on their individual sales; sometimes maximum per-farmer payout limits apply. For the Western Grain Stabilization Act (WGSA), payout is related to gross earnings (price times volume). This system is used because both prices and volumes sold fluctuate substantially.

The WGSA model was used to examine the potential impact of stabilization programs for the fishery. This was chosen because returns to fishermen, like grain farmers, can fluctuate with prices, volumes harvested, or both. Analysts, therefore, used landed values in the fisheries to model the potential impact of fisheries stabilization programs.

One of the arguments for considering the agricultural stabilization programs was certain similarities between fisheries and agriculture:

— both are based on biological (living) resources;
— both are subject to the vagaries of weather and markets;
— both depend on international markets;
— similarly to prairie grain incomes, fisheries incomes vary with prices, volumes or both and income downturns for one crop or species may be offset by improvements in others.

Unlike the agricultural models, national averages were not used in modelling potential fisheries stabilization programs. Instead, the Pacific and Atlantic coast fisheries were modelled separately. In addition, because of the considerable diversity among regions in the Atlantic, the TFAF areas were modelled separately. On the Atlantic, groundfish models were developed for each Task Force area and lobster models for eight areas. On the Pacific, one model was developed for all species combined and one for salmon.

Over the period 1973–1984, for Atlantic groundfish only 29 payouts were triggered in total for the 13 areas modelled. For Atlantic lobsters, there would have been no payouts in four of eight areas and only one payout in each of three others. In Eastern Nova Scotia, the worst stock collapse in the lobster fishery would have triggered seven payouts over the 12 years (DFO 1986g).

There would have been five payouts totalling $73 million for Pacific salmon over the 25 years to 1984. For all Pacific species, eight payouts totalling $95 million would have occurred over the same period.

Based on these analyses, it was concluded that "stabilization programs based on agriculture models would be of little benefit to Atlantic fishermen but may be worth further examination on the Pacific coast" (DFO 1986g). The reason for this was that there was insufficient variation in landed value and generally rising returns from fishing over time.

More recently, Grady and MacLean (1989) carried this analysis to a further level of detail within provinces. They concluded that the problem of income variability is difficult to assess:

> "At a provincial level fishermen's incomes only appear to be volatile in British Columbia. In Nova Scotia and Newfoundland fishermen's incomes appear no more variable than the average income across all sectors in these provinces. However, landed value data does suggest that variability is greater at the sub-provincial level. The direction and size of changes in landed values varies greatly from year to year, by species and areas."

Perhaps the most interesting aspect of the Grady and MacLean study was their comparison of fishermen's incomes with that of farmers. The average total income of farmers and fishermen from 1977 to 1985 is shown in Fig. 14-6. Total income included the income of farmers and fishermen from all sources, including for farmers, payments made under stabilization schemes. Even after stabilization payments, farmers' total income was more variable than that of fishermen. Similarly, net farming income was more variable than net fishing income even though net farming income included stabilization payments (Fig. 14-7). These comparisons indicated that farmers' incomes prior to stabilization payments were more variable than those of fishermen.

The bottom line is that the chief issue respecting fishermen's incomes appeared to be the inadequacy of total income. The problem of chronically low incomes was particularly apparent in some fisheries and areas (e.g. Newfoundland inshore fishery).

The 1986 studies suggested that an income stabilization fund would not provide a satisfactory replacement for the existing Fishermen's Unemployment Insurance Program. Stabilization schemes might provide some measure of assistance in moderating fluctuations in fishermen's incomes. This would be of greater benefit on the Pacific than the Atlantic coast because income tends to be more variable and there is not as great a degree of dependence on Unemployment Insurance for income supplementation. Despite the deficiencies of the Unemployment Insurance Program as presently constituted, it provided badly needed income support to fishermen with very low incomes. Fishermen, plant workers and fishing communities continued to rely on the much-maligned Unemployment Insurance system to supplement chronically low incomes.

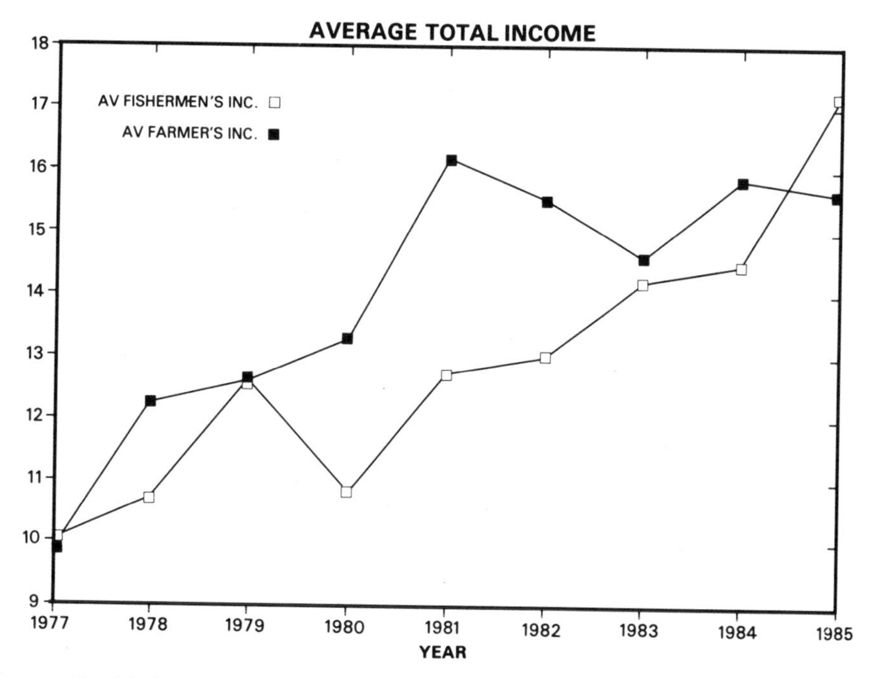

FIG. 14-6. Average total income (thousands of dollars) of fishermen and farmers (from Grady and Maclean 1989).

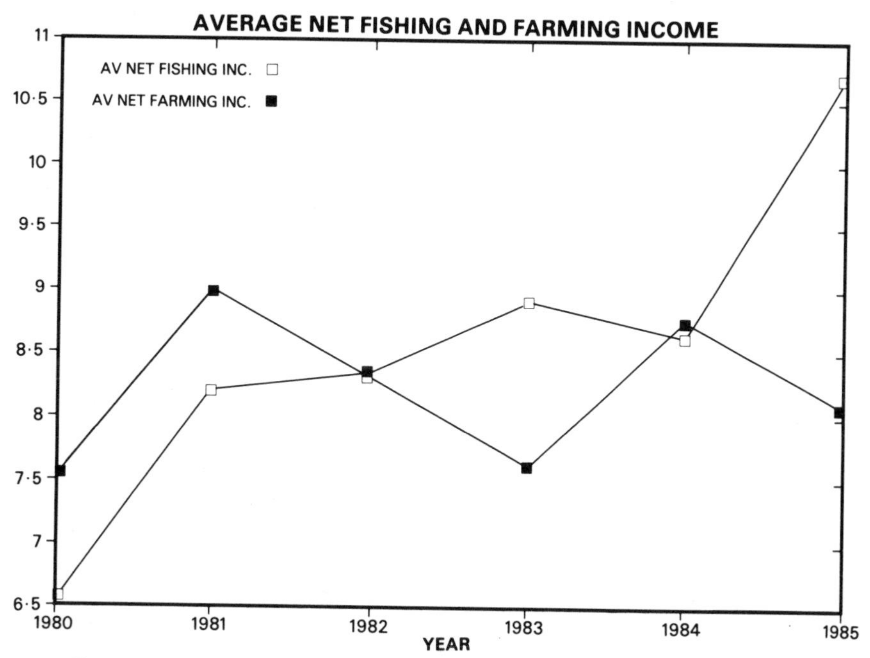

FIG. 14-7. Average net fishing and farming income (thousands of dollars) (from Grady and Maclean 1989).

In March 1992, Fisheries Minister Crosbie and Employment and Immigration Minister Valcourt announced the formation of a Task Force on Incomes and Adjustment in the Atlantic Fishery (DFO 1992f). The six-member group was chaired by Richard Cashin, President of the Newfoundland Fishermen, Food and

Allied Workers Union. The 1991 Atlantic catch declines had highlighted the need to address the problem of overcapacity and inadequate incomes that the fishing industry had faced for several years (see Chapter 13).

The purpose of the Task Force was to develop a comprehensive, long-term strategy for the industry. The long-term strategy would have as its goals:

"— to advise on the continued supply of the fishery resource;
"— to work towards stability and profitability in the industry;
"— to ensure stable, adequate incomes for those whose employment is sustainable by the industry; and
"— to find alternative training, employment and other adjustment possibilities for displaced fishermen, plant workers and affected communities." (DFO 1992f)

The group was to do its work in two stages. In phase one, it was to provide an interim assessment on the state of the resource, the industry and those employed in it. In the second phase, it would prepare a set of final recommendations for an adjustment strategy, and propose communities in Atlantic Canada in which to test this strategy through pilot projects. The federal government had set aside $45 million to test Task Force proposals for pilot projects in selected Atlantic communities.

The dramatic closure of the Newfoundland northern cod fishery for a 2-year period commencing in July 1992 highlighted the need for such a comprehensive adjustment strategy. To replace incomes lost because of this moratorium, the federal government put in place a multi-faceted program costing in the order of $800 million. This was to cover not only income replacement but also to facilitate reduction in capacity when this fishery reopened (see Chapter 13). It was anticipated that the Task Force on Incomes and Adjustment would develop specific proposals for adjustment in this and other Atlantic fisheries, recognizing the twin problems of overcapacity and income inadequacy.

6.0 NATIVES AND THE FISHERY

Canada's native peoples fished for food and barter long before Europeans arrived. In recent decades considerable attention has been focused on native peoples' rights to fish and, in particular, on the terms and conditions under which such fishing may be conducted.

Indian participation in the fishery is a high profile issue in British Columbia. Confrontation and violence has occurred on salmon rivers as fishery officers try to enforce regulations which Indians see as infringing on their aboriginal or treaty rights.

In 1986, more than 700,000 persons (representing 3% of the total population of Canada) reported at least one aboriginal origin. Approximately 286,000 gave a single North American Indian origin, 60,000 a single Métis origin and 27,000 reported a single Inuit origin. Approximately 330,000 reported both aboriginal and non-aboriginal origins (Canada 1990d). The fishing rights accorded these groups differed according to their traditional fishing practices before the Europeans arrived, the status and areas encompassed by any treaties signed, and other factors.

In this chapter I will deal primarily with issues concerning the Indian fisheries in British Columbia. For a discussion of native fisheries in the Arctic, see Clarke (1993). Native freshwater fisheries are dealt with by Pearse (1988). In terms of Canada's marine fisheries, the native fishery is particularly important in British Columbia where there are 72,000 status Indians in 196 bands on 1662 reserves. Many of these reserves border on or are adjacent to important salmon rivers. Much of what follows is based upon Pearse's 1982 and 1988 reports.

In British Columbia there are two components to fishing by Indians — participation in the commercial fishery and participation in the so-called Indian fishery or Indian food fishery. During the 1980's considerable controversy raged concerning the scope and nature of the Indian food fishery, with much of the debate focusing on whether Indians should be allowed to sell their catches.

As Pearse (1982) wrote, "The Indian fishery puts relatively light demands on the fish resources in the Pacific region but it involves issues of profound social, political and economic consequences."

6.1 Indians in the Commercial Fisheries

As discussed earlier, the early commercial salmon fishery in B.C. was carried on from numerous locations along the coast. Canneries were located near major salmon runs in areas where Indian communities were usually also situated. Indian men from those communities worked on the fishing vessels and Indian women and children worked in the canneries. In the early 1920's, the canneries employed more than 9,000 people, most of whom were Indians. More than one-third of all salmon fishermen were Indian (Sinclair 1971).

Opportunities diminished for Indians during the 1920's and 1930's with the building of larger and more expensive fishing vessels and the consolidation of canneries. Their fortunes improved temporarily during World War II but Indian employment in the fisheries declined sharply after the war (Hawthorn et al. 1958).

With the consolidation of canneries, by 1970 Indians

Catching salmon with a dip-net.

Preparing salmon for drying.

accounted for 1500 of a total of 3700 cannery employees. During the 1950's and 1960's the number of Indian-owned salmon fishing vessels declined by about 60% to 599 in 1971. At the beginning of the 1970's, the fishing and processing industries employed less than half the number of Indians that had been employed two decades before (Pearse 1982). By 1980 the number of Indian salmon licensees had declined to 452.

Pearse (1982) noted:

> "The erosion of employment in fishing and related occupations has had a devastating impact on dozens of Indian communities that offered no other employment opportunities and where unemployment was already chronic.... And it created severe economic and social problems beyond those normally attributed to unemployment. For example, vessels were displaced that had been depended upon for food fishing and for transportation links with other communities."

Various policies were adopted in the 1970's to promote Indian participation in the commercial fishery. When limited entry licensing was introduced for salmon in 1968, a number of measures were taken to halt the decline in Indian participation.

In 1970, the Department of Indian and Northern Affairs provided funds to purchase old vessels from the existing fleet to create a "tonnage bank" for Indian fishermen wanting to enter the salmon fishery. In 1971, an Indian licence, known as category "A1", was created. Vessels with these licences could not be purchased under the buyback program but could be transferred to other Indians. DIAND purchased some licensed vessels for the "tonnage bank", from which they were available to Indians who required assistance. In 1972, all temporary, category "B", licences held by Indians were converted to full fledged Indian, category "A1" licences.

In spite of these measures, the number of Indian participants in the salmon fishery continued to decline initially under the limited entry licensing regime. In the first 3–4 years, the number of Indian vessel-owners and crewmen declined by about 8% (Campbell 1974). After 1972, for a few years, aided by DIAND's Indian Fishermen's Assistance Program, Indians increased their share of the salmon fisheries (Pearse 1982)

When the roe-herring fishery developed, licences for Indians were not limited until 3 years after they were limited for non-Indians. Indians were charged only nominal fees for their licences. When the herring spawn-on-kelp fishery developed, individual Indians and band councils had priority for new licences. As of 1982, they held 18 of the 28 licences in this fishery. (In 1991, the total number of licences was increased by 10, with all of the increase going to native bands. As of 1991, bands held 15 licences and individual Indians 8).

Three federal programs were introduced from 1968 to the mid-1980's to improve Indian participation in the Pacific commercial fisheries. These were:

1. The Indian Fishermen's Assistance Program;
2. The Indian Fishermen' Emergency Assistance Program; and
3. The acquisition of the B.C. Packers northern fleet by the Northern Native Fishing Corporation.

The first of these, the Indian Fishermen's Assistance Program, was introduced in 1968 in conjunction with limited entry licensing for salmon. It had three objectives:

1. To arrest the decline in the number of Indian-owned vessels, and, if possible, to reverse it;
2. To improve the earnings of Indian-owned vessels so they equalled the average of the rest of the fleets; and
3. To improve the versatility and mobility of the Indian fleet to the level of the rest of the fleet.

By 1979, when the program ended, $163 million had been expended, half in grants and half in loans. The program appeared to arrest the decline in the proportion of the salmon fleet owned by Indians. The average gross earnings of Indian vessels increased from a low of 61% of the average for the salmon fleet as a whole in 1967 to a high of 109% in 1973, and averaged 84% during the last 5 years of the program. The total tonnage of Indian-owned vessels increased by 33% and their average value increased from 67 to 87% of the average for the entire fleet. However, while achieving its main objectives, the program benefited primarily those Indians who were already well-established and successful fishermen (Pearse 1982).

Despite the apparent initial success of the Indian Fishermen's Assistance Program, between 1977 and 1980 the Indian-owned fleet decreased by about 100 vessels. Given the financial difficulties in the salmon industry in 1980 and 1981, the Department of Indian and Northern Affairs became concerned that the displacement of Indians would intensify. Accordingly, an Indian Fishermen's Emergency Program was established late in 1980. Funded by DIAND, the program was administered by two boards controlled by the Native Brotherhood of British Columbia and the Nrun-chah-nulth Tribal Council. When the program ended early in 1982, $2 million in grants, $200,000 in direct loans and $700,000 in loan guarantees had been made. While this program did not have a major impact, it did enable some Indian fishermen to fish in 1981 who might otherwise have not been able to operate.

A much more significant initiative was a federal grant in 1982 of $11.7 million to fund the purchase of 243 vessels and 252 licences from B.C. Packers Ltd. by the Northern Native Fishing Corporation, a corporation formed by three tribal councils. This corporation leased the licences and sold the boats to Indian fishermen, most of whom had previously rented these vessels from B.C. Packers.

From his historical study, Pearse (1982) concluded that large numbers of Indians had been displaced from the commercial fishing industry in recent decades. He observed: "This displacement has generated serious economic and social distress in Indian communities, many of which offer no alternative employment. The relative immobility of Indian people has left them heavily dependent on unemployment insurance and welfare payments. This is costly to the taxpaying public and, at the same time, inflicts high costs on the Indians themselves in the form of idleness, dependency, demoralization and social and personal breakdowns."

Despite this, it was obvious that Indians could adapt and perform well in the commercial fisheries. Pearse concluded that the fisheries were an obvious base for Indian social and economic development policies. He saw a choice "between increasing subsidization to coastal Indian communities in the form of welfare funds and personnel needed to cope with the growing problems of unemployment, dependency and demoralization, on the one hand, and subsidizing fisheries programs that will provide productive employment and contribute to individual and community morale on the other."

In support of this second alternative, Pearse recommended that:

1. The federal government proceed with the Indian Fishermen's Economic Development Program, a proposal by the Native Brotherhood of British Columbia, as quickly as possible;
2. The Department of Indian and Northern Affairs provide staff and resources to monitor the financial performance of Indian fishing operations under the Indian Fishermen's Economic Development Program;
3. Licences held by Indian fishing corporations not be transferable to non-Indians and licensing policies be developed to enable such licences to be leased to individual Indians; and
4. The Department of Indian and Northern Affairs provide Indians and Indian corporations the financial assistance they need to compete successfully in the proposed periodic reissuing of licences of licences by competition (the 'auction' proposal discussed in Chapter 8).

As of 1985, 923 B.C. salmon licences were owned or operated by Indians. This represented 20.5% of the total licence entitlements in the fleet of 4,497. In that year 27.7% of the commercial salmon catch in pieces was taken by native participants. The Indian fleet accounted for 22.7% of the landed value. [This figure is higher than earlier estimates because it includes a more comprehensive list of native operations and joint owners of vessels]. For details, see James (MS 1987). The fishing industry accounted for an estimated 25–30% of total Native employment in British Columbia.

In July 1985, the federal government announced an $11 million B.C. Indian Fishermen's Economic Development Program. The program had three main elements:

1. Assistance to existing Native vessel owners for debt reduction and a fisherman's training program emphasizing basic business and financial management skills;
2. A Native-only buy-back program; and
3. Financial assistance to Native fishermen operating company rental boats to purchase their own vessels.

The overall intent of the program was to stabilize Native participation at existing levels, and to provide a bridging mechanism pending the settlement of the fisheries component of Native comprehensive claims in British Columbia (*The Vancouver Sun*, July 9, 1985).

6.2 The Indian Food Fishery Legal Framework

Much of the debate concerning Indians and the fishery in recent years has centered on the Indian Food Fishery (IFF) and the possible legalization of the sale of this catch. The debate reflects fundamental differences of view on Indian rights to participate in the fishery.

The present Indian food fishery is a continuation of traditional native fishing practices. Although the term "Indian food fishery" is now widely used, Indians reject the term because it implies a traditional dependence on fish for direct consumption only. They contend that fish were important in trade and barter as well as providing food. Pearse (1982) recognized this broader historical role:

> "The traditional importance of fish extended well beyond its food value.... Fish were also a major commodity of trade among Indian bands and tribal groups. The pattern of Indian settlement can be traced in large part to the accessibility of fish both on the coast, where permanent villages and seasonal camps were located near fishing grounds, and in the interior, where villages and fishing stations

were established on rivers and streams near places where salmon could be easily caught. Today, this pattern of Indian settlement remains in large part unchanged. Seasonal fishing established the annual routine of life, and the runs and catches of salmon were viewed with reverence since fish were the primary means of survival. The great social and cultural significance of fish, especially salmon, is reflected in the important role they play in elaborate traditions of feasts, ceremonies, myths and art."

The legal basis for native rights to fish in Canada have recently been clarified by court decisions. The courts have recognized certain aboriginal rights of native peoples to hunt and fish. Certain rights were stipulated in treaties signed by the British government and its representatives. However, in British Columbia only a few bands formally relinquished their claims to lands and resources under treaties. Also in British Columbia certain rights to fish were expressly guaranteed to Indians when allotments of reserves were made by the Reserve Commissioners.

Native representatives argue that these issues are linked because aboriginal rights were the underlying source of rights for both categories. From their perspective, the rights to hunt and fish stipulated in treaties acknowledged aboriginal rights to be protected as the whites proceeded to settle Crown land. Similarly, they argue that the reserved fishing rights attached to the allotments by the Reserve Commissioners represent aboriginal rights explicitly recognized for specific sites. In the United States, courts have held that "reserve" and "treaty" rights in the Pacific Northwest are recognized aboriginal rights.

Since most Indians in British Columbia never surrendered lands and resources through treaties, their claims to fish rest on the legitimacy and definition of their claim to aboriginal rights. The Nishga Band in the early 1970's sought clarification through a law-suit against the province. It based its claims on unsurrendered aboriginal rights and a 1763 British Royal Proclamation. The B.C. Court of Appeal declined to recognize aboriginal rights (Calder et al. v. A.G.B.C., B.C. Court of Appeal). The Supreme Court of Canada in the same case failed to resolve the issue, leaving the question of aboriginal title in British Columbia unsettled. In 1980, the Federal Court of Canada in the case of *Hamlet of Baker Lake v. Minister of Indian and Northern Affairs et al.* held that "the law of Canada recognizes the existence of an aboriginal title independent of *The Royal Proclamation* or any other prerogative act or legislation." Justice Mahoney listed four elements which must be proven to establish an aboriginal title recognizable in common law:

1. They and their ancestors were members of an organized society;
2. The organized society occupied the specific territory over which they assert the aboriginal title;
3. The occupation excluded other organized societies; and
4. The occupation was an established fact at the time England asserted sovereignty.

The nature, extent or degree of the aboriginal people's physical presence on the land they occupied was to be determined in each case by a subjective test.

The Supreme Court of Canada in 1984 in *Guerin v. the Queen* recognized an obligation by the Crown to deal with land in the best interest of the Indians when it has been surrendered to the Crown. Chief Justice Dickson stated that in the *Calder* case the Supreme Court had "recognized aboriginal title as a legal right derived from the Indians" historic occupation and possession of their tribal lands.

The first case which really dealt with the aboriginal right to fish was *Sparrow v. the Queen* (1986, the B.C. Court of Appeal). The B.C. Court of Appeal accepted the contention by the accused that he was exercising an aboriginal right to fish for food when he was fishing for salmon with a drift net in the Fraser River some 10 miles from his reserve.

In the *Sparrow* case the accused claimed the right to fish for food on the band's traditional fishing grounds. Whether there is an aboriginal right to fish for other than food purposes was not at issue. The Court decided that there was an aboriginal right to fish for food within the following context:

> "The breadth of the right should be interpreted liberally in favour of the Indians. So food purposes should not be confined to subsistence."

Elsewhere in the judgment, the breadth of the aboriginal right to fish was defined as the right to take fish "for food and for the ceremonial purposes of the band" and/or "for reasonable food and societal needs."

The questions whether there is an aboriginal right to fish for sale or barter and/or whether aboriginal fishing for commercial purposes has been extinguished were expressly left to be decided in another case.

Thus, the courts decided that native peoples have aboriginal rights distinct from any that may have been arranged under treaties, constitutions or laws. These include the rights to use the fruits of the land, including fish and wildlife.

The 1982 amendments to the Canadian Constitution in the form of the Canadian Charter of Rights and Freedoms provided that "the existing aboriginal and treaty rights of the aboriginal people of Canada are hereby recognized and affirmed."

In the case of *Sparrow v. the Queen* 1986, the B.C. Court of Appeal observed:

> "Any definition of the existing right must take into account that it exists in the context of an industrial society with all of its complexities and competing interests. The 'existing right' in 1982 was one which had long been subject to regulation by the federal government."

The court held that the federal government has the power to regulate the aboriginal right to fish but there are limitations on that power:

> "The general power to regulate the time, place and manner of all fishing, including fishing under an aboriginal right, remains. The essential limitation upon that power is that.... in allocating the right to take fish, the Indian food fishery is given priority over the interests of other user groups. What is different is that where the Indian food fishery is in the exercise of an aboriginal right, it is constitutionally entitled to such priority. Furthermore, by reason of section 35(1) [of the Constitution] it is a constitutionally protected right and cannot be extinguished."

The court concluded with the following principles:

> "1. regulations which do not infringe the aboriginal food fishery, in the sense of reducing the available catch below that required for reasonable food and societal needs, will continue to regulate the aboriginal right to fish.
>
> 2. regulations which do not infringe the aboriginal food fishery may still be valid, but only if they can be reasonably justified as being necessary for the proper management and conservation of the resource or in the public interest."

As mentioned, Indian treaties in British Columbia are confined to Vancouver Island and the northeast part of the province. Under the 14 Douglas treaties negotiated with various bands on Vancouver Island by James Douglas, of the Hudson's Bay Company, the bands formally surrendered claims to certain lands for cash but they retained their village sites and fields. In all of these treaties, they were assured that they were "at liberty to hunt over the unoccupied lands, and to carry on fisheries as formerly."

Pre-1990 the courts had held, notwithstanding the assurance of access to traditional fisheries provided for in these treaties, that Indians are required to comply with regulations under the Fisheries Act regarding permits, gear, fishing times and similar measures.

Another legal issue which has complicated the management of the Indian food fishery is the relative legal status of two federal statutes, the *Fisheries Act*, administered by the Minister of Fisheries and Oceans, and the *Indian Act*, administered by the Minister of Indian and Northern Affairs. Under section 81(1[o]) of the *Indian Act*, band councils are authorized to enact by-laws covering many activities including "the preservation, protection and management of...fish...on the reserve." The Minister of Indian and Northern Affairs may disallow such by-laws within 40 days; otherwise they take effect. A number of bands in British Columbia have adopted such fishing by-laws.

Under the *Fisheries Act*, the Minister of Fisheries and Oceans is authorized to issue licences to status Indians and to make regulations governing their fishing. These regulations differed from province to province. Generally, however, they required Indians to acquire food fish licences or permits and prohibited the sale or trade of fish caught under these licences. Some bands issued by-laws which conflicted with these regulations. These by-laws applied only to the reserves but there was considerable ambiguity concerning some reserve boundaries. Decisions in the lower courts indicated that such by-laws superseded federal fisheries regulations.

6.3 The Evolution of Policy and Practice Respecting the Native Fishery

Given this somewhat confused legal framework, we will now examine the evolution of federal policy for the Indian fishery.

Throughout the past century there has been a tug-of-war between the Indian traditions of fishing and hereditary fishing areas, the need to conserve fish stocks, and the demands of commercial and recreational fishermen. There has been a long history of conflict over Indian fishing rights between Indians and non-native users and between Indians and government regulators. This conflict has involved hundreds of charges for violations of fisheries regulations, violence resulting in criminal charges, civil litigation, injunctions against Indians and other injunctions against governments and third parties.

Before 1877 there was essentially no regulation of fisheries in British Columbia. Ware (1978) observed:

> "In this era there was no distinction between "food fishery" and commercial fishing. There were no regulations, no Proclamations, no

Orders-in-Council, no laws of any kind which specifically restricted or regulated Indian fishing in British Columbia."

The Dominion Fisheries Act (1867) authorized the federal Minister to issue licences to Indians to catch fish for their own use. A package of British Columbia Fishing Regulations was adopted in 1878 but regulations dealing specifically with the Indian fishery were not enacted until 1888. In that year, the Regulations stipulated that, although other users required a lease or licence to fish in all waters of British Columbia:

> "Indians shall, at all times, have liberty to fish for the purpose of providing food for themselves, but not for sale, barter or traffic, by any means other than with drift nets, or spearing."

In 1910, a regulation was enacted which required Indians to acquire a permit. This permit could specify the area and time when fishing could take place and the gear to be used.

These regulations were essentially unchanged until 1977 when the requirement was changed to a licence instead of a permit. Indians opposed this even though there was no substantive change at that time. In 1981, a new regulation required permits to specify both the species and the quantities of fish that could be taken. This stipulation that the quantity of fish be specified increased an already high level of tension between fisheries officials and Indians. In 1977, the Department began issuing permits to some Indian bands instead of to individuals, allowing the permits to be administered by band councils. By 1982, about 10% of the bands engaged in the Indian fishery did so under this arrangement.

The most authoritative written expression of federal policy respecting the Indian food fishery during the 1970's is contained in a February 15, 1974, letter from the then-Minister of the Environment, Jack Davis, to the Union of B.C. Indian Chiefs. Mr. Davis stated:

> "As stated at our meeting of November 16, 1973, Indians have a right to fish for food use but not for sale or barter. Furthermore, ...in the exercise of this right the Indian food fishery has priority second only to conservation.
>
> "I also advised you that I anticipated Indians would act responsibly in exercising this food fish right.... I consider the sale of fish caught under food permits to be a serious violation."

Indians have been very critical of the system used to regulate their fishing times and places. However, the system enabled Indians to fish in areas and using methods that non-Indians may not. Permits were the means to allow special exemptions for Indians. They also assisted in gathering needed statistics. Pearse (1982) summarized the Indians' dissatisfaction with the permit system:

> "While permits confer fishing privileges that are not available to non-Indians, they have also been the government's means of curtailing Indian fishing. But many Indians feel that their traditional access to fish is their right and not merely a privilege to be meted out by the authorities as they see fit.... This is the root of much of the discontent and friction that have erupted in the field and spilled over into the courts. Under current policy, Indians view their access to fish to be vulnerable to changes in Indian fishery regulations and the Department's policies, to catches by other, larger resource users, to pollution and other habitat damage and to the Department's difficulties in managing the resources. Thus, the permit system offers the Indians no security for their claim on the resource."

One of the most controversial issues concerned the illegal sales of fish from the Indian food fisheries and whether the commercial sale of such fish should be legalized. This was identified as a key issue by Pearse and remained on the front-burner throughout the 1980's.

Pearse (1982) concluded that "Indian fisheries policy cried out for reform." As a basis for a new policy, he recommended:

1. The Department should allocate a specific quantity of fish annually to each Indian band involved in the Indian fishery.
2. The quantity and kind of fish to be allocated to each band should be determined through negotiations with the bands, primarily with reference to their catches in recent years but also taking into account special circumstances relating to population trends and economic opportunities.
3. The Department should give the catch allocated to Indian bands priority over the commercial and sports fisheries.
4. No royalties should be levied on fish harvested by Indians under the allocations proposed above.
5. Each band should be allowed to choose whether its entitlement to fish would be allocated through an Indian fishing agreement.
6. Indian fishing permits should be issued annually to individual fishermen through band councils. Permits should authorize Indians to take fish for

food and ceremonial purposes only. They should specify the quantity and composition of the authorized catch, and the location, time and method of fishing as required for management.

7. The Department should enter into Indian Fishery Agreements with Indian bands. These agreements should —
   (i) Carry terms of 10 years with provision for renewal 1 year before the term expires;
   (ii) Specify the band's allocation of fish;
   (iii) Authorize the band to harvest its allocation according to an annual fishing plan determined jointly by the band and the Department;
   (iv) Where appropriate, authorize the band to engage in enhancement activities on or near their reserves and to augment their allocated catch by a portion of the enhanced stocks, under fisheries management plans; and
   (v) Exempt the band from restrictions on the sale of fish under agreed monitoring and marketing arrangements.
8. Where they are willing and able to do so, band councils should be encouraged to take administrative responsibility for Indian fisheries, including apportioning the band's allocation among band members and issuing individual fishing permits under the umbrella of a general permit to the band.
9. The Minister of Fisheries and Oceans, the Minister of Indian and Northern Affairs and representatives of Indian organizations should attempt to reconcile band fishing by-laws with the paramount responsibility of the Department of Fisheries and Oceans for fish conservation and management.

Pearse's recommendations for the B.C. Indian fishery were generally enlightened and progressive. His proposals would have given Indian bands greater opportunities and at the same time placed greater responsibility on their shoulders to share in the management of the resource. These proposals offered hope of resolving the conflicting views of Indians and non-native users. However, these recommendations were not implemented.

The June 1984 federal Policy for the Pacific Salmon Fisheries recognized that native participation in the commercial Pacific salmon fishery was threatened by the general resource and financial difficulties faced by the fishing industry. The Policy promised a Native Fishermen's Economic Development Program to promote Native participation in the commercial fishery. This was implemented in July 1985 (see section 6.1).

The proposed Pacific Fisheries Policy stated that maintaining an equitable share of the salmon resource for the Indian food fisheries would remain an important priority:

> "Allocations for these fisheries will be established for the short-term by DFO in consultation with Indian communities. In the longer term, it is anticipated that allocations for the food fisheries will be determined in the context of comprehensive Native claim settlements."

Missing was any sense of Pearse's vision of changing the approach to management of the Indian food fisheries.

In a December, 1984, review of the status of the recommendations of the Pearse Report, DFO recognized that the Indian food fishery recommendations had not been implemented. This lack of action was linked to the comprehensive land claims issue. Basically, the pre-Pearse policy was still in effect.

Over the next 5 years the polarization between Indians and non-Indians on the food fishery issue increased. There was violence on the Fraser River in the summers of 1986 and 1987 and confrontation between DFO officers and Indians over enforcement of fisheries regulations. Non-Indian user groups formed the Pacific Fishermen's Defence Alliance to block attempts to legalize the sale of fish caught in the Indian food fishery and to seek court injunctions to overturn by-laws of the Gitksan Wet' suwet'en Tribal Council concerning fishing plans on the Skeena River.

The conflict intensified when the federal Ministers of Fisheries and Oceans and Indian Affairs and Northern Development released in March, 1986, the discussion paper "A Policy Proposal for a B.C. Indian Community Salmon Fishery" (Canada 1986). This was "intended to stimulate discussion on the development and management of an Indian community salmon fishery which would differ from an existing food fishery, conducted by status Indians under a licence, in that there would be an established catch limit and provision for sale."

The paper reiterated that "the current allocation policy of DFO places Indian food fisheries second only to conservation and before commercial and recreational fisheries. This policy will remain in force for Indian food fisheries and for those converted to Indian community fisheries, and will continue to be implemented in that from the expected returns of salmon the first commitment made will be those for spawning requirements, followed by Indian food fish/community fishery requirements." The two Ministers agreed that "the legitimate aspirations of native peoples must be addressed, within the context of sound management principles which recognize the paramountcy of conservation and the need for overall fishery management plans developed in consultation with all user groups under the provisions of the *Fisheries Act*."

The paper envisaged co-management of Indian community fisheries. To achieve this, (1) catch levels would have to be established and appropriate fishing plans developed at the band or tribal council level, in consultation with DFO; (2) these plans would have to be consolidated into a river management plan; and ultimately (3) river management plans would have to be integrated into the overall plan developed in consultation with all user groups.

These proposals reflected the intent, if not the detail, of Pearse's 1982 recommendation on the Indian food fishery. They advanced the concept of Indian community fisheries, involving co-management and legalization of the sale of fish caught in such fisheries. Non-native user groups vigorously opposed these measures. The B.C. government in 1986 secured an injunction on behalf of non-Indian users against four band fishing by-laws. Once the extent of opposition became clear, the proposal was shelved for the time being.

Meanwhile, court rulings were confirming aboriginal fishing rights and asserting that these had priority over fishing by other groups. Court judgments also indicated that these rights should be interpreted liberally and that the federal government can apply reasonable controls on native fishing only for conservation purposes. The government faced a dilemma given these court rulings and non-native groups' opposition to change. The situation was exacerbated by the lack of agreement at the First Ministers' Conference on Aboriginal Rights in March 1987. This led many natives to think that they must provoke the government to recognize their rights. DFO was confronted with protest fishing by Indians and noncompliance with fisheries regulations. (This was an harbinger of the national unrest among natives during the summer of 1990).

In July 1987, the fisheries component of a proposed Nishga land claims settlement was leaked to the public. This resulted in a very negative reaction from the fishing industry and derailed any attempts to resolve the conflict. The Indians blamed DFO for leaking the information and setting back the land claims negotiations.

By this time there were many links between Indian fishery management initiatives and the possible settlement of land claims in British Columbia. There were renewed efforts to move toward a negotiated settlement of land claims as well as Indian fishery issues. Recognizing that these discussions would take time, DFO in the spring of 1988 adopted a framework for management of the Indian food fishery. This provided interim guidelines for day-to-day management of the fishery, pending the clarification of native rights through legal, constitutional and land claim processes. It also provided for regulations prohibiting the sale and barter of food fish.

DFO also undertook more consultation with natives. Specifically, the new 12-member Pacific Regional Council (PARC) included three native representatives (see Chapter 15). Native representatives were appointed to the International Pacific Salmon Commission and its Panels. This paralleled existing arrangements on the east coast where natives had previously been appointed to the Atlantic Regional Council (ARC) and the Atlantic Salmon Board.

By the summer of 1988, British Columbia Indians were clearly dissatisfied with what they saw as lack of progress on their legitimate aspirations. August 20 became a day of protest on which Indians throughout British Columbia fished without a food fish permit to demonstrate their aboriginal right to fish. Richard Watts, acting vice chairman of the Aboriginal Fisheries Commission, stated that the province-wide protest fishery was to educate the public about aboriginal fishing rights, not to provoke confrontations with fisheries officials (*The Vancouver Sun*, August 3, 1988).

Meanwhile, the policy framework of the spring of 1988 continued to apply, pending the outcome of further discussion among these stakeholders. This national policy covered Indian food fisheries on the east coast as well as the west coast. There are approximately 35,000 Indians living on 110 reserves in Atlantic Canada and Québec in 66 bands. In 1987, 14 of these bands had authorized food fishery licences, one in Newfoundland, seven in New Brunswick, one in Nova Scotia and five in Québec. Other bands pushed for access to a food fishery licence for Atlantic salmon. Similar problems with band fishery by-laws were emerging. There was evidence of quota violations. In Newfoundland and the Maritimes, 63 cases had come before the courts during the previous 2 years. Enforcement was becoming increasingly difficult. Considerable pressure was also building for government to allow Natives access to existing commercial fisheries for other species.

The situation in the Arctic differs from that on the Atlantic and Pacific coasts because settlement of aboriginal land claims is assigning specific fishery management responsibilities to aboriginal peoples. Two claims have been settled in the area, the James Bay-Northern Québec Agreement with the Cree and Inuit of Northern Québec, and the Inuvialuit Final Agreement with the Inuit of the western Arctic. Agreement has also been reached on the Tungavik Federation of Nanuvut claim which covers the remaining coastal area in the Northwest Territories. This includes an offshore component, the first claim to do so.

The Inuit and the Cree covered by the first two agreements have greater access to fisheries and a greater voice in their management. Under the James Bay-Northern Québec Agreement, a Hunting, Fishing and Trapping Committee was established. Under the Inuvialiut Final Agreement, a Fisheries Joint Management

Committee was established. The native peoples appoint members to these committees which have significant responsibility for fisheries management. A Nunavut Wildlife Management Advisory Board was also established as a precursor to the settlement of the Tungavut Federation of Nunavut land claim. These committees all involve various degrees of cooperative management.

Under the James Bay-Northern Québec Agreement and the Inuvialuit Final Agreement, the Inuit and Cree have gained exclusive subsistence and commercial fishing rights to all species on some lands and exclusive rights to certain species on others. The James Bay Agreement guaranteed harvests of all species equal to the 1976 levels, subject to conservation requirements. The Inuvialuit and TFN rights are also limited only by conservation requirements.

Thus arrangements for cooperative management with native peoples in the Arctic were far advanced compared with the Indian food fishery on the west coast, in central Canada and, to a lesser extent, on the east coast. Pearse (1988) documented a progressive arrangement in Québec with the Micmac Indians of the Maria Reserve, who have traditionally depended on Atlantic salmon in the Cascapedia River. This has been a clear example of successful co-management. These examples show that cooperative management is possible.

In his 1988 review of the freshwater fisheries in Canada, sponsored by the Canadian Wildlife Federation, Peter Pearse again called for negotiated fishing agreements to address native fishing issues. His recommendations were very similar to those he made in 1982 for the Indian fishery in British Columbia.

Meanwhile, in May, 1990, the Supreme Court of Canada rendered a decision in *R. v. Sparrow*. This was the first attempt by the Supreme Court to explore the scope of s.35(1) of the *Constitution Act, 1982*. The Court decided that "existing aboriginal and treaty rights" means "unextinguished" rather than exercisable at a certain time in history. It concluded that these "must be interpreted flexibly so as to permit their evolution over time" and stated that "those rights are affirmed in a contemporary form rather than in their primeval simplicity and vigour."

The Court established a justificatory standard that must be met for any infringement of an aboriginal right to fish for food to be valid:

> "The constitutional entitlement embodied in s.35(1) requires the Crown to ensure that its regulations are in keeping with that priority. The objective....is not to undermine Parliament's ability and responsibility with respect to creating and administering overall conservation and management plans regarding the salmon fishing. The objective is rather to guarantee that those plans treat aboriginal peoples in a way ensuring that their rights are taken seriously."

The Court determined that the Musqueam Band had established an "existing right to fish for food and social and ceremonial purposes," and that the right "may be exercised in a contemporary manner." The Court did not try to establish the quantities of fish that might be associated with that right. It also did not deal with the issue of commercial rights to fish.

6.4 A New Aboriginal Fisheries Strategy

Even though the Supreme Court did not deal with the commercial rights to fish, it was becoming apparent that the sale of food fish, under the auspices of Indian community fishery agreements, was a necessary element of a new approach to the aboriginal fisheries issue. Unless the prohibition on the sale of fish was removed, relations with aboriginal groups would continue to be characterized by confrontation and conflict. The challenge was to put aside the differences of the past and negotiate workable fisheries arrangements with Canada's native peoples.

The years 1991 and 1992 represented a major turning point in the federal government's approach to aboriginal fisheries policy. In 1990, $1.1 million was spent on pilot cooperative management programs with native groups in British Columbia. In June 1991, Fisheries Minister Crosbie announced an expanded multi-year Aboriginal Cooperative Fisheries and Habitat Management Program (DFO 1991f).

Aboriginal people became much more involved in the design and delivery of projects to manage their fisheries, to collect information needed for effective management of fisheries and to increase fish production through habitat restoration and fish enhancement. One hundred and fifty agreements worth more than $11 million were signed by DFO with Aboriginal groups in the Pacific, Atlantic and Arctic regions. These programs created approximately 2,000 jobs for Aboriginal people in 1991 (DFO 1992g).

The Deputy Minister of Fisheries and Oceans, Mr. Bruce Rawson, held a major series of meetings with British Columbia commercial, recreational and Aboriginal representatives during 1991 and 1992. In addition to direct consultations and negotiations with many Aboriginal groups, DFO provided funding for the B.C. Fisheries Commission, a representative third-party group, to ensure that commercial and recreational fishing groups were informed and consulted and their interests protected (DFO 1992g).

On June 29, 1992, Fisheries Minister Crosbie announced a new Aboriginal Fisheries Strategy, as the basis for a new social contract among government,

Aboriginal people and non-Native fishing groups. The aim was to increase economic opportunities in Canadian fisheries for Aboriginal people while achieving "predictability, stability and enhanced profitability for all participants." (DFO 1992g).

The Strategy provided for negotiated agreements with First Nations, covering a spectrum of fisheries management activities, including:

- fixed, numerical harvest levels;
- enhanced self-management of Aboriginal fishing;
- demonstration projects to test the sale of fish caught by Natives;
- fish habitat improvement and fishery enhancement;
- research; and
- fisheries-related economic development and training.

This Aboriginal fisheries strategy was expected to cost approximately $140 million over the initial 7 years, about 70% of which would be spent in British Columbia. It was estimated that $73.5 million would be spent from 1992 to 1997 for fisheries-based economic development, on-the-job training and Native participation in fisheries management activities.

The government had concluded agreements on fish harvesting and cooperative management plans with three of the First Nations based on the Fraser River. The agreements established allocations of salmon to the Sto:lo Nation and Tribal Council, the Musqueam Nation and the Tsawwassen Nation.

These three First Nations, represented in a new Aboriginal fishing agency, called the Lower Fraser Fishing Authority, were allocated 395,000 sockeye salmon as part of the overall plan to manage the 1992 Fraser salmon harvest. Minister Crosbie indicated that these were the first of more than 30 such agreements that DFO hoped to conclude with B.C. First Nations. The intent was to negotiate communal harvesting plans for all rivers and all groups.

Demonstration projects to test the sale of fish were a major element of the new strategy. Interest in selling fish appeared to be concentrated in a limited number of watersheds and geographic areas where fish were abundant and Native bands had traditionally harvested large quantities of fish.

The sales would take place out of the total catch permitted under negotiated agreements. The agreements would specify numerical limits of fish, within the levels of historical catches and DFO-Native accommodations on harvests in recent years. The negotiated quantities would encompass all uses, including sale and food, social and ceremonial requirements. The demonstration projects would take place under 1-year agreements, subject to evaluation and review by third parties (DFO 1992g).

In the past, Native groups fished to plans under which their fishery was open for a period of time and they kept whatever fish they caught during the opening. This led to unpredictable numbers of fish being taken in some cases. Under the new agreements, the number of fish taken would be counted and the fishery closed when the limit was reached.

DFO would allow sales only within the context of written agreements. The sale of fish caught other than under commercial licence and these agreements would remain illegal.

Beyond negotiated baseline allocations, further increases in allocations to Aboriginal fisheries would be accomplished by voluntary buyout of existing commercial licences. Minister Crosbie announced a pilot program to assess options for voluntary licence retirements in the commercial salmon fishery. The program was designed to neutralize impacts that might result from the reallocation of salmon resources under the Aboriginal fisheries strategy.

To begin, $7 million was committed for the 1992 pilot program. The Minister indicated that the Department intended to ensure that licence retirements were structured in such a way as to maintain the salmon fishery's existing ratio of allocations between fleet sectors – gillnet, seine and troll — and to match as closely as practical licence retirements to those areas where agreements had been reached with Aboriginal groups (DFO 1992g).

This new Aboriginal fisheries strategy of June 1992 was a major departure from the Indian food fishery policy and practice of previous decades. It represented fundamental reform of the federal government's approach to the Aboriginal fisheries issue.

7.0 FISHERMEN AND THEIR ORGANIZATIONS

7.1 The Context

As can be seen from the preceding sections, the term "fisherman" (or more recently "fisher") is used generally to describe those who fish. That umbrella designation in Canada encompasses a wide variety of people who earn their living from fishing. As pointed out by Clement (1986):

> "The designation 'fisherman' is an omnibus term covering many class relations ranging from the captain of a giant trawler, who is part of a fish processor's management structure, to the crew of sixteen working under his command, to the 'independent' boat owner-operator who works alone. As an activity, fishing is heterogeneous".

Because of this heterogeneity, fishermen's attempts at organizing to represent their interests, whether at the bargaining table with management or lobbying governments for changes in policy, have often been fragmented. Hanson and Lamson (1984) observed:

> "Atlantic fishermen have never found a 'collective voice' nor have they successfully mobilized their potentially powerful political will to serve their collective interests. In part this can be explained by structural divisions within the industry."

Certainly, fishermen have not had a collective voice on either the Atlantic or Pacific coasts. Numerous groups have sprung up over the past century purporting to "represent fishermen". Numerous organizations have been born, struggled and often died due to limited support.

Fisheries Minister Romeo LeBlanc was popularly regarded as the "Minister of Fishermen". He urged fishermen in the 1970's to organize and associate. They would gain strength by speaking with a common voice. In an October, 1974 speech to the Atlantic Provinces Economic Council he said:

> "I intend that our fisheries officials will consult the fishermen to the utmost. But I would remind the fishermen we can't consult every single one of them. In a word: organize. Be sure your voice is heard, and be sure that your spokesmen are properly mandated and accountable to you" (DOE 1974c).

Hanson and Lamson (1984) described fishermen's organizations as having emerged in five major forms on the Atlantic coast:

1. Cooperatives often linked to a central cooperative responsible for processing and/or marketing;
2. Unions often affiliated with larger unions and ultimately with the Canadian Labour Congress (CLC);
3. Associations of independent businessmen established to influence government and fish processors on an organized and sustained basis;
4. Associations organized on a situational rather than a sustained basis; and
5. Confederations of existing organizations.

These five types can be consolidated nationally in three categories — unions, cooperatives and associations. There is considerable heterogeneity within these categories. Clement (1986) noted:

> "Associations, unions and cooperatives are not mutually exclusive forms, nor do they always have mutually exclusive objectives."

Fishermen's struggle to organize in Canada has been at times a difficult and perilous one, fraught with uncertainty. It has faced resistance from fishermen themselves and from corporations and governments. Wallace Clement chronicled this struggle in a fascinating 1986 book *The Struggle to Organize — Resistance in Canada's Fishery*. Much of what follows is based on his study.

Clement described the fishermen's organizations that have emerged in Canada as outcomes of economic, ideological and political struggles and as "organized resistance". This resistance is not confined to fighting the companies but includes struggling with the cooperatives, the state, and other fishing organizations. He described this resistance as "the struggle to fight back and attempt to shape the lives of those who directly produce fish for sale".

Clement traces the history of the struggle to organize on both the Atlantic and Pacific coasts. Space precludes me from dealing with this historical context here. Rather, I will make a few brief remarks on the earlier organizations and concentrate primarily on the major ones emerging since World War II, and particularly since 1970 on the Atlantic coast.

Fishing in Canada has become industrialized. This has dramatically changed the relations of production. Merchant capitalists of the saltfish trade have largely been replaced by industrial capitalists of fresh/frozen/canning production. Simple commodity producers have generally become industrial workers. However, the ideologies of those engaged in fishing, and labour laws governing fishing within various provinces, have not kept pace with these changes. For example, the concept of fishermen independence in the sense of "working for oneself" or "being your own boss" is still very prevalent in some sectors of the fishery.

Clement observed in an earlier (1984) paper:

> "When a fisherman says he stays fishing because he loves the 'independence' he means independence with respect to his immediate labour process (not his dependence on processors or banks). The work itself is long, hard, and dangerous (including long hours in repairs, maintenance and rigging — a boat being 'a hole in the sea into which you pour money'), but it also taxes the skills of fishermen: knowledge of fish, catching techniques, navigation, mechanics, electronics, etc. This does not apply to industrial fishermen aboard trawlers or offshore scallopers who are

> subjected to a detailed division of labour and have limited discretion, being heavily supervised and directed in their work."

Trawler crews have no ownership of the means of production and are directed by the representatives of management — the captain and his officers. They have none of the rights of property and all of the characteristics of labour. Despite this, for decades they were designated by law as "co-adventurers". This traditional view was historically one of three legal barriers to unionization for Canadian fishermen.

Steinberg (1973, 1974) identified these three barriers as follows:

1. The first was a 1947 Nova Scotia Supreme Court ruling, the so-called "Zwicker Decision", which designated fishermen as "joint-adventurers" rather than as "employees". The court invalidated a 1946 War Labour Board certification of Nova Scotia fishermen. Steinberg described the implications thus:

   "Zwicker was echoed and reechoed in subsequent fishermen's collective bargaining cases on both coasts. This formidable legal precedent not only excluded fishermen from certification under existing labour laws, but it also discouraged both the federal, and the several coastal provincial governments, from any fresh positive enactments until 1971."

2. The second legal problem was fishermen's liability under anticombines statutes. Fish prices are an important earnings variable and a prime bargainable item. Since fish is regarded as an "article of commerce", it was not exempt from the effect of anticombines law, as was labour. Collective bargaining by fishermen was therefore held by the courts to be in restraint of trade. The United Fishermen, Food and Allied Worker's Union (UFFAWU) in British Columbia secured voluntary collective bargaining arrangements with B.C. processors after a prolonged struggle but this resulted in innumerable legal snarls in the courts.

3. The third legal barrier to collective bargaining for fishermen was, as Steinberg puts it, "uniquely Canadian". For decades there was a continuing legal tussle over whether the federal or provincial governments had jurisdiction over the labour relations of fishermen.

The first legal obstacle was finally overcome when the share system on trawlers was recognized as "an old incentive system". Because of UFFAWU efforts to unionize fishermen in Nova Scotia in the late 1960's, National Sea Products extended "voluntary" recognition to the Canadian Brotherhood of Railway, Transport and General Workers Union (CBRT) in 1969. Steinberg (1974) observed:

> "The reputation of the UFFAWU in the West Coast industry as a militant, perhaps leftist-led union, and the prospects of possibly dealing with it, apparently fostered a reevaluation by the Nova Scotia fish processors of their traditional and strong antiunion sentiments."

The UFFAWU signed up fishermen in the Mulgrave-Canso area and pressed Booth and Acadia Fisheries for an agreement. This resulted in an 8-month strike, a federal-provincial Industrial Inquiry Commission, mediation attempts, a request to the Nova Scotia Supreme Court for a constitutional ruling and the eventual shutdown of Booth Fisheries and the move of much of Acadia Fisheries to Newfoundland. On March 18, 1971, Nova Scotia amended its Trade Union Act to allow unions of "shoresmen" to certify as bargaining agents. The Canadian Food and Allied Workers Union contested the UFFAWU jurisdiction and won certification from the Nova Scotia Labour Relations Board.

Real progress came with legislation in Newfoundland during 1971–73 which permitted the unionization of independent fishermen as well as trawlermen. In 1982, New Brunswick also passed legislation permitting collective bargaining by fishermen. In British Columbia, the UFFAWU arrangements are still "voluntary", even though the UFFAWU has been the most aggressive fishermen's union in the country.

Clement (1986) described the Canadian fishing industry in terms of three social categories. These categories are:

1. *Small-Scale* (one to three people) — Small-scale fishermen work alone or in de facto partnerships, occasionally employing a helper. On the Atlantic, these include inshore fishermen who catch lobster, cod by line or trap, herring by gillnet or weir. They usually fish daily from their home port and in Québec are referred to as "la flotte artisanale". In British Columbia, almost all gillnetters and trollers fit this category.
2. *Intermediate Scale* (four to 10 people) — These include purse and Danish seiners who fish herring on both coasts, salmon on the west coast and some groundfish and mackerel on the Atlantic. Halibut longliners on both coasts are included as are shrimpers, scallopers, crab boats and small draggers on the east coast. These vessels range

from 46 to 75 feet long, with most under 65 feet. They include a mixture of company-owned and fishermen-owned vessels. Quite often processors have shares in these vessels. In Québec these are called "la flotte côtière."

3. *Large Scale* (more than 10 people) — On the Atlantic, the most obvious examples are the large groundfish trawlers and scallop draggers. Nearly all are company-owned vessels which fish year-round. They average about 150 feet long with crews of approximately 18. The length of trips tends to range from 10 to 20 days. On the west coast, this category includes some vessels used for groundfish and halibut. In Québec they are known as "la flotte hauturière."

Small-scale producers have a low division of labour and control their immediate labour process. Intermediate-scale producers have a clear division of labour between the captain and the crew but low division of labour within the crew. Large-scale producers have a clear hierarchical division of labour and clearly differentiated roles involving captain, mate, engineer, bosun, cook, trawlerman, deckhand.

Clement observed:

> "Unionization has tended to be strongest among trawlermen and small-scale fishermen (gill-netters on the West Coast and inshore fishermen on the East Coast) who face industrial processors as markets. Small-scale fishermen whose markets are more diverse, such as trollers, have not been as forceful in drives to unionize, although political and cultural/ideological factors have at times made them union supporters. Medium-scale fishermen, especially seiners (South-west Seiners and also Association Professionelle des Pêcheurs du Nord-Est), have been more successful financially, closer to the skipper and less forceful than other sectors for unionization. Political and cultural/ideological factors on the West Coast and in Newfoundland, however, illustrate these barriers can be overcome."

Against this background, I will now describe the major organizations within each of the three categories identified previously — unions, cooperatives and associations.

7.2 Plant Worker Unions

Most large-scale fish processing plants can be considered industrial factories. There are also many small, seasonal plants, whose workers are unorganized. Seasonal plant workers are generally drawn from a reserve labour force of women who engage in "domestic labour" for the rest of the year.

On the West Coast, the United Fishermen Food and Allied Workers Union was formed in 1945 by a merger between the Fish Cannery, Reduction Plant and Allied Workers Union and the United Fishermen's Federal Union. The shore workers section of the UFFAWU has legal bargaining rights. The fishermens' section has used this as leverage to force processors to negotiate "voluntary" agreements on minimum prices and working conditions through the Fisheries Association of British Columbia (FCBC). The UFFAWU represented shore workers at the Prince Rupert Fishermen's Cooperative until 1968 when they were decertified following a fiercely contested strike. Shore workers there were subsequently represented by the Prince Rupert Amalgamated Shoreworkers' and Clerks' Union.

On the Atlantic Coast, the Canadian Seafood Worker's Union (CSW) represents workers in various plants in Nova Scotia and New Brunswick. The CSW was formed in 1957 through a merger of the Canadian Fish Handlers and the United Fisheries Workers of Canada. It represents about 4,500 plant workers in the summer and 2,500 year-round. In the 1980's, the CSW was threatened by the NFFAWU, which made in-roads into Petit-de-Grat and Canso. Predecessors of the CSW were involved in actions such as the 1952 organization of the National Sea plant at Lockeport and the CSW had a 1970 strike by 1200 members against five plants of National Sea. However, Clement notes that "Recently, membership has been diminished by the weak economy, and the Union has stagnated to the point where its autonomous future is threatened."

In Newfoundland, the Canadian Fishermen and Allied Workers Union (CFAWU) was granted jurisdiction over fish plant workers in 1967 by the CLC and absorbed six locals which had previously been direct charters of the CLC. In 1970, the CFAWU merged with the newly formed Northern Fishermen's Union to form the Newfoundland Fishermen, Food and Allied Workers Union. The NFFAWU soon represented all of the province's major fish plants, with two locals, one for plant workers and one for fishermen.

7.3 Fishermen's Unions

There are currently five major fishermen's unions in Canada: the UFAWU in British Columbia, the Newfoundland Fishermen's Union (formerly the NFAWU), the CBRT representing trawlermen in Nova Scotia, the Cooperative Fishermen's Guild in British Columbia, and the Maritime Fishermens' Union (MFU) representing small-scale fishermen in New Brunswick and southwest Nova Scotia.

7.3.1 UFAWU (United Fishermen and Allied Workers' Union)

The 8,000 – member UFAWU represents most gillnetter and seine crews in British Columbia, as well as shore workers. It negotiates share agreements and working conditions with intermediate and large-scale vessel owners through the Vessel Owners' Association. It negotiates annual price agreements for salmon and herring. While some trollers are members of the UFAWU, in general trollers belong to the Pacific Trollers' Association (PTA), which has long opposed the UFAWU.

The UFAWU restricts its membership to labour. Anyone who regularly employs more than two people is ineligible. The UFAWU has been a militant and aggressive force on behalf of fishermen and plant workers. It has led numerous strikes, fought other fishermen's organizations in British Columbia, and been accused of being communist-led. During the 1950's and 1960's, the UFAWU was attacked from many quarters. Legal action under the Combines Investigation Act was launched against it in 1952, 1955, 1956, 1958 and 1960. As the result of a bitter strike in 1967, the union lost part of its membership. Two leaders, President Steve Stavenes and Secretary Homer Stevens, were jailed for a year for contempt of court.

The UFAWU tried in the late 1960's to organize fishermen in parts of Nova Scotia. This attempt failed but laid the groundwork for successful organizational drives by other unions. Clement described the UFAWU as:

> "The most powerful force in the West Coast fishery. All other organizations, including the cooperative and various associations, define their place within the industry in terms of the UFAWU. Its militant practices and constant political/educational work have it a force with which to reckon."

7.3.2 Newfoundland Fishermen's Union

By the mid-1970's, the most comprehensive fishermen's union in Canada had emerged as the Newfoundland Fishermen, Food and Allied Workers Union (NFFAWU). This union was formed from a merger of the Northern Fishermen's Union and the Canadian Food and Allied Workers Union in April 1971. Within a decade it had become one of the two most powerful fishermen's unions in Canada, the other being the UFAWU.

It may seem surprising that such a comprehensive union emerged in an area where fishermen were scattered among hundreds of isolated communities and earning only a marginal existence from the fishery. However, it must be remembered that Newfoundland had earlier been the scene of an innovative and, for a time, powerful organization of fishermen. Led by William Coaker, the Fishermen's Protective Union (FPU), was formed in 1908 in Herring Neck on Twillingate Island on the northeast coast of Newfoundland to fight the stifling credit system and the St. John's elite. This populist movement became a political party which was a dominant force from 1910 to the mid-1920's. It formed the opposition party in 1913 and was a junior partner in a coalition government in 1919.

While sealers and the fishermen who participated in the Labrador floater fishery formed the base of the FPU, it grew to include loggers, farmers and other labour groups. Coaker attempted to reform the export trade by establishing marketing boards and cooperatives to import supplies wholesale. The intent was to free the people from the iron grip of the merchants. By 1914, the FPU had 20,000 members. Coaker also advocated such social measures as compulsory education, local hospitals and universal pensions. Opposed by the St. John's merchants and the Catholic Church because its strength lay on the Protestant northeast coast, the FPU was undermined by wartime conditions and diminished in influence in the mid-1920's. For more about the rise and fall of the FPU, see Panting (1963), McDonald (1974), and Brym and Neis (1978).

Soon after Newfoundland entered Confederation with Canada in 1949, the government passed a Labour Relations Act which prevented collective bargaining for fishermen. Until 1970, the Labour Relations Board refused to recognize any group of fishermen as employees. From 1951 onward, fishermen were purportedly represented by the Newfoundland Federation of Fishermen, a weak-kneed organization which did little to improve their lot.

In November 1969, Father Desmond McGrath assisted a group of Port au Choix longline fishermen to band together. Father McGrath enlisted the help of a former classmate at St. Xavier University, Richard Cashin, to draft a constitution. In May 1970, the Northern Fishermens' Union was formed. The constitution allowed membership both by fishermen and plant workers. Later that year McGrath and Cashin met with representatives of the Canadian Food and Allied Workers' Union which represented workers at eight Newfoundland fish plants. The CFAWU was an international affiliate of the Amalgamated Meat Cutters and Butcher Workmen of North America.

These discussion resulted in formation of the NFFAWU in April 1971 as a division of the CFAWU. The approach was comprehensive. From the beginning the following were eligible for membership in the Fishermen's Section of the Union:

1. The owner or part-owner of a vessel actively pursuing the fishery;
2. Any person who fishes with others for shares of the catch or wages; and
3. Any person below the rank of officer on a trawler, dragger or seiner. (This was later amended to include officers below the rank of captain).

The NFFAWU enjoyed remarkable success. A history of the union up to the early 1980's can be found in *More Than Just a Union — The Story of the NFFAWU* by Gordon Inglis (1985). By the mid-1980's, the NFFAWU represented almost all 10,000 full-time inshore fishermen, 12,000 plant workers and 1,050 trawler crew in Newfoundland plus some shore workers, trawler crew and scallop draggers crew in Nova Scotia. On behalf of inshore fishermen, the NFFAWU negotiated price agreements for various species and sizes of fish with the Fisheries Association of Newfoundland and Labrador (FANL). After 1975 a combination of price of fish and a per diem were negotiated for trawler crews.

Two events aided the NFFAWU's success in its early period, the Burgeo strike of 1971 and the trawler crews' strike of 1974–75. The Burgeo strike broke the paternalistic hold of one of Newfoundland's fish merchants, Spencer Lake, and gave the Union momentum to organize successfully across the province. The 1974–75 trawlers' strike tied up all plant workers and fishermen in the province in July for 6 weeks. A Board of Inquiry was established, headed by Leslie Harris of Memorial University.

Harris recommended that "Trawlermen should negotiate with companies not the price of fish but rather the income level that will be attainable for full-time work." He also recommended an increase in the level, as well as a change in the method, of payment. At first the companies rejected the recommendations. Trawlermen refused to sail. In March, 1975, a settlement was reached which abolished the co-adventurer system but provided less generous wage increases than those Harris had proposed. The crews won a per diem of $20 a day plus a payment based upon the amount of fish caught. For details of this strike and its impact, see MacDonald (1980).

In the early 1980's, the NFFAWU made in-roads in Nova Scotia. By 1983, it represented a significant portion of the crew on offshore trawlers and scallopers, particularly at Canso and Riverport. In Newfoundland, bargaining became more difficult during these years of crisis. Up to 1980 the Union had focused primarily on the incomes of crews and plant workers. A province-wide disruption in Newfoundland in 1980 was partly a strike, partly a lock-out. The Union won from the province a workers' compensation scheme for fishermen with the fish buyer designated as the employer.

The Union also had a significant impact on fisheries policy when it advocated priority for full-time fishermen to receive limited entry licences. It got involved in over-the-side sales to provide an alternative market for areas and species where the onshore market was weak.

When asked to enter a Social Compact during the restructuring crisis of 1983, the Union refused, recognizing that the proposal would probably depress wages to workers and prices to fishermen. In October 1984, the Union struck against the restructured Fisheries Products International. Following a 6-month strike, the Union made some gains in pensions, wage levels and shore leave schedules and successfully resisted attempts to take away rights it had achieved prior to restructuring. The name of the union was changed to the United Food and Commercial Workers Fishermen's Union in 1984.

Several commentators have speculated on the reasons for the Newfoundland Union's success. I believe the major factor was the leadership of Richard Cashin and Father McGrath. Substantial funding from the international union obviously helped a great deal in the formative stages. The charismatic leadership of Cashin brought the union into conflict with its national and international leadership in 1987. Cashin withdrew members from the existing structure and reconstituted under another name, Fishermen, Food and Allied workers, in affiliation with the Canadian Auto Workers. Under Bob White, the Auto Workers had broken away from their U.S. parent only a year or two previously. A bitter struggle ensued with the newly constituted Newfoundland Fishermen's Union having to fight for certification, group by group, plant by plant. But ultimately it prevailed, again largely due to Cashin and McGrath and their track record. Under a new label it remains one of Canada's two most powerful fishermen's unions.

7.3.3 Canadian Brotherhood of Railway, Transport and General Workers (CBRT)

This union has its major activities outside fishing. Following the attempt by the UFAWU to organize companies in the Canso Strait area, CBRT was voluntarily recognized for some trawler crew and scallopers by National Sea Products, Riverside Seafoods, and Nickerson's. By 1971, it represented the bulk of trawler crews and some scallopers in Nova Scotia. In 1982 and 1983 it lost some of its membership to the NFFAWU, with the NFFAWU gaining most of the scallopers and the CBRT holding a substantial proportion of the trawler crews. Generally, the CBRT has not been as militant nor has it had the impact of the UFAWU or the Newfoundland Fishermen's Union.

7.3.4 The Maritime Fishermen's Union (MFU)

Unionization of small-scale, inshore fishermen in the Maritime provinces has proven a difficult task. While there is a multiplicity of fishermen's organizations throughout the Maritimes, there is only one organization which functions as a union for inshore fishermen — the MFU. The MFU was established in Baie Ste. Anne, New Brunswick, in 1977. Its original membership was primarily inshore Acadian fishermen. It has waged a long and difficult struggle for collective bargaining rights for inshore fishermen throughout the Maritimes, achieving some success in New Brunswick in 1982. It has a union ideology, viewing fishermen as labourers. Following passage of the collective bargaining legislation in New Brunswick, the MFU was certified in 1984 to represent fishermen in two areas: the Caraquet Peninsula and the southeast. It has not yet obtained bargaining rights for fishermen in Prince Edward Island or Nova Scotia. It has some support on the west end of Prince Edward Island and in Cape Breton but initially met considerable resistance in traditionally anti-union southwest Nova Scotia. More recently, it has concentrated on organizing small-boat fishermen in southwest Nova Scotia, with greater success. It became involved in over-the-side sales for herring gillnetters in the Bay of Fundy and used this as leverage to gain some momentum. But, as a union, its collective bargaining function is confined to the Gulf shore of New Brunswick.

The MFU has obtained moral, and some financial, support from the Newfoundland Union which has not attempted to organize inshore fishermen in the Maritimes.

The MFU has not tried to organize the crews on intermediate size vessels, who on the Gulf shore are represented by a very strong fishermen's association, the Association Professionelle des Pêcheuries du Nord-Est (APPNE).

Many fishermen's associations in the Maritimes have strongly resisted the MFU's union approach.

7.3.5 Co-operative Fishermen's Guild (CFG)

The CFG grew out of the controversial 1967 UFAWU strike. It represents crew members on intermediate and large-scale vessels fishing for the Prince Rupert Fishermen's Co-operative. The CFG does not negotiate prices since these are set by the cooperative structure. Its members also belong to the Co-op. It was really set up to allow the Prince Rupert Co-op to continue fishing during UFAWU-led strikes.

7.4 Fishermen's Cooperatives

Fishermen who belong to cooperatives have a different view of the world from union members. They see themselves as outside private corporate relationships. The fishermen cooperatives in Canada tended to be formed for cooperation in selling and marketing rather than in fishing. Cooperatives compete with private processors for markets. Fishermen continue to own their vessels and fish individually.

There were three major cooperatives in Canada's fishing industry: the Prince Rupert Fishermen's Cooperative, the United Maritime Fishermen, and Les Pêcheurs-Unis du Québec. There are now only two, Pêcheurs-Unis having disappeared in the 1984 restructuring of the Atlantic fishery.

7.4.1 The United Maritime Fishermen's Cooperative (UMF)

The UMF was founded in Halifax in 1930. It was organized by Moses Coady of St. Francis Xavier University. UMF started as an educational body of regional associations. In 1934, the central structure began to market fish. There was a clear division of functions from 1934 to 1961. Fishermen fished, local cooperatives processed the fish and UMF (the central cooperative) marketed the fish and provided supplies. During this period the local co-ops were community based. In 1961, UMF central broadened to include production and direct ownership, ultimately owning six facilities. Starting in 1979, UMF accepted individual direct members from areas without local co-ops.

The UMF represents primarily owners of small-scale vessels but also some crew members. There are 29 member co-ops, each with different membership criteria. Through these member co-ops the UMF represents 2,000 fishermen; it has an additional 500 direct members of the central co-op. The UMF's areas of operation overlap significantly those of the MFU. The UMF itself opposed the extension of bargaining rights to the MFU, seeking exemptions for co-op fishermen. However, the MFU appears to be winning the struggle since many of the UMF fishermen are now MFU members.

The UMF had financial problems in the early 1980's and received assistance from the federal government in the form of $5.6 million in preferred shares and the "forgiving" of $2.5 million in interest payments on loans.

7.4.2 Les Pêcheurs Unis du Québec (PUQ)

PUQ was formed in 1939 as a federation of 31 co-ops in Québec. At that time it had eight branches and 1,500 members. This increased to 35 branches and 3,000 members in 1944. By the early 1980's PUQ had about 600 fishermen members, and owned six plants. Seven other plants were affiliated with the PUQ for marketing. The majority of the membership consisted of small-scale fishermen, but intermediate-sized vessels caught most of the fish.

During the industry crisis of the early 1980's, PUQ was plunged into financial crisis. It disappeared as a result of restructuring, with the provincial and federal governments gaining ownership of various groups of plants. The federal government formed Pêcheries Cartier as a government-financed company with some participation from the banks. This was subsequently privatized in 1985.

7.4.3 The Prince Rupert Fishermen's Co-operative (PRFC)

The PRFC was also founded in 1939 assisted by St. Francis Xavier University. At first it only marketed fish caught by trollers, but in 1944 it became involved in processing. Unlike the UMF, the PRFC is a direct co-operative with 1450 members. Members must sell all fish through the co-op which generally does not accept fish from non-members (especially during strikes involving the UFAWU).

Initially, the membership consisted solely of trollers. Later some halibut and gillnet fishermen joined, and more recently some dragger fishermen have become members. By the mid-1980's the Co-op represented 25 seiners, 175 trollers, 134 gillnetters and 138 combination vessels (gillnet and troll gear). The PRFC has become the second largest processor in British Columbia. It has its own unions, the CFG and the PRASWGU.

Boyd (1968) observed that only certain types of fishermen are attracted to the PRFC:

> "To be a member, a fisherman must be financially independent or be able to obtain financing independently. He must be able to invest share capital in the cooperative. These requirements tend to restrict membership to the better fisherman who is in an independent financial position."

Over the years the Prince Rupert Co-op has had a stormy relationship with the major fishermen's union, the UFAWU. The story of the PRFC is told by A.V. Hill (1967) in the *Tides of Change: A Story of Fishermen's Co-operatives in British Columbia.*

7.4.4 Other Cooperatives

Besides these three, there were other smaller cooperatives. On the west coast, there was the Central Native Fishermen's Co-operative (CNFC), formed in 1975 and affiliated with the Native Brotherhood. The CNFC included boat owners, crews, shore workers, and administrative personnel. Its core consisted of 40 seiners (about 200 members), 40 trollers and 10 gillnetters, primarily crewed by Natives. It had financial trouble beginning in 1981 and ceased operations in 1984.

On the Atlantic coast, there are numerous other small cooperatives. These include the Atlantic Herring Fishermen's Marketing Co-operative (AHMC) and the Southwest Seiners Association, which represent the owners of the herring seiner fleet in southwest Nova Scotia-Bay of Fundy. Their formation was linked to the restructuring of the Bay of Fundy herring fishery in the late 1970's. Kearney (1984) provides a history of the transformation of this fishery and the role of these groups in it.

Perhaps the best-known of the smaller cooperatives on the Atlantic coast is the Fogo Island Cooperative in northeast Newfoundland. It increased in membership from 127 in 1967 to 1500 in 1983, including 750 fishermen, 400 plant workers and 350 members-at-large. Its story captured the imagination of the National Film Board and has recently been described by Carter (1984). The Newfoundland Fishermen's Union has not attempted to prevail upon the Co-op's turf.

7.5 Fishermen's Associations

Associations are the third major category of fishermen's organizations. These tend to oppose the unionization of fishermen. Their main function is to lobby governments on fisheries policy.

There is a multiplicity of individual associations, many representing gear or types of vessels, and several groupings of associations into federations. Such associations are not generally involved in labour-management issues but rather concentrate on presenting fishermen's views to government. They do this through membership on advisory committees, and direct representations to fisheries officials, Members of Parliament and the Minister of Fisheries and Oceans.

Individual associations have existed on the east coast for decades. In 1979, an attempt was made to bring together many of these under an umbrella organization, the Eastern Fishermen's Federation (EFF). The EFF was formed with government encouragement and funding through a squid allocation of 2,000 tons, worth $1 million in the squid boom at that time. It started with six fishermen's associations, increased to 13 in 1980 and reached a high of around 25 in 1982. During the 1980's the core support came from the Prince Edward Island Fishermen's Association and associations in Nova Scotia.

Theoretically, any fishermen's association with 100 or more members could join. In practice the associations which joined represented primarily small — and intermediate — scale vessel owners. For a while it included one crew association (Scotia-Fundy Seiners), several associations representing both captains and crew (Nova Scotia Fishermen's Association and the APPNE), and a large cooperative (UMF).

While the EFF's primary function was lobbying, over time and in competition with the MFU it began some marketing of fish and over-the-side sales. By the mid 1980's, the EFF had lost several of its most powerful associations. The Atlantic Herring Fishermen's Marketing Cooperative and the South-West Seiners withdrew in 1982; UMF withdrew in 1983, and in 1984 the APPNE left the EFF. This considerably weakened the EFF's effectiveness. The residual membership consists primarily of small vessel owners in Prince Edward Island and Nova Scotia and on the Fundy shore of New Brunswick.

Apart from the Herring Seiner Associations which have been influential lobbyist groups for seiner owners, two other individual associations merit some comment. One of the strongest associations in the Maritimes is the APPNE, headed for years by lawyer Gastien Godin. It is an association of captains and crew of intermediate-sized vessels based on the Gulf (Acadian) shore of New Brunswick. Virtually all vessels are fishermen-owned. These vessels fish for shrimp (100 members), crab (350 members), herring (100 members) and groundfish (150 members). Altogether, members fish from a mobile fleet of around 130 vessels. Members of the APPNE participate in lucrative fisheries such as shrimp and crab as well as groundfish. Although not a union, they have sometimes forced prices up. They have been a major lobbying force.

Another relatively strong and well-organized association is the Fundy Weir Fishermens' Association, formed in 1973. Its original purpose was to lobby government to save the weir operations when stocks were declining and markets poor. In 1978, the Association hired a former Area Manager of Fisheries and Oceans, Walter Kozak as Executive Director. The Association for a time experimented, with government assistance, with a Catch Insurance Scheme which ultimately collapsed. It has negotiated prices with the major buyer in the area, Connors Brothers. In many respects this association has functioned like a union.

In British Columbia, there are also individual associations focused on gear, area or species groupings. The most notable are the Pacific Trollers' Association (PTA), the Gulf Trollers Association, and the Pacific Gillnetters Association. These groups represent owners of small and intermediate-size vessels who have traditionally opposed the UFAWU and, in particular, its use of strikes. They are influential voices for the groups they represent in lobbying on fisheries management policies, fish allocations and related matters.

The EFF was instrumental in organizing a national convention of fishermen's associations in 1981. Out of this came the Western Fishermen's Federation (WFF) led by the Pacific Trollers Association, in January 1982. Other members included the Native Brotherhood, the Gulf Trollers' Association, the Pacific Coast Salmon Seiners' Association, the Deep Sea Trawlers' Association, the Fishing Vessel Owners' Association, and the Pacific Gillnetters' Association. The WFF was led by John Sanderson of the Pacific Trollers' Association.

As Clement (1986) put it, "the Federation never really got off the ground." The UFAWU would not join; the seiner groups withdrew. In November 1983, the EFF and WFF met in Ottawa to discuss the possibility of a national organization for lobbying. Shortly after, the EFF lost several of its key members, as did the WFF. In February 1984, the Fishing Vessels Owners' Association withdrew from the WFF which declined thereafter.

Clement (1984) explained the failures of the EFF and the WFF:

> "Federations of fishermen's associations are inherently unstable because they can seldom 'speak with one voice' and unanimity reduces the scope of issues they can address. Membership loyalty tends to be low, except around specific issues. Each association tends to have its particularistic concerns and they can only be unified through opposition, such as the EFF's struggle with the MFU or WFF's with UFAWU, not by solidarity.... They tend to be initiated by outside problems rather than internal unity. Generally they struggle to maintain the status quo rather than see their vested interests eroded since there is little common basis for directing social change. Whereas unions tend to be inclusive of various factions, sectors, and regions, associations thrive on area, gear, and species divisions."

Despite this, some individual associations have played influential roles in shaping government fisheries policy.

7.6 The Native Brotherhood

One organization influential in the B.C. fishery, which does not fit into any of the previous three categories is the Native Brotherhood. It is an umbrella organization for various categories of activity. It acts as a union when it negotiates alongside the UFAWU with the Fisheries Council of British Columbia. It also functions as an association, for example, it was a member of the WFF. Members of its executive led the Central Native Fishermen's Cooperative. It acted as the catalyst to purchase B.C. Packers Northern Fleet, through the North-West Tribal Council Corporation, to supply vessels to native fishermen.

The Brotherhood was formed in 1931 as a fraternal association of coastal Indian fishermen. Later Natives from the Interior of British Columbia joined, and until 1969, it represented all British Columbia Native people. At that time the Brotherhood decided to confine itself to fishery-related matters. The membership is diverse, including Native people who are intermediate-scale vessel owners (mainly seiners), crew on these vessels, and small-scale vessel skippers (primarily gillnetters). The leadership has come primarily from seiner captains. While the Brotherhood has included some shore workers, its influence in this sector has been marginal. The UFAWU has been the dominant force in representing shore workers.

7.7 General

It is clear from this review that fishermen in Canada are heterogeneous and so are the organizations which represent them. There is a wide diversity of structure, practice and ideology which make generalizations about the interests and views of Canadian fishermen meaningless. The lack of a collective voice on either the east or the west coasts has sometimes weakened fishermen's representations in the corridors of power. On the other hand, organizations such as the Fishermen's Union in Newfoundland and the United Fishermen and Allied Workers' Union in British Columbia have brought together diverse interest groups. They wield sufficient power to change significantly the wages and working conditions of fishermen and plant workers.

Cooperatives generally are not a major force in the fishery. One notable exception is the Prince Rupert Fishermen's Co-operative which wields considerable influence in British Columbia. The United Maritime Fishermen, which along with Les Pêcheurs Unis du Québec, arose from cooperative idealism of the 1930's, has declined in influence in recent years. Locally-based associations can be effective in lobbying governments for policy changes.

Given the complexity and diversity of the fishing industry, it is not surprising that a complex web of fishermen's organizations has evolved. The influence of these organizations is similarly complex. The lack of a collective voice does not mean that fishermen are voiceless. Far from it! Over the past quarter century there have been dramatic advances in the power and influence of fishermen in shaping their way of life and government policies which impact on that way of life.

8.0 CONCLUSIONS

For centuries fishermen in hundreds of coastal communities have earned their living from the sea. Modern technology has changed the ways in which the fishery is conducted. Still, in most of these communities fishermen battle with the elements to eke out a living for themselves and their families, buffeted by the vagaries of nature and a volatile resource. Many of these small, scattered and isolated coastal communities are almost totally dependent upon the fishery with little in the way of alternative employment opportunities. This dependence upon the fishery is not restricted to small, inshore-fishery based communities but also includes larger, trawler-supplied towns where major fish plants are threatened with closure after decades at the heart of the deep-sea fishery.

The majority of fishermen who live in these fishery-dependent communities earn inadequate incomes from fishing. There is a wide disparity in the incomes of the poorest and the wealthiest fishermen. There is also considerable geographic diversity in incomes earned, depending upon the species fished and whether the fishery is year-round or seasonal. However, most fishermen on the Atlantic coast live close to or below the poverty line. Income inadequacy, not income instability, is their major problem. Since 1957, the Special Fishermen's Benefits Program of Unemployment Insurance has evolved into the major means of income supplementation for these fishermen. It contributes as much as 40% to the total incomes of fishermen in some areas of the country. Fishermen, plant workers and the communities in which they reside became dependent upon the various categories of Unemployment Insurance to supplement inadequate incomes.

Numerous studies in the 1970's and 1980's identified deficiencies in the Unemployment Insurance program, particularly the Benefits for Self-Employed Fishermen. Various alternative basic income security programs were proposed. None of these alternatives was found politically or socially acceptable. Meanwhile, fishermen and plant workers continued to depend upon Unemployment Insurance to supplement inadequate incomes.

The Task Force on Incomes and Adjustment was tasked by the federal government in March, 1992, with devising a comprehensive long-term strategy to address the problem of income inadequacy in many areas of the Atlantic fishery and developing specific proposals for adjustment. Whether this Task Force will succeed where others have failed remains to be seen. A major downturn in the abundance of Atlantic groundfish resources, coupled with a continuing overcapacity problem, at the beginning of the 1990's heightened the need for a comprehensive but achievable adjustment strategy.

Canada's Native people have a traditional dependence upon the fishery for food and other societal purposes, predating the arrival of the white man in North America. This relationship to the fishery is particularly evident in British Columbia where band reserves are often adjacent to major salmon rivers. Over the past century there was conflict between the traditional dependence of Native peoples on fish and the development of the industrial fishery of the twentieth century. In the 1970's and 1980's this conflict was manifested in confrontation between Native peoples, other user groups and government authorities.

Recent court decisions confirmed aboriginal rights to fish, subject to conservation requirements. Traditional fisheries policy placed Natives' needs second only to conservation.

Proposals to provide for Indian community fisheries and the sale of fish taken in these fisheries were strongly resisted by other user groups for many years. Progress toward a more realistic Indian fishery policy was thwarted by the often-blanket resistance of commercial and sports fishing groups. The new Aboriginal fisheries strategy of June, 1992, was a major step forward in addressing the legitimate fisheries aspirations of Canada's Aboriginal peoples. Demonstration projects to test the sale of fish taken in Aboriginal fisheries represented a radical departure from previous policies.

The mythical image of a fisherman is that of a rugged individualist who has little "truck or trade" with his fellow fishermen. Common-property mythology suggests that they are all out there on the fishing grounds competing in a mad rush for the fish, with little thought for cooperation. Although fishermen have a strong, rugged, individualistic streak, they have banded together to achieve mutually beneficial goals. While Canadian fishermen as a whole lack a collective voice, this is not surprising given the complexity and wide geographical diversity of Canadian fisheries. Despite this, some strong and effective organizations of fishermen have emerged. Prominent among these are the United Fishermen and Allied Workers' Union in British Columbia and the Fishermen's Union in Newfoundland. These unions have achieved significant changes in wages and working conditions, benefiting fishermen and plant workers.

It has long been perceived that social factors have dominated fisheries policy in Canada over the past century. Numerous economists over recent decades have called for fisheries rationalization, meaning reductions in the numbers of fishermen and, in some instances, onshore processing capacity. Michael Kirby in his 1982 Task Force Report recognized this social dimension but stated that the first objective of government policy should be economic viability. He saw the 1983 restructuring debate with Newfoundland as a disagreement over objectives:

> "A debate over the 'economic vs the social-fishery'....was at the root of the disagreement between the federal and provincial governments" (Kirby 1984).

This categorization of the positions of two governments is somewhat simplistic. In the end, little "rationalization" resulted from the 1983–84 restructuring. Several smaller companies were merged into two larger ones but there was virtually no reduction in harvesting or processing capacity.

During the boom of the mid-1980's, onshore processing capacity again increased significantly. As the industry confronted another Atlantic crisis in the early 1990's, there was insufficient fish to feed expanded harvesting and processing capacity. It appeared that some plants would close. Reduction of harvesting and processing capacity will mean some loss of jobs. However, this is necessary if the industry is to become economically viable.

Despite these factors, the social dynamics of the fishery will continue to play a major role in shaping fisheries policy as we enter the next century. The notion of the fishery as the employer of last resort remains a major obstacle to meaningful change. Some rationalization of overcapacity is needed. This will, however, mean fewer jobs in isolated coastal communities. The consequent dislocation and social adjustment will be traumatic. While some rationalization is needed, this cannot ignore the human cost of disrupting a centuries-old way of life.

CHAPTER 15
RECONCILING CONFLICTING INTERESTS

"Conflict is the name of the game."

– G. Brewer, 1983

1.0 INTRODUCTION

Fisheries are characterized by conflict. From angry fishermen picketing fisheries offices to the burning of fisheries patrol boats, conflict often runs rampant. These are examples of conflict directed at the management agency and its representatives. But conflict in the fishery runs deeper. Group is often pitted against group in the struggle for their "rightful" share of the resource. Values and beliefs on how the resource should be managed and shared frequently clash.

2.0 THE NATURE OF CONFLICT

For the first few decades of this century, conflict was viewed as negative or destructive. Early sociologists saw conflict as "a force destroying stability and endangering the structure of ....society" (Loewen 1983). The famous sociologist Talcott Parsons saw conflict as "having primarily disruptive, dissociating, and dysfunctional consequences" (Coser 1956).

Since the 1950's, sociologists and social psychologists have paid increasing attention to conflict. It is now recognized that conflict can have positive as well as negative impacts. Many different definitions of conflict have been suggested.

Coser (1968) proposed one of the most influential definitions:

> "Social conflict may be defined as a struggle over values, or claims to status, power, and scarce resources, in which the aims of the conflicting parties are not only to gain the desired values but also to neutralize, injure, or eliminate their rivals."

Conflict occurs between individuals, between groups, or between an individual and a group. Coser contended that institutionalization of conflict often functions as a safety-value, helping to maintain the integrity and stability of social systems. Some conflicts, instead of tearing the social order apart, "function to sew it together" (Himes 1980).

Boulding (1962) defined conflict as a form of competition in which the competing parties recognize that they have mutually incompatible goals. Kriesberg (1982) observed:

> "Conflict is related to competition, but the two are not identical....In the case of competition, parties are seeking the same ends whereas conflicting parties may or may not be in agreement about the desirability of particular goals....
>
> "The term social conflict...refers to a situation in which parties believe they have incompatible goals."

Some researchers hold that conflict is subjective; others that it can be based on objective factors. Bercovitch (1984) observed:

> "Subjective approaches to conflict assert that, at the most basic level, conflicts are about values and values are ultimately dependent upon perceptions."

Under this interpretation, conflicts are subjective and the parties' perception of the values in conflict is all that counts. Such perceptions can create conflict and can also transform a conflict from violence and coercion into one with mutually beneficial outcomes.

Under the subjective view of conflict, a conflict exists only if the adversaries perceive that they are in conflict. This harks back to Kriesberg's observation that the parties *believe* that they have mutually incompatible goals.

The alternative view is that conflict exists whenever there are incompatible interests, regardless of whether the actors are aware of these interests. The subjective approach emphasizes motivational and attitudinal factors whereas the objective approach stresses structural factors. These different approaches envisage different methods of conflict resolution. Under the subjective approach, conflict resolution involves a change in perceptions; the objective approach requires a fundamental restructuring of social situations.

Himes (1980) defined social conflict as:

> "*purposeful* struggles between collective actors who use social power to defeat or remove opponents and to gain status, power, resources, and other scarce values."

Fundamental to Himes' model is the view that in conflict the contending parties struggle for scarce values. The structurally inherent potential for conflict is heightened when one party believes the other blocks access to such values.

Social psychologist Morton Deutsch (1973) provided a different perspective with one of the simplest definitions of conflict:

> "A conflict exists whenever incompatible activities occur. From this perspective, an action that is incompatible with another action prevents, obstructs, interferes, injures, or in some way makes the latter less likely or less effective."

Conflicts may be *intra*personal, *intra*group, *intra*national or they may be *inter*personal, *inter*group or *inter*national. Deutsch also commented on the confusion in the use of the terms *competition* and *conflict*. He observed:

> "Although competition produces conflict, not all instances of conflict reflect competition....Conflict can occur in a cooperative or a competitive context, and the processes of conflict resolution that are likely to be displayed will be strongly influenced by the context within which the conflict occurs."

According to Deutsch, a conflict is usually about one or another of five basic types of issues:

1. Control over resources;
2. Preferences and nuisances;
3. Values;
4. Beliefs; or
5. The nature of the relationship between the parties.

Deutsch also distinguished between destructive and constructive conflicts:

> "A conflict clearly has destructive consequences if its participants are dissatisfied with the outcomes and feel they have lost as a result of the conflict. Similarly, a conflict has productive consequences if the participants are satisfied with their outcomes and feel that they have gained as a result of the conflict. Also, in most instances, a conflict in which the outcomes are satisfying to all the participants will be more constructive than one that is satisfying to some and dissatisfying to others."

It is clear from the foregoing that, although the term conflict drops loosely from our lips, it is an ambiguous and complex concept. Galtung (1971) conceived a triangle relating conflict to attitude and behaviour (Fig. 15-1). Bercovitch (1984) describes it:

> "A conflict situation, corresponding to the wider or objective approaches to conflict, refers to a situation which generates incompatible goals or values among different parties. Conflict attitudes are closer to the subjective approach to conflict, consisting of the psychological and cognitive processes which engender conflict or are consequent to it. And conflict behaviour consists of actual, observed activities undertaken by one party and designed to injure, thwart or eliminate its opponent."

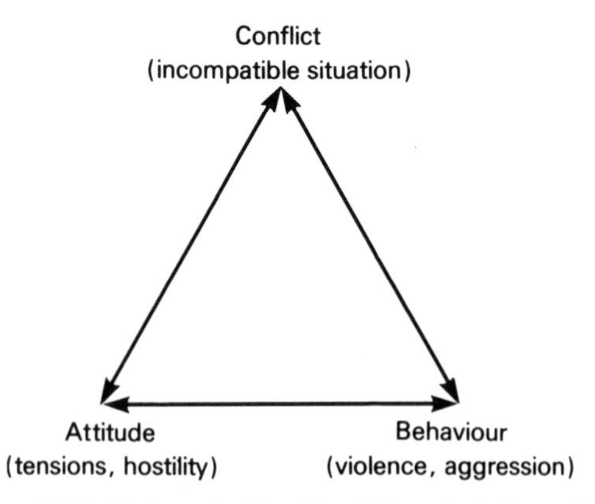

FIG. 15-1. Galtung's Conflict Triangle (adapted from Bercovitch 1984).

The consequences of social conflict are captured in the concept of *function*. Merton (1957) described three basic types: manifest functions, latent functions and dysfunctions. *Manifest functions* are the intended and recognized "objective consequences" of social action that contribute to the adaptation or adjustment of a social system. *Latent functions* are the unintended and often unrecognized objective consequences of organized social activity. *Dysfunctions* are the objective consequences of social activity that militate against the adjustment or adaptation of a system.

Himes (1980) concluded that there is a significant scholarly consensus regarding the *manifest* functions of conflict:

> "They declare that groups engage in struggle for the express purpose of gaining values that either are in fact or are believed to be in scarce supply, or that are incompatible. Further, there is impressive agreement that power, status, and resources are the leading categories of scarce resources for which people engage in conflict. Also...the definitions imply that change is a manifest aim of social conflict. The status, power, resources and other values to which the scholars allude can be gained only at the price of change."

The *latent* functions of conflict are the unexpected developments, occurrences, or changes that arise in the course of conflict action.

Dysfunctions occur when social conflicts attack or threaten core values and basic concerns of social systems. The consequences of such conflict may be maladjustments and disintegration of the system. Such a dysfunction may be the price a system has to pay for a functional gain. Dahrendorf (1958) suggested that one of the consequences of social conflict is further conflict.

Destructive conflict has a tendency to expand and escalate. As Deutsch (1973) noted, expansion occurs along the various dimensions of conflict: the size and number of the immediate issues involved; the number of motives and participants implicated on each side of the issue; the size and number of the principles and precedents that are perceived to be at stake; the costs that the participants are willing to bear in relation to the conflict; the intensity of negative attitudes towards the other side.

There also tends to be an increasing reliance upon a strategy of power and upon tactics of threat, coercion and deception. There is a shift away from a strategy of persuasion and from tactics of conciliation, minimization of differences, and enhancement of mutual understanding and goodwill. There is pressure on each side towards uniformity of opinion and a tendency for leadership and control to be taken over by militant elements (Deutsch 1973).

This tendency towards conflict escalation results from the combination of three factors:

1. Competitive processes involved in the attempt to win the conflict;
2. Processes of misperception and biased perception; and
3. Processes of commitment stemming from pressures for cognitive and social consistency.

The competitive process tends to produce the following effects which help to perpetuate and escalate a conflict (Deutsch 1973):

1. Communication between the conflicting parties becomes unreliable and impoverished;
2. It fosters the view that the solution of the conflict can only be imposed by one side or the other by means of superior force, deception or cleverness; and
3. It fosters a suspicious, hostile attitude which increases sensitivity to differences and threats and minimizes awareness of similarities.

Misjudgment and misperception are also at work to distort the situation. Such misperceptions can transform a conflict into a competitive struggle. Two individuals will interpret the same act quite differently. Most people are strongly motivated to maintain a favourable view of themselves and less strongly motivated to view others favourably. Thus, people are biased to perceive their behaviour towards others as more benevolent and more legitimate than others' behaviour towards them. Under such circumstances, conflicts will often spiral upward in intensity.

Deutsch (1973), however, stressed that conflict can also be productive. The social science literature on this aspect is sparse. Intense inner conflict is often the prelude to major emotional and intellectual growth in individuals. Deutsch postulated that a creative function of conflict is its ability to motivate solutions to problems that might otherwise not be dealt with. As an example, those who have authority and power and who are satisfied with the status quo may be aroused to recognize problems and be motivated to work on them when opposition from the dissatisfied makes the customary relations and arrangements unworkable. Such acceptance of the need for change is more likely, however, when the circumstances bringing new motivation suggest action involving minimal threat to those who would have to change.

From a *cooperative* perspective, a conflict can be viewed as a common problem in which the conflicting parties have a joint interest in a mutually satisfactory solution. Deutsch suggested the following reasons why a cooperative process is likely to lead to productive conflict resolution:

1. It aids open and honest communication between the participants;
2. It encourages recognition of the legitimacy of the other's interest and of the necessity to search for a solution that responds to each side's needs.

It tends to limit rather than expand the scope of conflicting interests and thus minimizes the need for defensiveness; and

3. It leads to a trusting, friendly attitude, which increases sensitivity to similarities and common interests, while minimizing the significance of differences.

Cooperative processes also generate certain forms of misperception and misjudgment. Cooperation, for example, tends to minimize differences and enhance the perception of the other's benevolence. This can dampen conflict or prevent conflict. Such benevolent misperceptions limit the frequency and intensity of opposition in conflict.

Deutsch pondered the factors which may determine whether a conflict becomes constructive or destructive. He formulated "Deutsch's crude law of social relations." The strategy of power and the tactics of coercion, threat and deception result from, and also result in, a competitive relationship. Similarly, the strategy of mutual problem solving and the tactics of persuasion, openness, and mutual enhancement result from, and also result in, a cooperative relationship. In short, cooperation breeds cooperation, and competition breeds competition.

More recently, Carpenter and Kennedy (1988) emphasized the spiral of unmanaged conflict. They observed:

> "Conflict is dynamic. Unmanaged conflicts seldom stay constant for long. Simple solutions that might have worked in the beginning may be ineffective and even cause more damage if they are attempted when the conflict is fully developed....
>
> "The following sequence is typical of public disputes: one or more parties choose not to acknowledge that a problem exists. Other groups are forced to escalate their activities to gain recognition for their concerns. Eventually everyone engages in an adversarial battle, throwing more time and money into 'winning' than into solving the problem."

The changes in activities, issues, and psychological perceptions that occur as a conflict escalates are shown in Fig. 15-2.

Carpenter and Kennedy observed:

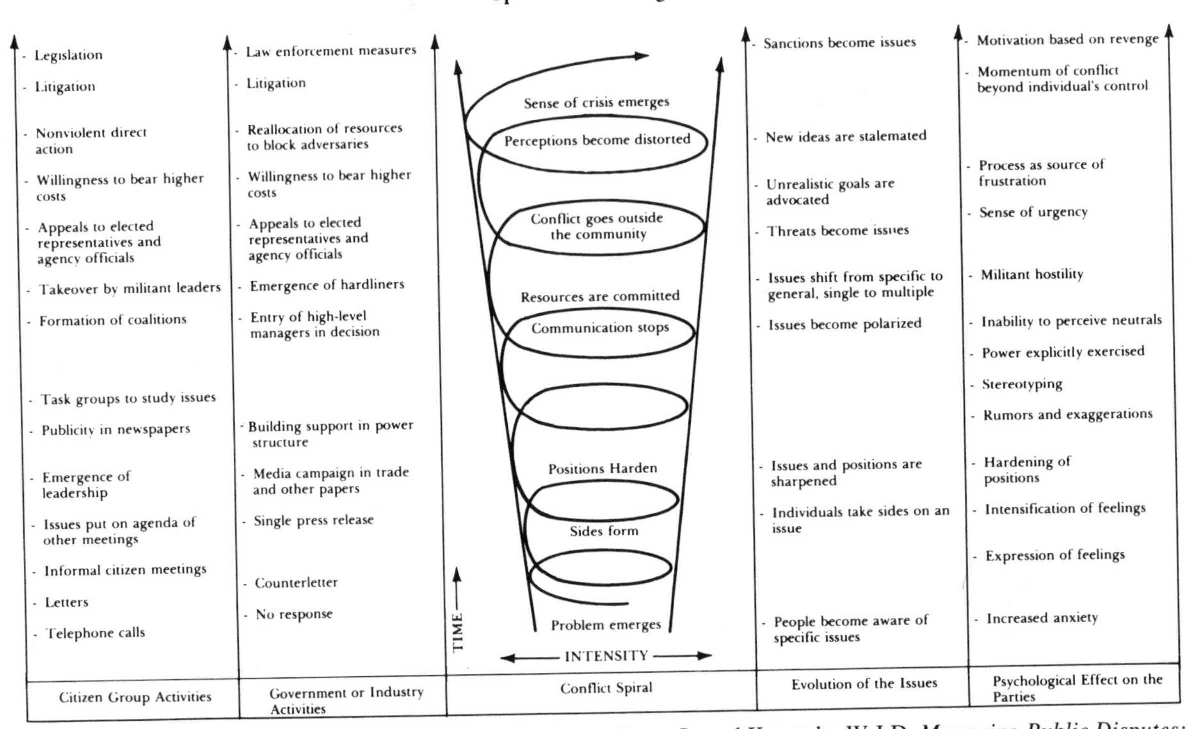

FIG. 15-2. Spiral of Unmanaged Conflict (from Carpenter, Susan L. and Kennedy, W.J.D. *Managing Public Disputes: A Practical Guide to Handling Conflict and Reaching Agreement*, Figure 1, p. 12. Copyright 1988 by Jossey-Bass Inc., Publishers. Reprinted by permission of the publisher.)

"Many conflicts start with a resolvable problem and grow beyond hope of resolution because they are not dealt with early. It is sometimes said that the conflict manager should let a situation 'ripen' or polarize before attempting to handle it. This suggestion seems tantamount to telling a doctor that a bad cold should be allowed to develop into pneumonia before he or she prescribes treatment. On the contrary, the great value of taking a hard look at where the dispute is on the spiral is that one can then choose an interim strategy that will slow down or stop expansion of the conflict. The purpose of conflict management activities, such as establishing communication, defining issues, and facilitating effective meetings, is to interrupt the spiral of conflict....The lesson of the conflict spiral is not that its progress is inevitable but that it is predictable when nothing is done to manage the conflict."

Their model of the spiral of conflict owes much to Boulding (1962), Deutsch (1973) and Kriesberg (1982). Their focus is on the destructive aspects of conflict.

A summary of the sources of conflict, conflict consequences and outcomes is depicted in Fig. 15-3.

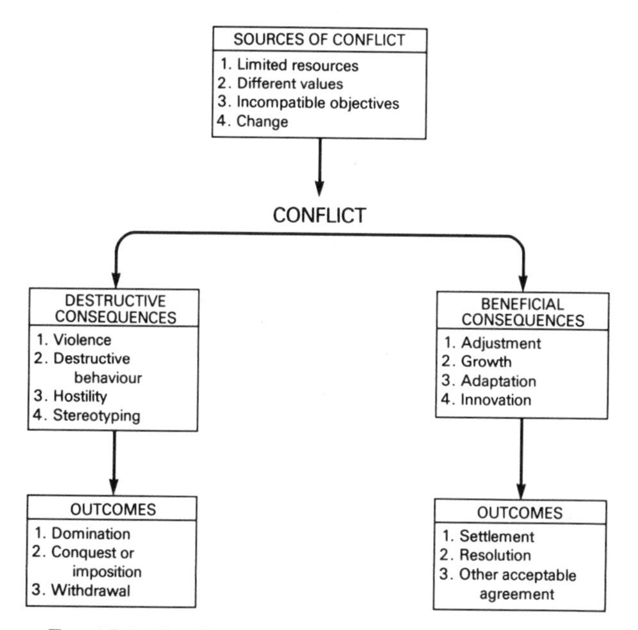

FIG. 15-3. Conflict and Conflict Consequences (adapted from Bercovitch 1984).

3.0 CONFLICT MANAGEMENT

Deutsch (1973) noted that "the point is not how to eliminate or prevent conflict but how to make it productive." Bercovitch (1984) observed that the purpose of conflict management "is not to eliminate, prevent, or control conflicts, but rather to increase their values and benefits, and decrease costs and dissatisfaction.... The purpose of conflict management is to influence the course of a conflict."

Returning to Galtung's Triangle (Fig. 15-1), recall that conflicts may start with any of the three integrated components — conflict situation, attitudes and behaviour. They may start with incompatible goals, negative attitudes, or destructive behaviour. Similarly, conflict management may attempt to influence the conflict situation or alter conflict attitudes and behaviour.

Conflict management can be endogenous or exogenous. Endogenous conflict management is that undertaken by the parties to a conflict, for example, negotiation. Exogenous conflict management involves the activities of a third party. These efforts of third parties may involve voluntary efforts such as mediation or compulsory (binding) arbitration or adjudication.

Ruble and Thomas (1976) postulated that a party to conflict will exhibit one of five orientations to resolve the conflict, depending on the outcome desired (see Fig. 15-4). The two dimensions of their model of conflict resolution behaviour are *cooperativeness* (attempting to satisfy the other party's concerns) and *assertiveness* (attempting to satisfy one's own concerns). Various combinations of assertiveness and cooperativeness produce five conflict management approaches which were summarized by Cooze (1989):

Forcing— where one party is unwilling to deal with the other party, and tries to subdue it, using power to win at the other's expense.

Compromise— this approach partially satisfies both parties since each gains as well as loses something.

Accommodating— this involves willingness to put the concerns of the other person or group ahead of one's own.

Collaborating— this involves a desire to fully satisfy the concerns of both parties. The two parties are open and respect each other.

Avoiding— this involves refusing to deal with the conflict, thus exhibiting indifference to the concerns of the other party.

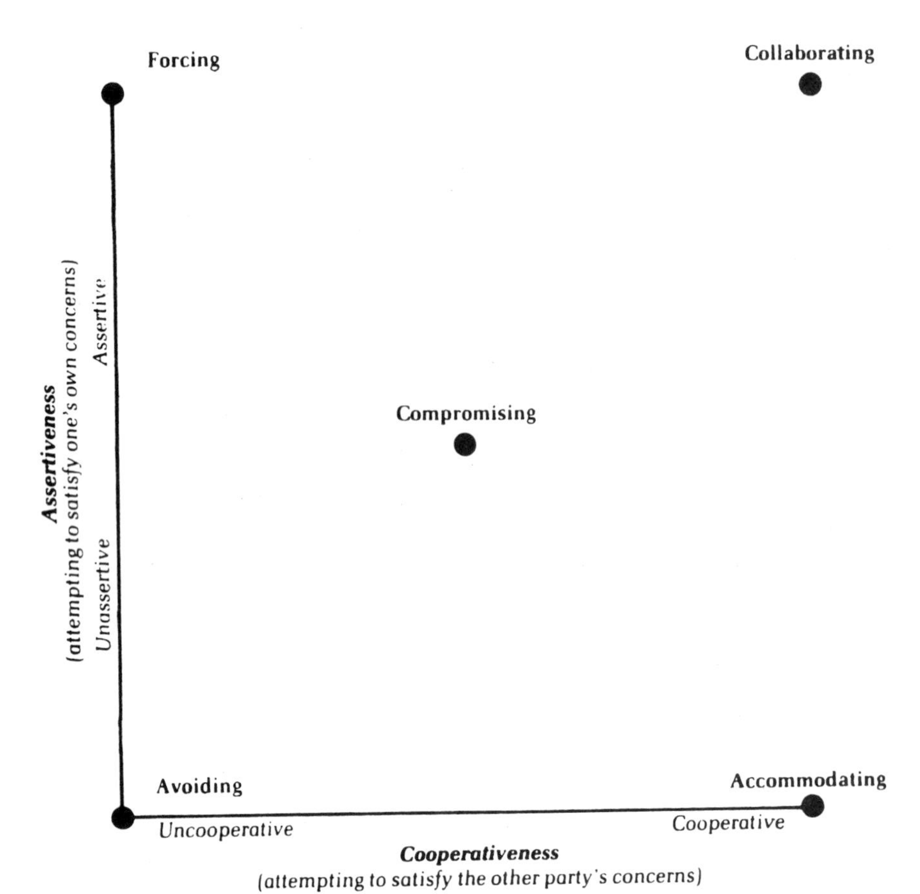

FIG. 15-4. Two-Dimensional Model of Conflict Resolution Behavior (adapted from Cooze 1989).

Conflict resolution techniques that work well in one situation may be quite ineffective in another.

A wide variety of procedures have been developed to manage conflict. These include institutional forms (e.g. collective bargaining), social roles (e.g. third parties), or social norms. Among the best known methods are legal regulation, bargaining and negotiation. There is now a vast literature on these various forms of conflict management, particularly negotiation.

Himes (1980) examined the process by which societies develop mechanisms and processes for regularizing emotional expression and social relations. These mechanisms and processes he termed *institutionalization*, a conflict-regulatory strategy. He defined "institutionalization" as a set of binding social norms and a pattern of collective action that endorse, regularize and reward "legitimate" social struggle to prevent the resort to "nonlegitimate" conflict. This concept involves the process of subjecting conflict to binding rules, the cluster of those rules in effect at any given time, and the pattern of collective action required by those rules.

Himes (1980) suggested that a strategy of institutionalizing social conflicts requires four kinds of actions by a governing body or other authoritative organization:

1. Recognition as a matter of policy that various groups and sectors of the community are likely to have divergent, incompatible or clashing interests;
2. Guarantees to the several groups and sectors of the right to pursue their interests, with the understanding that this right is limited by the rights of other groups and sectors and by the requirements of the inclusive social system;
3. Establishment and enforcement of [binding] institutions to regularize and regulate the resultant struggles to ensure satisfaction to the adversaries and consider the reciprocal rights of participants and the welfare of the society or community; and
4. Establishment and support of an agency to supervise and enforce the institutional rules, and to attempt to solve disputes and conflicts that issue from the pursuit of divergent and clashing interests.

Conflicting interests are more likely to be settled or resolved harmoniously to the extent that they take place in a context of cooperative relations. As Deutsch (1973) pointed out, harmonious relations are less likely to occur if either or both sides: (1) feel that their existence or their rights are threatened by the other side; (2) think that their survival is endangered by external competition; (3) are torn by internal factionalism; (4) have little local autonomy so that agreements cannot be responsive to local conditions; (5) are constantly subjected to changing conditions.

Managers tend to follow one or more of four conventional approaches to conflict (Carpenter and Kennedy 1988):

1. *Avoiding the Issue*
   It often appears easiest to pretend that everything is all right and hope that the issue will go away.
2. *Leaping into Battle*
   Under this scenario, the immediate reaction is to take an adversarial approach. Potentially satisfactory solutions are overlooked and all energies are diverted to a fight.
3. *Finding a Quick Fix*
   There is a temptation to produce an immediate solution. Such a solution appears before there is a clear understanding of the issue.
4. *The Solomon Trap*
   This involves four phases. The person who has to resolve a conflict seeks to identify all affected parties and to obtain their views. Secondly, this person examines the comments, weighs the trade-offs, and attempts to play King Solomon by devising a solution that comes closest to addressing everyone's interests and that is in harmony with the agency's goals and priorities. Thirdly, the decision is communicated to the parties, who are upset when they find that key issues have not been addressed exactly as they wished. The parties lash out at the decision-maker and reject the decision. In the fourth phase the decision-maker spends a great deal of time defending the merits of the decision to each of the interested parties. No one is willing to support it. This leads to a situation where the decision gets tossed out and the process begins anew.

There are alternative approaches for managing conflicts. Carpenter and Kennedy (1988) identified several common characteristics of these approaches:

1. The decision maker is a facilitator;
2. The focus is on solving a problem;
3. The parties meet face to face to work out differences;
4. The parties help to shape the process; and
5. Decisions are made by consensus.

Carpenter and Kennedy summarized it thus:

> "The goal of consensus decision making is to reach a decision that all parties can accept. The parties reach agreement by gathering information, discussing and analyzing it, and convincing each other of its merits. They combine or synthesize proposals or develop totally new solutions. Not everyone will like the solution equally well or have an equal commitment to it, but the group recognizes that it has reached the best decision for all parties involved."

Based on their experience in attempting to deal with conflict productively, Carpenter and Kennedy suggested the following ten principles for an effective program of conflict management:

Principle 1. Conflicts are a mixture of procedures, relationships and substance.
Principle 2. To find a good solution, you have to understand the problem.
Principle 3. Take time to plan a strategy and follow it through.
Principle 4. Progress demands positive working relationships.
Principle 5. Negotiation begins with a constructive definition of the problem.
Principle 6. Parties should help design the process and solution.
Principle 7. Lasting solutions are based on interests, not positions.
Principle 8. The process must be flexible.
Principle 9. Think through what might go wrong.
Principle 10. Do no harm.

Their ten principles constitute a productive pragmatic approach to deal with conflict. They are a fitting introduction to conflict in the fisheries arena.

4.0 CONFLICT AND THE FISHERIES

Day after day, week after week, year after year the news media report countless examples of fisheries-related conflict. Throughout this book the subject of conflicting interests keeps popping up. The examples of fisheries-related conflict are endless; they range from the trivial to the major controversies which involve most of the significant players. There is occasional violence (usually limited to property destruction). Conflict is all-pervasive in the fisheries arena.

Conflict may be as minor as a dispute between fishermen over the deleterious effects of a particular type of fishing gear. Major conflicts occur in the context of allocating scarce fish resources among different fleet sectors (see Chapter 7). Conflict between countries over fish quotas and appropriate management of fish resources can adversely affect international relations (e.g. the Canada-France dispute, the Canada-USA dispute over Pacific salmon interceptions). Some of these conflicts can be resolved locally through the intervention of fishery officials closest to the scene. Others, for example, the annual Atlantic groundfish resource allocation process, work through consensus-seeking and ultimately a decision by the federal Minister of Fisheries and Oceans, advised by his provincial counterparts. Interprovincial disputes over the allocation of resources such as northern cod require a similar decision-making process. Matters such as the Canada-France dispute over fish quotas and boundary matters have required a mediator and an arbitration tribunal. Canada and the U.S. went to the International Court of Justice at the Hague to seek a judgement on their conflicting views of where the boundary line should be drawn in the Gulf of Maine. The Pacific Salmon Treaty was arrived at by negotiation between Canada and the USA. Thus, it is clear that a variety of conflict management techniques are employed, depending upon the nature and extent of the fisheries conflict and the players involved.

4.1 The Nature of Fisheries Conflicts

Most fisheries-related conflicts have their roots in the common property character of the fish resource, internationally beyond 200 miles and nationally within 200 miles. The common property aspects of the resource were discussed at length in Chapter 7. To briefly recap, since no one owns the fish, each individual fisherman is motivated to increase his catching capacity to maximize his share of the available catch. Inevitably, this leads to overcapacity in harvesting and in processing. Chapters 8 and 9 deal with methods to combat this tendency. Use of these methods has not removed the inherent competition and resulting conflict stemming from the common property phenomenon.

Most of the definitions of conflict cited previously include the concept of struggle over claims to scarce resources. Control over resources was one of the five basic types of conflict issues listed by Deutsch (1973). Conflict over access to and allocation of limited fish resources is the fundamental conflict in fisheries. There is often also conflict over objectives and methods of resource management. This sometimes involves a conflict of values, as distinct from a conflict over how the resource is to be allocated. Often these forms of conflict become intertwined, as, for example, in the recurrent disputes over the level at which the Gulf of St. Lawrence redfish TAC should be established. This dispute was in fact motivated by the underlying struggle over how to share access to the resource between Gulf and non-Gulf-based interests.

Consensus on how the resource is to be managed is sometimes more easily resolved than conflict over how to share access to the resource. The matter is often complicated by interjurisdictional disputes which lead to interprovincial rivalry and conflict on distribution of the social and economic benefits of exploitation. Concern about maintaining certain levels of employment and economic activity can bring conflict over the appropriate level (target fishing mortality rate) of exploitation. One of the most dramatic examples has been the recent dispute over the appropriate level of the TAC for northern cod. The federal government initially chose to reduce the TAC slowly to minimize social and economic disruption, even though the Departmental scientific advice and an independent Panel report called for a more rapid reduction. Ultimately, following a sudden, drastic and unexpected decline in the abundance of the northern cod stock in 1991, the federal government imposed a 2-year moratorium commencing in July, 1992.

In this situation there was consensus neither about what level of harvest was appropriate, nor the allocation of whatever level was chosen. Normally, consensus about the criteria of allocation is less likely than consensus about the desirable level of harvest. This leads to conflict and high levels of dissatisfaction within the fishing industry.

The reason for this widespread conflict lies in the zero-sum nature of the issue. People are dealing with a scarce resource, in most cases insufficient to meet existing demands. Because the valued resource is less than the desired amount and is (usually) not increasing, the government must divide a fixed pie which is insufficient to satisfy all the appetites. Increasing one party's share means diminishing that available to others. The parties are involved in a zero-sum relationship. This inevitably produces conflict and high levels of dissatisfaction.

Recently, Charles (1992) has presented a typology of fisheries conflicts, arguing that the wide range of fishery conflicts can be organized into a relatively small number of categories. He proposed four interrelated headings:

(1) *Fishery jurisdiction* — involving fundamental conflicts over who 'owns' the fishery, who controls access to it, what is the optimal form of fisheries management, and the appropriate role for governments in the fisheries system.

(2) *Management mechanisms* — concerning relatively short term issues arising in the development and implementation of fisheries management plans.
(3) *Internal allocation* — involving conflicts arising within the specific fishery, between different user groups and gear types, as well as between fishermen, processors and other players.
(4) *External allocation* — incorporating the wide range of conflicts arising between internal fisheries players and 'outsiders', including foreign fleets, aquaculturists, non-fish industries, such as tourism and forestry, and the public at large.

Many examples of these types can be found throughout this book. Earlier I argued that conflict over access to and allocation of limited fish resources is the fundamental conflict in fisheries. Charles (1992) also concluded that fishery conflicts tend to be dominated by allocation issues. As I noted, however, there is often also conflict over objectives of resource management (see Chapter 4). Charles (1992) translated these conflicting policy objectives into what he described as three fisheries *paradigms* — conservation, rationalization and social/community. He saw these paradigms as forming a paradigm triangle (Fig 15-5). He suggested that fisheries conflicts can be viewed as reflecting tensions between the triangle's three corners, with extreme policy proposals lying relatively close to one of the corners, and attempts at conflict resolution typically aiming at the "middle ground". Charles' paradigms correspond closely to the conflicting objectives which I have described in Chapter 4 and elsewhere throughout this book.

Government's challenge is to manage this conflict and minimize the dissatisfaction by instituting a process for management and allocation which is considered fair and equitable by the parties concerned.

Peyton (1987) concluded that one or more of three important components contribute to intergroup conflicts over natural resources. These are: the state of the science or technology, what the public believes to be true (public beliefs), and the public values and priorities assigned to those beliefs. He suggested that the extent to which resource conflict is avoided or satisfactorily resolved is determined by how well each of these components is managed.

The question of the nature and adequacy of fisheries science and its use in resource management is examined in Chapters 17 and 18. In many cases, user groups' response to scientific uncertainty is to question the credibility of managers and the scientific basis for management. In such situations, user groups often become disruptive and prefer their own experiences, observations and intuition as a basis for management.

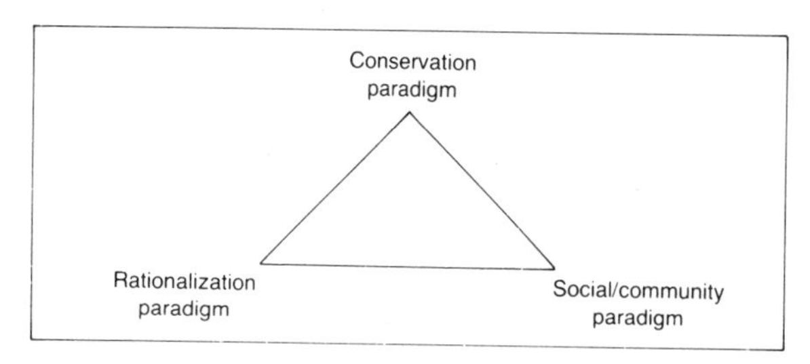

Fig. 15.5. The paradigm triangle (Charles, 1992).

This generates conflict in the resource management process. A dramatic example of this is the 1989–90 controversy throughout Atlantic Canada, but particularly in Newfoundland, over the scientific advice and management of northern cod (Lear and Parsons 1993).

A major component of fisheries conflict is the divergent values held by various groups involved in the fishery. Examples are the conflict between natives and other user groups in the Pacific salmon fishery and that between recreational and commercial users, as illustrated by salmon management on the east and west coasts. Fisheries managers find it difficult to resolve value-related conflict. These conflicts center on resource allocation. Managers have the difficult task of priorizing values of competing groups and dividing a fixed or declining resource "pie". The result often is considerable bitterness towards the resource managers unless, perchance, compromises can be achieved through the use of appropriate conflict management techniques.

The base of Galtung's Conflict Triangle consists of attitudes and behaviour. In managing fisheries, both the attitudes and behaviour of groups on particular issues must be addressed. Attitudes may be expressed in the group's position on an issue. Behaviours may be manifested as active opposition or support, compliance or noncompliance. Clearly, an individual or a group's response to an issue is a function of many factors. Behaviours are frequently inconsistent with the apparent attitudes. Groups may express agreement with the need for particular regulations but fail to comply with such regulations (Peyton 1987).

A complex variety of factors can be involved in attitude formation. While values are enduring standards or preferences which influence choice and action, attitudes are more ephemeral and may change as new priorities are assigned to values or as beliefs are modified by new information (Peyton 1987). Ladd (1985) demonstrated from fifty years of public opinion research that the public holds firmly to an underlying structure of core values, while expressing great confusion and instability on the details of public policy.

It is obviously desirable, from a resource management perspective, for user group attitudes to be formed through thorough rational processes. If groups quickly form attitudes linked to strongly held values, there is a strong tendency to defend those attitudes and values instead of attempting to resolve a conflict. Often, individuals or groups become overexcited and, as a result, are incapable of innovative or constructive actions. Under such circumstances, individuals or groups tend to revert to old patterns of response. They accept information which supports their position and reject that which runs counter to it.

As Peyton (1987) noted, a common response by groups to a proposed management plan is to opt for the status quo and to reject any new proposals. This occurs because humans tend to resist change and avoid risk. One way of dealing with this is to present the user groups with not one plan but alternative plans to compare. This fosters a fuller understanding of the benefits and risks involved in each alternative. Unfortunately, quite often groups and individuals who oppose a particular course of action become involved in greater numbers than those who support it. This can generate further conflict as the majority subsequently attempts to overturn a decision taken as the result of lobbying by a vocal minority.

To deal effectively with conflicts arising from differing values, attitudes and beliefs requires a structured framework for the regular input of views from all groups involved in fishery issues. It also requires the patience of Job to sort through those views and attempt to find a compromise which will satisfy the majority, minimize the dissatisfaction of the minority and meet the basic objectives of the management agency. In the next section I describe the fisheries decision-making process that has evolved in Canada and, in particular, the consultative structures and processes for input by the multitude of publics concerned with fisheries issues.

4.2 The Framework for Fisheries Decision-Making in Canada

4.2.1 Fisheries Legislation

In Chapter 2, I described the constitutional framework for fisheries management in Canada. I outlined the courts' interpretations of the federal government's authority to manage the seacoast and inland fisheries under Section 91(12) of the Constitution Act, 1867. I also examined the courts' interpretation of the extent this authority was limited by provincial constitutional authority over property and civil rights. Essentially, the federal government has authority over fish in the water until it is caught, the provinces have jurisdiction over land-based activities such as fish processing, and the federal government has authority over interprovincial and international trade and export. From this latter provision, the federal government also derives the right to set fish inspection standards and, with the cooperation of the provinces, has developed a nationwide fish inspection system.

This federal mandate to manage Canada's fisheries has been exercised primarily through two statutes — the *Fisheries Act*, which is the primary source of authority for managing domestic fisheries, and the *Coastal Fisheries Protection Act*, which is the primary authority for managing foreign fisheries under Canada's jurisdiction within its 200-mile fisheries zone.

The *Fisheries Act* (R.S.C. 1985, c. F-14, as amended) grants the federal Minister responsible for fisheries and the Governor General in Council considerable discretion to regulate the fisheries and to change rules and regulations without further reference to Parliament. One of the most powerful sections of the *Fisheries Act* is Section 7 which allows the Minister to issue fishing leases and licences "in his absolute discretion". The use of this discretion and appeal processes were discussed in Chapter 8. Section 36 of the *Fisheries Act* prohibits the disposal of deleterious substances into water frequented by fish. While the Act defines "deleterious substance" (section 34), the actual determination of deleterious substances is left to the Governor General in Council (in essence the federal Cabinet). Section 43 of the *Fisheries Act* grants the Governor General in Council discretion to regulate:

(a) the proper management and control of seacoast and inland fisheries;
(b) the conservation and protection of fish;
(c) the catching, loading, landing, handling, transporting, possession and disposal of fish;
(d) the operation of fishing vessels;
(e) the use of fishing gear and equipment;
(f) the issue, suspension and cancellation of fishing licences and leases;
(g) the conditions under which a licence and leases may be issued;
(h) the obstruction and pollution of any waters frequented by fish;
(i) the conservation and protection of spawning grounds;
(j) the export of fish;
(k) the interprovincial transport or trade of fish;
(l) the duties of federal employees; and
(m) the delegation to federal officials of the administration of the authority to vary any close time or fishing quota.

These provisions, taken together, give the federal Minister of Fisheries and Oceans and his Cabinet colleagues enormous discretion to manage fisheries. There has been no mandatory obligation to seek public input before exercising these powers, except for the short-lived reference to consultation in the Sunset Amendment of 1986 (see Chapter 2).

Similarly, the *Coastal Fisheries Protection Act* (R.S.C. 1985, c. C-33 as amended) authorizes the Governor General in Council to establish the conditions for foreign fishing vessels to fish in Canadian waters and authorizes protection officers to board and search foreign fishing vessels in Canadian waters. Under this Act, the government has made regulations pertaining to mesh sizes, area closures and species quotas.

These two Acts provide the basic underpinning of the Canadian fisheries management system. The widespread discretionary powers under these Acts essentially permit the federal Minister and the federal Cabinet to establish a flexible fisheries management system and modify it almost at will. These acts do not contain fisheries policies as such. The objectives of Atlantic fisheries policy were embodied in the preamble to the *Atlantic Fisheries Restructuring Act* of 1983. Apart from that, the statutes are generally silent on the direction of fisheries policy and leave a wide latitude as to how these policies may be implemented. With the government-wide regulatory reform of the mid-1980's, there are now requirements for advance notice of proposed regulations to allow time for public comment. This comment primarily addresses the scope of fisheries policies and specific fishing plans as these are being developed rather than focusing on the regulations themselves.

For a long time there were four statues authorizing direct financial assistance to the fishing industry — the *Fisheries Development Act*, the *Fisheries Improvement Loans Act*, the *Fisheries Price Support Act*, and the *Saltfish Act*. The *Fisheries Improvement Loans Act* was allowed to lapse in June 1987. For a discussion of these Acts and programs undertaken under their authority, as well as the *Atlantic Fisheries Restructuring Act*, see Crowley et al. (1993). These statutes in general do not specifically address fisheries management. The *Fisheries Development Act* is related to fisheries management in that programs under this Act at times have fostered overcapacity in the industry and hence conflicted with fisheries policy objectives.

4.2.2 The Organization of Fisheries Management

The organization of fisheries management in Canada has gone through numerous statutory changes since 1867 (Table 15-1). In 1971, the management of fisheries was subsumed in the Fisheries and Marine Service component of a new multi-faceted Department of the Environment. This Department was established as a result of growing concern about the need for environmental protection at the beginning of the 1970's. From 1971 to 1974, the Minister of the Environment, Jack Davis, was also the Minister responsible for fisheries. The fisheries constituency in Canada was very dissatisfied with this and lobbied to restore a full-fledged fisheries department. Roméo LeBlanc was appointed Minister of State for Fisheries in 1974, as a junior minister under the Minister of the Environment. In 1976, the Department's name was changed to Fisheries and the Environment and Roméo LeBlanc was appointed Minister of Fisheries and the Environment.

TABLE 15-1. The organizational management of fisheries in Canada — 1867 to present.

Years	Administrative Body Responsible for Fisheries Management
1867–1884	Marine and Fisheries — Fisheries Branch
1884–1892	Department of Fisheries
1892–1914	Marine and Fisheries — Fisheries Branch
1914–1920	Naval Services — Fisheries Branch
1920–1930	Marine and Fisheries — Fisheries Branch
1930–1969	Department of Fisheries
1969–1971	Department of Fisheries and Forestry
1971–1976	Environment — Fisheries and Marine Service
1976–1979	Fisheries and the Environment — Fisheries and Marine Service
1979–	Fisheries and Oceans

Source: DFO Factbook, 1989.

In 1979, Parliament split the Department of Fisheries and the Environment into two departments — a new Department of Fisheries and Oceans, and the Department of the Environment. After many changes, a single Minister again spoke with authority on fisheries (and oceans) issues. From 1979 onward the Minister of Fisheries and Oceans had the statutory responsibility to "coordinate the policies and programs of the government of Canada with respect to oceans." This mandate was not exercised until Fisheries and Oceans Minister Tom Siddon began to consult with representatives of the oceans industry at an Oceans Forum, in Sydney, B.C. in September 1986. This culminated in the announcement of a new Oceans Policy for Canada in September, 1987, and a National Marine Council to advise the minister on a broad spectrum of oceans-related issues (DFO 1987k).

From 1979 to 1986, the Department of Fisheries and Oceans consisted of four primary components, each headed by an Assistant Deputy Minister: Atlantic Fisheries, Pacific and Freshwater Fisheries, Economic Development and Marketing, and Ocean Science and Surveys. Under this set-up, the Atlantic Fisheries Service was responsible for fisheries management in the four Atlantic provinces and Quebec. There were four regional fisheries offices (Newfoundland, Scotia-Fundy, Gulf and Quebec Regions), headed by Directors General of Fisheries Management. The Newfoundland Region was responsible for fisheries along the south and east coasts of Newfoundland and in Labrador. The

Scotia-Fundy Region was responsible for fisheries on Georges Bank, in the Bay of Fundy and on the Scotian Shelf. The Gulf and Quebec Regions shared management of fisheries in the Gulf of St. Lawrence. Under the ADM, Pacific and Freshwater Fisheries, who was responsible for fisheries in central and western Canada and the Arctic, there were three Regions — Ontario, Western (the Prairie Provinces and the Northwest Territories) and Pacific (British Columbia and the Yukon). Ocean Science and Surveys, responsible for oceanographic and hydrographic programs, had four regional science centres, headed also by Directors General — the Bedford Institute of Oceanography at Dartmouth, Nova Scotia; the Champlain Centre for Marine Science and Surveys in Quebec City; the Bayfield Laboratory in Burlington, Ontario; and the Institute of Ocean Sciences at Sydney, British Columbia. Altogether there were 11 regional Directors General within the Department of Fisheries and Oceans, whose geographic, but not functional, mandates overlapped.

The ADM, Economic Development and Marketing, ran a headquarters operation with four main functions: (1) promotion and fish marketing services, (2) economic and statistical information and policy, (3) directing financial assistance programs, and (4) international fisheries relations.

In the fall of 1985, the Neilson Task Force Study Team on Natural Resources had proposed streamlining and reducing the size of the Department of Fisheries and Oceans (Canada 1985a). The new Minister of Fisheries and Oceans, Tom Siddon, appointed in November 1985, decided to make major changes in the structure of the Department. In February, 1986, the Minister announced that the regional operations of the Department would be consolidated into six regions, by merging the seven fisheries and four ocean science and surveys regions. The headquarters staff of the department was reduced by 23%, amounting to 200 person years. The fisheries research components of the former Atlantic Fisheries and Pacific and Freshwater Fisheries Services were merged with the oceanographic and hydrographic programs of the former Ocean Science and Surveys to form a new integrated science sector, headed by a new Assistant Deputy Minister for Science (DFO 1986h).

Subsequently, the Department's regional structure was consolidated (DFO 1986i). The new Regional Directors General were to manage all of the Department's operations within the six new geographic regions — Newfoundland, Scotia-Fundy, Gulf, Quebec, Central and Arctic, and Pacific. They would report directly to the Deputy Minister but were also accountable to the Sector ADMs, Atlantic Fisheries, Pacific and Freshwater Fisheries, and Science for the delivery of sector programs in their regions (Fig. 15-6).

Under this model, the respective responsibilities of the Deputy Minister, the Sector ADMs, and the Regional Directors General, were as follows:

MANAGEMENT LEVEL	RESPONSIBILITIES	AUTHORITIES
Deputy Minister	• overall results of the program • implementation of Ministerial priorities • effective management of Departments' operations	• approval of sector plans and budgets • major resource adjustments affecting: —Sectors —Parliamentary Votes, TB allotments —Commitments to the Minister, Parliament or Central Agencies
Sector Head	• achievements of sector results • sector policies, priorities, program proposals, plans and budgets • advice to the Deputy on performance against sector plans and adjustments to sector plans	• approval of regional sector plans and budgets • direction, coordination implementation and monitoring of sectoral programs, plans and budgets • authority to act on all operational matters pertaining to the sector • resource adjustments between or within work activities and/or regions within sector • review and approval of human resource plans, classification and staffing
Regional Director General	• general manager of region and custodian of Departmental resources within region • delivery of all programs per approved plans and budgets • provision of support services to regional programs • regional input to sector and departmental policies, plans and budgets	• preparation of regional sector plans and budgets and approval of subordinate plans and budgets • adjustments to plans and budgets within work activities subject to regular reporting and concurrence of sector head

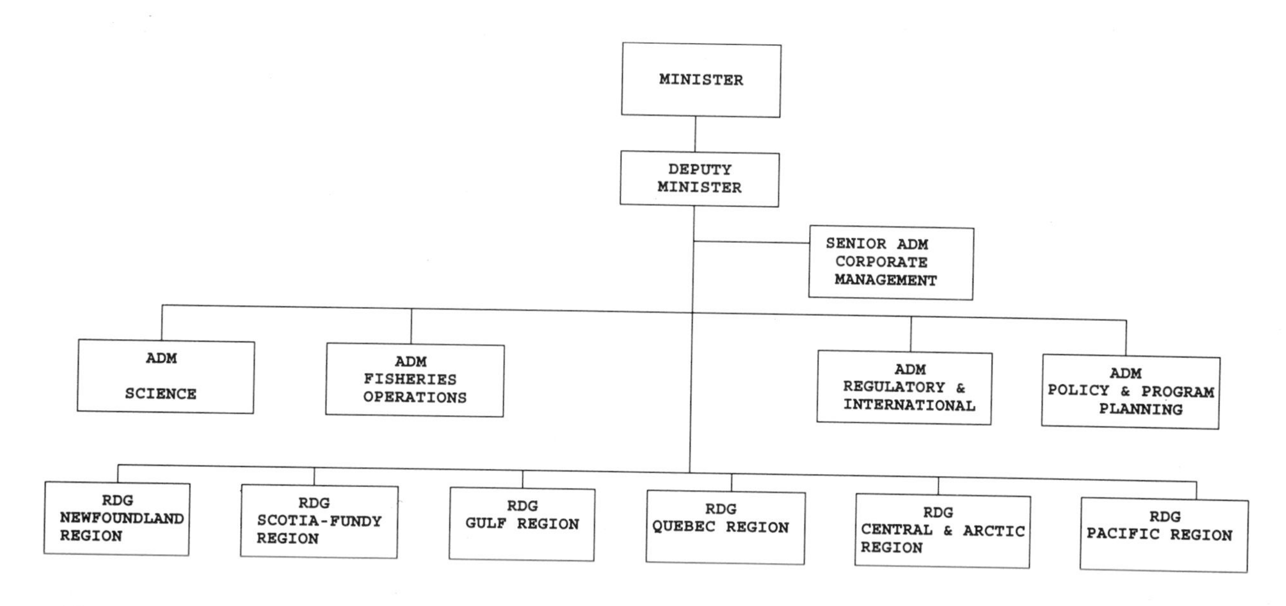

FIG. 15-6. Organization Chart for the Department of Fisheries and Oceans (December 1992).

The new organization also included a Senior ADM and two staff ADMs, Policy and Program Planning, and International, whose responsibilities differed from those of the Sector ADMs. There was further departmental reorganization at the ADM level in the early 1990's, with the establishment of an ADM, Fisheries Operations combining the functions of the previous ADMs of Atlantic Fisheries and Pacific and Freshwater Fisheries. The responsibilities of the former ADM, International, were enlarged to include responsibility for the inspection and enforcement functions. The relationship between the departmental organization, the program activity structure and 1993-94 resources (dollars and person-years) is shown in Fig. 15-7.

To function effectively, the new organization would have to rely upon a team approach, management discipline and sharing of information. This required new departmental management structures. This was accomplished through a new Departmental Management Committee (DMC) consisting of the ADMs and RDGs, chaired by the Deputy Minister. This met several times annually to discuss departmental administration issues and major policy proposals with nationwide implications. Another committee, the Departmental Executive Committee (DEC), consisting of the ADMs and the Deputy Minister, met weekly and helped to shape the Department's direction and response on a myriad of fronts. The purpose of these two committees was to promote a more cohesive, integrated approach to major policy issues.

But day-to-day fisheries management issues tended to be handled through mechanisms such as the Atlantic Directors General Committee, chaired since 1978 by the ADM, Atlantic Fisheries. This Committee had existed since the early 1970's and survived several departmental reorganizations. It had both regional and headquarters representatives and played a key role in the annual development of fishing plans and the development of Atlantic-specific fisheries policies.

However, the Minister of Fisheries and Oceans is the ultimate decision-maker. As the Department became more heavily involved in fisheries regulation during the 1970's and 1980's, there was an increasing tendency for the Minister to become involved not just in policy direction for the Department but also in decisions on politically sensitive operational issues. On the one hand, the departmental bureaucracy is heavily decentralized with more than 90% of the department's staff located outside Ottawa. On the other hand, because the Minister was seen as accountable for every fisheries management decision big or small, a major component of even micro-level decision-making was centralized. The Minister and his staff took most key decisions, taking into account the advice of the senior bureaucrats channelled through the ADMs and the Deputy Minister. This occurred in spite of the increasing complexity of the fisheries management system, particularly since the extension of fisheries jurisdiction. McCorquodale (1988) hinted at the reason for this when she observed:

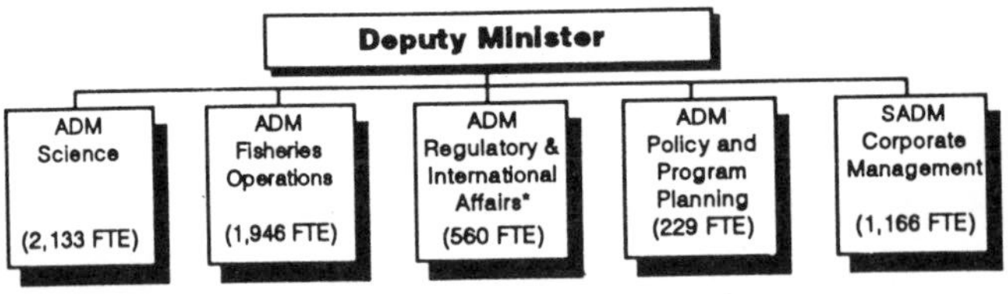

Deputy Minister

	ADM Science (2,133 FTE)	ADM Fisheries Operations (1,946 FTE)	ADM Regulatory & International Affairs* (560 FTE)	ADM Policy and Program Planning (229 FTE)	SADM Corporate Management (1,166 FTE)	Activity Totals
Science	204, 913 (2,133 FTE)					204, 913 (2,133 FTE)
Fisheries Operations		436,461 (1,946 FTE)				436,461 (1,946 FTE)
Inspection			35,188 (525 FTE)			35,188 (525 FTE)
International			6,779 (21 FTE)			6,779 (21 FTE)
Corporate Policy and Program Support			2,552 (14 FTE)	41,341 (229 FTE)	225,611 (1,166 FTE)	269,504 (1,409 FTE)
Sub-total	204,913	436,461	44,519	41,341	225,611	952,845 (6,034 FTE)
Corporate Executive					2,947 (41 FTE)	2,947 (41 FTE)
Total	204, 913	436,461	44,519	41,341	228,558	955,792 (6,075 FTE)

* For the purposes of this Expenditure Plan, resources devoted to enforcement activities are displayed under Fisheries Operations. During 1993-94, enforcement activities will be consolidated under one sector reporting to the Assistant Deputy Minister of Regulatory and International Affairs. Corresponding changes will be incorporated in the 1994-95 edition of the Department's Expenditure Plan.

FIG. 15-7. 1993–1994 DFO Resources by organization/activity.
FTE = numbers of staff expressed as Full Time Equivalents. Expenditures are in $ 000s.

"What makes fisheries different is that this is one of the few... policy areas wherein the federal government is a direct deliverer of services, and where the economic consequences of policy hit the policy community immediately. To allocate fish is to allocate wealth, or more exactly, to allocate the opportunity for wealth."

But DFO and the Minister do not make these decisions in a vacuum. The Department has to respond to a multitude of "publics", including formal and informal groups of stake-holders. Conflicting values bombard the Department from commercial, recreational and native fisheries representatives, provincial governments, foreign governments and international agencies, to cite just a few. At the national level the Department has to function in a bewildering array of checks and balances imposed by central agencies and the Cabinet Committee system. Unlike many departments, DFO cannot ignore the claims and counterclaims of fishermen, processors, provincial governments and other interest groups. Because it is allocating the opportunity for wealth and the right to earn a livelihood from fishing, it must continually reconcile conflicting interests. The Minister, in particular, under the institutional arrangement in effect until 1992 had to try to render decisions which would be perceived as fair and reasonable by that vast multitude of interested parties. Given the wide diversity of those interests, the challenge has been to design structures which allow for meaningful input from all groups with a stake in fisheries management, while at the same time avoiding policy paralysis. In a later section I discuss major changes to this institutional structure being implemented in 1993 and 1994.

Before reviewing the processes and structures for public input to decisions in effect until 1992, I will digress briefly to discuss some general approaches to policy making and decision making and to situate fisheries in that wider context.

4.3 Approaches To Policy Making

4.3.1 The Nature of Policy and the Decision-making Process

"Policy" can be defined in many different ways. Dye (1978) offered a rather simple definition: "Public policy is whatever governments choose to do or not to do." This recognizes that public policy sometimes involves maintaining the status quo or doing nothing. Brooks (1989) defined public policy as: "the broad framework of ideas and values within which decisions are taken and action, or inaction, is pursued by governments in relation to some issue or problem."

Wilson (1981) pointed out that decision-making is crucial to public policy making and that, in the vast number of instances *a series of decisions extending over time* is the crucial distinction between policy-making and decision-making.

Similarly, many different models of the policy making process have been advanced. Anderson (1984) offered the following distinction between decision making and policy-making:

> "Decision-making involves the choice of an alternative from among a series of competing alternatives. Theories of decision-making are concerned with how such choices are made.... Policy-making typically involves a pattern of action, extending over time and involving many decisions, some routine and some not so routine.... It is the course of action that defines policy, not the isolated event."

There is a multiplicity of theories of decision-making. For a discussion of the historical evolution of these theories, see Wilson (1981).

For simplicity, I will concentrate here upon three main theories of decision-making. Probably the best known and most widely accepted is the *rational-comprehensive theory*. This theory has the following elements (Anderson 1984):

1. The decision maker is confronted with a given problem that can be separated from other problems or at least considered meaningfully in comparison with them.
2. The goals, values, or objectives that guide the decision-maker are clarified and ranked according to their importance.
3. The various alternatives for dealing with the problem are examined.
4. The consequences (costs, benefits, advantages and disadvantages) that would follow from the selection of each alternative are investigated.
5. Each alternative, and its attendant consequences, can be compared with the other alternatives.
6. The decision-maker will choose that alternative, and its consequences, that maximizes the attainment of his goals, values, or objectives.

This, in theory, will produce a rational decision, i.e. one that most effectively achieves a given end. This theory, although widely espoused, has been criticized on several grounds. First of all, decision makers are rarely confronted with concrete, clearly defined problems. Defining the problem accurately is often a significant challenge for the decision maker. A second criticism of the rational-comprehensive approach is the multiplicity of information demands it places on the decision maker. The decision maker faces barriers such as lack of time, difficulty in collecting information and predicting the future, complexity of measuring and calculating costs and benefits.

From the perspective of its utility as a decision-making model for fisheries, the rational-comprehensive approach has another significant deficiency. The fisheries decision maker is confronted with conflicting values, rather than value agreement, and these conflicting values are not easily compared or weighted.

The second major theory is the *incremental* approach. Incrementalism has the following features (Anderson 1984):

1. The selection of goals or objectives and the empirical analysis of the action needed to attain them are closely intertwined with, rather than distinct from, one another.
2. The decision maker considers only some of the alternatives for dealing with a problem, and these will differ only incrementally (i.e. marginally) from existing policies.
3. For each alternative only a limited number of "important" consequences are evaluated.
4. The problem confronting the decision maker is continually redefined. Incrementalism allows for countless ends-means and means-ends adjustments that have the effect of making the problem more manageable.
5. Incremental decision making is essentially remedial and is geared more to the amelioration of present, concrete social imperfections than to the promotion of future social goals.

Charles Lindblom has been a major proponent of incrementalism, advocating it in his famous paper *The Science of "Muddling Through"* (Lindblom 1959). Anderson (1984) summarized the arguments in favour of incrementalism as follows:

> "Decisions and policies are the product of "give and take" and mutual consent among numerous participants ("partisans") in the decision process. Incrementalism is politically expedient because it is easier to reach agreement when the matters in dispute among various groups are only modifications of existing programs rather than policy issues of great magnitude or an "all or nothing" character. Since decision-makers operate under conditions of uncertainty with regard to the future consequences of their actions, incremental decisions reduce the risks and costs of uncertainty. Incrementalism is also realistic because it recognizes that decision-makers lack the time, intelligence, and other resources needed to engage in comprehensive analysis of all

> alternative solutions to existing problems. Moreover, people are essentially pragmatic, seeking not always the single best way to deal with a problem, but more modestly, 'something that will work.' Incrementalism, in short, yields limited, practicable, acceptable decisions."

Incrementalism has been attacked on the grounds that it is too conservative and too focused on the existing order, thus impeding innovation. Furthermore, in crises, incrementalism is of no value. Another deficiency is that it may discourage the search for alternatives. Despite these criticisms, incrementalism has gained acceptance in recent years as being more in tune with reality. One of the best examples of incrementalism is the federal budgeting process in the United States. For an analysis of this, see Wildavsky (1979).

While incrementalism is a useful theory, frequently there are nonincremental, or fundamental, policy decisions which depart sharply from past practice or require large increases or decreases in the commitment of resources to given policies. In the broader Canadian political arena, examples include Medicare, Constitutional changes, bilingualism, the Free Trade Agreement, and privatization.

For fisheries management, one of the most useful distinctions is that made by Manzer (1984) between elitist planning and pluralist exchange models of public policy-making. The elitist planning paradigm is essentially the rational decision-making model. As Manzer describes it, this approach involves "collective decisions as deliberate choices from available options made by designated decision-makers on behalf of a group. In this view public policies are, or perhaps, should be the result of anticipatory problem-solving, synoptic planning and rational choice."

The pluralist exchange paradigm involves "collective decisions as [related] outcomes of decisions made by many individuals or groups interacting with one another. Public policies in this view are the result of reactive problem-solving, strategic planning and incremental decisions."

Manzer (1984) described the pluralist exchange model thus:

> "Decision-makers wait for the pressure of events to establish their priorities. Then they fight fires or prescribe cures, dealing with pressing issues and political crises as the need arises."

From my observations of the fisheries management scene in Canada, it appears that the government bureaucracy adopted a variety of decision-making approaches based in theory on the rational approach to problem-solving. In practice, however, actual decision making in the fisheries management arena has been largely characterized by incremental adjustments to policy, based on interactive processes. There is intellectual analysis (e.g. the 1976 Policy for Canada's Fisheries, the 1982 Report of the Task Force on Atlantic Fisheries and the 1982 Commission on Pacific Fisheries Policy, the 1986 Fish Habitat Management Policy) but actual policies applied are shaped to a large extent by the interaction of conflicting interest groups. As Brooks (1989) observed:

> "There can be little doubt that the process of fragmented decision-making and incremental change described by the pluralist exchange model is a much closer characterization of reality than the elitist planning model."

Nonetheless, in the Canadian fisheries instance, there are clear examples of fundamental policy shifts which depart sharply from past practice. Recent examples include the Aboriginal Fisheries Strategy of June 1992 and initiatives currently underway to reform the licensing and allocation process through the establishment of quasi-judicial Boards, under legislation, at arms-length from the Minister.

4.3.2 The Institutional Framework for Policy Making

Various theoretical approaches have been developed to explain the behaviour of political systems. These include systems theory, group theory, elite theory, functional process theory, and institutionalism. The following description of these approaches is based largely on Anderson (1984).

Systems Theory

This views public policy as a political system's response to demands arising from its environment. Easton (1957) defined the political system as being composed of those identifiable and interrelated institutions and activities that make authoritative decisions or allocations of values that are binding on society. Demands and supports comprise the inputs into the political system from the environment. The environment consists of those conditions and events external to the political system. Demands are the claims made by individuals and groups on the political system for action to satisfy their interests. Support occurs when groups and individuals accept the results of elections, pay their taxes, comply with the laws. In a process of feedback, public policies may alter the environment and the results may generate new demands, leading to further policy outputs in a continuing flow of public policy.

Group Theory

This views public policy as the product of group struggle. It regards interaction and struggle among groups as the central facts of political life. A group becomes a political interest group "when it makes a claim through or upon any of the institutions of government." According to this scenario, "public policy, at any given time, will reflect the interests of dominant groups. As groups gain and lose power and influence, public policy will be altered in favour of the interests of those gaining influence against the interests of those losing influence" (Anderson 1984).

Anderson observed that group theory, while focusing attention on one of the major dynamic elements in policy formation, especially in pluralist societies, overstates the importance of groups and understates the role that public officials play in the policy process.

Elite Theory

This approach perceives public policy as reflecting the values and preferences of a governing elite. Elite theorists argue that it is not the people who determine public policy through their demands and action but rather a ruling elite which decides on policy which is then implemented by public officials and agencies. Elite theory is less relevant than group theory to explain public policy formation in pluralist democracies such as Canada.

Institutionalism

This theory contends that political life revolves around governmental institutions such as legislatures, executives, courts, and political parties. The institutional approach focuses on the more formal and legal aspects of governmental institutions.

Each of these theories explains facets of how particular policies evolve. None is all-encompassing or explains fully the public policy-making process in Canada. From the fisheries perspective, there is considerable evidence to support the view that policy is shaped by group struggle, the struggle of conflicting interests. Nonetheless, government institutions and officials strongly influence policy as they attempt to reconcile these conflicting interests.

In the Canadian context, Wilson (1981) described the institutional model as the "hierarchical responsibilities paradigm" (Fig. 15-8). This does not recognize sufficiently the interaction of actors and forces in a pluralist society. Doern and Aucoin (1971) described the most important locus of political action as the "executive-bureaucratic arena." Wilson (1981) defined this as the sphere of interaction between the mandarins, the members of the cabinet and the powerful interest groups who have access to the corridors of power. Wilson (1981) depicted these various relationships in what he termed the pragmatic paradigm of political responsibility (Fig. 15-9). In this figure, the reality of power in the Canadian political system is seen as lying within the circle. Note that this includes interest groups but shows only a weak linkage to the general public.

Hartle (1979) put a slightly different perspective on this pragmatic paradigm:

> "What is the role of government? Is it nothing more than a highly complex means of resolving, through an endless series of compromises, the conflicting interests of those subject to its authority in such a way as to avoid the open use of force: 'the war of all against all.'.... The answer is usually yes, but occasionally no. There is a role for government leadership by which we mean the adoption of policies that are unpopular in prospect but come to be accepted perhaps even with enthusiasm and pride, after they have proven themselves."

This sums up very well the task of fisheries policy makers. They are preoccupied day-to-day with reconciling conflicting interests of the myriad of stake holders in the fisheries. But, if progress is to be made, they must, and do, on occasion rise above the art of compromise, to chart the way to more fundamental change. Sometimes these attempts at leadership succeed; sometimes they fail. The extent to which they succeed usually depends on the interaction between the executive, the bureaucracy and the interest groups, and the extent to which the interest groups can be persuaded to set aside their differences in pursuit of a common goal. This interaction with the conflicting interest groups is a key feature of a fisheries manager's daily life.

Wilson (1981) noted that interest groups can be differentiated into "public interests" and "special interests". Public interests he defined as those which transcend the special interests of groups or individuals. For example, a public interest in the case of fisheries would be the conservation and protection of the fishery resources for future generations. Special interests, on the other hand, are usually exclusive issue groups, identifiable and readily recognizable. These interests may or may not be well organized.

4.3.3 Pressure Groups and Public Policy

Paul Pross has written extensively on the subject of pressure groups and public policy in Canada (e.g. Pross 1986). Pross sees pressure groups as playing a key role in the policy process, with specialized publics dominating decision-making in fields of policy where they have competence.

Pross terms these specialized publics "policy communities". He defines a policy community as "that part of a political system that — by virtue of its functional responsibilities, its vested interest, and its specialized knowledge — acquires a dominant voice in determining government decisions in a specific field of public policy."

A generalized model of the policy community for a field in which the federal government is prominent is shown in Fig. 15-10. This model can apply to fisheries. At the center of this community are the key federal organizations involved: the department or agency primarily responsible for formulating policy and carrying out programs (DFO); Cabinet, with its co-ordinating committees and support structure; the Privy Council office, Treasury Board, etc. None of these are located at the very center of the figure because no single agency is ever consistently dominant in the field. Because so much policy-making is routine, the lead agency tends to be most influential over time. In the case of fisheries, DFO generally has the lead in setting federal fisheries policy. However, in times of crisis, e.g. the Northern Cod controversy, the central agencies (e.g. PCO) come to play a key role, special coordinating structures may be established (the Ad Hoc Cabinet Committee on Northern Cod), and other agencies become involved, e.g. Employment and Immigration; Industry, Science and Technology; External Affairs.

Clustered around, watching closely federal activities and forming part of the *sub-government* are certain

THE HIERARCHIAL RESPONSIBILITIES PARADIGM

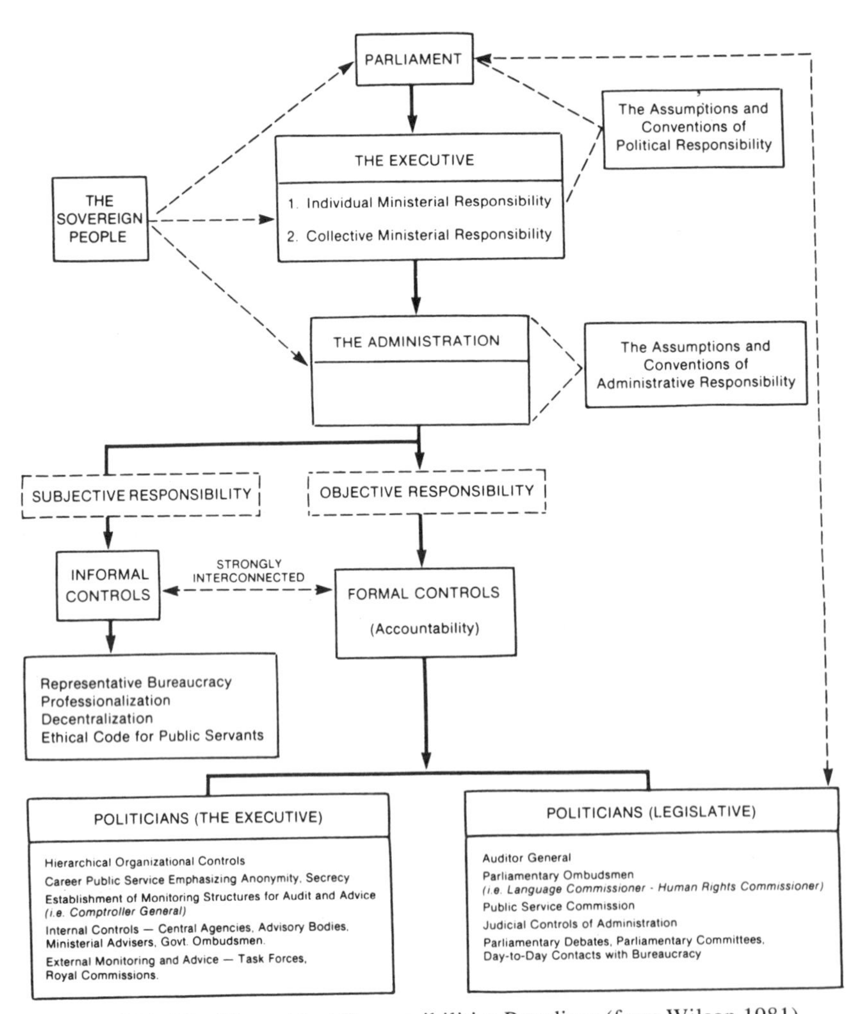

FIG. 15-8. The Hierarchical Responsibilities Paradigm (from Wilson 1981).

provincial government agencies and key pressure groups, e.g. the Fisheries Council of Canada, the NFFAWU, the MFU and the EFF. Pross suggests that such sub-governments consist of very small groups of people.

The second component of Pross's model is what he terms "the attentive public." This is neither tightly knit nor clearly defined:

> "It includes any government agencies, private institutions, pressure groups, specific interests, and individuals — including academics, consultants and journalists — who are affected by, or interested in, the policies of specific agencies and who follow, and attempt to influence, those policies, but do not participate in policy-making on a regular basis. Their interest may be keen but not compelling enough to warrant breaking into the inner circle."

Thus, pressure groups can play a key role in policy formulation. Governments have encouraged the formation of special interest groups in many fields. One of the first things Fisheries Minister Roméo LeBlanc did, upon assuming office, was to encourage fishermen to organize and to speak, where possible, with a single voice. Such groups, through special consultative structures established to receive their input, can persuade policy makers that changes in policy are desired, and worthwhile.

On the opposite side of this coin, special interest groups can mobilize opposition to policies which they consider inimical to the interests of their members. Through effective lobbying and public protests, they can undermine new policies and force governments to reverse or abandon policies in mid stream. A dramatic example of this was the banding together of diverse interest groups on the Pacific coast to block Peter Pearse's proposals to reform the licence access system and institute auctions and individual transferable quotas in certain fisheries.

Occasionally, special interest groups administer policy. A fisheries example is the involvement of the Atlantic Herring Marketing Co-operative in the efforts to transform the Bay of Fundy herring fishery in the late 1970's.

Pross has classified interest groups along a continuum, ranging from the loosely-organized *issue-oriented* group to the more highly-organized *institutionalized* group (Fig. 15-11).

He defined institutionalized pressure groups as:

> "Groups that possess organizational continuity and cohesion, commensurate human and financial resources, extensive knowledge of those sectors of government that affect them and their clients, a stable

THE PRAGMATIC PARADIGM OF POLITICAL RESPONSIBILITY

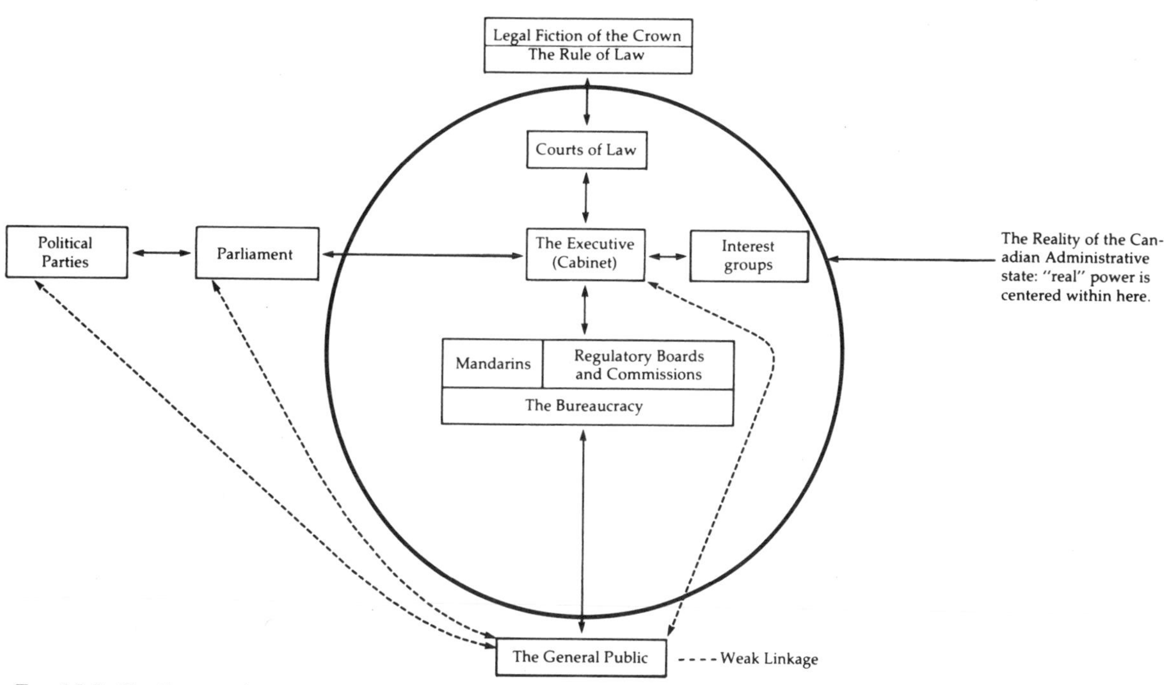

FIG. 15-9. The Pragmatic Paradigm of Political Responsibility (from Wilson 1981).

membership, concrete and immediate operational objectives associated with philosophies that are broad enough to permit [them] to bargain with government over the applications of specific legislation or the achievement of particular concessions and a willingness to put organizational imperative ahead of any particular policy concern."

At the other extreme are "issue-oriented groups". These are:

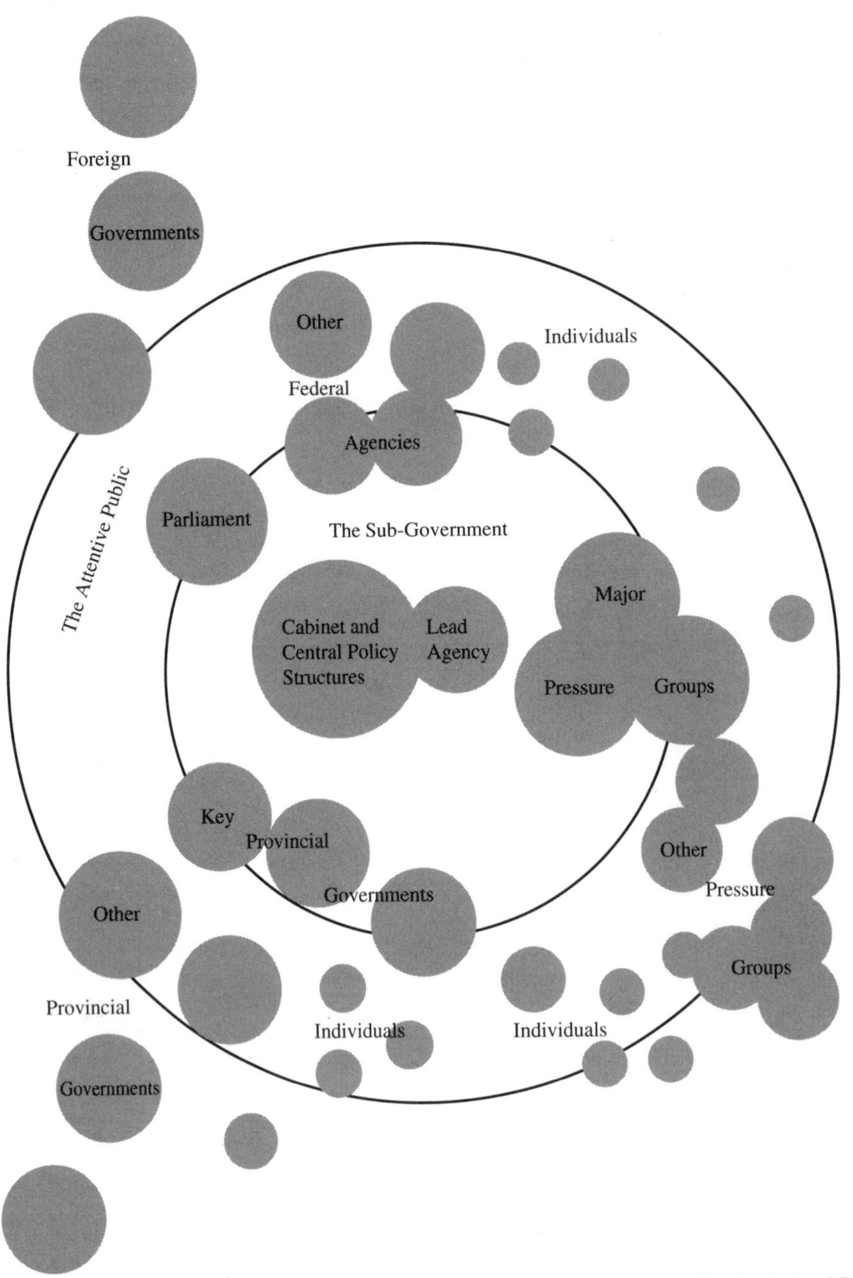

FIG. 15-10. The Policy Community (from *Group Politics and Public Policy* by A. Paul Pross, Fig. 4.1. Copyright © Oxford University Press Canada 1986. Reprinted by permission of the publisher.)

CATEGORIES	GROUP CHARACTERISTICS							
	Objectives				*Organizational Features*			
	single, narrowly defined	multiple but closely related	multiple, broadly defined & collective	multiple, broadly defined, collective & selective	small membership/ no paid staff	membership can support small staff	alliances with other groups/staff includes professionals	extensive human and financial resources
Institutionalized								
Mature								
Fledgeling								
Issue-oriented								

CATEGORIES	LEVELS OF COMMUNICATION WITH GOVERNMENT					
	Media-Oriented:			*Access-Oriented*		
	publicity-focused protests	presentation of briefs to public bodies	public relations; image-building ads, press releases	confrontation with politicians, officials	regular contact with officials	regular contact, representation on advisory boards, staff exchanges
Institutionalized						
Mature						
Fledgeling						
Issue-oriented						

FIG. 15-11. The Continuum Framework for Interest Groups (from *Group Politics and Public Policy* by A. Paul Pross, Fig. 5.1. Copyright © Oxford University Press Canada 1986. Reprinted by permission of the publisher.)

> "governed by their orientation toward specific issues...[and have] limited organizational continuity and cohesion, minimal and often naive knowledge of government, fluid membership, a tendency to encounter difficulty in formulating and adhering to short-range objectives, a generally low regard for the organizational mechanisms they have developed for carrying out their goals, and, most important, a narrowly defined purpose, usually the resolution of one or two issues or problems, that inhibits the development of 'selective inducements' designed to broaden the group's membership base."

There are many diverse interest groups in the fisheries field. These range from the issue-oriented (e.g. a group formed to save a local fish plant from imminent closure) to fledging (e.g. various local habitat preservation groups) to mature (e.g. EFF), to institutionalized. Examples of the institutionalized model include the Fisheries Council of Canada, the Atlantic Salmon Federation, the FFAW, the UFFAWU, the MFU and the APPNE. The Canadian Wildlife Federation, with some selective interest in fisheries issues, is another example of this type.

In the case of fisheries, successive ministers have encouraged fishermen to organize and other interests to band together to speak with a common voice. The clash of conflicting interests in fisheries matters is so great that the federal government has historically taken other initiatives to provide a more structured input from these conflicting interest groups into the policy-making process. These many groups speak with a thousand voices. To deal with this, government found it necessary during the 1970's and the 1980's to create an elaborate consultative framework in an attempt to reconcile these interests and to bridge the wide gap between differing views.

4.4 The Fisheries Consultative Framework

4.4.1 Atlantic

4.4.1.1 Evolution of the Consultative Process

Policies for management of Canada's Atlantic fisheries evolved rapidly during the 1970's and 1980's. Management-consultative processes developed at a similar pace. As policies changed, consultative and decision-making structures also changed. The evolution of the consultative process occurred in four phases to the mid-1980's.

Phase I

Prior to 1971, the main emphasis in marine fisheries policy was on the development and expansion of Canada's capability to harvest and process offshore fish resources in competition with foreigners. The only structured consultative mechanism was the Federal/Provincial Atlantic Fisheries Committee (FPAFC) of Deputy Ministers. This was established originally to coordinate federal and provincial efforts to "modernize" the fleets. There was no formal domestic framework to feed industry and provincial views into fisheries management decisions. The Minister made policy and management decisions on the advice of the department. As this was prior to the introduction of TACs and limitation of access, this advice was largely focused on conservation and protection activities within Canada's narrow fisheries waters.

Phase II

During the period 1972–76 there was a dramatic shift in the approach to management. ICNAF adopted the approach of setting TACs and national quotas and Canada took its first steps toward limited entry. New consultative mechanisms were established in the form of fisheries advisory committees (species specific). These had federal, provincial and industry representatives. Examples included the Offshore Groundfish Advisory Committee (OGAC) and the Atlantic Herring Management Committee (AHMC). Because of declining fish abundance during that period, these committees focused primarily on stock management issues rather than on resource allocation or industry viability. There was a general lack of organization among inshore fishermen which resulted in an ineffective input from that sector into the decision-making process.

In contrast, fish processors were well represented by the Fisheries Council of Canada (FCC). The FCC frequently had the ear of the Fisheries Minister and was quite influential in shaping fisheries policy. An example of this was the FCC's brief on extension of Canada's fishing zones. It largely shaped the government's proclamation of fisheries closing lines on both coasts in the early 1970's.

Also during this period much consultation occurred in the context of ICNAF, where the battle over management approaches was being fought. Industry representatives also became heavily involved on an ad hoc basis in Canada's delegations to the Law of the Sea Conference (Johnston 1978).

Phase III

The post-extension period from 1977 to 1981 witnessed the proliferation of consultative committees, as government assumed a greater regulatory role with the proclamation of the 200-mile fisheries zone. Major stocks were declining, e.g. Atlantic groundfish. The TAC controls introduced by ICNAF had proven ineffective and Canadians were forced to tighten their belts. The first Atlantic groundfish fishing plan was introduced to share among the fleets an insufficient supply of fish. With the introduction of resource allocation and the more widespread adoption of entry controls, the clash of conflicting interests became apparent. So, too, did the need for structured consultation. OGAC evolved into AGAC, the Atlantic Groundfish Advisory Committee, which became a major forum for debate of stock management and resource allocation issues.

The process began to focus on the contentious question of allocating among fleet sectors and achieving a balanced input by all participants to allocation decisions. The fisheries consultative committee system expanded. So did the federal government's attempts to deliver local services. Decentralization of services arrived with the Area Manager system. Policy decisions, however, remained centralized. While large companies and the processing sector in general were organized for significant input to policy decisions, the inshore sector had still not acquired the analytic and communication capabilities to represent its members, with certain notable exceptions (e.g. the NFFAWU). During this period the Minister of Fisheries, Roméo LeBlanc, encouraged and the Department assisted fishermen to organize. An example is the establishment of the Eastern Fishermen's Federation, funded by a special squid allocation.

Phase IV

From 1981 to 1985, the Department attempted to strengthen the consultative mechanisms to provide for a decision-making system involving annual consultations on TACs, allocations and regulations for all fisheries. This included efforts to ensure a balanced membership on species and other advisory committees.

In addition to the established advisory/consultative committees, the Department convened special policy seminars which focused on the longer term strategies for certain fisheries (e.g. northern cod, Atlantic herring) or addressed particular policy issues (e.g. quality improvement). Representation was drawn from all sectors of the industry as well as the provinces. As well, there were meetings with industry representatives which focused on strategies/positions to be taken in annual bilateral discussions with foreign countries fishing in Canadian waters.

In addition, the Federal-Provincial Atlantic Fisheries Committee continued to meet regularly to discuss not only resource management questions but an array of fisheries policy issues of mutual concern to both levels of government. The FPAFC also coordinated programs between the two levels of government.

The Atlantic Council of Fisheries Ministers (ACFM), established in 1978, met several times a year to discuss resource management and fisheries policy issues.

In 1985, another consultative body, the Atlantic Regional Council (ARC), was added to the consultative structure. It was followed some time later by the establishment of a similar Pacific Regional Council (PARC). I deal with the mandate of these Councils later in this chapter.

The structure of consultative/advisory committees had evolved differently in different regions, influenced by the differences in the fisheries and the structure of fishermen and processors' organizations.

4.4.1.2 The Advisory Committee Structure in the late 1980's.

As of the late 1980's there were eight multi-regional species advisory committees. The largest and best known of these was the Atlantic Groundfish Advisory Committee (AGAC) which had regional and sub-regional committees. The Atlantic Salmon Board, representing the recreational, commercial and native user groups, as well as the provinces, was another major multi-regional species consultative committee. Other species-based consultative committees dealt with seals and sealing, northern shrimp, squid and bluefin tuna.

There was a series of regional and sub-regional licence appeal committees plus a large number of species oriented committees. The Newfoundland Region had 11 committees, including the licensing appeal committees. All committees were regional, except the licence appeal committees and the local inshore groundfish advisory committees which were subregional. There were also advisory committees for the following species or groups of species: groundfish, small pelagics, salmon, seals, lobsters and crabs.

The Scotia-Fundy Region had 34 regional advisory committees, apart from the licensing appeal committees. There were regional or sub-regional (area) advisory committees for the following species or groups of species: groundfish, crab, shrimp, scallops, clams, lobster (nine district committees), marine plants, herring, swordfish, tuna, mackerel, salmon (five zonal management committees), gaspereau and shad.

The Gulf Region had 46 advisory committees, plus five joint committees with the Quebec Region, in addition to its licence appeal committees. There were regional or sub-regional committees for the following species or groups of species: groundfish, snow crab, shrimp, trout, oysters, scallops, lobsters (five district committees), marine plants, gaspereau, mussels, tuna, herring, mackerel, smelt, eels, seals and salmon (five zonal committees).

The Quebec Region had 14 advisory committees. These covered the following species or groups of species: groundfish, lobster, scallops, tunas, crabs, shrimp, herring, mackerel, seals and whales. There was also a committee to develop management plans for marine mammals (seals and whales) for Northern Quebec which met once a year in a northern location.

The species advisory committees included DFO officials and representatives designated by provincial deputy ministers, fishermen's organizations and fishing companies including the processing sector. Each interregional management advisory committee was supported by a Working Group of DFO officers who consolidated biological, economic (including marketing), and other input into draft fishing plans for committee consideration. Terms of Reference for each committee focused primarily on TACs or alternative management measures, allocations and regulations which were expressed in an annual fishing plan.

4.4.1.3 The Advisory/Consultative Process

This section describes the process in effect until 1992. Chapter 17 describes the scientific advisory process. Advice from the Canadian Atlantic Fisheries Scientific Advisory Committee (CAFSAC) was channelled to the Assistant Deputy Ministers of Atlantic Fisheries (Operations) and Science. From there:

(a) the advisory document proceeded to an interregional management advisory committee for interregional consultations leading to a draft management plan; or

(b) in the case of stocks of concern to one region only, the advice proceeded to the appropriate single region advisory committee for consultation and development of a plan;

(c) draft management plans were forwarded to the Atlantic DGs Committee for key recommendations on strategies and TACs;

(d) the DGs Committee sent recommendations to the Minister, via the ADM and the Deputy Minister, for approval. The Deputy Minister consulted with his provincial colleagues through the FPAC and the Minister with his provincial colleagues through the Atlantic Council of Fisheries Ministers; and

(e) following the Minister's decision, or the decision of cabinet in exceptional cases, regulations were drafted, finalized and promulgated, and the fishing plan announced.

For stocks managed by the Northwest Atlantic Fisheries Organization (NAFO), the NAFO Scientific Council advised the NAFO Fisheries Commission. The Atlantic DGs Committee also used the Council's advice, in consultation with interregional advisory committees, to develop the Canadian negotiating position. This included estimating Canada's harvesting requirements. Following NAFO decisions on TACs and allocations, interregional advisory committees were consulted regarding development of a Canadian management plan, following the process described in (a) above.

This process for the Atlantic groundfish fishery is summarized in Fig. 15-12 and 15-13.

Management plan preparation took place over a number of months preceding the fishing season to which a plan applied. The process started with the receipt of the scientific advice on the fish stocks. Given this advice and the experience of the current fishing

Meeting of a Small Pelagics Advisory Committee

ATLANTIC GROUNDFISH MANAGEMENT PROCESS

GROUNDFISH	STOCKS/NAFO DIVISION	JAN.FEB.MAR.APR.	MAY	JUNE	JULY/AUG.	SEPT.	OCT.	NOV.	DEC.
a) Canadian stocks (Inside 200 mi.)	Cod 2GH, 2J3KL, 3Ps, 4RS, 3Pn, 4Vn,4T, 4VsW, 4X Haddock - 4VWX+5 Pollock 4VWX+5 Redfish - 2 + 3K, 3O, 3Ps, 4RST, 4VX American Plaice 2 + 3K, 3Ps, 4T Witch 2J, 3KL, 3Ps, 4Rs Greenland Halibut 2J+3KL, 3Ps, 4RS Silver Hake 4VWX White Hake 3LNO, 4T Argentine 4VWX Round Nose Grenadier 2 + 3 Other Groundfish 5 + 6	Bilaterals with Foreign Countries on Fishing Plans and Allocations (ongoing)	CAFSAC Groundfish Sub-committee	Atlantic Region DGs Committee			OVOWG* and IOG**	Federal Provincial Atlantic Fisheries Committee	Atlantic Fisheries Ministers Conference
			CAFSAC Steering Committee	Atlantic Groundfish Advisory Committee			Atlantic Groundfish Advisory Committee		Management Plan Announced for Next Year
				NAFO Scientific Council					
b) Canada/NAFO (Stocks overlapping and outside 200 mi.)	Cod 3M, 3NO Redfish 3M, 3LN American Plaice 3M, 3LNO Yellowtail 3LNO Witch 3NO Squid 3 + 4					NAFO General Council & Fisheries Commission Annual Meeting			

* Offshore Vessel Owners Working Group (OVOWG)
** Independent Offshore Group (IOG)

FIG. 15-12. Atlantic Groundfish Management Process (until 1992).

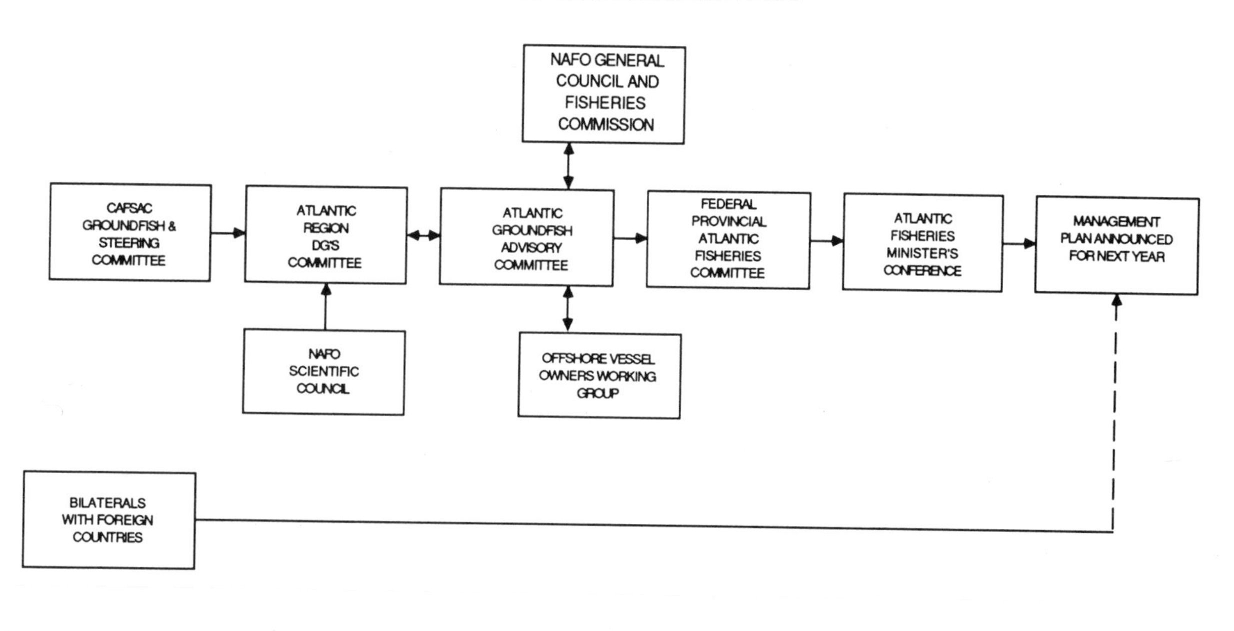

FIG. 15-13. Atlantic Groundfish Management Information Flow (until 1992).

season, DFO then prepared draft plans for subsequent consultations with representatives of the fishing industry and the governments of the five eastern provinces. The next stage was a series of government-industry consultations through the advisory committees at local, area, regional and inter-regional levels. Intergovernmental consultations took place within FPAFC and ACFM.

DFO (1986j) has described the purpose of these widespread and often prolonged consultations as essentially twofold:

1. To advice the principal user groups and the provincial governments on the basic direction and content of the proposed management plans; and
2. To arrive at a broadly-based consensus on the major elements of these plans, particularly with respect to the sharing of the fish quotas among the different user groups.

Consensus is not easily achieved. The consultative committees bring to the table the diverse views and conflicting interests of those involved in a fishery. They compel representatives of the various groups to express their demands and views in front of other user groups. This motivates the various interests to articulate and support a position, rather than just criticize government proposals. Every year for every fishery there are contentious issues on which there is no consensus. Under such circumstances, the Committee Chairman works to narrow the gap between conflicting views, puts alternatives on the table and guages the degree of support for each alternative. Conflicting views are noted. If there is an impasse, the Chairman, usually a federal official, ensures that these conflicting views are taken into account when the decisions are made.

The consultative committees, in effect, institutionalize the conflict between different user groups. They channel it into an orderly discussion of the pros and cons of various alternatives, bridge the gap to the extent possible, and finally assure the participants their views will be considered. One negative effect of the process is that it can polarize views, as parties stake out extreme positions expecting a compromise in the final decision.

With goodwill among participants, a more genuine attempt to bridge the differences is possible. Sometimes differences can be narrowed.

Frequently, a great deal of posturing takes place in the consultative committees. Progress often depends on the genuine extent of the conflict, and, in particular, on previous relations among the parties. If these have been acrimonious, then the Chairman will have great difficulty. Horse-trading between the parties will sometimes occur outside the formal consultative structures.

The Task Force on Atlantic Fisheries recognized that it is difficult to consult the myriad interests in the Atlantic fishery. It acknowledged there was a great deal of formal and informal consultation about the Atlantic fisheries. However:

> "There remains widespread criticism that such consultation is not as effective as it should be and that, on some important issues, it does not take place at all. The Department of Fisheries and Oceans is sometimes accused of being secretive and arbitrary and of taking decisions without adequate knowledge or advice from knowledgeable sources (that is, processors and fishermen). Provincial governments complain that they are treated as merely one of many interest groups rather than as another level of government" (Kirby 1982).

The Task Force concluded that the processing sector of the industry was well organized in each province, as well as under the umbrella of the FCC. There was, however, a considerable problem of fragmentation among Atlantic Canada fishermen which the Task Force perceived to be "a major stumbling block to effective communication and consultation." As a consequence, "DFO appears to adopt a paternalistic approach and attempts to do for fishermen what they cannot do for themselves — that is, represent their own interests. The result is occasionally commendable, sometimes pathetic, and always awkward, if not inappropriate."

Given this assessment of the situation, the Task Force recommended that:

> "Steps be taken by DFO, in concert with industry organizations, provincial departments of fisheries, and other federal departments as appropriate, to formalize and streamline the process for consultation and communication on fisheries policy and programs."

The Task Force recommended that the government:

1. Review membership on management advisory committees and, in the interest of effective communication and serious pursuit of consensus, reduce numbers to the minimum necessary to ensure that essential interests are represented. Delegate greater responsibility to sub-committees to provide for greater efficiency and more effective representation (R52).
2. Encourage organization of fishermen generally, as well as umbrella groupings that can represent the

fishermen's viewpoint on region-wide and Atlantic-wide issues (R53).

3. Make greater use of the Federal-Provincial Atlantic Fisheries Committee to develop policy, to harmonize programs and to resolve conflicts. This will probably require the creation of a network of subcommittees on a continuing or ad hoc basis (R54).
4. Create an Atlantic Fisheries Consultative Group of knowledgeable and experienced individuals. The maximum size should be 10 to 12, with occasional rotation of members. The Group would operate informally, with a mandate to advise the Minister and senior officials on major strategic issues (R55).

The Task Force argued for an ongoing mechanism to identify and develop strategic issues; hence, its proposal for an Atlantic Fisheries Consultative Group. The Task Force rejected a proposal based on the U.S. Regional Council system.

These Task Force recommendations had little immediate impact on the Atlantic fisheries consultative process. A serious effort to change the number and structure of consultative committees did not occur until 1986–1987. At that time, DFO brought before the FPAFC and ACFM a proposal to restructure the consultative committee process (DFO 1986j). That proposal examined existing and alternative consultative committee structures based on *nine criteria*:

1. *Clear Client Input.* All clients must have an opportunity to advise the Department, normally in a common and public forum.
2. *Fairness.* The system must be perceived to be fair and to consider all points of view prior to the final decision.
3. *Timeliness of Decisions and Announcements.* Decisions and announcements must be made sufficiently in advance of the normal fishing season to permit clients to prepare and participate.
4. *Client Satisfaction.* The clients must feel that the system is useful, meaningful, relevant and takes their representation and concerns into account.
5. *Cost Effectiveness.* The costs for consultation borne by the department must be in reasonable proportion to the remainder of the implementation costs.
6. *Effective Provincial Input at All Levels.* The provinces must have the opportunity for meaningful input at various levels in the process, from working groups to the Atlantic Council of Fisheries Ministers.
7. *Equitable Representation.* Membership must be representative, relatively compact, articulate, and accountable. There must also be a mechanism for regular turnover in membership.
8. *Consistency.* There must be interregional consistency, unless circumstances clearly dictate different approaches.
9. *Availability of Most Recent and Relevant Scientific Advice.* Up-to-date and relevant scientific advice must be available to support management decisions.

This document concluded that the existing consultative system met most of these criteria. However, it had some major weaknesses, particularly for anadromous species, and in the timeliness of the consultative process in general.

A more systematic system was proposed which would establish a system of multi-regional, regional and subregional committees for groundfish, anadromous species, pelagics and shellfish. Each region would have had a similar consultative structure, with local and regional committees for the four main species groups. The regional committees would feed into multi-regional committees, which in turn would advise the Minister. Essentially, the system for other species groups would be modelled on that in effect for Atlantic groundfish. The Atlantic Salmon Board would have formed the basic structure for the Anadromous Advisory Committee, with a revised membership and new regional committees in some regions. The multi-region Shellfish and Pelagic Advisory Committees would have been new.

The stated intent was "to give industry a stronger voice across the board, with a greater and more systematic influence on decisions. More information would flow between governments and industry. Management decisions would be more timely."

FPAFC and ACFM considered this proposal. Initial reaction appeared favourable, with all provinces expressing support. However, questions were raised about the proposed representation on the committees. As well, some provinces questioned the adequacy of AGAC as a model and the appropriateness of Atlantic-wide committees for species such as shellfish and pelagics. In September 1987, the Atlantic Council of Fisheries Ministers decided to maintain the existing consultative structures and process.

During the 1980's governments continued to encourage fishermen to organize. A new alliance of fishermen was formed in Québec, with government assistance. The MFU became more firmly established in New Brunswick through collective bargaining legislation, and strengthened its representation in southwest Nova Scotia.

The TFAF recommendation to make greater use of the FPAFC was implemented. Virtually all major Atlantic fisheries policy issues were brought before this committee and, where appropriate, to its ministerial counterpart, ACFM. FPAFC failed, however, to address seriously the issue of fish processing capacity. This is borne out by the dramatic expansion in processing capacity during the 1980's. The FPAFC also failed to harmonize the various federal and provincial subsidy and loan programs for fishermen, as proposed by the TFAF.

The proposal to establish an Atlantic Fisheries Consultative Group was acted upon in 1985 when a new Atlantic Regional Council was established.

4.4.2 Pacific

The evolution of consultative arrangements on the Pacific coast is less clear than on the Atlantic coast. As on the Atlantic, over time numerous consultative committees have been put in place. The fisheries consultative processes have changed significantly as the complexity of management has increased. As late as the mid 1970's, much of the public advisory process was informal, occurring between ad hoc groups or individual users and fisheries staff, especially fishery officers in the field.

A 1975 brief from the Vessel Owners Association of British Columbia to Fisheries Minister Roméo LeBlanc stated:

> "Government regulations are frequently made without any industry consultation, and especially without any input from fishermen's groups. It is hardly surprising, therefore, if such regulations turn out to be ineffective and inequitable....Fisheries Service, Pacific Region, has, in the past, for all practical purposes, ignored our presence" (*Western Fisheries*, June 1975).

The Association called for input into the decision-making process and improved mechanisms for consulting the fishing industry.

By the late 1970's a formal consultative structure had evolved. In 1982, this consisted of 13 formal advisory groups (Table 15-2).

The fishing industry at this time was also involved as advisors to the Canadian delegations to three international commissions, the International North Pacific Fisheries Commission, the International Pacific Halibut Commission, and the International Pacific Salmon Fisheries Commissions.

According to DFO's Submission to the Pearse Inquiry:

> "These groups play a very important role in ensuring industry's participation in important decisions which shape the conduct and future of the fisheries. As management problems grow more complex, the importance of these groups becomes more significant" (DFO 1982d).

TABLE 15-2. Formal fisheries advisory groups on the Pacific coast in 1982.

1. Pacific Region Fisheries Management Advisory Council to the Minister of Fisheries and Oceans (This became known during the mid-1980s as the Minister's Advisory Council (MAC)).
2. Pacific Region Fisheries Management Advisory Council. (This council was essentially a duplication of the Minister's Advisory Council and soon became defunct).
3. Central Coast Advisory Committee.
4. Fraser River Advisory Committee.
5. Groundfish Advisory Committee.
6. Herring Advisory Committee.
7. Herring Spawn-on-Kelp Advisory Committee.
8. Johnstone Strait Chum Advisory Committee.
9. Queen Charlotte Islands Advisory Committee.
10. Skeena River Advisory Committee.
11. Sport Fishing Advisory Board.
12. Stikine River Advisory Committee.
13. The Yukon River Advisory Committee.

Despite this emphasis on industry participation, an examination of the mandates and composition of these various groups suggests a patchwork of area and species advisory groups, with no consistent structure. The only global group was MAC which became increasingly used as the major consultative body until it was replaced in 1987 by a new Pacific Regional Council (PARC).

Pearse's 1982 Report devoted a chapter to consultation. He noted the increasing importance of participation by the public and special interest groups in public agencies' decisions. He observed that formal structures and informal channels for consultation and advice had profilerated in wide variety. This was attributed in part to the growing complexity of governmental regulation, which created a need for outside advice.

Pearse, like the Task Force on Atlantic Fisheries, recognized the need for effective consultative processes and their potential role in moderating conflict and reconciling conflicting interests. Pearse acknowledged that the Department had responded to this growing need by "creating a host of consultative committees, advisory boards, task groups and other channels for liaison with the interested public." Pearse pointed out that these processes consumed a great deal of time and effort but had come under "heavy and widespread criticism at my public hearings and are now being undermined by a lack of confidence."

Pearse added:

> "With a few exceptions, most commentators are distressingly critical of the consultative

process, describing it in such terms as an "exercise in frustration", "window dressing" and a "dialogue of the deaf". Although specific criticisms vary, many who have served on advisory committees complain that they lack direction, clear terms of reference and orderly procedures. Insufficient advance notice of issues to be discussed and inadequate information for informed discussion are also common complaints. Others have charged that consultations are a public relations exercise on the part of the Minister or the Department, only rubber-stamping decisions already made. And most worrisome...is the widespread perception that advice is not seriously sought or listened to.

Pearse expressed concern about the representations and lobbying that circumvent the consultative system, through delegations to the Minister, meetings and interviews with senior officials, and endless phone calls with demands and complaints. He noted:

"The Department apparently tries to follow an open-door policy, accommodating all these representations, but the appeal of this approach is superficial. The Department should, of course, respond to private concerns, but by tolerating and encouraging all these informal representations, which are usually not public and are often between acquaintances, the consultative structure is undermined. It also exhausts the time of senior officials, who seem to spend an inordinate proportion of their time in meetings.

"These methods cannot provide the Department with balanced advice. Clearly we need a consultative system that will relieve officials of the flurry of unstructured lobbying so they can attend to their responsibilities in the context of publicly articulated advice from interested private groups. For this to work, interested individuals and groups must have confidence that the channels provided for this purpose offer the most effective means of exercising influence."

Pearse concluded fundamental reform of the consultative structure, process and procedures was needed.

He recommended that MAC be replaced with a new Pacific Fisheries Council with the following provisions:

(i) The council should be provided for in legislation.
(ii) The council's terms of reference should embrace all Pacific fisheries matters within the responsibility of the Minister of Fisheries and Oceans.
(iii) It should consist of not more than eight members, appointed by the Minister for staggered 3-year terms. They should be appointed in their personal capacities and selected for their knowledge, experience and judgement, and not for their affiliations.

Pearse intended that this council have a high status, and become the central forum for consultations between the Minister and public interests. It should be the channel for coordinating communication with more specialized advisory committees. Pearse recognized the need for more specialized advisory committees "to deal with the narrower, but often complicated, problems associated with particular fisheries, regions and interest groups."

He proposed three distinct categories of advisory committees:

1. Fisheries advisory committees;
2. Fisheries conservation committees; and
3. Special regional management committees.

He envisaged a special fisheries advisory committee for each of the significant fisheries that had special regulatory policies, including the sport and Indian fisheries, the separately licensed commercial fisheries and mariculture.

Compared with the existing fisheries advisory committees, Pearse proposed restricting the functions of the proposed new committees:

"First, they should not concern themselves with the fractious question of catch allocations among competing groups. The general policy on this issue should be established at a higher level in consultation with the Pacific Fisheries Council, and specific arrangements should be laid out in pre-season fishing plans.... However, these committees should be involved in setting objectives for resource management and appraising the results achieved. Second, these committees should not concern themselves with day-to-day, in-season management, but rather with policy, planning and results."

Pearse proposed abolishing the Salmonid Enhancement Task Group and establishing three regional fisheries conservation committees, one each for the north, south and Fraser River administrative areas. These would address enhancement and habitat management matters.

In addition, Pearse envisaged local advisory committees for consultation on special fisheries habitat or management issues in areas where these problems could not be adequately dealt with by the fisheries advisory committees or the fisheries conservation committees.

Ironically, it was to MAC that the Minister of Fisheries and Oceans turned for advice on the Pearse Report. MAC endorsed in principle the proposal for new policies and procedures for more effective public consultation. The specifics of any new arrangements were deferred pending advice from MAC. As of 1986, the advisory process had not changed significantly. The one significant change came late in 1986 — early in 1987 when MAC was replaced by PARC (see Pacific Regional Council, section 4.4.3.2). Few of Pearse's specific proposals for changes to the rest of the consultative structure were acted upon.

Since PARC was established in 1987, there have been some changes in the advisory/consultative process in the Pacific Region. As of 1989, there were 18 advisory bodies. Five of these reported to the Minister of Fisheries and Oceans directly and 13 were domestic boards or committees advising the Department directly. Major changes included the new committee structure serving the Pacific Salmon Commission, the establishment of PARC, the establishment of a B.C. Aboriginal Peoples Fisheries Commission, and the establishment of a Commercial Fishing Industry Council following the disbanding of MAC.

Bodies advising the Minister included:

1. *Pacific Salmon Commission*
   This was established under the Canada-U.S. Pacific Salmon Treaty of March 1985. The Commission advises each country on matters pertaining to the Treaty and serves as the forum for consultation and negotiation of annual management plans for the major intercepting fisheries.
   The Commission meets formally on three occasions annually:
   (i) Consultation involving the Panel Chairs/Vice-Chairs and Joint Technical Committee Chairs — October.
   (ii) Post fishing season meeting — November.
   (iii) Annual meeting of the Commission — February.
   The Fraser Panel, because of its responsibilities concerning in-season management of the fisheries for Fraser River sockeye, meets frequently throughout the year. (In 1988 the Fraser Panel met 28 times).
2. *Pacific Regional Council (PARC)*
   This is now the senior advisory board to the Minister on Pacific fishery matters.
3. *Commercial Fishing Industry Council (CFIC)*
   This Council was formed in April, 1987, following the disbanding of MAC, to provide a forum in which general issues affecting the commercial fishery would be discussed.
   Each CFIC member organization is represented on the Council by a single individual chosen by that organization. The Council is co-chaired by two members elected from the 14 council members. CFIC's mandate includes the development of annual domestic commercial salmon allocation plans.
   From 1987 to 1990 CFIC was a focus for discussion of commercial salmon allocations.
4. *The Sportfish Advisory Board*
   This advisory board has existed since the 1970's. As of 1990, it had 28 representatives. It advises DFO on tidal and non-tidal sport fishing matters and assists in publishing information about those fisheries.
5. *B.C. Aboriginal Peoples Fisheries Commission*
   This Commission was formed in March 1984 for consultation on Indian fisheries issues. The BCAPF represents 16 Native Bands.

There are 13 domestic advisory committees and/or boards advising the Department at the Pacific Region level. Seven of these are concerned with salmon management. These are area- and gear-specific and include advisory committees for the Queen Charlotte, Skeena River, Central Coast, South Coast and Fraser River areas and for the Inside Troll and Outside Troll fisheries.

These committees are chaired by Pacific Region officials, with representation from the diverse groups involved in the fishery in particular areas. Generally, these committees meet once a year formally, with ad hoc meetings as required to discuss in-season management.

Two bodies advise on Pacific herring management — the Herring Industry Advisory Board (HIAB) and the Herring Spawn-on-Kelp Committee. Stocker (1993) has described the role they play in developing management plans for Pacific herring. The Herring Industry Advisory Board, comprised of 22 representatives from non-federal organizations, is chaired by DFO's Herring Coordinator. Many in the department and industry consider this to be the most successful and effective consultative body in the Pacific Region (Pacific Region Headquarters staff, pers. comm.).

The Herring Spawn-on-Kelp Committee consists of the approximately 28 (now 38) licence holders. It advises on the planning and development of this fishery.

A Pacific Groundfish Advisory Committee provides a consultative forum for all matters relating to planning, policy development and development of annual management plans for groundfish. It also contributes to longer term development plans for groundfish. A Sablefish Advisory Committee has similar functions.

There is also an umbrella Shellfish Advisory Committee and five Shellfish Sector Committees. The Shellfish Advisory Committee provides input to DFO regarding management and research. This committee deals with broad issues not addressed by the Shellfish Sectoral Committees. Such issues include overall shellfish stock status and impacts from other fisheries, e.g. aquaculture.

The five Shellfish Sector Committees are: Shrimp (Prawn) Trap, Crab Trap, Net Fisheries, Dive Fisheries, and Intertidal Fisheries. These report directly to the Department and provide information reports to the Shellfish Advisory Committee.

These Sectors Committees:

(i) Make recommendations and provide industry input to DFO regarding annual management plans and short term operational needs;
(ii) Provide recommendations on issues such as regulation requirements, licensing and long term planning;
(iii) Identify research needs; and
(iv) Increase awareness and understanding among industry and other publics by assisting DFO to disseminate information concerning shellfish stocks and fisheries.

The other major advisory group is the Salmonid Enhancement Task Group (SETG) representing the general public. This has been around for many years. Until about 1986 it advised a Salmonid Enhancement Board, which was recently disbanded, its functions being assumed by PARC.

Figure 15-14 shows the relationships among these various consultative bodies for the Pacific fisheries.

From the foregoing, it is clear that the consultative structure has evolved since the Pearse Report of 1982. While the new structure is not exactly as Pearse envisaged, it comes closer to his criteria for effectiveness than the structure existing in 1982. One of the most significant changes has been the emergence of a Pacific Regional Council in 1987, paralleling the Atlantic Regional Council established in 1985.

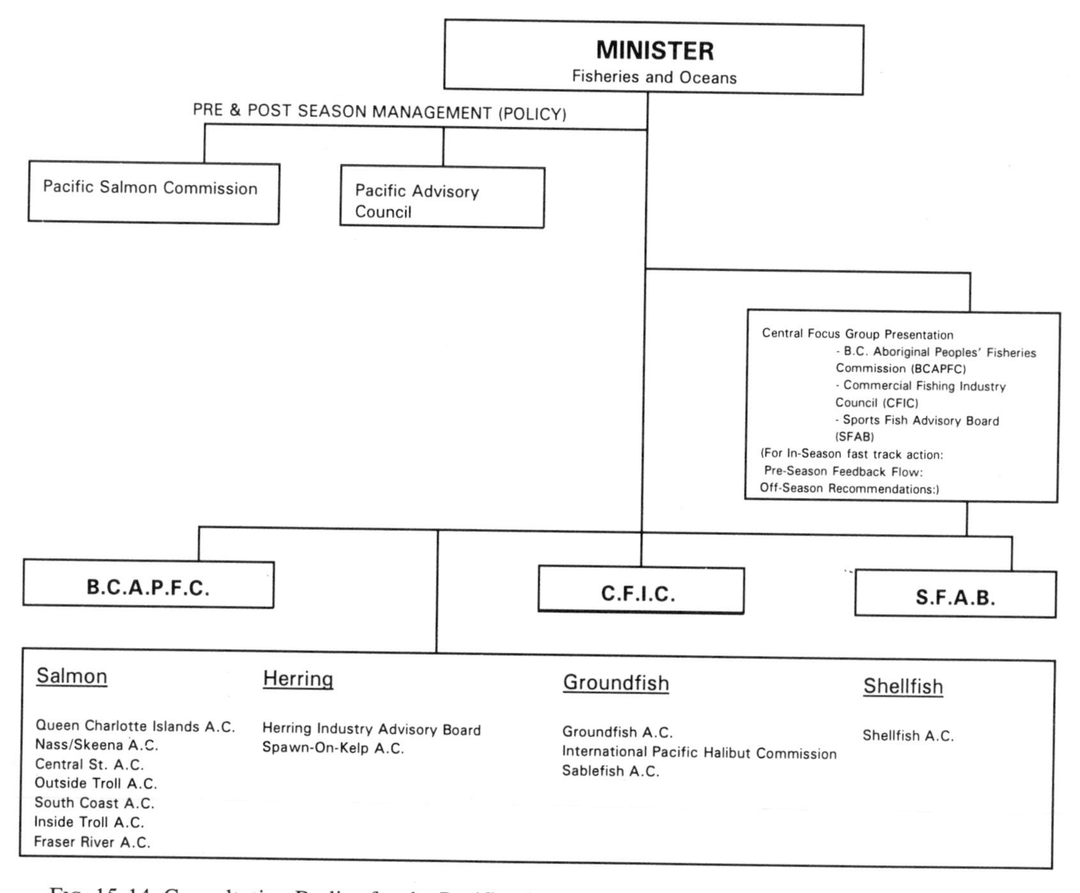

FIG. 15-14. Consultative Bodies for the Pacific Fisheries (AC = Advisory Committee).

4.4.3 The Atlantic and Pacific Regional Councils

4.4.3.1 The Atlantic Regional Council

In a December, 1981, Brief to the Department of Fisheries and Oceans entitled "Is Our Groundfish Being Managed?", the Eastern Fishermen's Federation (EFF) contended that "industry participation in formulating management policy has not, and cannot be achieved through 'advisory committees'".

The EFF stated:

> "Although many such committees have been created to discuss particular issues they have not had real authority nor impact, other than to advise senior bureaucrats who are not bound to act on these recommendations. It is the feeling of the directors of the EFF, many of whom have been members of various advisory committees, that these committees have been generally a public relations effort [while] the Departmental planners continued to formulate and implement policies on their own. This experience has increased mistrust of the Department rather than achieved better communication and response to issues" (EFF 1981).

The EFF proposed restructuring the existing consultative system:

> "The way to achieve effective collaboration between fishermen and managers of the resource is to make the industry participants, in fact, the managers. Three Regional Councils, representing the fishery participants and governments involved could have as their prime function the responsibility to formulate and recommend to the Minister of Fisheries, policies to guide the management of the Regional Fisheries."

The Executive Director of the EFF, Mr. Allan Billard, became an outspoken advocate of this Regional Council proposal, based largely on a U.S. system. The Task Force on Atlantic Fisheries rejected this suggestion, concluding:

> "We were unable to generate any enthusiasm for a consultative and decision-making model based on the U.S. Regional Council system. It is apparently a system that is appropriate, if not inevitable, in the U.S. jurisdictional context, where states exercise fisheries management authority within three miles. It is of dubious relevance to the Canadian scene, where it would make the decision-making process more complicated than it is now, without contributing anything more to the substance of the issues or any better means for special interests to be heard."

The TFAF proposed a new Atlantic Fisheries Consultative Group to advise the Minister and senior officials on major strategic issues. On the Pacific coast, Pearse recommended a new Pacific Fisheries Council, with a policy advisory rather than a management function. Nothing was done in the 2 years following these recommendations to create such structures on the Atlantic and Pacific coasts.

A new government took office in September, 1984. The new Minister of Fisheries and Oceans, John Fraser, in the initial months of his term, travelled extensively and consulted widely on both the Atlantic and Pacific coasts on possible changes to the consultative process. On the Atlantic, it soon became clear that, except for the EFF, no other major fishery group wanted a U.S. type system, i.e. Regional Councils with management authority.

A DFO analysis made public in the spring of 1985 (DFO 1985k) identified the following weaknesses in the existing structure:

1. The existing advisory committee structures were set up along species lines. These committees were considered effective for designing and developing annual fishing plans and dealing with other practical operational management issues. However, this mechanism's structure prevented it from accomplishing certain things. Multispecies approaches were difficult to initiate or discuss within the existing committee structure. A need was identified for a continuing forum to coordinate and pursue cross-fertilization and multispecies issues.
2. Existing management-consultative structures tended to focus primarily on stock management and allocation issues. Other major concerns were dealt with either in special 'one-time' seminars, ad hoc annual meetings, or most often in direct appeals to the Minister. There were no continuing broadly representative fora to discuss such things.
3. Existing management-consultative structures were very pragmatic, issue-oriented institutions. They tended to focus on short term problems — trying to reconcile disputes over this year's or next year's allocations of specific species and stocks or tactics for upcoming discussions with foreigners. There was little room to discuss or debate longer term policy.

4. Without continuing fora with balanced membership to review and debate many kinds of key issues and policies, there was no structured context for progress. Instead, individual proposals or ideas representing a particular point of view were presented to the department or Minister. This made the department or Minister look paternalistic or arbitrary when deciding between competing proposals or differing views.

The DFO paper concluded that the TFAF recommendations ignored some fundamental weaknesses in the species advisory committee approach, as discussed above. It criticised the Task Force recommendation to create an informal Atlantic Fisheries Consultative Group of 10 or 12 knowledgeable, experienced individuals. The recommendation did not address the broad need for regular, open and established fora with balanced representation from all groups to input into all aspects of the policies and programs affecting the harvesting and processing sectors.

The paper concluded that the regional council system as it had developed in the American context was "widely recognized to be an ineffective mechanism for consultation and decision-making.... The system has largely failed to overcome the problems implicit to the unique jurisdictional situation between the states and the federal government."

The cumbersome and time-consuming fisheries management procedures implicit in the U.S. system were identified as the primary reason for its ineffectiveness. DFO suggested that a simpler regional council structure would suit Atlantic Canada.

The paper proposed three main alternative organizational approaches:

1. A number of councils could be established to coincide with each DFO Region. Also there could be one Atlantic-wide Council made up of representatives of the regional councils.
2. A similar arrangement to the first alternative but structured along provincial boundaries rather than DFO regional lines.
3. One Atlantic-wide council with geographically and structurally balanced representation from fishermen and processor groups.

The perceived drawbacks to the first two alternatives were that there would be many different councils dealing with common fisheries issues, possibly recommending considerably different approaches. Council positions could be "solidified" before the Atlantic-wide council could consider their interregional implications. This could paralyze the Atlantic-wide council with little prospect of consensus on major issues.

Mandate was one key issue. Two models were considered: (1) A Policy Instrument, and (2) A Management Instrument.

Under Model One, the regional council(s) would review and advise the Minister on major policy objectives for their particular regions within a framework of objectives for the Atlantic fishery as a whole. The regional councils would have a mandate to review all aspects of policies and programs affecting the fisheries in their regions to ensure their overall consistency with policy objectives.

Under this model, they would operate in parallel to the existing species advisory committee structure. The reporting relationships of the existing species advisory committees would remain the same.

Under Model Two, regional councils would be integrated with the existing regional species advisory committee structure. These latter committees would in essence become subcommittees of the regional councils. All proposed annual fishing plans and attendant regulations would be forwarded to the regional councils for review and further refinement according to regional policy objectives. Atlantic-wide species advisory committees would report to the Atlantic-wide Fisheries Council to vet their annual plans for transmittal to the Minister.

DFO suggested starting with one Atlantic-wide council. Experience would then determine the need for a more disaggregated system. Regarding representation, the paper suggested that initially, the council have only industry representation, excluding provincial and federal officials as members. DFO believed that Regional Councils as a management instrument might duplicate the weaknesses of the American system. The consultative/decision-making process could become encumbered with so many steps that the system would become paralysed. Annual fishing plans would take too long to produce. Regional councils would be so busy that they could not go beyond the existing short-term focus and the emphasis on species-by-species allocation issues. The longer term policy review function would be sacrificed for more pragmatic and immediate management concerns.

Minister Fraser put these alternatives to an Atlantic-wide meeting of representatives of the harvesting and processing sectors in March 1985, in Halifax. They generally disapproved of the Management Instrument approach. There was consensus on the following points:

— one Atlantic-wide council;
— no need for provincial participation;

— two or three members from outside the fishing industry;
— focus on policy rather than management issues;
— report directly to Minister and senior executive of the department; and
— other advisory committees and consultative structures to be left intact.

In August, 1985, Minister Fraser appointed the first Atlantic Regional Council (ARC). Composed of 18 individuals the Council's purpose was to advise the Minister and his senior officials on key policy issues affecting the Atlantic fishing industry (DFO 1985l).

ARC elected its own Chairman and Vice-Chairman and began to address key policy issues. One of these was the factory freezer trawler issue (see Chapter 8). This proved one of the most contentious. The Council did not achieve consensus but gave the Minister informed comment on the various factors to consider in his decision. Between 1985 and 1990 ARC discussed and offered advice on a wide range of issues (see Table 15-3).

From conversations with DFO officials and certain ARC members, it appears that ARC was influential for a while. The interaction among people of deeply opposing views helped members understand the concerns and priorities of those from other sectors of the industry.

However, by the early 1990's ARC had become inactive. It was eliminated in the federal Budget of April 1993.

4.4.3.2 Pacific Regional Council

Meanwhile on the west coast the Minister's Advisory Council (MAC), which had been regularly consulted about policy in the post-Pearse era (1982 to 1985), became immersed in debate over intersectoral allocation issues. Relations within MAC became fractious. Some groups, particularly the Pacific Trollers Association, withdrew from MAC, charging that the process was ineffective and unrepresentative.

In 1985 and 1986, MAC wrestled with the structure for the Pacific consultative process. The committee agreed that some changes were required but could not agree on the nature of these changes.

After soliciting views from the Pacific fishing industry, Fisheries and Oceans Minister Tom Siddon in October, 1986, announced plans for a new structure for the Pacific fisheries consultation process:

> "This new consultation structure will be an integrated system consisting of the Minister's Pacific Regional Council (PARC), a number of coastwide fisheries allocation and management advisory committees and local advisory groups. Through this new structure, all fisheries user groups will be assured of a vehicle for providing me with their advice and input on policy as well as operational management issues. It will, in turn, provide a channel for efficiently communicating government policy to the industry" (DFO 1986k).

TABLE 15-3. List of issues considered by the Atlantic Regional Council, between August 1985 and June 1991.

MAJOR POLICY ISSUES — DOMESTIC	*MAJOR POLICY ISSUES — INTERNATIONAL*
— Factory Freezer Trawler Policy	— Overfishing
— Amendments to Fisheries Act	— NAFO Structure
— Licensing and Regulations	— Canada/U.S. and Canada/France
— Overfishing	— U.S. Countervail Action
— Quality Improvement Program	— Port Access
— Income Stabilization	— Foreign Investment in Fishery
— Consultative Process	— Canada/U.S. Free Trade
— Underutilized Species	— Catch Transshipments
— Direct Sales	
— Short/Long-Term Vessel Charters	
— Seals-Parasite Infestation	
— Offshore EA Review	
— Resource Sharing Arrangements	*GOVERNMENT INITIATIVES*
— Management of Northern Cod	— Forget commission of Inquiry on unemployment Insurance
— Aquaculture	— Malouf Royal Commission on Seals and Sealing
— Vessel Replacement Guidelines	— Regulatory Reform
— Overcapacity in Groundfish Sector	— Cost Recovery/User Pay
— Observer Program	— Report of Nielsen Task Force
— Government Restraint and Department Cutbacks	— Northern Cod Task Force
— Native Fisheries	

The new Pacific Regional Council (PARC) would consist of 12 knowledgeable and experienced persons appointed by the Minister. Members would be chosen to achieve a balance of interests. PARC was to advise the Minister on policy development and issues such as fleet rationalization, licensing and trade policy and interrelationships within the consultative process itself (DFO 1986k).

In April, 1987, Fisheries Minister Tom Siddon announced the membership of PARC.

> "The members of PARC are people with a broad view of the fisheries who do not come to PARC as advocates for any one group, but as advisors on all the fisheries and all aspects of that subject... What we learned through MAC is that individual groups found it difficult to do battle for individual shares of the fishing resource pie, and, at the same time, give advice designed to serve the common interest of all the fisheries" (DFO 1987l).

Mr. Siddon named Ms. Wendy McDonald, a North Vancouver business person and community worker, as Chairperson of PARC (ARC had selected its own Chairperson).

During its first 2 years of operation, PARC addressed a number of major issues. It met approximately 10 times during that period, and advised the Minister on the following issues:

- Policy on Disposal of Surplus Fish
- Chinook Tagging Program
- Policy on Roe Herring Licensing
- Impact of the GATT Ruling
- Native Fisheries Co-Management Proposals

PARC experienced some difficulty during this period in coming to grips with its policy mandate. PARC members disagreed over their role and terms of reference, specifically, what issues to consider and in what detail. This was manifested in a tendency to focus on the short term rather than on long term policy development.

There was some evidence that PARC was not functioning as intended, as a council of wise persons. PARC members reportedly divided along sector lines and had trouble dealing with intersector issues effectively.

4.5 Consultation and Conflict

4.5.1 The Role of Consultation

Conflict is endemic in the common property fisheries. The extent of conflict ranges from differing views moderately expressed at committee meetings to confrontation, hostage-taking and occasional violence, usually involving destruction of property. Through elaborate consultative structures developed in the 1970's and 1980's, the government attempted to channel conflicting views into the policy-making process. These processes also moderated conflict by institutionalizing it. The interest groups were less likely to fight when sitting across the negotiating table from each other. However, the nature of fishing and fisheries management means that consensus is often elusive.

Writing about the Canadian political system, Faulkner (1982) observed:

> "Pressure from a variety of sources is one of the dominant features of political life. Pressure from interest groups is just one source, though it is a source that in recent years has become increasingly prominent. The accommodation or rejection of interest group pressure reflects an enormously complex and subtle process of interaction of multiple interests."

Faulkner noted that; "an interest group that is related to his portfolio is, for any cabinet minister, a primary constituency: the bankers for the Minister of Finance; the Indian associations for the Minister of Indian Affairs; the Canadian Legion for the Minister of Veterans Affairs". Similarly, for the Minister of Fisheries and Oceans, the various fisheries interest groups are his primary constituencies. As Faulkner observed:

> "Their backing is often essential to maintaining ministerial clout within cabinet, even of surviving in the job. Their support in the implementation of programs and policies is usually important. A minister who loses the confidence of his primary constituency is in trouble."

These primary constituencies are also a principal alternative source of information for a minister. They provide ministers with "grass-roots feed-back on issues and concerns of [their] membership". Ministers need "very good communications with these primary constituencies to provide intelligence on what programs are working and on how they are working. They need their views on what policies should be changed. In short, they depend on the primary constituency as a source of policy countervail to departmental advice. Likewise, those primary constituencies, those interest groups, very much need the politician. From the interest group's point of view, the cabinet minister directly responsible for their interest is their single most important ally. None of the issues of concern to them will make it through the decision-making system without the assiduous attention and support of their minister" (Faulkner 1982).

Despite this interdependence, few such groups provide the public and political support that would ensure or reinforce their cause. This occurs because the nature of the relationship is perceived to be adversarial. As Faulkner notes, "Leaders of interest groups seem compelled to posture as fighters, not negotiators. If they appear to be supportive of the Minister, they risk losing their credibility as negotiators. The dynamics of the interaction reflects the lack of maturity, and sophistication in many Canadian interest groups and their leadership" (Faulkner 1982).

This adversarial relationship is particularly evident in the fisheries arena. The representatives of various fisheries interest groups tend to regard departmental officials and the Minister as whipping boys for every flaw of fisheries management, whether imagined or real. Most of these groups have experienced difficulty in making the transition from an adversarial system to one which seeks consensus. This is not surprising. The stronger an interest group is, the less it need to compromise, and the less attractive is consensus.

In general, the federal government provided greater access to decision makers through increased use of consultative committees, by providing assistance to groups to enable them to better input into the policy process, and by creating or enhancing consultative structures.

There was a proliferation and an evolution of fisheries consultative structures during the 1970's and 1980's. These bodies brought the conflicting interest groups together. Sometimes these processes brought consensus on advice to the department and the Minister. Other times the Minister had to intervene and choose among the opposing viewpoints. This occurred frequently in disputes over fish allocations.

Nonetheless, the process put opposing viewpoints on the record, and frequently people came to understand other's viewpoints. In many instances where consensus was achieved, it helped shape policy.

When interest groups are unsuccessful in the consultative process, they resort to lobbying. Well-organized groups such as the FCC, the FCBC and the various unions, recreational associations and native groups are quite effective in reaching decision-makers. They are persuasive and eloquent. But they may overshadow less effective groups with an equal stake in the outcome. Such behind-the-scenes lobbying threatens the process of seeking a fair and equitable outcome.

If consultation appears ineffective or insincere, it loses credibility. As a result, senior bureaucrats and the Minister are inundated with special pleadings from the multiplicity of actors on the fisheries stage. This is precisely the scene which the consultative process was intended to avoid. In the final analysis, such end runs of the established consultative process are counterproductive even for those groups which might gain a temporary advantage. However, groups which can afford to take both approaches will likely continue to do so.

The federal government has sometimes been accused of failing to consult adequately those affected by fisheries policies. Snow (1977) commented:

> "Consultation with Canadian fishermen is on an informal, individual level and consists of managers contacting union and company leaders when they want their advice or cooperation."

That was certainly true in the 1960's and the early 1970's but from the mid 1970's on the situation changed significantly. In the 1980's an inordinate proportion of the time of senior fisheries officials was consumed by the structured consultative process. While that process was not established in legislation, it was nonetheless formal and well understood by the major fisheries interest groups.

The formal structuring of consultation was a formal adaptation, namely, the *institutionalization* of mechanisms to deal with conflict, as described by Himes (1980).

A 1991 internal DFO review of the Atlantic Fisheries consultative and advisory processes identified various benefits and shortcomings. It concluded that consultation can generate both short and long term benefits for the Department and its clients. It can:

- allow clients to be kept abreast of departmental programs and policies;
- provide opportunity to participate in the resource management process;
- provide early warnings as to whether the Department's policies and new initiatives will be well-received and successfully implemented;
- act as a sounding board for the evaluation of plans, programs and policies as well as the emergence of trends and issues;
- provide clients with a clearer insight into the constraints facing the Department as well as the role of the bureaucracy in the decision-making and policy formulation processes;
- foster client-to-client interaction;
- assist the Department in focusing its activities where the public demand and need are greatest;
- assist the Department in developing new or refining existing approaches relative to its mandate, in resolving conflict and in reconciling the "public interest".

The existing process had a number of shortcomings. The process was:

- labour intensive, time consuming and expensive to administer in times of increasing fiscal restraint;
- so heavily structured that it was at times dysfunctional and lead to complacency of the membership;
- by and large, ineffective in achieving consensus amongst all parties on sensitive issues;
- without guidelines to deal with such issues as representation and the participation of stakeholders;
- prone to re-directing further representations to the Minister's office to a much greater extent than was necessary or desirable;
- not easily amenable to considerations of scientific advice in the discussion of fisheries management initiatives given the time constraints under which the advice is generated, reviewed and released;
- often lacking in mechanisms which forge accountability for acting upon accepted courses of action.

4.5.2. Co-management

The government has also experimented with alternative approaches to involving fishermen in developing novel approaches to fisheries management. One of the best known is the Bay of Fundy Herring Project, which involved transforming the Bay's herring fishery from a reduction to a food fishery in the late 1970's. This experiment involved an alternative model in which fisheries participants themselves had substantial decision-making powers. This approach has been termed "co-management" (Kearney 1984). Government, purse seiners and other fishermen's groups cooperated to transform this fishery. A catalyst was the formation of the Atlantic Herring Fishermen Marketing Co-operative (AHFMC) or the Club as it was called. The AHFMC sub-allocated the fleet quota and introduced, for the first time, voluntary individual annual and weekly boat quotas. Initially these quotas were not formalized in regulations. Rather, they were self-imposed by the purse seiner captains and administered by the AHFMC. The fishermen became for a while, at least in part, the managers of this fishery. Although there was a history of underreporting, the quotas had to be self-policed. Over-the-side sales to Polish factory ships played a key role in the project's economic viability. Kearney (1984) saw substantial progress towards the project's objectives:

> "As a result of the Bay of Fundy project, the purse seine fishermen, to an extent almost undreamed of in 1975, owned their boats, ran their business, negotiated prices and working conditions, and had become partners fully equal to the processors."

The vessel quota allocation scheme synthesized the government's objective to improve the incomes and bargaining position of independent purse seine fishermen with the fishermen's objective to reduce the competition among themselves.

The AHFMC assumed substantial management responsibilities. It played the lead role in policing vessel quotas, negotiating prices, allocating nightly markets, supervising over-the-side sales, and collecting statistics.

As Kearney (1984) recognized, "the successes of the Bay of Fundy Project did not persist for long." A reduction in herring catches and a downturn in the European markets because of increased landings from European herring stocks split the AHFMC. By the end of 1981 the purse seine fleet's survival was doubtful.

The Bay of Fundy Project was integrated into the mainstream of fisheries decision-making in the late 1970's. Kearney gave this part of the blame for the decline of the project. He contended that the consultative process "greatly dampened the innovative trends of the Bay of Fundy Project, namely, the evolution of joint policies/strategies and fishermen management, and eventually put almost a complete end to them."

Kearney recognized a problematic side to co-management in that its success depends to a large extent on a supportive government and favourable government measures. Nonetheless, he suggested that it could still be a model for decision-making:

> "The Bay of Fundy Project, however, offers an alternative vision for fisheries management. It is a vision of people collectively formulating their own objectives and deciding their own future, of regulation without a morass of regulations, and of the complementariness of social and economic goals rather than their contradiction."

These are laudatory sentiments. In fact, however, the Bay of Fundy Project broke down from the pressure of the conflicting interests of the participants. Circumstances were briefly favourable and cooperation flourished. The available evidence suggests its initial success was due to an unusual combination of circumstances. Once the situation changed, the Project withered on the vine.

The *Policy for Canada's Atlantic Fisheries in the 1980's* (DFO 1981a) suggested co-management might be a way to delegate certain responsibilities to fishermen, and a mechanism to tailor resource management to the needs and objectives of local communities. What was really envisaged was a decentralized form of the advisory/consultative model rather than substantial delegation of decision-making to fishermen.

The Task Force on Atlantic Fisheries paid scant attention to the concept:

> "The idea of 'co-management' has not been developed in detail by those who advocate it and appears for the moment to be more of a catch-phrase than a well thought out proposal of substance."

One of the options in the DFO (1985k) paper examining the Regional Management Council concept, the Council as a management instrument, would have represented true co-management. Fishermen and processors would have shared in the fisheries decision-making process. This alternative was rejected by the industry and the government at that time in favour of the Regional Council as a policy instrument. The Atlantic Regional Council established in 1985 was an evolution of the advisory/consultative decision-making model, rather than co-management. At that time all major participants in the fishery favoured the approach adopted.

Charles (1992) described "co-management" as an "increasingly attractive option" to resolve fisheries conflicts. He described co-management as "shared responsibility and shared decision-making between fishers and government, providing an efficient mechanism for conflict resolution, or avoidance, under a variety of circumstances". The major institutional changes being implemented in 1993 and 1994, for example, the Fisheries Resource Conservation Council for the Atlantic fisheries, and the proposed quasi-judicial Atlantic and Pacific licensing and allocation Boards (see section 6.0), will give the fishing industry a much greater say in the decisions which so dramatically influence their livelihoods. These initiatives represent a move towards greater sharing of responsibility with the industry.

5.0 OTHER PERSPECTIVES

Others, outside the fisheries mainstream, have criticized the consultative process. Davis and Kasdan (1984) described relations between fishermen, fish processors, and government in Atlantic Canada as "rife with conflict. These range from mild disagreements over policy to violent confrontations concerning policy enforcement practices. This situation is particularly evident in the relations between small boat fishermen and the federal government's Department of Fisheries and Oceans."

Based on case studies of two incidents in southwest Nova Scotia, they generalized that DFO policy was initiated and implemented with minimal consultation, placing fishermen in the position of passive respondents. They suggested that fishermen's frustrations with the consultative process frequently resulted in confrontation, which was ultimately resolved through more meetings and compromise.

Davis and Kasdan reached this conclusion based on two incidents in southwest Nova Scotia. One was an April, 1977, mass demonstration in Shelburne County during which fishermen vandalized the property of fish buyers in the offshore lobster fishery. The second was an intense confrontation between DFO and some southwest Nova Scotia lobster fishermen over DFO enforcement of trap limitations. As the conflict escalated, about one hundred fishermen burned and sunk two DFO patrol boats beside a wharf in East Pubnico.

Earlier I distinguished between constructive and destructive conflict. This latter incident is certainly one of the most destructive conflicts witnessed in Canadian fisheries in the modern era. The lack of adequate communication and consultation in these two incidents produced violent outcomes. Both incidents involved lobster fishermen in southwest Nova Scotia. The Task Force on Atlantic Fisheries found that fishermen in this area earn the highest fishing incomes in Atlantic Canada. They are hardly typical of the east coast or of Canadian fishermen as a whole.

These incidents revealed flaws in the consultative processes for the management of the lobster fisheries of this area. Since there was no representative association of fishermen, the system was changed to allow fishermen to elect their own representatives to the lobster advisory committees. Subsequently, the silent majority of fishermen, who wanted trap limits enforced, prevailed and a compromise was achieved.

Hennessy and LeBlanc (1982) compared the Canadian and American fisheries management systems:

> "[The] relatively closed and centralized [Canadian] system has two major advantages as well as a number of disadvantages. The major advantages lie in the speed or dispatch with which the minister and his staff can respond to changing fishery circumstances. In this sense the Canadian system of administration and management is highly flexible because decision-making costs are extremely low; a very small proportion of the affected individuals are required to make a decision. Indeed,... only one decision maker is required — the minister himself....flexibility is high but accountability low."

Regarding the American system, Hennessy and LeBlanc (1982) concluded:

> "By attempting to maximize accountability, the management process becomes very complex and lengthy thereby reducing the flexibility necessary to respond in a timely manner to changing fishery conditions."

Apostle et al. (1984) concluded that both systems lead to frustration and alienation, the Canadian one because of the lack of decision-making power on the part of fishermen, and the American because its elaborate participatory decision-making process is slow and inflexible.

Dubinsky (MS 1987) examined the Canadian consultative structures from a slightly different perspective. He concluded that inshore fishermen and members of the fisheries bureaucracy belong to different and conflicting cultures.

He argued that the formal organizational structure "prevents those unable to master the skills of such an environment from having a direct impact... Given the unique occupational environment of fishing communities... inshore fishermen are not equipped to deal with the formal organizational structure in terms of both decision-making (e.g. advisory committees) and structures which try to represent the concerns of fishermen to government i.e. various pressure groups."

He contended that DFO had attempted to design structures which are organizationally pluralistic, but which had little or no relevance to fishermen, given their unique norms and values. He argued that the consultative models "assume the ability of the particular constituency to give its allegiance to a particular organizational representative, as well as an ability to determine particular policy platforms which can thus be transmitted to government in an expeditious and timely fashion and without having to grab media attention, in a low key manner. These skills have never been present among small-scale fishermen."

There is some merit to Dubinsky's arguments about cultural conflict between the norms and values of inshore fishermen and the administrative model which emphasizes structure, process and representation. However, his analysis also is primarily rooted in the dynamics of the southwest Nova Scotia fisheries. Fishermen there have been slow to organize and in particular have resisted the unionization/representation model of organization (see Chapter 14). This, however, is changing. The MFU is making inroads in the local milieu and bridging the gap between individualistic fishermen and the bureaucracy.

Dubinsky, Davis and Kasdan, and Apostle et al. demonstrate a lack of faith in the ability of small-scale fishermen to participate in consultation and to shape and influence the resulting decisions and public policy. There is abundant evidence in recent years to suggest that inshore fishermen are key players in the fisheries system. Bureaucrats or politicians ignore these players at their own peril.

The advisory/consultative model of decision making has been instrumental in shaping fisheries policy in Canada, both on the micro and macro scale. Keating (1983–84) noted this for international fisheries policy:

> "The conduct of bilateral fisheries relations has been complicated over time by the activities of politically important groups on both sides of the border. However trite a statement, it is important to reiterate that the domestic fishing constituency in Canada, while small in size and economic significance from a national perspective, has a substantial regional economic importance and considerable political salience at both the regional and national levels. The views of these constituents are thus an important factor in policy deliberations, and policy makers have displayed a willingness to consult with domestic fishing interests almost as great as the latter's penchant for voicing their demands."

6.0 TOWARDS INSTITUTIONAL CHANGE

The Public Service 2000 Task Force on Service to the Public, chaired by Deputy Minister Bruce Rawson, pointed out that governments consult with groups and individuals in the private sector for many reasons: to collect information needed for policy making, to involve external groups in the policy development process so that they are comfortable with the outcome, to gauge the impact of public policy decisions on a particular group, or to determine the level of support a proposed initiative might enjoy among the public. It suggested that a shift toward a substantially more active and open consultative relationship with the public "is singularly important for the future effectiveness of the public service" (Canada 1990e).

The Task Force concluded that a formula approach to consultation is neither desirable nor feasible–different circumstances require different approaches. It recognized that consultations should not always be expected to end in consensus — some views are irreconcilable, or time constraints may not allow agreement to be reached. The Task Force suggested that, at a minimum, consultation should be authentic and accessible; its objectives and limitations should be clear to participants from the outset; and it should make provision not just for clients to be heard but for various stakeholders and

TABLE 15-4. Principles for Consultation.

1. Consultation between government and the public is intrinsic to effective public policy development and service to the public. It should be a first thought, not an after-thought.
2. Mutual respect for the legitimacy and point of view of all participants is basic to successful consultation.
3. Whenever possible, consultation should involve all parties who can contribute to or are affected by the outcome of consultation.
4. Some participants may not have the resources or expertise required to participate and financial assistance or other support may be needed for their representation to be assured.
5. The initiative to consult may come from inside government or outside - it should be up to the other to respond.
6. The agenda and process of consultation should be negotiable. The issues, objectives and constraints should be established at the outset.
7. The outcome of consultation should not be pre-determined. Consultation should not be used to communicate decisions already taken.
8. A clear, mutual understanding of the purpose and expectations of all parties to the consultation is essential from the outset.
9. The skills required for effective consultation are: listening, communicating, negotiating and consensus building. Participants should be trained in these skills.
10. To be effective, consultation must be based on values of openness, honesty, trust and transparency of purpose and process.
11. Participants in consultation should have clear mandates. Participants should have influence over the outcome and a stake in implementing any action agreed upon.
12. All participants must have reasonable access to relevant information and commit themselves to sharing information.
13. Participants should have a realistic idea of how much time a consultation is likely to take and plan for this in designing the process.
14. Effective consultation is about partnership. It implies a shared responsibility and ownership of the process and the outcome.
15. Effective consultation will not always lead to agreement; however, it should lead to a better understanding of each other's positions.
16. Where consultation does lead to agreement, wherever possible, participants should hold themselves accountable for implementing the resulting recommendations.

Source: Public Service 2000. Service to the Public. Task Force Report

interest groups to learn from each other,preferably in a common setting.

The Task Force proposed a statement of principles for consultation (Table 15-4). These principles can serve as valuable guidelines for improving the fisheries management consultative process.

Fisheries management, to a large extent, involves managing conflict. A major challenge of the fisheries manager is to reconcile the myriad conflicting interests he or she confronts almost daily. The fisheries manager functions in a complex web of demands for time, energy and attention. The Auditor General's 1986 Report recognized that the operational context for fisheries decision making was complex (Fig. 15-15):

> "The Pacific fishery is characterized by many groups competing for a scarce resource. The Region operates in a complex environment composed of recreational and commercial groups that demand a share of the resource, Indian people who have traditionally fished salmon and herring, and a variety of other interested parties including industry and local communities....other federal departments....the government of British Columbia and other countries.... Because of the difficulties of obtaining consensus among the various commercial fishing groups, and among the commercial, recreational and Indian interests, a considerable amount of senior management attention has been devoted to obtaining agreement on how to manage the fisheries each season....A major feature of the in-season fishery is the interaction between fisheries managers and fishing groups. Groups pressure the Region and the Minister for additional fishing time or a greater share of the catch. The onus is then on fisheries managers to demonstrate that conservation of the salmon will be endangered by additional fishing. On several occasions in the past few years, fisheries groups have taken the Department to court or occupied departmental offices to protest its decision. As a result of these influences, a considerable amount of effort is required by fisheries managers to obtain consensus with fishing groups and to keep the various fishing interests informed. This has sometimes made it difficult for fisheries managers to give sufficient time and energy to the main objective of the Department — the protection and conservation of fish" (Auditor General 1986).

Put more simply, the life of a fishery manager is demanding and stressful. Ability to tolerate abuse is a requirement of the job. Managers must be sensitive to a variety of interests as they search for workable solutions to complex problems.

Fisheries managers live in a practical world. Thus they are likely to question general solutions, recognizing that these are normally less useful than highly specific, realistic ones. Brewer (1983) observed:

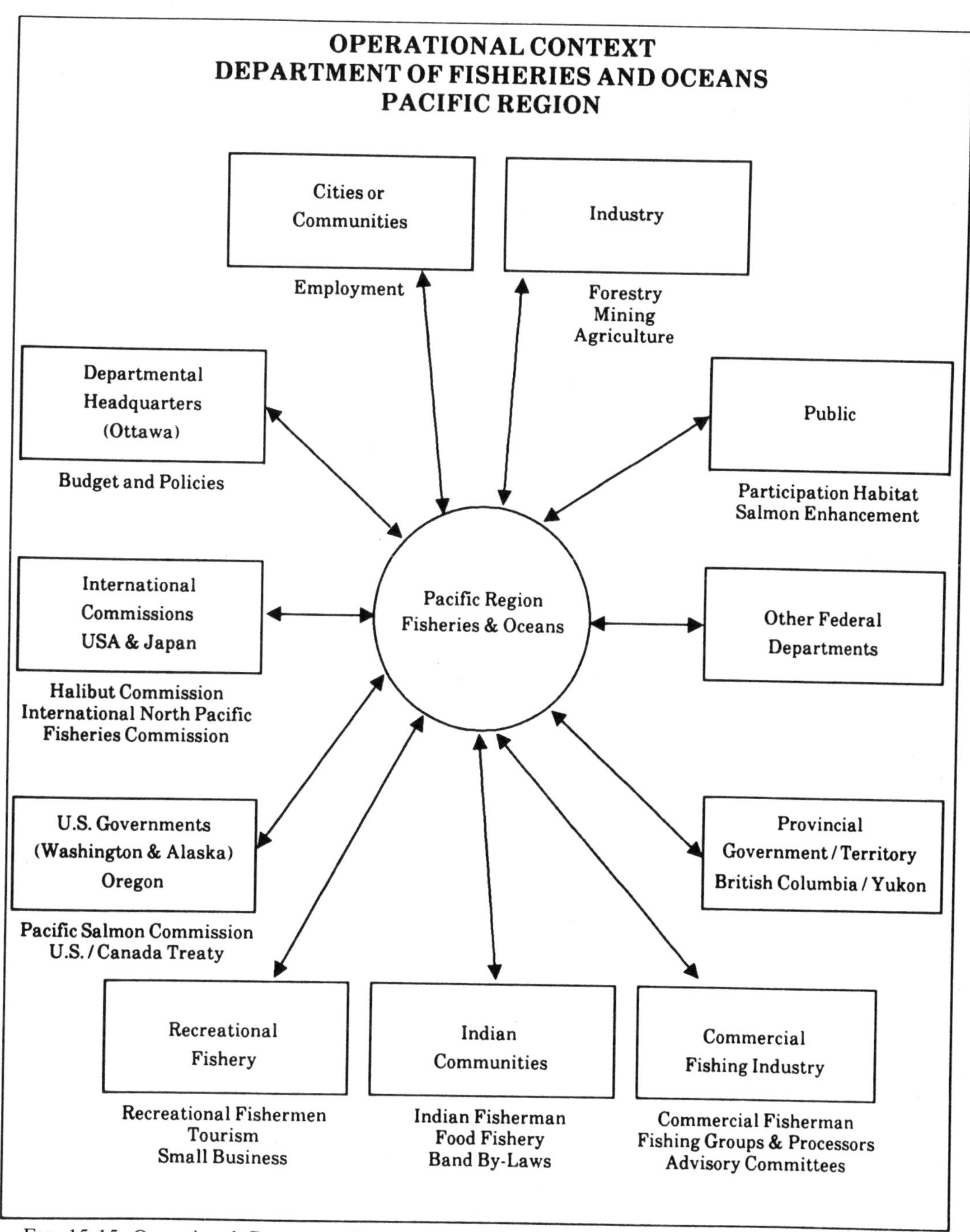

FIG. 15-15. Operational Context — Department of Fisheries and Oceans (from the Auditor General's Report for 1986).

"From the manager's perspective, there are no general solutions, just oceans of nasty particular details."

The task of balancing diverse and conflicting interests is frequently a thankless one, and absolute consensus is not a realistic goal. As more people and interest groups become involved in fisheries policy, the greater the likelihood that some will be disappointed by a decision.

In the end, fisheries managers and user groups must realize that no two individuals will see the same event the same way. Thus, there will inevitably be conflicting views about the appropriate course of action. The goal is not to eliminate such conflict but to manage it.

The success of any consultative process depends upon demonstrating its effectiveness to user groups. This requires structures which facilitate fairness, compromise, and mutual respect among groups.

Two major studies of Canadian fisheries policy (Kirby 1982; Pearse 1982) recommended that decision making on fisheries licensing/allocation issues be depoliticized through new independent agencies or boards, at arms length from the federal Minister for fisheries. The government at that time rejected these proposals. Instead, through the 1980's the federal government concentrated on strengthening and streamlining the consultative/advisory structure.

By the late 1980's a number of problems with the consultative/advisory structure became apparent:

— Decision-making was seen by many industry participants as too centralized and not sufficiently sensitive to regional and local needs;
— Decision-making rules were perceived as confusing and inconsistent;
— Fisheries management was, therefore, perceived to lack stability or continuity from year to year;
— Consultative processes had, in many cases, come to be viewed as ineffectual;
— There was widespread dissatisfaction with the handling of licensing appeals and sanctions for violations of fisheries management plans.
— The industry did not know how and why final decisions on licensing and allocation matters were made.

In the fall of 1991, Fisheries Minister John Crosbie in a speech to the Fisheries Council of Canada (DFO 1991b) announced the government's intent to establish independent Boards for the Atlantic and Pacific fisheries. The new Boards would take over some key powers currently exercised by the Minister and DFO under the *Fisheries Act*, namely, licensing and allocation for the marine commercial fisheries.. The existing system requiring the Minister to make all the decisions was "simply archaic" and "too political."

Mr. Crosbie stated that the existing allocation/licensing system was overwhelming:

> "No human being can take into account representations on over 400 allocation decisions made in over 100 advisory committees. Like the dog that walked on its hind legs, the system is remarkable not because it walked poorly, but that it walked at all... Instead of concentrating on broad policies to strengthen the fisheries industry, leading minds in the department spend most of their working hours in a politically-charged swamp of individual decision making."

The responsibility of the new Boards would be:

> "to apply clearly defined policies on licensing and allocations to decisions in individual cases. Decision making by this institution must be open to public view and provide for direct participation by those whose interests are affected. And the decisions should be made by panels that reflect in a fair and balanced way those who rely on various fisheries resources for their livelihood" (DFO 1991b).

The basic rationale was the perceived need to replace an anachronistic system of decision-making based on ministerial discretion with a fairer, more impartial system responsive to the needs and views of industry. It was envisaged that the creation of these Boards would lead to a more productive working relationship between the federal government, as steward of the resource, and the fishing industry.

The perceived problems listed above would be addressed by changing the way decisions are made. This change would eliminate any appearance of unfair and closed decision-making, permit the industry to participate directly in decision-making in its own region, and provide better service to clients by conferring benefits impartially on all applicants. It was also perceived that these Boards would offer more consistency and continuity in decision-making.

In March, 1993, Fisheries Minister Crosbie released a public discussion paper, preparatory to the introduction of legislation to establish the two Boards (DFO, 1993a). The proposal described how the Boards would be established, their mandate and their powers to make decisions on licensing and allocation. Under this proposal, the Boards would be mandated to administer, under ministerial policy direction, licensing and allocation for the Atlantic and Pacific marine commercial fisheries.

There would be no change to federal responsibility for the international aspects of marine fisheries management, and no change to the federal responsibility for the Aboriginal fishery. The proposed legislation would not affect federal and provincial responsibilities for freshwater fisheries or co-management arrangements in the Northwest Territories and Northern Quebec. Biological resource assessment and surveillance and enforcement would continue to be the responsibility of the Department of Fisheries and Oceans. The Minister of Fisheries and Oceans would remain responsible for conservation and would continue to set the overall levels of harvest for the marine commercial fisheries.

Overall, it was envisaged that the system would work as follows:

1. The framework would be established by law; hence the mandate and decision-making structure of the Boards would not change with each Minister.
2. The Minister would guide by setting policy but would have no direct say in specific decisions.
3. The Board would decide, within the policy framework set by the Minister, who would get the licences and allocations.
4. DFO would carry out the Board's decisions through the day-to-day management of the fishery and routine licence administration.
5. The Board would hear appeals on licence decisions taken by DFO or Board staff. There would be no subsequent appeal to the Minister.
6. The Board would penalize Canadian commercial fisheries violators brought before it by departmental enforcement personnel.

Thus, the Minister would permanently give up the power to decide on individual cases. The Board, operating under Ministerial policy and conservation directions, and within the limits of the overall harvest levels set by the Minister, would decide exactly who gets the fish and how many. Although a judicial review of the Board's decisions by the Federal Court would be possible, such a review would not consider the substance of the Board's decision. Rather it would examine whether the Board had exceeded its jurisdiction, or ignored some fundamental principle of natural justice.

The Boards would be composed of appointed members supported by Board employees. Board members would have to be knowledgeable about and have experience related to the fishing industry but they could not directly or indirectly be engaged in a fisheries business. Each Board would have a core membership, the Executive Board, consisting of up to 7 members on the Atlantic and up to 5 on the Pacific. Additional members would be appointed to serve on Board panels. The principal role of the Panels would be to recommend allocations.

It was envisaged that the Boards would commence operations in 1994. There were provisions to ensure no change in licence holders or existing allocation arrangements during the transition period.

In addition, in December 1992, the Minister announced his intention to create, beginning in 1993, a new Fisheries Resource Conservation Council for the Atlantic fisheries. This Council would subsume portions of the functions previously performed by CAFSAC and AGAC. The Council, to be composed of scientists and persons knowledgeable of the fishing industry, would integrate industry experience and scientific expertise by involving industry more directly in the scientific process. The Council would advise the Minister on research and assessment priorities, conduct public hearings on stock assessments and proposed conservation measures, and make public recommendations to the Minister on TACs and other conservation measures.

Commenting on the DFO Reform initiative, Deputy Minister Bruce Rawson said:

> "The Reform initiative and the creation of the two agencies is a very logical evolution in the history of our Department. Management and enforcement by ministerial edict is now an anachronism.
>
> "Stakeholders demand their say about matters that affect their lives. Bureaucratic structures are giving way, more and more, to arms-length operational systems. Politicians can't be expected to continually play Solomon between conflicting elements of society. Criminal law can't be expected to police economic boundaries." (PISCES, October-November, 1992).

These proposed changes would substantially alter the institutional and decision-making structure for fisheries management. At first glance, these could be the most profound changes in the management of Canada's marine fisheries since the introduction of the 200-mile fisheries zone.

CHAPTER 16

HABITAT MANAGEMENT

"The protection of aquatic habitat is considered by many to be the 'first and foremost' problem of fisheries policy."

– Peter Pearse, 1982

1.0 INTRODUCTION

The aquatic habitat is the essential resource base for fish populations. Concern about humans' impact on that resource base has been evident ever since the passage of Canada's first Fisheries Act and long before that. That concern has deepened in recent years as civilization's encroachment upon fish habitat has intensified. The impact of other natural resource industries, e.g. forestry, mining, upon fish habitat has long been significant. In recent years offshore oil and gas developments have been perceived as a threat to fish habitat and fish populations. The detrimental consequences of acid rain on freshwater fish habitat and on some Atlantic salmon populations are well documented. There is growing concern about contaminants in the freshwater and marine environments and their deleterious effect on fish habitat, fish and fishing. In the late 1980's the "greenhouse effect" and global warming emerged as global environmental issues. The potential overall impact upon the oceans and fisheries is uncertain. Overall, competing uses for and impact of other industries upon fisheries habitat are a major source of fisheries-related conflicts.

2.0 LEGISLATIVE BASIS FOR FISH HABITAT MANAGEMENT

Fish habitats are the natural systems or environments where fish live. The *Fisheries Act* (Section 34[1]) defines fish habitat as:

> "spawning grounds and nursery, rearing, food supply and migration areas on which fish depend directly or indirectly in order to carry out their life processes."

The very first Canadian Fisheries Act (1857) required the construction of fishways to permit fish, specifically salmon, passage around dams. Since that time, the habitat provisions of the *Fisheries Act* have been strengthened considerably. A major overhaul of these provisions occurred in 1977. This included broadening the definition of habitat. The current Act contains several sections which deal with the protection of fish habitat.The latest amendments involving increased emphasis on the protection of fish habitat were approved in 1991.

Section 20 contains provisions for safe fish passage around an obstruction in a waterway. Section 22 states that the Minister may require the owner to provide a sufficient flow of water to permit the safe and unimpeded descent of fish. Section 28 of the Act prohibits destruction of fish by explosives.

Section 30 authorizes the Minister to require that every water intake used for conducting water from any "Canadian fisheries waters" for irrigating, manufacturing, power generation, domestic or other purposes, be provided at its entrance with a fish guard or screen, to prevent the passage of fish into such water intake, ditch, channel or canal.

A key habitat provision of the *Fisheries Act* is Section 35 which stipulates:

> "35. (1) No person shall carry on any work or undertaking that results in the harmful alteration, disruption or destruction of fish habitat.
>
> (2) No person contravenes subsection (1) by causing the alteration, disruption or destruction of fish habitat by any means or under any conditions authorized by the Minister or under regulations made by the Governor in Council under this Act."

In effect, this Section prohibits anyone from altering fish habitat without the Minister's authorization. Given the definition of fish habitat in Section 34(1), this gives the Minister very broad habitat protection powers.

The other key habitat-related provisions of the Act are in Section 36 which deals with injury to fishing grounds and water pollution. Only the most relevant of this

Section's numerous provisions will be mentioned here.
Section 36(3) stipulates:

> "Subject to subsection (4), no person shall deposit or permit the deposit of a deleterious substance of any type in water frequented by fish or in any place under any conditions where such deleterious substance or any other deleterious substance that results from the deposit of such deleterious substance may enter any such water."

36(4) stipulates:

> "No person contravenes subsection (3) by depositing or permitting the deposit in any water or place of
> (a) waste or pollutant of a type, in a quantity and under conditions authorized by regulations applicable to that water or place made by the Governor in Council under any Act other than this Act, or
> (b) a deleterious substance of a class, in a quantity or concentration and under conditions authorized by or pursuant to regulations applicable to that water or place or to any work or undertaking or class thereof, made by the Governor in Council under subsection (5)."

Pearse (1988) pointed out that although these various provisions appear to be powerful tools for habitat management, they have proven difficult for governments to use. He identified two main reasons: "the inflexibility of the law, including its failure to accommodate competing water users, and a complex administrative structure."

The administrative structure is complex in two senses. When the Department of the Environment was formed in 1970, it included a Fisheries and Marine Service and an Environmental Protection Service. An internal memorandum of agreement assigned the responsibility for the control of physical alterations to the Fisheries and Marine Service and pollution control responsibilities went to the Environmental Protection Service. When a separate Department of Fisheries and Oceans was established in 1979, administration of the Fisheries Act pollution control provisions (most of the former Section 33 — now 36) was delegated to the Department of the Environment. Working relationships between the two Departments were set forth in a Memorandum of Understanding of May, 1985, signed by the Deputy Ministers of the two departments (DFO - DOE 1985). This Memorandum acknowledged that then-Section 33 (now Section 36) was "intended to protect fish and fish habitat from actions that cause or may cause the pollution of waters and is therefore of interest to both Parties in fulfilling their respective mandates." Under this Memorandum, DFO reserved the right to act directly in circumstances where the fisheries resource is being affected by the deposit of a deleterious substance and where DOE "is unable or unwilling to take such action."

Other federal laws affect fish habitat management. One of the most significant is the recent *Canadian Environmental Protection Act* (1988) (CEPA)(R.S.C., 1985, C.16 as amended). This replaced the *Environmental Contaminants Act*, which had been widely critized as ineffective. Not one prosecution had been undertaken under it. Most prosecutions for pollution had been made under the *Fisheries Act* or the *Canada Shipping Act*.

Part IV of CEPA also replaced the *Ocean Dumping Control Act*. This Part is to protect human health, marine life and legitimate uses of the sea by controlling the dumping of waste and other substances into the ocean. Under the Act, disposal of wastes at sea is regulated by permit. The terms and conditions of permits vary with the type of substance dumped. These conditions govern such things as timing, handling, storage, loading and placement at the disposal site.

The *Navigable Waters Protection Act* (R.S.C., 1985, C.N.-19, s.1 as amended) prohibits anyone from constructing anything in "navigable waters" without the permission of the Minister of Transport. The *Northern Inland Waters Act* (R.S.C., 1985, c.28, s.1 as amended) provides the authority to regulate water quality and use in the Yukon and Northwest Territories.

In addition to this federal legislation, there are many provincial laws deriving from provincial ownership of land and water under the *Constitution Act*. Provinces regulate the sale and lease of land and water, activities that use land and water and construction projects such as dams, and roads and water diversions. There is also provincial legislation regulating water discharges, waste and chemical disposal and pollution control. Regulatory activities under such provincial legislation can have a significant impact on fish habitat. Considerable controversy erupted in the early 1990's concerning the federal environmental role in major projects such as dams.

Historically, the federal government acted to protect fish habitat only at the request of a province in those areas where it does not directly manage the fisheries itself. It directly administers habitat protection in Newfoundland and in the territories and in marine areas where it manages fisheries (e.g. British Columbia and the maritime provinces). It has of course been heavily involved in cross-boundary issues and those involving international waters.

There are numerous informal federal-provincial referral arrangements. Pearse (1988) pointed out certain positive and negative aspects of these arrangements:

> "Most referral procedures are informal rather than required by law. This gives managers leeway to develop contacts and procedures suited to their needs. Referral procedures are an effective way of integrating regulatory activities and keeping the lines of communication open. But such arrangements also mean that other federal and provincial agencies are not legally obliged to follow the comments and recommendations of the Department of Fisheries and Oceans. Another difficulty is the sheer burden of processing thousands of proposals annually through the referral system."

The extent to which the *Fisheries Act* and federal-provincial environmental regulation duplicate or complement each other has been questioned in the courts. In a 1980 judgement, Fowler v. the Queen (1980), the Supreme Court concluded that the former section 33(3) of the *Fisheries Act* was *ultra vires* because it made no attempt to link the proscribed conduct to actual or potential harm to fisheries:

> "It is a blanket prohibition of certain types of activity, subject to provincial jurisdiction, which does not delimit the elements of the offence so as to link the prohibition to any likely harm to fisheries" (R. v. Fowler [1980]).

The former section 33(3) was thereby deemed beyond the constitutional scope of the federal government. The former section 33(2) (now 36 [3]) was also challenged in the case of Northwest Falling Contractors Ltd. v. the Queen (1980). This judgement concluded that Section 33(2) was within the constitutional powers of the federal government. The judges observed:

> "The definition of 'deleterious substances' ensures that the scope of S.33(2) is restricted to a prohibition of deposits that threaten fish, fish habitat or the use of fish by man" (Northwest Falling Contractors Ltd. v. The Queen [1980].

In 1989 and 1990, two additional court decisions, relating to the federal application of environmental guidelines, criticized the federal government for leaving fish habitat management to the inland provinces. The first of these decisions, in 1989, concerned the Rafferty Dam in Saskatchewan. The judges held that the federal environment minister did not follow his own environmental regulations. The second decision by the Federal Court of Appeal, in 1990, concerned the environmental review of the Oldman Dam in Alberta. Basically, the judges concluded that the federal Transport and Fisheries ministers were bound by federal environmental guidelines and had failed to live up to their responsibilities regarding the controversial $353 million dam. The ruling seemed to extend federal responsibilities and powers in environmental matters. It nullified an approval from the federal Transport Department, granted under the *Navigable Waters Protection Act*, because the Transport Minister had failed to do an environmental review.

This court decision raised important questions about the federal role in those areas where the federal government had essentially left the management of fish habitat to the provinces. Stephen Hazell, a lawyer specializing in environmental issues, stated operations as simple as driving logs down a river, upsetting spawning fish, could now come under federal environmental review laws instead of being left to the provinces. Environmentalists speculated that hundreds of more projects — anything that could affect fish habitat — would now fall under federal review even if they were provincial projects (*The Ottawa Citizen*, March 14 and March 15, 1990).

In January 1992, the Supreme Court of Canada dismissed an appeal from Alberta against federal intervention in the Oldman River Dam project. The high court ruled that federal environmental assessment rules were valid and that the federal government has both the right and the responsibility to conduct environmental assessments for projects for which it is a decision-making authority. This meant that DFO would be increasingly called upon to advise on impacts on fish and fish habitat for projects initiated by provincial governments.

3.0 ACTIVITIES AFFECTING FISH HABITAT

3.1 General

Pressures on fish habitat are many and diverse. From a marine fisheries perspective, habitat problems over recent decades have assumed greater prominence on the Pacific than on the Atlantic coast. This is because the primary commercial species there — Pacific salmon and Pacific herring — are particularly sensitive to habitat disturbances. Pacific salmon spawn in freshwater and spend critical stages of their life cycle there. Pacific herring spawn in the intertidal zone. For salmon in particular, activities affecting the freshwater habitat some considerable distance from the ocean can threaten the survival of particular stocks. The watersheds which are the nursery areas for Pacific salmon also support a wide range of human activities based on abundant timber, mineral, agricultural and hydroelectric resources. The estuaries important to salmon are the centres of human population and industrial activity. Thus, there is inevitable conflict between utilization of these other resources and protection of fish habitat.

Pearse (1982) identified the greatest potential threats to the habitat of salmon and related species on the Pacific coast as dams and diversions, forestry, mining and foreshore development. He also recognized potential damage from pollution, oil spills and certain specific effects of urbanization.

A 1983 review of fish habitat management on the Atlantic coast (DFO 1983m) classified habitat issues by the ecotype affected: freshwater, coastal/estuary and marine.

For freshwater, most of the major fish habitat issues in eastern Canada were rooted in the development of primary resources: forestry, mining and agriculture, and hydroelectric development. Secondary manufacturing industries outside of these sectors appeared to be relatively minor contributors. The major freshwater habitat issues involved energy development, particularly hydroelectic development, forestry, mining agriculture, public and municipal works, and environmental contaminants.

Stream bank Erosion.

Priority fish habitat issues in the coastal/estuarine ecotypes were quite different from those in freshwater. Four major categories of issues were identified as unique problems in estuaries and coastal zones: shorefront developments, land-based industrial pollution, municipal pollution and shipping pollution.

Only two major categories of fish habitat issues were considered unique to the offshore marine environment: the oceans' function as a "sink" for all environmental contaminants, and offshore oil and gas developments. Sea-bed mining was not yet an issue although the mining of silica sand deposits around the Magdalen Islands was considered a threat.

The multitude of diverse types of impacts upon fish habitat on both the Atlantic and Pacific coasts, and in the Arctic, are too numerous to catalogue in this chapter. An in-depth examination of the array of activities which impact adversely on fish habitat would require a book in itself. In this chapter I will provide only a broad overview and deal selectively with certain significant issues chosen to illustrate the diversity of the problems of fish habitat management and the nature of the government's response.

3.2 Forestry

Canada's forests are its most valuable natural resource. Among the primary industries, forestry ranks third in terms of its contribution to Canada's GDP and forest products rank first among the country's export commodities. Forests occur side by side with abundant fish resources, particularly in the case of salmon and freshwater fish. Thus conflict occurs when certain forestry activities damage fish habitat. The area of potential conflict occurs in streams and rivers, estuaries and nearshore coastal waters.

Under natural conditions forests can play a key role in maintaining fish habitat. Trees influence the amount of water that reaches a stream. Roots stabilize soil and reduce erosion. The reduction of tree cover can accelerate the melting of snow and influence the timing and rate of stream flows. Forest debris falling into streams can create a diversity of habitats necessary for high productivity. Hence, forest harvesting and management practices can impact significantly upon fish habitat.

Pearse (1982) noted that the adverse impact of forest development and harvesting operations had received a great deal of attention in his Commission's public hearings. He concluded:

"Logging and related activities are now widely agreed to have had a greater overall impact on salmon stocks than any other single source of habitat damage."

Similarly the 1983 Atlantic Fish Habitat Review concluded:

"No other industry has the capability of causing as widespread and significant physical and chemical impacts on the freshwater aquatic environment and fish habitat specifically as does forestry" (DFO 1983m).

The impacts of forestry are felt in many different ways, ranging from harvesting practices to effluents from pulp and paper processing.

Logging can affect stream flow and run-off, accelerate erosion, impede spawning through deposit of logging debris in streams, or eliminate streamside vegetation. Stream flow changes can shift or displace gravel used by spawning fish. Animals or plants dwelling on the bottom can be swept away or reduced in numbers. On the other hand, increased water run-off during summer can increase stream flows at rearing times and benefit the fish population.

Erosion can result in the transport of more sediment into streams. This can reduce the flow of clear oxygen-rich water necessary for egg survival. Accumulation of debris can block the passage of migrating fish and impede their spawning. Reduction of streamside vegetation can elevate water temperatures to levels unsatisfactory or even lethal to fish.

Forest roads can have a more adverse impact on fish habitat than logging practices. Road construction can result in increased deposition of sediments in streams. Water drainage patterns can be altered. The Atlantic Fish Habitat Review concluded:

"Forest access roads have been built with more of a mind towards short-term economy than long-range benefits (least of all those of an environmental nature). Serious damage has been done to fish habitat in the form of siltation, barriers to migration and destruction or removal of spawning or nursery substrate" (DFO 1983m).

Log transportation and storage can also produce adverse effects on fish habitat. Dumping and storage in shallow areas can destroy food vegetation. Decomposition of accumulated sunken bark and logs and leaching of chemicals can contaminate surrounding waters and reduce dissolved oxygen levels.

Fertilization of forest areas can increase nutrient concentrations in streams. This can lead to rapid algal growth, again reducing dissolved oxygen concentrations. The use of pesticides or biocides to control insects and unwanted vegetation can poison streams, killing fish and food organisms and reducing stream productivity.

The processing of wood into pulp and paper can result in the discharge into the aquatic environment of effluents toxic to fish. This is particularly true of chlorinated organic compounds produced during the pulp bleaching process. These chemicals can accumulate within fish and produce physiological or behavioral damage. They can render fish unfit for human consumption. In the late 1980's the problem of dioxins and furans produced by pulp and paper mills gained national attention. Extensive cross-Canada sampling was conducted and areas around some mills were closed to fishing as a result (see Dioxins and Furans, section 3.7.2).

Pearse (1982) noted that logging in the early decades of the 20th century was extremely destructive to Pacific anadromous fish. This occurred because of a lack of controls to protect streams from road and railroad construction, log jams and debris, log driving, siltation, reduction of streamside vegetation and many other effects of cutting, storing and transporting timber. Sediments disturbed spawning gravel and log jams and debris blocked fish access to spawning and rearing waters. Pearse contended that the causes of some of these early losses, such as log-driving and dams built for log transport, had since been eliminated. Many coastal streams appeared to have recovered following years of natural rehabilitation and forest regrowth.

Forestry in British Columbia was widely considered to be a major cause of declines in salmon stocks. In response, Pearse argued that the available scientific evidence contradicted some superficial impressions. He referred to results of the Carnation Creek project on Vancouver Island and other studies in the United States. These, he stated, showed that clearcutting does not necessarily reduce run-off; that stable large debris in streams is normal and created the pools needed for overwintering fry; and that higher stream temperatures after forest clearing do not always impair fish productivity. He noted, however, that such studies also revealed the possibility of destabilizing streams through poorly planned streambank activities. Pearse took a somewhat sanguine view of present day practices:

"Logging operations today are undoubtedly less damaging than they were in the past; and we now know a great deal more about how to reduce their detrimental effects on fish habitat; they can be dispersed to avoid total removal of forest cover over entire watersheds; unstable slopes can be avoided; streambank vegetation can be preserved and the streambeds left undisturbed; denuded areas can be quickly reforested; and logging and roadbuilding methods can be modified in a host of ways."

Counterbalancing this is the fact that more trees are being harvested than in the past and there is also more clearcutting.

Recently, the results of the Carnation Creek experience have been documented more extensively. This project covered the period from 1971 to 1986 in three phases: (1) pre-logging, 1972–76; (2) logging, 1977–81; and (3) post-logging, 1982–86. Hartman and Scriviner (1990) documented the effects observed during this study. Waldichuk (1993) has summarized the results:

> "Very briefly and simply, the effects of logging appeared to have a positive effect initially on the production of coho, but there was a later decline because recruitment of coho to the adult population decreased. In more detail, again highly simplified, removal of the forest canopy over the creek by logging increased solar insolation and water temperature. The fish-egg incubation period was reduced, leaving a longer period for the fry and smolts to feed in the creek. Coho fry reached a larger size earlier than before logging and the proportion of 1-to-2-year-old smolts leaving Carnation Creek increased. Because 1+ smolts were smaller than 2+ smolts, and smaller smolts survived poorly in the ocean compared with larger ones, overall survival in the ocean declined. This would have reduced spawning escapement unless increased numbers of smolts were sufficient to offset the increased mortality. Coho escapement after logging (1982–87) was lower than before logging (1972–76), but whether this was due to increased marine mortality or increased exploitation was unknown."

The effects of both logging and ocean conditions were especially severe on chum salmon. Waldichuk (1993) concluded that 74% of the negative impact on chums could be attributed to recent ocean conditions and 26% to logging.

The Carnation Creek project, the only one of its kind in Canada, demonstrated in a concrete fashion the adverse impacts of logging on salmonids.

3.3 Hydroelectric Development, Dams and Diversions

Rivers and streams have been harnessed for hydroelectric development for generations. Natural landslides, dams and diversions have over the years significantly affected productivity of anadromous species (Pearse 1982). Some of the most famous landslides were the Hells Gate slides of 1913 and 1914 which reduced Fraser River runs of sockeye salmon as well as certain runs of pink salmon. The natural Babine River slide of 1951 reduced significantly the Babine River sockeye runs.

Dams were constructed with little knowledge or regard for their effects on fish. Pearse (1982) cited the examples of dams on the Adams River, built for logging purposes, and on the Quesnel River for placer mining. Both contributed to a decline in Fraser River salmon stocks. A permanent dam on the Puntledge almost wiped out unique runs before hatcheries were built. Despite this, Pearse contended that hydroelectric dams had caused "less permanent loss of fish on Canada's Pacific coast than have other causes of environmental damage." He acknowledged that dams and diversions posed the greatest potential threat to natural salmon stocks. Whether this materialized would depend on political decisions on flood control and hydroelectric development.

The Atlantic Fish Habitat Review (DFO 1983m) concluded that the impacts of hydroelectric development on fisheries and fish are manyfold. It noted that many older dams were not equipped with even rudimentary fish passage devices (despite the Fisheries Act requirement for such devices).

In British Columbia, and more recently in Alberta and Saskatchewan, Manitoba and Quebec, there has been widespread concern about the environmental impacts of hydroelectric projects. One of the most controversial recent cases pertaining to Pacific salmon was Phase II of the proposal by Alcan Smelters and Chemical Ltd. (Aluminum Company of Canada Ltd.) to generate additional hydroelectric power for their aluminum smelters at Kitimat.

British Columbia issued a conditional water licence to Alcan in 1950, authorizing it to develop hydroelectricity in north-central British Columbia. This licence did not contain any requirement to protect fish.

Phase I, completed in 1957, involved the Kenny Dam on the Nechako River to form the Nechako Reservoir and a tunnel through the Coast Range mountains to divert the water to a powerhouse on the Kemano River (Fig. 16-1). During the 1950's salmon runs in the Nechako River were affected adversely both during the construction phase and for a while thereafter.

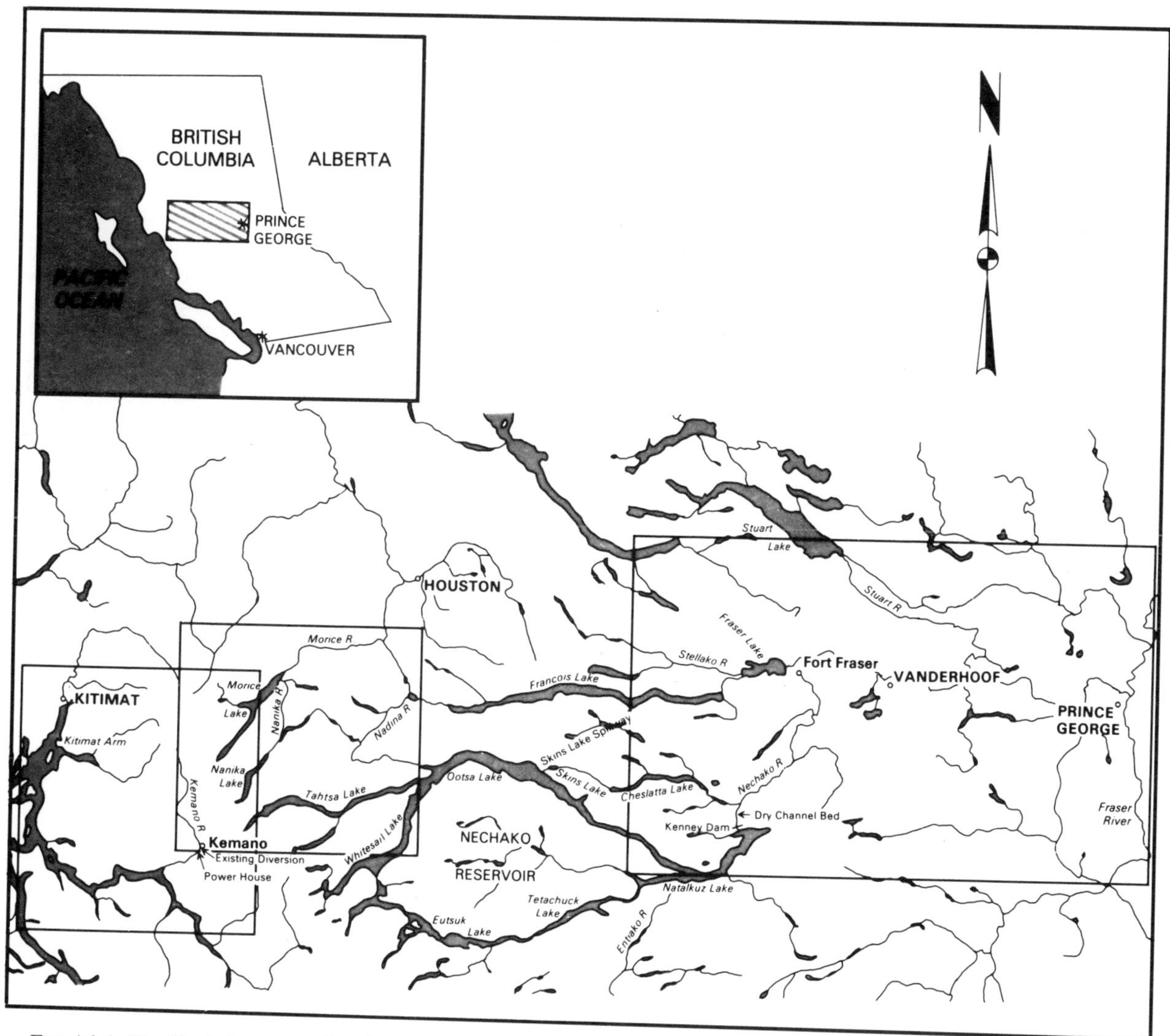

FIG. 16-1. The Nechako reservoir, after the building of Kenney Dam and the lower Nechako River and its main tributaries, the Nautley and Stuart rivers.

During the 1960's and 1970's there was some improvement in salmon production. This occurred because Alcan did not use all the available water and allowed high volumes to flow into the Nechako River. In 1978, Alcan began to sell excess electric power to B.C. Hydro. About that time also a series of dry weather years commenced. DFO became concerned about the impact of the reduced water flows in the winter of 1979–80. These averaged 12.7 cubic metres per second, only about one-third of that necessary for chinook salmon production. In 1980, the Minister of Fisheries and Oceans directed Alcan to release a minimum of 31.1 cubic metres per second for spawning and incubation from September to March and 56.6 cubic metres per second for rearing from April to August.

Alcan refused to comply on the grounds that the former section 20(10) (now 22[3]) of the *Fisheries Act* was unconstitutional and that the Minister's decision was "uninformed." Court action resulted in the Supreme Court of British Columbia granting a mandatory interlocutory injunction ordering Alcan to comply with the Minister's directions. Alcan complied with the injunction which was renewed annually until 1985, when Alcan challenged it. The company decided to take the matter to the Supreme Court of British Columbia for a permanent settlement. The province of British Columbia joined Alcan as a co-defendant. This became a classic case of development of water resources versus protection of fish habitat.

To place the issue in context, the Nechako River and its tributaries contributed about 19% of the average annual sockeye run and 5% of the average annual chinook run of the entire Fraser River system. At that time these Nechako contributions had an estimated annual value of $42 million. The recreational fishery contributed close to another $1 million. Alcan projected that it would create an additional 780 permanent jobs from the planned additional smelter. To achieve this, Alcan argued it required most of the Nechako River flows.

DFO scientists estimated that flows from the reservoir to the Nechako River were reduced to 38% of historical values (1930–1954) when Kemano I commenced operations and to 20% after 1980 under the injunction flow regime. Over the 30 years up to 1985, Alcan's Nechako Reservoir releases had contributed approximately 36% of the average annual flow of water into the Nechako River. The Nechako River in turn contributed about 28% of the average annual flow of the Fraser River at Prince George.

Alcan proceeded to challenge the powers of the federal government under the former section S.20(10) of the *Fisheries Act*. The contested section stated:

> "The owner or occupier of any slide, dam or other obstruction shall permit to escape into the river bed below the said slide, dam or other obstruction, such quantity of water, at all times, as will, in the opinion of the Minister, be sufficient for the safety of fish and for the flooding of the spawning grounds to such depth as will, in the opinion of the Minister, be necessary for the safety of the ova deposited thereon."

In February, 1986, DFO registered a statement before the Supreme Court of British Columbia containing its "pleading flows" compared with those proposed by Alcan. The DFO flows would have used 20% of the average annual reservoir inflow, while the Alcan flows would have used 8%. The DFO flows were derived from empirical data based on actual field measurements of salmon spawning and rearing habitat in the Nechako River. Alcan's proposed flows were derived from a computer simulation model which was used to estimate the area of the Nechako River falling within certain depth, velocity, substrate and cover criteria. There was scientific controversy over whether this latter method could predict actual fish habitat. DFO maintained that Alcan's proposed flows were insufficient to protect fish habitat.

The court case, scheduled to commence on March 31, 1987, was postponed several times to allow the parties to negotiate. A settlement was reached among Alcan, the Minister of Fisheries and Oceans and the B.C. Minister of Energy, Mines and Petroleum Resources in September, 1987. A key element in reaching this settlement was the formation of a Nechako River Working Group, composed of scientists and engineers from Alcan, DFO and the province of British Columbia, headed by Dr. David Strangway, President of the University of British Columbia. This group concluded:

1. That the principal and guiding concern should be for chinook conservation. To this end, it recommended protection of habitat for an average annual spawning population of 3,100 chinook salmon in the Nechako River;
2. That conserving and protecting the chinook stocks of the Nechako River could be achieved with an acceptable level of risk through a specific program of remedial measures (Anon. 1987a).

This group's recommendations led to a three-part agreement. First, an agreement was concluded with Alcan regarding short and long term water flows in the Nechako River, and an associated program of fisheries protection measures and applied research to provide an "acceptable" level of certainty that fish populations would be conserved. Second, agreement was reached with the province of British Columbia regarding measures to ensure that resident trout managed by the province would be protected. Third, Alcan consented to abandon forever its rights to develop the Nanika river system.

Alcan agreed to construct a Kenney Dam release facility and dredge certain areas to increase the operating efficiency of the Nechako reservoir. During this period, water flows in the river would be maintained at the levels DFO proposed to protect fish. With these measures, water released from the reservoir would be reduced to 19.6 cubic metres per second, approximately 50% of the release volume with short term flows. Actual flows in the river were anticipated to be greater than this due to inflows from tributary systems and the periodic release of cooling water. This reduction was to coincide with the implementation of other remedial measures. In return, Alcan was given some measure of certainty regarding long term flows in the Nechako River. The federal government passed an order-in-council pursuant to the *Fisheries Act* which confirmed the limits of Alcan's obligation to release water for the protection of fish.

The Minister of Fisheries and Oceans announced that the recommended program of fish habitat protection measures would be implemented (DFO 1987m). This was to consist of a hierarchy of measures from instream modifications to increase the productivity of the river for fish, to artificial spawning and rearing areas. If these measures proved inadequate, a hatchery would be constructed. In addition, a comprehensive program of monitoring would evaluate fish stocks, habitat and the effectiveness of remedial measures.

In return for Alcan's abandoning the rights from their 1950 water licence to develop the hydroelectric potential of the Nanika watershed, the province would provide rights of comparable value elsewhere.

Public reaction to this agreement was mixed. Local governments and the Chambers of Commerce saw it as a step toward removing a substantial impediment to economic development in northern B.C. Others, including environmental groups, natives and certain segments of the fishing industry, denounced the reduction of flows in the Nechako River as a failure of the Minister of Fisheries and Oceans to carry out his legislative responsibilities.

In response, Minister Siddon argued that it was better to have an agreed solution than an imposed solution. A court decision would have confirmed one of two flow regimes in the Nechako River, Alcan's or the Minister's, but would have taken years and considerable expense to achieve. Minister Siddon contended that the Agreement protected the Nechako and eliminated the threat to the Nanika.

As a result of the Settlement Agreement, the DFO Minister issued an opinion regarding water flow needs pursuant to the *Fisheries Act* and the Minister of Transport issued an exemption under the *Navigable Waters Protection Act*. Following court decisions regarding the environmental assessment process for the Rafferty and Oldman dams in 1989, the federal government's responsibilities under the Environmental Assessment Review Process were interpreted more broadly than before. Various interest groups argued that EARP should be applied to the Kemano Completion Project. An October, 1990, Order-In-Council exempted the project from one federal Environmental Assessment Review Process. The Carrier-Sekani Tribal Council and the Save the Bulkley Society et al. initiated legal action against the Ministers of DFO, DOT, DOE and DIAND.

On May 14, 1991, Justice Walsh of the Federal Court quashed the Kemano Completion Project Guidelines order, the Kemano Settlement Agreement and the DFO Minister's decisions on water flows, and the exemptions and approval pursuant to the *Navigable waters Protection Act*. She directed the Federal Ministers to apply the EARP Guidelines order to the Kemano Completion Project. In response, the federal government appealed the Court's decision but began the preliminary steps to apply EARP.

The federal government was successful in its appeal. In May 1992 a Federal Court panel overturned the ruling by Justice Walsh. In February 1993 the Supreme Court confirmed that the EARP Guidelines did not have to be applied to the Kemano Completion Project. Meanwhile, the Province of British Columbia decided to conduct its own public hearings on the matter.

The negotiation of the Settlement Agreement and the subsequent court decisions provide an interesting example of the difficulties in reconciling and accomodating conflicting resource development and fish habitat protection objectives.

3.4 Mining

Mining is another potential conflict area between industrial activity and fish habitat, this time between the utilization of nonrenewable and renewable natural resources. Mining's impact on fish resources can be either chemical or physical. Water plays a key role in the mining process, whether it be placer mining in streams, seabed mining in marine coastal waters or hardrock mining in underground or open pit mines.

In the case of hardrock mining, ores are mined, crushed, ground-up and concentrated in mills. Large quantities of water are used in these processes. Milling separates the desired metal from waste material using water in the process of grinding and concentrating the desired metals. Chemicals may also be added during milling. Mine tailings, a waste mixture of water, chemicals and finely pulverized rock, usually contain traces of metals. Tailings and chemical wastes from mining operations, if not properly treated, can pollute water, kill fish, and damage habitat.

As one example, the discharge of untreated nitrates and ammonia into fish habitats can cause immediate mortality of some fish and rapid growth of aquatic vegetation, leading to reduced oxygen levels killing still more fish. In another well-known example, sulphur dioxide and nitric oxide emissions from metal smelters can be carried long distances by atmospheric transport. When these gases mix with water vapour, they form sulphuric and nitric acid and reach the earth as *acid rain*. The detrimental effects of acid rain have been well documented (see Acid Rain, section 3.7.1).

There can also be physical damage to fish and fish habitat from mining operations when waste rock, particulate material, sand or gravel is dumped or washed into streams and renders the water turbid. The removal of large amounts of streamside cover can increase water temperatures, affecting the survival of fish and organisms lower in the food chain on which they feed.

The potential environmental effects of marine disposal of mine tailings are summarized in Table 16-1.

The disposal of mine tailings has posed problems for fish. In older mining operations, for example, adequate precautions were not taken to prevent leaching. Sprague (1964) and Sprague et al. (1965) and Saunders and Sprague (1967) documented the problems created by the leaching of metals such as copper, zinc and lead from tailings ponds.

Waldichuk (1993) has described some examples of the disposal of mine tailings into the ocean. Of particular interest was a molybdenum mine in Kitsault, B.C., which, from 1968 to 1972, discharged its tailings into Lime Creek, and thence into Alice Arm on the northern British Columbia coast. During the late 1970's environmentalists became concerned about plans to expand the mine and discharge the tailings directly into Alice Arm. This operation lasted only 18 months.

Waldichuk (1993) has described the operations and impact of the one remaining mine which discharges tailings directly into the sea. This is the open-pit Island Copper Mine, which has been operated since 1971 near Port Hardy on the northeast coast of Vancouver Island. Tailings from this mine have covered the deepest parts of the Rupert Inlet – Holberg Inlet portion of Quatsino Sound. According to Waldichuk, the bottom habitat, particularly of the deeper parts of this inlet system, has been essentially obliterated by the mine tailings.

Waldichuk and Buchanan (1980) concluded that the deposit of these mine tailings on the Rupert Inlet – Holberg Inlet ecosystem had the following impacts:

1. Deep-water benthic habitats were drastically modified by the tailings;
2. Intertidal and shallow subtidal habitats on the northeast side of Rupert Inlet were eliminated by the waste rock;
3. There has been no significant bioaccumulation of metals by marine organisms from the tailings or the waste rock;
4. The prawn habitat in Rupert Inlet was undoubtedly severely damaged by the tailings, and the deeper habitat of Dungeness crabs was probably also adversely affected; and
5. There was no evidence of a significant effect on salmonid runs to tributaries of the Rupert – Holberg Inlet system.

In summary, Waldichuk (1993) concluded that there had been physical impact locally but no indication of a human health hazard from contaminants in seafood, and the impact on commercial fisheries of the area has been small, if any.

Placer Mining

Placer mining, which involves washing gravel to extract gold, has been controversial at times. During the mid-1980's placer mining in the Yukon was a high profile issue. There were then over 300 licensed placer mining operations in the Yukon. Negative impacts on fish and their habitats can be caused by excessive sediment discharges to fish spawning beds, removal of streamside vegetation, oxygen depletion, construction of dams and berms in streams, diversion of streams and instream use of heavy equipment.

Throughout the 1980's there had been continuing attempts to develop an agreed set of regulations to protect fish habitat while providing legal certainty for the placer mining industry. In May, 1987, the Ministers of Indian Affairs and Northern Development and Fisheries and Oceans made a public commitment to resolve long-standing issues associated with placer

TABLE 16-1. Potential environmental impacts of marine disposal of mine tailings.

Category	Environmental impacts
Physical	— Loss of habitat, especially for bottom-dwelling organisms.
	— Increased turbidity around the point of discharge.
	— Spread of tailings and other associated contaminants.
Chemical	— Increased concentration of trace elements in the water column, particularly metals and certain organic compounds (mill reagents)
	— Changes in the redox potential and pH of seawater.
	— Increase in chemical oxygen demand and possibly in biomechanical oxygen demand.
	— Creation of a resevoir of contaminants in the sediments and/or tailings which may be remobilized through time into the water column.
Biological	— Lethal and sublethal toxic effects on plants and animals.
	— Loss of diversity and possible disappearance of life in the severely affected areas.
	— Bioaccumulation and bioconcentration of certain trace elements in biota.
	— Change in the benthic community from sensitive species to more pollution-tolerant species.
	— Decreased photosynthesis due to high turbidity levels.

Source: Kay (1989).

mining. In June, 1987, they established a government-industry Yukon Implementation Review Committee (IRC) to develop a workable regime for the placer mining industry.

As a result of the work of this Committee, in May, 1988, Ministers Siddon and McKnight released a Yukon Fisheries Protection Authorization, and a Policy Directive for the Protection of Fish and Fish Habitat in the Yukon from Placer Mining Activity (Canada 1988g). These documents specified a new operational regime for placer mining, including stream classification into four types, monitoring responsibilities, legal compliance procedures, methods of measurement and site-specific authorizations.

The new regime provided for:

1. The protection of salmonid species as a first priority, with less stringent standards applied for freshwater fish species.
2. The continuation of historic placer mining activities, applying minimal protection standards in streams where fish habitat had been disrupted, and in streams where there are fish of little or no value.
3. The periodic reclassification of streams, through reference to the IRC, which in turn makes recommendations to the Minister of Fisheries and Oceans.
4. The maintenance of the productive capacity of fish habitat.
5. A commitment by the miners to restore fish producing streams that had been disrupted by placer mining activities.

3.5 Foreshore and Estuary Developments

Estuaries are often critical areas for fish. Adult salmon assemble there before commencing their upstream migrations. Young fish feed there en route from freshwater to the ocean. Estuaries are often nursery areas for other species which depend on habitat such as estuarine marshes.

These areas are often also the chief centres of human settlement, ports, marinas and industrial development. Because of the heavy development of such areas, fish and fish habitat can sometimes be imperiled.

On the Pacific coast, the most important estuary is that of the Fraser, one of the most important salmon rivers in the world. Here salmon congregate both en route to the ocean and during their return spawning migration. The Fraser estuary is also surrounded by considerable urban development, including Vancouver, one of Canada's largest cities.

Early in the 20th century much of the wetland delta area was dyked for agriculture use. In the early 1980's plans to expand a coal port terminal led to a major review by a Federal Environmental Assessment Panel. The Panel concluded that eelgrass beds provide the prime habitat for both juvenile salmonids and birds. It called for design measures to minimize the impact on eelgrass. Waldichuk (1993) reported that follow-up studies after the terminal was enlarged indicated that protecting eelgrass beds was a wise decision.

In June 1991, the federal government announced a $100-million program to clean up the Fraser River, restore salmon populations to historical levels of abundance, and restore the river to environmental health. The Fraser River Basin Action Plan includes:

— working in partnership with the province to identify and control contaminants entering the rivers from industrial and domestic point and non-point sources;
— identification and clean-up of waste sites through the application of the "polluter pays" principle and a National Contaminated Sites Remediation Program;
— enforcement of environmental protection laws;
— restoration, enhancement and conservation of fish and wildlife habitats, including doubling of sockeye salmon stocks within 20 years and increasing other salmon stocks (Canada 1991a).

Regarding foreshore development on the Atlantic coast, the 1983 Atlantic Fish Habitat Review criticized the lack of an integrated Coastal Zone Management approach anywhere in Atlantic Canada. The Review Team observed:

> "The result has been a continuing series of conflict-ridden harbour developments in areas considered to be of high value to shipping, the fishing industry, recreational development and municipal expansion. It is difficult to judge what percentage of the limited coastal resource of fish habitat has been eaten up by senseless and ill-conceived harbour expansion through infilling and unzoned development. Uncontrolled development of this sort has led to a proliferation and dispersal of pollution sources affecting coastal resources in a 'crazy-quilt' pattern which is impossible to manage or monitor. Harbours have been developed in areas where unsuitable oceanographic conditions, aggravated by harbour structures such as breakwaters and wharves, have led to a never-ending cycle of

Fraser River Esturary (Photo courtesy of Otto Langer).

> remedial dredging and dumping.... The need for a strong system of Coastal Zone Management *must* be addressed, and soon" (DFO 1983m).

This plea is equally valid today. Other examples of problems of shorefront development include the construction of coastal roads and bridges. Road builders have traditionally driven causeways, rather than bridges, across coastal marshes and estuaries and have destroyed the habitat responsible for a significant amount of marine production.

For years there was considerable controversy over the impact of the Canso Causeway linking Cape Breton to mainland Nova Scotia. The Causeway made possible the creation of one of the finest, ice-free deepwater ports in Atlantic Canada, which stimulated growth of industry and population there. Various groups in the 1970's blamed the Causeway for pollution and for failures in the lobster and herring fisheries.

The Fisheries and Marine Service of DOE convened Canso Marine Environment Workshops in November, 1977, and February, 1978. The results of those Workshops were summarized by McCracken (1979). The purpose was to assess the effects of the Canso Causeway on the marine ecology and fisheries of the area. This task was difficult because the Causeway had been constructed during the early 1950's without any impact statement or collection of baseline data. Furthermore, during the two decades following construction, various ecologically significant events had confounded the situation, obscuring causal relationships. Despite these limitations, the Workshop reached the following conclusions:

1. With the possible exception of lobsters and the disappearance of species or small fisheries from extremely local areas, no adverse changes in biota could be identified which could be attributed to the building of the Causeway.
2. For lobsters, it was agreed that there had been a recruitment failure on the Chedabucto Bay side of the Strait. The workshop was unable to distinguish among the possible causes of this recruitment failure. It may have been due to one or more of these factors: construction of the Causeway and prevention of lobster larvae from St. Georges Bay entering Chedabucto Bay, recruitment overfishing, non-Causeway related environmental effects or changes in the marine environment.

There was a subsequent resurgence of the lobster population in the 1980's because of a major recruitment pulse Atlantic-wide. This tends to confirm the earlier conclusion that the Causeway had had no significant effect on marine fish and shellfish populations in the area.

The coastal zone in extensive portions of Atlantic Canada has deteriorated considerably through private projects of infilling, wharf construction, shoreline protection, breakwater construction and gravel removal. Such developments and activities are so extensive that they have added up to millions of dollars of lost production potential through simple thoughtlessness (DFO 1983m).

3.6 Offshore Hydrocarbon Exploration and Development and Oil Spills

3.6.1 General

More than one quarter of world oil production in recent years has come from offshore areas. The portion increased at a rate of nearly 10% per year from the mid-1970's to the mid-1980's (Ulfstein 1988). Exploration for offshore hydrocarbons and the subsequent exploitation of proven reserves pose conflicts with other uses of the oceans, particularly fisheries. Transporting oil by tankers sometimes results in spectacular spills which foul beaches and kill marine life. Spills attract considerable media attention and public concern abut the impact on the marine environment and marine life. Notable examples include spills from the tanker *Arrow* in Chedabucto Bay, Nova Scotia, in February 1970, and more recently the barge *Nestucca* off the southwest coast of Washington State in December, 1988, and the *Exxon Valdez* spill in Prince William Sound, Alaska, in March, 1989,

During the 1970's and 1980's there was considerable exploration for offshore hydrocarbons in the Beaufort Sea and off Canada's east coast. The discovery of significant oil and gas reserves off Canada's Atlantic coast, possibly large enough to make the region one of the world's major oil producers, has promised economic development for a disadvantaged region of Canada. But this is a region which has traditionally depended on the fishery. Development of these reserves of oil and gas can only proceed at a rate and in a manner which does not jeopardize the renewable fish resource base or otherwise harm the fishing industry. Similar concerns have arisen about the petroleum reserves in the Beaufort Sea. Both off the east coast of Newfoundland and in the Beaufort Sea development is complicated by the fact that these reserves are in waters which are ice-covered, either seasonally or year-round.

This has generated important questions about the impact of oil and gas spills upon the marine environment and marine living resources. The extent to which offshore drilling and exploitation activities will interfere with traditional fishing operations is unknown. Major environmental reviews were conducted for the Beaufort Sea development proposals and the Hibernia proposal off Newfoundland. These reviews resulted in decisions to permit controlled development provided adequate safeguards are taken to protect living marine resources and the fishery.

Shoreline fouled with oil.

In the late 1980's considerable controversy arose over proposals to permit exploration drilling for oil and gas on Canada's portion of Georges Bank, which supports the rich scallop fishery so important to the economy of southwest Nova Scotia. Scientific review of the risks in this instance plus lobbying from fishery and environmental interests led the federal government to establish a moratorium to the year 2000 on exploration activities in that area.

Offshore oil and gas exploitation and fishing coexist in several areas of the world. One striking example is the North Sea. Nonetheless, Canadians pose three sets of legitimate questions which have to be addressed before permitting such development. The key questions are:

1. Will the ocean's ability to produce fish be affected by the oil spills which, even with most vigilant management, are certain to occur eventually?
2. Even if numbers of fish stay at existing levels, could their quality be reduced by oil spills or by drilling operations?
3. What will be the effects on fishing? Will vessels be obstructed or crowded out by rigs or pipelines? Will the fouling of nets and traps be a major problem? Will the dumping of debris from rigs and supply vessels be a problem? How will spills be dealt with?

3.6.2 Impact on Fish

There is a vast literature on the effects of petroleum hydrocarbons on aquatic organisms. For a partial listing, see Waldichuk (1993). There are five major ways in which oil can affect fish (DFO 1987n):

Toxicity — Under certain conditions, most oils are lethal. The power to kill fish increases with the oil's solubility in water. This means that light refined oils like gasoline and diesel fuels, which mix easily, can be very dangerous. On the positive side, these oils evaporate quickly. Heavier oils such as "Bunker C" may be less of a threat, though they are harmful in other ways. Crude oil is composed of numerous chemical compounds ranging from light to heavy in molecular weight. "Bunker C" is a particularly heavy component remaining after the refining of crude oil.
Juvenile fish are particularly sensitive to oil-related toxicants. Some of the oil exploration activities occur in the areas where juvenile fish congregate, for example, on the Grand Banks and the Scotian Shelf.

Physiological and behavioral effects — In very small concentrations oil which does not kill fish may trigger changes in their growth rate, their metabolism and their ability to reproduce. These oil traces may also cause changes in the pattern of fish migrations.

Incorporation into tissue — Most aquatic organisms will absorb fractions of oil. This absorption could taint the fish, rendering them unsaleable as food.

Changes in biological communities — If oil spills kill certain species, the community structure of the aquatic ecosystem could be altered, resulting in declines in fish stocks and economic losses. It is possible that the aquatic community may never return to its original state.

Incorporation of oil into sediments — Oil may become mixed into mud or gravel on the sea bottom and remain there. From these pockets, oil may leach out over the years, contaminating an area and resulting in ongoing damage to fish and other aquatic life.

The first victims of a marine oil spill are usually sea birds. For example, following the *Exxon Valdez* oil spill, more than 30,000 dead seabirds were counted. The impact on fish and invertebrates is not as clear cut. The fisheries for herring and salmon in Prince William Sound were closed following the *Exxon Valdez* oil spill and those for molluscan shellfish and crabs on the west coast of Vancouver Island following the *Nestucca* spill. In both instances this occurred because of concern about tainting of seafood (Waldichuk 1993). In August, 1991, scientists from the Bedford Institute visited Prince William Sound. They witnessed once heavily oiled shorelines that bore few traces of oil. Intertidal plant and animal communities appeared healthy and productive. Marine mammal and fish populations were at normal levels. They concluded that the Prince William Sound ecosystem recovered from this accident in just over two years (DFO 1991g). Other studies indicated that the cleanup treatments apparently had a greater deleterious impact on shoreline communities than did oil alone (*The Vancouver Sun*, January 8, 1993). At a Symposium in January 1993 some scientists suggested that some Prince William Sound wildlife would suffer for decades (*The Ottawa Citizen*, February 7, 1993).

The extent of damage to fishery resources will vary among species, depending on the nature of the petroleum product. Shellfish are likely to be the most seriously affected. Refined petroleum products such as diesel fuel are more toxic than Bunker C which will not easily penetrate the substrate. Herring, which spawn in the intertidal zone, may be seriously affected if an oil spill coincides with spawning and hence damages the herring eggs or the vegetation to which the eggs are attached. Species with pelagic eggs could be affected if an oil slick forms offshore. Waldichuk concluded that there are not many known examples in the marine environment where there has been permanent or long-term damage from an oil spill, with the possible exception of the very large *Amoco Cadiz* crude oil spill off France in 1978. In that instance coastal oyster beds were severely polluted by oil. This persisted for several years.

In Canada one of the best documented studies of the long-term effects of an oil spill is that from the tanker *Arrow* in February, 1970. After the *Arrow* went aground in Canso Strait, scientists doubted the summer-long manual clean-up of the coast would be of much benefit. The final report on the accident stated:

> "The *Arrow* disaster revealed dramatically and at substantial cost the woefully inadequate measures in place for contending with oil spills of this magnitude" (Anon. 1970).

Recently, Dr. John Vandermeulen of DFO observed that "after two decades, cleanup technology still hasn't progressed very far, and perhaps never will" (*The Ottawa Citizen*, March 4, 1990). Research at the Bedford Institute indicates that the early work done on approximately 300 km of Chedabucto Bay beachline may have made it easier for the natural systems to cleanse themselves. Vandermeulen observed that, "Today the beaches of Chedabucto Bay are 99.9% clean, although there are still some detectable hydrocarbons at some beaches." There was no obvious impact on fish and lobster at the time and there has not been any since.

On a related point, Vandermeulen noted that only about 12% of ocean oil pollution can be directly traced to tanker spills. He speculated that the gross impact on the ocean shoreline of numerous, small spills occurring each day can cause far more damage over time than an *Arrow*-type accident.

Recent oil spills in Washington State and Alaska raised concerns again in Canada about the rules and procedures governing tanker passage through Canadian waters. The *Nestucca* spill off Washington State, which would be considered a large spill (more than 1,000 barrels or 159,000 litres), dumped 875,000 litres into the ocean. Although only a fraction of that drifted into Canadian waters, it was sufficient to foul with oil more than 150 km of the west coast of Vancouver Island. The *Exxon Valdez* spill which dumped 261,000 barrels in Alaska fell into the category of a catastrophic spill (more than 150,000 barrels or 24 million litres). A 1989 study conducted for Environment Canada and the Canadian Coast Guard calculated that a "large" oil spill can be expected to hit in or near Canadian waters "every one or two years". It concluded that the chance of a catastrophic spill is "reasonably remote...but probable enough to be cause for concern considering the potential for damage" (*The Ottawa Citizen*, December 8, 1989).

As a result of these spills, the federal government appointed a Public Review Panel on Tanker Safety and Marine Spills Response Capability, chaired by Mr. David Brander-Smith. In its Report of September 1990, the Panel noted that it had not learned of any agency that could handle a major oil spill under any conditions. The *Nestucca* and *Exxon Valdez* experiences had demonstrated that spill response equipment and organizations had serious limitations in containing or cleaning up marine spills. The Panel considered spill prevention to be the highest priority for protecting the coast and marine environments (Brander-Smith 1990).

In addition to the potential impact of oil spills on fish stocks, the fishing industry has a number of other concerns about offshore hydrocarbon exploration and development. These include: potential loss of access to fishing grounds; damage to fishing gear by oil-related bottom debris, seafloor installations, and vessel traffic; movement of skilled labor and vessels from fisheries to petroleum-related industries; competition for shoreline facilities, space and service; and damage to nets and gear by oil pollution.

3.6.3 Loss of Access to Fishing Grounds

Fishermen's concerns include both temporary loss around exploratory wells and the potential for long-term loss at the site of a producing platform or a pipeline. The situations experienced in various countries differ on this aspect. It appears that loss of access to their fishing grounds by fishermen in the United States has been relatively minor. In Norway and the United Kingdom, however, the situation has been more serious. In the North Sea, there has been extensive exploration and production activity, with significant loss of access to fishing grounds. Pipelines interfere with fishing patterns as fishermen avoid them through fear of damaging their gear.

3.6.4 Damage to Nets and Gear by Oil Pollution

Since exploratory drilling began off Canada's east coast in 1966, a considerable number of wells have been drilled. No serious pollution problems have developed to date from this activity. Government and industry adopted a conservative preventive approach to reduce the probability of a serious blow out. Under Canada's oil and gas regulations, detailed drilling plans had to be presented to the Canada Oil and Gas Lands Administration (COGLA)[1] before a company could begin operations. There were elaborate provisions regarding operating and emergency procedures, including specification of drilling rig and support facilities; special equipment at the site; details on the geology of the area; an oil spill contingency plan; and proof of pollution liability coverage.

The *Canada Oil and Gas Act* (1981, c.81, s.1, as amended) contains far-reaching provisions regarding liability and compensation. One key feature is the establishment of an absolute liability regime whereby no proof of fault or negligence is required to establish liability. This liability scheme applies to spills, whether authorized or unauthorized, as well as to discharges, emissions or escape of gas or oil, and to damage resulting from debris. The definition of damage recoverable includes loss of income, including future income.

3.6.5 Hibernia and Georges Bank — Different Responses to Potential Conflict Between Fisheries and Hydrocarbon Development

The fishing industry and government response to offshore hydrocarbon exploration and development in two areas off Canada's east coast has differed considerably in recent years. Considerable exploratory activity has taken place on the Grand Banks, off Labrador and on the Scotian Shelf. With petroleum companies the federal and provincial governments established conditions for development of the Hibernia reserves on the Grand Banks. This occurred after a Hibernia Environmental Assessment Panel review. The Panel concluded that the major impact resulting from an oil spill would be disruption of fishing rather than effects on fish stocks. It noted:

> "Delineation of the area from which fishing would be excluded in the event of a spill will be a difficult task, given that the present knowledge of the exposure required for fish tainting is limited and that the oil spill may extend in an irregular manner over large areas. In the event that an oil spill were to occur in the inshore area, severe localized damage to fish populations, especially shellfish, might result" (Canada-Newfoundland 1985).

The Panel also made recommendations about loss of access to fishing areas, debris and compensation. Following the Panel Review, the federal and provincial governments committed themselves to encouraging development of the Hibernia fields, while ensuring adequate safeguards to minimize the project's negative impact. In October 1990, development of the Hibernia reserves commenced.

In the late 1980's, concern emerged in Nova Scotia about proposals for exploratory hydrocarbon drilling on Georges Bank. Fishing industry groups lobbied the federal and provincial governments for a ban on exploratory drilling, arguing that Georges Bank was a particularly sensitive marine ecosystem. The Gulf of Maine Advisory Committee requested that DFO scientists provide background information on the potential impact of exploratory drilling. The Science Sector of DFO's Scotia-Fundy Region, in January, 1988, released an assessment of the possible environmental impacts of exploratory drilling on Georges Bank fishery resources (Gordon 1988). In 1986, COGLA had conducted an environmental evaluation of potential exploratory drilling on Georges Bank using information provided by DFO and other agencies. At that time COGLA concluded:

> "The proposed exploratory drilling program would not be expected to have any significant environmental impacts, and, therefore, no further environmental review is required under the federal environmental assessment review process" (COGLA 1986).

The 1988 scientific report stated that Georges Bank has many physical and biological features which have the potential of magnifying the environmental impacts of possible blowouts and major spills compared to other continental shelf areas. The report suggested that exploratory drilling could have many environmental impacts causing conflict between the fishing and oil industries.

The report concluded:

> "To date, it has not been possible to demonstrate conclusive evidence of long-term impacts on highly variable resource populations in case studies of actual offshore events such as the *Argo Merchant* spill on Nantucket Shoals (1976), the Uniacke blowout near Sable Island (1984) and drilling/oil production in the North Sea. Ten exploratory wells were recently drilled on the American sector of Georges Bank without demonstrated environmental

[1] COGLA was disbanded in April, 1991. Its activities were reassigned to the Minister of Indian Affairs and Northern Development and to the Minister of Energy, Mines and Resources.

Offshore Oil Drilling Rig.

effects or mishap. This does not rule out, however, that long-term impacts might have occurred but were undetectable at the population level by available sampling methods.

> "There is a range of opinion among DFO scientists as to the probability of damage to fisheries resources from hydrocarbon activities and this reflects in part a deficiency in the amount of site-specific information available for Georges Bank. Additional research is necessary to increase knowledge of key environmental processes and to define more precisely what parts of Georges Bank, both spatially and temporarily, should be classified as critical habitat in order to protect, maintain and restore valuable renewable fishery resources.
>
> "Proposed drilling sites should continue to be reviewed on an individual basis. If a decision is made to allow exploratory drilling at a particular site, it should proceed only under stringent regulations which reflect the environmental and habitat concerns" (Gordon 1988).

On April 18, 1988, the Honourable Marcel Masse, federal Minister of Energy, Mines and Resources, called for an environmental moratorium to at least the year 2000 on drilling on Georges Bank. An environmental review would be carried out during the moratorium. Mr. Masse explained the decision:

> "Georges Bank has one of the richest, most economically important fisheries in the world. The unique characteristics that have endowed it with such wealth also make it particularly sensitive from an environmental perspective.... I have concluded that by law no drilling should be permitted on Georges Bank to at least the year 2000 given the current state of uncertainty with respect to potential impacts. Moreover, the Canada-Nova Scotia offshore legislation should provide for a thorough and independent public review of petroleum exploration on Georges Bank prior to the year 2000" (EMR 1988).

Provisions for the moratorium and a future environmental review were incorporated into legislation passed that year. Fisheries and environmental groups welcomed this decision. It is debatable whether the scientific uncertainty about drilling on Georges Bank was much different from that for the Grand Banks. Certainly, the support for offshore hydrocarbon development was much greater in the Newfoundland instance.

Miles and Geselbracht (1986) reviewed various conflicts between fisheries and other uses of the sea. For oil and gas exploration and exploitation, they concluded that the conflicts which were predicted between fishing operations and the offshore oil and gas industry had materialized. However, conflicts predicted between oil and gas development and fisheries resources or the marine environment had not always emerged in the manner foreseen. In the North Sea, damage to fishing gear and vessels had occurred due to contact with pipelines and debris from oil development activities. On the other hand, predictions about adverse impacts on fishery resources, spawning grounds or nursery sites had not been fulfilled.

Despite these observations, recent oil spills have renewed concerns about the potential detrimental effects of blowouts and accidental spills on fish and the marine environment. Vigilance is required to minimize the occurrence and negative effects of such events.

3.7 Contaminants

In recent decades the impact of contaminants on aquatic ecosystems, and potentially upon human health has become a matter of increasing concern not only to environmentalists but also to the general public. The number of contaminants affecting the freshwater and marine environments is large and the potential effects diverse and difficult to measure. A more widespread preoccupation with the problem of chemical contaminants at the end of the 1980's reflected a more heightened consciousness of man's negative effects upon the environment. Summarizing these diverse impacts is beyond the scope of this chapter. Here I will attempt merely to illustrate the range and magnitude of the problems with a few examples.

3.7.1 Acid Rain

Acid rain was one issue which captured popular attention during the 1970's and 1980's. This became a major bilateral irritant between Canada and the United States. Although the term "acid rain" was popularized during the 1970's, it originated in the 19th century. Certain features of the acid rain phenomenon were first described by an English chemist named Robert Smith. In 1852, he published a report on the chemistry of rain around Manchester. In 1872, he first used the term "acid rain" in a prophetic book entitled "Air and Rain: The Beginnings of Chemical Climatology" (Smith, 1872). His work was, however, quickly forgotten.

Not until the 1970's did the phenomenon of acid rain become the focus of widespread scientific and public attention. Within a decade it became identified as one of the major environmental problems facing North Americans. The term "acid rain" refers to precipitation in the form of rain, snow or hail which results when oxides of sulphur and nitrogen react chemically with oxygen and moisture in the atmosphere. Unpolluted rain is already slightly acidic because of the presence in air of carbon dioxide, which combines with water to form carbonic acid. This rain usually has a pH value of 5.6. When rain is contaminated with sulphuric and nitric acids the pH falls below 5.6; this precipitation is referred to as acid rain. To put this in perspective, acidity is measured on a logarithmic scale of 0 for maximum acidity, 7 for neutral, and 14 for maximum alkalinity. A change of one unit on this scale represents a 10-fold change in acidity or alkalinity of a solution.

The acidic pollutants in acid rain originate as emissions of sulphur dioxide (SO_2) and oxides of nitrogen (NO_x). They are converted in the atmosphere to acidic compounds (sulphates and nitrates) and return to earth in precipitation. Most of these pollutants in acid rain come from three main sources:

1. burning of fossil fuels (coal and oil) in thermal power plants;
2. burning of gasoline in motor vehicles; and
3. smelting operations of plants which refine non-ferrous metal ores (Fig. 16-2).

Most of the sulphur dioxide (SO_2) pollutants originate in power plants and industrial sources. Nitrogen oxides (NO_x) in the atmosphere originate primarily in the emissions of motor vehicles.

The effects of acid rain often materialize in areas far from the actual sources of emission. This phenomenon is referred to as long-range transport of atmospheric pollutants or LRTAP. The groundwork for this understanding of the acid rain phenomenon was established through research by Gorham and his colleagues in England and Canada during the 1950's and 1960's (e.g. Gorham 1976). This research established the following principles:

1. Much of the acidity of precipitation in industrial regions can be attributed to atmospheric emissions produced during combustion of fossil fuels;
2. Progressive loss in alkalinity of surface waters and increase in acidity of bog waters can be traced to atmospheric deposition of acidic substances through precipitation;
3. Much of the free acidity in soils receiving acid precipitation was due to sulphuric acid;

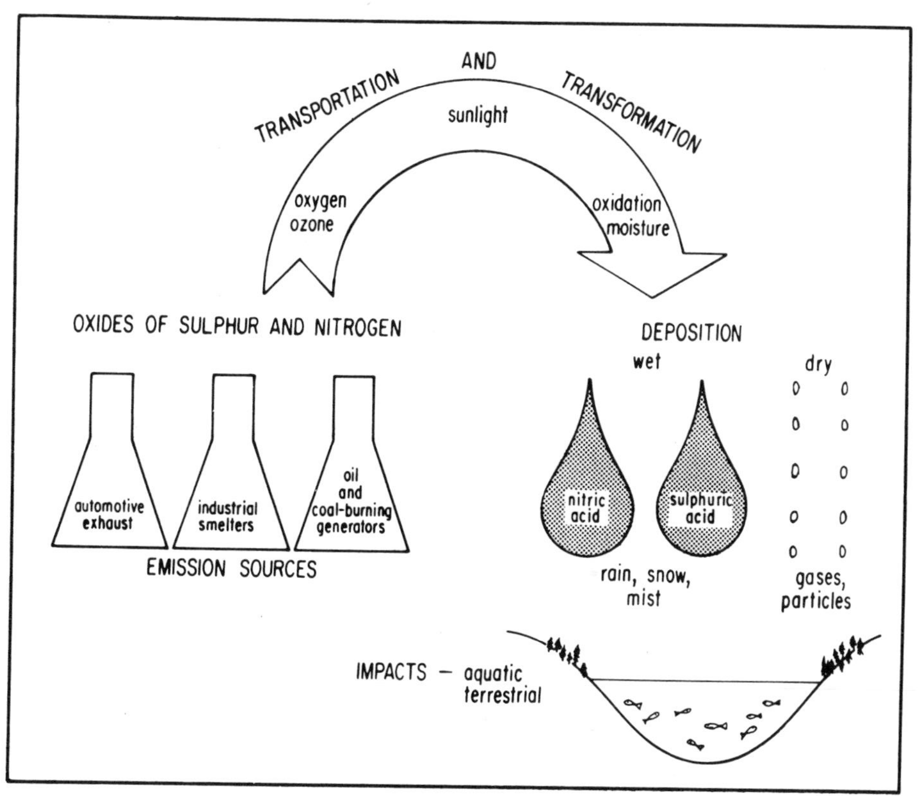

FIG. 16-2. The origins of acidic deposition.

4. Fumigation with sulphur dioxide and the resulting acid rain contributed to the deterioration of vegetation, soils, and lake water quality around metal smelters.

Thus, Gorham by the early 1960's had established a major part of the present understanding of the sources and the limnological and ecological significance of acid rain (Cowling 1980). By the late 1960's Svante Odén, a soil scientist at the Agricultural College near Uppsala in Sweden, conducted analyses of the trajectories and chemistry of air masses which showed that:

1. Acid precipitation was a large scale regional phenomenon in much of Europe with well-defined source and sink regions;
2. Both precipitation and surface waters were becoming more acidic; and
3. Long-distance (100–2000 km) transport of both sulphur and nitrogen-containing air pollutants was taking place among the various nations of Europe (Odén 1968).

Odén also predicted the following ecological consequences of acid rain:

1. decline of fish populations;
2. decreased forest growth;
3. increased plant diseases; and
4. accelerated damage to materials.

Odén's hypothesis generated a storm of debate. The Swedish government responded by presenting to the 1972 UN Conference on the Environment a case study entitled: "Air Pollution Across National Boundaries: The Impact of Sulphur in Air and Precipitation."

In North America concern about acid rain developed first in Canada and subsequently in the United States. Initial attention focused on the effects of sulphur dioxide and related acid rain and heavy metal deposition near metal smelting, especially near Sudbury, Ontario. In the late 1960's, Harold Harvey and Richard Beamish discovered the detrimental effects of acid rain on fish in several lakes in the La Cloche Mountains of Killarney Provincial Park. Gorrie (1986) credited Harvey and Beamish with being the first Canadians "to alert the public to the invisible danger in gentle summer rain and soft swirling snowflakes." Their investigations revealed that, where fish were dying, the lakes were very acid. Beamish and Harvey (1972) concluded that the most likely sources of this were the Inco and Falconbridge nickel smelters near Sudbury, about 100 km northeast of their study area. Subsequently, declining fish populations were discovered in numerous other lakes, in southern Ontario and Nova Scotia, remote from local sources of atmospheric sulphur.

In the years since then much has been learned about the effects of acid rain, both on the aquatic and terrestrial ecosystems. Scientists discovered that acid rain in lakes or streams harms fish in several ways. Minns et al. (1986) summarized five ways in which fish may be affected:

1. Enhanced bioaccumulation of some metals that may render fish unsafe for human consumption;
2. Declines in fish production;
3. Alteration of community structure;
4. Imposition of mortalities from episodic changes in chemistry; and
5. Extinction of fish populations and aquatic communities.

Large areas of North America were threatened by the negative effects of acid rain because of the sensitivity of the terrain downwind of emission sources. The major concern originated with fish, e.g. freshwater fish in lakes and streams and Atlantic salmon in Nova Scotia, but the concerns then broadened to include effects on forests, agricultural crops, buildings and human health.

The ability of lakes or streams to neutralize acid rain, i.e. their buffering capacity, differs depending on the type of surrounding rock and soil. If a lake is surrounded by limestone or other rock containing carbonate, the water can neutralize much of the acid. These are acid-tolerant lakes. Lakes surrounded by rocks like granite or others containing little alkaline material are acid-sensitive. In large parts of eastern North America, lakes and streams are acid-sensitive. In particular, Canadian Shield lakes on the Precambrian bedrock areas of central and eastern Canada are sensitive and are also subject to large amounts of acid precipitation. Kelso et al. (1990) estimated that approximately 21% of Canada's total lake surface area (and 38% of Canada's total number of lakes) occurs in the region susceptible to acid precipitation.

The lakes near Sudbury were drastically acidified. Lakes in and near the La Cloche Mountains experienced pH declines averaging 0.16 units/year for lakes in and east of the Mountains. The pH in Lumsden Lake, where Harvey and Beamish first stumbled on the acid rain phenomenon, declined from 6.8 in 1969 to 4.4 in 1971, indicating a 200-fold increase in acidity in 10 years (Beamish and Harvey 1972). Beamish (1976) found that lakes in this region were much more acidic than northwest Ontario lakes in similar geological terrain but not receiving acidic precipitation. In south-central Ontario, lakes receive acidic precipitation, with the pH in some being as low as 5.0.

Of 19 lakes surveyed in Nova Scotia, pH declined in all of them between 1955 and the late 1970's (Watt 1981) Similarly, pH had declined in 14 Nova Scotia rivers.

In the 68 La Cloche Mountains lakes studied by Harvey and Beamish, the number of species of fish present decreased as lake pH decreased. Several species of fish disappeared from lakes in this region (see Beamish and Harvey 1972; Harvey 1975). Harvey (1982) reported that fish had been lost from around 200 Ontario lakes. As of 1980, seven rivers in southwestern Nova Scotia were believed to be unsuitable for Atlantic salmon reproduction and another eight threatened (Farmer et al. 1980).

Although some mortalities result from acidification, the most common reason for the decline in fish populations and the extinction of certain species in certain lakes is reproductive failure. Recruitment declines and eventually a species may disappear from an acidified lake or stream (Haines 1981).

There is another detrimental effect of acidification of lakes or streams. Thompson et al. (1980) reported that fish from remote lakes in Ontario and southeastern New Brunswick had elevated levels of mercury apparently associated with declining pH. It has been speculated that a continuing input of acid rain can release metals such as mercury, aluminum, lead and manganese from the surrounding soils or lake beds.

The 1986 State of the Environment Report for Canada (Bird and Rapport 1986) presented information on trends in the emissions of SO_2 and NO_x in the United States and Canada from 1955 to 1980. In eastern Canada and the United States (i.e. east of the Manitoba-Saskatchewan border and the northeastern and midwestern states), sulphur dioxide emissions did not change greatly from 1955 to 1980. There was, however, a substantial increase in nitrogen oxide emissions. Electrical utilities were the main source of SO_2 in the United States and non-ferrous smelters the most significant Canadian sources. In both countries, motor vehicles were the major source of nitrogen oxide emissions.

The relative contribution of Canadian and U.S. sources to wet sulphate deposition in various parts of eastern Canada is shown in Table 16-2. Depending on the area, from 40 to 70% of the wet sulphate depositions in eastern Canada came from emissions originating in the United States. Sulphate deposition was identified as the dominant factor contributing to the long-term acidification process in eastern Canada (Fig. 16-3).

The 1986 SOE Report concluded that, for regions receiving sulphate depositions of less than 20 kg per hectare per year ($kg{\cdot}ha^{-1}{\cdot}yr^{-1}$), detrimental or biological effects had not been observed. For regions receiving sulphate deposition of between 20 and 30 $kg{\cdot}ha^{-1}{\cdot}yr^{-1}$, there was evidence of chemical alteration and acidification. For regions experiencing loading greater than 30 $kg{\cdot}ha^{-1}{\cdot}yr^{-1}$, serious long-term chemical and/or biological effects, as well as short-term effects on non-alkaline surface waters, have been observed. A broad band stretching across southern Ontario, Quebec, the Maritimes and Newfoundland received sulphate depositions between 20–40 $kg{\cdot}ha^{-1}{\cdot}yr^{-1}$.

Including the well-known case of lakes in the La Cloche Mountains, over 300,000 hectares of 11,400 lakes in Ontario were considered to be at moderate to high risk, representing a potential loss of 30% or more of the total provincial resource of at least 5 major fish communities.

Apart from the documented loss of fish populations in Ontario, one of clearest impacts of acid rain has been on Atlantic salmon populations in Nova Scotia. Watt (1986, 1987) summarized acid rain impacts on Atlantic salmon in eastern Canada. An extensive survey of water chemistry conducted in 1979-80 revealed that the only severely acidified area in the Maritimes was the Southern Upland zone of Nova Scotia. The area contains 60 rivers, of which 44 had a mean annual pH less than 5.4.

Watt's analyses indicated that, in 13 rivers with pH less than 4.7, salmon runs no longer existed. For rivers in the pH range 4.7 to 5.0, the angling catch had declined to about 10% of levels prevalent during the 1936–1953 period. Rivers in pH categories 5.1 to 5.4 and greater than 5.4 showed no signs of an impact of acidification on angling returns. Watt estimated that the loss of Atlantic salmon production capacity attributable to acidification was 49.8% in Nova Scotia's Southern Upland, 33.3% for all of Nova Scotia but only 3.2% for eastern Canada as a whole. Although the local impact of acidification on the Atlantic salmon resource in Nova Scotia had been substantial, acidification did not threaten the resource as a whole.

DFO has undertaken experiments to test the feasibility of mitigating acidification by adding limestone or other substances to lakes or streams. Watt (1987) estimated the total cost for a 20 year project of de-acidifying the Atlantic salmon habitat of the Southern Upland of Nova Scotia at $95 million (1984 dollars). Given that this would amount to $400 per restored salmon and that the value per landed salmon to the eastern Canadian economy was, on average, less than $100 per fish, he concluded that a major liming operation was not economically justified. Watt suggested that approximately 6 of the 18 stocks threatened with extinction should be preserved by creating deacidified refuges through headwater lake liming.

Reducing the threat posed by acid rain required a reduction in the emissions producing acid rain. In 1985 the federal government and the seven easternmost provinces agreed to reduce the pollution causing acid rain by 50% from 1980 levels by 1994. The agreed goal was to reduce the amount of deposition to 20 $kg{\cdot}ha^{-1}{\cdot}yr^{-1}$. Reduced acid-loading was expected to protect all but the most sensitive lakes — the latter amounting to 20,000 of the 2,000,000 in the seven provinces involved.

This target has been criticized by some because it deals only with the average amount of acidity in a lake during the year and ignores periodic heavy doses when as much as 25% of the annual deposition can occur in a few weeks. "The environment is not being asked to accept the annual average," said Harvey, "but a small number of rather severe acid pulses. Fish have to be able to survive the most extreme event they encounter" (Gorrie 1986).

Despite the Canadian federal-provincial initiative, the damage from acid rain could not be halted unless U.S. emissions were reduced, since more than half the emissions producing acid rain in eastern Canada originated in the U.S. For more than a decade Canada lobbied the U.S. for a bilateral treaty on acid rain. The Reagan administration steadfastly maintained that the cause of the effects was not proven and that more research was required. In a 1985 report, special envoys appointed by Canada and the U.S. concluded that acid rain was a serious problem and recommended that the U.S. spend $5 billion over 5 years to research new methods to control emissions.

TABLE 16-2. Percentage Contribution to Wet Sulphate Deposition.

Source	North-central Ontario %	South-central Ontario %	Southern Québec	Southern Nova Scotia
Canada	29–32	32–46	35–58	30–34
United States	68–71	54–68	42–65	66–70

Source: Bird and Rapport (1986).

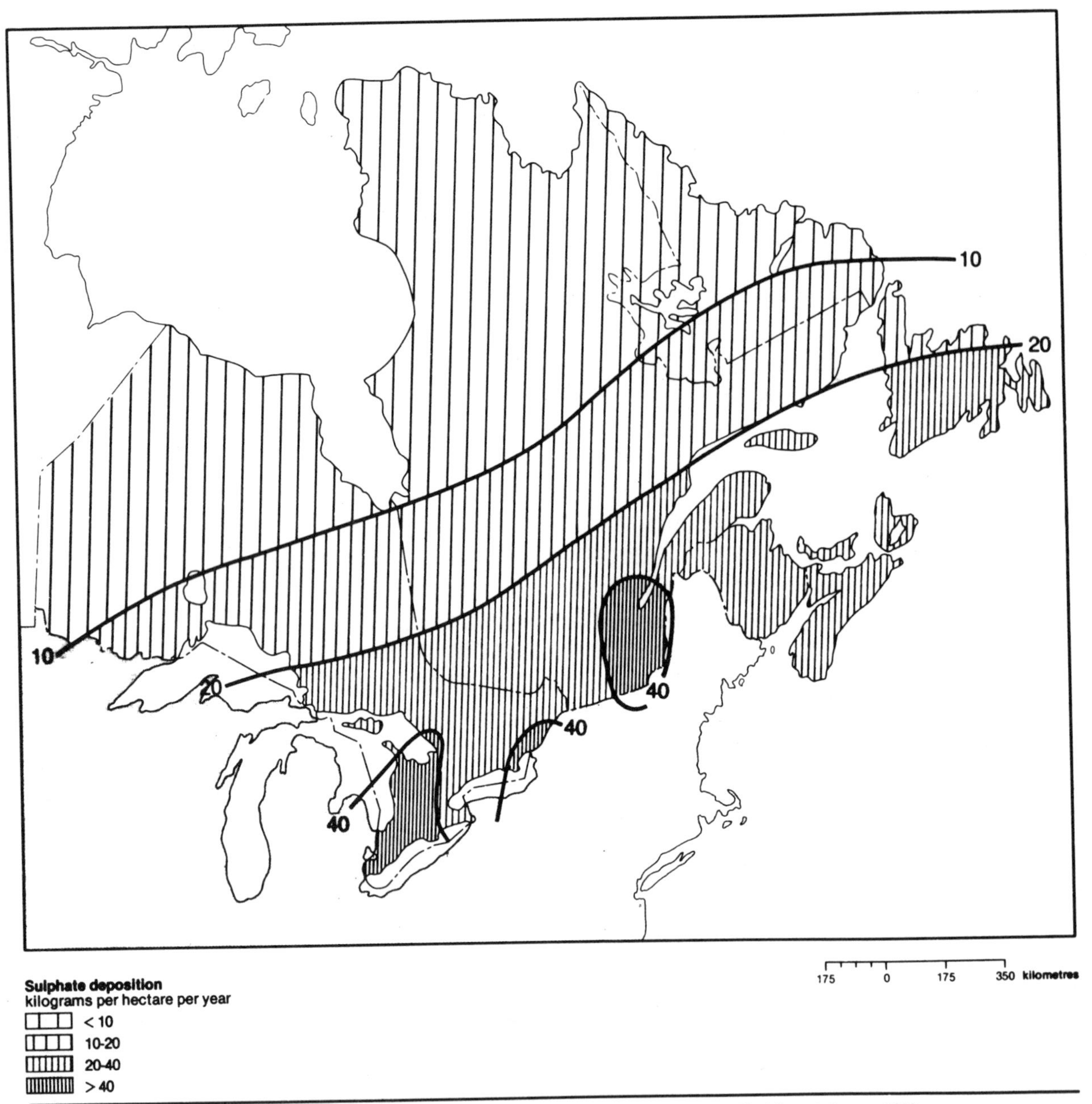

FIG. 16-3. Annual Sulphate Deposition Regime for Eastern Canada Based on 1980 Data (from Bird and Rapport 1986).

When President Bush took office in 1989, he pledged environmental action. In June, 1989, he outlined a clean air proposal, including acid rain legislation. In the latter half of 1989, debate began in the U.S. Congress. An intensive lobby was mounted by representatives of the polluting industries. In the end, some aspects of the proposed legislation were strengthened.

On November 15, 1990, President Bush signed the Clean Air Act of 1990 into law. The Act proposed a 40% reduction in acid rain emissions. On March 13, 1991, Canada and the U.S. signed the Canada–U.S. Air Quality Agreement which commits both countries to reduce emissions. This was good news for Canada which had lobbied hard for U.S. controls.

Meanwhile, some hope can also be garnered from the recent observations by Kelso et al. (1990). They concluded that, for most Canadian lakes, evidence indicates that damage due to acidification peaked in the late 1970's. Recently, recruitment has resumed for at least some brook trout and lake trout populations near Sudbury. There is also evidence of re-occupation by white sucker in two lakes. Repopulation of these lakes appears to be associated with improvement of water quality to pHs greater than 5.0 in the early 1980's. But future progress hinges on the success of the recent Canadian and U.S. initiatives to lessen emissions causing acid rain.

3.7.2 Dioxins and Furans

In recent years there has been growing concern about the effects of persistent toxic chemicals on the ecosystem and potentially on human health. These persistent pollutants include trace metals, and organic compounds present in the environment as a result of human activities. Examples of the latter include polychlorinated biphenyls (PCBs), chlorophenals, polycyclic aromatic hydrocarbons (PAHs) and polychlorinated dibenzodioxins (PCDDs) and polychlorinated dibenzofurans (PCDFs). These persistent organic compounds, with the exception of PAHs, are known as organochlorines or chlorinated hydrocarbons. These are mostly synthetic compounds with unique chlorine-carbon bonds that are not readily broken down by natural processes. The chlorinated hydrocarbons can cause biological damage at low concentrations. They persist in the environment, and they have the potential to accumulate in individual species (bioaccumulation) or to concentrate through entire food webs (biomagnification) reaching the highest concentrations in top predators, including humans (DOE 1989a).

About six million chemicals have been produced worldwide and about 100,000 are in commercial use globally, more than 20,000 in Canada. Some 1,000 new chemicals are prepared annually. Chemicals derived from resource extraction and manufacturing, from industrial, commercial and home use, and from waste treatment and disposal, have entered virtually every aquatic ecosystem within Canada's jurisdiction. No plant, fish, or mammal, including man, is free from exposure to chemicals or contamination (Piuze 1989 and pers. comm.).

Some contaminants kill fish directly. Toxic chemicals may render fish unfit to eat. Commercial fisheries have been closed because of pollution by mercury (a trace metal), e.g. the fishery for Atlantic swordfish and the walleye fishery in the English-Wabigoon river system in Northwestern Ontario.

I will focus here on the recent concern over the effects of dioxins and furans on aquatic life and potentially on human health. Polychlorinated dibenzodioxins (dioxins) and polychlorinated dibenzofurans (furans) constitute a family of 210 chemically-related compounds, some of which are known to be extremely toxic to laboratory animals. They are highly persistent compounds which accumulate in bottom sediments and biological tissues. They have been found in all parts of the ecosystem including air, water, soil, sediments, animals and foods. All animals and humans in Canada are exposed to these substances.

Unlike PCBs, dioxins and furans have never been manufactured deliberately. They are the byproducts of some manufacturing or industrial processes that use chlorine and from incineration of certain wastes. In 1983, the major environmental inputs of dioxins to the Canadian environment were ranked as follows:

1. Municipal and industrial incineration sources;
2. Chlorophenols, particularly in the wood preservation and treatment industries and pesticide uses;
3. Landfills containing liquid organic wastes and precipitated fly ash from municipal incinerators;
4. Other combustion sources, such as wood burning and fires involving electrical equipment containing chlorinated organics;
5. Pesticides synthesized from chlorophenol precursors, including phenoxy herbicides and others;
6. Pharmaceutical and domestic products, including some disinfectants (Health and Welfare Canada/ Environment Canada 1983).

Pulp mills were suspected as the most important source of these compounds for West Coast fish and shellfish. Bleached kraft mills were known to discharge dioxins, furans and other organochlorines. These are produced through the use of chlorine in the pulp bleaching process and through use of wood chips contaminated with chlorophenols.

Prior to 1987 there was little testing for the presence of dioxins and furans in the marine environment. Tests conducted in 1986 did not detect dioxins in the edible portions of crab, shrimp, and fish from Boundary Bay, Burrard Inlet and Vancouver Harbour in B.C. Levels of 2–3 parts per trillion (ppt) were detected in other crab tissue. This compared with a "regulatory tolerance" of 20 ppt in the edible portion of fish for the most toxic dioxin, established by Health and Welfare Canada (DOE 1989a).

Early in 1988 the results of a study in the United States by the Environmental Protection Agency aroused concern in Canada. That study found elevated levels of dioxins and furans in the wastewater from five mills producing bleached kraft paper. In January, 1988, in response to growing concern over the production of dioxins and furans by chlorine bleaching pulp mills, Fisheries and Oceans Minister Tom Siddon announced the acceleration of the federal government sampling program to assess the situation near Canadian pulp and paper mills.

From January to April 1988, DFO and DOE conducted a dioxin/furan screening survey in finfish and shellfish as well as in sediments at 14 of the 47 Canadian chlorine bleaching pulp mills. Health and Welfare's evaluation of the results indicated that the levels found in edible portions of fish samples did not pose a health hazard to consumers. Because higher levels of dioxins and furans had been found in the digestive gland of crab samples at four British Columbia locations, additional samples were examined from those sites.

On May 16, 1988, the Ministers of Fisheries and Oceans, Environment and Health and Welfare jointly announced the results of the preliminary sampling program. They indicated that the three Departments would extend the fish and sediment sampling program to include all the Canadian pulp and paper mills using chlorine in the bleaching process (Canada 1988h).

An expanded sampling program was implemented in the latter half of 1988. On November 23, 1989, results for three of the four problematic B.C. mills were released. Minister Siddon announced the closure of prawn, shrimp and crab fishing in the immediate vicinity of these three mills. Health and Welfare had advised that these species from those sites could pose a health hazard if eaten frequently. While there was no apparent health hazard if these species were eaten only occasionally, Health and Welfare was concerned that frequent consumption could have adverse health effects (Canada 1989b).

Despite these closures, the Government of Canada informed the public that no long term effects had yet been documented in fish, wildlife or domestic animals exposed to levels of dioxins and furans typically found in the environment. Recent scientific studies had, however, indicated that elevated levels of dioxins and furans can significantly damage the health of laboratory animals. Health and Welfare Canada observed:

> "Despite the lack of direct evidence that current exposures to dioxins and furans in Canada contribute to health problems in humans, the Government of Canada recognizes that these compounds are undesirable environmental contaminants and that, where possible, their unintentional production should be limited" (Health and Welfare Canada 1990).

For a discussion of the effects of dioxins and furans on fish, see Waldichuk (1990, 1993).

DFO's nationwide sampling program of fish and shellfish in the vicinity of all Canadian pulp mills using chlorine bleaching was completed in 1991. Close to 1000 fish sample results were submitted to Health and Welfare Canada for human health risk assessment. Following the initial May 1988 release of information there were 13 subsequent public releases (as of January 1992). Based on Health and Welfare's assessment of the data, various areas, stocks and associated fisheries were closed. Advisories to limit consumption were issued where this was deemed appropriate (DFO 1992h).

On the regulatory front, in December, 1991, the federal Ministers of the Environment, Fisheries and Oceans and Health and Welfare announced a Pulp and Paper Regulatory Package aimed at ensuring improvements in water quality in Canada. This included new regulations under the Canadian Environmental Protection Act and amendments to Regulations under the Fisheries Act. The new regulations would require the industry to introduce changes to pulp and paper processing to prevent the formation of dioxins and furans, reduce organochlorine levels overall and control conventional pollutants (suspended solids and oxygen demand) in pulp mill affluents (Canada 1991b).

3.7.3 Marine Pollutants

Oil and gas spills and the closure of certain fisheries because of elevated levels of dioxins in certain species are but examples of a growing problem of marine pollution. As Piuze (1989) noted, many people believe that the oceans assimilative capacity for waste is virtually endless. They treat the oceans as a bottomless sink for civilization's garbage. This ignores two important factors:

1. Man's activities are concentrated along estuaries and coastal areas; and
2. Coastal waters, which constitute about 10% of the total area of the oceans, are the sites of most biological activity.

Piuze (1989) presented an overview of the major sources and types of marine pollutants. The types are many and the sources diverse (Table 16-3).

Major sources of marine pollution include ocean-based activities (navigation, ocean dumping, fishing, offshore hydrocarbon exploitation) as well as land-based industrial, municipal and domestic activities. Pollutants generated by the latter make their way to the oceans through direct coastal input of effluents, river runoff or long-range atmospheric transport. In the next section I discuss contamination in the Arctic occurring through atmospheric transport.

TABLE 16-3. Major Types and Sources of Marine Pollutants.

CATEGORY	TYPES	SOURCES
PHYSICAL	Persistent debris (plastics, glass, metals)	Coastal effluents; Navigation; Fishing; Offshore activities
	Oil, grease, tar	Marine transportation; River runoff; Coastal effluents
	Dredge spoils, sludge	Ocean dumping
	Suspended solids	Coastal effluents; River runoff; Coastal engineering
	Heat	Thermal generation of electricity; Industrial processes; Municipal sewage
CHEMICAL	*Synthetic Organic Chemicals*	
	Organochalogen Compounds	
	DDT, aldrin, toxaphene, lindane, chlordane, mirex, HCB, 2,4-D, etc	Biocides (pesticides, herbicides, fungicides) for agriculture, forest spraying, antimalaria programmes and other uses
	PCBs	In industrial products (as insulators, plasticizers)
	Dioxins and furans	Combusion processs; chemical dumps; pulp and paper effluents
	Chlorophenols (PCP)	Wood preservatives
	Freons	Propellants; Aerosol cans; Refrigerants
	Solvents	Dry cleaning; Industrial use
	Alphatic chlorinated hydrocarbons	Vinyl chloride for PVC production
	— Non-halogenated compounds	
	Organotins	Marine antifouling agents
	Organophosphates (e.g. fenitrothion)	Agriculture and forest spraying pesticides
	Organoarsenicals	Pesticides; Manufacturing and processing; Combustion of fossile fuels
	Phthalic acid esters	Plasticizers (found in effluents from textile, pulp and paper, oil refineries, electrical manufacturing)
	Aromatic amines	Textile industry (dyes)
	Carbamates (e.g. aminocarb, carbaryl)	Forest spraying; Agricultural pesticides
	Alkyldinitrophenols	Agricultural pesticides
	Petroleum hydrocarbons	
	Crude oils	
	Refined petroleum products	Marine transportation; River run-off; coastal effluents; Atmosphere (from cars, planes, industries); Naturalseeps (account for about 10%)
	Polycyclic aromatic hydrocarbons (PAHs)	Fuel combustion; Coal power generation
	Inorganic chemicals	
	Metals (lead, mercury, cadmium. copper, zinc, etc.)	Industrial effluents; domestic sewers, River runoff; Ocean dumping
	Nutrients	Sewage; River discharges; Agricultural runoff
	Acids and bases	Fossil fuel burning; Industrial processes; Domestic uses
	Carbon dioxide	Fossil fuel combustion
	Artificially produced radionuclides (e.g. Caesium-137, Strontium-90, Plutonium Carbon-14, Tritium)	Nuclear weapons testing; Nuclear reactors and reprocessing of fuel
	Drugs, antibiotics	Aquaculture
BIOLOGICAL	Pathogens (e.g. bacteria, viruses, fungi)	Sewage
	Biochemical oxygen demand (BOD) material	Sewage; Plant growth from excess nutrients

Source: Puize (1989).

Wilson and Addison (1984), in a volume entitled *Health of the Northwest Atlantic*, concluded that the offshore Northwest Atlantic was relatively uncontaminated. Among the reasons for this:

1. Population density along the eastern Canadian coast is relatively low, and there are relatively few industrial sources of pollution along the Canadian Atlantic coast. Thus sources of man-made contaminants are limited.
2. The Northwest Atlantic is a very dynamic environment, and any pollutants which are introduced to it tend to be dispersed rapidly and widely.

They also observed that, although offshore contamination is low, there are examples of local inshore pollution, usually associated with local industrial discharges. Much of the pollution found in the offshore Northwest Atlantic arises from diffuse or non-point sources, such as by atmospheric transport from the Northeastern U.S.

Wilson and Addison concluded:

> "In most instances, levels of contaminants in the offshore are not high enough to cause any obvious or acute toxic effects, but these effects may sometimes be seen in a few cases of localised inshore populations. Levels of contaminants in fish products from the Northwest Atlantic are usually well within ranges acceptable for public health reasons, although in some instances, inshore fisheries may be closed because of unacceptably high contaminant levels."

Piuze (1989), speaking more generally, concluded that toxic chemicals are widespread in the oceans. While there are no more pristine areas, open-ocean levels are not elevated. Piuze noted that coastal and estuarine areas are much more contaminated and, in some cases, polluted enough to affect the resident animals and fish. Levels of some toxic chemicals in this fauna often exceed human health guidelines set to protect consumers, thereby necessitating fisheries closures, e.g. the closures due to dioxins and furans in British Columbia.

Some of the most comprehensive data on pollutants in the marine environment were presented by Kay (1989) for British Columbia (see also DOE 1989b). This report quantified some of the major stresses to the marine environment of the Pacific coast, based on data available from the 1970's to the end of 1986. It also identified the environmental changes or effects that had been observed.

Three major sources of pollution of British Columbia marine waters were identified:

1. Municipal and other sewage wastes;
2. Pulp and paper industry wastes;
3. Mining wastes (Discussed earlier).

In 1988, 393 marine discharges were authorized by provincial permits, up from 85 in 1973. The majority of permits were issued to allow the discharge of municipal sewage, but the greatest volume of effluent was disposed of under pulp mills' permits (DOE 1989b).

Sewage can produce bacteriological contamination of swimming beaches and shellfish harvesting areas. This can result in closures of these areas to protect public health. Of particular concern is the transmission of disease through human consumption of molluscan shellfish (oysters, clams and mussels). Other impacts of sewage discharge include nutrient enrichment, oxygen depression, accumulation of solids on the ocean floor and effluent toxicity. These effects are usually confined to the area immediately around sewage outlets.

We have already discussed some of the impacts of pulp and paper industry wastes. Effluents from these mills contain chemicals from the pulping and bleaching processes, as well as wood fibres and other wood wastes. Effluent from coastal pulp mills has been shown to reduce the dissolved oxygen available to fish over an area ranging from 0.3 to 10 km^2. Pulp mills resulted in dramatic reductions in Biochemical Oxygen Demand (BOD) and TSS (Total Suspended Solids) between 1975 and 1980 but levels remained stable in the 1980's (DOE 1989b).

The dumping of specific materials at sea is controlled under federal legislation (formerly the *Ocean Dumping Control Act* and now Part VI of the *Canadian Environmental Protection Act*). These provisions ban or limit the dumping of substances that can seriously harm the environment, including mercury, cadmium, oil and gas, high-level radioactive wastes, persistent plastics and various other toxic materials. Disposal is carried out at designated locations under a permit system. On the Pacific coast, most dumping occurs at six major sites in the waters off the lower mainland of British Columbia. The quantity disposed of each year varies. The bulk of the material is the product of dredging operations. The potential environmental impacts of ocean dumping are summarized in Table 16-4.

Persistent contaminants are a problem. Trace metals such as cadmium, lead and mercury are all commonly reported as occurring in the marine environment. All of these are known to affect human health. Levels of trace metals exceeding background levels have been observed in marine sediments at various locations (Fig. 16-4). It is not yet clear whether these accumulations are affecting B.C.'s marine environment or human health.

We have already discussed the effects of certain organic contaminants, i.e. dioxins and furans. Other organic contaminants include PCBs, chlorophenols and PAHs. PCBs are well known as a long-term hazard to the environment. They have not been manufactured in North America since 1978 but are still in use in closed circuit electrical systems. Surveys in 1973, 1980 and 1985 of PCBs in edible tissues of fish and shellfish indicated amounts well below the acceptable level established by Health and Welfare Canada. PCB levels in the eggs of Great Blue Herons from the Fraser River declined from 20 ppm in 1977 to about 5 ppm in 1988.

TABLE 16-4. Potential environmental impacts of ocean dumping.

Category	Environmental impacts
Physical	— Loss of habitat through smothering — Creation of new habitat (artificial reefs) — Alteration of bottom currents — Attenuation of light in the water column through fractionation of wastes — Dispersal of sediment-associated contaminants
Chemical	— Increased concentration of elements (e.g. metals) in water column and sediments — Introduction of wastes containing various chemical forms of elements compared with those occurring naturally — Oxygen depletion resulting from biodegradation of waste products
Biological	— Lethal and sublethal toxic effects on plants and animals — Bioconcentration, bioaccumulation and/or biomagnification of trace elements or organic substances

Source: Kay (1989).

Chlorophenols have been used in the forest industry in fungicides to protect freshly cut lumber. These compounds are toxic to fish and other aquatic life. Many forest companies have recently ceased using them. The impact on levels in the environment is not yet clear.

Polycyclic aromatic hydrocarbons (PAHs) are associated with an increased incidence of some forms of cancer in humans and animals. PAHs are produced by fires and also by fossil fuel combustion, aluminum smelting, and refuse burning. They are transported to the ocean by the atmosphere in rain. Levels of the most carcinogenic PAH, benzopyrene, appear to be very low or nondetectable in commercially harvested shellfish from British Columbia. Levels are higher in harbour areas. This chemical is suspected of causing liver lesions in English sole in Vancouver Harbour (DOE 1989b). It has also been found in high levels in the St. Lawrence belugas (see section 3.8)

3.7.4 Contaminants in the Arctic

An emerging environmental issue of considerable concern, both nationally and internationally, is the discovery of chemical contamination of the Arctic ecosystem. The Arctic was long thought to be relatively pristine. Recent studies have, however, found contaminants in the Arctic environment, contaminants which threaten the Arctic ecosystem and are potentially harmful to humans.

Initial concerns about contaminants in native diets were raised by the clean-up of PCBs and other wastes at Canada's abandoned Distant Early Warning (DEW line) sites in 1985. The results of surveys carried out then revealed only limited contamination at some of the sites. That same year the Department of Indian and Northern Affairs (DIAND) established an inter-agency working group on contaminants with representatives from Health and Welfare, Environment, Fisheries and Oceans and the Government of the Northwest Territories.

This group concluded early on that contamination of the North was serious and widespread, but that the small quantities of PCBs at the DEW line sites were not contributing to the problem to any significant extent. Attention has focused on four classes of contaminants — organochlorines, acids, metals and radionuclides. Initial efforts were directed towards studying the occurrence of these contaminants in the Arctic.

The contaminants of the greatest concern have been the organochlorine compounds. These compounds reach the Arctic through long range atmospheric transport and are also carried there by water currents. They accumulate in the food chain, thus threatening health of humans, particularly Inuit, a substantial portion of whose diet still consists of Arctic animals. Organochlorines persist in the environment, resist biological metabolization and therefore tend to bioaccumulate. They are highly soluble in animal fat and tend to have high chronic toxicity. Because of their high solubility in fat, they bioaccumulate in Arctic animals such as whales, polar bears and seals which have relatively long lifespans. These animals comprise the major component of Inuit diets.

Muir et al. (1992) have summarized the results of scientific studies on Arctic marine ecosystem contamination. Lockhart et al. (1992) have similarly documented the presence and implications of chemical contaminants in the freshwaters of the Canadian Arctic. Organochlorine contamination is widespread throughout the Arctic but is generally lower than that observed in marine, freshwater and terrestrial environments in the mid-latitudes. Nearly all organochlorine contaminants occurring in southern parts of Canada have been detected in Arctic biota. Concentrations of these contaminants in fish, marine and terrestrial wildlife have been found to be similar over wide areas of the Arctic.

Studies indicate that metals and acids deposited in the North have originated primarily at Eurasian sources. Annually, over 90% of the sulphur reaching the North originates in Eurasia (DIAND 1989). The source and movement of organochlorine contaminants to the Arctic is still largely unknown. It is thought that organic contaminants derived from agricultural and industrial sources are transported to the Arctic via the atmosphere, ocean currents and river run-off. There have been very few measurements. Hence, exact sources are not known but are thought to be global, with some sources in the Northern Hemisphere.

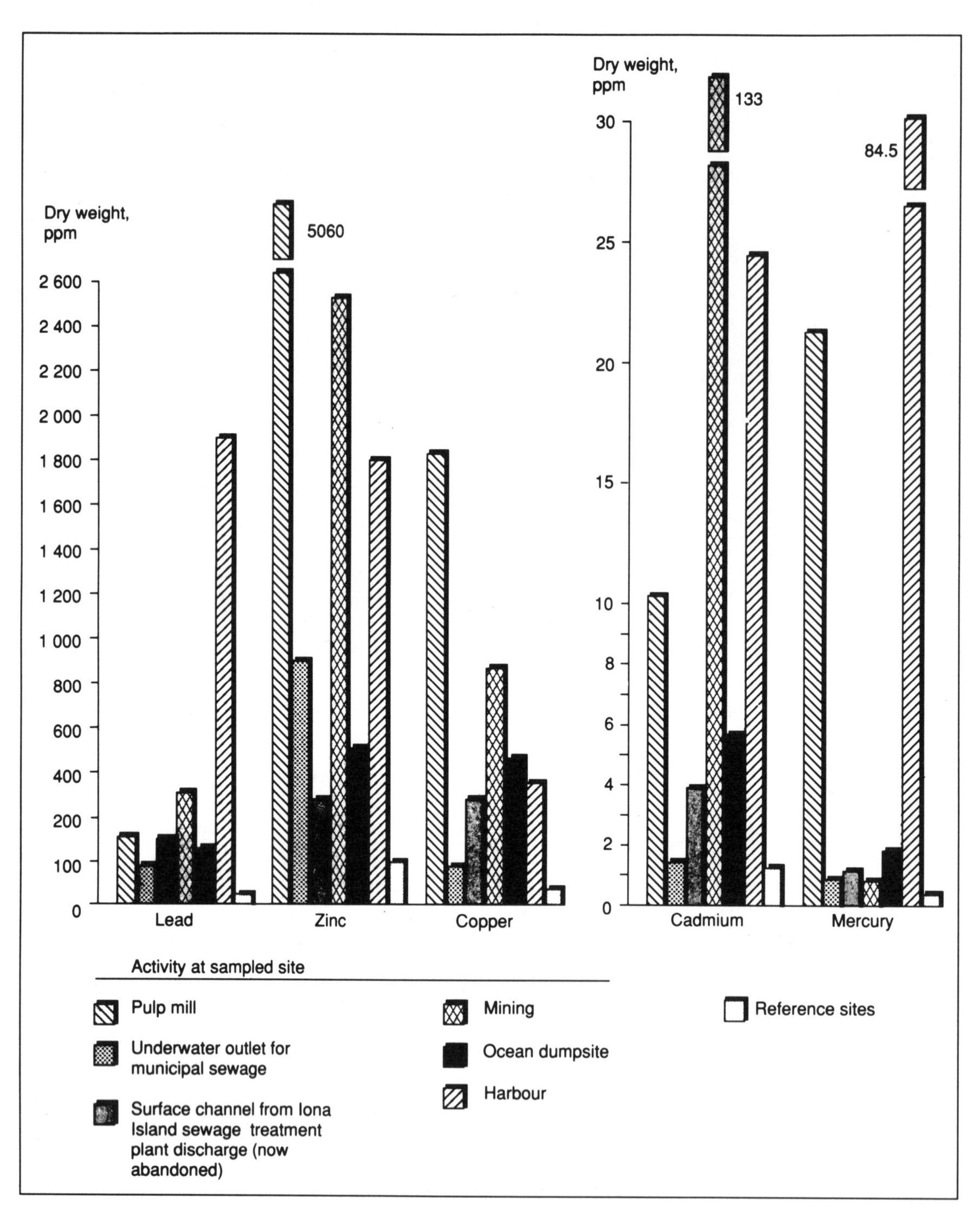

FIG. 16-4. Maximum trace metal concentrations in sediments between 1975 and 1985, at contaminated and reference sites on the British Columbia coast (from DOE 1989b).

Major organic contaminants detected in Arctic biota include PCBs, DDT and its metabolites, chlordane-related compounds, toxaphene and dioxins/furans. Levels of chlordane compound residues in polar bear fat were reported to be four times higher in 1984 than in 1969. Levels of DDT did not change. Other organochlorines measured were twice as high in 1984 as in 1969. Dioxins and furans have been detected in ringed seal blubber samples from seven Arctic communities in the range from 2 to 37 parts per trillion.

The levels of some contaminants appear to have declined. PCB and DDE (a metabolite of DDT) residue levels in Arctic ringed seal blubber collected at Holman Island in the Northwest Territories were lower in 1981 than in 1972. PCB levels in ringed seals and seabirds from the central and eastern Arctic also appeared to have declined (DIAND 1989).

The highest average concentrations of DDT and PCBs in marine mammals have been found in narwhal blubber from Pond Inlet and Pangnirtung (5 ppm). Levels were somewhat lower in beluga (3 ppm in the blubber) and ringed seals (0.8 ppm). These latter levels are much lower than in marine mammals in the Gulf of St. Lawrence.

Muir et al. (1992) reported that measurements of organochlorines in air, snow, ice and sea water from an Ice Island indicated that atmospheric supply was a major route for transport of these contaminants to surface layers of the Arctic ocean. Muir et al. (1992) also reported a lack of agreement on trends in organochlorine residues over time between various studies of marine biota. DDT residues declined in polar bear fat and in ringed seals over intervals of 14 and 9 years respectively. In the case of seals this could be explained by body fat and age differences in the samples analyzed. PCBs definitely declined in seals and seabirds but appeared to increase in polar bears. Limited evidence suggests that toxaphene and chlordane have increased relative to PCBs and DDT in beluga whale fat over the past 20 years. Muir et al. (1992) called for further studies of temporal trends of contaminants in marine mammal tissue because of the implications for the health of wildlife populations and for human consumers of these animals.

With respect to metals, high cadmium and mercury levels have been found in Arctic marine mammals, mainly in the kidney and liver. Cadmium levels in narwhal kidney averaged 63.5 ppm, among the highest levels reported for marine mammals. These levels appear to originate from natural rather than industrial sources. Lead in the Arctic apparently originates from industrial sources. Greenland glacial ice deposits indicate higher concentrations in recent years than in the preindustrial period.

Muir et al. (1992) concluded that most heavy metals have increased in the Arctic relative to historical times. Some local sources exist, but differentiating between natural and anthropogenic sources is still a problem. With the possible exception of lead, the existing data base on heavy metals in arctic biota is too limited to indicate spatial or temporal trends.

Despite these data deficiencies there is considerable concern about the implications of Arctic contamination for both the marine ecosystem and human health. Early in 1989, DIAND convened a meeting to evaluate the potential impact of contaminants in northern ecosystems upon native diets (DIAND 1989). The results of studies of Inuit foods and diet carried out at Broughton Island since 1985 were evaluated. These studies focused on PCBs.

Results of the initial study, conducted in September 1985, indicated a high intake of Inuit foods and an associated intake of PCBs which exceeded, for 18.9% of the study participants, the amount considered "tolerable" by Health and Welfare Canada. Blood PCBs exceeded the HWC "tolerable" levels in 63% of the children and 39% of the women of childbearing age.

The intake of PCBs from food, blood levels for PCBs, and the organic mercury blood levels observed during the pilot survey were not considered to present any immediate threat. Later studies indicated that Inuit food was used by nearly all Broughton Island residents and constituted a major part of the community's diet.

All Inuit foods tested contained some PCBs, with the largest levels found in narwhal and beluga blubber and in polar bear fat. Overall, about 10% of female participants and 15% of male participants consumed more than the "tolerable" amounts of PCBs. The study concluded, nonetheless, that a change in diet was not advisable.

No change was recommended for a combination of reasons. First, the nutritional value of Inuit foods was shown to be high, superior to the marketed foods used in the community. Second, substitution of Inuit foods by marketed foods currently available and consumed in the community would result in poorer diet, with risk of damage to health. Overall, the benefit of Inuit foods and of breast feeding to Broughton Island residents was considered to be greater than the risk from the consumption of PCBs in Inuit foods or in breast milk.

Despite these results, the pattern of consumption of Inuit foods is largely unknown for the Arctic as a whole. The proportion of the Inuit population consuming more than the "tolerable" level of organochlorines is also unknown. Similarly, spatial and temporal trends of the concentrations of many organochlorines in the Arctic are unknown. It was impossible to determine whether the potential risk is increasing or decreasing and whether contaminant hot spots exist. Further research was suggested:

(1) to define the contaminant sources and transport mechanisms;

(2) to document the spatial and temporal trends to identify the magnitude, geographic extent and duration of the problem; and
(3) quantify the risks of contaminants to the Arctic ecosystem and the relative risks and benefits to humans from consuming harvested animals.

3.8 The Plight of the St. Lawrence Beluga

In recent years the belugas or white whales of the St. Lawrence estuary have become widely known as the "miner's canaries", contaminated by pollutants discharged into the St. Lawrence River from the industrial heartland of central Canada. Numbering in the thousands in the 1800's, they have declined in the latter half of this century to approximately 500 animals. There is some debate whether the population has stabilized or is continuing to decrease. There is no evidence that the population is increasing.

The beluga is essentially an Arctic species. The population in the St. Lawrence is at the southern limit of its range in North America. It inhabits a 145 km stretch of the St. Lawrence estuary between Petite-Rivière-Saint-François and Pointe-des-Monts on the north shore and between Ile d'Orléans and Cap-Chat on the south shore. The belugas occur in the Saguenay River at its confluence with the St. Lawrence (Fig. 16-5).

The means of asessing the number of belugas is inexact because they are in constant movement and never evenly distributed. Exploitation of the St. Lawrence beluga population began in the 18th century, with the most extensive hunting occurring between 1860 and 1945. Peak kills occurred in 1889 (850 animals), 1915 (900 animals) and 1935 (692 animals). There were three peak periods of hunting: 1886–1890, 1915–1921, and 1932–1938.

Attempts have been made to estimate the earlier population levels. Reeves and Mitchell (1984) concluded that the St. Lawrence beluga population had decreased dramatically during the present century. They estimated that over 16,000 belugas were killed between 1886 and 1960. From an analysis of trends in the hunt, they calculated the population was above 5,000 animals near the end of the 1800's. A substantial hunt into the 1940's reduced the population from about 5,000 to several hundred. Commercial hunting ended around 1955. Since the early 1970's several surveys suggested that the population was either stable or decreasing, numbering approximately 500 animals. The range of survey estimates was 368–715 in 1982, 187–773 in 1984 and 250–740 in 1985 (Sergeant 1986). This indicates the current St. Lawrence beluga population is scarcely 10% of that in 1885 (Reeves and Mitchell 1984).

CAFSAC (1988) reported, however, that estimates of the earlier population levels could not be verified. Despite the uncertainties surrounding recent population estimates, CAFSAC concluded that the population in the late 1980's was within the range 350–700, and was probably about 500 animals.

Beached beluga.

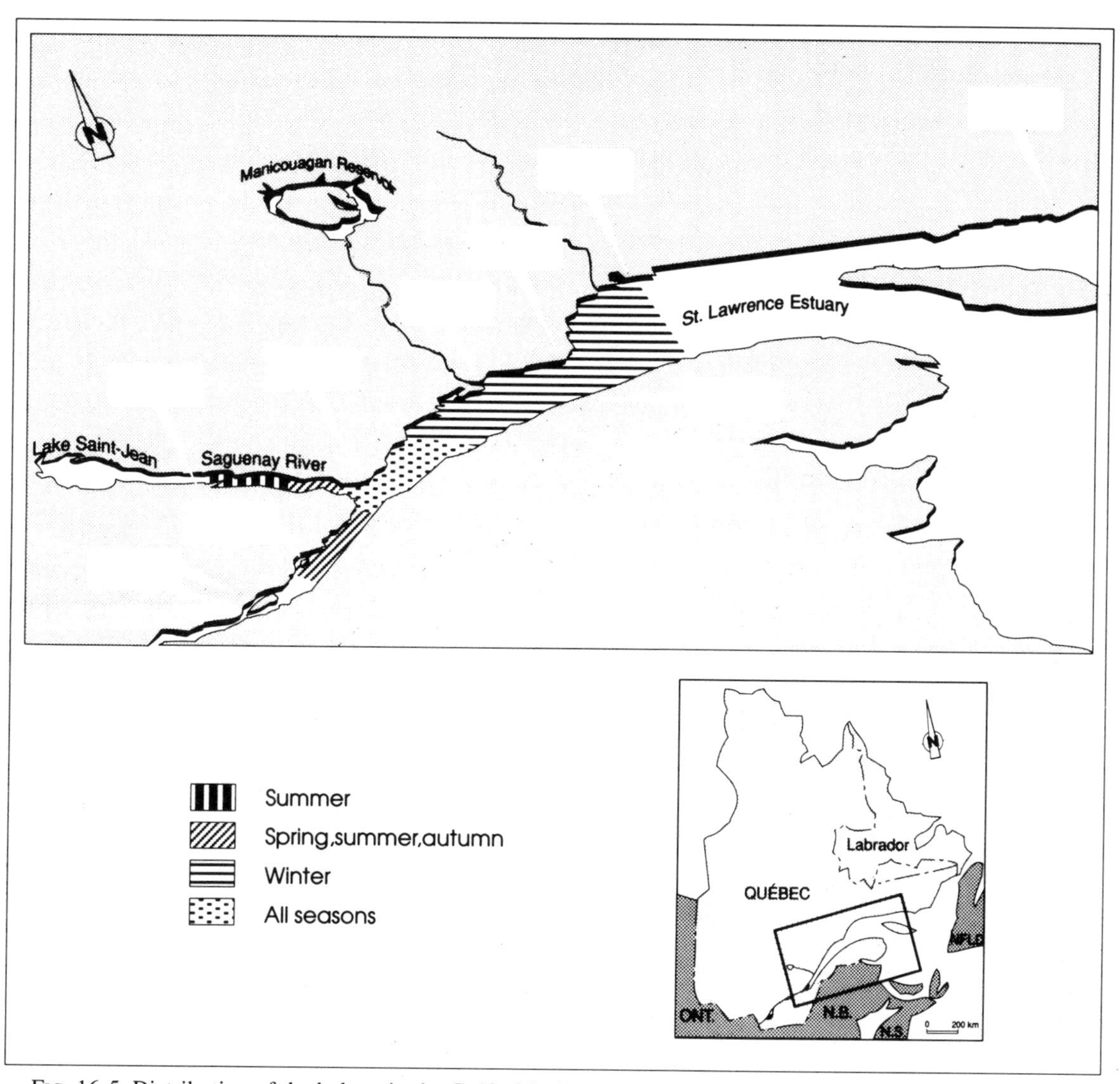

FIG. 16-5. Distribution of the beluga in the Gulf of St. Lawrence (adapted from Canada 1991c).

In 1979, the federal government listed the St. Lawrence beluga as a protected species (population). In 1983, the Committee on the Status of Endangered Wildlife in Canada (COSEWIC) placed it on the endangered list. Several observers questioned the failure of the population to increase following the cessation of commercial hunting at the end of the 1940's. Earlier in this century overexploitation appears to have led to a decline in the population. But this does not explain its failure to recover since the 1950's. In the 1970's, several hypotheses were put forward involving overexploitation, habitat loss through dredging and hydroelectric projects, and pollution. In support of this latter hypothesis, Sergeant and Brodie (1975) reported high levels of PCBs and DDT in a very young beluga.

In the early 1980's, Béland and colleagues commenced a program of autopsies on stranded belugas. This revealed a large number of contaminants in the dead belugas (Béland 1988). During the 1980's considerable attention focused on contamination by toxic chemicals. Sampling of dead belugas stranded along the St. Lawrence revealed heavy organochlorine loads in all animals. Martineau et al. (1987) reported high levels of organochlorine chemicals in the blubber of 26 stranded carcasses of beluga. Over 25 individual contaminants were identified. In their 1987 analysis, Martineau et al. dealt with contamination by PCBs, DDT and their metabolites. The blubber concentrations of total DDT were two orders of magnitude higher than in beluga whales from the MacKenzie Delta. PCB and DDT levels were one order of magnitude greater than those found in grey seals from Sable Island, in harp seals from the Gulf of St. Lawrence and in beluga whales from the Baltic.

Martineau et al. (1987) noted that the proportions of calves and juvenile belugas in the St. Lawrence were lower than in Alaskan populations. They hypothesized that, if this difference is real, it would indicate a decreased birth rate and/or increased juvenile mortality. They suggested that the degree of organochlorine contamination found in the population could explain a decreased birth rate, as comparable or lower concentrations have had such an effect in other animals.

Subsequently, Martineau et al. (1988) examined evidence of diseases in stranded carcasses of St. Lawrence belugas and attempted to determine whether a relationship between these conditions and the toxicity of organochlorines and potential carcinogens could be established. They concluded that various observed conditions in the dead belugas were consistent with chronic organochlorine toxicity. They suggested that contamination by organochlorines might be causing diseases in the St. Lawrence belugas, and might explain the population's failure to recover.

A variety of chemical contaminants have been found in the St. Lawrence beluga population (Table 16-5). As Béland (1988) observed, every St. Lawrence beluga sampled so far has been highly contaminated with PCBs and DDT. The Saguenay fjord which is central to the beluga habitat is home also to one of the largest aluminum-producing complexes in the world. This complex has introduced over the years large amounts of mercury, fleurons and PAHs into the environment. It is a major source of Benzo(a)pyrenes (BaPs). Large quantities of BaPs have been detected in sediments of the Saguenay and the blue mussels in the St. Lawrence at the mouth of the Saguenay. Béland (1988) summed up the available evidence thus:

> "In the case of the St. Lawrence beluga, many lines of evidence point in the same direction: the presence of high levels of toxic products; a high incidence of lesions, many of which are known consequences of intoxication with these products; a long-term decreasing trend in the population; and a low percentage of juveniles. Such indicators have led us to believe that toxic contaminants are threatening the population."

TABLE 16-5. Chemical contaminants found in the St. Lawrence beluga population.

Organic contaminants		Heavy metals,
(PaHs)	Polycyclic aromatic hydrocarbons	Total mercury
(PCBs)	Polychlorinated biphenyls	Cadmium
	Mirex	Cobalt
(DDT)	Dichlorodiphenyltrichloroethane	Chromium
(HCB)	Hexachlorobenzene	Copper
	Dieldrin	Manganese
	Heptachlor	Nickel
	Lindane	Lead
	Aldrin	Zinc
	Chlordane	
	Endrin	

Source: Anon. (1987b).

In 1986, DFO organized an interdepartmental federal-provincial Ad Hoc Committee on the Conservation of the St. Lawrence Belugas. In February, 1988, this committee produced a report which concluded:

> "Despite all the controversy in determining the number of individuals making up the St. Lawrence beluga population, everyone agrees that the population has reached a critical level, requiring an energetic response from the authorities. Not only is the number of individuals disturbing, but their health has been greatly impaired, as shown by the autopsies performed since 1982....
>
> "The constant pressure on the beluga population from industrial pollution and disturbing influences, as well as the responsibility of governments toward a population that is recognized as being threatened, suggest it is urgent that we take the necessary steps to enable the St. Lawrence belugas to find an environment in which they can increase their numbers and stay healthy" (Anon. 1987b).

In June, 1988, Fisheries and Oceans Minister Tom Siddon announced an action plan to protect the St. Lawrence belugas. Proposed actions included halting harassment of the population by shipping activities, legislative and regulatory changes, greater control of activities that might be destroying the habitats of this population and creating a Saguenay Marine Park. Six million dollars were allocated to protect the belugas (DFO 1988i).

Elements of this Action Plan were discussed at an International Forum for the Future of the Beluga held in Tadoussac in the fall of 1988. The final Plan had three major components: (1) controlling disturbances; (2) controlling toxic chemicals; and (3) increasing scientific knowledge about the status of and factors affecting the St. Lawrence beluga. Details of actions taken on these fronts are given in federal annual reports on the Interdepartmental Action Plan to Ensure the Survival of the St. Lawrence Beluga (Canada 1991c).

To prevent disturbances of beluga in their natural habitat, DFO in 1988, issued a set of Guidelines Applying to Beluga and Cetacean Watching in the St. Lawrence River. DFO prepared beluga habitat distribution maps and used these in an awareness campaign. It also introduced a whale-watching permit system. Finally, the federal government took steps to create a Saguenay Marine Park.

Reducing industrial pollution was recognized to be a key element in restoring and maintaining the beluga's habitat. The federal government developed targets to reduce, by 1993, 90% of the liquid toxic substances discharged by 50 priority industries. Negotiations with Alcan reportedly resulted in reductions in PAH emissions. Additional efforts were aimed at reducing discharges by pulp and paper companies of dioxins, furans and organochlorines.

DFO intensified its research to determine population size and understand the dynamics of the beluga population. Studies were stepped up or initiated on contaminants in carcasses, in the beluga's food chain, and in the environment.

Whether these initiatives will be sufficient to stabilize the St. Lawrence beluga population and permit it to increase will not be clear for many years to come. Meanwhile, the beluga has become a very visible symbol of the impact of man's activities on the aquatic ecosystem.

3.9 Plastic Debris

3.9.1 An Emerging Problem

The problem of toxic chemicals is now recognized as a major threat to the world biosphere ecosystem, including the aquatic ecosystem. However late and however inadequate, some initiatives are being taken to address the problem. An emerging problem which has not yet captured world attention to the same extent is that of plastic debris. Plastic is a miracle substance of the 20th century – strong, durable and versatile. These qualities, which make it attractive for a wide variety of uses in today's consumption-oriented society, also make it particularly dangerous when it is thrown into our oceans, rivers, lakes and streams. Plastic endures for years. That makes it a problem for sea life and for fishermen. The amount of plastic debris in the oceans is accumulating so rapidly that scientists and environmentalists now rate plastic debris as potentially one of the major ocean pollution problems.

Not much is known about the extent of the plastic debris problem in Canada's oceans or on its beaches. Nonetheless, the federal government's 1987 Oceans Policy for Canada (DFO 1987k) identified plastic trash as one of the major marine pollution problems of the 1980's. The federal government promised action plans to address the problem of discarded fishing nets and gear at sea, and to deal with the more general problem of plastic debris in the oceans. To follow up on that commitment, an interdepartmental working group was established through the federal Interdepartmental Committee on the Oceans. This working group commissioned a background study of the problem and sponsored a workshop on plastic debris in the aquatic environment, held in Halifax in May, 1989.

There had been growing international concern about marine debris, particularly plastic debris, during the previous decade. The issue has been discussed at meetings of the London Dumping Convention and the International Maritime Organization. Major international conferences to examine the issue were held in 1984 and 1989.

A background study presented at the 1989 Halifax workshop (Buxton 1989) drew largely on research conducted in the United States over the previous decade. This indicated that the production of plastics had increased dramatically. For example, the U.S. production of plastic resins in 1985 was about 10 times greater than the production in 1960. However, there are only crude estimates of the amount of plastic debris entering the oceans.

Buxton (1989) observed:

> "The damage has not yet reached a crisis point. Compared with other environmental problems, persistent debris in the sea does not rank high in many people's minds. No one knows, let alone has proved, how much harm is caused. However, apart from limited efforts to clean up beaches and retrieve lost fishing nets, the debris can but accumulate, unless steps are taken to deal with the problem."

Plastics persist in the environment and have harmful effects over fairly long periods of time. It is their improper disposal which makes them a substantial contributor to the debris problem.

3.9.2 Sources of Plastic Debris

The plastic materials found in the oceans and washed up on beaches come from many different sources.

Some major ones are:

Commercial Fishing — Most commercial fishing gear is now made from plastic. Plastics are used in the manufacture of fixed gillnets and driftnets, trawls, fishing line, and traps for crab, lobster and some fish. Plastic is also used for buoys, bags and containers. This fishing gear and related equipment becomes debris when it is dumped as garbage from fishing vessels or is lost during fishing.

Shipping and Boating — Cargo ships are a major source of plastic debris, much of which is domestic garbage dumped at sea. Similarly, cruise ships dump bags of garbage overboard. Military ships also contribute. One U.S. study revealed that plastics as a component of garbage on U.S. Navy vessels increased about 20 fold between 1977 and 1988 (Alig et al. 1989).
A major source of plastic debris in the aquatic environment is recreational boating. Boaters freely discard large quantities of plastic bottles, six-pack rings, food wrap, and plastic bags. It has been estimated that more than 50% of garbage dumped in U.S. waters originates with recreational boaters (Crampton 1989).

Plastics Manufacturing and Processing — The small plastic pellets which are the first stage in making plastics can enter the environment as losses from manufacturing plants (discharged with plant waste water), during transport and at processing plants.

Sewers and Sewage Treatment Plants — Normal sewage includes plastic wastes such as condoms, disposable diapers and tampon applicators. Where sewage treatment systems do not filter sewage properly, these plastic wastes end up in the aquatic environment.

Solid Waste Dumping and Littering — Other debris which originate on land may end up in the ocean or on beaches. Accidental releases from coastal landfills, littering on beaches, release of helium balloons and deliberate dumping in coastal waters all contribute to the problem.

It is often difficult to determine the specific sources of particular items of aquatic debris. The source of materials such as fishing gear is obvious. But the origin of other frequently-occurring items such as strapping bands, sheeting, bottles and containers, six-pack yokes, cups and balloons is less obvious. The composition of aquatic debris changes significantly from one region to another, reflecting the variation in regional distribution of urban areas, shipping, fishing and recreational activities. Debris can of course be carried from one area to another by ocean currents.

3.9.3 Effects of Plastic Debris

Buxton (1989) observed:

> "Effects depend on where debris ends up, how long it takes to get there, and what it does while getting there."

Effects range from entanglement and ingestion resulting in death or injury for marine mammals, seabirds, fish, sea turtles and crustaceans to the aesthetic and economic effects of litter washed up or discarded on beaches.
There are four major categories of effects:

- Effects on fish and wildlife;
- Effects on humans on shore;
- Effects on humans who use the ocean;
- Economic effects of these other impacts.

The full effects are not known because there has been little research on the subject.

3.9.3.1 Effects on Fish and Wildlife

Persistent debris can affect fish and wildlife in two ways. Individual animals may become entangled in it or eat it. Either of these can result in injury or death. There is abundant evidence of effects on individual animals but the effects at the population level are generally unclear. There are only a few cases of documented damage by debris to populations of a species. These include the northern fur seals and Hawaiian monk seals and even in these cases population decline may not be solely due to damage from debris (Buxton 1989).
One of the best known effects is that of "ghost fishing" on fish and crustaceans. Studies in Newfoundland (Way 1976) demonstrated that ghost nets may continue to fish effectively for groundfish for at least one to two years. There was no apparent adverse effect on crab stocks.

3.9.3.2 Shore Effects on Humans

The effects on humans can include shoreside aesthetic and possibly health and safety effects. Beaches covered with litter can deter tourists and large amounts of time and money are required in some instances to keep beaches clean. In recent years the potential negative health effects of medical wastes washing up on beaches has raised concerns.

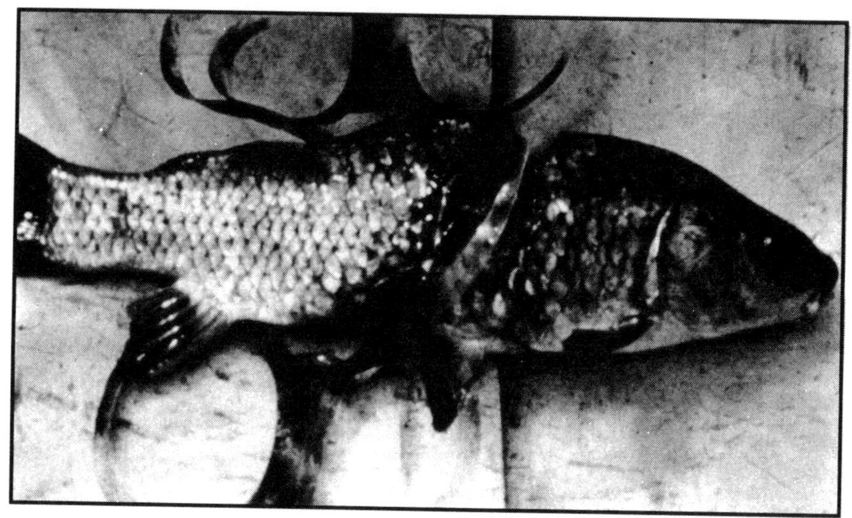

Plastic debris victim.

3.9.3.3 Damage to Ocean Users

Various types of adverse effects occur for those who depend on the ocean for their livelihood or who use it for recreation. Discarded or lost nets and ropes can foul propellers. Plastic bags and sheeting can clog water intakes.

3.9.3.4 Economic Effects

Economic impacts can include lost fishing opportunities, damage to markets for fish products resulting from concern about the "polluted ocean", the costs of beach clean-ups, loss of tourism as a result of littered beaches, costs and time involved in vessel and gear repairs.

The Interdepartmental Committee on Oceans Working Group on Plastic Debris concluded that more knowledge about these effects is necessary to establish priorities. However, enough was known "to state clearly that we have a significant and growing problem that warrants action".

The 1989 Halifax Workshop (DFO 1989l) made the following recommendations:

1. The federal government should lead national action on the problem of plastic and other persistent debris in the aquatic environment by developing and articulating a clear national policy on the issue. This should include:
   (a) Recognition that plastic and other persistent debris in the aquatic environment are a serious issue which needs to be addressed immediately, before it reaches crisis proportions;
   (b) Recognition that this is an issue where the federal government can be most effective by leading and coordinating other levels of government, private groups, and individuals, by sparking initiative and spurring public action;
   (c) Recognition that there are legal and policy measures that can only be taken by the federal government; and
   (d) Recognition that the problem can probably be dealt with adequately by modest direct funding at the federal level.
2. In support of this policy, the federal government should establish an office to coordinate and act as a focal point for Canadian action on this issue.
3. Funding should be provided for appropriate information, education and publicity campaigns designed to stimulate public response.
4. Research should be sponsored to obtain specific information on the debris problem in Canada, with a high priority being research into the extent and impact of ghost fishing.
5. Proper disposal facilities should be provided in those ports and harbours where it is the federal government's responsibility to do so; in other cases the responsible parties should be required to provide such facilities.
6. Canada should accede to Annex V of the MARPOL Convention[2] as a matter of priority.

[2] The MARPOL Convention (International Convention for the Prevention of Pollution from Ships) covers different types of vessel pollution through its various Annexes. Annex V (MARPOL 73/78) prohibits vessels from dumping garbage except in restricted circumstances. Dumping of plastics is prohibited anywhere.

7. Regulations giving effect to Annex V should be promulgated.
8. A coherent information, compliance, and enforcement policy supported by all concerned federal government departments should be developed.
9. Ports which would be allowed to receive garbage originating in foreign countries should be designated.
10. All vessels operated by the Canadian government should set an example by complying fully with the government's policies and regulations pertaining to persistent debris.

The workshop recognized three categories of solutions to the problem of persistent debris — cleanup, better use and disposal practices, and reduction at source. Cleanup action can be fairly straight forward. Better use and disposal practices require reception facilities, recycling and regulation of some aspects. Reduction at source involves issues such as packaging, alternative materials and biodegradation. Each of these categories of action involves public education and public action.

Debris in the aquatic environment is a component of a much larger solid waste problem which society is only now recognizing. A change in public attitudes is an essential element of any action plan to deal with debris in the aquatic environment. For effective action, the public must first recognize the nature and dimensions of the problem. Our "throwaway" society is the root of the problem. Improved disposal practices must be seen to be in the best interests of society in general and individuals in particular. Public education must be accompanied by the adequate reception facilities and recycling programs.

While the full extent of the persistent debris problem in Canada is not known, we know enough to realize that Canadian waters are adversely affected and that action now will help to prevent a costly future crisis. Effective action will require a number of different measures. More importantly, it will require a wide range of people, from various levels of government, from industry, from public interest organizations across the country. In particular, concerned individuals must combat this emerging problem in the aquatic environment.

4.0 CANADA'S FISH HABITAT MANAGEMENT POLICY

Human activities threaten the aquatic environment, fish habitat and fisheries in a myriad of ways. The issues discussed in the preceding sections are not a catalogue of all the threats to fish habitat and fish. They do, however, illustrate the wide range of ways in which we affect the habitat upon which fish depend for survival.

Addressing this multiplicity of threats and potential threats to fish habitat requires a policy and process for dealing with emerging habitat management issues.

Some steps in this direction were taken with the 1977 amendment of the Fisheries Act to strengthen its habitat protection provisions. As a result, the federal government can now take legal action to ensure that work done on or near fish habitats is deferred pending a thorough examination by fisheries experts. Under the revised Act, fisheries authorities were empowered to demand all information about a proposed development necessary to make a decision about the possible impacts on fish habitat. This provided the authority, if development projects are judged likely to cause damage to habitats, to delay or halt them until necessary changes are made. The 1981–1988 Nechako experience illustrated the application of these powers in practice. The Nechako experience also illustrated the conflicts that can arise between the federal and provincial governments and between the federal government and developers over the use of Canada's aquatic resources.

Pearse (1982) described the division of resource ownership and jurisdiction between the two levels of government in Canada as the root of much of the difficulty associated with habitat protection. Conflict arises because forest, mining and other operators on provincial Crown land must often serve two masters. They must abide by the terms of their resource agreements with the provincial government but at the same time ensure that their activities do not violate the *Fisheries Act* and its regulations. Developments on privately owned land also pit private interests against the public interest in protecting fish habitat.

At that time (1982), the Department's objective for habitat management was "to conserve and develop habitat of federally managed aquatic species in a manner that will serve fisheries management goals." This diffuse objective provided no guidance on how this might relate to specific habitat use conflicts. Pearse argued for more explicit objectives to strengthen the Department's hand in dealing with other resource industries that threaten habitat. At the same time, he argued for flexibility on the question of whether all habitat must be protected at all costs.

Pearse recommended that:

(R3.2) The policy of the Department should be to ensure that total fish production capacity will not be diminished as a result of industrial and other activities that impinge upon fish habitat. Identifiable and measurable harm to fish habitat should be tolerated for any particular development only if the damage is fully compensated through expanded fish production elsewhere.

(R3.3) The Department should adopt an explicit policy for assessing proposed developments that threaten fish habitat and for determining compensation where required, based on the following precepts:

(i) In considering proposals for new developments, the Department should investigate their impact on fish habitat and all feasible means of avoiding or minimizing harm to fish.

(ii) Developers should be required to adopt all reasonable measures to avoid or mitigate damage to fish habitat.

(iii) If such measures are insufficient to prevent habitat damage, the Department should be authorized (but not required) to approve the development, but only if the loss in fish production capacity is fully compensated through increased fish production capacity elsewhere. The compensation should take the form of new fish production capacity created by the developer, or cash sufficient to enable the Department to replace the equivalent of the lost productive capacity.

In effect, Pearse's habitat proposals amounted to a *no-net-loss* approach. Pearse also proposed a Pacific Fisheries Conservation Fund, holding cash from compensation for damage to fish habitat. The proposed fund would only be used for habitat improvement and other fish-production measures.

During the early to mid-1980's, DFO undertook extensive consultations on various drafts of a proposed national fish habitat management policy with all interested parties, including fishermen, environmental groups, major developers and provincial governments.

On October 9, 1986, Fisheries and Oceans Minister Tom Siddon tabled in the House of Commons a new Policy for the Management of Fish Habitat (DFO 1986l). Mr. Siddon described the new policy as "an explicit recognition by Canada that fish habitats are important national assets." Going beyond Pearse's 1982 recommendations, the specific objective of the new policy was the achievement of an overall *Net Gain* of habitat productivity. The Department would strive to balance unavoidable habitat losses with habitat replacement on a project-by-project basis, to prevent further reductions to Canada's productive fish habitats. A net gain would result from habitat restoration and enhancement. The policy incorporated the concept of integrated resource planning, to reconcile the interests of the many sectors which compete for the use of fish habitat.

During the consultations leading up to the 1986 Policy, it became clear that an improved approach was needed to manage fish habitat and to consider opposing views before habitat decisions are taken. In particular, integrated resource planning needed to be more widely applied in fish habitat management.

DFO expected to apply this policy primarily in freshwaters, estuaries and coastal situations where most damage to fish habitats has taken place and where the risk of future damage is highest. The policy also applied to the marine waters on Canada's continental shelves, the main areas of interest being: (1) the surveillance and control of chemical hazards introduced, or that might be introduced, by human activities, and (2) managing the potential adverse effects of ocean dumping, shipping and oil and gas exploitation activities. DFO indicated that the policy would be applied to projects and activities of any scale, large or small, to avoid cumulative losses of habitats that support Canada's fisheries resources.

A net gain in habitat productivity was to be achieved through the active conservation of the current productive capacity of habitats, the restoration of damaged fish habitats, and the development of habitats. Underpinning the "net gain" objective were three goals:

— Fish Habitat Conservation
— Fish Habitat Restoration
— Fish Habitat Development (Fig. 16-6).

The fish habitat *Conservation* goal is to maintain the current productive capacity of fish habitats supporting Canada's fisheries resources. Where there is a risk of potential damage to habitat, the Department strives to prevent losses of natural fish production areas, to produce fish in perpetuity and to help maintain genetic diversity. The habitat provisions of the *Fisheries Act* are utilized to control the negative impacts of existing and proposed projects and activities that have a potential to alter, disrupt or destroy habitats. DFO recognized that there are limitations on the use of the *Fisheries Act* to control widespread activities on an ecosystem-wide basis, such as land use developments and the release of air pollutants (e.g. acid rain).

To control ocean pollution and chemical contamination of fish and fish habitat, the Department pledged to cooperate with and provide criteria for fisheries protection to provinces, territories and other federal departments. Examples of such cooperation include lobbying the United States on the acid rain issue and multi-departmental cooperation on the clean-up of oil spills and dioxins from pulp mills. Future regulations under the new *Canadian Environmental Protection*

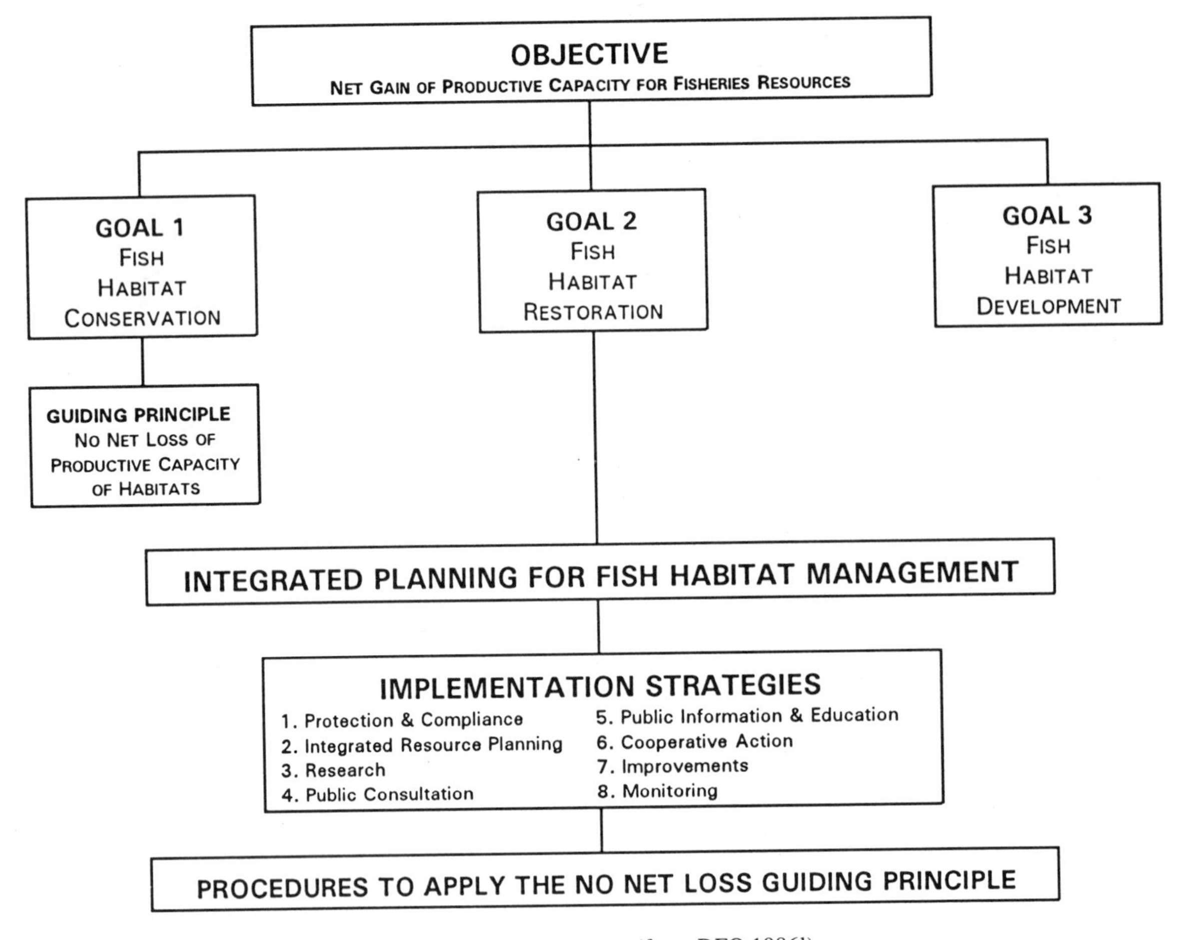

FIG. 16-6. Policy Framework for Fish Habitat Management (from DFO 1986l).

Act would play an important role in controlling chemical contamination of fish habitat.

The fish habitat *Restoration* goal is to rehabilitate the productive capacity of fish habitats in selected areas where the fisheries resource can provide economic or social benefits. This complements the preventive approach of the conservation goal.

The third goal of fish habitat *Development* involves improving and creating fish habitats in selected areas where the production can be increased. This can involve manipulating naturally occurring chemical, physical and biological factors, and creating, or providing access to, new spawning, rearing and food-producing areas.

Fundamental to this overall approach is the *guiding principle of no net loss* of the productive capacity of habitats. Under this principle, the Department strives to balance unavoidable habitat losses with habitat replacements project-by-project to prevent further reductions due to habitat loss or damage.

A key facet of the 1986 Habitat Policy was its emphasis on integrated resource use planning. The Policy recognized that other natural resource interests such as the forest, mining, energy and agricultural sectors make legitimate demands on water resources, and that ways must be found to reconcile differences of opinion on their best use. Regional or local fish habitat management plans were to be developed to allow the involvement of other stakeholders.

Eight Implementation Strategies were identified for the policy:

1. Protection and Compliance
2. Integrated Resource Planning
3. Research
4. Public Consultation
5. Public Information and Education
6. Cooperative Action
7. Improvement
8. Monitoring

The protection and compliance strategy involves protecting fish habitats by administering the *Fisheries Act* and incorporating fish habitat protection requirements into land and water use activities and projects. Frequently, potential adverse effects on fish habitats can be avoided by modifying the plans, designs and operating procedures for projects and activities and by incorporating mitigation and compensatory measures. Other instances require collaboration with other federal departments and/or other levels of government.

Under Section 37.(1) of the *Fisheries Act*, the Minister or his officials may ask proponents to provide a statement of information so that the Department can assess the potential impact of existing or proposed works on the resource. Special procedures apply for major projects. The policy defines major projects as: "Those works, undertakings and activities that could potentially have, or be perceived to have, significant negative impacts on the habitats supporting Canada's important fisheries resources." Examples given include: large-scale aerial biocide spraying of forest and agricultural lands, deep-draft marine terminals, hydroelectric dams and diversions, integrated mining operations, offshore oil and gas exploration and development, large industrial and municipal waste discharges, large pipelines, roads and transmission lines, and large dredging operations.

The 1986 Policy provided special procedures for major projects. The Department would conduct detailed reviews, preferably as a participant in a provincial or federal environmental review process, of major proposed industrial undertakings that could potentially harm fish habitats. For such development projects, a senior level Habitat Policy Steering Committee, chaired by an Assistant Deputy Minister of Fisheries and Oceans, would be established to direct the Department's actions.

Regional Project Committees would be established for each major project for detailed review and interaction with the proponent(s). The Department followed this general process for the review of the proposed Northumberland Strait Fixed Link Project.

DFO realized that applying the policy's no net loss principle would be difficult. Accordingly, it committed to evaluate each development, whether major or minor, in the planning phase to determine if it would reduce the capability of habitat to sustain fisheries resources. If this proved to be true, the Department would adhere to the following hierarchy of principles to achieve no net loss of productive capacity:

1. The first preference of the Department is to maintain without disruption the natural productive capacity of the habitat in question by avoiding any loss or harmful alterations at the site of the proposed project or activity. This could be achieved by encouraging the proponent to redesign the project, to select an alternate site, or to mitigate potential damages using other reliable techniques.
2. Only after it proves impossible or impractical to maintain the same level of habitat productive capacity using the approached outlined above would the Department accede to the exploration of compensatory options. First, the possibilities of like-for-like compensation would be assessed. This would involve replacing natural habitat at or near the site. With respect to chemical contamination, compensation options would not be considered. The Department would insist on the installation of reliable control techniques to mitigate such problems from the outset.
3. In those rare cases where it is not technically feasible to avoid potential damage to habitats, or to compensate for the habitat itself, the Department would consider proposals to compensate in the form of artificial production to supplement the fishery resource.
4. The costs associated with providing facilities or undertaking measures to mitigate and compensate for potential damages to the fisheries resource will be the responsibility of proponents, along with the costs to operate and maintain such facilities.

To apply the no net loss guiding principle, the Department's reviews would follow the six steps illustrated in Fig. 16-7.

The 1986 Policy was a major step to improve the Department's processes for habitat management. Most groups with an interest in habitat issues reacted positively to this policy framework. Pearse (1988), in his review of freshwater fisheries in Canada, described the 1986 Policy as "commendable and much-needed." Pearse called on DFO to press ahead to give effect to the policy through habitat management plans for areas under federal jurisdiction. These should clearly set out fish production targets to measure potential losses from development. He also called for agreements with other levels of government to define resource management goals and guidelines for habitat replacement, restoration and development. Agreements should establish habitat steering committees where required for major development projects.

Court judgements during the period 1989–1991 introduced a new element into the equation by questioning the way in which the federal government was discharging its environmental assessment responsibilities in the inland provinces. The court decisions suggest that the federal government cannot abdicate its

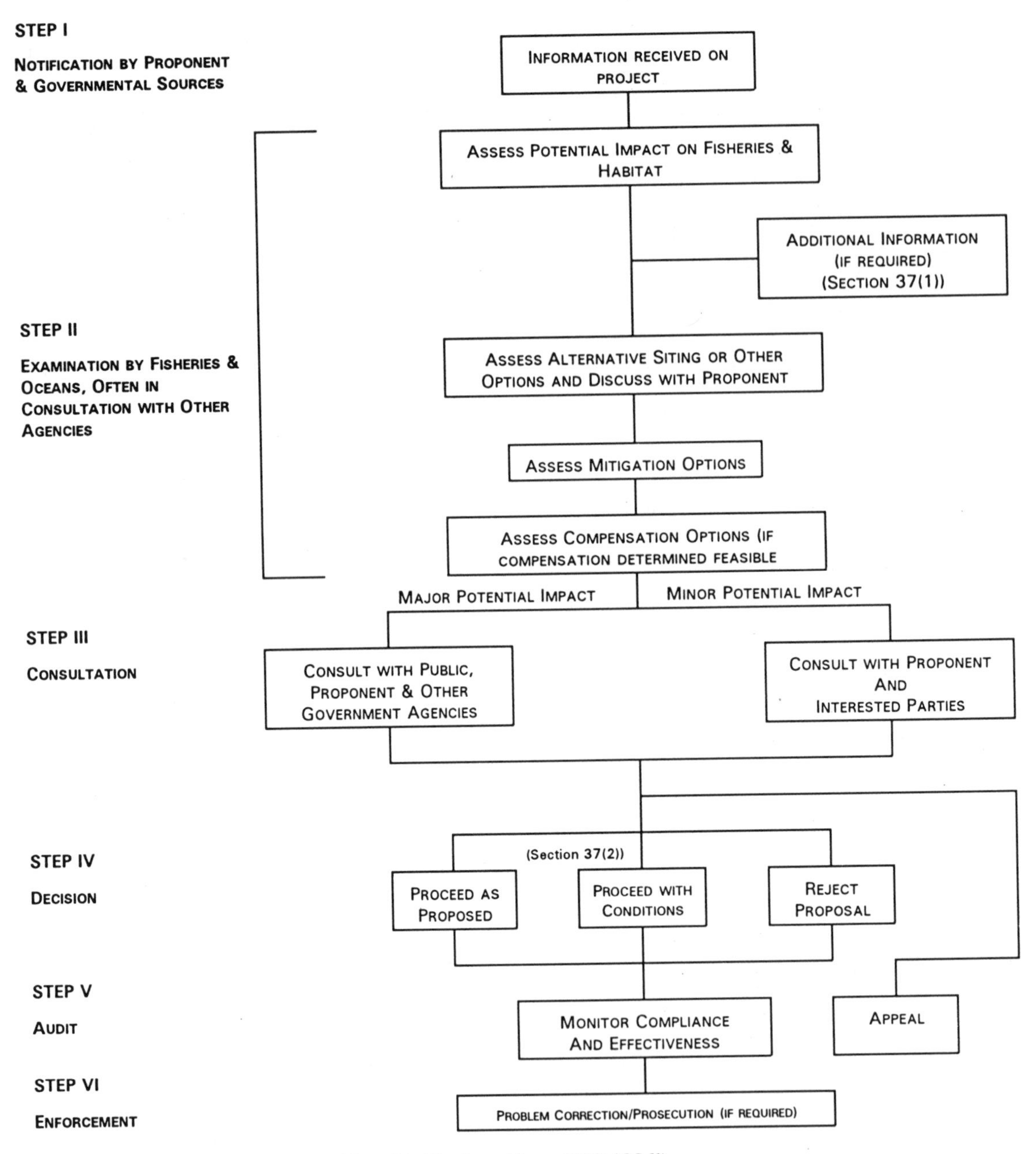

FIG. 16-7. Procedural Steps to Achieve No Net Loss (from DFO 1986l).

responsibilities in this area by delegating them to the provinces. Meanwhile, the government must address effectively the challenges to the habitat of anadromous and marine species.

Chemical contamination, in particular, poses a major threat to the habitat of certain stocks and the fisheries for those stocks. This has implications for human health. It has become increasingly difficult to separate fish habitat issues from the general concerns about man's impact on the environment, concerns which loomed large as we entered the last decade of the 20th century.

These concerns are global. They have widespread implications for fish and fisheries but far more profound implications for humanity as a whole.

5.0 GLOBAL WARMING

During the 1980's the issue of climatic change and global warming moved from backroom scientific discussions to prominent attention in newspapers and on the television screens of the industrial world.

Scientists gathered in Villach, Austria, in October, 1985, in a meeting organized by the World Meteorological Organization (WMO), the UN Environmental Programme (UNEP), and the International Council of Scientific Unions (ICSU), concluded that climate change had to be considered as "a plausible and serious probability." They estimated, if recent trends continued, the combined concentration of CO_2 and other greenhouse gases in the atmosphere would be equivalent to a doubling of CO_2 from pre-industrial levels, possibly as early as the 2030's. This could lead to a rise in global mean temperatures greater than any in man's history. Modelling studies indicated an increase in globally averaged surface temperatures, associated with a doubling of CO_2, of between 1.5°C and 4.5°C. The warming would become more pronounced at higher latitudes during winter than at the equator (WMO 1986).

The Conference speculated that a global temperature increase of 1.5–4.5°C, with as much as two to three times this at the poles, could lead to a sea level rise of 25 to 140 centimetres. The 1987 Brundtland Commission on Environment and Development commented:

> "A rise in the upper part of this range would inundate low-lying coastal cities and agricultural areas, and many countries could expect their economic, social, and political structures to be severely disrupted. It would also slow the 'atmospheric heat-engine', which is driven by the differences between equatorial and polar temperatures, thus influencing rainfall regimes. Experts believe that crop and forest boundaries will move to higher latitudes. The effects of warmer oceans on marine ecosystems or fisheries and food chains are also virtually unknown" (Brundtland 1987).

Canada has been at the forefront in research on climate change and assessing the impact of such changes. In June, 1988, Canada sponsored a World Conference on "The Changing Atmosphere: Implications for Global Security" in Toronto. More than 300 scientists and policy makers from 46 countries, United Nations organizations, other international bodies and non-governmental organizations participated. The Conference reached the following conclusions (DOE 1988):

> "Humanity is conducting an unintended, uncontrolled, globally pervasive experiment whose ultimate consequences could be second only to a global nuclear war. The Earth's atmosphere is being changed at an unprecedented rate by pollutants resulting from human activities, inefficient and wasteful fossil fuel use and the effects of rapid population growth in many regions. These changes represent a major threat to international security and are already having harmful consequences over many parts of the globe.

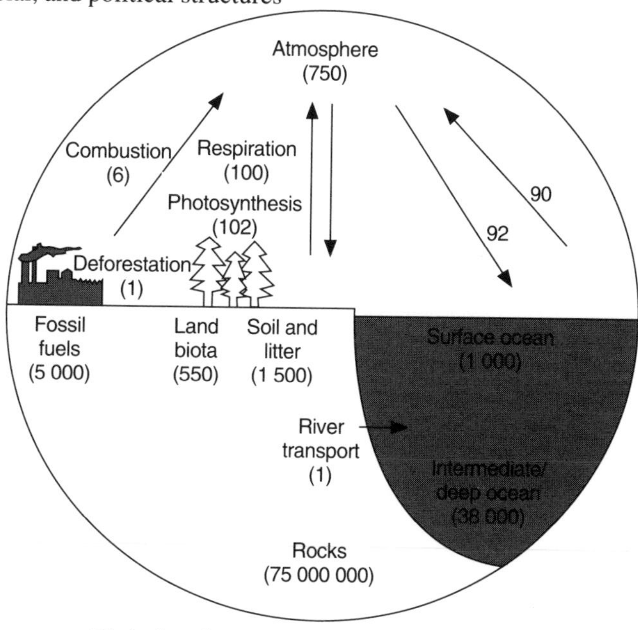

Global carbon cycle.

"Far-reaching impacts will be caused by global warming and sea-level rise, which are becoming increasingly evident as a result of the continued growth in atmospheric concentrations of carbon dioxide and other greenhouse gases. Other major impacts are occurring from ozone-layer depletion resulting in increased damage from ultra-violet radiation. The best predictions available indicate potentially severe economic and social dislocation for present and future generations, which will worsen international tensions and increase risks of conflicts between and within nations. It is imperative to act now."

Regarding climate warming, the Conference Statement observed:

1. There has been an observed increase of globally-averaged temperature of 0.7°C in the past century which is consistent with theoretical greenhouse gas predictions. The accelerating increase in concentrations of greenhouse gases in the atmosphere, if continued, will probably result in a rise in the mean surface temperature of the Earth of 1.5 to 4.5°C before the middle of the next century.
2. Marked regional variations in the amount of warming are expected. For example, at high latitudes the warming may be twice the global average. Also, the warming would be accompanied by changes in the amount and distribution of rainfall and in atmospheric and ocean circulation patterns. The natural variability of the atmosphere and climate will continue and be superimposed on the long-term trend, forced by human activities.
3. If current trends continue, the rates and magnitude of climate change in the next century may substantially exceed those experienced over the last 5,000 years. Such high rate of change would be sufficiently disruptive that no country would likely benefit in total from climate change.
4. The climate change will continue so long as the greenhouse gases accumulate in the atmosphere.
5. There can be a time lag of the order of decades between the emission of gases into the atmosphere and their full manifestation in atmospheric and biological consequences. Past emissions have already committed planet Earth to a significant warming.
6. Global warming will accelerate the present sea-level rise. This will probably be of the order of 30cm but could possibly be as much as 1.5m by the middle of the next century. This could inundate low-lying coastal lands and islands, and reduce coastal water supplies by increased salt water intrusion. Many densely populated deltas and adjacent agricultural lands would be threatened. The frequency of tropical cyclones may increase and storm tracks may change with consequent devastating impacts on coastal areas and islands by floods and storm surges.

What will be the implications of these changes for the oceans and for fish, fish habitat, and fisheries? In a 1989 Submission to the House of Commons Standing Committee on the Environment, on *Global Warming, the Oceans, and Canada's Fisheries*, the Department of Fisheries and Oceans addressed this question (DFO 1989m):

"It is clear that global warming will affect the oceans, aquatic living organisms in them, the abundance, distribution and quality of freshwater, and the people who exploit the oceans, lakes, and rivers or who live near their shores. Global warming could have major impacts on Canada's fisheries. It is recognized that the distribution and abundance of many fish species are sensitive to small changes in ocean conditions. Other coastal and inland species, such as whitefish and Pacific salmon, are vulnerable to changes in precipitation and lake and stream temperatures. Weather pattern changes will affect marine transportation. The predicted rises in sea level could have dramatic consequences. Changes in the thickness and duration of ice cover or the frequency of icebergs could be particularly important to Canada."

The DFO Submission offered the following assessment:

"Climate change will have a profound impact on Canadian fisheries. Climate is a powerful determinant of the distribution, abundance and species mix of fishes in coastal and inland fisheries. The lakes of Canada owe their origin to glacial action and their persistence to the modern excess of precipitation over evaporation. The distribution of species and productivity of fisheries has a strong climate component through the interplay of currents and coastal geology. For example, the marine mammal and fish fauna of the Arctic has evolved in intimate association with the extensive ice cover and cold waters imposed by the extreme Arctic climate. The natural southern limit of salmonids is determined by climatic factors as is the species composition of east coast fisheries. It is thought that environmental changes played a significant, role in the recent variations in the Newfoundland inshore cod fisheries. Bottom water temperatures on the Atlantic continental shelf may influence the ratio of valuable groundfish to less valuable pelagic fish such as herring and mackerel....

"Some important species, notably salmon, gain most of their weight in the open ocean far from Canadian shores and are thus sensitive to ocean climate and productivity changes over the whole area of their range.

"The effects of climate change on Canadian fisheries could be both positive and negative, but even positive changes require sometimes painful adjustments. A warming of our waters would affect the northern limit of the range in Canadian waters of species such as squid and mackerel on the Atlantic coast and hake and tuna on the Pacific, leading to new or expanded fisheries. Some northern salmonid populations might experience increased production through more favorable conditions. The effects of global warming on freshwater fisheries could be more striking: it has been estimated that 31 new species might successfully invade the Great Lakes from the south with a climate change resulting from a doubling of atmospheric carbon dioxide. In central and northern Canada, where the greatest projected changes in annual temperatures [would] occur, melting of permafrost would destroy fish habitat and change the whole character of the geographic landscape. If precipitation decreases as projected by some models Lake Manitoba might dry out and Lake Winnipeg shrink dramatically" (DFO 1989m).

These comments are largely conjecture. There are considerable uncertainties surrounding the predicted climate changes. Similar uncertainties surround any prediction of marine populations' reactions to environmental changes of this nature and scale.

The oceans play a significant role in influencing climate. They provide a stabilizing influence over the more volatile atmosphere because of their role as a source or sink for the fluxes of gases, moisture and energy across the ocean/atmosphere interface. At mid-latitudes the oceans transport as much heat poleward as the atmosphere and play an equal role in moderating climate (Stoddart and Clarke 1988). There is, however, considerable uncertainty about the degree to which physical, chemical and biological processes in the atmosphere and in the oceans are interlinked. There is also uncertainty whether the oceans will continue to behave as in the past. If the predicted global warming occurs, the oceans might react quite differently, setting up major new circulation patterns and upwelling processes that could totally disrupt weather patterns, climate, fisheries and other uses of the sea (Stoddart and Clarke 1988).

Projections for future climate are based on complex numerical simulations which include the atmosphere, the land surface and water, and the oceans. The greatest deficiency in these computer simulations lies in inadequate knowledge of oceans' influence on climate. We cannot even simulate present day climate with confidence, let alone accurately predict future climate (DFO 1989m).

It is not generally recognized that the oceans are important components in the system controlling the levels of greenhouse gases in the atmosphere. The widely-publicised projections of long-term trends in global temperature depend on predictions of atmospheric concentrations of CO_2 and other gases, and hence on understanding the ways in which they are removed or produced by the oceans. For example, the atmospheric increase of CO_2 has been only about half as great as expected from rates of deforestation, fossil fuel use and other human activities. The discrepancy could be accounted for by the oceans uptake of CO_2.

The present projections of global warming cannot be accepted with certainty because the large scale models on which they are based have serious deficiencies. The most important deficiency relates to inadequate knowledge about the transport of heat poleward from the tropics by the oceans. The ocean's role in moderating the build-up of greenhouse gases must also be better understood for reliable long term projections of global warming. Increased research is required to fill this knowledge gap (DFO 1989m).

Meanwhile, despite these uncertainties, there have been calls for action to halt the increase in greenhouse gases before it is too late. The 1988 Conference on the Changing Atmosphere urged "immediate action by governments, the United Nations and their specialized agencies, other international bodies, non-governmental organizations, industry, educational institutions, and individuals to counter the ongoing degradation of the atmosphere." The Conference called on governments and industry to take the following actions to reverse the deterioration of the atmosphere:

— *Ratify the Montreal Protocol on Substances that Deplete the Ozone Layer.* The Protocol should be revised to ensure nearly complete elimination of the emissions of fully halogenated CFCs by the year 2000.
— *Set energy policies to reduce the emissions of CO_2 and other trace gases in order to reduce the risks of future global warming.* Stabilizing the atmospheric concentrations of CO_2 is an imperative goal. It was estimated (1988) that this required reductions of more than 50% from existing emission levels.
— *Reduce CO_2 emissions by approximately 20% of 1988 levels by the year 2005 as an initial global goal.*
— *Set targets for energy efficiency improvements* that are directly related to reductions in CO_2 and other greenhouse gases. A challenging target would be to achieve a 10% energy efficiency improvement by 2005.
— *Initiate the development of a comprehensive global convention as a framework for protocols on the protection of the atmosphere.*

On June 30, 1988, at the close of the Changing Atmosphere Conference, Canada ratified the Montreal Protocol for the Protection of the Ozone layer. On that same day, Canada proclaimed the new *Canadian Environmental Protection Act.* The Montreal Protocol established a timetable to control major industrial substances that deplete the ozone layer. In the short term, it established phasedown requirements for chlorofluorocarbons and halons. In announcing Canada's ratification of the Protocol, Environment Minister McMillan said that it represented "an important step towards Canada's goal of an International Law of the Atmosphere" (DOE 1988a).

Just over a year later, Environment Minister Lucien Bouchard addressed a Ministerial Conference on Atmospheric Pollution and Atmospheric Change in the Netherlands (November 1989). In that speech, Mr. Bouchard acknowledged that reducing emissions of greenhouse gases would be difficult. Canadian emissions had grown consistently over the past century. Projections suggested that, without intervention, CO_2 emissions could rise by 50% by the year 2005. Mr. Bouchard noted that Canada had already taken certain actions:

"— We are committed to phasing out controlled CFCs by 1999.
— We have established automobile fuel efficiency guidelines.
— We have announced our intention to reduce automobile CO_2 emissions by 20% by 2005.
— We have also announced our intention to reduce automobile NO_x emissions by 30% by 1994.
— Federal and Provincial Energy Ministers are currently considering actions to further reduce CO_2 emissions through energy efficiency and conservation" (DOE 1989b).

In April, 1990, Canada's energy ministers decided to drop national targets for cutting air pollution. The federal, provincial and territorial representatives agreed unanimously that proposed country-wide restrictions on greenhouse gases were "premature, given the extensive consultations now underway in most jurisdictions." The energy ministers agreed each province should do its best within its own boundaries and participate in forthcoming public consultations on the costs and benefits of reducing air pollution. They said: "The place to resolve the question of targets is in the context of international negotiations" (*The Ottawa Citizen*, April 3, 1990).

Apparently, a key turning point in the talks came at a luncheon at which Canada's leading climate guru, Dr. Kenneth Hare of the University of Toronto, pointed out to the Ministers that there was greater conflict among scientists than among politicians on global warming. Hare was quoted as saying that another 10 years might pass before the scientists reached agreement on climate change (*The Financial Post*, April 5, 1990).

However, in August, 1990, the Intergovernmental Panel on Climate Change reported that global mean surface air temperatures had increased by 0.3 to 0.6°C over the previous 100 years, with the five global-average warmest years being in the 1980's. Over the same period global sea-level increased by 10 to 20cm. Subsequently, it was recognized that 1990 was the warmest year on record worldwide, with six of the seven warmest years in more than a century occurring between 1981 and 1990 (IPCC 1990).

In November, 1990, the Second World Climate Conference concluded that a clear scientific consensus had been reached on the range of global warming to be expected as a consequence of the emissions of greenhouse gases resulting from human activity. The Conference concluded:

> "Without actions to reduce emissions, global warming is predicted to reach 2 to 5 degrees C over the next century, a rate of change unprecedented in the past 10,000 years. The warming is expected to be accompanied by a sea level rise of 65 cm ± 35 cm by the end of the next century....climate change and sea level rise would seriously threaten low-lying islands and coastal

zones. Water resources, agriculture and agricultural trade, especially in arid and semi-arid regions, forests and fisheries are especially vulnerable to climate change" (Anon. 1990).

The Conference also concluded that technically feasible and cost-effective opportunities exist to reduce CO_2 emissions in all countries. It considered such opportunities to be sufficient to allow many industrialized countries to stabilize CO_2 emissions from the energy sector and to reduce these emissions by at least 20% by 2005. The Conference also recommended that nations launch negotations on a convention on climate change "with the aim of signing such a convention in 1992." At the United Nations Conference on Environment and Development held in Brazil in June 1992, 154 countries signed a Framework Convention on Climate Change. This Framework Convention, which was to enter into force after 50 ratifications, established a blue print for starting the international response to global warming. The objective was to stabilize greenhouse gas concentrations in the atmosphere at a level preventing dangerous anthropogenic interference with the climate system and allowing ecosystems to adapt naturally and development to proceed sustainably.

Whether this convention will precipitate action to address the predicted global warming remains to be seen. Currently, Canada is committed to stabilize net carbon dioxide emissions at the 1990 level by the year 2000.

6.0 SUSTAINABLE DEVELOPMENT

The recent debate in Canada about targets for reducing CO_2 emissions and the implications for energy usage and development emphasize the difficulty that Canada, along with most other countries, is having in implementing the concept of "sustainable development" advanced by the Brundtland Commission in *Our Common Future* (Brundtland 1987). "Sustainable development" has become the buzz-phrase for the 1990's. Everyone pays lip service to the concept but nations wrestle with the practical implications. The Brundtland Commission defined *sustainable development* as:

> "development that meets the needs of the present without compromising the ability of future generations to meet their own needs. It contains within it two key concepts:
> — the concept of 'needs', in particular the essential needs of the world's poor, to which overriding priority should be given; and
> — the idea of limitations imposed by the state of technology and social organization on the environment's ability to meet present and future needs."

The Report identified the following critical objectives for environment and development policies that flow from the concept of sustainable development:

— reviving growth;
— changing the quality of growth;
— meeting essential needs for jobs, food, energy, water and sanitation;
— ensuring a sustainable level of population;
— conserving and enhancing the resource base;
— reorienting technology and managing risk; and
— merging environment and economics in decision making.

An in-depth discussion of the national or global implications of the sustainable development concept is beyond the scope of this chapter. But clearly the objective of "conserving and enhancing the resource base" is directly relevant to habitat and fisheries management in Canada. In effect, Canada's 1986 Habitat Management Policy attempted to set forth an approach for the management of fish habitat which involved striking a balance between sustainability and development. Sustainable development is the goal of fish habitat management.

The Canadian government's 1990 Green Plan set out additional actions, one effect of which would be to protect and restore fish habitat. These include measures to control toxic substances, protect and enhance water quality, reduce the risk of ocean spills, and enhance Canada's ocean spill response capability. The government also introduced increased penalties for habitat violations through amendments to the *Fisheries Act* in 1991. It pledged to update and strengthen pollution prevention regulations made under the *Fisheries Act*, beginning with the Pulp and Paper Effluent Regulations and the Metal Mining Liquid Effluent Regulations.

Expanded scientific assessment, monitoring and research on toxic substances and their effects on fish and fish habitat were promised.

The government also pledged to improve the level of habitat protection across the country through a more consistent application of the *Fisheries Act*.

Sustainable fisheries depend on both *productive fish habitats* and *sound harvesting practices*. The concept of sustainable fisheries has broader implications for fisheries management. As part of the Green Plan, the federal government committed to develop a National Sustainable Fisheries Policy and Action Plan, in cooperation with the provinces, territories, the commercial, recreational and native fishing communities and other interested parties. The stated purpose was to focus national attention on the importance of the fisheries and aquatic ecosystems, identify key issues and establish a national framework for co-operation to achieve sustainable fisheries.

7.0 CONCLUSIONS

Too often when we think of fisheries management we think only of managing the level of harvest. But fisheries management in the broadest sense includes the management of the habitat upon which fish depend. If we pollute the rivers and streams, if we alter the water flow patterns, if we contaminate the marine environment, we threaten the survival of fish by reducing or contaminating the habitat which sustains them.

The examples given in this chapter show that the threats to fish habitat are many, diverse, and sometimes subtle. Oil spills have a readily apparent impact upon seabirds and upon the shoreline. Contaminants in the marine environment are often detectable only with sophisticated analytical equipment but over time they can devastate the aquatic ecosystem and harm humans who eat fish. Cleaning up an oil spill can be an achievable though demanding task. Cleaning up contaminants in the aquatic environment is exceedingly difficult.

As Piuze (1989) put it:

> "There is only one lasting solution to the problem of marine pollution. It is neither dilution, nor more research and monitoring. It is not continually trying to clean up. The only permanent solution is to turn off the tap, all the others being mitigating measures."

He listed numerous obstacles to this, including:

- the magnitude and complexity of the problem;
- economic considerations, including human greed;
- a long-standing attitude which pits economy against environment;
- fragmented intergovernmental jurisdictional wrangling;
- the complexity of the international situation, with controls differing from country to country;
- lack of scientific understanding of the total picture;
- resistance to change.

An effective solution to these problems requires cooperative action on a variety of fronts. Fundamental to any solution is a change in attitude and lifestyles. Polluting, altering or destroying fish habitat must be rejected by society. Everyone involved must share responsibility for the problem and for finding solutions. There must be a change in the waste and throw-away philosophy which pervades modern society to one where reduced use and recycling are the norm. On the energy front there must be movement toward less wasteful practices and cleaner energy sources. Stringent effluent regulations and marine environmental guidelines must be developed and applied. Those who violate these regulations must pay dearly for their actions. There is a need for increased research on means of preventing pollution at the source. Prevention rather than clean-up, mitigation or rehabilitation must become the norm.

Progress is being made. It appears that emissions in the U.S. contributing to acid rain in Canada are finally being addressed as a result of the 1990 U.S. Clean Air Act. It was a long and divisive battle but reason finally prevailed over inertia. Chemical contamination of the aquatic environment is now at the forefront of public consciousness. Increased monitoring and research is revealing the extent of the problem in the freshwater and marine environments. Action is being taken to reduce the sources of pollution through tougher effluent regulations for pulp and paper mills and other polluting industries.

But not all of the encroachments on fish habitat are major. Small-scale diversions of streams, dumping of wastes and littering also have a negative cumulative effect by altering or destroying valuable habitat. This aspect of the problem requires attention by concerned citizens. Public information and education is a vital link in the chain leading to corrective action.

Vigilance in the interest of a cleaner environment is essential to a successful program of fisheries and habitat management.

CHAPTER 17

SCIENCE AND FISHERIES MANAGEMENT I — THE DEVELOPMENT OF FISHERIES RESEARCH AND THE SCIENTIFIC ADVISORY PROCESS

"The aim of fisheries science becomes maximization of understanding so that whatever the undefined and shifting social objectives, there will always be some notion of how to get there. Our real obligation to the future is to let others know how it all works, leaving them enough options to do whatever they wish in the future."

– Peter Larkin, 1978

1.0 INTRODUCTION

Fisheries science is a relatively recent development in the history of science. It had its beginnings in the work of some European investigators in the last part of the 19th century, and in the first two decades of this century, e.g. Petersen and Hjort, but did not really crystallize until the interval between the two World Wars. One of the by-products of the First World War was that it demonstrated the impact of fishing on fish stocks. Some Northeast Atlantic stocks which had been reduced by intensive fishing before the war recovered with the cessation of fishing during the war years. Catch rates and size of fish increased. This provided one of the first practical demonstrations of man's impact on nature through fishing.

During the twentieth century the scope of research on fish has expanded considerably. Taxonomy, life history, physiology, disease and behaviour studies all form part of the wide array of research on fish. Some of these studies have stimulated the development of alternatives to wild harvesting, e.g. aquaculture. In this chapter I use the term *fisheries research* to describe research, primarily in fish biology, aimed at providing a scientific basis for managing wild fish stocks. I use the term *fisheries science* to encompass other disciplines as well as biology, e.g. economics and sociology, which have an important contribution to make to the rational management of fisheries in support of societal objectives.

Central to the attempts to provide a scientific basis for fisheries management has been the study of fisheries population dynamics. These studies proliferated as the century advanced. They have provided models which have been used as the basis for management of fisheries, first in certain international commissions and later by coastal states following the proclamation of 200-mile fisheries zones. Certain of these models have been discussed in Chapter 3. In this chapter I examine the historical development of the science of fish population dynamics internationally and in Canada. I also describe the evolution of processes to ensure the orderly provision of scientific advice to managers. Certain limitations and inadequacies in the existing processes are identified, and ways of improving the credibility of scientific information for the fishing industry and fisheries managers are suggested.

2.0 BRIEF HISTORY OF FISHERIES RESEARCH RELATED TO MANAGEMENT

2.1 The Initiation of Fisheries Research in the International Context

Ritchie-Calder (1978) suggested that marine biology dates back to Aristotle. Modern fisheries research had its origins in the nineteenth century (Smith 1988). One of the first major events was the recognition that fish species are subdivided into self-sustaining populations. This was demonstrated for herring in the Northeast Atlantic by Heincke during the last quarter of the nineteenth century. Sinclair and Solemdal (1988) concluded that Heincke's studies on herring populations strongly influenced subsequent developments in fisheries biology and management by leading to the definition of management units based on geographic populations or population complexes.

The International Council for the Exploration of the Sea (ICES) was founded in Copenhagen in 1902. Three committees were established to address migration, overfishing and hydrography. The mandate of the migrations committee included the question of geographic origins of fish of different species appearing in the coastal waters of various European nations at different times. At that time, it was thought that variations in migration patterns generated the interannual and decadal variability in catches in particular fishing areas. Thinking changed following the discovery that populations of fish are restricted to specific geographic areas throughout their life cycles, and Johann Hjort's demonstration that variability in the abundance of populations was due to year-class size variability. These discoveries indicated that interannual fluctuations in catches were in some cases largely due to population fluctuations within a fishing area rather than changes in migration patterns (Sinclair and Solemdal 1988).

Meanwhile, the overfishing committee of ICES wrestled with the impact of fishing on fish populations (Smith 1988). As noted in Chapter 3, Petersen, as early as 1894, distinguished between *growth overfishing* and *recruitment overfishing*, even though he did not use these terms. Much of the early research encouraged by ICES focused on the problem of growth overfishing. The British fished small plaice close to the European coasts and discarded six times as many fish as they kept. Petersen suggested that this problem could be resolved if fishermen restricted themselves to fishing the larger fish.

Following the First World War, the abundance of plaice was substantially higher than in the pre-war years and there was a significant increase in the average size caught (Smith 1988). A minimum size and a closed area were implemented in the 1933 Convention for the protection of plaice and flounder in the Baltic (Cushing 1972). Russell, in his 1931 paper "Some theoretical considerations on the 'overfishing problem'", put forward a model of overfishing. This was further developed by Graham (1935), who modelled the maximization of yield as fish grew in weight despite the loss in numbers (see Chapter 3). In 1937, an International Convention on the Protection of Undersized Fish was signed by ten European countries. This was not implemented because of the Second World War.

These developments in Europe during the 1930's were influenced to some extent by measures taken in the International Pacific Halibut Commission involving Canada and the United States (Smith 1988). W.F. Thompson, the first Director of the Halibut Commission, identified a relationship between catch per unit of effort and the amount of fishing effort (Thompson and Bell 1934). This led him to suggest that some restriction of fishing effort would improve the state of the halibut stocks. Some management measures were taken by the Halibut Commission and the halibut stocks increased in abundance. There was later controversy about the extent to which this was due to the appearance of a strong year-class at the same time the measures were introduced.

The apparent success of Thompson's attempts to apply some tentative scientific findings in management spurred the further development of population models in the Northeast Atlantic (Smith 1988). Thompson's observations on the relationship between CPUE and effort were extended by Russell, Hjort et al. (1933) and Graham to model the relationship between catch and fishing. This led subsequently to the idea of maximum sustainable yield as a management objective (see Chapters 3 and 4). As described earlier, this set the stage for much of applied fisheries research today.

2.2 The Development of Fisheries Research in Canada (1900–1972)

Canada had developed a fisheries administration soon after Confederation and passed a Fisheries Act as one of the federal government's first legislative initiatives. However, there was no scientific investigation on any organized basis before the early 1900's. Until then, research on fish in Canada was largely restricted to describing and identifying the various species of fishes and determining their distribution. There was some research into the life history of certain species. Research on ecology, physiology and population dynamics were still to come.

There was extensive activity in the area of fish culture, with hatcheries being established in Ontario, Quebec and the Maritimes. However, there was little scientific basis for the activities undertaken and the need for objective scientific research was just beginning to be recognized. In 1893, a specialist in fish embryology, Dr. E.E. Prince was appointed Canada's Commissioner of Fisheries. One of Prince's first proposals was to establish a marine scientific station for Canada. Parliament established in 1898 a Board of Management and appropriated $15,000. This represented a turning point for Canadian marine biological research. A station was built which could be transported on a scow from one location to another. This "station" was located at St. Andrews from 1899 to 1901, at Canso from 1901 to 1903, at Malpeque, PEI from 1903 to 1904, at Gaspé from 1905 to 1906 and was finally abandoned at Sept Iles in 1907.

The Board of Management in 1904 also assumed partial responsibility for a freshwater station at Go Home Bay on an island in Georgian Bay, Ontario. In 1908, it decided that a permanent Atlantic Biological Station should be established at St. Andrews, New Brunswick, and that same year a station was established at Nanaimo, British Columbia. In 1912, the Board of Management became the Biological Board of

First Biological Station of Canada — Barge built in 1899.

Biological Station Nanaimo, British Columbia, established in 1908.

Biological Station St. Andrews, New Brunswick, established in 1908.

Canada, operating under a special Act of Parliament. It continued to manage the two marine stations but the Georgian Bay Station was abandoned in 1913. The Board's membership was broadened in 1924 to include representatives of the fishing industry and a wider spectrum of academic expertise.

In 1937, the name of the Board was changed to the Fisheries Research Board of Canada. It continued to manage Canada's federal fisheries research effort until 1973 when the laboratories and personnel were integrated with the Department of Fisheries, leaving the Board an advisory role. By the end of that decade it was disbanded.

Johnstone (1977) detailed the activities of the Fisheries Research Board and its predecessors in his entertaining book *The Aquatic Explorers*. He drew upon interviews with many of the Board's leading figures to present a colourful, anecdotal history. Ricker (1975) summarized the Board's 75 years of achievements in many fields of aquatic science.

The transition from a Board of Management to the Biological Board of Canada in 1912 occurred partly because of conflict with the administrative section of the Department of Marine and Fisheries regarding the Board's finances (Huntsman 1943). Another major factor was the question of objectives. The opening of the first marine biological station provided a great opportunity for on-the-spot research and hence attracted many university researchers. Although some of these scientists became involved with fisheries problems, the majority pursued goals that were mainly 'curiosity-driven' rather than 'mission-oriented' research.

Johnstone (1977, p.72) observed:

> "The two objectives which both boards undertook to achieve, one of independent aquatic research and the other of providing answers to the practical problems of the fisheries, required that they do a nice balancing act, with the pole tilted now one way, now the other.... When science was harnessed to the service of fisheries, it proved a mettlesome and high-spirited steed that must have caused deputy ministers to wonder who was smuggling in the oats. In their own university departments the members of the Board were a law unto themselves, respected for their scholarship and achievements, and not at all prepared to have their decisions reviewed by 'bureaucrats' unfamiliar with biological matters."

This tug-of-war between basic and applied research continued throughout the history of the Fisheries Research Board and its predecessor. At various times, under different leaders it tilted one way or the other. For the most part, the view prevailed that basic and applied research went hand-in-hand.

In any event the 1912 Act to Create the Biological Board of Canada gave the Board the administrative independence it sought. Shortly thereafter Canada became involved in collaborative research efforts with other countries. Johann Hjort, who had established the variability in abundance of particular year-classes of fish, was enticed to come to Canada to lead a major survey called the Canadian Fisheries Expedition. This involved Norwegian, Canadian and United States researchers. The objectives were: to determine whether Canadian herring were all of one race or several different races; to look for variations in growth rate in different waters; to obtain information on year-class strengths; and to study the regime of ocean temperatures, salinities, currents, and plankton.

Although the Expedition as such was not continued in 1916 because of the First World War, efforts continued by Canadian investigators along the practical lines suggested by Hjort. Huntsman carried out a major field investigation in the Gulf of St. Lawrence. Applied scientific work was also initiated on lobsters. Researchers from St. Andrews became involved in meeting with fishermen to persuade them to return egg-bearing lobsters to the water. Work was also begun to study the circulation of Bay of Fundy waters and to relate this to the seasonal movement of fish and their spawning and migration patterns.

In 1921, another international collaborative initiative was launched, an international committee on marine fishery investigations, with members from Canada, Newfoundland and the United States. This Committee, which later became known as the North American Council on Fishery Investigations, also had occasional representation from France. Over time the committee became more and more a forum for scientific exchanges of plans and results of fisheries, including oceanographic, research. Between 1921 and the beginning of the Second World War there were about two dozen meetings. The Council encouraged the collection of more complete statistics on the offshore fisheries and to some degree coordinated investigations of major fisheries of common interest. It developed some collaborative projects with ICES.

On the west coast, from 1912 to 1921, there was a similar trend toward applied research. The ageing and determination of the growth rate of salmon from scales was undertaken. Efforts were initiated to restore salmon runs on the Fraser River reduced by the construction of the Canadian Northern Railway in 1912–13. The Pacific Biological Station became involved in 1918 in a cooperative tagging program between Canada and the United

States to ascertain the routes and rates of movement of sockeye during their migration, the percentage being caught, and where the survivors went to spawn.

At all its stations during the first 10 years of the Biological Board there was a trend towards more applied research. However, the biological stations continued to serve mainly as research facilities for university scientists. At St. Andrews, for example, there was no permanent scientific staff other than the Director until 1925. The Department of Fisheries, particularly the Assistant Deputy Minister, W.H. Found, was not satisfied, regarding the Biological Board as an obstacle to building scientific capability within the Department. In 1919, a Bill was passed in the House of Commons which in effect would have taken control of the stations out of the hands of the Board and restricted the work to "such investigations...as may be assigned to the Board by the Minister."

Board members were successful in having this legislation blocked in the Senate. In 1921, Dr. A.P. Knight became Chairman of the Board and relations between the Board and the Department improved considerably. Knight was more inclined toward applied research. The year 1925 marked a significant change in the policies of the Board. Two new technological stations were established at Halifax and Prince Rupert. Their mission was to investigate the technological problems of the fishing industry. At the same time a start was made towards acquiring a full-time scientific staff at the two biological stations. Both initiatives were steps towards research of more direct value to fisheries. Fishing industry members were appointed to the Board. By 1930 specific investigations were established at St. Andrews on oysters, lobsters, groundfish, salmon, trout and fish disease. In 1928, a resident oceanographer, H.B. Hachey, was appointed. At Nanaimo, studies were launched on salmon propagation and migration, herring and pilchards, oysters and other shellfish, trout propagation and oceanography.

One of the most significant results of the Nanaimo work in the late 1920's was the Board's adoption of a conclusion that, in areas where a natural run of salmon occurs with a reasonable expectancy of successful spawning, artificial propagation was unnecessary. Any results over and above natural spawning would not be worth the cost. This convinced the Department of Fisheries to close all the salmon hatcheries in British Columbia.

Meanwhile, work in Europe was advancing understanding of the nature of fish populations (Smith 1988). Needler (1987) has indicated that the population concept did not immediately take hold in Canada. Referring to the 1920's and 1930's, he stated:

> "Many concepts very familiar to us had not yet emerged. The existence of more or less distinct populations or 'stocks' of the same specie and many concepts of population dynamics were not yet imagined."

Professor A.T. Cameron, who was appointed Chairman in 1934, was strongly oriented towards applied research and quickly moved to restrict the relationship with volunteer workers which had characterized the Board's activities from 1898 to 1930.

Two former Station Directors, W.A. Clemens and A.G. Huntsman, criticized this move as short-sighted. Clemens (1958) said:

> "The Depression of the early 1930's with its severe curtailment of funds brought the volunteer investigator arrangement to an end in 1934. That this or some similar system was not reinstated seems to me unfortunate, because the association of university and Station personnel was mutually stimulating and beneficial."

Huntsman (1953) stated:

> "I still believe, in spite of the very great material success of the Board and its phenomenal growth since that time, that its action twenty years ago in doing away with volunteer research was unwise and that organized research is decidedly expensive and unproductive scientifically in comparison with volunteer research."

In addition to phasing out the volunteer workers, one of Cameron's major objectives was to organize activities on both the east and west coasts with full-time workers devoting their efforts to solving the problems of the fishing industry (Johnstone 1977).

During the 1930's a more intensive study of the groundfish fishery in the Bay of Fundy was undertaken by the St. Andrews' Biological Station. The investigations included the life history of cod and haddock, including growth, distribution, migrations, temperature preferences, spawning and egg distribution, and the collection of statistics on the fisheries. Huntsman conducted extensive studies on Atlantic salmon and elucidated certain aspects of the freshwater and ocean life of Atlantic salmon.

On the west coast, one of the major areas of investigation involved pink salmon and, in particular, why there was no overlapping or interbreeding between fish of alternate years. Large-scale field transplantation experiments were carried out. Attempts to establish runs in "off" years were unsuccessful.

In response to industry pressures to expand the herring fishery, the Pacific Biological Station initiated a more intensive investigation of British Columbia herring under the leadership of A.L. Tester. This study encompassed the delineation of herring populations, their abundance, fluctuations and response to fishing. Tester used differences in vertebral counts to establish the existence of about 12 stocks. Tester also attempted to determine the abundance and exploitation of these stocks from surveys of the quantity of spawn deposited each year.

Methodology became available to use internal metal tags and recover them with magnets in reduction plants. This made possible extensive tagging experiments. These experiments suggested that herring homed to specific sections of the coast but with significant 'wandering' among stocks. Thus the groundwork was laid for understanding the population dynamics of Pacific herring.

In 1937, the Biological Board became the Fisheries Research Board of Canada. A.T. Cameron chaired the new Board until 1947 and steered it through the Second World War years, fighting for its survival in the face of a restricted budget. By the end of the War the scope of the Board's mandate had been increased and new funding was available for new tasks.

On the west coast, studies were launched on salmon in the Skeena River system, to complement the study of Fraser River sockeye launched by the International Pacific Salmon Fisheries Commission in 1937. Population trends were monitored using counting fences. Tagging provided some information on migration patterns and exploitation rates. This study concluded that the commercial fishery was mainly responsible for the decline in sockeye salmon populations.

On the east coast, herring were a major underutilized resource. An Atlantic Herring Investigation Committee was established, involving the Atlantic Provinces and Newfoundland as well as the federal government. The age composition of samples collected by research vessels showed that exploitation was relatively light.

Lobster studies demonstrated the value of increasing the minimum size limit in certain areas to increase the catch of "market" lobsters.

For groundfish, information was being accumulated which would later be the basis for the initial regulations under ICNAF. The establishment of ICNAF in 1949 stimulated increased effort in assessing commercial fish species.

The Board's activities in the post-war years increased significantly in scope. Two factors contributed to this: increased demands from the fishing industry for information, and the need for accurate information to support Canada's position in international fisheries negotiations.

Alfred Needler, who served as Director at both Nanaimo and St. Andrews and later as Deputy Minister of Fisheries and Canada's chief international negotiator in fora such as ICNAF, summed up the situation (see Johnstone 1977, p.185):

> "I think you might say from the years 1945 to about 1960, and even to 1963 and 1969, research was the magic word in government finance. Research, on the whole, received more assistance than anything else. It was allowed a higher rate of increase.... "

Another significant event during this period was the addition of a Biological Station in St. John's to the Fisheries Research Board system. Fisheries research in Newfoundland had been initiated in 1931 when a laboratory was set up at Bay Bulls under Harold Thompson as Director. This laboratory was destroyed by fire in 1937. Staff moved to St. John's where they shared space with the Public Health and Analytical Chemistry Laboratories in a new Newfoundland government laboratory on Water Street. In 1944, Dr. Wilfred Templeman, who had in the 1930's been involved in lobster research at St. Andrews, became Director. He served until the various Biological Stations were integrated with the Department of Fisheries.

Templeman was active in groundfish research, utilizing the 82- foot research vessel the *Investigator II* to explore offshore fishing grounds in weather conditions for which it was never designed. He concentrated on mapping the distribution of potentially valuable species such as American plaice and redfish as well as the more familiar cod and haddock. He discovered new fishable concentrations and thereby assisted the development of an offshore Canadian fishery by trawlers.

Templeman also initiated oceanographic work by establishing an annual series of oceanographic observations across the Grand Bank. He focused much of the work of the St. John's Station in the areas necessary to provide the background information required for the initial discussion of regulations of groundfish in ICNAF. As an example, extensive experiments were conducted on the selectivity of various mesh sizes and types of trawls and trawl materials.

G.B. Reed, Chairman of the Board from 1947 to 1953, wanted to restore links with the universities and to encourage fundamental research. In his 1949 Annual Report, he wrote:

> "In the highest interests of fisheries research, it is necessary that fundamental research in

aquatic biology be adequately supported. Some of this should be done in the Board's Biological Stations but most of it will have to be done in universities. It is desirable too that university personnel be encouraged to undertake research in the Board's Stations. Unless fundamental research is more adequately encouraged, the quality of applied research will suffer" (Johnstone 1977, p.193).

In 1953, Dr. J.L. Kask was appointed the first full-time Chairman of the Board. Prior to Kask's chairmanship, the board was a loosely-knit group of stations run by Directors who worked relatively independently in geographic isolation. During his 10- year term he forged it into an integrated national organization. There were major changes in Board policy and significant expansion. The staff more than doubled during this period (Johnstone 1977).

The international commissions of interest to Canada such as ICNAF, the International Whaling Commission and the North Pacific Fur Seal Commission increased their activities considerably. The International North Pacific Fisheries Commission was established. This generated increased requirements for information on salmon migrations and intermingling on the high seas.

Kask had two aims as Chairman. One was to establish objectives for the Board's programs: "The purpose of the Fisheries Research Board of Canada, in my view, is to enhance the value of the Canadian fisheries through scientific research...one of my first jobs was to see that the objectives of each investigation were very clearly defined." Kask's second goal was administrative. He took measures to create a national organization and succeeded in curbing the powers of the individual Station directors.

Kask argued for a development branch in the Fisheries Research Board. Instead, an Industrial Development Branch (IDB) was created in the Department of Fisheries. Kask opposed this because he believed that research and development belonged together, to give research groups responsibility for fisheries development. Separating research and development set the two groups up as competitors, a trend which continued into the 1970's.

Another significant event during the Kask years was the acquisition of two large (177-foot) research ships modelled after commercial side trawlers. These ships, named the *A.T. Cameron* (1958) and the *G.B. Reed* (1963) after former Board Chairmen, added significantly to the Board's research capability in offshore waters.

By the early 1960's a large proportion of the Board's biological and oceanographic research was carried out to support the international fisheries and marine mammal commissions to which Canada belonged.

In 1963, the Royal Commission on Government Organization (The Glassco Commission) commented on scientific research and development activities in government (Glassco Commission 1963). Regarding fisheries, it noted that research activities were shared by the Fisheries Research Board and the Department of Fisheries. Research expenditures ($6.4 million and a staff of 218) represented about 25% of the total costs of the department (including the Board). It commented that "the work carried out is on the whole of high scientific quality, particularly in oceanography."

The Commission noted that scientific research and development in fisheries was conducted not only by the Fisheries Research Board but also by two units of the Department of Fisheries — the Fish Culture Development Branch (later known as the Resource Development Branch) of the Conservation and Development Service, and the Industrial Development Service.

The greatest proportion of the expenditures on these various programs was for Fisheries Research Board biological research. From 1952 to 1962, the Board's biological research budget had increased fivefold (Fig. 17-1).

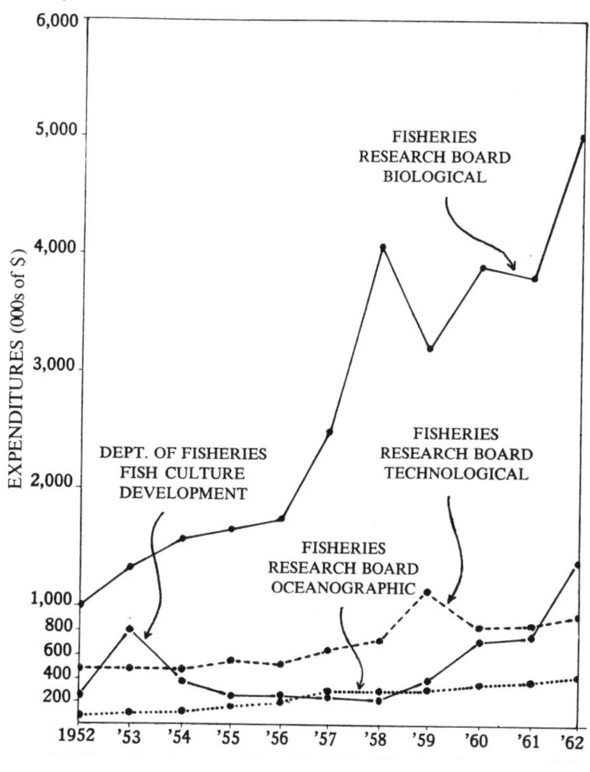

FIG. 17-1. Expenditures on fisheries research from 1952 to 1962 (from the Glassco Commission).

The Glassco Commission concluded that the wide geographic dispersal of research activities had brought about various organizational problems. Research projects developed in widely separated units of the Department of Fisheries were administered by its regional offices. This made coordination of the research program difficult. Furthermore, biological and technological units of the Fisheries Research Board were generally not located in the same geographic areas and, where they were, there was little coordination of work.

The Commission saw a similar lack of coordination in research planning, with little contact and discussion even at senior levels of administration.

It proposed combining within a single authority all research activities relating directly to Canadian fish resources. Specifically, it recommended that:

1. The areas of research assigned to branches of the Department of Fisheries and to the Fisheries Research Board be brought under a single Research Branch of the Department.
2. The Fisheries Research Board, renamed the Fisheries Research Advisory Board, assume an advisory role and maintain a continuing scrutiny of all programs of research.

These recommendations were not implemented immediately but hung over the Board like the sword of Damocles through the 1960's. Despite this, Dr. F.R. Hayes, who chaired the Board from 1964 to 1969, took the Board in directions which in some ways reinforced the concerns raised by the Glassco Commission. Hayes' inclination was to support more fundamental or basic research rather than the applied end of the spectrum. He took action to increase university participation in Board activities — by establishing a grants program to develop centers of excellence in aquatic science in Canadian universities, by encouraging universities to use Board facilities, and promoting graduate student and postdoctoral fellowships at Board stations.

Water pollution and, in particular, lake eutrophication, were becoming visible public issues. The Board's research was diversified and greater emphasis placed on experimental investigation and the ecosystem approach to study of aquatic life. Research in the areas of productivity, effects of heavy metals and organochlorine pesticides on aquatic organisms, fish diseases and the role of hormones in fish was encouraged. The former Atlantic and Pacific Oceanographic Groups were incorporated respectively into the newly established Marine Ecology Laboratory in Dartmouth (part of the Bedford Institute of Oceanography) and the Pacific Environmental Institute at West Vancouver. The Freshwater Institute was beefed up at Winnipeg (Johnstone 1977).

At these latter three locations, ecological research including oceanography and limnology was emphasized. Hayes had a certain disdain for the applied research at St. John's, St. Andrews and Nanaimo in support of domestic fisheries management and Canada's role in the international fisheries commissions (see Johnstone 1977, p.249). This is reflected in Hayes' comments on classical versus experimental science:

> "Templeman was doing the systematics of the fishes of the North Atlantic, and general population studies without a quantitative basis. This costs a great deal of money. You have to keep ships at sea. And I felt that the case history method of repetitive observation is of very, little value in science beyond a limited number of years.... I have felt by analogy with medicine and everything I know about elsewhere that you don't make progress by repetitive observations, except, of course, for immediate use. So I was interested in experimental work and getting it going and I was pushing for the kind of thing that Dickie (i.e. the Marine Ecology Lab) was doing" (Johnstone 1977, p.249).

Fortunately for Canada, the necessary background information on the fish stocks and the fisheries continued to be collected. These "repetitive observations" were to be the basis for the bold management interventions on both coasts in the 1970's.

Hayes' views were not shared by many senior managers at the Board. Dr. Bill Ricker, one of the Board's most distinguished scientists and interim Chairman in 1963–64, commented:

> "Ron Hayes and I differed in respect to the emphasis that should be given to systematic collection and interpretation of data. Fish populations are not static entities that can be studied thoroughly for a year or two and then you know how to manage them for all time. Each and every one of them changes continuously from year to year, in response to a variable environment and in response to the fishery. The more important a population is commercially, the more exposed it is to changes; in abundance, rate of growth, age composition and so on. As I see it one of the Board's main responsibilities is to analyze the state of our fish stocks on a continuing basis, and particularly to sound a clear warning if

overexploitation is threatened. Ron seemed to think that this was someone else's job, if it needed doing at all, and that the Board should restrict itself to short-term experimental work. Fortunately the needs of the international commissions prevented any major erosion of Board involvement in this area" (Johnstone 1977, p.253).

Although the necessary fish population research was continued, Hayes' attitude was clear to those around him. While it may not have been the only or the major factor, it certainly contributed to the Board's demise. At the end of the 1960's expansion slowed down. The handwriting was on the wall for the Board.

The Board attempted to head off its impending demise. In a 1968 brief to the Senate Special Committee on Science Policy, the Board sought a broader mandate in the water resources area. However, there seemed to be little recognition of the need to expand conventional management-related research on species such as groundfish which were being intensively fished by foreign fleets and for which international scientific committees (e.g. STACRES) were already calling for more direct management action. The Board wanted to increase the emphasis on improving the efficiency of Canada's fishing fleet at a time when fisheries managers were beginning to take steps through limited entry to tackle the problem of harvesting overcapacity (Canada 1968). This suggests that the Board was out of tune with the Department's research needs.

The plea in this brief for an expanded mandate and increased resources was the Board's swan song. The Science Council of Canada undertook a study on

Canadian fisheries research vessel, A.T. Cameron, built in 1958.

Canadian fisheries research vessel, G.B. Reed, built in 1963.

Scientific Activities in Fisheries and Wildlife Resources, published in 1971 as Special Study No. 15 (Pimlott et al. 1971). The results of the study were reflected in an earlier Science Council Report (No. 9) titled, *This Land is their Land* (Science Council 1970).

Unlike the 1968 FRB brief, the Science Council Study emphasized the importance of groundfish research in the Atlantic area by the FRB Stations at St. John's and St. Andrews. It noted that recent studies by ICNAF showed that data available from many years of research were "barely adequate for even the roughest assessment of basic population parameters." The Study Team observed:

> "The uncertainty about the future of these fisheries makes it imperative that high priority be given to research aimed at providing adequate biological assessment of both exploited and unexploited groundfish stocks to ensure optimum utilization by Canada in the face of increasing foreign competition" (Pimlott et al. 1971).

The Study Team concluded that Canada had gained a wide reputation in fisheries science. It noted that the emphasis in fisheries research had shifted from background biology to population dynamics. Similarly, the demands of modern food technology had shifted research from the simple essentials of canning and preserving to the complex biochemistry of marine products. Multiple resource use had created needs for knowledge of mechanisms of natural productivity and the effects of a multitude of natural and unnatural factors on the physiology and behaviour of aquatic organisms. The scope of fisheries research had been broadened to reflect a diversity of species, environments, technologies and resource use problems.

It recognized that the development of a strong federal research organization, the Fisheries Research Board, had placed Canada among the leaders in biological research related to fisheries. However, there was a lack of balance between research, management and development. The Atlantic herring fishery was cited as an example of the inadequacies existing in research, management and development programs.

The Science Council itself in *This Land is Their Land* suggested that "a stable and healthy environment of high ecological quality, maintained over the long term, should be defined as a new national goal." It proposed creating an Environmental Council of Canada and a Department of Renewable Resources.

The Science Council proposed that a new Department of Renewable Resources should include the existing federal groups concerned with fisheries, forestry, parks and wildlife. To some extent, this proposal was implemented with the creation in the early 1970's of a new federal Department of the Environment, including at that time fisheries and forestry. In 1979, fragmentation occurred again with the establishment of the new Department of Fisheries and Oceans. In 1990, a separate Department of Forestry was established.

Regarding fisheries research, the most significant recommendation in this Science Council Report was for the integration of research, management and development within fisheries. The Council observed:

> "The relationship between the Fisheries Research Board and the Services of the Department of Fisheries is not as effective as it could be; there are problems where overlapping functions have been difficult to resolve. The development of a large biology group in the Resource Development Service, for example, is at least in part due to the fact that the Service considers that the Board does not provide the research it needs to support its activities. Similarly, it is apparent that there are areas where much can be done to improve co-ordination between activities of the Fisheries Research Board and the Industrial Development Service."

By now the momentum for change was growing. Johnstone (1977, p. 269) noted:

> "Hayes must have known that time was running out for the Board. The Glassco Commission recommendations were suspended over its head like an axe and needed only the weight of a strong parliamentary majority to have them implemented. This majority was to be provided in the election of 1968 which saw (Trudeaumania) sweep the country."

In 1969, Dr. J.R. Weir was appointed to succeed Hayes as Chairman of the Fisheries Research Board and also as adviser on renewable resources to the Minister of Fisheries and Forestry. Weir had been a consultant to the Glassco Commission and Director of the Government's recently established Science Secretariat. With the transition from a Department of Fisheries and Forestry to the Department of Environment in 1971, Weir also took on the responsibility of Assistant Deputy Minister of Fisheries.

The environmental issue loomed large during Weir's term as Chairman. He emphasized that the FRB was the agency chiefly concerned with providing the research and technical background to manage aquatic renewable resources.

However, 1972 was the last year for the Board as an operational entity actively managing the governments' fisheries research program. In 1973, it lost its direct control over research programs and facilities and became solely an advisory body. This marked a major change in the organizational structure for fisheries research in Canada. Before discussing the new arrangements which governed fisheries research for the next decade, it is worth noting some of the major accomplishments of the Board in research related to fisheries management. Despite the huge influence of individual approaches of Board Chairmen and the Station Directors, the necessary work was still being done. It was the Board that failed, not the individual scientists.

2.3 Management-related Research Achievements of the Fisheries Research Board

Ricker (1975) summarized the accomplishments of the Board over its 75 years of active research. He noted:

> "A major aspect of the Board's work over the years has been investigations to support or to improve the management of our fisheries. Ideally, this includes estimating the size of each stock, the fraction of it that is taken each year by Canada and by other nations, its rate of replenishment from growth and from new recruits, effects of utilization on the age and size of fish caught, variations in recruitment from year to year, and forecasts of abundance and of catches in the case of stocks that fluctuate rapidly. When fishing becomes intensive it is necessary to consider also the interactions between different stocks and different species either as competitors for food or as predator and prey. A further problem is to know how many adult fish are needed to obtain adequate recruitment, averaging out the years of favourable and unfavourable environmental conditions. To all of these aspects of population dynamics Board scientists have made major contributions."

Huntsman in 1918 had demonstrated how fishing reduces the average age and size of fish in a population and the abundance of the stock. Board scientists had demonstrated that the number of recruits added to a stock each year depends partly on climatic conditions in the water at time of spawning and during early life. Some progress had been made in identifying conditions favourable for each species, and hence making predictions of future catches "but much remains to be done." The Board had also investigated the relationship between the number of recruits each year and the size of the spawning stock. Studies had revealed a variety of recruitment patterns, including the discovery (by Ricker himself) that at intermediate stock sizes the absolute numbers of recruits can be much greater than when spawners are either abundant or scarce. It had been shown that, if a stock is reduced too much, recruitment could collapse.

Such analyses required basic information on growth, natural mortality, and fishing mortality. During the Board's first two decades methods for estimating growth were developed. Ricker noted that methods of estimating fish populations were developed or adapted for use in Canada by D.B. DeLury and others, particularly the "change-in-fishing-success" and "mark-and-recapture" methods.

Extensive tagging of cod, haddock, herring, flatfish and salmon had provided information on the abundance of many of the major stocks. Ricker had authored a handbook on problems and techniques in fish population dynamics which is used throughout the world.

On the Atlantic Coast, in the pre-ICNAF era research scientists had developed methods of ageing, improved understanding of life histories and methods of distinguishing between different stocks. Most of the research was carried out by scientists from Canada and the United States, with discussion of results occurring in the North American Council on Fishery Investigations. In addition to biological research, steps had been taken to collect basic data that would be required for management such as catch and fishing effort statistics.

During the 1950's, much of the Board's research effort was centered on the groundfish stocks following the formation of ICNAF. As noted by Regier and McCracken (1975), major efforts were made to improve the "tools of the trade" for both managers and scientists, i.e. landings, catches, sizes, catch per unit effort. Much of the effort sought to determine the distribution of various groundfish species. Research on pelagic fish had a lower priority. Most of the research had an exploratory or regulatory function. This began to establish the capability for management by species and stock.

During the 1960's, more direct mapping of the distribution of traditional groundfish species was made possible by the Board's acquisition of two new research vessels, the *A.T. Cameron* and the *E.E. Prince*. Under the umbrella of ICNAF, scientists intensified efforts to develop a thorough statistical series, including information on the species and quantities caught, where, when and how they were caught and the amount of fishing effort. Otter trawl surveys were launched to

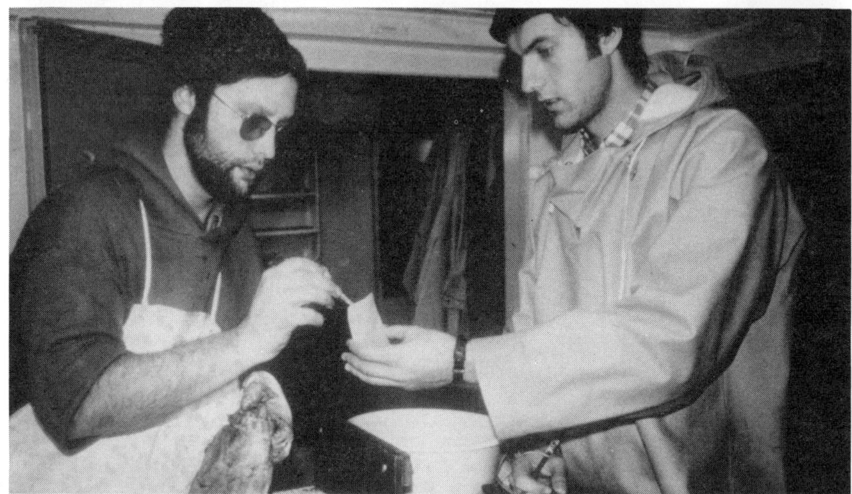

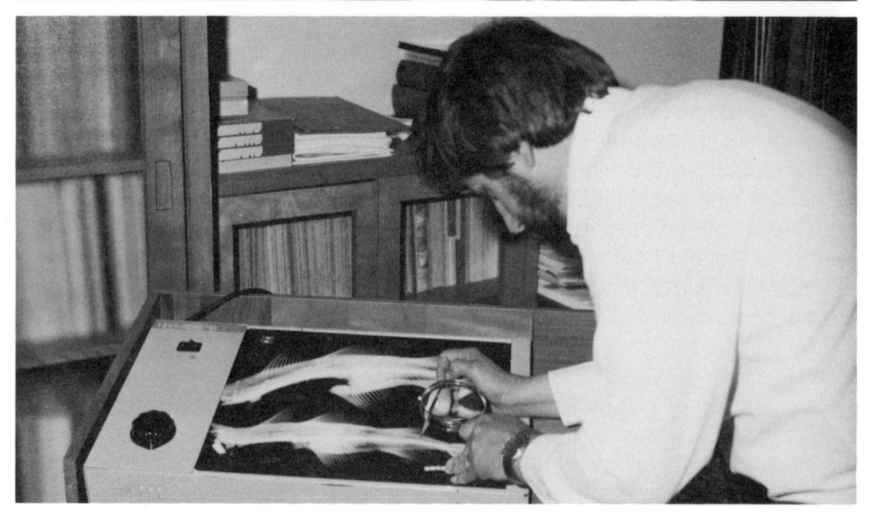

Scientific activities: Tagging Atlantic Cod (Top); Sampling fish for size, age and sexual maturity (Center); Scientist examining spinal x-rays of cod to determine stock origin (Bottom).

provide a measure of relative abundance from year to year and to provide some idea of pre-recruit abundance.

By 1970, Board scientists' investigations had laid the groundwork for ICNAF's establishment of Total Allowable Catches in the early 1970's (see Chapters 5 and 6). These quotas placed increasing demands upon the scientific community to estimate and predict stock abundance and provide management advice. Research vessel surveys began to make it possible to predict periods of abundance and scarcity in species such as haddock and redfish which exhibit considerable year-to-year variability in recruitment.

Tagging and morphometric and meristic analyses had established the migration patterns and stock divisions of Atlantic herring. Estimates of abundance were beginning to be used as the basis for catch quotas.

Tagging experiments had also showed that the rate of commercial utilization of salmon larger than grilse was 75% or more before the fishery at West Greenland commenced, with anglers catching about 50% of the remainder. The number of spawners needed in major streams had been defined.

For Atlantic lobsters, the Board's first management-related work had established the best minimum size for maximum economic return in various areas. This had enabled the development of regulations for lath spacing on traps to permit small lobsters to escape (see Chapter 6).

On the Pacific coast, both for salmon and for herring, the Board had directed considerable effort to determining how many spawners were required to produce a maximum harvest. During the 1920's and 1930's, the Board's research on Pacific salmon centered on migration, artificial propagation and enhancement. Larkin (1979) reviewed the history of research in relation to the management of Pacific salmon. He divided research from the early 1900's to the mid-1970's into four phases. The first phase from 1900 to the 1930's — *Discovery* — established some of the simple life history facts of salmon biology. By the 1930's what Larkin termed the "standard religion" of Pacific salmon research and management was developed. This held that "the adequate husbandry of the Pacific Salmon as a resource requires :

1. research on the biology of salmon;
2. regulation of the fisheries to ensure sustained yield;
3. protection of the environment to ensure the means of production; and
4. enhancement of natural conditions to augment natural production and to mitigate the effects of imperfect regulation and protection."

Studies during this phase focused on obtaining basic qualitative life history information on Pacific salmon. As Larkin noted, they did not speculate on where salmon go when they go to sea, nor did they develop any notion of population dynamics that could be used as a basis for management.

The second phase of Pacific salmon research from 1940 to the mid 1950's — *Gaining Understanding* — involved getting a handle on "the arithmetic of salmon abundance." Various researchers from Nanaimo, including Clemens, Forester, Ricker, Pritchard and Neave, studied the freshwater aspects of sockeye, pink and coho salmon. Larkin observed:

> "By the early 1950's it was clear that though the freshwater part of the life cycle had an indispensable role to play in determining production, it was not the sole determinant of abundance. Marine conditions were also at work in varying survival from one generation to the next."

Meanwhile, investigators tried, from the mass of statistical information that had been accumulated by sampling catches and estimating (by eye) escapement to spawning grounds, to develop some systematic basis for management. Ricker (1954), in his famous paper on stock and recruitment, brought some order out of these analyses and provided a theory of salmon population dynamics. During this period "management based on numerical assessments of catch and escapement of each race became firmly entrenched" (Larkin 1979). As mentioned earlier, fish culture had been essentially abandoned in the late 1930's.

The third phase from the mid 1950's to 1965 — which Larkin termed *Impacts from Outside* — was heavily influenced by events imposed from outside the natural fish production system. In particular, there was substantial pressure for hydroelectric development:

> "The 1950's were characterized by a major emphasis on research concerned with the engineering and biology associated with getting adult salmon up over dams, and getting young salmon back down again in fit condition to go to sea. A great deal was learned about the physiology and behaviour of salmon, the work of Brett and his associates at Nanaimo...being most notable" (Larkin 1979).

With the establishment of INPFC in 1954, an intensive investigation of the high seas migrations of Pacific salmon was undertaken. Understanding the ocean migrations of salmon was the highlight of salmon biology of the late 1950's and early 1960's. These scientific discoveries set the scene for attempts by Canada and the United States to limit Japanese interceptions of North American-origin salmon on the high seas (see Chapter 10). The discoveries also raised further questions: "What influences marine survival?" and "How do salmon navigate to find their way home?" Some progress was made on understanding the process of mortality and growth at sea but much was still unknown.

As Larkin put it, "Without question, the period from 1950 to 1965 was marked by major advances in understanding the biology of Pacific salmon." During the latter part of the 1960's, as demonstrated in earlier chapters, economics became an important factor influencing salmon management. This aspect is examined in Chapter 7 and 8.

Larkin termed the fourth phase of salmon research from 1965 to the late 1970's — *Inspiration*. The advent of computers led to computer-projected scenarios of various alternative courses of action. Economic factors, particularly fleet overcapacity, generated pressures for even more intensive management. Larkin observed: "Research on salmon population dynamics became typically to be more of the same, concerned largely with how to fine-tune an evidently cumbersome and antiquated system."

Meanwhile, environmental pressures on salmon habitat were growing. Increased fishing pressure and habitat deterioration led managers to turn again to fish culture techniques to provide a solution — produce more salmon. This was the genesis of the Salmonid Enhancement Program (SEP), which became a major component of Pacific salmon management and development in the 1970's and 1980's. Ricker (1975) described the considerable advances in fish culture techniques that had occurred since the 1950's.

Overall, on both the Atlantic and Pacific coasts Fisheries Research Board scientists had developed a considerable understanding of the biology and life history of the major commercial species. Through advances in population dynamics, they had laid the groundwork for the more interventionist management regime of the 1970's and 1980's.

2.4 Organizational Arrangements for Fisheries Research — 1973–1985

In December, 1972, Environment Minister Jack Davis announced a new organizational structure for the Department of Environment, which had been established in 1971. The new organization had two principal components: a Fisheries and Marine Services and an Environmental Service. The Board's research establishments were integrated into the new Fisheries and Marine Service organization. The new structure replaced the important research of the Board which had lasted for several decades. In July, 1974, the Deputy Minister of Environment, Robert Shaw, announced the details of the new organizational structure for the Fisheries and Marine Service (DOE 1974d). Administration of the line operations of the Fisheries and Marine Service were brought under two assistant deputy ministers: one responsible for Fisheries Management and the other for Ocean and Aquatic Affairs (OAA), later called Ocean Science and Surveys (Fig. 17-2).

The new ADM, Fisheries Management, was responsible for resource management and conservation and enforcement of fisheries regulations. Fisheries research was part of those responsibilities.

The ADM, Ocean and Aquatic Affairs, however, was responsible for physical and chemical oceanographic research, biological research related to the quality of the marine environment and environmental assessments of activities affecting freshwater and marine life, as well as hydrography.

The regional activities grouped under Fisheries Management were consolidated under five Regional Directors-General located at Vancouver, Winnipeg, Quebec City, Halifax and St. John's. These Regional Directors-General were responsible for fisheries research as well as fisheries operations. The Biological Stations at Nanaimo, St. Andrews, and St. John's, for example, reported to these Fisheries Management RDGs.

The Department's Annual Report for 1974–75 indicated that the fisheries research program (within Fisheries Management) covered two broad areas: renewable resources and environmental quality. The primary objective of the renewable resources research was "to possess the capacity for effective development and management of the fisheries resources, including all aspects of the primary fishing industry, and utilization by the secondary industry" (DOE 1975i). The environmental quality programs involved developing the capability to maintain or restore the aquatic environment to "acceptable conditions to permit it to support desirable life forms." This reflected the growing public concern over damage to fish habitat caused by other resource industries (DOE 1975i).

By 1975–76 the former research activities of the Fisheries Research Board and the Resource Development and the Industrial Development Branches of the former Fisheries Department had been consolidated. This consolidation occurred more smoothly and completely on the Atlantic than on the Pacific coast. There was a new functional structure at Headquarters and in the Regions. Fisheries Management and Research activities were organized under three major categories — Resource Services, Fishing Services and Industry

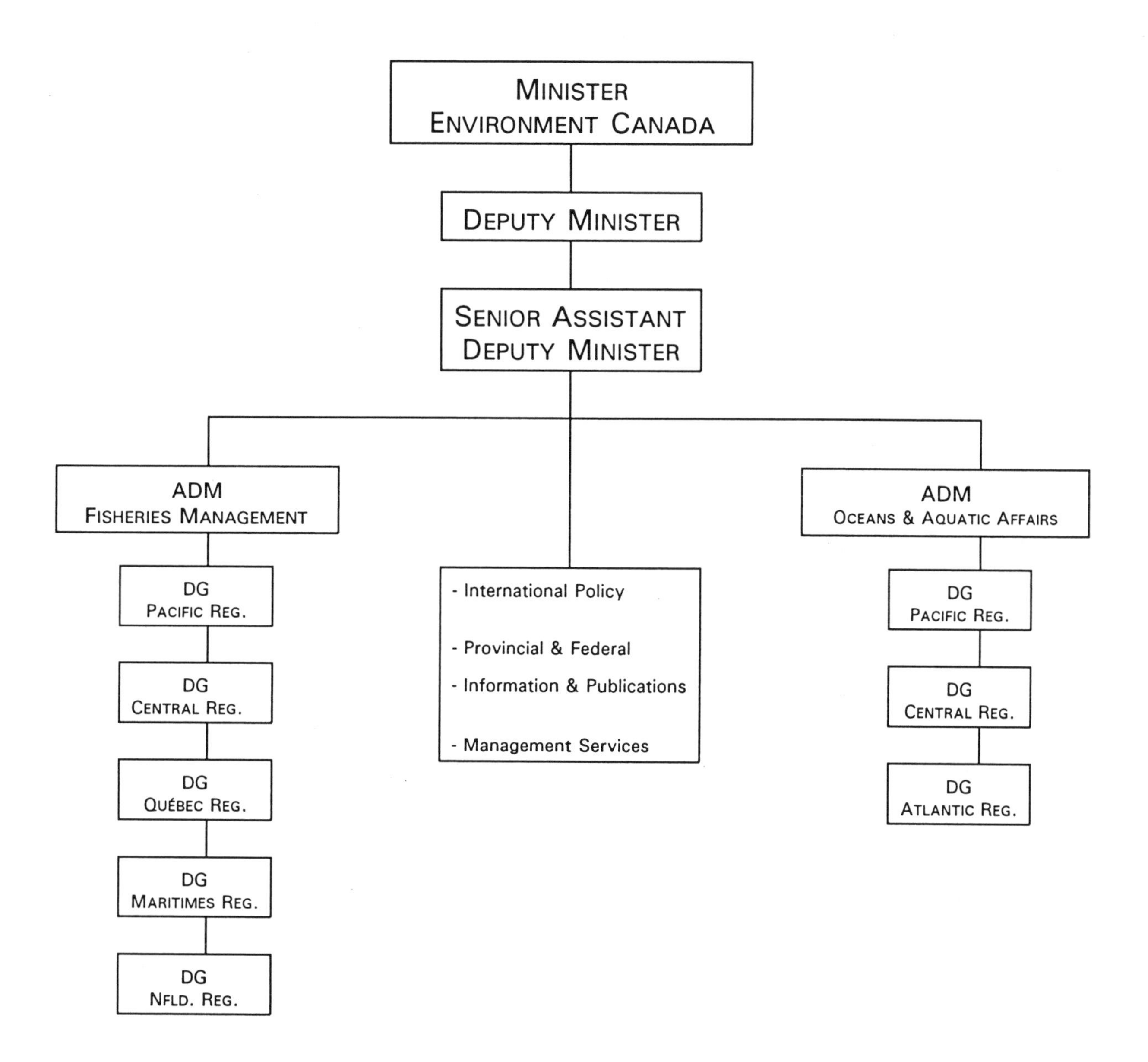

FIG. 17-2. Organization of the Fisheries and Marine Service in 1974.

Services. The biological work of the former FRB Biological Stations and the Resource Development Branch had been brought together under the new organizational grouping of Resource Services. Resource Services activities were directed "toward improving the management and sustained economic use of Canada's marine and aquatic renewable resources *compatible* with a concern for the quality of the environment."

The 1975–76 Annual Report emphasized the importance of scientific advice for management of the marine resources on the Atlantic and Pacific coasts. Scientists provided background information and advice to a number of national and international bodies for establishment of quotas and other management measures for various fisheries (see Chapter 10) (DOE 1976l).

2.4.1 Preparing for the 200-mile Zone

A major component of the Canadian response to the 200-mile limit was to strengthen the Canadian capability for fisheries resource assessment research. In order to rebuild the resource, it was recognized that a more sophisticated Canadian capability was required for estimation and prediction of fish stock abundance. The role of resource assessment was now the responsibility of Canada as the coastal state. The failure of many international commissions to manage the fisheries effectively was partly the result of the failure of member countries to devote significant efforts to assessing the resource. Analyses revealed that the level of sampling of the Atlantic offshore commercial fish catch in the mid-1970's was only 40% of that required to meet the minimum level of sampling identified by ICNAF's Scientific committees (DOE 1976i). Similarly, resource survey activity had been inadequate, far less than the minimum required. The data presented by certain countries was suspect and appeared to have been manipulated by those countries to suit a particular end.

The DOE Task Group on Extended Fisheries Jurisdiction concluded that Canada as the coastal state must develop an independent capability to assess adequately the fish resources off its coasts without having to rely upon information or analyses of other countries. To achieve this, it was necessary to increase significantly the levels of scientific activities aimed at assessing the resource potential, monitoring resource fluctuations and predicting trends in resource abundance.

In 1976, considerable scientific effort was devoted to the preparations for an orderly transition to the 200-mile fishing zone in January 1977. This included identifying the research that would be required to manage the new zone effectively, acquiring the necessary additional funding, and hiring new staff.

Proposals were made to double commercial catch sampling capability. It was also estimated that 1,500 ship days were required for various kinds of resource survey activity on the Atlantic coast. The Department at that time operated only two research trawlers capable of offshore fishing, neither of which was ice-strengthened. Overall, it was calculated that the equivalent of an additional five vessel years were required: (1) to conduct the required resource inventory surveys; (2) to conduct surveys with specialized gear to determine the abundance of precommercial-age fish for fish species which recruit to the fishery at an early age but whose young are not taken in regular inventory surveys; (3) to conduct egg and larval surveys to provide: (a) a basis for describing the early life history of several major fish species and enable the construction of population models combining early life history stages with adult production, (b) and insights on stock recruitment relationships; and (4) to conduct specialized surveys to evaluate and improve existing survey techniques and research methodology (DOE 1976i).

A need for laboratory experimentation was also identified. Recent initiatives in multispecies fisheries modelling had produced hypotheses about the nature and functioning of fish population control mechanisms. It was necessary to test hypotheses derived from analyses of field observational data through laboratory experimentation under controlled conditions. This might elucidate important links between the basic mechanisms by which a fish community controls and stabilizes its growth, reproduction and survival, and population dynamics theory as applied by resource assessment scientists (DOE 1976i).

Additional resources were required for the processing of samples collected in the field, age determination, and ichthyo-zooplankton sorting, identification and quantification. None of these additional data would be of much value unless the results could be analyzed, interpreted and advice provided. About 29% of existing manpower was involved in analysis, interpretation and advice. The analytical capability required ranged from data summarization, plotting and updating of standard calculations to development of theory, methodology and analysis of the dynamics and interrelationships of fish stocks. Three main types of analytical activities were distinguished:

1. Distribution, life history and stock discrimination studies;
2. Single species stock assessments;
3. Multispecies fisheries modelling.

Only about half of the TACs imposed by ICNAF were derived from analytical single species stock assessments. The remainder were based on even cruder models or approaches. The Task Group proposed that the number of scientific person-years devoted to that activity should be doubled (DOE 1976i).

The existing single species stock assessment approach did not adequately reflect the ecological interaction between species as affected by fishing and the problem of incidental catches in mixed fisheries. The very simple single stock assessment models in use were considered useful for forecasting allowable catches a year or two in advance, given an early direct estimate of the abundance of recruiting year-classes, but were not satisfactory for longer term prediction. A major thrust in multispecies fishing modelling was required.

By 1979–80 the government had allocated 216 person-years and $23.0 million for the required additional research (W.G. Doubleday, DFO, Ottawa, pers. comm.).

This more than doubled Canadian resource assessment activities in the offshore area. A large proportion of this increase was used to expand Canadian direct-survey capability at sea. This included the long term charter of a 263-foot ice-strengthened offshore trawler, the *Gadus Atlantica*, capable of conducting research off Newfoundland-Labrador during the winter, and another large stern trawler, the *Lady Hammond*, capable of bottom and midwater trawling.

2.4.2 A New Department of Fisheries and Oceans

In December, 1978, the emerging Fisheries and Oceans Department was restructured. This was a prelude to the split of the Fisheries and Marine Service from the Department of the Environment and the establishment of a separate Department of Fisheries and Oceans in 1979. Associate Deputy Minister of Fisheries and Environment (later Deputy Minister of Fisheries and Oceans) Donald D. Tansley announced that the former functions of the ADM, Fisheries Management, would be carried out by three ADMs in the new Department. The new ADMs were responsible for Pacific and Freshwater Fisheries, Atlantic Fisheries and Fisheries Economic Development and Marketing (DOE, 1978g) (Fig. 17-3).

One of the consequences of the new organizational structure was that the line authority for the management of the Department's marine fisheries research program was split between the ADMs of Atlantic Fisheries and Pacific and Freshwater Fisheries. To overcome this fragmentation, functional leadership was provided by a restructured Resource Services Directorate, reporting to the ADM, Atlantic Fisheries. In practice, this proved difficult because research managers in the Pacific and Freshwater Fisheries organization were reluctant to

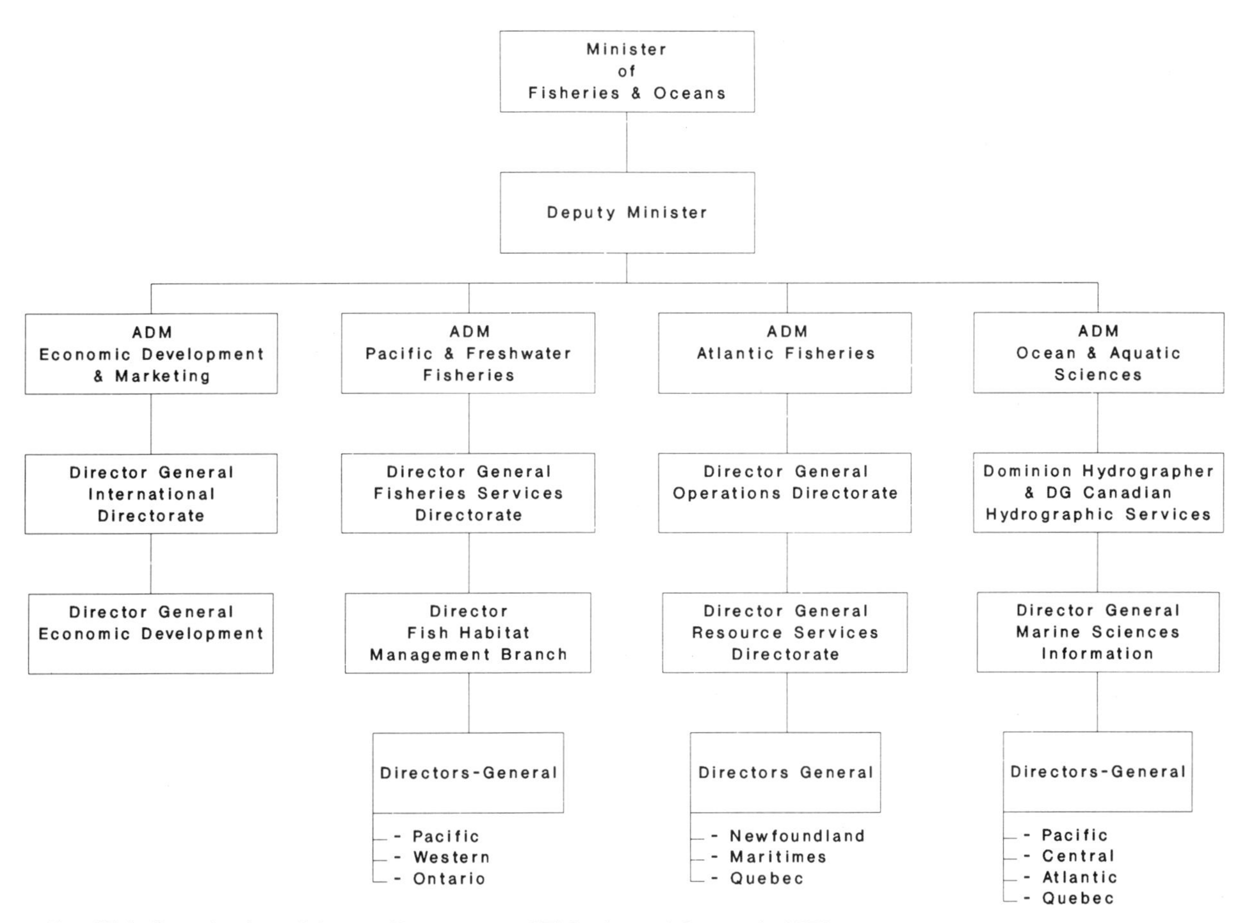

FIG. 17-3. Organization of the new Department of Fisheries and Oceans in 1979.

accept direction from a Director General under the ADM, Atlantic Fisheries.

The Department of Fisheries and Oceans was established on April 2, 1979, following proclamation of the *Department of Fisheries and Oceans Act*. Under the new Act, the duties, powers and functions of the Minister included:

(i) seacoast and inland fisheries;
(ii) fishing and recreational harbours;
(iii) hydrography and marine sciences; and
(iv) coordinating the policies and programs of the Government of Canada respecting oceans.

The stated objectives of the new Department were to ensure:

"— the comprehensive husbandry and management of Canada's fisheries resource base, through the protection, rehabilitation and enhancement of individual fish stocks and the aquatic habitat upon which these resources depend;
— the 'best use' of fisheries resources, through a variety of measures affecting when, where, how and by whom these resources are harvested, processed and marketed to obtain optimal social-economic benefits;
— an adequate hydrographic survey and chart production program to enable hydrographic charts and other publications to be produced for safe navigation in Canadian waters;
— the acquisition of the necessary knowledge base pertaining to oceanic processes and environments to support activities related to defence, marine transportation, the exploitation of offshore energy resources, and the management of the fishery resource and its aquatic habitat;
— the provision of a national ocean information system; the provision and administration of a national system of harbours in support of commercial fishing vessels and recreational boating" (DFO 1980h).

The objectives of fisheries research were subsumed in the phrase "the acquisition of the necessary knowledge base pertaining to...the management of the fishery resource and its aquatic habitat." This was linked to the acquisition of knowledge on "oceanic processes and environments" even though structurally fisheries research continued to be carried out separately from oceanographic research.

Despite the organizational fragmentation, the Resource Services Directorate launched some initiatives to provide a national focus for the Department's fisheries research program. In May, 1981, it produced a National Plan for Fisheries Research for the period 1981–1985 (DFO 1981n).

DFO fisheries research was organized along the following lines. A Research Director in each Fisheries Region reported directly to a Regional Director-General. Research conducted by regional establishments was directed primarily to the resources within the region managed by that Director-General. The headquarters Resource Services Directorate was responsible for "planning and recommending broad priorities and goals of major departmental fisheries research programs, evaluating programs, coordinating integration of activities between Regions and Services, representing fisheries management interests in government-wide science policy initiatives, and managing the development and functioning of mechanisms to integrate fisheries science with other elements of fisheries management". The line responsibility for the conduct of programs rested with the regional Research Directors.

The Fisheries Research programs were multidisciplinary. The staff involved a wide range of scientific disciplines, including population dynamics, ecology, bioenergetics, physiology, biochemistry, systematics, nutrition, microbiology, genetics, chemistry, limnology, oceanography and toxicology. The integration of disciplines was considered fundamental to develop a basic understanding of the aquatic ecosystem and to move towards multispecies management.

The Fisheries Research programs were subdivided into three categories:

— Resource Assessment and Related Research
— Aquaculture and Resource Development Research
— Habitat Assessment and Related Research

Extensive differences existed in the balance and scale of programs among regions. For instance, the largest marine finfish assessment program was on the Atlantic coast where annual assessments were being provided for some 50 stocks over the major continental shelf. On the Pacific coast, approximately 30 marine finfish stocks (excluding anadromous fish) were the subject of a much smaller research program on a narrow continental shelf. Assessment activities on the Pacific coast focused on the major salmon resources and on herring. Research to support enhancement was most developed on the Pacific coast for the Salmonid Enhancement Program initiated in 1977. Habitat-related research was most highly developed in the regions where major habitats were degraded or threatened by such manmade problems as eutrophication, energy-related development, logging, and mining, and acid rain (e.g. Pacific and Western Regions).

Resource assessment and related research was recognized as a major and essential input to fisheries management. The major components of resource assessment included:

1. Regular monitoring of resource stock sizes and yields, including:
   (a) periodic assessments of stocks (including surveys of abundance and composition of commercial catches)
   (b) continuous refinement of methodologies for monitoring and prediction;
2. Advising fisheries managers, including:
   (a) developing mechanisms to ensure objectivity and quality of advice;
   (b) ensuring effective mechanisms to integrate resource advice with economic data and analyses and with the consultative processes (see Chapter 15) which bring in the social and political dimensions;
   (c) generating and analyzing alternative management options and strategies.
3. Continued basic research on population ecology and fisheries ecosystems, including field and laboratory experimentation and systems modelling, to improve the overall basis for resource management (DFO 1981n).

These components of resource assessment continue to be essential elements among which resources must be balanced. The short term and advisory functions are essential to support legislation and regulations administered by the Department. The longer term research programs are just as essential to maintain and enhance the capacity for reliable short-term advice and to maximize long-term benefits from the fisheries system.

One interesting section in the 1981–85 Fisheries Research Plan dealt with the interaction between Fisheries Research and Ocean Science and Surveys. It suggested that Fisheries Research and OSS were linked at several levels. The Plan listed a number of committees and administrative arrangements to facilitate interaction. It glossed over the very real strains developing between the essentially mission-oriented Fisheries Research program and the largely separate oceanographic programs of Ocean Science and Surveys. The need to link environmental variability to changes in fish abundance and distribution was identified but no clearcut mechanisms were established to foster the necessary interaction.

In 1984 and 1985, fragmentation of the Department's scientific research efforts became a topic of considerable debate within the Department's senior executive. Ways were discussed of forging a closer working relationship between the fisheries research and oceanographic research programs of the Department. A Task Force headed by Mr. G. Duclos from the Office of the Comptroller General recommended in 1985 that all of the scientific research programs be combined under a single Assistant Deputy Minister of Science.

2.5 Organizational Arrangements for DFO Research — 1986–1992 — Science Integration

In November, 1985, the Honourable Tom Siddon was appointed Minister of Fisheries and Oceans. In late December, 1985, Dr. Peter Meyboom, who had served as Secretary to the Neilsen Task Force on Program Review, was appointed Deputy Minister. One of Dr. Meyboom's first tasks was to consider how best to meet the reduction targets for Fisheries and Oceans established in the Government's November 1984 Economic Agenda (Department of Finance 1984). The number of Fisheries and Oceans headquarters staff would be reduced by approximately 25%, with consolidation of certain departmental functions.

Part of this restructuring of DFO involved the creation of a new Science sector. On February 13, 1986, Minister Siddon announced that all fisheries research and oceans science programs would be consolidated under a new Assistant Deputy Minister for Science "to raise the profile and improve the focus of the Department's world-class science effort, to increase the coherence and visibility of the Department's science effort and to ensure the full integration of science with the daily operations of the Department and the government's priorities" (DFO 1986h).

Immediately thereafter, a series of studies of DFO's scientific research programs were initiated, to provide a basis for decisions on how to integrate the former Fisheries Research and Ocean Science and Surveys Organizations. These studies included reviews of:

— Pacific and Freshwater Fisheries Research
— Physical and Chemical Oceanography
— Marine Ecology and Biological Oceanography, and
— An Atlantic Fisheries Research review already underway.

These reviews included client consultations. For the Atlantic Fisheries Research review, for example, nearly fifty meetings were held in seventeen locations to consult with fisheries research clients.

Atlantic Fisheries Research

In the Atlantic Research review there was no consensus among clients on overall priorities or on how such priorities should be established. Most interest focused on resource assessment and related research. Representatives of the fishing industry called for more emphasis on assessing major stocks. Clients expected a high degree of accuracy in the provision of stock assessment advice. They were not satisfied with the current level of accuracy, nor with the communication of advice to clients (DFO 1986m).

Research on anadromous stocks was controversial. Representatives of recreational fishermen considered that every dollar invested in research and enhancement of Atlantic salmon was well spent. Commercial fishermen, on the other hand, considered that expenditures on Atlantic salmon were far out of proportion with those on commercially valuable species such as cod, lobster and crab.

There was widespread support for fisheries ecological research, particularly the study of seals' effects on fisheries through predation and parasites. Much interest was expressed in aquaculture and in science to support it.

The Atlantic review revealed that about two-thirds of Atlantic fisheries research funding was devoted to resource assessment and related research. Most of the remainder was allocated to aquaculture and resource development research, and a small amount (about 9%) supported fish habitat research. Within these categories, most effort and expenditure was devoted to the provision of information and advice to external clients, and the production of fish in the case of salmon enhancement. Targetted basic research was being carried out to improve the basis for future advice, but very little basic research without an identifiable future payoff was being conducted.

The balance of the program among subject areas and among regions reflected historical evolution. There was a regional concentration of aquaculture and resource development and habitat capabilities in the Scotia Fundy Region (about 60% and 55–60% respectively). In addition, departmental programs in the Ocean Science and Surveys area were focused in its Atlantic (Bedford Institute) and, to a lesser degree, Québec regions. There appeared to be a significant overlap of capabilities and functions between OSS and Fisheries Research regarding fisheries ecology, fish habitat and chemical contaminants research.

The resource assessment program was spread more evenly among Atlantic regions and showed a more even distribution of attention to species and stocks than would result from a direct correspondence to the commercial and recreational importance of the stocks. Funding for diadromous species and marine species such as silver hake and squid with limited Canadian fisheries at that time attracted criticism from clients in the fishing industry. The review identified a need to strengthen marine stock assessment capabilities in the Gulf Region.

The diadromous stock assessment activity was paradoxical in that it had substantial funding, but not enough to provide reliable assessments stock by stock due to the very large number of relatively small stocks (e.g. over 1,000 for Atlantic salmon). The non-vessel person-years devoted to salmon were almost equal to those for cod. Study of one stock or river was considered to have limited potential for extrapolation elsewhere.

During the client consultations, the fishing industry had strongly criticized the accuracy and reliability of the assessments of various major fish stocks. Review of historical stock assessments revealed substantial inaccuracies. Available funding would not permit a high level of service for all stocks. The study, therefore, suggested that the emphasis on resource assessment of major stocks be increased at the expense of minor ones, and that the capability to estimate stock abundance and forecast yields be improved.

Pacific, Freshwater and Arctic Research

The review of Pacific and Freshwater Fisheries Science (including the Arctic) also reached a number of conclusions which influenced the organization of and priorities for the new Science Sector. In the Pacific, the Department's research activities were scattered among three branches. The Fisheries Research Branch studied salmon habitat, fish populations and ecology, lake enrichment, fish culture and health, and various aspects of groundfish, shellfish and herring. The Field Services Branch, which managed fisheries and fish habitat, did stock assessment and monitoring (mostly salmon) and investigations of salmon habitat. Some research was also done by the Salmonid Enhancement Program to develop techniques and evaluate and improve operations. There appeared to be a heavy emphasis on the freshwater phases of salmon in the Region and insufficient investigation of the marine phases (DFO 1986n).

This diffusion of science roles in the Pacific Region could be traced back to the mid-1960's when the Fisheries Research Board under Chairman Hayes was placing greater emphasis on basic rather than applied research. Various fisheries developments and crises resulted in ad hoc science initiatives by the Department directed at specific fisheries management issues. To address this growing demand, a number of biologists were added to fisheries management staff.

These developments led to a lack of coordination and cooperation among the different science based units. Amalgamation of the Fisheries Research Board and the former Fisheries Department in the early 1970's did little to alter the balance of science roles in the research and operations branches, and the lack of coordination persisted.

The Pacific and Freshwater Fisheries Science review concluded that the highly dispersed responsibility for science activities resulted in gaps for which no one was clearly responsible, overlapping activities from lack of adequate coordination, and inadequate quality control of studies performed in isolation from DFO's mainstream science activity. Various options to address the weaknesses of the dispersed responsibilities were considered, but none were implemented.

This problem was finally addressed in 1993 when the Regional Director, Science, was given budgetary control over all resource assessment activities within the Pacific Region, even though the personnel remained split among the various Branches.

Regarding the Arctic, the review concluded that the Department's research mandate was clear. However, the state of knowledge was inadequate in many areas, and the demands for advice on the marine fish and mammal resources were increasing and becoming more specific. This contrasted with the Department's uncertain mandate and lack of direct management responsibility south of 60° north latitude. Despite this, the former Western Region had committed to the Arctic less than half of its A-base science person-years. Of the 62 research person-years devoted to the Arctic, only about 16 were involved in actual stock assessment work on anadromous fish, marine fish and marine mammals, with 18 person-years devoted to process oriented Arctic research. Overall, it appeared that DFO's capability in Arctic resource assessment research was inadequate to meet the demands for specific advice on the status of fish and marine mammal stocks. These demands were increasing with the shared management evolving as a result of native land claims (DFO 1986n).

DFO decided to approximately double the existing effort in Arctic marine resource assessment by an increase of 15 person- years. This would bring DFO closer to meeting its mandate and client needs.

Ecological and Biological Oceanographic Research

The review of Marine Ecology and Biological Oceanography concluded fisheries ecology studies utilized less than 4% of the department's scientific resources. Clients viewed this as inadequate. The review recommended that greater attention be given to the linkages between ichthyoplankton and zooplankton to improve long-term forecasting of species production. Also regarded as very important were accurate abundance estimates of year-classes not yet recruited to the fishery.

Clients were not satisfied with the scope of the Department's resource assessment research. While increased accuracy in catch projections was of critical concern, demands for broader and more comprehensive perspectives on fish stock distribution, abundance and variability were considered equally important. This included study of the reasons for in-season catch rate variations, explanation of past abundance trends, increased ecological understanding of predator/prey interactions, and the influence of oceanographic factors on changes in distribution, migratory patterns and recruitment mechanisms of marine fish populations (DFO 1986o).

Physical and Chemical Oceanographic Research

The review of Physical and Chemical Oceanography considered the requirements for physical oceanographic research for fisheries. It concluded that the linkages between resource assessment-related research and physical oceanography within DFO were weak. New initiatives were required to develop adequate linkages. More cooperative projects involving a multidisciplinary approach to tackle fisheries problems were proposed. The review also suggested that more attention be given to fisheries needs in planning for oceanographic programs (DFO 1986p).

A second issue was the balance between targetted basic, applied research and monitoring in physical and chemical oceanography. Interviews with departmental, federal and non- federal clients highlighted the need for a strong data collection, archival, interpretive and dissemination function. Approximately 5% of existing program effort was devoted to monitoring and related activities. The bulk of existing research effort in this program area was targetted basic rather than applied, with the ratio being approximately 75:25. The review proposed that the proportions of effort devoted to applied research be increased significantly. A more systematic approach was necessary in the design and application of monitoring systems (DFO 1986p).

Another major issue identified in the review was the need to consolidate contaminants chemistry within an integrated science organization. Contaminant chemistry was defined to include toxicology and biological effects chemistry. This definition encompassed the total departmental science activity dealing with contaminants and pollution impacts. Science activities resulting in advice on habitat issues existed in both the previous Fisheries Research and OSS organizational components, yet neither had given the necessary priority to this responsibility. Research relevant to habitat issues needed focus. Contaminant chemistry within the department was capital poor and spread among too many locations. There was no regional or departmental plan. Adequate monitoring was not being done, departmental clients were not being satisfied and the department's mandate was not being met. The review concluded that the various components should be consolidated into one activity.

A New Organizational Structure for DFO Science

Based on these reviews of the major research programs conducted by the former Fisheries Research and OSS organizations, options were developed to address deficiencies in existing programs and to integrate them into a cohesive and effective departmental program. Organizational proposals were based on the principle of organizing DFO Science programs along disciplinary lines and deploying staff in multidisciplinary teams to address important issues. This matrix approach to program management was considered to be the most effective means to integrate the variety of scientific expertise within the Department. This would enable strategic needs for specialized staff to be systematically defined and met. It would help maintain high professional standards while responding to changing priorities by establishing and disbanding project teams.

Accordingly, DFO decided to bring all Biological Sciences research together in one "Work Activity" and to combine Physical and Chemical Sciences, including contaminants chemistry, in another. The Hydrography Work Activity remained unchanged. The Chemical Science sub-activity included the responsibility to carry out research on chemical aspects of fish habitat, which was formerly shared between the Fisheries Research and Oceanographic Work Activities. Biological ecosystem and primary and secondary productivity research was brought under the new Biological Sciences Work Activity. These changes eliminated the former boundaries between Fisheries Research and Oceanography, to encourage greater interdependence and teamwork and foster integrated management of the Department's Science programs.

Science Integration Initiatives

New objectives were adopted for the Science Sector of DFO:

- To ensure that scientific information of high international standards is available to the Government of Canada for use in developing policies, regulations and legislation regarding the oceans and aquatic life, and to other government departments, private industry and the public for use in planning and carrying out aquatic activities;
- To provide and communicate a reliable scientific basis for the management of fisheries and fish habitat and for aquaculture;
- To acquire and communicate scientific information on the impact of deleterious substances on fish, fish habitat and aquatic ecosystems;
- To describe and understand the climate and processes of the ocean, influence on fish stocks and their interaction with the atmosphere;
- To describe and quantify marine environmental parameters relevant to marine engineering, transportation and other activities;
- To chart Canadian waters for the purpose of safe navigation;
- To facilitate fishing activities and to assist coastal and offshore development and other interests;
- To develop and refine methodology and technology necessary to carry out the Department's scientific role and to transfer relevant technology to Canadian industry to develop the private sector's capability;
- To facilitate and coordinate the government's marine science programs in collaboration with interested departments through the Interdepartmental Committee on Oceans (ICO) (Canada 1987b).

Resource reallocations were also made, shifting program resources from areas of low priority to high priority areas not being adequately addressed. Examples of lower priorities included resource assessment of squid and diadromous species on the Atlantic coast. High priority areas requiring additional funding included studies of Atlantic groundfish parasites, the predation of marine mammals on fish stocks, and resource assessment of Arctic marine species, as well as commercially important species such as lobster, cod and snow crab on the Atlantic coast. For Pacific salmon, increased emphasis on studies required to implement the new Canada-U.S. Pacific Salmon Treaty, and study of the marine phase of the life cycle were recommended. This latter priority resulted in a multidisciplinary project team, including physical oceanographers as well as fisheries research biologists, to study the marine aspects of salmon survival (Project MASS).

In keeping with the theme of reallocation from lower to higher priority areas, the original program plan for the newly established Maurice Lamontagne Institute in Mont Joli, Québec, was reexamined. Adjustments were made to reflect the general thrust of proposals for Science Programs in the Atlantic, to increase the emphasis on major species and stocks, and to ensure that research programs at MLI and the Gulf Fisheries Center in Moncton were complementary.

DFO consultations with the fishing industry had revealed very serious concerns about the spread and increasing severity of parasite infestations in groundfish and the potential impact of increased abundance of seals on the parasite problem and on the abundance of fish stocks. There appeared to be an adequate capability to assess the abundance of grey seals in the Scotia Fundy Region but estimates of grey seal abundance in the Gulf of St. Lawrence were less reliable. This species is the major primary host for the sealworm. Harp seals play a minor role in the transmission of groundfish parasites but were very abundant and increasing because of low harvest rates. Fishermen were concerned about harp seals' consumption of fish.

DFO concluded that the existing departmental response had to be augmented. To this end, the Department established a Centre of Disciplinary Expertise for parasitology at the Maurice Lamontagne Institute, involving 6 person-years, to study the biology and population dynamics of fish parasites of significance to the Atlantic commercial fisheries. In addition, units were set up in the Gulf, Scotia-Fundy, and Newfoundland regions to carry out a program of applied research related to parasite problems in these regions. This was later buttressed in 1988 by the addition of additional person-years for this purpose in the Scotia-Fundy Region and the funding of a collaborative multi-year program involving DFO, the fishing industry and Dalhousie University.

Overall, these initiatives redirected a significant portion of the new Science Sector's efforts.

The other major initiative was the establishment of seven national centres of Disciplinary Expertise in existing departmental facilities to address critical long-term research issues. The initial Centres of Disciplinary Expertise were:

- An Atlantic Resource Assessment and Survey Methodology Centre at the Northwest Atlantic Fisheries Centre in St. John's, Newfoundland;
- A Biological Oceanography Centre and a Marine Contaminants and Toxicology Centre at the Bedford Institute in Dartmouth, Nova Scotia;
- A Parasitology Centre at the Maurice Lamontagne Institute in Mont Joli, Québec;
- A Freshwater Fisheries Contaminants Centre at the Freshwater Institute in Winnipeg, Manitoba and the Bayfield Laboratory in Burlington, Ontario;
- A Centre on Genetics and Biotechnology for Aquaculture at the West Vancouver Laboratory in Vancouver, and at the Pacific Biological Station in Nanaimo, British Columbia; and
- An Ocean Climate Chemistry Research Centre at the Institute of Ocean Sciences in Sidney, British Columbia.

Overall, one of the more significant impacts of the Science Integration initiative was the decision to broaden the scope of and augment the Department's scientific effort in Newfoundland. This involved establishing the Resource Assessment CODE and adding physical oceanographic and hydrographic program components to that region, as well as bolstering the capability for fisheries ecological research. Altogether the Region's Science Sector was strengthened by adding approximately 55 scientists and support staff. This was done because the former Ocean Science and Surveys, Atlantic Region, which had an Atlantic-wide mandate, did not devote as much effort to addressing the problems of the Newfoundland fishery as was needed.

Within 3 years, these initiatives were bringing new understanding of the dynamics and abundance of northern cod, and the recognition that even more resources were required to address these issues to the extent required for rational fisheries management. Increased resources for the study of northern cod were provided through the Atlantic Fisheries Adjustment Program in May 1990 (see Chapter 13).

Minister Siddon announced the Science Integration and other Departmental Management initiatives on September 24, 1986 (DFO 1986i). Between February and September of 1986, the Department's management structure had been consolidated such that there were now only six Fisheries and Oceans Regions headed by Regional Directors General (Fig. 15-5). The Regional Director General would serve as the general manager and custodian of Departmental resources within the region. Reporting to each RDG was a Regional Director of Science responsible for all Science programs within a Region. The RDG was responsible for the delivery of all programs in accordance with approved plans and budgets. The Sector Heads (ADM, Science, ADM, Atlantic Fisheries and ADM, Pacific and Freshwater Fisheries and the Senior ADM, Corporate and Regulatory Management with respect to Inspection) were responsible for developing, planning, directing, and coordinating sector policies and programs. Sector Heads also had the authority to initiate and/or authorize adjustments to plans and budgets within the sector, including adjustments between or within work activities and/or regions within the sector.

To facilitate communication between the various Sectors and between Headquarters and the Regions, a National Science Committee was established, chaired by the ADM, Science, with the Regional Directors General, Headquarters Science Directors General and the Science Sector Director of Policy and Program Coordination as members. This committee would meet several times a year to direct Science Sector priority setting and planning. The ADMs of Atlantic Fisheries and Pacific and Freshwater Fisheries[1] (or their representatives) participate in the deliberations of the National Science Committee for priority setting where fisheries and habitat management needs are concerned. Reporting to the National Science Committee is a National Science Directors Committee composed of Regional Science Directors, Headquarters Science Directors General and the Director of Policy and Program Coordination. The National Science Directors Committee addresses issues in each of the three Science Sector Work Activities (Biological Sciences, Physical and Chemical Sciences and Hydrography), in three subcommittees chaired by the corresponding Headquarters Science Director General. Overall, this committee structure deals with the development and approval of policies, strategic plans, workplans, capital plans, vessel acquisition and deployment strategies, and decisions on organization and human resource planning issues for the Science Sector.

It was a long way from the first mobile station on the Atlantic coast, manned by visiting university staff, to a fully integrated Science effort within the Department of Fisheries and Oceans. The new Science Sector represented approximately 31% of the total 1987–88 department operating expenditures and 36% of the total departmental person-years. With the new structure in place, it was time to address the priorities and scientific challenges of the coming decades. With the new Centres of Expertise (in Resource Assessment and Survey Methodology, Parasitology, Marine Contaminants and Toxicology, Freshwater Fisheries Contaminants, Biological Oceanography and Ocean Climate Chemistry Research, and Aquaculture and Biotechnology), the Department was positioned to meet emerging priorities. Priorities included resource assessment (improving methodologies for abundance estimation and understanding recruitment mechanisms and multispecies interactions), habitat and environmental concerns, the role of the oceans in climatic change, and the needs of the emerging aquaculture industry. These scientific challenges are addressed in Chapters 16 and 18.

[1] In October 1991 a new position of ADM, Fisheries Operations, was established bringing together fisheries operations nationally under one ADM. The Science Sector structure in December 1992 was largely the same as that implemented in 1986.

Before turning to the examination of emerging scientific challenges related to fisheries management, I will first consider the evolution of the process for the provision of scientific advice to resource managers. Demands on resource assessment scientists had increased enormously since the mid-1960's as fisheries management interventions became increasingly complex. The systems for transforming data collection and scientific analyses into advice on alternative management measures evolved and adapted considerably between 1965 and 1992.

3.0 THE SCIENTIFIC ADVISORY PROCESS

3.1 International Fisheries Scientific Advisory Bodies

3.1.1 General

At the beginning of the 20th century ICES began to lay the foundations for fisheries science which would later support fisheries management. Early on, however, it concluded that the high variability of catches "precluded the possibility of any reliable combination of the trawling records" (Garstang 1904). From the conclusion that catches were too variable flowed a second conclusion that management was impossible. Cushing (1974) pointed out that this led to a change in the name of the original "overfishing" committee of ICES to that of "investigating the biology of the Pleuronectidae and other trawl-caught fish." ICES did not again consider the problem of overfishing until after World War I. Intergovernmental action did not blossom until the Overfishing Convention of 1946 and the subsequent establishment in the Northwest Atlantic of ICNAF (1949) and in the Northeast Atlantic of the Permanent Commission in 1953, later to become the Northeast Atlantic Fisheries Commission (NEAFC).

The establishment of these commissions, and other international fisheries commissions, for example in the North Pacific, led to formal processes for scientific advice to these management bodies. If these international bodies were to become involved in the management of fisheries as provided in their Conventions, scientific advice would be a key requirement. The scientific advisory processes differed among Commissions.

Over time three different mechanisms evolved for the conduct of research and the provision of scientific advice to these international commissions:

1. Use of an independent research staff as part of the functions of the commission;
2. Creation of a committee of scientists drawn primarily from the scientific ranks of member nations; and

3. Naming a completely independent group of scientists as the body for the provision of scientific advice.

The first two mechanisms predominated in the North Pacific (Miles et al. 1982). In the North Atlantic, ICNAF took the second approach while NEAFC used ICES as its advisory body. This latter arrangement could be considered a variation of the second and third approaches.

The use of research staff as part of the functions of an international fisheries commission first occurred in the North Pacific in the International Pacific Halibut Commission. The International Pacific Salmon Fisheries Commission later used the same approach. Both commissions continued to operate this way until Canada and the United States extended fisheries jurisdiction in 1977. Under the revised Convention for the Halibut Commission which took effect in 1979, this arrangement for research and scientific advice was maintained. The International Pacific Salmon Fisheries Commission was replaced in 1985 with the negotiation of the new Canada-USA Pacific Salmon Salmon Treaty. This Treaty provided that the new Canada-USA Pacific Salmon Commission would use the services of bilateral scientific committees involving the nationals of the two countries. Miles et al. (1982) concluded that the research operations of the IPHC and the IPSFC were highly successful in providing the scientific basis for management.

In the North Pacific, the second approach to scientific research and the provision of advice was used in the International North Pacific Fisheries Commission and the International North Pacific Fur Seal Commission. Under these arrangements, the parties carried out national research and planned, coordinated, interpreted and reported on their work during the annual commission meetings.

Arguments have been advanced for and against the commission research staff approach. Miles et al. (1982) summarized these:

> "Within the specific management contexts involved, including the shared experiences of two neighbouring and culturally similar nations, the research staff has had undeniable advantages for the parties, as evidenced in part by their willingness to continue support for so many years. Such an arrangement facilitates the development and implementation of a sustained research program directed at meeting needs identified as important for realizing management goals. The independent staff also avoids the sometimes substantial transaction costs involved in arranging for the continual series of meetings of groups of scientists that are necessary to meet management needs.
>
> "But even in this favourable situation, the staff approach may have problems. It has been suggested that there is a tendency for research staffs to become ingrown and lose communication with other scientists. Staff attitudes about the correctness of their methodology and analyses may occasionally discourage further research and lead to discounting responsible external criticism."

When the Canada-USA Pacific Salmon arrangements were negotiated, the commission research staff approach was abandoned. This indicates some dissatisfaction had arisen over the years with this approach. I believe that the Commission research staff approach fosters insular thinking and analyses, and does not allow for broad peer review of the advice to management. Such peer review must be an integral element of any effective scientific advisory process.

The other type of arrangement in the North Pacific was exemplified in the International North Pacific Fisheries Commission. The original INPFC Agreement provided that scientific investigations were to be a key function of the Commission. The initial major research emphasis was placed on investigating the offshore distribution of salmon, and determining areas of intermingling of salmon of Asian and North American origin. Somewhat later, research effort switched to studies to determine whether particular stocks continued to qualify for abstention under the conditions established in the INPFC Convention (see Chapters 9 and 10). In the later years of the original agreement, the research activities were diversified to include groundfish in the Bering Sea and the Gulf of Alaska. Over the years the Commission also conducted research into oceanographic conditions in the North Pacific that might be related to the extent and variability of intermingling of salmon stocks.

The Commission's main research coordinating mechanism was its Committee on Biology and Research (CBR). This was divided into subcommittees dealing with different aspects of the salmon question, crabs, groundfish, oceanography and research planning. The CBR emphasized the need for coordination of research among the members and examined means to this end. The CBR does not seem to have evolved into a specific mechanism for scientific advice on management, to the extent witnessed in the North Atlantic during the 1960's, 1970's and 1980's.

3.1.2 International Council for the Exploration of the Sea (ICES)

Canada has been a member of ICES since 1967. The present membership of ICES includes 17 countries. Fourteen of these are coastal states in the Northeast Atlantic, seven are members of the EC. Although ICES on the surface appears to provide a forum to coordinate and promote scientific research in the North Atlantic, it is primarily a regional organization whose principal focus is on the Northeast Atlantic, the North Sea and the Baltic Sea.

ICES' objectives are:

— to promote, encourage, develop and coordinate marine research;
— to publish and otherwise disseminate results of research; and
— to provide scientific advice to member governments and regulatory commissions.

It is noteworthy that the formal objectives of ICES, as provided in Article I of its Convention, do not explicitly include scientific advice to international fisheries bodies or member countries. In practice, over the past two decades this advisory role has become one of the major components of the Council's activities.

Miles et al. (1982) summarized the functioning of ICES:

> "Scientific activities of the Council arise in its standing committees. A review of contributed scientific papers reveals a need for a concerted joint effort on a certain problem. The Committee decides to develop a project which is then designed by an ad-hoc working group. Members determine the extent of their interest and the nature and magnitude of their contribution by ships and personnel. A working group of the Standing Committee coordinates the development and implementation of the project and the analysis of its results. The final step may be a symposium and the publication of a symposium volume and/or an atlas of the results.... As a formality, committee recommendations are reviewed by the Consultative Committee and approved by the Delegates."

This is a very brief description of the way that ICES functions, but it depicts reasonably accurately the basic structure and process.

The advisory functions that ICES has assumed in recent decades are discharged differently. Initially, ICES provided scientific advice to NEAFC through its Liaison Committee whose chairman sat in the Commission meetings. The Liaison Committee was advised by a number of working groups, one for each stock or group of stocks. These groups met separately over a period of several weeks at the ICES headquarters in Copenhagen in winter or early spring.

With the proliferation of requests for advice on management, ICES instituted a more formal structure for stock assessments and advice on management in 1977. This involved establishing the Advisory Committee on Fisheries Management (ACFM). ACFM includes the chairmen of three standing committees (Demersal Fish, Pelagic Fish, and Baltic Fish) together with a number of scientists nominated by member countries (one each) and appointed by the Council. The Chairman of ACFM is proposed by the Consultative Committee to the Council for approval. On behalf of the Council, ACFM provides fisheries management advice directly to member governments and the EC on request, to NEAFC, the International Baltic Sea Fisheries Commission (IBSFC), and the North Atlantic Salmon Conservation Organization (NASCO). These three commissions formally recognize ICES as a statutory advisory body and contribute to its funding. In the late 1980's, Council approved an arrangement whereby ACFM could provide advice to the EC. In return the EC helps fund ICES.

In the late 1980's, there were some 25 working groups involved in assessment of various fish stocks. ACFM reviews the conclusions and recommendations of these working groups to formulate advice on stock abundance and alternative management measures for the appropriate body (e.g. NEAFC, NASCO, the EC or member governments).

Canada participates in ICES and in ACFM. Its main interest has been in ICES as a forum to discuss advances in fisheries science. The only advice that ICES provides for stocks of Canadian interest is in its capacity as the scientific advisor for NASCO. Occasionally, Canada has referred special questions to ICES, e.g. the status of harp seals in the Northwest Atlantic in the 1970's when the EC was under internal pressure to ban import of whitecoat pelts.

3.1.3 International Commission for the Northwest Atlantic (ICNAF)

ICNAF and NAFO have been the primary international commissions of interest to Canada in the Northwest Atlantic. Chapter 10 describes ICNAF's origins. Article VI of the International Convention for the Northwest Atlantic Fisheries (1949) provided that:

"The Commission shall be responsible in the field of scientific investigation for obtaining and collating the information necessary for maintaining those stocks of fish which support international fisheries in the Convention area."

The Commission was to achieve this through or in collaboration with agencies of the Contracting Governments or other public or private agencies and organizations, or, when necessary, independently (ICNAF 1974b).

Although the Commission was empowered to conduct independent scientific investigations, in fact its studies were done through Contracting Governments' agencies. Representatives of those scientific agencies met in committees established by ICNAF to plan such research, report on findings and advise the Commission.

The chief scientific committee of ICNAF was the Standing Committee on Research and Statistics (STACRES). The Commission's Rules of Procedures, adopted in June, 1969, stipulated that STACRES shall:

"(a) develop and recommend to the Commission such policies and procedures in the collection, compilation, analysis, and dissemination of fishery statistics as may be necessary to ensure that the Commission has available at all times complete, current, and equivalent statistics on fishery activities in the Convention area and adjacent waters;

"(b) shall keep under continuous review the research programs in progress in the Convention Area and adjacent waters, and shall develop and recommend to the Commission from time to time such changes in existing programs, or such new programs as may be deemed desirable; and

"(c) keep under review the state of exploited fish stocks and the effects of fishing on these and provide the Commission and Panels with regular assessments" (ICNAF 1974b).

STACRES became very active throughout the life of ICNAF. It developed a system for collecting statistics on the fisheries of the Northwest Atlantic. It planned coordinated research programs, and, particularly during the last decade of ICNAF's existence, advised ICNAF on fisheries management.

In the 1950's and 1960's, STACRES coordinated studies on mesh selectivity as input to the Commission's decisions on minimum mesh size regulations. In the mid-1960's (see Chapter 6), it warned that some stocks were becoming overfished and from 1969 to 1976 it advised on TACs for stocks throughout the convention area. It brought together respected fisheries scientists from many countries, including coastal states such as Canada and the United States and distant water fishing countries from the Eastern European Bloc and Western Europe. STACRES was chaired at various times by individuals who became well known international fisheries scientists, e.g. Wilfred Templeman, John Gulland, Basil Parrish and Arthur May.

During the period when catch quotas became the primary management tool (the early 1970's), most of the supporting analyses were carried out in the Assessments Subcommittee of STACRES. This committee functioned by peer review. Underlying data and assumptions were examined and challenged, and the advice to the Commission was the best available scientific opinion at that time. It used "state of the art" stock assessment methodologies. The scientists were there as representatives of their governments, and national interest was sometimes evident in the discussions. However, scientific integrity was usually dominant, and the advice offered was largely free of national bias. The distant-water nations had sufficient fishing interests in the Northwest Atlantic up to the mid-1970's to send their best fisheries scientists to participate in STACRES and advise national delegations during the Commission meetings. The scientific debate was of high calibre. However, the measures taken by ICNAF were too little too late. This was not due to scientific intransigence. It was more the fault of the limited data available and the models in use, combined with the gradual adaptation to a TAC regime and lack of adequate international enforcement. In fact, STACRES prodded the Commission to adopt progressively more stringent conservation regulations.

3.1.4 The Northwest Atlantic Fisheries Organization (NAFO)

With the extension of fisheries jurisdiction by Canada and the United States in 1977, it became necessary to negotiate a new international management regime for stocks beyond 200 miles and stocks straddling the 200-mile limit. This led to the formation of NAFO in 1979 (see Chapter 11). Regarding scientific advice, Canadian representatives involved in negotiating this successor organization to ICNAF briefly considered using ICES as the scientific advisor. They rejected this option because ICES seemed preoccupied with Northeast Atlantic matters and dominated by scientists whose countries were distant water states in the Northwest Atlantic.

The situation was complicated by the fact that Canada wanted the new scientific advisory body to provide scientific advice, on request, to the coastal states, not just to the new Fisheries Commission responsible for stocks beyond 200 miles. These considerations stimulated the creation of a three-headed organization, NAFO. This is comprised of two main bodies, the Fisheries Commission and the Scientific Council. The administrative activities of these two bodies are coordinated by a General Council (see Chapter 10).

Articles VI–VIII of the 1978 International Convention on Future Multilateral Cooperation in the Northwest Atlantic Fisheries establish the functioning of the Scientific Council (NAFO 1984b). The net effect of these provisions is that, for the Fisheries Commission, the Scientific Council functions much as STACRES did in ICNAF, providing advice on request. The difference is that, unlike STACRES, it is not a subsidiary body but is, according to the Convention, equal in status but with different functions. The relationship of the Scientific Council to coastal States is more complex. The Council provides advice to a coastal State only on request and within the terms of reference stipulated by the coastal State (see Article VII.2).

Initially, the Scientific Council functioned relatively smoothly providing advice on the stocks managed by the Fisheries Commission, usually in terms of the $F_{0.1}$ catch, following procedures established during the last days of ICNAF. Canada, as a coastal State, requested advice from the Scientific Council on those stocks within its 200-mile zone which involved a significant foreign fishery for quantities surplus to Canadian requirements (e.g. silver hake, roundnose grenadier, Greenland halibut). Initially, because there were some foreign allocations of northern cod, it also requested advice on the status of that stock and the catch at various levels of fishing mortality ($F_{0.1}$ and less) until 1986. At that time Canada excluded northern cod from the stocks referred to the NAFO Scientific Council for advice. (Canada subsequently requested the Scientific Council to review the status of northern cod in 1992 — see Chapter 7).

The relative harmony of the NAFO Scientific Council was shattered when the EC began to demand advice for a range of fishing mortality options. Canada rejected this demand (see Chapter 11). However, from this point on relations within the Scientific Council degenerated. The polarization between Canada and the EC in the Fisheries Commission spilled over into the Scientific Council, even to the point where the EC representative demanded votes on the scientific advice as well as procedural matters. There were attempts to stimulate the Scientific Council to provide, on its own initiative, advice on matters which the Fisheries Commission had explicitly chosen not to request advice.

Another significant factor which hampered the Council's effectiveness was certain countries' diminished participation in Council deliberations. Compared with the last days of STACRES, the quantity and quality of representation from non-coastal States decreased. The distant water fishing countries had significantly diminished access to fish allocations in the Northwest Atlantic (see Chapter 10) and, consequently, little incentive to send their best fisheries scientists to NAFO Scientific Council meetings. Instead, their attention turned to fisheries assessment issues within ACFM and its working groups. This meant that the NAFO Scientific Council failed to meet expectations of the drafters of the Convention. It continues to advise the Fisheries Commission and coastal States. But it is not the vigorous forum for the discussion and advancement of fisheries science that was originally envisaged.

3.2 New Domestic Scientific Advisory Mechanisms Following Extended Jurisdiction

3.2.1 General

While formal advisory structures were being developed in the international context, on the domestic front in the pre-extension era advice to managers generally came from individual scientists. Peer review had become an accepted and necessary operating procedure in the international fora, but the mechanisms were much more fluid for those coastal fisheries under Canadian management. Scientific advice on the management of particular fisheries could often pass directly from an individual scientist to fisheries managers without being peer reviewed internally. This meant that the quality of the advice depended heavily on the idiosyncrasies of particular individuals. There was no assurance that the scientific methodology being applied was the best available. Individual scientists would often consult colleagues but this was not required.

This situation was untenable as Canada moved to extend its fisheries jurisdiction over the major fish stocks of the continental shelf. For fisheries beyond 200 miles international scientific advisory bodies would continue to provide advice. Canadian managers needed a new mechanism for scientific advice on the management of fisheries within the extended zone.

3.2.2 The Canadian Atlantic Fisheries Scientific Advisory Committee (CAFSAC)

Canadian scientists had realized for several years that a formal, continuing mechanism was required for scientific advice on all fisheries management matters on the Atlantic coast. They envisaged the fundamental purpose as examining, debating and arriving at a consensus to advise fisheries managers on biological matters. In the year prior to extension of fisheries jurisdiction, the Canadian Atlantic scientific community took action to insure that a suitable mechanism was put in place. To this end it proposed that a scientific advisory body be established immediately (A. Pinhorn, DFO, St.John's, pers. comm.). It was envisaged that the new committee would function along the lines of STACRES. DOE established the Canadian Atlantic Fisheries Scientific Advisory Committee (CAFSAC) in January 1977. Under its Terms of Reference, CAFSAC is responsible for providing advice to the Atlantic Fisheries Management Committee on the management, including the full range of conservation measures taking into account economic objectives, of all stocks of interest or potential interest to Atlantic coast fishermen. This advice was to be provided in accordance with specific fisheries management objectives and strategies and published as a matter of routine (CAFSAC 1978).

The Steering Committee of CAFSAC was comprised of the Directors of Resource branches in the Maritimes, Newfoundland and Québec regions, the Director of the Fisheries Research Branch in Ottawa, chairmen of the Subcommittees and a few appointed expert members. The Steering Committee became the focal point of CAFSAC although the actual stock assessment work was done in various Subcommittees. It reviews the Subcommittee reports to ensure all relevant information is submitted to senior management for consideration. It was also "vested with the responsibility for identifying weak areas in the scientific data base and methodology used by the Subcommittee to reach conclusions." Scientific advice was submitted to senior fisheries managers by means of Advisory Documents. These Advisory Documents were made public so that clients and others were fully aware of the advice.

Initially, eight Subcommittees were established. These covered Pelagics; Groundfish; Invertebrates and Marine Plants; Marine Mammals; Marine Environment and Ecosystems; Anadromous, Catadromous and Freshwater Fisheries; Aquaculture; and Statistics, Sampling and Surveys. Three of the initial Subcommittee Chairmen later served as Chairmen of CAFSAC.

The initial Chairman's Report by Dr. B.S. Muir, in the first Annual Report of CAFSAC, gave the context in which CAFSAC was established, its modus operandi and suggested areas of future activity. The Chairman observed:

> "The key to the present and the future success of CAFSAC is that it functions as a peer group. On scientific matters, the collective judgement of the group takes precedence over individual scientists' considerations....
>
> "CAFSAC ensures: *peer review* — All assessments and analyses are reviewed in detail by other specialists and new methods, if more appropriate, are made readily available for reanalyses of data.
>
> "*Collective responsibility* — Biological advice on each stock is prepared by a group of specialists who have examined the analysis, re-analyzed it, tried new methods and made a judgement concerning adequacy of the data and the analysis. Group responsibility does not make the individual less responsible. It does increase the overall competence of all assessment work, ensures a uniformity of methodology and encourages new methods to be tried. Above all else, it ensures a high degree of scientific objectivity which is mandatory in the business of providing scientific advice.
> "*documentation* — The basic theory of fisheries science is reasonably old, but much of it remains untested theory or special cases developed for a particular species. Considerable effort is made by Canadian scientists to develop new methods of analysis and to bring new dimensions to the theory under the demanding tests of practical application. Careful documentation is essential for future reference and to facilitate incorporation of new information" (CAFSAC 1978).

These three pillars — peer review, collective responsibility and documentation — were CAFSAC's operating philosophy throughout its existence.

Over the next several years CAFSAC settled into the routine of advising on Total Allowable Catches and other management measures for all of the major fish and invertebrate stocks within the Canadian Atlantic fisheries zone. By its fifth year, its activities had grown to the extent that DFO managers decided that a full-time chairman was required. The demands on the time of Subcommittee chairpersons and members also increased significantly. CAFSAC produced a steady flow of Advisory Documents on the management of particular species and stocks which were discussed in the appropriate fisheries management advisory committee and formed the scientific basis for each year's management measures. For the groundfish stocks the advice was generally offered in terms of the $F_{0.1}$ catch. The implications of alternative management options were described on request.

For the first few years the CAFSAC advisory process worked relatively well and was reasonably well accepted by the industry despite cautious advice for several major groundfish stocks. But problems arose in the early 1980's for various Atlantic herring stocks. In 1981, CAFSAC warned of potential stock collapse for the southern Gulf of herring (CAFSAC 1982).

CAFSAC Steering Committee, Fifth Anniversary, 1982.

The CAFSAC Report for 1983 reported a sharp decline in the Atlantic salmon fishery in 1983 which led to the first 5-year Atlantic salmon management plan beginning in 1984 (see Chapter 7 and Lear 1993). The low abundance of herring in the southern Gulf was again a matter for concern. After several years of advice that herring catches in that area be kept to the lowest practical level, CAFSAC found it necessary to recommend closure of the spring spawner fishery in 1984. It informed fisheries managers that the fall spawners were also "at a low level of abundance relative to historical values" (CAFSAC 1984c).

This produced one of the first major crises of confidence in the credibility of the CAFSAC advice. Herring fishermen in the southern Gulf claimed that the stocks were much more abundant than CAFSAC had estimated. CAFSAC stuck to its assessment. However, TACs in 1985 and 1986 were set higher than CAFSAC advised. Ultimately, CAFSAC revised its assessment to recognize that the stocks had bottomed out and were beginning to recover.

During the first years of CAFSAC's existence the groundfish stocks had exhibited a general upward trend (see Chapter 10). The Chairman's Report for 1984 noted that the steady increase in reference catch levels of groundfish since 1977 seemed "to have paused or come to an end and some stocks, such as both cod stocks in the Gulf of St. Lawrence, now show slight declines in forecast yield, towards long-term expected levels. Haddock in Division 4X is experiencing a sharp decline following a period of above average catches" (CAFSAC 1985).

This marked the beginning of more difficult years for CAFSAC. Various sectors of the fishing industry began to question its credibility, often motivated by the fact that their fleet sector's allowable catch would be reduced if the CAFSAC advice were followed. The Chairman's Report for 1985 recognized a communication gap between CAFSAC scientists and fishermen:

> "The problem of adequate communication of the scientific advice is vexing. On the one hand there is the difficulty of oversimplification without adequate background information and on the other hand the danger of the use of complete but technical information to the point of obscurity" (CAFSAC 1986b).

The 1985 Report also recognized a problem in reconciling conflicting sources of information used in the assessment process:

> "The evidence from the experience of fishermen in one gear component is not always consistent with that of fishermen in another component or, apparently, with the advice provided by CAFSAC.... A working group of CAFSAC was set up and will meet in early 1986 to review the accuracy of assessments" (CAFSAC 1986b).

The Chairman's Report for 1986 gave concrete examples of the results of this problem of reconciling conflicting abundance indices. It also indicated that previous assessments for several groundfish stocks had been overly optimistic:

> "The Groundfish and Pelagic Subcommittees were particularly active during the year because of the need to examine further the status of certain stocks. This was on occasion because inherent contradictions in the available data necessitated additional data after the initial assessment, but more often was because a number of stocks have not only ceased the increase in abundance generally seen since extension of fisheries jurisdiction, but are now estimated to be declining or to *have been fished harder than had been intended*" (Emphasis mine) (CAFSAC 1987b).

In particular, cod stocks reviewed by CAFSAC were estimated to be less abundant, except for the stock in the southern Gulf of St. Lawrence. Catches in 1987 at the $F_{0.1}$ level implied a reduction in the TAC as a result of the conclusion that past fishing mortalities had been much higher than estimated previously. The Labrador-eastern Newfoundland (2J3KL) stock was "continuing to rebuild, but more slowly than had been envisaged." The traditional haddock stocks were "declining in response to heavy fishing and low recruitment" (CAFSAC 1987b).

The early 1980's crisis for herring stocks was now over. The various stocks were generally increasing with a particularly marked recovery of the spring spawning stocks along the east and south coasts of Newfoundland as the very strong year-class of 1982 had reached harvestable size. The stock complex in the southern Gulf had increased about eight fold since the beginning of the 1980's (CAFSAC 1987b).

Despite the recovery of herring stocks, signs for the groundfish stocks were disturbing. Of particular concern was the indication that past levels of fishing mortality for many stocks had been overestimated. To address this, DFO established a Centre of Expertise for Atlantic Resource Assessment and Survey Methodology (fully realizing that it would be some years before this initiative would bear fruit). Also, in 1986, DFO established a special Ad Hoc Working Group to:

1. Identify the existing strengths and weaknesses of the CAFSAC system.
2. Develop proposals to remedy any weaknesses identified.

This Working Group, chaired by A.T. Pinhorn of the Newfoundland Region, reported to the ADM, Science, in November, 1986. It identified certain perceived problems with CAFSAC, summarized the strengths and weaknesses of the CAFSAC-type system, analyzed how CAFSAC had performed in relation to its terms of reference, and proposed improvements (Anon 1986).

The Working Group stated:

> "There is renewed questioning by industry clients of CAFSAC's competence to accurately calculate TAC levels because of wide variations in advised catch levels for important stocks between years. There has also been the suggestion that CAFSAC advice is, in some way, tampered with by Departmental fishery managers prior to release. A long-standing complaint is that CAFSAC is remote from the fishery and that fishermen's practical observations of the fishery are not communicated to, and do not adequately influence, CAFSAC advice.
>
> "Departmental clients of CAFSAC, in broad measure, share the industry's doubts about CAFSAC competence to determine TAC levels, and all clients agree that there is a need for greater clarity in the advice given. Advisory Documents are not understandable. Fishery managers have consistently been dissatisfied with the time frames within which CAFSAC provides advice, and have had trouble on occasion in eliciting from CAFSAC the advice they perceived they needed."

On the positive side, the Group recognized that CAFSAC had provided a clearly-defined institution for scientific advice on fisheries management. It had spoken authoritatively with a single voice on such matters on the Atlantic coast. The potential dangers of regionalization of scientific opinion had been avoided. The CAFSAC system allowed all the relevant Departmental scientific expertise to be brought to bear on the advisory process. The Working Group concluded:

> "It has lived up to its three original basic principles of peer review, collective responsibility, and documentation. Peer review of stock assessments and other analyses is the core of CAFSAC work. In taking collective responsibility for advice, it has provided an environment for objective scientific evaluation and advice formulation."

The Working Group summarized CAFSAC's overall effectiveness:

> "CAFSAC has proven to be effective in serving the ongoing needs of fisheries management programmes for resource management advice. This is no small achievement. However, CAFSAC has not been a particularly progressive organization. It has not been able to integrate itself into the more general process of fisheries management and contribute to development of new fisheries management approaches. It has not been a major stimulant in the development of new scientific approaches to management, or had much direct impact on research conducted... The high degree of scientific association which it facilitates, however, must have a significant positive effect on scientific progress."

The Working Group proposed a number of actions to strengthen CAFSAC. It dealt with questions of scientific credibility, competence and communication at length. Most of the points made are dealt with in a later section (3.3 — The Credibility of Scientific Advice). Here I will address only a couple of the key recommendations.

The Working Group noted that large fluctuations in TACs will always occur because of either large fluctuations in recruitment for some short-lived species or a changed perception of stock status resulting from an improved data base or improved analytical techniques. The first step to solve this problem was for CAFSAC clients to "accept the fact that large fluctuations in calculated catch at $F_{0.1}$ will continue to occur." The next step was to deal with these fluctuations. One of the first concrete actions to address this was the adoption in 1986 of the so-called 50% rule for setting groundfish TACs in cases where there were large annual fluctuations in advised catch at $F_{0.1}$ (for details, see Chapter 7).

The Working Group recommended that CAFSAC revert to precautionary TAC advice when major changes in calculated $F_{0.1}$ catch levels occur, unless there is a high degree of certainty associated with the calculation.

In light of the conclusion that CAFSAC scientists were slow to adopt new scientific techniques, the Working Group suggested establishing an assessment methodology working group to develop new techniques or ideas for use in future assessments. For more general aspects of fisheries science, it suggested the NAFO Scientific Council organize more of the formal types of scientific consultation such as symposia.

In January, 1987, DFO adopted the vast majority of the Ad Hoc Working Groups proposals. Urgent action was required to address some of the perceived shortcomings of the existing system (see The Credibility and Scientific Advice, section 3.3).

Meanwhile, by mid-1987 CAFSAC's credibility was again under attack as inshore fishermen in Newfoundland questioned the accuracy of its assessment of northern cod. Despite scientific analyses which indicated that the stock was increasing, inshore catches had been on the decline since 1982. The Minister of Fisheries and Oceans appointed the Alverson Task Group, to review the status of this fishery. They essentially confirmed the CAFSAC assessment of stock status and the hypothesized reasons for the "failure" of the inshore fishery. CAFSAC analyses indicated that the stock had been increasing at a rate of 15% per year since 1977. The Alverson group indicated a rate of increase of about 13% per year (see Chapter 13 and Lear and Parsons 1993). Inshore fishermen remained unconvinced by these analyses.

The Auditor General for Canada's 1988 report examined the Resource Management process for the Atlantic fisheries, including the scientific input to the process. Regarding the scientific advisory process and CAFSAC, he observed:

> "[CAFSAC] provides a review and challenge of [scientific] advice designed to ensure that it is as good and stable as the state of the art in stock assessment will allow it to be. Estimating the number, age and size of the fish in the sea is a difficult task, and the estimates have relatively large and inescapable error bands. We are concerned that the scientific community has not always been successful in making resource managers and industry aware of the process behind developing these estimates and of the estimates' inherent limitations.... Canada has a good reputation for its management of stocks in the Atlantic, with most stocks either rebuilding or stable, and the fishing industry experiencing significant growth. The stock assessment process contains controls designed to ensure that the scientific advice on fish stocks is both consistent from year to year and reliable" (Auditor General 1988).

This pronouncement that the scientific advice was consistent and reliable was shattered less than a year later. In January, 1989, CAFSAC produced its assessment for northern cod indicating that stock abundance had been consistently overestimated in recent years. In particular, the calculated catch at $F_{0.1}$ in 1989 was less than half the TAC in effect for 1988 (125,000 tons versus 266,000 tons). Because of the importance of northern cod to the Atlantic fishing economy, this new assessment shocked the scientific community, the fishing industry and both levels of government. The Minister of Fisheries and Oceans established a Special Panel on Northern Cod, led by President Leslie Harris of Memorial University. The panel was to undertake an intensive review of the scientific basis for the management of this stock. Studies were also commissioned on how to deal with the social and economic impact of a reduced Total Allowable Catch (see Chapter 13 and Lear and Parsons 1993).

The results of the Harris Review and its implications for the scientific advisory process and future research are dealt with in Chapter 18.

3.2.3 The Pacific Fisheries Scientific Advisory Process (PSARC)

As described earlier, the applied research effort on the Pacific coast was more diffuse and uncoordinated. Ever since the introduction of herring quotas in the 1940's, there had been informal mechanisms for scientists to advise managers. The Pacific salmon fishery tended to be managed species by species, stock by stock, and by gear-type, with in-season estimates of the size of the returning spawning runs. Estimates of required escapement were based on scientific models developed by Ricker and others. Fishing was permitted each year for the surplus above the required spawning escapement. Fisheries on mixed stocks during migration complicated the picture.

The Committee on Biology and Research of INPFC provided a forum to discuss the problem of high seas interceptions of salmon. The IPHC staff advised the Halibut Commission management of halibut stocks. Prior to extension of jurisdiction the groundfish fishery on the Pacific coast (except for halibut) was of minor significance.

Because of the narrow continental shelf off Canada's Pacific coast, extended jurisdiction did not mark such a dramatic change in the requirements for scientific advice and the structures for channelling this advice to managers. Hence, DFO did not immediately recognize the need for a formal scientific advisory structure (like CAFSAC). This was compounded by the rivalries among the various applied research groups within the Pacific region. The biologists of the former Resource Development Branch and the Fisheries Research Board had not been integrated into one scientific research arm of the Department during the pre-extension period from 1973 to 1976.

Following the first five successful years of CAFSAC, senior fisheries research managers in Ottawa began to promote a CAFSAC- type mechanism for the stocks of the Pacific coast. In August of 1983 they proposed a Pacific Stock Assessment Review committee based on the CAFSAC model. In February, 1985, the Regional Director General announced a new stock assessment review process. The first series of reviews of groundfish, herring and shellfish stock assessments took place in the fall of 1985. There was no examination of salmon assessments in that year (Stocker 1987).

In 1986, with the establishment of an integrated national Science Sector within DFO, attention again focused on the need for a formal Pacific stock assessment peer review process. The 1986 Review of Pacific and Freshwater Fisheries Science noted that the review process initiated in 1985 was "up and running" only for herring, groundfish and some shellfish "in that a full stock assessment review encompassing all the regional biological information is completed and passed on to an identifiable next step for processing into management plans." The Review observed that "stock assessment work for salmon is still diffuse and much effort is required to develop a PSARC model for this species group" (DFO 1986n).

The review concluded that the PSARC model should be developed for salmon as quickly as possible. During 1986 the Regional Director General appointed Dr. Max Stocker as Chairman and instructed him to develop PSARC formally to improve stock assessment methodology and capability in the region, with special emphasis on salmon. In 1986, PSARC again reviewed groundfish, herring and shellfish stock assessments. Three salmon programs were also reviewed: the chinook "key stream" program, coho stock assessments, and Barkley Sound sockeye. The first five formal advisory documents resulting from these reviews are contained in the PSARC Annual Report for 1986 (Stocker 1987).

PSARC was formally launched in 1987. It was to review biological advice on the status and management of Pacific fisheries resources. The Committee was tasked with reviewing methodologies and criteria employed in the stock assessment process, and formulating and evaluating methodologies employed to establish management plans and to assess local fisheries. PSARC encompasses the stock assessment community of DFO in the Pacific Region responsible for biological advice to senior management in the Region. PSARC is administered by a Steering Committee, with a Chairman who reports to the Regional Director General. The Steering Committee of PSARC had 16 members (see Table 17-1). The technical work of PSARC is performed by subcommittees, organized by species or subject.

The Steering Committee reviews the Subcommittee reports to ensure all relevant information has been analyzed, and formulates appropriate biological advice on management questions. It also identifies weak areas

TABLE 17-1. Membership of the Steering Committee for the Pacific Stock Assessment Review Committee.

- — The Chairman
- — Past-Chairman
- — Biological Sciences Section Head for Salmon
- — Biological Sciences Section Head for Groundfish, Herring and Shellfish
- — Five Subcommittee Chairmen
- — Head of Biological Services, Fisheries Operations
- — Chief, Resource Allocation and Industry Liaison, Fisheries Operations
- — A representative of the regional Planning and Economics Branch
- — A representative of the Salmon Enhancement Plan
- — A representative from the Fisheries Research Branch in Ottawa

in the scientific database and methodology used by the Subcommittees and recommend improvements.

Five subcommittees were established initially:

- — Salmon Stock Assessment
- — Herring Stock Assessment
- — Groundfish Stock Assessment
- — Shellfish Stock Assessment
- — Stock Assessment Data Systems

Thus, to a large extent, PSARC followed the CAFSAC model. There was one exception, namely, the inclusion of some representatives from the Fisheries Operations Branches. Given the dispersion of applied research within the Pacific Region, this was a necessary feature.

Perhaps one of the best examples of how PSARC has been integrated into Pacific fisheries management is the case of Pacific herring. Stocker (1993) has described the herring management structure which consists of four interacting groups: (1) PSARC Herring Subcommittee, (2) Herring Management Working Group, (3) Herring Industry Advisory Board, and (4) In-season DFO Management Teams (Fig. 17-4). The following description is drawn from Stocker's paper.

The annual cycle begins with collection of data from the current fishery. Stock assessment analyses are undertaken and forecasts of potential catches made. The PSARC Herring Subcommittee reviews these analyses, identifies trends in the stock and determines the harvestable surplus. Based on this and other information, the Herring Management Working Group drafts fishing plans. The Herring Industry Advisory Board reviews the results of the previous fishing season, proposed management plans, and provides input to the final fishing plans. In-season management of the herring fishery includes assessment with echo sounding, test sets for roe content, tagging sometimes, and biological sampling for stock characteristics. In-season management controls include restrictions on fishing periods, enforcement of mesh sizes, and managing to quota for each gear type.

For stock assessment specifically, the PSARC Herring Stock Assessment Subcommittee's biological advice must be "consistent with sound conservation principles and a view to optimizing production of herring stocks." The chairmanship alternates every 2 years between members of the Biological Sciences Branch and the Fisheries Operations Branch. The Subcommittee consists of scientific staff from the Herring Section of the Pacific Biological Station, fisheries management biologists and Fisheries Officers from Districts, the Coordinator for the Herring Fishery, an economist from the Program Planning and Economics Branch, and a representative of the Fisheries Research Branch in Ottawa. Representatives from industry are also invited to participate as observers. The Subcommittee meets in early September to review stock assessment analyses, to discuss stock status and recommend TACs. The PSARC Steering Committee reviews the Subcommittee report which is by then forwarded to the Regional Fisheries Resource Management Committee as an advisory document for approval.

The scientific advisory process for salmon is more diffuse. The Salmon Stock Assessment Subcommittee of PSARC provides only a overview of trends. It focuses on certain key stocks and issues. The ongoing management of the many diverse stocks of the various species is complex. It relies heavily on "in-season" data collection and management. Perhaps one of the best descriptions of this process was provided by Sprout and Kadowaki (1987), who commented on the management of the Skeena River sockeye salmon fishery in northern British Columbia. They described the management of this sockeye fishery as an "intricate task complicated by biological uncertainties and multiple objectives." The Skeena sockeye run is composed of over 50 stocks, with two enhanced stocks producing most of the run.

The major components of the modern-day management process are:

1. developing concurrent biological, economic, social, and legal objectives;
2. forecasting run size and formulating fishing plans;
3. consulting with user groups;
4. finalizing fishing plans;
5. executing fishing plans; and
6. evaluating management impact on the resource (Fig. 17-5).

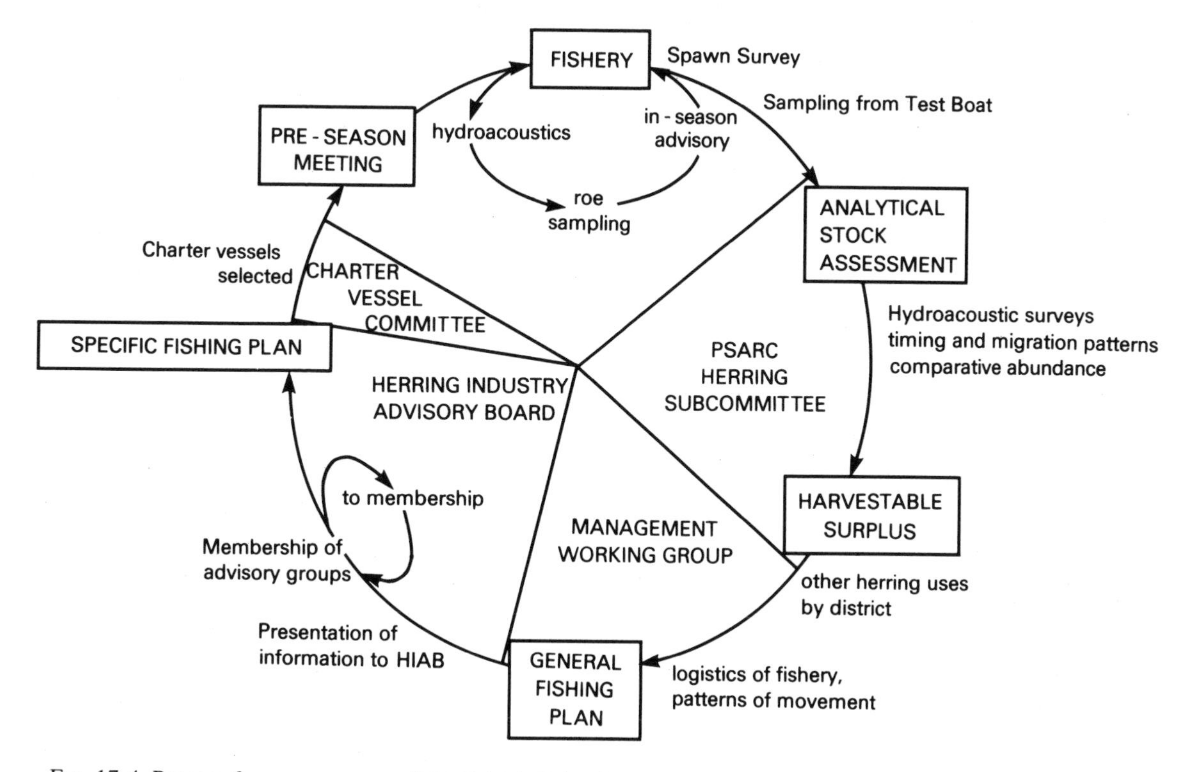

FIG. 17-4. Process for management of the British Columbia Herring Fishery (adapted from Stocker 1993).

One of the few explicit objectives for managing this fishery is an escapement target of 0.8 to 0.9 million spawners to the Skeena system. This target was derived from a stock-recruit relationship which suggested the maximum sustainable yield. The forecasts of adult returns to the Skeena River are developed using three methods : average rate of return per spawner, smolt-to-adult survival, and the age 3 to age 4, or age 4 to age 5 ratios. The average rate of return forecast is based on the average number of recruits per spawner for all complete brood years from 1940 to the present. The total spawners for a given brood year are multiplied by this average rate of return to forecast total return. These three methods often yield a wide variability between estimates. A best estimate is chosen by comparing the predictions using these forecasting methods with actual returns in past years.

The pre-season run forecasts by the biologists are used to formulate a fishing plan. This is done by calculating the number of days fishing each week using historical data to estimate the relationship between run size and effort. DFO discusses this fishing plan with user groups through the Skeena River Salmon Management Committee. Once the final fishing plan is determined, decisions are through "in-season management". This is the process for actions during the fishing season to meet the pre-season objectives. This iterative decision-making process includes monitoring catch and escapement and achieving weekly harvest rates by manipulating openings, fishing areas, and gear types.

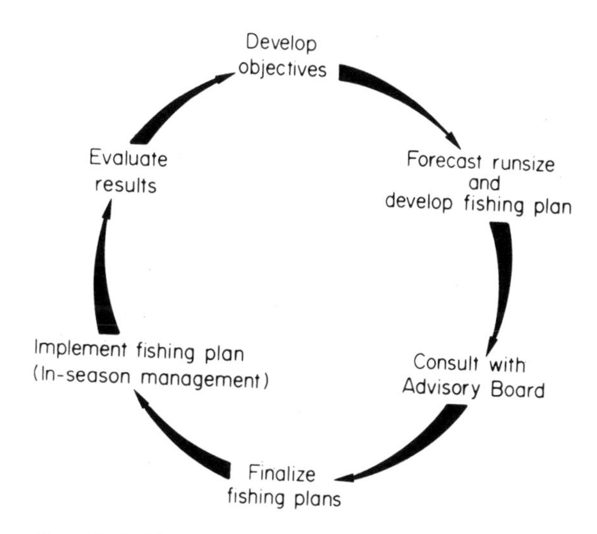

FIG. 17-5. The annual process for managing the Skeena River sockeye fishery (from Sprout and Kadowaki 1987).

The in-season management process has four components : (1) data collection, (2) run assessment, (3) decision-making, and (4) implementation.

A gillnet test fishery provides a daily in-season estimate of fish escaping the fishery into the Skeena River. Commercial catches are a second important source of data on run strength and timing. The test fishery and commercial catches are sampled for age, size and sex composition.

DFO monitors the runs during the season and, where required, modifies the fishing plans. Because the accuracy of the pre-season forecast is highly variable, once in-season information is available the pre-season estimate becomes progressively less important.

Sprout and Kadowaki (1987) noted that in-season decision- making is driven by the rate of fishery exploitation required to produce the desired spawning escapement. Measurement error and uncertainties in the relationships between indices and actual abundance lead to extensive discussion. Deciding when and how long to allow fishing is very difficult. Managers are usually faced with a variety of uncertain biological and non-biological data as a basis for decisions.

A group decision-making process is used on the Skeena to reconcile uncertainties and evaluate inputs. This involves daily meetings to decide the fishing pattern for the following week and determine if in-season adjustments to fishery plans are necessary. Table 17-2 shows the participants and their roles in the in-season decision-making process.

Sprout and Kadowaki (1987) described the process:

> "Prior to the meeting each participant decides how much weight should be assigned to specific factors. The fishery officer and management biologist outline their initial position as to the appropriate management decision and explain their rationale for the relative weight they have assigned to the available information. Colleagues discuss these positions and may offer views of their own. More discussion occurs if the positions and the emphases differ. Differences are likely to occur because: (1) regional objectives are open to differing interpretations, (2) there are contrasting perspectives on acceptable risk, (3) there are differing gear expectations, and so on. This discussion usually produces a decision supported by all participants. It is very much like a negotiation but rather than being adversative, it usually brings cooperating individuals to a shared decision."

This process for input of biological advice to in-season management of the Pacific salmon fisheries is quite different from the formalized CAFSAC preseason process used for Atlantic groundfish, for example, or even the mixed system used for Pacific herring. It lacks the peer review dimension but does have an element of collective responsibility. Decision-making is subjective rather than analytical.

A bilateral peer review process is used for Pacific salmon in the various scientific committees established to advise the Canada- USA Pacific Salmon Fisheries Commission. Article II.17 of the Pacific Salmon Treaty provides for a Standing Committee on Research and Statistics. The various Annexes also provide for several Joint Technical Committees. As of March, 1989, these were:

1. Joint Chinook Technical Committee
2. Joint Coho Technical Committee
3. Joint Chum Technical Committee
4. Joint Northern Boundary Technical Committee
5. Joint Transboundary Technical Committee
6. Joint Technical Committee on Data Sharing.

TABLE 17-2. Participants and their roles in the in-season decision-making process for Skeena River sockeye.

Participant	Role
1) District supervisor	— the senior officer in the District, acts as chairman
2) Sub-district officer	— fishery officer in charge of the fishing area, has the legal authority to regulate fisheries and is usually most familiar with the actual prosecution of the fishery. He provides data on current and historical catch and effort patterns.
3) Management biologist	— an advisor to the fishery officer, responsible for stock and fishery assessment. This involves the analysis of catch and escapement data and projections of the run to come.
4) Senior management biologist	— ensure consistency in stock and fishery assessment between fishing areas. Not always present at routine management meetings but is usually present for critical and very difficult decisions to lend his experience and expertise.
5) Area manager	— has final authority at the field level for all fishing plans. Attends crucial meetings but reviews all decisions before they are implemented to ensure consistency with objectives.

Source: Sprout and Kadowaki (1987).

The Standing Committee on Research and Statistics consists of two Commissioners and two technical advisors from each Party plus the Executive Secretary of the Pacific Salmon Commission or his designee. The primary functions of the Committee are to:

— develop and review research plans, data collection and analyses
— provide scientific advice on research and statistics.

This Committee serves many of the same functions as STACRES did in ICNAF, or the NAFO Scientific Council does, but on a bilateral rather than a multilateral basis. The difficult bilateral management issues involved ensure vigorous scientific review and debate of all analyses being brought forward for the Commission's consideration.

3.2.4 The Arctic Fisheries Scientific Advisory Process (AFSAC)

In a parallel initiative to the establishment of PSARC, in May, 1986, DFO established an Arctic Fisheries Scientific Advisory Committee (AFSAC). According to Clarke et al. (1989), AFSAC was intended to ensure biologically sound advice for fisheries management. Original terms of reference were established in July 1986 and modified later to reflect the revised departmental organization.

The stated objective of AFSAC was:

> "To provide sound biological advice to Fisheries Management (the Director-General and the Director of Fisheries Operations of the Central and Arctic Region) on Arctic fish and marine mammal stocks so that management decisions result in the long term well-being of these stocks" (Clarke et al. 1989).

The responsibilities of AFSAC are to:

1. assess the status of fish and marine mammal stocks and make "biologically sound recommendations" on harvest levels;
2. assess other biological aspects of management (such as size restrictions, closed seasons, mesh size, etc.) and make "biologically sound recommendations" on such operational requirements for management;
3. assess the adequacy of biological information available for making fisheries management decisions and make recommendations for its improvement;
4. review and assess biological aspects of general management plans for fish and marine mammals; and
5. respond to any other requests for biological advice on fisheries management and habitat issues (Clarke et al. 1989).

AFSAC consists of a Chairperson, an Executive Committee, a Fish Subcommittee, and a Marine Mammals Subcommittee. The peer review of scientific advice occurs in the Subcommittees. The AFSAC Executive Committee consists of the Chairperson of AFSAC, the Subcommittee Chairpersons and the Manager of the Regulatory and Native Affairs Division. Each Subcommittee consists of:

1. A chairperson selected from Regional staff;
2. Three research scientists or biologists from the Central and Arctic Region;
3. One expert from outside the DFO Central and Arctic Region; and
4. Additional experts on specific topics being considered, and the people involved in the preparation of background papers.

One noteworthy aspect of AFSAC's terms of reference is the explicit reference to marine mammals. This reflects the social, cultural and economic importance of marine mammals such as beluga, narwhal, bowhead whales, walrus and seals to residents of Canada's north.

Because of the geographic overlap of responsibilities within DFO and the migratory nature of several fish and marine mammal species, other jurisdictions or review processes (e.g. CAFSAC) are involved in developing scientific advice for the Arctic. Clarke (1993) provides a general description of Canada's Arctic fisheries and the management processes.

Clarke et al. (1989) described the activities of AFSAC during its first 2 years of operation (1986/87 and 1987/88). The main results were management recommendations and identification of research requirements for twelve fish stocks, three fish stock complexes and commercial quotas for anadromous Arctic charr, six marine mammal stocks and one marine mammal species.

Reviews during this period identified some problems common to management of many of the fish and marine mammal stocks. The problems result primarily from the size of the area, the scattered nature of the resource, different stocks having very different management problems, and an inadequate data base. Some general requirements identified included:

- better harvest statistics (e.g. species, numbers, size, data, location, biological samples), especially for "domestic" fisheries;
- collection of long-term data series so that techniques such as cohort analysis could be applied to improve stock estimates;
- development, testing and application of models appropriate for the management of Arctic fish and marine mammal stocks; and
- continued improvement of data on stock identity, stock size and vital parameters of stocks.

During the 1986 Science Integration review, the Department recognized that the Arctic marine resource assessment effort was inadequate. Steps were taken to bolster this applied research effort by doubling (approximately) the existing level of activity.

3.3 The Credibility of Scientific Advice

3.3.1 Fisheries Scientists and the Fishing Industry

Scientific assessment of fish stocks and advice on resource management have been subject to criticism ever since fisheries scientists began to produce estimates of abundance and the impact of fishing upon fish stocks. In particular, scientists' estimates often do not coincide with fishermen's anecdotal observations for particular stocks in particular locations. This has frequently resulted in denunciations by the fishing industry of scientific advice and, in particular, criticism of the failure of scientists to take fishermen's observations and knowledge into account in the resource assessment process. It is not uncommon for fishermen to suspect the validity of scientific methodology and to distrust the motives of the scientists in offering specific advice on management options.

Criticism by the fishing industry of the credibility of scientific advice depends on the perspective of vested interests of specific components. Offshore trawler companies criticized the scientific advice when non-Gulf based trawlers were excluded from the Gulf of St. Lawrence redfish fishery in 1976–77. During the early 1980's scientists advised that the stock was increasing because of improved recruitment. Gulf-based fishermen then criticized the advice as overly optimistic while, on the other hand, the representatives of the offshore trawler companies argued that the scientists had got it right this time. The results were similar in the most dramatic recent instance where the credibility of the scientific advice has come under assault: the 1989 revised assessment of the $F_{0.1}$ catch for northern cod. Inshore fishermen felt vindicated and argued that the scientists had finally realized what they had known for years. The offshore companies, on the other hand, argued that this cod stock was increasing and demanded that scientists come to sea on offshore trawlers to observe the high catch rates in the offshore fishery.

On the Atlantic coast, the questioning of the scientific advice began to intensify in the mid-1980's. While scientists were advising that the upsurge in groundfish abundance which had been witnessed since extension of jurisdiction was plateauing and, in some instances, stock abundance was declining, competition for the available resource intensified. Overcapacity in the harvesting and processing sectors stimulated both fishermen and processors to attack the scientific advice in an attempt to maintain or increase Total Allowable catches in the face of scientific advice to the contrary.

The 1986 Ad Hoc Working Group examining CAFSAC concluded:

> "A credibility problem will always occur with the fishing industry whenever CAFSAC provides a piece of advice that is contrary to the observations by the fishing industry or that prevents some sector of the industry from fully utilizing all of its fishing effort on an annual basis. This has happened in the past, is a large part of the credibility problem with the industry at present, and will continue in the future. While in some cases there is genuine doubt about the accuracy of advice, this is not always so. It has been a standard ploy the world over to try to discredit the scientific advice as a first defense against the implementation of rational fisheries management schemes" (Anon 1986).

The Working Group noted that one problem experienced in relations with industry clients is that advice provided by CAFSAC is normally well researched and peer reviewed. On the other hand, industry officials and fishermen often make contentions which are diametrically opposed to the findings of CAFSAC but rarely, if ever, supported by appropriate documentation. The Working Group urged that these clients be required to support their contentions with data that CAFSAC could analyze when these appear to conflict with CAFSAC analyses and advice.

One problem which has become evident to the author in recent years is the expectation developed over the years since extended jurisdiction that the resource could be stabilized. The 1976 Policy for Canada's Commercial Fisheries suggested that "the fishery economy of the future would be a vigorous and *stable one*" (DOE 1976a). The 1981 Policy for Canada's Atlantic Fisheries in the 1980's stated: "In the 1980's resource harvesting policies will aim at increasing the economic *viability and stability* of the Atlantic commercial fisheries" (DFO 1981a). It suggested that setting TACs at or below the $F_{0.1}$ reference level would provide for "stock rebuilding, larger average size of fish in the catch, improved catch rates and *greater stability* of catches." The first Annual Report of DFO for 1979–80 stated:

> "During the year under review the department continued to concentrate its efforts on *achieving stability* and increased viability in the fishing industry across Canada, taking optimum advantage of the opportunities created by the extension of fisheries jurisdiction by 200 miles" (DFO 1980h) (Emphasis mine).

The theme of fostering stability was a major Departmental objective in the decade following extension of jurisdiction. To some extent, the scientific community lent credence to this objective. For example, the document Resource Prospects for Canada's Atlantic Fisheries 1979–1985 (DOE 1978a) stated:

> "Fish stocks can be managed to give stable average catch rates over the long-term at various levels within the biological limits of the species, taking into account fishing costs and market prices. Within biological limits, the supply can be managed up or down in response to social and economic factors including market prospects."

DFO scientists were required to project trends in the Atlantic fish stocks. This requirement arose because of the pressure from the provinces and the fishing industry to expand the offshore fleet to take advantage of the anticipated growth resulting from the 200- mile limit. The original Resource Prospects documents were prepared to dampen these expectations and to demonstrate that the anticipated growth resulting from stock recovery could, for the most part, be harvested by the existing fleet. Resource projections included important caveats:

> "These projections...should be viewed only as a general guide to likely events. While 1979 predictions are based largely on formal calculations, and actual events should not differ widely from those predicted, projections of stock status in the 1980's *are to a considerable extent best guesses*. The precision of these estimates varies greatly depending on whether the assessment of the stock is based on known age composition, fishing mortality rates and predicted levels of recruitment, or on generalized production models relating overall landings and fishing effort, or on "best estimates from the scientists and managers concerned" based upon recent catch trends.... it must be borne in mind that the actual TAC for a particular stock in any year may differ widely from those projected here" (DOE 1978a).

These resource projections were updated annually and the caveats repeated. Unfortunately, there was a tendency both within the fishing industry and government circles to focus on the numbers and ignore the fact that these projections had their foundation in the shifting sands of natural variability. The result was that over the post-extension decade it was widely assumed that stability in TACs and catches could be achieved. Thus, the downturn in certain stocks in the latter part of the 1980's came as a major shock to the industry. The fishing industry, government administrators and, to some extent, the scientific community had forgotten that fish populations are subject to large fluctuations because of natural variability in the environment and the factors influencing year-class strength. In retrospect, it is clear that greater emphasis should have been placed on the importance of natural variability and continuing fluctuations in fish resources, despite the best human efforts to impose order through management interventions in this natural system.

The wide inter-annual variations in TACs which became pronounced for some stocks in the late 1980's resulted from two factors:

1. large changes in recruitment or growth rates due to natural variability, or
2. a changed perception of stock status resulting from improvements in the database or improved analytical techniques.

The credibility of the scientific advice has been questioned because of TAC fluctuations resulting from both factors. In the first instance, the scientific community and fisheries managers must hammer home the realization that fish resources will always be subject to natural variability and TACs can be expected to fluctuate because of this. Fluctuations in TACs for the second set of reasons, however, raises questions in the public mind about the competence of the scientists. It is not generally recognized that changed perceptions of stock status are an integral part of the progress of science. Collection of new data and development of new analytical techniques will often invalidate previous hypotheses. The accuracy of projections by Canadian advisory bodies such as CAFSAC is no worse than that of other advisory bodies such as ICES. As noted by the Ad Hoc Working Group on CAFSAC, "the predictive accuracy which can be attained is low in relation to the demand for accuracy by the present management system" (Anon 1986).

Clients of the scientific advisory process must recognize that large fluctuations in calculated catches at $F_{0.1}$ (or whatever other biological reference point is being used) will continue. The solution lies in devising ways to deal with these fluctuations through management planning. Steps need to be taken to improve our knowledge of the marine ecosystem, recruitment fluctuations, multispecies interactions and the influence of the environment on fish abundance. We must also improve the methods for estimating and projecting fish abundance. The nature of the required initiatives is discussed in Chapter 18.

There is another dimension to the problem of scientific credibility. Often there is a large communication gap between fishermen and scientists involved in resource assessment. The Ad Hoc Working Group on CAFSAC acknowledged the need to improve communication between these groups. It called for more readable scientific advisory documents and a broader range of communication vehicles. These documents should include clear, simple statements on the status of the stock and the changes in the fishery. In particular, they should clearly explain dramatic changes in advice from one year to the next and should always link this year's advice about a stock to that of previous years.

Around this time the need became clear to allow fishermen more input to the scientific advisory process, e.g. the Science Integration reviews. The industry wanted more involvement in the assessment process, complaining that DFO scientists knew little about the fishery. This called for action to better explain scientific programs to fishermen, and more frequent meetings with fishermen to obtain their views on the past fishing season before assessments were performed. The mechanisms for gathering pre- assessment information on the past fishing season and improving communication of the scientific results would vary from area to area, depending upon the nature of the fisheries. They might be stock specific or regional in scope.

DFO's Science Sector identified several avenues for increasing fishermen's involvement (B.S. Muir, DFO, Ottawa, pers. comm.):

1. *Tagging experiments.* Ways to improve participation would include: (1) increasing the reward to fishermen to ensure a higher return of tags; (2) increasing feedback to fishermen returning tags to encourage them to provide details on each return; and (3) improving communication on such experiments. The feedback already involves in some regions a personalized card showing on a map the location of the release and of the recapture, as well as some statistics on the total number of tags released and the number of returns to date.
2. *Monitoring fisheries performance through the log-books.* Participation could be improved by simplifying the information requested and by informing fishermen of the actual use of this data in the annual assessments.
3. *Monitoring the performance of index fishermen.* In certain fisheries (e.g. southern Gulf herring), some fishermen had been participating in a special program to monitor their fishing. The approach involves two components: the Log Book programs in which fishermen provide data collected from their fisheries, and SITE programs, which use small representative geographic sites to identify trends in larger geographic areas. Log Book programs have been used to derive abundance indices for Atlantic herring and gaspereau assessments while SITE programs have contributed most to Atlantic salmon assessments (Claytor et al. 1991). This approach was expanded to other fisheries.
4. *Special research and exploratory surveys.* In some instances, commercial vessels are chartered for special surveys. Because such surveys involve commercial fishermen with significant experience, the results which they generate tend to be more credible to the industry. Commercial charters have some drawbacks (calibration problems, scheduling difficulties and high cost, particularly during the peak of the fishing season) and cannot be used as a general means for surveying fish populations. However, they are an acceptable means for exploratory work and for certain specialized surveys.
5. *Inviting fishermen along on research vessels as observers.* This allows fishermen to learn how research is actually conducted at sea and to discuss with scientists the rationale for the research.
6. *Workshops on special topics.* Regions were being encouraged to hold workshops with fishermen on special topics to communicate the results of scientific research, e.g. workshops on Gulf herring stock assessments and lobster research results.
7. *Provisions for input by fishermen into the assessment process.* This involved scientists meeting with fishermen to discuss results of the fishing season as input to upcoming assessments.
8. *Encouraging better explanation of the stock assessment process to fishermen.* This included the

development of a video presentation (One Fish, Two Fish) explaining what stock assessment is, how it is done, and what role fishermen can play in improving it. Other initiatives included more meetings with fishermen to explain the stock assessment process and assessment results.

The Atlantic Fisheries Crisis of 1989–1993 intensified the pressures for such initiatives. The Panel on Northern Cod (Harris 1990) recognized the need for greater communication between fishermen and DFO scientists to encourage mutual understanding and to ensure that the collective wisdom and experience of the fishing community can be fully integrated into the stock assessment process (Recommendation 14). In response, DFO expanded its initiatives to open up the scientific process to the fishing community. The *Enhanced Scientific Dialogue Initiative* was intended to cover all Atlantic fisheries but, because of the Northern Cod controversy, it initially focused on that fishery. The intent was to increase direct contact between and mutual understanding of fisheries scientists and fishermen. This was to be accomplished through three activities:

— *Verification* of the scientific interpretation of information provided by fishermen and the results of assessments by meetings with fishermen before assessments were finalized in CAFSAC Advisory Documents. These Documents were circulated to industry in draft form first. (This extended the earlier idea of post-season meetings between scientists and fishermen);

— *Personal contact* between scientists and fishermen who now go to sea on each other's vessels;

— *Extension of the index fishermen programs* to allow scientists to understand more fully events in the fishery through discussion of detailed observations with small groups of experienced fishermen.

This Enhanced Scientific Dialogue Initiative began with discussion of the new scientific estimates of the Northern Cod stock in the spring of 1990. In the summer of 1990 it was extended to other stocks and areas. Index fishermen programs were expanded to cover all major cod and haddock fisheries.

The other major issue involved opening up CAFSAC. We saw earlier that the PSARC process has involved personnel other than scientists in the Steering Committee and that fishermen are invited to participate as observers in the meetings of the Herring Subcommittee. On the Atlantic coast, two major steps were taken in 1991 to strengthen the CAFSAC process. Four professors from different Canadian Atlantic universities were appointed to the CAFSAC Steering Committee. Also a number of non-DFO experts were to participate in subcommittee meetings for some of the major CAFSAC assessments.

These initiatives to address the credibility issue were intended to bridge the gap between scientists and fishermen and increase the objectivity of the CAFSAC review process. But they were probably too little and too late. Peyton (1987) pointed out the importance of resource users' perception of science: science provides an important basis for decision-making in Western society, yet the information from science is always tentative and rarely complete. The scientific capability to predict outcomes of management strategies is frequently limited by an inadequate data base and by variables missing from the resource management models. This creates an important dilemma for the fisheries manager and can colour users groups' perceptions of scientific credibility.

Pringle (1985) commented on the human factor in fisheries management. He observed that scientist/resource manager/fisherman relations are an often-ignored variable in the resource management equation. He noted that, if little confidence exists among these parties, the best of management plans will fail. Pringle appealed to scientists and fisheries managers, as a first priority, to look at government's resource management performance from fishermen's perspective:

> "One might then excuse the cynicism of fishermen. Secondly, that government personnel approach fishermen with the philosophy that the latter care for their resource. Thirdly, that a forum be instituted where government personnel and industrial sector delegates...can meet, at least annually, to develop a management plan.... It takes time for each (fishermen and scientists) to gain the confidence of the others. The scientist must present data in a format understandable to the layman, he must 'come clean' and admit a lack of knowledge where appropriate. Government personnel should realize that fishermen are neither 'the salt of the earth' nor fools, but people who enjoy their vocation, are knowledgeable about harvesting technology, but who tend to know little about biology or ecology."

The basic gist of Pringle's analysis was that, once a mutual understanding is developed, good resource management can follow.

The public's perception of science is as important in resolving resource issues as the adequacy and nature of the science itself. Peyton (1987) noted that often the public does not understand the scientific process nor the proper use and limitations of science. He observed that a common response by resource managers is to "educate" users about the informational products of science. These products are often oversimplications which lead to credibility problems when the simplified concept does not apply.

Peyton (1987) pointed out:

> "We have traditionally ignored the process side of science in our education attempts. The public is not exposed to the complexities of the scientific process, nor to the limitations imposed by difficulties of data collection or incomplete experimental design. User groups often face inconvenience and/or economic hardships resulting from fisheries regulations. If they are to maintain a sense of credibility in management programs, they must be helped to develop realistic expectations of the scientific basis of management."

In the case of the Atlantic groundfish fishery, bridging of the gap between the analyses of fisheries scientists and the perceptions of the fishing industry occurred in the early 1990's. The sudden, drastic decline of the northern cod stock during 1991 was verified by research vessels surveys, special exploratory charters and the experiences of the various sectors fishing that stock. Commercial catch rates declined dramatically and there was a general consensus that there were few large cod to be found. When the 2-year moratorium on fishing for northern cod was introduced in July, 1992, based on scientific evidence of an abrupt stock decline, the fishing industry concurred with the need for this dramatic conservation action.

Similarly, when CAFSAC in the autumn of 1992 advised that several other cod stocks were in decline and recommended drastic reductions in TACs for these stocks (see Chapter 6), the fishing industry, almost without question, accepted the scientific assessments as accurate and accepted the need for draconian reductions in the TACs.

However, by this time it had become apparent that it was necessary to forge a better partnership between scientists and the fishing industry. Accordingly, in December 1992, when announcing the 1993 Atlantic Groundfish Plan, Fisheries Minister Crosbie also announced his intent to create a new Fisheries Resource Conservation Council for the Atlantic fisheries. The purpose was to "make the fishing industry a full partner with scientists in the process that generates resource assessments and translates these assessments into conservation actions." The Council would replace CAFSAC and AGAC for the purpose of setting TACs and other related conservation measures.

Speaking in Clarenville, Newfoundland on January 21, 1993, Minister Crosbie announced that the Council would be chaired by Herb Clarke, former Executive Vice-President of Fishery Products International and former head of the Newfoundland public service. Minister Crosbie indicated that the Council would act as"a board of directors for fisheries science. It will also

TABLE 17-3. Problems and advice needs at different stages of fishery development.

Stage	Problems	Stock Assessment Advice
Under-developed	How can the fishery be developed? Is the resource big enough to justify large efforts to develop the fishery?	Rough estimates of the possible annual yield
Growth	Apparently few. In reality how to slow growth down as the limit set by the resource is approached and so reduce or eliminate risk of later problems of over-development.	More precise estimates of sustainable yield, and of the effort (e.g. number of boats of different types) needed to take this yield.
Over-development	Over-capacity, declining catch rates (and sometimes failing total catch). Economic losses. Conflict between different groups of fishermen.	Explicit advice on the specific measures (mesh size; catch quota; length of closed season) needed to achieve the manager's objectives (which themselves need to be clearly understood)
Management (a)	Adjustment of management measures (e.g., size of quotas) to take account of natural fluctuations in recruitment, or developments in fisheries on related species.	1) Precise and explicit advice on adjustments to, e.g., annual catch quotas, based largely on single-species models. 2) More strategic, and less quantitative advice on modifications to policy to take account of, e.g., species interactions.

Note (a): By the time a fishery reaches this stage it is likely that the complexities outside the simple description, e.g., natural fluctuations, or interactions between fisheries on different species, will probably have become relatively important. This is reflected in the nature of the problems and the advice.

Source: Gulland (1983).

review resource assessments and will recommend total allowable catches and other conservation measures (e.g. concerning selectivity of harvesting technologies) following public hearings." Its recommendations would be set out in written, public reports to the Minister (DFO, 1993b).

At the time this book was going to press, the exact terms of reference and the membership of the Council had not been finalized. The Council was expected to commence operations in May 1993, initially with assessments and conservation measure for groundfish and later to include other species. Meanwhile, DFO was revising the process for peer review of scientific estimates and information to be presented to the new Council. One fundamental change was already apparent. DFO scientists would no longer *recommend* TACs and other conservation measures for Atlantic groundfish. Instead, they would provide assessments of stock status and trends and the implications of alternative actions, with conservation recommendations being formulated jointly by scientists and industry through the new Council, with public hearings on proposed conservation measures as necessary.

It was envisaged that this new initiative would improve fisheries clients' understanding of the scientific process and the limitations, as well as the potential, of fisheries science. Partnership in the formulation of conservation recommendations with the fishing industry offered the potential to improve the scientific basis for fisheries management in Atlantic Canada.

3.3.2 Relations Between Fisheries Scientists and Fisheries Managers

Equally important is the understanding and perceptions of fisheries managers of the nature and limitations of fisheries science. Effective fisheries management requires timely and effective scientific information and/or advice.

Gulland (1983) distinguished four main phases in the development of a fishery. These are: an initial phase of under- utilization, in which there is either no fishing, or a small traditional fishery which takes much less than that potentially available, a phase of development and rapid growth, a phase of over-development characterized by overcapacity, and a final phase of management.

The management problems and advice needs at different stages of development differ (see Table 17-3). Most of Canada's marine fisheries are in the "over-development" or "management" phases. Thus managers require precise and explicit advice on adjustments to management measures, e.g. catch quotas. These are derived largely from single-species stock assessment models. There is a growing requirement for more strategic advice on modifications to policy to take account of multispecies interactions, spawning stock considerations, and the impact of environmental variability on the fishery (see Chapter 18).

The nature of the problems fishermen and fisheries managers face, and consequently the nature of the questions posed to scientists, vary with the state of development of the fishery. In all fisheries there is a necessary linkage and a complex set of interactions between fisheries scientists and fisheries managers. These linkages can range from weak to strong (Fig. 17-6a and 6b). Where the linkages are weak, the interactions between the two groups are considered peripheral to the main activities within each group. Where the linkages are closely linked, questions by managers and advice by fisheries scientists are regarded as key activities of the two groups.

Gulland (1983) pointed out that much of the success of fisheries management depends on how the two linkage stages — posing questions for scientific research and providing the answers and advice on management — are handled. If they are treated as marginal, the research may be irrelevant. If scientists and managers work closely together, the result can be useful and timely advice for ongoing management of the fishery.

Earlier I discussed the formal mechanisms which have been developed in Canada in recent years to ensure that managers' questions are addressed and that the results of stock assessments are made regularly available to managers.

As early as 1971, Gulland pointed out that there was frequently confusion about the roles, and methods of work, of science and management:

> "Management is a matter of taking decisions, and often it is at least as important to take a decision in time as to make precisely the best decision. Management has to resolve a wide range of often conflicting objectives, of a political, social or economic nature. Science has to provide evidence on the likely results, within its field of competence, of possible management actions, and so enable more rational decisions to be made" (Gulland 1971).

Science's role is to project the effects and interactions of particular management measures so that the management agency can make an appropriate choice of measures within the context of its fisheries objectives. This requires ongoing dialogue between scientists, the fishing industry and fisheries managers. The problem of adequate communication between scientists and managers

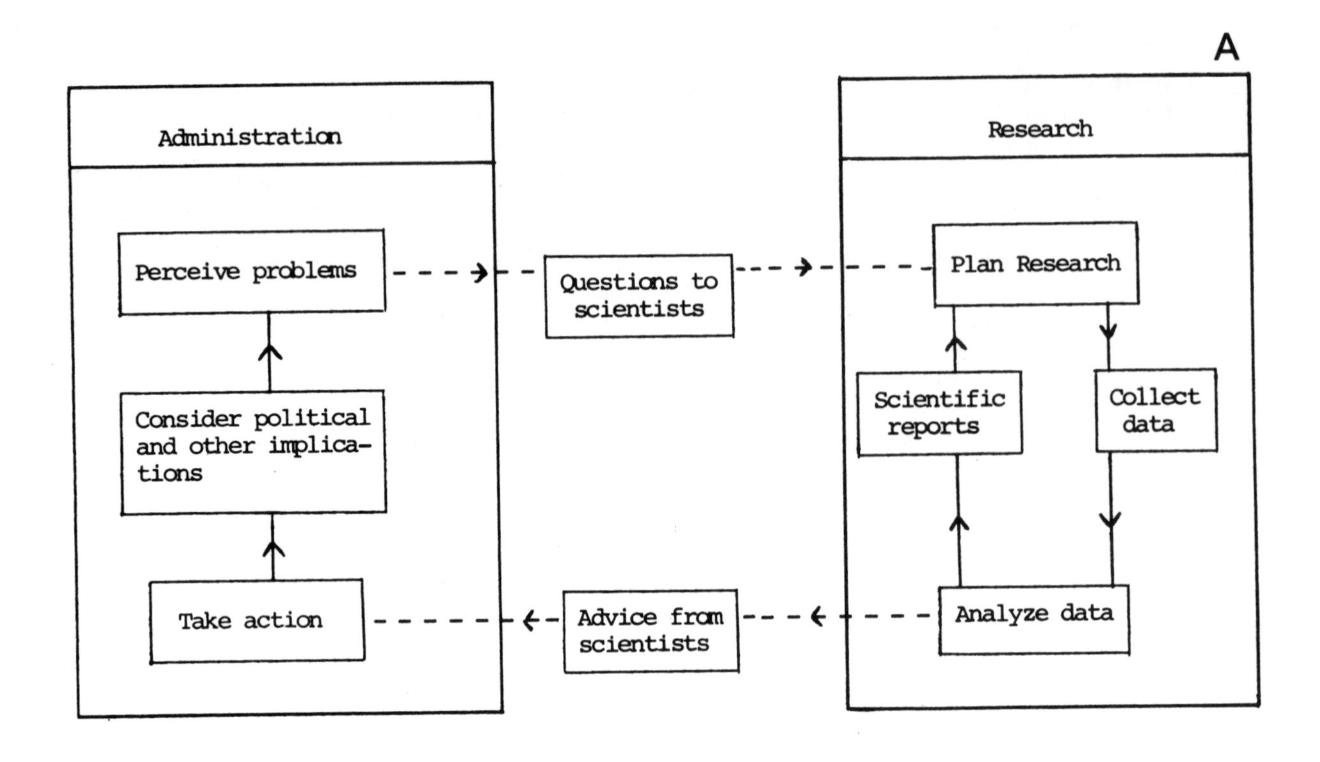

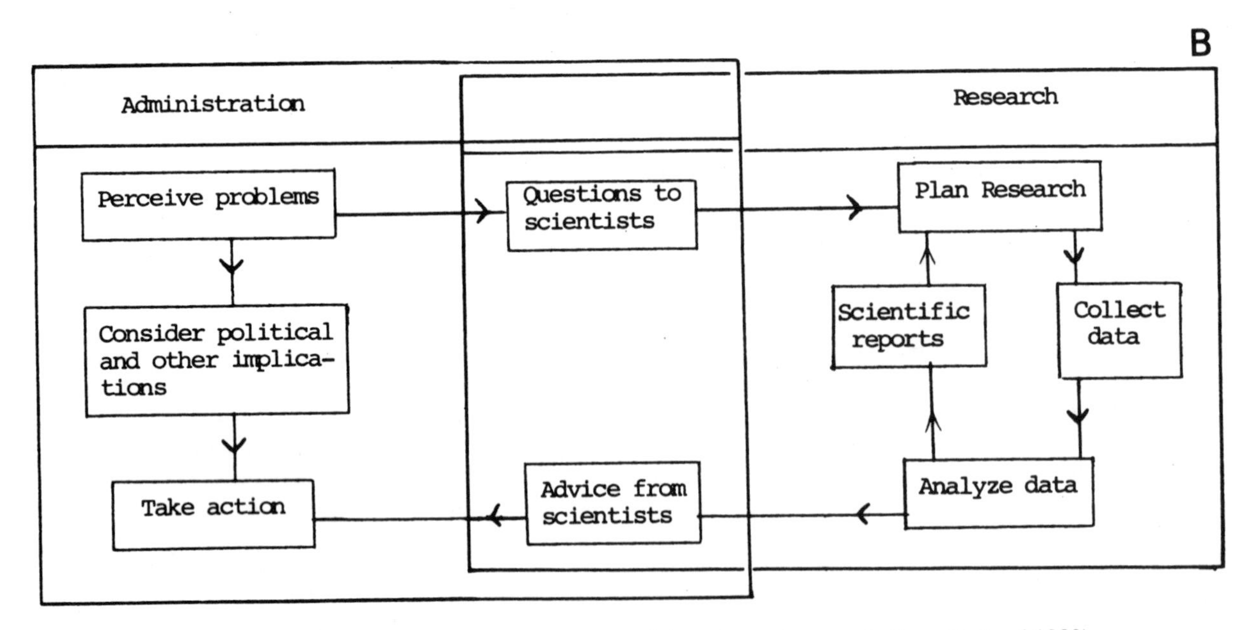

FIG. 17-6. A. Pattern of activity with weak links between administration and research (from Gulland 1983).
B. Pattern of activity with close links between administration and research (from Gulland 1983).

is a world wide problem. Mechanisms for integrating scientific information into the management process, as in the case of the Skeena River sockeye, can help to ensure that this is not a "dialogue of the deaf." Effective dialogue will help to clarify and define the respective views of the scientist and the manager and help each group to understand both the capabilities and the limitations of the other. As Edwards (1980) observed:

> "The day has passed wherein the manager can casually scapegoat the scientist because he uses strange words, and wherein the scientist can sulk in his value-neutral box because the world at large won't adapt to him. All sides have to make an effort to communicate."

4.0 CONCLUSIONS

While much has been learned about Canada's marine fish stocks and the aquatic ecosystem over the past century, to a large extent we have only seen the tip of the iceberg. Managers, scientists and the fishing industry must be prepared to act in the absence of certainty. It is a fallacy to think that effective management can only occur where there is complete scientific understanding of a fishery. It is also a mistake to think that fisheries science can provide the complete answer to management problems and to delay action pending complete knowledge.

Gulland (1971) put this succinctly:

> "Scientific finality cannot be achieved. Science advances by disproof rather than proof; a succession of hypotheses are put forward capable of explaining the observed facts, and have to be abandoned or revised as further observations show them to be inadequate. Complete and final scientific advice cannot therefore be provided. All that can, or should, be provided is advice that is sufficiently accurate and detailed for the immediate purpose."

Because absolute knowledge is impossible to achieve, scientific assessments should be couched in terms of probabilities. It should forecast the probable results of a wide range of possible actions. The reliability of scientific advice will improve as understanding of fish stocks, the marine ecosystem, and the impact of fishing on stocks and complexes of stocks improves.

While much has been accomplished by applying scientific information in the management of marine fisheries in Canada, we have only begun to understand the complex interactions in the marine ecosystem. Man, as predator, is but one part of this complex system. Immense scientific challenges remain to be addressed if the scientific basis for the management of fisheries is to be improved. Meeting those challenges will put fisheries science to the test in the decades ahead. In the next chapter I address some of the most urgent and difficult of those challenges.

CHAPTER 18

SCIENCE AND FISHERIES MANAGEMENT
II — FISHERIES SCIENCE: CHALLENGE AND OPPORTUNITY

"The true abundance of a fish stock will always be uncertain."

– M.K. Lapointe et al. 1989

1.0 INTRODUCTION

For far too long "fisheries science" has been regarded as synonymous with "fisheries biology". This reflects the dominant role fisheries biology has played as the scientific basis for fisheries management (Wooster 1988). I use the term "fisheries science" to encompass other disciplines as well as biology, e.g. economics and sociology. These have an important contribution to make to the rational management of fisheries in support of societal objectives.

Fisheries systems are exceedingly complex, characterized by interactions among the biological, environmental, economic and social dimensions. To meet the challenge of understanding even some of those dimensions will require a much more effective holistic approach than has been the practice in the past.

Fisheries biology, physical oceanography, fisheries economics, sociology and social anthropology must all be a part of a truly integrated approach to fisheries science. In this chapter I address some of the major scientific challenges in each of these areas where progress is required to improve the scientific basis for fisheries management.

One of the central problems of fisheries science continues to be estimating the number of animals in a fish stock, documenting the history of its abundance and predicting future abundance at various levels of fishing. This problem is complicated by inadequate data, methodologies for estimation which are still relatively primitive, and the fact that any fish stock is part of a vast marine ecosystem subject to complex interactions and considerable natural variability.

While significant progress has been made over the past century in developing an understanding of fish population dynamics, the contemporary fisheries scientist confronts immense research challenges. Science and management function in a world where uncertainty is the norm rather than the exception.

Although fisheries science through the 1970's and 1980's continued to wrestle with the problems of uncertainty, demands for short-term advice for use in management increased exponentially. Despite the caveats attached to such advice, managers and the fishing industry often treated it as absolute. For a while fisheries scientists applied existing models routinely to generate advice. They failed to emphasize the limitations of these models.

By the mid-1980's, however, it became apparent to many scientists that their estimates of population numbers had wide confidence limits. There was also a growing recognition that single species stock assessments were inadequate for management, given the diverse interactions among species within the marine ecosystem. Single species stock assessment models used for abundance estimation needed to be improved. A greater understanding of the dynamics of the marine ecosystem and, in particular, multispecies interactions, was also required. Understanding of the mechanisms determining recruitment, including the relative influence of parent stock and natural variability of the marine environment, was still limited.

Cushing (1983) speculated that over the next decade fisheries science would develop in three directions:

1. Stock estimation independent of catches;
2. Interactions between species, biological or operational; and
3. The understanding of recruitment and its dependence on parent stock.

Gulland (1988) similarly identified stock and recruitment, multispecies interactions and improved estimates of natural mortality as the basic problems in fisheries assessment science.

These issues identified by Cushing and Gulland are as problematic in the early 1990's as at the beginning of the 1980's. I would add to their lists the need to understand better the impact of environmental variability upon the availability and abundance of marine fish stocks and the more general question of climatic change. There are other equally important issues to be addressed in the social-economic research sphere if progress is to be made in forging a truly integrated fisheries science. Before addressing this latter category of issues, I will examine the biological-oceanographic issues raised earlier.

2.0 IMPROVING RESOURCE ABUNDANCE ESTIMATION

2.1 Methodology

2.1.1 Analytical Stock Assessment Methodology

In Canada over the past two decades the management of marine finfish stocks (anadromous species excluded) has been generally based on a catch quota system. Establishing catch quotas involves determining Total Allowable Catches for particular stocks. Such Total Allowable Catches have been based usually either on surplus production or age-structured models. Most commonly, for stocks where catch-at-age data exist, TAC estimation involves reconstructing historical population levels from a virtual population model, pioneered by Gulland (1965) and Pope (1972), and a projection of future catches. For some stocks, e.g. certain redfish stocks for which historical catch-at-age data sets do not exist, Schaefer-type general production models (see Chapter 3) are still used.

Management by catch quotas generally requires short-term (one to three years) forecasts of catch at designated levels of fishing effort. Shepherd (1988) noted that such forecasts need to be rather precise to be useful and are very demanding of data and analysis. The standard procedure for short-term assessments is to use virtual population analysis (VPA). The basic data used for the assessment of fish stocks include catch, age and weight information, biomass estimates from research vessel surveys, and information on commercial catch rates. Generally, VPA or cohort analysis is used to reconstruct historical population levels.

The abundances of each age-group within a cohort are estimated sequentially, starting with the oldest age-group for that cohort in the catch-at-age matrix, and working backwards to the youngest ages (Fig. 18-1). To initiate the VPA calculations, estimates of F for the oldest age group for each cohort (*terminal fishing mortality*) and an estimate of natural mortality M are required. Terminal F is selected to give the best fit between the estimates of temporal trends in stock size and trends in independent indices of abundance, e.g. from Research

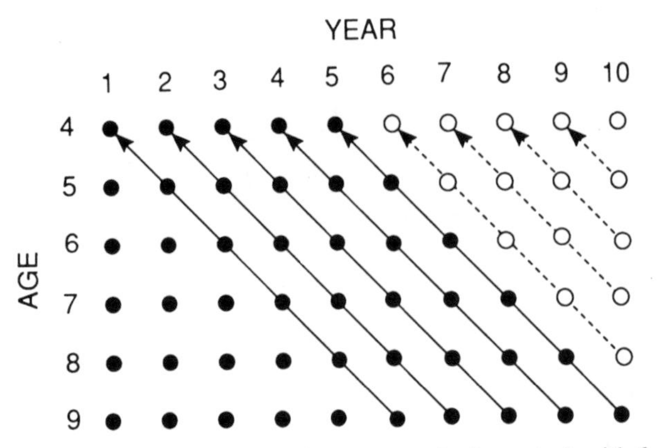

FIG. 18-1. Idealized catch at age matrix for a stock with 6 fished ages, beginning at age 4 (from Bradford and Peterman 1989).

Vessel Surveys or acoustic estimation. Errors in estimated abundances due to an erroneous choice of terminal F for the oldest age group diminishes as the reconstruction proceeds backwards to the younger ages, for cohorts in which all age groups are present in the catch matrix (see the solid diagonals of Fig. 18-1) (Pope 1972). In the recent years of any set of catch data there are many year-classes or cohorts for which one or more age-classes are not yet present in the catches (see the dashed diagonals in Fig. 18-1).

Estimates of historical population levels are usually obtained by comparing research vessel (RV) and catch per unit effort data (CPUE) from the commercial fishery to the trends in stock size estimated from VPA using different assumed fishing mortalities.

It follows, of course, that such estimates of recent population size are only as good as the trends indicated by the RV survey and commercial CPUE data. Tuning with RV survey or CPUE indices have different strengths and weaknesses. The RV survey is normally based on a stratified-random sampling approach. The survey vessel carries out trawling, and takes samples, at various depth strata within particular statistical areas at a randomly selected group of fishing stations (locations). The age and weight samples taken at these locations are used to estimate the age composition of the population as a whole and to provide an estimate of minimum trawlable biomass and population numbers in the area surveyed.

These RV survey estimates represent an index of population trends rather than estimates of absolute biomass or numbers. An underlying assumption is that the RV index represents a reasonably constant proportion of the true population size over time. Because the research vessels use small-meshed fishing gear, estimates are also obtained of the relative abundance of younger age groups which are not taken in the commercial fishery. RV survey data can be biased by year to year changes in the seasonal availability of particular species. Commercial CPUE data are also used to provide another index of population trends. Normally it is assumed that there is a direct linear relationship between CPUE and population size. In other words, catch is related to fishing effort by a constant *catchability coefficient*, q.

Experience has shown that for schooling species, particularly pelagic species such as herring, the catchability coefficient is not linearly related to population size. Catch rates can be maintained as the population size is reduced by fishing. It has also long been recognized that catch rates in the early years of a fishery can be influenced by a learning factor, as fishermen learn better where the fish are located and how to fish them most efficiently. Unnoticed changes in fishing technology or different types of fishing gear can also complicate the interpretation of apparent trends in CPUE data.

Thus, tuning of VPA estimates using RV survey or commercial CPUE is not as simple as it might seem at first glance. Conflicting trends in RV and CPUE data raise important questions as to which data source is most likely to reflect population trends in a given situation.

Despite these and other limitations, VPA and related tuning techniques were widely used to estimate and project population size and provide the basis for advice on Total Allowable Catches during the 1970's and 1980's. Scientists were aware that their estimates of population size and projected catch at particular levels of fishing mortality (e.g. $F_{0.1}$) have large error limits. Until recently, however, these shortcomings of the abundance estimation methodology were not communicated clearly to the fishing industry and fisheries managers.

2.1.2 Independent Measures of Stock Size

Other measures of stock size are sometimes used to provide independent estimates of population abundance.

Egg and Larval Surveys

Production of eggs and larvae are a function of the size of a spawning stock. Thus egg and larval surveys are sometimes used to estimate spawning stock size. The objective is to obtain production estimates at the earliest life history stages possible. With estimates of fecundity and mortality in the egg and larval stages, estimates of spawning stock size can be made. For some stocks, in the Northeast Atlantic (e.g. mackerel west and south of Britain), this has become a standard procedure. Such surveys, like groundfish trawling surveys, are very expensive and hence are used only when particularly appropriate.

In the Canadian Atlantic, extensive ichthyoplankton surveys were conducted off the coast of Nova Scotia for the decade following extension of jurisdiction (Scotian Shelf Ichthyoplankton Program). While this series of surveys yielded valuable information on the life history of various species, it could not be used to provide independent abundance estimates of major fish populations because it was not designed for that purpose.

Juvenile Fish Surveys

This type of survey can be employed to develop relative indices of prerecruit abundance. Since natural mortality rates appear to be somewhat stable following the egg and larval stages for many species, abundance at the juvenile stages prior to recruitment to the fishery can be a useful predictor of imminent recruitment. They can be most profitably utilized when the areas and times at which juveniles concentrate are well known and change relatively little from year to year.

Tagging

Tagging experiments have been frequently used for a variety of species to determine migration patterns and assist in determining appropriate management units. Tagging can also be used to produce estimates of stock size provided there is a reliable reporting mechanism. Tagging was used extensively during the "reduction" fisheries for herring on both the Pacific and Atlantic coasts, using internal metallic tags which were recovered by magnets in the reduction plants.

Ricker (1975) discusses at length the methodology of using marking (tagging) experiments to estimate population size and mortality rates. Jones (1979) discusses the theoretical methods for and the practical difficulties of tagging experiments.

Tagging is rarely used as a routine means of updating estimates of population size and mortality rates. However, it has proved useful for deciphering migration patterns and providing initial estimates of these parameters in situations where ongoing estimates from other methods are not readily available.

Acoustic Surveys

Scientists have been working for several decades on developing acoustic technology to the point where it can be used for routine abundance estimation. Acoustic surveys have been used for some time to provide estimates of abundance for pelagic species such as capelin and herring, both in the Northeast and Northwest Atlantic. There have also been attempts to employ this technique for demersal species.

Perhaps the most striking example of its application in the Canadian fishing zone has been for yearly estimates of capelin abundance in the Northwest Atlantic. These have provided the basis for managing the major capelin stocks off Newfoundland and Labrador. For this species it has become the primary tool for abundance estimation.

Acoustic techniques are now routinely used in Newfoundland for three capelin stocks, five herring stocks and for redfish. They are being explored for Northern cod. They are also used in Scotia-Fundy and in the Gulf for herring stock assessment, and on the Pacific coast for salmonids in freshwater.

Acoustic techniques are also widely used for studying fish behaviour, establishing the temporal and spatial distribution of fish in a given area, and in studying migration patterns. It appears that a number of countries are applying acoustic techniques to studying populations of demersal fish. However, no country is obtaining absolute biomass estimates for demersal stocks from acoustic surveys, with the possible exception of redfish (B.S. Muir and S.B. MacPhee, DFO, pers. comm.).

These so-called "independent" measures of abundance are unlikely to supplant but can enhance the use of methodology such as VPA or next generation models based on commercial catch and research vessel survey information.

2.1.3 Natural Mortality

Gulland (1988) pointed out that natural mortality was likely to give rise to significant uncertainties in long-term stock assessment using traditional approaches. Normally in the stock assessment process for marine finfish it has been customary to assume that the natural mortality, M, is constant for the exploited part of the fish population. Assuming that $M=0.1$ or $M=0.2$ (or some other value depending on the species) has become accepted practice in doing virtual population analysis or other assessment calculations. Yet there is no reason to believe this to be the case. Indeed, there is increasing evidence in recent years that M varies significantly with age.

One of the reasons why M has so often been assumed to be constant is that it is difficult to measure directly for fish stocks. Cushing (1975) demonstrated natural mortality rates decreasing dramatically with age for certain Northeast Atlantic stocks for which he was able to estimate total numbers from the egg to the adult stages. The most dramatic changes were in the egg and larval stages. Compared with this, changes during the exploited phase of the stock appeared negligible.

Various researchers, (e.g. Agger et al. (1973), Ulltang (1977), Mesnil (1980), Sims (1984)), have assessed the effect of errors in the natural mortality rate upon the values of the fishing mortality rates and stock sizes calculated using VPA. Sims' analyses indicated that under certain conditions the percentage error in stock-size estimates due to an error in the assumed natural mortality rate can build up to quite high levels as the VPA proceeds backward in time. Also, it appeared that an overestimate of the natural mortality rate leads to considerably higher percentage errors in stock-size estimates than does an underestimate.

Vetter (1988) reviewed the estimation of natural mortality in fish stocks. He assessed the sensitivity of fishery models to changes of a given magnitude in M by comparing percent change reported in model response (output) to percent change in M (input). The results showed that errors in estimates of M propagate into roughly equal errors in estimates of maximum yield per recruit (Y/R_{max}) but with sign reversed. For example, a 10% overestimate in M will lead to an approximately 10% underestimate of Y/R_{max}. The actual magnitude of the effect depends not just on the error in M but on the values chosen for the other parameters in the model. Various comparisons between age-structured versus constant M, or between different constants, demonstrated that effects on results can be large for some combination of parameters yet small for others. Vetter (1988) concluded that specific amounts of change depend strongly not only on the values chosen for M, but also on the value of M relative to values chosen for the other interacting parameters in the yield models. Errors in natural mortality rate along with trends in fishing mortality rate can cause spurious time trends in fish stock abundances estimated by VPA (LaPointe et al. 1989).

Vetter described three methods used to estimate M in fish populations:

1. Analysis of catch data, usually from commercial fisheries but also from sampling programs specifically conducted for stock assessment;
2. Correlation of M with other life history parameters; and
3. Estimation of deaths due to predation.

Of the major methods available for estimating M, the catch curve method is the only one which has been utilized extensively to generate estimates of M for use in stock assessments.

Although M is usually assumed to be constant for the exploited age groups in fish stocks, there is abundant evidence indicating that this is incorrect. Various studies have shown natural mortality to vary with age, density, disease, parasites, food supply, predator abundance, water temperature, fishing pressure, sex and size (Vetter 1988).

Pauly (1980) showed that M varies considerably between groups of fish (Fig. 18-2). Based on a literature review, Vetter (1988) noted that estimated rates of natural mortality are not particularly constant for either unexploited or exploited groups, and are only slightly less variable within stocks than they are within species. None of the studies from either unexploited or exploited fish stocks indicated that M is constant for any given stock or species. The within-stock ranges were often considerable.

Overall, Vetter concluded that natural mortality is far from constant for many fish stocks and that the variability is sufficiently extensive that it should not be ignored in fish stock assessments. Assuming constant M, in the face of evidence to the contrary, could seriously distort fish stock abundance estimation with serious consequences for fisheries management. The natural mortality rate is probably the single most important but least well-estimated parameter in fishery models. Fisheries scientists have not chosen to ignore the variability in natural mortality but rather have no way as yet to include it realistically in stock assessments.

2.2 Some Errors in Estimating Stock Abundance

2.2.1 Some Early Revisions of Stock Abundance Estimates

As early as 1981, Rivard (1981) studied the performance of assessments carried out on eight groundfish stocks of the Northwest Atlantic. He observed a maximum difference of more than 33% (on average) among estimates of stock size produced in three consecutive assessments. A relative error of 15% was calculated for the southern Gulf cod stock, with more than 60% of the variance due to the uncertainties associated with the estimates of mean weights-at-age.

Pope and Gray (1983), in a simulation of three North Sea fish stocks, estimated coefficients of variation for TAC estimates of 12–25% for North Sea plaice, 27–33% for North Sea cod, and 39–53% for North Sea sprat. Brander (1987) assessed how well ICES working groups performed in forecasting catches. The average prediction error for "year ahead" forecasts of the nine stocks examined was 21% (see Table 18-1) but ranged from 11–34%. He found no evidence of any change in forecasting performance over the years 1979 to 1984.

Sinclair et al. (1985), in a study of stock assessment problems for Atlantic herring in the Northwest Atlantic, examined the accuracy of the biomass estimates generated by sequential population analysis for three of

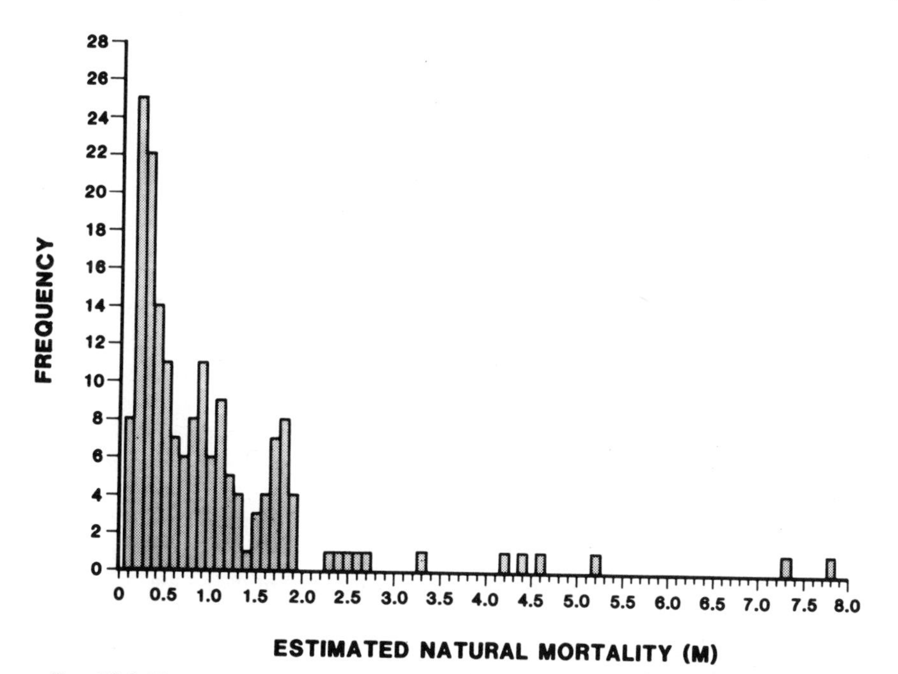

FIG. 18-2. Frequency of estimated instantaneous annual rates of natural mortality (M) in 175 different fish stocks, populations, or species. Estimates include both freshwater and marine species. Data from Pauly 1980. Adapted from Vetter 1988.

TABLE 18-1. Root mean square prediction error of "year ahead", "current year" and "precautionary" catch forecasts, all adjusted for change in F in the forecast year and expressed as percentages for various ICES stocks from 1977–1984.

Stock	ICES Division	Years	Forecasts: Year ahead	Forecasts: Current year	Forecasts: Precautionary	Coefficient variation of actual landings
Cod	VIIa	1979–1984	34	16	23	25
	IV	1979–1984	17	11	25	16
Plaice	VIIa	1979–1984	29	25	14	9
	IV	1977–1984	21	11	11	11
Sole	VIIa	1979–1984	11	8	15	23
	VIIf,g	1980–1984	13	12	17	9
	VIIe	1980–1984	22	17	23	9
	IV	1980–1984	18	8	22	24
W. mackerel		1979–1984	22	16	32	6
	Mean		21	14	20	15

Source: Brander (1987).

the 10 management units for the Canadian Atlantic herring fisheries (Scotian Shelf, southern Gulf of St. Lawrence, and Fortune Bay in Newfoundland). The accuracy of the annual stock assessments was evaluated retrospectively using the subsequent assessments.

There were major changes in the assessments of these stocks over time (Table 18-2). For example, the 1977 biomass estimate for 4WX adult herring in the 1978 stock assessment was 551,000t. The 1983 assessment indicated, however, that the revised value was about 288,000t. On this basis, it appeared that the 1978 assessment overestimated stock size of the most recent year by about 90%. The quota was too high by the same margin.

For the southern Gulf of St. Lawrence stock, the 1978 biomass estimate in the 1979 stock assessment was 361,000t. The 1982 assessment indicated that the correct value was closer to 52,000t, an overestimate of almost 600%. The quotas were so high during this period in which the stock size was grossly overestimated that they were never reached. Sinclair et al. (1985) concluded that the stock assessment biologists had tended to overestimate stock size.

Rivard and Foy (1987) evaluated the past performance of catch projections from a retrospective analysis of the assessments for 18 groundfish and pelagic stocks of commercial importance in the Northwest Atlantic. The study covered six stocks of cod, two stocks of haddock, five stocks of herring, and one stock each of redfish, yellowtail flounder, American plaice, pollock and mackerel.

To assess the performance of past projections, they compared the 1980 TACs calculated in 1979 (using 1978 data) and the 1982 TACs calculated in 1981 (using 1980 data) with the catch projected at $F_{0.1}$ using the "most recent" information on stock size, mean weight, and partial recruitment coefficients. The result of these retrospective analyses (summarized in Table 18-3), indicated considerable differences between projected catch levels and actual $F_{0.1}$ catches for the two years studied. There was a general tendency to overestimate the $F_{0.1}$ catch such that the actual fishing mortality rate was much higher than $F_{0.1}$. The overall error was +26.1% for 1980 and +33.1% for 1982.

The major sources of error which contributed to the difference between the advised TAC and the "true" catch at $F_{0.1}$ were differences in exploitation pattern (partial recruitment coefficients), mean weight-at-age, stock size estimates for year 2 of the projections, and the $F_{0.1}$ value utilized (see Fig. 18-3). Sources of error tended to vary considerably between stocks and between species. The sources of error in these projections were, in order of importance:

1. *The estimation of stock size for the last year of the projection period.* This can be influenced by several sources of error: (i) the estimation of population abundance from the VPA; (ii) difference between the TAC in the first year of the 2-year projection and the actual catch in that year; (iii) difference between the catch composition in the first year of projection and the actual catch composition in that year; and (iv) differences in estimations of recruiting cohorts during the projection period.
2. *The changes or variations in partial recruitment coefficients.* These may be due to (i) estimation errors, or (ii) external changes in fleet behaviour, gear types, seasonality of the fishery, etc.;

TABLE 18-2. Estimates of population biomass (age 4 +) in 000s tons for the 4WX, 4T, and Fortune Bay herring management units.

	Year																	
Assessment	1965	1966	1967	1968	1969	1970	1971	1972	1973	1974	1975	1976	1977	1978	1979	1980	1981	1982
4WX																		
1977	—	—	523	534	502	478	319	220	161	510	449	410						
1978	336	408	523	539	505	481	319	223	170	571	523	505	551					
1979	354	439	569	586	553	528	361	254	188	538	466	414	384	245				
1980	337	410	527	541	506	481	318	221	164	504	430	366	314	179	106			
1981	343	418	537	551	510	481	319	220	163	495	419	356	301	169	104	439		
1982	338	412	530	544	504	475	314	216	159	484	406	340	285	154	95	407	356	
1983	341	416	534	548	516	480	317	219	160	488	411	343	288	156	92	355	310	219
4T																		
1978	—	—	—	—	1125	684	461	333	260	220	182	161	157					
1979	—	—	—	—	1117	700	475	346	257	242	209	188	214	361				
1980	—	—	—	—	1213	742	514	380	292	247	209	186	189	304	295			
1981	—	—	—	—	1014	584	396	257	179	138	108	74	59	77	54	26		
1982	—	—	—	—	989	616	401	272	184	142	111	77	63	52	59	33	36	
Fortune Bay																		
1977	—	8	44	35	20	29	24	21	10	5	4	4						
1978	—	—	—	—	20	29	24	22	8	7	5	4	3					
1979	—	9	43	34	19	27	23	19	8	4	2	2	1	6				
1980	—	8	43	34	19	27	23	19	7	4	2	2	1	4	3			
1981	—	—	—	—	19	27	23	19	7	4	2	2	1	4	3	2		

Source: Sinclair et al. (1985).

TABLE 18-3. Estimates of the 1980 and 1982 catch at $F_{0.1}$ for commercial fish stocks of the Northwest Atlantic, as initially projected and as back-calculated from the results of the 1985 assessments.

		1980			1982		
Species	Stock	Projected	Back-calculated	Relative difference (%)	Projected	Back-calculated	Relative difference (%)
Cod	2J-3KL	212,000	144,350	46.9	270,000	207,410	30.2
	3NO	16,500	22,760	−27.5	20,000	23,857	−16.2
	3Ps	28,000	18,834	48.7	33,000	17,814	85.2
	3Pn-4RS	59,000	58,310	1.2	105,000	75,540	39.0
	4TVn[a]	63,000	20,699	204.4	65,000	32,945	97.3
	4VsW	45,000	31,338	43.6	53,000	23,613	124.5
Haddock	4VW	15,000	8,461	77.3	24,000	7,127	236.7
	4X[b]	28,000	17,436	60.6	32,000	12,655	152.9
Redfish	4RST[c]	18,000	59,094	−69.5	31,000	76,571	−59.5
Yellowtail flounder	3LNO	18,000	8,509	111.5	23,000	5,413	324.9
American plaice	3LNO	43,400	41,419	4.8	45,000	15,581	188.8
Pollock	4VWX-5	41,750	47,403	−11.9	55,000	32,589	68.8
Herring	4WX[d]	83,000	68,514	21.1	92,000	74,344	23.7
	4T, spring[e]	50,102	8,992	457.2	20,000	31,831	−37.2
	4T, fall[e]	39,271	11,332	246.5	—	—	—
	4R, spring[f]	13,160	23,497	−44.0	8,202	11,547	−29.0
	4R, fall[f]	4,911	5,841	−15.9	4,542	4,058	11.9
Mackerel	2–6[g]	197,000	176,710	11.5	118,000	97,700	20.8
Overall		975,094	773,499	26.1	998,744	750,595	33.1

[a] The TAC was revised to 54,000 t for 1980 and to 60,000 t for 1982.
[b] A 1-year projection was done for 1980.
[c] The projections for 1979 were considered to be indications only because of the unreliability of the data.
[d] The 1980 projection was revised to 65,000 t.
[e] The $F_{0.1}$ projection were not the recommended options for 1980; instead, a strategy to reduce the decline of the stock was recommended.
[f] A 1-year projection was done for both 1980 and 1982.
[g] The value of 118,000 t for 1982 corresponds to one of the many options explored.

Source: Rivard and Foy (1987).

3. *The estimation of $F_{0.1}$ for the projection years.*

The variance of catch projections, expressed in terms of coefficient of variation, for nine key stocks, are shown in Fig. 18-4. The coefficients of variation of the catch projected at $F_{0.1}$ were as follows: 15–20% for cod stocks; 25–50% for haddock; 28% for pollock; 16% for redfish, and 35–42% for herring.

From their analysis, the following factors emerged as key issues in projecting TACs: dependence on abundance estimates for prerecruited age groups; dependence on stock size estimates for partially recruited and fully recruited age-groups; and dependence on partial recruitment and $F_{0.1}$ estimates for the projection period.

2.2.2 Northern Cod

While scientists recognized these pitfalls in the assessments, the fishing industry and managers were generally not aware of the large error limits.

On Canada's Atlantic coast this blissful ignorance was shattered by the widespread publicity surrounding CAFSAC's January 1989 revision of its estimate of the status of the northern cod stock and the advised TAC for northern cod fishing at $F_{0.1}$. The estimated catch at $F_{0.1}$ was revised downwards by more than 50%, from 266,000 to 125,000 tons. Because of the widespread perception that the northern cod stock was increasing and would continue to increase in abundance, this revised CAFSAC estimate sent a shockwave through the Atlantic fisheries system and shattered expectations about future growth. Because of the relative size of the northern cod stock and its importance particularly to inshore fishing communities and trawling ports in Newfoundland and, to a lesser extent, in Nova Scotia, Fisheries Minister Tom Siddon established an Independent Review of the State of the Northern Cod Stock, headed by Dr. Leslie Harris, President of Memorial University (For details, see Chapter 13 and Lear and Parsons [1993]).

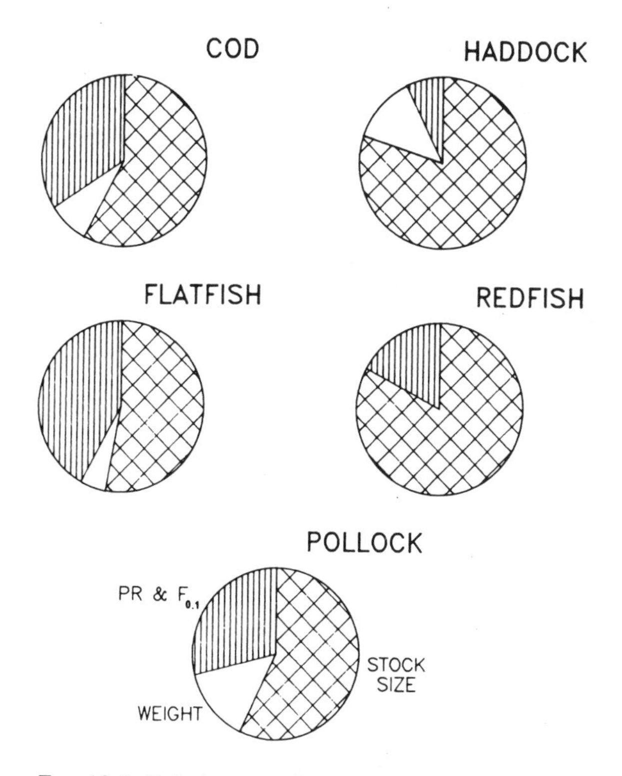

FIG. 18-3. Relative contribution of each input variable (partial recruitment, $F_{0.1}$, weight, and stock size) to error in catch projections for cod, haddock, flatfish, redfish and pollock (from Rivard and Foy 1987).

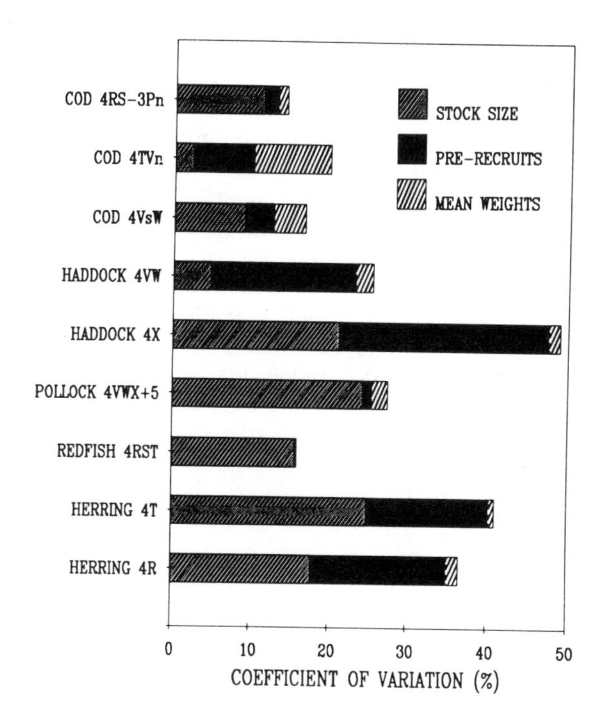

FIG. 18-4. Components of variance for catch projections of various Atlantic Canada fish stocks. The variance components are expressed in terms of their contribution to the coefficient of variation. The coefficient of variation for catch-at-age was less than 1% for all stocks and is not illustrated (from Rivard and Foy 1987).

The Northern Cod Panel (Harris 1990) concluded that the 1989 CAFSAC assessment of the status of the northern cod stock was essentially correct. CAFSAC had indicated that the northern cod stock increased between 1977 and 1984, subsequently stabilized and, depending on harvesting strategies, could decline over the next several years because of poor recruitment. Although the population biomass remained rather static between 1984 and 1988 (with maybe a slight decline), there was a more obvious downward trend in the number of fish and the population. This reflected a decline in the abundance of younger age groups entering the population (Fig. 18-5).

The Panel concluded that "the state of the stock measured by the biomass trends does not support a conclusion that anything drastic or threatening has occurred to the northern cod stock to date." However, the Panel expressed concern that the decline in recruitment, coupled with the continued catch levels experienced during 1986, 1987 and 1988, "could sharply erode the gains made in rebuilding the northern cod stock during the late 1970's and early 1980's."

The Panel made recommendations for future management of this stock and for strengthening the scientific basis for management. I will return to its proposal for improving the scientific basis later in this chapter. The Panel's management proposals are addressed in Chapter 13. At this point I wish to explore the Panel's explanation for the difference between the 1989 and earlier scientific advice on the overall state of the 2J3KL cod stock. The Panel noted that up until the time Canada extended its fisheries jurisdiction to 200 miles the main index of abundance was derived from the catch per unit effort of foreign fishing vessels. From 1978 onward the CPUE series was based upon the developing fishery by offshore Canadian trawlers. There was an overlap of only two years (1978 and 1979) between these two series of abundance indices. This made it impossible to derive a long-term CPUE index. These were the initial years of development of a significant Canadian offshore fishery for this stock. The Panel commented:

> "With the wisdom of hindsight, it is possible to see that the efficiency of the Canadian vessels increased quite sharply after these two earlier years."

Offshore fleet efficiency for the fully exploited part of the population appeared to have increased quite sharply between 1980 and 1985. After that it declined (Fig. 18-6), perhaps in response to the enterprise allocation regime. This increase in efficiency was interpreted as an increased stock size. The error became apparent only as the longer time series of CPUE data became available and as the RV survey data series, which started in 1981, became available for use.

The significant difference in the 1989 scientific advice from that of earlier years was attributed in part to the addition of a new analytical method of handling the data inputs, in part to the changes in the state of stock which had occurred since 1986, and in part to a significant adjustment in the 1986 RV survey abundance estimates. Two additional years of data added to a reasonably short series of observations showed a marked decline in recruitment over that observed in years prior to 1985. The 1986 survey values which were incorporated into the earlier RV survey calibration were shown to be an artifact of resource availability, probably brought about by a change in the timing of the 1986 RV survey.

Thus, complications with the RV and CPUE data for tuning the VPA and the manner in which the VPA

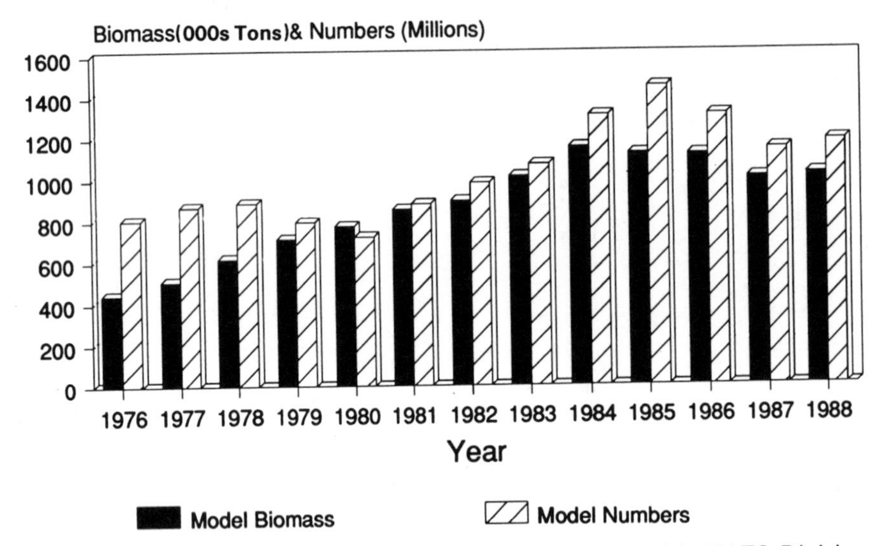

FIG. 18-5. Biomass and Numbers Estimates of Age 3+ cod in NAFO Divisions 2J3KL (from Harris 1990).

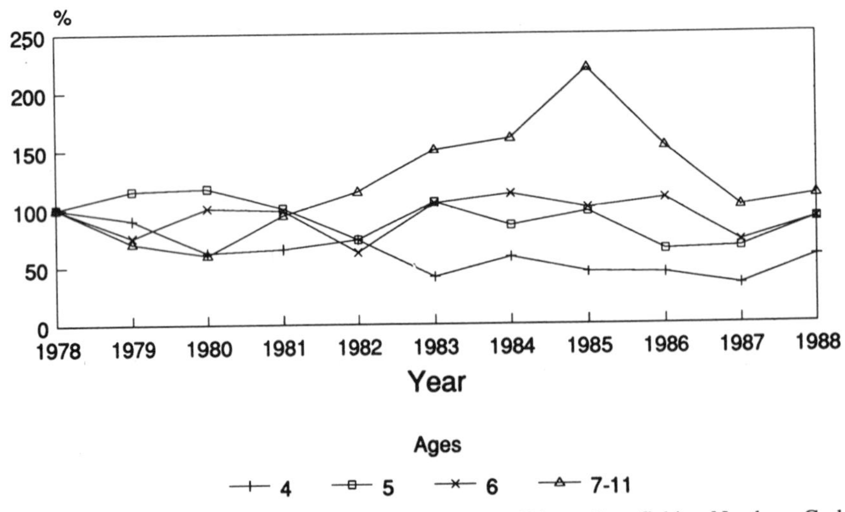

FIG. 18-6. Changes in Efficiency of the Canadian offshore fleet fishing Northern Cod (from Harris 1990).

itself was applied contributed to the underestimation of fishing mortality and overestimation of population size in the earlier years. The actual fishing mortality rates since 1977 had in fact been at least double those projected in the earlier assessments. As the assessments were revised annually, the estimates of F for the earlier years were revised upwards because of the nature of the VPA approach but until 1989 scientists kept underestimating the current F and overestimating the projected catch at $F_{0.1}$.

The Panel attributed part of the blame for this on the tuning methodology used for the VPA. Up until 1989 the northern cod population estimates were tuned using a "bulk biomass" method. This method used the overall (all ages combined) catch rate estimates from RV surveys and/or the commercial fishery. These indices of biomass change were related to historical exploitable biomass estimated from the VPA. This relationship was then used to tune the population estimates for the most recent years so that the exploitable biomass showed an equivalent trend to the CPUE index.

The Panel observed that this approach was probably reasonably effective prior to 1978 when the foreign fishery was removing a large proportion of the stock and, hence, large year-to-year changes in the size of the biomass occurred. The method was less effective during the years following extension of jurisdiction partly because of the slower rate of biomass change but was probably also affected by increasing levels of uncounted bycatch and systematic increases in the efficiency of the Canadian fleet. The Panel noted that, apart from the problems inherent with the data set, the bulk biomass approach had several fundamental drawbacks which had been recognized in recent years, e.g. the exploitable biomass estimated from the VPA can be distorted by using incorrect estimates of the exploitation pattern or partial recruitment (the proportion of the full fishing morality acting on each age of fish). In the 1989 CAFSAC assessment a new approach was employed, using data from the RV surveys on an age-by-age basis. This method minimizes the discrepancies between the VPA population estimates for each age and the equivalent survey age-specific indices of abundance. The Panel's bottom line was that the more recent CAFSAC methodology "seems reasonable in light of the species involved and in view of the characteristics of the area and of the funds available" but "the independent estimates of population trends require much closer scrutiny."

The Panel suggested that the CAFSAC scientists were not sufficiently sceptical of the methodology early on because they strongly believed that the stock was indeed increasing in abundance:

> "It is possible that if there had not been such a strong emotional and intellectual commitment to the notion that the $F_{0.1}$ strategy was working, the open and increasing scepticism of inshore fishermen might have been recognized as a warning flag demanding more careful attention to areas of recognized weakness in the assessment process."

While the northern cod reassessment has been the object of considerable attention because of the size and importance of this stock and the "growth" mystique surrounding it in the post-extension years, it was by no means unique as an example of the difficulties of the abundance estimation process. Several examples of stock assessment errors were already mentioned in Section 2.2.1. In addition, the Statistics, Sampling and Surveys Subcommittee of CAFSAC in February 1989 carried out retrospective analyses for several stocks (Sinclair et al. 1990). These analyses indicated that northern cod was by no means the exception in terms of overestimation of stock size and underestimation of fishing mortality among Canadian Atlantic groundfish stocks. The changes in assessments for Scotian Shelf cod, southern Gulf cod and the two Scotian Shelf haddock stocks were more variable than those for northern cod, with each exhibiting a tendency to overestimate population numbers and underestimate fishing mortality in the earlier years.

It is perplexing, in light of these various studies, that the Harris Panel took a myopic view by treating Northern Cod as a special case and did not situate the reassessment of this stock in the broader context of a general pattern of revisions of stock estimates for Northwest Atlantic stocks and Northeast Atlantic stocks. By failing to do so, the Panel, to some extent, misled fisheries managers and federal Ministers into thinking that the northern cod overestimation was a special case which could be fixed by pouring more resources into the study of that particular stock. An opportunity was lost to make the more general point that abundance estimation for finfish stocks is still a primitive science which requires research on a broad array of fronts. The Panel did acknowledge that improvements in the data base and more biologically-based modelling were required.

The Panel recommended bolstering the Department's research effort on northern cod in several areas. The Minister of Fisheries and Oceans, Bernard Valcourt, responded to the Report stating, "I accept the basic principles of the report as well as its major recommendations aimed at rebuilding fish stocks through intensive scientific research by the Department of Fisheries and Oceans and conservation management measures, as well as direct communications with fishermen and the fishing industry." On May 7, 1990, in the context of a broader Atlantic Fisheries Adjustment Program, Minister Valcourt announced that $150 million would be devoted over the next 5 years to rebuilding fish stocks, including northern cod.

This expanded program built on the addition of 31 researchers to the Newfoundland Region's Science Branch during 1987 and 1988 in areas supporting cod research, as part of the Science Integration initiative. These new resources reassigned in 1987 and 1988 had permitted new research initiative in areas of longstanding concern to fisheries managers and the industry:

— research on cod had increased with the Newfoundland Region designated the lead Region for cod research in Atlantic Canada;
— an oceanographic research component had been added to examine ocean climate and environmental influences on fish distribution and abundance;
— an Atlantic centre for Resource Assessment and Survey Methodology had been established in St. John's to improve the estimation of Total Allowable Catches, catch rates, stock forecasts, and other outputs from stock assessment in support of the fishing industry;
— increased attention had been focused on fisheries ecology research; i.e. the interaction of fish with their environment and processes affecting survival and recruitment.

As part of the Atlantic Fisheries Adjustment Program, $42.8 million over 5 years and 18 person-years were allocated to undertake new initiatives in the following areas:

— a major study of ecosystem relationships in the Newfoundland area;
— increased study of the possible impact of harp and hooded seals on commercial fish stocks;
— additional studies on stock structure of Northern Cod relevant to more refined management measures;
— oceanographic and other studies relevant to both inshore migration of Northern Cod and the survival of young stages of commercial groundfish species;
— measures to enhance stock assessment through additional research vessel survey activity, the utilization of new techniques such as hydroacoustics, and more research on trawl fishing characteristics;
— studies of the impact of fishing gear on the environment and on fish behaviour;
— greater involvement of the fishing industry not only in interpreting data from commercial operations but also in collecting the data and discussing the results of the assessments (see Chapter 17).

Many of the recommendations made by the Harris Panel to improve abundance estimation are similar to

TABLE 18-4. Recommendations for improving catch projections (from Rivard and Foy 1987).

FACTOR	RECOMMENDATION	PROGRAM ACTIVITY OR FUNCTION AFFECTED
Estimates of stock size	Improve estimation of stock size obtained from virtual population analysis. Here, there is a wide variety of actions possible:	
	(a) Improve independent measure of stock size (catch rate indices, research survey estimates); related questions are how to combine independent measures when more than one is available and how to resolve conflicting trends in time series.	Research surveys; logbook analysis; observer program
	(b) Improve catch composition estimates.	Sampling
	(c) Improve methodology; calibration process; need way to cope with variability in survey data; more objective approach.	Analysis
Estimates of prerecruited age-groups and of recruitment	Improve estimation of prerecruited age-groups and forecasts of recruitment:	
	(a) Obtain independent measure of abundance.	Research surveys
	(b) Better forecasts of recruitment from independent index.	Analysis
	(c) Reduce dependence of projections on recruitment by reducing forecast horizon.	Scheduling of advice
	(d) Develop stock recruitment relationships.	Analysis
	(e) Measure partial recruitment independently of virtual population analysis.	Analysis; surveys
Forecasts of fishing mortalities for the projection period	Improve forecasts of partial recruitment coefficients for year 2 of the projection:	
	(a) Develop a model taking into consideration fleet behaviour.	Analysis
	(b) Do 1-year projections.	Scheduling of advice
	(c) Do multi-year projections which accommodate known levels of error.	Analysis
	(d) Other methods/approaches.	Analysis

recommendations made earlier by Rivard and Foy (1987) for improving catch projections (Table 18-4). As Rivard and Foy observed, the objective of recommendations to improve catch projections should be to reduce errors which contribute significantly to the accuracy or precision of projections. Because there is no single source of error, there is no simple solution for improving catch projections. Bradford and Peterman (1989) pointed out that better estimates of abundance would likely result if a larger number of independent and reliable sets of information are included in an analysis. A bias in any one type of input information is more likely to be offset by an opposite bias in another set than in VPA, which uses relatively few types of information. They advocated the use of sensitivity analyses.

The bottom line is that managers will have to learn to live with the uncertainties inherent in these methodologies. Fisheries managers and the fishing industry must understand the limitations of existing abundance estimation procedures. The confidence limits for estimates of stock size and TACs at a particular reference fishing mortality can be large. In the past, Canadian scientists have provided population estimates and advised catches in terms of numerical estimates. They believed that providing confidence limits for these estimates would tempt managers to base management measures on the upper margin of the confidence limits for any particular estimate. The time has come for scientists to provide advice in a manner which displays the extent of uncertainty involved. Experience has shown the tendency to overestimate stock size and underestimate fishing mortality. This suggests that managers would be wise to err on the side of caution to maintain resource productivity. However, the scientific estimates will continue to exhibit unpredictable changes, because of the shortcomings of the methodologies available and because of the inherent variability of the marine ecosystem.

3.0 STOCK, RECRUITMENT, AND ENVIRONMENTAL VARIABILITY

3.1 Some General Considerations

Interannual variability in recruitment is a phenomenon characterizing the majority of marine fish stocks. The quest to understand and predict this variability has bedeviled fisheries scientists since the turn of the century. There have been two broad approaches to the problem. One, which has been vigorously pursued since the work of Beverton and Holt (1957) and Ricker (1954, 1958) attempting to relate parental stock size and subsequent recruitment, has involved the fitting of models to existing data and the further development of models, both functional and empirical. The second approach has involved an attempt to understand the mechanisms determining interannual variability in recruitment.

There is a large volume of literature on this subject. For recent reviews, see Rothschild (1986), Cushing (1988), Larkin (1989) and Wooster and Bailey (1989).

Johann Hjort first vividly demonstrated the existence of year-class variability. He proposed two hypotheses to account for recruitment variability. Hjort's first hypothesis stated that variability in year-class size is a result of between-year shifts in the precise timing of spawning and of the phytoplankton bloom, which initiates the seasonal zooplankton cycle upon which the fish larvae feed. Laboratory studies had indicated that first feeding during the few days following final resorption of the yolk sac were critical to larval survival. Sinclair (1988b) has pointed out that Hjort considered the early life history stages as a whole (eggs, larvae, post-larvae and early juvenile stages) to be the "critical period" rather than just the first few days of feeding, which has been frequently assumed in the literature.

Hjort's second hypothesis suggested that interannual differences of advection of eggs and larvae away from appropriate geographic areas for feeding and life cycle continuity generated subsequent recruitment variability.

3.1.1 Stock and Recruitment Models

Stock-recruitment relationships expressed as mathematical models came to prominence in fisheries biology in the mid-1950's with the work of Ricker and Beverton and Holt. Larkin (1989) noted that both models were derived from notions of density dependent predation, Ricker's with mortality dependent on initial abundance, and that of Beverton and Holt with densities over a sequence of stages. Both Sinclair (1988b) and Larkin (1989) noted that these models arose in part from a much broader literature concerned with the regulation of abundance of natural population. For a discussion of this context, see Larkin (1989).

In subsequent decades there have been numerous attempts to fit various forms of these models to data on recruitment and stock size for numerous fish populations, with varying degrees of success. Generally there was such a wide scatter that it was impossible to distinguish between the validity of various models or determine whether there was really any clear evidence of a particular stock-recruitment relationship. This became evident at the 1973 ICES Symposium on Fish Stocks and Recruitment (Parrish 1973).

During the 1980's new modelling approaches were applied to the stock and recruitment question. Garrod

(1982) noted that a number of stocks had suffered recruitment collapse during the 1960's and 1970's. He acknowledged that environmental change had no doubt contributed to the collapse. But in each case it occurred under continued high levels of exploitation and at a time when the spawning stock had been reduced to relatively low levels. It was desirable to base a definition of the stock and recruitment relationship upon a definition of the biological mechanisms involved. However, he concluded that this did not appear to be practical for any stock in the foreseeable future. He, therefore, suggested focusing on extracting as much information as possible from the empirical analysis of time series data, bearing in mind that any relationship so determined can only represent the average outcome of a very complex biological process.

Regarding models, it has been generally recognized that any stock and recruitment relationship reflects the variation in pre-recruit mortality with stock size. That mortality consists of density dependent and density independent components, however these might be defined. It is generally assumed that the density dependent component includes a compensatory mortality where increased egg production is offset by increased prerecruit mortality, causing a less than proportional increase in recruitment. Dome-shaped S-R curves can be generated by hypothesizing more complex biological models (e.g. Ricker (1954), Cushing and Horwood (1977) and Ware (1980)). Any of these models is possible in a given situation. However, the use of any particular model necessitates imposing the underlying biological rationale as an assumption for any data set. This assumption may not be valid.

Shepherd (1982) proposed a more general model which made no implicit assumption concerning the form of any dependent compensation. The Shepherd model can take any form dictated by the data.

In this model, b is a compensation coefficient. If $b < 1$, the curve rises continuously, if $b=1$, it is asymptotic, and if $b > 1$ it forms a dome. Shepherd suggested that for pelagic fish $b < 1$, for flatfish $b=1$ and that if $b > 1$ the dome was generated by cannibalism (e.g. gadoids).

Garrod (1982) applied a log transformation of the Shepherd equation to fit data for 12 stocks. He concluded that the data used for individuals or in grouped form did not allow a convincing estimation of the appropriate parameters of stock and recruitment curves. The resulting S-R curves for Northeast Atlantic stocks are shown in Fig. 18-7.

Another approach to estimating recruitment without the constraints of a particular model was suggested by Getz and Swartzman (1981). Their approach was to let the data themselves determine the probability density function (pdf) of recruitment at a given stock size without determining that it fit some preconceived form. This involved partitioning the stock and recruitment data into intervals based on historical stock and recruitment data. For any stock interval, these data then determine the probability that recruitment will be in each recruitment interval. Overholtz et al. (1986) modified this approach to project recruitment of Georges Bank haddock. This method has been used to determine fisheries management strategies in New England.

Evans and Rice (1988) concluded that algorithms that use stock and recruitment data directly, in raw form, may project future recruitment of fish more directly than algorithms that use estimated parametric summaries. One difficulty with these new methods is that it takes a long time to obtain enough data to be confident about predictions (Larkin 1989).

Recruitment overfishing has usually been considered to have occurred when a population has been fished down to a point where recruitment is substantially reduced or fails. This definition is vague and has not been used traditionally to provide a biological reference point for fisheries management. Levels of spawning biomass below which recruitment seems to be reduced have been used, but their determination from available data is usually difficult and controversial.

Sissenwine and Shepherd (1987) proposed an alternative definition of recruitment overfishing in terms of the level of fishing pressure that reduces the spawning biomass of a year class over its lifetime below the spawning biomass of its parents on average.

For situations where the spawner-recruit relationship is indeterminate, Sissenwine and Shepherd (1987) defined a new biological reference point, F_{rep}. F_{rep} is the fishing mortality rate that corresponds to average replacement.

They recognized that, while compensation must exist at some level of population size, it need not apply over the range of spawning biomass for which there are quantitative observations. For the population to persist where compensation is inoperative, F should not exceed F_{rep}. If compensation is operative, but obscured by the effects of a fluctuating environment, then F_{rep} could be considered a conservative biological reference point. ICES in 1984 suggested using spawning biomass per recruit as the basis for another biological reference point. The guideline of allowing enough spawning biomass per recruit so that year classes replace the spawning biomass of their parents on average was adopted as an objective of the Multispecies Fisheries Management Plan for New England in 1985.

Walters and Ludwig (1981) pointed out that large errors in measuring spawning stocks can make recruitments appear independent of spawning stock size.

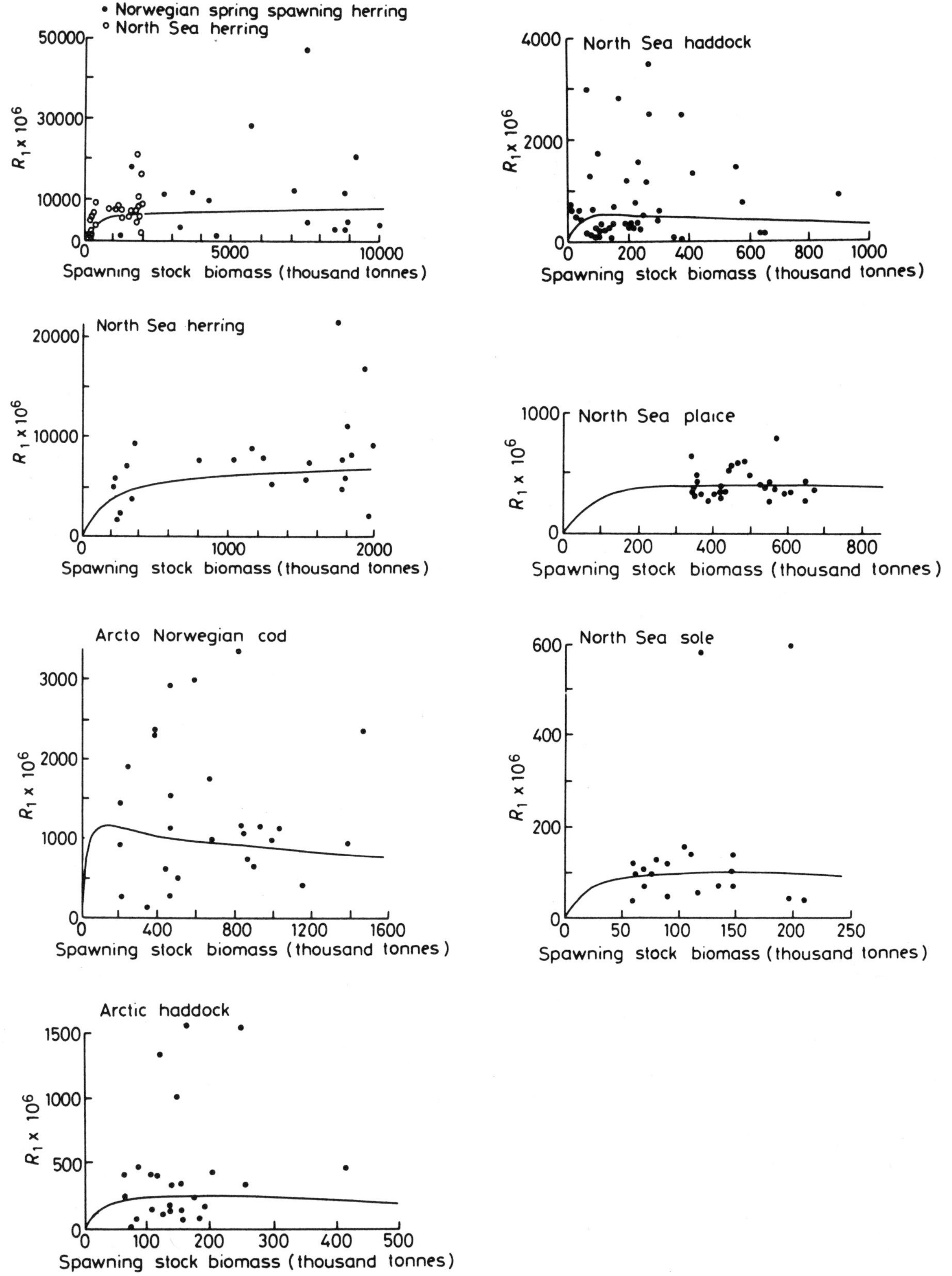

FIG. 18-7. Stock and recruitment curves for selected stocks (© British Crown Copyright, 1982; reproduced from Garrod, 1982.)

Unfortunately, this effect promotes overexploitation rather than simply making the relationship "noisier" and harder to measure. They noted that much fisheries theory and practice are based on the assumption that recruitment is largely independent of spawning stocks. They warned that fisheries managers should not trust models based on this assumption, unless it can be clearly demonstrated that spawning stocks have been measured "accurately" (i.e. errors less than ± 30% or so).

Progress requires a better understanding of the mechanisms determining recruitment. Essentially, attempts to fit S-R models to data on spawning stock size and numbers of recruits have yielded as much information as is likely without further understanding of the factors involved. This leads us to the question: What progress has been made in understanding mechanisms determining recruitment variability?

3.1.2 Recruitment Processes

Two key contributions to the question of mechanisms determining recruitment variability were the work of R. Harden-Jones (1968) in his book *Fish Migration* and that of David Cushing summarized in a series of papers and his book *Marine Ecology and Fisheries* (Cushing 1975). Harden-Jones's migration triangle (see Chapter 3) concept for population persistence conceptualized larval drift from spawning grounds to juvenile nursery areas via the surface layer residual currents. It provided a geographic framework to consider recruitment variability. He considered events during the larval drift phase to be critical to determining variability in year-class size. Cushing developed the match-mismatch theory. This linked Hjort's first hypothesis concerning the variability in timing of spawning and time of blooming with the "critical depth" concept of oceanographer Sverdrup. The principle of the match-mismatch hypothesis is illustrated in Fig. 18-8. The hypothesis linked oceanographic processes, including between-year differences in wind events at the time of spawning and the resulting differences in thermal stratification, with fish stock fluctuations. As a result, food-chain events during the early larval drift phase became the focus of study, in a series of field studies on the first stages of the life cycle concerned with the first feeding of larvae.

Some studies (e.g. Californian sardine and the northern anchovy) indicate an increase in mortality rate at hatching (Figs. 18-9, 18-10). Numerous other observations do not show such an increase. Cushing (1988) found no evidence that such steps in mortality affect the subsequent recruitment. Density dependence in the egg and larval stages has been shown for cod (Daan 1981), mackerel (Ware and Lambert 1985) and, to some extent, herring (Burd 1985).

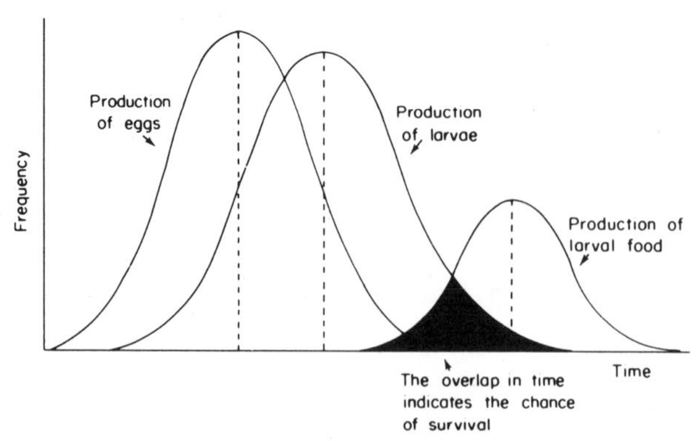

FIG. 18-8. Illustration of Cushing's match/mismatch hypothesis (adapted from Cushing 1982).

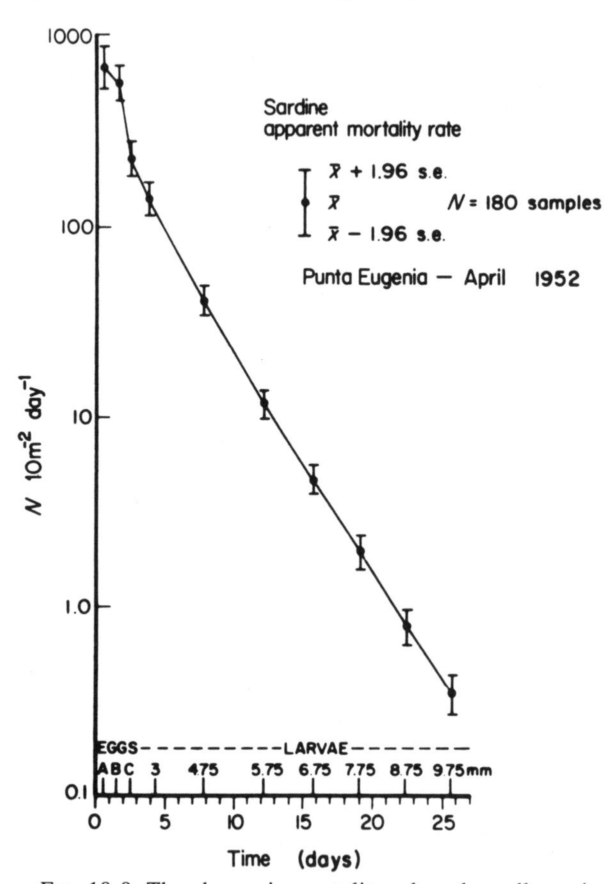

FIG. 18-9. The change in mortality when the yolk sac is exhausted for the California sardine (D.H. Cushing [ed]. *Fish Population Dynamics (second edition)*. p107 Copyright 1988. Reprinted by permission of John Wiley & Sons Ltd.)

Three processes could generate density dependent mortality: cannibalism, predation and starvation (Cushing 1988). While the opportunity for cannibalism

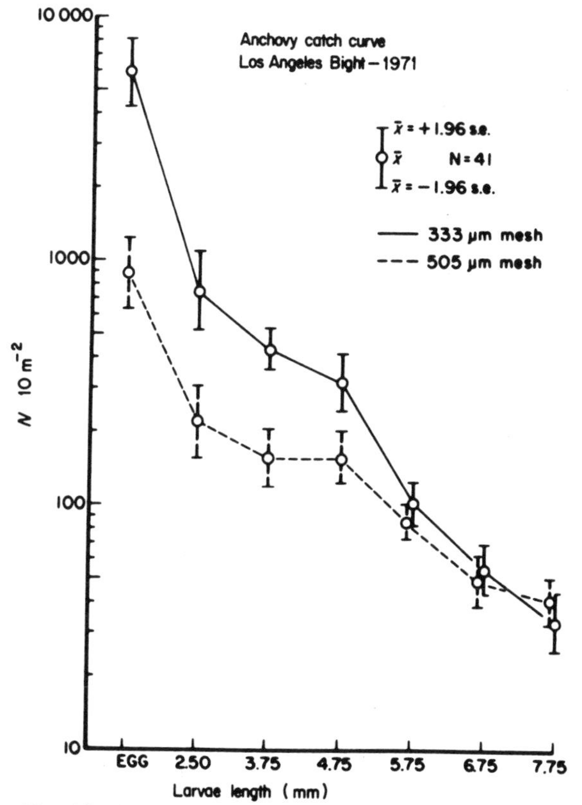

FIG. 18-10. The change in mortality when the yolk sac is exhausted for Northern anchovy (D.H. Cushing [ed]. *Fish Population Dynamics (second edition)*. p108 Copyright 1988. Reprinted by permission of John Wiley & Sons Ltd.)

may be low for most species, cannibalism has been shown to affect recruitment in some species (e.g. the work of Cook and Armstrong (1986) showing North Sea haddock and whiting recruitment being reduced by prior abundant year classes).

Lasker (1975, 1978, 1981) hypothesized that larval fish need a "threshold" concentration of prey particles (proper-sized food) to feed and survive. He suggested (his "stability" hypothesis) that, for northern anchovy, necessary concentrations occur in small-scale layers that form in inshore waters during periods of water column stability. These layers could be dispersed by storms or by intense upwelling. From this, he concluded that anchovy year-class strength depended on the stability of the environment.

Peterman et al. (1988) rejected Hjort's and Lasker's hypotheses that the abundance of recruits in fish is determined at an early life history stage. They suggested that attempts to understand interannual variability in recruitment might have to rely on abundances estimated closer to the age at recruitment. Methot (1986) emphasized the importance of understanding the sources of mortality during the last three months prior to recruitment as age 1 fish.

Although starvation mortality is important, it does not appear to be the single controlling mortality factor under normal ranges of prey density (Sissenwine 1984a).

Peterman et al. (1988) summarized reported correlations between abundances at early and later life stages (Table 18-5). They concluded that the diversity of results emphasized the varied effect of different physical and biological processes on the relative variability in mortality at early and late life stages.

Predation is a recognized factor causing variability in mortality in the early life stages of marine fish. However, it has been among the most difficult areas to study.

TABLE 18-5. Summary of reported correlations between abundances at early and later life stages. Size of fish, when reported, are given in parentheses below the age. References: A. Cushing (1974); B. Bannister (1978); C. Rauck and Zijlstra (1978); D. Tormosova (1980); E. Postuma and Zijlstra (1974); F. Burd (1985).

Species	Beginning stage	Ending stage	N (yr)	Correlation	P	Reference
Baltic cod	30-d-old larvae	Age 1 group	6	0.81	0.05	A
	Age 0 group	Age 1 group	6	0.90	0.015	A
North Sea plaice	Eggs	Age 2	10	0.25	0.49	B
	Age 0 group	Age 2	14	0.71	0.0005	C
North Sea sole	Age 0 group	Age 2	—	Low	—	C
	Age 1 group	Age 2	12	0.68	0.13	C
North Sea haddock	First-feeding larvae (≥6mm)	Age 2	7	0.60	0.15	D
North Sea Herring						
Downs Stock	30-d-old larve	Age 3	15	0.34	0.22	E
	30-d-old larvae	Age 0	13	0.76	<0.01	F
Doggerbank stock	Small larvae (<11mm)	Age 3	9	0.58	0.10	E
	Large larvae (≥11mm)	Age 3	9	0.71	0.025	E
Bank stock	20-d-old larvae	Age 0	11	0.66	0.027	F

Source: Peterman et al. (1988).

Wooster and Bailey (1989) cited two studies that attempted to partition larval mortality into predation and starvation (Hewitt et al. 1985; Leak and Houde 1987). These indicated that predation is the major source of larval mortality but did not show a direct correction between predation on eggs and larvae and subsequent recruitment.

Frank and Leggett (1985), based on a reanalysis of published studies, concluded that failure to consider the importance and influence of environmental data had contributed to incorrect conclusions about the causes of changes in predator-prey numbers reported. Other studies have also demonstrated the important effect of environmental variability in determining recruitment.

3.1.3 Environmental Variability and Recruitment

Hjort (1914), in his second hypothesis, had suggested that recruitment variability was related to transport of eggs and larvae away from favourable nursery areas. Various studies have shown a net movement of eggs and larvae by currents (see Wooster and Bailey 1989).

Correlations have been demonstrated between recruitment and environmental conditions associated with transport (Nelson et al. 1977; Bailey 1981; Sinclair et al. 1985).

A number of recent studies have emphasized the importance of environmental variability in determining recruitment. During the 1960's and early 1970's several authors explored the role of ocean climate, specifically large-scale changes in physical oceanographic processes, on recruitment fluctuations. Iles (1973) drew attention to correlations between year-class variability in the North Pacific and in western Canadian lakes and large-scale ocean climate changes.

Cushing (1978) extended Iles's suggestion of the effect of global climatic variations on year-class variability. By the time of the ICES Symposium on Marine Ecosystems and Fisheries Oceanography in 1976 there was growing evidence linking oceanographic conditions to recruitment fluctuations (Parsons et al. 1978).

That climatic change affects fisheries has long been recognized. Cushing (1982) provided an extensive overview in his book *Climate and Fisheries*. He dealt briefly with the dependence of recruitment upon environmental factors. He cited a number of early studies correlating year-class strength to environmental variables — usually temperature. Both positive (e.g. Johansen 1927; Dickie 1955; Hermann et al. 1965) and negative correlations (Uda 1952; Marr 1960; Ketchen 1956; Martin and Kohler 1965; Dickson et al. 1973) have been shown between temperature and year-class strength. One of the more striking examples was the apparent dependence of recruitment to the West Greenland cod stock upon temperature (Hermann et al. 1965; Fig. 18-11).

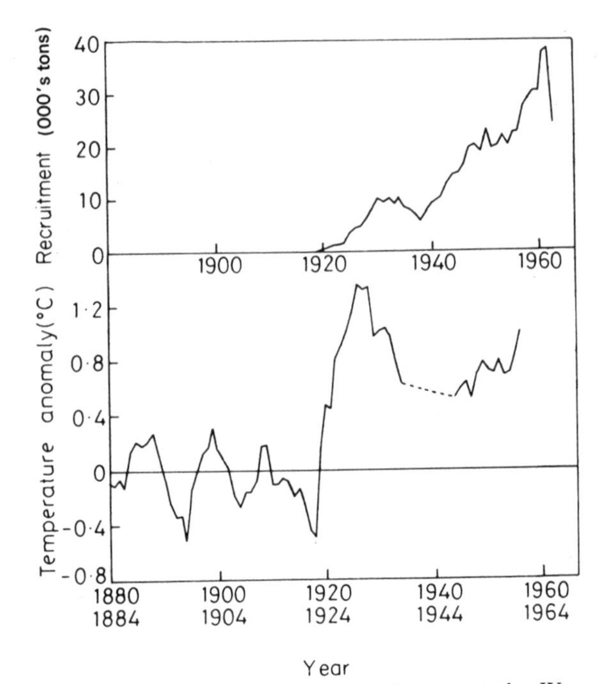

FIG. 18-11. Dependence of recruitment to the West Greenland cod stock upon temperature during the early years (from Hermann et al. 1965).

Not all of these relationships to temperature could be explained in the same way. Cushing (1982) suggested that differences in recruitment from year to year were affected by differences in wind strength and direction.

Templeman (1972) reviewed data on year-class success for the previous 30 years for most of the main stocks of cod and haddock in the North Atlantic. There appeared to be area relationships for year-class success of these species, between success of haddock on Georges and Browns Banks and of year-class success either of cod or haddock or both on Sable Island Bank, St. Pierre Bank, and the Grand Bank one year later or sometimes in the same year. Success in the Icelandic, Greenland and Norwegian-Spitzbergen-Barents Sea areas appeared to occur in the same year but usually one year later than good year-classes in the Sable Island to southern Grand Bank area. Templeman suggested that some of the factors affecting year-class strength were: *temperature*, but temperature in itself might be of secondary rather than of direct importance; *larval drift* in currents to unfavourable situations; *mixtures of water* by winds and currents, the resulting high plankton abundance being favourable. He also suggested that density dependent factors played a role.

Shepherd et al. (1984) acknowledged that major long-term changes of abundance and/or geographical range of fish populations could be attributed to the effects of climatic change. They suggested that, if it were possible to allow for the effects of short-term environmental fluctuations, "one could "clean up" the stock-recruitment plot, and clarify whatever relationships there might be."

They suggested that environmental changes are likely to affect fish stocks via four principal processes:

1. Direct physiological effects;
2. Disease;
3. Feeding, with successful feeding being affected through:
   a) Food abundance,
   b) Food quality,
   c) Temporal match/mismatch of food production and feeding,
   d) Spatial distribution of food relative to fish,
   e) Local concentrations of food,
   f) Competition,
4. Predation.

Shepherd et al. (1984) noted that each species of fish has its own unique set of relationships with environmental variables. A simple change in one environmental factor could produce a complex change in the web of competition and predation interactions between species that would be extremely difficult to predict. They observed:

> "Looking for a single responsible mechanism for high and low recruitment in a particular year, as has often been done, is thus likely to be misleading and unrewarding."

Many relationships have been reported over the years between recruitment and various environmental factors. Examples of some of these are presented in Tables 18-6 to 18-9. From further analyses of the data from these studies, Shepherd et al. (1984) concluded that there is strong evidence, from the long history of fluctuations of abundance and the existence of fairly well-defined geographical ranges, that an important relationship must exist between recruitment and climate. Koslow (1984) and Koslow et al. (1987) provided evidence of this from the Northwest Atlantic.

There have been several recent studies demonstrating the influence of oceanographic and meteorological processes on recruitment of certain Pacific coast stocks, particularly Pacific halibut and Pacific hake (Parker 1989; Hollowed and Bailey 1989; Francis et al. 1989).

Francis et al. (1989) illustrated the importance in fisheries management of incorporating linkages between environment and stock production. Over the range of stock sizes observed to date, Pacific hake recruitment appeared to be independent of spawning stock size and strongly correlated with environmental conditions occurring during early life history. Francis et al. (1989) argued that a failure to take into account the role of the physical environment in the recruitment process, and incorporate this into long-term fisheries production estimates, can provide results that are misleading to fisheries managers.

They also noted that the large-scale temporal pattern in the environment, hypothesized to drive hake production, is not unique to the California Current or eastern boundary current system in general. Francis et al. (1989) had linked the production dynamics of several important fisheries (walleye pollock, king crab, sockeye salmon) of the Eastern Bering Sea with large-scale environmental dynamics. It was suggested that the major dynamics of many fish resources are strongly related to oceanographic and ecological processes that are large-scale (regional or global), apparently periodic and climatic driven.

Mann and Lazier (1991), in their book *Dynamics of Marine Ecosystems*, examined relationships between changes in fish stock abundance and large-scale processes in the ocean for both the North Atlantic and the North Pacific. They traced the effects of the "great salinity anomaly" of the 1960's and 1970's in the North Atlantic, which retained its identifiable characteristics for nearly two decades, while travelling around the sub-Arctic gyre. There were marked effects in the plankton community north of Iceland during the formation of the great salinity anomaly in the early 1960's and later on the Grand Banks in 1973. Cushing (1988) attempted to relate the spawning success of several stocks to the passage of the great salinity anomaly. He found that year-classes produced during the low salinity years were significantly below normal for 11 stocks. However, other stocks in the area were apparently not affected.

Despite many unanswered questions about why some stocks were affected and others not, Mann and Lazier (1991) concluded that "the case for the formation of the great salinity anomaly, its journey of 10,500 km around the subarctic gyre, and its depressing effect on the recruitment of fish stocks in its path is a persuasive one." They postulated a link to meteorological events.

The link to meteorological changes is clearer for events in the North Pacific. The effects of El Niño off Peru are well known. Mann and Lazier (1991) cited evidence of meteorological effects on fish stocks in the North as well as the South Pacific. They identified that deepening and intensification of the Aleutian

TABLE 18-6. Temperature as a recruitment factor.

Species	Reference	Proposed mechanism	Hypothetical ultimate cause
Plaice	Bannister et al. (1976)	Reduction of predator nos. or whereabouts or food requirements of predators	—
Atlanto-Scandian herring	Benko and Seliverston (1971)	—	—
North Sea plaice	Beverton and Lee (1965)	Duration of pelagic phase → drift	—
Manx herring	Bowers and Brand (1973)	—	—
Herring, East Anglian autumn fishery	Carruthers and Hodgson (1937)	—	—
Arcto-Norwegian cod	Cushing (1972)	Match/mismatch	Changes in wind strength and direction
North Sea cod	Dickson et al. (1974)	Physiological effect of temperature on larval size, density dependence; ability of larvae to consume food. Match/mismatch	—
West Greenland cod	Elizarov (1965)	—	St. Lawrence discharge
South Newfoundland and Iceland cod	Garrod and Colebrook (1978)	—	—
West Greenland cod	Hermann et al. (1965)	—	—
North Sea cod	Holden (1970)	—	—
Plaice	Johansen (1927)	Physiological effect of survival; reduction of food supply. Predation	Baltic run-off
Norwegian and Barents Sea cod	Kislyakov (1961)	Larval drift and development time	—
Flemish Cap cod	Konstantinov (1975, 1977, 1980 1981)	—	—
Pacific sardine	Marr (1960)	Physiological effect on maturation. Match/mismatch. Competition with anchovy.	Upwelling/advection
Southern ICNAF cod	Martin and Kohler (1965)	Larval transport	—
Downs and Dogger herring	Postuma (1971)	Physiological effect on egg mortality	—
North Sea herring	Postuma and Zijlstra (1974)	—	—
Northeast Baltic herring	Rannak (1971)	Physiological effect on mortality	Inflow of North Sea water
Baltic spring herring	Rannak (1973)	—	Winds
Pacific sardines	Sette (1958/1959)	—	—
Yellowtail flounder, Southern New England	Sissenwine (1974)	—	Overall warming in 40's
Sand-eel (*A. marinus*) North Sea	Hart (1974)	Reduction in either food or predation	Atlantic penetration into North Sea
General, off Plymouth	Russell (1973)	Increased survival, due to more productive water	Water flow up channel

Source: Shepherd et al. (1984).

TABLE 18-7. Salinity as a recruitment factor.

Species	Reference	Proposed mechanism	Hypothetical ultimate cause
Plaice	Johansen (1927)	Ova/larva viability (physiological); reduction of food supply; predation	Baltic outflow
North Sea sand-eel	Hart (1974)	Reduction in either food or predators	Penetration of Atlantic water
Pacific sardine	Walford (1946)	—	Upwelling/winds
Arcto-Norwegian cod	Cushing (1972)	Match/mismatch	Wind strength/direction

Source: Shepherd et al. (1984).

TABLE 18-8. Wind, pressure gradients, and upswelling as recruitment factors.

Species	Reference	Related to	Proposed mechanism	Hypothetical ultimate cause
Northern anchovy	Bakun and Parrish (1980)	Upwelling	—	—
Pacific hake	Bakun and Parrish (1980)	Offshore transport	—	—
Pacific bonito	Bakun and Parrish (1980)	Upwelling	—	—
Pacific makerel	Bakun and Parrish (1980)	Wind stress curl, temperature, upwelling	—	—
Atlantic menhaden	Nelson *et al.* (1977)	Ekman transport	—	—
Pacific sardine	Bakun and Parrish (1980)	Upwelling, wind curl	—	—
Dover sole	Bakun and Parrish (1980)	Upwelling, winds	—	—
North Sea herring, haddock and cod	Carruthers (1938) Carruthers and Hodgson (1937) Carruthers *et al.* (1951)	Pressure gradients	—	—
North Sea plaice and Arctic cod	Garrod and Colebrook (1978)	Wind	Turbulence	Pressure over Europe
Bear Island cod	Hill and Lee (1958)	Wind	Water transport	—
Sardines (Pacific)	Sette (1958/1959)	Wind	—	—
North Sea haddock	Rae (1957)	Wind	Movement of food	—

Source: Shepherd et al. (1984).

TABLE 18-9. Miscellaneous recruitment factors.

Species	Reference	Related to	Proposed mechanism	Hypothetical ultimate cause
Plaice (North Sea Kattegat)	Johansen (1927)	Days of ice in Danish waters	Physiological effect of larval survival. Reduction of food. Predation.	Freshwater outflow from Baltic
Norwegian and Scottish herring	Beverton and Lee (1965)	Ice cover north of Iceland	Temperature affects duration of development and pelagic phase, hence affects distance drifted and thereby how close they get to location of polar front.	Change in water temperature
Cod (Norwegian)	Sund (1924)	Tree-ring width	Heavy spring floods drive larvae out to sea, hence poor settlement.	Snowfall
Atlantic halibut	Sutcliffe (1972, 1973)	St. Lawrence discharge	—	—
English Channel herring	Cushing (1961)	PO_4 winter maximum	Affects/results from balance of competition with pilchard	
Northern anchovy	Lasker (1975, 1981)	Stability of water	Stable water allows local concentrations of food to build up; these enable better feeding for larvae	Storms
Labrador cod	Borovkov (1980)	Index of meridional atmos. circ.	Distance drifted by larvae	—
Haddock (Québec landings)	Sutcliffe (1972)	St. Lawrence run-off	—	—
Mackerel and herring, Pacific	Skud (1982)	Temperature and competition	Competition	Temperature changes
Sardines, Pacific	Skud (1982)	Temperature and salinity and competition	Competition	Upwelling
4WX herring	Sinclair et al. (1980)	Wind and sea level	—	Barometric pressure
Skagerrak and Kattegat sprat	Lindquist (1978)	Ratio of temperature and wind force	Fungal infection of eggs	—
Georges Bank haddock	Chase (1955)	Temperature and wind strength	—	—

Source: Shepherd et al. (1984).

atmospheric low pressure system, which often follows an El Niño — Southern Oscillation event, is associated with a rise in temperature and sea level along the west coast of North America and a reduced flow of the California Current. They suggested that interannual variations in the strength of the northward flowing coastal current are correlated with changes in the survival of cod and in the migration routes of salmon.

Subsequently, Beamish and Bouillon (1993) examined the relationship between salmon production in the North Pacific and climate. They suggested that trends in salmon production from 1925 to 1989 were not primarily a result of fishing effort, management actions, or artificial rearing, but rather that the trends in survival were caused by the environment. Total annual all-nation catches of three Pacific salmon species — pink, chum and sockeye — from 1925 to 1989 exhibited long-term parallel trends. The strong similarity in the pattern of catches indicated that common events over a vast area affect the production of salmon in the North Pacific. They noted that the climate over the northern North Pacific is dominated in the winter and spring by the Aleutian Low pressure system. Their analyses indicated that the long-term pattern of the Aleutian Low pressure system corresponded to the trends in salmon catch. Beamish and Bouillon (1993) concluded that climate and the marine environment may play an important role in salmon production.

All of these studies provide further evidence of the important influence of environmental variability on marine fish production.

3.2 The Canadian Experience

One of the most interesting recent Canadian theoretical contributions to the recruitment question is that by Sinclair and Iles. For Atlantic herring, Iles and Sinclair (1982) suggested that the richness of stocks of Atlantic herring is determined by the number of areas that are suitable for larval retention. Further, the abundance of a stock was considered to be determined by the size of its larval retention area.

Sinclair (1988a) and Sinclair and Iles (1989) extended this hypothesis to situate recruitment as a component of the broader question of the regulation of animal populations. For marine fish species with complex life histories, they considered the specific patterns in spawning location and the associated numbers of populations. These they interpreted as a function of the requirement for coherence at the early life history stages in the face of the diffusive and advective characteristics of the oceans. They suggested an integrated conceptual framework within which both physical oceanographic and food-chain processes play a direct role in regulating abundance.

They suggested four aspects of populations may be involved in the regulation of abundance: spatial pattern, richness, absolute abundance, and temporal variability (Fig. 18-12). *Spatial pattern* was defined as the geographical distribution of self-sustaining populations of the same species. *Population richness* was defined as the number of discrete self-sustaining populations within a species. The range of *absolute abundance* between component populations of marine fish species may cover several orders of magnitude. The fourth characteristic is *temporal variability* of numbers within populations due to year-class variability. Sinclair and Iles argued that progress in understanding the causes of temporal variability in abundance within populations may be hindered by the lack of understanding of the processes regulating pattern, richness and absolute abundance.

Sinclair and Iles proposed a population-regulation hypothesis (the member/vagrant hypothesis) to account

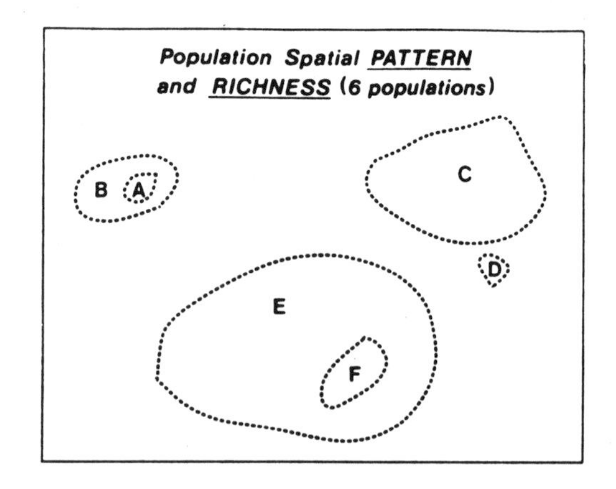

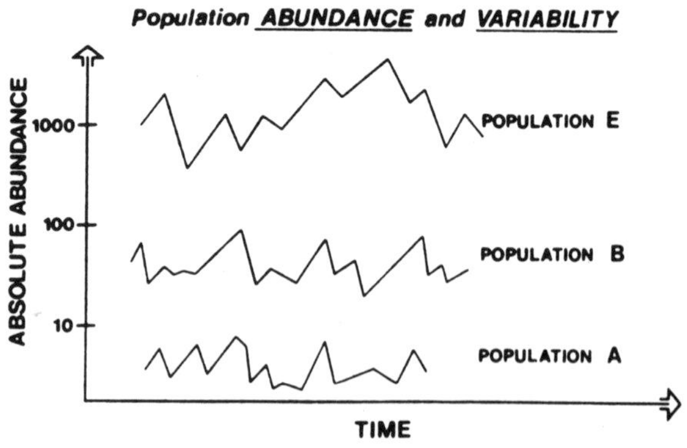

FIG. 18-12. Schematic representation of the four components of the population regulation question (from Sinclair and Iles 1989).

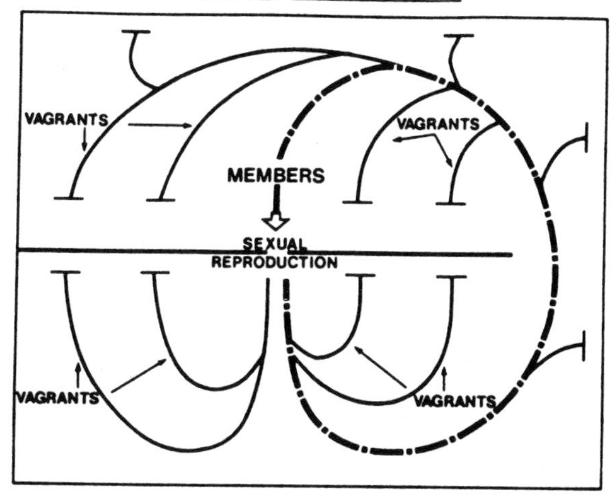

FIG 18-13. Schematic representation of life-cycle closure of a marine population with a complex life history in relation to spatial constraints (from Sinclair and Iles 1989).

for these four population characteristics. This concept of the complex life histories of populations is shown in Fig. 18-13. The member/vagrant hypothesis has three components:

1. Population pattern and richness are functions of the number and location of geographic settings (within the overall distributional area of the species) within which the particular life cycle is capable of closure.
2. Absolute abundance is scaled according to the size of the geographic area in which there is closure of the life cycle of the free-crossing population.
3. Temporal variability is a function of the inter-generational losses of individuals (vagrancy and mortality) from the appropriate distributional area that will ensure membership within a given population.

Both density-dependent and density-independent factors are implicated in the member/vagrant hypothesis as it relates to the determination of temporal variability. This hypothesis envisages an important role for environmental variability in helping to determine variability in year-class strength of marine fish populations. Space precludes a full examination of this complex hypothesis. For more details, see Sinclair (1988a).

Sinclair (1988b) summarized the field studies on the recruitment question involving Canadian scientists during the 1970's and 1980's. Most of the Canadian field studies on recruitment during this period took place in the Atlantic. Two major initiatives were the ICNAF/NAFO-sponsored multinational studies on Georges Bank herring and Flemish Cap cod and redfish. For reviews of these studies, see Grosslein (1987), Lilly (1987b) and Grosslein and Lilly (1987).

To understand recruitment variability of the Georges Bank herring population, it was considered necessary to investigate the relevant physics, biological oceanographic production processes as well as larval behaviour. For various reasons, including the collapse of the herring population, the program was terminated. Sinclair (1988b) noted that the descriptive data collected and its analysis contributed to:

1. a hypothesis on Atlantic herring population patterns and abundance for the North Atlantic as a whole;
2. space- and time-scale characterization of circulation and mixing over submarine banks in general; and
3. the conclusion that events during the early life stages as a whole (rather than the first few weeks of the larval stage) are important in defining year-class strengths.

This study was not a test of Cushing's match/mismatch hypothesis. The Flemish Cap study on cod and redfish, however, was more influenced by Cushing's food-chain interpretation of year-class variability. The central hypothesis of the project was that "The year-class strength of the Flemish Cap cod stock varies as a result of specific biological and environmental conditions." This central hypothesis was divided into four main parts: physical environmental conditions, predation conditions, food abundance conditions, and characteristics of the spawning stock. This study did not advance the understanding of mechanisms determining recruitment, but the empirical observations clarified hypotheses and descriptions of the relevant physics (Sinclair 1988b).

A third study, conducted between 1977 and 1982, was the Scotian Shelf Ichthyoplankton program (SSIP). The aim was to describe the seasonal distribution of eggs and larval distributions for all fish species on the Scotian Shelf. This was preparatory to a study on the recruitment dynamics of a particular population. The observations suggested persistence of egg and larval distributions for some species over the banks rather than drift with the residual surface layer circulation. The results led to a recruitment study on Brown's Bank haddock.

Parallel to these three expensive, large-scale field programs, two classes of inexpensive small-team studies were carried out in the Atlantic (Sinclair 1988b). The first was a series of statistical studies initiated by W. Sutcliffe and his colleagues in the former Marine Ecology Laboratory (MEL). These studies involved the analysis of data on year-class sizes (or landings) and physical oceanographic parameters. The second category included the nearshore field studies by William Leggett and his colleagues on the spawning and early life history of capelin in both the St. Lawrence Estuary and Newfoundland bays.

More recently, studies have been initiated on the Pacific coast. In 1985, the Pacific Biological Station and the Institute of Ocean Sciences (both part of the Science Branch of the Pacific Region of DFO) began a joint, decade-long investigation on the effects of oceanic variability on fish stocks on La Pérouse Bank (Thompson and Ware 1988). In 1986, an initiative was also undertaken to improve our understanding of the mechanisms controlling marine survival and thus recruitment of Pacific salmon. It was hoped that this would allow forecasts of subsequent abundance to be more accurate than is currently the case and lead to improved management of the salmon resource. The Marine Survival of Salmon Project (MASS) began in 1987, with the objective of understanding the mechanisms that determine variation in recruitment of Pacific salmon. The MASS program, like the La Pérouse Bank study, and the Canadian Atlantic projects, is a multidisciplinary fisheries and oceanography study. It is attempting to relate physical and biological oceanographic studies to the distribution and survival of salmon during their few months of marine residence.

Several Canadian studies have attempted to relate recruitment variability to environmental fluctuations.

Koslow (1984) analyzed recruitment time series for 14 stocks of Northwest Atlantic fish. He found consistent positive correlations in recruitment among stocks within such species as cod, haddock and herring. Significant positive correlations were also found in recruitment among demersal, offshore-spawning species (cod, haddock and redfish). Recruitment in these groups tended to be negatively correlated with that of pelagic species, which spawn inshore (herring) or in restricted waters (mackerel). These patterns spanned the region from West Greenland to Georges Bank. Koslow suggested that their spatial extent indicated that large-scale physical forcing, rather than local biological interactions, predominately regulates recruitment to northwest Atlantic fisheries. His results were largely consistent with Templeman's (1972) qualitative analysis of recruitment records. The latter indicated widespread year-class synchrony among Northwest Atlantic gadoid stocks.

In a follow-up study, Koslow et al. (1987) explored the most likely causes of large-scale coherence in recruitment for gadoid stocks in the Northwest Atlantic. Four distinct mechanisms were examined: wind-driven advective losses, run-off regulated nutrient dynamics, slope water inputs of nutrients, and predation. They considered also temperature and large-scale, but less well-understood, variables represented by principal components of sea surface temperature (SST) over the Northwest Atlantic and basin-wide atmospheric pressure patterns.

Koslow et al. (1987) also examined recruitment in relation to stock size. There was generally no clear linear or nonlinear relationship between egg production and recruitment for these stocks. They concluded that endogenous population mechanisms are unlikely to be the significant factors producing the apparent periodic fluctuations in these populations.

Several physical and biological variables in the region were found to exhibit similar large-scale coherence and apparent periodicity. Multiple regression analysis indicated that year-class success in Northwest Atlantic cod stocks tends to be associated with large-scale meteorological patterns and offshore winds. Recruitment to most haddock stocks from the Scotian Shelf to Georges Bank was negatively associated with abundance of 0-group mackerel, which could be due to predation during the winter and/or to a combination of environmental features including SST, large-scale atmospheric pressure systems, and freshwater outflows. Statistical analyses often did not define a unique set of variables that best predicted fishery recruitment. This was due to widespread intercorrelations among environmental processes and the likelihood that not all relevant processes entered directly into the analyses.

Koslow et al. (1987) suggested that the evidence of large-scale, apparently periodic, behaviour in the meteorology and physical and biological oceanography of the Northwest Atlantic demonstrated the need for improved understanding of regional climatic processes and their interactions. They called for detailed field study to determine the actual physical and biological mechanisms underlying apparent statistical relationships.

Cohen et al. (1991) reexamined the northwest Atlantic cod and haddock recruitment data, including additional years of recruitment data not used in Koslow's analyses and some stocks for which data were not previously available. They used the mathematical technique of first differencing to obtain time series data that were stationary. Their results indicated that, while large-scale effects probably do affect recruitment, local-scale processes operating over 10 to a few hundred kilometres, rather than large-scale physical forcing over several hundreds to thousands of kilometres, dominate the recruitment patterns of cod and haddock in the northwest Atlantic. Recruitment correlations were strongest for neighbouring stocks.

Winters and Wheeler (1987) found the recruitment patterns of the seven major spring-spawning herring populations in the Northwest Atlantic to be synchronous. They were determined largely by annual variations in overwintering temperatures and salinities associated with the Labrador Current. Removal of the environmental component of recruitment variability revealed dome-shaped stock-recruit curves with sharply ascending left limbs and relatively high degrees of compensatory density dependence in recruitment. The degree of such density dependence was positively correlated with recruitment variability which was negatively associated with mean stock size. This suggested that large herring stocks tended to have less recruitment variability but lower levels of density dependence in recruitment than smaller stocks.

Environmental variability was positively associated with recruitment variability. For stocks of the same size, the more northern populations of spring-spawning herring, inhabiting a more variable environment, had a higher level of density dependence on recruitment. Winters and Wheeler attributed the apparent high level of density dependence of recruitment in herring stocks to several features:

1. large egg size and herring's adaptability to a wide range of spawning conditions;
2. a density-dependent mortality associated with egg hatching success; and
3. a strong homing tendency, ensuring that changes in spawning populations sizes are spread evenly throughout the spawning area, thereby optimizing larval production.

Stocker et al. (1985) concluded that variation in recruitment of the Strait of Georgia stock of Pacific herring could be explained to a reasonable degree by spawning stock size and environmental factors that affect critical periods in their early life. They used a multiplicative, environmental-dependent Ricker spawn-recruitment model to identify significant environmental variables. The model indicated a significant dome-shaped relationship between temperature and spawning success with an optimal temperature during larval stages resulting in maximum production of recruits. Also, increased spawning success was associated with increased summer river discharge. Moderate density dependence was indicated for the Strait of Georgia herring stock.

Leggett and his colleagues have studied recruitment mechanisms in Atlantic capelin. Leggett et al. (1984) demonstrated the influence of meteorological and hydrographic factors in regulating year-class strength in eastern Newfoundland stocks of capelin. Year-class strength from 1966 to 1978 was found to be strongly influenced by onshore wind frequency during the period immediately following hatching and by water temperatures experienced during the subsequent period of larval drift. Wind conditions subsequent to hatching influenced both the timing of larval emergence from the beach spawning beds and the physical condition of larvae at emergence. The influence of wind on early larval survival appeared strong. The authors suggested that the role of water temperature during drift may be indirect, operating via its influence on food production. They speculated that this could operate by altering mortality rates due to starvation or by altering growth rates which could influence swimming performance, predator avoidance, and time to metamorphosis. The strong influence of these abiotic variables, independent of parent stock size, added support to their hypothesis that abiotic factors operating at critical periods in larval development may be more important than spawning stock biomass as regulators of year-class strength.

Frank and Carscadden (1989), for an offshore spawning capelin population on the Southeast Shoal of the Grand Bank, found that the formation of the dominant cohort during spawning season coincided precisely with a storm and sharp increases in both bottom temperature and currents in the area. This appeared to reflect destratification due to in site mixing. Emergence timed to periods of destratification was argued to be beneficial to larval survival.

The NAFO Scientific Council and CAFSAC in 1992 concluded that recruitment to the northern cod stock appears correlated with environmental factors such as the volume of the Labrador Current and salinity (NAFO 1992a; CAFSAC 1992b). In particular, similarities were noted between the time series of the areal extent of the Cold Intermediate Layer (<0° C) of the Labrador Current and of northern cod recruitment. A negative relationship was found between the area of the CIL waters along a transect off Bonavista and the number of age 3 cod for the years 1978 to 1988 (NAFO 1992a).

3.3 Some Broader Perspectives

Rothschild (1986) reviewed the extensive literature on recruitment variability and concluded:

> "The specific causes of fluctuations in fish-population abundance are supported by a long list of speculations and a short list of facts. The facts are: (1) the abundance of individual stocks has fluctuated for centuries; (2) evident human interactions with most

marine stocks is relatively recent; and (3) fish stocks have continued to fluctuate in abundance in the presence of increased toxic-chemical loading, eutrophication, habitat modification and fishing."

Rothschild (1986) proposed a population-dynamics-process model as an alternative to the traditional views of the stock and recruitment problem. This process model included a stability-inducing component associated with the species of interest and its trophic status, and a stability deteriorating component that is more a function of the environment, external to but affecting the trophic interaction of the species of interest. Rothschild's model is complex but so is recruitment variability. It attempts to provide a unifying model to explain fluctuations in population abundance on the grand scale. (For details, see Rothschild (1986)).

Others, however, take a more pragmatic view of the study of recruitment variability. Leggett (1988) observed:

> "I am convinced that there is no one explanation for, or solution to, the recruitment problem. New approaches, and a new willingness to consider alternative hypotheses will therefore be required. There are, undoubtedly, common themes. These provide a framework for a coordinated attack. It is clear, however, that many factors, both biotic and abiotic, operate to regulate the success of a year class. The hierarchy of importance in this framework is likely to vary through time and space, and between species. Identifying the important variables and understanding how they interact in response to key forcing functions is the challenge."

Wooster and Bailey (1989) concluded that there is "no simple, unifying hypothesis to explain recruitment variations for all species in all circumstances." Individual populations respond to their local environments in different ways.

In 1988, the Science Sector of DFO convened a national workshop on the factors determining recruitment variability in marine fish populations. The Steering Committee for the DFO National Recruitment Workshop (Sinclair 1988b) reached the following conclusions:

1. The characteristics of recruitment problems vary in important ways among species and areas. Thus, the preferred approach to research on recruitment questions is generally considered to be unique to the specific problem under consideration.
2. The rate of increase in understanding of the mechanisms controlling recruitment variability has not, in recent decades, been characterized by major breakthroughs that increase our understanding of recruitment variability of different species and geographic areas. It is probable that future increases in understanding will be incremental rather than dramatic.
3. There has been a shift in emphasis from the critical period concept to the view that recruitment may be more influenced by significant events and processes in the adult and juvenile life history stages. The current approach to solving the recruitment problem has become more balanced by placing some emphasis on later life history stages.
4. There is a paucity of knowledge of the biology and life history of many marine species. The result is that many hypotheses remain untested or poorly defined, especially for juvenile stages.
5. Advances in recruitment research depend upon repeated measurements of important biological and oceanographic features. This information is necessary on inter-annual and inter-decanal time scales. Although there are often time series data on fish stock abundance, there is a lack of comparable time series for physical and biological oceanographic features such as currents, temperature, salinity, and zooplankton abundance.
6. There is a growing acceptance that large-scale environmental forcing is of fundamental importance to recruitment of marine species. These environmental processes can operate simultaneously over large areas within one year and can also unfold slowly over many years.

On balance, I conclude that integrated physical-biological studies, combined with in-situ experimental studies, while expensive, show promise of advancing our understanding of recruitment variability. However, there will be no "quick fix" solution to the problem. Progress will require substantial funding and collaboration among biologists, and physical oceanographers on the one hand, and government and university scientists on the other. This is unlikely to provide an improved basis for fisheries management in the short term but could improve the basis for management sometime in the decades to come.

Meanwhile fisheries scientists, managers and the fishing industry will have to live with natural resource variability. The experience of the past several decades suggests that such variability will continue to confound man's attempts to manage the wild fisheries.

4.0 MULTISPECIES INTERACTIONS

In Chapter 3, I touched briefly on the subject of multispecies interactions. Scientists have been intrigued by the nature and implications of such interactions for quite some time. As early as 1972, ICNAF attempted to address the complexities of managing multispecies fisheries through the two-tier catch quota system implemented in the southern part of the ICNAF area. From the late 1970's onward, this topic has been of considerable interest to fisheries scientists.

There have been many attempts to categorize types of multispecies interactions. Two broad categories of species interrelationships are generally recognized: biological interactions and technological interactions. Biological interactions (such as predator-prey interactions, competition, habitat overlap) occur among the species present. The complexity of these interactions ranges from the relatively simple two-species krill and baleen whale systems in the Antarctic (see May et al. 1979) to systems involving a wide range of species with more complex trophic relationships. Biological interactions occur regardless whether any of the species are fished.

4.1 Technological Interactions

With technological interactions, the link between species is due to the presence and nature of fishing. The complexity of technological interactions increases as the number of species involved increases. These interactions may be direct, e.g. when two or more species are caught together, or indirect, e.g. when the gear used for one species alters the suitability of the habitat for another.

Examples of technological interactions are numerous. Murawski et al. (1989) analysed interactions among Gulf of Maine mixed-species fisheries. Pikitch (1989) examined technological interactions in the U.S. west coast groundfish trawl fishery and their implications for management.

Closer to home are studies of by-catch problems in the Scotian Shelf foreign fishery and their impact on domestic fisheries (Waldron and Sinclair 1985). These types of technological interactions do not pose any great scientific challenge. If adequate data are collected from the fisheries involved, the assessment of technological interactions is largely straight forward. The problem comes more with devising appropriate management regimes and enforcing them.

4.2 Biological Interactions

4.2.1 1977 FAO "Expert Consultation"

Biological interactions are, however, quite a different story. Here the scientific challenge is considerable. This is exemplified by four major scientific conferences on multispecies fisheries problems held from 1977 to 1989. FAO in 1977 (FAO 1978) convened an "expert consultation" on the management of multispecies fisheries. This session concluded that the theory of single species, existing and exploited independently of any other species, was not sufficient. It did not consider technological or biological interactions. Furthermore, most single species methods required extensive data. As more species are exploited, it becomes increasingly impractical to collect enough data to apply single species methods to each and every species of interest.

4.2.2 1979 Canadian Multispecies Workshop

In 1979, the Canadian Department of Fisheries and Oceans convened an international workshop on "Multispecies Approaches to Fisheries Management Advice" (Mercer 1982).

There were widely diverging views on the efficacy of existing single species approaches to management. However, the workshop concluded that there were examples of successful management actions based on single species concepts and that single species assessment models would continue to be useful for short-term prediction and be used for management advice indefinitely. It was also agreed that multispecies approaches to management would become increasingly important.

The workshop noted the following situations where multispecies approaches appeared to be particularly important:

1. For some fish stocks interaction effects appeared to play such a significant part that behaviour of the stocks, when standard single species management methods were applied, was totally unpredictable. Such an effect was seen in the events that had taken place in the North Sea since 1960. Recruitment and yield of many species increased when herring and mackerel stocks were reduced. The species composition of the area changed dramatically while the total fish biomass seemed to remain relatively constant.
2. Sometimes a single dominant predator-prey relationship can have such a significant influence that management advice for exploiting predator and prey might be erroneous. The interactive effects between cod and capelin off northeast Newfoundland and Labrador could fall into this category.

The Workshop concluded that, if the interaction effects of major predators were ignored, useful advice on large-scale changes in predator stocks might be impossible. Simultaneous advice on prey stock management might be unreliable because the natural mortality of prey species is affected by the biomass and age (size) structure of predator stocks. When single species assessments are used, management agencies should be advised that decisions affecting the abundance of predator stocks might influence the abundance of other stocks even if the effects could not yet be quantified.

4.2.3 1984 Dahlen Workshop on Exploitation of Marine Communities

The multispecies interactions issue was again addressed at the international Dahlen Workshop on Exploitation of Marine Communities in Berlin in 1984 (May 1984). In a paper on the observed patterns in multispecies fisheries, Gulland and Garcia (1984) stated:

> "The question of interaction and stability in multispecies fisheries has therefore become recognized as a matter deserving of increased practical and theoretical attention, and a number of studies have addressed this problem...but these have done little more than open the door on a complex maze of interrelated problems."

The authors considered the relevance of existing arguments about objectives of fisheries management to the multispecies situation. They concluded that there was a new element — the degree to which the balance between species should be an objective in itself. In this connection they noted that the problem of species balance was particularly prominent for marine mammals. The best formal attempt to deal with it was in Article 2 of the new Convention for the Conservation of Antarctic Marine Living Resources.

Gulland and Garcia (1984) distinguished two approaches to analyzing biological systems:

a) To look at each species separately and add, within the analysis of each species, appropriate terms for the interaction with other species (bottom-up approach); or
b) To look at the whole system in terms of energy flow and total catch and separate production into species only to the extent that it is possible and desirable (top-down approach).

The report of the Ecosystem Dynamics Group at this workshop noted:

> "Notwithstanding its potential shortcomings, the single-species approach has met with remarkable success, especially in managing certain long-lived species, and is the basis for most of our present-day management decisions. It was perhaps this lack of clear necessity that kept the ecosystems perspective from advancing in a field whose pragmatic concern is fishery management."

The Group considering Strategies for Multispecies Management suggested that research be concentrated on the problem of predicting major changes in species composition, particularly when there are large changes in fishing mortality. It recognized that predictions at the multispecies level might be unrealistic, except for communities with few dominating species. It argued that the essential requirement is for understanding the mechanism of interaction among groups.

Larkin (1989) recently described the Dahlen Conference as "a comprehensive but discouraging account of present knowledge of multispecies fisheries."

4.2.4 Modelling Multispecies Interactions

CAFSAC sponsored a Workshop in 1984 on the inclusion of fishery interactions in management advice (Mahon 1985). Halliday and Pinhorn (1985), in a review of existing management strategies for Canadian Atlantic marine fisheries, noted that biological interactions had been extensively debated in the scientific advisory process. This had influenced advice but had, as yet, seldom resulted in management strategies which took biological interactions explicitly into account. For cod and capelin, this had taken the form of setting low exploitation rates for capelin because of its importance as food for cod. DFO attempted a more quantitative approach for the southern Gulf cod stock where the effect of mackerel biomass on cod recruitment was modelled. Management strategies for cod from 1976 to 1978 were based on this model. These models, however, proved to have little predictive capability and in 1979 the $F_{0.1}$ management strategy was again pursued for this stock.

During the 1980's a variety of approaches were developed to model multispecies interactions internationally. Andersen and Ursin (1977) made one of the first large-scale attempts to describe a complicated fisheries systems with several interacting species. This ecosystem approach was unsuitable for fish stock assessment purposes but it stimulated the development of multispecies virtual population analysis (MSVPA) methods (Sparre 1980; Pope and Knights

1982; Gislason and Helgason 1985). This has become one of the most visible attempts to develop a multispecies approach for fisheries management. Before tracing the evolution of this methodology, I would first like to situate this in the broader context of alternative approaches.

Kerr and Ryder (1989) reviewed approaches to multispecies analyses of marine fisheries. They grouped the various approaches into four categories:

1. Descriptive Multivariate
2. Dynamical Multivariate
3. Multivariate System
4. Integral Systems

The *Descriptive Multivariate* category refers to a variety of statistical and other techniques used to discern and describe patterns in ecological data, usually obtained from commercial catch reports or research vessel surveys (e.g. Gabriel and Murawski 1985; Mahon 1985).

The *Dynamical Multivariate* category of analyses builds upon knowledge of single-species processes to deduce the behaviours of multispecies systems. Included in this category is the ICES MSVPA approach, discussed later.

The *Multivariate System* category includes models which depend upon aggregations of data derived from sources such as conventional catch statistics. One of these is the multispecies Schaefer model which treats the total fish biomass of all species as if it were a single stock (see Chapter 13). Other models have been formulated at different levels of aggregation (e.g. Ralston and Polovina 1982; Silvert and Crawford 1989).

The *Integral Systems* category operates at the larger aggregation scale of community or ecosystem behaviour. Sheldon et al. (1972) observed that the average particle-size spectrum of offshore pelagic communities tends to be smooth and orderly. In other words, the distribution of biomass concentration is a relatively flat but slightly decreasing function of organism body size. From this, efforts have been made to estimate the biomass of a size interval that is difficult to estimate directly, e.g. fish, from the biomass of a lower trophic level, e.g. phytoplankton. It is a considerable distance from this observation however, to estimating fish biomass. Some attempts have been made. Kerr and Ryder (1989) recognized that "although the evidence accumulates, much careful work remains to be done before the biomass-spectrum can be routinely used as a fisheries forecasting procedure." This is probably an understatement of the difficulties involved.

Another example of the Integral Systems approach is the mordoedaphic index (MEI) which has been often applied to lakes and reservoirs and more recently to rivers. Community production is predicted based on one or more controlling or limiting abiotic factors. The MEI consists of a nutrient value in its numerator and an energetic (morphometric) variable in the denominator. Kerr and Ryder (1989) suggested that it may be possible to extend the MEI concept to marine systems.

Of the four categories described by Kerr and Ryder, the greatest effort is being applied in northern temperate waters to develop the *dynamical multivariate* type of model. Of these, the most visible have been multispecies virtual population analyses in the Northeast Atlantic. Andersen and Ursin (1977), in their attempt to construct an ecosystem model of the North Sea, pointed out that predation can be a substantial part of the natural mortality of young fish. Sparre (1979) and Ursin (1982) noted that the single-species yield-per-recruit models generate severe inconsistencies when applied to predator and prey stocks simultaneously. Daan (1973, 1987) noted that species interactions may affect fish population dynamics in several ways. The most obvious is predation of one species on another. Predation may occur in the prerecruit stage with a consequent effect on recruitment to the prey stock, or in the postrecruit stage, in which case there may be direct competition between predation and the fishery for a given species. The interactions between two species may be complex. For example, herring larvae have been shown to contribute to the food of juvenile cod and juvenile herring to the food of adult cod. Conversely, cod eggs have been found in herring stomachs.

Existing assessment models could not capably account for these interactions in the prerecruit stages. At the 1979 ICES meeting Helgason and Gislason (1979) and Pope (1979) independently presented a method to calculate predation mortalities. This method led to Multispecies Virtual Population Analysis. This in turn led to the establishment in 1980 of an ICES Ad Hoc Working Group on multispecies assessment model testing in 1980. This Working Group stimulated a large international fish stomach sampling program in 1981 for the predators — cod, whiting, saithe, mackerel and haddock. Hence, 1981 is known in ICES circles as the "year of the stomach". The results of this sampling program were reported to the 1983 and 1984 ICES meetings. This led to the establishment of the ICES Multispecies Working Group which first met in 1984. Through its activities, the approach to MSVPA has evolved. For details of the Working Group's activities, see Pope (1991).

Whereas in the single species VPA, natural mortality, M, is input as a constant, in MSVPA it is partitioned into M1 (Mortality due to disease, starvation, spawning stress, senility, etc.) and M2 (Mortality due to predation by other species and to cannibalism). VPAs for each stock are done simultaneously with M1 fixed and M2 calculated according to the size of the relevant predator populations. Upon completion of the first set of VPAs, new population size conditions are calculated and the M2 values for each population are recalculated. The analysis is repeated with changing M2 values until certain convergence criteria are met.

The MSVPA requires information on the annual food composition for the various predators by age group. The MSVPA estimates simultaneously the number of each species at each age in the sea, the fishing mortalities, and the predation mortalities (M2). For a description of the model and its assumptions, see Daan (1987).

Daan (1987) reported that one main conclusion from the trial runs with MSVPA so far is that for adult fish the results agree well with traditional single-species assessments. It seemed unlikely that MSVPA would affect the short-term catch predictions on which the advice on catch quotas is based, when recent levels of exploitation are projected forward. This occurs because stock sizes and fishing moralities change in proportion. However, the MSVPA yields different results when management aims at changing short-term exploitation levels on particular stocks because of the variable impact of the predator stocks.

The multispecies approach has not replaced the single-species assessment as the basis for fish stock management in the Northeast Atlantic. As Daan (1987) observed:

> "There are still too many uncertainties in the underlying assumptions of the model to be generally accepted. More importantly, however, it represents only a tool that assists in evaluating past changes, and as long as sound multispecies management approaches have not been formulated, the scope of this work is bound to remain limited."

One important result is the estimate of high predation mortality on juvenile fish. The age of first capture in the food fisheries lies generally within the stage of life when predation mortality is still considerably higher than had been assumed formerly. This means that, particularly for North Sea groundfish, past mesh assessments have undoubtedly overestimated long-term gains associated with increased mesh sizes. The adverse effects of industrial (reduction) fisheries on the human consumption stocks through the by-catch of juvenile fish appear to have been overestimated.

Pope (1991) noted that the most important insight gained by the ICES Multispecies Working Group is: due to predation mortality, natural mortality rates on the younger ages are higher than previously estimated and are variable from year to year. This means that the numbers of young fish in the North Sea were previously underestimated. Conflicting advice emerges from the multispecies model versus single species models. Single species models predict increases in long term yield if effort is decreased in the roundfish and saithe fisheries or if mesh sizes are increased. The results from the MSVPA predict the opposite. Pope noted that these differences arise not because the multispecies natural mortality rate is high but because it varies with predator density and hence increases if management measures generate higher stocks of predators such as cod, whiting and saithe. This difference between single and multispecies models only seems to apply for long term yield. In the short term the yield seems to be much the same for either approach. Apparently, this is because short term yield calculations are little affected by the use of any average level of natural mortality. Systematic shifts in the level of predation mortality might, however, cause differences.

O'Boyle et al. (1985) pointed out that the North Sea MSVPA model appears inappropriate for Canada's east coast because of the relative lack of exploitation of age one to three fish in these waters. It is in these age groups that predation mortality appears to be highest. The fisheries in the North Sea generally exploit populations at substantially younger ages than is the case off Canada's east coast. For instance, North Sea cod and haddock are fully recruited to the fishery at age three whereas in Canada recruitment occurs more commonly at ages of five or older. In addition, there are large industrial fisheries in the North Sea for Norway pout, sandeel and sprat, all of which exploit fish of ages zero to four. Because Canadian fisheries generally exploit older, larger fish, the MSVPA approach is likely to be less useful. The authors recommended pursuit of other models more suitable to available data sets and the exploitation patterns in the fisheries.

4.2.5. Utility of Multispecies Models

In 1989, ICES sponsored a Symposium on Multispecies Models Relevant to Living Marine Resources, in the Netherlands. The first session of the Symposium was devoted to MSVPA. The Symposium concluded that multispecies models have had little influence on fisheries management, but it would be only a matter of time before widespread multispecies management becomes a reality (ICES 1991).

A session which considered the usefulness of multispecies models was most controversial. Some participants believed that such models did not help because they were too uncertain and difficult to understand. Others argued that it was better for long-term management decisions to use uncertain multispecies models than single-species models which were likely to be wrong.

Two of the most interesting papers at this Symposium were those by Brugge and Holden (1991) and Gulland (1991). Both questioned the utility of multispecies models for management. Brugge and Holden (1991) addressed three types of questions about species interaction models and fleet interaction models (the biological interactions and technological interactions referred to earlier):

1. How credible are the models?
2. What solutions do they provide to present problems?
3. What new problems do they present?

The authors concluded that species interaction models will make management more difficult because the credibility of the results which they produce can be easily questioned and because they produce far more problems than solutions. On the other hand, fleet interaction models were seen as far more convincing and likely to provide managers with a very useful decision-making tool.

Gulland (1991) posed the question, "Do managers want multispecies models?" He noted a widespread belief: if scientists could devise models that described adequately multispecies interactions and modified their advice accordingly, managers would welcome this. It would be a significant step towards the Holy Grail of perfect fishery management. Gulland's thesis was that this belief is largely misplaced. Scientists should recognize that multispecies advice will not necessarily be welcomed, and should take this into account in giving their advice. He suggested that managers are well aware that the widespread use of multispecies models would complicate their lives, and doubt that they would lead to better decisions.

Gulland identified several conditions under which managers might welcome multispecies advice:

1. It implies changes in current management practices that are straightforward and preferably minor.
2. The science involved is transparent to a non-specialist such that the inclusion of new information clearly suggests changes in the advice on management of the resource.
3. The new results are scientifically uncontroversial, i.e., the provision of the new advice will not be accompanied by substantial debate over the science itself.
4. Suggested new measures will reduce or at least not increase conflicts between different interest groups.
5. There will be no clear losers that cannot easily be justified and dealt with.

Judged against his five criteria, Gulland concluded that most analyses of technological interactions would be welcomed. He considered the science to be transparent and the results likely to be clear. Managers could easily take the required actions: adjustments to TACs to account for incidental catches, closures of areas or times of very high by-catch.

Regarding the conclusions of the ICES Multispecies Working Group, Gulland opined that the science was transparent. The conclusions generally called for less dramatic changes to the status quo (less dramatic reductions in fishing effort, smaller mesh sizes) than the single-species models. While managers might welcome these (Brugge and Holden suggest they would not), they were unlikely to result in much benefit. Gulland suggested that the effect could be negative. By throwing doubt on the single species models, the multispecies results could be used to argue against any movement toward a lower fishing mortality.

Gulland cited the interactions between krill and whales in the Antarctic as one fishery in which the interaction between marine mammals and fisheries could become the most serious management issue. If quantitative models are developed which show that the krill harvest is affecting marine mammals, great uncertainty will probably surround the conclusions. Gulland observed:

> "Managers are...unlikely to welcome advice that...such a threat is present unless the proof is strong enough to overcome the opposition of the harvesters. Such proof will be difficult to obtain for the Antarctic ecosystem and the manager's position will undoubtedly be awkward."

Another example where multispecies interactions could have significant management implications is the hypothesized interaction between pelagic and demersal fish stocks in the North Sea. From the early 1900's to 1960 the catches of the chief demersal species, except during the two world wars, was generally stable around 200,000 tons. After 1960 the catches surged to the 400,000–500,000 ton range. Year-class strengths of the demersal species increased considerably. This occurred at a time when the herring and mackerel stocks in the North Sea had collapsed. Various authors have suggested that diminished predation by adult pelagic fish on larval cod and other demersal species could have caused the abundance of herring or mackerel to affect demersal recruitment.

Walters et al. (1986) demonstrated an interaction between Pacific cod and herring in Hecate Strait, British Columbia. The abundances of these species were found to be inversely correlated. The authors speculated that peak cod abundances in northern British Columbia during the late 1950's may have been partly responsible for the collapse of the herring reduction fishery of the 1960's. In this instance, they hypothesized that cod predation strongly influences herring recruitment rates. They suggested the need to search for cod-herring "co-management" strategies that might prevent the two fisheries from limiting each other's fishing opportunities.

For the North Sea demersal-pelagic interaction hypothesis, Gulland (1981) suggested that the sensible management policy would be to keep the herring stocks at a level no higher than needed to supply the relatively small human consumption market. He speculated that the declared policy of fully rebuilding the pelagic stocks could significantly reduce the total value of North Sea landings. Gulland (1991) noted that this observation had been made on several occasions but managers had found it difficult to act because of the uncertainty surrounding this hypothesis.

Overall, Gulland observed that the conclusions reached so far looked rather discouraging for those who believe that building better multispecies models will bring better management. He suggested that the best course of action was for scientists to take the conditions under which managers have to operate more fully into account in their analysis and presentation of results. This could involve:

1. Paying more attention to policies which reduce risk rather than maximizing expected catches;
2. Where uncertainties are inevitable, doing more to explain them and their implications (e.g. giving confidence limits or estimating probabilities of alternative outcomes);
3. Giving advice that looks ahead to facilitate predictive rather than reactive management; and
4. Improving communications between biologists, managers and fishermen.

Despite the misgivings expressed by Brugge and Holden (1991), Gulland (1991), and others, the Co-Convenors of the ICES Symposium were optimistic that multispecies models would be used appropriately for multispecies management. They noted that during the 1980's multispecies models had become part of the thinking of the community of scientists that formulate management advice. In their view, it was only a matter of time before multispecies management becomes a reality (ICES 1991).

4.2.6. A Canadian Example — Cod/Capelin/Seals

Meanwhile, multispecies interactions keep surfacing in Canada as a very real and pressing issue. The 1990 Harris Panel Report on Northern Cod raised important questions concerning predator/prey relationships and their impact on the abundance of northern cod. The Panel was repeatedly confronted in the course of its public hearings with the issue of the growth of the seal herds and the possible impacts upon cod abundance. The Panel examined three sets of predator-prey relationships: Cod/Capelin; Seal/Capelin; and Cod/Seal.

The fact that capelin are an important prey species for cod and other fish, for seals and other marine mammals, and for sea birds had contributed to the conservative strategy for capelin management. This strategy set the TAC at 10% of the estimated spawning biomass. The Panel concluded:

> "On the basis of current knowledge....a large capelin population may be an essential precondition of a large cod population. Until we are convinced that we have an adequate understanding of cod/capelin interaction, we should err, if at all, on the side of underexploitation of capelin rather than on the other side."

For seals and capelin, the Panel noted that, while cod may be the most important predator of capelin, the harp seal is also a significant competitor for the same source of food. Capelin appeared to be a major prey of harp seals. The Panel observed that a harp seal herd of several million animals possibly consuming 6% of its biomass daily requires fish of whatever species in millions of tons. If a considerable portion of this total is capelin, the question arises as to how big the herd can grow before its predation precipitates a collapse of capelin stocks. If capelin decline, would the seal herd decline or concentrate more heavily upon other prey?

Regarding the relationship between cod and harp seals, many in the fishing industry contend that the harp seal is a significant predator of cod. Recognizing that the available scientific evidence was limited, the Panel was convinced that seal predation is "a matter of concern and clearly unresolved", notwithstanding the earlier views of the Malouf Commission that harp seals were not a significant predator on cod (Malouf 1986). More recently, the NAFO Scientific Council reviewed the sudden, drastic and unexpected decline in abundance of northern cod in 1991. It concluded that the increase in the harp seal population probably had an effect on cod either directly by predation or indirectly by competition and could have contributed to, but not accounted for, the decrease in abundance of northern cod (NAFO 1992a).

The issues the Northern Cod Panel raised concerning the interactions among cod, capelin and seals illustrate the importance that multispecies interactions could play in influencing management of particular species. It also illustrates the considerable uncertainties which continue to surround all questions of multispecies interactions. The 1980's saw some advances in tackling the complexities of modelling multispecies interactions. But multispecies management remains a distant and elusive objective. The evidence from modelling in the North Sea suggests that it is also an objective which managers in many circumstances may prefer to avoid. Nonetheless, improved knowledge of multispecies interactions is required in order to take informed single species management decisions.

5.0 FISHERIES ECONOMICS

5.1 Early Development and Impact

As described in Chapter 3, the foundation of fisheries economics was laid in the mid 1950's at a time when Beverton and Holt, and Ricker, were establishing the basis for fish population dynamics. The fundamental tenets of fisheries economics theory flow from the seminal articles by Canadian economists Scott Gordon (1953, 1954) and Anthony Scott (1955). They established that, in an open access fishery, economic rent would be dissipated by the inevitable rush of fishermen to maximize their individual share of the catch, leading to an excess input of labour and capital into the fishery. Following Hardin (1968), this became known as the Tragedy of the Commons (see Chapters 7–9).

Up until the early 1970's, the scientific input to fisheries management, in Canada and elsewhere, had been dominated by natural scientists, particularly fisheries biologists. They played a key role in assisting Canadian fisheries managers in the international commissions (e.g. ICNAF). They emphasized the need to control and reduce fishing effort from a biological perspective. Fishing effort controls in the open-access international fishery were impractical. This led to the widespread introduction of catch quotas (TACs) as the regulatory control mechanism.

While TACs could control the overall level of fishing mortality, they did nothing to deal with the economists' preoccupation, the need for efficiency. Leading fisheries economists have all emphasized economic efficiency as the appropriate objective for fisheries management. This is evident not only in the early literature of the 1950's but also in more recent expositions by Clark (1976), Anderson (1977, 1986) and Hanneson (1978). Economists argued that management advice based on the biological approach almost always led to economic inefficiencies (Crutchfield 1961; Pearse 1979b).

Economic theory began to affect management in the late 1960's for Canada's domestic fisheries such as Pacific salmon and lobster (see Chapter 7). Perhaps the chief impact was the recognition that, without management intervention to alter the common property "race for the fish", Canada would have too many fishermen and boats chasing too few fish. Economists proposed limitation of entry to deal with this problem. At the end of the 1960's, Canada began to experiment with limited entry. By the mid 1970's, limited entry licensing had become an established tool of Canadian marine fisheries management.

Anderson (1987) portrayed the evolution of thinking from the early biological models to the fisheries economic models (what he termed 'bioeconomic' models) in the following terms. Fisheries biology concentrated on fishing mortality as the control variable but considered it exogenous to the biological model (Fig. 18-14). However, fisheries economics theory as advanced by Gordon, Scott and Crutchfield considered fishing mortality as endogenous.

The 'bioeconomic model' got this name because it applied economic concepts to the Schaefer general production model. It viewed the fishing industry as a combination of individual firms or boats which react to market prices, input costs, and the production function and exert fishing effort as long as it is to their financial advantage (Fig. 18-15). The production function is the relation between fishing effort and yield. This model portrays profits as the driving force of the whole system. It assumes that fishermen do not fish for fish but rather for the profits they can earn. Some economists have questioned this assumption recently based on studies which show that nonmonetary factors also motivate fishermen to fish (e.g. Karpoff 1985; Apostle et al. 1985). However, it is still a central tenet of fisheries economics.

Also central to this model is the assumption of an open-access equilibrium, at which the stock will not change because the catch is equal to natural growth and fleet size does not change because profits are sufficient to keep existing vessels operating but not high enough

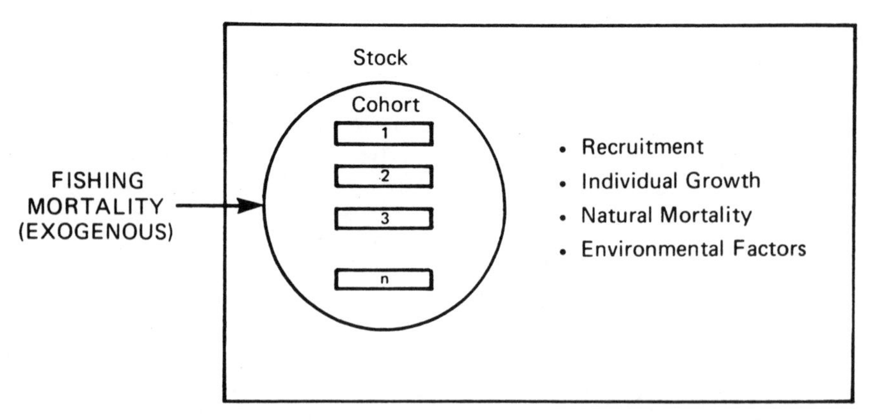

FIG. 18-14. The biological approach to fisheries management. Fishing mortality is viewed as an exogenous variable which acts on the stock of fish to determine the equilibrium stock size and composition (from Anderson 1987).

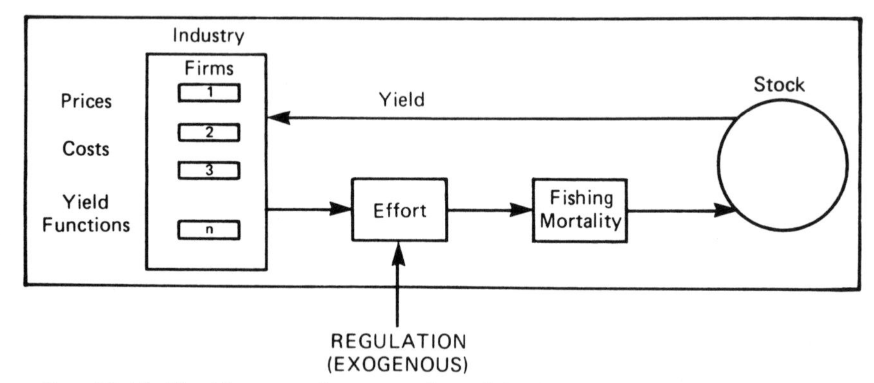

FIG. 18-15. The bioeconomic approach to fisheries management. Regulation is viewed as an exogenous variable which, given industry and fish stock characteristics, determines the equilibrium fleet size and level of fishing effort, and hence determines fishing morality and equilibrium stock size and composition (from Anderson 1987).

to attract new entrants (see Chapter 3). At this open-access equilibrium point, excess effort is being employed in the fishery. Hence, it is economically inefficient. As an alternative, economists postulated an optimum point called Maximum Economic Yield (MEY) (see Chapter 4).

Although other objectives were acknowledged to exist (e.g. income distribution, employment and balance of trade), the pursuit of economic efficiency became the primary obsession of fisheries economists. Economists proposed policies to change the behaviour of fishermen in the 'desired' way. Policies which had been favoured on 'conservation' grounds, e.g. total quotas, gear restrictions, closed seasons (see Chapter 5), were rejected as economically inefficient.

Limited entry licensing was initially favoured as the means to achieve economic efficiency. Economists persuaded fisheries managers to use this approach widely during the 1970's. By the end of the 1970's, however, it had become apparent that limited entry licensing was insufficient to achieve their desired objective of economic efficiency. Economists then promoted taxation and individual transferable quotas as alternative regulatory tools. Taxation on landings has been generally ignored as a regulatory tool, but individual quotas were increasingly embraced during the 1980's as more promising to achieve economic efficiency. From a small beginning in Canada in the Bay of Fundy herring fishery in the late 1970's, individual quotas were implemented during the 1980's in many east coast fisheries (for details, see Chapter 9). In general, these have been implemented without a provision for free transferability. This violated one of the cardinal tenets of recent fisheries economics theory. At the beginning of the 1990's, however, individual transferable quotas were being tried in some fisheries (e.g. Gulf of St.

Lawrence small dragger groundfish fishery). This approach will likely become more widespread during the 1990's.

5.2 Fisheries Economics Within Canada

A small group of Canadian academics, along with their U.S. counterparts, have been the major developers and contributors to fisheries economics theory. Such names as Gordon, Scott, Pearse, Copes, Clark and Munro are well known in the field. Within the federal fisheries department, the resources devoted to economic research have been very small relative to those devoted to the natural sciences. By the late 1960's the Department recognized that economic rationalization of the Canadian fishing industry was an important objective (DOF 1968e). The Department's organization included an Economics Service, divided into two sections, one dealing with resource management and the other with regional development. Most of the field work was undertaken by Economics units attached to each administrative region. Most of the professional staff, however, were at the national headquarters. The Economic Service's research unit then consisted of 17 professionals (The number of economics professionals in DFO in 1991 was 51). During the 1960's expenditures on economics declined from 5.8% of the Department's scientific expenditures in 1962–63 to 3.9% in 1968–69. The expenditures on biology remained relatively stable in percentage terms during this period (around 23–24%). A large proportion of the so-called scientific expenditures during the 1960's were directed toward engineering and fishing technology reflecting the developmental emphasis of that era (Table 18-10).

University and management consultants were being utilized "to investigate such aspects of the fisheries economy as the underlying motivation of those individuals who enter the salmon sport fishery on the West Coast; the development of the fisheries of the Northwest Territories, as related to the employment of native peoples; the impact of recent technological advances and other factors on the administration of the Atlantic oyster fishery; the rationalization requirements of the Atlantic ground fishery in relation to the market ability of its products; and the productivity implications of the program by which certain Newfoundland fishermen and their families are being re-located in more viable communities" (DOF 1968e).

Other investigations during the previous 5 years included "a cost-and-earnings study of the Atlantic fishing fleets, which had been updated each year since 1952, rationalization studies of such diverse fisheries as those of the Northwest Territories, and that of Pacific salmon, Atlantic lobster, and groundfish off Newfoundland; and benefit-cost studies of the Great Lakes – lamprey control program and the Newfoundland household-resettlement program" (DOF 1968e).

This snapshot is sufficient to illustrate that the activities of the small band of Departmental economists reflected the preoccupation of academic economists at that time — the pursuit of economic efficiency and "fisheries rationalization."

The Science Council of Canada in its 1970 Report "This Land is their Land" recognized that there was a

TABLE 18-10. Canadian (Department of Fisheries) Expenditures by discipline, for fisheries related scientific activities during fiscal years 1962–63 to 1968–69.

	($000)						
Discipline	1962–63	1963–64	1964–65	1965–66	1966–67	1967–68	1968–69
Bacteriology	118	122	128	144	155	177	192
Biology	489	478	506	763	956	1130	1397
Chemistry	59	61	64	72	78	87	96
Economics	119	127	140	151	206	236	271
Engineering	432	424	578	991	1249	1366	1421
Fishing Technology	291	280	563	1159	1490	1540	1366
Food Technology	318	330	345	388	421	473	517
Naval Architecture	64	62	124	254	326	336	299
Sociology	38	36	73	153	196	202	178
Statistics	19	20	22	23	32	37	42
Administration	105	105	154	266	329	345	322
TOTAL	2052	2045	2697	4364	5438	5929	6101

Source: The Senate of Canada. Proceedings of the Special Committee on Science Policy. 1st session — 28th Parliament. 1968.

need to "develop social and economic research related to fisheries and wildlife."

Limited entry programs lacked a sufficient background of social and economic research. Much of the research that had been done was not being applied (Science Council 1970).

Although the Science Council judged the economic research effort by the Department as inadequate, some economists played a prominent role in shaping fisheries policy during the 1970's. The 1976 Policy for Canada's Commercial Fisheries was shaped largely by the Department's chief economic guru for many years, W.C. (Bill) MacKenzie. This policy enunciated the "best use" concept as the guiding principle for future fishery management. The "best use" concept incorporated biological, economic and social considerations (see Chapter 4).

While this policy document emphasized economic and social factors, there is little evidence that economic and social research was bolstered following its publication. In the post-extension era after 1977, the Department hired a handful of additional economists while marine fisheries stock assessment research was doubled and surveillance and enforcement strengthened considerably. The emphasis was on stock protection and stock recovery. In the post-extension euphoria many in the industry and provincial governments felt that with the displacement of foreign fishing effort and stock recovery there would be a need for more, not less, fishing effort (see Chapter 6). While the federal government did not share this view, it did not significantly strengthen the capability for economic and social research, except for economics staff hired for the British Columbia Salmonid Enhancement Program (SEP).

SEP represents an instance where economists made a significant contribution to Canadian fisheries policy. In general, in fisheries there has often been a wide gap between the theory of economics models and management practice. This may be the result of the lack of suitable models. SEP is an exception to this general pattern. In May 1977, Canada and British Columbia completed a $6 million two-year planning phase for the proposed Salmonid Enhancement Program. The general objective was to double the existing level of catches to historic levels, and to maximize the net social and economic benefits subject to technical and budgetary restraints. Economists developed a planning framework to measure the impact of the proposed expenditure of $150 million over 5 years. The framework was multi-objective. It utilized what was called a "5-account" system to evaluate the impact in the following areas:

1. National Income
2. Regional Development
3. Native People
4. Employment
5. Environmental and Resource Preservation.

Pepper and Urion (1979) described this model. Details of implementation are given in the SEP Annual Reports.

5.2.1 Some Shortcomings in Fisheries Economics

The 1976 Policy specified the incorporation in resource management models of "not only biological and environmental, but also major social and economic components of the system." However, there appear to be no fully developed models in the Canadian context that meet that goal.

To some extent, this reflects the broader world situation with respect to the application of economics in fisheries management. Huppert (1982) pointed out that the established economic theory of fisheries examines the fundamental economic issues under severely simplified assumptions and produces "conclusions of great generality." He suggested that for specific, applied analysis the empirical content of the theory needed to be improved. Work was required to specify underlying bio-physical relationships, assumptions about industry, structure and market power and uncertainty.

Regarding the biological underpinning, we have seen that the general production models, which were the basis for general fisheries economics theory, are not satisfactory from an ecological perspective. Characteristics of a single species often fail to meet the assumptions of the models and the models fail to consider multispecies interactions explicitly.

Another shortcoming is the use of fishing effort as an "omnibus variable" (Scott 1955). Early on, economists adopted the neoclassical assumptions of the theory of the firm and industry which produce short run adjustments to input and output prices. Anderson (1986) has integrated the neoclassical theory of the firm with the fishing effort model (see Chapter 3).

Huppert (1982) suggested that applied economic research must find tractable, yet realistic, specifications of the cost functions and gross value functions and tie these to biological relationships. He noted that some recent studies had examined how to incorporate risk (e.g. Smith 1980; Dudley and Waugh 1980) and how to design a harvest strategy incorporating adaptive control for learning (e.g. Walters and Hilborn 1976). An important point is that virtually all such studies overlook or assume away some important feature of the fishery being studied.

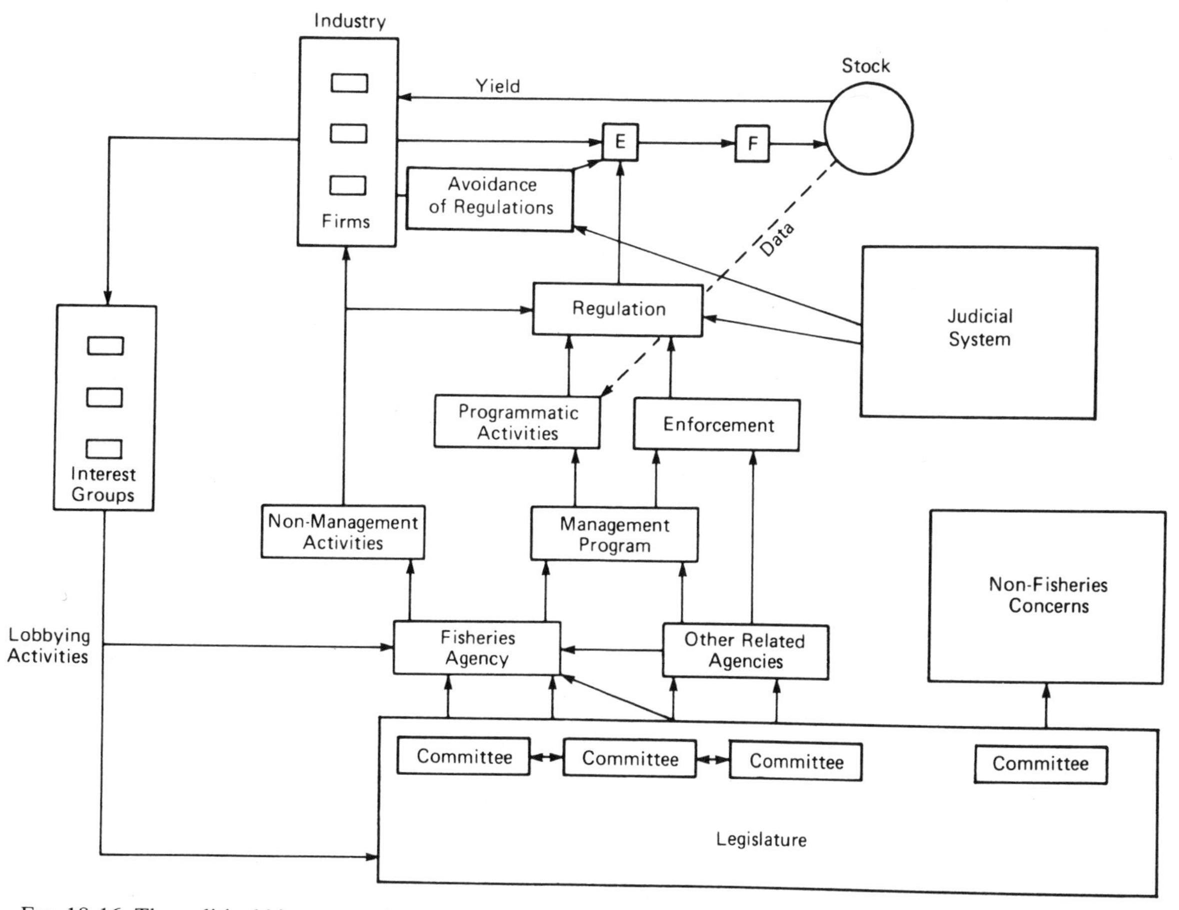

FIG. 18-16. The political bioregunomics approach to fisheries management. All major elements are endogenous to the model. Regulations and their actual effect on fishing effort are determined by the interactions of industry participants, interest groups, enforcement officers, and personnel from various agencies, the judicial system, and the legislature (from Anderson 1987).

5.3 Future Research Directions for Fisheries Economics

Anderson (1987) argued that the existing paradigms of fisheries management, which had been broadened from the biological to the bioeconomic, were insufficient. He proposed expansion of the fisheries management paradigm to include institutional structure and function. He termed this the "political bioregunomics" approach. Anderson noted that fisheries managers often ignore the advice resulting from biological and bioeconomic models. Actual programs often contradict the advice of either the biologists or the economists. This is sometimes seen as a failure by the biologists and economists to communicate clearly their message to decision makers.

Anderson contended it was more likely that the decision makers understood the arguments very well but considered other issues more important. From this he concluded that it is also important to study why officials implement the type of regulations they do. His expanded paradigm (Fig. 18-16) encompasses aspects of economics, political economy and biology as they apply to regulation:

> "Just as the bioeconomic approach considers the interactions of the fish population and of the industry, the political bioregunomic approach looks at the three-way interaction among fish stocks, industry and government entities."

Anderson describes the components of this model in the context of the U.S. political and fisheries management system. The legislative/political arrangements differ in Canada. Nonetheless, the approach suggested has considerable merit for future research.

Such system models tend to be generalized and academic, rather than specific and practical. It would be difficult now to construct a meaningful model describing in an integrated manner all the elements concerned, but work should be pursued on submodels of the individual elements. But, we must not let the pursuit of this attractive systems approach divert us from pragmatic and immediate economic issues of concern to fisheries managers.

Managers have immediate and practical requirements for economic analyses concerning issues such as allocation of the resource among competing groups. Economic analysis can be used to examine trade-offs among economic gains and losses at different points in time, resulting from specific fisheries management decisions and regulations (e.g. Rettig 1987).

Economists have preferred to set aside "messy and emotional" distributional issues or to leave them to others such as political scientists and sociologists (Crutchfield 1985). Yet these distribution issues are the most vital concern to many in the industry.

Crutchfield (1985) criticised a wide split in the discipline of fisheries economics between those engaged in high level theoretical modelling and those doing applied research:

> "In some instances the practitioners of mathematical bioeconomic modelling have strayed far beyond the reasonable limits of our biological data and reasonable assumptions as to the nature of the industry.... As the models themselves become more elaborate the likelihood of obtaining adequate data for practical implementation of the models decreases (probably at an exponential rate).... The danger here is that the pursuit of the theoretical at the expense of the applied can lead to fundamental mistakes when the analyst loses touch with the industry."

This damning indictment of theoretical fisheries economics appears justified. The theoretical economists, and indeed some of the theoretical fisheries biologists, seem to have lost touch with the realities of the industry they are studying.

Crutchfield also took issue with the traditional fisheries economist's slavish devotion to economic efficiency:

> "Current management practices seem to be undertaken to minimize conflict between user groups, to protect the livelihood of those in the industry and to minimize the political discomfort of the regulators. Economic efficiency as a goal of management is of secondary importance, verging on the inconsequential".

Crutchfield went beyond criticism to offer suggestions for a fisheries economics research agenda. While his comments were particularly aimed at the fisheries management scene in the United States, they are also relevant to Canada. In particular, he suggested that further research is required in a number of areas, especially for certain key assumptions of the traditional bioeconomic models:

1. Fishing effort can be measured by a well-defined unit ("E").
2. The fleet consists of a large number of homogeneous vessels.

These assumptions are contradicted by some recent studies which show that skill, experience and knowledge are important but are difficult to measure, e.g. Crutchfield (1985). These indicate that the fleet is more heterogeneous than had been hitherto assumed in economic models.

Recent studies also indicate that fishermen do not just pursue profit. They fish for a variety of reasons, some of which have no apparent economic rationale. Fishermen receive satisfaction from fishing which is not measurable in monetary terms (Anderson's Worker's Satisfaction Bonus concept). Further research is required to determine how fishermen respond to profits and losses.

Another important area of research involves applying stochastic modelling to fisheries economies. This gets around the assumption of certainty because stock size and other variables are assumed to be random with a finite distribution around some expected value. For a review of basic methods and results of stochastic bioeconomics, see Andersen and Sutinen (1984). They found that, in several cases, optimal policy under stochastic conditions was qualitatively different from optimal policy under deterministic conditions. However, such differences were ambiguous. Lewis (1982) and Smith (1980) concluded that deterministic policies are normally reasonably good substitutes for stochastic policies.

With respect to future research, Bromley (1977) argued that fisheries economics and policy are concerned with the study of alternative institutions pertaining to who may fish, what technology will be employed, and what will happen to the harvest. This suggested that fisheries economics should be approached from the intellectual framework of welfare economics. He rejected the notion of Anthony Scott (1977) that economists should press for the economic management of fisheries "as far as possible on allocative or efficiency criteria before modification, if absolutely necessary, by equitable or distributional criteria." From this perspective, Bromley suggested that economists should model the level and incidence of transaction costs relevant to bargaining about who gets what. He also suggested desegregating "fishing effort" into its component parts rather treating it as some homogeneous and standard input. He also proposed building fisheries models that recognize there are individuals of a wide variety of skills and capital in the fishery and with a wide disparity of economic opportunity outside of the fishery.

Lane (1988) noted that until recently there has been little research into the decision making processes and behaviour of fishermen. This was pointed out by Hilborn (1985) who argued that most fisheries problems arise from failure to understand fishermen's behaviour. He contended that many fisheries failures could be ascribed to "poor understanding of the dynamics of fishermen, how they fish, and how they invest." He suggested that the study of fishermen and fleet dynamics should be a major element of fisheries science.

Hilborn proposed a conceptual framework for four elements of fisheries — population dynamics, fleet dynamics, processing and marketing — and how they interact (Fig. 18-17). He deplored the paucity of research on fleet dynamics, processing, or marketing, observing that such studies were not considered an integral part of fisheries science. He took issue with the traditional barriers between fisheries biology, economics, and the social sciences.

Lane (1988) followed up on Hilborn's suggestions by developing and applying an investment decision-making model for fishermen. Lane emphasized the need to verify the implications of such models by interviewing individual fishermen and by collecting more practical empirical data.

One of the major benefits of Hilborn's and Lane's work is its illustration of the need to bridge the gap among fisheries biologists, fisheries economists and other social scientists. I return to this later under section 7.0 *The Need for Collaboration*. At this point, I will examine social impact research needs for fisheries

6.0 FISHERIES SOCIOECONOMICS AND SOCIAL IMPACT RESEARCH

Although few in number, fisheries economists have been in the vanguard of the social science examination of fisheries. The involvement of sociologists, social anthropologists and other social scientists in the study of fisheries has been very limited until recently.

Charles (1988) pointed out that there is a wide variety of socioeconomic factors which determine the validity and effectiveness of regulatory instruments

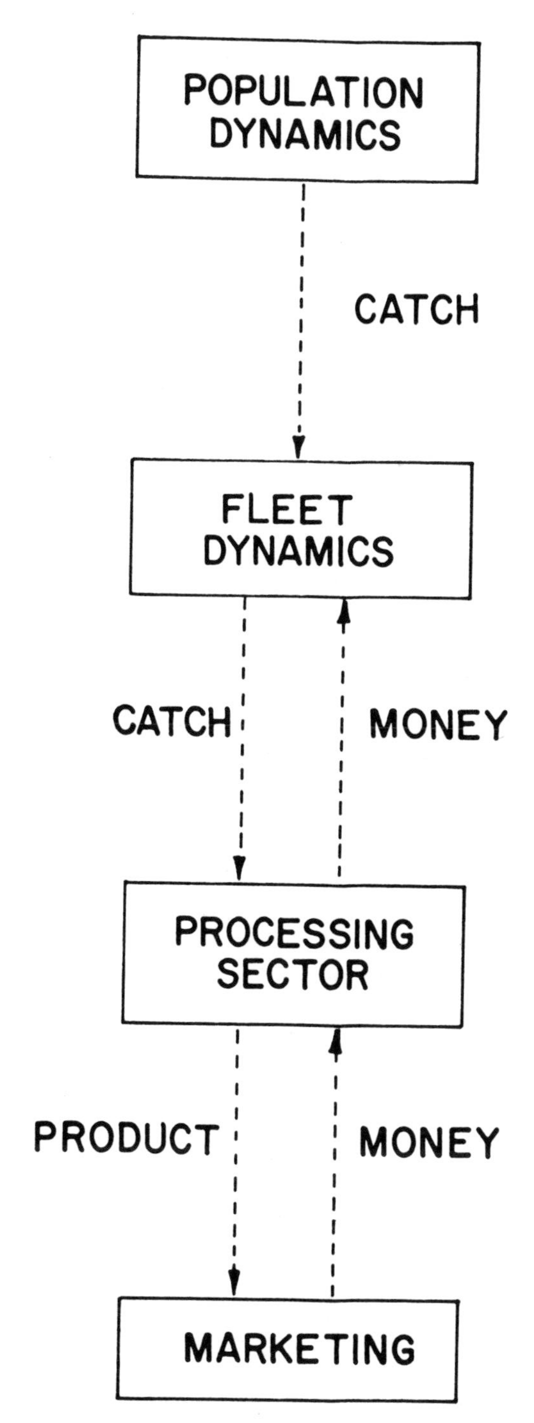

FIG. 18-17. Interaction among the four major elements in a fishery (from Hilborn 1985).

for fisheries managers. These factors included: distributional issues, labour market structure, social and opportunity costs, fishing community dynamics, and fishermen decision-making processes.

The distributional issue has already been discussed. I begin with comments on the social cost and opportunity cost of labour. There are often differences between private, financially motivated actions and those deemed socially optimal. Sinclair (1983) studied the effects of limited entry licensing in Northwest Newfoundland. He argued: "Even advocates of limited entry and enterprise allocation sometimes concede that the social consequences may be excessive and the economic gains illusory." Crutchfield (1979), for example, argued that unemployment is seldom a serious problem as a result of limited entry. However, he noted: "An important exception would be isolated fishing communities where other employment opportunities are severely limited. It is quite possible that spreading employment in the fishery would still be the least cost method of achieving some desired minimum economic and social standards." For most of rural Newfoundland, this appears to be a valid argument.

In such isolated communities, where alternative employment possibilities are very limited, the opportunity costs of labour are low. In such circumstances social factors play a major role in shaping fisheries policy.

Fisheries decision makers in Canada frequently have to consider the extent to which the opportunity cost of fishing for inshore fishermen can be raised by providing non-fishing employment alternatives. The Atlantic Fisheries Adjustment Program of 1990 involved a number of initiatives aimed at providing such alternatives.

Panayotou (1982) examined the modifications required to incorporate a variety of socioeconomic factors into the traditional static bioeconomic model. He considered problems of income distribution, lack of mobility, and of alternate employment, and conflicts between industrial and small-scale fisheries. Panayotou's model is illustrated in Fig. 18-18. He divided fishing costs into two components, private labour costs (wages) and other capital and operating costs. Where there are high levels of unemployment in the economy, the opportunity cost of labour is essentially zero. As he defined it, the "social yield" consists of resource rents plus wages. Panayotou identified five possible equilibrium points at different levels of fishing effort:

1. Maximum Economic Yield — (MEY) — the private optimum
2. Maximum Social Yield — (MScY) — the social optimum
3. Maximum Sustainable Yield — (MSY) — maximum physical yield
4. Zero Resource Rent — (OAE) — open access equilibrium

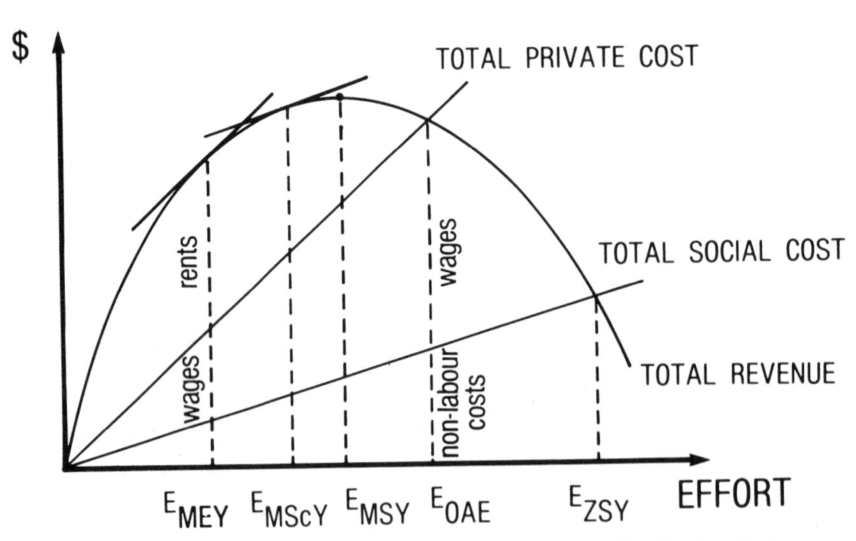

FIG. 18-18. The division of fishery revenues into (1) non-labor fixed and variable costs, (2) private labor costs, and (3) fishery rents. From a societal perspective, if the opportunity cost of labor is zero, net social benefits comprise the sum of fishery rents and payments to labor. The social optimum (MscY), which may be determined through a balancing of multiple fishery objectives, is unlikely to coincide with either the Maximum Economic Yield or the Maximum Sustainable Yield (from Charles 1988).

5. Zero Social Yield — (ZSY) — the employment optimum

Panayotou's social optimum (MScY), which represents a balancing of multiple objectives, is unlikely to coincide with either MEY or MSY. This was an interesting attempt to model factors which are often regarded only as qualitative modifiers to the traditional bioeconomic analysis. It suggests that social costs and opportunity costs of fisheries labour can be taken into account in determining the target fishing effort or catch level.

The process of labour adjustment in fisheries is related to the overall environment in which labour functions. This issue has been examined previously for the Newfoundland inshore fishery. Copes (1972) argued that per capita incomes could be increased by reducing the numbers of inshore fishermen. His analyses helped shape the federal - provincial resettlement program in the 1960's. Antler and Faris (1979), among others, criticized the resettlement program as misguided. Among the impacts they identified were:

1. a dramatic transformation from self-employed fishing to wage labouring;
2. a shift in the type of fish produced; and
3. a change in the organization of processing activity from family based, fishermen-recruited production units to privately owned fish plants.

They also suggested that it meant the end of women's participation in the fishery. This latter conclusion is questionable in light of the later employment of women extensively in fish processing plants.

Ferris and Plourde (1980, 1982) concluded that (1) unemployment insurance plays a major role in maintaining the number of inshore fishermen; and (2) the inshore fisherman is more responsive to market incentives than might have been expected (see Chapter 14).

Panayotou and Panayatou (1986), from a study of the Thailand fishery, concluded that labour is mobile between occupations but less so between locations. Terkla et al. (1985) examined the interaction of labour mobility and fisheries management in two New England fisheries. They identified difficulties in transferring labour out of fishing in small isolated ports where there are few alternative sources of employment and where labour outmigration is low because of individuals' strong attachment to the community and family.

Apostle et al. (1985) posed the questions: "Do people continue to live in.... small villages by choice, or from lack of alternatives? Is work satisfaction a prime reason for wishing to remain within the community, or is the work secondary to other factors related to place?"

Based on surveys carried out in southwest Nova Scotia and New England, they identified three general classes of fishing communities characterized by: (1) a high level of job satisfaction and strong community attachment in southwest Nova Scotia; (2) a high level of job satisfaction but weak community ties as in Point Judith, Rhode Island; and (3) a weak commitment to fishing, but a strong community attachment and/or interest in maintaining of kin and status positions, as in New Bedford, Massachusetts.

In general, fisheries management agencies in Canada, the United States and elsewhere have concentrated traditionally on collecting and analysing biological data. Insufficient attention has been paid to the need for economic and social information. Earlier I pointed out that the fisheries economics research capability of the Canadian Department of Fisheries and Oceans has been inadequate. There has been historically even less capability for sociological or social anthropological studies. Pimlott et al. (1971), in their background study for the Science Council of Canada, noted that the sociological aspects of the use of natural resources had been almost completely neglected in Canada and the United States. They observed that the expertise to tackle the problems did not exist in Canada:

> "There are very few sociologists who are practising their science in ways that have much relevance to resource management."

Maiolo and Orbash (1982), in their book on *Modernization and Marine Fisheries Policy*, observed:

> "As in so many cases when social science breaks through, the discipline of economics has had greater success than other social sciences in permeating the management arena, and has done so more quickly. Economics has a language, a method, and, for many, a track record that are more impressive to the nonsocial scientist than sociology, anthropology or other social disciplines."

As we have seen, Canadians have been prominent contributors to the development of fisheries economics theory and practice. The track record for the other social sciences has been less impressive.

Andersen (1978), in a review of fisheries social science needs, argued that "the human and nonhuman sides of the equation require systematic monitoring...much of our knowledge of the human factors — fishermen adjustment, their communities, industry, and government — in the fisheries management and development process is fragmentary and, because of extended jurisdiction, obsolescent." He identified the limited resources being deployed to the scientific study of fisheries social issues and proposed that these be expanded significantly. To a large extent, this plea fell on deaf ears. The Institute for Social and Economic Research at Memorial University has conducted some useful studies in the interim (for example, the work by Andersen, Sinclair and colleagues). Patricia Marchak and colleagues at the Department of Anthropology and Sociology at the University of British Columbia conducted a three year Fish and Ships research project which resulted in a useful study incorporating social and economic aspects of the British Colombia fishing industry. One product was a volume entitled *Uncommon Property: The Fishing and Fish Processing Industries in British Columbia*. Hanson, Lamson and colleagues at Dalhousie University produced a book (Hanson and Lamson 1984) which focused on the fisheries decision-making process, and in particular, how it impacted on fishermen. But these have been rare exceptions to a general pattern of inattention to the non-economic social science dimensions of the Canadian fisheries.

In the United States the situation has not been appreciably better. Acheson (1981) reviewed the literature on the anthropology of fishing. He synthesized disparate studies on various topics from around the world. This ranged widely covering such topics as:

- institutional response of fisherman to uncertainty in the marine environment (e.g. crew organization, access to fishing rights);
- the relationship between fishermen and fish buyers;
- cooperatives;
- information management by fishermen;
- competition (politics and conflict, the importance of skill in fishing, switching among fisheries, capital management, innovation and technical change);
- the psychological characteristics of fishermen (commitment to fishing, psychological adaptation and personality traits); and
- the role of women in the family life of fishermen.

One of his more significant conclusions was a limitation on the assumption that fishermen are motivated to catch as many fish as fast as possible for monetary reasons alone and operate completely independently. This may describe the situation in certain fisheries but is highly inaccurate for others. He cited studies which show that fishermen interact a great deal, and depend on each other for information about the location of concentrations of fish and to evaluate innovations. He also cited studies which suggest that income may be relatively unimportant in selecting fishing as an occupation and the degree of commitment to it.

Acheson also noted that, except for economists, social scientists had played a very small role in fisheries management planning in the U.S. Some social scientists had participated in the work of the Scientific and Statistical Committees (SSCs) established by the U.S. Regional Councils. Paredes (1985) commented on the role of social scientists in such committee work. Other social scientists who had also participated (Acheson, Leary, McCay, Orbach, Peterson, Spoehr and Langdon) responded to and amplified on Paredes' comments.

Acheson (1981, 1985) gave reasons why social and cultural issues were ignored in fisheries management plans even though they were mandated in the Magnuson Fishery Conservation and Management Act. First, little social and cultural data on fisheries exist and very few social scientists had undertaken the kinds of studies of coastal communities that would produce those data. Second, there is no agreed-upon way to operationalize Optimum Sustainable Yield (OSY), nor any agreement on which social, economic and biological data are required. Nor is there any model which would allow the SSCs to integrate the data from these diverse fields. Given the difficulties with OSY, the Regional Councils had ignored the concept and the social data requirements. Third, a great deal of social and cultural information was deemed politically inexpedient. Acheson observed:

> "Too much social and cultural data threatens to bring to light the 'downstream effect' of management — the winners, the losers and the effects on the distribution of income — with all that indicates about possibilities for political opposition."

Fricke (1985) compared the use of sociological data in the allocation of common property resources in fisheries and forestry in the United States. He identified three areas of importance of social science information to the resource manager — in determining the allocation of resources, in avoiding conflicts over resources, and in effective management of the resources. These three uses of sociological data he saw as complementary to the use of ecological and economic information and not as substitutes (Fig. 18-19).

He noted that the National Marine Fisheries Service had only one staff sociologist. Thus review and advice on social impact analysis occurred only after the plan had been submitted for formal consideration by the agency. The SSCs normally had at least one university sociologist or anthropologist as an appointed

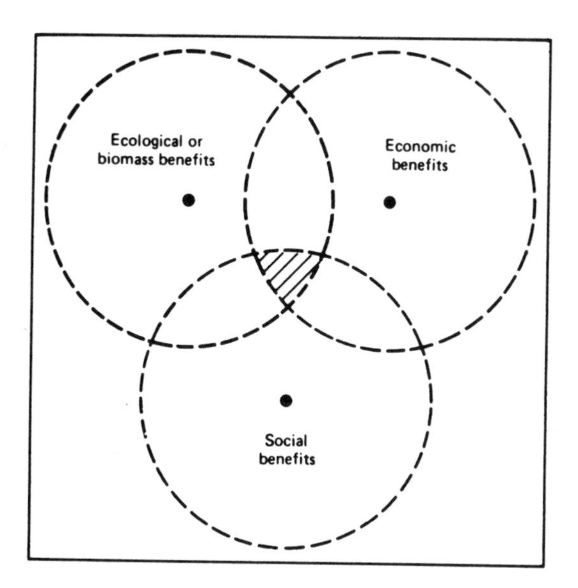

● indicates the point of maximum sustainable yield or benefits in this ideal situation.

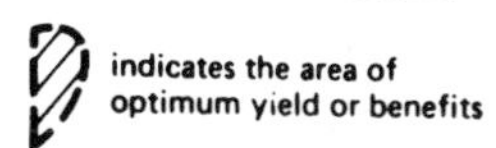

FIG. 18-19. The elements of MSY and optimum yield (from *Marine Policy*, Vol. 9, No 1, January 1985, p44, P. Fricke Reproduced with permission of Butterworth-Heinemann, Oxford, UK).

member. As Peredes and the others observed, these had had little impact in incorporating social considerations into the planning process.

Fricke suggested two reasons why sociological data were not being used. One was the organizational climate in which resource planning decisions were taken, and the second was the problem of uncertainty. In general, the councils and the Fisheries Service had access to biological knowledge and knew that it would be used in decision making. The availability and scope of economic and sociological information was uncertain.

Agreement that knowledge about the biological aspects of fisheries was needed for resource management led to automatic incorporation of such information in the Fishery Management Plans. The Regional Councils and the Fisheries Service also agreed on the need to include economic data in the plans, even though knowledge of the economics of fisheries was uncertain. There was disagreement on the incorporation of social impact analysis in the plans. Given the uncertain level of sociological information, the solution was to not include such analysis. Social impact analysis had not been included in the majority of plans.

Pollnac and Littlefield (1983) concurred that, although there was a general awareness of the need for sociocultural information, its acquisition and use had been, at best, superficial. They suggested this was due to the lack of a general theoretical framework to deal with the multivariate interrelationships between the fishery and its sociocultural matrix. They identified five categories of negative social consequences which fisheries management can have on a fishery or fishermen: (1) decreases in income; (2) changes in the structure of the fishery; (3) displacement from the fishery; (4) negative impacts on job satisfaction levels resulting from the previous three impacts; and (5) perceptions of the rules as "bad" or "unfair" in terms of their potential impacts. They identified job satisfaction as a key sociocultural variable associated with fisheries management.

In Canada, the consultation process (advisory committees) does address social issues in the development of fishing plans and the provision of advice to the Minister of Fisheries and Oceans. Much of the resource allocation analysis that is done in DFO is back-of-the-envelope socioeconomics. However, this is usually not backed up by systematic socioeconomic impact analysis. Resource allocation staff tend to be more involved in crisis management than in focused long-term analysis and research.

An FAO Expert Consultation on the Acquisition of Socio-Economic Information in Fisheries (FAO, 1985) dealt primarily with small-scale fisheries. The report distinguished between the role of socioeconomic information in managing fully exploited fisheries and in developing new fisheries. For the former, employment, efficiency, food supply and demand and distribution of income and wealth are primary concerns. For the latter, improving the standard of living, increasing fish production, and expansion of markets for fish are often the chief concerns.

An Annex to this report provides an extensive "indicative list" of information requirements for a selected set of government measures affecting the fisheries sector. This was grouped into three categories:

1. Information that should be routinely acquired by Departments of Fisheries or fisheries research institutes;
2. Information that should be routinely acquired by non-fisheries agencies; and
3. Information that commonly requires special studies, either to collect basic data or to analyze data available within fisheries and non fisheries agencies.

This list of information requirements was intimidating. It is doubtful that all of the identified requirements are being met for any fishery anywhere in the world.

7.0 THE NEED FOR COLLABORATION

The research challenges in all disciplines pertaining to fisheries biology, the physical environment, economics and the other social sciences are immense. Fisheries systems are exceedingly complex, characterized by interactions among the biological, environmental, economic, social and regulatory dimensions. To meet the challenge of understanding even some of the interactions will require much more effective interdisciplinary collaboration than has been the practice in the past.

Even fisheries biology, which has provided the scientific basis for fisheries management for most of the past century, can no longer be studied in isolation. Variability in the physical environment can have a profound effect on year-class success as well as migration and patterns of availability of fish stocks to fishermen in particular areas and seasons. The relative impact of parental stock size and environmental variability upon recruitment success remains to be determined. The influence of the various factors probably differs among species and among fish populations. The need for integrated biological and physical oceanographic studies is now recognized as a priority. Past efforts have been sporadic and ad hoc.

Since the integration of its scientific activities in 1986, DFO has launched a number of new interdisciplinary initiatives. Examples include the La Pérouse Bank project and the Marine Aspects of Salmon Survival (MASS) projects in British Columbia. In these, biologists and oceanographers are working together in interdisciplinary teams to plan and implement major research projects aimed at improving our understanding of the relationship between variability in the marine environment and marine fish stocks. The La Pérouse project also addresses species interactions. On the Atlantic coast, a similar interdisciplinary effort is being mounted to study the interaction between environmental variability and the population abundance and availability of the major northern cod stock off Newfoundland. Here also species interactions are being addressed on a more extensive scale than previously, with the help of additional funding under the Atlantic Fisheries Adjustment Program.

Off Nova Scotia there have been two large scale federal fisheries ecology field programs on the Scotian Shelf since 1978: the Scotian Shelf Ichthyoplankton Program from 1978 to 1982 and the Southwest Nova Scotia Fisheries Ecology Program from 1982 to 1985. Both programs involved collaboration between fisheries biologists and physical oceanographers. In response to one of the recommendations of the Scotia-Fundy Groundfish Task Force, DFO initiated a new project in the early 1990's on the Eastern Scotian Shelf. This too is focused on seeking an increased understanding of key biological parameters. These include stock boundaries and reproductive affinities within and between stocks, annual and interannual cycles in spatial distribution and the environmental factors which drive them, and the productive dynamics of the resources, including recruitment and growth. This multi-institutional program is also interdisciplinary, bringing together biologists and oceanographers.

Cod is presently the focus of several multidisciplinary research initiatives in Canada and abroad. The Canadian Government's Network of Centres of Excellence program has recently funded the Ocean Production Enhancement Network (OPEN) which has initiated studies on cod recruitment and distribution. ICES has started a study group on cod population fluctuations and climate (the Cod and Climate Change Program) that proposes to examine all historical data on cod population dynamics and associated physical and biological parameters influencing reproduction and growth at several life history stages. Initial planning has centered on research questions that clearly require a multidisciplinary approach.

These studies recognize the need for a multidisciplinary collaborative effort if there is to be any prospect of answering some of the fundamental questions concerning variability in resource abundance and availability. Clearly, interdisciplinary collaboration in the natural sciences is being harnessed more effectively than in the past to tackle questions of fundamental importance to fisheries management.

The track record in bridging the gap between the natural and social sciences, and within the social sciences, is less impressive. Fisheries economists describe their models as bioeconomic models. In reality, they are for the most part economic models superimposed on a basic and, to a large extent, little used, model of fisheries biology, the general production yield/effort relationship. Complex theoretical mathematical models have been erected but these are not widely applied in fisheries management. Crutchfield described some of these models as inestimable and largely irrelevant to fisheries managers.

There is also a gap on the applied fisheries economics end of the spectrum. There is insufficient dialogue between fisheries biologists and applied fisheries economists. There is insufficient systematic monitoring of basic economic data. For example, we saw in Chapter 14 that there has been no sustained program to monitor the economic/financial pulse of the fishing industry. Apart from basic ongoing cost and earnings studies of certain fleets, specific in-depth studies seem to be undertaken only when there is a crisis in one fishery or another. Such economic analyses as are conducted are carried out in isolation from the biological analyses. The latter are subjected to regular peer review. No such process exists on the economics research side.

We need to strengthen the applied fisheries economics research capability in Canada and develop better linkages between fisheries biologists and economists. More concerted economic analyses should be brought to bear on a wide variety of fisheries management issues, including the ongoing development of fishing plans. Copes (1984) speculated that the early interest of Canadian economists in fisheries questions was stimulated by the evidently severe fisheries problems in Canada. He remarked: "The persistence of those problems in Canada, as elsewhere, suggests that the solutions offered by economists have been politically or socially unacceptable or that, where they have been tried, they have been flawed in design or application."

Copes' 1984 conclusion seems equally valid today. Despite the lack of sufficient resources devoted to applied fisheries economics research, fisheries economics appears relatively strong in comparison with the other social sciences. The paucity of resources devoted to sociological and social anthropological research on fisheries questions is surprising. As an example, the federal Department of Fisheries and Oceans does not employ a single sociologist. This is puzzling given the high visibility of social issues, and the continual tug-of-war between economic and social factors in fisheries management. There is social impact analysis but this is done on an ad-hoc basis. A more systematic approach to social impact analysis as part of the fisheries management planning process is required.

Charles (1989) has attempted to broaden existing fisheries models in a move towards bio-socio-economic fishery models. He has argued that the joint dynamics of the fish stocks and fishermen must be taken into account in determining appropriate management policies. He developed a bio-socio-economic modelling approach to incorporate these effects within a multi-objective optimization framework.

As we saw in Chapter 4, fisheries management must balance multiple objectives. These include the following: (1) resource conservation, (2) food production, (3) generation of economic wealth, (4) reasonable incomes for fishermen, (5) maintaining employment for fishermen, and (6) maintaining the well-being and viability of fishing communities.

In his bio-socioeconomic models, Charles (1989) focused on objectives (2) to (6). He attempted to develop a framework for analysing the joint ecological and socioeconomic dynamics inherent in fisheries systems. His approach involved: (1) determining appropriate adjustment processes to predict the response of fish stocks and of fishermen to changing conditions in the fisheries, and (2) using these dynamics to undertake multiobjective management of fisheries harvests. Charles included in the optimization process the size and stability of the labour force, the level of employment in the fishery-dependent economy, and the per capita income level. The data necessary to fit such models are lacking. There is a requirement to collect time series of data on fisheries labour forces, fishing community populations, and fisheries participation rates as well as the more traditional data on fish stock dynamics and economic parameters.

This sort of modelling effort is a step in the right direction. But we must learn to walk before we can run. If we begin now to increase our capability to undertake and apply socioeconomic research and to collect the necessary data over time, perhaps several decades hence we will have truly interdisciplinary models of the fisheries. Meanwhile, the contribution of economists, sociologists and other social scientists to the fisheries management process can be improved even in the absence of such models.

There is another area where more effective collaboration is required between government and university scientists. We saw in the previous chapter that such collaboration in the natural sciences waxed and waned during the 75-year history of the Fisheries Research Board and its predecessor, often depending on the whims of the Chairman of the day. Following the demise of the Fisheries Research Board, relations between government and university scientists weakened during the late 1970's - early 1980's but were strengthened in the latter half of the 1980's. This renewal of collaboration between government and universities applies to fisheries and oceanographic research generally. The Canadian plans for participation in the Joint Global Ocean Flux Experiment (JGOFS) and the World Ocean Circulation Experiment (WOCE) are excellent examples of harnessing the scientific talents and resources of government and academia. There are numerous other examples at the individual project level.

Another significant opportunity for collaboration is offered by the new Centre of Excellence, the Ocean Production Enhancement Network (OPEN). This is a university-led initiative, involving a network of Canadian universities with an interest in fisheries and marine science. But the initiative draws upon government facilities and resources, for example, ship-time for at sea surveys and experiments. Government scientists are also participating in various aspects of the multidisciplinary studies.

Testifying before the Senate Standing Committee on Fisheries on May 15, 1990, Dr. William Leggett, then

Program Leader for OPEN, described the program's approach:

> "We are not just studying fish. We are studying fish in great depth and we are studying every aspect.... from the parents of the egg through the egg and larval stages, right up to the adults. We are also studying the dynamics of the food chain that intersects with those organisms, both what they eat and what eats them. We are studying, in real detail and simultaneously with the biological observations, the physical oceanography associated with them, such as changes in currents, temperatures and the way those currents and temperatures influence distribution and survival. We will have molecular biologists working with us who are looking into the genetic structure of these populations and using that genetic knowledge to help us to assess which of the individuals in the parent community are successful in producing offspring and whether there are characteristics of females, either in terms of their own biology or when they spawn, in relation to other cycles of production such as temperature and currents in the ocean which lead to particularly successful survival and production of fish.
> "...what it will not do — and we should be realistic about this — is to allow us to solve all of the problems of the fisheries in three or four years, but we strongly believe that we will be in a position to provide serious answers to some of the questions and good insights into where we should go next in terms of solving these problems in the medium term."

As these remarks attest, OPEN is a multidisciplinary, collaborative program in many respects. It illustrates the approach required across the natural and social sciences for improving the scientific basis for fisheries management as we enter the next century.

8.0 CONCLUSIONS

Despite nearly a century of research in the natural sciences, our understanding of the dynamics of marine fish populations, their interactions within the food web, and with the marine environment, and the role of man as predator is still weak. Fishing theory as devised in mid-century has provided the basis for attempts to manage fisheries to produce sustainable yields. But this theory generally assumes steady-state equilibrium conditions. In reality, marine fish populations and the ecosystem in which they live can be extremely variable.

Recent experience has demonstrated that our methods of estimating marine fish population abundance require improvement. The techniques which have been widely used since the introduction of catch quota regimes in the early 1970's (Virtual Population or Cohort Analysis) have tended to underestimate fishing mortality and overestimate the abundance of fish populations. Priority in research must be directed towards improving our methods of estimation, including more accurate determination of the natural mortality rate of fish. We cannot count fish directly as forest scientists can do for trees or wildlife biologists for birds or mammals. Hence, our estimates will likely continue to have wide variances. Fisheries managers must be aware of these variances and must understand that fluctuations in advice on allowable yields at particular target fishing mortalities are normal. This is particularly so in situations where the existing level of fishing effort is excessive in relation to the sustainable yield.

Understanding the relative influence of parental stock size and environmental variability in determining the strength of year-classes remains a challenge for fisheries scientists. While it is accepted that there are minimum spawning stock sizes below which the probability of successful year-classes is low, there is a growing awareness that widespread fluctuations in year-class strength are largely environmentally induced. This does not, however, obviate the need for precaution in setting the level of harvest so as to maintain "adequate" spawning stock. Additional research into the mechanisms determining recruitment success is necessary. No simple solution to the recruitment problem is likely to emerge. Different combinations of factors probably act at different times in different places to produce the observed fluctuations in fish productivity. Resource variability will continue to be an inherent component of the fisheries system.

Recent attempts to model multispecies interactions indicate that natural mortality on the younger ages is higher than was previously thought. The development of Multispecies Virtual Population Analysis in the Northeast Atlantic has suggested implications for management which differ significantly from single species models. There is a legitimate question whether fisheries managers will want multispecies advice. There are, however, specific instances where a better understanding of multispecies interactions is clearly required. It is necessary, for example, to obtain a better understanding of the interactions among the capelin — cod — seals complex off Newfoundland to guide future management decisions for this multispecies complex.

From the 1950's onwards, complex economic models of fishing have been developed. Economists have had a significant influence on fisheries management policy. Their focus on economic efficiency and the common property "race for the fish", resulting in excess input of labour and capital into the fishery, has led to various attempts to combat this phenomenon. Limited entry licensing, first promoted as the solution, has proved ineffective in many cases. Taxation on landings has been largely rejected as an option. Individual quotas became an accepted management tool in the 1980's and are being introduced in a wide variety of fisheries. The economist's prescription of transferable quotas has been largely ignored to date. However, it is likely that we will see individual transferable quotas introduced in a number of fisheries over the next decade. But individual quotas will not be the panacea for the common property problem. Problems of enforcement loom large and the approach is clearly unsuitable in many small boat fisheries.

There appears to be little prospect that many of the complex biomathematical models recently developed by academic economists can be applied by fisheries managers. While mathematically attractive, they require volumes of data which are unlikely to be forthcoming. There is a need for a greater emphasis on applied research in fisheries economics. The existing fisheries economics research capability in Canada is inadequate.

There has been little research on the social aspects of fisheries. Social impact analysis is generally done only on an ad hoc basis. Given the prominence of issues such as maintaining employment, reasonable incomes and community dependence in shaping fisheries management policy, this paucity of research on the social dimensions of the fisheries is clearly untenable. Steps need to be taken to foster an adequate capability for such social science research and its application in fisheries management.

Biologists, economists and other social scientists all have a role to play in determining the "best use" of fisheries resources, i.e. in defining the Optimum Sustainable Yield (Fig. 18-19). There is a need to array for decision makers the implications of alternative management options. Those options have biological, economic, and social dimensions. Currently, the economic and social implications are ill defined in the fisheries management planning process.

For too long "fisheries science" has been regarded as synonymous with "fisheries biology." This reflects the dominant role fisheries biology has played as the scientific basis for fisheries management. It is time to change that perspective. Physical oceanography, fisheries economics, sociology and social anthropology must also be part of a truly integrated approach to fisheries science. As we move toward the next century, we must also move toward an integrated bio-socioeconomic approach to fisheries science as the scientific input to fisheries management. This will not be achieved overnight or without difficulty. It will require a recognition by fisheries biologists that the other scientific disciplines have a legitimate role. It will also require improving the capability for social and economic research on fisheries in Canada. A broader view of fisheries science is necessary if we are to continue improving the scientific basis for fisheries management.

CHAPTER 19

ENSURING COMPLIANCE: FISHERIES ENFORCEMENT

"Fisheries law enforcement has traditionally been treated as the bastard child of fisheries management"

– Margaret Frailey and Robert Taylor, 1987

1.0 INTRODUCTION

Although enforcement is an essential component of fisheries management, the extensive world literature on fisheries management has given it little attention. This is slowly changing with the growing recognition that effective fisheries enforcement is crucially important to achieve management objectives. Without enforcement, the most complex array of regulations is likely to be ineffective, given the well known incentive for fishermen to ignore such regulations. This incentive is powerful if there is little likelihood of being detected, apprehended and convicted, or if the penalties upon conviction are low compared to the gains from illegal fishing.

Recently, various authors (e.g. Sutinen and Andersen 1985; Anderson and Lee 1986; Milliman 1986; Sutinen and Hennessey 1987; and Anderson 1989) have begun to examine how noncompliance and the costs of enforcement affect the economics of fishing and the relative efficiency of various fisheries management controls. Blewett et al. (1987) analyzed Canada's experience in measuring the deterrent effect of fisheries law enforcement. The results of their study are summarized later in section 5.

Meanwhile, fisheries managers are confronted daily with practical questions of how and where to enforce fisheries regulations. They wrestle with choices as to the appropriate mix of monitoring, control and surveillance modes and how to achieve maximum compliance at minimum cost.

Within Canada, efforts to enforce fisheries regulations date from pre-Confederation days. In the 18th and 19th centuries Britain, France and the United States fought over fishing rights in what subsequently became Canadian waters. Treaties were enforced by warships and examples are numerous of New England fishermen's attempts to defy bans on fishing. Fishermen in Nova Scotia, New Brunswick and Newfoundland were demanding that the British government enforce the exclusion of Americans from the inshore fishery.

Moses Perley, in a report titled *The Sea and River Fisheries of New Brunswick* (Perley, 1850), commented that the laws to regulate the inland fisheries appeared well founded. However, there had been a total failure to enforce them. He lamented "the decay of these once valuable and prolific fisheries now hastening rapidly to their termination". He called for cooperation with the Government of Lower Canada to police the Restigouche for salmon violations.

The first vessel specifically dedicated to fisheries patrol was the Coast Guard schooner *Alliance*. This was used in the early 1850's by a young medical doctor from McGill, Pierre Fortin, who was employed by the Canadian government to patrol the fisheries of the Gulf of St. Lawrence. Fortin's annual reports from 1852 to 1867 make fascinating reading. By 1855 Fortin had acquired a faster schooner, *La Canadienne*. During his 16 year patrol of Gulf fisheries, he earned the title "Le Roi du Golfe". Johnstone (1977) summarized his accomplishments:

> "During that time he had brought law and order to an area that had been terrorized by the depredations of foreign fishermen. With his handful of well-drilled seamen he didn't hesitate to board vessels whose masters had violated the law and hold them to account, or appear at stormy town meetings when a show of authority was required to restore order. He also made sure his men attended Sunday mass 'to set an example' for the local population. He tracked down poachers and violators of the new Fishery Act, handed out penalties, and administered an impartial justice. And above all, he showed a deep concern for the preservation and development of a fishery which, he felt sure, could mean a

better life to the hard working people who pursued it."

It is fitting that a modern day Quebec Region fisheries patrol vessel is named in honour of Fortin. In many respects, today's Canadian fishery officer is following the path blazed by Pierre Fortin. Since the Second World War, and particularly during the 1970's and 1980's, a much more sophisticated fisheries enforcement system has been established. This encompasses a broad range of Monitoring, Control and Surveillance (MCS) activities (e.g. air surveillance, sea patrols, an observer program and dockside and quota monitoring). These are coupled with penalties for violations which include fines levied by the courts under the Coastal Fisheries Protection Act and the Fisheries Act, forfeiture of illegally caught fish and fishing gear and vessels used to commit an offence, and suspension of fishing licences.

Fisheries enforcement can be a controversial, difficult and dangerous undertaking. One of the most dramatic illustrations of this in the modern era involved the May, 1983 burning of a fisheries patrol vessel at West Pubnico, Nova Scotia by fishermen protesting the enforcement of lobster trap-limit regulations in southwest Nova Scotia. Enforcement difficulties brought this fishery to national attention. In the 1980's angry fishermen also occupied fisheries offices and held staff hostage in Cape Breton, Nova Scotia, and Memramcook, New Brunswick. There have been armed confrontations between fishery officers and salmon poachers in New Brunswick and British Columbia. In resisting arrest for fishing illegally in the Canadian 200-mile zone, Spanish fishing vessels in 1986 ended up transporting Canadian fishery officers half way across the Atlantic before they were released. In July, 1989, the U.S. scallop dragger *Bountiful* rammed the Canadian patrol vessel *Cygnus*, when ordered to stop fishing on the Canadian side of the Hague line on Georges Bank and subsequently evaded arrest by fleeing into the U.S. zone. These and other incidents attest to the strong emotional reaction of some fishermen to government officials' attempts to enforce fisheries regulations.

These incidents also show that some fishermen are strongly motivated to violate regulations and evade arrest. To find the reasons for this and the rationale for fisheries enforcement it is necessary to examine the economics of crime and the concept of deterrence.

2.0 FISHERIES CRIME AND THE CONCEPT OF DETERRENCE

The objective of fisheries enforcement is to achieve compliance with fisheries management regulations by deterring illegal fishing activity. Enforcement programs attempt to achieve this by influencing fishermen's behaviour. Fisheries enforcement/deterrence models involve a combination of the economic theory of criminal behaviour with the bioeconomic theory of fisheries management.

The underlying assumption is that criminals are "rational" decision makers. In economic parlance, this merely means that the individual makes choices based upon the costs and benefits of participating in illegal activities (Becker 1968). As Furlong (1985) has noted, these benefits and costs of committing a crime could be monetary and/or psychological. Money is the primary motivation in most violations of fisheries regulations. In deciding whether to commit a crime, an individual considers the potential gain if the crime is successful, the potential punishment if unsuccessful, and the likelihood of success. Becker (1968) pointed out a number of factors which will influence individuals in making the compliance/noncompliance choice. These include:

1. the probability of avoiding detection;
2. the benefits gained from undetected illegal activities;
3. the probability of being detected but avoiding sanctions;
4. the benefits gained in such a situation; and
5. the probability of being detected and sanctioned and the costs of such sanctions.

According to the economist's definition of rationality, the individual acts to maximize individual expected revenues rather than social welfare (Arrow 1951). Under such circumstances he will choose that course of action which appears to offer the greatest expected revenue. He will prefer to fish illegally if the potential gain outweighs the potential loss. To determine whether gain outweighs loss, the individual has to weigh the potential gain against: the probability of being apprehended and arrested if detected; the probability of conviction; and the likely penalties if convicted. Since the fisherman does not know these probabilities, he acts on his subjective perceptions of these probabilities and the likely penalties. Economists tend to ignore the fact that, for moral reasons, many individuals would not commit a crime even if the perceived gains outweighed the perceived losses.

Blewett, Furlong and Toews (1987) elaborated on this general concept. They pointed out the uncertainty surrounding these factors. The magnitude of the gain and the likelihood of success are not known with certainty prior to actually committing the offence. The penalty that could result from conviction is also uncertain. The individual's attitude towards risk is also a significant factor. If the potential losses exceed the potential gains, risk prone individuals might still fish illegally.

Since the objective of fisheries enforcement is compliance with regulations, managers try to structure compliance-inducing-mechanisms to achieve "acceptable" compliance levels. A mix of enforcement mechanisms is used in an attempt to influence the implicit gain/loss calculations of the individuals making the choice whether to comply with fisheries regulations.

Hennessey and Sutinen (1987) pointed out that the extent of overall compliance is nearly impossible to measure. The actual number of offences in the fishery is unobservable (Blewett et al. 1987). Data can be collected on the extent of detected noncompliance but this is only a part of the picture. Hennessey and Kaiser (1987) observed that levels of detected noncompliance cannot be extrapolated to the entire population to estimate the overall levels of noncompliance. This is because monitoring and surveillance — especially boarding and inspections — are not random. Normally enforcement efforts are targetted towards vessels suspected of violations.

Blewett et al. (1987) estimated the extent of noncompliance with regulations and the probabilities of apprehension, arrest and conviction based on interview surveys of fishermen in selected Canadian fisheries (See section 5).

Before discussing the Canadian fisheries enforcement system, it is perhaps useful to focus on the definition of enforcement. Quite often, enforcement is thought of as the physical activity associated with detecting and apprehending fisheries violators. However, this is only part of the process. A comprehensive enforcement program has three basic elements:

1. ability to detect infractions when they occur;
2. capability to apprehend the violators; and
3. imposition of penalties or sanctions as deterrents.

I consider sanctions to be an integral part of the enforcement process. For the purposes of this chapter the enforcement process can be subdivided into two spheres of activity:

1. Monitoring, Control and Surveillance (MCS)
2. Sanctioning

Sanctioning includes the various forms of penalty imposed for violations (e.g. fines, seizures, forfeitures, licence suspensions or imprisonment). Fishermen's perceptions of the nature and level of the penalty likely to be imposed are equally as important in achieving deterrence as the perceived chances of detection and apprehension.

Stigler (1970) suggested that there are four basic means to improve compliance:

1. Minimize the chances that violations will go undetected;
2. Maximize the probability that sanctions will follow the detection of violations;
3. Speed up the process from time of detection of a violation to sanction; and
4. Make the sanctions large.

Hennessey and Kaiser (1987) noted that there are differing views about the best mix of alternatives among the four Stigler identified. Becker (1968) and Tullock (1974), for example, argued that the probability of being sanctioned is more important than the size of the sanction. Other experts have argued that imposing large sanctions is necessary to increase the deterrent effect. There is another means of improving compliance. This involves an effective information education program. If fishermen do not understand or support a particular regulation, they are less likely to comply with it (D. Brock, DFO, Ottawa, pers. comm.).

3.0 MONITORING, CONTROL AND SURVEILLANCE

Apart from the Royal Canadian Mounted Police, DFO has one of the largest enforcement programs in the Canadian government. DFO employs more than 700 full-time and seasonal enforcement officers. These officers utilize a fleet of 50 Departmental patrol vessels larger than 15 tonnes, and 34 smaller vessels. The total DFO expenditure on enforcement was approximately $65 million in 1988-89. The cost and distribution of fishery officers by region is shown in Table 19-1. Table 19-2 shows the cost by region of the vessels and associated crew.

TABLE 19-1. Department of Fisheries and Oceans: Cost by Region of fishery officers for fiscal year 1987–1988.

Region	Salaries ($000s)	Person-years
Québec	2,099.4	53
Scotia-Fundy	5,696.24	154
Gulf	6,348.55	184
Newfoundland	4,759.55	125
SUB-TOTAL	18,903.80	516
Pacific	6,598.74	175
Central & Arctic	583.37	17
SUB-TOTAL	7,182.11	192
TOTAL	26,085.90	708

The total O&M cost associated with these enforcement officers was approximately $15 million.

Fishery enforcement officers boarding a fishing vessel.

Checking otter trawl to ensure compliance with mesh size regulations.

TABLE 19-2. Department of Fisheries and Oceans: Cost by Region of enforcement vessels and associated crew for fiscal year 1987–1988.

		$000s			
Region	Number of Vessels	Salaries	Operating & Maintenance	Capital	Person-years
Québec	20	1,066.09	786.67	187.07	22.90
Scotia-Fundy	10	4,273.20	2,793.75	54.66	117.50
Gulf	13	539.34	184.40	6.84	16.70
Newfoundland	9	3,524.36	3,100.75	55.17	108.60
SUB-TOTAL	52	9,402.99	6,865.57	303.73	265.70
Pacific	31	5,097.90	2,198.94	197.85	142.78
Central & Arctic	1	—	21.54	—	—
SUB-TOTAL	32	5,097.90	2,220.47	197.85	142.78
TOTAL	84	14,500.89	9,086.04	501.57	408.78

356 of the ships' crew were also designated as fishery officers.

The bulk of enforcement costs are associated with the fishery officers and the patrol vessels. Approximately $5 million was allocated for air surveillance and the observer program in 1987–88. The funding for these programs was later increased significantly (see sections 3.2 to 3.5).

DFO's enforcement program has four distinct yet interrelated components:

1. Inland, Coastal and Offshore Sea Surveillance;
2. Observer Program;
3. Air Surveillance; and
4. Dockside and Quota Monitoring.

The mix of these enforcement modes varies by fishery, from season to season, and from year to year in a given fishery.

3.1 Sea Surveillance

3.1.1 General

Patrol vessels provide platforms from which fishery officers inspect fishing vessels and take direct enforcement action to apprehend violators. Patrol vessels also check fishing gear for compliance with mesh size regulations and patrol closed areas. They enforce size limits in fisheries such as those for crab and lobster.

DFO operates five offshore patrol vessels on the Atlantic. These required 210 person-years to operate in 1990–91, with a total operating cost of approximately $13 million (Table 19-3).

These vessels provided a total of 1297 sea days, 545 by the two Newfoundland based vessels and 752 by the three based in Scotia-Fundy.

TABLE 19-3. Department of Fisheries and Oceans: Cost of Atlantic offshore patrol vessels for fiscal year 1990–91.

		$000s			
Resources	Person-years	Salaries	Operating & Maintenance	Capital	TOTAL
Newfoundland					
Cape Roger	40.00	1,602.00	1,407.00	54.00	3,063.00
L.J. Crowley	40.00	1,602.00	1,200.00		2,802.00
Scotia Fundy					
Cygnus	39.50	1,463.00	895.00	2.00	2,360.00
Chebucto	37.70	1,405.10	786.50	1.00	2,192.00
Louisbourg	22.50	871.00	656.40	10.00	1,537.40
SUB-TOTAL	179.70	6,943.10	4,945.60	67.00	11,955.70
Fishery Officers:					
Offshore	30.00	1,290.00	—	—	1,290.00
TOTAL OFFSHORE:	209.70	8,233.10	4,945.6	67.00	13,245.70

356 of the ships' crew were also designated as fishery officers.

On the Atlantic, DFO also had, in 1990, 16 inshore and nearshore patrol vessels which operate in areas up to 30 to 40 miles from shore. They range from 45 to 85 feet long and are normally crewed by 2 to 4 people. They operate seasonally with approximately 150 operating days per year per vessel. The costs amount to 78 person years and approximately $5 million. These vessels provided the following sea days for patrol (in 1990–91): Newfoundland — 1295; Scotia-Fundy — 993; Québec — 338.

The Gulf Region had no patrol vessels other than small craft operated by fishery officers, e.g. eleven 42 foot long patrol vessels manned by 2 people. Patrol vessels from the Québec Region assisted in surveillance for Gulf Region waters. Vessels from Newfoundland and Scotia-Fundy also assisted as required.

On the Pacific coast, DFO operates two patrol vessels considered "offshore," the *James Sinclair* and the *Tanu*. In 1990–91, these provided 454 sea days, at a total cost of $3.6 million (D.A. Good, DFO, Ottawa, pers. comm.).

Also on the Pacific coast, DFO operates a fleet of 25 inshore patrol vessels which usually operate in areas that are within 25 miles of the shore. These vessels are normally crewed by 2 to 6 people. They operate seasonally averaging approximately 145 operating days per year per vessel. The costs amounted to approximately 80 person years and $5.4 million in 1990–91.

3.1.2 Offshore Sea Patrols

At the time of extension of fisheries jurisdiction, Canada faced the challenge of demonstrating that it could police effectively the 200-mile zone. It had been preparing for these responsibilities for some time. In March 1976, the federal Cabinet had approved a 5-year plan for fisheries surveillance and enforcement involving the coordinated deployment of vessels of the Department of Fisheries and the Environment, aircraft and ships of the Department of National Defence and vessels of the Department of Transport. An additional $4 million per year was diverted to offshore surveillance and enforcement for a total of $12 million per year commencing in fiscal 1976–77.

The additional funds were intended to enable:

1. An increase in Tracker aircraft patrols from 2000 to 4000 hours per year;
2. A doubling of offshore ship time to 1,650 sea days on the Atlantic and 495 days on the Pacific;
3. Air surveillance once weekly of sensitive fishing ares; (Flying hours were to total 3,750 per year on the Atlantic coast and 480 hours on the Pacific); and
4. The boarding of vessels offshore, with foreign vessels to be boarded at least four times per year, and Canadian vessels at least twice per year. The objective was to board and inspect one third of the foreign fleet and one-sixth of the Canadian fleet each month.

At that time the Department of Fisheries and the Environment had eight major vessels deployed on the Atlantic coast and three on the Pacific, but not all of these were capable of offshore patrols. Sixteen Tracker aircraft plus three squadrons — six aircraft each — of long range Argus were used on the Atlantic coast and three Trackers plus another squadron of six Arguses on the Pacific.

Canada was able to maintain a credible presence in enforcing the 200-mile limit in the early years of extended jurisdiction. During the first year 900 inspections were conducted on foreign vessels and 170 on Canadians. Fourteen convictions were obtained against foreign captains for violating Canadian regulations. The major challenges to Canadian authority came later in the 1980's, with incursions of foreign vessels, particularly Spanish, into the Canadian zone on the Grand Banks. This resulted in increased surveillance and new enforcement approaches. There were also problems with incursions of American vessels into the Canadian zone following the boundary delimitation of Georges Bank by the International Court of Justice in 1984.

An analysis of the effectiveness of offshore fisheries surveillance operations in 1980 resulted in recommendations for increasing penalties under the Fisheries Act and the Coastal Fisheries Protection Act. It also called for revised operational goals for offshore fisheries surveillance and proposed additional resources to meet these revised goals (Clough 1980).

DFO adopted the following revised operational goals:

1. that observers be placed on board 100% of non-USA foreign vessels and 50% of USA and domestic trawlers over 65 feet long;
2. that 3% of all offshore fishing vessels should be sighted daily on the east coast and 2% on the west coast;
3. that 15% of all offshore vessels not carrying observers should be boarded monthly; and
4. that a presence should be maintained over fishing grounds that are intersected by the fishing zone closure lines or the 200-mile limit.

The 1980 optimization analyses (Clough 1980) indicated that a mix of approximately 1200 offshore sea-days, with observers on all foreign vessels and at least

20–30% of domestic vessels over 65 feet would provide a credible deterrent on the Atlantic, excluding Flemish Cap and the far north (Davis Strait). The patrol sea-day requirement for Flemish Cap and Davis Strait was estimated to be 150 sea-days, for an overall Atlantic offshore total of 1350 patrol sea-days.

The offshore patrol effort averaged 1225 sea-days from 1977 to 1979, compared with the 1980 estimated requirement of 1350 sea-days for the Atlantic coast.

As a result of Clough's 1980 study, DFO decided to establish a basic offshore fleet of four large offshore patrol vessels of the Cape Roger class to provide about 1000 sea days. The balance was to be supplied by DND or Coast Guard vessels. In 1990–1991, DFO operated four Cape-Roger class vessels and one intermediate-size vessel, the *Louisbourg*, providing 1297 offshore patrol sea-days. On the surface, these met the requirements first estimated in 1980.

Since 1980, however, there have been dramatic changes in the Atlantic offshore fishery. The 1980's saw a growing presence on the Nose and Tail of the Grand Bank just outside the 200-mile limit of foreign vessels not licensed to fish in the Canadian zone. This increased the requirement for almost constant patrol vessel coverage of these sensitive boundary areas. Thus, the two existing offshore patrol vessels based in Newfoundland were required to patrol these areas, at the expense of coverage of the traditional offshore fisheries within the Canadian zone. In 1990, DFO concluded that a third offshore vessel was required in Newfoundland to ensure regular boardings of the domestic fleet (E.B. Dunne, DFO, St. John's, pers. comm.). This was achieved by obtaining 100 sea days from the Canadian Coast Guard and 150 additional days from DND (L. Strowbridge, DFO, St. John's, pers. comm.).

For the Pacific coast, the offshore surveillance requirements estimated in 1980 were significantly less than those for the Atlantic because of the different nature of the fisheries on the two coasts. An annual requirement of 490 patrol sea-days was estimated to permit boardings of the many small domestic vessels that fished in "offshore" areas (e.g. trollers off the West Coast of Vancouver Island). The patrol vessel *Tanu* was to provide 200 sea-days. The remainder was to be made up by deploying the smaller DFO vessels *Laurier* and *Howay* (or their replacements), by some DND support, and by charters from the private sector.

3.1.3. Inshore/Nearshore Patrols

No explicit operational goals have been established for inshore/nearshore sea patrols on either the Atlantic or Pacific coasts. The increasing complexity of fisheries regulations through the 1980's led to increased attempts by inshore fishermen to evade regulations. Examples include the sometimes controversial efforts to enforce trap limits in the Atlantic lobster and crab fisheries.

The 1990 Harris and Haché reports (Harris 1990; DFO 1989i) identified a need for improved enforcement in the inshore/nearshore fisheries. DFO concluded that chartering vessels was the most cost effective means of providing additional inshore patrol days to address these recommendations. Funds were initially allocated under the Atlantic Fisheries Adjustment Program to charter five vessels for 7 months of the year at an annual cost of approximately $3 million. These vessels were to be distributed as follows: Newfoundland (3); Scotia-Fundy (1); and Gulf Region (1). However, this need was ultimately met by increasing the utilization of existing DFO vessels and obtaining additional support from the Canadian Coast Guard and DND.

The combination of these measures provided additional inshore patrol vessel coverage along the Labrador coast and northeastern and eastern Newfoundland, on the Scotian Shelf and in the Gulf of St. Lawrence. The increased effort was to concentrate on increased patrols on spawning grounds, enforcement of minimum fish and mesh sizes and closed areas.

Data are not available to analyse the effectiveness of such inshore/nearshore patrols or their contribution to the overall enforcement effort for either the Atlantic or Pacific coasts.

3.2 Observer Program

3.2.1 Origin and Development

One of the key components of the DFO enforcement process since the mid-1970's has been an observer program on both the Atlantic and Pacific coasts. It originally targetted foreign fleets but, over time, was extended to the Canadian offshore fleet.

The Canadian observer program was first conceived at a special ICNAF meeting in Tenerife, Spain, in 1976, as a bilateral program involving fisheries in the then Maritimes Region under the auspices of ICNAF (ICNAF 1977). The impetus came from disputes about the extent of by-catch of other species in the Scotian Shelf foreign silver hake fishery. ICNAF introduced a silver hake "Small-Mesh Gear" box to address the problem (see Chapter 5). ICNAF decided that scientific information should be collected to monitor the efficacy of this measure in reducing the by-catch of small fish of valuable species such as cod and haddock.

The first bilateral accords covering observers on foreign vessels in the Canadian fisheries zone were signed with the USSR and Cuba in 1977, and a small observer program was initiated in the Maritimes Region. By 1978 the program had also been extended to the Newfoundland Region. The Atlantic Foreign Observer Program grew from a small start in 1977 to 5520 observer sea-days in 1978, and 11,235 sea days in 1979–80 (D. Kulka, DFO, St. John's, pers. comm.).

In the late 1970's DFO began an observer program on the Pacific coast. It was much smaller in scale (270 observer-sea-days in 1979 and 170 observer-sea-days in 1980).

These observer programs had two objectives: (1) collecting data for resource assessment, and (2) promoting deterrence through the presence of the observer on the fishing vessel. The observer was required to collect certain data, maintain records, report violations, and if necessary call for a boarding inspection and further action by a DFO fishery officer.

An onboard observer can be a deterrent, provided he or she is required to report violations and to maintain records at a specified standard of accuracy. An observer could, however, be intimidated or bribed. Accordingly, they have to be checked. Thus, even vessels carrying observers need to be boarded occasionally (see Section 3.3.2).

Clough's 1980 analysis indicated an offshore observer requirement of 54,000 to 58,000 sea-days. This assumed coverage of the full foreign fleets and 20 to 30% of the domestic offshore fleet. On the Pacific coast, the study suggested that the observer coverage should increase to 2200 sea-days annually.

The budget for the Atlantic observer program was increased from $1.6 million to $2.7 million in 1981–82 and thereafter remained static until 1986. By 1986, DFO estimated that observer coverage was about 46–50% on licensed foreign vessels and 8% on domestic vessels, far short of the targets adopted in 1980. The static budget for the observer program had eroded coverage.

In 1986, facing growing concern about foreign overfishing, the government adopted a cost-recovery approach to the foreign observer program. This was part of a package of measures to enhance Canada's surveillance and enforcement capability. It was intended to provide full observer coverage on licensed foreign vessels fishing in the Canadian Zone at no cost to the Treasury. Observers were employed by a third-party contractor engaged by DFO. Foreign countries (or companies) were required to pay the observer company directly for the costs of mandatory observer coverage. A variation of this was utilized in the case of the USSR, whereby DFO instituted fees to recover observer program costs with the funds going to the Treasury Board. The latter increased the Department's budget accordingly.

From 1987 onward, coverage was extended to 100% for both the foreign fleet and certain domestic fisheries on the Atlantic. Observer-sea-days approximately doubled from 1986 to 1988. The Pacific Region implemented a dedicated observer program for foreign vessels as part of the national initiative in 1987. Previously the Pacific program had been ad hoc, utilizing student employment programs, term employees and trainee fishery officers. There was no observer program for the domestic Pacific fishery until EAs were introduced.

A review of the national observer program early in 1990 suggested that emerging demands required increased coverage. However, the data already being collected by observers was not being processed and utilized in a timely and effective manner.

The existing program covered vessels longer than 45 feet but targetted mainly vessels exceeding 100 feet. The sea-day coverage in 1987–88 amounted to 20,500 for fleet components whose sea-days totalled 180,500. Thus the overall level of coverage was about 11% for the fleets on which observers were being utilized. Of this total, about 11,000 observer sea-days were directed towards foreign vessels and 9,500 sea-days toward Canadian vessels. The level of coverage of the foreign fleets was close to the target of 100%. Coverage of the domestic fleets ranged from 1–2% in the Gulf of St. Lawrence, to 7% in Scotia-Fundy, to a high of 12% in Newfoundland. The higher percentage coverage in Newfoundland reflected increased coverage of the controversial offshore northern cod fishery.

The 1988 level of coverage (20,500 sea-days) was less than half of the target level for the offshore fleet recommended in 1980 (Clough 1980). The actual observer sea-days had, however, doubled since 1980, with the increase occurring from 1987 onward. Changing circumstances in the fishery and enforcement practices during the 1980's had led to a downward reassessment of target levels of coverage.

Despite this, pressure intensified to expand further the observer program. The Scotia-Fundy Groundfish Task Force called for an industry-funded observer program for vessels greater than 100 feet on enterprise allocations, fishing on the Scotian Shelf (DFO 1989i). Also, the Panel on Northern Cod recommended that DFO expand the observer program to the inshore fleet and expand services for analyzing observer data (Harris 1990).

The 1990 Atlantic Fisheries Adjustment Program provided additional resources to achieve 100% observer coverage of the winter northern cod fishery and to provide for more rapid analysis of observer data from all Atlantic regions.

3.2.2 The Bribery/Intimidation Factor

From the beginning of the observer program there has been concern about the possible intimidation or bribery of observers. Observers are usually not enforcement officers or permanent DFO employees but instead work for private companies contracted by DFO. Vessels with observers onboard are sometimes boarded by officers from a fisheries patrol vessel as a check on observer activities.

Potential dishonesty in the observer program became a public issue in the spring and summer of 1985. Reporter Stephen Thornhill, in a series of articles for Canadian Press, reported allegations by a former observer of bribery and corruption in the program (Former Fish Observer Took Foreigner's Bribes, *The Evening Telegram*, May 8, 1985, and related stories). In response, DFO launched an Internal Audit of the program and requested the RCMP to investigate the allegations.

The Internal Audit was conducted in May and June 1985. Fisheries and Oceans Minister John Fraser released the report that September. The results of the audit indicated that management generally had satisfactory systems for administering and controlling the Observer Program and for taking action on suspected violations reported by observers or DFO Fishery officers (DFO 1985m). Some weaknesses were identified. For example, there was frequently not an adequately documented trail of actions taken on observer reports.

Regarding the specific allocations of bribery, the Audit concluded that the systems the Department had in place did not eliminate the possibility of observers taking bribes. The Audit noted:

> "In order to provide assurance against bribery the Department would have to place two observers on each assigned vessel and while this would further decrease the probability of observers accepting bribes it would not eliminate it. The Department does, however, have checks which would identify situations where significant irregularities may be occurring" (DFO 1985m).

The Audit concluded:

> "We feel that the effort made by the Department to help prevent the taking of bribes is reasonable, considering that it is expensive to have two observers per vessel, the incremental surveillance supplied by the additional observer is not significant, and prevention of acceptance of bribes would still not be assured" (DFO 1985m).

On January 9, 1987, DFO announced that the observer program had been given "a clean bill of health" following an extensive RCMP investigation into the allegations of bribery and corruption. In its investigation, the RCMP had found no evidence to support criminal charges. DFO concluded that the combination of the RCMP and internal audit reports had proven the soundness of the management systems and the integrity of departmental surveillance officers and observers (DFO 1987o).

The observer program has proven a useful adjunct to the more traditional methods of fisheries enforcement, sea patrols and aircraft surveillance. A number of other countries have adopted observer programs in the years following the widespread adoption of the 200-mile limit. Observers' presence on fishing vessels helps deter illegal fishing and provides data for resource assessment. French et al. (1982) reported on the benefits of the program established by the Northwest and Alaska Fisheries Center, National Marine Fisheries Service, for observers on foreign fishing vessels. They concluded that this program had provided an important source of information for managing foreign fisheries in the U.S. Fishery Conservation Zone and basic biological data on important species taken by foreign vessels.

Recently, some misgivings have been expressed about the utility of the observer program as an enforcement tool. These are based on anecdotal evidence which suggests that observers who come from the same fishing communities as vessel captains and crew are unlikely to report violations because of perceived community pressures.

3.3 Air Surveillance

3.3.1 General

Air patrols are the third major component of DFO's enforcement program. Air surveillance is effective for tracking fleet movements and helps deter illegal fishing. Aircraft can survey large areas, locate and track fishing fleets, detect violations and are a visible enforcement presence. Aircraft often provide the first indication of illegal incursions by foreign vessels in boundary areas. Routine sightings information can be used to ensure that patrol vessels are deployed effectively.

Air patrols provide photographic and eye-witness evidence of fishing in areas where vessels are not licensed to operate. Frequent air patrols provide a visible deterrent, particularly in the case of short-term air-detectable violations. As an example, a vessel may be able to increase revenues substantially on a single day by dodging back and forth across closure lines. Air patrols can discourage such violations. An aircraft has

TABLE 19-4. Aerial surveillance on the Atlantic Coast 1978/79 — 1988/89 fixed wing aircraft.

			TOTAL HOURS		HOURS PER REGION			
FISCAL YEAR	DFO BUDGET (MILLION)	COST/HR TO DFO ($)	DFO FUNDED	DND[1] FUNDED	TOTAL	NFLD.	SCOTIA-FUNDY	GULF/ QUÉBEC
1978/1979	1.200	542	2,227	1,440	3,667	2,536	1,131	
1979/1980	1.500	660	2,295	1,440	3,735	2,580	1,155	
1980/1981	2.000	807	2,464	1,440	3,904	2,688	1,216	
1981/1982	1.900	961	1,967	1,440	3,407	2,369	1,038	
1982/1983	2.000	1,147	1,776	1,440	3,216	2,085	881	250
1983/1984[2]	1.900	1,183	1,616	1,360	2,976	1,902	824	250
1984/1985	2.245	1,305	1,686	1,360	3,046	1,963	857	226
1985/1986	2.245	1,274	1,727	1,360	3,087	1,989	872	226
1986/1987	2.245[3]	1,536	1,367	1,360	2,727	1,790[4]	737	200[5]
1987/1988	2.245[6]	1,599/2,750	1,157	1,360	2,517	1,536[7]	801	330
1988/1989	2.245[6]	1,716/2,750	1,022	1,360	2,382	1,429[8]	756[9]	330

(1) Includes Tracker/Argus (Aurora) allotment
(2) Converted Argus to Aurora (500 h to 420 h)
(3) DND Temporary Duty Allowance increased from 45K to 150K
(4) Converted 113 hours to cover Atlantic Airways Pilot Project (100 h)
(5) Portion of hours converted to charter aircraft
(6) Qué/Gulf provided with 150K annually to charter aircraft(s)
(7) Converted 565 hours of 1346 Tracker hours to cover Atlantic Airways Charters (325 hrs)
(8) Converted full Nfld. Region to portion (1.250) of air surveillance budget to Atlantic Airways Charter(s)
(9) Converted 52 hours of 386 Tracker hours to cover Atlantic Airways Pilot Project (30 hrs)

Source: Aerial Surveillance Review for Atlantic Canada (Atlantic Operations Branch, Atlantic Fisheries Service) November, 1988.

to fly at low altitude to read the name of a vessel off its hull; this provides a clear deterrent threat.

The deterrence goal for air patrols adopted in 1980 translated into an operational requirement for the east coast of an average of 1.5 sightings per vessel per month. The operational requirement for the west coast was an average of 0.9 sightings per vessel per month (Clough 1980).

To meet these air surveillance goals, about 5900 air-hours per year were required on the east coast and 1600 air-hours per year on the west coast. The actual air patrol level was about 5000 air-hours on the east coast in 1977 and 1978. The west coast air patrol level declined from 1344 air-hours in 1977 to 826 air-hours in 1979 (Clough 1980).

Initially, air surveillance was provided exclusively by the Department of National Defence (DND). DFO purchased tracker aircraft hours to provide short-range patrols while Argus aircraft (subsequently replaced by Aurora) met long-range requirements. Up until 1990 DND provided both the Tracker and Aurora through a combination of DFO – purchased and free air hours.

By 1986, new approaches were required. Although DFO was entitled to purchase up to 2250 hours annually of Tracker air time, the rapid escalation in the cost to DFO of DND incremental Tracker aircraft time had eroded the number of purchased air hours by about 25% since 1980. Air coverage was further eroded by a relocation of the permanent Tracker base from Greenwood, Nova Scotia, to Summerside, Prince Edward Island in 1989.

The changing profile of the air surveillance resources over the 1978–79 to 1988–89 period is shown in Table 19-4. Substantial increases (approximately 15% annually) in the incremental rates charged by DND were a major factor impelling DFO to explore alternative delivery mechanisms. The incremental rate increased from $542 per air hour in 1978/79 to $1716 in 1988–89.

In the latter half of the 1980's DFO diverted portions of the fixed Atlantic air surveillance budget of $2.2 million to private sector companies in an effort to obtain more cost-effective air surveillance. During the 1987–89 period DFO chartered a twin-engined VFR helicopter as a pilot project to provide additional offshore (shipborne) capability with some coastal patrols. In 1989 the government provided additional funding ($7.5 million over 3 years) to enable DFO to acquire a long-range IFR rotary wing aircraft to conduct air patrols in Southwest Nova Scotia during the 1989–92 period. This aircraft was to provide 800 patrol hours annually.

In addition, the escalating cost of DND aircraft led DFO to experiment with an Atlantic Airways Ltd. twin-engine Beech King Air-B-200 aircraft with a state-of-the-art Litton search radar system. The cost effectiveness of the Beechcraft, operating from St. John's, was four times greater than the Tracker aircraft supplied by the Department of National Defence for the equivalent search capability (Auditor General 1988). In 1988–89 the Newfoundland Region converted its full portion of the Atlantic air surveillance budget to purchase Beechcraft air hours from Atlantic Airways Ltd. The Scotia-Fundy Region also began to experiment with this.

At the same time as DFO's air surveillance purchasing power had been declining, the demand for air surveillance had been increasing. Boundary areas such as the nose and tail of the Grand Banks, the area of boundary dispute in 3Ps between Canada and France, and the Hague Line between Canada and the U.S. had become the scene of unauthorized cross-boundary incursions by foreign fishing vessels. This was particularly true in the Gulf of Maine and on the Grand Banks. Detected unlicensed foreign fishing within the Canadian zone increased dramatically between 1984 and 1985 from 13 to 24 incidents in the Gulf of Maine and 5 to 63 incidents on the Grand Banks.

By 1989–90 the total DFO annual air surveillance budget had been increased substantially from the previous $2.2 million to $5.8 million. The government's decision to close the Summerside Air Base, announced in 1989, meant that the Tracker fisheries air surveillance capability would terminate. In October, 1989, the government provided $28 million over 5 years to maintain the air surveillance capability previously provided by DND, and to augment air surveillance activities to more adequately monitor the foreign and domestic fisheries. The annual budget from 1990–91 to 1993–94 was established at $6.5 million (DFO 1989n).

In April, 1990, DFO contracted with Atlantic Airways Ltd. for approximately 3500 hours annually over 4 years for offshore surveillance on the Atlantic. DND would continue to provide 420 free Aurora air hours each year. Helicopters and smaller fixed wing aircraft would continue to be chartered for patrols in the Gulf of St. Lawrence.

On the Pacific coast, DFO replaced the air hours provided previously by DND Tracker aircraft with chartered aircraft at an annual cost of $1.3 million in 1990–91. The Pacific Region also continued its use of helicopters and other smaller aircraft.

3.3.2 Goals for Air Surveillance

The operational goals for air-hours established following Clough's analyses in 1979 and 1980 were not met during the 1980's. In the late 1980's the air surveillance program was augmented to deal with unauthorized incursions into the Canadian zone. It is questionable whether the operational goals derived in 1979 and 1980, using 1977 and 1978 data on offshore fishing activity by foreign and domestic vessels, are still applicable. Clearly, these goals had not guided the deployment of aircraft patrols through the 1980's.

In practice, the actual deployment of aircraft patrols, as well as sea patrols and observers, has tended to be based on subjective decisions on the relative priorities of various areas, vessel nationalities, fishing practices and species. These priorities shift over time.

3.4 Dockside and Quota Monitoring

The Monitoring, Control and Surveillance program also includes dockside monitoring and quota management systems. Since all domestic vessels cannot be inspected at sea or carry observers, vessels are frequently inspected upon landing to ensure compliance with fisheries regulations, particularly quotas. Dockside monitoring involves enforcing regulations for misreporting, trip limits, gear restrictions, minimum fish size limits, and bycatch limits. These tasks are performed by land-based fishery officers who must visit more than a thousand landing sites on the Atlantic coast where there is not a formal dockside monitoring program. On the Pacific similar enforcement functions are performed by fishery officers.

Quota monitoring includes:

- collecting catch and effort information;
- assessing stock harvest status;
- issuing variation orders and closure notices for day to day control of fishing activity; and
- producing weekly quota reports.

Both the Harris and Haché reports recommended additional conservation measures. These required improving and expanding the existing system for inspecting vessels at the point of landing. In-port verification of fish landing must be coordinated with air and sea surveillance to ensure that the origin of all fish landed is accurately reported against the appropriate quotas and TACs.

4.0 SANCTIONING (PENALTIES)

The activities described so far are intended to increase the probability of detecting violations and apprehending violators. Catching violators is of little consequence unless they are sufficiently penalized to deter further illegal activity. If the perceived penalties and probabilities of detection, apprehension and/or arrest are low in relation to the perceived gains from illegal fishing, then the enforcement system is unlikely to deter further violations.

Penalties can take the form of fines levied under the *Fisheries Act* or the *Coastal Fisheries Protection Act* (CFPA), and forfeiture of illegally caught fish or fishing gear and equipment (including vessels) used to commit an offence.

The primary statute used to control the domestic (Canadian) harvesting sector is the *Fisheries Act*, particularly sections 7 and 43. Section 7 authorizes the Minister "in his absolute discretion" to issue licences or leases for fishing. Section 43 authorizes Cabinet to make regulations on a wide variety of matters pertaining to fishing. The *Coastal Fisheries Protection Act* (CFPA) provides the primary authority to control the harvesting of fish by foreigners in Canadian waters.

Fines levied under the *Fisheries Act* have generally been low. The use of licence suspension and forfeiture provisions in the *Fisheries Act* has varied over time. For a while during the 1980's there was increasing use of the Ministerial authority to suspend and forfeit. However, once the implications of the Charter of Rights and Freedoms became clearer, questions arose whether an individual was being subjected to double jeopardy for the same offence. The use of licence suspensions and forfeitures decreased from 1985 onward, after the *Fisheries Act* was amended to limit the Minister's suspension and forfeiture powers.

4.1 Penalties under the Coastal Fisheries Protection Act

Specific penalties respecting violations of Canadian fisheries laws by foreigners are established in the *Coastal Fisheries Protection Act*. Foreign fishing is governed both by the Coastal Fisheries Protection Regulations made under the CFPA and the Foreign Vessel Fishing Regulations made under the *Fisheries Act*.

The Coastal Fisheries Protection Regulations govern the licensing of foreign vessels to fish in the Canadian zone, the terms and conditions of such licences, requirements of the masters of foreign fishing vessels for notification of entry to and exit from the zone, taking of observers on board and boarding and inspection procedures.

The Foreign Vessel Fishing Regulations contain the detailed fisheries management provisions governing foreign fishing, including close times, size limits, incidental catch limits, mesh size, closed areas and seasons. Relevant provisions of these regulations are also generally stipulated in the licence issued to a foreign fishing vessel.

Penalties established in the CFPA relate only to offences stipulated in that Act and violations of the Coastal Fisheries Protection Regulations made under that Act. Fisheries Act penalties apply to violations of the Foreign Vessel Fishing Regulations. In addition to the general provisions of the CFPA, Section 13 of the Coastal Fisheries Protection Regulations allows the Minister to suspend or cancel any licence or permit.

The CFPA provides for two categories of fines for *summary conviction* and for *indictment* and different fines depending upon the offence. This Act as of the early 1950's (1952–53, C.15, s.I) provided for a fine of up to $5,000 on summary conviction or $25,000 on indictment for:

1. entering Canadian fisheries waters without authorization or fishing or transhipping without authorization.

A lesser fine of $2,000 on summary conviction or $10,000 on indictment applied for:

2. failure to "bring to" when required by a fishery officer or upon signal of a government vessel; and
3. failure to answer a question under oath when asked by an officer.

A violator was also liable to imprisonment for up to 3 months on summary conviction and 2 years on indictment, depending upon the nature of the offence.

In the early 1980's it became apparent that these penalty provisions were low relative to the potential gains a foreign vessel could obtain by fishing illegally. The Act was amended in June 1984 and the maximum penalties increased to $25,000 on summary conviction and $100,000 on indictment for illegal entry into the Canadian zone or unauthorized fishing, and $5,000 and $25,000 respectively for failing to "bring to", or resisting or obstructing a protection officer in the execution of his duty. The imprisonment option was retained only for resisting or obstruction offences.

With the growing concern about foreign overfishing and the call by First Ministers for tougher action against violators, the Act was again amended in December, 1986, to increase the maximum penalties to $100,000 on summary conviction and $500,000 on indictment for most categories of offences and $150,000 and $750,000 respectively for unauthorized fishing in the Canadian zone (DFO 1987p).

Thus, between 1983 and 1986, maximum penalties under the CFPA were increased substantially (thirty fold for incidents of unauthorized fishing).

Section 14 of the CFPA (R.S.C. 1985, c. 39, (2nd Suppl.)) provides for forfeiture of any fishing vessel or "goods" used in the commission of an offence. Forfeiture of a fishing vessel is a potentially powerful sanction. Canadian courts have, however, been reluctant to forfeit fishing vessels for fisheries violations. The number

TABLE 19-5. Number and type of prosecutions under the Coastal Fisheries Protection Act (CFPA) — 1975–84.

(A).

1975	1976	1977	1978	1979	1980	1981	1982	1983	1984
9	9	6	14	66	25	11	14	12	7

(B).

Offence	1979	1980	1981	1982	1983	1984
Unauthorized Entry	31	1	3	1	4	1
Illegal Gear	—	—	—	—	—	—
Licence Infractions	1	7	—	—	—	—
Illegal Fishing	34	15	8	12	8	—
Miscellaneous	—	2	—	1	—	6
TOTAL	66	25	11	14	12	7

TABLE 19-6. The disposition of charges laid under the Coastal Fisheries Protection Act — 1985 to 1989.

Year	Charges Laid	Withdrawn/Stay	Conviction	Dismissal	Outstanding
1985	70	5	38	10	17
1986	38	nil	28	5	5
1987	18	3	13	nil	2
1988	14	1	2	nil	11
1989	32	nil	3	1	28
TOTAL	172	9	84	16	63

and type of the charges laid under the CFPA in the early 1980's are shown in Table 19-5.

The relatively large number of prosecutions under the CFPA in 1979 related to the Canada-U.S. tuna dispute on the west coast. From 1979 to 1984, the majority of prosecutions were for fishing without authorization in Canadian waters.

From 1985 onward the number of charges laid was significantly higher than in the early 1980's (see Table 19-6). About 37% of the charges laid from 1985 to 1989 were still outstanding in 1990. Of the 109 charges dealt with by the courts, conviction was secured in 77% of the cases, 14.7% were dismissed or acquittal granted, and 8% were withdrawn by the prosecutors.

4.2 Penalties under the Fisheries Act

4.2.1 Fines

Under the *Fisheries Act* from 1977 to 1990, most fishing violations were subject "on summary conviction to a fine not exceeding $5,000 or to imprisonment for a term not exceeding 12 months or to both" (section 79.8 of the Act as amended in 1977). The 1977 version of the Act contained 14 sections setting out fisheries and habitat penalties.

More stringent penalty provisions were available for pollution of fish habitat (deposit of deterious substances), up to $50,000 on summary conviction and up to $100,000 for second and subsequent offences.

Various studies over the years (eg. Clough 1980; Blewett et al. 1987) concluded that these penalties, in particular, the $5,000 maximum fine for general offences under which most fish harvesting prosecutions fall, were much too low to deter violations of the *Fisheries Act* or regulations made under the Act.

Since 1984, the total number of convictions for violations of the *Fisheries Act* and regulations increased from 2588 in 1984 to 3830 in 1986 and thereafter stabilized in the 3300–3500 range annually (Table 19-7). The average fine was incredibly low, in the order of $160–$240 for an offence. The total number of convictions was relatively stable in the Atlantic Regions but almost doubled between 1984 and 1986 in the Pacific Region. The average fine for habitat charges was generally more than 10 times the national average for all

Table 19-7. Convictions, fines, licence suspensions and forfeitures made under the Fisheries Act — 1984–1988.

All Offences	1984	1985	1986	1987	1988
Total number of convictions	2,588	3,137	3,830	3,459	4,913
Total amount of fines	$424,839	$581,039	$677,922	$764,428	$859,242
Average fine	$ 164	$ 165	$ 177	$ 221	$ 175
Licence suspensions	26	7	13	211	79
Forfeitures	534	608	669	672	626
All Offences — Atlantic Region					
Total number of convictions	1,333	1,162	1,318	1,422	1,366
Total amount of fines	$288,323	$375,039	$378,223	$471,997	$489,759
Average fine	$ 216	$ 323	$ 287	$ 332	$ 358
Licence suspensions	26	3	11	21	74[1]
Forfeitures	199	237	304	291	291
All Offences — Pacific Region					
Total number of convictions	1,255	1,975	2,512	2,037	3,547
Total amount of fines	$136,516	$206,000	$299,699	$292,431	$369,483
Average fine (excluding Habitat charges and B.C. Sportfishing)	$ 154	$ 177	$ 212	$ 234	$ 250
Average fine — Sportfishing charges	$ 67	$ 80	$ 65	$ 81	$ 88
Average fine — Habitat charges	$ 3,785	$ 1,642	$ 1,780	$ 1,050	$ 2,400
Licence suspensions	—	4	2	17	5
Forfeitures	335	371	365	381	335

[1] 68 of the 74 suspensions were in the Scotia-Fundy region, mostly lobster licences.

fisheries-related violations. These fines were, however, still relatively low in relation to the nature of the offences and the maximum penalties available for habitat violations.

Two factors appear to have influenced the level of fines. First, the courts are reluctant to levy maximum penalties for "routine" offences. Second, the courts appear to perceive violations of the *Fisheries Act* and regulations as minor offences compared with other charges which come before the courts daily.

As a result, fines levied under the *Fisheries Act* became a minor cost of doing business and in no way served to deter illegal fishing. The fines were inconsequential in the context of the potential gains from illegal fishing.

4.2.2 Licence Suspensions and Forfeitures

The *Fisheries Act* also provides for another class of penalty, licence suspensions and forfeitures of gear, equipment and vessels. Section 9 of the Act (1977 version) provided that:

> "The Minister may cancel any lease or licence issued under the authority of this Act, if he has ascertained that the operations under such licence were not conducted in conformity with its provisions".

This gave the Minister complete discretion to suspend or cancel a fishing licence or lease, regardless of whether the licence holder had been charged for committing an infraction of the Act or regulations made under the Act.

Section 58 (5) of the Act also granted the Minister or the court the power to order forfeiture of "any vessel, vehicle, article, goods or fish seized" or "proceeds" in cases where a person was found guilty.

In 1985, the Minister's power to cancel a licence was altered with the passage of the *Statute Law (Canadian Charter of Rights and Freedoms) Amendment Act* (R.S.C. 1985 c.31 (1s Suppl.)) such that:

> "9. The Minister may suspend or cancel any lease or licence issued under the authority of this Act, if
> (a) the Minister has ascertained that the operations under the lease or licence were not conducted in conformity with its provisions, and
> (b) no proceedings under this Act have been commenced with respect to the operations under the lease or licence."

This wording clarified the Minister's power to suspend as well as cancel a licence or lease and attempted to foreclose arguments of double jeopardy under the Charter of Rights and Freedoms. In other words, the Department had an option of seeking Ministerial suspension or cancellation of a licence only if a case had not been brought before the courts.

The 1985 Amendment also removed the Minister's authority to order forfeiture, leaving this power to the court:

> "72(1), where a person is convicted of an offence under this Act or the regulations, the convicting court or judge may in addition to any punishment imposed, order that anything seized pursuant to subsection 71(1), or the whole or any part of the proceeds of a sale referred to in subsection 71(3), be forfeited."

Enforcement data are available for the Atlantic coast for the period immediately prior to and following the passage of the *Statute Law (Canadian Charter of Rights and Freedoms) Amendment Act* in October 1985. An analysis of forfeitures and suspensions during the periods January, 1984, to October, 1985, and October, 1985, to May, 1987 is enlightening. Prior to October 1985, both the court and the Minister had the power to order forfeiture. From January, 1984 to October, 1985, the courts ordered forfeitures in 451 cases. Between October, 1985, and May, 1987, when only the courts had the authority to order forfeitures, the courts ordered forfeiture in 291 cases. In absolute terms this represented an approximate decrease of 25% in court forfeitures in the period immediately following the Amendment. However, in relative terms, comparing what DFO sought and obtained in each case, the percentages did not change appreciably. A review by DFO's Atlantic Surveillance and Enforcement Committee (June 1989) concluded there had been no noticeable change in court forfeitures because of the 1985 Amendment.

From January 1984 to October 1985, the Minister only ordered forfeitures where the courts either failed to order forfeiture or where DFO considered it futile to ask the courts to do so. This situation occurred only in the Gulf Region where the Minister ordered 157 forfeitures, mostly for violations in the lobster fishery.

The Minister suspended licences in 29 cases in the period immediately prior to the 1985 Amendment. For all practical purposes, the Minister stopped suspending licences early in 1985 to conform to the spirit of the Amendment Bill. Thus the 29 cases were from the calendar year 1984. Following the Amendment the court suspended licences in 22 cases, a decrease of approximately 25%.

Larger discrepancies were noted, however, when post-Amendment suspensions were compared with Ministerial suspensions during 1982–83. Reports compiled by the Regulations and Enforcement Branch (DFO, Ottawa) indicate that, between 1980 and 1983, the total number of suspensions per year for the Atlantic varied from a low of 116 to a high of 146. Compared with these numbers, there was a significant drop in the number of licence suspensions in the period following the 1985 Amendment.

In an attempt to persuade the courts to use the licence suspension and forfeiture powers, DFO, in conjunction with the Department of Justice, developed a set of guidelines in 1986 for prosecutors handling fisheries violations. The guidelines were an attempt to encourage uniformity in approach across the country. They were developed for the use of crown prosecutors in recommending to the courts the suspension or cancellation of licences, or forfeitures of property, as penalties for fisheries offences.

More recent data (Table 19-7) indicate that the number of licence suspensions remained low from 1984 to 1987 but increased significantly to 79 in 1988 (from 21 in 1987). The sudden increase was the result of 68 suspensions in the Scotia-Fundy Region, mostly in the lobster fishery. When one considers that the total number of convictions from 1985 to 1987 was nationally in the 3100 to 3800 range, the number of licence suspensions was obviously very low.

The number of forfeitures from 1985 to 1988 was relatively stable (in the 600–670 range), applied in about 20% of convictions. Since a large number of these forfeitures involved only the fishing gear and illegally caught fish, it is not clear to what extent the exercise of the forfeiture powers by the courts was deterring illegal activity.

Meanwhile, the Minister retained powers under Section 9 of the *Fisheries Act* to suspend or cancel licences, provided that no proceedings had been commenced under the Act with respect to the operations under the lease or licence. From 1986 to 1988 this power went essentially unused, with the Minister leaving the question of licence suspensions to the courts. However, in the face of increasing pressure for a tougher stance against fisheries violations, DFO developed a new procedure for seeking Ministerial suspension, cancellation or refusal of a licence. This procedure was issued as national enforcement directive (NDE 1/89) on July 21, 1989. This was "intended to provide a procedure for use in instances of repeated non-compliance by a fisherman with the Fisheries Act and the regulations made thereunder."

This Directive incorporated procedures to ensure due process reflecting the principles of natural justice. The three essential elements were that:

— the licence holder be advised of the action being proposed;
— the licence holder be allowed reasonable time to present his side of the case; and
— his/her views be truly and impartially considered in making a decision.

The potential to suspend or cancel licences under Section 9 of the Act appeared to be limited to situations where there is a violation of the terms and conditions of the licence itself. This represents a fairly narrow range of offences, depending on the fishery. It does not, for example, include violations or regulation governing minimum fish size or mesh size limits.

Provisions of the *Fisheries Act* for penalties (both fines and suspensions and forfeitures) appeared in the late 1980's to offer little incentive to comply with the Act and its regulations. The fines levied by the courts were extremely low. The Minister's powers to suspend and order forfeiture were severely constrained by the October 1985 Amendments to conform to the Charter of Rights and Freedoms. The courts appeared not to have chosen to utilize their suspension and forfeiture powers to augment significantly the fines levied. Regardless of the probability of detection and apprehension for violations, the penalties (fines, suspensions and forfeitures) did not outweigh the potential significant gains from illegal fishing.

5.0 MEASURING THE DETERRENT EFFECT OF CANADA'S FISHERIES LAW ENFORCEMENT

From 1982 to 1984, DFO evaluated its fisheries enforcement programs in five regions to investigate the extent to which the Department's programs were deterring violation of the *Fisheries Act* and regulations. The results are contained in a series of DFO Evaluation Reports and are summarized in a 1987 paper by Blewett et al.

These evaluations were premised on the economic model of crime discussed earlier. The investigators assumed that the extent of non-compliance with fisheries regulations, and the associated probabilities of apprehension, arrest and conviction, cannot be directly measured but rather must be estimated.

Questionnaires were used to gather data on fishermen's perceptions of enforcement, the extent of non-compliance, and the probabilities, gains and losses associated with fisheries violations. Questions were asked about the probabilities of arrest, prosecution, conviction and punishment. Details about the questions can be found in Blewett et al. (1987).

TABLE 19-8. Department of Fisheries and Oceans Enforcement Program Evaluation Formula for Computing Probability of Conviction and Expected Net Return: Glossary of Variables and Variable Names.

Glossary of Variables and Variable Names	
individual violation rate	IVR
daily violation rate	VR
participation rate	PR
Pr(arrest)	P_A
Pr(prosecution/arrest)	$P_{PR/A}$
Pr(conviction/prosecution)	$P_{C/P}$
Pr(punishment/conviction)	$P_{PN/C}$
Pr(fine/conviction)	$P_{F/C}$
value of the fine	F
Pr(catch forfeiture/conviction)	$P_{CA/C}$
value of forfeited catch	CA
Pr(gear forfeiture/conviction)	$P_{GE/C}$
value of forfeited gear	GE
Pr(licence suspension/conviction)	$P_{L/C}$
value of lost fishing time	L
perceived penalty	PEN
illegal gain share	IGS
illegal gain	G

Source: Blewett, Furlong and Toews, 1987.

The initial formula used to compute the overall probability of conviction and calculate the expected net return was:

$$P_C = P_A * P_{PR/A} * P_{C/P} * P_{PN/C}$$

For a glossary of the variables and variable names see Table 19-8.

In this formula, used in the Pacific and Atlantic Evaluation studies, Pc included the probability of being punished if convicted. In the Quebec study (Furlong 1985) the formula was modified such that the probability of conviction was defined as above and the punishment probabilities were then entered into an equation for the perceived penalty.

$$PEN = (P_{F/C} * F) + (P_{C/A} * CA) + (P_{GE/C} * GE) + (P_{L/C} * L)$$

For further details on the model, see Furlong (1985).

For the Pacific Region, the illegal gains, penalties and probabilities as perceived by fishermen are summarized in Table 19-9. The perceived gain, G, and the perceived penalty, P, associated by the fisherman with a particular violation, varied among violations and among fishermen for the same violation. Some violations were perceived as very lucrative while others were seen to offer only a small return. The range of perceived

TABLE 19-9. Canadian fisheries law enforcement: illegal gains, penalties and probabilities as perceived by fishermen on the Pacific Coast.

	Value of Perceived			*Value of Expected*		
Violation	Illegal Gain (G)	Penalty (P)	Perceived Probability of Apprehension & Punishment	Illegal Gain[1]	Penalty[2]	Net Return[3]
GN: Creek Robbing	$ 1,500	$ 2,500	0.00	$1,500	$ 0	$ 1,500
GN: Net Violations	1,000	10,250	0.009	991	92	899
	200	4,200	0.01	198	42	156
GN: Area Violations	500	5,300	0.045	478	239	239
S: Net Violations	1,000	20,300	0.0099	990	201	798
	6,000	60,350	0.0097	5,942	585	5,357
	40,000	160,500	0.0095	39,620	1,525	38,095
S: Area Violations	5,000	20,250	0.0099	4,951	200	4,751
	60,000	100,000	0.009	59,460	905	58,555
T: Barbed Hooks	25	525	0.0099	25	5	20
	25	1,700	0.01	25	17	8
T: Undersized Fish	200	2,100	0.0098	198	21	177
T/GN: Area Violations	200	2,300	0.0098	198	23	175
H: Tab Violations	6,000	15,300	0.0098	5,941	150	5,791
	2,400	10,500	0.00	2,400	0	2,400

1. $E(G) = (1-p)G$
2. $E(P) = p(P)$
3. Net Return $= E(G)-E(P) = (1-p)G-p(P)$

GN = Gillnetter S = Seiner T = Troller T/GN = Troller/Gillnetter Combination H = Herring

Source: Pacific Region Evaluation Study, DFO.

Illegal Gains ranged from $25 for trollers fishing with barbed hooks to $60,000 for seiners fishing in closed areas or during closed times.

Perceived penalties exceeded perceived gains but the relative magnitude of the perceived penalties varied considerably. As Blewett et al. (1987) reported, the ratio of penalty to gain ranged from less than 2:1 to almost 70:1, with the ratio varying inversely with the size of the illegal gain. They observed:

> "The larger the potential gain from non-compliance, therefore, the more difficult it is to create a credible deterrent effect."

Fines were perceived to be an insignificant part ($100 to $500) of the total penalty, generally accounting for less than 10% of the total perceived penalty. The largest share of the perceived penalty were losses incurred from confiscated catch, forfeited fishing gear, and lost fishing time.

Probabilities of apprehension and punishment are the product of four probabilities: apprehension, prosecution, conviction, and penalties. Table 19-10 shows the perceptions of these probabilities. The probabilities of prosecution, conviction and punishment were all perceived to be at or very near 100% but the perceived probability of apprehension was around 1%. The fishermen interviewed believed their chances of being caught and punished for a violation were less than one in 100.

DFO carried out a similar study in three Atlantic Regions — Scotia-Fundy, the Gulf and Newfoundland. This study focused on the lobster fishery. Table 19-11 shows the perceived illegal gains, penalties and probabilities. Four basic types of violations are most common in the lobster fishery:

TABLE 19-10. Canadian fisheries law enforcement: components of the perceived probabilities of apprehension and punishment for fishermen on the Pacific Coast.

COMPONENTS OF THE PERCEIVED PROBABILITIES OF APPREHENSION AND PUNISHMENT (FISHERMEN)

Violation		Perceived Probability of Being			
		Apprehended	*Prosecuted*	*Convicted*	*Penalized*
Creek Robbing		0.00	1	1.00	1
GN:	Net Violations	0.01	1	0.90	1
		0.01	1	1.00	1
GN:	Area Violations	0.05	1	0.90	1
S:	Net Violations	0.01	1	0.99	1
		0.01	1	0.97	1
		0.01	1	0.95	1
S:	Area Violations	0.01	1	0.99	1
		0.01	1	0.90	1
T:	Barbed Hooks	0.01	1	0.99	1
		0.01	1	1.00	1
T:	Undersized Fish	0.01	1	0.98	1
T/GN:	Area Violations	0.01	1	0.98	1
H:	Tab Violations	0.01	1	0.98	1
		0.00	1	1.00	1

Source: Pacific Region Evaluation Study, DFO.

1. Taking undersized lobsters;
2. Taking egg-bearing females;
3. Fishing with an excess number of traps; and
4. Fishing in closed areas.

Perceived illegal gains were generally low, ranging from $59 in Scotia-Fundy to $254 in the Gulf. Perceived penalties were significantly higher in the Gulf ($8,000 to $13,000 range) compared with Scotia-Fundy and Newfoundland ($2,000 to $3,000 range). The perceived probability of apprehension and punishment ranged from 2% in the Gulf to 4% in Newfoundland. The expected net returns were low in all three regions for violations in the lobster fishery.

Perceived penalties were significantly higher in the Gulf because of the anticipated cost and probability of a licence suspension. In the Gulf, licence suspensions accounted for 50 to 90% of the total perceived penalty. Fishermen believed the duration of licence suspensions averaged about 3 weeks in the Gulf compared with 2 weeks in Scotia-Fundy. The perceived cost of a licence suspension was about $9,200 in both the Gulf and Scotia-Fundy but only $2,700 in Newfoundland. There was a significant difference in the perceived probability of a licence suspension being imposed — only 40% in Scotia-Fundy compared with 75% in Newfoundland and 90% in the Gulf. Overall, there was a perception of higher penalties through licence suspensions in the Gulf.

A subsequent study in Québec Region found a significant difference in the perceived probability of apprehension in the Gaspé Peninsula (20 to 30%) compared with the Magdalen Islands (2 to 9%). However, the perceived probability of prosecution given apprehension was considerably higher in the Magdalen Islands (75–85% compared with 47 to 51%). The perceived probability of apprehension in the earlier Atlantic study was 5 to 7% and the perceived probability of prosecution 87 to 95% (Table 19-12). Further analyses of actual data on arrests and prosecutions supported these regional rankings of perceived probabilities. When all of the perceived probabilities are factored in, violations in both the Gaspé and the Magdalen Islands were expected to produce a positive net return (Table 19-13).

The bottom line was that fishermen expected a positive net return from illegal activities in virtually all of the examples examined in these studies. With fishermen expecting to end up winners as a result of fishery violations ("crime pays"), Blewett et al. (1987) concluded that the existing enforcement system did not provide a

TABLE 19-11. The Atlantic lobster fishery: perceived illegal gains and penalties, and expected values of net returns.

	Value of Perceived			*Value of Expected*		
Region	Illegal Gain	Penalty	Perceived Probability of Apprehension & Punishment	Illegal Gain	Penalty	Net Return
Scotia/Fundy	$ 59	$ 2,215	0.029	57	64	(7)
Gulf	254	10,978	0.021	249	230	19
Nfld	110	2,394	0.04	105	96	9
All Regions	139	5,792	0.029	135	168	(33)

Source: Blewett et al. (1987).

TABLE 19-12. Perceived probability of arrest, prosecution and conviction in the Québec lobster fishery.

	Probability of Arrest	Probability of Prosecution Given Arrest	Probability of Conviction Given Prosecution	Probability of Conviction
Gaspé				
Trap Limit	0.212	0.477	0.614	0.062
Undersized	0.302	0.512	0.522	0.081
Iles-Madeleine				
Trap Limit	0.025	0.863	0.854	0.018
Undersized	0.093	0.751	0.838	0.059
Atlantic				
Trap Limit	0.057	0.877	0.893	0.045
Undersized	0.073	0.947	0.933	0.064

Source: Blewett et al. (1987).

sufficient deterrent. The particular Atlantic fishery studied most intensively, the lobster fishery, is, however, not a firm foundation for generalizing to other Atlantic fisheries because:

1. it is managed quite differently from the finfish which are generally under catch quotas;
2. at the time of the study the use of licence suspensions and forfeitures was much more prevalent in the Atlantic lobster fishery than in the Atlantic finfish fisheries or indeed other invertebrate fisheries.

Misreporting of area of capture, discarding of small fish, and underreporting of catch are prevalent violations in these other fisheries. Peacock and MacFarlane (1986) reported that the herring catch by Bay of Fundy purse seiners was underreported for several years by as much as half. Fraser and Jones (1989) cited instances of significant misreporting and discarding in the west Newfoundland small dragger fishery. These examples tend to reinforce the conclusion that fishermen perceive that crime pays and that the existing enforcement system could not ensure compliance.

Blewett et al. (1987) suggested that a small probability of apprehension coupled with large probabilities of prosecution and conviction is likely to be the most effective enforcement mechanism. Similarly, a small likelihood of apprehension and conviction coupled with large monetary penalties would be effective. Detecting, arresting, prosecuting and convicting offenders is a costly process (witness the overall cost of the DFO enforcement program). Payment of penalties, on the other hand, is a cost borne entirely by the offender. Blewett et al. (1987) concluded that, to the extent that greater severity of punishment could be substituted for greater likelihood of apprehension and conviction, it would make sense to do so.

The economic model of crime deterrence suggests that when the likelihood and severity of punishment are optimally set, the expected net return from commission of a crime is negative so that, on average, crime does

TABLE 19-13. Expected penalties and expected gains from non-compliance in the Québec lobster industry.

	Value of Perceived			*Value of Expected*		
Region	Illegal Gain	Penalty	Probability of Punishment	Illegal Gain	Penalty	Net Return
Gaspé Peninsula						
Trap Limit	$ 94	$ 172	0.062	$ 86	$11	$75
Undersize	$127	$ 203	0.081	$116	$18	$98
Iles de la Madeleine						
Trap Limit	$130	$1882	0.018	$128	$34	$94
Undersize	$112	$1438	0.059	$105	$85	$20

Source: Blewett et al. (1987).

not pay. When this point is reached, only risk preferrers are likely to fish illegally. Since the evaluation studies indicated that the combination of penalties and likelihood of conviction was too low, either the probability of apprehension and conviction or the penalties levied, or both, needed to be increased so that a potential violator would expect a negative return from illegal fishing.

Socio-economic variables also influence the propensity to fish illegally. Blewett et al. (1987) concluded that increases in the share of family income derived from the fishery, or in the proportion of the family employed in the fishery, were likely to decrease the violation rate. Conversely, an increase in the household unemployment rate was likely to increase illegal fishing activity. They suggested that greater family dependence on the fishery leads to a greater interest in a sustainable fishery and hence a greater respect for resource protection regulations. On the other hand, high unemployment creates a need for more income and provides a stronger incentive to violate fisheries regulations. Also, in some communities it appears to be more socially acceptable to violate fisheries regulations.

Overall, Blewett et al. (1987) concluded that the economic model of criminal behaviour was applicable to the fishery. Their results supported the assumption that individuals rationally decide to violate a fishery regulation by implicitly weighing the costs against the benefits.

They concluded that illegal fishing could be more effectively controlled by altering the perceived gains and losses. The relative effects of altering the various instruments of deterrence were less clear, due to small sample sizes.

6.0 RECENT CANADIAN INITIATIVES

6.1 Demands to Enhance the Surveillance/ Sanctioning System

In the latter half of the 1980's the problem posed by significant noncompliance with fisheries regulations became increasingly apparent. Various studies (the Report of Fisheries Ministers to First Ministers in 1986, the 1989 Scotia-Fundy Groundfish Task Force Report, and the 1990 report of the Harris Panel on Northern Cod) focused public and government attention on the problem of foreign and domestic overfishing. They called for improvements in the enforcement system (both surveillance and sanctions) to provide increased deterrence of fisheries violations.

The Report of Fisheries Ministers to the 1986 First Ministers Conference (DFO 1986g) drew attention to the escalating problem of foreign overfishing outside 200 miles on the Atlantic coast. It recommended that Canadian surveillance and enforcement capabilities be enhanced by:

> "a) amending legislation to allow for increases in maximum fines for illegal activities by foreign vessels;
> b) providing full observer coverage on all licensed foreign vessels fishing in Canadian waters by requiring foreign interests to pay the observer company directly for the costs of mandatory observer coverage, and with the existing resources of the foreign observer program redeployed to other offshore surveillance and enforcement activities;
> c) arming Department of Fisheries and Oceans offshore patrol vessels and boarding parties, commencing with the five Atlantic vessels;
> d) providing helicopter surveillance;
> e) developing an electronic identification system to be carried on fishing vessels; and
> f) denying port privileges to foreign fishing vessels or fleets which consistently disregard conservation rules established by international or bilateral agreement, subject to obligations under international law and relevant treaties, and prohibiting the use of such vessels in direct sales; i.e. for transporting Canadian fish exports."

The Scotia-Fundy Groundfish Task Force (DFO 1989i) reported, following extensive consultations with fishermen:

> "Fishermen tended to want a system of enforcement that was stronger and simpler. They wanted higher penalties for offenders, and more extensive use of licence suspensions and cancellations. The sentiment among the inshore fishermen seemed to favour a policeman on every offshore vessel."

During these consultations the quality of the commercial catch data provided by the existing monitoring system was condemned. Fishermen regularly claimed to misreport the area of their catch and to underreport their total catches. The Task Force examined the catch monitoring system and made recommendations to improve it.

The Task Force identified certain deficiencies in the *Fisheries Act*. The Act did not require industry participants (fishermen, buyers or processors) to maintain systematic records of any kind:

"The system places the onus on DFO to continually solicit data essential for management and there is little discipline or rigour in the system. Consequently, log and purchase slip forms are often incomplete. Catch and sales information can be withheld or misreported with relatively low probability of detection. Indeed, few charges have been successfully laid for failure to submit a report, logbook or slip, and weak penalties have been applied in instances when misreporting has been established."

To address these deficiencies in the commercial catch monitoring system, the Task Force recommended:

"7. Amend the Fisheries Act and regulations to place responsibility on fishermen and fish buyers to keep auditable up-to-date records as required for a comprehensive commercial catch monitoring system and to complete and submit periodic reports on their activities.

"8. Develop over four years an industry funded observer program for EA fleets (greater than 100 feet) fishing on the Scotian Shelf.

"9. Support studies into the development of electronic surveillance techniques for "at sea" monitoring and provide the legal authority to require the use of such devices.

"10. Modify the Region's catch monitoring program, define the data to be maintained, identify the reports to be submitted, improve dockside control through the use of designated ports and legal weigh-out sites, and develop the audit and analysis expertise required by the new system."

The Task Force proposed two approaches for improving enforcement. Both were based on the principle that enforcement can be made more effective and thus less costly by focusing on the consequences of breaking the rules. The proposals involved increased penalties for violations, and changes to the licence suspension procedures.

Specifically, the Task Force proposed (R.11) amending the *Fisheries Act* "to increase maximum fines and broaden the range of penalties by providing courts with more flexibility and discretion in order that a more effective deterrence is provided." It acknowledged that many in the industry regarded the low level of existing penalties as "a cost of doing business." It called for significantly increased penalties.

The Task Force also proposed increasing the use of Ministerial suspensions. To accomplish this, it suggested a new administrative approach:

"Infractions which are to be treated administratively should be removed from the regulatory scheme and imposed directly as a condition of licence. Guidelines must be established which define the consequences for violating the conditions. Finally, a formal review and appeal process or tribunal must be set up in order to assure that alleged violators are given a fair hearing before penalties are imposed."

The Panel on Northern Cod (Harris 1990) focused primarily on scientific and management issues, but also made some recommendations about enforcement. The Panel supported the need for enhanced aircraft surveillance and encouraged the deployment of additional patrol vessels to support aircraft surveillance.

The Panel expressed concern that the Department lacked sufficient personnel to implement properly the observer program, particularly to utilize fully the data gathered through the program. It identified a need for more training of observers, regular audits or "spot checks" in the field and analysis of observers' logs.

Noting that potential penalties for illegal entry by a foreign vessel into the Canadian zone and illegal fishing in the zone were by this time (1990) substantial, the Panel observed that penalties levied by the courts had not even approached those levels:

"We cannot escape the feeling that the violations of fishery regulations, the poaching of fish within the Canadian zone, and such like activities are regarded as mere peccadillos; as if the whole matter of enforcement were a game in which a few tons of illicitly taken fish were of no greater significance than the crabapples stolen by naughty boys from a neighbour's tree....such attitudes must change. Canada must convince both her own fishermen and the international community that conservation and proper management are matters of vital concern. In no cases should the potential gain from defiance of regulations exceed the penalty for being apprehended in violation of them."

Specifically, the Panel recommended:

"18. ....that the Government investigate the use of satellite or other advanced technologies for the purposes of surveillance; and that arrangement be imposed or negotiated as appropriate for fitting all vessels involving the Canadian shelf fisheries with transducers (sic) for ease of monitoring their movements and location.

"21. That DFO should expand the observer program to include observation on the inshore sector of the fleet and to expand support services for analyzing observer data.

"22. That the Government....undertake the provision of additional patrol vessels for offshore surveillance to provide adequate on-site action in respect of violations reported by aircraft or by observers and that helicopters be employed in conjunction with smaller patrol boats for inshore surveillance.

"24. That the Government....should urge the appropriate authorities to treat violations of fisheries regulations aimed at conservation as serious offences and to ensure that penalties imposed upon convicted violators be sufficiently onerous as to fully offset any potential gain from violations."

These three reports (Fisheries Ministers in 1986, the Scotia-Fundy Task Force in 1989, and the Harris Panel in 1990) each called for strengthening of Canada's surveillance capability and increased penalties for violators. Various initiatives have been undertaken since 1986 to address their major recommendations respecting enforcement.

6.2 The Government's Response

Many of the recommendations of the 1986 Report to Fisheries Ministers were acted upon shortly following submission of the Report. Penalties under the *Coastal Fisheries Protection Act* were increased substantially in December 1986 (see section 4.2). DFO increased observer coverage on foreign vessels to 100%, with the costs being paid by the foreign countries or companies involved. This freed up resources which were redeployed to bolster the observer coverage of the domestic fleet (see section 3.3). DFO adopted a policy of denying port privileges to foreign violators in an attempt to deter overfishing beyond the Canadian zone and illegal entry into and fishing in the Canadian zone. Helicopter surveillance was provided as suggested, and DFO bolstered air surveillance with the switch to private aircraft in the late 1980's (see section 3.4).

Other major initiatives involved arming DFO offshore patrol vessels and, more recently, experimentation with an electronic identification system.

6.3 Arming of Offshore Patrol Vessels on the Atlantic

Up until 1986 DFO was unable to independently implement the arrest of offshore vessels which chose to flee. DFO patrol vessels were generally unarmed. If a foreign vessel was detected fishing illegally in Canadian waters but resisted arrest by DFO fishery officers, DFO had to call on DND or the RCMP for armed support. From June, 1985, to April, 1986, DFO had to rely on such support four times. There were also at least four other instances when resisting vessels had avoided DFO arrest and fled the Canadian zone before the arrival of armed assistance.

In the summer of 1986 the Government proceeded with the arming of five DFO Atlantic offshore patrol vessels with 0.50 calibre portable machine guns, and the arming of boarding parties (DFO 1986g). This would mean significant savings since a DND destroyer, at that time, cost some $125,000 a day, compared with about $7,000 per day for DFO patrol vessels. One incident in 1986 involving the fishing vessel *Peonia #7* had cost more than $500,000 for a DND destroyer. The cost for that one incident exceeded the cost of arming and training all crews on the five DFO Atlantic offshore patrol vessels.

DFO envisaged that arming the patrol vessels would permit a timely reaction to violations and minimize the risk of foreign violators fleeing, on occasions with DFO officers onboard. Arming vessels should reduce DFO patrol days lost awaiting the arrival of armed assistance. A further consideration was that armed DFO patrol vessels provide a lower authority response to a violation than the executive power represented by DND. Other coastal states already had armed vessels and/or armed boarding parties to enforce fisheries legislation, e.g. Iceland, Norway, France, U.S., New Zealand and Australia.

The five Atlantic patrol vessels have been armed since December 30, 1987. Armed boarding parties are deployed only when a foreign vessel is suspected of illegal entry into Canadian waters, or illegal fishing, or when the vessel has failed to comply with orders to stop and receive a boarding party. They are not used to arrest Canadian vessels, although fishery officers may wear sidearms when carrying out boardings of Canadian vessels in potentially dangerous situations.

Armed boardings are carefully controlled. Approval authority levels vary according to the force used, but the final decision to use a disabling force rests with the Minister of Fisheries and Oceans in consultation with the Secretary of State for External Affairs.

6.4 Bolstering Atlantic Surveillance Activity under the 1990 Atlantic Fisheries Adjustment Program

In response to the recommendations of the Scotia-Fundy Groundfish Task Force and Harris Reports, the Government allocated resources under the Atlantic Fisheries Adjustment Program to increase the use of existing patrol vessels, to charter one offshore patrol vessel and additional inshore patrol vessels, and to bolster dockside monitoring capability. Funds were also provided to investigate available technologies (e.g. transponder devices) to identify fishing vessels and monitor fishing activity, including fishing outside the 200-mile limit by foreign vessels. The intent was to introduce this technology beginning with the offshore sector. DFO ended up increasing the utilization of its own patrol vessels and securing additional support from the Canadian Coast Guard and DND.

6.5 Amendments to the Fisheries Act

On June 6, 1990, Fisheries Minister Bernard Valcourt introduced in the House of Commons a *Bill to Amend the Fisheries Act*. This followed years of debate within government and the fishing industry on the need to increase the penalty provisions of the Act. The proposed amendments focused on increased penalties to provide greater deterrence for fisheries violations and for fish habitat offences, and to strengthen the effectiveness of fishery management through improvements to the fisheries statistical collection process. While the Bill was national in scope, the government linked it to the recently announced Atlantic Fisheries Adjustment Program. These amendments became law in 1991 (S.C. 1991, c.1). The three major thrusts of the Amendments are discussed below.

6.5.1 Penalties

Penalties in the *Fisheries Act* had not kept pace with those of related statutes, particularly the *Coastal Fisheries Protection Act* (amended in 1986) and the *Canadian Environmental Protection Act* (CEPA), passed in 1988. Nor were Canadian penalties for domestic fishing violations comparable to those in some other countries. A Six-Country Review of Legislation Pertaining to Fisheries Management (Gardner Pinfold Consulting Economists Ltd. 1988) had revealed significant differences in penalty provisions among Canada, New Zealand, the United States, the United Kingdom, Iceland and Australia. The differing provisions at that time are summarized in Table 19-14.

In all of these countries, violations were considered criminal offences, but the United States also has a category of civil offences. In all cases offences were punishable by fine, with prison terms specified in the United States and Norway. There were wide differences in the fines levied for general offences. Canada and Iceland levied the lowest fines, with maximum

TABLE 19-14. Penalty provisions for fisheries violations in various countries (prior to 1991 Amendments to *Fisheries Act*).

	Canada	New Zealand	United States
Offences and Penalties	Offences punishable by fine on summary conviction	Offences punishable by fine on summary conviction	Offences are divided into civil and criminal
	Fines up to $5,000 for general offences	Fines up to $8,000 for general offences	Fines up to $60,000 for general offences
	Certain specific offences identified	Certain specific offences identified	Civil offences are punishable by fine and subject to review by district court
	Forfeiture may be ordered at descretion of judge	Act provides for strict liability for offences	Criminal offences punishable by fine and/or prison term
	Suspension and cancellation of licences may be imposed	Court has discretion to order forfeiture	

Norway	United Kingdom	Iceland	Australia
Offences punishable by fine and prison term (for repeat offences)	Offences punishable by fine on summary conviction/indictment	Offences punishable by fines	Offences punishable by fine on summary conviction/indictment
Level of fine not specified	Fines up to $100,000 for general offences	Fines up to $3,500 for general offences	Fines up to $50,000 for general offences
Vessels, gear and catches may be confiscated	Court may also order forfeiture of fish and gear and also disqualify person from holding licence	Ministry empowered to revoke licence	Court has discretion to order forfeiture of vessel, fish and proceeds of catch

Source: Gardner Pinfold Consulting Economists Ltd. (1988).

levels of $5,000 and $3,500 respectively. New Zealand was next with fines of up to $8,000. Fines of up to $50,000 and $60,000 could be levied in Australia and the United States. The highest penalties were in the United Kingdom, where the Act provided for maximum fines of up to $100,000.

Most Acts gave the courts the discretion to order the forfeiture or confiscation of vessels, gear, catch and proceeds from the catches. Suspension and cancellation of licences were also provided for in the Acts of the United Kingdom, Iceland and Canada. Only in the New Zealand legislation were the courts given guidance on procedural matters.

With the 1991 Amendments to the *Fisheries Act*, Canada acquired one of the most stringent penalty regimes for offences of this kind of any major fishing nation. The changes in the penalty provisions for fisheries and habitat offences are summarized in Table 19-15.

Another major feature of the 1991 Amendments involved additional options for the courts. For all fisheries and habitat offences, the courts now had the discretion to make orders:

— prohibiting a person from committing any act or engaging in any activity that may result in the continuation or repetition of an offence;
— directing a person to take action the court considers appropriate to remedy or avoid any harm to any fish, fishery, or fish habitat that resulted, or might result from an offence;
— directing a person to publish, in any manner the court considers appropriate, the facts relating to an offence;
— directing a person to perform community service;
— directing a person to pay an amount of money the court considers appropriate to promote the proper management and control of fisheries or the conservation and protection of fish.

These Amendments also included a provision to increase the fine for all ticketable offences proscribed in regulations from an existing maximum of $100 to a new maximum of $1,000. There was an additional proviso for automatic forfeiture of fish and gear seized.

6.5.2 Stronger Statistical Powers

The Scotia-Fundy Groundfish Task Force had noted that the existing provisions of the Act respecting the need to provide statistics were clearly inadequate to deal with the complexities of today's fisheries (DFO 1989i).

The previous version of Section 61 of the Act required an individual request by the Minister or a fishery officer for the filing of a "true return". There was no obligation on fishermen, buyers or processors to keep records. Revisions to this Section were extensive, detailing:

— who could be required to provide information and keep records;
— the type of information that could be required;
— the responsibility to keep records as prescribed by regulations or licence; and
— how such information could be sought.

In addition, Section 49 of the Act was revised to provide inspection powers for fishery officers to examine records and books of account to verify the accuracy of information.

6.5.3 Stronger Enforcement and Management Powers

The 1991 Amendments also included:

1. Broader liability for offenders
   — Licence holders could be held liable for offences;
   — Officers and directors of corporations could be held liable for offences.
2. New powers for fishery officers
   — fishery officers were authorized:
      (i) to take fish samples,
      (ii) to seize books and records under warrant.
3. Regulation – Making Changes
   Powers were refined and clarified concerning:
   — orders to vary area closures and fish weights and sizes;
   — fishing vessel identification and tracking
   — the requirement to carry observers.

In addition to these major changes, the legislation repealed certain archaic and redundant provisions of the *Fisheries Act*.

From the onset, there was widespread support in the fishing industry for the proposal to increase penalties. The Atlantic Council of Fisheries Ministers had previously supported publicly the need for the amendments. Responding to the tabling of the legislation, Randy Baker, President of the Eastern Fisherman's Federation said that the Minister's bid for increased fines was a "very positive step". Mr. Baker said the majority of fishermen feel overfishing is wrong and must be stopped: "It's been accepted in the past in the sense that we know it's going on. But the majority of fishermen know that it's taking away from everyone, and it's

TABLE 19-15. Changes in the Penalty Provisions for Fisheries and Habitat Offences.

Previous	1991
FISHERIES OFFENCES	
General Fisheries Offences	
Previous:	*1991:*
Up to $5,000 or 12 months imprisonment, or both on summary conviction.	Up to $100,000 on summary conviction.
No provision for indictment.	Option for indictment. Up to $500,000 on indictment. Court Options: For second and subsequent convictions: Up to one year imprisonment on summary conviction. Up to Two Years on indictment.
Obstruction of a Fishery Officer	
Previous:	*1991:*
Up to $100 or six months of "hard labour" on summary conviction.	Up to $100,000 on summary conviction.
Up to Two Years imprisonment on indictment.	Up to $500,000 on indictment. Court Options: For second and subsequent convictions: Up to One Year imprisonment on summary conviction. Up to Two Years on indictment.
HABITAT OFFENCES	
Alteration of a Fish Habitat	
Previous:	*1991:*
Up to $5,000 on a summary conviction.	Up to $300,000 on summary conviction.
Up to $10,000 for second and subsequent offences.	Up to $1,000,000 on indictment.
Up to Two Years imprisonment.	Court Options: For second and subsequent offences: Up to six months imprisonment on summary conviction.
	Up to Three Years on indictment.
Pollution of Fish Habitat (deposit of deleterious substances)	
Previous:	*1991:*
Up to $50,000 on summary conviction.	Up to $300,000 on summary conviction.
Up to $100,000 for second and subsequent offences.	Option for indictment.
No provision for indictment.	Up to $1,000,000 on indictment. Court options: For second and subsequent offences: Up to six months imprisonment on summary conviction.
	Up to Three Years on indictment.
Failure to Provide Habitat Information	
Previous:	*1991:*
Up to $5,000 on summary conviction.	Up to $200,000 on summary conviction.
For second and subsequent offences: Up to $10,000.	Court options: For second and subsequent offences: Up to $200,000 and/or six months imprisonment.

hurting them by the shortage of fish" (*The Halifax Chronicle Herald*, June 7, 1990).

Gastien Godin, President of the Association of Acadian Professional Fishermen (APPNE), said that fishermen were completely in favour of the move to stiffer fines. Noting that his group, among others, had been pressing for stiffer fines and better enforcement of the regulations for several years, Godin stated that it was a necessity. He acknowledged that in the past there had been arrangements between fishermen and processors which often encouraged fishermen to underreport or misreport their catch. Godin also noted that fines for infractions in other sectors of society had increased dramatically in recent years and said it was time for similar fines in the fishing industry (*The Moncton Times Transcript*, June 7, 1990).

In an editorial, *The Halifax Chronicle Herald* characterized the proposed amendments as "tough, but necessary." It observed:

> "Frankly, the existing fine structure for illegal fishing by Canadians is something of a joke. Mr. Valcourt is moving in the right direction by proposing penalties which are a real deterrent to serious overfishing in waters off Atlantic Canada" (*The Chronicle Herald,* June 11, 1990).

The penalty provisions of the *Fisheries Act* were amended as proposed (S.C. 1991 c.1). The impact of the modified sanctions will ultimately depend upon actual fines levied by the courts in the coming years.

7.0 THE FISHERIES ENFORCEMENT PROCESS — OTHER PERSPECTIVES

In recent years a few interesting studies have been done on the fisheries enforcement process in other jurisdictions, particularly the United States. A Workshop on Fisheries Law Enforcement, held at the University of Rhode Island in October, 1985, focused on the programs, problems and evaluation of fisheries law enforcement in the United States (Sutinen and Hennessey 1987).

In an overview of U.S. fisheries law enforcement presented at this workshop, Hennessey and Sutinen (1987) observed that enforcement is one of the most costly components of U.S. federal fisheries management. In 1985, nearly 60% of all expenditures to carry out the Magnuson Fishery Conservation and Management Act were for enforcement. They noted that the high costs and problems associated with enforcement were seriously complicating management programs. No management plans had been implemented for certain U.S. fisheries because of the high costs of enforcement. In other fisheries there was a persistent problem of devising regulations that were biologically suitable, politically feasible and enforceable.

Overall, detected violations in the U.S. had been on the increase among domestic vessels while the number of foreign fishery violations exhibited no clear pattern after an initial decline following extension of fisheries jurisdiction. There had been no marked increase in detected foreign violations since the introduction of an observer program.

Implementation of the MFCMA in 1977 had more than doubled enforcement expenditures. The average annual expenditure from 1977 to 1983 was approximately $90 million. The bulk of fisheries enforcement expenditures were by the U.S. Coast Guard, with the National Marine Fisheries Service having only a small enforcement capability.

Hennessey and Sutinen (1987) concluded that overall compliance was nearly impossible to measure given existing enforcement practices. Actual levels of compliance were unknown. The available data simply measured the extent of detected noncompliance. They observed:

> "We cannot say whether the current levels of expenditure are justified nor how well the program is working. It is not clear whether society would be better off with more or with less enforcement."

Enforcement is expensive and imperfect. Some types of management regulations are more costly than others to enforce. For example, at sea enforcement operations are significantly more expensive than dockside operations. Hennessey and Sutinen concluded:

> "Management regulations restricting how, when, and where fishing is conducted at sea may not be economically justified in some fisheries. Similarly, other regulations, while desirable in a costless, perfect enforcement context, may not be when the realities of enforcement are accounted for."

Margaret Frailey (1987), Assistant General Counsel for Enforcement and Litigation in NOAA, picked up on this theme. She suggested that enforcement had often been an after-thought in fisheries management, a postscript to Fisheries Management Plans. Rather than making enforceability an important criterion in choosing among alternate management strategies, some managers seemed (to her) to assume that fishermen would automatically comply with whatever regulations were promulgated.

Pallozi and Springer (1987) argued that program effectiveness in fisheries law enforcement should be assessed in terms of "reasonable compliance." They defined "reasonable compliance" as a situation where violations in a fishery under regulation are occurring at a rate which:

1. is much lower than that at which they would occur with no enforcement;
2. is acceptable to the industry and the public; and
3. contributes to the conservation goals established by the U.S. Fishery Management Plans.

They were, however, unable to translate the concept of "reasonable compliance" into tangible measurements. They examined alternative enforcement strategies and concluded that management measures and regulations requiring costly enforcement should be limited, and where possible, eliminated. This could mean reducing at-sea enforcement. They tempered their conclusion with the realization that measures and regulations that are both effective and inexpensive cannot always be found. While dockside enforcement modes are less costly than at-sea modes, they are not suitable for ensuring compliance with certain types of management measures, e.g. closed areas.

In a paper presented to a 1986 European Workshop on the Regulation of Fisheries (Ulfstein et al. 1987) P.J. Derham commented on enforcement of EC fisheries legislation. He criticized the legislative process and the resulting regulations:

> "Quite a few of the regulations adopted can be seen as excessively complex, and fishermen, who are well aware for example, that a sizeable sole will get through a small mesh net, condemn them as impractical. Certainly from a policing point of view, many have proved to be so difficult if not impossible to enforce that they seem to have but presentational value" (Derham 1987).

Derham discussed various types of management measures from the perspective of enforceability. He argued that quotas were largely unenforceable and that regulation of fishing effort was equally difficult to apply effectively.

In a later paper, Frailey and Taylor (1987) argued that the actual impact of enforcement sanctions is difficult to determine. The primary sanction used against violaters of the Magnuson Act is the administrative assessment of civil penalties. Frailey and Taylor analysed required penalty assessments and concluded that, depending on the assumptions, the penalty level necessary to offset illegal gains could be many times the statutory maximum under the MFCMA. In response to the question, "why not seek higher statutory maximums and higher penalty assessments to make our penalties more effective?", they advanced several arguments:

1. A high-penalty case is less likely to be settled before the hearing stage and will drag on through the justice system;
2. The higher the level of the assessed penalty the greater the likelihood that the fisherman will exert whatever political influence over the process and level of sanction he can muster;
3. Lack of data on the U.S. fisherman's perception of detection; and
4. Fishermen might be motivated by other than economic factors.

Frailey and Taylor noted an increasing use of licence suspensions or cancellations as sanctions for domestic fisheries violations. However, because only a few fisheries are under limited entry licensing in the U.S., the utility of this approach is limited there.

Finally, Frailey and Taylor argued that the probability of detection is critical to the enforcement process. Since increased enforcement dollars are unlikely, the answer, she contended, must be to make regulations more easily enforceable. Where possible, management measures should be enforceable at dockside.

What Derham, Frailey and Taylor failed to consider was the relative effectiveness of particular types of management measures in achieving management objectives (see Chapter 5). This is tied to the question of their enforceability. Perfect enforcement of ineffective management measures is not the solution. Management plans and enforcement plans must be linked. Greater attention must be given to the proper mix of enforcement instruments, including sanctions.

Anderson (1989) presented a net gains model for a regulated fishery. To calculate the net gains from a regulated fishery, it is necessary to subtract from the value of fisheries output a multiplicity of costs — the cost of fishing effort, compliance costs, avoidance costs, initial government implementation costs and government enforcement costs. In the most general case, the problem for a management agency is to select the appropriate combination of policy instruments and to allocate implementation and enforcement inputs so that the net gains from the regulated fishery are maximized. However, as noted by Anderson, the problem is more than a simple maximization problem. Agencies seldom know the exact nature of the gains function to be maximized. Also, there are often other maximums, mostly dealing with income distribution and other politically sensitive issues (see Chapter 4), that are often driving forces in determining how and for what purpose a fishery will be managed.

Anderson identified the following issues as important in the practical application of fisheries policy: (1) the type of monitoring (dockside versus at-sea), (2) ease of government implementation, (3) duration of the period at risk when in noncompliance, (4) ease and cost of compliance, (5) ease of distinction between honest mistakes, sloppy practices and deliberate cheating, (6) initial versus continued compliance, (7) ease of communicating requirements, (8) ease of disguising noncompliance, (9) ease of detecting noncompliance that is admissible as evidence, (10) demonstrating personal or social benefits from compliance, (11) potential for citizen cooperation in identifying offenders, (12) detection of illegal activities under various conditions, (13) efficacy of enforcement with respect to management objectives, and (14) identification of enforcement priorities.

These are all important issues which have a bearing on the choice of the mix of enforcement instruments and can influence how enforcement is conducted. See Anderson (1989) for an interesting discussion of how these issues can affect the enforcement process.

Anderson concluded:

> "When comparing fisheries regulations programs, it is necessary to go beyond the efficiency effects on the production of effort. The absolute and relative effects on industry compliance and avoidance costs as well as government implementation and enforcement costs can be very important in determining which program produces the largest net benefits."

8.0 CONCLUSIONS

Enforcement is a neglected but vital component of the fisheries management process. The objective is to influence the behaviour of fishermen (and other industry participants) in such a manner as to deter illegal fishing and secure compliance with fisheries management measures and regulations.

Deterrence and compliance are not directly measurable. However, studies of the Canadian fisheries enforcement program indicate that the economic model of crime provides a framework for considering how best to achieve deterrence. Apart from moral considerations, fishermen are motivated to maximize their gains from fishing. Implicitly, they calculate whether the gains from illegal fishing outweigh the costs of detection and punishment.

The Canadian studies, and studies elsewhere, suggest that, generally, in the fisheries situation, crime pays. The probability of detection is generally perceived as low (less than 10%). Penalties have been historically low. Therefore, the combined probabilities of detection, apprehension and arrest, and the likely level of penalty to be levied are so low that fishermen perceive a positive net return from fisheries violations.

The appropriate mix of sea patrols, observer programs, air surveillance and dockside monitoring depends upon the fishery and the types of management measures being enforced. Air surveillance, for example, is especially useful in detecting illegal fishing by foreign vessels in the Canadian zone. At-sea boardings are necessary to monitor compliance with gear regulations, minimum fish size, prohibited species, etc, and to provide the platforms to effect arrests of suspected violators. Observers provide useful data for scientific resource assessment and to facilitate the detection of violations on offshore vessels. The presence of observers on vessels can in some instances deter illegal fishing activity.

Since the mid-1980's Canadian monitoring, control and surveillance activities have been strengthened considerably. Full observer coverage of the foreign fleets, increased observer coverage on the domestic offshore fleet, increased offshore sea patrol capability, the arming of Atlantic offshore patrol vessels, increased air surveillance, and improvements to the dockside commercial catch monitoring process, have all contributed to increasing the probability of detection, apprehension and arrest of violators. However, many fishermen still perceive the probability of detection and apprehension as low.

Amendments to the *Coastal Fisheries Protection Act* in 1986 strengthened considerably the penalty provisions applicable to violations under this Act. Amendments to the *Fisheries Act* in 1991 should provide a stronger deterrent to violations of the *Fisheries Act* and regulations. Increased penalty provisions do not by themselves instill a deterrent effect. The extent to which the sanctioning component of the enforcement process is strengthened will depend upon the actual fines levied by the courts. If the courts levy higher penalties, then these will affect fishermen's perception of the gain versus potential loss from illegal fishing. It will be several years before the effect of increased penalties on compliance can be assessed. By significantly increasing the penalty provisions, Parliament has sent a signal to the courts and to the fishing industry that penalties for fisheries violations can no longer be regarded as merely a cost of doing business.

Changes to the sanctions process were being proposed in early 1993, linked to the establishment of new Atlantic and Pacific Licensing and Allocation Boards. It was proposed that breaches of conditions of licences and regulations by licensed commercial fishermen, under the new Act establishing these Boards, would be handled by the Boards rather than the criminal court system. Other crimes under the *Fisheries Act*, e.g. habitat destruction, and criminal code offences, e.g. fraud and assault, would continue to be handled by the criminal courts, as would unlicensed fishing.

The proposed Boards would have the authority to impose a range of administrative sanctions. These would include: forfeiture, quota reduction, suspension, non-renewal or cancellation of a licence, or monetary penalties. The onus of proof would become a "balance of probabilities" rather than the criminal law requirement for proof beyond a reasonable doubt.

Also, minor violations would be made a ticketable offence. Tickets would result in financial penalties. Contested tickets would lead to an oral or a paper hearing at the option of the violation.

The intent was to allow these proposed Boards to function as a knowledgeable fisheries court, handing down appropriate penalties swiftly and fairly. Retention of a licence would be more directly linked to a fisherman's willingness to abide by the rules.

The combined effect of increased and improved monitoring, control and surveillance activities and increased penalties should over time change fishermen's perceptions about the net return from fisheries violations. The economic disbenefits of fisheries violations, resulting from detection, apprehension, conviction and punishment, should outweigh the economic gains from illegal activity. If this can be achieved, there will be greater compliance with fisheries regulations. This will only occur if fishermen perceive that it no longer pays to commit a fisheries offence.

The optimal mix of enforcement instruments for particular fisheries situations is still largely unknown. Further research is warranted on the enforcement process and on the effectiveness of recent enforcement initiatives in providing an incentive for increased compliance.

Most fisheries enforcement is reactive rather than proactive. Another dimension to fisheries enforcement, too often ignored, is education. Greater emphasis should be placed on educating/informing fishermen, processors and the judiciary of the rationale for particular fisheries regulations and the serious effects of illegal fishing. There should be a greater focus on crime prevention. Crime prevented is the best end product.

CHAPTER 20

SOME OBSERVATIONS ON MARINE FISHERIES MANAGEMENT IN OTHER COUNTRIES

1.0 INTRODUCTION

So far we have examined the evolution of fisheries management practice for marine fisheries in Canada over the past three decades. This included particular emphasis on events since the transition from expansion-oriented development activities to government attempts to arrest declining resources and address the common property problem. During this period other countries have confronted similar problems and adopted divergent approaches. Before drawing conclusions about the Canadian experience, it seems prudent to first situate that experience in a world context. This chapter attempts to do that. It examines the major trends in fisheries management in certain other countries, particularly since the worldwide move in the mid-1970's to enclose the common property resource by 200-mile fishing or exclusive economic zones. This is a cursory description of the accomplishments and shortcomings of management regimes in these countries. However, it illustrates some of the lessons learned from extended jurisdiction and should assist us in drawing conclusions about the relative success of Canada's initiatives.

Enclosure of the ocean commons had considerably different impacts upon the fishing industry and the fisheries management systems of the various countries examined. Some countries have benefited significantly. These include Iceland, the U.S. (in the North Pacific) and New Zealand. While there have been extensive changes to the fisheries management system in certain countries (the U.S. and New Zealand, for example), very few have established clear, specific objectives for management. Resource conservation has been a concern for most countries. Only New Zealand and, more recently, Australia, have embraced maximization of economic return as an explicit fisheries management objective. Several countries, including the U.S., Australia and New Zealand, have attempted to involve industry more extensively in fisheries management. This ranges from the traditional advisory structures in most countries, to a more significant role for Management Advisory Committees in Australia, to the almost complete decentralization of fisheries management decision-making in the U.S. to Regional Councils dominated by industry representatives.

Miles (1989) reported the results of a 1985 workshop which looked at the early post-extension experience in various parts of the world. Case studies presented at this workshop indicated that variations among countries and in different regions of the world made it difficult to draw overall conclusions.

While many coastal states clearly benefited from the new 200-mile zones, Miles (1989) concluded that the case studies generally illustrated five kinds of management failures:

1. failures in conservation;
2. avoiding basic allocation decisions and, as a consequence;
3. failure to deal with overcapitalization;
4. developing countries' failure to establish adequate management systems because they lack the capability and resources; and
5. failures in fisheries development (either incomplete or inappropriate development).

Miles also concluded that there is no necessary connection between extended jurisdiction and fisheries management performance. Extended jurisdiction only provided an opportunity to increase control over foreign and domestic fishing. Coastal states differed significantly in exercising that control.

The following sections describe some successes and failures of post-extension fisheries management in selected countries.

2.0 UNITED STATES

2.1 Overview

The United States is one of the world's chief fishing nations, ranking in the top six in terms of landings consistently since 1970. With extension of fisheries jurisdiction to 200 miles in 1977, the U.S. moved from 6th place in the early 1970's to 4th place through most of the 1980's. It has been surpassed consistently only by

Japan, the USSR and China, which were ranked one, two and three, respectively, through the 1980's. The U.S. share of world fish production increased from 5 to 6% during the 1980's. Over the past three decades it has generally been in the 4–6% range.

Concern about the effects of foreign fishing in the late 1960's-early 1970's led to pressure from various groups for the U.S. to join those states seeking extended fisheries jurisdiction. However, the Executive Branch of the U.S. government resisted the worldwide move to a 200-mile limit in UNCLOS III. It was more concerned with security and other related aspects of the Law of The Sea discussions. Nevertheless, fishing interests secured the support of key members of Congress for a U.S. 200-mile limit (Hollick 1981).

2.2 The Magnuson Fishery Conservation and Management Act (MFCMA) and Its Application

2.2.1 Provisions of the Act

After much debate, and attempts by the Executive Branch to block the proposed legislation, Congress in March 1976 passed the Fishery Conservation and Management Act (FCMA). This legislation, later renamed the Magnuson Fishery Conservation and Management Act (MFCMA), ushered in a new era in U.S. marine fisheries management.

For a discussion of events leading up to the MFCMA, see a legislative history of the Fishery and Conservation and Management Act of 1976 (USA 1976). Here I will describe the management regime resulting from the MFCMA and some of its accomplishments and shortcomings.

The MFCMA appears to be unique in the detail and specificity of the fisheries management regime it supports.

Apart from the purposes and definition section, the Act is divided into four parts or Titles:

Title I describes the fishery management authority of the United States, including the proclamation of the new Fishery Conservation Zone and U.S. rights within that zone;

Title II, Foreign Fishing and International Fishery Agreements, sets out the terms, conditions, and mechanisms to govern foreign fishing in the new zone;

Title III, National Fishery Management Program, sets out National Standards for fishery conservation and management, establishes a mechanism for fisheries management — Regional Councils — a process for and the contents of fishery management plans, and various provisions relating to enforcement;

Title IV, Miscellaneous Provisions, contains housekeeping provisions, including amendments to other fisheries related legislation;

Here I focus on the purposes and objectives, the national standards, the Regional Council framework, and the fishery management plans process.

The MFCMA established a comprehensive framework for managing fisheries in a Fishery Conservation Zone (FCZ) that extended from 3 to 200 nautical miles. The Act extended the legal jurisdiction of the U.S. over fisheries from an area of about 545,000 square nautical miles to over 2,000,000 square nautical miles, the largest fishing zone in the world.

The purposes of the Act (section 2(b)) were:

1. To take immediate action to conserve and manage the fishery resources off the coasts of the U.S., and the anadromous species and Continental Shelf fishery resources of the U.S. by (A) establishing a fishery conservation zone within which the U.S. would assume exclusive fishery management authority over all fish, except highly migratory species, and (B) exclusive management beyond the FCZ over anadromous species and Continental Shelf fishery resources;
2. To support and encourage international fishery agreements for highly migratory species;
3. To promote domestic commercial and recreational fishing "under sound conservation and management principles";
4. To provide for fishery management plans "which will achieve and maintain, on a continuing basis, the optimum yield from each fishery";
5. To establish Regional Fishery Management Councils; and
6. To encourage development of fisheries which were currently underutilized or not utilized by U.S. fishermen.

Title II of the Act authorized foreign fishing under certain conditions. It provided for Governing International Fishery Agreements (GIFAs) which would govern foreign fishing in the U.S. FCZ. It provided for the establishment annually for particular stocks of a Total Allowable Level of Foreign Fishing (TALFF), which was defined as that portion of the optimum yield (OY) of a fishery which could not be harvested by U.S. vessels.

Title II authorized the Secretary of State to allocate TALFFs, in cooperation with the Secretary of

Commerce, taking into account whether nations:

(i) had traditionally engaged in a particular fishery;
(ii) had cooperated with the U.S. in research and the identification of fishery resources; and
(iii) had cooperated with the U.S. in enforcement.

Under Title III, Section 301 established certain National Standards for fishery conservation and management.

Section 302 of the Act established eight Regional Fishery Management Councils. Each council was to manage the species within its geographical area in the FCZ (later Exclusive Economic Zone) adjacent to the 3-mile territorial sea which remained under jurisdiction of the States. Where species were important in more than one area, they were managed jointly by the relevant councils, usually with one council designated the lead.

The majority of voting members on each council were selected by the Secretary of Commerce from lists submitted by the governors of the States in a given region. Other mandatory members included the respective Regional Director of the National Marine Fisheries Service and the official from each State who had principal responsibility for marine fisheries management. There were also nonvoting members such as regional directors of the U.S. Fish and Wildlife Service, the commander of the Coast Guard district, the executive of the interstate Marine Fisheries Commission for the region concerned and a representative of the Department of State.

According to the Act (Section 302.h.) the functions of each Council were to:

1. Prepare and submit to the Secretary of Commerce a management plan for each fishery within its geographical area of authority and such amendments as become necessary;
2. Prepare comments on any application for foreign fishing and any fishery management plan or amendment transmitted to it;
3. Conduct public hearings to allow all interested persons to be heard in the development of FMPs and amendments; and
4. Submit reports to the Secretary of Commerce.

Section 303 of the Act established certain requirements for Fishery Management Plans (FMPs). Section 304 of the Act laid out the process and timetable for review by the Secretary of Commerce of any FMP submitted by a Council. Section 304(c) provided for the Secretary to prepare a FMP if:

(a) the appropriate Council fails to develop and submit to the Secretary, "after a reasonable period of time," a FMP for a particular fishery; or
(b) the Secretary disapproves or partially disapproves of any FMP or amendment, and the Council involved fails to change the Plan or amendment.

Section 304(e) required the Secretary of Commerce to develop a comprehensive program of fishery research to apply the Act.

The most important provisions of the MFCMA were those setting the management objective as "optimum yield", establishing the Regional Councils as the primary management authority in the FCZ (EEZ), and provisions regarding the preparation of the FMPs.

In the context of subsequent attempts to evaluate the implementation of the MFCMA, it is important to note that "optimum yield" was defined (Section 3(18)) as:

"the amount of fish —
"(a) which will provide the greatest overall benefit to the Nation, with particular reference to food production and recreational opportunities; and
"(b) which is prescribed as such on the basis of the maximum sustainable yield from such fishery, as modified by any relevant economic, social, or ecological factor."

In the years since the MFCMA became law, various authors have analyzed it and commented on the strengths and deficiencies of the resulting management system.

Christy (1977) criticized the objective of *optimum yield*, as defined in the Act:

> "The nebulous nature of this standard...renders it ineffective in providing a basis for decision-making. "Optimum yield" becomes merely a "best" yield, to be defined on an ad hoc basis by decision-makers. Thus, in view of the current dominance of industry representatives on the Regional Fishery Management Councils, "optimum" is likely to be interpreted as that yield best for industry, while other valid interests are de-emphasized or even ignored."

Christy noted that the goals of consumers and taxpayers are mentioned only indirectly in the Act and that the goal of maintaining employment opportunities for fishermen was absent. He concluded:

> "In short, there is a significant inadequacy in the Act's stated management objectives. In the

preparation of management plans, the Regional Councils will be left to their own devices, resolving policy objectives on an ad hoc basis in the context of the objectives for the particular fishery involved."

Pontecorvo (1977) identified four main shortcomings of the Council system:

1. The Councils did not appear to have sufficient expertise in the natural or social sciences to accommodate the complexities of the management task they confronted;
2. The Act did not favour the "general welfare", defined in terms of the largest output for the lowest price, constrained by the need for conservation;
3. Fishing industry representatives would dominate the Councils; and
4. The structure of representation by state and individuals suggested that the Councils would support the status quo in their approach to management.

Pontecorvo concluded:

> "The Act was intended to protect United States interests from unregulated depredation of our resources by other nations. Unfortunately, it appears to have been constructed and effectuated in a way that is likely to be detrimental to the general welfare and of doubtful value in advancing fisheries management."

Given these gloomy predictions, what happened in practice? There are a number of ways of addressing that question. I will illustrate the range of opinion on the efficacy of the Councils and the Fishery Management Plans process, and offer some personal observations based on an examination of the track record of two Councils.

2.2.2 Some Critiques of the MFCMA System

Spencer Appollonio, who served for a while as Chairman of the New England Council, offered some insightful observations on the initial difficulties experienced by some of the Councils (Appollonio 1983). He described "optimum yield" as "an ill-defined and, as it turned out, elusive concept that allows social and economic considerations." Despite the wide range of available options, the initial experience indicated that Councils had trouble defining management objectives. FMPs contained very general objectives. Appollonio suggested that many of these were so general that they amounted to no clear objective at all. The MFCMA gave extraordinary latitude to the Regional Councils in setting policy. In practice, optimum yield became "whatever the regional managers say it is."

Anderson (1982) described two significant impacts of the MFCMA:

1. It created the "potential for strong and unified management where previously international agreements...were the only source of control"; and
2. The right to manage carried with it the right to utilize the stocks. This offered considerable potential to develop the U.S. fisheries by displacing foreign fishing.

Finch (1985) identified three principal problems with the management process when it was first implemented:

- the long time it took to put a FMP or amendment in place (initially, approving and implementing a FMP took as much as a year);
- the central-regional tug-of-war; and
- uncertainty about responsibility for managing the U.S. fisheries (the Councils frequently saw the National Marine Fisheries Service (NMFS) as second-guessing their decisions contrary to the purposes of the Act).

Finch concluded that management mechanisms under the Act, although more complex than desirable, were working reasonably well.

Fullerton (1987) attributed the following changes in U.S. fisheries to the MFCMA:

1. The Act had provided a climate in which domestic commercial harvesting and processing activities had expanded, sometimes contributing to overcapitalization and gear conflicts.
2. Foreign fishing in the U.S. zone had declined.
3. U.S. harvesters had benefited from joint ventures with foreign nations.
4. A new U.S. catcher-processor vessel fleet was being acquired by fishermen in New England and Alaska.
5. Foreign and domestic interest in underutilized species had encouraged new product development, such as surimi.
6. Competition within domestic fisheries had intensified. This had made allocation decisions more difficult for the Regional Councils.
7. Settlement of sensitive international negotiations, for example the Canada-USA Pacific Salmon Treaty, had been spurred.

8. As national standards in the Act had become more accepted, intergovernmental relations in the U.S. had improved. Although the federal government had twice preempted a state's authority to regulate its fisheries, it had generally adopted the fishery regulations of states when these had conformed with the Act. Cooperative agreements on exchange of fisheries data and enforcement were now in place.

A 1986 study sponsored by the National Oceanic and Atmospheric Administration (NOAA) identified significant defects in the way the Act had been implemented and called for major conceptual and operational changes. The deficiencies included continued overfishing of some stocks, a lack of coordination between the Councils and NMFS in setting research agendas, conflicts among users, the vulnerability of the fishery management plan process to delays and political influence, lack of accountability, inconsistency in management measures in federal and state waters, and adoption of unenforceable management measures (NOAA 1986).

In general, the NOAA Study suggested that the Regional Councils had tended to sacrifice resource conservation for short-term economic gains. The Study contended that the Councils were too close to the user groups and consequently had neglected the future interests of society as a whole. The Study concluded that fishery management would be markedly improved by a clear separation between conservation and allocation decisions. It considered the goal of conservation to be maintaining resource productivity for future generations. Allocation, on the other hand, involved distributing the opportunity to participate in a fishery.

The NOAA Study proposed that the conservation decision, expressed as an acceptable biological catch (ABC) determination, be made before deciding who the users would be. Combination of the two processes in one decision under the existing system resulted in the addition of more fishing effort to serve the interests of more users.

The Study considered an array of alternative institutional arrangements for conserving and allocating U.S. fishery resources. These ranged from a modification of the existing Council system to entirely private, entirely federal or entirely state systems. It endorsed management by a partnership of State, Federal, Council and private entities.

Specifically, the Study recommended that NOAA set the ABCs for regional fisheries at the national level, based on the best scientific information available. This determination would become the ceiling for a Council's allocations in regional fisheries. Council allocations could not exceed an ABC, would be based on a redefined optimum yield, and would be subject to a revised Secretarial review process.

2.2.3 The MFCMA System in Practice

So far I have concentrated on the U.S. fishery management structure and process in the post-extension era, including critiques of the existing system. The real issue is: what have been the results in terms of the state of the stocks and the health and general development of the fishing industry? I address this by looking at events in New England and the North Pacific as particular examples of two ends of the continuum of fisheries management success under the MFCMA.

2.2.3.1 The New England Groundfish Experience

Anthony (1990) reviewed the management of the New England groundfish fishery in the first 10 years under the MFCMA. Much of what follows is drawn from his analysis.

Extended jurisdiction brought great expectations about the benefits for New England. This was exemplified by a very rapid expansion of U.S. fishing effort off New England during the late 1970's (see Fig. 20-1). The number of vessels increased from 825 in 1977 to 1423 in 1983 and then declined to 1334 by 1987. The number of fishing trips increased by about 47% from 1977 to 1983, followed by a slight decline. The total number of days fished by otter trawlers increased 73% overall from 1976 to 1986; there was a 100% increase in the Gulf of Maine, a 57% increase on Georges Bank, and an 82% increase in southern New England. There was also an increase in the average size of vessels in the groundfish fleet. Anthony (1990) concluded that the New England fisheries moved from being generally underfished to overfished in a very few years.

The FMPs for New England groundfish have had a chequered history. The first FMP was a complicated and unenforceable combination of catch quotas, closed spawning areas, mesh-size regulations, minimum fish size and trip limitations. Fishermen were severely critical of this approach.

Following protests by fishermen, the quota system was subsequently dropped. This left a plan with no direct controls on fishing mortality. Under intense fishing pressure, groundfish stocks continued to decline. Fishing mortality on many stocks reached record levels generally doubling from 1977 to 1986 (Fig. 20-2). Overall, there was a 65% decline in abundance of principal groundfish off the New England coast from 1977 to record lows in 1987 (Fig. 20-3). The decline in abundance was reflected in a dramatic decline in the

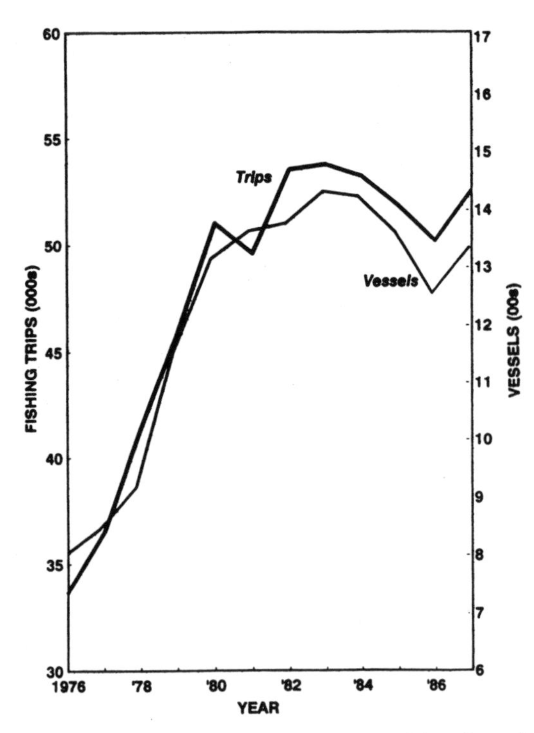

FIG. 20-1. The numbers of trips (thousands) and numbers of vessels (hundreds) larger than 5 GRT for all gear types fished off New England: 1976–1987 (from Anthony 1990).

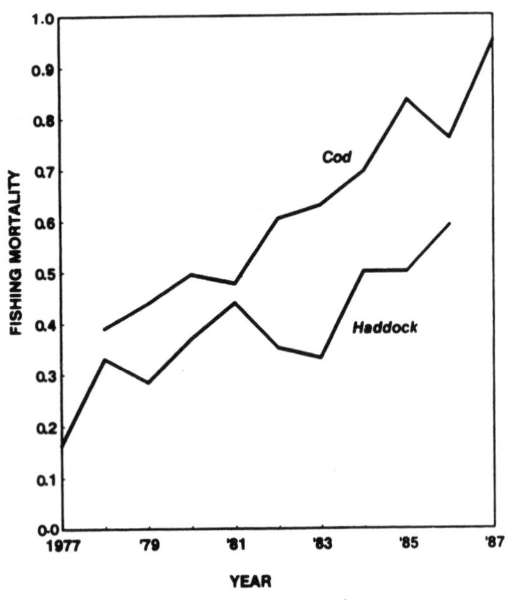

FIG. 20-2. Fishing mortality for Georges Bank stocks of Atlantic cod and haddock: 1977–1987 (from Anthony 1990).

catch per unit of effort by otter trawlers after 1978, in the range of 50–60% (Fig. 20-4). The gross revenue per vessel also declined steadily from 1978 to 1985.

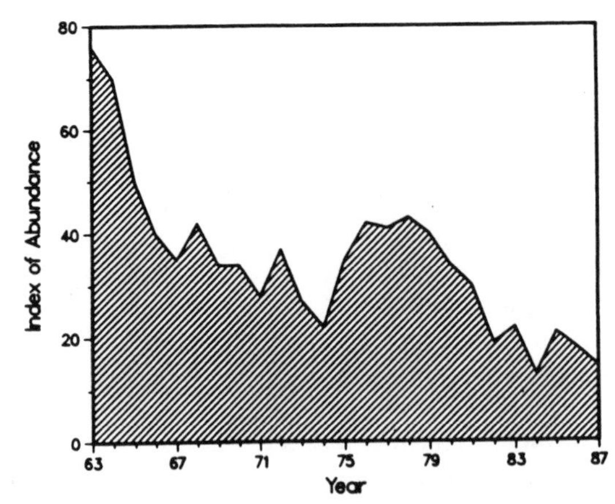

FIG. 20-3. Multispecies index of abundance of principal groundfish species determined from research vessel surveys off the New England coast: 1963–1987 (from Anthony 1990).

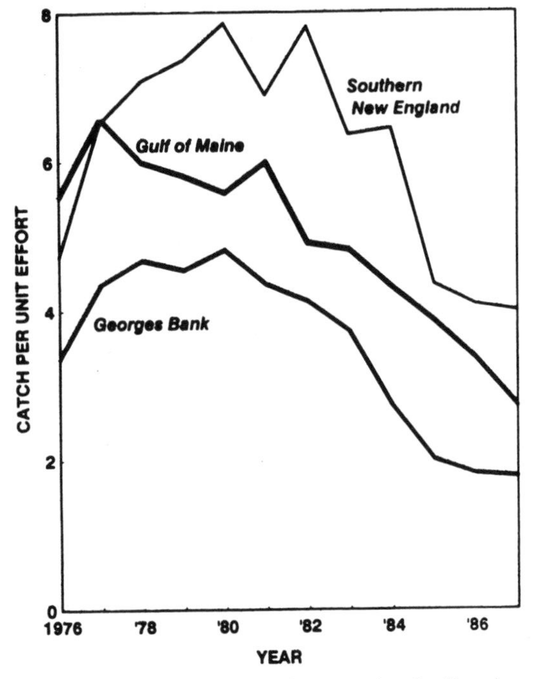

FIG. 20-4. Standardized catch-per-unit of effort in tons per day in the groundfish otter-trawl fishery of the Gulf of Maine, Georges Bank and southern New England: 1976–1987 (from Anthony 1990).

Anthony (1990) was extremely critical of the effectiveness of the Council's management efforts during the first 10 years of extended jurisdiction:

"The effective (real) fishing effort in 1987 was, in fact, more than twice as great as it was in 1977 because of the increase in both size and number of vessels and the improvements in fishing technology. This is the problem that the New England Fishery Management Council has had to face during the last decade. The council did not anticipate this great increase in fishing power and apparently failed to recognize its significance. The control of such increases in fishing potential with indirect control measures, such as mesh sizes, minimum fish sizes, and spawning area closures, simply was insufficient. In the case of New England groundfish, therefore, the MFCMA did not result in successful management during its first decade of implementation."

2.2.3.2 The Alaskan Groundfish Experience

Following implementation of the MFCMA in 1977, the new North Pacific Fisheries Management Council (NPFMC) adopted Interim Fishery Management Plans for the Bering Sea-Aleutian Islands and Gulf of Alaska regions. Through these FMPs the NPFMC attempted to remedy two decades of overfishing through catch reductions and a catch-quota management system.

In some instances, catch levels were reduced from pre-MFCMA levels to protect stocks such as Pacific halibut, which was being adversely affected by incidental catches in other fisheries. In other instances, species formerly harvested by both foreign and U.S. fishermen (such as snow [tanner] crabs and Pacific herring) were allocated exclusively to U.S. fishermen. Although the foreign catch for groundfish was reduced, the U.S. fishing industry initially showed little interest in the large groundfish resource. Foreign fisheries were allowed to continue, harvesting that part of the catch quota above the U.S. harvesting and processing capacity.

A significant event occurred in 1980 which would later shape development of a U.S. groundfish fishery in the North Pacific. Under the American Fisheries Promotion Act (Public law 96–561), commonly referred to as the "fish and chips" policy, foreign allocations were based on the extent to which a nation contributed to the development of the U.S. fishing industry. These contributions could be in the form of tariff reductions on U.S. fishery products, purchases of U.S. fishery products, or participation in joint ventures. These "joint ventures" essentially consisted of sales over-the-side of fish caught by U.S. vessels to foreign processing vessels. This was somewhat similar to the "allocations for access" policy pursued by Canada for a while in the late 1970's-early 1980's, with poor results (see Chapter 11).

The "joint venture" aspect of the U.S. policy hastened transformation of the groundfish fisheries off Alaska from a foreign to a domestic operation. In the Bering Sea and Gulf of Alaska, foreign fisheries dominated until the early 1980's, when U.S. vessels began to fish groundfish as a result of joint-venture agreements. The joint-venture catch grew from a small percentage to about 80% of the Alaskan groundfish harvest in 1987 (Fig. 20-5). The changeover occurred rapidly in the Alaska pollock fishery. Before 1980, the catch was taken almost exclusively by foreign trawlers. By 1981, joint-venture fisheries began to displace the foreign trawlers. By 1983, joint-venture operations took approximately 76% of the total trawl catch of pollock. By 1988, foreign fishing had been eliminated completely from the Gulf of Alaska, and domestic fisheries began to quickly displace joint ventures (Fig. 20-6).

The rapid growth of joint ventures was stimulated by two factors. The decline of the fishery for red king crab led many crab fishermen to convert their vessels to groundfish trawlers. The "fish and chips" policy linked foreign allocations to foreign purchases through joint ventures. The transition from a foreign to a domestic groundfish fishery took place more rapidly than the government anticipated. There was no foreign allocation in the Gulf of Alaska in 1987 and by 1989 no joint-venture operation. The last foreign groundfish allocation in the Bering Sea was in 1987. By 1988 the entire northeast Pacific groundfish fishery was fully utilized by the U.S. domestic fisheries.

Regarding groundfish abundance, Megrey and Wespestad (1990) concluded:

> "The effect of the MFCMA on groundfish resource abundance off Alaska has been to reverse or arrest the decline in abundance of some major exploited species. The resulting increase in abundance of some species has been due to a combination of catch reductions or stabilization. Also contributing to the observed increase in abundance has been the coincident high levels of recruitment in the mid-to-late 1970's. Several stocks have returned to lower levels of abundance because strong year-classes have been replaced by less abundant year-classes. If management practices instituted under the MFCMA continue, we expect all stocks to remain at healthy levels and depleted stocks to continue to rebuild."

Overall, the NPFMC appears to have helped conserve most groundfish stocks off Alaska.

In short, this fishery was successfully Americanized between 1977 and 1990. Unlike the New England

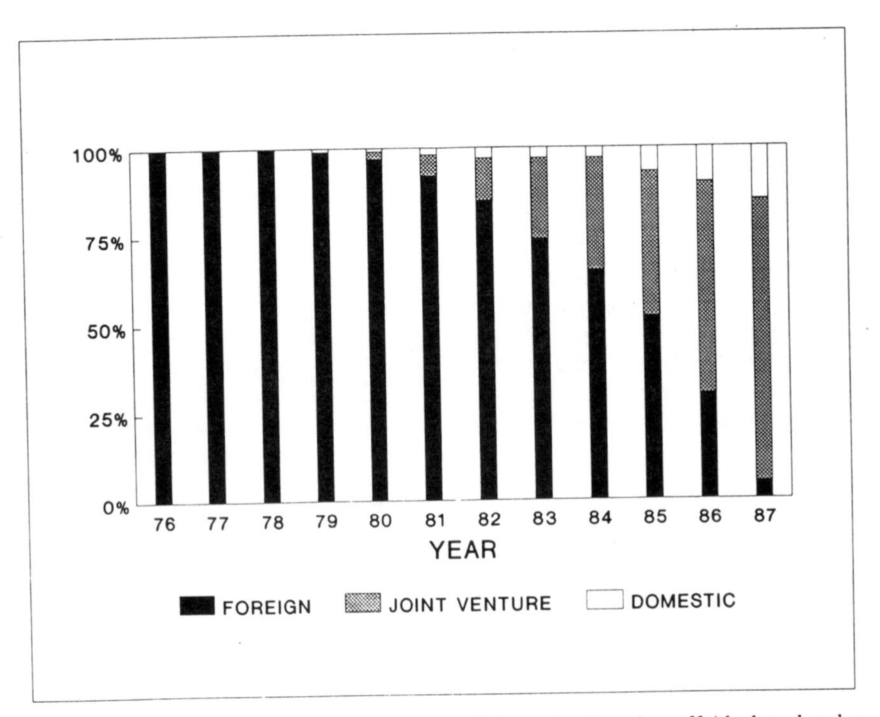

FIG. 20-5. Percent of the total annual groundfish catch allocation off Alaska taken by foreign, joint venture and domestic fisheries: 1976–1987 (adapted from Megrey and Wespestad 1990).

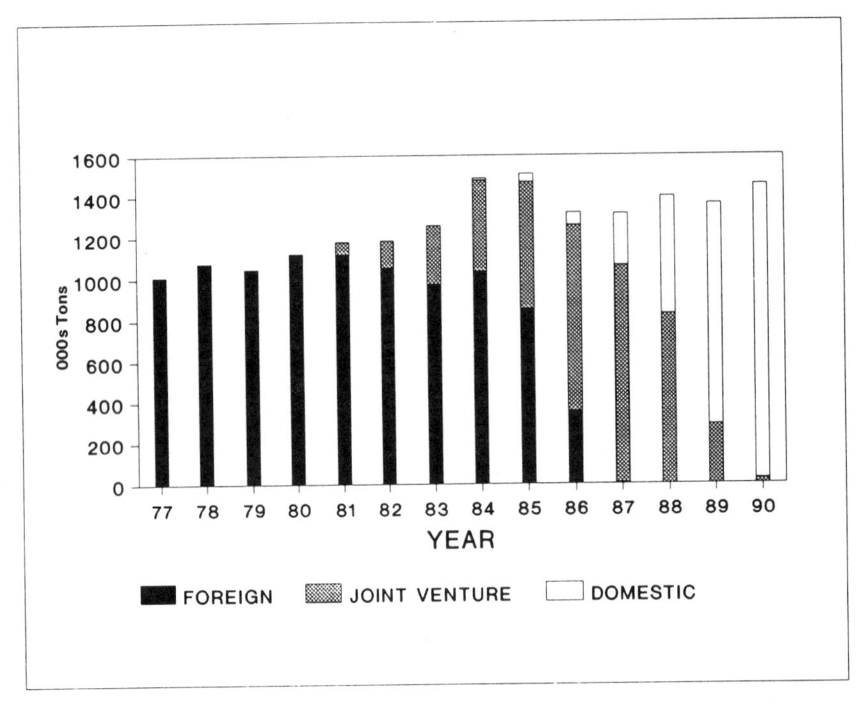

FIG. 20-6. Catches of Alaskan Pollack by U.S. domestic, joint venture and foreign countries in the Gulf of Alaska, Eastern Bering Sea and Aleutian Islands area (Catches in metric tons).

experience, the U.S. industry was able to grow by displacing the massive foreign fishery, particularly for pollock. Now that the fishery has been Americanized, the traditional common property "race for the fish" is beginning to emerge within the U.S. fleets. New conflicts have emerged between onshore processors and the new U.S. factory trawler fleet. The overall catching capacity has not diminished (Fig. 20-7). The danger is that the apparent economic benefits from displacing foreigners may be quickly dissipated by overinvestment and overcapacity. The challenge for the NPFMC is to prevent this.

2.2.3.3 Some Lessons from the New England and Alaskan Groundfish Fisheries

These case studies of the New England and Alaskan groundfish fisheries illustrate the widely divergent approaches to fisheries management among the Regional Councils. The New England Council chose the path of open access and no direct controls on fishing mortality. The measures to control age at first capture have by themselves been ineffective. In stock conservation and economic terms, the results have been disastrous. Groundfish stocks are at historic low levels and fishing capacity is far more than required to harvest the available resource profitably.

The NPFMC instituted catch quotas to control fishing mortality but did not limit the amount of fishing effort. To date, the Alaskan groundfish story is one of successful displacement of foreign fishing effort. The groundfish stocks appear to be in relatively good condition compared with their pre-extension status. The problem of the unregulated growth of domestic capacity only began to surface at the end of the 1980's. Unless the common property issue in this fishery is addressed, the success of the late 1980's may be succeeded by the Tragedy of the Commons in the 1990's.

One lesson which emerges from the different management approaches to the New England and Alaskan groundfish fisheries is that the objectives and National Standards of the MFCMA are so vaguely worded that they are being interpreted in diametrically opposite ways by these two Councils. Other authors have suggested a similar lack of consistency in other Councils' application of the Act. What has emerged in the U.S. in the post-MFCMA era is extremely decentralized management of the fisheries by the various Councils.

The checks and balances exercised by the Department of Commerce under the MFCMA appear to have been

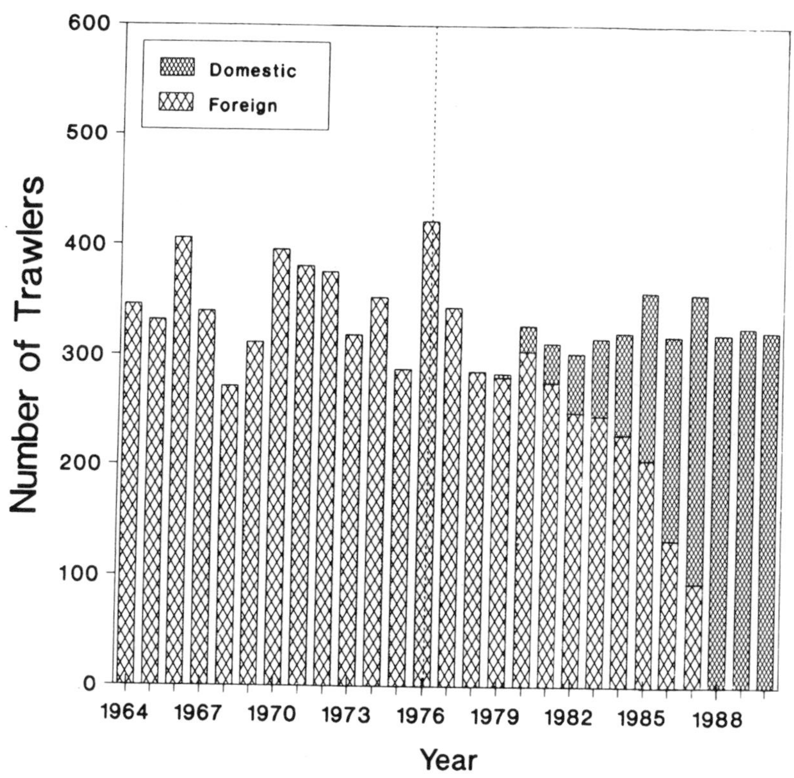

FIG. 20-7. Total number of trawlers fishing the waters off Alaska: 1964–1990 (adapted from Megrey and Wespestad 1990).

more procedural than substantive. The 1986 NOAA Fishery Management Study identified certain deficiencies of the existing Act and its application. However, Congress largely ignored the study's recommendations when it reauthorized the Magnuson Act in 1986.

2.3 Conclusions from the U.S. Experience

The total fish and shellfish catch off the U.S. coast, from 0 to 200 miles, increased from just under 4 million tons in 1977 to more than 6 million tons in 1987 and 1988. Since the catch within 3 miles remained relatively stable at just under 2 million tons for most of 1977–1988, the growth occurred in the EEZ. The U.S. catch in the EEZ increased from about 700,000 tons in 1977 to almost 3 million tons in 1988 (Fig. 20-8). Foreign catches in the U.S. EEZ declined from about 1.7 million tons in 1977 to less than 50,000 tons in 1989. These figures illustrate the successful Americanization of the fishery in the U.S. EEZ. The most dramatic example is the Alaskan pollock fishery. The "fish and chips" policy of 1980, particularly its joint ventures aspect, was instrumental here.

To a large extent, the growth made possible by displacement of foreign fishing effort has masked some deep-rooted problems in the U.S. fishery. These are likely to hit home with a vengeance in the 1990's. There are two fundamental problems. One is the lack of a concerted, consistent and effective management policy related to stock conservation. Some stocks are already severely overfished. Many others could be overfished in the 1990's. The second major flaw, which contributes to the first, is the general failure to address the problems arising from the common property nature of the fisheries. Numerous economists have lamented the failure of the Regional Councils to come to grips with this aspect of fisheries management.

Anderson (1987) summarized the problem as follows:

> "The fisheries management institutional structure as set up by the MFCMA, the operational guidelines, and formal and informal standard operating procedures, is not very conducive to introducing economic efficiency reasoning into management plans. There are too many places where individuals, groups, or agencies can step in and stop, slow down, or reroute the process if it does not appear to be producing results favourable to a particular point of view. Many can hinder progress, but no one is held accountable for not achieving it."

A Massachusetts Offshore Groundfish Task Force examined the precipitous decline in groundfish landings in New England and concluded that "the primary cause

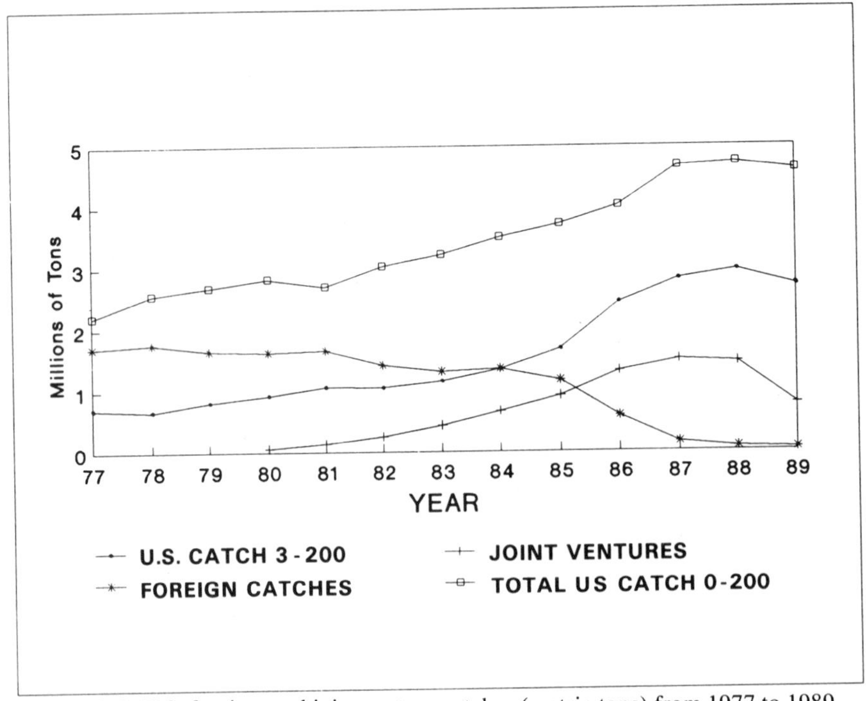

FIG. 20-8. U.S. foreign, and joint venture catches (metric tons) from 1977 to 1989.

is overfishing by American fishermen due to mismanagement of the resource." Major management problems identified included:

1. The FMP for groundfish contained *no direct limits on fishing mortality* and no limit on the number of fishermen.
2. The number of domestic groundfish vessels increased dramatically after the passage of the Magnuson Act, with strong encouragement by federal financial assistance programs and liberal tax laws.
3. Compliance with fisheries regulations at sea had been seriously compromised by poorly designed regulations, so cheating had become a way of life (Massachusetts 1990).

The Task Force recommended the following actions:

1. Directly controlling fishing mortality by reinstituting species-specific quotas;
2. Reducing wasteful fishing mortality for both juvenile and adult groundfish through area and seasonal closures in areas with high concentrations of juveniles;
3. Improving compliance with regulations through shifting the enforcement burden from at-sea to dockside and using sanctions as the primary deterrent against violations.
4. Strengthening the biological basis for management, including increased funding for stock assessment.
5. Allocating the groundfish catch among fishery components.

Overall, the Task Force concluded:

> "Groundfish management has failed to ensure the future of New England's traditional groundfish fishery because short term economic considerations were allowed to prevail. The Council and the Secretary (of Commerce) have allowed American fishermen to take tomorrow's fish yesterday."

The success of U.S. marine fisheries management under the MFCMA is thus mixed. There has been growth in some sectors, particularly Alaska, through the displacement of foreign fishing effort. For the most part, however, the Regional Councils have not dealt with the problems of stock conservation and the common property nature of the resource. Whether the Regional Council system can respond effectively to these problems depends on Council members' willingness to set aside parochial interests and act for the common good.

In New England the U.S. appears to have replaced an ineffective international fisheries management regime with an equally ineffective national regime. It is thus difficult to offer an overall prognosis for U.S. fisheries in the next decade or two. Progress will be made on some fronts but it is likely that in other areas overfishing and overcapacity will continue to be major problems.

A central lesson from the U.S. experience is the danger inherent in an extremely decentralized system of fisheries management. While there are some advantages in allowing objectives to be set on a region-by-region or fishery by fishery basis, the failure to apply a national framework of objectives and standards can result in an extremely fragmented and ineffectual system of fisheries management. Canada, with its vast geography and great diversity of fisheries, has maintained a more balanced fisheries management system, with central direction but allowance for tailoring fishing plans to local circumstances. In general, this has proven more effective than the extremely decentralized U.S. fisheries model.

3.0 AUSTRALIA

3.1 Overview

While a small player on the world fisheries scene in volume and value terms, Australia has been on the cutting edge of fisheries management.

Australia has one of the largest coastlines of any country in the world (37,000 km) and (since 1979) a large fishing zone of 9 million square km, including areas surrounding offshore island territories. The Australian fishing zone is approximately the same size as the U.S. zone. Despite this, Australia is a relatively small player in world fish production. Landings in the late 1980's were in the 160,000 to 200,000 ton range, approximately 0.2% of total world production. These landings, however, consist primarily of relatively high valued shellfish such as prawns and rock lobsters. There are few areas of coastal upwelling. Consequently, demersal and pelagic species, which contribute to the higher productivity in northern temperate waters, are relatively insignificant.

The landed value of the commercial fishery was around $900 million Australian dollars in recent years ($951 million in 1989–90). At least 90% of Australians live within 50 km of the coast and Australia's seas are an important recreational resource. Approximately $2.2 billion was estimated to have been spent on recreational fishing in Australia in 1983–84 (Australia 1989). The Australian fishing industry ranks about fifth among Australia's primary industries well behind the major rural industries such as wool, beef, wheat and dairying. Rock lobster and prawns account for more than 50% of both the Gross Value of Production and exports. About half the domestic consumption of seafood is supplied by imports (Australia 1991).

Approximately 21,000 persons work in the harvesting sector. They operate about 9,000 boats, most of which are less than 20 m in length. There was, however, an increase in the number of larger vessels in the 1980's (Bain 1985). Another 4000–5000 people are employed in the processing and wholesaling sector. The fishing industry is the main source of economic activity for many towns along the coast.

3.2 Jurisdictional Arrangements

Australia is a federation of 6 States and one Territory, forming the Commonwealth of Australia (Fig. 20-9). States traditionally exercised fisheries jurisdiction in coastal waters extending to 3 miles. The Commonwealth government had jurisdiction beyond State coastal waters of 3 miles. Although fisheries had been an area of concurrent jurisdiction from federation in 1901, it was more than a half-century before the Commonwealth government focused on fisheries matters in any significant way. The national government in 1952 passed the Commonwealth Fisheries Act to regulate fishing beyond the 3-mile limit (Herr and Davis 1986). While this signified a growing increase in fishing by the national government, it had little immediate impact since most fishing continued to occur within state-managed territorial waters.

Australia extended its fisheries jurisdiction to 200 miles in November 1979. This gave Australia authority to control foreign fishing in the expanded zone. It also led to a greater involvement in marine fisheries management by the Commonwealth government. Meanwhile, efforts had been made to coordinate national and state policies through the establishment in 1968 of the Australian Fisheries Council (AFC). The AFC was created to provide a mechanism for the joint coordination of fisheries policy through annual meetings of

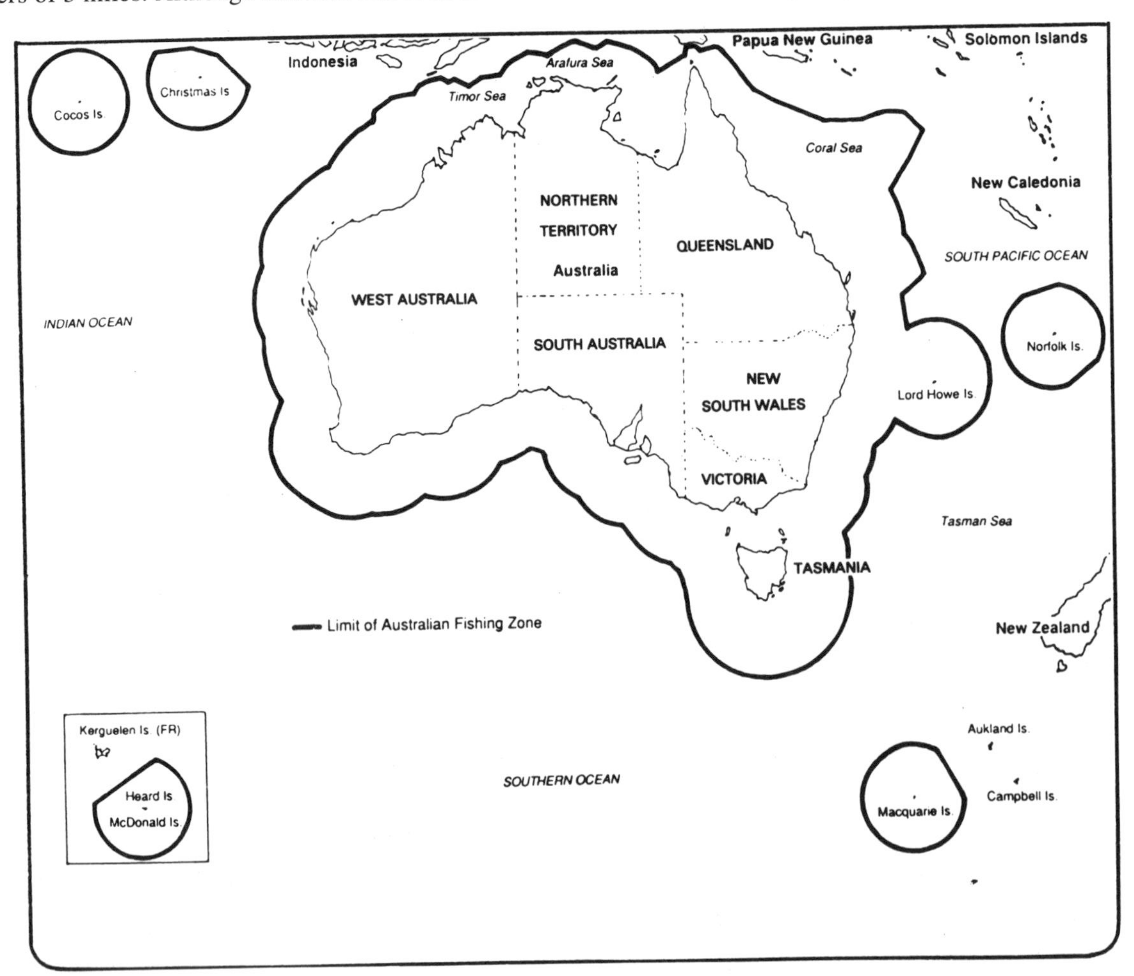

FIG. 20-9. The Australian Fishing Zone.

State and Commonwealth Ministers responsible for fisheries. This council was chaired by the Minister of Primary Industries of the Commonwealth.

Virtually all major fisheries in Australia overlap the boundary between state and national fisheries jurisdiction. Commonwealth and state legislation implementing the Offshore Constitutional Settlement of 1979 allowed for establishment of a single jurisdiction for particular fisheries.

In May, 1987, the Commonwealth Minister for Primary Industry announced new arrangements, effective June 1, 1987, under which Queensland, Tasmania, South Australia and Western Australia would exercise exclusive jurisdiction over a total of 16 fisheries. These were previously managed by the Commonwealth outside State waters (*Australian Fisheries*, May 1987). As well the Commonwealth would exercise jurisdiction over parts of the tuna fishery in all waters off Queensland, South Australia and Western Australia. Previously those States controlled the tuna fishery in State waters.

Legislation provided for the establishment of four Joint Authorities, each comprised of the Commonwealth Ministers and the appropriate Ministers of the member State or States, with the authorities varying in size from one to five State(s). The Joint Authorities were responsible for the management of particular fisheries in waters adjacent to one or more of the States represented on an Authority.

3.3 Post-extension Initiatives

For some considerable time overcapacity has been a recognized problem in most of the traditional fisheries. Some state governments, the South Australian government in particular, became concerned about economic efficiency and adopted a cautious approach to management at a relatively early stage. The introduction of limited entry in the Spencer Gulf prawn fishery in South Australia in 1969 is an example. There was a rapid escalation of fishing effort in the rock lobster and abalone fisheries in the 1960's because of the high unit value of these species. The government introduced limited entry programs to these fisheries in the early 1970's. Limited entry was later introduced in the major northern prawn fishery.

Following the establishment of Australia's 200-mile zone in 1979 there was some euphoria about the potential for expansion of the fishery. The government encouraged expansion, causing a rapid build-up in fishing capacity (Lilburn 1986). Even for the higher unit-value established fisheries, governments and industry were slow to accept the potential economic benefits of rigorous limited entry programs. Consequently, fishing capacity soon became excessive in all these fisheries, despite implementation of limited entry. Lilburn (1986) stated that "controls on the total amount of capacity and effort have often been loose". Restrictions on the quantity and size of fishing gear and the capture of small fish provided some protection for the stocks.

By the early 1980's, there was a growing realization that the common property race for the fish led to low or negative economic returns from the fishery and declines in total catches as well as catch per unit of effort. Various groups called for improved management plans for Australia's fisheries.

The election of a new federal government in 1983 and a Senate committee of inquiry into the fishing industry later that year led to a National Fishing Industry Conference in February, 1985. One of the conclusions of this Conference was that uncontrolled access to fisheries eventually depletes the resource and leads to overcapitalization and low or negative returns on investment. The Conference concluded that fishermen should have clearly defined fishing rights and that catches must not exceed the productivity of the resources (*Australian Fisheries*, February, 1985).

The Conference also concluded that industry's role in management decision-making should be expanded. It recommended that decisions on the management of a fishery be taken as early as possible in its development, taking into account biological, environmental, economic and social factors. Management objectives should be determined fishery-by-fishery.

The Conference proposed establishing a system of fully transferable fishing rights agreed to by industry and Government. It suggested that these rights allow access to a restricted area or a defined share of the catch and be defined through legislation to afford maximum protection under the law.

The results of this Conference shaped the Commonwealth government's approach to fisheries management for the remainder of the 1980's. The government, through the Australian Fisheries Service of the Department of Primary Industry, assumed a more proactive role in fostering new management approaches.

One of the initiatives resulting from the Conference was a greater role for industry in decision making. This was addressed in two ways. First, an independent national organization was established to represent the commercial fishing industry. The National Fishing Industry Council (NFIC) was intended as a forum for all sectors of the commercial fishing industry to discuss national issues and put forward their views to the federal government. This Council did not become as important or effective as originally envisaged. Early in 1989 it was restructured. In the interim, State Fishing Industry Councils had become more powerful voices for industry interests (*Australian Fisheries*, March, 1989).

Another initiative resulting from the 1985 Conference, which had more lasting impact, was the government's establishment of management advisory committees, to promote industry-government consultation on the development of new management regimes. Regional committees of Commonwealth/state/territory officials were replaced by industry/government task forces and management advisory committees to advise governments about particular fisheries. These task forces and advisory committees played a key role in developing innovative approaches to the management of Australia's fisheries in the second half of the 1980's. Industry representatives were usually in the majority. The committees reported to Ministers through a national standing committee of senior government advisers.

Lilburn (1986, 1987) observed that these industry/government committees had been very successful in resolving conflicting interests. Over time they contributed to a cooperative approach to management among operators, whereas previously fishermen had concentrated exclusively on individual objectives.

Following the 1985 Conference, Australia moved to introduce more clearly defined regimes of fishing rights in certain fisheries. These took the form of either a capacity-units approach under the general umbrella of limited entry (e.g. the northern prawn fishery) or the introduction of individual transferable quotas. The government first introduced ITQs in the southern bluefin tuna fishery and later extended them to abalones. These fishing rights were incorporated into fishery management plans which became law when signed by the Minister.

3.4 Defining Fishing Rights

Many fishery management plans were introduced for various Australian fisheries in the late 1980's. Here I discuss the Northern Prawn Fishery Management Plan and the Southern Bluefin Tuna Plan to illustrate the innovative approaches to define fishing rights. Both of these plans have been extensively analyzed (e.g. Wesney 1989; Franklin 1987; Geen and Nayar 1989).

3.4.1 Northern Prawn Fishery

This has been for some time Australia's single most valuable fishery. In 1989–90, 7000 tons were landed valued at $70 million. The major species fished are tiger, banana and endeavour shrimp. The fishery extends from Western Australia east across northern Australia to Queensland. The northern prawn fishery developed rapidly to full exploitation despite its distance from the more settled parts of Australia.

Total catch has fluctuated around an average of about 9,500 tons. The fishery is conducted by a fleet of specialized trawlers, generally steel-hulled, 19 to 23 m freezer vessels. A total of 215 vessels operated in this fishery in 1989–90.

The government first introduced limited entry in this fishery in 1977. The number of vessels eligible to be licensed to fish in this area was limited to 292. By the early 1980's profitability in this fishery was declining significantly, despite a vessel replacement policy that was intended to be restrictive. This proved ineffective in controlling the size and fishing power of vessels.

As the total number of days fished increased, the catch per vessel declined. The fleet became overcapitalized. An economic study in 1984 revealed that total annual fleet costs exceeded total returns by about $10 million in 1980–81. About 55% of the boats did not yield a positive return to capital (Wesney 1989).

The problem in this fishery was overcapacity. The drive to revise the management system originated with economic rather than biological concerns. Biological concerns did, however, emerge later in the process. The objectives of the new Northern Prawn Fishery (NPF) Management Plan were:

> "(a) To conserve and reduce the fishing pressure on the stocks of prawns in the area of the fishery;
> (b) To promote the economic efficiency of the fishery" (Australia 1986 and 1987).

The measures to attain these objectives were to include:

(a) protecting the juvenile phases of the prawns' life cycle;
(b) controlling the number and total fishing capacity of boats authorized for the fishery;
(c) determining the manner of measuring the fishing capacity of the fishery;
(d) determining a permitted fishing capacity for the fishery; and
(e) facilitating withdrawal of units from the fishery (Section 7.2 of the 1986 NPF Plan).

The withdrawal of units from the fishery was to be facilitated by a Voluntary Adjustment Scheme (VAS), and a vessel replacement policy.

During 1986 scientific evidence indicated that the tiger prawn stocks were declining due to fishing pressure. The government amended the NPF plan in 1987 to include additional measures to address the problem of declining stocks. Additional objectives were specified:

1. the reduction, by 1993, of the total number of registered "Class A" units to not more than 70,000;
2. the reduction, by 1993, of the number of registered Class B units by approximately 40%; and
3. the prevention of an increase in the number of vessels in the fishery through the assignment of Class B units to a vessel. (For a description of this classification system, see below).

The NPF plan was based on input controls, despite the known limitations of input controls and the ineffectiveness of a previous input-oriented regime in controlling capacity growth. The government considered a quota-based system but rejected it. The reasons for the choice of input-related controls lay in the nature of the northern prawn fishery. The fishery takes place across the top of Australia. A majority of the vessels are capable of remaining at sea for months, catching prawns on their way to and from the main area of the fishery and transferring product at sea to carriers. Prawns from NPF vessels caught within and outside the NPF area could be landed at ports extending from the southwest of the continent to the northeast coast. This made timely monitoring of catches against individual quotas difficult and expensive (Wesney 1989).

In addition, the year-to-year availability of prawns is highly variable. This would make the task of setting an annual TAC and sharing it among individual fishermen difficult. This has been done in the case of the Northern Shrimp fishery in Atlantic Canada but the number of vessels involved is much smaller.

Two main components of the plan included a new vessel replacement policy and an adjustment scheme to assist in the removal of excess fishing capacity. The NPF plan was based on a ceiling on the total number of "boat units" allowed to participate in the fishery. These units are transferable. "Boat units" are in effect an indirect measure of the fishing power of a vessel. They were arrived at by adding together the measured under-deck-volume and the manufacturer's specified continuous kilowatts brake power of the boat's engine. The total number of boat units allowed in the fishery was derived from the number for the fleet of up to 292 allowable licensed boats under the previous management scheme. This number was 131,769 so-called "Class A" units in 1984. In addition to these "Class A" units, one "Class B" unit was allocated to recognize each fisherman's historical right to an endorsement on his Commonwealth licence to operate in this fishery (Wesney 1989).

This merely defined the existing number of boat units, which was excessive in relation to the objective of reducing capacity. As one means to decrease the number of units, the fishing industry proposed a voluntary unit buyback scheme (the voluntary adjustment scheme or VAS). Fishermen who wished to leave the fishery could sell their units to a buyback authority, and be paid out of a fund financed by an annual compulsory buyback levy on all NPF fishermen.

A VAS was established and managed by the NPF Trading Corporation Ltd., by agreement with the Australian government. Industry established this Corporation to foster removal of boats under the VAS to meet the target reduction of 40% of the Class A units by 1993.

Wesney (1989) indicated that the management measures might be working. Progress was being made in reducing the number of potential and operational boats by means of the VAS, together with the unit surrender provisions of the boat replacement policy. As of March, 1988, 11.3% of the original number of allowable boats had been removed from the fishery by the VAS. When combined with the units surrendered through replacement, the number of potential boats had declined by 13%. In addition, 13% of the original number of Class A units had been purchased under the VAS. This level of reduction had been achieved at a total cost to fishermen of about A $4.8 million.

Wesney (1989) noted that the NPF input control management plan was complex and involved considerable government intervention. Constraints such as gear restrictions, seasonal closures and bans on daylight fishing, plus the unit surrender provisions impose additional costs on fishermen.

In 1987, the NPF harvested its total catch in about six months with fewer vessels than previously. Fishermen indicated that their profitability was good despite additional costs resulting from the Management Plan. Wesney (1989) suggested that the Plan was yielding positive results. Most of the potential additional capacity had been removed along with some vessels which had been active in the fishery.

The adjustment mechanism (the VAS) was voluntary and worked only if fishermen chose to leave the fishery in return for compensatory payments funded by the rest of the industry. Such a mechanism was likely to reduce capacity only slowly and probably would be most effective during downturns in the fishery.

As part of a broader move to reduce overcapacity through wider use of ITQs, the Commonwealth Government in October, 1990, approved a new buy-back scheme to reduce the number of Class A units in the NPF to 50,000 by 1993. A fixed price of $450 was offered for each active and surplus unit purchased. All remaining units would be freely tradeable on the market. For each boat withdrawn from the fishery before April 1, 1991, there was also a redeployment payment to compensate fishermen for the financial loss they would experience when they sold their boats. Government assisted this restructuring through a guarantee for the commercial loans necessary to fund the scheme (about $40 million) and a $5 million grant over 3 years from the National Fisheries Adjustment Program (*Australian Fisheries*, October, 1990).

The intent was to reduce the NP fleet by half. The first phase of the restructuring program ended in April, 1991, with the number of units being reduced from 96,300 to 76,700 at a cost of $18.3 million. A further compulsory reduction of 35% would take effect before April 1993. The government also initiated a study of the feasibility of introducing ITQs in the NPF (*Australian Fisheries*, June, 1991).

3.4.2 Southern Bluefin Tuna Fishery

Southern bluefin tuna (SBT) is a single stock, highly migratory and long-lived (up to 20 years). It spawns in the waters south of Java in the Indian Ocean and then migrates through the southern oceans from east of South Africa to east of New Zealand. Juveniles are fished at 2–3 years of age but do not reach sexual maturity until 7 or 8 years old. Exploitation rates on the immature tuna can be quite high.

This stock has been fished since the early 1950's, mainly by Japanese longliners, Australian purse seiners and pole and line vessels and, to a much lesser extent, by New Zealand fishermen. The Australian fishery occurs primarily along the southern coast, with the age of fish exploited ranging from 2 years in western Australia to 3 years along the southern coast and 4–5 years off the southeastern part of the continent.

Japan has been a major player in this fishery. The Japanese catch peaked at 77,500 tons in 1961 and then declined to 17,000 tons in 1983. The Australian catch ranged between 10,000 and 15,000 tons through the 1970's, then peaked at 21,000 tons in 1982. At the beginning of the 1980's there were about 140 Australian vessels involved in this fishery, in two components. In Western Australia there was a small-scale inshore fishery which harvested very young SBT-less than 3 years old. Ninety vessels, less than 15 m long, caught about 6000 tons in 1982–83. The dominant sector was a fleet of about 35 pole and line vessels based in South Australia. Almost all of the Australian catch was landed at five or six main ports.

Profitability in this fishery fluctuated considerably during the 1970's due to varying catches and prices. Despite a freeze on entry into the eastern sector from 1976 to 1981, investment in vessels increased. In the early 1980's, the profitability of the fleet deteriorated. The average operating surplus of vessels in the eastern area dropped from $120,000 in 1980–81 to less than $10,000 per vessel in 1981–82 (Bureau of Agricultural Economics 1986).

The situation facing the fleet was worsened by scientific advice that the stock was in trouble and the catch should be reduced. The spawning stock had declined from about 600,000 tons prior to fishing to about 220,000 tons by 1975, then remained stable for the next 5 years. Scientists from Australia, Japan and New Zealand advised that the spawning stock would probably decline to dangerously low levels if the existing high exploitation rates were maintained. Administrators from the three countries agreed in 1982 on the urgent need to restrain the total level of fishing to maintain the spawning stock at the 1975–80 level. The mechanism adopted was a global quota shared among the three countries.

Australia decided that its excess fleet capacity must be reduced. Input capacity controls were rejected in light of previous experience and the need for an adjustment mechanism to reduce capacity. Because of concern about the future status of the stock, a quick-action adjustment mechanism was required.

Fisheries officials chose a system of ITQs as the basis for management partly because they thought such a system would be relatively easy to control and monitor. The landing of the catch at only five or six ports would facilitate this. Also the domestic tuna market was limited and this limited the possibility for "black market" sales.

The SBT Management Plan incorporated the following objectives (Australia 1985):

- "stabilisation of the parental biomass of the stock of southern bluefin tuna in the fishery at, or about, a level consistent with scientific assessments of the status of the fishery" and
- "optimum utilisation of the resource in proclaimed waters".

Measures to attain the objectives were to include:

- determining the TAC in proclaimed waters;
- enforcing the TAC; and
- enforcing individual quotas.

The interim management plan put in place in October 1983 did not include individual quotas. It established a national quota of 21,000 tons for the 1983–84 season. Minimum sizes of 54 and 70 cm were introduced for the western and eastern sectors. A lower size limit was set in the west because a national limit of 70 cm would likely have closed the fishery off Western Australia.

The Australian government introduced ITQs in the longer term management plan implemented in October,

1984. The national quota for 1984–85 was set at only 14,500 tons. Size limits were abolished. Wesney (1989) described the formula used for calculating the ITQs. For 1984–85 and 1985–86 each quota unit was 2,712 tons of SBT. The plan also included a levy to recover part of the costs of management.

Wesney (1989) credited the management plan with reducing the catch consistent with scientific advice. He also reported a rapid adjustment of the fleet and a major improvement in the efficiency and profitability of operators who remained in the fishery.

The number of active participants declined from 136 in 1984 to 63 in 1987. This was accomplished largely by the buyout of quota from fishermen in Western Australia and New South Wales by the larger operators based in South Australia. High priced sashimi markets in Japan provided the incentive for the quota purchases. As a result of a change in the marketing of the tuna, the value of each ton of fish averaged four or five times the pre-management price level. Despite a decline in catch of 50%, the value of the catch doubled. The per-ton quota value increased from $1000 to more than $5000.

In 1986, the Japanese agreed to reduce their catch of SBT to 19,500 tons. Also, the Tuna Boat Owners' Association of Australia and the Japanese Tuna Federation agreed to an arrangement whereby Australian quota holders would not catch 3000 tons of their annual quota of 14,500 tons for a 3-year period ending in September, 1989. These arrangements reduced the tri-national TAC to 32,000 tons.

Wesney (1989) concluded that the changes in the size and structure of the fleet and the reduction in the catch of SBT were directly associated with the introduction of ITQs. The speed of fleet restructuring was facilitated by the sale of quotas by those choosing to leave the fishery. Overall, the number of quota holders was reduced by 54% between 1984 and 1987.

Geen and Nayar (1989) suggested that the introduction of ITQs into the SBT fishery greatly improved the profitability of the fleet, compared to even the most optimistic profit expectations under alternative management schemes.

Despite the favourable experience with ITQs, the southern bluefin tuna resource situation deteriorated after these initial assessments of the ITQ system were written. In August, 1988, scientists from Japan, Australia and New Zealand drew attention to a rapidly declining resource. They recommended immediately reducing global SBT catch limits in all sectors of the fishery substantially below existing catch levels. They suggested that the only safe catch level was zero (*Australian Fisheries*, August, 1989).

After months of negotiations and a failed attempt by Australia to introduce a global moratorium, Australia, Japan and New Zealand finally agreed in November, 1989, to cut the 1989–90 quota by almost 25% to 11,750 tons. This was allocated 6065 tons to Japan, 5265 tons to Australia, and 420 tons to New Zealand. This was a cut of about 31% for Japan and 15% for Australia, thus representing a significant increase in the proportion allocated to Australia (*Australian Fisheries*, December, 1989).

This quota reduction forced the Australian fleet to concentrate on catching larger fish and to maximize the proportion going to the sashimi market. The industry faced a rocky road, which led to further rationalization of fleet capacity. There were only 10 vessels active in 1989–90 (Australia 1991).

3.4.3 Other Fisheries Adjustment Mechanisms

Until about 1985, governments in Australia had confined their fisheries management objectives largely to biological and social concerns. Controls on fishing activity generally involved area and seasonal closures, mesh size restrictions, and other restrictions on fishing gear and on the size and types of vessels allowed to operate in a particular fishery.

The limited entry schemes introduced in the 1970's were generally not intended as tight controls on fishing capacity (Lilburn 1986). Only in the 1980's did the government consider incorporating automatic adjustment mechanisms in limited entry management plans. Without such mechanisms, capacity tended to increase under most limited entry regimes. As an example, during 20 years of limited entry in the Western Australia rock lobster fishery, the efficiency of fishing operations increased despite strict controls on boat replacement. In 1986, economists estimated that the same number of boats were taking a similar total catch but with more than twice the fishing effort (Lilburn 1987). Controls on the minimum size at first capture had provided some stock protection. Nonetheless, the fishery was acknowledged to be overexploited both biologically and economically. Attempts to reduce the number of lobster pots per boat were resisted because many fishermen had entered the fishery in recent years paying large premiums for the pot authorities.

In the late 1980's, the Australian government attempted to introduce some form of adjustment program in various fisheries under limited entry. Examples included surrender of units of capacity:

1. on the introduction of a new boat or engine in the Southeastern Trawl Fishery;

2. on transfer of licences in the New South Wales and Victoria abalone fisheries; and
3. periodically, as in the lobster fishery in south-eastern South Australia where all licence holders were required to surrender a percentage of their lobster pot holdings.

The most visible adjustment mechanism was that incorporated in the NPF Management Plan. ITQs have acted as a means to reduce fleet size in the SBF. ITQs were also introduced in the abalone fishery in Tasmania, the western sector of the South Australia abalone fishery and the Bass Strait scallop fishery.

During the late 1980's, the government experimented with both the unit capacity approach and ITQs in various fisheries. The choice of management approach depended on the circumstances of the particular fishery. However, ITQs became more popular with the government towards the end of the 1980's. The government also adopted the "user pays" concept for fisheries management, as well as in other sectors of the economy.

3.5 Policy Directions for the 1990's

In December, 1989, the Government released a policy statement titled "New Directions for Commonwealth Fisheries Management in the 1990's" (Australia 1989). It identified three objectives for fisheries management policies:

1. ensuring the conservation of fisheries resources and the environment which sustains them;
2. maximizing the economic efficiency with which these resources are exploited; and
3. ensuring there is an appropriate return to the community from those permitted to exploit community-owned resources for private gain.

Minister Kerin identified overcapacity as the greatest single problem facing Australia's fisheries managers:

> "The excessive number of boats puts undue pressure on what are finite fishery resources and this in turn leads to the over-exploitation of fish stocks. As the catch is shared between all the fishermen involved, the greater the number of boats fishing, the smaller the catch per boat. As fishing costs do not decrease as catch falls, this means that profitability is reduced" (*Australian Fisheries*, February, 1990).

The Government established a new statutory authority — the Australian Fisheries Management Agency (AFMA). With a separate governing board, the new Authority was aimed at flexible, open and less bureaucratic fisheries management. It was intended to allow greater industry and community participation in management. The Agency was to give individual advisory committees increased responsibility for day-to-day management. The AFMA's operating budget was to be met by the industry and the Government in proportion to the benefits each receives.

The Government indicated that it would amend the Fisheries Act to recognize the "ongoing nature" of fishing rights in established fisheries. It also indicated that ITQs would be the preferred option unless it could be shown that other controls would be better in a particular fishery. Transferable quotas would allow the market, rather than direct Government intervention, to determine an individual fishermen's level of operation. This was a situation the Government wanted to encourage.

The Minister announced that the Government would also set up a formal register akin to a land title register to record the ownership of fishing rights. Access rights to developed fisheries would be allocated to fishermen entitled to operate in those fisheries.

For new fisheries, the Government would invite people to bid for the right to carry out trial commercial fishing in developing fisheries. If new resources proved profitable, a management plan would be developed and fishermen invited to bid for access rights to the new fishery.

The Government would provide funds for restructuring as long as it was convinced that this would lead to a more efficient industry. As fisheries became more profitable, fishermen would be expected to pay a share of their profits to the community in exchange for the right to exploit a public resource for private gain. One key element of the new policy was a requirement that fishermen pay most of the costs of fisheries management. Various studies indicated that fishermen received about 90% of the financial benefits of management. The Policy Statement suggested that, subject to review, fishermen pay up to 90% of the costs of fisheries management by a management levy. This would be a payment for services rendered and would not give industry ownership of the resource.

This Policy Statement established an ambitious agenda for Australian fisheries management in the 1990's. Industry members strongly supported the statement, hailing it as a major step forward and a blueprint for future fisheries management.

In June, 1991, the House of Representatives passed legislation to replace the 1952 Fisheries Act, and put in place the Australian Fisheries Management Authority and other aspects of the new policy for Australia's fisheries.

3.6 Conclusions from the Australian Experience

Although Australia gained a vast fishing zone with its extension of fisheries jurisdiction in 1979, this had little effect on fisheries development because of the relatively low biological productivity offshore. Some new fisheries have been developed but in general the Australian fisheries continue to be predominantly inshore or nearshore, with high-value, low-volume invertebrate species such as prawns and rock lobsters contributing most in value.

The 1980's brought major changes in fisheries management policies and practice. These changes included innovative approaches to the common property problem through experimentation with fishing access rights. Experiments with ITQs in the Southern bluefin tuna fishery and unit capacity controls and the Voluntary Adjustment Scheme in the northern prawn fishery led to the extension of ITQs to other fisheries. The inclusion of adjustment mechanisms for fleet reduction in other limited entry fisheries helped reduce capacity in some fisheries.

Despite these initiatives, overcapacity remained the central problem. The Australian Government endorsed the idea of applying ITQs in all fisheries where it is feasible to do so. The auctioning of access to new fisheries is breaking new ground in practice, if not in theory.

Another important aspect of the Australian approach to fisheries management in the 1980's was industry's very extensive involvement in decision-making. The management advisory committees, with majority industry representation, played a crucial role in developing innovative approaches such as ITQs, and fleet adjustment schemes. The Policy Statement on New Directions for the 1990's envisaged an even more prominent role for these committees.

Underlying the recent approach of the Commonwealth Government to fisheries management has been the concept that market forces, not government intervention, should dictate the fortunes of those participating in the fishery. At the same time, the Government contended that, because the resource is community-owned, those who gain from using the resource should pay for the privilege. It adopted a system whereby the beneficiaries, the fishermen, pay the major portion of the costs of fisheries management.

Australia's system of involving industry in decision-making through management advisory committees is closer to the Canadian than the U.S. model. As Canada implements proposed changes to the institutional structures for fisheries decision making, it may find some guidance in the new Australian Fisheries Management Agency approach. Australia's cost recovery initiatives may also offer some lessons for Canadian proponents of such an approach.

4.0 NEW ZEALAND

4.1 Overview

Adjacent to Australia, New Zealand has a much smaller land mass but also has a large exclusive economic zone, the fourth largest in the world. New Zealand has benefited more than Australia from extension of jurisdiction to 200 miles because of the existence of several important deepwater species, e.g. hoki and orange roughy.

The New Zealand 200-mile EEZ at 1.3 million square nautical miles is large, more than fifteen times its land mass. Although the size of the zone is extensive, much of the area is not productive in fisheries terms because of the limited continental shelf. The water is deeper than 1000 m in 72% of the zone, and shallower than 200 m in only 6% of the zone (Clark et al. 1989) (Fig. 20-10).

During the late 1960's foreign countries commenced fishing around New Zealand. By 1971 the Japanese finfish catch was about 44,000 tons, equivalent to the New Zealand catch. Japan was subsequently joined by vessels from the USSR, Korea and Taiwan. Overall, fishing effort and total catch increased rapidly, peaking at more than 500,000 tons in 1977.

The expansion of the catch from New Zealand waters between 1971 and 1977 was due almost entirely to increased foreign fishing effort. The New Zealand industry remained relatively static during this period, following some increase in the late 1960's.

New Zealand introduced a 200-mile Exclusive Economic Zone in April, 1978. The government embarked upon a program of encouraging expansion into the deepwater trawl fishery by New Zealanders, with the aim of displacing the foreign effort over time.

4.2 Immediate Post-extension Initiatives

Following an initial period of limited licensing, the fishery was open access from 1963 to 1978. However, by the mid-1970's problems began to emerge in the traditional New Zealand inshore fishery. It appeared that many stocks were overfished and the industry was becoming overcapitalized.

In 1977, the Fisheries Act was amended to give the Minister of Fisheries the power, when necessary, to declare a "controlled fishery" regulating the species, quantity and size of fish that could be taken from a fishery, the type of fishing method to be used, the areas that

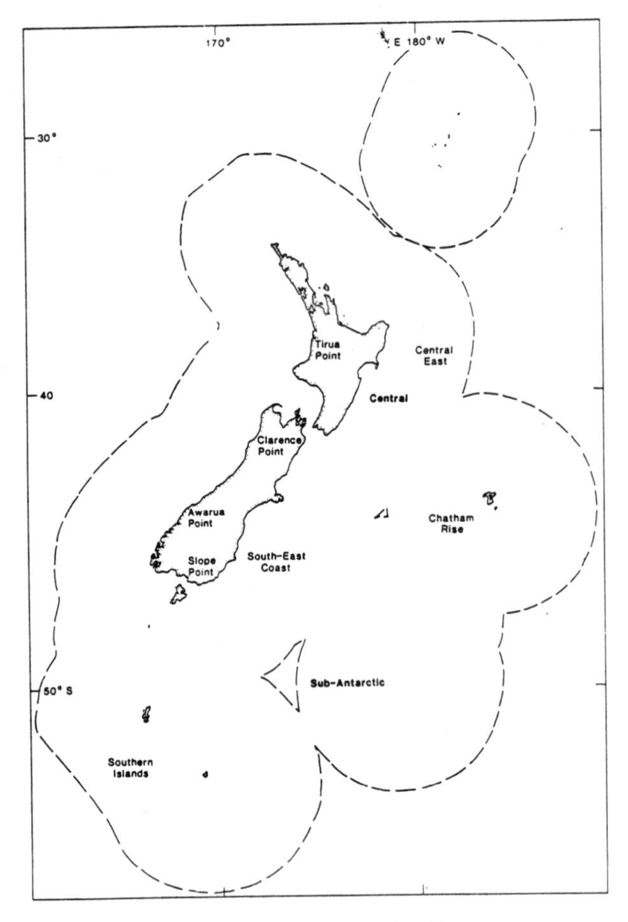

FIG. 20-10. The New Zealand Fishing Zone.

could be fished and the persons who could engage in such a controlled fishery (Fisheries Amendment Act No. 7, Part III, 1977). Several inshore fisheries were designated as "controlled fisheries" over the next several years.

A new Fisheries Act in 1983 introduced the concept of fishery management plans. The first FMP introduced under this Act was for the Deepwater Trawl Fishery. This incorporated a system of company quotas.

The success of and positive response to company quotas for the deepwater trawl fishery, combined with persistent overcapacity and conservation problems in the inshore fishery, prompted a new government in 1984 to move to a system of Individual Transferable Quotas for virtually the entire New Zealand fishery.

At the onset of extended jurisdiction it was calculated (OECD 1979) that a TAC of 405,000 tons of wetfish was a realistic estimate of the safe biological yield from New Zealand waters. To this could be added 91,000 tons of squid and 30,000 tons of tuna (OECD 1981). Given that the New Zealand industry had the capacity to harvest only 80,000–100,000 tons of wetfish, the government decided to use joint ventures to foster New Zealandization of the deepwater fishery resources.

In 1977, the foreign fleet caught 82% of the total catch off New Zealand. By 1988 its share was less than 6%. In 1989–90, less than 2% of the total catch quota was allocated to foreign fishing fleets.

4.2.1 The Deepwater Trawl Policy

Prior to 1983 the New Zealand deepwater fleet was regulated by gear restrictions, species quotas in various areas and limits on the number of joint venture vessels allowed in the fishery. Despite the expansion of the deepwater trawl fleet, its profits had been marginal (New Zealand Fishing Industry Board 1986).

During 1982 the Government worked out with the deepwater trawl operators a new policy for the management of this fishery. Effective April 1, 1983, company quotas were introduced for this fishery. This involved the allocation of rights to catch deepwater species to individual companies with a commitment to and dependence on deepwater species in both the catching and processing sectors. The rights were transferable between eligible companies. To encourage New Zealandization of this fishery, quota allocations were restricted to firms with at least 75.1% New Zealand ownership. Also quota holders were required to process 35% of the total deepwater catch onshore. No company was permitted to hold more than 35% of the allocated resources.

Allocations were based on a weighting of three criteria: investment in onshore processing facilities; investment in fishing vessels; and quantities of product processed. Species allocations were made to nine companies, including two consortia of smaller operators. These allocations include hoki, orange roughy, oreo dory, trawl ling, silver warehou, hake and squid. These quotas were issued for a period of 10 years.

The reaction of the deepwater companies to this system of company quotas was generally favourable.

4.2.2 Policy for the Inshore Fishery

The inshore fisheries continued to experience substantial fishing pressure. As a result of declining catches by inshore vessels combined with rising operating costs, fishermen were leaving the inshore sector of the industry.

Following consultation with the industry, consensus was reached in 1983 concerning the need for a reduction of fishing effort in the inshore fishery to achieve long-term biological and economic viability. The first step agreed was to more precisely define commercial fishermen and to eliminate part-time fishermen from the fishery effective October 1, 1983.

In 1984, the plight of the inshore fisheries continued to dominate fisheries management discussions. Despite a moratorium on new entrants effective March, 1982, and the departure of some part-time fishermen from the fishery, economists estimated that the harvesting sector remained overcapitalized by about NZ $28 million. They argued for measures to reduce catches to a level which would enable fish stocks to return to their previous abundance and "to provide a management regime whereby sustainable yields are harvested by the most efficient means" (OECD 1985).

The New Zealand Cabinet decided in May, 1985, to manage the inshore fisheries under ITQs beginning in October, 1985. The process of quota allocation delayed the effective date of implementation until October, 1986. The Fisheries Amendment Act of 1986 provided a legislative base for the new system. Company quotas in the deepwater trawl fishery were subsumed as part of the general ITQ program.

4.3 The ITQ System

4.3.1 Initial Features of the ITQ System

Descriptions of the ITQ system can be found in Clark et al. (1989). MacGillivray (1990) and Sissenwine and Mace (1992) have summarized the key features of the system. The following description is drawn from these sources.

Establishment of TACs and Initial Allocations

Assessments of stock status were conducted for 153 management units, and TACs established for the major finfish species in each of 10 management areas. These initial TACs were rather crudely estimated. Most TACs for the troubled inshore species were set between 25 and 75% lower than 1983 catch levels.

The deepwater company allocations were converted directly to ITQs. In the inshore sector, provisional maximum allocations were determined separately for each permit holder as the average catch of the best 2 out of 3 years (1981/82, 1982/83 and 1983/84). The initial quota allocations were based on these catch histories. These were expressed as a fixed volume in weight, granted as a fishing right in perpetuity.

Transferability

Relatively unrestricted transferability was considered an essential element of the ITQ system. Quota could be leased or traded in perpetuity or for a fixed term. Foreign ownership of quota was prohibited and no person or company could own more than 20% of a species TAC.

To facilitate trading and the operation of a market in quotas, a computerized quota exchange was established. Face-to-face trades were also permitted but all transactions had to be registered.

Enforcement

The nature of enforcement changed dramatically. Clark et al. (1989) described it as a change from the traditional game-warden approach to an auditing approach tracing the flow of money and product through the system. This meant that the government pulled back significantly from its traditional at-sea enforcement activities. This has been criticized by some participating in the ITQ system (Muse and Schelle 1988).

TAC Adjustments

If the TAC increased or decreased in the future, the actual increase or decrease was to be achieved by having the government sell or buy quotas. This proved to be a major drawback in the system.

4.3.2 Factors Facilitating the Introduction of ITQs in New Zealand

Clark and Duncan (1986) cited four characteristics of the New Zealand fisheries which assisted in the management of ITQs. These were:

1. Geographical isolation;
2. Ease of monitoring;
3. One jurisdiction; and
4. Nature of the resource.

Because of New Zealand's geographical isolation, poaching by foreign vessels is not overly attractive. Monitoring of landings is facilitated by the largely export orientation of the industry, with 80% of the output, by value, being exported. This limits the potential for the creation of an unpoliced domestic black market. The fact that the New Zealand central government has authority to monitor product flow both at sea and onshore facilitates enforcement. The nature of some of the resources, e.g. the schooling nature of orange roughy and hoki, and low bycatch, makes the application of ITQs for these species simpler than in the case of some inshore species, at least in theory.

4.3.3 Impact of the ITQ System

New Zealand's implementation of an ITQ system in most of its fisheries has been widely hailed as a bold and innovative step forward in fisheries management (e.g. Neher et al. 1989). By 1990 more than 33 species were included in the ITQ System.

Clark et al. (1989) cited two fundamental principles underlying the ITQ system: "to protect the resource and, within that constraint, to maximize efficiency." Initially, most of the attention focused on the second objective, maximizing economic efficiency. Clark et al. (1989) stated:

> "It is our view that the scheme can be judged already to be successful, and that it effectively addresses both the economic and biological aspects of fisheries management."

Other authors have also described the system as successful (e.g. Muse and Schelle 1988; MacGillivray 1990) but they have presented little in the way of data to support this assertion.

Under the ITQ system, foreign vessels under charter continued to form a substantial part of the catching sector. They accounted for 55.9% of the so-called "New Zealand" catch in 1986, 66.8% in 1987 and 57.5% in 1988 (Fig. 20-11). It was anticipated that this would continue under the ITQ system as the foreign presence enabled domestic operators to harvest in the most efficient manner by choosing between domestic and foreign catching capacity. They chose to rely on "cheaper" foreign catching capacity, to a large extent. Clark et al. (1989) estimated the contribution by the foreign charters to the total value of the New Zealand catch. The estimated value of the 1986 catch was $427.1 million. The domestic (noncharter) fleet's contribution to this total was estimated at $257 million, or 60% of total value; the foreign charter contribution at $170 million or 40% of the total. The major species caught by the charter fleet were hoki, orange roughy, squid and oreo dories.

Despite this reliance on charter capacity, there was an expansion in deepwater capacity through investment in the domestic fleet, which increased by $81 million between 1983 and 1986. Investment in processing facilities during the same period increased by around $50 million.

The fishing industry claims that harvesters have modified their fishing practices to reduce the costs and/or increase the market value of their catches. There is also some evidence that quota holdings have been consolidated. From October 1986 to April 1988, there were 15,580 quota sales involving 453,000 tons and 3,417 leases of quota involving 253,000 tons (Sissenwine and Mace 1992). Bevan et al. (1989) reported that the total number of quota holders decreased by 5.7% during the first 2 years of ITQs. Also the top 10

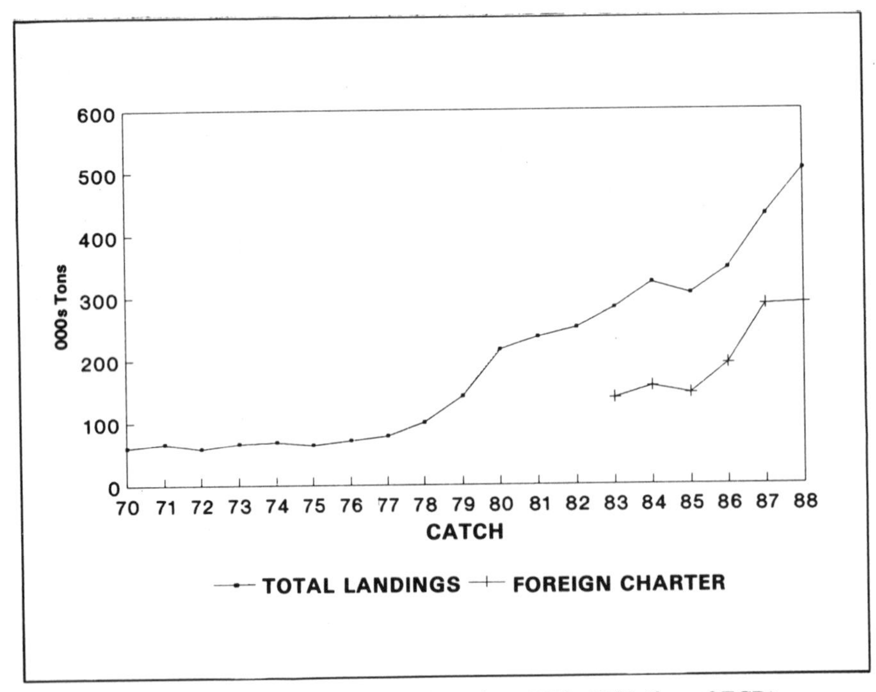

FIG. 20-11. Trend in total New Zealand catches: 1970–1988 (from OECD).

quota owners increased their share of the quota from 57 to 80% of the total. However, some vessel owners who sold their quota continued to fish it under lease.

Thus, the economic effects of ITQ management were not clear as of 1990. According to Bevan et al. (1989) the industry experienced only a 4.3% return on investment, before income taxes, during 1987–88. Other factors — lower prices in export markets and unfavourable exchange rates — had an adverse effect on the economic situation of the industry.

Virtually all of the economics literature on the New Zealand ITQ system assumes that there would be a beneficial effect on stock conservation. Sissenwine and Mace (1992) examined the conservation aspect and suggested that the ITQ system had had a detrimental effect on the resource management system. They found little evidence of improvement in the condition of the fisheries resources.

Because the knowledge of these resources is limited and the amount of fisheries research was inadequate, it is difficult to know exactly what has been happening to the stocks. There have been some increases in TACs but these resulted from adjustment to the stock assessments rather than increases in abundance. Some stocks have declined. Orange roughy, in particular, is now estimated to be much less productive than previously thought (Annala 1990). Sissenwine and Mace (1992) reported that "the current TAC for the largest stock of orange roughy exceeds even the most optimistic estimates of long-term sustainable yield by a factor of three."

Assessments in 1989 and 1990 (Annala 1989; 1990) indicated that there were other species/stocks for which the TACs should be reduced. It appeared that TACs were too high for species such as hoki, squid, paua, and rock lobster.

At the 1989 Fisheries Assessment Meeting, in 36 cases yield estimates were less that 50% of the existing TAC.

Reductions in the TACs were recommended for several species. Overall, Sissenwine and Mace (1992) concluded:

> "TACs are not closely tied to the best available assessments of the resources, nor are catches strongly controlled by the TACs. To date, the track record of ITQ management with respect to conservation is not good."

Part of the problem lies in the fact that, following the initial buy-back/rationalization, the government failed to enter the market to buy quota to reduce TACs in line with the scientific advice.

It also appears that the ITQ system had not reduced government intervention to the extent envisaged. The moratorium on new licences was removed. In effect, this was replaced by a requirement to own a quota. Input controls such as minimum fish size and closed areas or seasons have generally been maintained. To minimize conflict between the large FFTs and smaller vessels, the large FFTs were prohibited from fishing within 25 miles of the coast. To this has been added a complicated system of recordkeeping and reporting requirements. Overall, there is still a considerable amount of government intervention in the fishery, even in the absence of restrictions on the number and size of fishing vessels (Sissenwine and Mace 1992).

4.4 Changes to the Original ITQ Program

The definition of an ITQ as a fixed volume of a particular species in perpetuity led to a reluctance by the government to intervene in the market to purchase quota to reduce TACs. If maintained over time, this could have had catastrophic consequences.

A major change in the ITQ system came when in 1989 the government announced its intention to convert from a volume-based to a percentage-based system. The need for this redefinition resulted from government's failure to purchase quota on the market to reduce TACs when necessary. A proposed "Revolving Fund", incorporating resource rentals and revenues from the sale of quota, was never established. Sissenwine and Mace (1992) estimated that, if government had entered the marketplace to purchase quota to implement the TAC reductions suggested by the 1989 scientific assessments, the cost would have far exceeded the revenue to date from the ITQ system. They estimated that the reductions for orange roughy alone would have cost in the range of NZ $60–150 million.

Why did this problem arise? Sissenwine and Mace (1992), based on discussions with government officials, concluded that the likelihood of needing to purchase quotas to reduce TACs was underestimated. It was thought that TACs had been established at a conservative level. Insufficient attention was paid to the inherent variability in abundance of fish stocks.

In any event, effective April 1, 1990, the ITQ system was changed to one where the ITQ entitlement was expressed as a percentage of the TAC for particular species/stocks. To secure agreement to the new approach, the government agreed to freeze resource rental rates for 5 years. Funds thus made available were to be redistributed to compensate ITQ holders for the necessary quota reductions.

4.5 Conclusions from the New Zealand Experience

Following the declaration of its 200-mile Exclusive Economic Zone in 1978, New Zealand began to develop fisheries for the resources of the new zone and to restructure its fisheries management system. In fisheries management terms, New Zealand is now regarded as one of the most innovative countries in the world because of its widespread experimentation with ITQs.

The resources of the New Zealand 200-mile zone offered greater potential for fisheries development than did those of its neighbour, Australia. Even though large areas of the zone are unproductive, New Zealand has succeeded in developing domestic fisheries for deepwater species such as orange roughy, hoki and squid, which had been exploited by foreigners prior to extension of jurisdiction. To develop these deepwater fisheries, it relied extensively on chartered foreign vessels which in 1990 continued to contribute approximately half of the New Zealand catch.

The 1980's brought dynamic changes in the approaches to management of New Zealand's fisheries. The government embraced an explicit objective of maximizing economic efficiency. After a brief trial with input controls from 1978 to 1982, New Zealand experimented with company quotas in the deepwater trawl fishery. The successful implementation of this experiment coincided with the 1984 election of a new government committed to bring market forces to bear in fisheries management. This resulted in a decision to adopt ITQs as the preferred method of management for most fisheries. The government introduced ITQs after securing industry's acceptance.

The initial ITQ system expressed fishing rights in terms of fixed volumes of particular species guaranteed in perpetuity. These quotas were freely transferable and saleable, within certain constraints respecting foreign ownership and maximum shares of a species for an individual quota holder. The government facilitated transition to the ITQ regime through buy-back of quotas to align the sum of the ITQs with the TACs for various species. Most of the reduction occurred in overfished species such as snapper. Subsequently, the government more than recouped its buy-back costs through sale of additional quotas and resource rentals.

There was, however, a major flaw in the initial ITQ approach New Zealand adopted. The fixed-volume approach assumed a certain stability in TACs. Although managers thought that TACs had been set conservatively, scientific assessments soon indicated a need for TAC reductions. Under the ITQ system as originally established, the government was supposed to buy quota to reduce the TACs. This did not happen because of the substantial costs. In 1989, the government recognized that the fixed-volume approach was not the best way to structure the ITQ system. It took a new approach which expressed ITQs as a percentage of TACs.

New Zealand's ITQ approach to fisheries management has been widely praised. However, it is premature to judge its long term success in terms of conservation and economic efficiency. In the short term, the system has not improved, and has perhaps hindered, stock conservation. The economic impact is unclear. However, the fishing industry supports the system and argues that it has reduced harvesting costs.

The nature of the change to the fisheries management system was revolutionary rather than evolutionary. The new regime met virtually all of the criteria advocated by economist proponents of property rights in the late 1970's and early 1980's. The change occurred as part of a large scale movement to market-driven management of the New Zealand economy. It will be perhaps a decade before a proper evaluation can be done.

Whatever the ultimate judgement on the ITQ system, New Zealand clearly established itself as a leading innovator in fisheries management. As Canada moves towards a wider implementation of individual quotas, it can learn from the New Zealand experience. Percentage-based rather than fixed-volume approaches to individual quotas appear preferable. A closer examination of the impact of transferable quotas might assist Canada in future decisions on whether ITQs are appropriate for particular fisheries.

5.0 NORWAY

5.1 Overview

The fishery has long been important in Norway but has declined in relative economic significance in recent years. In the early 1980's, the fishing industry employed about 3% of the working population (Laursen 1985). By 1989 only 2% of Norwegians were directly involved in fishing. The fishing industry contributed less than 1% of Norway's Gross National Product (*World Fishing*, July, 1989).

Despite the low overall significance to the nation's economy, the fishing industry is still very important along large stretches of the Norwegian coast, particularly in the north.

The total Norwegian catch increased from 1.6 million tons in 1955 to 2.3 million tons in 1975. In 1981, the catch reached 2.5 million tons or 3.4% of the world's total catch. In that year Norway ranked as the world's seventh largest fishing nation. In the late 1980's Norway experienced declining catches in the Barents Sea because of declines in the abundance of key species such as cod and capelin. The total catch declined from 2.5 million tons in 1984 to 1.9 million tons in 1986 and 1987. Norway dropped from seventh place in terms of the world fish catch in 1981 to 9th in 1984 and 12th in 1986 and 1987. (These figures do not include its growing aquaculture industry which recently has equalled the traditional fishery in value). This decline in catch occurred despite Norway's extension of fisheries jurisdiction in 1977 over some of the world's richest fishing grounds. More recently, the Barents Sea cod and capelin catches have increased again.

5.2 Fisheries Policy and Management Approaches

Brochman (1983) distinguished two phases of Norwegian fisheries policy during the period 1920–82. There was a phase of expansion in catch volume and catching capacity which lasted until about 1970. During the expansion phase the government considered fishery policy part of regional policy aimed at industrializing the northern part of the country.

After 1970 the emphasis in Norwegian fisheries policy shifted to curbing overfishing of fish stocks and to limiting the growth of excessive harvesting capacity. This led to a comprehensive system of Total Allowable Catches, vessel and gear sector allocations, and in some fisheries a form of individual vessel quotas.

In 1977, the government submitted a Long-term Plan for the Norwegian Fisheries to the Storting (Parliament). Although it focused particularly on the 1978–81 period, the Plan incorporated certain long term considerations. It stated that the goals for fishery policy must be viewed in the light of overall community development goals. It identified the following paramount societal objectives for fisheries policy:

1. Preserving the pattern of settlement;
2. Conserving and developing fish resources; and
3. Providing secure and attractive employment opportunities.

Brochman (1983) noted that the political justification for growing governmental support to the fisheries had increasingly been based on the need for jobs in the sparsely populated regions. The fishery was relied upon to preserve the pattern of settlement. To achieve this, the fisheries structure would consist of small decentralized production units. Unprofitable activities within the fisheries would be supported by the government.

Implicit in the third objective was an assumption that secure employment in the fisheries sector could be maintained by continuing government support. Conspicuous by its absence was any objective concerning profitability or productivity. Brochman (1983) noted, however, that the Plan stressed the need to adjust both the catching and processing capacity to the resource base, with an implied reduction in both sectors.

Recently, the Norwegian government submitted a "white paper" on fisheries management to Parliament (Norway 1991–92). This emphasized resource conservation as the primary objective of fisheries management but there was a renewed emphasis on economic efficiency. The report diagnosed fleet overcapacity as the main problem of the industry (K. Mikalsen, Tromso, Norway, Personal Communication).

5.3 Trends in Stock Abundance and Management

The Norwegian fisheries are diverse and exploit stocks over a wide region. Two primary areas of importance are the Barents Sea and the North Sea. Since the European Community (EC) is heavily involved in the management of the North Sea stocks, I deal with these in the next section on the EC. In the Barents Sea, Norway experienced a reduction in its total catch in the late 1980's. To a large extent, this was due to trends in certain key stocks. Declines in capelin and cod in the late 1980's were, to some extent, offset by the partial recovery of the Atlanto-Scandian herring stock and increased catches of blue whiting, mackerel and northern prawns. In the early 1990's, the capelin and cod stocks recovered.

5.3.1 Atlanto-Scandian Herring

The fishery on this stock expanded considerably during the 1950's and 1960's as the herring's migration patterns became better known. Catches peaked at 1.9 million tons in 1966 (Fig. 20-12A). Thereafter, catches declined precipitously to 700,000 tons in 1968 and to only 20,000 tons by 1970 (Jakobsson 1985).

The spawning stock biomass was reduced from 10 million tons in 1957 to about 2.5 million tons in 1963, increased to about 3.7 million tons in 1965 and subsequently collapsed. This occurred because of poor recruitment to the adult stock and a rapid escalation of fishing mortality (Fig. 20-12B).

The spawning stock recovered to the 400,000–500,000 tons range during 1978 to 1982. This was only 5–10% of the spawning stock in the 1950's.

In 1965, when this stock had all but collapsed, scientists advised that the exploitation was at a level where no benefit for landings could be expected from regulation. No restrictions on fishing were recommended until 1970. A total ban on fishing was not advised until 1975, following recognition that the stock had collapsed (Saetersdal 1989). In 1979, ICES advice to cease commercial fishing was accepted.

The stock recovered somewhat during the 1980's, with the occurrence of some strong year-classes. Trends in yield, fishing mortality and spawning stock biomass in recent years are shown in Fig. 20-12 A and B. The ICES Advisory Committee on Fishery Management in

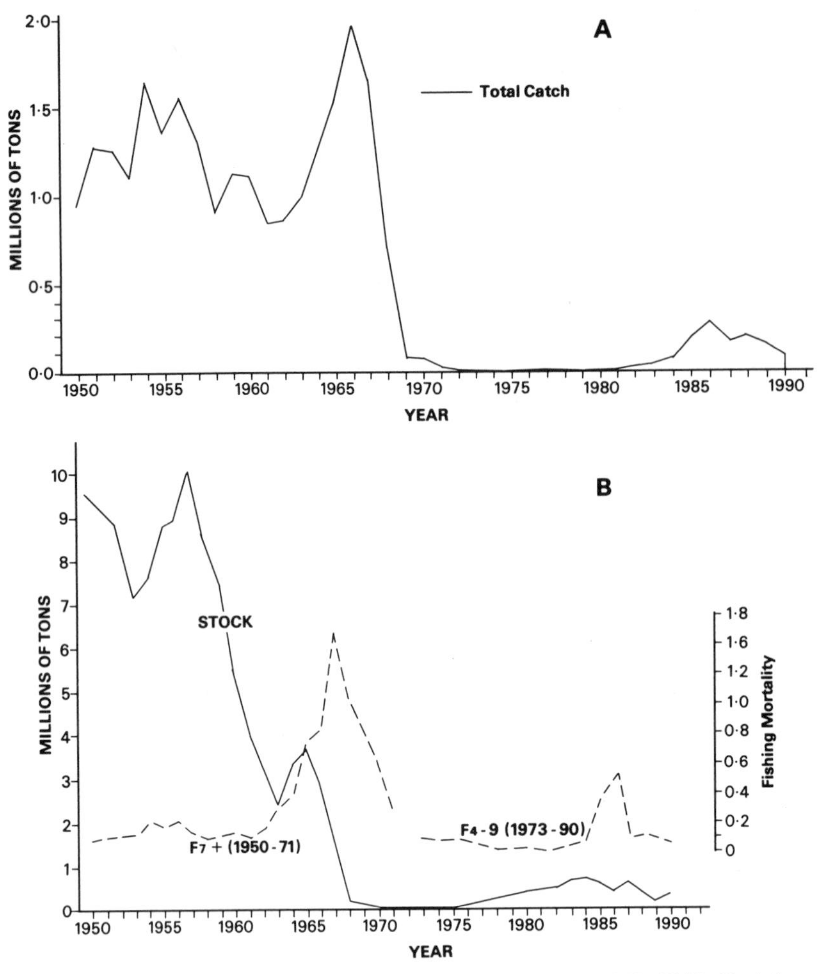

FIG. 20-12. A. Atlanto-Scandian herring — catches from 1950–1990; B. Atlanto-Scandian Herring — trends in spawning stock biomass and fishing mortality (F) from 1950–1990. (Adapted from Jakobsson, 1985 and ACFM Reports, various years).

1991 concluded that the spawning stock was about 60 % of the lowest level (2.5 million tons) known to have given good recruitment in the period prior to the collapse of the stock and was still very much below the historic level in the 1950's. It recommended that no fishing should take place until a substantial increase in recruitment and biomass was evident.

5.3.2 Barents Sea Capelin

The Barents Sea capelin is the largest capelin stock in the world. Spawning mortality is considered total. The primary management target in recent years has been the maintenance of a minimum spawning stock of 500,000 tons.

During the 1960's, catches increased rapidly to 680,000 tons by 1969. Catches further increased during the 1970's and peaked at 2.9 million tons in 1977. This increase was partly the result of diversion of fishing effort from the collapsed herring fisheries. Catches dropped to 1.6 million tons by 1980 but then subsequently increased to 2.4 million tons in 1983. Thereafter catches decreased rapidly to 123,000 tons in 1986, due to collapse of the stock (Fig. 20-13).

In 1985, ACFM abruptly revised its assessment of this stock. The total stock biomass in 1985 was estimated to be 0.82 million tons, compared with an estimate of 2.9 million tons in 1984. A target spawning stock of 300,000–400,000 tons had been used. Even without any fishing, the estimated spawning stock in the winter of 1986 would be 200,000 tons. Hence, ACFM recommended that catches during the winter season of 1986 should be reduced to the lowest practicable level.

ACFM concluded that the decrease in the stock

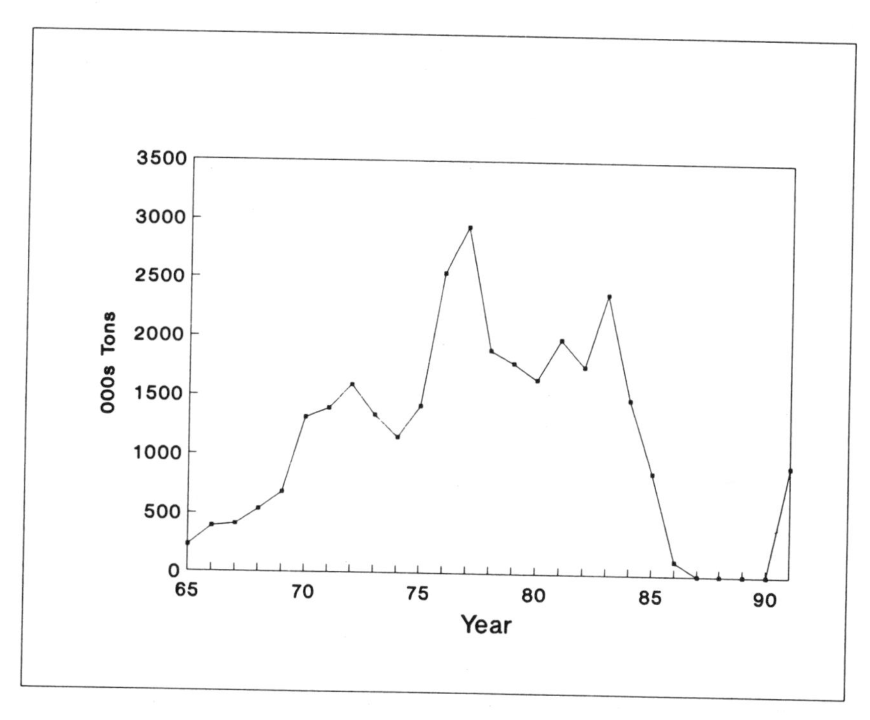

FIG. 20-13. Trends in catches of Barents Sea Capelin from 1965–1991.

exceeded by far what could be explained by the fishery and "is most likely connected to changes in the ecosystem." ACFM suggested that the rapidly increasing abundance of cod in 1984 and 1985 might have caused an increase in the natural mortality of capelin due to predation by cod. It also suggested that the presence of juvenile herring in the Barents Sea since 1983 had influenced the survival of 0-and 1-group capelin (ICES 1985). These suggestions were all attempts to find an explanation for what could only be regarded as a sudden catastrophic decline of the capelin stock.

From May 1986 until January 1991 the fishery was closed. The fishery was re-opened in January 1991 with a TAC for the winter (January–April) fishery of 1.1 million tons. In 1991, ACFM advised that the spawning stock biomass was recovering rapidly, dominated by the very strong 1989 year-class. In October, 1991, the maturing stock was estimated to be 2.1 million tons.

Whatever the future of this stock, its sudden catastrophic collapse in just 1 year, without warning, is testimony to the vulnerability of the marine ecosystem and to the lack of scientific understanding of the nature and scale of multispecies interactions within the ecosystem. The predictive capability of fisheries science in this instance was nonexistent.

5.3.3 Northeast Arctic Cod

The Barents Sea cod, or Northeast Arctic cod as it is more commonly known, has long been considered to be the largest cod stock in the North Atlantic. In the 1980's, catches plunged well below the long-term average and the estimated sustainable yield. Historically, the fishery occurred in three main areas: (1) at the beginning of the year on the approaches to the spawning area around the Lofoten islands; (2) in the feeding grounds to the north — from Bear Island to Spitzbergen; and (3) to the east in the Barents Sea from the Soviet coast northwards (Gulland 1983c). Catches increased rapidly before World War II, reaching a peak at around 900,000 tons in 1937. During the war, effort declined and catches dropped to around 200,000 tons. But catches recovered rapidly after the war, reaching almost 900,000 tons in 1947. In the mid-1950's the fishing effort and the total catch increased significantly with a peak catch in 1956 of almost 1.4 million tons (Fig. 20-14A). As late as 1974, catches were in excess of 1.1 million tons. From 1974 to 1984 catches declined substantially to a low of about 275,000 tons in 1984. There was a brief recovery to 500,000 tons in 1987 following which TACs had to be lowered substantially. The 1990 catch of about 200,000 tons represented the lowest catch level since World War II. This compares with an average catch of 750,000 tons during the decade 1961–1970.

TACs have been in effect for this stock since 1976 but the TACs established were often set in excess of the levels recommended by ICES. The actual catches exceeded the quotas because of the discarding of considerable quantities of small fish at sea (Gulland 1983c). The result was that F remained very high, more that 0.6 from 1968 to 1976, and around 0.8, on average, from 1977 to 1988 (Fig. 20-14A). In the late 1980's, the stock was reduced to low levels. F on this stock increased significantly following extension of fisheries jurisdiction and persisted at this higher level through the 1980's. In 1985, ACFM calculated that the stock biomass had declined from a level of about 3 million tons in 1974 to 0.8 million tons in 1984, the lowest figure on record to that time. The 1983, 1984 and 1985 year-classes appeared to be well above the average for the period 1960–1980. Accordingly, ACFM predicted that the spawning stock biomass would reach its lowest level in 1986 and increase to more than 1 million tons by 1989 (ICES 1985). ACFM suggested that the managers take advantage of the improved recruitment to rebuild the spawning stock biomass while increasing catch quotas and, at the same time, reducing fishing mortality towards F_{max}. It suggested a management strategy which would have seen the TACs increasing, with F supposedly being reduced.

In 1986, it stated that survey results "confirm that there will be a vast improvement in the recruitment to the stock in the period 1986–1989 as a result of the strong contribution from the 1983 to 1986 year-classes." It observed that "a further significant increase in spawning stock biomass is expected in 1990, mainly due to contributions from the 1983 year-class" (ICES 1986).

By 1987, however, ACFM was revising its estimates of recruitment downward and its estimates of recent Fs upwards. The revised assessment implied higher fishing mortalities in recent years than previously estimated because of reductions in the estimates of recent recruitment and growth rates. Meanwhile, managers had increased the TAC from 400,000 tons in 1986 to 560,000 tons for 1987 in line with the earlier scientific optimism.

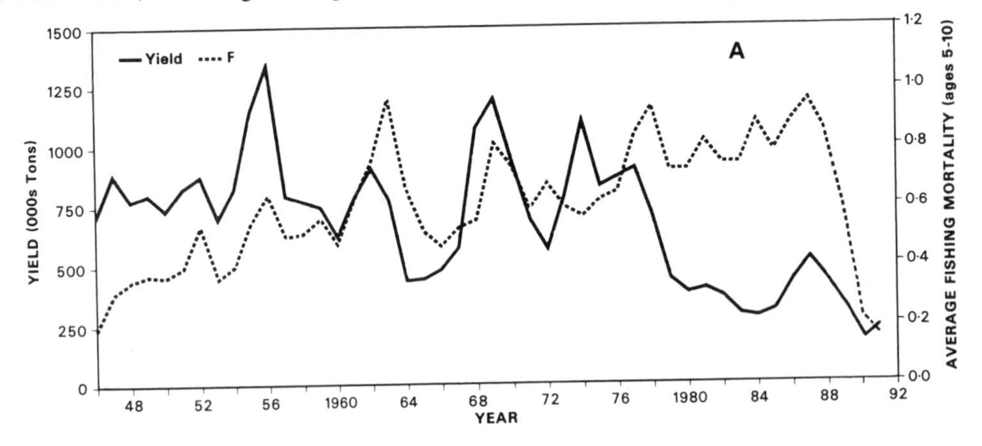

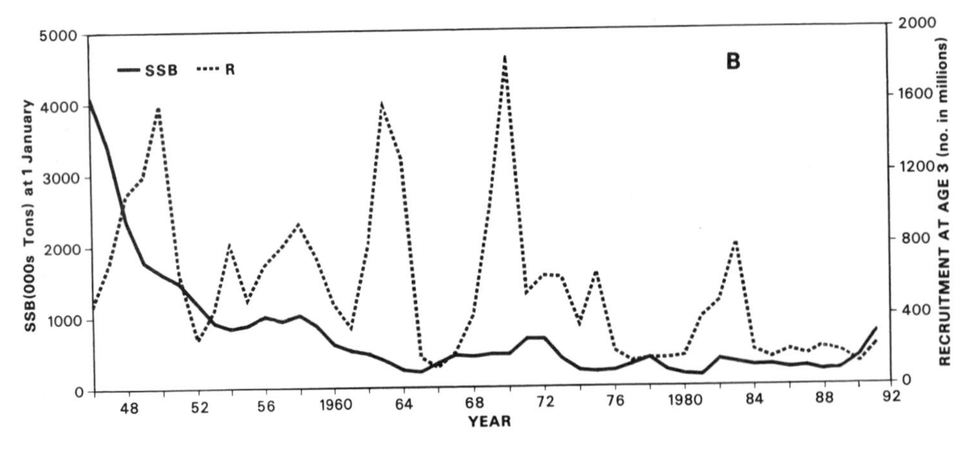

FIG. 20-14. A. Northeast Arctic Cod — Trends in catch and fishing mortality from 1946 to 1991; B. Northeast Arctic Cod — Trends in spawning stock biomass (SSB) and recruitment (R) from 1946 to 1991.

ACFM advised a reduction in F to avoid a decline back to the very low biomass levels of the late 1970's and mid 1980's. Despite these revisions to the assessment, ACFM still projected a F_{max} catch of 530,000 tons in 1989 and 646,000 tons in 1990.

Actual landings increased from 308,000 tons in 1985 to 518,000 tons in 1987. Discard rates were estimated to be high in 1986 and 1987. In 1988, landings decreased to 432,000 tons, considerably less than the TAC of 590,000 tons. The 1988 spawning fishery in Lofoten was the worst on record for more than 100 years. In 1988, ACFM estimated that the spawning stock biomass had fallen to a low level of 187,000 tons (Fig. 20-14B). It predicted that the 1990 level might be the lowest in the history of this stock and that continued fishing at current levels of F would lead to a rapid decline in spawning stock biomass after 1990 (ICES 1988). A rapid rebuilding of the spawning stock was urgently called for. ACFM recommended that F be reduced substantially by 1990.

A TAC of 300,000 tons was established for 1989. In 1989, the spawning fishery in Lofoten was still at a historic low level. By the time of the 1989 assessment, ACFM estimated that all the year-classes after 1983 were well below the long-term average. With no evidence of good year-classes recruiting to the stock in the near future, ACFM suggested that "the stock may be in danger of recruitment overfishing" (ICES 1989). ACFM recommended that the TAC be reduced to 172,000 tons. This was to include all cod caught except the Norwegian coastal cod. Norway and the USSR responded by setting a TAC for 1990 of 160,000 tons. The Norwegian small boat coastal fishery was placed on individual quotas for the first time, with the exception of about 500 of the smallest boats which fished only up to 300 m from shore.

Given the dramatic change in the prognosis for this stock, and contradictory advice on management over 3 years, ACFM felt compelled in 1989 to offer a detailed explanation for the changes. Estimates of the strength of the 1983, 1984 and 1985 year-classes from surveys were revised drastically downwards from 1986 to 1989. Possible explanations for the reduction in numbers of these three year-classes included higher-than-usual cannibalism, lower growth rates and discarding rates higher than previously estimated (ICES 1989).

Although no explicit reference was made to the collapse of the capelin stock from 1986 onward, Norwegian scientists were of the view that the lack of capelin as food for cod led to the decreased growth rates and hence underestimation of effective fishing mortality (Odd Nakken, Bergen, Norway, pers. comm.). In any event, ACFM failed both in the case of the Barents Sea capelin and the Northeast Arctic cod to provide reliable and timely advice for management of these stocks. Nakken (pers. comm.) has attributed the collapse of these two stocks not to failure of managers to respond to scientific advice, but rather to the failure of scientists to predict accurately the size of the stocks and the interactions between species.

Whatever the reason, the record for this stock in the post-extension era is a sad one. Catches declined from the 750,000–1,000,000 ton level in the 1970's to 278,000 tons in the early 1980's. After a brief resurgence to 500,000 tons in 1987, based on one large year-class (1983), catches again declined to the lowest level on record in 1990. The spawning stock biomass in the latter part of the 1980's, at less than 200,000 tons, was only 5% of the post-World War II level and only 10% of the 2,000,000 ton level of the late 1950's (Fig. 20-14B). The maintenance of high F_s overall through the 1980's and high exploitation rates on younger fish contributed to the collapse of the stock in the late 1980's.

In 1992, ACFM concluded that the recent severe measures taken by the management bodies had succeeded in decreasing fishing mortality from a record high of 1.02 in 1987 to 0.19 in 1991, the lowest level since 1946. As a result, the stock biomass had started to increase. The spawning stock biomass was estimated to be at the highest level since 1972 (about 680,000 tons in 1991), and the total stock biomass had reached 1.8 million tons in 1992. This was due to low fishing mortality, early maturation, especially of the 1983 year-class, and improved individual growth.

5.4 Measures to Address Overcapacity

As early as 1977, the Norwegian government's Long-Term Plan recognized that there was an overcapacity in nearly all sectors of the fishing industry. The Plan indicated that, even with an anticipated increase in the Barents Sea cod quota, there was an overcapacity in the Norwegian cod fisheries. A reduction in the factory trawler fleet was proposed. No new fresh fish trawlers were envisaged, beyond the replacement of existing capacity. The Plan did not envisage room for any substantial increase in the capacity of coastal vessels, beyond the increase associated with technological progress. The capacity of the purse seine fleet was too large. The Plan proposed that an attempt be made to reduce it. There was an overcapacity in the fish-oil and fish-meal industry. The Plan suggested some factories would have to be closed (Brochman 1983).

Entry to particular fisheries has, since 1972, been limited by means of specific licences. By the late 1980's entry to virtually all fisheries was limited, except for those in northern Norway. Under the Fisheries Act, the Minister has the authority to introduce special quota arrangements in an attempt to reduce harvesting capacity. This allows for an individual vessel quota system.

An example of the vessel quota system is that in effect in the capelin fishery. The purse seiners were allocated about 90% of the quota. The purse seine quota was then divided into specific vessel quotas. Each vessel was assigned a quota according to its cargo capacity. The vessel quota increased digressively in relation to the cargo capacity. The purpose of this system was to ensure equality of income distribution among fishermen (Brochman 1983).

In the cod fishery, vessel quotas were first introduced for the deep-sea trawler fleet in 1976. The coastal cod fishery was placed on individual quotas for the first time in 1990. About 500 boats, which fish only up to 300 m from shore, were placed on a fleet quota. The remainder, which fish beyond 300 m, were assigned individual quotas. This was the first attempt to apply this system to the small boat fishery. It was recognized that this might prove difficult to enforce (Odd Nakken, Bergen, Norway, pers. comm.). Prior to this, the coastal fishery was exempt from the TAC and fishing continued even after the Norwegian share of the TAC was reached.

In 1979, the government introduced a scrapping program for the purse seiner fleet. The aim was to reduce catching capacity by 20%. In 1980, a scrapping and refinancing program was adopted for trawlers in the cod fishery. At that time, a combined reduction of 25% was suggested for purse seiners and trawlers. Fourteen out of 80 fresh fish trawlers were scrapped or sold. It has been estimated that the fleet reduction program helped to reduce capacity in the purse seine and trawler fleet by approximately 25% by 1983 (OECD 1983).

From 1985 to 1988 approximately 384 million Norwegian kroner were spent on a new fleet scrapping program introduced in 1984. More than 472 vessels were withdrawn from the fishery by this means. Approximately 143 million Nkr were spent to reduce capacity in the processing sector. More than 24 plants were closed as a result of this program (OECD 1986 to 1989).

Despite these measures, there is little evidence to suggest that the real harvesting overcapacity of the late 1970's was reduced to any significant extent. Technological improvements and resulting increases in efficiency have probably more than offset whatever reduction was achieved initially. It appears that the number of vessels and fishermen who fish year round has increased slightly in recent years, despite a slight decline in the total number of fishing vessels and fishermen over the years.

Because of the lack of effective controls on capacity, the average size of vessels increased. The total capacity of the fleet in the late 1980's was still too large in comparison to available resources (*World Fishing*, July, 1990). This situation was exacerbated by the late 1980's declines in the Barents Sea capelin and Northeast Arctic cod and haddock stocks.

On the processing side, the number of shore-based plants was reduced substantially in the 1980's. The total capacity of the fish processing industry is still thought to be excessive in relation to the amount of raw material available. Although there has been a decline in the number of plants, the size and productivity of the remaining plants has increased.

The 1991–92 Norwegian "white paper" on fisheries management considered various ways of dealing with the overcapacity problem. The vast majority of those consulted opposed the use of Individual Transferable Quotas, on the grounds that ITQs would lead to a concentration of quotas and a centralization of fishing operations. The report proposed subdividing the inshore fleet into a number of groups, with quotas allocated on a group basis and seasonal quota breakdowns. The individual vessel quotas adopted in 1990 would be abandoned. For the offshore fisheries, vessel quotas would be retained. Transferability would be permitted within enterprises, but individual quotas would not be freely transferable. The general approach was one of incremental rather than comprehensive adjustment. (K. Mikalsen, Tromso, Norway, Personal Communication).

5.5 Conclusions from the Norwegian Experience

The traditional Norwegian fishing industry experienced increasing problems in the late 1980's. These resulted from a combination of declining resources and already excessive capacity in the harvesting and processing sectors. Norway's shrimp and fishing industries lost huge amounts of money in 1987, according to a study by the Institute of Fishery Technology Research. The year 1987 was one of the worst on record. The shrimp industry experienced its fourth consecutive year of "red ink" with losses totalling $10 million (U.S.). Three of every four frozen fish plants and every other conventional fish processor ended up in financial difficulty. Firms in northern Norway were particularly hard hit (*Fishing News International*, December, 1988). It is unlikely the situation improved with the substantial quota reductions in 1989 and 1990. At the beginning of the 1990's the resource supply (cod and capelin) improved.

Despite attempts by Norway and the USSR to manage the fish resources of the Barents Sea, on which the fishery of northern Norway depends, things took a turn for the worse in the late 1980's. The sudden collapse of the capelin stock in 1985–86 combined with the virtual collapse of the Northeast Arctic cod at the end of the 1980's, threatened the future of important segments of Norway's fishery. Meanwhile, the once-great Atlanto-Scandian herring stock had been rebuilt to only a modest fraction of its former abundance. Fishing mortality on the Northeast Arctic cod was excessive through the 1980's, greater than the levels exerted in the pre-extension days. Emerging problems with certain North Sea stocks (see the section on the EC), combined with the Barents Sea stock declines, did not augur well for Norway's fisheries at the beginning of the 1990's. There were, however, some positive signs with the reopening of the capelin fishery and rebuilding of the cod and haddock stocks underway.

Norway's fisheries policy has been driven largely by equity rather than economic efficiency considerations. Hannesson (1985) reported that the Norwegian fisheries stagnated in terms of output after 1972. Input of capital continued to increase until 1978, while labour input, which had been declining through the 1960's, stopped declining after 1975.

Hannesson (1985) concluded that the Norwegian fishery was caught in a classic policy dilemma. The income support policy, intended to redistribute income, was also adversely affecting the allocation of resources. Too many boats and fishermen were chasing too few fish in an economy where there was nearly full employment and excess demand for investment funds. Furthermore, he suggested, the income support policy was self-perpetuating. Subsidization leads to increased fishing effort, which depletes fish stocks, generating pressure for higher subsidies to maintain incomes. Recent events in the fishery suggest that this analysis is correct.

Norwegian policy makers were trapped in a vicious circle. At the beginning of the 1990's, they faced a major challenge of rebuilding depleted fish stocks and reducing fishing effort to a level more in balance with the likely sustainable catch. Given the events of the 1980's in the Barents Sea, the sustainable catch from that ecosystem is a major question. The answer will help shape the future of Norwegian fisheries. Norway's experience underlines the importance of understanding multispecies interactions (e.g. cod/capelin). This has particular relevance for Canada which has faced similar questions on its Atlantic coast in recent years.

6.0 ICELAND

6.1 Overview

Fisheries are the most important sector of Iceland's economy. Marine products in the 1980's accounted for 70–80% of the value of the country's export commodities. Iceland provides about 5% of the world's fish exports.

The Icelandic fisheries are based principally upon several groundfish species, of which cod, haddock, saithe and redfish are the most important, and two pelagic species, herring and capelin. There are also important invertebrate fisheries for shrimps, lobsters and scallops. The groundfish fisheries produce between 75 and 80% of the total value of the Icelandic fisheries in a typical year (Arnason 1986). Pelagics contribute about 15% of the total catch value and the other fisheries about 5%. The single most important fishery is that for cod, which usually accounts for about 50% of the total value of the catches.

6.2 Fisheries Policy and Management Approaches

Iceland had long been preoccupied with two primary fisheries goals: (1) a wider fisheries limit, to secure a greater share of the fish catch; and (2) conservation and management policies "to secure optimum sustainable yield of the various stocks of fish and other living marine resources." It pursued these goals on many fronts: within ICES, the Permanent Commission, NEAFC, the Third Law of the Sea Conference and through several unilateral extensions of its fisheries limit (Elisson 1981).

Iceland's fisheries limit was extended in stages, from 3 to 4 nautical miles in 1951, to 12 miles in 1958, to 50 miles in 1972 and to 200 miles in 1975. The last move anticipated the growing consensus at the Law of the Sea Conference on extended coastal state fisheries jurisdiction. Iceland's actions, including two well-publicized Cod Wars with Britain, were prompted in large part by concern about the state of the fish stocks off Iceland and the desire to secure a greater share of the available catch.

For decades until 1972 Iceland's share of the important groundfish stocks averaged only 50%. With the near collapse of the herring stocks in the late 1960's, securing a greater share of the groundfish catch became vitally important.

By the late 1950's fishing effort (Icelandic and foreign) was already excessive. This trend continued into the 1970's, prompting Iceland to act unilaterally. Because many of the stocks important to Iceland are local, coastal state jurisdiction offered the opportunity to control those resources. There were other stocks, however, such as the Atlanto-Scandian herring, capelin and redfish which extended beyond Iceland's jurisdiction, requiring cooperation with other countries.

In 1973, the Althing (Parliament) enacted a new law empowering the Minister of Fisheries to adopt conservation and management measures to restore overexploited stocks to "optimum yield" levels and to secure an optimum utilization level for all types of fish, crustacea and shellfish. This legislation was revised in 1976.

Policies adopted under it included an increase in the minimum mesh size of trawls for bottom fishing from 120 to 135 mm, and later to 155 mm, except for redfish, for which trawlers are allowed to use 135 mm meshes in specified areas. For Danish seines the minimum mesh size was increased to 155 mm. The mesh size of all other gear is also regulated, including shrimp trawls, gillnets, driftnets and purse seines.

Another management measure involved special closed areas during the spawning seasons of several species of fish. Closed areas for immature fish have also been established either permanently or temporarily. Closed seasons are used for capelin, herring, lobster and shrimp.

In the post-200-mile limit era TACs were set annually for herring, lobster, and shrimp and then were extended to capelin and cod. Elisson (1981) expressed the hope that a "resort to catch quotas will be necessary only over a short time." By 1990, however, TACs had become an established part of the Icelandic fisheries management system.

By the early 1980's the government was considering means to restrict entry into the fishing industry. However, concrete action did not come until 1984 when Iceland adopted a system of individual quotas for the demersal fisheries.

By the late 1980's fisheries objectives incorporated economic considerations. The revised Fisheries Act adopted in January 1988 was intended to promote protection of fish stocks and "their cost-effective utilization and thereby ensure steady employment and settlement in the country" (Article I). Iceland's fisheries minister, Halldor Asgrimsson, in an interview with *World Fishing* (March 1988) commented:

> "In formulating fisheries policy, the long-term goal is first of all to ensure a sound utilisation of the most important fish stocks with the aim of attaining maximum sustainable yield. Secondly, the quota system is intended to improve the economic performance of the fishing industry by cutting down unnecessary investment and fishing effort. Thirdly, the system is intended to take into account, as far as possible, regional equity among the main fishing communities around the country."

These are very general objectives. According to Jakob Jakobsson, Director of Iceland's Marine Research Institute, managers and industry have been reluctant to establish more specific objectives for management of the fishery (Jakobsson, Iceland, pers. comm., 1990).

6.3 Trends in Stock Abundance and Management

Three stocks have been crucially important to the Icelandic fishery during the past two decades — cod, and, to a lesser extent, capelin and summer spawning herring.

6.3.1 Icelandic Cod

This stock was brought under Icelandic control when Iceland claimed a 200-mile zone in 1975. During the period 1961–70 the total catch from this stock averaged 395,000 tons. Total catches declined from a peak of 470,000 tons in 1970 to 330,000 tons in 1978, following extension, but subsequently increased to 469,000 tons in 1981. The catch dropped quickly back to a low of 284,000 tons in 1984 but increased to 370,000–390,000 tons during 1986-88 (Fig. 20-15A). The foreign share of the total catch decreased dramatically from 45% in 1971–72 to 3% by 1977 and less than 1% in the 1980's.

Foreign fishing effort was replaced by Icelandic fishing effort as purse seiners were converted to fish for cod and additional units were added to the fleet. The fully exploited fishing mortality (on age groups 7 to 11) had been in excess of 1.0 during 1973, 1974 and 1975 (Fig. 20-15A). There was a sharp reduction in fishing mortality post-extension to a low of around 0.45 in the 1978–1980 period. This reduction occurred at a time when the exploited stock size was somewhat above average (Fig. 20-15B) as a result of good 1970, 1972 and 1973 year-classes. This offset any obvious effect of reduced effort on the catches although there was a decline to a slightly lower level in 1978. The change in F was accompanied by an increase in the mesh size for bottom trawls and in the minimum landing size and the closure of certain grounds where small fish concentrated. The spawning stock increased up to 1980 but subsequently declined to the lowest level on record in 1983, 1984 and 1988 (Fig. 20-15B). With a lower fishable biomass during the 1980's than in the late 1970's and the late 1960's, catches were maintained in excess of 300,000 tons only by an increase in fishing mortality to 0.8 and higher.

Iceland's aim was to rebuild the fishable stock to the 1.5-1.7 million ton range but this was not achieved. Recent catches of 350,000–390,000 tons, (primarily 4, 5 and 6 year-old fish) have been as large as the spawning stock, which fluctuated between 300,000 and 400,000 tons during the 1980's (Fig. 20-15B). Large year-classes are less frequent now than in previous decades. There used to be 4 to 6 large year-classes each decade but there were only two in the 1980's (Jakobsson, Iceland, pers. comm., 1990).

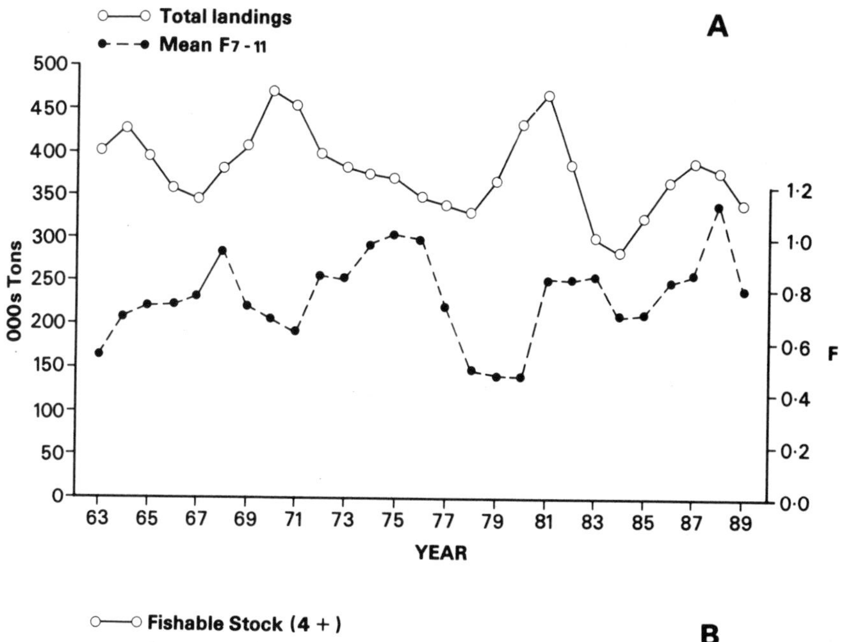

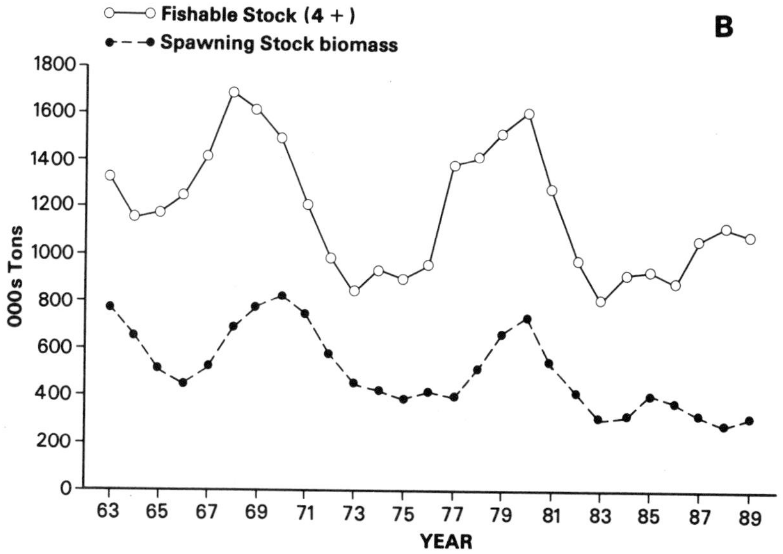

FIG. 20-15. A. Icelandic Cod — Total landings and average fishing mortality (F) from 1963 to 1989; B. Icelandic Cod — Trends in fishable stock and spawning stock biomass from 1963 to 1989 (from Iceland 1989).

Iceland has not been as successful as might be generally regarded in managing its cod stock. At the end of the 1980's fishing mortality was high and the spawning stock was low. The government was setting TACs for cod above the levels proposed by the Marine Research Institute. For 1987 and 1988, the MRI proposed cod TACs of 300,000 tons. The actual catches were 390,000 tons and 380,000 tons. A similar pattern of exceeding the scientific advice was evident for certain other stocks (Iceland 1988).

In 1989, the Institute advised that "if the cod stock is to be kept on the present level the catch should not be more than 250,000 tons in 1990 or 1991." The government established the target as 260,000 tons for 1990 with the catch pegged at a maximum of 300,000 tons.

In 1992, ACFM concluded that the fishing morality on this stock was high at 0.8. The spawning stock biomass was close to the lowest level on record. Recruitment of year-classes from 1985 onwards had been poor,

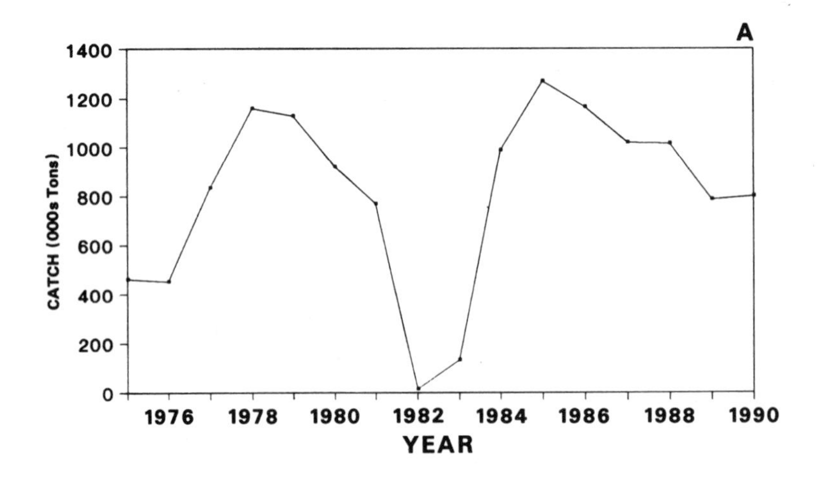

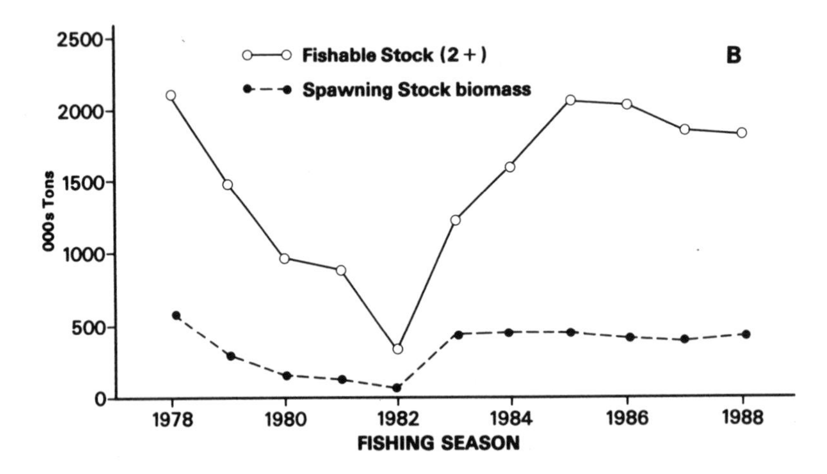

FIG. 20-16. A. Icelandic capelin — Trends in catch; B. Icelandic Capelin — Abundance of the fishable stock at the beginning of the 1978/79 – 1988/89 seasons and the remaining spawning stock biomass (courtesy of H. Vilhjalmsson, Iceland).

following the good 1983 and 1984 year-classes. ACFM advised that this stock had reached a level of exploitation which would bring it below the minimum biologically acceptable level. Therefore, ACFM recommended an immediate substantial reduction in fishing mortality in 1993, corresponding to a catch in 1993 of no more than 154,000 tons. Catches had exceeded national advice and national TAC levels considerably for the previous decade.

6.3.2 Icelandic Capelin

The Icelandic stock of capelin inhabits the area between Iceland, Greenland and Jan Mayen. Catches surged upwards through the 1970's, exceeding 1 million tons in 1978 and 1979 (Fig. 20-16A). This occurred partly as a result of the diversion of fishing effort from the depleted Atlanto-Scandian herring stock. During the 1978/79–1988/89 period this capelin stock exhibited dramatic changes in abundance. Adult stock abundance decreased from 2.35 million tons in 1978 to around 320,000 tons in 1982 but increased rapidly thereafter to a level of more than 2 million tons in 1985 and 1986 (Fig. 20-16B). After 1983 the stock was rebuilt rapidly as a result of a most remarkable increase in recruitment and a more conservative fishing policy.

Vilhjalmsson (1983,) has described the management measures implemented during the 1970's and 1980's for this fishery. Because most of the capelin spawn only once and then die, the primary management objective since the early 1980's has been to maintain an "adequate" spawning stock. Following the collapse of the capelin stock in the early 1980's and the apparent detrimental impact on the growth of cod, it has also been considered important to leave enough capelin for cod to feed upon.

In 1980, it was decided that it would be inadvisable to reduce the spawning stock to less than 2/3 of the 1979 level or 400,000 tons. This minimum spawning stock size target has been maintained since then.

This target spawning biomass was not maintained during the period of low recruitment in the early 1980's. The 1980 spawning stock was reduced to about 300,000 tons and the 1981 and 1982 spawning stocks to extremely low levels (Fig. 20-16B).

A fishing ban was recommended and accepted for the 1982/83 season. Following the stock collapse, scientists took a cautious approach in providing advice on management. The countries concerned — Iceland, Norway and Greenland (EC) — set TACs somewhat above the levels advised by ICES but the minimum spawning stock size was met. With a rapid rebuilding of the stock, catches again exceeded 1 million tons annually from 1985 to 1989. The stock collapse in the early 1980's was due to recruitment failure and not to overexploitation. (H. Vilhjalmsson, Iceland, Personal Communication).

This stock again declined in the late 1980's-early 1990's. The TAC for the 1990–91 season was set at 600,000 tons, higher than the 500,000 tons suggested by ICES. The catch in 1990–91 was approximately half the TAC. Recent year-classes had been below average strength. In 1991, ACFM concluded that the spawning stock had been below the "minimum safe level of 400,000 tons during the last two years."

6.3.3 Icelandic Summer-spawning Herring

Historically, three stocks of herring were fished off Iceland-the Atlanto-Scandian herring, the Icelandic spring spawners, and the Icelandic summer spawners. The Icelandic spring spawners should probably be considered as a component of the Atlanto-Scandian herring living at the outer limits of their distribution (Jakobsson 1985). This stock collapsed in the late 1960's before there was any attempt at management. I deal here only with the Icelandic summer-spawning herring.

Catches from this stock were between 20,000 and 30,000 tons during the 1950's (Fig. 20-17A). Catches rose rapidly in the early 1960's, with the introduction of new technology, and peaked at 130,000–140,000 tons during 1963–65. Catches declined dramatically in the late 1960's. A ban on purse seining was introduced in 1972. This remained in effect until the last quarter of 1975. Since then the stock has recovered and catches increased to 50,000–60,000 tons by the early 1980's and rose sharply to more than 90,000 tons during 1988–1990.

Fishing mortalities in the 1950's were low, fluctuating between 0.1 and 0.2. In the early 1960's, F increased sharply to 1.17 in 1965 and remained at this high level until purse seine fishing was banned in 1972. Since the recovery of the stock in the mid-1970's, F averaged about 0.2 from 1965–1982 (Fig. 20-17A). Since 1984, F has averaged about 0.25, slightly above $F_{0.1}$.

Spawning stock biomass plummeted from 300,000 tons in 1960 to a low of around 10,000 tons by 1972 (Fig. 20-17B). The stock recovered rapidly because of very strong year-classes in 1971 and 1974, produced when the spawning stock was at a low level. From 1979 to 1988, three large year-classes were produced (1979, 1983 and 1986). These, combined with a relatively low F, helped to rebuild the spawning stock biomass to more that 500,000 tons by 1988.

Early scientific advice on desired management actions from 1965 to 1970 was largely ignored, with the exception of a minimum landing size regulation. Following the reopening of the fishery in 1975 it has been managed through TACs and individual allocations to purse seiners. TACs are divided between driftnetters, which usually have been allocated 30–40% of the total, and purse seiners. In addition to the TACs, a minimum landing size of 27cm was implemented, combined with a closed season for about nine months each year.

ACFM calculated the long term average yield to be in the order of 75,000 tons for this stock. The TAC advice for 1988 to 1990 was in excess of this because of above-average recruitment. Despite some significant shifts in the scientific advice in recent years, the rebuilding of the Icelandic summer-spawning herring is a scientific/management success story, with the cooperation of nature. A collapsed stock was rebuilt and F has been held close to the target level of $F_{0.1}$. Above-average recruitment has helped to make this possible. But certainly additional fishing pressure would have been exerted on the stock during the 1980's if it were not for prudent management.

6.4 Measures to Address Fleet Overcapacity

During the late 1970's and the 1980's, Iceland went from an open access fishery to a situation where free access competitive fisheries are virtually nonexistent. When the herring fishery was reopened in the mid-1970's, an individual vessel quota system with limited eligibility was introduced. In 1979, relatively unrestricted transfers of quotas between vessels was permitted.

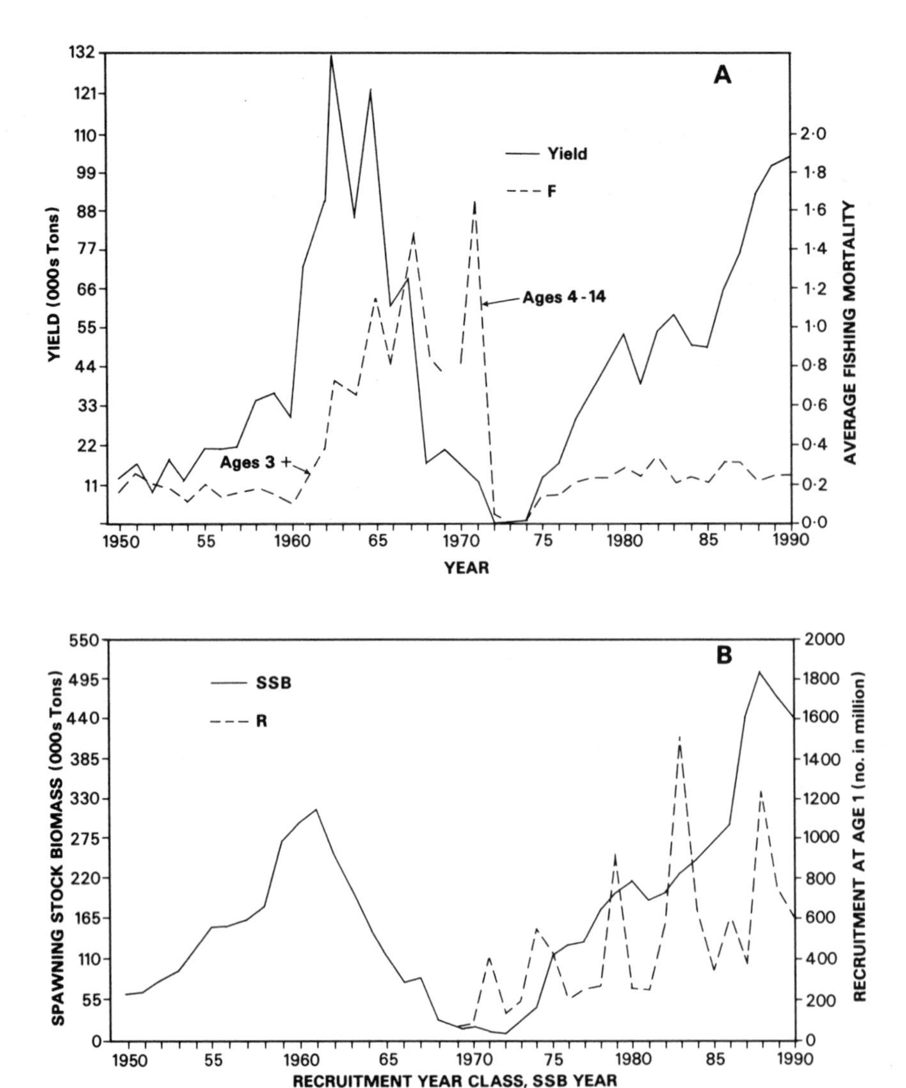

FIG. 20-17. A. Icelandic summer spawning herring — Catches and fishing mortality (F) from 1950 to 1990; B. Icelandic summer spawning herring — Spawning stock biomass (SSB) for ages 3+ and recruitment (R) at age 1 from 1950–1990. (Adapted from Jakobsson 1985 and ACFM Reports, various years).

Limited entry and individual vessel quotas were introduced to the capelin fishery in 1981. In the groundfish fishery, individual vessel effort restrictions were introduced in 1978. Because entry was not limited, the fleet continued to grow. However, as a result of a crisis in the groundfish industry in the early 1980's, a major new approach to management of that fishery, involving individual vessel quotas, was introduced in 1984. For details on the introduction of this system, see Arnason (1986).

Following the initial trials with individual quotas for the demersal fisheries, in January, 1988, the Althing passed a new Act covering the period 1988–1990, which embodied social and economic objectives. It required the Minister to set TACs. It also established limited entry licensing and legislated a transferable vessel quota system.

No one is permitted to participate in the major fisheries without having been issued a specific licence. Licences are issued to vessels, not individuals. The vessels that may be considered for licences are those (or their replacements) that were permitted to fish in 1985. Eligibility was also open to those vessels that were issued special fishing licences under fisheries management legislation passed in 1985.

Two kinds of fishing permits continued to be available, those with catch quotas and those with effort quotas. The choice of control mechanism was left up to the vessel operator who had to make a decision each year. However, vessels on effort quotas were allowed to fish only 10% above the set catch quota of cod instead of 20% as before. Catch quotas were made freely transferable among vessels belonging to the same firm or based in the same port. Transfers that would affect the regional distribution of landings and impact on employment were prohibited unless approved by the Minister. However, permanent transfers of quota are not really feasible, since quotas continue to be issued annually. Under the 1988 Act, effort quotas and allowable catches under the effort quota system are not transferable.

Vessels under 10 GRT were required to obtain a license, but entry was not limited. However, licenses were not to be granted to new vessels over 6 GRT unless a comparable vessel ceased fishing.

According to OECD statistics, nominal fishing effort increased in Iceland from 1985 to 1988. According to the Icelandic government, this enlargement of the fishing fleet was due to two factors. Firstly, fishing vessels below 6 GRT had not been limited by the individual quota system which meant that they enjoyed free entry to the fishery. Secondly, a favourable cost/earnings ratio during the previous couple of years (1985 and 1987) had led to significant vessel construction. The measures adopted in 1988 with respect to vessels less than 10 GRT were intended to halt this expansion (OECD 1990).

As of 1988, profitability of the groundfish industry had improved considerably since the introduction of the vessel quota system. How much of the improvement was due to the vessel quota system was not clear. The new system made it possible for firms to reduce aggregate fishing effort and to utilize the remaining effort more cheaply. The Icelandic Economic Institute estimated that in 1985 the system improved profitability in the groundfish industry by $15 million (U.S.).

Arnason (1986) suggested that some overall redistribution of landings had occurred to the western part of the country, primarily from the southwest and the east. No information was available on the effects of the quota system on the distribution of personal income in the fishing industry.

Arnason (1986) argued that the quota system also halted the previous growth in fishing capital. Aggregate fishing effort, measured in terms of ton-days at sea, decreased by 15% in 1984 and by an additional 6% in 1985. Arnason (1986) estimated that economic rents, on the basis of quota values, yielded some $24 million (U.S.) in 1984 and $35 million in 1985.

In 1987 and 1988 the economy experienced a dramatic downturn. The nation's fishery experienced crisis conditions by the end of 1987. In its summary of the position during 1987, the National Economic Institute estimated an average profit of 5–8% on groundfish fishing, while the profitability of processing varied considerably. By the end of 1987, the entire processing sector was 4–5% in the red and the freezing plants were losing 7–8% (Iceland 1988).

Iceland's Fisheries Minister, Halldor Asgrimsson, suggested that the harvesting sector had performed relatively better than the processing sector because of the system of individual vessel quotas.

6.5 Conclusions from the Icelandic Experience

Iceland's fisheries policy in recent years has been very much shaped by the fact that the fleet is still too large for the available stocks. Overinvestment in vessels from 1971 to 1983 led the government to introduce an individual vessel quota system. The objective is to gear investment and fishing effort as closely as possible to estimates of the sustainable yield. It remains to be seen whether that will be achieved.

On the conservation front, Iceland, through prudent management and with good recruitment, successfully rebuilt the Icelandic summer-spawning herring stock. The capelin stock, again with favourable recruitment, was brought back rapidly from a very low level in the early 1980's. By 1990, however, it was again declining. The cod stock, the backbone of the Icelandic fishery, experienced two downturns in the 1980's. Fishing mortalities (above 0.8) were considerably beyond even the F_{max} reference point. This meant that the fishery was very vulnerable to downturns in recruitment. The government consistently set the TAC at a level higher than Icelandic scientists recommended. As Iceland enters the 1990's, it remains to be seen whether the recent catch level of 300,000–400,000 tons can be sustained. A succession of poor year-classes could lead to a severe downturn in the fishery.

Like other countries, Iceland has had to grapple with the problems of overfishing and overcapacity. It has had some success but problems remain. The chief lesson for Canada is that unless capacity can be brought into line with the available resource, key fisheries will continue to be vulnerable to periodic downturns.

7.0 EUROPEAN COMMUNITY

7.1 Overview

The European Community has become a principal player in fisheries management in the Northeast Atlantic only since 1977. Before that, the Northeast Atlantic

Fisheries Commission (NEAFC) was the international agency responsible for stocks outside national jurisdiction, which was then generally 12 miles or less.

On January 1, 1977, all member states of the EC extended their fishery limits to 200 miles or to the appropriate median line. At that time the EC decided to take over NEAFC's responsibility for managing fish stocks within EC waters. The individual member states of the EC resigned from NEAFC. According to Holden (1984), this reflected increasing disillusion with NEAFC's decision-making processes. Although TACs and quotas were frequently agreed at NEAFC meetings, use of the objection procedure meant that often these were not applied. The EC countries thought that the EC framework of Community law decision-making "gave the possibility of effective fisheries management" (Holden 1984).

Approximately 11% of the ocean's known fish resources occur within the 200-mile economic zones of EC member countries. Despite the curtailment of distant-water fishing in the post-extension era, nine of the current 12 member states of the EC rank among the top 50 fishing nations in the world (FAO 1989). Four of these (Denmark, Spain, the United Kingdom and France) ranked among the top 23 fishing nations. Thus the EC is a major player in fisheries, not just in the Northeast Atlantic, but on the world scene. In 1985, (the last year prior to the accession of Portugal and Spain) the EC's total reported catch was 5.9 million tons valued at 4.4 billion $U.S.

7.2 Fisheries Policy and Management Approaches

With the establishment of a single EC fishing zone, the EC became actively involved in matters of conservation and resource management. The first proposals for a new Common Fisheries Policy (CFP) tabled in November 1976 suggested that the principal objectives should be:

(i) to ensure "the optimal exploitation of the biological resources of the Community Zone" in the medium-and long-term interests of fishermen and consumers; and
(ii) to ensure "the equitable distribution of these limited resources between member states" while maintaining "as far as possible the level of employment and income in coastal regions which are economically disadvantaged or largely dependent on fishing activities" (European Community 1976).

Holden (1984) concluded that the management objective of the Community in the immediate post-extension years was to stabilise the level of fishing effort at its existing level. The Council of Ministers had never taken an explicit decision on the management objectives for stocks occurring exclusively within Community waters. Until and including 1981, the Commission appeared to base its proposals on the following objectives:

1. closure of fisheries on stocks which were in danger of or had suffered a recruitment failure; and
2. achievement of exploitation at F_{max} on all stocks by a reduction of fishing mortality by 10% each year for stocks exploited at greater than F_{max} (Holden 1984).

This in effect was the approach that was being suggested in ICES advice at that time.

By 1981, however, it became clear that the second objective was not likely to be achieved. This objective was modified in 1982 to "setting TACs which would achieve a stabilisation of the fisheries at the existing level of fishing mortality. Once this was achieved, then the objective would be to set TACs which would reduce fishing mortality by about 10% per year on stocks which were overexploited, but not necessarily to the F_{max} level" (Holden 1984).

Holden (1984) indicated that the management policy of the EC in theory incorporated four elements:

(i) indirect limitation of the amount of fishing effort by setting TACs;
(ii) regulating the methods and technical aspects of fishing (e.g. minimum mesh sizes);
(iii) control and enforcement; and
(iv) limitation of access.

In practice, in the immediate post-extension era the emphasis was placed, in the absence of a comprehensive Common Fisheries Policy, on setting TACs and allocating these among member states.

In May, 1980, the Foreign Ministers of EC member states proposed the following criteria for the allocation of quotas: "Fair distribution of catches having regard, most particularly to fishing activities, to the special needs of regions where the local populations are particularly dependent upon fishing and the industries allied thereto, and to the loss of catch potential in third country waters" (European Community 1980).

There was understandable jockeying by member states on the relative importance that should be attached to these various criteria. After years of debate, with interim rollovers of existing arrangements from one year to the next, the Council agreed on January 25, 1983, on a Common Fisheries Policy which included provisions for (1) an overall percentage allocation between member states of the seven main species (cod, haddock, saithe, whiting, plaice, redfish and mackerel) in terms of a cod equivalent, and (2) the percentage allocation for each stock between member states.

The intricate negotiations and trade-offs leading up to this compromise are detailed in Wise (1984) and Farnell and Elles (1984). Canada's negotiations with the EC on a Long-Term Bilateral Fisheries Agreement (see Chapter 11) occurred during this tumultuous period.

Koers (1989) undertook a comprehensive examination of the EC's fisheries management objectives as articulated in the CFP and subsidiary regulations. He concluded that, although the various regulations supposedly included objectives, the real objectives of EC fisheries policy could not be identified.

Despite this, a complex system of TACs and national quotas was implemented. The system operates largely as follows: (1) TACs are fixed for certain stocks or groups of stocks in EC waters; (2) the share of these TACs to be allocated to EC members is set; (3) the share so determined, plus any catches allocated to the EC in the waters of nonmember states, is allocated among member states; and (4) specific conditions for fishing stocks subject to a TAC are established (e.g. bycatch regulations) (EC Regulations 170/83 and 172/83). The actual TACs and allocations are the product of time-consuming and often tough negotiations.

Recently, senior European scientists have confirmed that no explicit management objectives had been set. The basic approach has been to muddle through, with whatever compromise can be achieved. It appears that the EC's overall objective has been to maintain relative stability, i.e. maintenance of the status quo prevailing at the time of extension of jurisdiction. This is borne out by an examination of what has happened to various key stocks which fall under EC management.

7.3 Stock Abundance and Management

The North Sea has proved to be a very productive area over many decades. It is bordered by seven countries, six of which are members of the EC. More than 20 different species of commercial fish occur in the North Sea, 10 of which are abundant enough to support distinct fisheries targeted on that species (Gulland 1987).

Because of the location of numerous major fisheries laboratories around the North Sea, the North Sea and its commercial fish stocks are better studied than those in almost any other area of similar size in the world. For a general review of the fisheries for and status of the North Sea stocks pre-extension, see Hempel (1978).

Despite these studies, there is much that is not known, particularly about multispecies interactions and how to manage such a complex ecosystem. In the 1960's and 1970's, there were two major recruitment changes in the North Sea. Herring recruitment collapsed and many demersal species experienced a recruitment surge. While the total weight taken from the North Sea increased greatly in the 1960's and 1970's, the total value (in constant dollars) did not. The rise of demersal species has been attributed to the decline of the herring stocks by some investigators, but the interspecific mechanism by which this may have been triggered is still unclear. Increasing attention has been paid by ICES to multispecies interaction questions in recent years (see Chapter 18).

The basic management tool in the North Sea continues to be single stock TACs, accompanied by technical measures aimed at reducing bycatches and discards. I have selected certain key stocks to illustrate the trends in stock status and management approaches.

7.3.1 North Sea Herring

Like the Atlanto-Scandian herring, the North Sea herring was overfished and collapsed in the 1970's. The stock subsequently recovered, following closure of the fishery. Annual catches of North Sea herring were relatively stable from 1947 to 1963, ranging between 500,000 and 700,000 tons (Fig. 20-18A). Catches increased rapidly in 1964 and 1965 to about 1.2 million tons. By 1972, catches had dropped to 500,000 tons. From 1973 onward catches decreased sharply until the fishery was closed in 1977.

Catches were very low from 1977 to 1979 but from 1980 onward they increased steadily from 61,000 tons in 1980 to about 235,000 tons in 1982 and thence to 700,000 tons by 1988.

The total North Sea herring stock was at a relatively high level of 1.5-3.0 million tons during the 1950's to the mid-1960's. From 1963 onward the spawning stock biomass declined from 2.2 million tons to about 400,000 tons in 1968 and then declined further to a low of about 200,000 tons in 1977 (Jakobsson 1985).

Fishing mortality on 2-ringers and older herring, which had been 0.4 or less in the 1950's, rose to 0.7 in 1965 and 1966. From 1968 to 1976 the fishing mortality rose dramatically to very high levels (in excess of 1.0). With the closure of the fishery in 1977, fishing mortality dropped to a low level in 1978 and 1979. By 1980, however, it was back up to 0.3. It remained in the 0.3 to 0.4 range from 1980 to 1984 but thereafter increased to about 0.5 on average (Jakobsson 1985).

Jakobsson (1985) concluded that the fishing ban in force from 1977 to 1979 helped to reduce the exploitation rate and thus preserve sufficient spawning potential of the North Sea herring to make possible the production of large year-classes in 1979 to 1982.

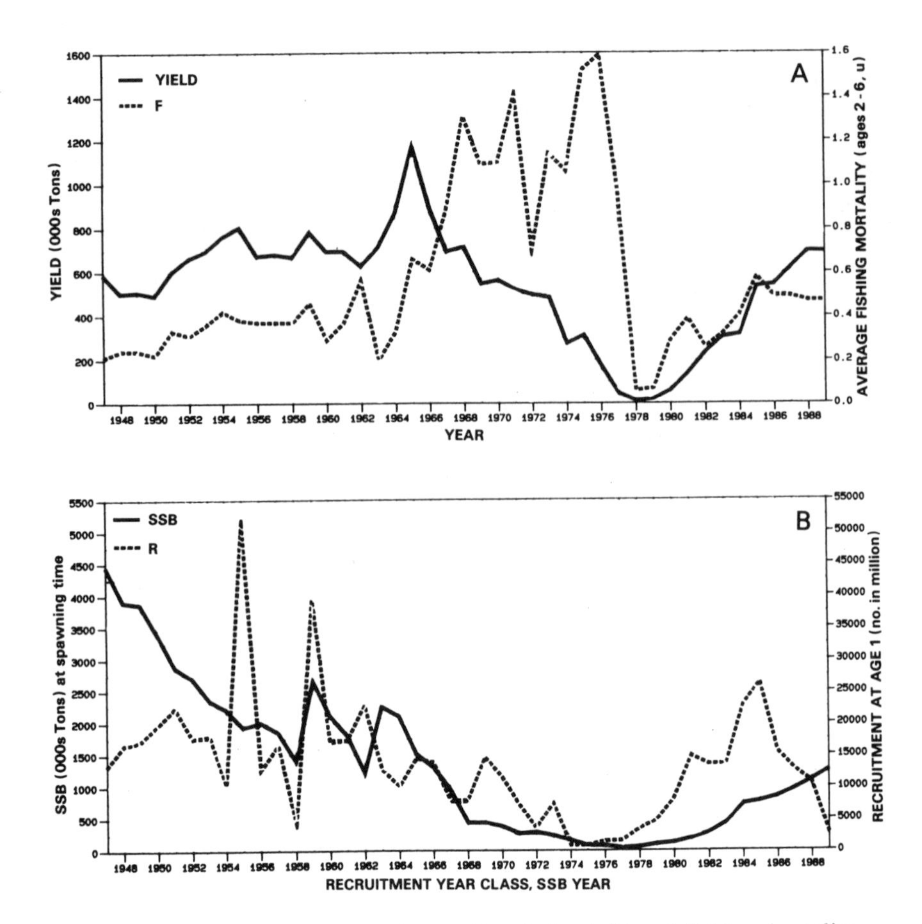

FIG. 20-18. A. North Sea Herring — Trends in yield and fishing mortality — 1947–1989; B. North Sea Herring — Trends in spawning stock biomass (SSB) and recruitment (R) — 1947–1989.

In 1980, ACFM suggested that the policy should be to rebuild the spawning stock to at least 800,000 tons. It described this as "the estimated minimum spawning stock size required to avoid a risk of recruitment failure with a reasonable degree of probability" (ICES 1980).

In 1983, herring fishing was allowed in all parts of the North Sea for the first time since 1977. The total catch was about 308,000 tons compared with a TAC of 145,000 tons. In both 1982 and 1983, approximately half of the catches were not officially reported. However, despite illegal fishing, the prospects for the stock improved because of the recruitment of two strong year-classes in 1981 and 1982. ACFM recommended fishing at $F_{0.1}$, to allow a rapid rebuilding of the spawning stock.

From 1983 to 1986 the TACs established by the EC and Norway were about double the TACs advised by ACFM. By 1986 ACFM was advising that, since the level of F was already considerably higher than $F_{0.1}$, it would be advisable to reduce it while recruitment was at a high level. The spawning stock biomass had reached 800,000 tons by 1986 and 1987 (Fig. 20-18B).

In 1991, ACFM advised that, because of reduced recruitment, overshooting of TACs, and high catches of juveniles, the increase in this stock appeared to have halted. It predicted that continued fishing at existing levels of fishing mortality would lead to a decrease in spawning stock biomass because of the recent decrease in recruitment (ICES 1991b).

While on the surface the recovery of the North Sea herring stock in the 1980's is a success story, in reality managers were fortunate that nature provided good recruitment. Management actions left a lot to be desired. Illegal fishing and misreporting were major problems throughout the period. Up until 1987, agreed TACs were established considerably above the level of scientific advice. Actual catches in many years significantly exceeded these agreed TACs. With fishing

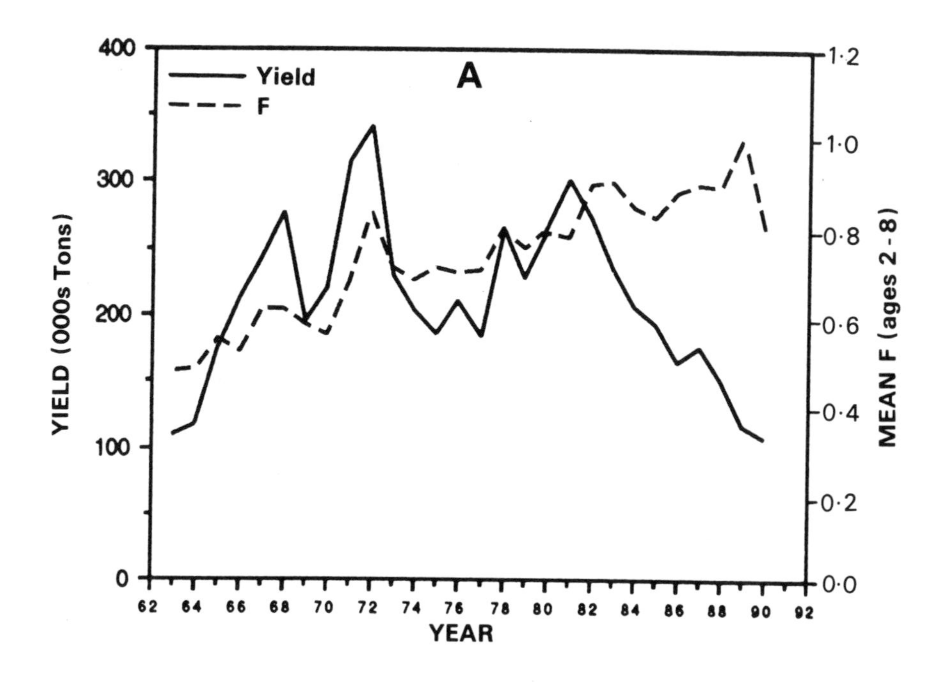

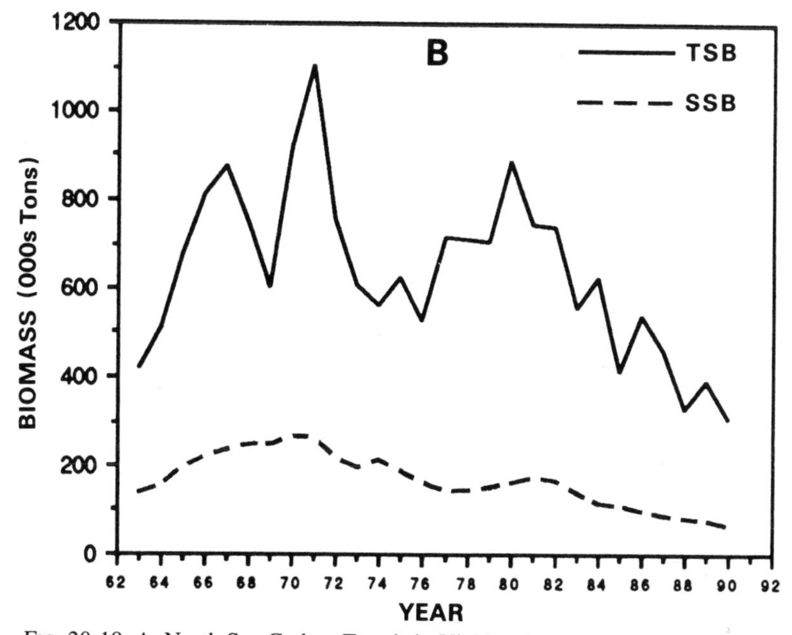

FIG. 20-19. A. North Sea Cod — Trends in Yield and mean fishing mortality for ages 2–8 during 1963 to 1990; B. North Sea Cod — Trends in Total Stock biomass (TSB) and Spawning Stock biomass (SSB) during 1963 to 1990.

mortality at 0.5 or more since the mid-1980's, managers were failing to take advantage of the opportunity offered by good recruitment. With declining recruitment in the early 1990's combined with a relatively high level of fishing mortality (for herring), it is possible that stock abundance could decline rapidly.

7.3.2 North Sea Cod

In recent years it has been generally accepted that pelagic stocks are vulnerable to recruitment overfishing but groundfish stocks have been considered to be much more resilient and able to withstand higher fishing pressure. The North Sea groundfish stocks have often been cited as examples to support this view. However, recent events in the North Sea have cast doubt on the validity of this accepted wisdom.

As early as 1978, ACFM was advising that F values were greatly in excess of F_{max} for cod, haddock and whiting in the North Sea. A high proportion of the catch consisted of young fish. Also, for all three species, annual yields and spawning stock biomasses had declined significantly since the end of the 1960's. ACFM suggested that these stocks were severely overexploited, that fishing effort should be considerably reduced, and that effective action should be taken to improve the exploitation pattern. However, none of these stocks was considered to be in immediate danger of recruitment overfishing. ACFM recommended, on the basis of yield per recruit considerations, that fishing mortality should be reduced in steps of 10% per year.

In 1980, ACFM took a look at what had happened over the period 1976–79. For North Sea cod and haddock F was estimated to be in excess of 1.0 during those years. ACFM concluded that in practice little progress had been made in improving the situation. In 1981, ACFM noted that the existing level of exploitation for cod was 3 times F_{max}, for haddock 4.5 times F_{max}, and for whiting 3 times F_{max}. ACFM recommended that fishing mortality on all three North Sea stocks be reduced as much as possible towards F_{max}.

The turning point in the assessment of the North Sea cod came in 1985. ACFM noted a trend of steadily increasing overall fishing mortality. Spawning stock biomass had been declining since 1971. It was estimated to have been 107,000 tons in 1984, the lowest value on record. A further decline to 94,000 tons in 1985 was forecast, with the spawning stock expected to remain at this level through to 1987 (Fig. 20-19A).

F was now four times F_{max}. The spawning stock size was at a historically low level and the 1984 year-class was estimated to be the smallest ever. Faced with this situation, ACFM called for a substantial reduction in fishing mortality in 1986 (ICES 1985).

In 1986, ACFM noted that the lowest stock level at which consistently good recruitment had been observed was about 150,000 tons. Such a biomass level could only be obtained in 1988 by a complete closure of the fishery in 1987. ACFM expressed "a very strong preference in favour of a TAC not exceeding 100,000 tons, which would permit a more rapid recovery of the spawning stock biomass."

The TAC for 1987 was established at 125,000 tons. At its May 1987 meeting ACFM revised its estimates of year-class strengths upwards. As a result, the TAC was revised upwards in mid-season to 175,000 tons.

By November, 1987, ACFM had revised the estimates of the strength of the 1985 and 1986 year-classes downwards again. Landings in 1986 of 157,000 tons were the lowest in a 20-year period. ACFM recommended a 1988 TAC of 148,000 tons.

In 1988, ACFM estimated that spawning stock biomass had fallen to a record low level of 95,000 tons in 1987, compared with a suggested minimum level of 150,000 tons (Fig. 20-19B). F continued to be very high (0.86). Recent year-classes all appeared to be below average. ACFM predicted that the stock size and sustainable catches would remain at low levels unless recruitment returned to the historically high levels observed in the 1970's (Fig. 20-19B).

The dramatic downturn in the North Sea cod in the late 1980's was attributed to several factors:

1. Excessive fishing mortality (generally 0.8 to 0.9 throughout the 1980's and at least those levels previously);
2. Exploitation of very young, immature fish;
3. Misreporting and underreporting of catches;
4. Failure to recognize that stock recruitment could be a problem for this stock;
5. Failure to take sufficiently severe and quick action when the decline in spawning stock size was first recognized as a problem; and
6. The TACs established, although lower in the late 1980's, had maintained F at too high a level (ICES 1989).

In 1991, ACFM noted that F remained near historically high levels. Fishing effort had remained above the levels necessary to harvest the reduced TACs with the result that quantities of fish in excess of the TACs were often caught, with high discards. For 1992, ACFM did not recommend a TAC. Instead, it recommended that fishing effort in the directed fisheries on North Sea roundfish stocks in 1992 be limited to 70% of the 1989 level.

7.3.3 North Sea Haddock

In the mid-1970's NEAFC was establishing TACs for North Sea haddock around 200,000 to 275,000 tons, almost twice the level recommended by ICES. By 1978 and 1979 ICES was recommending TACs of 105,000 and 83,000 tons, less than half the actual catch in 1976. An estimated F of 1.00 in 1979 compared with a F_{max} estimate of 0.33. By 1978, 1979 and 1980, TACs were being set at approximately the level of the recommended TAC. The recommended TAC was increased to 160,000 tons by 1982.

ACFM noted in 1982 that the TACs adopted by agreement between the EC and Norway were not being effectively enforced. In addition to the human consumption fishery, significant quantities of haddock were being taken in the industrial fishery. For example, in 1983 landings included 16,000 tons taken as bycatch in the industrial fishery. In addition, 65,000 tons were estimated to have been discarded. This compared with reported landings of 168,000 tons.

Agreed TACs increased to 207,000 tons and 230,000 tons in 1985 and 1986. Actual landings fell short of the agreed TACs from 1983 onward, with the shortfall increasing in 1985 and 1986. For 1987, ACFM recommended a TAC of 120,000 tons. Fishing mortality appeared to have almost doubled from 0.65 in 1982 to 1.14 in 1986. With these high F_s the fishery was catching mainly juvenile fish. The spawning stock decreased significantly in 1987 and 1988 and was forecast to be close to the historical minimum by 1989.

ACFM warned that the stock was very heavily exploited and that fishing mortality should be reduced. Reducing discards was suggested as the most effective conservation measure, since the industrial bycatch had declined to a low level.

The recommended TAC decreased substantially to 68,000 tons in 1989. Recruitment since 1983 had been generally poor. In 1989, ACFM concluded that continued fishing at recent levels of fishing mortality would lead to a reduction in the spawning stock biomass "well below any acceptable level." ACFM recommended that F in 1990 be reduced to 0.8, corresponding to a TAC of 50,000 tons.

ACFM warned:

> "The stock is very heavily exploited and fishing mortality must be greatly and permanently reduced to allow recovery and subsequent maintenance of stock size when good recruitment occurs" (ICES 1989).

The status of whiting, which is caught in a mixed fishery involving cod and haddock, was relatively good. This highlighted the problems in regulating fishing mortalities in mixed fisheries by the application of TACs. ACFM estimated that, to allow spawning stock biomass to recover to the lowest acceptable values (150,000 tons for cod, 100,000 tons for haddock) by the start of 1991 would have required that F on cod in 1990 be reduced by 80% and that on haddock by 60% from the values in 1988. This would have meant TACs of 35,000 tons for cod and 31,000 tons for haddock.

ACFM concluded that enforcement of such low TACs in one step would have very serious socioeconomic consequences for the fisheries. It adopted a policy of recommending TACs which would reduce F by relatively small amounts (usually 20%) from the most recent level. ACFM recognized that problems would be encountered in applying this approach to the type of mixed fishery in which North Sea cod, haddock and whiting are exploited.

ACFM found it difficult to propose a package of TACs which would ensure rebuilding of the North Sea cod and haddock stocks while allowing whiting to be exploited at a higher level of F than the other two species. For 1993 ACFM proposed dealing with this problem by limiting the fishing effort in the directed fisheries on North Sea roundfish stocks, except saithe, to 70% of the 1989 fishing effort.

7.4 Overcapacity and Measures to Address It

It is evident from the very high fishing mortalities (0.8 to 1.0) being exerted on many stocks in the North Sea that the EC has a massive overcapacity problem. Extension of jurisdiction left the EC with a legacy of overcapacity.

The total number of vessels registered in the Community changed only slightly from about 53,000 in 1973 to about 51,000 in 1980. This, however, masked important changes in the fleet. The distant-water fleets of the United Kingdom and West Germany, in particular, were hit hard by their displacement from fishing grounds elsewhere in the North Atlantic. Over a 10-year period the United Kingdom and West Germany lost about 190 vessels greater than 500 GRT, i.e. about three quarters of the number registered in 1971. The trend was more complicated for vessels between 150 and 500 GRT. There was a substantial reduction in the number of such vessels in Germany, France and the U.K. but an increase in Belgium, Ireland and the Netherlands. Over a 10-year period the offshore fleet of Denmark expanded by 75% (Koers 1989).

Looking at the member states fishing in the North Atlantic, Koers (1989) calculated that the number of full-time and part-time fishermen declined from about 90,000 in 1970 to just over 70,000 in 1980. There were significant differences in the trend among member states: France, West Germany and the Netherlands experienced the biggest drop in the numbers of fishermen. In Denmark and the United Kingdom the numbers remained more or less stable. In Ireland the number increased by about 50%. In 1986, Spain brought an additional 105,000 fishermen into the EC and Portugal about 35,000.

It is difficult from the available information to draw any conclusions about the extent of overcapacity. Gulland (1987) suggested that overcapacity was very high. He noted that this overcapacity had, in the case of the major groundfish stocks, existed for most of this century and had been widely recognized for almost as long. Overcapacity in the pelagic fisheries was more recent but had had dramatic effects, e.g. the collapse of the herring stocks. In his view, little effective action had been taken to reduce this overcapacity.

Despite short-term pressures to maintain the status quo, there were some attempts in the 1980's to reduce overcapacity. These generally took the form of financial assistance for mothballing or decommissioning vessels. There was, however, no comprehensive attempt to limit entry and restrict the growth of the fleet.

In December, 1986, the Community adopted a 10-year program of measures "to improve and adapt structures in the fisheries and aquaculture sector" (Council Regulation No 4028/86). This Regulation allowed the Community to grant financial aid for the restructuring, renewal and modernization of the fleet.

The Regulation called for the submission by member states of Multiannual Guidance Programmes. One of the requirements was that such Programmes be designed "to establish a viable fishing fleet in line with the economic and social needs of the regions concerned and the foreseeable catch potential in the medium term". An overall target reduction in the EC fleet of 4.3% was set for 1991, with the greatest reductions targeted for Ireland (17.2%), Denmark (9.5%) and the Netherlands (16.4%).

Not much progress was made in the first years of this program. In 1987, most countries were still increasing their fleets and others were not successful in complying with the fleet reduction targets. The EC Commission decided, in December, 1988, to reschedule the target to a 3% reduction in gross registered tonnage and a 2% reduction in engine power in the 5 years ending in 1991. Ten percent of this reduction was to be achieved by 1989, 60% by the end of 1990 and 100% by the end of 1991 (*Eurofish Report*, January 19, 1989).

In practice, little progress was made towards achieving these targets by 1990. Many countries did not submit the required Multiannual Guidance Plans and the Community suspended grants temporarily. Some countries (e.g. U.K.) chose not to accept Community financial assistance. The U.K. government chose to leave the question of fleet reduction to market forces. This decision angered an industry desperately trying to adjust to the extremely low quotas in the late 1980's-early 1990's (*Fishing News International*, January 5, 1990).

Some 14 years after extended jurisdiction, the EC continued to face a massive overcapacity problem. With a few exceptions, its funding programs aimed at reduction of capacity had had little impact on the problem.

7.5 The Misreporting and Enforcement Problem

The problems of declining stocks and substantial overcapacity were compounded by an acknowledged massive problem of misreporting and cheating on catch quotas. ACFM Reports are replete with estimates of substantial quantities of catch going unreported and significant misreporting of area of capture. The lack of accurate catch statistics undoubtedly contributed to errors in stock assessments. The nature of the EC system has rendered ineffective attempts to introduce rigorous enforcement of quotas and other controls on fishing. Member states were responsible for ensuring adherence of their fleets to control measures. The EC had acquired a small group of inspectors whose task is to monitor and verify that member states have in place systems which enforce the Community's fisheries legislation fairly, effectively and without discrimination. Member state are obliged to inform the Commission of the number of vessels inspected, their nationality, any infringements observed and any action taken. For a description of the EC system, see Koers (1989).

Over the years there have been frequent allegations of widespread misreporting, underreporting and cheating by fishermen in some member states. One of the more spectacular misreporting incidents is the so-called "grey market" case in the Netherlands. This matter erupted into a major issue during the summer of 1990 when the Dutch Fish Inspectorate was accused of fraud. In mid-September, 1990, Fisheries and Agriculture Minister Gerrit Braks resigned from his post because of evidence that government inspectors falsified fisheries landings data to allow Dutch fishermen to exceed their cod and sole quotas. Both quotas had been exceeded by 50% or more in 1989 but figures officially transmitted to Brussels had not given any warning of the impending exhaustion of the quota, which should have led to closure of these fisheries (*Eurofish Report*, September 27, 1990).

A survey carried out by the European Commission late in 1990 concluded that fisheries inspection facilities in EC member states were "woefully inadequate" (*Eurofish Report*, January 31, 1991). Figures released by the Commission indicated that the problem was especially serious in certain member states, with Spain, France and Belgium all identified as particularly ill-equipped to enforce EC conservation laws. The number of port-based inspectors in the Community — just over 600 — was considered scarcely half the number required. Aerial surveillance was either inadequate or totally lacking in nearly all EC countries. The inadequate enforcement resources make it "comparatively easy for fishermen to break EC rules with impunity."

7.6 Conclusions from the EC Experience

In the summer of 1990 the EC began to debate the future direction of the Common Fisheries Policy. In a background paper on "Conservation and Management of Fisheries Resources," Commission officials stated that the existing fisheries management system was not working satisfactorily. They noted that an excessively large fishing fleet had depressed stock levels, decreased profitability of the fishing industry and reduced the reliability of market supplies (*Eurofish Report*, June 21, 1990).

The Commission stated that at least 75% of all Community fish stocks were being exploited at rates significantly above the optimum for long-term economic development of the fishery. It observed that "none of the fish stocks in Community waters are being exploited at a rate which generates the maximum economic yield." The Commission acknowledged that keeping TACs at unsuitably high levels to offset short-term economic hardship to fishermen only perpetuated the "vicious circle" of low returns and ever-increasing levels of fishing effort. The only way to break this cycle was to reduce the size of the Community fleet. This objective, the paper acknowledged, was being realized "as yet very slowly" (*Eurofish Report*, June 21, 1990).

The North Sea stocks have been fished very intensively since extension of jurisdiction. The EC had opted to maintain the status quo. This resulted in very high exploitation rates, concentrated on very young fish. While catches were maintained at a high level in the short-term, in the long term stocks have become vulnerable to downturns in recruitment. This became evident in the late 1980's in the cases of the North Sea cod and haddock. In stock conservation terms, the EC's fisheries management approach appears unsuccessful.

The central problem confronting the EC fisheries is chronic overcapacity. Despite various programs aimed at reducing the EC fleet, little progress had been made in this direction.

Neither conservation nor economic benefits are being achieved under the existing fisheries management regime. The TAC and national quota system has been subverted by widespread cheating. Recently, there have been some initiatives to strength enforcement mechanisms. But there is still some considerable distance to go before achieving a reasonable level of compliance.

Koers (1989) concluded that for the EC member states the extension of fishing limits produced few benefits and many costs. Part of the costs were paid by nonmember states who were displaced from EC waters. A large part was paid by the fishing industry. The shift from high-value to low-value species, the loss of employment opportunities, scrapping larger vessels, building smaller, more sophisticated vessels — all were costs that industry had to absorb. The consumer paid part of the costs through a steep rise in prices. The taxpayer has also paid. Koers estimated that over a 15-year period the EC alone spent about 190 million ECUs under the structural policy. Individual member states also contributed large sums: in 1981 member states spent about 85 million ECUs on their fishing industries, often through subsidies to reduce operating expenses or invest in new vessels.

Koers (1989) concluded:

> "In no EC member state does employment in fisheries exceed three percent of the total working population, and in no member state is the fishing industry's contribution to the gross national product greater than 0.7 percent. Yet fisheries management has absorbed much of the EC's energy, and on occasion it has developed into a major political issue. It seems to me that there is a lesson here — in any fisheries management system and in any discussion on these systems great allowances should be made for the irrationality of man."

Gulland (1987) pointed out that, unless the beneficiaries were clearly identified, the short term pressures against significant reduction in fishing effort are likely to be stronger than pressure for change. He estimated that the current values of fishing mortality for each of the main demersal stocks in the North Sea were far above the optimum, whether defined as that giving the maximum yield per recruit (F_{max}) or the greatest economic return (approximately $F_{0.1}$). He estimated the first-hand sale value of the North Sea fisheries production of over half a million tons at one-half billion pounds sterling. The costs of catching it were probably similar. With a conservative assumption that fishing mortality could be halved and the costs halved, the potential benefits would be around 250 million pounds sterling annually. However, Gulland observed:

> "Unfortunately, North Sea fishing policies are in a vicious circle. Until there is effective pressure for action to reduce excess effort, there is little substantive discussion of the form the benefits from reduction should take or how the reduction should be achieved. Effective pressure, however, is unlikely until substantive discussions have identified the beneficiaries, and they can start generating pressure."

There seemed to be a consensus among several senior European scientists whom I interviewed in October, 1990, that the situation in the North Sea is essentially the same as that before extension of fisheries extension in 1977. Little or no progress had been made in stock conservation or achieving economic benefits. With the recent decline in certain key stocks, prospects appear grim for the EC's Northeast Atlantic fisheries in the first half of the 1990's.

A group of independent international fisheries experts convened by the EC to recommend improvements for the EC's fisheries policy concluded that the capacity of the Community fishing fleet was seriously out of balance with available resources. The group calculated that fishing mortality was at least four times higher than it should be for North Sea cod, haddock, plaice and herring. They recommended a two-stage approach to resolve the overcapacity problem:

> "1. Capacity should be brought into equilibrium with current quota levels within the first two years of the next Multiannual Guidance Programme (1992–96);
>
> "2. Capacity should be further reduced to a level corresponding to an agreed long-term target fishing mortality. This step should start within the period of the next MGP and should be completed by 2002 at the latest" (*Eurofish Report*, December 6, 1991).

The group suggested, as long-term targets, mortality levels of F_{max} for demersal species and $F_{0.1}$ for pelagic stocks.

In November, 1990, Fisheries Commissioner Manuel Marin unveiled a "white paper" indicating the direction in which the Commission intended to steer the CFP from 1992 onwards. This document stated that the capacity of the EC fishing fleet must be reduced by at least 40% over the next 10 years "to save key fish stocks from extinction".

Commissioner Marin stated that the next set of Multiannual Guidance Programmes, to run from 1992 to 1996, would have to bring fleet capacity in line with sustainable patterns of exploitation. This would imply cutbacks of around 70% for North Sea cod, haddock and herring, and lesser, but still substantial, reductions for other stocks. He said:

> "Conservation policy has to be the central element of the reform of the policy after 1992. We simply have to conserve our resources" (*Eurofish Report,* December 6, 1990).

In December 1992, EC Fisheries Ministers agreed on a compromise proposal to reduce EC demersal fleets by 20% and bottom/beam(benthic) trawlers by 15% during the 1993–1996 Multi-Annual Guidance Programme. The proposal allowed for nearly half of the reduction to take the form of effort restriction, e.g. tie-up schemes. This fell far short of the European Commission's previous fleet reduction proposals (*Eurofish Report*, December 3, 1992).

The Council committed itself, as part of an agreement on the framework for the Common Fisheries Policy for the next ten years, to setting a multi-annual programme for attaining a long-term balance between stocks and fishing effort by January 1994. It was not clear how much the 1993–1996 Multi-Annual Guidance Programme would be considered part of this goal. The new framework agreement called for the introduction by 1995 of a Community regime for fishing licences managed by member states (*Eurofish Report*, December 20, 1992).

The key question is: will the revision of the CFP come to grips with the underlying problems and lead to fundamental change? There appears to be a slowly growing awareness in parts of Europe of the nature and magnitude of the problem. However, should the Community opt for fundamental change, the task ahead is daunting.

Canada can take some solace from the EC's experience. While Canada has had to wrestle with stock management and overcapacity problems, these pale in comparison to those encountered by the EC in the post-extension era. The EC's track record in attempting to maintain the status quo vindicates Canada's initiatives in setting conservative TACs, even though these have failed to maintain key cod stocks at desired levels. Canada's attempts to deal with the common property problem through limited entry licensing and individual quotas, while imperfect, are light years ahead of the EC's feeble efforts at structural adjustment.

The traditional wisdom that fisheries managers need not worry about the possibility of recruitment overfishing of demersal stocks is also called into question by the late 1980's overfishing of the North Sea cod and haddock stocks. ACFM called for substantial reductions in fishing effort because of its concern that the spawning biomass of these stocks was being reduced to a level which could impair future recruitment. This has particular relevance for Canada, which has experienced declines in a number of major groundfish stocks in recent years. Fishing mortality rates here, however, while above the target of $F_{0.1}$, are considerably less than the levels exerted over the past two decades in the North Sea.

8.0 CONCLUSIONS

Despite experimentation with new approaches, two fundamental problems continue to confront fisheries managers in most of the countries examined here. These are overfishing and overcapacity.

Attempts to conserve and increase fisheries resources have been only partially successful. Norway experienced major resource problems with its Barents Sea stocks. The EC, in attempting to maintain the status quo, reduced many of its fish stocks to levels where maintaining an adequate spawning stock became a major concern. The groundfish stocks off New England were reduced to historic lows.

New Zealand's industry grew through displacement of foreign fishing activity and utilization of joint ventures. Even there, however, TACs for some key species had to be reduced in 1989 and 1990.

Iceland, on the other hand, managed to rebuild several important stocks for a while by being more conservative in setting levels of fishing mortality. Stocks were also maintained or increased in the U.S. North Pacific through conservative TACs and restrictions on foreign fishing. Exploitation of this major resource has now been Americanized. It is doubtful that this resource conservation approach can be maintained without effective controls on domestic fishing effort.

Equity and regional distribution concerns have been major forces shaping fisheries policy in Norway and the EC. Iceland, through its system of individual quotas in some fisheries, has implicitly recognized the importance of economic efficiency. The U.S. Regional Councils have generally failed to accept the economic need to limit access to the fishery. New Zealand transformed its fisheries management system to maximize economic efficiency. Australia has adopted similar economic efficiency goals in its fisheries policy for the 1990's.

Overcapacity remains the second fundamental problem in most of the countries examined. Overcapacity, which has its roots in the common property character of the resource, impedes progress towards a viable fishing industry except for the New Zealand deepwater fishery and the U.S. North Pacific offshore fishery. In the latter two instances, growth of the domestic industry occurred in the post-extension era at the expense of foreign fishing. New Zealand's new ITQ system should prevent overcapacity in the deepwater fishery and reduce existing capacity in the inshore fishery. Although the U.S. North Pacific offshore fishery has been successfully Americanized, it has no adequate mechanisms to prevent overcapacity from occurring.

Norway is wrestling with overcapacity which was exacerbated by a dramatic downturn in its major resources in the late 1980's. The resource situation improved at the beginning of the 1990's. Overcapacity remains a problem. It is being addressed through incremental rather than comprehensive adjustment.

The EC has a monumental overcapacity problem. There have been some feeble efforts to address this through vessel scrapping programs but these have been generally ineffectual. Indeed, excessively high fishing mortalities were maintained on most North Sea fish stocks through the 1980's. Stock declines by 1990 were forcing a reexamination of the EC's approach. The situation in the North Sea has been described by many senior scientists and administrators as "a mess". There was essentially no change from conditions prior to extension of jurisdiction up to 1990. The major question is whether the EC in its current revision of the Common Fisheries Policy will begin to deal with the problems of overfishing and overcapacity which prevail in the Northeast Atlantic.

Iceland, while still having trouble reducing the exploitation rate on its major cod stock to a more desirable level, has made some progress. Its individual quota system offers some promise of reducing capacity over time.

The record low abundance of the New England groundfish stocks are the result of a lack of control of fishing mortality and massive overcapacity, in the absence of any form of limited entry. It is perhaps the most extreme example of what has occurred in the absence of an effective fisheries management regime. Without new approaches to management, the situation is likely to worsen.

New Zealand during the 1980's embarked upon an ambitious new experiment in fisheries management. Its adoption of individual transferable quotas for most of its fisheries represented a major departure from the traditional path. Other countries, such as Iceland and Canada, have experimented with individual quotas. These have generally been introduced in specific fisheries on a trial basis. The intent has been to extend the concept to other fisheries gradually if these experiments succeed. New Zealand's broadscale adoption of ITQs was a radical step forward. It is testing a concept which had been widely advocated by fisheries economists but had not yet been generally accepted by the world fisheries management community.

The New Zealand ITQ system incorporated essentially all of the elements economists advocated: divisible, transferable and saleable individual quotas guaranteed in perpetuity. Generally, radical change in fisheries management is made a crisis atmosphere, which provides the catalyst. In New Zealand, however, this was not the case. Instead, Wilen (1989) has argued that innovative leadership played the critical role.

The New Zealand ITQ approach was certainly bold. Although economists have widely hailed this initiative, it is premature to judge its success. There is some evidence that the system as initially implemented may have been counterproductive in resource conservation terms. The assignment of ITQs as fixed volumes in perpetuity meant that when TACs had to be reduced, government had to purchase quota in the market. Government did not in fact do this. Hence, TACs were not reduced when needed. Because the cost of purchasing the necessary quota would have been very high, government decided to change to a system based on percentages of TACs. This is more likely to succeed. One of the problems was that the initial TACs were based on inadequate scientific information. The introduction of fixed volume ITQs in a situation where knowledge of the resource base was limited was a flaw in the design of the ITQ system.

The real question is whether the New Zealand system will achieve the expected improvements in minimizing costs and increasing economic returns. The economic impact of the program is still unclear. Much hinges on the realization of the projected economic benefits in the 1990's.

Even if the New Zealand experience were wildly successful, conditions in many other countries do not favour such an ambitious ITQ system. New Zealand has essentially ignored equity and distributional issues. Elsewhere, these remain key concerns of fisheries managers and their political masters. Many other countries committed to the free enterprise market-driven approach have not embraced ITQs. In the U.S., for example, there is continued widespread opposition within the fishing industry to limited entry. In New England, overfishing and overcapacity are acute. There, the Regional Council has concentrated on indirect controls aimed at regulating the age or size at first capture rather than regulation of fishing mortality. These efforts were doomed to failure.

The 200-mile zones brought tremendous optimism in many countries which acquired large fishing zones. In most cases, that optimism was misplaced because the fisheries management system did not utilize the opportunities created by the new 200-mile regimes.

Overall, the world record in marine fisheries management is not particularly inspiring. The problems of the pre-extension period have continued, in many instances unaltered, into the new era. The EC and some Regional Councils in the U.S., as examples, have shown a reluctance to grapple with the fundamental problems.

The EC adopted maintenance of the status quo as the policy objective. This meant perpetuating overcapacity and excessively high fishing mortalities. Eventually, even as resilient a fisheries production system as the North Sea showed signs of stress from continued excessive fishing pressure. Agreement on the necessary reductions in fishing mortality/capacity is a difficult challenge, given the diverse conflicting interests of the Community's member states.

Under the decentralized Regional Council management system in the U.S., management of the fishing industry was essentially turned over lock, stock and barrel to industry. So far, with some modest exceptions, these Councils have appeared unwilling to address the overfishing/overcapacity/common property problems.

There is also a middle road. Iceland and Australia have made progress on a course somewhere between the laissez-faire policies of the EC and the U.S. and the radical economic-efficiency theology of New Zealand. Recent experience in these countries suggests that the problems of overfishing and overcapacity can be addressed provided the political will is forthcoming to tackle the underlying common property problem, the nub of the fisheries management dilemma.

CHAPTER 21

CONCLUSIONS

"[Sustainable development] means, among other things, passing on to those who follow us renewable resources at least as abundant as those passed on to us."

– Fisheries Minister John Crosbie, January 1993.

1.0 DECADES OF CHANGE

The 1960's, 1970's and 1980's were decades of dynamic change in the management of Canada's marine fisheries. During this period management interventions proliferated.

From the Second World War up until the early 1960's, development and modernization had been emphasized. In the late 1960's, the emphasis shifted to the pursuit of conservation and economic/social objectives. Prior to the mid-1960's, the regulatory regime had been relatively laissez-faire. In the late 1960's, it became evident that major stocks were being threatened by intense fishing pressure, from both the massive build-up of foreign fishing effort (e.g. Atlantic groundfish) and the build-up of domestic fishing effort (e.g. B.C. herring). Domestic harvesting overcapacity was dissipating potential economic benefits.

From 1967 to 1976 major changes occurred in the way Canada's marine fisheries were managed. Domestically, the federal government moved to limit entry into the fisheries, commencing with the Atlantic lobster and Pacific salmon fisheries in 1967 and 1968. By the mid-1970's limited entry licensing had been extended to virtually all major fisheries in Canada. The goal of this form of licensing was to limit fishing effort. Federal officials expected this to reduce fishing pressure on threatened fish stocks and foster a more profitable industry.

The other major change during this period was Canada's push for direct controls on the amount of catch in order to limit fishing mortality. The government had already introduced catch quotas in the B.C. herring fishery, following a ban on fishing during 1968–70. Canada was instrumental in securing agreement in ICNAF in the early 1970's to introduce Total Allowable Catches accompanied by national allocation of these TACs. Because of limited scientific information and international pressure to maximize the amount of fish available for capture, these initial TACs were set too high.

Recognizing that existing international control mechanisms were inadequate, Canada proclaimed a 12-mile territorial sea in 1971 and declared the Gulf of St. Lawrence, the Bay of Fundy and certain areas on the Pacific coast to be Canadian waters. It entered into phase-out agreements with various countries which had fished historically in the Gulf of St. Lawrence. This, of course, was not sufficient. On the east coast, several major groundfish stocks extended to the edge of the continental shelf. On the west coast, Canadian salmon migrated vast distances and were intercepted in fisheries on the high seas. During 1972–77, Canada worked within international commissions such as ICNAF to reduce fishing pressure, in an attempt to arrest the decline in the resource. At the same time it became a major proponent, in the context of UNCLOS III, of extended jurisdiction for coastal states over renewable resources adjacent to their coasts. By 1976, international consensus had been achieved on the concept of a 200-mile exclusive economic zone for coastal states. Recognizing that it would be some time before a new Law of the Sea Treaty text would be finalized, Canada, in concert with many other countries, proclaimed a 200-mile fishing zone effective January 1, 1977.

This was part of a worldwide movement to enclose what had been the global commons. The Canadian government anticipated that jurisdiction over fisheries resources within the 200-mile zone would enable Canada to control the harvest from its fish stocks in a manner which better met its fisheries management objectives.

Resource decline and a market downturn had culminated in a major crisis in the Atlantic groundfish fishery during 1974–76. The federal government stepped in with major financial assistance (close to $200 million over 3 years). The immediate objective following the 200-mile limit was to rebuild fish stocks which had been reduced to low levels. To this end, Canada abandoned the pursuit of Maximum Sustainable Yield (MSY) as a management objective and adopted a more diffuse concept of "best use", which was to incorporate biological, economic and social objectives. The closest Canada came to replacing MSY with a more explicit objective, incorporating both biological and economic considerations, was the adoption of $F_{0.1}$ as the reference level of fishing mortality for setting TACs for most

Atlantic finfish stocks. On the Pacific coast, a constant spawning escapement strategy continued to be pursued for Pacific salmon, leaving the residual available for harvest.

The 200-mile limit permitted stricter controls over foreign fleets fishing in the Canadian zone. The Canadian fleet on the Atlantic coast was faced with conservative TACs in a series of Atlantic Groundfish Fishing Plans aimed at rapid stock recovery. Meanwhile, several Atlantic provinces and large fishing companies pushed for an aggressive fleet expansion program to take advantage of the resources previously exploited by foreigners. The federal government resisted, arguing that the existing fleet was capable of harvesting the likely available catch of traditional groundfish species. Strict controls largely held the line on the number of units in the offshore fleet. However, effective fishing effort increased somewhat because of technological improvements in fishing vessels, fish-finding equipment and fishing gear. Because of delayed implementation of limited entry in certain sectors of the groundfish fishery, and "leakage" once measures were adopted, it was less successful in constraining harvesting capacity of the inshore and midshore fleets.

In British Columbia, the federal government attempted to moderate the natural fluctuations in Pacific salmon resource abundance through a major Salmonid Enhancement Program. This aimed at doubling the production of B.C. salmonids. Meanwhile, limited entry controls introduced in 1968 (the so-called Davis Plan) had failed to constrain growth in fleet capacity. By 1980, the B.C. fleet was significantly larger than required to harvest the available resource economically.

First on the Atlantic coast, and later on the Pacific, major battles occurred over the allocation of access to a limited resource. This started as a fight over how to divide a shrinking pie but later degenerated into bitter disputes over how to share future growth.

In the euphoria generated by the 200-mile limit, processing capacity on the Atlantic expanded considerably through debt financing. When the recession of the early 1980's hit, the fishing industry on both the Atlantic and Pacific coasts was vulnerable and plunged into crisis. The federal government appointed a Commission of Inquiry into the Pacific Fisheries and a Task Force on Atlantic Fisheries. These reported in 1982 and 1983. Most of the recommendations of the Atlantic Task Force were adopted by the federal government, including its proposed objectives for fisheries policy. During 1983 and 1984, the federal government restructured the five large fishing companies on the Atlantic coast into two giants, Fishery Products International and a revamped National Sea Products. This cost the federal treasury nearly $300 million.

Pearse's proposals for a major revamping of the licensing system on the Pacific coast through introduction of 10-year quota licenses and an auctioning system to determine access were repudiated by the British Columbia fishing industry. Industry proposed a fleet buy-back to reduce the overcapitalized fleet which faced bankruptcy. The federal government proposed a new Pacific fisheries policy involving quota licenses. This died with the change of government in 1984. An upward surge in British Columbia salmon abundance in the mid-1980's pushed the fleet overcapacity problem to the back burner where it remains. The question, however, is not whether there will be another crisis in the B.C. fisheries but when.

On the Atlantic coast, the industry rebounded in response to buoyant market conditions from 1985 to 1988. High prices masked a levelling off in the recovery of the major fish stocks and a downturn in some. The dramatic results of a reassessment of the northern cod stock in January 1989 rapidly dissipated the euphoria of the mid-1980's. A downturn in some other groundfish stocks (e.g. northern Gulf cod, Scotian Shelf haddock) and a massive overcapacity problem in the Scotia-Fundy dragger fleet prompted another spate of special studies (the Stein Task Force on Northern Cod, the Harris Panel, and the Haché Task Force on Scotia-Fundy Groundfish). These studies recommended more conservative TACs for northern cod, a reduction of onshore processing capacity, and measures to reduce fleet overcapacity, particularly in Scotia-Fundy.

The federal government responded in May 1990 with the Atlantic Fisheries Adjustment Program. The government committed $597 million over 5 years to bolster scientific research, increase enforcement capability, assist workers displaced by plant closures, and support development of alternative industries in one-industry fishing communities whose survival was threatened by plant closures.

And then in 1992 new scientific advice indicated that there had been a sudden, drastic and unexpected decline in the abundance of the northern cod stock during 1991. This triggered a 2-year moratorium on fishing for northern cod by Canadian fishermen, commencing in July, 1992, and lasting until the spring of 1994. The federal government again stepped in with massive financial assistance (in the order of $800 million) to assist fishermen and plant workers affected by the northern cod closure and to facilitate a reduction in harvesting and processing capacity before the fishery reopened.

Later in 1992 the Canadian Atlantic Fisheries Scientific Advisory Committee reported that certain other Atlantic cod stocks, particularly in the southern Gulf of St. Lawrence and on the Scotian Shelf, were at a low level. CAFSAC advised substantial reductions in TACs for 1993. The government responded with draconian reduction in TACs for these stocks. It abandoned the gradualist approach and adopted an immediate move to $F_{0.1}$. The government also introduced tough measures to reduce the capture and discarding of small fish. This deepened the most recent crisis in the Atlantic groundfish industry which had commenced in 1989.

Given these events and, in particular, the boom-and-bust pattern evident in three major crises on the Atlantic coast in the past 20 years, and at other intervals in the Pacific fisheries, one may well question the effectiveness of Canada's management of its marine fisheries. It is difficult to assess the degree of success because of (1) the lack of a consistent, weighted framework of specific management objectives, and (2) the wide diversity of fisheries and circumstances involved.

The three Atlantic coast crises were due to different combinations of factors. The 1974–1976 crisis resulted from a resource decline due to foreign overfishing combined with a market downturn. That of the early 1980's was primarily the result of debt-financed over-expansion which crippled the industry when interest rates rose to 20% plus. The 1989–1993 Atlantic crisis was the result of a reduction in TACs due to a decline in some stocks, an initial reassessment of the status of the major northern cod stock and then a sudden, drastic and unexpected decline in northern cod abundance in 1991, probably due to environmental factors. This was combined with further growth in onshore processing capacity in the mid-1980's and continued fleet over-capacity in some areas. Severe recessionary conditions in Canada in the late 1980's — early 1990's also contributed to this crisis.

The one common feature of these crises was the fishing industry's inability to withstand downturns, whether precipitated by resource, market or financial factors. A major contributing factor to this vulnerability is the continued trend toward overcapacity in both the harvesting and processing sectors. Excess capacity is a feature of most fisheries. This overcapacity has its roots in the multiple conflicting objectives of fisheries management and the common property character of the resource.

2.0 CONFLICTING OBJECTIVES

In 1976, Canada, like the United States, embraced the concept of optimum yield, or "best use" as it was called in the Policy for Canada's Commercial Fisheries, as the objective of fisheries management. "Best use" was intended to encompass biological, economic and social considerations. This was to replace the traditional reliance on Maximum Sustainable Yield as "the" objective of fisheries management. MSY had come under assault, particularly by fisheries economists who argued that the appropriate management objective was Maximum Net Economic Yield (MEY). Both MSY and MEY were subsumed in the diffuse optimum yield concept.

In 1982, the Task Force On Atlantic Fisheries proposed an explicit, ranked set of objective for Atlantic fisheries policy which was endorsed by the government of the day (Kirby 1982). These objectives, in order of priority, were: (1) economic viability; (2) maximization of employment with an acceptable income constraint; and (3) Canadianization of the fishery. Pearse (1982) promoted economic rationalization but emphasized resource conservation as the paramount federal government fisheries obligation. The 1985 Study Team on Natural Resources recommended reemphasizing conservation and enhancement of fisheries resources in the wild as the chief objective of fisheries management (Canada 1985a).

Many observers of the fisheries scene have argued that, in the absence of a clear, systematic, hierarchial set of objectives, there can be no rational fisheries policy. The reality is that in Canada, as elsewhere, fisheries managers are faced with multiple, conflicting objectives which shift over time. Official objectives tend to be stated as a general framework, incorporating biological, economic and social dimensions, with the balance and relative weights attached to component subobjectives shifting from time to time and from fishery to fishery.

Considerable lip service has been paid in recent decades to the economic rationalization/viability objective. In reality, however, social factors have played a large role in shaping fisheries policy. Maximizing employment within a constraint of "reasonable" incomes from fishing has been a dominant consideration. The tug-of-war between economic and social concerns has waxed and waned.

There has, however, been general agreement that resource conservation should have precedence over economic and social considerations when there is a threat to the future of the resource. This is not because fishermen, managers or politicians are ichthyocentric but because conservation is like motherhood — nobody can be against it. Conservation considerations led to the July, 1992 decision to impose a 2-year moratorium on fishing for northern cod, despite the profound social and economic impacts.

When the survival of the resource is not at stake, economic/social factors will often take precedence over achieving some arbitrary biological reference point, whether it be MSY, $F_{0.1}$, or something else. This was the case when the TAC for Gulf snow crab was increased to lengthen the season and provide more weeks of work for processing plant workers. Another example is the attempt in Atlantic Groundfish Management Plans in the late 1980's to minimize the disruptive effects of dramatic adjustments in TACs suggested by the stated objective of fishing at $F_{0.1}$, by applying the so-called 50% rule to cushion the effects that would flow from sudden downward adjustments in TACs. This gradualist approach was abandoned in 1992 when the government decided to implement draconian TAC reductions for 1993, to arrest the decline in certain key stocks and facilitate stock rebuilding.

Attempts to forge a clear, national, hierarchial set of fisheries management objectives, with relative weights to the various components, are probably doomed to failure in a country with fisheries as diverse as those in Canada.

Ultimately, society through the political system determines the objectives for fisheries management. Thus, the interpretation of "best use" is subject to change. What is "best" will vary from time to time and from fishery to fishery. National objectives for fisheries management are likely to remain vague. This does not mean that we should abandon all efforts to develop clear, explicit, ranked and measurable objectives for fisheries management. It does mean, however, that such efforts are more likely to succeed if they are fishery-specific. Sophisticated tools exist to assist managers in this process (e.g. Healey 1984). These should be tested to determine the optimum yield for particular fisheries. The weighing of biological, economic and social factors will vary from fishery to fishery.

3.0 HAVE MANAGEMENT OBJECTIVES BEEN MET?

3.1 Conservation Objectives

In terms of conservation, Canada's track record for marine fisheries has been generally good. The *Fisheries Act*, which provides the legal basis for most of the fisheries management activities of the Department of Fisheries and Oceans, refers explicitly to "conservation and protection" (Section 43). Almost all of the Fishing Plans which have been developed in the past couple of decades or so refer to conservation and protection as a primary objective or basic principle but very few define "conservation".

Since DFO has never formally adopted a definition of conservation, it is useful to examine the definition adopted by the Department of the Environment in 1986, based on the UNEP World Conservation Strategy. It defined *conservation* as:

> "That aspect of renewable resource management which ensures that utilization is sustainable and which safeguards ecological processes and genetic diversity for the maintenance of the resources concerned. Conservation ensures that the fullest sustainable advantage is derived from the living resource base and that facilities are so located and conducted that the resource base is maintained".

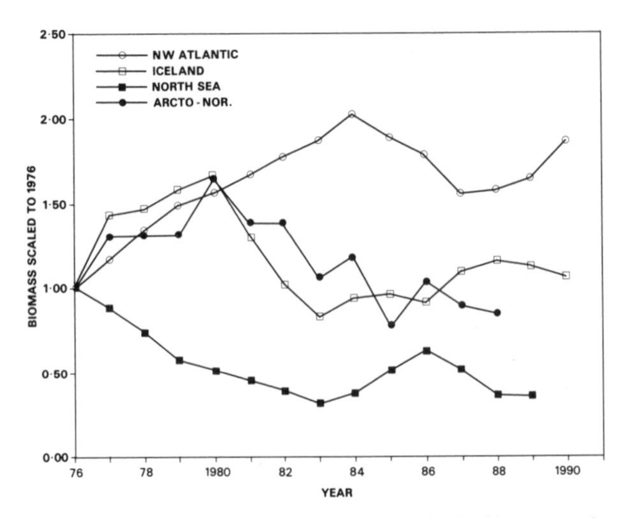

FIG. 21-1. Comparison of relative trends in the biomass estimated for the cod stocks in the Northwest Atlantic with that for cod stocks around Iceland, in the North Sea and in the Arcto-Norwegian area (from Rivard and Maquire 1993).

Conservation is sometimes viewed as synonymous with preventing overfishing. As I explained earlier (Chapter 3), generally two types of overfishing are distinguished — *growth overfishing and recruitment overfishing*. Both growth overfishing and recruitment overfishing have sometimes been used as diagnostic tools to measure the degree to which fishery management plans meet conservation objectives. I believe this defines "conservation" too broadly. Growth overfishing is really an economic rather than a biological or resource concern. It is more appropriate to evaluate whether the conservation objective has been met in terms of recruitment overfishing or, more specifically, whether the survival of a stock or group of stocks is threatened.

Viewed in this light, Canada's management efforts have generally, but not always, prevented recruitment overfishing. The most likely exceptions are three haddock stocks on the Scotian Shelf and Georges Bank and certain anadromous stocks on both the Atlantic and Pacific coasts. Compared with most other countries, Canadian marine fisheries management has been relatively successful in meeting the conservation objective(s), even though the latter has been poorly defined (M. Sinclair, DFO, Halifax, pers. comm.).

As an example, the rebuilding of Atlantic cod stocks in Canada during the 1980's contrasts sharply with the trends in cod stocks in the Northeast Atlantic. The major Northeast Atlantic cod stocks have shown either fairly consistent declines or, at best, variations with no trend during this period (Fig. 21-1). Also the cod and other groundfish stocks off New England declined dramatically in the post-extension era (see Chapter 20). Stocks having faster individual growth rates and younger ages of maturity, would be expected to increase at a faster rate than cod in Canadian waters and to respond faster to reductions in fishing pressure. The fact that these stocks experienced fishing mortality levels much higher than those exerted on cod stocks in Canadian waters may explain, at least in part, the differences in stock trends (Rivard and Maguire 1993).

Although specific management targets, e.g. the $F_{0.1}$ reference fishing mortality for Atlantic groundfish, have been exceeded in many instances, there are only a few examples where the survival of a marine fish stock in Canadian waters has been jeopardized by fishing. Nonetheless, there is sometimes a perception of failure because of the divergence between expectations and "reality", e.g. the reassessment of Northern Cod in 1989. This divergence is a feature of the inherent inadequacies of the scientific data base and existing models.

In June and July, 1992, the NAFO Scientific Council and CAFSAC concluded that there had been a sudden, drastic and unexpected decline in the abundance of northern cod during 1991. The total biomass had been reduced by half and the spawning stock biomass by three-quarters. The spawning stock biomass, estimated to be in the range of 48,000 to 108,000 tons, was near or at the lowest level ever observed and probably only 10% of the average spawning biomass from 1962 to 1990 (Fig. 13–18). The Canadian government reacted swiftly to the CAFSAC advice that catches be kept to the lowest level possible by imposing a 2-year moratorium on fishing for northern cod. It was hoped that, by protecting the 1986 and 1987 year-classes from fishing, this would rebuild the spawning stock rapidly and permit resumption of fishing at some level in 1994. This was probably the most dramatic conservation action in Canadian fisheries history.

It appeared that the sudden, unexpected decline in northern cod during 1991 was the result not of high fishing mortality, but rather abnormal environmental conditions (e.g. extremely cold water, extensive ice coverage) which, through mechanisms not yet understood, may have led to an abrupt increase in natural mortality in the early 1990's. Research is underway to clarify whether the decline was as sudden and abrupt as had been suggested by CAFSAC and the NAFO Scientific Council and to determine the cause of the decline.

In the early 1990's several key Atlantic groundfish stocks were at a very low level. It will probably be several years before these stocks rebuild to more satisfactory levels even if there is a return to environmental conditions which favour higher levels of stock productivity. If environmental conditions remain unfavourable, stock rebuilding is likely to take even longer.

3.2 Economic Objectives

Since the late 1960's, Canada has placed considerable emphasis upon creating a more economically efficient fishing industry. Although economic objectives have rarely been clearly articulated, from the late 1960's onward they have been influential in shaping fisheries policy and management initiatives. Canada's track record in meeting economic efficiency/viability/profitability objectives has been less impressive than fisheries managers would have desired.

The management tool favoured by economists in the late 1960's- early 1970's to achieve a more efficient fishery was limited entry licensing coupled with restrictions on vessel capacity. Canada adopted limited entry licensing for most of its major fisheries during the period 1967 to 1973.

Despite numerous attempts to bring capacity more in line with the available resource, excess capacity continues to be a problem in many Canadian fisheries. This failure to meet economic objectives in many instances is rooted in the common property nature of fisheries resources. The race for the fish has led inexorably to excess capacity in most fisheries. The use of limited entry licensing in Canada to curb overcapacity and the tendency to overinvest has met with a mixture of success and failure. By the late 1970's, it was clear that limited entry licensing and capacity controls were not having the desired impact in constraining or reducing fleet overcapacity. Based on the B.C. Salmon licensing experience, many analysts of fisheries management became increasingly pessimistic about the utility of limited entry licensing as a means to curb overcapacity. This view is supported by the experience in the inshore-nearshore groundfish fishery on the Atlantic coast, particularly the overcapacity problem in the southwest Nova Scotia dragger fleet.

The experience in the Atlantic lobster fishery has been more positive. In this fishery, limited entry licensing has helped by constraining additional entry during a resource resurgence. Similarly, limited entry licensing and stringent vessel replacement controls restrained the growth of capacity in the offshore groundfish fleet on the Atlantic coast up until 1984 when a 5-year trial of enterprise allocations was introduced. Thereafter, the larger offshore companies reduced their harvesting capacity.

Many of the perceived shortcomings of limited entry can be traced to the fact that it is usually introduced late in the game, after the overcapacity problem has become nearly uncontainable. One of the fundamental challenges is what to do in such overcapacity situations. Licence buy-back schemes have been tried in certain fisheries. The Atlantic Lobster Licensing Buy-Back

Program in the early 1980's was successful in reducing the number of participants in the Atlantic lobster fishery. The Atlantic Salmon Licence Buy-Back program in 1984, 1985 and 1986 succeeded in reducing the number of active participants in the Atlantic salmon fishery during a resource crisis. A more extensive Atlantic Salmon Licence Buy-back program implemented in Newfoundland and Labrador in 1991 and 1992 coupled with a closure of the commercial fishery for a least 5 years, reduced the number of commercial salmon licences by 86%. Properly designed buy-back programs remain a viable option for fisheries where significant overcapacity exists.

Two other approaches in the form of output control, taxation and ITQs, have been suggested as more effective rationalization schemes. Although advocated by some economists, taxes or royalties have found little or no acceptance in practice as a fisheries management tool. During the 1980's, the concept of individual quotas was widely debated and tested in several major fisheries on Canada's Atlantic coast, and, to a lesser extent, on the Pacific coast. Based on that experience, it seems that individual quotas are an effective management tool to foster economic viability in some fisheries. Reduction of gluts, the landing of better quality fish, the reorientation from a volume-driven to a market-focused fishery, and fleet rationalization (i.e. reduction) are all tangible benefits which are being achieved in Canadian experiments with individual quotas.

Individual quota management seems best suited to foster economic viability in fisheries where:

1. The resource is relatively stable.
2. The number of enterprises is relatively small, perhaps tens or a hundred participants rather than thousands.
3. The number of landing points is relatively small and easily accessible to enforcement personnel.
4. There is formal organization of the enterprises into effective associations than can speak for, and negotiate on behalf of, the members.
5. The participants recognize the negative effects of "the race for the fish" under open access and are willing to try new approaches; and
6. There is a voluntary commitment to comply with an individual quota regime and to assist in its enforcement.

Individual quotas will not work in all fisheries. Nonetheless, they are a promising tool to counter the incentive for a fisherman to maximize his share of the catch and hence build bigger and better boats to attain it. Like all fisheries management tools, individual quotas have drawbacks, the most serious of which is the incentive to cheat. Hence, there is a need to design adequate and cost-effective mechanisms to achieve compliance.

3.3 Social Objectives

The fisheries are a vital source of employment and income in approximately 1500 coastal communities. Although social objectives for fisheries policy have rarely been clearly articulated, social considerations have often strongly influenced fisheries management decisions.

In many instances, community survival is closely linked to the fate of the fishery in a particular region. Closing a fish plant can put hundreds of people out of work and, in one-industry towns, threaten a community's future. It is for this reason that rationalization of excessive onshore processing capacity was bitterly opposed by community groups in the Atlantic fisheries restructuring debates of the early 1980's and again in the 1989–1993 Atlantic fisheries crisis.

Not only does the fishery provide jobs. It is the "ticket of entry" to the unemployment insurance social safety net. Fisheries policy and management initiatives have often been called upon to serve social goals of maximizing employment and supporting community maintenance in areas where there are few alternative employment opportunities. This has often distorted management initiatives to conserve stocks or improve the economic viability of particular fisheries. Numerous regional economic development programs and initiatives have been undertaken over the past several decades in an effort to diversify the economy of the Atlantic provinces and reduce regional disparity. Regrettably, most of these programs have failed to foster meaningful economic diversification. This has generated a continuing demand that the fisheries support those who are unable to find alternative employment within the region. This demand has resulted in too many fishermen chasing too few fish. It has also created a self-perpetuating cycle of low or poverty-level incomes in fishing communities. Reliance on the fisheries as the employer of last resort has stymied attempts to achieve economic viability in particular fisheries.

3.4 Stability

Although stability has never been explicitly adopted as a Canadian fisheries management objective, it was a recurrent theme through the 1970's and 1980's. The 1976 policy for Canada's Commercial Fisheries suggested that "the fishery economy of the future would be a vigorous and stable one" (DOE 1976a). The 1981 Policy for Canada's Atlantic Fisheries in the 1980's stated: "In the 1980's resource harvesting policies will aim at increasing the economic viability and stability of the Atlantic commercial fisheries" (DFO 1981a). It suggested that setting TACs at or below the $F_{0.1}$ level would provide for "stock rebuilding, larger average size of fish in the catch, improved catch rates and greater stability of catches." The first Annual Report of DFO for 1979–80 stated:

> "During the year under review the department continued to concentrate its efforts on achieving stability and increasing viability in the fishing industry across Canada" (DFO 1980h).

The theme of fostering and achieving stability was clearly a major goal in the decade following extension of jurisdiction.

It is clear in hindsight that attempts to achieve stability were doomed to failure. While there can be some relative stability, fisheries are prone to fluctuations because of diverse, complex factors which lie beyond the control of fisheries managers. Over the first post-extension decade, it was widely assumed that stable TACs and catches could be achieved. Thus, downturns in certain stocks came as a major shock to certain segments of the industry in the latter part of the 1980's and the early 1990's. Such fluctuations are, however, but one feature of the fisheries management dilemma.

4.0 THE FISHERIES MANAGEMENT DILEMMA

The crux of the fisheries management dilemma can be found in the nature of the fisheries system which encompasses the resource, the harvesting, processing and marketing sectors, and the people involved in all facets of the fishing industry. This system is characterized by:

1. Natural resource variability, often environmentally induced;
2. The common property character of the resource which leads to overcapacity in both harvesting and processing;
3. Fluctuations in market conditions similar to, but not necessarily in tandem with, resource fluctuations;
4. Heavy dependence on the fisheries in isolated coastal communities with few alternative employment opportunities;
5. Heavy dependence on government support programs such as Unemployment Insurance;
6. Tremendous diversity among fisheries;
7. Recurrent conflict among competing user groups.
8. Conflicting objectives for fisheries management (see above discussion);
9. The migratory nature of fish stocks necessitating bilateral and multilateral management of transboundary stocks; and
10. The necessity to manage despite uncertainty in scientific advice.

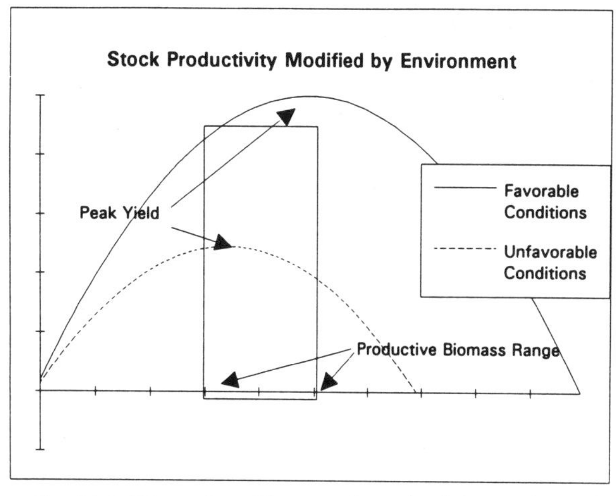

FIG. 21-2. Illustration of how stock productivity can be modified by environmental conditions (courtesy of Dr. W.G. Doubleday, DFO, Ottawa).

4.1 Resource Variability

In the increasing attempts to manipulate or manage fisheries over the past three decades, managers have relied upon scientific models. These models generally assume steady-state equilibrium conditions, conditions which in fact are rarely found in fisheries systems. The role of natural resource variability was generally underestimated or downplayed as scientists strived to provide simple answers to complex questions. Although it has long been known that large-scale fluctuations in fish stocks can occur because of environmental factors, there is now increasing recognition that environmental perturbations can overshadow the effects of fishing (see Chapter 18). Attempts to stabilize yield through regulation of fishing will often be foiled by natural events beyond man's control.

A recent dramatic example is the sudden, drastic and unexpected decline in the abundance of northern cod during 1991. While no single factor was identified as the cause of the decline, the primary factors suggested were ecological. Extreme environmental conditions in 1991 (the worst ice coverage and the coldest ocean temperatures off northeast Newfoundland-Labrador in 30–40 years) were thought to be the primary factors contributing to the abrupt decline.

In Chapter 3, I described the simple general production model, which has been the chief underpinning of fisheries theory for decades. This assumes that the environment is stable or fluctuating without trend. In reality, changes in environmental conditions can have a profound impact on the abundance of marine fish populations. Stock productivity can be substantially modified by the environment (Fig. 21-2). Under favourable conditions, a stock can be more productive at a higher stock size. There is a range of stock sizes which takes best advantage of varying conditions. A prudent fisheries management strategy is to build up a high biomass as a buffer against periods of unfavourable environmental conditions.

4.2 The Common Property Problem

The effects of the common property nature of fisheries resources are well documented. The race for the greatest share of the available pie leads inexorably to excess capacity in harvesting and processing. Fisheries managers around the world are struggling to find ways to remove this incentive for destructive competition. Canadian attempts to address this problem through limited entry licensing have met with only limited success. Recent experiments with individual quotas offer greater promise of success.

Onshore processing falls under provincial jurisdiction with provinces competing for the available resource in the same way that individual fishermen do. They want to secure the maximum economic and employment benefits for their own people. Hence, they have been slow to halt the build-up in excess onshore processing capacity.

On the Atlantic coast, two major surges in processing capacity occurred in the late 1970's and in the mid-1980's. Only recently have some provinces, e.g. Nova Scotia, even begun to licence processing operations. Attempts to reduce excess processing capacity have so far foundered on the shoals of provincial rivalries and imperatives to maximize employment and maintain existing settlement patterns.

Canada's fisheries are heavily export-oriented and dependent on conditions in foreign markets, particularly the United States. While generally trending upwards over the past couple of decades, market demand for fishery products fluctuates. When market conditions are favourable, e.g. from 1985 to 1988, the fishing industry tends to overinvest, forgetting previous downturns. When market corrections occur or market demand stagnates or decreases because of general economic conditions, the effect is felt throughout the fisheries system.

Excess capacity makes the fisheries system particularly vulnerable to both resource and market fluctuations, general recessionary conditions and shifts in national monetary policy. The dependence on the fishery to provide employment and to maintain communities in isolated coastal regions exacerbates this vulnerability. It is, therefore, not surprising that there is a heavy dependence on government support programs even in so-called "normal" times and particularly in times of crisis.

The results of this fisheries management dilemma have been a recurrent boom-and-bust pattern, recurring crises, and recurring demands for government financial assistance. Governments have repeatedly stepped in with financial assistance to minimize the impact of increased unemployment, plant closures and communities thrown into turmoil. Often the full range of social and economic policy instruments of the federal government are called upon to prevent a crisis from becoming a social and economic catastrophe. What to do under these circumstances becomes a major government policy issue and can no longer be confined to fisheries management in the traditional context.

4.3 Reconciling Conflicting Interests

Another major feature of the fisheries system is the all-pervasive nature of conflict. Group is often pitted against group in the struggle for the greatest share of the available resource. Values and beliefs over how the resource should be managed and shared frequently clash. Conflict among the multiple objectives of fisheries management is predominant, followed by disputes over access to and allocation of limited fish resources.

A major component of conflict is the differing values held by various groups involved in the fishery. Examples are disputes between natives and other user groups in the Pacific salmon fishery and between recreational and commercial users nationally. Because DFO has been allocating the opportunity for wealth and the right to earn a livelihood from fishing, it has continually had to reconcile these conflicting interests.

In matters involving fisheries a multiplicity of groups speak with a thousand voices rather than a single voice. To deal with this, the federal government during the 1970's and 1980's created an elaborate consultative framework to bring these differing views to the same table. Sometimes these processes resulted in consensus on advice to the department or the Minister. On many occasions, consensus could not be achieved; the Minister had to arbitrate between opposing viewpoints. Where consensus was achieved, this helped to shape fisheries policy.

Fisheries managers, both bureaucrats and politicians, cannot expect to be loved. The task of balancing diverse and conflicting interests is frequently a thankless one. Absolute consensus is not a realistic goal in fisheries management. As more people and interest groups become involved in the formulation of fisheries policy, the greater is the likelihood that at least some of them are going to be disappointed by a specific decision. The goal is not to eliminate conflicting views about the appropriate course of action in any given situation but to manage conflict constructively. While the government can suggest dimensions for possible compromise, ultimately the goodwill of the users themselves determines the possibility for compromise and accommodation of conflicting interests.

In an attempt to provide better mechanisms for reconciling the conflicting interests of groups involved in the fisheries, the Canadian government is in the process (1993 and 1994) of making significant modifications to the institutional structures for decision-making on licensing and resource allocation matters and for formulating recommendations on conservation measures. The proposed new Atlantic and Pacific Boards to deal with licensing and allocation matters will distance micro-level decision-making on licensing and allocation matters from the Minister. These Boards, proposed to be established under new legislation, would have the authority to make licensing and allocation decisions, within a framework of policy and conservation directives issued by the Minister of Fisheries and Oceans.

Also, in December 1992, Fisheries Minister Crosbie announced the formation of a Fisheries Resource Conservation Council for the Atlantic Fisheries. The purpose of this Council, consisting of DFO scientists, external scientists and persons knowledgable about the fishing industry, was to integrate industry experience and scientific expertise by involving industry more directly in the scientific process. It was envisaged that the new Council would hold public hearing to review DFO stock assessments and solicit industry views on proposed conservation measures. Recommendations to the Minister on conservation measures would come in the future from this new Council rather than directly from DFO scientists and fisheries managers. The Minister would retain his decision-making role on conservation matters. The Council was to begin with Atlantic groundfish and later extend its operations to other finfish and invertebrate species.

4.4 Migratory Stocks — The International Dimension

The proclamation of a 200-mile fisheries zone in 1977 brought valuable fish resources under Canadian jurisdiction. However, important stocks of interest to Canada straddle the 200-mile limit on the Nose and Tail of the Grand Banks and migrate into the waters of adjacent coastal states. Canada led the way in putting in place a new international mechanism to manage stocks beyond its 200-mile zone on the east coast, i.e. NAFO. Despite this, overfishing of stocks beyond 200 miles became a matter of increasing concern in the late 1980's.

Initially, Canada's international fisheries policy post-extension focused on securing both conservation and market benefits for Canada from the allocation of surplus fish in the Canadian zone. There were attempts to negotiate long-term arrangements for improved market access in return for assured allocations in the Canadian zone. By the early 1980's, it became evident that these initiatives were of dubious benefit. The problem of foreign overfishing intensified with the accession of Spain and Portugal to the European Community in 1986.

Canada increased its efforts to bring the foreign overfishing problem under control by attempting to persuade the European Community to accept and implement NAFO TAC and national quota allocations. A concerted Ministerial-level campaign to persuade the Community to change its approach began to yield some benefits by mid-1992. The European Community withdrew its fleet from fishing northern cod after Canada imposed a 2-year moratorium in its own waters. At the September 1992 NAFO meeting, agreement was reached on TACs and national allocations for 1993. Then, in December 1992, Canada and the EC initialed an agreement which, when ratified, could resolve the dispute with the EC about its fishing outside 200 miles on the Nose and Tail of the Grand Banks. The problem of fishing by non members of NAFO remained. Canada continued to pursue a legal initiative, under the auspices of the United Nations, to develop an improved framework for the management of fisheries on the high seas beyond coastal state jurisdiction. Continued uncertainty about the evolution of the international fisheries management regime in this area introduces another element of instability into the Atlantic fisheries management picture. Continuing declines in the abundance of key transboundary stocks have emphasized the need for a new approach to the management of straddling stocks.

Canada's attempts to negotiate maritime boundaries and transboundary fisheries management agreements have met with only limited success. Canada's fisheries relations with Denmark (Greenland), France and the United States in the post-extension era have demonstrated that the goal of cooperative management of transboundary resources is more elusive than the drafters of the 1982 Law of the Sea Convention envisaged. Where the fisheries management systems of two countries differ radically, as with Canada and the United States on the Atlantic coast, this task is even more difficult. Because fish migrate across man-made boundaries, these migrations will continue to complicate the task of fisheries management.

4.5 Science and Management

4.5.1 Managing under Uncertainty

The complex fisheries management system that has evolved over the past several decades depends heavily on fisheries science. Effective management requires timely and reasonably reliable scientific estimates of fish abundance.

The role of fisheries science is to project the likely effects and interactions of particular management measures so that the management agency can make an appropriate choice of measures to meet its objectives. Significant advances have been made since the 1950's in scientific understanding of the marine ecosystem, fish stock distribution and abundance, and in modelling the effects of fishing on fish abundance. Increasing demands are being placed upon scientists to generate routine advice for the year-to-year adjustment of catch quotas or prediction of the size of spawning runs of salmon, and to take into ecological and environmental trends in forecasting fish abundance and availability.

The 200-mile limit facilitated the introduction of a more conservative management regime which assisted stock recovery. When stocks and allowable catches were on the increase, the scientific advice tended to go largely unchallenged, with some local exceptions. In the late 1980's, however, the post-extension stock recovery on the east coast levelled off and some stocks experienced dramatic downturns. The scientific perception of the status of other stocks also changed quite radically (e.g. northern cod). The models being used tended to overestimate stock abundance and underestimate fishing mortality. Also, environmental changes were affecting the productivity of fish stocks, e.g. recruitment, growth rates, in ways that could not be foreseen.

Scientific assessments of fish stocks and advice on resource management have come under increasing criticism in recent years. This has been manifested in denunciations by the fishing industry of the scientific advice and, in particular, criticism of scientist' failure to consider fishermen's observations and knowledge in the resource assessment process. It is not uncommon for fishermen and other sectors of the fishing industry to criticize the validity of the scientific methodology.

A changed perception of stock status, as the result of improvements in the data base or improved analytical techniques, is an integral part of the scientific process. Collection of new data and development of new techniques will often invalidate previously held hypotheses. The accuracy of projections by Canadian scientific advisory bodies is no worse, and probably no better, than that of other advisory bodies such as ICES. In all cases the predictive accuracy which can be attained is low in relation to the demand for accuracy by the present management system.

The clients of fisheries science must recognize that large fluctuations in calculated catches at a particular target level of fishing mortality, e.g. $F_{0.1}$, will continue to occur. The solution lies in devising ways to deal with these fluctuations through the management planning process. Reduction of excess capacity in the harvesting and processing sectors would make such changes in catch levels less stressful for the industry. Managers must, however, be prepared to adjust fishing plans rapidly if there is evidence that current levels of fishing are excessive and will thwart the achievement of long-term objectives.

Another aspect of scientific credibility is a major communications gap between fishermen and scientists involved in resource assessment, and sometimes between scientists and fisheries managers. Recent initiatives to increase the input of fishermen to the scientific assessment process should be encouraged and expanded. These include: monitoring the performance of index fishermen, meetings with fishermen to discuss the results of the fishing season and assessment results prior to finalizing scientific assessments, and personal contact between fishermen and scientists. The new Fisheries Resource Conservation Council for the Atlantic fisheries will provide the opportunity for a more meaningful dialogue between DFO scientists, external scientists and the fishing industry. DFO scientific assessments will be subject to scrutiny in public hearings to review stock assessments and proposed conservation measures.

Ensuring adequate communication between scientists and managers is a worldwide problem. In some respects, Canada is ahead of other countries in developing mechanisms to ensure that managers' questions are addressed and that the results of stock assessments are made available to managers in a timely and systematic manner. Inadequate dialogue between the two groups sometimes leads to confusion of roles. The role of the fisheries scientist is to analyze, assess and project the likely impacts of alternative management approaches. It is not to advocate solutions which favour a particular objective considered most important by the scientist. Deciding the management approach is the role of the fisheries manager.

This does not mean that scientists should not act as catalysts for change. The scientist can seek clarification of management objectives and stimulate managers to consider alternative approaches which might have a higher probability of attaining their objectives.

The fisheries manager functions in a complex web of demands for his time, energy and attention. He is bombarded by conflicting views and representations from various interest groups. He often has to rely on inadequate information to make quick decisions with major implications for fishermen or for the resource.

Managers must be sensitive to a variety of interests while also searching for workable solutions to complex problems. The manager deals in a very practical world and is likely to question theoretical or general solutions, recognizing that they are less likely to be useful than highly specific, realistic ones. But, more importantly, managers face uncertainty every day and frequently turn to fisheries scientists seeking certainty. Increasingly, scientists realize that they cannot provide that certainty.

In Canada scientific assessments have long been couched in terms of a specific projection of the likely response of fish stocks to a particular level of exploitation. Scientists have realized for some time that there are large margins of uncertainty surrounding their estimates of stock abundance, fishing mortality and other vital parameters. However, they have been reluctant until recently to make explicit the tremendous uncertainty surrounding these estimates. That reluctance, particularly in Atlantic Canada, was grounded in the experience of the international commissions. When international fisheries managers were offered a range of management options and their likely implications, they inevitably chose the option allowing the highest catch possible in the short term.

In the system which evolved following extended jurisdiction, Canadian scientists and managers agreed that biological advice would be offered for Atlantic marine finfish stocks as one number, corresponding to the catch at a particular target fishing mortality, $F_{0.1}$. This became the standard practice. The tendering of advice in the form of a specific number created an atmosphere of certainty, even though the scientists involved were fully aware that the margin of uncertainty could be considerable. However, a margin of uncertainty around $F_{0.1}$ gave more comfort (at least to the scientists) than a similar margin around F_{max}. As the data bases gradually improved in the decade after extended jurisdiction, it became clear that early estimates of stock abundance were often wide of the mark by a considerable margin, the tendency being to overestimate stock abundance. If the management objective had been F_{max} the consequences could have been catastrophic. In this sense, $F_{0.1}$ worked.

But the Atlantic fishing industry was jolted when the full extent of uncertainties involved was revealed by the dramatic reassessment of the major northern cod stock in January 1989. The uncertainties involved in the northern cod assessment were not unique. Similar instances of radical revision of scientific advice had occurred in the case of several major Northeast Atlantic stocks (see Chapter 20). The irony is that in the latter instances ICES advice had been offered in the form of a range of options for many years, with managers generally choosing "Status Quo" TACs.

Recently (September 1990), the NAFO Scientific Council convened a Special Session on Management under Uncertainty (NAFO 1990c). This brought together scientists from Canada, Denmark (Greenland) the EC, Japan, New Zealand, the USA and the USSR. This session concluded that almost everything about management advice is uncertain to some extent. The effect of those uncertainties on management advice is quite variable.

Analyses presented at this Session indicated that traditional (constant mortality or constant escapement) policies may not be optimal when judged by objective criteria. However, constant low fishing mortality strategies may be close to optimal under a wide range of conditions, so that the $F_{0.1}$ rule of thumb for quota setting is often an appropriate choice in practice.

The NAFO Scientific Council Session focused on how uncertainties of various sorts in assessments could best be portrayed for fisheries managers. One approach suggested would summarize the effects of these uncertainties in terms of risk, i.e. the probability of "something nasty" happening. It was suggested "something nasty" be defined in terms of something immediately comprehensible to managers and the fishing industry, e.g. economic loss, numbers of boats forced out of operation or some other relevant factor. Advice to managers should display the results for a range of risk levels, and not prejudge the acceptable level of risk.

Placing confidence limits around the advised level of catch would also emphasize, for managers, the extent of uncertainty involved. Whether this would lead to management options being selected from the upper range of the confidence interval remains to be seen. The desire to maximize short-term catch would probably lead initially to some risk-prone decisions, particularly if the stock in question appeared robust.

The Session concluded that, according to various criteria, compound management strategies which interpolate between classical strategies (F_{max}, $F_{0.1}$) and "status quo" management are superior and yield lower variability of catch and effort. It proposed that managers seriously consider such strategies especially when current F values are far from the long-term target level. In effect, this had been Canada's approach in the late 1980's with the so-called "50% rule" for adjustment of

Atlantic groundfish TACs. This, however, contributed to the trend of tolerating fishing mortalities in excess of the target of $F_{0.1}$ and was abandoned in 1992.

The explicit recognition of the degree of uncertainty associated with scientific assessments is likely to be useful. The bottom line, however, is that managers must be prepared to take decisions on fisheries issues in the absence of certainty. It is a fallacy to think that effective management can only occur where there is complete and accurate scientific understanding of the fishery being managed. Fisheries science can rarely specify exactly the implications of a particular management strategy.

Honest mistakes will be made. Fish populations are generally sufficiently resilient that corrections or adjustments can be made in time to avert catastrophe. This is not always the case; sometimes resources can be reduced to economic extinction. The failure to take corrective action in such cases usually results not so much from uncertainty as from paralysis in the decision-making process, often because of the tug-of-war between conflicting objectives.

4.5.2 Research Priorities and Interdisciplinary Collaboration

Despite the progress of recent decades, our understanding of the dynamics of marine fish populations, their interactions within the food web and with the marine environment, and the role of man as predator is still weak. The fisheries models which are the foundation of current management approaches generally pertain to steady-state equilibrium conditions when, in reality, marine fish populations and the environment in which they live can be extremely variable. Fish resources will always be subject to natural variability and catches will fluctuate because of this, despite the best efforts to impose order through management interventions.

Improving our methods of abundance estimation must be a research priority. Understanding the relative influence of parental stock size and environmental variability in determining the relative strength of year-classes is another priority. Despite the growing awareness that widespread fluctuations in year-class strength are largely environmentally-induced, additional research into the mechanisms determining recruitment success is necessary. A greater ability to forecast changes in physical oceanographic parameters and the marine environment and their effects on fish populations is also crucial. This will require a substantial research effort.

There are specific instances where a better understanding of multispecies interactions is clearly required. It is necessary, for example, to obtain a better understanding of the interactions among the capelin-cod-seals complex off Newfoundland to guide future decisions on management of this multispecies complex.

Fisheries systems are exceedingly complex, characterized by interactions among the biological, environmental, economic, social and regulatory dimensions. To meet the challenge of understanding even some of these interactions will require much more effective interdisciplinary collaboration than has been the practice in the past.

Since the 1950's economics has also significantly influenced fisheries management policy. Various approaches to address the problems arising from the common property aspect of the resource, e.g. limited entry licensing, individual quotas, were first advocated by fisheries economists and then gradually adopted as management tools.

There is little prospect that many of the more advanced models recently developed by academic economists can be applied by fisheries managers. While mathematically elegant, they require enormous volumes of data which are unlikely to be forthcoming. There must be greater emphasis on applied research and monitoring in fisheries economics.

There has been relatively little research on the social aspects of fisheries. Social impact analysis is relatively rare and sociological research relevant to fisheries management has been extremely limited. Given the prominence of issues such as employment maximization or maintenance, "reasonable" incomes and community dependence in shaping fisheries policy, the paucity of research on the social dimensions of the fisheries is puzzling.

Steps should be taken to improve the capability for such social research, so that fisheries decision-makers can have access to a full array of biological, economic and social information and analyses.

The need for greater collaboration within the natural sciences related to fisheries is being addressed both within government and between government and university scientists. There is a similar need to foster greater collaboration between biologists and economists and to move towards a bio-socio-economic scientific framework of information and analysis for fisheries management.

There is a need to forge new linkages among biologists, economists and other social scientists to bring together the combined efforts of the natural and social scientists to provide for more effective scientific input to the fisheries management process.

Biologists, economists and other social scientists all have a role to play in determining the options for "best use" of fisheries resources. There is a need to array for decision makers the implications of alternative management options. These options have biological, economic and social dimensions. Currently, the economic and social implications are ill-defined in fisheries management planning.

For too long "fisheries science" has been regarded as synonymous with "fisheries biology". This reflects the dominant role biology has played as the scientific basis for fisheries management (Wooster, 1988). It is time to change that perspective. Physical oceanography, fisheries economics, sociology and social anthropology must also be part of a truly integrated approach to fisheries science. This is necessary to improve the scientific basis for fisheries management to meet the challenges ahead.

5.0 THE FUTURE

I have described in general terms some conclusions which emerge from this study of recent Canadian marine fisheries management. Behind these generalizations lies the incredible richness and diversity of Canada's fisheries.

While there has been a recurrent boom-and-bust syndrome and recurring crises, the extent varies among fisheries. Some have enjoyed prolonged periods of prosperity, yielding good economic returns to participants. Many of the shellfish fisheries, for example, have avoided the periodic downturns which have plagued the Atlantic groundfish fishery. But even the shellfish fisheries are subject to variability. The virtual collapse of the Gulf snow crab fishery in the late 1980's contrasts with the dramatic surge in Atlantic lobster catches through the 1980's. The Pacific salmon fishery, which harvests a variable resource, experienced a prolonged period of prosperity in the mid to late 1980's. It faced a market downturn at the beginning of the 1990's. Even within the Atlantic groundfish fishery, there are enormous differences between the relative prosperity of midshore dragger fishermen in southwest Nova Scotia and the poverty of small-boat inshore fishermen in eastern and northern Newfoundland and Labrador.

Significant changes in Canada's marine fisheries will occur over the next several decades. Aquaculture will become increasingly important as we enter the 21st century. There will be growing competition between aquaculture enterprises and harvesters of wild fish resources. Aquaculture operators and other users of nearshore habitat are already in conflict. This may intensify. While aquaculture will undoubtedly comprise a greater share of the value obtained from Canada's fisheries resources, it is unlikely to displace the predominant role of the wild fisheries in the foreseeable future.

The use of individual quotas as a fisheries management tool will probably spread to many other fisheries by the beginning of the next century. Transferable quotas will probably be introduced in a greater number of these fisheries. This offers promise of reducing the common property tendency to invest in excess capacity. But, because individual quotas do not suit all fisheries, more traditional methods will continue to be employed. Individual quotas will only work if more effective means are found to minimize the tendency to cheat, which is the Achilles' Heel of this approach.

Major changes in the institutions and mechanisms of fisheries management are being proposed for implementation in 1993 and 1994. The most significant is the move to distance the Minister from micro-level decision-making on licensing and allocation matters through the proposed establishment of Atlantic and Pacific Licensing and Allocation Boards. Also significant is the initiative to forge a new partnership between scientists and the fishing industry through the establishment of a Fisheries Resource Conservation Council for the Atlantic fisheries. Scientists will now provide assessments of stock status and predict the effects of alternative management options, instead of advising one-number TAC recommendations as in the past for Atlantic groundfish. This means that industry will have to share the responsibility for formulating recommendations on conservation measures and, hence, for its future.

Whatever the shape of these changes, the essential fisheries management dilemma will remain. The combination of natural resource variability, fluctuations in market demand, the common property problem leading to excess capacity, and the continued dependence on the fishery in isolated coastal communities, will result in a repetition of the boom-and-bust pattern. Governments will continue to be called upon for assistance during these crises.

The past three decades were a period of dynamic change in marine fisheries management. The next several decades will undoubtedly also be a period of considerable change and challenge. Change will occur as the result of factors which can only be dimly perceived today. The challenge will be:

(1) to ensure sustainable fisheries;
(2) to foster a more resilient industry better able to withstand the inevitable downturns; and
(3) to lessen dependence on the fishery in the coastal regions of Canada, particularly parts of Atlantic Canada.

These three elements of challenge are interrelated. Sustainability of the resource base is achievable unless global change, e.g. global warming, occurs at a pace and in a manner as yet unforeseen. Assuming that market demand for fishery products will increase, rather than lessen, this means that fisheries can be sustained. This does not, however, mean maintaining the fisheries as they exist today.

To foster a more resilient industry, it is essential to address the problem of excess capacity (both capital and labour). In areas where alternative economic opportunities exist, this may be achieved, at least partially and in some fisheries, through the more widespread adoption of individual transferable quotas. For other fisheries, industry-funded buy-backs of excess capacity may become an option linked to individual quotas. In still other fisheries, this may mean more rigorous enforcement of existing or improved limited entry licensing (e.g. terminal licences) and fleet capacity controls, with a gradual reduction in capacity over time. Because of the diversity of Canada's fisheries, no one solution will suit all circumstances.

It is impossible to devise a uniform set of fisheries objectives and management measures which fit all the circumstances of the Canadian marine fisheries. There is as much difference among the fisheries of Atlantic Canada as there is between the Atlantic, Arctic and Pacific fisheries.

What is required is a national framework of management objectives and policies which can be tailored to the specific circumstances of particular regions and fisheries. Management initiatives must respond to local circumstances and local priorities without violating national precepts and standards. Examples of such national standards are these generally-agreed priorities in resource allocation: resource conservation, native food fisheries, and communities adjacent to the resource. Economic viability will be the dominant objective for certain fisheries, and social objectives will predominate in other areas. Where fish resources are being threatened by habitat deterioration or overfishing, resource conservation will be the main concern. In other instances, resource restoration will be emphasized. Objectives have to be implemented on a fishery by fishery basis.

Economic downturns in the fisheries are often attributed specifically to stock failures, market failures, price increases, ice conditions, weather, etc. The reality is that fishery downturns are the inevitable consequences of an overcapitalized industry which has also assumed the role of employer of last resort and means of access to the social safety net. The fishing industry would be as viable as any other industry if it were not located in rural or remote settings and if the social conscience of the country were not so highly developed.

Because of this, a more resilient fishing industry cannot be achieved through fisheries management initiatives alone. It will be necessary to harness the collective policy instruments of governments, both federal and provincial. The creation of a resilient industry in parts of Atlantic Canada, for example, Newfoundland and Labrador, will mean fewer people employed in the harvesting sector. This will only be achievable if alternative employment opportunities become available. This in turn depends upon diversifying the economy in those coastal regions. Thus, fisheries policy is inextricably linked to social policy and to regional economic development policy. Without alternative economic opportunities, it will be difficult to create a resilient fishing industry able to prosper in good times and survive without government handouts in bad times.

If harvesting and processing capacity can be brought in line with the capability of the resource to sustain economically viable fisheries, the next century could offer considerable promise for Canada's fisheries. Governments and the fishing industry must work together to achieve that brighter future.

REFERENCES

ACFM. (Various years). Reports.

ACHESON, J. A. 1985. Comments. *In* Human organization. J. Soc. Appl. Anthropol. 44: 182–183.

ACHESON, J. M. 1981. Anthropology of fishing. Annu. Rev. Anthropol. 10: 275–316.

AFDC. 1979. Statement of Leigh S. Ratiner Counsel and Washington Representative of the American Fisheries Defense Committee before the Subcommittee on Fisheries and Wildlife Conservation and the Environment of the Committee on Merchant Marine and Fisheries. U.S. House of Representatives. June 22, 1979.

AGAC. 1980. Summary Minutes of the Atlantic Groundfish Advisory Committee Meeting. Montreal, Quebec. December 9, 1980.

AGGER, P., I. BOETIUS, AND M. LASSEN. 1973. Error in the virtual population analysis: the effect of uncertainties in the natural mortality coefficient. J. Cons. int. Explor. Mer 35: 93.

AIKEN, D. 1993. Aquaculture in Canada. *In* L.S. Parsons and W.H. Lear [ed.]. Perspectives on Canadian marine fisheries management. Can. Bull. Fish. Aquat. Sci. (In press)

AKENHEAD, S. A., J. CARSCADDEN, H. LEAR, G.R. LILLY, AND R. WELLS. 1982. Cod-capelin interactions off northeast Newfoundland and Labrador, p. 141–148. *In* M.C. Mercer [ed.] Multispecies approaches to fisheries management advice. Can. Spec. Publ. Fish. Aquat. Sci. 59.

ALEXANDER, D. 1977. The Decay of Trade. An Economic History of the Newfoundland Saltfish Trade, 1935–1965. Newfoundland Social and Economic Studies No. 19. Institute of Social and Economic Research, Memorial University of Newfoundland, Nfld. 174 p.

ALEXANDER, L. M., AND R.D. HODGSON. 1975. The impact of the 200-mile economic zone on the Law of the Sea. The San Diego Law Rev. 12: 569–599.

ALIG, C., L. KOSS, T. SCARANO, AND F. CHITTY. 1989. Control of Plastic Wastes Aboard Naval Ships at Sea. Proceedings of the Second International Conference on Marine Debris, Honolulu, April 1989.

ALVERSON, D. L., AND G.J. PAULIK. 1973. Objectives and problems of managing aquatic living resources. J. Fish. Res. Board Can. 30: 1936–1947.

ALVERSON, L. G. [Chairman]. 1987. A study of trends of cod stocks off Newfoundland and factors influencing their abundance and availability to the inshore fishery. Submitted by the Task Group on the Newfoundland inshore fisheries. 125 p.

ANDERSEN, K. P., AND E. URSIN. 1977. A multispecies extension to the Beverton and Holt theory of fishing, with accounts of phosphorus circulation and primary production. Medd. dan. Fisk. Havunders. 7: 319–435.

ANDERSEN, P., AND J.G. SUTINEN. 1984. Stochastic bioeconomics: a review of basic methods and results. Mar. Resour. Econ. 1(2): 117–136.

ANDERSEN, R. 1978. The need for human sciences research in Atlantic Coast fisheries. J. Fish. Res. Board Can. 35: 1031–1049.

ANDERSON, J. E. 1984. Public policy-making. Third Edition. Holt, Rinehart and Winston. New York, NY. 179 p.

ANDERSON, L. G. 1977. Economic impacts of extended fisheries jurisdiction. Ann Arbor Science Publishers, Ann Arbor, MI.

ANDERSON, L. G. 1977. The economics of fisheries management. John Hopkins Univ. Press, Baltimore, MD. 214 p.

ANDERSON, L. G. 1982. Marine fisheries, p. 149–178. *In* P.R. Portney [ed.] Current issues in natural resource policy. John Hopkins Univ. Press, Baltimore, MD.

ANDERSON, L. G. 1983. Economics and the fisheries management development process, p. 211–227. *In* B.J. Rothschild [ed.] Global Fisheries: Perspectives for the 1980's. Springer-Verlag, New York, NY.

ANDERSON, L. G. 1986. The economics of fisheries management. Revised edition. John Hopkins Univ. Press, Baltimore, MD. 296 p.

ANDERSON, L. G. 1987. Expansion of the fisheries management paradigm to include institutional structure and function. Trans. Am. Fish. Soc. 116(3): 396–404.

ANDERSON, L. G. 1987. Bridging the gap between economic theory and fisheries management: can the MFCMA produce economically rational management? Mar. Fish. Res. 49: 13–25.

ANDERSON, L. G. 1989. Enforcement issues in selecting fisheries management policy. Mar. Res. Econ. 6: 261–277.

ANDERSON, L. G., AND D.R. LEE. 1986. Optional governing instrument, operation level, and enforcement in natural resource regulation: the case of the fishery. Am. J. Agric. Econ. 68: 678–690.

ANNALA, J. H. 1989. Report from the Fishery Assessment Plenary, July 1989: stock assessments and yield estimates. Fisheries and Research Centre, New Zealand Ministry of Agriculture and Fisheries, Wellington, NZ. 158 p.

ANNALA, J. H. 1990. Report from the Fishery Assessment Plenary, May 1990: stock assessments and yield estimates. Fisheries and Research Centre, New Zealand Ministry of Agriculture and Fisheries, Wellington, NZ. 165 p.

ANON. 1952. Texts for the Santiago agreements were reproduced in Resita Peruana de Derecho Internacional, XIV, No. 45/Jan–June. 1945: 104–113.

ANON. 1967. Proceedings Canadian Atlantic herring fishery conference. Can. Fish. Rep. No. 8, Dec. 1966. 268 p.

ANON. 1970. Report of the Task Force — Operation Oil. Vol. I, II, III, IV. Can. Minist. Transp. 1971.

ANON. 1974a. Law of the Sea and International Fisheries Regulation. International Canada. June 1974. 120 p.

ANON. 1974b. Law of the Sea and International Fisheries Regulation: Law of the Sea Conference, p. 129–132. *In* International Canada. July and August 1974.

ANON. 1980. Report of the Canada-EEC Scientific Working Group on Joint Stocks in NAFO Subareas 0 and 1. November 22–25, 1980.

ANON. 1982. Report of the Fleet Rationalization Committee. Vancouver, B.C. 113 p. + app.

ANON. 1986. CAFSAC. Proposals for a 10 year refit. Report of an Ad Hoc Working Group to the ADM, Science. 51 p.

ANON. 1987a. Summary Report of the Nechako River Working Group. Dr. David Strangway (Facilitator).

ANON. 1987b. Ad Hoc Committee for the Conservation of the St. Lawrence Belugas. Statement of problems of the St. Lawrence belugas. 28 p.

ANON. 1988a. Minutes, Georgia Strait Task Group February 1988.

ANON. 1988b. Report of the Scotia-Fundy Groundfish Capacity Advisory Committee. November 10, 1988. 16 p. + app.

ANON. 1989. In the matter of Canada's landing requirements for Pacific coast salmon and herring. Final report of the Panel established under the Canada-USA Free Trade Agreement. October, 1989.

ANON. 1990. Second World Climate Conference. 7 November 1990.

ANON. 1991. Agreement between Felogia Laksaskif and the Committee for the Purchase of Open-Seas Salmon Quotas, Reykjavik, 17 April, 1991.

ANTHONY, V. C. 1990. The New England Groundfish Fishery after 10 years under the Magnuson Fishery Conservation and Management Act. N. Am. J. Fish. Manage. 10: 175–184.

ANTHONY, V. C., AND J.F. CADDY [ed.]. 1980. Proceedings of the Canada-US Workshop on Status of Assessment Science for N.W. Atlantic Lobster (*Homarus americanus*) Stocks (St. Andrews, N.B., Oct. 24–26, 1978). Can. Tech. Rep. Fish. Aquat. Sci. 932: vii + 186 p.

ANTLER, E., AND J. FARIS. 1979. Adaptation to changes in technology and government policy: a Newfoundland example (Cat Harbour), p. 129–154. *In* R. Andersen [ed.] North Atlantic Maritime Cultures: anthropological essays on changing adaptations. Mouton Publishers.

APOSTLE, R. L. KASDAN, AND A. HANSON. 1984. Political efficacy and political activity among fishermen in southwest Nova Scotia: a research note. J. Can. Stud. 10(1): 157–165.

APOSTLE, R., L. KASDAN, AND A. HANSON. 1985. Work satisfaction and community attachment among fishermen in southwest Nova Scotia. Can. J. Fish. Aquat. Sci. 42: 256–267.

APPLEBAUM, B. 1990. The straddling stocks problem: The Northwest Atlantic situation, international law, and options for coastal state action. 23L Sea Int. Proc: 282–317. Alfred. H.A. Soons [ed.].

APPOLLONIO, S. 1983. Fisheries management. Oceanus 25(4): 29–38.

ARGENTINA. 1946. Presidential Decree No. 14708, Concerning National Sovereignty over the Epicontinental Sea and the Continental Shelf, October 1, 1946 Boletin oficial, December 5, 1946.

ARGUE, A. W., R. HILBORN, R.M. PETERMAN, A.J. STALEY, AND C.J. WALTERS. 1983. Strait of Georgia chinook and coho fishery. Can. Bull. Fish. Aquat. Sci. 211: 91 p.

ARMSTRONG, D. W. 1978. Management of fish stocks. Scottish Fish. Bull. 44: 4–7.

ARNASON, R. 1986. Management of the Icelandic demersal fisheries, p. 83–101. *In* Fishery access control programs worldwide: proceedings of the workshop on management options for the North Pacific longline fisheries. Orcas Island, Washington. April 21–25, 1986. Alaska Sea Grant Report No. 86-4.

ARROW, K. J. 1951. Social choice and individual values. Cowles Foundation Monograph 12. John Wiley and Sons, Inc., New York, NY.

ATKINSON, D. B., W.R. BOWERING, AND D.G. PARSONS. 1982. Review of the biology and fisheries for roundnose grenadier, Greenland halibut and northern shrimp in Davis Strait. NAFO Sci. Counc. Stud. 3: 7–27.

ATLANTIC GROUNDFISH MANAGEMENT PLANS. (Various Years).

AUDITOR GENERAL. 1986. Report of the Auditor General of Canada to the House of Commons. Fiscal Year Ended 31 March 1986.

AUDITOR GENERAL. 1988. Report of the Auditor General of Canada to the House of Commons. Fiscal Year Ended 31 March 1988.

AUSTRALIA. 1985. Plan of Management No. 1 Southern Bluefin Tuna Fishery. Cat. No 854820X. Australian Government Publishing Service.

AUSTRALIA. 1986 and 1987. Northern Prawn Fishery Management Plan. Cat No. 8740112. Australian Government Publishing Service.

AUSTRALIA. 1989. New Directions for Commonwealth Fisheries Management in the 1990s. A Government Policy Statement, December 1989. Australian Government Publishing Service. 114 p.

AUSTRALIA. 1991. Background Fisheries Statistics. Australian Fisheries Service. Department of Primary Industries and Energy, Canberra, Australia. May 1991. 42 p.

BAILEY, K. 1981. Larval transport and recruitment of Pacific hake, *Merluccius productus*. Mar. Ecol. Progr. Ser. 6: 1–9.

BAIN, R. 1985. Major Australian fisheries. Infofish Mark. Dig. Canberra, A.C.T. Australia. 14–17.

BARRETT, L. G. 1984. Capital and the state in Atlantic Canada: the structural context of fishery policy between 1939 and 1977, p. 77–104. *In* C. Lamson and A.J. Hanson [ed.] Atlantic Fisheries and Coastal Communities: fisheries decision-making case studies. The Institute for Resource and Environmental Studies, Halifax, N.S.

BARRETT, L. G., AND A. DAVIS. 1984. Floundering in troubled waters: the political economy of the Atlantic fishery and the Task Force on Atlantic fisheries. J. Can. Stud. 19: 125–137.

BATES, S. 1944. Report on the Canadian Atlantic sea-fishery. Halifax, N.S. King's Printer. 179 p.

BEAMISH, R. J. 1976. Acidification of lakes in Canada by acid precipitation and the resulting effects on fishes. Soil Pollut. 6: 501–514.

BEAMISH, R. J., AND D.R. BOUILLON. 1993. Pacific salmon production trends in relation to climate. Can. J. Fish. Aquat. Sci. 50. (In press)

BEAMISH, R. J., AND H.H. HARVEY. 1972. Acidification of the La Cloche Mountain lakes, Ontario and resulting fish moralities. J. Fish. Res. Board Can. 29: 1131–1143.

BECKER, G. S. 1968. Crime and punishment: an economic approach. J. Polit. Econ. 76: 169–217.

BEDDINGTON, J. R., AND R.B. RETTIG. 1983. Approaches to the regulation of fishing effort. FAO Fish. Tech. Pap. 39 p.

BÉLAND, P. 1988. Witness for the prosecution: scientific studies on beached belugas are strengthening the case against chemical pollutants. Nat. Can. 17(4): 28–36.

BELL, F. H. 1981. The Pacific halibut, the resource, and the fishery. Alaska Northwest Publishing Co. Anchorage, AK. 267 p.

BERCOVITCH, J. 1984. Social conflicts and third parties: strategies of conflict resolution. Westview Press, Boulder, C O. 163 p.

BERKES, F. 1985. Fishermen and The Tragedy of the Commons. Environ. Conserv. 12 (3): 199–206.

BEVAN, D. E. 1965. Methods of fishery regulation, p. 25–40. *In* J. A. Crutchfield [ed.] The fisheries: problems in resource management. University of Washington Press, Seattle, WA.

BEVERTON, R. J. H., AND S.J. HOLT. 1957. On the dynamics of exploited fish populations. U.K. Min. Agric. Fish., Fish. Invest. (Ser.2)19: 533 p.

BEVERTON, R. J. H., AND V.M. HODDER. [ed.]. 1962. Report of the working group of scientists on fishery assessment in relation to regulation problems. Int. Comm. Northw. Atl. Fish. Suppl. Annu. Proc. 11: 83 p.

BEVIN, G., P. MALONEY, AND P. ROBERTS. 1989. Economic review of the New Zealand fishing industry, 1987–1988. New Zealand Fishing Industry Board. 56 p.

BIRD, P. M., AND D.J. RAPPORT. 1986. State of the environment report for Canada. Department of the Environment. 263 p.

BISHOP, C. A. 1982. Cod trap mesh selection studies. Can. Tech. Rep. Fish. Aquat. Sci. No. 1075.

BISHOP, C. A., J.W. Baird and R. Wells. 1988. Yield-per-recruit analyses for cod in Div. 3NO. NAFO Sci. Doc. 88/20.

BISHOP, R. C., D.W. BROMLEY, AND S. LANGDON. 1981. Implementing multiobjective management of commercial fisheries: a strategy for policy-relevant research, p. 197–222. *In* L.G. Anderson [ed.] Economic analysis for fisheries management plans. Ann Arbor Science Publishers Inc., Ann Arbor, MI.

BLACKWOOD, C. M. 1974. Fish processing capability of plants in the Atlantic Region. *In* Environment Canada Proceedings Government-Industry Utilization of Atlantic Marine Resources. Montreal, P.Q. February 5–7, 1974.

BLEWETT, E. W. FURLONG, AND P. TOEWS. 1987. Canada's experience in measuring the deterrent effects of fisheries law enforcement, p. 176–212. *In* J.G. Sutinen and T.M. Hennessey [ed.] Fisheries law enforcement: programs, problems and evaluation. Proceedings of a workshop on fisheries law enforcement, the University of Rhode Island, October 21–23, 1985. NOAA/Sea Grant. The University of Rhode Island Marine Technical Report 93. 237 p.

BOGHEN, A. D. 1989. Cold-water aquaculture in Atlantic Canada. 404 p.

BOULDING, K. E. 1962. Conflict and defence: a general theory. Harper & Row, New York, NY.

BOYD, J. 1968. The industrial relations system of the fishing industry, task force on labour relations Project No. 55a (Ottawa, Privy Council Office).

BRADFORD, M. J., AND R.M. PETERMAN. 1989. Incorrect parameter values used in virtual population analysis (VPA) generate spurious time trends in reconstructed abundances, p. 87–99. *In* R.J. Beamish and G.A. McFarlane [ed.] Effects of ocean variability on recruitment and an evaluation of parameters used in stock assessment models. Can. Spec. Publ. Fish. Aquat. Sci. 108.

BRANDER, K. 1987. How well do working groups predict catches? J. Cons. int. Explor. Mer 43: 245–252.

BRANDER-SMITH, D. 1990. Protecting our waters. Final report September 1990. Public review panel on tanker safety and marine spills response capability. Minister Supply and Services Canada. 263 p.

BREEN, P. A. 1986. Management of the British Columbia fishery for northern abalone (*Haliotis kamtschatkana*), p. 300–312. *In* G.S. Jamieson and N. Bourne [ed.] North Pacific Workshop on stock assessment and management of invertebrates. Can. Spec. Publ. Fish. Aquat. Sci. 92.

LA BRETAGNE ARBITRATION. 1986. Award of July 17, 1986 by the Arbitral Tribunal established by agreement of October 23, 1985 between Canada and France on the dispute concerning filleting within the Gulf of St. Lawrence.

BREWER, G. D. 1983. The management challenges of world fisheries, p. 195–210. *In* B.J. Rothschild [ed.] Global fisheries: perspectives for the 1980's. Springer-Verlag, New York, NY.

BREWER, G. D., AND P. DELEON. 1983. The Foundations of Policy Analysis. Dorsey Press. Homewood, Ill.

BROCHMANN, B. S. 1983. Fishery policy in Norway — experiences from the period 1920–82 case study, p. 108–122. *In* Case studies and working papers presented at the Expert Consultation on strategies for fisheries development, Rome, 10–14 May 1983 FAO Fish. Rep. (295). Suppl.

BROMLEY, D. W. 1977. Distributional implications of the extended cconomic zone. Am. J. Agric. Econ. 59: 182–185.

BROOKS, S. 1989. Public policy in Canada: An introduction. McClelland & Stewart Inc., Toronto, Ont. 378 p.

BROWN, B. E., J.A. BRENNAN, M.D. GROSSLEIN, E.G. HEYERSDAHL, AND R.C. HENNEMUTH. 1976. The effect of fishing on the marine finfish biomass in the northwest Atlantic from the eastern edge of the Gulf of Maine to Cape Hatteras. ICNAF Res. Bull. 12: 49–68.

BRUGGE, W. J., AND M.J. HOLDEN. 1991. Multispecies management: a manager's point of view. ICES Marine Science Symposia. The Hague, 2–4 October 1989. 353–358.

BRUNTLAND, G. H. [Chairman]. 1987. Our common future. The world commission on environment and development. Oxford University Press. 400 p.

BRYM, R., AND B. NEIS. 1978. Regional factors in the formation of the Fishermens' Protective Union of Newfoundland. Can. J. Sociol. 3: 4.

BURD, A. C. 1985. Recent changes in the central and southern North Sea herring stocks. Can. J. Fish. Aquat. Sci. 42(1): 192–206.

BUREAU OF AGRICULTURAL ECONOMICS. 1986 and 1987. Southern Bluefin Tuna Survey 1980–81 and 1981–82 Canberra.

BURKE, W. T. 1988. Coastal state fishery regulation under international law: a comment on the La Bretagne award of July 17, 1986 (the arbitration between Canada and France). San Diego Law Review. 25: 495–534.

BURKENROAD, M. D. 1948. Fluctuations in abundance of Pacific halibut. Bull. Bingham Oceanogr. Collect. Yale Univ. 11: 81–129.

BURKENROAD, M. D. 1953. Theory and practice of marine fishery management. Rapp. P.-v. Reun. Cons. int. Explor. Mer 18(3): 300–310.

BUXTON, R. 1989. Plastic debris and lost and abandoned fishing gear in the aquatic environment — A background paper from a Canadian perspective for a workshop held in Halifax, May 1989. Report on a workshop held at the Citadel Inn, Halifax, N.S., May 1989. 62 p. + app.

BUZAN, B. G. 1982. Canada and the Law of the Sea. Ocean Devel. Intl. Law 11: 149–180.

BYNKERSHOEK, C. VAN. 1702. De Dominion Maris Dissertatio. Clarendon Press, Oxford. (1923).

CADDY, J. F. 1984. Indirect approaches to regulation of fishing effort, p. 63–75. *In* Papers presented at the Expert Consultation on the Regulation of Fishing Effort (Fishing Mortality) Rome, 17–26 Jan. 1983. FAO Fish. Rep. No. 289 (Suppl. 2).

CADIEUX, M. 1980. Address by M. Cadieux. The Oceans and Canada/US Relations. Dalhousie, University, 6 February 1980.

CAFSAC. 1978. Annual report of the Canadian Atlantic Fisheries Scientific Advisory Committee for 1977–1978. Vol. 1: 58 p.

CAFSAC. 1981. Advice on 4X Haddock closed area. CAFSAC Adv. Doc. 81/9, CAFSAC Annu. Rep. 4: 121.

CAFSAC. 1982. Annual report of the Canadian Atlantic Fisheries Scientific Advisory Committee 1981. Vol. 4: 156 p.

CAFSAC. 1984a. Advice on the management of scallop resources. CAFSAC Adv. Doc. 84/11, CAFSAC Ann. Rep. 7: 71–82.

CAFSAC. 1984b. Advice on the management of groundfish stocks in 1984. CAFSAC Adv. Doc. 83/19, CAFSAC Annu. Rep. Vol. 6.

CAFSAC. 1984c. Annual report of the Canadian Atlantic Fisheries Scientific Advisory Committee for 1983. Vol. 6: 166 p.

CAFSAC. 1985. Annual report of the Canadian Atlantic Fisheries Scientific Advisory Committee for 1984. Vol. 7: 164 p.

CAFSAC. 1986a. Advice on the status and management of the cod stock in NAFO Divisions 2J3KL. CAFSAC Adv. Doc. 86/25, CAFSAC Annu. Rep. Vol. 9.

CAFSAC. 1986b. Annual report of the Canadian Atlantic Fisheries Scientific Advisory Committee for 1985 Vol. 8: 237 p.

CAFSAC. 1987a. The status of Atlantic salmon stocks in Atlantic Canada and advice for their management in 1988. CAFSAC Adv. Doc. 87/24.

CAFSAC. 1987b. Annual report of the Canadian Atlantic Fisheries Scientific Advisory Committee for 1986 Vol. 9: 349 p.

CAFSAC. 1988. Present information on the status of the population of beluga whales in the Gulf of St. Lawrence, and research requirements. CAFSAC Adv. Doc. 88/16.

CAFSAC. 1989a. Advice for 1989 on the management of cod in Divisions 2J3KL. CAFSAC Adv. Doc. 89/1.

CAFSAC. 1989b. Advice on the management of scallop resources on Georges Bank. CAFSAC Adv. Doc. 89/7.

CAFSAC. 1989c. Advice on the management of groundfish stocks in 1990. CAFSAC Adv. Doc. 89/12.

CAFSAC. 1989d. Advice on the management of herring stocks on the Atlantic Coast of Canada in 1990. CAFSAC Adv. Doc. 89/11.

CAFSAC. 1991. Status of Atlantic salmon stocks. CAFSAC Adv. Doc. 91/4.

CAFSAC. 1992a. Advice on the management of cod in Divisions 2J3Kl. CAFSAC Adv. Doc. 92/2.

CAFSAC. 1992b. Advice on the management of groundfish stocks. CAFSAC Adv. Doc. 92/7.

CALDER V. A.-G. B.C. 1973. S.C.R. 313, (1973) 4 W.W.R. 1.

CAMPBELL, A. 1985. Application of a yield and egg-per-recruit model to the lobster fishery in the Bay of Fundy. N. Am. J. Fish. Manage. 5: 91–104.

CAMPBELL, A. 1990. The lobster (*Homarus americanus*) fishery off Lower Argyle, southwestern Nova Scotia, Can. J. Fish. Aquat. Sci. 47: 1177–1184.

CAMPBELL, B. 1974. Licence limitation in the British Columbia fishery. Department of Environment, Vancouver, B.C.

CANADA. (Various Years). Dominion of Canada. Department of Marine and Fisheries. Annual Reports (Various years). *In* Dominion of Canada, Parliament, Sessional Papers, Ottawa.

CANADA. 1852. Report of the Crown Lands Department.

CANADA. 1913. Report of the Dominion Shellfish Fishery Commission, Ottawa.

CANADA. 1949. Statement by Prime Minister St. Laurent in the House of Commons February 8, 1949. H.C. Debates (Can.) 1949, Vol. 1, p. 349.

CANADA. 1956a. Statement by Prime Minister St. Laurent in the House of Commons, July 30, 1956. H.C. Debates (Can.) 1956, Vol. VII 6702–6703 p.

CANADA. 1956b. Statement by Minister of Fisheries, James Sinclair, in the House of Commons, August 13, 1956. H.C. Debates (Can.) 1956, Vol. VII p. 7528.

CANADA. 1958a. Official records of UN Conference on Law of the Sea, Vol. III, A/Conf. 13/C.1/L.77, Rev.1.

CANADA. 1958b. Statement by the Hon. George Drew in the House of Commons cited in External Affairs, Statements and Speeches, 58/31 p. 4.

CANADA. 1959. The Law of the Sea: A Canadian Proposal, a pamphlet published by the Canadian government in December 1959.

CANADA. 1963. Statement by Prime Minister Pearson in the House of Commons, June 4, 1963. House of Commons Debates (Can.) 1963, Vol. 5, p. 621.

CANADA. 1964. An Act Respecting the Territorial Sea and Fishing Zones of Canada, 13 Eliz.2, Ch.22 (assented to 16 July 1964).

CANADA. 1968. The Senate of Canada. Proceedings of the special committee on Science Policy, Thursday, December 12th, 1968. Appendix 16: Brief submitted by the Fisheries Research Board of Canada.

CANADA. 1971. Statement to the Preparatory Committee for the 1973 Law of the Sea Conference Sub-Committee II, at Geneva, August 6, 1971 cited in The Canadian Yearbook of International Law. Vol. X. 1972: 295–297.

CANADA. 1972. House of Commons Debates, May 25, 1972 at 2542.

CANADA. 1975a. House of Commons Debates, May 9, 1975, at 5635.

CANADA. 1975b. House of Commons Debates, May 12, 1975, at 5674.

CANADA. 1975c. House of Commons Debates, July 22, 1975 at 7800.

CANADA. 1976. Statement by the Hon. A.J. MacEachen, Secretary of State for External Affairs concerning the extension of Canada's fishing zones, June 4, 1976. Reprinted in the Canadian Yearbook of International Law 1977: 326–328.

CANADA. 1977a. Statutory Orders and Regulations: 77–62. Registered January 1, 1977. Canada Gazette, January 12, 1977.

CANADA. 1977b. Statutory Orders and Regulations: 77–173. Registered February 25, 1977. Canada Gazette, March 9, 1977.

CANADA. 1977c. Office of the Prime Minister, Press Release, 27 July, 1977.

CANADA. 1979. Fishing Zones of Canada (Zones 4 and 5) Order, Amendment, Order in Council P.C. 1979–1984, 25 January 1979 (Canada Gazette, Part II, Volume 113, Number 3, 14 February 1979).

CANADA. 1981a. Canada-EEC Long-Term Fisheries Agreement. Government of Canada. News Release. December 30, 1981.

CANADA. 1982a. International Court of Justice. Delimitation of the Maritime Boundary in the Gulf of Maine Area (Canada\United States of America) Memorial Submitted by Canada September 1982.

CANADA. 1982b. Office of the Prime Minister. "Release" January 8, 1982.

CANADA. 1983a. House of Commons. Minutes of Proceedings and Evidence of the Standing Senate Committee on Fisheries and Forestry. Issue No. 92. Thursday, November 24, 1983. Respecting: Bill C-170 the Atlantic Fisheries Restructuring Act.

CANADA. 1983b. Atlantic Fisheries Restructuring Act Bill B-170 presented in the House by Minister De Bané. October 31, 1983.

CANADA. 1984a. Proposal to authorize investment in the Pacific Fisheries for the purpose of conserving, protecting and developing the fish resources and improving their management and utilization and to amend the Fisheries Act. Draft legislation tabled in the House of Commons. June 18, 1984 Ref. No. 322–7/19.

CANADA. 1985a. Economic Growth. Natural Resources. Natural Resources Program: From Crisis to Opportunity. A Study Team Report to the Task Force on Program Review. 155 p.

CANADA. 1985b. Arbitral Tribunal Established by Agreement of October 23, 1985. Dispute concerning filleting within the Gulf of St. Lawrence by the French trawlers referred to in article 4(b) of the Fisheries Agreement between Canada and France of March 27, 1972. Memorial Submitted by Canada. February 22, 1985.

CANADA. 1986. A policy proposal for a B.C. Indian community salmon fishery. Released for Discussion by the Minister of Indian Affairs and Northern Development. March, 1986.

CANADA. 1987a. Canada-France, Our Fish Our Boundaries. Brochure issued by the Government of Canada. February 1987.

CANADA. 1987b. Government acts against French overfishing. NR-HQ-87-25E. March 17, 1987.

CANADA. 1987c. Fisheries and Oceans 1987–88 Estimates. Part III Expenditure Plan.

CANADA. 1988a. Aquaculture in Canada. Report of the Standing Committee on Fisheries and Oceans. House of Commons. 129 p.

CANADA. 1988b. Canada receives formal reply from France. NR-HQ-88-31E. May 13, 1988.

CANADA. 1988c. International Scientific Report confirms effect of overfishing off the South Coast of Newfoundland. NR-HQ-88-47E. July 7, 1988.

CANADA. 1988d. Canada-France fisheries dispute: Mediation. NR-HQ-88-063E. September 21, 1988.

CANADA. 1988e. Parliament. Senate. Proceedings of the Standing Senate Committee on Fisheries. Issue No. 30. Tuesday, April 19, 1988. The examination of all aspects of the marketing of fish in Canada and all implications thereof.

CANADA. 1988f. Statement by the Hon. Pat Carney Minister for International Trade in House of Commons Debates, March 21, 1988 at pp. 13930–13931.

CANADA. 1988g. Federal Policy Addresses Long-Standing Yukon Placer Mining Issues. News Release by Ministers Siddon and McKnight. Whitehorse, May 11, 1988.

CANADA. 1988h. Preliminary results of fish and sediment sampling program for dioxins and furans announced. NR-HQ-88-032E. May 16, 1988.

CANADA. 1989a. Settlement reached in Canada-France fisheries and boundary disputes. NR-HQ-89-010E. March 31, 1989.

CANADA. 1989b. Fishing closures and consumption restrictions issued for eight B.C. sites. News Release. Vancouver. November 23, 1989.

CANADA. 1990a. Atlantic Fisheries Adjustment Program. News release. St. John's, May 7, 1990. Ministers Valcourt, Crosbie and MacKay.

CANADA. 1990b. Statement by the Hon. John C. Crosbie in response to announced plant closures by Fishery Products International. January 5, 1990.

CANADA. 1990c. Canada and U.S. reach agreement on salmon and herring trade dispute. NR-90-038. February 22, 1990. Government of Canada.

CANADA. 1990d. Canada Year Book 1990. A review of economic, social and political developments in Canada.

CANADA. 1990e. Service to the Public Task Force. Report for Discussion. October 12, 1990.

CANADA. 1991a. Federal Government announces $100 million program to clean up Fraser River. PR-HQ-091-23. June 1, 1991.

CANADA. 1991b. New federal regulations to control pulp mill pollution. PR-HQ-091-45. December 4, 1991.

CANADA. 1991c. Annual report for 1990–91 on the Interdepartmental action plan to ensure the survival of the St. Lawrence Beluga whale. 33 p.

CANADA 1992. Notes for a Statement by the Honourable John C. Crosbie, Minister of Fisheries and Oceans and Minister for the Atlantic Canada Opportunities Agency, on an Agreement with the European Community regarding fisheries, St. John's, Newfoundland, December 21, 1992.

CANADA-DENMARK. 1973. Agreement Between the Government of the Kingdom of Denmark and the Government of Canada Relating to the Delimitation of the Continental Shelf Between Greenland and Canada.

CANADA-EEC. 1980a. Agreement of Letters initialled at Brussels April, 1980.

CANADA-EEC. 1980b. Agreement on Fisheries between the Government of Canada and the European Economic Community. Initialled at Brussels, November 29, 1980.

CANADA-EEC. 1980c. Summary Record. Canada-EEC (0 Plus 1) Negotiations. Ottawa, November 27–28, 1980.

CANADA-EEC. 1981a. Agreement on Fisheries between the Government of Canada and the European Economic Community signed at Brussels, December 30, 1981.

CANADA-EEC. 1981b. Agreement in the form of an exchange of letters between the Government of Canada and the European Economic Community concerning their fisheries relations signed at Brussels, December 30, 1981.

CANADA-FRANCE. 1972. Agreement between Canada and France on their Mutual Fishing Relations, Ottawa, 27 March, 1972.

CANADA-FRANCE. 1977. Proces-Verbal. October 21, 1977.

CANADA-FRANCE. 1978. Proces-Verbal. November 30, 1978.

CANADA-FRANCE. 1980. Proces-Verbal. October 3, 1980.

CANADA-FRANCE. 1987a. Agreed Record of January 24, 1987 Negotiations in Paris.

CANADA-FRANCE. 1987b. Report of the ad hoc Working Group for the assessment of the cod stock in Subdivision 3Ps. May 28–30, 1987.

CANADA-FRANCE. 1989a. Agreement establishing a Court of Arbitration for the purpose of carrying out the delimitation of Maritime Areas between Canada and France. March 30, 1989.

CANADA-FRANCE. 1989b. Proces-Verbal. Interim Agreement on Fisheries Quotas 1989–1991. March 30, 1989.

CANADA-FRANCE. 1992. Court of Arbitration for the delimitation of Maritime Areas between Canada and France. Case concerning delimitation of Maritime Areas between Canada and the French Republic. Decision of June 10, 1992.

CANADA-NEWFOUNDLAND. 1983a. Canada-Newfoundland Ministerial Memorandum of Understanding on Fisheries Restructuring. May 17, 1983 signed by Ministers De Bané and Morgan.

CANADA-NEWFOUNDLAND. 1983b. Addendum to Canada-Newfoundland Ministerial Memorandum of Understanding on Fisheries Restructuring. May 18, 1983 signed by Ministers De Bané and Morgan.

CANADA-NEWFOUNDLAND. 1983c. Agreement between the Government of Canada and the Government of Newfoundland and Labrador concerning the restructuring of the Newfoundland fishery. September 26, 1983.

CANADA-NEWFOUNDLAND. 1985. Hibernia Development Project. Report of the Environmental Assessment Panel. 62 p.

CANADA-NOVA SCOTIA. 1983. Federal-provincial agreement to cooperate in restructuring two major Nova Scotia Deep-sea fishing companies. Department of Fisheries and Oceans. NR-HQ-083-070E. September 30, 1983.

CANADA-PORTUGAL. 1981. Summary Record of Portugal-Canada Fisheries Discussion. Signed at Lisbon, September 17, 1981.

CANADA-PORTUGAL. 1983. Summary Record of Portugal-Canada Fisheries Discussion. Signed at Lisbon, May 26, 1983.

CANADA-PORTUGAL. 1984. Summary Record of Portugal-Canada Fisheries Discussion. Signed at Aveiro, January 11, 1984.

CANADA-SPAIN. 1976. Agreement between the Government of Canada and the Government of Spain on Mutual Fisheries Relations. Signed at Madrid, June 10, 1976.

CANADA-SPAIN. 1979. Memorandum of Understanding between Canada and Spain. June 1, 1979.

CANADA-SPAIN. 1980. Memorandum of Understanding between Canada and Spain. Ottawa, March 7, 1980.

CANADA-SPAIN. 1982a. Draft Ad Referendum Memorandum of Understanding Between Canada and Spain. February 11, 1982. Ottawa.

CANADA-SPAIN. 1982b. Summary Record of Canada-Spain Fisheries Consultations. May 17–20, 1982, Ottawa.

CANADA-USA. 1970. Agreement between the Government of Canada and the Government of the United States of America on Reciprocal Fishing Privileges in certain areas of their Coasts. Ottawa, April 24, 1970.

CANADA-USA. 1971. Agreement between United States and Canada. Consultations on Salmon Problems of Mutual Concern. Seattle, Washington, June 17–18, 1971.

CANADA-USA. 1977. Reciprocal Fisheries Agreement between the Government of Canada and the Government of the United States of America, 24 February 1977.

CANADA-USA. 1978. Exchange of Notes between the Government of Canada and the Government of the United States of America constituting an Interim Agreement amending and extending the Reciprocal Fisheries Agreement of 1977, 11 April 1978.

CANADA-USA. 1979a. Agreement between the Government of Canada and the Government of the United States of America on East Coast Fishery Resources, 29 March 1979.

CANADA-USA. 1979b. Special Agreement between the Government of Canada and the Government of the United States of America to Submit to a Chamber of the International Court of Justice the Delimitation of the Maritime Boundary in the Gulf of Maine Area, Washington, D.C. March 29, 1979.

CANADA-USA. 1980. Agreed Summary Record of Canada/United States Discussions on a Comprehensive Agreement on the Management and Development of Pacific Salmon Stocks of Mutual Concern, Lynnwood, Washington, October 20–25, 1980.

CANADA-USA. 1985. Treaty between the Government of the United States of America and the Government of Canada concerning Pacific Salmon, Ottawa, January 28, 1985.

CARPENTER, S. L., AND W.J.D. KENNEDY. 1988. Managing public disputes. A practical guide to handling conflict and reaching agreements. Jossey-Bass Inc., San Francisco, CA. 293 p.

CARPENTIER, G. R., D.S. MACDONALD, AND P.R. HOOD. 1980. Atlantic Coast Groundfish Trawler Study. Resource requirements 1981–1985. Department of Fisheries and Oceans, Ottawa. August, 1980. 48 p.

CARTER, R. 1984. The Fogo Island co-operative: An alternative development strategy? Master's thesis, Memorial University, Nfld.

CHARLES, A. T. 1988. Fishery socioeconomics: a survey. Land Econ. 64: 276–295.

CHARLES, A. T. 1989. Bio-socio-economic fishery models: labour dynamics and multi-objective management. Can. J. Fish. Aquat. Sci. 46: 1313–1322.

CHARLES, A.T. 1992. Fishery Conflicts. A unified framework. Marine Policy: 379–393.

CHRISTY, F. R., JR., and A.D. Scott. 1965. The common wealth in ocean fisheries. Some problems of growth and economic allocation. John Hopkins University Press, Baltimore, MD. 281 p.

CHRISTY, F. T., JR. 1973. Fisherman quotas: a tentative suggestion for domestic management. University of Rhode Island, Law of the Sea Institute, Kingston, RI. Occas. Pap. 19: 17.

CHRISTY, R. T., JR. 1977. The fishery conservation and management act of 1976. Management objectives and the distribution of benefits and costs. Wash. Law. 52:

CICIN-SAIN, B., J.E. MOORE, AND A.J. WYNER. 1978. Limiting entry to commercial fisheries: some worldwide comparisons. Ocean Manage. 4: 21–49.

CLAIN, L. E. 1985. Gulf of Maine — A disappointing first in the delimitation of a single Maritime Boundary. Virg. J. Int. Law 25(3): 521–620.

CLARK, C. 1988. Rights to common-property fisheries resources: a theoretical review. Proceedings of the International Conference on Fisheries, University of Quebec, Rimouski, Canada. August 1986. Vol. I: 425–427.

CLARK, C. W. 1973. Profit maximization and the extinction of animal species. J. Political Econ. 81: 950–961.

CLARK, C. W. 1976. Mathematical bioeconomics: The optimal management of renewable resources. John Wiley & Sons, New York, NY. 352 p.

CLARK, C. W. 1980. Towards a predictive model for the economic regulation of commercial fisheries. Can. J. Fish. Aquat. Sci. 37: 1101–1110.

CLARK, I. N., AND A.J. DUNCAN. 1986. New Zealand's fisheries management policies — past, present and future: The implementation of an ITQ-based management system, p. 107–140. *In* N. Mollett [ed.] Fishery Access Control Programs Worldwide. Proceedings of the Workshop on Management Options for the North Pacific Longline Fisheries.

CLARK, I. N., P.J. MAJOR, AND N. MOLLETT. 1989. The development and implementation of New Zealand's ITQ management system. 117–149. *In* P.A. Neher, R. Arnason and N. Mollett [ed.] Rights based fishing. Kluwer Academic Publishers, Norwell, MA. 541 p.

CLARKE, R. MCV. 1993. An overview of Canada's Arctic marine fisheries and their management: with emphasis on the Northwest Territories. In L.S. Parsons and W.H. Lear [ed.] Perspectives on Canadian marine fisheries management. Can. Bull. Fish. Aquat. Sci. (In press)

CLARKE, R. MCV, L. JOHNSON, G.D. KOSHINSKY, A.W. MANSFIELD, R.W. MOSHENKO, AND T. SHORTT. 1989. Report of the Arctic Fisheries Scientific Advisory Committee for 1986/87 and 1987/88. Can. Manuscr. Rep. Fish. Aquat. Sci. 2015: iv + 68 p

CLAY, D. 1979. Mesh selection of silver hake (*Merluccius bilinearis*) in otter trawls on the Scotian Shelf with reference to selection of squid (*Illex Illecebrosus*). ICNAF. Res. Doc 79/II/3 39p.

CLAYTOR, R. R., E.M.P. CHADWICK, G.A. NIELSEN, G.J. CHAPUT, D.K. CAIRNS, S.C. COURTENAY, AND H.M.C. DUPUIS. 1991. Index Programs: Their value in southern Gulf of St. Lawrence fish stock assessments. ICES C.M. 1991. Department of Fisheries and Oceans, Science Branch, Moncton, N.B.

CLEMENS, W. A. 1958. Reminiscences of a director. J. Fish. Res. Board Can. 15: 779–796.

CLEMENT, W. 1984. Canada's coastal fisheries: formation of unions, cooperatives and associations. J. Can. Stud. 19: 5–33.

CLEMENT, W. 1986. The struggle to organize: resistance in Canada's fishery. McClelland and Stewart, Toronto, Ont. 219 p.

CLOUGH, D. J. 1980. Optimization and implementation plan for offshore fisheries surveillance. Unpublished study done under contract for the Department of Fisheries and Oceans. 206 p.

COGLA. 1986. An environmental evaluation of proposed exploratory drilling on Georges Bank. Unpublished report. 156 p. + app.

COHEN, E. B., D.G. MOUNTAIN, AND R. O'BOYLE. 1991. Local-scale versus large-scale factors affecting recruitment. Can. J. Fish. Aquat. Sci. 48: 1003–1006.

COLLIE, J. S., R.M. PETERMAN, AND C.J. WALTERS. 1990. Experimental harvest policies for a mixed-stock fishery: Fraser River sockeye salmon (*Oncorhynchus nerka*) Can. J. Fish. Aquat. Sci. 47: 145–155.

COLLINS, E. JR., AND M.A. ROGOFF. 1986. The Gulf of Maine Case and the future of ocean boundary delimitation. Maine Law Rev. 3(8): 11–48.

CONAN, G. Y., R.W. ELNER, AND M. MORIYASU. 1989. Review of literature on life histories in the genus *Chionoecetes* in light of the recent findings on growth and maturity of *C. opilio* in eastern Canada. 163–179. *In* Proc. Int. Symp. King and Tanner Crabs, Nov. 1989. Anchorage, Alaska.

CONNELLY, P., AND M. MACDONALD. 1983. Women's work: domestic and wage labour in Nova Scotia community. Stud. Polit. Econ. 10: 45–72.

COOK, B. A., AND P. COPES. 1987. Optimal levels for Canada's Pacific halibut catch. Mar. Res. Econ. 4: 45–61.

COOK, R. M., AND D.W. ARMSTRONG. 1986. Stock-related effects in the recruitment of North Sea haddock and whiting. J. Cons. int. Explor. Mer 42: 272–280.

COOPER, J. 1986. Delimitation of the maritime boundary in the Gulf of Maine area. Ocean Devel. Intl. Law 16: 59–90.

COOZE, J. 1989. Conflict management. Part I: Philosophical approaches to conflict handling.

COPES, P. 1972. The resettlement of fishing communities in Newfoundland. Canadian Council on Rural Development. Ottawa. 259 p.

COPES, P. 1980. The evolution of marine fisheries policy in Canada. J. Bus. Admin. 11(1/2): 125–148.

COPES, P. 1981. Fisheries on Canada's Pacific Coast: The impact of extended jurisdiction on exploitation patterns. Ocean Manage. 6: 279–297.

COPES, P. 1983. Fisheries management on Canada's Atlantic Coast: economic factors and socio-political constraints. Can. J. Reg. Sci. VI(1): 1–32.

COPES, P. 1984. Introduction: regional science and fisheries analysis in the Canadian context. Institute of Public Affairs. 145–151.

COPES, P. 1986. A critical review of the individual quota as a device in fisheries management. Land Econ. Vol. 62(3): 278–291.

COPES, P., AND B.A. COOK. 1982. Rationalization of Canada's Pacific halibut fishery. Ocean Manage. 8: 151–175.

COSER, L. A. 1956. The functions of social conflict. The Free Press, New York, NY. 188 p.

COSER, L. A. 1968. Conflict: social aspects, p. 232–236. *In* David Sills [ed.] International Encyclopedia of the Social Sciences. MacMillan, New York, NY.

COWLING, E. B. 1980. An historical resume of progress in scientific and public understanding of acid precipitation and its biological consequences. SNSF Project, Oslo, Norway. 29 p.

CRAMPTON, C. 1989. The Coast Guard's Annex V Compliance Report. A Case Study. Proceedings of the Second International Conference on Marine Debris, Honolulu, April, 1989.

CROUTER, R. A. 1984. Quotas by fishing gear for the herring fishery of the Bay of Fundy. *In* Expert consultation on the regulation of fishing effort (fishing mortality). FAO Fisheries Rep. No. 289 (Suppl. 3): 251–273.

CROWLEY, R. W., AND H. PALSSON. 1988. Enterprise allocations, transferable quotas and the Canadian approach to fisheries management. Econ. Comm. Anal. Direc. Dec. 29, 1988.

CROWLEY, R. W., AND H. PALSSON. 1992 Rights Based Fisheries Management in Canada. 1–21. *In* Marine Resource Economics. 7: 1–21.

CROWLEY, R. W., B. MACEACHERN, AND R. JASPERSE. 1993. A Review of Federal Assistance to the Canadian Fishing Industry 1945–1990. *In* L.S. Parsons and W.H. Lear [ed.] Perspectives on Canadian marine fisheries management. Can. Bull. Fish. Aquat. Sci. (In press)

CRUTCHFIELD, J. A. 1961. An economic evaluation of alternative methods of fishery regulation. J. Law Econ. 4: 131–143.

CRUTCHFIELD, J. A. 1965. Economic objectives of fisheries management, p. 43–64. *In* J.A. Crutchfield [ed.] The Fisheries: problems in resource management. University of Washington Press, Seattle, WA. 136 p.

CRUTCHFIELD, J. A. 1979. Economic, and social implications of the main policy alternatives for controlling fishing effort. J. Fish. Res. Board Can. 36: 742–752.

CRUTCHFIELD, J. A. 1981. The Pacific halibut fishery. (Case Study No. 2, Tech. Rep. No. 17) The Public Regulation of Commercial Fisheries in Canada. Economic Council of Canada. 69 p.

CRUTCHFIELD, S. R. 1985. Fishery economics: current status and outlook for the future. Fish. Econ. Newsl. 20: 8–20.

CUNNINGHAM, S. 1981. The evolution of the objectives of fisheries management during the 1970's. Ocean Manage. 6: 251–278.

CUNNINGHAM, S. 1983. The increasing importance of economics in fisheries regulation. J. Agric. Econ. 34(1): 69–77.

CUNNINGHAM, S., M. R. DUNN, AND D. WHITMARSH. 1985. Fisheries economics: an introduction. Mar. Policy Mansell Publishing, London (UK). 372 p.

CUSHING, D. H. 1968. Fisheries biology. University of Wisconsin Press, Madison, WI. 200 p.

CUSHING, D. H. 1971. The dependence of recruitment on parent stock in different groups of fishes. J. Cons. Perma. int. Explor. Mer 33: 340–362.

CUSHING, D. H. 1972. A history of some of the international fisheries commissions. The Royal Society of Edinburgh Proceedings (section B) 73: 362–390.

CUSHING, D. H. 1974. A link between science and management in fisheries. Fishery Bull. 72: 859–864.

CUSHING, D. H. 1975. Marine ecology and fisheries. Cambridge University Press, Cambridge. 278 p.

CUSHING, D. H. 1978. Biological effects of climatic change. Rapp. P. -v. Reun. Cons. int. Explor. Mer 173: 107–116.

CUSHING, D. H. 1982. Climate and fisheries. Academic Press, New York, NY. 373 p.

CUSHING, D. H. 1983. The outlook for fisheries research in the next ten years, p. 263–275. *In* B.J. Rothschild [ed.] Global fisheries: perspectives for the 1980's. Springer-Verlag, New York, NY.

CUSHING, D. H. 1988. The study of stock and recruitment, p. 105–128. *In* J. A. Gulland [ed.] Fish population dynamics (second edition): the implications for management. John Wiley & Sons, Ltd., Great Britian.

CUSHING, D. H., AND J.G.K. HARRIS. 1973. Stock and recruitment and the problem of density-dependence. Rapp. P. -V. Reun. Cons. Perma. int. Explor. Mer 164.

CUSHING, D. H., AND J.W. HORWOOD. 1977. Development of a model of stock and recruitment, p. 21–35. *In* J.H. Steele [ed.] Fisheries mathematics. Academic Press, London.

DAAN, N. 1973. A quantitive analysis of the food intake of North Sea cod, *Gadus morhua*. Neth. J. Sea Res. 6: 479–517.

DAAN, N. 1981. Comparison of estimates of egg production from the Southern Bight cod stock from plankton surveys and from market statistics. Rapp. P. -v. Cons. int. Explor. Mer 178: 242–243.

DAAN, N. 1987. Multispecies versus single-species assessment of North Sea fish stocks. Can. J. Fish. Aquat. Sci. 44: 360–370.

DAHRENDORF, R. 1958. Toward a theory of social conflict. J. Conflict Resolution 2: 170–183.

DAVIS, A., AND L. KASDAN. 1984. Bankrupt government policies and belligerent fishermen responses: Dependency and conflict in the southwest Nova Scotia small boat fisheries. J. Can. Stud. 19(1): 108–124.

DAVIS, A., AND V. THIESSEN. 1986. Making sense of the dollars: Income distribution among Atlantic Canadian fishermen and public policy. Mar. Policy 10(3): 201–214.

DAVIS, A., AND V. THIESSEN. 1988. Public Policy and Social Control in the Atlantic Fisheries. Canadian Public Policy 14(1): 66–77.

DE MESTRAL, A. L. C., AND L.H. LEGAULT. 1979–1980. Multilateral Negotiation — Canada and the Law of the Sea Conference. Intl. J. 35: 47–69.

DE VORSEY, L. 1987. Historical Geography and the Canada-United States Seaward Boundary on Georges Bank. Maritime Boundaries and Ocean Resources. 182–207 p.

DEPARTMENT OF FINANCE. 1984. A New Direction for Canada: An agenda for economic renewal. 115 p.

DERHAM, P. J. 1987. The implementation and enforcement of fisheries legislation. *In* G. Ulfstein et al. [ed.] The regulation of fisheries: legal, economic and social aspects. Proceedings of a European Workshop, University of Tromso, Norway, 2–4 June, 1986.

DEUTSCH, M. 1973. The resolution of conflict: constructive and destructive processes. Yale University Press, New Haven, CT. 420 p.

DEWOLF, A. G. 1974. The lobster fishery of the Maritime Provinces: economic effects of regulations. Bull. Fish. Res. Board Can. 187: 54 p.

DFO. (Various Years). Atlantic Fisheries Restructuring Act. Annual Reports for 1983–84 through 1985–86.

DFO. 1979a. Final Report Northern Cod Seminar. Glynmill Inn, Corner Brook, Newfoundland. August 28–30, 1979. 214 p.

DFO. 1979b. Toward a Policy for the Utilization of Northern Cod: A Discussion Paper. September, 1979. Department of Fisheries and Oceans. 30 p.

DFO. 1979c. Set Policy for Licensing Freezer Trawlers. Department of Fisheries and Oceans. NR-HQ-79-055E, November 30, 1979.

DFO. 1980a. Rapporteur's Report. (Final Version). Gulf Groundfish Seminar. Memramcook, N.B. September 23–25, 1980.

DFO. 1980b. 1981 Atlantic Groundfish Management Plan. NR-HQ-080-050E. December 29, 1980.

DFO. 1980c. Fisheries Minister announces changes for west coast salmon fisheries. News Release. Department of Fisheries and Oceans, Pacific Region. October 28, 1980.

DFO. 1980d. An address by the Hon. Romeo LeBlanc, Minister of Fisheries and Oceans at the Gulf Groundfish Seminar, Memramcook, N.B. September 25, 1980.

DFO. 1980e. Notes for an address by the Hon. Romeo LeBlanc, Minister of Fisheries and Oceans at the NFFAWU Annual Meeting, St. John's, Newfoundland. November 12, 1980.

DFO. 1980f. Canada and Spain resume fishing negotiations. NR-HQ-80-009E. February 29, 1980.

DFO. 1980g. Portugal to buy more Canadian fish. NR-HQ-80-007E. February 14, 1980.

DFO. 1980h. Department of Fisheries and Oceans Annual Report. 1979–80.

DFO. 1981a. Policy for Canada's Atlantic Fisheries in the 1980s. Atlantic Fisheries Service, Ottawa.

DFO. 1981b. Toward a Policy for the Management of Gulf Groundfish: A Discussion Paper. Atlantic Fisheries Service. Department of Fisheries and Oceans, Ottawa, February, 1981. 37 p. + app.

DFO. 1981c. Heavy Fishing May Result in Closure. NR-HQ-081-004E. February 9, 1981.

DFO. 1981d. 1982 Atlantic Groundfish Management Plan Announced. NR-HQ-081-065E. December 23, 1981.

DFO. 1981e. Conservation Measures Announced for Pacific Coast Chinook Salmon. NR-HQ-081-006E. February 11, 1981.

DFO. 1981f. Revisions to 1981 Pacific Fishing Plan. NR-HQ-081-022E. April 24, 1981.

DFO. 1981g. Inquiry into Pacific Coast Fishing Industry Announced. NR-HQ-081-001E. Department of Fisheries and Oceans, Ottawa. January 13, 1981.

DFO. 1981h. Vessel Buy-Back Program Announced for Pacific Coast. NR-HQ-081-007E. February 12, 1981.

DFO. 1981i. Lobster Licence Buy-Back Program Extended. NR-HQ-081-064. December 9, 1981.

DFO. 1981j. Notes for an address by the Hon. Romeo LeBlanc to the Annual Meeting of the Maritime Fishermen's Union. Halifax, Nova Scotia. February 26, 1981.

DFO. 1981k. New Guidelines for Fishing Vessel Assistance Program and Vessel Replacements. Department of Fisheries and Oceans. NR-HQ-081-032E. June 16, 1981.

DFO. 1981l. Heavy Fishing May Result in Closure. NR-HQ-081-004E. February 9, 1981.

DFO. 1981m. Rules for the Administration of Enterprise Allocations for the Atlantic Offshore Groundfish Fisheries during the 1982 Trial Period. Department of Fisheries and Oceans.

DFO. 1981n. National Plan for Fisheries Research 1981–1985. Resources Services Directorate. Department of Fisheries and Oceans. May. 1981.

DFO. 1982a. Fisheries Minister Announces Decision on Pearse Commission Recommendations. NR-HQ-082-005. March 1, 1982.

DFO. 1982b. A draft framework agreement between Canada and the United States for the management of Pacific salmon. Background Information.

DFO. 1982c. New Canada-USA Pacific Salmon Agreement One Step Closer. NR-HQ-082-063. December 30, 1982.

DFO. 1982d. Submission by the Department of Fisheries and Oceans to the Pearse Inquiry, Vancouver, April 27, 1982.

DFO. 1983a. Response of the Government of Canada to the Report of the Task Force Study on the Atlantic Fishery. News Release. February 17, 1983.

DFO. 1983b. 1984 Canadian Atlantic Groundfish Plan Announced. NR-HQ-083-095E, December 29, 1983.

DFO. 1983c. Evaluation study of the lobster vessel certificate retirement program (reference report). Program Evaluation Branch, Department of Fisheries and Oceans, Ottawa. July, 1983. 21 p. + app.

DFO. 1983d. Restructuring of Atlantic Herring Purse Seine Fleet to Begin. NR-HQ-083-053E. August 2, 1983.

DFO. 1983e. Minister welcomes progress in Restructuring Atlantic Herring Purse Seine Fleet, NR-HQ-083-093E. December 19, 1983.

DFO. 1983f. Enterprise Allocations for the Inshore Mobile Gear Groundfish Fleet 19.8m or less in NAFO Divisions 4R-3Pn. Department of Fisheries and Oceans Area office — Western Nfld. and Southern Labrador — Gulf Region (1984 First year Pilot Project). Department of Fisheries and Oceans, Gulf Region. December 31, 1983. 29 p.

DFO. 1983g. Reduced fish allocation for EEC. NR-HQ-083-007E. January 28, 1983.

DFO. 1983h. New deal with EEC on long-term fisheries agreement. NR-HQ-083-092E. December 19, 1983.

DFO. 1983i. Resource-Short Atlantic Fish Plant Program Announced. NR-HQ-83-086. December 8, 1983.

DFO. 1983j. Pacific Salmon Negotiators Submit Treaty Text. NR-HQ-083-010. February 21, 1983.

DFO. 1983k. No progress made on Pacific Treaty Discussions. NR-HQ-083-041. June 27, 1983.

DFO. 1983l. Statement by the Hon. Pierre De Bané Minister of Fisheries and Oceans in St. John's, Newfoundland July 4, 1983 on Fisheries Restructuring.

DFO. 1983m. Atlantic fish habitat task force: Report of the policy review committee. R.H. Cook (chairman). Fish Habitat Management in the Atlantic Fisheries Service: Policy Review. 51 + app.

DFO. 1984a. Sector Management of Canada's Atlantic Groundfish Fishery. Atlantic Fisheries Service Department of Fisheries and Oceans July, 1984. 7 p.

DFO. 1984b. 1984 Troll Season Announced. News Release. March 16, 1984.

DFO. 1984c. Strait of Georgia Chinook Conservation. NR-HQ-084-27E. April 19, 1984.

DFO. 1984d. New Policy for Pacific Salmon Fisheries Announced. NR-HQ-84-047, June 18, 1984.

DFO. 1984e. Atlantic Fisheries Licence Review Board Established. NR-HQ-84-053, June 25, 1984.

DFO. 1984f. Statement of the Minister of Fisheries and Oceans on the New Policy for the Pacific Fisheries. NR-HQ-084-59E. June 28, 1984.

DFO. 1984g. Statement by the Hon. Pierre De Bané on Restructuring National Sea Products. NR-HQ-84-011. Ottawa.

DFO. 1984h. Offers to Purchase the Assets of PUQ Made by Pêcheries Cartier Inc. NR-HQ-84-009. Ottawa.

DFO. 1985a. Trollers to share in Fraser sockeye run. NR-HQ-85-063. August 16, 1985.

DFO. 1985b. Pacific Region Salmon Resource Management Plan: discussion document. Canada. Department of Fisheries and Oceans.

DFO. 1985c. Atlantic Salmon Management Plan Announced. Department of Fisheries and Oceans. NR-HQ-85-026, April 25, 1985.

DFO. 1985d. Discussion Paper on Factory Freezer Trawlers. Department of Fisheries and Oceans, August, 1985.

DFO. 1985e. Three Factory Freezer Trawler Licences Approved. Department of Fisheries and Oceans NR-HQ-85-077, November 8, 1985.

DFO. 1985f. 1985 Bay of Fundy — Nova Scotia Herring Plan Announced. NR-HQ-085-028E. April 25, 1985.

DFO. 1985g. Developments in foreign allocations on the Atlantic coast of Canada since the extension of jurisdiction. Atlantic Fisheries Service. 118 p.

DFO. 1985h. Canada/Spain Fisheries agreement to lapse. NR-HQ-86-51E. June 28, 1985.

DFO. 1985i. Fisheries Minister satisfied that FRG vessels leave area. NR-HQ-085-036. May 10, 1985.

DFO. 1985j. 1985 Atlantic Salmon Management Plan announced. NR-HQ-085-0E. April 25, 1985.

DFO. 1985k. The Regional Council Concept as applied to Atlantic fisheries management. Discussion Paper. 21 p.

DFO. 1985l. Atlantic Regional Council Established to advise Minister on Fisheries Policy. NR-HQ-85-061E. August 14, 1985.

DFO. 1985m. Audit Report on Observer Program. Evaluation and Audit Branch. Department of Fisheries and Oceans. August, 1985.

DFO. 1986a. 1986 Pacific Salmon Allocations. NR-PR-086-04A. February 14, 1986.

DFO. 1986b. Canadian Atlantic Groundfish Management Plan announced. Department of Fisheries and Oceans. NR-HQ-87-098, December 29, 1986.

DFO. 1986c. Atlantic Fisheries Licence Appeal Board Appointed NR-HQ-86-053 July 7, 1986.

DFO. 1986d. 1986 Bay of Fundy/Nova Scotia Herring Plan Announced. NR-HQ-86-012E. March 11, 1986.

DFO. 1986e. Notes for an address by the Hon. Tom Siddon, Minister of Fisheries and Oceans to the St. John's Board of Trade, St. John's, Nfld. June 13, 1986.

DFO. 1986f. Statement by Fisheries and Oceans Minister Tom Siddon. NR-HQ-86-92E. November 26, 1986.

DFO. 1986g. Challenges Facing the Fishery Sector: A report to the annual conference of first ministers. November, 1986. 156 p. Department of Fisheries and Oceans, Ottawa.

DFO. 1986h. Minister announces streamlining of Fisheries and Oceans headquarters. NF-HQ-086-08E. February 13, 1986.

DFO. 1986i. Fisheries and Oceans Minister Announces major initiatives. NR-HQ-86-72E. September 24, 1986.

DFO. 1986j. Streamlining the Atlantic Fisheries Consultative Process. Discussion paper. June, 1986.

DFO. 1986k. New Pacific Regional Council Announced. News Release. October 31, 1986.

DFO. 1986l. Policy for the management of fish habitat. Department of Fisheries and Oceans. 30 p.

DFO. 1986m. Review of Atlantic Fisheries Research. Department of Fisheries and Oceans. July, 1986.

DFO. 1986n. Review of Pacific, Freshwater and Arctic Science. Department of Fisheries and Oceans. June, 1986.

DFO. 1986o. Review of Marine Ecology and Biological Oceanography. Department of Fisheries and Oceans. June, 1986.
DFO. 1986p. Review of Physical and Chemical Oceanographic Research. Department of Fisheries and Oceans. June, 1986.
DFO. 1987a. Canada's recreational fisheries: An overview and a description of Department of Fisheries and Oceans Programs. 18 p.
DFO. 1987b. 1988 Canadian Atlantic Groundfish Management Plan Announced. NR-HQ-087-109E. December 30, 1987.
DFO. 1987c. Chinook. *In* Pacific Region Salmon Stock Management Plan, DFO discussion document. Vol. K: 81 p.
DFO. 1987d. Review of Vessel Replacement Policy for the Atlantic Fisheries. April 1987. Department of Fisheries and Oceans, Internal Report. 28 p. + app.
DFO. 1987e. Minister of Fisheries and Oceans calls for review of moratorium on unused groundfish licences. Department of Fisheries and Oceans. NR-HQ-87-024.
DFO. 1987f. Minister lifts moratorium on unused groundfish licences in the Scotia-Fundy Region. Department of Fisheries and Oceans. NR-HQ-87-038. April 16, 1987.
DFO. 1987g. Canada and EC Consult on Future Fisheries Relations. NR-HQ-87-37E. April 13, 1987.
DFO. 1987h. Announcement on Resource-Short Plant Program. NR-HQ-87-96E. September 30, 1987.
DFO. 1987i. Address by the Hon. Tom Siddon, Minister of Fisheries and Oceans, at the Opening Plenary Session of the 34th Annual Meeting of the International North Pacific Fisheries Commission. Vancouver, November 3, 1987.
DFO. 1987j. Statement by the Hon. Tom Siddon, Minister of Fisheries and Oceans, issued in the House of Commons by the Hon. John C. Crosbie, Minister of Transport and Regional Minister for Newfoundland and Labrador, regarding port privileges of foreign fishing vessels. NR-HQ-87-16E.
DFO. 1987k. Ocean's Policy for Canada. A strategy to meet the challenges and the opportunities on the oceans frontier. 15 p.
DFO. 1987l. Pacific Regional Council Appointments Announced. NR-HQ-087-036. April 13, 1987.
DFO. 1987m. Fisheries Minister announces agreement on Nechako and Nanika River systems. NR-PR-87-12. September 14, 1987.
DFO. 1987n. Fish Habitat. The foundation of Canada's fisheries. 25 p.
DFO. 1987o. RCMP gives observer program clean bill of health. NR-HQ-087-001. January 9, 1987.
DFO. 1987p. Increased fines now in effect for foreign fishing violations. NR-HQ-87-04E. Ottawa, January 13, 1987.
DFO. 1988a. Government/Industry Workshop. Inshore/Offshore Sharing of the Atlantic Groundfish Resources (Final Report) April 20-22, 1988. Montreal, Quebec. 34 p. + app.
DFO. 1988b. 1989 Canadian Atlantic Groundfish Plan Announced. NR-HQ-88-073. December 30, 1988.
DFO. 1988c. 1989 Atlantic Groundfish Management Plan. Atlantic Fisheries Department of Fisheries and Oceans.
DFO. 1988d. Tough, Fair Conservation Program for Lower Strait of Georgia Chinook Announced. NR-PR-88-03. March 7, 1988.
DFO. 1988e. Continuation of Offshore Groundfish Enterprise Allocation (EA) Program. NR-HQ-088-072E. December 30, 1988.
DFO. 1988f. Notes for a Statement by Dr. P. Meyboom to the 1988 Annual Meeting of the General Council of NAFO.
DFO. 1988g. European Community continues to overfish NAFO stocks. NR-HQ-88-071E. December 13, 1988.
DFO. 1988h. The Canada-U.S. Free Trade Agreements and Fisheries. 29 p.
DFO. 1988i. Announcement of an Action Plan to ensure the survival of the St. Lawrence Belugas. C-RQ-88-025A. June 21,1988.
DFO. 1989a. Report on Northern Cod Released. NR-HQ-89-016E. May 26, 1989.
DFO. 1989b. The Scotia-Fundy Lobster Fishery Phase One: Issues and Considerations. Department of Fisheries and Oceans. May, 1989. Communications Branch, Scotia-Fundy Region, 111 p.
DFO. 1989c. Northern Cod TAC for 1989 Announced. NR-HQ-98-005. February, 8, 1989.
DFO. 1989d. Further measures regarding Northern Cod announced. NR-HQ-89-006. February 12, 1989.
DFO. 1989e. 1989 Salmon Allocation Plan Report. Document submitted to the Commercial Fishing Industry Council by P.S. Chamut, April 28, 1989.
DFO. 1989f. Backgrounder. Salmon Allocation Plan for 1989. June 14, 1989.
DFO. 1989g. 1989 Atlantic Salmon Management Plan. Guiding principles and major elements. Atlantic Fisheries Service. Department of Fisheries and Oceans. May, 1989. 33 p.
DFO. 1989h. Commercial Fisheries Licensing Policy for Eastern Canada. Department of Fisheries and Oceans, Ottawa. January 1989. 54 p.
DFO. 1989i. Report of the Scotia-Fundy Groundfish Task Force. J.E. Hache, Chairman, December, 1989. 86 p.
DFO. 1989j. Discussion paper on individual quotas in the halibut industry. Department of Fisheries and Oceans, Ottawa. September 11, 1989. 6 p. + app.
DFO. 1989k. Discussion paper on individual quotas in the sablefish industry. Department of Fisheries and Oceans, Ottawa. November 17, 1989. 6 p. + app.
DFO. 1989l. Plastic debris in the aquatic environment — Halifax Report on a workshop held at the Citadel Inn, Halifax, N.S., May 1989. 46 p.
DFO. 1989m. Global warming, the oceans, and Canada's fisheries. Submission to the House of Commons Standing Committee on Environment on behalf of the Department of Fisheries and Oceans. 13 p.
DFO. 1989n. $28 million for fisheries air surveillance. NR-HQ-89-037E. October 26, 1989.
DFO. 1990a. Aquaculture strategy for the 90s: cultivating the future. Department of Fisheries and Oceans. 40 p.
DFO. 1990b. Fisheries-A National Priority says Siddon. NR-HQ-90-01E. January 2, 1990.

DFO. 1990c. Minister Valcourt says new multi-year groundfish plan strikes balance for sustainable Atlantic fishery. NR-HQ-90-50E. December 14, 1990.
DFO. 1990d. Pacific Salmon Allocations for 1990 Announced. NR-PR-90-10E. June 8, 1990.
DFO. 1990e. Northern Shrimp Fishery: 1990–91 Management Plan. Atlantic Fisheries, Department of Fisheries and Oceans. May, 1990.
DFO. 1990f. Pacific Coast Commercial Fishing Licensing Policy: Discussion Paper. Department of Fisheries and Oceans. 83 p.
DFO. 1990g. Gulf of St. Lawrence Groundfish: An alternative approach to management. Communications Branch, Department of Fisheries and Oceans, Ottawa. 16 p.
DFO. 1990h. Individual Quotas and Catch Monitoring in Scotia-Fundy. Backgrounder to NR-HQ-90-050E. December 14, 1990.
DFO. 1990i. Closure to Protect Abalone Stocks. NR-PR-90-22. November 1, 1990.
DFO. 1990j. Individual Quotas Examined in Halibut Fishery. NR-PR-90-23. November 1, 1990.
DFO. 1990k. Report of the Implementation Task Force on Northern Cod. Atlantic Fisheries Adjustment Program. 104 p. + app.
DFO. 1991a. Factory Freezer Trawler Program to Continue. NR-HQ-91-028E. June 10, 1991.
DFO. 1991b. Notes for an address by the Hon. John Crosbie, Minister of Fisheries and Oceans to the Fisheries Council of Canada Annual Convention. October 3, 1991. Ottawa, Canada.
DFO. 1991c. Review of Canadian Enterprise Allocation Programs September, 1990. (Unpublished internal document). Department of Fisheries and Oceans, Ottawa. 88 p.
DFO. 1991d. Canada calls for emergency decision to improve controls on fishing. NR-HQ-91-08-E. February 28, 1991.
DFO. 1991e. NAFO adopts new measure to monitor fishing outside 200 miles. NR-HQ-91-018E. May 1, 1991.
DFO. 1991f. Crosbie announces new initiative on native co-management and west coast fishery. NR-PR-91-10. June 24, 1991.
DFO. 1991g. Weekly Scientific Briefing. Bedford Institute Vol. 10(34). August 23, 1991. EXXON VALDEZ revisited.
DFO. 1991l. Fisheries Management: a proposal for reforming licensing and allocation systems. 17 p.
DFO. 1991m. Atlantic Fisheries Adjustment Program. Backgrounder Department of Fisheries and Oceans. B-HQ-91-006E. June 1991.
DFO. 1992a. Crosbie announces first steps in Northern Cod (2J3KL) recovery plan. NR-HQ-92-58E. July 2, 1992.
DFO. 1992b. Crosbie seeks global cooperation to protect straddling stocks. NR-HQ-92-36E. May 6, 1992.
DFO. 1992c. Canada and Newfoundland offer $40 million to retire commercial salmon fishing licenses. NR-N-92-18E. March 6, 1992.
DFO. 1992d. Canada-France Boundary Decision is in Canada's Favour. NR-HQ-92-51E. June 11, 1992.
DFO. 1992e. Notes for a Statement by the Hon. John C. Crosbie, Minister of Fisheries and Oceans and Minister for the Atlantic Canada Opportunities Agency, on Income Support and Related Measures for Fishermen and Plant Workers affected by the Northern (2J3KL) Cod Moratorium, St. John's, Newfoundland, July 17, 1992.
DFO. 1992f. Fishery Incomes and Adjustment to be Studied. NR-HQ-92-023E. March 23, 1992.
DFO. 1992g. Notes for Remarks by The Hon. John C. Crosbie, Minister of Fisheries and Oceans, to a Media Conference to Announce the Aboriginal Fisheries Strategy, Ottawa-Vancouver. June 29, 1992. Crosbie Announces Aboriginal Strategy for B.C. Fishery. NR-HQ-92-55E. Crosbie Announces Agreements with Lower Fraser Aboriginal Groups. NR-HQ-92-56E. Voluntary Licence Retirement Program Announced. NR-HQ-92-57E. Aboriginal Fisheries Strategy — Backgrounder I: The Program — Backgrounder II: The Context — Backgrounder III: Native Groups Participate in Management, Enhancement — Backgrounder IV: Sale to be Tested in 1992.
DFO. 1992h. Dioxins and Furans. B-HQ-92-001E. January 1992.
DFO. 1992i. Notes for a Statement by the Honourable John C. Crosbie, Minister of Fisheries and Oceans and Minister for the Atlantic Canada Opportunities Agency. Atlantic Groundfish Management Plan for 1993. Halifax, Nova Scotia, 1992.
DFO. 1992j. Financial Performance of The British Columbia Salmon Fleet 1986–1990. Prepared by the Program Planning and Economics Branch July, 1992. Department of Fisheries and Oceans, Pacific Region, Vancouver, B.C. 59 p.
DFO. 1992k. Fishing Opportunities for St. Pierre and Miquelon conditional on conservation measures and reciprocal access to resources. News release October 9, 1992.
DFO. 1992l. Canadian salmon stock benefits to be vigorously pursued at negotiations with U.S. Department of Fisheries and Oceans. NR-PR-92-26E. November 27, 1992.
DFO. 1992m. Incomes and Adjustment in the Atlantic Fishery. A Discussion Paper by The Task Force on Incomes and Adjustment in the Atlantic Fishery. Department of Fisheries and Oceans, Ottawa. 19 p.
DFO. 1993a. Fisheries Management. A Proposal for Reforming Licensing, Allocation and Sanctions Systems. Department of Fisheries and Oceans, Ottawa. 37 p.
DFO. 1993b. Notes for an Address by 'The Honourable John C. Crosbie, Minister of Fisheries and Oceans and Minister for the Atlantic Canada Opportunities Agency, on Sustainable Development and the Atlantic Fishery, to the Annual Meeting of The Inshore Fishermen's Improvement Committee Clarenville, Newfoundland, January 21, 1993. Department of Fisheries and Oceans, Ottawa.
DFO AND DOE. 1985. Memorandum of Understanding between the Department of Fisheries and Oceans and the Department of the Environment on the subject of The Administration of Section 33 of the Fisheries Act. May, 1985.Ottawa, Ontario.
DFO/DREE. 1982. Report of the DFO/DREE Working Group on Processing Capacity in Atlantic Canada, Ottawa.
DIAND. 1989. Contaminants in Northern ecosystems and native diets: summary of an evaluation meeting held in Ottawa, Feb. 28–Mar. 2, 1989. Department of Indian and Northern Affairs. 7 p.

DICKIE, L. M. 1955. Fluctuations in abundance of the giant scallop (*Placopecten magellanicus*) (Gmelin) in the Digby area of the Bay of Fundy. J. Fish. Res. Board Can. 12: 797–856.

DICKSON, R. R., J.G. POPE AND M.J. HOLDEN. 1973. Environmental influences on the survival of North Sea cod, p. 69–80. *In* the early life history of fish. J.H. S. Blaxter [ed.] Springer-Verlag. Berlin.

DOE. 1973a. Opening Remarks by Ken Lucas, Senior Assistant Deputy Minister Fisheries Marine Service. *In* Licensing for Atlantic Coast Fisheries. Government/Industry Policy Development Seminar Citadel Motor Inn, Halifax, N.S. October 22–23, 1973.

DOE. 1973b. New Atlantic fishing vessel licensing program under consideration. News Release. August 13, 1973.

DOE. 1973c. State of the Resource and General Recommendations on Atlantic Licensing Policy. Remarks by Lorne Grant. *In* Licensing for Atlantic Coast Fisheries. Government/Industry Policy Development Seminar, Citadel Motor Inn, Halifax, N.S. October 22–23, 1973.

DOE. 1973d. Atlantic fishing fleet development policy announced. Department of the Environment. News Release. November 14, 1973.

DOE. 1973e. Speech by Dr. A. May to Fisheries Council of Canada. Charlottetown, April 25, 1973.

DOE. 1974a. $4 million extension to assistance program. News release. November 19, 1974.

DOE. 1974b. Minister of State for Fisheries Romeo LeBlanc today announced a new short term federal government plan for Canadian Fishermen. News Release. December 20, 1974.

DOE. 1974c. Notes for a speech by the Hon. Romeo LeBlanc to the Atlantic Provinces Economic Council, October 22, 1974.

DOE. 1974d. Fisheries and Marine Service Executive Appointments. July 19, 1974.

DOE. 1975a. Notes for a Speech by the Hon. Romeo LeBlanc Minister of State (Fisheries) for an address to the 30th Annual Meeting of the Fisheries Council of Canada. Halifax, Nova Scotia. May 5, 1975.

DOE. 1975b. Lobster Fishery Task Force Final Report, March, 1975. 180 p.

DOE. 1975c. Tighten controls on east coast lobster fishery. News Release. December 30, 1975.

DOE. 1975d. Atlantic Ports to be Closed to Soviet Fishing Fleet. News Release, Ottawa, July 23, 1975.

DOE. 1975e. Foreign Fishing to be Cut 40% off Atlantic Coast. News Release, Montreal. September 28, 1975.

DOE. 1975f. A 50 million dollar program leading to long-term measures to strengthen Canada's fishing industry. News Release. Ottawa. April 23, 1975.

DOE. 1975g. Notes for a speech by the Hon. Romeo LeBlanc Minister of State (Fisheries) for an address to the 30th Annual Meeting of the Fisheries Council of Canada, Halifax, Nova Scotia, May 5, 1975.

DOE. 1975h. Stricter Quality Requirements for Groundfish Assistance Program. News Release. October 8, 1975.

DOE. 1975i. Annual Report of the Department of the Environment for 1974–75.

DOE. 1976a. Policy for Canada's Commercial Fisheries. Department of the Environment, Ottawa. 70 p. + app.

DOE. 1976b. Announce Steps to Keep East Coast Groundfish Fleet Working. News Release. June 17, 1976. Environment Canada.

DOE. 1976c. Atlantic Coast Groundfish Fisheries Management Plan for 1977. December, 1976. Environment Canada.

DOE. 1976d. 1977 Fishing Plans for Atlantic Groundfish Fleet. Press Release. Environment Canada. December 21, 1976.

DOE. 1976e. Moonlighters to be phased out of Maritimes lobster fishery. News Release. September 11, 1976.

DOE. 1976f. Aid Program for Herring Fishermen. News Release. April 9, 1976.

DOE. 1976g. Government — Industry Plan to Increase Value of Bay of Fundy Herring Fishery. News Release. July 16, 1976.

DOE. 1976h. Notes for Statement by the Hon. Romeo LeBlanc, Minister of State (Fisheries) at the opening of 1976 Annual Meeting of ICNAF in Montreal, Quebec, June 8, 1976.

DOE. 1976i. Resource assessment requirements in relation to Canada's extension of fisheries jurisdiction. A study undertaken for the Task Group on Extended Fisheries Jurisdiction, Ottawa. 177 p.

DOE. 1976j. Canada/France Interim Fisheries Arrangement. News Release. December 30, 1976.

DOE. 1976k. Fisheries Minister announces $44 million assistance program. FMS. HQ. News Release. Ottawa. March 3, 1976. Department of Environment.

DOE. 1976l. Annual Report of the Department of the Environment for 1975–76.

DOE. 1977a. Notes for a Speech by the Hon. Romeo LeBlanc, Minister of Fisheries and the Environment to the Rotary Club Yarmouth, Nova Scotia. November 28, 1977.

DOE. 1977b. Notes for a Speech by the Hon. Romeo LeBlanc, Minister of Fisheries and the Environment to the Rotary Club St. John's, Newfoundland. May 19, 1977.

DOE. 1977c. Management of Canada's 200-mile Fishing Zone. An address to the Canadian Labour Congress by Dr. A.W. May, Acting Assistant Deputy Minister Fisheries Management, Fisheries and Marine Service, Department of Fisheries and the Environment. Ottawa, January 13, 1977.

DOE. 1978a. Resource Prospects for Canada's Atlantic Fisheries 1979–1985. 53 p. + app.

DOE. 1978b. Cod Quotas Increased under 1979 Atlantic Groundfish Fishing Plan. Press Release. 1979 Groundfish Plan. December 28, 1978.

DOE. 1978c. Federal aid to lobster fishermen. News Release. July 29, 1978.

DOE. 1978d. Impose Licensing Freeze in Atlantic inshore fishery. Department of the Environment. FMS-NR-78-36, November 30, 1978.

DOE. 1978e. Notes for a speech by the Hon. R. LeBlanc to The Fisheries Council of Canada, Quebec, P.Q. May 3, 1978.

DOE. 1978f. Licence allocations announced for new shrimp fishery. FMS-HQ-NR-17. June 2, 1978.

DOE. 1978g. Fisheries and Oceans ADM's Appointed. FMS-HQ-NR-37. November 30, 1978.

DOE. 1988. The Changing Atmosphere: Implications on global security. Conference Statement. Department of the Environment. Toronto, Ontario. June 27–30, 1988.

DOE. 1989a. Pollutants in British Columbia's marine environment. A state of the environment fact sheet #89-2. Environment Canada, Conservation and Protection.

DOE. 1989b. Notes for a Statement by the Hon. Lucien Bouchard, Minister of the Environment for Canada. Ministerial Conference on Atmospheric Pollution and Climatic Change. Noordwijk, The Netherlands, November 6–7, 1989.

DOERN, G. B., AND P. AUCOIN. 1971. The structures of policy making in Canada. MacMillan Co., Toronto, Ont.

DOF. 1966. Controlled lobster trap limitation, coupled with the registration of lobster boats. NR-HQ-66-20, March 31, 1966.

DOF. 1967a. Lobster trap limits announced. NR-HQ-67-005, January 31, 1967.

DOF. 1967b. Licences Limited. NR-HQ-67-1, February 27, 1967.

DOF. 1967c. Licensing restrictions for Northumberland Strait. NR-HQ-67-31, June 29, 1967.

DOF. 1968a. New regulations for B.C. Salmon fishing Industry. News Release. September 6, 1968.

DOF. 1968b. Registration of lobster fishing boats. NR-HQ-68-004, February 6, 1968.

DOF. 1968c. Limitation of lobster fishing licences. NR-HQ-68-007, February 28, 1968.

DOF. 1968d. Speech by Jack Davis, Minister of Fisheries and Forestry to the American Commercial Fish Exposition in Boston, Massachusetts, October 18, 1968.

DOF. 1968e. Submission of the Ministry of Fisheries to the Senate of Canada Special Committee on Science Policy. Part III: Department of Fisheries of Canada. December 1968. Appendix 17.

DOF. 1969a. Lobster licensing limitations in the Maritimes. NR-HQ-69-001, January 20, 1969.

DOF. 1969b. New licensing program in waters off N.S., N.B. and P.E.I. NR-HQ-69-007, February 27, 1979.

DOF. 1969c. Limitation of multiple licences in the lobster fishery. NR.-HQ-69-014, April 3, 1969.

DOF. 1970. Davis hikes Salmon Licence fees to fund Vessel Buy-Back Program. News Release. January 16, 1970.

DOUBLEDAY, W. G. 1976. Environmental fluctuations and fisheries and management. ICNAF Selected Pap.1: 141-150.

DOUBLEDAY, W. G., AND T.D. BEACHAM. 1982. Southern Gulf of St. Lawrence cod: a review of multispecies models and management advice, p. 133–140. *In* M.C. Mercer [ed.] Multispecies approaches to fisheries management advice. Can. Spec. Publ. Fish. Aquat. Sci. 59.

DOUBLEDAY, W. G., A.T. PINHORN, R.G. HALLIDAY, R.D.S. MACDONALD, AND R. STEIN. 1989. The impact of extended jurisdiction in the Northwest Atlantic, p. 33–73. *In* E. Miles [ed.] Management of world fisheries: implications of extended coastal state jurisdiction.

DOUCET, F. J., AND P.H. PEARSE. 1980. Fisheries policy for the Pacific coast: issues and advice. A report to the Hon. Romeo LeBlanc, Minister of Fisheries. 46 p.

DREE. 1974. Speech by Minister Jamieson to the Rotary Club, St. John's, Newfoundland, December 12, 1974.

DUBINSKY, W. E. 1987. Cultures in conflict and the problems of objective interpretation, as applied to fisheries management in Canada and the United States. Unpublished M.A. thesis, Carleton University, Ottawa, Ont. 308 p. + app.

DUDLEY, N., AND G. WAUGH. 1980. Exploitation of a single-cohort fishery under risk: a simulation-optimization approach. J. Environ. Econ. Manage. 7: 234–255.

DUNBAR, M. J. 1951. Resources of Arctic and Subarctic Sea. Trans. R. Soc. Can. 45, 3rd Ser., 5th Sec.

DUNBAR, M. J. 1970. On the fishery potential of the sea water of the Canadian North. Arctic 23: 150.

DYE, T. R. 1978. Understanding public policy. 3rd ed. Prentice-Hall, Englewood Cliffs, NJ. 3 p.

EASTON, D. 1957. An approach to the analysis of political systems. World Politics, IX (April, 1957) pp. 383–400. C. Easton, A Framework for Political Analysis.

EDWARDS, R. L. 1980. Problem of the determination of allocation principles in the northwest Atlantic region of the USA, p. 109–118. *In* J. H. Grover [ed.] Allocation of fishery resources, proceedings of the technical consultation on allocation of fishery resources held in Vichy, France, 20–23 April 1980.

EFF. 1981. Is our groundfish being managed? Brief to the Department of Fisheries and Oceans by Eastern Fishermens Federation, December, 1981.

EGAN, D., AND A. KENNEY. 1990. Salmon farming in British Columbia — an industry in transition. World Aquacult. 21: 6–11, 24–29.

EIC. 1989a. Notes for an Address by the Hon. Barbara MacDougall, Minister of Employment and Immigration and Minister responsible for the status of women, on the Labour Force Development Strategy. Employment and Immigration Canada.

EIC. 1989b. Success in the Works: a policy paper. A Labour Force Development Strategy for Canada. Employment and Immigration Canada.

ELISSON, M. 1981. Conservation and management. *In* Iceland 1981 Fisheries Yearbook. 8–9.

ELLIOTT, G. H. 1973. Problems confronting fishing industries relative to management policies adopted by governments. J. Fish. Res. Board Can. 30: 2486–2489.

EMR. 1988. Minister Marcel Masse calls for moratorium on Georges Bank drilling to at least the year 2000. News Release. April 18, 1988. Energy Mines and Resources. Halifax, N.S.

EUROPEAN COMMUNITY. 1976. Future external fisheries policy and internal fisheries system, Commission communication to the Council, COM (76) 500, 21.9.76.

EUROPEAN COMMUNITY. 1980. Official Journal of the European Communities, NoC 158 of 27/6/1980.

EVANS, G. T., AND J.C. RICE. 1988. Predicting recruitment from stock size without the mediation of a functional relation. J. Cons. int. Explor. Mer 44: 111–122.

EVENSEN, J. 1985. The effect of the law of the sea conference upon the process of the formation of international law: rapprochement between competing points of view, p. 23–40. *In* R.B. Krueger and S.A. Riesenfeld [ed.] The developing order of the oceans. Honolulu, HI.

EXTERNAL AFFAIRS. 1970. Promulgation of Fisheries Closing Lines: a statement tabled in the House of Commons by the Hon. Jack Davis, Minister of Fisheries and Forestry, December 18, 1970. Statements and Speeches. Department of External Affairs. No. 70/26.

EXTERNAL AFFAIRS. 1973. Third United Nations Conference on the Law of the Sea. Department of External Affairs, Ottawa, November 1973. 26 p.

EXTERNAL AFFAIRS. 1975a. Joint Communique on Canada-Spain Discussions of Fisheries Matters of Mutual Concern. Communique No. 67, August 6–7, 1975, Department of External Affairs, Ottawa.

EXTERNAL AFFAIRS. 1975b. Fisheries. Canada-USSR Agreed Record of Understanding, August 27, 1975. Communique No. 72, August 28, 1975. Department of External Affairs, Ottawa.

EXTERNAL AFFAIRS. 1975c. Canada-USSR Joint Communique on Fisheries. Communique No. 83, September 26, 1975. Department of External Affairs, Ottawa.

EXTERNAL AFFAIRS. 1975d. Canada-Norway Agreement on Mutual Fisheries, Relations. Communique No. 116, December 2, 1975, Department of External Affairs, Ottawa.

EXTERNAL AFFAIRS. 1977. Department of External Affairs, Communique no. 1,-1, 21 October, 1977.

EXTERNAL AFFAIRS. 1978. Department of External Affairs, Communique no. 39, 12 April, 1978.

EXTERNAL AFFAIRS. 1979. Department of External Affairs, Communique no. 29, 29 March, 1979.

FAO. 1978. Some scientific problems of multispecies fisheries. FAO Fish. Tech. Paper 181. 42 p.

FAO. 1979. Interim report of the ACMRR working party on the scientific basis of determining management measures. FAO Fish. Circ. No. 718.

FAO. 1980. Report of the ACMRR working party on the scientific basis of determining management measures: Hong Kong, 10–15 Dec. 1979, FAO Advisory Committee on Marine Resources Research, Rome (Italy), FAO Fish. Rep. No. 236.

FAO. 1984a. Report of the Working Party on the principles for fisheries management in the new ocean regime. Nantes, France, 14–18 March 1983. Advisory Committee on Marine Resources Research. FAO Fish. Rep. 299: 11 p.

FAO. 1984b. Expert consultation on the regulation of fishing effort (fishing mortality). Rome, 17–26 January 1983. FAO Fish. Rep. No. 289: 470 p.

FAO. 1985. Report of the eleventh session of the Advisory Committee of experts on marine resources research. Rome, 21–24 May 1985. FAO Fish. Rep. 338: 20 p.

FAO. 1989. FAO Yearbook in 1987 of Fishery Statistics Catches and Landings. FAO Yearbook, Vol. 64: 490 p.

FARIS, J. 1972. Cat Harbour: A Newfoundland fishing settlement (with appendices). Newfoundland Social and Economic Studies No. 3, Institute of Social and Economic Research, Memorial University of Newfoundland, Nfld.

FARMER, G. J., T.R. GOFF, D.O. ASHFIELD, AND H.S. SAMANT. 1980. Some effects of the acidification of Atlantic salmon rivers in Nova Scotia. Can. Tech. Rep. Fish. Aquat. Sci. 972: 13 p.

FARNELL, J., AND J. ELLES. 1984. In search of a common fisheries policy. Gower Publishing Co. Ltd. Brookfield, Vermont. 213 p.

FAULKNER, J. H. 1982. Pressuring the executive. *In* A.P. Pross [ed.] Governing under pressure. Can. Pub. Admin. 25(2): 240–253.

FCC. 1963. A Brief concerning Canada's National and Territorial Waters, submitted to Government of Canada by the Fisheries Council of Canada, January 28, 1963, Ottawa. 11 p.

FCC. 1971. A Submission from the Fisheries Council of Canada to the Standing Committee on External Affairs and National Defence Respecting a "Foreign Policy for Canadians". January, 1974.

FCC. 1974. Fisheries Council of Canada. Brief to the Parliamentary Standing Committee on External Affairs and National Defence with reference to United Nations Law of the Sea Conference. Ottawa, January 4, 1974.

FERRIS, J. S., AND C.G. PLOURDE. 1980. Fisheries management and employment in the Newfoundland economy. Economic Council of Canada. Discussion paper 173. 164 p.

FERRIS, J. S., AND C.G. PLOURDE. 1982. Labour mobility, seasonal unemployment insurance and the Newfoundland inshore fishery. Can. J. Econ. 15: 426–41.

FINCH, R. 1985. Fishery management under the Magnuson Act. Mar. Policy 9: 170–179.

FINKLE, P. Z. R. 1974. The International Commission for the Northwest Atlantic Fisheries: An experiment in conservation. Dalhousie Law J. 1: 526–550.

FISHERIES COUNCIL OF BRITISH COLUMBIA. 1989. Trends in the commercial fishing industry of British Columbia 1983–1987. 31 p.

FOGARTY, M. J., AND J.S. IDOINE. 1988. Application of a yield and egg production model based on size to an offshore American lobster population. Trans. Am. Fish. Soc 117: 350–362.

FORGET, C. E. [Chairman]. 1986. Commission of inquiry on unemployment insurance. Supply & Services, Canada. 516 p.

FOWLER v.R. 1980. 25. C.R. 213, 113 D.L.R. (3d) 513, revg. 93 D.L.R. (3rd) 724.

FPI. (Various Years). Fishery Products International, Annual Reports.

FPI. 1990a. Press Statement by Victor Young, CEO of Fisheries Products International, re: plant closures, January 5, 1990. St. John's. Nfld.

FPI. 1990b. Fishery Products International Annual Report. 20 p.

FPI. 1991. Third quarter report to the shareholders for the thirty-nine weeks ended September 28, 1991 by the Fishery Products International.

FPSB. (Various Years). Annual Reports Fisheries Prices Support Board.

FRAILEY, M. 1987. Problems of case management, p. 70–92. *In* J.G. Sutinen and T.M. Hennessey [ed.] Fisheries law enforcement: programs, problems and evaluation. Proceedings of a workshop on fisheries law enforcement, the University of Rhode Island, October 21–23, 1985. NOAA/Sea Grant. Univ. R.I. Mar. Tech. Rep. 93: 237 p.

FRAILEY, M., AND R. TAYLOR. 1987. Rationalizing sanctions for fisheries violations, p. 215–234. *In* J.L. Bubier and A. Rieser [ed.] East coast fisheries law and policy. Proceedings from the June 17–20, 1986 Conference on East Coast Fisheries Law and Policy. Marine Law Institute, Maine. 482 p.

FRANCE. 1977. Décret no. 77–169 du 25 février 1977 portant sur la création, en application des dispositions de la Loi du 16 juillet 1976, d'une zone économique au large des côtes du département de Saint-Pierre-et-Miquelon.

FRANCE-GREAT BRITAIN. 1904. Convention between France and Great Britain respecting Newfoundland and West and Central Africa Signed at London, 8 April 1904, Article I.

FRANCIS, R. C., S. ADLERSTEIN, AND R. BRODEUR. 1987. Biological basis for management of commercial fishery resources of the eastern Bering Sea, p. 187–209. *In* W. S. Wooster [ed.] Fishery science and management. Lecture notes on coastal and estuarine studies. Vol. 28. Springer-Verlag, New York, NY.

FRANCIS, R. C., S.A. ADLERSTEIN, AND A. HOLLOWED. 1989. Importance of environmental fluctuations in the management of Pacific Hake (*Merluccius productus*), p. 51–56. *In* R.J. Beamish and G.A. McFarlane [ed.] Effects of ocean variability on recruitment and an evaluation of parameters used in stock assessment models. Can. Spec. Publ. Fish. Aquat. Sci. 108.

FRANK, K. T., AND W.C. LEGGETT. 1985. Reciprocal oscillations in densities of larval fish and potential predators: a reflection of present or past predation? Can. J. Fish. Aquat. Sci. 42: 1841–1849.

FRANK, K. T., AND J.E. CARSCADDEN. 1989. Factors affecting recruitment variability of capelin (*Mallotus villosus*) in the Northwest Atlantic. J. Cons. int. Explor. Mer 45: 146–164.

FRANKLIN, P. 1987. Australian southern bluefin tuna fishery, p. 412–426. *In* Papers presented at symposium on the exploitation and management of marine fishery resources in southeast asia, IPFC meeting, Darwin, Australia, 16–19 February 1987. RAPA report: 1987/10.

FRASER, C. A., AND J.B. JONES. 1989. Enterprise allocations: the Atlantic Canadian experience, p. 267–288. *In* P.A. Neher et al. [ed.] Rights based fishing. Kluwer Academic Publishers, Norwell, MA. 541 p.

FRASER, G. A. 1978. Licence limitation in British Columbia salmon fishery, p. 357–381. *In* R.B. Rettig and J.J.C. Ginter [ed.] Limited entry as a fishery management tool. University of Washington Press, Seattle and London.

FRASER, G. A. 1979. Limited entry: experience of the British Columbia Salmon Fishery. J. Fish. Res. Board Can. 36: 754–763.

FRASER, G. A. 1982. Licence limitation in the British Columbia roe herring fishery: an evaluation, p. 117–137. *In* N.H. Sturgess and T.F. Meany [ed.] Policy and practice in fisheries management. Proceedings of the national fisheries seminar on economic aspects of limited entry and associated fisheries management measures, held in Melbourne, February 1980. Department of Primary Industry. Canberra, Australia.

FRENCH, R., R. NELSON JR., AND J. WALL. 1982. Role of the United States observer program in management of foreign fisheries in the Northeast Pacific Ocean and Eastern Bering Sea. N. Am. J. Fish. Manage. 2: 122–131.

FRICKE, P. 1985. Use of sociological data in the allocation of common property resources. Mar. Policy 9(1): 39–52.

FULLERTON, E. C. 1987. Regional fishery management councils: the federal perspective. Coastal Management. 15(4): 305–308.

FURLONG, W. J. 1985. An economic analysis of the deterrent effect of Quebec Regions surveillance, and enforcement program. A study prepared under contract by Dr. W.J. Furlong, University of Guelph, for the Program Evaluation Branch, Department of Fisheries and Oceans. 123 p.

GABRIEL, W. L., AND S.A. MURAWSKI. 1985. The use of cluster analysis in identification and description of multi-species systems, p. 112–117. *In* R. Mahon [ed.] Towards the inclusion of fishery interactions in management advice. Can. Tech. Rep. Fish. Aquat. Sci. 1347.

GALTUNG, J. 1971. Peace thinking. *In* A. Lepawski et al. [ed.] The search for world order. Appleton Century-Crafts, New York, NY.

GARDNER PINFOLD CONSULTING ECONOMISTS LTD. 1988. A six-country review of legislation pertaining to fisheries management. Prepared for the Department of Fisheries and Oceans. November, 1988. 68 p. + app.

GARDNER, M. 1988. Enterprise allocation system in the offshore groundfish sector in Atlantic Canada. Mar. Resour. Econ. Vol. 5: 389–454.

GARDNER, M. 1989. The enterprise allocation system in the offshore groundfish sector in Atlantic Canada, p. 293–320. *In* P.A. Neher et al. [ed.] Rights based fishing. Kluwer Academic Publishers, Norwell, MA.

GARROD, D. J. 1975a. Resource management, its objectives and implementation. Seventh Special Commission Meeting — September 1975. ICNAF Res. Doc. 75/IX/128. Ser. No. 3679. 12 p.

GARROD, D. J. 1975b. Specified stock size as an objective for the management of single stocks. ICNAF Res. Doc. 75/XI/124 Ser. No. 3675: 6 p.

GARROD, D. J. 1982. Stock and recruitment — again. Fish. Res. Tech. Rep. MAFF Dir. Fish. Res., Lowestoft (68). 22 p.

GARROD, D. J. 1987. The scientific essentials of fisheries management and regulations. Min. Agric. Food Fish. Lowestoft. Lab. Leafl. No. 60: 14 p.

GARSTANG, W. 1904. Report of the convenor on comparative trawling experiments in 1903. Rapp. P. -v. Reun. Cons. Perm. int. Explor. Mer 2: 47–57.

GASKILL, H. S., S. MAY, AND C.A. CLARK. 1987. Economic Strategies in Outport Fishing Communities in Newfoundland. Report prepared for Newfoundland and Labrador Department of Fisheries. 35 p.

GATT. 1987. Canada — Measures affecting exports of unprocessed herring and salmon. Report of the Panel, GATT Doc. L/6268, 20 November 1987.

GEEN, G., AND M. NAYAR. 1989. Individual transferable quotas in the southern bluefin tuna fishery: an economic appraisal, p. 355–381. *In* P.A. Neher, R. Arnason and N. Mollett [ed.]. Rights based fishing. Kluwer Academic Publishers, Norwell, MA. 541 p.

GETZ, W. M., AND G.L. SWARTZMANN. 1981. A probability transition model for yield estimation in fisheries with highly variable recruitment. Can. J. Fish. Aquat. Sci. 38: 847–855.

GIBSON, A. [Chairman]. 1982. Report of the Halibut Vessel Quota Implementation Committee. April 27, 1982.

GISLASON, H., AND T. HELGASON. 1985. Species interaction in assessment of fish stocks with special application to the North Sea. Dana 5: 1–44.

GLASSCO COMMISSION. 1963. The Royal Commission on Government Organization. Volume 4. Special Areas of Administration Report 23: Scientific Research and Development. The Queen's Printer. Ottawa, Canada.

GOMAC. 1984. The International Court of Justice Gulf of Maine Boundary Decision. A First Analysis of the Impact on Fisheries by Fisheries Research Branch Scotia-Fundy Region November 1984. Research Report No. 1 to the Gulf of Maine Advisory Committee.

GORDON, D. C. [ed.]. 1988. An assessment of the possible environmental impacts of exploratory drilling on Georges Bank fishery resources. Can. Tech. Rep. Fish. Aquat. Sci. 1633: 31 p.

GORDON, H. S. 1953. An economic approach to the optimum utilization of fishery resources. J. Fish. Res. Board Can. 10: 447–457.

GORDON, H. S. 1954. Economic theory of a common property resource: the fishery. J. Polit. Econ. 62: 124–142.

GORHAM, E. 1976. Acid precipitation and its influence upon aquatic ecosystems — an overview. Water, Air and Soil Pollut. 6: 457–481.

GORRIE, P. 1986. Acid rain fighter: the story of one man's persistent efforts to get us to stop polluting the environment (Harold Harvey). Canadian Geographic Oct/Nov 1986. 8–17.

GOTLIEB, A. E. 1964. The Canadian contribution to the concept of a fishing zone in international law. *In* The Canadian Yearbook of International Law 2: 55–76.

GOUGH, J. 1988. Fisheries history, p. 781–784. *In* The Canadian Encyclopedia (2nd ed.). Hurtig Publishers.

GOUGH, J. 1993. Fisheries management in Canada: a historical sketch. *In* L.S. Parsons and W.H. Lear [ed.] Perspectives on Canadian marine fisheries management. Can. Bull. Fish. Aquat. Sci. (In press)

GRADY, P., AND D. MACLEAN. 1989. An income stabilization program for fishermen: its need, feasibility and cost. Report by Global Economics Ltd. for the Department of Fisheries and Oceans. 73 p.

GRAHAM, M. 1935. Modern theory of exploiting a fishery, and North Sea trawling. J. Cons. int. Explor. Mer, 10(2): 264–274.

GRIGOR'EN, G. V. 1972. Reproduction of *Macrurus rupestris* Gunner of the northern Atlantic. Tr. Pinro, 28: 107–115. Fish. Res. Board Can. Transl. Ser. No. 2529.

GROSSLEIN, M. D., AND G.R. LILLY. 1987. Summary report of the special session on recruitment studies. NAFO Sci. Coun. Studies 11: 83–90.

GROSSLEIN, M. D. 1987. Synopsis of knowledge of the recruitment process for Atlantic herring (*Clupea harengus*) with special reference to Georges Bank. NAFO Sci. Coun. Stud. 11: 91–108.

GROTIUS, H. 1609. Mare Librum. *In* The Freedom of the Seas or The right which belongs to the Dutch to take part in the east Indian trade. Reprint 1916 J.B. Scott [ed.] Oxford University Press. 83 p.

GUERIN V. THE QUEEN (the "Musqueam" case). 1984. 55 N.R. 161 (S.C.C.).

GULF TROLLERS ASSOCIATION. 1984. Press Release. September 13, 1984.

GULLAND, J. A. 1965. Estimation of mortality rates. Annex to Rep. Arctic Fish. Working Group, int. Counc. Explor. Sea C.M. (3): 9 p.

GULLAND, J. A. 1968. The concept of the maximum sustainable yield and fishery management. FAO Fish. Tech. Pap. 70: 30 p.

GULLAND, J. A. 1968. The concept of the marginal yield from exploited fish stocks. J. Cons. 32: 256–261.

GULLAND, J. A. 1971. Science and fisheries management. J. Cons. int. Explor. Mer 33(3): 471–477.

GULLAND, J. A. 1974. The management of marine fisheries. University of Washington Press, Seattle, WA. 198 p.

GULLAND, J. A. 1977. Goals and objectives of fishery management. FAO Fish. Tech. Pap. No. 166.

GULLAND, J. A. 1981. Long-term potential effects from the management of the fish resources of the North Atlantic. J. Cons. int. Explor. Mer 40(1): 8–16.

GULLAND, J. A. 1983a. Stock Assessment: Why? FAO Fish. Circ. 759: 18 p.

GULLAND, J. A. [ed.]. 1983b. Fish stock assessment. John Wiley and Sons, London. 372 p.

GULLAND, J. A. 1983c. World resources of fisheries and their management, p. 839–1061. *In* O. Kinne [ed.] Marine ecology. John Wiley & Sons. New York, N.Y.

GULLAND, J. A. 1984. Control of the amount of fishing by catch limits, p. 119–127. *In* Papers Presented at the Expert Consultation on the Regulation of Fishing Effort (Fishing Mortality) Rome, 17–26 Jan. 1983. FAO Fisheries Report.

GULLAND, J. A. 1987. The management of the North Sea fisheries: looking towards the 21st century. Mar. Policy 11: 259–272.

GULLAND, J. A. 1988. The problems of population dynamics and contemporary fishery managementp, p. 383–406. *In* J.A. Gulland [ed.] Fish population dynamics (second edition): the implications for management. John Wiley & Sons, Ltd., Great Britian.

Gulland, J. A. 1991. Under what conditions will multispecies models lead to better fisheries management? 348–352. ICES Marine Science Symposia. The Hague, 2–4 October 1989. 358 p.

GULLAND, J. A., AND M.A. ROBINSON. 1973. Economics of fishery management. J. Fish. Res. Board Can. 30: 2042–2050.

GULLAND, J. A., AND K. BOEREMA. 1973. Scientific advice on catch levels. Fish. Bull. 71: 325–335.

GULLAND, J. A., AND S. GARCIA. 1984. Observed patterns in multispecies fisheries, p. 155–190. *In* R.M. May [ed.] Exploitation of marine communities. Springer-Verlag, Berlin.

GUPPY, N. 1987a. Labouring on shore: transforming uncommon property into marketable products, p. 199–222. *In* P. Marchak, N. Guppy, and J. McMullan [ed.] Uncommon property: the fishing and fish-processing industries in British Columbia. Methuen, Toronto, Ont. 402 p.

GUPPY, N. 1987b. Labouring at sea: harvesting uncommon property, p. 173–198. *In* P. Marchak, N. Guppy, and J. McMullan [ed.] Uncommon property: the fishing and fish-processing industries in British Columbia. Methuen, Toronto, Ont. 402 p.

HAGE, R. 1984. Canada and the law of the sea. Mar. Policy 8: 2–15.

HAIG-BROWN, R. 1974. The Salmon. Fisheries and Marine Service. Department of the Environment. Ottawa. 79 p.

HAINES, T. A. 1981. Acidic deposition and its consequences for aquatic ecosystems: a review. Trans. Am. Fish. Soc. 110(6): 669–707.

HALL, D. L., R. HILBORN, M. STOCKER, AND C.J. WALTERS. 1988. Alternative harvest strategies for Pacific herring (*Clupea harengus pallasi*) Can. J. Fish. Aquat. Sci. 45: 888–897.

HALLIDAY, R. G., AND W.G. DOUBLEDAY. 1976. Catch and effort trends for the finfish resources of the Scotian shelf and estimates of the maximum sustainable yield of groundfish (except silver hake). ICNAF Selected Pap. 1: 117–128.

HALLIDAY, R. G., AND A.T. PINHORN. 1985. Present management strategies in Canadian Atlantic marine fisheries. Their rationale and the historical context in which their usage developed, p. 10–33. *In* R.Mahon [ed.] Towards the inclusion of fishery interactions in management advice. Can. Tech. Rep. Fish. Aquat. Sci. 1347.

HALLIDAY, R. G., J. MCGLADE, R. MOHN, R.N. O'BOYLE, AND M. SINCLAIR. 1986. Resource and fishery distributions in the Gulf of Maine area in relation to the subarea 4/5 boundary. NAFO Sci. Counc. Stud. 10: 67–92.

HAMLET OF BAKER LAKE V. MINISTER OF INDIAN AFFAIRS AND NORTHERN DEVELOPMENT. 1980. 1 F.C. 518 (S.C.C.).

HANNESON, R. 1978. Economics of fisheries. Bergen, Norway: Universitetsforlaget.

HANNESSON, R. 1985. Inefficiency through government regulations: the case of Norway's Fishery Policy. Marine Resource Economics 2: 115–142.

HANSON, A. J., AND C. LAMSON. 1984. Fisheries decision making in Atlantic Canada, p. 1–13. *In* C. Lamson and A. J. Hanson [ed.] Atlantic fisheries and coastal communities: fisheries decision-making case studies. The Institute for Resource and Environmental Studies, Halifax, N.S.

HARBO, R. 1990. PSARC Working Paper 1–87–4. (1976–1985). DFO Statistical Division (1986–1990).

HARDEN JONES, F. R. 1968. Fish migration. Edward Arnold Ltd., London. 325 p.

HARDIN, G. 1968. The Tragedy of the Commons. Science 162: 1243–1248.

HARE, G. A., AND D. DUNN. 1993. A retrospective analysis of the Gulf of St. Lawrence snow crab (*Chionoecetes opilio*) fishery 1965–1990. *In* L. S. Parsons and W.H. Lear [ed.] Perspectives on Canadian marine fisheries management. Can. Bull. Fish. Aquat. Sci. (In press)

HARHOFF, F. 1983. Greenland's Withdrawal from the European Communities, 20 Common Market L.R. 13.

HARRIS, L. DR. [Chairman]. 1989. Independent review of the state of the northern cod stock. Interim report prepared for the Hon. Tom Siddon, May 15, 1989. 67 p.

HARRIS, L. DR. [Chairman]. 1990. Independent review of the state of the Northern Cod Stock. Final report prepared for the Hon. Tom Siddon, Minister of Fisheries and Oceans. 154 p. + app.

HARTLE, D. G. 1979. Public policy decision making and regulation. Institute for Research on Public Policy. 218 p.

HARTMAN, G. F., AND J.C. SCRIVENER. 1990. Impacts of forestry practices on a coastal stream ecosystem, Carnation Creek, British Columbia. Can. Bull. Fish. Aquat. Sci. 223: 148 p.

HARVEY, H. H. 1975. Fish populations in a large group of acid stressed lakes. Verh. Internat. Verein. Limnol. 19: 2406–2417.

HARVEY, H. H. 1982. Population responses of fishes in acidified waters, p. 227–244. *In* T.A. Haines and R.E. Johnson [ed.] Acid Rain/Fisheries. Proc. Int. Symp. Acid Precipitation and Fishery Impacts in Northeastern North America. Am. Fish. Soc. Bethesda, MD. 357 p.

HARVILLE, J. P. 1975. Multidisciplinary aspects of optimum sustainable yield, p. 59–64. *In* P.M. Roedel [ed.] Optimum sustainable yield as a concept in fisheries management. American Fisheries Society. Allen Press Inc. Lawrence, Texas.

HAWTHORN, H. B., C.S. BELSHAW, AND S.M. JAMIESON. 1958. The Indians of British Columbia: A study of contemporary social adjustment. University of British Columbia Press, Vancouver, B.C. 474 p.

HEALEY, M. C. 1993. The management of Pacific salmon fisheries in British Columbia. *In* L.S. Parsons and W.H. Lear [ed.] Perspectives on Canadian marine fisheries management. Can. Bull. Fish. Aquat. Sci. (In press)

HEALEY, M. C. 1982. Multispecies, multistock aspects of Pacific salmon management, p. 119–126. *In* M.C. Mercer [ed.] Multispecies approaches to fisheries management advice. Can. Spec. Publ. Fish. Aquat. Sci. 59.

HEALEY, M. C. 1984. Multiattribute analysis and the concept of optimum yield. Can. J. Fish. Aquat. Sci. 4 (9): 1393–1406.

HEALTH AND WELFARE CANADA/ENVIRONMENT CANADA. 1983. Report of the Joint Health and Welfare Canada/Environment Canada Expert Advisory Committee on Dioxins, November 1983. 57 p.

HEALTH AND WELFARE CANADA. 1990. Issues. Dioxins and Furans. Health and Welfare Branch. September 20, 1990.

HELGASON, T., AND H. GISLASON. 1979. VPA analysis with species interaction due to predation. ICES CM 1979/G:52.

HEMPEL, G. 1978. North Sea fish stocks — recent changes and their causes. Rapp. P -v. Cons. int. Explor. Mer 172.

HENNESSEY, T., AND M. LEBLANC. 1982. Fishery Administration and Management in Canada and the United States: Some Preliminary Comparisons. Paper presented at the FMG Forum, Digby, Nova Scotia, October, 1982.

HENNESSEY, T., AND D. KAISER. 1987. Fisheries law enforcement: indicators of system performance, p. 201–214. *In* J.L. Bubier and A. Rieser [ed.] East coast fisheries law and policy. Proceedings from the June 17–20, 1986 Conference on East Coast Fisheries Law and Policy. Marine Law Institute, Maine. 482 p.

HENNESSEY, T., AND J. SUTINEN. 1987. Fisheries Law Enforcement in the United States, p. 1–31. *In* J.G. Sutinen and T.M. Hennessey [ed.] Fisheries law enforcement: programs, problems and evaluation. Proceedings of a workshop on fisheries law enforcement, the University of Rhode Island, October 21–23, 1985. NOAA/Sea Grant. Univ. R.I. Mar. Tech. Rep. 93: 237 p.

HERMANN, F. HANSEN, P.M. HANSEN, AND S. HORSTED. 1965. The effect of temperature and currents on the distribution and survival of cod larvae at West Greenland. ICNAF Spec. Publ. 6: 389–395.

HERR, R. A., AND B.W. DAVIS. 1986. The impact of UNCLOS III on Australian federalism. Intl. J. 41: 674–693.

HEWITT, R. P., G.H. THEILACKER, AND N. LO. 1985. Causes of mortality in young jack mackerel. Mar. Ecol. Prog. Ser. 26: 1–10.

HILBORN, R. 1985. Fleet dynamics and individual variation: why some fishermen catch more fish than others. Can. J. Fish. Aquat. Sci. 42: 2–13.

HILBORN, R., AND C.J. WALTERS. 1992. Quantitative fisheries stock assessment. Choice, dynamics and uncertainty. 570 p. Routledge, Chapman & Hill, Inc. New York, N.Y.

HILL, A. V. 1967. Tides of Change: A story of fishermen's co-operatives in British Columbia Prince Rupert: Prince Rupert Fishermen's Co-operative Association, 1967 .

HIMES, J. S. 1980. Conflict and conflict management. University of Georgia Press, Athens, GA. 329 p.

HJORT, J. 1914. Fluctuations in the great fisheries of northern Europe. Rapp. P. -v. Cons. int. Explor. Mer 20: 228 p.

HJORT, J., G. Jahn, and P. Ottestad. 1933. The optimum catch. Hvalrad. Skr. 7: 92–127.

HJUL, P. 1972. The stern trawler. Fishing News (Books) Ltd.

HOAG, S. H., AND R.R. FRENCH. 1976. The incidental catch of halibut by foreign trawlers. International Pacific Halibut Commission. Sci. Rep. 60: 24 p.

HODDER, V. M., AND A.W. MAY. 1964. The effect of catch size on the selectivity of otter trawls. ICNAF Res. Bull. 1:28–35

HODGETTS, J. E. 1955. Pioneer public service. University of Toronto Press, Toronto, Ont. 292 p.

HOLDEN, M. J. 1984. Management of fishery resources: the EEC experience, p. 113–120. *In* Experiences in the management of national fishing zones. Organization for Economic Co-operation and Development.

HOLLICK, A. 1977. The origins of 200-mile offshore zones. Am. J. Int. Law 71: 494–500.

HOLLICK, A. L. 1981. U.S. foreign policy and the law of the sea. Princeton University Press, Princeton, NJ. 496 p.

HOLLOWED, A. B., AND K.M. BAILEY. 1989. Introspectives on the relationship between recruitment of Pacific hake, *Merluccius productus*, and the ocean environment, p. 207–220. *In* R.J. Beamish and G.A. McFarlane [ed.] Effects of ocean variability on recruitment and an evaluation of parameters used in stock assessment models. Can. Spec. Publ. Fish. Aquat. Sci. 108.

HORWOOD, J. W. 1976. Interactive fisheries: a two species Schaefer model. ICNAF Selected Pap. 1: 151–155.

HOURSTON, A. S. 1978. The decline and recovery of Canada's Pacific herring stocks. Can. Fish. Mar. Serv. Tech. Rep. 784: 22 p.

HOURSTON, A. S. 1980a. The biological aspects of management of Canada's west coast herring resource, p. 69–90. *In* B.R. Melteff and V.G. Wespestad [ed.] Proceedings of the Alaska Herring Symposium, February 1980. Alaska Sea Grant Rep. 80–4.

HOURSTON, A. S. 1980b. The decline and recovery of Canada's Pacific herring stocks. Rapp. P.-v. Reun. Cons. int. Explor. Mer 177: 143–153.

HOUSE OF COMMONS DEBATES. 1987. Statement by the Hon. Don Mazankowski. January 30, 1987, p. 2912.

HOUSE, D., M. HANRAHAN, AND D. SIMMS. 1986. Fisheries policies and community development: proposal for a revised approach to managing the inshore fisheries in Newfoundland. A study undertaken for the Royal Commission on Employment and Unemployment, Newfoundland and Labrador. 232 p.

HUNTSMAN, A. G. [Chairman]. 1943. Report of the committee on research. Trans. Can. Conserv. Assoc. for 1942. 2: 71–75.

HUNTSMAN, A. G. 1953. Fishery management and research. J. Cons. Perm. int. Explor. Mer 19(1): 44–45.

HUPPERT, D. 1982. Living marine resources, p. 44–66. *In* G. M. Brown Jr. and J. A. Crutchfield [ed.] Economics of ocean resources: a research agenda, proceedings of a national workshop sponsored by Office of Ocean Resources Coordination and Assessment, National Oceanic and Atmospheric Administration, September 13–16, 1981, Orca Island, Washington.

ICELAND. 1982. Yearbook of trade and industry '82. Iceland Review. 74 p.

ICELAND. 1988. Yearbook of trade and industry '88. Iceland Review. 98 p.

ICELAND. 1989a. Nytjastofnar Sjavar og Umhverfishpaettir. Aflahorfur, 1990. State of marine stocks and environmental conditions in Icelandic water, 1989. Fishing Prospects, 1990. Hafrannsoknastofnun Fjolrit NR. 19.

ICELAND. 1989b. Yearbook of trade and industry '89. Iceland Review.

ICES. Reports of the Advisory Committee on Fishery Management. International Council for the Exploration of the Sea. (Various Years).

ICES. 1984. Report to the North Atlantic Salmon Conservation Organization. ICES Cooperative Research Report 131.

ICES. 1986. Report to the North Atlantic Salmon Conservation Organization. ICES Cooperative Research Report 146.

ICES. 1990. ICES/CIEM Information Bulletin, April 1990. Issue no. 15.

ICES. 1991a. ICES Marine Science Symposia. Multispecies Models Relevant to Management of Living Resources. The Hague, 2–4 October 1989. 358 p.

ICES. 1991b. Extract of the Report of the Advisory Committee on Fishery Management to the North-East Atlantic Fisheries Commission on Herring stocks north or south of 62°N. May 1991.

ICNAF. 1964a. Proceedings of the 14th Annual Meeting of the International Commission for the Northwest Atlantic Fisheries, June, 1964. Dartmouth, Canada.

ICNAF. 1964b. Report of the Standing Committee on Research and Statistics. Redbook, Part I. Proceedings Annual Meeting of the International Commission for the Northwest Atlantic Fisheries. Dartmouth, Canada.

ICNAF. 1967. Report of working group on joint biological and economic assessment of conservation actions. Part 4 Annual Proceedings of the International Commission for the Northwest Atlantic Fisheries. Vol.17 (1966–67).

ICNAF. 1968a. Report of the Working Group on Joint Biological and Economic Assessment of Conservation Actions. Annual Proceedings Vol. 17 for the year 1966–67. 48–84. International Commission for the Northwest Atlantic Fisheries. Dartmouth, Canada.

ICNAF. 1968b. Proceedings of the 18th Annual Meeting of the International Commission for the Northwest Atlantic Fisheries. Dartmouth, Canada.

ICNAF. 1969a. Report of the Standing Committee on Research and Statistics. Proceedings of the 1968 Annual Meeting of the International Commission for the Northwest Atlantic Fisheries. Dartmouth, Canada.

ICNAF. 1969b. Proceedings of the 19th Annual Meeting of the International Commission for the Northwest Atlantic Fisheries. Redbook, Part I. June, 1969. Dartmouth, Canada.

ICNAF. 1970. Proceedings of the Twentieth Annual Meeting of the International Commission for the Northwest Atlantic Fisheries. June, 1970. Dartmouth, Canada.

ICNAF. 1971. Report of the Standing Committee on Research and Statistics. Redbook, Part I. Proceedings of the 1971 Annual Meeting of the International Commission for the Northwest Atlantic Fisheries. Dartmouth, Canada.

ICNAF. 1972a. Report of Assessment Sub-Committee to 1972 Annual Meeting. int. Comm. Northwest Atl. Fish. Annual. Proc. 17- 42.

ICNAF. 1972b. Proceedings of the Special Meeting on Herring of the International Commission for the Northwest Atlantic Fisheries. Dartmouth, Canada.

ICNAF. 1972c. Proceedings of the 22nd Annual Meeting of the International Commission for the Northwest Atlantic Fisheries. Dartmouth, Canada.

ICNAF. 1973a. Report of the Standing Committee on Research and Statistics. Redbook. 1973 Part I. Proceedings of January 1973 Special Meeting of the International Commission for the Northwest Atlantic. Dartmouth, Canada.

ICNAF. 1973b. Proceedings of the 23rd Annual Meeting of the International Commission for the Northwest Atlantic Fisheries. Dartmouth, Canada.

ICNAF. 1974a. Proceedings of the 24th Annual Meeting of the International Commission for the Northwest Atlantic Fisheries. June 1974. Dartmouth, Canada.

ICNAF. 1974b. Handbook of the International Commission for the Northwest Atlantic Fisheries. Dartmouth, Canada. 78 p.

ICNAF. 1975a. Proceedings of the 25th Annual Meeting of the International Commission for the Northwest Atlantic Fisheries, June, 1975. Dartmouth, Canada.

ICNAF. 1975b. Report of the Standing Committee on Research and Statistics. 1975 Redbook. Proceedings of the May–June 1975 Annual Meeting of the International Commission for the Northwest Atlantic. Dartmouth, Canada.

ICNAF. 1975c. Proceedings of the Seventh Special Meeting of the International Commission for the Northwest Atlantic Fisheries. September, 1975. Dartmouth, Canada.

ICNAF. 1976a. Proceedings of the 26th Annual Meeting of the International Commission for the Northwest Atlantic Fisheries, June, 1976. Dartmouth, Canada.

ICNAF. 1976b. Report of the Standing Committee on Research and Statistics. Redbook, 1976. Proceedings of the Seventh Special Meeting of the International Commission for the Northwest Atlantic Fisheries. September, 1975. Dartmouth, Canada.

ICNAF. 1976c. Report of the Standing Committee on Research and Statistics. Redbook, 1976. Proceedings of the May–June, 1976. Annual Meeting of the International Commission for the Northwest Atlantic Fisheries. Dartmouth, Canada.

ICNAF. 1977. Proceedings of the Ninth Special Commission Meeting of the International Commission for the Northwest Atlantic Fisheries, December, 1976. Dartmouth, Canada.

ICNAF. 1979. Proceedings of the Tenth Special Meeting of the International Commission for the Northwest Atlantic Fisheries. March, 1979. Dartmouth, Canada.

ICNAF. 1985. Statistical Bulletin Vol. 28 for the year 1978 (2nd Revision) for the International Commission for the Northwest Atlantic Fisheries. Dartmouth, Canada.

ILES, T. D. 1973. Interaction of environment and parent stock size in determining recruitment in the Pacific sardine as revealed by analysis of density-dependent O-group growth. Rapp. P. -v. Reun. Cons. int. Explor. Mer 164: 228–240.

ILES, T. D., AND M. SINCLAIR. 1982. Atlantic herring: Stock discreteness and abundance. Science 215: 627–633.

INGLIS, G. 1985. More Than Just a Union: The Story of the NFFAWU. St. John's, Newfoundland. Jesperson Press. 331 p.

INNIS, H. A. 1978. The cod fisheries — The history of an international economy. Revised edition. University of Toronto Press, Toronto, Ont. 522 p.

INPFC. Various Years. International North Pacific Fisheries Commission.

INTERNATIONAL COURT OF JUSTICE. 1984. Delimitation of the Maritime Boundary in the Gulf of Maine Area (Canada v United States of America), Judgement of the Chamber of the International Court of Justice, 12 October 1984.

IPCC. 1990. Report of the Intergovernmental Panel on Climate Change. First Assessment Report. Overview. 31 38August 1990.

IPHC. 1989. Annual Report. International Pacific Halibut Commission Celebrating the Pacific Halibut Fishery Centennial 1988–1989, 62 p.

IPHC. 1990. Annual Report. International Pacific Halibut Commission 1989, 39 p.

JACKSON R.I., AND W.F. ROYCE. 1986. Ocean Forum: an interpretative history of the International North Pacific Fisheries Commission. Fishing News Books Ltd. Farnham, England. 239 p.

JAKOBSSON, J. 1985. Monitoring and management of the Northeast Atlantic herring stocks. Can. J. Fish. Aquat. Sci. 42(1): 207–221.

JAMES, M. 1987. Review of Indian participation in the 1985 commercial salmon fishery. Unpublished Manuscript, Department of Fisheries and Oceans, Vancouver.

JAMIESON, G. S., AND W.D. MCKONE [ed.]. 1988. Proceedings of the International Workshop on Snow Crab Biology, December 8–10, 1987, Montreal, Quebec, Can. Ms. Rep. Fish. Aquat. Sci. 2005: 163 p.

JENSEN, T. C. 1986. The United States — Canada Pacific Salmon Interception Treaty: An historical and legal overview. Environ. Law. 16 (3): 363–422.

JOHANSEN, A. C. 1927. On the fluctuations in the quantity of young fry among plaice and certain other species of fish, and causes of the same. Rep. Danish Biol. Sta. 33: 1–16.

JOHNSON, B. 1977. Canadian foreign policy and fisheries, p. 53–95. In B. Johnson and M.W. Zacher [ed.].Canadian Foreign Policy and the Law of the Sea.

JOHNSON, B., AND M.W. ZACHER [ed.]. 1977. Canadian Foreign Policy and the Law of the Sea. University of British Columbia Press, Vancouver, B.C. 387 p.

JOHNSTON, D. M. 1978. The administration of Canadian fisheries. Halifax, N.S. Dalhousie University, Institute for research and environmental studies. 45 p.

JOHNSTON, D. M. 1988. The theory and history of ocean boundary-making. McGill-Queen's University Press. 445 p.

JOHNSTONE, K. 1977. The aquatic explorers: a history of the Fisheries Research Board of Canada. Toronto, Ont. University of Toronto Press, Toronto, Ont. 342 p.

JONES, R. 1961. The assessment of long-term effects of changes in gear selectivity and fishing effort. Mar. Res. (Scotland) (2): 1–19.

JONES, R. 1979. Materials and methods used in marking experiments in fishery research, FAO, Fish Tech. Pap. T190 134 p.

KARPOFF, J. M. 1985. Non-pecuniary benefits in commercial fishing: empirical findings from the Alaska Salmon Fisheries. Econ. Inquiry 23: 159–174.

KAY, B. H. 1989. Pollutants in British Columbia's marine environment. A Status Report. A state of the environment report. Department of the Environment. 59 p.

KEARNEY, J. F. 1984. The Transformation of the Bay of Fundy Herring Fisheries 1976–1978: An experiment in fishermen-government co-management, p. 165–203. *In* C. Lamson and A.J. Hanson [ed.] Atlantic fisheries and coastal communities: fisheries decision-making case studies. Dalhousie Ocean Studies Programme. The Institute for Resource and Environmental Studies, Halifax, N.S.

KEATING, T. 1983–4. Domestic groups, bureaucrats, and bilateral fisheries relations. Int. J. 39: 146–170.

KELSO, J. R. M., M.A. SHAW, C.K. MINNS, AND K.H. MILLS. 1990. An evaluation of the effects of atmospheric acidic deposition on fish and the fishery resource of Canada. Can. J. Fish. Aquat. Sci. 47: 644–655.

KERR, S. R., AND R.A. RYDER. 1989. Current approaches to multispecies analyses of marine fisheries. Can. J. Fish. Aquat. Sci. 46: 528–534.

KETCHEN, K. S. 1956. Climatic trends and fluctuations in yield of marine fisheries in the North Pacific. J. Fish. Res. Board Can. 13: 357–374.

KIRBY, M. J. L. [Chairman]. 1982. Navigating troubled waters — a new policy for the Atlantic fisheries. Report of the Task Force on Atlantic Fisheries. Supply and Services, Ottawa. 379 p.

KIRBY, M. J. L. 1984. Restructuring the Atlantic Fishery: A Case Study in Business — Government Relations. Paper. Dalhousie Univ. School of Business 1 March 1984. 55 p.

KOERS, A. W. 1989. What Trends and Implications? The Northeast Atlantic: EEC, p. 77–120. *In* Management of World Fisheries: Implications of Extended Coastal State Jurisdiction, E.L. Miles (ed), 318 p. University of Washington Press, Seattle, Washington.

KOSLOW, J. A. 1984. Recruitment patterns in northwest Atlantic fish stocks. Can. J. Fish. Aquat. Sci. 41: 1722–1729.

KOSLOW, J. A., K.R. THOMPSON, AND W. SILVERT. 1987. Recruitment to northwest Atlantic cod (*Gadus morhua*) and haddock (*Melanogrammus aeglefinus*) stocks: influence of stock size and climate. Can. J. Fish. Aquat. Sci. 44: 26–39.

KRIESBERG, L. 1982. The sociology of social conflicts. Prentice-Hall Inc. Englewood Cliffs, New Jersey. 300 p.

LADD, E. C. 1985. The American Polity. The people and their government. W.W. Norton & Co. Ltd. New York, NY. 678 + app.

LAMSON, C. 1986. On the line: women and fish plant jobs in Atlantic Canada. Industrial Relations. 41(1): 145–156.

LANE, D. E. 1988. Investment decision making by fishermen. Can. J. Fish. Aquat. Sci. 45: 782–796.

LAPOINTE, M. F., R.M. PETERMAN, AND A.D. MACCALL. 1989. Trends in fishing mortality rate along with errors in natural mortality rates can cause spurious time trends in fish stock abundances estimated by virtual population analysis (VPA). Can. J. Fish. Aquat. Sci. 46: 2129– 2139.

LARKIN, P. A. 1971. Simulation studies of the Adams river sockeye salmon, *Oncorhynchus nerka*. J. Fish. Res. Board Can. 28: 1493–1502.

LARKIN, P. A. 1977. An epitaph for the concept of maximum sustained yield. Trans. Am. Fish. Soc. 106(11): 1–11.

LARKIN, P. A. 1978. Fisheries management — an essay for ecologists. Annual. Rev. Ecol. Syst. 57–73.

LARKIN, P. A. 1979. Maybe you can't get there from here: a foreshortened history of research in relation to management of Pacific salmon. J. Fish. Res. Board Can. 36: 98–106.

LARKIN, P. A. 1980. Objectives of management, p. 245–262. *In* R.T. Lackey and L.A. Nielsen [ed.] Fisheries management.

LARKIN, P. A. 1988. Pacific salmon, p. 153–184. *In* J.A. Gulland [ed.] Fish population dynamics (second edition): The implications for management. John Wiley & Sons, Ltd., Great Britian.

LARKIN, P. A. 1989. Before and after Ricker (1954) and Beverton and Holt (1957), p. 5–9. *In* R.J. Beamish and G.A. McFarlane [ed.]. Effects of ocean variability on recruitment and an evaluation of parameters used in stock assessment models. Proceedings of the international symposium held at Vancouver, B.C., October 26–29, 1987. Can. Spec. Publ. Fish. Aquat. Sci. 108.

LASKER, R. 1975. Field criteria for survival of anchovy larvae: the relation between inshore chlorophyll maximum layers and successful first feeding. Fish. Bull. US 73: 453–462.

LASKER, R. 1978. The relation between oceanographic conditions and larval anchovy food in the California current: identification of factors contributing to recruitment failure. Rapp. P. -v. Reun. Cons. int. Explor. Mer 173: 212–230.

LASKER, R. 1981. Factors contributing to variable recruitment of the northern anchovy (*Engraulis mordax*) in the California current: contrasting years, 1975 through 1978. Rapp. P.-v. Reun. Cons. int. Explor. Mer 178: 375–388.

LAUBSTEIN, K. 1990. Canadian Atlantic groundfish management: Introduction of enterprise allocation (EA) program, p. 641–668. *In* H. Frost and P. Andersen [ed.] Seafood trade, fishing industry structure and fish stocks — the economic interaction. The Institute, Corvallis, Ore.

LAURSEN, F. 1985. Norwegian marine policy. Mar. Policy Rep. 8: 1–6.

LEAK, J. C., AND E.D. HOUDE. 1987. Cohort growth and survival of bay anchovy *Anchoa mitchilli* larvae in Biscayne Bay, Florida. Mar. Ecol. Prog. Ser. 37: 109–122.

LEAR, W. H. 1993. Management of the Canadian Atlantic Salmon Fishery. *In* L.S. Parsons and W.H. Lear [ed.] Perspectives on Canadian marine fisheries management. Can. Bull. Fish. Aquat. Sci. (In press)

LEAR, W. H. 1985. Atlantic Cod. Underwater World. DFO.

LEAR, W. H., AND L.S. PARSONS. 1993. History and management of the northern cod fishery. *In* L.S. Parsons and W.H. Lear [ed.]. Perspectives on Canadian marine fisheries management. Can. Bull. Fish. Aquat. Sci. (In press)

LEGAULT, L. H. 1974. Maritime Claims: 377–397 *In* Canadian perspectives on international law and organization. Macdonald, Morris and Johnston [ed.].

LEGAULT, L. H. 1977. Notes for statement at Canadian Labour Congress Fisheries Meeting. Ottawa, January 13, 1977.

LEGAULT, L. H. 1985. A line for all uses: the Gulf of Maine boundary revisited. Intl. J. 40: 461–477.

LEGAULT, L. H., AND D. M. MCRAE. 1984. The Gulf of Maine Case. Notes and Comments. Canadian Yearbook of International Law. 22: 267–290.

LEGAULT, L. H. S., AND B. HANKEY. 1985. From sea to seabed: the single maritime boundary in the Gulf of Maine case. Am. J. Intl. Law. 79: 961–991.

LEGGETT, W. C. 1988. Comments: 222–225. *In* Report from the National Workshop on Recruitment. M. Sinclair et al. [ed.]. Can. Tech. Rep. Fish. Aquat. Sci. 1626.

LEGGETT, W. C., K.T. FRANK, AND J.E. CARSCADDEN. 1984. Meteorological and hydrographic regulation of year-class strength in capelin (*Mallotus villosus*). Can. J. Fish. Aquat. Sci. 41: 1193–1201.

LETT, P. F. 1977. A preliminary discussion of the relationship between population energetics and the management of southern Gulf of St. Lawrence cod. Can. Atl. Fish. Sci. Adv. Comm. Res. Doc.77/78.

LETT, P. F. 1978. A multispecies simulation for the management of the southern Gulf of St. Lawrence cod stock. Can. Atl. Fish. Sci. Adv. Comm. Res. Doc. 78/21.

LEVELTON, C. R. 1979. Toward an Atlantic Coast Commercial Fisheries Licensing System: a report prepared for the Department of Fisheries and Oceans Government of Canada. 95 p. + app.

LEWIS, T. R. 1982. Stochastic modelling of ocean fisheries resource management. University of Washington Press, Seattle, WA. 109 p.

LILBURN, B., AND N. MOLLETT [ed.] 1986. Management of Australian fisheries: broad development and alternative strategies, p. 141–187. *In* Fishery access control programs worldwide: proceedings of the workshop on management options for the North Pacific longline fisheries, Orcas Island, Washington, April 21–25, 1986.

LILBURN, B. 1987. Formulation of fisheries management plans. Indo-Pacific Fishery Commission (FAO): symposium on the exploitation and management of marine fishery resources in south east Asia, Darwin, Australia, 16–17 Feb. 1987. 507–527.

LILLY, G. R. 1987a. Interactions between Atlantic cod (*Gadus morhua*) and capelin (*Mallotus villosus*) off Labrador and eastern Newfoundland: a review. Can. Tech. Rep. Fish. Aquat. Sci. 1567. 37 p.

LILLY, G. R. 1987b. Synopsis of research related to recruitment of Atlantic cod (*Gadus morhua*) and Atlantic redfishes (*Sebastes sp.*) on Flemish Cap. NAFO Sci. Counc. Studies 11: 109–122.

LINDBLOM, C. E. 1959. The science of "Muddling Through". Public Admin. Rev. 19(2): 79–88.

LOCKHART, W. L., R. WAGEMANN, B. TRACEY, D. SUTHERLAND, AND D.J. THOMAS. 1992. Presence and implications of chemical contaminants in the freshwaters of the Canadian Arctic. 165–243. Science of Total Environment 122(1/2). Elsevier Science Publishers B.V., Amsterdam. 278 p.

LOEWEN, R. D. 1983. Tactics Employed by Senior Educational Administrators When Engaged in Conflict. Unpublished doctoral dissertation, University of Alberta, Edmonton, Alberta.

LYONS, C. 1969. Salmon: our heritage. Mitchell Press, Vancouver, B.C.

MAC. 1983. General Discussion Paper: A first step towards rationalizing the West Coast Fisheries. Minister's Advisory Council. September 28, 1983.

MACDONALD, D. 1980. 'Power Begins at the Cod End', The Newfoundland Trawlermans' Strike, 1974–75. Social & Econ. Studies No. 26. Inst. of Social and Econ. Research, Mem. Univ. of Nfld. 158 p.

MACDONALD, J. D. 1981. The Public Regulation of Commercial fisheries in Canada. Case Study No. 4. The Pacific Salmon Fishery. Econ. Counc. Can. Tech. Rep. 19. 168 p.

MACE, P. M. 1985. Catch rates and total removals in the 4WX herring purse seine fisheries. CAFSAC Res. Doc. 85/74. 31 p.

MACFARLANE, D. A. 1989. An economic assessment of the herring purse-seine fleet restructuring plan of 1983, the ten-year management plan at mid-term. Econ. Comm. Anal. Rep. No. 31, Department of Fisheries and Oceans. 49 p.

MACGILLIVRAY, P. 1986. Evaluation of Area Licensing in the British Columbia Roe Herring Fishery: 1981–1985, p. 251–274. *In* N. Mollett [ed.] Fishery Access Control Programs Worldwide: Proceedings of the Workshop on Management Options for the North Pacific Longline Fisheries, Orcas Island, Washington, April 21–25, 1986.

MACGILLIVRAY, P. B. 1990. Assessment of New Zealand's individual transferable quota fisheries management. Econ. Comm. Anal. Rep. No. 75: 19 p.

MACLEAN, A. K. 1928. Report of the Royal Commission investigating the fisheries of the maritime provinces and the Magdalen Islands. Ottawa, King's Printer. 125 p.

MAHON, R. [ed.]. 1985. Towards the inclusion of fishery interactions in management advice. Proceedings of a workshop at the Bedford Institute of Oceanography (October 30 – November 1, 1984). Can. Tech. Rep. Fish. Aquat. Sci. 1347: iii + 221 p.

MAIOLO, J. R., AND M.K. ORBACH. 1982. Modernization and Marine Fisheries Policy. Ann Arbor Science, Ann Arbor, MI.

MALOUF, A. H. [Chairman]. 1986. Report of the Royal Commission on Seals and Sealing in Canada. Supply and Services Canada. Ottawa. Volumes 1 to 3.

MANN, K. H., AND J.R.N. LAZIER. 1991. Dynamics of Marine Ecosystems. Biological-Physical Interactions in the Oceans. Blackwell Scientific Publications. Oxford. 466 p.

MANZER, R. 1984. Public policy-making as practical reasoning. Can. J. Pol. Sci. XVII(3): 577–594.

MARCHAK, P., N. GUPPY, AND J. MCMULLAN [ed.]. 1987. Uncommon property: the fishing and fish-processing industries in British Columbia. Methuen, Toronto, Ont. 402 p.

MARR, J. C. 1960. The causes of major variations in the catch of the Pacific sardine (*Sardinops caerulea*). Proc. World Sci. Meet. Sardines and Related Sp. 3: 667–791. FAO, Rome.

MARTIN, W. R., AND A.C. KOHLER. 1965. Variation in recruitment of cod (*Gadus morhua*) in southern ICNAF waters, as related to environmental changes. ICNAF Spec. Publ. 6: 833–846.

MARTINEAU, D., P. BELAND, C. DESJARDINS, AND A. LEGACE. 1987. Levels of organochlorine chemicals in tissues of beluga whales *Delphinapterus leucus* from the St. Lawrence estuary. Arch. Environ. Contam. Toxicol. 16(2): 137– 148.

MARTINEAU, D., A. LEGACE, P. BELAND, R. HIGGINS, D. ARMSTRONG, AND L.R. SHUGART. 1988. Pathology of stranded beluga whales *Delphinapterus leucas* from the St. Lawrence estuary Quebec Canada. J. Comp. Pathol. 98: 287–312.

MASSACHUSETTS. 1990. New England Groundfish In Crisis—Again. The Report of the Massachusetts Offshore Groundfish Task Force. 33 p.

MAY, R. M., J. R. BEDDINGTON, C.W. CLARK, S.J. HOLT, AND R.M. LAWS. 1979. Management of multispecies fisheries. Science 205: 267–277.

MAY, R. M. [ed.]. 1984. Exploitation of marine communities. Springer-Verlag, Berlin.

MCCAUGHRAN, D. A., AND R.B. DERISO. 1988. Effect of biological assumptions on halibut production estimates. p. 5–28. *In* W.S. Wooster [ed.] Lecture notes on coastal and estuarine studies No. 28. fishery science and management: objectives and limitations. Springer-Verlag.

MCCAUGHREN, D. A. 1989. Director's Report in Annual Report for 1988. International Pacific Halibut Commission. Seattle, Washington.

MCCORQUODALE, S. 1988. Fisheries and oceans: 1977–1987, p. 139–163. In How Ottawa Spends. 1988/89. The conservatives heading into the stretch. Carleton University Press Inc. Toronto, Ont.

MCCRACKEN, F. D., AND R. D. S. MACDONALD. 1976. Science for Canada's Atlantic inshore seas fisheries. J. Fish. Res. Board Can. 33: 2097–2139.

MCCRACKEN, F. D. 1979. Canso marine environment workshop part 1 of 4 parts: executive summary. Fish. & Mar. Serv. Tech. Rep. 834: 17 p.

MCDONALD, I. 1974. W.F. Croaker and the Balance of Power Strategy: The Fishermens' Protective Union in Newfoundland Politics. Atlantic Canada Studies Conference, Fredericton, 1974.

MCDONALD, J. 1981. The stock concept and its application to British Columbia salmon fisheries. J. Fish. Res. Board Can. 38: 1657–1664.

MCDORMAN, T. L. 1992. The Canada-United States Free Trade Agreement and the Canadian Fishing Industry. 433–462. *In* Canadian Ocean Law and Policy. D. VanderZwaag (ed), 1992, 546 p.

MCDORMAN, T. L., P. M. SAUNDERS, AND D.L. VANDERZWAAG. 1985. The Gulf of Maine boundary: dropping anchor or setting a course? Mar. Policy 9: 90–107.

MCHUGH, P. D. 1985. International law: delimitation of maritime boundaries. Nat. Res. J. 25: 1025–1038.

MCRAE, D. M. 1983. The Gulf of Maine Case: The Written Proceedings. The Canadian Yearbook of International Law 1983: 266–283.

MEGREY, B. A., AND V.G. WESPESTAD. 1990. Alaskan groundfish resources: 10 years of management under the Magnuson fishery conservation and management act. N. Am. J. Fish. Man. 10: 125–143.

MENSINKAI, S. S. 1969. Plant location and plant size in the fish processing industry of Newfoundland. DFO, Ottawa, Ont. 92 p.

MERCER, M. C. [ed.]. 1982. Multispecies approaches to fisheries management advice. Can. Spec. Publ. Fish. Aquat. Sci. 59: 169 p.

MERTON, R. K. 1957. Social theory and social structure. Rev. Ed. Fress Press. Glencoe. Ill.

MESNIL, B. 1980. Theorie et pratique de l'analyse des cohorts. Rev. Trav. Inst. Pêch. Marit. 44: 119–155.

METHOT, R. D. 1986. Synthetic estimates of historical abundance and mortality for northern anchovy, *Engraulis mordax*, NOAA Natl. Mar. Fish. Serv. Southwest Fish. Cent. Admin. Rep. LJ-86-29: 1–85.

MEXICO. 1945. UN Legislative Series: Laws and Regulations on the Regime of the High Seas, 1951, ST/LEG/SER.B/1, p. 13.

MEYER, P. A. 1976. A review of the herring roe licensing program — 1975. Social Science Unit, Fisheries and Marine Service, Pacific Region (Internal Report). 14 p.

MILES, E. L. [ed.]. 1989. Management of world fisheries: Implications of extended coastal state jurisdiction. University of Washington Press, Seattle, WA. 318 p.

MILES, E. L., AND L. GESELBRACHT. 1986. Fisheries and conflicting uses of the sea, p. 146–178. *In* G. Ulfstein, P. Andersen and R. Churchill [ed.].The regulation of fisheries: legal, economic and social aspects. Proceedings of a European Workshop, University of Tromso, Norway, 2–4 June, 1986.

MILES, E. L., S. GIBBS, D. FLUHARTY, C. DAWSON, D. TEETER, W. BURKE, W. KACZYNSKI, AND W. WOOSTER. 1982. The management of marine regions: the north Pacific. An analysis of issues relating to fisheries, marine transportation, marine scientific research, and multiple use conditions and conflicts. University of California Press, Berkeley, CA. 656 p.

MILLIMAN, S. R. 1986. Optimal fishery management in the presence of illegal activity. J. Environ. Econ. Manage. 13: 363–381.

MINNS, C. K., J.R.M. KELSO, AND M.G. JOHNSON. 1986. Large-scale risk assessment of acid rain impacts on fisheries: models and lessons. Can. J. Fish. Aquat. Sci. 43: 900–921.

MITCHELL, C. L., AND P. LENNON. 1975. An inquiry into the east coast herring industry with special reference to the Bay of Fundy and southwestern Nova Scotia areas. Department of the Environment. 98 p.

MOCKLINGHOFF, G. 1973. Management and development of fisheries in the North Atlantic. J. Fish. Res. Board Can. 30: 2402–2418.

MOHN, R. K., AND G. ROBERT. 1984. Comparison of two harvesting strategies for the Georges Bank scallop stock. CAFSAC Res. Doc. 84/10: 35 p.

MOLONEY, D. G., AND P.H. PEARSE. 1979. Quantitative Rights as an Instrument for Regulating Commercial Fisheries. J. Fish. Res. Board Can. 36: 859–866.

MOORES, R., D. BOLLIVAR, AND P. NICHOLSON. 1981. Enterprise Allocations. A Working Paper Prepared by Officials of: The Lake Group Limited, National Sea Products Limited and H.B. Nickerson and Sons Limited. November 10, 1981. 17 p.+ app.

MUIR, D. C. G., R. WAGEMANN, B.T. HARGRAVE, D.J. THOMAS, D.B. PEAKALL, AND R.J. NORSTROM. 1992. Arctic marine ecosystem contamination. 75–134. Science of Total Environment 122(1/2). Elsevier Science Publishers B.V., Amsterdam. 278 p.

MUNRO, G. R. 1980. A Promise of Abundance: extended fisheries jurisdiction and the Newfoundland economy. 111 p. Economic Council of Canada.

MUNRO, G. R. 1982. Fisheries, extended jurisdiction and the economics of common property resources. Can. J. Econ. 15(3): 405–425.

MURAWSKI, S. A., A.M. LANGE, AND J.S. IDOINE. 1989. An analysis of technological interactions among Gulf of Maine mixed-species fisheries. ICES CM. No. 7 1989.

MURPHY, G. I. 1965. A solution of the catch equation. J. Fish. Res. Board Can. 22: 191–202.

MUSE, B., AND K. SCHELLE. 1988. New Zealand's ITQ Program CFEC 88-3. Alaska Commercial Fisheries Entry Commission, Juneau, Alaska. 46 p.

NAFO. 1979. Scientific Council Report of the Northwest Atlantic Fisheries Organization. Dartmouth, Canada.

NAFO. 1981a. Proceedings of the Third Special Meeting of the Fisheries Commission. June, 1981. Dartmouth, Canada.

NAFO. 1981b. Proceedings of the Third Annual Meeting of the Fisheries Commission. September, 1981. Dartmouth, Canada.

NAFO. 1982. Scientific Council Report of the Northwest Atlantic Fisheries Organization. Dartmouth, Canada.

NAFO. 1983. Proceedings of the Fifth Annual Meeting of the Fisheries Commission. September, 1983. Dartmouth, Canada.

NAFO. 1984a. Proceedings of the Sixth Annual Meeting of the Fisheries Commission. September, 1984. Dartmouth, Canada.

NAFO. 1984b. Handbook of the Northwest Atlantic Fisheries Organization. Dartmouth, Canada. 103 p.

NAFO. 1985a. Proceedings of the Seventh Annual Meeting of the Fisheries Commission. September, 1985. Dartmouth, Canada.

NAFO. 1985b. Report of the NAFO Scientific Council, January 1985. Dartmouth, Canada.

NAFO. 1986a. Proceedings of the Eighth Annual Meeting of the Fisheries Commission. September, 1986. Dartmouth, Canada.

NAFO. 1986b. Report of the NAFO Scientific Council. Redbook. September 1986. Dartmouth, Canada.

NAFO. 1987. Proceedings of the Ninth Annual Meeting of the Fisheries Commission. September, 1987. Dartmouth, Canada.

NAFO. 1988a. Proceedings of the Tenth Special Meeting of the Fisheries Commission. February, 1988. Dartmouth, Canada.

NAFO. 1988b. Report of the NAFO Scientific Council. Redbook. September 1988. Dartmouth, Canada.

NAFO. 1988c. Proceedings of the Tenth Annual Meeting of the Fisheries Commission. September, 1988. Dartmouth, Canada.

NAFO. 1988d. Proceedings of the Tenth Annual Meeting of the General Council of NAFO. September, 1988. Dartmouth, Canada.

NAFO. 1988e. Report of the NAFO Scientific Council, June, 1988. Dartmouth, Canada.

NAFO. 1989a. Proceedings of the Eleventh Annual Meeting of the General Council of NAFO. September, 1989. Dartmouth, Canada.

NAFO. 1989b. Proceedings of the Eleventh Annual Meeting of the Fisheries Commission. September, 1989. Dartmouth, Canada.

NAFO. 1990a. Proceedings of the Twelfth Annual Meeting of the Fisheries Commission. September, 1990. Dartmouth, Canada.

NAFO. 1990b. Proceedings of the Twelfth Annual Meeting of the General Council of NAFO. September, 1990. Dartmouth, Canada.

NAFO. 1990c. Report of the NAFO Scientific Council Special Meeting. September, 1990. Dartmouth, Canada.

NAFO. 1992a. Report of the Special Meeting of the NAFO Scientific Council, June 1992. NAFO SCS. Doc. 92/20.

NAFO. 1992b. Report of Scientific Council Meeting, June, 1992. NAFO SCS Doc. 92/23.

Naidu, K. S. 1984. An analysis of the scallop meat count regulation. CAFSAC Res. Doc. 84/73. 18 p.

NASCO. 1984a. North Atlantic Salmon Conservation Organization. West Greenland Commission. Report of the First Annual Meeting. 23–25 May, 1984 and 18 July, 1984. Edinburgh, U.K.

NASCO. 1984b. North Atlantic Salmon Conservation Organization. North American Commission. Report of the First Annual Meeting. 3–4 May, 1984, Ottawa and 22–25 May, 1984. Edinburgh, U.K.

NASCO. 1985a. North Atlantic Salmon Conservation Organization. West Greenland Commission. Report of the Second Annual Meeting. 3–7 June, 1985, Edinburgh, U.K.

NASCO. 1985b. North Atlantic Salmon Conservation Organization. North American Commission. Report of the Second Annual Meeting. 21–22 February, 1985, Boston, USA and 3–7 June, 1985, Edinburgh, U.K.

NASCO. 1986a. North Atlantic Salmon Conservation Organization. West Greenland Commission. Report of the Third Annual Meeting. 23–27 June, 1986, Edinburgh, U.K.

NASCO. 1986b. North Atlantic Salmon Conservation Organization. North American Commission. Report of the Third Annual Meeting. 5–6 February, 1986, Quebec City, Canada. 23–27 June, 1986, Edinburgh, U.K.

NASCO. 1988a. North Atlantic Salmon Conservation Organization. Handbook. Basic Texts. Edinburgh, U.K. 135 p.

NASCO. 1988b. North Atlantic Salmon Conservation Organization. West Greenland Commission. Report of the Fifth Annual Meeting. 13–17 June, 1986, Edinburgh, U.K.

National Fisheries Institute. 1979. Statement of Gordon D. Murphy, President, National Fisheries Institute before the House Subcommittee on Fisheries and Wildlife Conservation and the Environment of the Merchant Marine and Fisheries Committee. June 22, 1979.

Needler, A. W. H. 1979. Evolution of Canadian fisheries management towards economic rationalisation. J. Fish. Res. Board Can. 36: 716–724.

Needler, A. W. H. 1987. The St. Andrews heritage. St. Andrews Biol. Sta. Newsl. 1(1): 1–2.

Neher, P. A., R. Arnason, and N. Mollett. 1989. Rights based fishing: proceedings of the NATO Advanced Research Workshop on Scientific Foundations for Rights Based Fishing, Reykjavik, Iceland, June 27–July 1, 1988. 541 p.

Nelson, W. M., M. Ingham, and W. Schaaf. 1977. Larval transport and year class-strength of Atlantic menhaden, *Brevoortia tyrannus*. US Fish. Bull 75: 23–41.

New Zealand Fishing Industry Board. 1986. Economics Division. "Economic Review of the New Zealand Fishing Industry 1985–85" Wellington: July, 1986.

Newfoundland. 1946. Newfoundland Fisheries Board. Report of the Fisheries Post-War Planning Committee, St. John's.

Newfoundland. 1979a. The Position of the Government of Newfoundland and Labrador on the Harvesting of the 2J + 3KL Cod Stocks. Presented at the Government/Industry Seminar on Northern Cod. August 28–30, 1979. Corner Brook, Newfoundland. Report of the Northern Cod Seminar: 207–214.

Newfoundland. 1979b. Discussion Paper on Fisheries Issues. Presented by Premier A.B. Peckford to The Hon. J. McGrath, Minister of Fisheries and Oceans. November 5, 1979.

Newfoundland. 1983a. Restructuring the Fishery. A detailed presentation by the Government of Newfoundland and Labrador to the Government of Canada, May 6, 1983.

Newfoundland. 1983b. Statement by the Hon. A. Brian Peckford, Premier of Newfoundland and Labrador. June 30th, 1983.

Newfoundland. 1983c. Restructuring the Newfoundland Fishery: the story has not been told. Statement by the Hon. A. Brian Peckford, Premier of Newfoundland and Labrador. August 18, 1983.

Newfoundland. 1985a. Appropriate offshore fish harvesting technology: an assessment of the detrimental effects of the use of factory freezer trawlers. Discussion paper. 32 p.

Newfoundland. 1985b. Statement to the House of Assembly by the Hon. A. Brian Peckford, Concerning Factory Freezer Trawlers, November 8, 1985.

Newfoundland. 1986. Building on Our Strengths. Final Report of the Newfoundland Royal Commission on Employment and Unemployment. 515 p.

Newfoundland. 1987. Canada/France Fisheries Agreement. A Secret Sellout of Northern Cod. Booklet issued by the Government of Newfoundland and Labrador, February 17, 1987.

Newfoundland. 1988. Northern Cod Under Attack: The Newfoundland and Labrador Perspective on the Allocation of Northern Cod. 11 p.

Newfoundland. 1992. Presentation by Honourable Walter C. Carter, Minister of Fisheries to House of Commons Standing Committee on Forestry and Fishing. December 8, 1992.

Newton, C. H. B. 1978. Experience with limited entry in British Columbia fisheries, p. 382–390. *In* R.B. Rettig, and J.J.C. Ginter [ed.] Limited entry as a fishery management tool. University of Washington Press, Seattle and London.

Nickerson, and National Sea Products. 1978. Where now? A Discussion of Canada's Fishery Opportunity and the Considerations Involved in Realizing It. Financial Post, April 29, 1978.

NOAA. 1986. National Oceanic and Atmospheric Administration fishery management study. 63 p.

Northwest Falling Contractors Ltd. v. R. 1980. 25.C.R. 292, 113 D.L.R. (3d) 1.

Norway. 1991–92. st. Meld nr. 58 (1991–92): Om struktur- og reguleringspolitikk overfor fiskeflaten. (Strukturmeldingen).

Nova Scotia Fish Packers Association. 1979. Freezing at sea — a Canadian opportunity. Discussion paper prepared for a seminar on northern cod, sponsored by Canada Department of Fisheries and Oceans. Corner Brook, August, 1979.

NPASC. 1992. Convention for the Conservation of Anadromous Stocks in the North Pacific Ocean. Signed at Moscow February 11, 1992.

NSP. 1985. An application for a licence to operate a factory freezer trawler. National Sea Products. June, 1985.

NSP. 1990. National Sea Products Limited. Annual Information Form and Management's Discussion and Analysis of Financial Condition and Results of Operation.

NTA. 1989. 1989 Northern Trollers Association Salmon Allocation Plan. May, 1989.

O'Boyle, R., J. Rice, and J.J. Maquire. 1985. The current approach of ICES to multispecies assessment, p. 201–208. *In* R. Mahon [ed.] Towards the inclusion of fishery interactions in management advice. Can. Tech. Rep. Fish. Aquat. Sci. 1347.

O'Boyle, R. N. 1985. A description of the two-tier catch quota system of ICNAF, with commentary on its potential usefulness in current fisheries management on the Scotian shelf, p. 196–200. *In* R. Mahon [ed.] Towards the inclusion of fisheries interactions in management advice. Can. Tech. Rep. Fish. Aquat. Sci. 1347.

Oden, S. 1968. The acidification of air precipitation and its consequences in the natural environment. Ecology Committee Bulletin No. 1. Swedish National Science Research Council, Stockholm. Translation Consultants Ltd., Arlington, VA, USA. 117 p.

OECD. 1979. Review of Fisheries in OECD Member Countries in 1978. Organisation for Economic Co-operation and Development. Paris 1979.

OECD. 1981. Review of Fisheries in OECD Member Countries in 1980. Organisation for Economic Co-operation and Development. Paris 1981.

OECD. 1983. Review of Fisheries in OECD Member Countries in 1982. Organisation for Economic Co-operation and Development. Paris 1983.

OECD. 1985. Review of Fisheries in OECD Member Countries in 1984. Organisation for Economic Co-operation and Development. Paris 1985.

OECD. 1986. Review of Fisheries in OECD Member Countries in 1985. Organisation for Economic Co-operation and Development. Paris 1986.

OECD. 1989. Review of Fisheries in OECD Member Countries in 1988. Organisation for Economic Co-operation and Development. Paris 1989.

OECD. 1990. Review of Fisheries in OECD Member Countries in 1989. Organisation for Economic Co-operation and Development. Paris 1990.

Overholtz, W. J., M.P. Sissenwine, and S.H. Clark. 1986. Recruitment variability and its implication for managing and rebuilding the Georges Bank haddock (*Melanogrammus aeglefinus*) stock. Can. J. Fish. Aquat. Sci. 43: 748–753.

Ozere, S. V. 1973. Needed: sea law to protect sea resources. Can. Geograph. J. 87: 4010.

Pacific Salmon Commission. 1986. Report Technical Committee: Chinook (86) 1. Final 1985 Report of the Chinook Technical Committee. 54 p.

Pacific Salmon Commission. 1988. Brochure describing the provisions of the Pacific Salmon Treaty and the structure of the Pacific Salmon Commission, Vancouver, B.C.

Pallozzi, M., and S. Springer. 1985. Enforcement costs in fisheries management: The alternatives. Workshop on Fisheries Law Enforcement, Oct. 21–23, University of Rhode Island, RI.

Panama. 1946. Amendment to the Constitution of Panama, Article 209, Decree No. 9938 March 1, 1946. UN Legislative Series: Laws and Regulations on the Regime of the High Seas, 1951, ST/LEG/SER.B/1, p. 15.

Panayotou, T. 1982. Management concepts for small-scale fisheries: economic and social aspects. FAO Fish. Tech. Pap. No. 228.

Panayotou, T., and D. Panayotou. 1986. Occupational and geographical mobility in and out of Thai fisheries. FAO Tech. Pap. No. 271.

Panting, G. E. 1963. The Fishermen's Protective Union of Newfoundland and the Farmers' Organizations in Western Canada, Canadian Historical Association Report, 1963.

Paredes, J. A. 1985. "Any comments on the sociology section, Tony?" committee work as applied anthropology in fishery management. Human Organization. 44(2): 177–182.

Parker, K. S. 1989. Influence of oceanographic and meterological processes on the recruitment of Pacific halibut, *Hippolglossus stenolepis*, in the Gulf of Alaska, p. 221–237. *In* R.J. Beamish and G.A. McFarlane [ed.] Effects of ocean variability on recruitment and an evaluation of parameters used in stock assessment models. Can. Spec. Publ. Fish. Aquat. Sci. 108.

Parrish, B. B. 1973. Fish stocks and recruitment. Proceedings of a Symposium held in Aarhus 7–10 July 1970. Cons. Int. Explor. Mer 164: 372 p.

Parsons, L. S. 1976. Distribution and relative abundance of roundnose, roughhead and common grenadiers in the Northwest Atlantic. ICNAF Sel. Papers 1:73–88.

Parsons, L. S. 1980. Fisheries management and regulatory measures, p. 158–166. In Selected lectures from the CIDA/ FAO/CECAF seminar on fishery resource evaluation, Casablanca, Morocco, 6–24 March 1978. FAO, Rome.

PARSONS, L. S. 1983. Enterprise Allocations for the Atlantic Offshore Groundfish Fisheries. Department of Fisheries and Oceans, Ottawa, Ont. 41 p.

PARSONS, L. S., AND V.M. HODDER. 1975. Biological characteristics of southwest Newfoundland herring. 1965–71. ICNAF Res. Bull. 11: 145–160.

PARSONS, T. R., B.O. JANSSON, A.R. LONGHURST, AND G. SAETERSDAHL. 1978. Marine ecosystems and fisheries oceanography. Rapp. P. -v. Reun. Cons. int. Explor. Mer 173: 240 p.

PAULY, D. 1980. On the interrelationships between natural mortality, growth parameters and mean environmental temperature in 175 fish stocks. J. Cons. int. Explor. Mer 39(2): 175–192.

PEACOCK, F. G., AND D.A. MACFARLANE. 1986. A review of quasi-property rights in the herring purse seine fishery of the Scotia-Fundy region of Canada, p. 215–230. *In* N. Mollett [ed.] Fishery Access Control Programs Worldwide: Proceedings of the Workshop on Management Options for the North Pacific Longline Fisheries, Orcas Island, Washington, April 21–25, 1986.

PEARSE, P. H. 1971. National regulation of fisheries. Canada Department of Fisheries and the Environment, Fisheries and Marine Service (Mimeo.)

PEARSE, P. H. 1972. Rationalization of Canada's west coast salmon fishery: an economic evaluation. 172–202. *In* Organization for Economic Co-operation and Development, Economic aspects of fish production. Paris: OECD.

PEARSE, P. H. 1979a. Property rights and the regulation of commercial fisheries. J. Bus. Admin. 11: 185–209.

PEARSE, P. H. [ed.]. 1979b. Symposium on Policies for Economic Rationalization of Commercial Fisheries. J. Fish. Res. Board Can. 36: 711–866.

PEARSE, P. H. (Commissioner) 1982. Turning the Tide: A New Policy for Canada's Pacific Fisheries. The Commission on Pacific Fisheries Policy. Final Report. 292 p.

PEARSE, P. H. 1988. Rising to the Challenge. A new policy for Canada's freshwater fisheries. 180 p.

PEARSE, P. H., AND J.E. WILEN. 1979. Impact of Canada's Pacific Salmon Fleet Control Program. J. Fish. Res. Board Can. 36: 764–769.

PEPPER, D. A., AND H. URION. 1979. Public expenditure and cost-recovery in fisheries: modelling the B.C. salmon fishery for policy analysis and government investment decisions. 393–407. *In* NATO Symposium on applied operations research in fishing, at the Marine Technology Center in Trondheim, Norway, August 14–17, 1979. Norwegian univ. of Fisheries. Trondheim (Norway).

PERLEY, M. H., ESQ. 1850. Report on the sea and river fisheries of New Brunswick, within the Gulf of Saint Lawrence and Bay of Chaleur. Fredericton, N.B. J. Simpson, Printer. 137 p.

PERU. 1947. Presidential Decree No. 781 of August, 1947. UN Legislative Series:High Seas, 1951, ST/LEG/SER.B/1, p. 16.

PETERMAN, R. M., M.J. BRADFORD, N.C.H. LO, AND R.D. METHOT. 1988. Contribution of early life stages to interannual variability in recruitment of northern anchovy (*Engraulis mordax*). Can. J. Fish. Aquat. Sci. 45: 8–16.

PEYTON, R. B. 1987. Mechanisms affecting public acceptance of resource management policies and strategies. Can. J. Fish. Aquat. Sci. 44(Suppl. 2): 306–312.

PFEIFFER, W. C., AND H. JORJANI. 1986. Investment analysis of commercial aquaculture in Central Canada. Can. Ind. Rep. Fish. Aquat. Sci. 160: 52 p. + app.

PHARAND, D. 1984. Delimitation of Maritime Boundaries: Continental shelf and exclusive economic zone, in light of the Gulf of Maine Case, Canada vs USA. Revue Generale de Droit 16: 363–386.

PIKITCH, E. K. 1989. Technological interactions in the U.S. West Coast groundfish trawl fishery and their implications for management. ICES MSM Symp. No. 22.

PIMLOTT, D. H., C.J. KERSWILL, AND J.R. BIDER. 1971. Scientific activities in fisheries and wildlife resources. Background study for the Science Council of Canada. Special Study No. 15: 191 p.

PINHORN, A. T. 1976. Catch and Effort Relationships of the Groundfish Resource in ICNAF Subareas 2 and 3. ICNAF Selected Pap. 1: 107–115.

PINHORN, A. T. 1979. The northern cod resource, p. 53–72. *In* Report of the Northern Cod Seminar, August 28–30, 1979. Corner Brook, Newfoundland. Department of Fisheries and Oceans. 214 p.

PINHORN, A. T., AND R.G. HALLIDAY. 1990. Canadian versus International Regulation of Northwest Atlantic Fisheries: Management Practices, Fishery Yields, and Resource Trends, 1960–1986. N. Am. J. Fish. Manage. 10: 154–174.

PINKERTON, E. 1987. The fishing-dependent community, p. 293–325. *In* Marchak et al. [ed.] Uncommon property.

PIUZE, J. 1989. Ocean pollution: an unavoidable threat? Paper presented at 7th Rotary Peace Forum, Sept. 21–23, 1989, Toronto, Ont. 11 p.

POETSCHKE, T. 1984. Community dependence on fishing in the Atlantic provinces. Can. J. Regional Sci. 7(2): 211–226.

POLLNAC, R. B., AND S.J. LITTLEFIELD. 1983. Sociocultural aspects of fisheries management. Ocean Devel. Intl. Law. 12: 209–246.

PONTECORVO, G. 1973. On the utility of bioeconomic models for fisheries management, p. 12–22. *In* A.A. Sokoloski [ed.] Ocean Fishery Management: Discussions and Research. NOAA Tech. Rep. NMFS Circ. 371: 174 p.

PONTECORVO, G. 1977. Fishery management and the general welfare: implications of the new structure. Wash. Law Rev. 52: 641–656.

POPE, J. G. 1972. An investigation of the accuracy of virtual population analysis using cohort analysis. int. Comm. Northwest Atl. Fish. Res. Bull. 9: 65–74.

POPE, J. G. 1976. The effect of biological interactions on the theory of mixed fisheries. ICNAF Selected Pap. 1: 157–162.

POPE, J. G. 1979. A modified cohort analysis in which constant natural mortality is replaced by estimates of predation levels. ICES Doc. CM 1979/H: 16 (Mimeo).

POPE, J. G. 1982. Background to scientific advice on fisheries management. Lab. Leafl. (New Series) Dir. Fish. Res. 54: 26 p.

POPE, J. G. 1991. The ICES Multispecies Assessment Working Group: evolution, insights and future problems. 22–33

ICES Marine Science Symposia. The Hague, 2–4 October 1989. 358 p.

POPE, J. G., AND D.J. GARROD. 1975. Sources of error in catch and effort quota regulations with particular reference to variations in the catchability coefficient. ICNAF Res. Bull. 11: 17–30.

POPE, J. G., AND B.J. KNIGHTS. 1982. Simple models of predation in multi-age multispecies fisheries for considering the estimation of fishing mortality and its effects, p. 64–69. *In* M.C. Mercer [ed.]. Multispecies approaches to fisheries management advice. Can. Spec. Publ. Fish. Aquat. Sci. 59.

POPE, J. G., AND D. GRAY. 1983. An investigation of the relationship between the precision of assessment data and the precision of total allowable catches, p. 151–157. *In* W.G. Doubleday and D. Rivard [ed.]. Sampling commercial catches of marine fish and invertebrates. Can. Spec. Publ. Fish. Aquat. Sci. 66.

PRICE WATERHOUSE. 1990. The economic impacts of fishing in British Columbia in 1988 and 1989. A study undertaken under contract for the Fisheries Council of British Columbia. 100 p.

PRINCE, E. E. 1902. Special Report on the Aim and Method of Fishery Legislation (III). 20–30. *In* Suppl. No. 1 to the Thirty-Fourth Annual Report of the Dept. of Marine and Fisheries for 1901.

PRINGLE, J. D. 1985. The human factor in fishery resource management. Can. J. Fish. Aquat. Sci. 42: 389–392.

PRINGLE, J. D., D.G. ROBINSON, D.G. ENNIS, AND P. DUBE. 1983. An overview of the management of lobster fishery in Atlantic Canada. Can. Manuscr. Rep. Fish. Aquat. Sci. 1701: vii + 103 p.

PRITCHARD, G. I. 1984. Proceedings of the National Aquaculture Conference. Strategies for aquaculture development in Canada. Can. Spec. Publ. Fish. Aquat. Sci. 75: 131 p.

PROSS, A. P. 1986. Group politics and public policy. Oxford University Press, Toronto, Ont. 343 p.

PTA. 1986. The troll salmon fishery: a time for new direction in management. Brief to the Minister of Fisheries and Oceans by the Pacific Trollers Association, January, 1986.

PTA. 1989. A Study of Significant Criteria Relevant to Equitable Allocation of all Pacific Salmon Species among participating geartypes. Brief by the Pacific Trollers Association, May 1, 1989.

R. v. SPARROW. 1990. 70 D.L.R. 94th) 385 (S.P.C.).

RALSTON, S., AND J.J. POLOVINA. 1982. A multispecies analysis of the commercial deep-sea handline fishery in Hawaii. Fish. Bull. 80: 435–448.

REEVES, J. E. 1974. Comparison of long-term yields and catch quotas and effort quotas under conditions of variable recruitment. ICNAF Res. Doc. 74/31 Ser. No. 3178: 16 p.

REEVES, R. R., AND E. MITCHELL. 1984. Catch history and initial population of white whales (*Delphinapterus leucas*) in the River and Gulf of St. Lawrence, eastern Canada. Naturaliste can. 111: 63–121.

REGIER, H. A., AND F.D. MCCRACKEN. 1975. Science for Canada's Shelf-seas fisheries. J. Fish. Res. Board Can. 32: 1887–1932.

RETTIG, B. R. 1973. Multiple objectives for marine resource management, p. 23–27. *In* A.A. Sokoloski [ed.] Ocean fishery management: discussions and research. NOAA Tech. Rep. NMFS Circ. 371: 174 p.

RETTIG, B. R. 1987. Bioeconomic models: do they really help fishery managers? Trans. Am. Fish. Soc. 116: 405–411.

RETTIG, B. R., AND J.J.C. GINTER [ed.]. 1978. Limited Entry. As a Fishery Management Tool. Univ. Washington Press, Seattle, WA. 455 p.

RHEE, S. M. 1980. The application of equitable principles to resolve the United States-Canada dispute over East Coast fishery resources. Harvard Int. Law J. 21: 667–683.

RICKER, W. E. 1950. Cycle dominance among the Fraser sockeye. Ecology 31: 6–26.

RICKER, W. E. 1954. Stock and recruitment. J. Fish. Res. Board Can. 11: 559–623.

RICKER, W. E. 1958. Handbook of computations for biological statistics of fish populations. Bull. Fish. Res. Board. Can. 119: 330 p.

RICKER, W. E. 1972. Hereditary and environmental factors affecting certain salmonid populations, p. 27–160. *In* R.C. Simon and P.A. Larkin [ed.] The stock concept in Pacific salmon. H.R. MacMillan Lectures in Fisheries. University of British Columbia, Vancouver, B.C.

RICKER, W. E. 1975. The fisheries research board of Canada — seventy-five years of achievement. J. Fish. Res. Board Can. 32: 1465–1490.

RICKER, W. E. 1975. Computation and interpretation of biological statistics of fish populations. Bull. Fish. Res. Board Can. 191: 382 p.

RIDDELL, B., AND P. STARR. 1987. Status of naturally spawning chinook salmon (*Oncorhynchus tshawytscha*) stocks in British Columbia through 1986. PSARC working paper no. S87-1:37. 22 p. + app.

RITCHIE-CALDER, LORD. 1978. Perspectives on the sciences of the sea. Ocean Yearbook 1: 271–292.

RIVARD, D. 1981. Catch projections and their relation to sampling error of research survey. *In* W.G. Doubleday and D. Rivard [ed.] Bottom trawl surveys. Can. Spec. Publ. Fish. Aquat. Sci. 58: 93–109.

RIVARD, D., AND J.J. MAQUIRE. 1993. Reference points for fisheries management: the Eastern Canadian Experience, Proceedings of the Workshop on Fish Evaluation and Biological Reference points for Fisheries Management. Nov. 19–22, 1991. (In press)

RIVARD, D., AND M.G. FOY. 1987. An analysis of errors in catch projections for Canadian Atlantic fish stocks. Can. J. Fish. Aquat. Sci. 44: 967–981.

RIVARD, D., W.D. MCKONE, AND R. W. ELNER. 1988. Resource Prospects for Canada's Atlantic Fisheries, 1989–1993. Department of Fisheries and Oceans.

ROBICHAUD, H. J. 1966. Atlantic fisheries: blueprint for an orderly revolution. The Atlantic Advocate 57(2): 14–16.

ROBICHAUD, H. J. 1967. From inshore to offshore: fundamental changes for coastal fishermen. The Atlantic Advocate 57(2): 18–20.

ROEDEL, P. M. 1975. A summary and critique of the symposium on optimum yield. Optimum sustainable yield as a concept in fisheries management. Am. Fish. Soc. Spec. Publ. 9: 79–89.

ROTHSCHILD, B. J. 1983. On the allocation of fisheries stocks, p. 85–91. *In* J. W. Reintjes [ed.] Improving multiple use of coastal and marine resources. Proceedings of a Symposium, Hilton Head, South Carolina, Sept. 22, 1982. American Fisheries Society.

ROTHSCHILD, B. J. 1986. Dynamics of marine fish populations. Harvard University Press, Cambridge, Mass. 277 p.

RUBLE, T., AND K. THOMAS. 1976. Support for a two-dimensional model of conflict behaviour. Organizational behaviour & human performance. 16: 143–155.

RUGMAN, A. M., AND A. ANDERSON. 1987. A fishy business: the abuse of American trade law in the Atlantic groundfish case of 1985–86. Can. Public Policy 13: 152–164.

RUSSELL, F. S. 1931. Some theoretical considerations on the "overfishing" problem. J. Cons. int. Explor. Mer 6: 3–27.

RUSSELL, F. S. 1942. The overfishing problem. Cambridge University Press.

RUTHERFORD, J. B., D.G. WILDER, AND H.C. FRICKE. 1957. An economical appraisal of the Canadian lobster fishery. Bull. Fish. Res. Board Can. 157: 127 p.

RYAN, S. 1986. Fish out of water. The Newfoundland Saltfish Trade 1814–1914. 320 p. + app. Breakwater Books, St. Johns, Nfld.

SAETERSDAL, G. 1989. Fish resources research and fishery management: A review of nearly a century of experience in the Northeast Atlantic and some recent global perspectives. J. Cons. int. Explor. Mer 46: 5–15.

SANGER, C. 1987. Ordering the oceans: the making of the Law of the Sea. University of Toronto Press, Toronto, Ont. 225 p.

SAUNDERS, R. L., AND J.B. SPRAGUE. 1967. Effects of copper-zinc mining pollution on a spawning migration of Atlantic salmon. Water Res. 1: 419–432.

SAVILLE, A. [ed.]. 1980. The assessment and management of pelagic fish stocks — a symposium held in Aberdeen, 3–7 July 1978. Rapp. P.-V. Reun. Cons. int. Explor. Mer 177: 517 p.

SAVVATIMSKY, P. I. 1969. The grenadier of the North Atlantic. Trudy Pinro, p. 3–72. Fish. Res. Board Can. Transl. Ser. No. 2879.

SCARRATT, D. J. 1964. Abundance and distribution of lobster larvae (*Homarus americanus*) in Northumberland Strait. J. Fish. Res. Board Can. 21: 661–680.

SCARRATT, D. J. 1979. The life and times of the Atlantic lobster. Nature Canada 8(3): 46–53.

SCHAEFER, M. B. 1957. Some considerations of population dynamics and economics in relation to management of the commercial marine fisheries. J. Fish. Res. Board Can. 14: 669–681.

SCHNEIDER, J. 1985. The Gulf of Maine case: the nature of an equitable result. Am. J. Intl. Law 79: 539–577.

SCIENCE COUNCIL OF CANADA. 1970. This land is their Land... a report on fisheries and wildlife research in Canada. Report No. 9. Ottawa. 41 p.

SCOTT, A. D. 1955. The fishery: the objectives of sole ownership. J. Polit. Econ. 63: 116–124.

SCOTT, A. D. 1962. Economic effects of regulation fisheries. *In* R. Hamlisch [ed.] Economic effects of fishery regulation. Report 45 of 1961 FAO Expert Meeting, Ottawa; FAO, Rome.

SCOTT, A. D. 1977. Commentary on Chapter 15. Economic aspects of extended fisheries jurisdiction. L.G. Anderson [ed.] 409–414, Ann Arbor, Mich. Ann Arbor Science Publishers, 1977.

SCOTT, A. D. 1979. Development of economic theory on fisheries regulation. J. Fish. Res. Board Can. 36: 725–741.

SCOTT, A. D., and P.A. Neher [ed.]. 1981. The public regulation of commercial fisheries in Canada. 9–20.

SCOTT, A. D., AND M. TUGWELL. 1981. The Maritime lobster fishery. The public regulation of commercial fisheries in Canada. Econ. Counc. Can. Tech. Rep. 16. Case Study no. 1, The Maritime Lobster Fishery. 67 p.

SCOVAZZI, T. 1985. Explaining exclusive fishery jurisdiction. Mar. Policy 9(2): 120–125.

SELDEN, J. 1635. Of the Dominion or Ownership of the Sea, Nedham translation (London, 1652); originally entitled Mare Clausum sev De Dominio Maris (1635).

SERGEANT, D. E., AND P.R. BRODIE. 1975. Identity, abundance, and present status of populations of white whales, (*Delphinapterus leucas*) in North America. J. Fish. Res. Board Can. 32: 1047–1054.

SERGEANT, D. E. 1986. Present status of white whales (*Delphinapterus leucas*) in the St. Lawrence Estuary. Naturaliste can. 113: 61–81.

SHELDON, R. W., A. PRAKASH, AND W.H. SUTCLIFFE JR. 1972. The size distribution of particles in the ocean. Limnol. Oceanogr. 17: 327–340.

SHEPHERD, J. G. 1982. A versatile new stock-recruitment relationship for fisheries, and the construction of sustainable yield curves. J. Cons. int. Explor. Mer 40(1): 67–75.

SHEPHERD, J. G. 1988. Fish stock assessments and their data requirements. 35–62. *In* Fish population dynamics (second edition): the implications for management. J.A. Gulland [ed.]. Great Britain. John Wiley & Sons, Ltd.

SHEPHERD, J. G., J.G. POPE, AND R.D. COUSENS. 1984. Variations in fish stocks and hypotheses concerning their links with climate. Rapp. P.-v. Reun. Cons. int. Explor. Mer 185: 255–267.

SILVERT, W., AND R.J.M. CRAWFORD. 1989. The periodic replacement of one fish stock by another. Proc. Int. Symp. on long term changes in marine fish populations. Vigo. Spain. Nov. 1986 .

SIMS, S. E. 1984. An analysis of the effect of errors in the natural mortality rate on stock-size estimates using virtual population analysis (cohort analysis). J. Cons. int. Explor. Mer 41: 149–153.

SINCLAIR, A., D. GASCON, R. O'BOYLE, D. RIVARD, AND S. GAVARIS. 1990. Consistency of some northwest Atlantic groundfish stock assessments. NAFO SCR Doc. 90/96: 26 p.

SINCLAIR, M. 1988a. Marine populations. An essay on population regulation and speciation. University of Washington Press, Seattle and London. 252 p.

SINCLAIR, M. 1988b. Historical sketch on recruitment research, p. 8–30. In M. Sinclair et al. [ed.] Report from the National Workshop on Recruitment. Can. Tech. Rep. Fish. Aquat. Sci. No. 1626.

SINCLAIR, M., AND P. SOLEMDAL. 1988. The development of "Population Thinking" in fisheries biology between 1878 and 1930. Aquatic Living Resources 1: 189–213.

SINCLAIR, M., AND T.D. ILES. 1989. Population regulation and speciation in the oceans. J. Cons. int. Explor. Mer 45: 165–175.

SINCLAIR, M., M.J. TREMBLAY, AND P. BERNAL. 1985. El Nino events and variability in a Pacific mackerel (*Scomber japonicus*) survival index: support for Hjort's second hypothesis. Can. J. Fish. Aquat. Sci. 42: 602–608.

SINCLAIR, M., V.C. ANTHONY, T.D. ILES, AND R.N. O'BOYLE. 1985. Stock assessment problems in Atlantic herring (*Clupea harengus*) in the northwest Atlantic. Can. J. Fish. Aquat. Sci. 42: 888–898.

SINCLAIR, P. R. 1983. Fishermen divided: the impact of limited entry licensing in northwest Newfoundland. Human Organization. 42(4): 307–313.

SINCLAIR, P. R. 1987. State Intervention and the Newfoundland Fisheries: Essays on Fisheries Policy and Social Structure. Aldershot, Hants, England: Brookfield, Vt. Gower Pub. Co. 155 p.

SINCLAIR, S. 1960. License Limitation — British Columbia. A Method of Economic Fisheries Management. 13–38. The Salmon Fishery. Department of Fisheries and Oceans, Ottawa.

SINCLAIR, S. 1978. A Licensing and Fee System for the Coastal Fisheries of British Columbia. Department of Fisheries and Oceans. Vol. I and Vol. II (statistical appendix): 313 p. + 115 p. statistics.

SINCLAIR, W. F. 1971. The importance of the commercial fishing industry to selected remote coastal communities of British Columbia. Dept. of the Environment, Fisheries Service, Pacific Region. 126 p.

SISSENWINE, M. P. 1984a. Why do fish populations vary? p. 59–94. In R.M. May [ed.] Exploitation of Marine Communities. Dahlem Konferenzen, Springer-Verlag, Berlin.

SISSENWINE, M. P. 1984b. The uncertain environment of fishery scientists and managers. Mar. Resour. Econ. Vol. 1(1). 1–30

SISSENWINE, M. P., AND P.M. MACE. 1992. ITQs in New Zealand: The era of fixed quota in perpetuity. Fishery Bulletin, U.S. 90: 147–160.

SISSENWINE, M. P., AND J.G. SHEPHERD. 1987. An alternative perspective on recruitment overfishing and biological reference points. Can. J. Fish. Aquat. Sci. 44: 913–918.

SKUD, B. E. 1973. Management of the Pacific halibut fishery. J. Fish. Res. Board Can. 30: 2393–2398.

SKUD, B. E. 1975. Revised estimates of halibut abundance and the Thompson-Burkenroad Debate. IPHC Sci. Rep. 56: 36 p.

SKUD, B. E. 1985. The history and evaluation of closure regulations in the Pacific Halibut fishery, p. 449–456. *In* FAO, 1985 Papers presented at the Expert Consultation on the regulation of fishing effort (fishing mortality). Rome, 17–26 January 1983. A preparatory meeting for the FAO World Conference on fisheries management and development. FAO Fish. Rep. (298) Suppl. 3: 215–470.

SLOAN, N. A., AND P.A. BREEN. 1988. Northern abalone, *Haliotis kamtschatkana,* in British Columbia: fisheries and synopsis of life history information. Can. Spec. Publ. Fish. Aquat. Sci. 103: 46 p.

SMITH, J. B. 1980. Replenishable resource management under uncertainty: a re-examination of the U.S. Northern Fishery. J. Environ. Econ. Manage. 7: 209–219.

SMITH, R. A. 1872. Air and rain: the beginnings of chemical climatology. Longmans, Green, London. 600 p.

SMITH, T. D. 1988. Stock assessment methods: the first fifty years, p. 1–33. *In* J.A. Gulland [ed.] Fish population dynamics (2nd edition). John Wiley & Sons, London.

SNOW, R. 1977. Lobster fishery licensing: injustice and muddling through? Dalhousie Law J. 4(1): 119–150.

SPARRE, P. 1979. Some remarks on the application of yield/recruit curves in estimation of maximum sustainable yield. ICES Doc. C.M. 1979/G:41 (Mimeo).

SPARRE, P. 1980. A goal function of fisheries (Legion Analysis). ICES CM 1980/G:40 81 p.

SPARROW V. R. 1986. 9 B.C.L.R. (2d) 300 (C.A.).

SPRAGUE, J. B. 1964. Avoidance of copper-zinc solutions by young salmon in the laboratory. J. Water Pollut. Contr. Fed. 36: 990–1004.

SPRAGUE, J. B., P.F. ELSON, AND R.L. SAUNDERS. 1965. Sublethal copper-zinc pollution in a salmon river — a field and laboratory study. Int. J. Air Water Pollut. 9: 531–543.

SPROUT, P. E., AND R.K. KADOWAKI. 1987. Managing the Skeena River sockeye salmon (*Oncorhynchus nerka*) fishery — the process and the problems, p. 385–395. *In* H.D. Smith, L. Margolis, and C.C. Wood [ed.] Sockeye salmon (*Oncorhynchus nerka*) population biology and future management. Can. Spec. Publ. Fish. Aquat. Sci. 96.

STAUFFER, G. 1988. Comments on the effect of biological assumptions on halibut production estimates, p. 29–30. *In* W.S. Wooster [ed.] Lecture notes on coastal and estuarine studies No. 28. Fishery science and management: objectives and limitations. Springer-Verlag, New York, NY.

STEINBERG, C. 1973. Collective bargaining rights in the Canadian sea fisheries: a case study of Nova Scotia. Ph.D. dissertation, Columbia University, New York. 436 p.

STEINBERG, C. 1974. The legal problems in collective bargaining by Canadian fishermen. Labour Law Journal. 25: 643–654.

STIGLER, G. J. 1970. The optimum enforcement of laws. J. Polit. Econ. 78: 526–536.

STILLMAN, P. G. 1975. The tragedy of the commons: a re-analysis. Alternatives 4(2): 12–15.

STOCKER, M. 1987. Pacific stock assessment review committee (PSARC) Annual Report for 1986. Can. MS Rep. Fish. Aquat. Sci. 1951: 97 p.

STOCKER, M. 1993. Recent management of the British Columbia herring fishery. *In* L.S. Parsons and W.H. Lear [ed.] Perspectives on Canadian marine fisheries management. Can. Bull. Fish. Aquat. Sci. (In press)

STOCKER, M., V. HAIST, AND D. FOURNIER. 1983. Stock assessment for British Columbia herring in 1982 and forecasts of the potential catch in 1983. Can. Tech. Rep. Fish. Aquat. Sci. 1158: ix + 53 p.

STOCKER, M., V. HAIST, AND D. FOURNIER. 1985. Environmental variation and recruitment of Pacific herring (*Clupea harengus pallasi*) in the Strait of Georgia. Can. J. Fish. Aquat. Sci. 42(1): 174–180.

STODDART, R., AND A. CLARKE. 1988. Oceans: the unknown factor. Can. Research. August, 1988. 16–17.

STOKES, R. L. 1979. Limitation of fishing effort — an economic analysis of options. Mar. Policy 3(4): 289–301.

STRAND, I. E., JR., AND V.J. NORTON. 1980. Some advantages of landings taxes in fishery management, p. 411–417. *In* J. H. Grover [ed.] Allocation of fishery resources. Proceedings of the technical consultation on allocation of fishery resources held in Vichy, France, 20–23, April 1980.

STROUD, R. H., G.C. RADONSKI, AND R.G. MARTIN. 1980. Evolving efforts at best-use allocations of fishery resources, p. 418–431. *In* J. H. Grover [ed.] Allocation of fishery resources, Proceedings of the Technical Consultation on Allocation of Fishery Resources held in Vichy, France, 20–23 April 1980.

SUTINEN, J. G., AND P. ANDERSON. 1985. The economics of fisheries law enforcement. Land Econ. 64: 387–397.

SUTINEN, J. G., AND T.M. HENNESSEY [ed.]. 1987. Fisheries law enforcement: programs, problems and evaluation. Proceedings of a workshop on fisheries law enforcement, the University of Rhode Island, October 21–23, 1985. NOAA/Sea Grant. Univ. R.I. Mar. Tech. Rep. 93: 237 p.

TAYLOR, F. H. C. 1964. Life history and present status of British Columbia herring stocks. Fish. Res. Board Can. Bull. 143: 82 p.

TEMPLEMAN, W. 1962. Division of cod stocks in the Northwest Atlantic. ICNAF Redbook 1962, Part 3.

TEMPLEMAN, W. 1972. Year-class success in some North Atlantic stocks of cod and haddock. ICNAF Spec. Publ. 8: 223–239.

TEMPLEMAN, W., AND J.A. GULLAND. 1965. Review of possible conservation actions for the ICNAF area. ICNAF Annual Proceedings 15 for the year 1964–65. Part 4: 47–56.

TERKLA, D. G., P.B. DOERINGER, AND P.I. MOSS. 1985. Common property resource management with sticky labour: the effects of job attachment on fisheries management. Discussion Paper No. 108. Department of Economics, Boston University, ME.

TERRES, N. T. 1985. The United States/Canada Gulf of Maine Maritime Boundary Delimitation. Maryland J. Int. Law and Trade 9: 135–180.

THIESSEN, V., AND A. DAVIS. 1986. A further note to making sense of the dollars: income distribution among Atlantic Canadian fishermen and public policy. Mar. Policy 10(4): 310–314.

THOMPSON, M. F. ELDER, A. DAVIS, AND S. WHITLOW. 1980. Evidence of acidification of rivers of Eastern Canada, p. 244–265. *In* D. Drablos and A. Tollan [ed.] Proceedings of the international conference on the ecological impact of acid precipitation. Acid Precipitation — Effects on Forest and Fish Project.

THOMPSON, R. E., AND D.M. WARE. 1988. Oceanic factors affecting the distribution and recruitment of west coast fisheries, p. 31–65. *In* M. Sinclair et al. [ed.] Report from the national workshop on recruitment. Can. Tech. Rep. Fish. Aquat. Sci.

THOMPSON, W. F. 1950. The effects of fishing on stocks of halibut in the Pacific. Fish. Res. Inst., Univ. Wash., Seattle, WA. 60 p.

THOMPSON, W. F. 1952. Condition of stocks of halibut in the Pacific. Rapp. P.-v. Reun. Cons. int. Explor. Mer 18: 141–166.

THOMPSON, W. F., AND H. BELL. 1934. Biological statistics of the Pacific halibut fishery. 2. Effect of changes in intensity upon total yield, and yield per unit gear. Rep. Int. Fish. Comm. No. 8.

TRUMAN, PRESIDENT HARRY. 1945a. Presidential Proclamation No. 2667, Concerning the Policy of the United States with respect to the natural resources of the subsoil and sea-bed of the Continental Shelf of 28 September 1945. 59. United States Statutes at Large 884.

TRUMAN, PRESIDENT HARRY. 1945b. Presidential Proclamation No. 2668, Concerning the Policy of the United States with respect to coastal fisheries in certain areas of the High Seas of 28 September 1945. 59 United States Statutes at Large 885.

TULLOCK, G. 1974. Does punishment deter crime? The Public Interest 36: 103–111.

TURRIS, B. R. 1988. Management Options for the Geoduck Fishery: A Discussion Paper, August, 1988. Department of Fisheries and Oceans, Pacific Region. Economics and Commercial Analysis Division. 49 p.

UDA, M. 1952. On the relation between the variation of the important fisheries conditions and the oceanographical conditions in the adjacent waters of Japan 1. J. Tokio Univ. Fish. 38: 364–389.

UFAWU. 1952. Submission by the United Fishermen, Food and Allied Workers Union, June 21, 1963 to the Minister of Fisheries.

ULFSTEIN, G. 1988. The conflict between petroleum production, navigation and fisheries in international law. Ocean Dev. Int. Law. 19: 229–262.

ULFSTEIN, G., P. ANDERSEN, AND R. CHURCHILL. 1987. The regulation of fisheries: legal, economic and social aspects. Proceedings of a European Workshop, University of Tromso, Norway, 2–4 June, 1986.

ULLTANG, O. 1977. Sources of errors and limitations of virtual population analysis (cohort analysis). J. Cons. Cons. int. Explor. Mer 37: 249–260.

UN. 1953. Canadian reservation to the ICNAF Convention. United Nations Secretariat, Treaty Series, 1953. 157:158.

UN. 1967. Speech by Ambassador Pardo to the UN General Assembly, 22nd Session, September 21, 1967, Summary Records (A/FOR/SR.166/p.3.)

UNDERWATER HARVESTERS ASSOCIATION. 1988. Proposal from the UHA to the Department of Fisheries and Oceans dated April 5, 1988. 4 p.

URSIN, E. 1982. Multispecies fish stock and yield assessment in ICES, p. 39–47. *In* M.C. Mercer [ed.] Multispecies approaches to fisheries management advice. Can. Spec. Publ. Fish. Aquat. Sci. 59.

USA. 1976. A Legislative History of the Fishery Conservation and Management Act of 1976. Senate Committee on Commerce and National Ocean Policy Study. 94th Cong. 2d Sess.

USA. 1978. Final Environmental Impact Statement. Renegotiation of the International Convention for the High Seas Fisheries of the North Pacific Ocean. Department of State. 147 p. + app.

USA. 1979. Background and Analysis: East Coast Maritime Boundary and Fisheries Treaties with Canada. March 28, 1979. U.S. Dept. State.

USA. 1981. Hearing on the Maritime Boundary Settlement Treaty with Canada before the Senate Committee on Foreign Relations, 97th Cong., 1st sess., 2–3 (March 18, 1981) (Letter of President Reagan).

USA. 1982. International Court of Justice. Case concerning the delimitation of the Maritime Boundary in the Gulf of Maine Area (Canada/United States of America) Memorial Submitted by the United States of America 27 September 1982.

USA. 1988. Report to the New England fishery management council' demersal finfish committee. An assessment of the effectiveness of the Northeast Multispecies FMP with recommendations for plan and management system improvements by Technical Monitoring Group. June 22, 1988.

VANDERZWAAG, D. L. 1983. The fish feud. Lexington Books, D.C. Heath & Co., Toronto, Ont. 135 p.

VETTER, E. F. 1988. Estimation of natural mortality in fish stocks: a review. Fish. Bull. 6: 25–42.

VILHJALMSSON, H. 1983. On the biology and changes in exploitation and abundance of the Icelandic Capelin. Proceedings of the expert consultation to examine changes in abundance and species composition of neritic fish resources, San Jose, Costa Rica, 18–29 April 1983. FAO Fisheries Report No. 291 Vol. 2: 508–520.

VON BERTALANFFY, L. 1938. A quantitative theory of organic growth. Human Biol. 10: 181–213.

WADEL, C. 1969. Marginal adaptations and modernization in Newfoundland: A study of strategies and implications of resettlement and redevelopment of outport fishing communities. Institute of Social and Economic Research, Memorial University of Newfoundland.

WALDICHUK, M. 1993. Fish habitat and the effect of human activity with particular reference to Pacific salmon. *In* L.S. Parsons and W.H. Lear [ed.] Perspectives on Canadian marine fisheries management. Can. Bull. Fish. Aquat. Sci. (In press)

WALDICHUK, M. 1990. Dioxin pollution near pulpmills. Mar. Pollut. Bull. 21: 365–366.

WALDICHUK, M., AND R.J. BUCHANAN. 1980. Significance of environmental changes due to mine waste disposal into Rupert Inlet. Canada Department of Fisheries and Oceans and British Columbia Ministry of Environment, West Vancouver and Victoria, B.C. 56 p.

WALDRON, D. E., AND A.F. SINCLAIR. 1985. Analysis of by-catches observed in the Scotian shelf foreign fishery and their impact on domestic fisheries, p. 60–91. *In* R. Mahon [ed.] Towards the inclusion of fisheries interactions in management advice. Can. Tech. Rep. Fish. Aquat. Sci. 1347.

WALLACE, D. H. 1975. Keynote address, p. 8. *In* P.M. Roedel [ed.] Optimum sustainable yield as a concept in fisheries management. American Fisheries Society.

WALTERS, C. J. 1986. Adaptive management of renewable resources. Macmillan, New York, NY. 374 p.

WALTERS, C. J., AND R. HILBORN. 1976. Adaptive control of fishing systems. J. Fish. Res. Board Can. 33: 145–159.

WALTERS, C. J., AND D. LUDWIG. 1981. Effects of measurement errors on the assessment of stock-recruitment relationships. Can. J. Fish. Aquat. Sci. 38: 704–710.

WALTERS, C. J., AND M.J. STALEY. 1987. Evidence against the existence of cyclic dominance in Fraser river sockeye salmon (*Oncorhynchus nerka*), p. 375–384. *In* H.D. Smith, L. Margolis, and C.C. Wood [ed.] Sockeye salmon (*Oncorhynchus nerka*) population biology and future management. Can. Spec. Publ. Fish. Aquat. Sci. 96.

WALTERS, C. J., M. STOCKER, A.V. TAYLOR, AND S.J. WESTRHEIM. 1986. Interaction between Pacific cod (*Gadus macrocephalus*) and herring (*Clupea harengus pallasi*) in the Hecate Strait, British Columbia. Can. J. Fish. Aquat. Sci. 43: 830–837.

WANG, E. B. 1981. Canada-United States Fisheries and Maritime Boundary Negotiations: Diplomacy in Deep Water. Behind the Headlines 38(6): 47 p.

WARD, F. J., AND P.A. LARKIN. 1964. Cyclic dominance in Adams River sockeye salmon. Int. Pac. Salmon Fish. Comm. Prog. Rep. No. 11: 116 p.

WARE, D. M. 1980. Bioenergetics of stock and recruitment. Can. J. Fish. Aquat. Sci. 43: 1028–1213.

WARE, D. M., AND T.C. LAMBERT. 1985. Early life history of Atlantic mackerel (*Scomber scombrus*) in the southern Gulf of St. Lawrence. Can. J. Fish. Aquat. Sci. 42: 577–592.

WARE, R. M. 1978. Five Issues Five Battlegrounds — An introduction to the history of Indian Fishing in British Columbia 1850–1930. Prepared for Coqualeetza Educational Training Centre, 1978. 14 p.

WARRINER, G. K., AND L.N. GUPPY. 1984. From Urban Centre to Isolated Village: Regional Effects of Limited Entry in the British Columbia Fishery. J. Can. Stud. 19: 138–156.

WATT, D. C. 1979. First steps in the enclosure of the oceans — the origins of Truman's proclamation on the resources of the continental shelf, 28 September, 1945. Mar. Policy 3(3): 211–224.

WATT, W. D. 1981. Present and potential effects of acid precipitation on the Atlantic salmon in eastern Canada. *In* Acid rain and the Atlantic Salmon. Int. Atl. Salmon Found. Spec. Pub. Ser. 10: 39–45.

WATT, W. D. 1986. The case for liming some Nova Scotia salmon rivers. Water, Air, and Soil Poll. 31: 775–789.

WATT, W. D. 1987. A summary of the impact of acid rain on Atlantic salmon (*Salmo salar*) in Canada. Water, Air and Soil Poll. 35: 27–35.

WAY, E. 1976. Lost Gill Net Retrieval Experiment. Environment Canada, Fisheries and Marine Service, Industrial Development Branch. St. John's, Nfld.

WELCH, D. W., AND D.J. NOAKES. 1990. Cyclic dominance and optimal escapement of Adams River sockeye salmon (*Oncorhynchus nerka*) Can. J. Fish. Aquat. Sci. 47: 838–849.

WENT, A. E. J. [ed.]. 1980. Atlantic Salmon: its future. Proceedings of the second International Atlantic Salmon Symposium, Edinburgh 1978, sponsored by The International Atlantic Salmon Foundation and the Atlantic Salmon Research Trust. Farnham, Surrey, England. Fishing News Books Ltd. 249 p.

WESNEY, D. 1989. Applied fisheries management plans: individual transferable quotas and input controls, p. 153–181. *In* P.A. Neher, R. Arnason, and N. Mollett [ed.]. Rights based fishing. Kluwer Academic Publishers, Norwell, MA. 541 p.

WILDAVSKY, A. 1979. The politics of the budgetary process. Little, Brown, Boston, MA.

WILDER, D. G. 1958. Regulation of the lobster fishery. Can. Fish. Cult. 22: 13–16.

WILDER, D. G. 1965. Lobster conservation in Canada. Fish. Res. Board Can. Stud. Ser. 1039: 237–245.

WILDSMITH, B. 1982. Aquaculture: the legal framework. Emond-Montgomery Ltd., Toronto, Ont. 313 p.

WILEN, J. E. 1988. Limited entry licensing: a retrospective assessment. Mar. Resour. Econ. 5: 313–324.

WILEN, J. E. 1989. Comments on: I.N. Clark, P.J. Major and N. Mollett's "The development and implementation of New Zealand's ITQ management system", p. 150–151. *In* P.A. Neher, R. Arnason, and N. Mollett [ed.]. Rights based fishing. Kluwer Academic Publishers, Norwell, MA. 541 p.

WILSON, R. C. H., AND R.F. ADDISON [ed.]. 1984. Health of the Northwest Atlantic: a report to the interdepartmental committee on environmental issues. Department of the Environment/Department of Fisheries and Oceans/ Department of Energy, Mines and Resources. 174 p.

WILSON, V. S. 1981. Canadian public policy and administration: theory and environment. McGraw-Hill Ryerson Ltd., Toronto, Ont. 442 p.

WINTERS, G. H., AND J.P. WHEELER. 1987. Recruitment dynamics of spring-spawning herring in the Northwest Atlantic. Can. J. Fish. Aquat. Sci. 44: 882–900.

WISE, M. 1984. The Common Fisheries Policy of the European Community. 316 p. Methuen Inc. New York, N.Y.

WMO. 1986. A Report of the International Conference on the Assessment of Carbon Dioxide and Other Greenhouse Gases in Climate Variations and Associated Impact, Villach, Austria, 9–15 October 1985. WMO No. 661 (Geneva: WMO/ICSU/UNEP, 1986).

WOLFE, E. E. 1987. East coast fisheries relationship, p. 161–170. *In* J.L. Bubier and A. Reiser [ed.] East Coast Fisheries Law and Policy. Proceedings from the June 17–20, 1986 Conference on East Coast Fisheries Law and Policy. Marine Law Institute, Maine.

WOODWORTH, B. 1983. Battling the storm. Atlantic Business. Halifax. April 1983.

WOOSTER, W. S. 1988. Biological objectives of fishery management, p. 1–4. In W.S. Wooster [ed.] Lecture notes on coastal and estuarine studies No. 28. Fishery science and management: objectives and limitations. Springer-Verlag, New York, NY.

WOOSTER, W. S., AND K.M. BAILEY. 1989. Recruitment of marine fishes revisited, p. 153–159. In R.J. Beamish and G.A. McFarlane [ed.] Effects of ocean variability on recruitment and an evaluation of parameters used in stock assessment models. Can. Spec. Publ. Fish. Aquat. Sci. 108.

YANAGIDA, J. A. 1987. The Pacific Salmon Treaty. Am. J. Intl. Law. 577–592.

ACRONYMS AND ABBREVIATIONS

ACFM	Advisory Committee on Fisheries Management
ACFM	Atlantic Council of Fisheries Ministers
AFSAC	Arctic Fisheries Scientific Advisory Committee
AGAC	Atlantic Groundfish Advisory Committee
ARC	Atlantic Regional Council
CAFSAC	Canadian Atlantic Fisheries Scientific Advisory Committee
CAW	Canadian Auto Workers
CBRT	Canadian Brotherhood of Railway, Transport and General Workers
CFG	Cooperative Fishermens Guild
CPUE	Catch Per Unit Effort
DFO	Department of Fisheries and Oceans
DOE	Department of the Environment
DREE	Department of Regional Economic Expansion
EA	Enterprise Allocation
EC	European Community
F	Fishing Mortality
FAO	United Nations Food and Agriculture Organization
FCC	Fisheries Council of Canada
FFT	Factory Freezer Trawler
FPAFC	Federal Provincial Atlantic Fisheries Committee
FPI	Fishery Products International
FPSB	Fishery Prices Support Board
ICES	International Council for the Exploration of the Sea
ICNAF	International Commission for the Northwest Atlantic Fisheries
ICO	Interdepartmental Committee on Oceans
INPFC	International North Pacific Fisheries Commission
IPHC	International Pacific Halibut Commission
IQ	Individual Quota
ITQ	Individual Transferable Quotas
JGOFS	Joint Global Ocean Flux Experiment
LOS	Law of the Sea
M	Natural Mortality
MAC	Minister's Advisory Committee
MASS	Marine Survival of Salmon Project
MEY	Maximum Economic Yield
MFU	Maritime Fishermen's Union
MScY	Maximum Social Yield
MSVPA	Multispecies Virtual Population Analysis
MSY	Maximum Sustainable Yield
NAFO	Northwest Atlantic Fisheries Organization
NASCO	North Atlantic Salmon Conservation Organization
NFFAWU	Newfoundland Fishermen, Food and Allied Workers Union
NIFA	Newfoundland Inshore Fishermen's Association
NSP	National Sea Products
OPEN	Ocean Production Enhancement Network
OSS	Ocean Science and Surveys
OSY	Optimum Sustainable Yield
OY	Optimum Yield
PARC	Pacific Regional Council
PRFC	Prince Rupert Fishermen's Cooperative
PSARC	Pacific Stock Assessment Review Committee
RSPP	Resource Short Plant Program
SEP	Salmon Enhancement Program
SOFA	Save Our Fisheries Association
STACRES	Standing Committee on Research and Statistics of ICNAF
TAC	Total Allowable Catch
UFAWU	United Fishermen and Allied Workers Union
UMF	United Maritime Fishermen's Cooperative
UNCLOS	United Nations Conference on the Law of the Sea
VPA	Virtual Population Analysis
WOCE	World Ocean Circulation Experiment
Y/R	Yield-per-recruit

APPENDIX I

Extracts from the 1982 Law of the Sea Convention

Article 61, dealing with the conservation of the living resources, provides that:

"1. The coastal State shall determine the allowable catch of the living resources in its exclusive economic zone.

"2. The coastal State, taking into account the best scientific evidence available to it, shall ensure through proper conservation and management measures that the maintenance of the living resources in the exclusive economic zones is not endangered by over-exploitation. As appropriate, the coastal State and competent international organizations, whether subregional, regional or global, shall co-operate to this end.

"3. Such measures shall also be designed to maintain or restore populations of harvested species at levels which can produce the maximum sustainable yield, as qualified by relevant environmental and economic factors, including the economic needs of coastal fishing communities and the special requirements of developing States, and taking into account fishing patterns, the interdependence of stocks and any generally recommended international minimum standards, whether subregional, regional or global.

"4. In taking such measures the coastal State shall take into consideration the effects on species associated with or dependent upon harvested species with a view to maintaining or restoring populations of such associated or dependent species above levels at which their reproduction may become seriously threatened."

Article 62, dealing with the utilization of the living resources, provides that:

"1. The coastal State shall promote the objective of optimum utilization of the living resources in the exclusive economic zone without prejudice to Article 61.

"2. The coastal State shall determine its capacity to harvest the living resources of the exclusive economic zone. Where the coastal State does not have the capacity to harvest the entire allowable catch, it shall, through agreements or other arrangements and pursuant to the terms, conditions, laws and regulations referred to in paragraph 4, give other States access to the surplus of the allowable catch, having particular regard to the provisions of articles 69 and 70, especially in relation to the developing States mentioned therein.

"3. In giving access to other States to its exclusive economic zone under this article, the coastal State shall take into account all relevant factors, including, inter alia, the significance of the living resources of the area to the economy of the coastal State concerned and its other national interests, the provisions of articles 69 and 70, the requirements of developing States in the subregion or region in harvesting part of the surplus and the need to minimize economic dislocation in States whose nationals have habitually fished in the zone or which have made substantial efforts in research and identification of stocks.

"4. Nationals of other States fishing in the exclusive economic zone shall comply with the conservation measures and with the other terms and conditions established in the laws and regulations of the coastal State. These laws and regulations shall be consistent with this convention and may relate, inter alia, to the following:

(a) licensing of fishermen, fishing vessels and equipment, including payment of fees and other forms of remuneration, which in the case of developing coastal States, may consist of adequate compensation in the field of financing, equipment and technology relating to the fishing industry;

(b) determining the species which may be caught, and fixing quotas of catch, whether in relation to particular stocks or groups of stocks or catch per vessel over a period of time or to the catch by nationals of any State during a specified period;

(c) regulating seasons and areas of fishing, the types, size and amount of gear, and the types, sizes and numbers of fishing vessels that may be used;
(d) fixing the age and size of fish and other species that may be caught;
(e) specifying information required of fishing vessels, including catch and effort statistics and vessel position reports;
(f) requiring, under the authorization and control of the coastal State, the conduct of specified fisheries research programmes and regulating the conduct of such research, including the samples of catches, disposition of samples and reporting of associated scientific data;
(g) the placing of observers or trainees on board such vessels by the coastal State;
(h) the landing of all or any part of the catch by such vessels in the ports of the coastal State;
(i) terms and conditions relating to joint ventures or other co-operative arrangements;
(j) requirements for the training of personnel and the transfer of fisheries technology, including enhancement of the coastal State's capability of undertaking fisheries research; and
(k) enforcement procedures.

"5. Coastal States shall give due notice of conservation and management laws and regulations."

SUBJECT INDEX